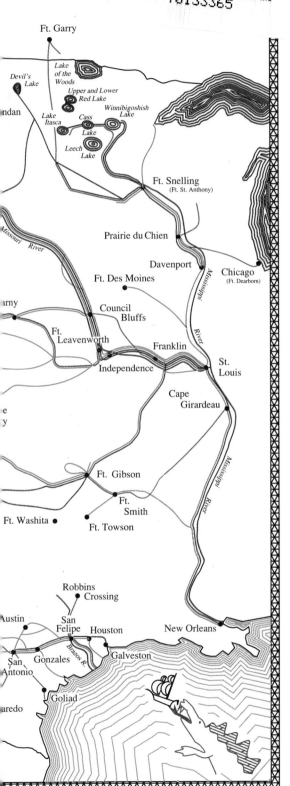

Map labels:
Ft. Garry
Devil's Lake
Lake of the Woods
Upper and Lower Red Lake
Winnibigoshish Lake
Lake Itasca
Cass Lake
Leech Lake
ndan
Lake
Ft. Snelling (Ft. St. Anthony)
Missouri River
Prairie du Chien
Davenport
Chicago (Ft. Dearborn)
Ft. Des Moines
Council Bluffs
arny
Ft. Leavenworth
Franklin
Independence
St. Louis
Cape Girardeau
Mississippi River
e
y
Ft. Gibson
Ft. Smith
Ft. Washita
Ft. Towson
Robbins Crossing
San Felipe
Houston
New Orleans
Austin
Gonzales
Galveston
San Antonio
Brazos R.
Goliad
aredo

Legend:

1790-1800
Haenke, 1791. Monterey
Menzies, 1792-94. Pacific coast

1800-1810
Lewis & Clark, 1804-06. St. Louis to Pacific and back
Langsdorff, 1806. Grays Harbor, San Francisco
Bradbury & Nuttall, 1809. St. Louis

1810-1820
Nuttall, 1810. St. Louis. Prairie du Chien down Mississippi R.
Bradbury & Nuttall, 1811. St. Louis to Ft. Mandan and back
Bradbury & Nuttall, 1811-12. St. Louis to New Orleans
Bradbury, 1812. New Orleans
    1818-20. Mississippi River
Eschscholz & Chamisso, 1816. San Francisco Bay
Baldwin, 1819. St. Louis to Franklin
Nuttall, 1819-20. Arkansas River

1820-1830
James, 1820. St. Louis to Rockies to Cape Girardeau
Douglass, 1820. Cass Lake, Prairie du Chien
Say, 1823. Lake of the Woods
Scouler, 1825. Lower Columbia, north Pacific coast
Douglas, 1825-27. Columbia River and Umpqua Valley
Collie, 1826-27. Santa Clara, Santa Cruz
Lay, 1827. Monterey
Botta, 1827-28. California coast
Berlandier, 1828-29. Texas and Mexico

1830-1840
Douglas, 1830. Oregon and Washington
    1830-32. San Francisco. Solano to San Buenaventura
    1832-33. Columbia River, Fort Vancouver to Fraser River
Coulter, 1831-32. Monterey to Yuma and back
Pitcher, 1831-34. Fort Gibson and vicinity
Houghton, 1832. Cass Lake to Ft. Snelling
Tolmie, 1833. Ft. Vancouver and vicinity
Gairdner, 1833-35. Columbia River to Ft. Walla Walla
Drummond, 1831-32. St. Louis to New Orleans
    1833-34. Texas
Wyeth, 1832-33. St. Louis to Ft. Vancouver and back
Maximilian, 1833-34. St. Louis to Ft. McKenzie and back
Leavenworth, 1833-40. Ft. Towson and vicinity
Beyrich, 1834. St. Louis to Cross Timbers and back
Nuttall & Wyeth, 1834. St. Louis to Ft. Vancouver & vicinity
Nuttall, 1836. Monterey to San Diego
McLeod, 1837. Ft. Vancouver to Wyoming
Hinds, 1837. Nootka Sound, San Francisco, Sacramento River, Monterey
    1839. Columbia River, San Francisco, California coast
Geyer, 1838. Ft. Snelling to Coteau du Prairie
    1839. St. Louis to Ft. Snelling

1840-1850
Brackenridge, 1841. Ft. Nisqually to Ft. Colville and Ft. Walla Walla; Puget Sound to San Francisco
Vosnesensky, 1841. California coast
Gambel, 1841. Santa Fe to Pacific coast
Fremont, 1842. St. Louis to Wind River Mts. and back. St. Louis to Ft. Laramie
Geyer & Fremont, 1841. Des Moines River
Geyer, 1843-44. St. Louis to Pacific coast
Burke, 1844-46. Northern Rockies, Great Basin, and Pacific Northwest
Gordon, 1845-48. New Orleans
Spalding, 1843. Lapwai
Audubon, 1843. St. Louis to Ft. Union on Missouri River
Fremont, 1843-44. St. Louis to Ft. Vancouver, Pyramid Lake, California, back to St. Louis
Lindheimer, 1843-45. Texas
Fremont, 1845-46. Bents Fort to Klamath Lake
Abert, 1845. Bents Fort to St. Louis
Wislizenus, 1846. St. Louis to Mexico
Hartweg, 1846-48. California coast, Sacramento River
Abert, 1846. Ft. Leavenworth to Bents Fort
Emory, 1846-47. Ft. Leavenworth to San Diego
Fendler, 1846-47. Ft. Leavenworth to Santa Fe and back
    1849. Ft. Leavenworth to Ft. Kearny
Parry, 1847. Des Moines
    1848. Davenport to Lake Superior
    1849. San Diego to Yuma
Lindheimer, 1847-48. Texas
Fremont, 1848-49. Bents Fort to California
Trecul, 1848. Eastern Missouri
    1849. Kansas, Mississippi River
    1849-50. Texas
Wright, 1848-49. Texas
Stansbury, 1849-50. Ft. Leavenworth to Great Salt Lake

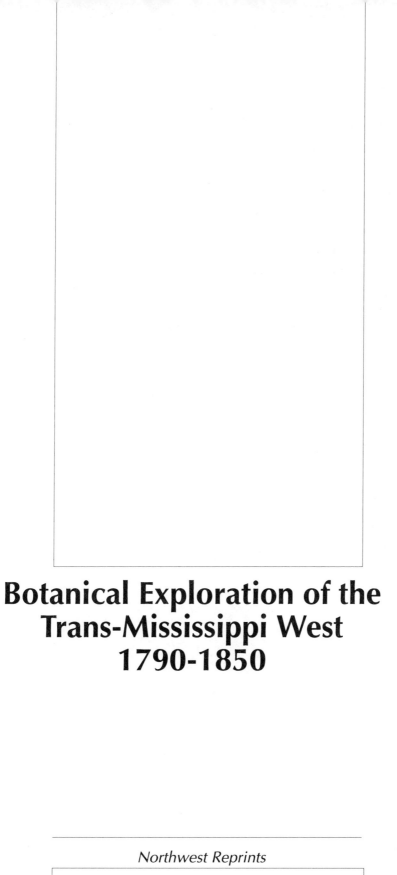

Botanical Exploration of the Trans-Mississippi West 1790-1850

Northwest Reprints

Northwest Reprints
Series Editor: Robert J. Frank

Other titles in the series:

Botanical Exploration of the Trans-Mississippi West 1790-1850

Susan Delano McKelvey

Foreword by Joseph Ewan
Introduction by Stephen Dow Beckham

Oregon State University Press
Corvallis, Oregon

The paper in this book meets the guidelines for permanence and durability of the Committee on Production Guidelines for Book Longevity of the Council on Library Resources and the minimum requirements of the American National Standard for Permanence of Paper for Printed Library Materials Z39.48-1984.

Library of Congress Cataloging-in-Publication Data
McKelvey, Susan Delano
        Botanical exploration of the trans-Mississippi west, 1790-1850 / Susan Delano McKelvey : introduction by Stephen Dow Beckham.
                p. cm. — (Northwest reprints)
        Reprint. Originally published: Jamaica Plain, Mass. : Arnold Arboretum of Harvard University, 1956.
        Includes bibliographical references and index.
        ISBN 0-87071-513-5 (alk. paper)
        1. Botany--West (U.S.)--History. 2. Plant collecting--West (U.S.)--History. 3. Plant collectors--West (U.S.)--History. 4. Scientific expeditions--West (U.S.)--History. 5. West (U.S.)--Discovery and exploration. I. Title. II. Series.
QK5.M3 1991
581.978—dc20   91-24466
                                                                      CIP

PREFACE

but there were things
That covered what a man was, and set him apart
From others, things by which others knew him. The place
Where he lived, the horse he rode, his relatives, his wife,
His voice, complexion, beard, politics, religion or lack of it,
And so on. With time, these things fall away
Or dwindle into shadows: river sand blowing away
From some long-buried old structure of bleached boards
That appears a vague shadow through the sand-haze,
*   and then stands clear,*
Naked, angular, itself.
*   from "Trial and Error," H.L. Davis*

People new to a region are especially interested in what things might set them apart from others. In works by Northwest writers, we get to know about the place where we live, about each other, about our history and culture, and about our flora and fauna. And with time, some things about ourselves start to come into focus out of the shadows of our history.

To give readers an opportunity to look into the place where Northwesterners live, the Oregon State University Press is making available again many books that are out of print. The Northwest Reprints Series will reissue a range of books, both fiction and nonfiction. Books will be selected for different reasons: some for their literary merit, some for their historical significance, some for provocative concerns, and some for these and other reasons together. Foremost, however, will be the book's potential to interest a range of readers who are curious about the region's voice and complexion. The Northwest Reprints Series will make works of well-known and lesser-known writers available for all.

RJF

FOREWORD

As a human document, the 1,184 pages of this book are history, although frequently interrupted in its writing for an aside or incident. Testy bits of philosophy are mortised between the tenons of fact. For this reason, the account is rather more narrative than encyclopedic. From an intimate field experience in the American West, Mrs. McKelvey composed this omnibus volume with verve and verdict. But once her hopes of publication were nearly abscised. Writing to me March 19, 1953, she sank in denigration that her "eight large fat loose leaf folders" assembled were "neither 'fish, flesh nor fowl' as far as a reading public (and hence a publisher) is concerned . . . I may do nothing about publication." However, Elmer Drew Merrill, Harvard botanist who read the manuscript from "cover to cover," turned dismay into print.

The inimitable Alice Eastwood botanized with Mrs. McKelvey in Arizona. Miss Eastwood presented her copy of volume two of the *Botany of California* to her companion, and it was used in the writing of this book. Since 1965, copiously annotated, it has been shelved with the Ewan floras. In my review of *Botanical Exploration of the Trans-Mississippi West* in *Rhodora* in 1957 I noted that some readers, like Joseph Grinnell who wrote on William Gambel's bird records, first read the footnotes and "if they proved tasty trimmings from the main dish" would "try the textual entree."

Museums and workers in natural history, in both botanical and zoological fields, will find the book of constant reference value. Nor will they be disappointed for the finer points of documentation and careful quotation. For the historian devoted to Western Americana—Mississippi River to the Pacific—and for the historically minded biologist, Susan McKelvey's opus is thick and rich. As a compendium of our knowledge of the lives and travels of botanical explorers (who were often also engaged in collecting insects, birds, or other "natural curiosities") this book ". . . is not a dull enumeration of facts. It might also be compared to the best type of fiction, were it not better still, reality."

Joseph Ewan
Missouri Botanical Garden

INTRODUCTION

The vast western lands of North America beckoned in 1790. The region beyond the Mississippi and inland from the shores of the Pacific was a place of mythical geography and unknown peoples and resources. Thomas Jefferson, author of a natural history, *Notes on the State of Virginia* (1787), wondered, for example, whether mammoths and mastodons might yet stalk the interior of the continent. His curiosity drove him several times to attempt to mount expeditions to penetrate the region. The third venture, the notable trek of Meriwether Lewis and William Clark, not only met his expectations but catapulted the young United States into the competition of Enlightenment era science.

The approaches to the Trans-Mississippi West came by sea and land. The Spanish voyages of the mid-1770s to the northwest coast yielded information which fed a burgeoning competition for colonies and trade. Great Britain sent Captain James Cook for a third voyage of discovery to the Pacific, and in 1778 he passed along the shoreline of North America on a solitary cruise toward the Bering Sea. Succeeding expeditions grew more ambitious with complements of specialists: cartographers, artists, and botanists (or natural historians) who collected "curiosities" and penned descriptions of exotic flora, fauna, and natives in their journals.

The contest grew. In 1786 the Lapérouse expedition, financed by France, explored coastal California. In 1790 the Spanish Malaspina expedition also visited that coastline. And in 1792, in a valiant effort to find the nonexistent Northwest Passage, Great Britain sent Captain George Vancouver to mount a minute examination of inlets, estuaries, straits, and seas from California to Alaska. These ventures, competitive and driven by geopolitical and economic motives, nevertheless were in the spirit of the eighteenth century. They were part of an intense desire to collect, describe, and organize the world. Participants in these events and scholars who examined their collections, logs, diaries, and charts sought to find order in diversity.

The commitment to examine nature closely, discern apparent and presumed relationships, and posit systems to describe observable phenomena were part of the spirit of the Enlightenment. Some of these figures dared to dream new dreams. Bernard De Fontenelle, the French scientific essayist, for example, authored *A Week's Conversation on the Plurality of Worlds* (1686). In these essays he said: "I have chosen that Part of Philosophy which is most likely to excite Curiosity; for I think nothing concerns us more, than to enquire how this World, which we inhabit, is made; and whether there be any other Worlds like it, which are also inhabited as This is." The prospect of speculating about other worlds and coming to terms with their own world inspired many.

Carolus Linnaeus (1707-78) helped fix the use of binomials to identify genus and species in his many works, including *Systema naturae* (1735), *Genera plantarum* (1737), and *Species plantarum* (1753). While a number of figures wrestled with systems of plant classification and wrote regional floras, others examined minerals, weather, diseases, and clouds. Luke Howard, for example, proposed a typology for identifying and naming clouds which gained sufficient currency to be used, in part, to the present.

These projects involved essentially two types of laborers: collectors, who increasingly after 1790 went into the field to search for and collect undescribed species; and the descriptive botanists in study centers, who worked with dried plant material and occasionally with fresh specimens grown from seeds. The two were willing partners in an era of descriptive science which flourished from the mid-1770s until the end of the nineteenth century. The Trans-Mississippi West became one of the important sources of new materials for these studies. Like China, India, South Africa, and parts of South America, it was a great storehouse of unexpected finds and opportunities.

Botanical Exploration of the Trans-Mississippi West, 1790-1850, provides a sweeping overview of the work of the plant collectors and, at times, their ties to and dependence upon the descriptive botanists who helped underwrite, order, and interpret what they found. The volume is an ambitious examination of the lives, labors, travels, and trials of often unheralded men who risked everything in their wandering to find new species for patrons and purchasers in distant cities and countries. Some of these plant collectors, like Archibald Menzies, were surgeons or staff members of maritime or military expeditions. Others, like David Douglas and Karl Theodor Hartweg, represented horticultural societies seeking exotic plants and new ornamentals for introduction into the gardens of Europe. A few, like Thomas Nuttall, had scholarly connections and sought to expand the herbaria of the institutions where they taught and wrote monographs. And still others, the professional collectors, like Ferdinand Jakob Lindheimer, collected plants as a living through subscriptions from those wanting Texas materials.

Often the work done by the plant explorers in the American West was unfinished. Some lost their specimens to accidents in the field or in transit to the herbaria or botanical gardens. Few kept detailed notes. The lack of journals and precise locations for plant collections tantalized later generations interested in species distribution, extinction, or ecological settings. Almost none of the collectors displayed interest in or awareness of the ethnobotanical potentials everywhere at hand. They did not seek information from the Native Americans living throughout the Trans-Mississippi West about the medicinal and practical uses of the species they and their ancestors had used for countless generations.

The botanical collectors were, however, the first "men of science" in much of the western United States. Their "science" was narrowly conceived but part of a tradition: seek, find, collect, and share with others. The plant collectors set an example subsequently nurtured by the Smithsonian Institution, colleges, universities, and fledgling scientific societies. By the 1850s the flow of curiosities from the American West mounted steadily. A new generation of field workers collected minerals, fossils, and Indian artifacts. In the 1850s Spencer Fullerton Baird began four decades of solicitation to pack the Smithsonian with a vast array of objects. In thousands of letters Baird encouraged military officers, travelers, civilian government employees, and others to send Indian word lists, botanical specimens, pickled fish and snakes—indeed almost anything to add to the holdings of the national museum in Washington, D.C.

The botanical exploration of the American West between 1790 and 1850 was thus the forerunner of increasingly institutionalized collecting and scientific inquiry. In many instances the explorer and plant collector were synonymous; in other cases the collector was part of a larger expedition with broader goals: furthering the national interest, military reconnaissance, or bartering and trapping for furs. The plant collectors coped with loneliness, inclement weather, impatient associates, and a host of dangers. Some labored for months only to receive chiding letters and criticism from their patrons or employers. Few saw the forest for the trees and fewer still in this period had interest in any but flowering plants. In spite of their limited vision and sometimes mercenary motives, these workers in the field made singular contributions to increasing botanical collections and enhancing the understanding of the American West.

Botanical Exploration of the Trans-Mississippi West, 1790-1850, explores the work of major collectors, lifts up a number of obscure figures, documents with as much precision as possible the travels and connections of these men, and provides fascinating information about the disposition and location of their specimens. The volume's detailed sources trace letters, diaries, catalogs, and identities of those who worked in the field and those who received the collections. The book is a fitting tribute both to those who collected specimens in the American West and to their collaborators who attempted to organize all nature and knowledge.

Susan Delano McKelvey did nothing halfway. A poised, dedicated researcher and writer, she spent forty-five years as patron and associate of the Arnold Arboretum at Harvard University. Little in her early life suggested such a course. Susan Delano was born on March 13, 1883, in Philadelphia. Her family had connections: wealth, education, and social prominence. She grew up in New York, attended Bryn Mawr, and majored in English and French. She graduated in 1907 and that fall married Charles Wylie McKelvey, an attorney. The McKelveys moved to Oyster Bay on Long Island, residing close to her brother, William Adams Delano, an architect of growing reputation. William Delano and Moreau Delano, another brother, and Charles McKelvey were all Yale graduates and probable acquaintances through their membership in the Scroll and Key society.

In the 1910s the Delano brothers and their sister, Susan McKelvey, embarked upon a lifetime of philanthropic support for the Arnold Arboretum. In 1919, her marriage at an end and one of her two sons dead, McKelvey moved to Boston and, under the nurturing of the aging Charles Sprague Sargent, then in his fifth decade as director of the arboretum, began rebuilding her life. Edmund A. Schofield, her biographer, remarked: "Once Sargent had given her the initial nudge, McKelvey threw herself wholeheartedly into mastering the various aspects of botany, maintaining her zeal for the subject virtually until her death in 1964."

McKelvey moved with increasing deliberation into a study of lilacs. Sargent made the suggestion, but McKelvey, possibly because of her interest in gardening and the use of *Syringa* as an ornamental, embraced the project and devoted most of a decade to carrying it out. Possessing ample means, she financed her own research. Her travels in pursuit of lilacs and information about their collection, introduction to Europe and America, and hybridization took her to Rochester, New York; to Theodore A. Havemeyer's gardens at Glen Head, New York; to the Jardin des Plantes, Paris; to the Royal Botanic Gardens at Kew in Great Britain; and to Nancy, France. As the project deepened, McKelvey explored herbals and botanical literature to trace the lilac from 1554 to the 1920s. She sought recondite but useful catalogs from nurserymen as further documentation.

The Lilac: A Monograph appeared in 1928 as a remarkably handsome, oversized volume of 581 pages containing 171 plates. McKelvey drew upon associates and included four contributions from them on "Diseases and Pests," "Culture," "Description of the Genus," and "History and Distribution." The great bulk of the work, however, was her own. She proceeded species by species to discuss diagnostic features, the history of its assessment in botanical literature, and notes on its collection and distribution. Within these discussions, she laid out many pieces of the history of botanical collecting in China by Europeans.

By the late 1920s, as *The Lilac* worked its way through the press at Macmillan Company in New York, McKelvey turned in another direction. Her field was the Southwest and her chosen genus, the yucca. A generation of Americans had discovered a land of enchantment in the watersheds of the Rio Grande and Gila. Unlike the artists and writers who became members of the Taos community, however, McKelvey kept her own counsel and found inspiration in journeys of weeks and months to the far edges of the region. With her chauffeur and associate in the field, O.E. Hamilton, touring car, and trailer for botanical presses and equipment, McKelvey explored the border of Mexico, the margins of the Pacific in southern California, the edge of the Grand Canyon, indeed all of the areas where she could find new species or varieties. In addition McKelvey worked in thirteen major herbaria, including the Englemann Herbarium at the Missouri Botanical Garden, the Torrey Herbarium at the New York Botanical Garden, and the National Herbarium in Washington, D.C.

Between 1928 and 1936 McKelvey mounted eight expeditions to the Southwest, five solely for yucca studies. Gradually she expanded her field to reach from Texas to Nevada and western Colorado to the Pacific. As in her previous work on the lilac, McKelvey listed the species, described their diagnostic features, identified their range (often with a map of the collecting sites), reviewed collectors and dates, and provided a summary of the notice of the species or variety in the literature.

Her *Yuccas of the Southwestern United States* appeared in two parts, 1938 and 1948, and was published by the Arnold Arboretum. The first volume contained 80 plates; the second had 55. McKelvey and Hamilton shared duties as photographers and, on occasion, one or the other appeared in the views to give scale to the towering desert plants which so challenged her. In fact, McKelvey confessed: "The writer admits that both temperamentally and physically she has been unqualified to uproot the yuccas encountered on her travels." The photographs and detailed descriptions served adequately for most readers and researchers.

McKelvey's work on the yuccas introduced her to the scattered but fascinating accounts and collections of those who had preceded her in pursuing flowering plants in the Trans-Mississippi West. During her southwestern travels she also met and became a friend and correspondent with Alice Eastwood, a longtime botanist and historian of botanical exploration in California. The history of plants, their collection, and disposition of specimens consistently gained attention in McKelvey's writing. So, too, did fascinating tangents and footnotes such as her discussion of the labors of Benedict Roezel in 1869-70 in California. McKelvey traced Roezel's work through plant catalogs advertising his materials. To her dismay she discovered he had collected 850 plants of the orchid *Talipogon Croesus*, not one of which had survived. She then wrote: "Although Roezel shipped great quantities of material, in one instance '10 tons of plants,' one may rest assured that he did not find it pleasurable to gather yucca in such numbers, with the result that the two San Diego species still survive."

The Great Depression did not deflect McKelvey's work or zeal. While her cousin, Franklin Delano Roosevelt, launched his New Deal and coped with events which erupted into World War II, McKelvey turned to historical studies. She also shifted from her work as a longtime Research Associate at the Arboretum at Jamaica Plain to the Botanical Museum in Cambridge. By 1944 she was deep into her research on her third, and greatest, book, *Botanical Exploration of the Trans-Mississippi West, 1790-1850* (1955). "Now I have begun on something else and am thrilled about it," she wrote to Eastwood, and added: "In fact so interested that I wish I had begun years ago." In a sense, though, her work of years ago became the foundation for crafting the volume. Her extensive field experience had contributed to a clear understanding of what frontier plant collectors had to endure.

McKelvey crafted a massive typescript of 1,853 pages, distilled into 1,184 pages when set in type and weighing 5 pounds, 14 ounces. The book brought together the scattered but fascinating literature of botanical work in the western United States. Joseph Ewan remarked appropriately in his review: "only one book of its kind is expected in a century." The reviewers uniformly praised the work and, though a few found McKelvey's footnotes tiresome, even they had to admit the information therein appended was useful and often suggestive of future research possibilities. The notes, as well, often revealed the good humor of a master of botany who saw the figures about whom she wrote as human beings with strengths, weaknesses, and passions.

"Days without time are needed to gather the facts. Nor is compilation alone the making of the book, for only a careful knitting of the skeins of facts will present a finished work," wrote Ewan. McKelvey had the time, resources, will, and ability to gather and knit. She lived purposefully from 1919 until her death in 1964. She remained until her death a patron of the Arnold Arboretum and willed her papers, books, photographs, and collections to the place which helped her bring together the loose threads of her life. Her works are valued today as when they were published decades ago. Librarians, in fact, have sequestered *The Lilac, Yuccas*, and *Botanical Exploration* in rare book cases, contemporary testimony to how well McKelvey's labors have withstood the test of time.

Stephen Dow Beckham
Lewis and Clark College

Bibliography

The most comprehensive biographical assessment of McKelvey is Edmund A. Schofield's "A Life Redeemed: Susan Delano McKelvey and the Arnold Arboretum," *Arnoldia,* the Magazine of the Arnold Arboretum, 47 (4), Fall 1987: 9-23. Schofield based his account on the manuscripts and published materials at Harvard and other institutions, and on interviews.

McKelvey, Susan Delano

1925 *"Syringa rugulosa,* a new species from western China." *Journal of the Arnold Arboretum,* 6 (3) (July): 153-154.

1927 "A new hybrid lilac." *Horticulture,* 5 (15) (August 1): 302.

1928 *The Lilac: A Monograph.* Macmillan Company, New York. xvi + 581 pages.

1933 "Taxonomic and cytological relationships of *Yucca* and *Agave.*" *Journal of the Arnold Arboretum,* 14 (1) (January): 76-81. [Written with Karl Sax.]

1934 *"Arctomecon californicum."* National Horticultural Magazine, 13 (4) (October): 349-350.

       "A verification of the occurrence of *Yucca Whipplei* in Arizona." *Journal of the Arnold Arboretum,* 15 (4) (October): 350-352.

1935 "Notes on *Yucca." Journal of the Arnold Arboretum,* 16 (2) (April): 268-271.

1936 "The Arnold Arboretum." *Harvard Alumni Bulletin,* 38 (15) (January 17): 464-472.

1938 *Yuccas of the Southwestern United States. Part One.* Arnold Arboretum, Jamaica Plain. 150 pages.

1947 *Yuccas of the Southwestern United States. Part Two.* Arnold Arboretum, Jamaica Plain. 192 pages.

1949 "A new *Agave* from Arizona." *Journal of the Arnold Arboretum,* 30 (3) (July): 227-230.

1955 *Botanical Exploration of the Trans-Mississippi West, 1790-1850.* Arnold Arboretum, Jamaica Plain. xl + 1144 pages.

1959 "A discussion of the Pacific Railroad reports as issued in the quarto edition." *Journal of the Arnold Arboretum,* 40 (1) (1959): 38-67.

Bibliographic Supplement: selected titles

The following selection of titles published since the publication of *Botanical Exploration of the Trans-Mississippi West* was made by Joseph Ewan, who also supplied the annotations.

Anderson, Bern. *Life and Voyages of Captain George Vancouver.* Seattle: University of Washington Press, 1960.

Beidleman, Richard G. "William Gambel—frontier naturalist." *Pacific Discovery* 11 (6) (1958): 10-15.

Berlandier, Jean Louis. *Journey to Mexico During the Years 1826 to 1834.* Austin: Texas State Historical Association, 1980. 2 vols. Transl. from French. Botanical notes by C.H. and K.K. Muller. [See *Sida* 9 (1982): 375-381.]

Blomkvist, E.E. "A Russian scientific expedition to California and Alaska, 1839-1849." *Oregon Historical Quarterly* 73 (1972): 101-170. [On I.G. Vosnesensky at Fort Ross, California, 105-116.]

Brosse, Jacques. *Great Voyages of Discovery. Circumnavigators and Scientists, 1764-1843.* New York: Facts on File, 1983. Transl. from French. [Excellent summaries of Cook, La Pérouse, Kotzebue, Barclay, Wilkes.]

Cutright, Paul R. *Lewis and Clark, Pioneering Naturalists.* Urbana: University of Illinois Press, 1969. [Essential reference.]

Davis, Goode P. Jr. *Man and Wildlife in Arizona. The American Exploration Period, 1824-1865.* Scotsdale: Arizona Game and Fish Department, 1982. [Annotated quotations. Popular plant names.]

Dupree, A. Hunter. *Asa Gray, 1810-1880.* Cambridge: Harvard University Press, 1959. [Acknowledged biography of the master of ceremonies.]

Ewan, J. "San Francisco as a mecca for nineteenth century naturalists. With a roster of biographical references to visitors and residents." In *Century of Progress in the Natural Sciences, 1853-1953.* San Francisco: California Academy of Sciences, 1955. 1-63.

Fraser, Esther. *Canadian Rockies: Early Travels and Exploration.* Edmonton, Alberta: Hurtig, 1969.

Geiser, Samuel W. "Men of science in Texas, 1820-1880." *Field and Laboratory* 26 (1958) 86-139; 27 (1959) 20-48, 81-96, 111-160, 163-256. [Essential supplement to his *Naturalists of the Frontier,* 1937, 1948.]

Goetzmann, William H. *Army Exploration in the American West, 1803-1863*. New Haven: Yale University Press, 1959. [See also *Exploration and Empire*. New York: Knopf, 1966. Historian's panorama on impact of exploration on culture and economy.]

Jackson, Donald. *Letters of the Lewis and Clark Expedition with Related Documents, 1783-1854*. Urbana: University of Illinois Press, 1962. [Detailed commentary on natural history.]

Lada-Mocarski, Valerian. *Bibliography of Books on Alaska Published Before 1868*. New Haven: Yale University Press, 1969. [Complements McKelvey, Chapter IV.]

Ledyard, John. *Journal of Captain Cook's Last Voyage*. Corvallis: Oregon State University Press, 1963. [Richly annotated; botanical notes by Helen M. Gilkey. Complement to Hakluyt Society edition, Cook's *Journals*, edited by J.C. Beaglehole, vol. 3, 1967.]

McClintock, Elizabeth. "Early plant explorers in the West." *Journal of the California Horticultural Society* 28 (1967): 114-121, 152-160.

McDermott, John Francis. *Audubon in the West*. Norman: University of Oklahoma Press, 1965. [Supplement to his *Up the Missouri with Audubon*, 1951.]

Mitchell, Howard T. *Journals of William Fraser Tolmie, Physician and Fur Trader*. Vancouver, B.C.: Mitchell Press, 1963. [See Mitchell, 232, and McKelvey, 475, on ascent of "Tolmie's Peak."]

Mozino, José Mariano. *Noticias de Nutka*. Seattle: University of Washington Press, 1970. Transl. from Spanish. [Fully documented. Catalogs of plants and animals.]

Norwood, William. *Traveler in a Vanished Landscape. The Life and Times of David Douglas, Botanical Explorer*. New York: Clarkson Potter, 1973. [Norwood's "probabilities" are undocumented. See *Plant Science Bulletin* 20 (no. 1) (1974): 19.]

Nuttall, Thomas. *A Journal of Travels into the Arkansas Territory During the Year 1819*. Edited by Savoie Lottinville. Norman: University of Oklahoma Press, 1980. [Valuable commentary.]

Paul Wilhelm, Duke of Württemberg. *Travels in North America, 1822-1824*. Transl. from German. Edited by Savoie Lottinville. Norman: Univer sity of Oklahoma Press, 1973. [Fully annotated.]

Pursh, Frederick. *Flora Americae Septentrionalis (1814)*. Reprint. Introduction by J. Ewan. Vaduz: Cramer, 1979. [Notes on Bradbury, Lewis, Menzies, Nelson, Nuttall, etc.]

Raven, Peter H. "George Barclay and the 'California' portion of the Botany of the *Sulphur.*" *Aliso* 5 (1969): 469-477. [McKelvey's "Mr. Barclay," 636-658, collected 4,000 specimens!]

Reveal, James L. "Botanical exploration in the intermountain region." In *Intermountain Flora. Vascular Plants of the Intermountain West, U.S.A.* New York: Hafner, 1972. 1:40-76. [Illustrated account, extensive bibliography.]

Thomas, John Hunter. "Botanical exploration in Washington, Oregon, California, and adjacent regions." *Huntia* 3 (1969): 5-62. [Excellent illustrated summary with references.]

BOTANICAL EXPLORATION
OF THE
TRANS-MISSISSIPPI WEST
1790–1850

Botanical Exploration of
the Trans-Mississippi West

1790-1850

by

Susan Delano McKelvey

Published by

The Arnold Arboretum of Harvard University

Jamaica Plain, Massachusetts

1955

PRINTED BY THE ANTHOENSEN PRESS, PORTLAND, MAINE

CONTENTS

CONTENTS

CONTENTS

CONTENTS

CONTENTS

INTRODUCTION

THIS story has for its locale the immense region extending westward from the Mississippi River to the Pacific Ocean and from the Canadian boundary southward to Mexico—the region now included within the boundaries of the United States. It covers, after an introductory chapter telling of an event which took place in 1786, the sixty years from 1790 to 1850. And it concerns the men who went into that region within those years and collected the plant material which, after careful study, classification and published description by qualified botanists, was to form the foundation of the reliable and comprehensive floras which enable persons of the present day to identify with a great degree of accuracy the plants which they see about them.

From the year 1607, when the first permanent British settlement was established on our Atlantic coast, to the year 1790 was slightly less than two hundred years. During this near bicentenary the frontier of settlement had spread gradually westward from the Atlantic seaboard until by 1790 (the year of the first census of the United States) it had reached the valley of the Mississippi River and in the persons of adventurers, traders and trappers was beginning its slow advance up some of the major tributaries entering that river from the west. But in 1790 what lay between Mississippi River and Pacific Ocean was virtually unknown to those advancing westward and any thought of that region's acquisition and settlement existed only in the visions of a few men in the nation's capital. Nonetheless between 1790 and 1850 (to be exact, 1853)— or within the sixty years chosen for my story—all the lands west of the Mississippi River were to become part of our national domain with exterior boundaries essentially what they are to-day. Assuring this territorial aggrandizement were: the Louisiana Purchase in 1803; the settlement of the "Old Northwest" in 1812 (although this mainly affected lands east of the Mississippi River); the annexation of Texas in 1845; the extinguishment of British claims in the Oregon Territory in 1846; the Mexican Cession in 1848; and the Gadsden Purchase in 1853. These treaties were consummated only after years of competition and of conflict as well as of prolonged negotiation between our nation and foreign powers whose individual claims involved great sections of the trans-Mississippi west—claims which, considered as a unit, represented a coverage of the entire area.

The activities of the plant collectors whose travels are told here were closely interlocked with these historical events. For it was only by associating themselves with any available trend in exploration, settlement, commerce or military movements that they were able to enter the then dangerous wildernesses where they wished to gather plants. I shall enumerate some of the more important of these men and indicate their relationship to the ever-changing historical picture of time and place which forms the background of my story.

Although to Americans advancing westward the lands beyond the Mississippi River were unknown, the first plant therefrom meriting inclusion in its future floras had already been collected before this story opens: the great scientific expedition sent out from France under command of the brave Lapérouse had, in 1786, anchored off Monterey, California, and a gardener, Collignon, had collected seeds of a little sand verbena which is a common coastal species for a long distance northward; these seeds, shipped to France before the ships and all aboard perished in the South Pacific, germinated, flowered and fruited, and the plant in orthodox manner was assigned to the new genus *Abronia* and given the specific name *umbellata* by two of the great botanists of France. And *Abronia umbellata* (Jussieu) Lamarck it triumphantly remains to this day!

Before the Louisiana Purchase, Spain had sent around the world an important scientific expedition commanded by Malaspina, and it was when its ships halted off Monterey in 1791 that the botanist Haenke was able to make the first collection of dried plants of California to reach Europe, and that the botanist Née was enabled to describe—from branches brought to him in Mexico by members of the party—two Californian oaks, the first trees from the trans-Mississippi west judged worthy of a place in the floras yet to be written of that region.

Still before the Louisana Purchase, or in the years 1792, 1793 and 1794, a Scottish gentleman, Menzies, attached in the dual capacity of naturalist and surgeon to the British surveying expedition commanded by Vancouver, had gathered many plants along the shores of Puget Sound and its neighboring waters and a lesser number along the coast of California. Although he was unable to penetrate any distance inland, all that Menzies found was new, and it was because of the floral wealth which he pictured on his return to England that the botanists and horticulturists of Great Britain sent in later years a succession of plant collectors—of whom Douglas was undoubtedly the most important—to what is now our west coast.

The extensive territory included in the Louisiana Purchase was geographically unknown to its purchaser, and the establishment of its ill-defined boundaries to the satisfaction of the United States and of the nations with abutting territorial possessions involved years of exploratory investigation. Lewis and Clark collected a fair number of plants, all things considered, on what was our first governmental expedition across the northern portion of these just-acquired lands and, as was to be expected in a region which was botanically untouched,[1] their collections included new genera and species. When, following the War of 1812, it was at length determined that the 49th parallel of north latitude should—as far west as the summit of the Rocky Mountains—

1. On April 7, 1805, the first collection of plants made within the area of the Louisiana Purchase—or between St. Louis and Fort Mandan, North Dakota—was shipped down the Missouri River just as the Lewis and Clark party left their winter quarters to start westward. Its safe arrival in Philadelphia was recorded in the Donation Book of the American Philosophical Society under date of November 16 of the same year.

constitute the dividing line between British and American possessions, the United States government had still to learn whether, for example, the sources of the Mississippi River and of some of its tributaries lay within its territory. Therefore, in the 1820's and 1830's, it sent out trained engineers and topographers to explore and to map the geography of these lands and to determine points along their northern boundary. Included in the expeditions were such men as Douglass, Say and Houghton whose partial duty it was to collect plants and to report upon the vegetation. The botanist Geyer, who later made important collections in the Rocky Mountains and westward, received some of his early training in field work when accompanying the famous topographer Nicollet.

As early as 1819 the botanist Nuttall entered southern portions of the lands acquired by the Louisiana Purchase. Traveling for the most part alone, he ascended the Arkansas River from its confluence with the Mississippi to Fort Smith, Arkansas, and from there made two trips into eastern Oklahoma, then set apart for occupancy by certain Indian tribes and officially termed the Indian Territory. This was a famous journey and was only possible because of the scattered settlements and military posts which were already in existence along the Arkansas waterway.

With the exception of the areas acquired by the Louisiana Purchase, all the territorial acquisitions of the United States in the trans-Mississippi west resulted from American expansion into regions where Spain—and after Spain, Mexico—held possession; or into regions where Great Britain through its fur companies had become the controlling power. When American fur traders (from 1807 to 1843 our outstanding pathfinders) found their way into the Rocky Mountains, and when, in their wake, American settlers turned first to Oregon and soon after to California, it is not surprising that, even from the start, such demonstrations of interest should have become a matter of concern to those already established there. With Great Britain the boundary question in the Oregon region, although a bitter one, was eventually and for the most part peaceably adjusted, with the 49th parallel the dividing line between British and American possessions from the Rocky Mountains to the Pacific Ocean. With Mexico, in the California region, war and subsequent negotiation left the United States in possession not only of California but of extensive and little-known lands to the east and, nearly simultaneously and by mere good fortune, in possession of gold fields which determined for some years the trend of westward migration. I mention some of the plant collectors who, as observers or even participants, shared in these events.

The first of a long series to avail themselves of the protection of the American fur traders and their outposts were the botanists Bradbury and Nuttall who made a journey up the Missouri River as far as southern North Dakota in 1811. When, sixteen years after the Louisiana Purchase, the United States government finally sent out an expedition initially intended to explore the upper Missouri River, one contingent included scientists, among them a surgeon-botanist who for the first year (1819) was

Baldwin, for the second year (1820) James. Although, just as this scientific section was to start on the more important part of its journey, the government—which then took financial expenditure into careful, even penurious, consideration—demanded its return, James, homewardbound, was able to make a famous climb and to gather alpine plants in the central Rocky Mountains of Colorado and desert plants in the arid regions of northern Texas and Oklahoma.

In 1834 the botanist Nuttall traveled overland to the Columbia River with an expedition organized by Wyeth, a Massachusetts resident with many plans for promoting American settlement in the rich lands of Oregon; before he sailed back to the United States Nuttall was able to make collections along the California coast. On his long journey Nuttall acquired much that was then unknown in the plant world and in the realm of ornithology as well.

When, in 1841, the ships of the grandiosely planned United States Exploring Expedition reached the shores of the Pacific northwest, its botanists—notably Brackenridge in my connection—were able, thanks to the facilities and advice of the Hudson's Bay Company, to make some inland trips into the state of Washington and to travel overland from the Columbia River to California by what was then a rarely used and dangerous route. Although the botanists did not return empty-handed, much if not most of the country which they traversed had already been intensively explored by the collector Douglas, many of whose plant exportations were already growing in the gardens of England. By the year of the expedition's arrival on the Pacific coast, authorities in Washington were well aware that the Oregon country offered rich possibilities for American settlers and the commander Wilkes did not fail to ask searching questions of the local representatives of the Hudson's Bay Company.

It was only belatedly, in 1843 or nearly forty years after the journey of Lewis and Clark, that the United States government sent a second overland expedition as far west as the Pacific coast. Americans were turning more and more to California as a place of settlement. The route to Oregon from the Missouri frontier went westward to Fort Hall in southeastern Idaho and thence northwest to the Columbia River. Although Indians and trappers had reached California overland from the east and although a few courageous men had even led parties thither, their routes were unmapped and, passed on by word of mouth, were subject to misinterpretation. Therefore, in the 1840's, the United States government—with emigrants and, one must believe, the possibility of military movements in mind—sent Frémont into the field to explore and to map direct routes to California. Such crossings, in more southern latitudes than the one leading to Oregon, necessitated the passage of the Great Basin, where knowledge of the location of water was essential, as well as the discovery of safe passes over the Sierra Nevada, that appalling termination of the long westward journey.

One of the most rapid explorers of all time, Frémont accomplished his assignment in the space of a few years, mapping two routes between Great Salt Lake and California—one into the northern, one into the southern part of that state—crossing the Sierra Nevada at more than one latitude, and journeying from the Columbia River southward by way of Klamath Lake, the valleys of the Sacramento and San Joaquin rivers and the Mojave Desert, to the latitude of Los Angeles. Frémont was not a botanist but he offered to collect plants from the time he began his western explorations in 1842 and, since he crossed many regions where the flora was unknown, the professional botanists who encouraged him in this avocation occasionally reaped a rich reward. Frémont's later activities—his participation in the war with Mexico, his gold, his railroad and his political interests—eventually superseded the topographical work which was probably his greatest contribution and ended, as well, his collecting of plants. Frémont's exploration of 1845-1846 brings us down to the period of the Mexican War in the California region. The routes which he had mapped from the east to Great Salt Lake and westward to California were put to prompt service—the Mormons are said to have used his reports and maps on their westward migration into Utah in 1847. By the time the gold seekers were moving west two years later the Mormon "State of Deseret" was well established and did a profitable business with the many persons bound for California. Stansbury, in 1849, was full of admiration for what the colony had accomplished in less than three years.

As yet I have not mentioned the vast Spanish possessions lying west of the southern portions of the lands acquired by the Louisiana Purchase, regions which, roughly speaking, extended southward from about Santa Fe to what is now Mexico, which were traversed throughout their western length by the waterway of the Rio Grande and which embraced eastward to the western boundary of the Purchase, a portion of western Oklahoma and the huge acreage now included in the state of Texas. Traders, operating from the Mississippi westward, had long desired entry to Santa Fe, not only for exchange of commodities but as a base from which they might extend their operations. But not until 1822, or a year after Mexico had declared its independence from Spain, was any American trader able to take his wagons into that city and to make an exchange of goods. Although this marked a beginning of such intercourse, all contacts with Santa Fe remained difficult, even dangerous, for some time after New Mexico came under American control in 1846. American colonists had been admitted to Texas in 1820 by the Spanish government and by 1831 had a well-established settlement at San Felipe de Austin on the Brazos River; other Americans, not all desirable additions, were also entering Texas, and although the later Mexican authorities made an attempt to check this influx, they were unable to do so and friction on both sides mounted rapidly, eventually culminating in the Texas revolution. After Texas in 1836 had become an independent republic, began the movement for annexation which, in its turn, brought on the Mexican War.

Persons other than traders and settlers had long been desirous of obtaining entry to these Spanish- or Mexican-held lands. Famous botanists in such secure regions as

Switzerland, Britain and the United States had been persistently eager to send their plant collectors thither. In 1828 Berlandier, emissary of the elder De Candolle in Geneva, was able to enter Texas, and in 1833 Drummond, working for the elder Hooker, got there from England; they were the first botanists to gather some of its central and southeastern flora. When, in the decade of 1840-1850, the United States government began to send troops into Texas to protect its frontiers and to make reconnaissances of various sorts, the able collector Wright visited Eagle Pass on the Rio Grande and made his famous journey from San Antonio to El Paso and back. But except for the hasty passage of Gambel in 1841, no botanist reached Santa Fe before the year of the Mexican War. However, before that conflict the United States government had sent topographers on a succession of exploratory expeditions over the several routes leading to Santa Fe from the east and on some of these trips plants were collected and reports published upon the flora. The younger Abert made two such journeys, one in 1845, one in early 1846,[2] and although on neither of these did he enter Santa Fe, he did, on the first, enter botanically unexplored regions in the "Panhandle" sections of present Oklahoma and Texas. In 1846, and just before war came to Santa Fe, the German botanist Wislizenus, a member of a private caravan traveling with United States troops, spent a short time in Santa Fe and then traveled south along the Rio Grande and crossed to El Paso which at that time was located in Mexico as now understood. This was a famous journey—even before entering Mexico where Wislizenus' most important work was accomplished—for he was the first to make a collection of plants in the Rio Grande valley and the first to describe its flora. War was well started when he made the crossing to El Paso.

In mid-August of 1846 Kearny and the "Army of the West," with some speeches and a show of force, took Santa Fe without firing a shot and, after arranging for control of that conquered region, started for California. With this army moved a contingent commanded by Emory. After descending the Rio Grande to the latitude of the Mimbres Mountains, this section turned west across that range, passed through "the deserted copper mines" of Santa Rita (where some plant collecting was done), reached the headwaters of the Gila River, and followed that stream to its junction with the Colorado. The Gila River route to California had already been traveled, at least in part, by traders, but Emory's party was the first to map its intricacies and the first, too, to collect plants from the headwaters of the Rio Grande westward to the California border, or across western New Mexico and the breadth of Arizona. Emory's "Reconnoissance" is a botanical classic. This route, a still more southern one than any yet mapped by Frémont, was followed in some measure by that explorer in 1849. Emory arrived in time to play a part in the conquest of California; Frémont at the crossing of the Colorado met Mexicans proceeding to the gold fields.

2. Abert's journey of *late* 1846 and 1847 (when he entered Santa Fe and explored portions of New Mexico) resulted, as far as I know, in no botanical report.

Thus, in brief, did history and plant collecting move hand in hand in my region and period. By 1850 routes to the Pacific coast had been recorded across the trans-Mississippi west at northern, central and southern latitudes and many of its important plants had been discovered.

Botanical work is, however, most productive when pursued deliberately, at seasons when plants are in flower or in fruit, and when halts can be made in rich floral regions, and but few of the early collectors had worked under such happy conditions. Most had moved perforce with great rapidity, often at the worst season of the year for acquiring plant material and without regard to their own regional preferences. But, shortly before my story's termination, began a type of collecting which was more deliberate and more concentrated in character. When Fendler, for example, was able to enter Santa Fe only a short time after its surrender to the Americans, his work of several months duration was restricted—for reasons of safety—to areas adjacent to that city. Intensive collecting, which was to become ever more important as botanists took up residence in various localities west of the Mississippi River, had its true inception, however, in the person of the German botanist Lindheimer who, for the nine years from 1843 to 1851 gathered Texan plants for Asa Gray.

Finally, in the year 1849, the last of my sixty-year period, were presaged—in a trip made by Parry from San Diego to the Colorado Crossing and in another made by Stansbury from Fort Leavenworth to Great Salt Lake—the government surveys which, in the decade of 1850 to 1860, were to cross the trans-Mississippi west at many latitudes and, with famous collectors participating, were to contribute vastly to the knowledge of its plants. The most important of these were surveys along the United States and Mexican boundary—an aftermath of the Mexican War—and others having to do with the selection of routes for the railroads which were ultimately to assure communication between the Mississippi River and the Pacific Ocean and, in addition, were to make settlement possible in the Great Plains and Rocky Mountains, two regions which those hastening farther west had passed over as rapidly as possible.

By 1850, when my story ends, we have reached not only a new era in United States history but also a new phase in botanical work west of the Mississippi River.

Even in the brief six decades considered here we observe not only the rapid exploration and settlement of a region which was an unknown wilderness when those sixty years began but also, as my decade maps depict, the amazing increase in the numbers of plant collectors who entered that area and, starting at a few accessible points along its Pacific coast, spread their investigations little by little into most of its major zones of vegetation. Had the story been continued for another forty years—or to cover the full centenary of 1790 to 1890—we should have reached the year in the historical picture when the "frontier-line" of the trans-Mississippi west had become non-existent or when, on the basis of population per square mile, the census authorities could report the region to have been settled. And we should have reached the point in the botanical picture where the main features of its vegetation were understood, where the bold outlines had been drawn up and only the elaboration of details remained to be filled in.

The first question asked by friends kind enough to take an interest in this book invariably concerns the contributions made to its content by those who antedated Americans in our west—Spaniards by two hundred and fifty years, Frenchmen by more than a century. I answer that query now. What their records tell of plants has a great historical interest, but is valueless from the point of view of scientific botany which lays down certain rules to which all plant descriptions must conform. And this story concerns those who contributed to scientific botany.

A certain parallel exists in the early maps which are also of historical interest but are not, all will admit, sufficiently accurate or complete to inspire confidence in a modern traveler. It was not until trained geographers had, little by little, determined the origins and courses of rivers, the placement of mountain ranges, their passes and infinitely more, and only after the accurate findings of these field-workers had been transferred to paper by skilled topographers in the draughting room, that the reliable maps now available to all came into being.

This story concerns the making of *botanical maps* in the sense that the books (commonly called "floras") enumerating the aggregate plants of a region may be regarded as such. As in the case of the geographical maps, the first step in their preparation was the obtainment of accurate data and to acquire this was the assignment of the plant collectors whose stories are told here. The data which they accumulated took the form of specimens of the plants found on their journeys. After these specimens and the important records of where they grew,[3] had been studied by qualified professionals on the home front, their findings were gradually assembled and published in the technical botanical literature which eventually and little by little became incorporated into

3. It may be well to explain why it is of the utmost importance to record where a plant was collected. In the early decades of my period the publications of the professional botanists were often, and perforce largely, mere descriptive lists of the plants gathered along the collector's route—what might be called "samples" of what grew along his way. But by the 1840's a marked change is noted in the content of such publications.

Not all plants are endemic, that is to say confined to a few localities or to a very limited region; most range over a moderate or even very large area; some even occur around the globe. Just as the geographer's perfected map indicates not only the outlet of a river but its entire course, so the botanist's account of the flora should supply what is known of a plant's distribution within, of course, the bounds of the area which they purport to cover. Undoubtedly, from the inception of their work, the descriptive botanists had envisioned the goal which they hoped to attain; but it was only after they had been supplied with plant material from many different regions that they were able to draw broad conclusions and to reason from the particular to the general. By the 1840's the number of specimens cited as examples of a plant's distribution had noticeably increased. Gray's paper on Fendler's New Mexican collections, published in 1849, is one excellent example.

the books which now make it possible for interested persons to identify the inhabitants of what to them may be a strange plant world.

It is unnecessary to comprehend the technicalities which must be observed by the cartographers in order to interpret or appreciate the value of their finished product. Similarly, it is unnecessary to understand the extremely complex rules laid down by the botanists in nomenclatorial congresses assembled—rules which determine the inclusion or exclusion of certain plant names and plant descriptions—in order to appreciate the usefulness of the descriptive floras now available.

Nor were the old Spanish and French descriptions of plants the only ones disregarded by taxonomic botanists. In their narrative journals many later explorers wrote of the plants which interested them, sometimes first descriptions and sometimes sufficiently good ones for identification. But unless these conformed to the nomenclatorial rules, they too were relegated to the category of mere historical interest.[4] There is a purpose behind such discriminatory rulings and this has, or so I surmise, eventual uniformity, in the sense of consistency, as its ultimate goal. In any event such decisions are explanatory of why much that seems pertinent thereto is not found in these pages.

Not many of the plant collectors mentioned in this volume were trained botanists and, as such, qualified to describe their own collections; even of the qualified, few to any noteworthy extent undertook the task of publication. Some who did so had assistance, and certain names which appear as coauthors of botanical works were undoubtedly embodied on a courtesy basis. Indeed, if I estimate the importance of the collectors included here on the three counts of scientific qualifications, breadth of knowledge of the living plant derived from actual field experience, and publication of personal discoveries, I find that Nuttall rates high on each count and stands virtually alone as the possessor of all three *desiderata*.

Few of the field workers undertook collecting on an independent basis. Some were sent by scientific or horticultural institutions—their sponsors were from many categories. By far the greater number went at the behest of professional botanists living in proximity to the essentials of herbaria and libraries and, in distinction to their emissaries, amid safe and comfortable surroundings. When the enormity—and costly aspect—of the task of amassing material from such a region as the trans-Mississippi west became evident, those sponsoring such work conceived a plan by which what came to be called the "profession" of plant collecting might be placed on a self-supporting

---

4. In a book such as *The Silva of North America,* Sargent is careful to distinguish between the individual who first saw or first described a particular tree and the individual who wrote its first valid description. But few of the descriptions found in the field-records of the explorers, or even of the collectors, appear in the scientific literature of plants. The man in the field could not transport the books from which to learn whether what he thought was new had already been described, nor was he near the herbaria in which he could study comparable material.

basis: collectors were to gather each plant in quantity and sell the duplicates in sets (the exsiccatae of the botanists) to subscribers. Because the initial financial outlay was thereby reduced, purchasers of the sets soon were asked to invest in the collector's wares before he went into the field. Written by those initiating such schemes, the prospectuses had much the ring of modern advertisements. It was a period when not only professionals but amateurs[5] were assembling herbaria and to the amateur such collecting had quite as strong an appeal as philately. However, I have yet to discover the field-worker who made more than a subsistence income, if that, from the arrangement. The backers of the scheme—often called "closet-botanists" for the reason that, working in offices, they may never have seen the living plants which they described— were engaged for the most part in descriptive botany, writing botanical papers or compiling floras of small or large scope. Since plant material constitutes the basis of all such work, they were obliged to send men into the field to acquire it. Very naturally, the less traveled the region, the greater the likelihood that the collector would discover new species or, preferably, new genera. In the early days of my period those first in a field invariably brought back something new, even the "tyros." But it was not long before the ability to discriminate became an important asset in a collector for, although his field might still be unexplored, its plants might already have been found in a belt of similar vegetation. There must have been complaints from subscribers regarding duplication, for the prospectuses soon stress that the field worker had promised to collect only "rarities" and avoid all "common plants."

During my sixty years the most important of the descriptive botanists concerned with plants of western North America were: the elder (Augustin-Pyramus) de Candolle in Geneva; Sir William Jackson Hooker in Britain who fell heir to the exalted position once held by Sir Joseph Banks; in the United States, John Torrey, Asa Gray who began his botanical career as Torrey's pupil, and George Engelmann who eventually turned in large measure from the successful practice of medicine to the work which had long had its appeal. Living in St. Louis, which in his day was the focal

5. Subscribers to the sets were promised not only specimens but determinations—in other words named specimens—supplied by the proponents of the scheme and there was, therefore, no necessity to be a "botanist" in order to enjoy this form of collecting! Many privately owned herbaria eventually found their way into botanical institutions and proved to be of great value. Dr. M. L. Fernald tells an amusing story of how one reached the Gray Herbarium:

"... one of the most fastidious and scrupulously groomed of amateur botanists, the late Nathaniel Thayer Kidder, walking up Chestnut Street, on Beacon Hill in Boston, was unceremoniously tempted by a familiar-appearing package on top of a barrel of ashes, ready for collection by the city ash-trucks. Poking at it with his stick, he verified his suspicion. His conventional propriety further overruled by curiosity, he rang, in the early forenoon, at the door opposite. The two elderly ladies living there explained that the package was discovered in clearing out the attic, and that Kidder might have it. The presentation slip in the Gray Herbarium does not mention Kidder. It reads: *Plants of Greenland* collected by *Benjamin Vreeland, M.D.*, on the *Kane 'Grinnell' Expedition, 1850* (Presented *1927*, by Misses Edith and Frances Fisher) ..."

I might add that, to the knowing, a bundle of "specimens" is not only "familiar-appearing," but unmistakable.

point of entry to and exit from all regions west of the Mississippi River, Engelmann had, in respect to location, considerably the advantage of his botanical associates, for he was consulted by every plant collector bound for or returning from that field of operation.[6]

Although each of the five men mentioned was keenly eager to acquire new plants and to publish their descriptions before anyone else did so, their competition was extraordinarily amicable and highly ethical insofar as encroachment on each other's prerogatives was concerned. Moreover, it was not long before enough material was coming in to keep everyone fully occupied—the universal plaint was "over" occupied. Partly because of such stress, but largely because it was the intelligent approach, it soon became customary for the recipient of a collection to turn over certain plant groups to specialists, "mosses to Sullivant," and so on. To the apparent satisfaction of all concerned, Engelmann became authority on the cacti, yuccas and century plants, spiniferous families which American botanists had hitherto been satisfied to admire from afar. To Gray went for identification all composites, a family in which he had a consuming interest. Including such plants as dandelions, daisies, etc., this family, although pleasanter to handle than Engelmann's specialties, was virtually limitless in number and of astonishing complexity—its members have a high birth rate and are not averse to intermarriage. Gray was in the habit of losing interest in a collection after he had completed in the botanical sequence the classification of his favorites. Torrey's interest appears to have been more catholic as far as plant groups was concerned. And it was under his leadership that American botanists changed from the Linnean System to the Natural System of plant classification, a change which Jussieu had inaugurated abroad.

Also, since no one person however capable could, with so much material coming in, describe all the plants of the North American continent, there arose what may be regarded as spheres of influence, although boundaries were not always held inviolate. Hooker who as successor to Sir Joseph Banks, sponsor of Menzies, had an incontestable head start on the northwest coast and in northern and arctic regions, seems to have been accepted as supreme authority on all that grew from Hudson Bay to the Pacific, an area which lies in large part outside my locale. But because arctic plants may descend to more southern latitudes at high elevations, one finds Hooker's interest reaching southward into our own Rocky Mountain region.[7] Torrey, Gray and Engel-

6. I have omitted all reference to Engelmann's field-work in Missouri and in Arkansas, considering him only in his rôle of descriptive botanist. Neither Torrey nor Gray traveled west of the Mississippi until after the period of this story. Neither the elder De Candolle nor the elder Hooker ever visited North America.

7. Hooker's connection with Texan botany is not explainable on such a basis for his collector Drummond selected that field after a visit to St. Louis. Interestingly enough it had been De Candolle in faraway Switzerland who sent out Berlandier, the first plant collector to enter Texas; but after his emissary's supposed fiasco the Genevese botanist seems to have been content to subscribe to the "sets" gathered by the envoys of others.

mann appear to have taken over as their sphere all that remained of the continent of North America, for there are indications that Mexico proper beckoned! These three men were responsible for describing most of the important plants which were brought out of the trans-Mississippi west in the years from 1840 to 1850, a decade prolific in collections.

De Candolle, Hooker, Torrey, Gray and Engelmann, as their accomplishments and fame betoken, dedicated their lives to botany and there are good reasons for supposing that they were not averse to having their collectors do likewise—as some, even literally, did do. Even before my sixty-year period has ended we hear of a new generation of botanists. Whether students of later decades will observe in the successors to these five great men the same all-consuming interest and the same invincible driving-force for accomplishment I cannot predict, but it seems unlikely; for is it not rare to find in those who fall heir to any cause the overwhelming devotion that motivated its first champions? The period of descriptive botany in my region was to continue to the end of the nineteenth century, after which other trends in botanical research were to come to the fore, with enthusiastic proponents for *their* new approach.

William Gilson Farlow—when writing of the first fifty years of the Smithsonian Institution, a period which began at much the time my story terminates and which ended in 1896—described the botanical trend which characterized his period, and mine, as well as trends which were to follow:

"... The first botanical problem to be solved in a new country is of necessity the exploration of its different parts and the description of the native species. As the systematic knowledge of the native flora increases, the important question as to the causes of the distribution of the different species, the effects of soil, temperature, and other climatic and biological conditions, assume a greater significance, and when a general knowledge of the flora has become widely diffused throughout a country, the stage is reached where the more general and abstract problems belonging to the domain of vegetable physiology and the minute investigations in cytology and the study of life-histories attract the attention of the rising generation of botanists. In the early years of the Institution the main object of botanists was to find out what plants grew in North America ..."

And Frederick Brendel's epitome of botanical accomplishment in the period of which I write is significant; it appears in his "Historical sketch of the science of botany in North America from 1635 to 1840":

"A history of the science of Botany in North America means not in this sketch a history of that science in all its branches, but rather the history of traveling and local collectors, and of descriptive botany so far as it concerns American plants. For until Prof[essor] A[sa] Gray's popular book *How Plants Grow* appeared in 1858, not a single work of any importance was published in this country, either on anatomy or on the physiology of plants, not even a single one of the many systems ever proposed had its origin in America. And yet the labors of American and foreign scientists in America contributed their large share to the advancement of science. They furnished the material for the work in all other branches of botany, and particularly in the geography of plants . . .''

To seasoned collectors what I now tell must be a familiar story. But to the un-schooled I believe that some explanation of what the task involved may make the var-ied problems which the field-workers faced and their successes and their failures more interesting and more understandable.

It is not difficult to learn some satisfactory system of keeping daily records.[8] And it is simple to learn, in theory at least, the technique of preparing the dried specimen destined for the herbarium—the form of material with which most of the collectors were concerned[9] and which was veritably "meat and drink" to the awaiting botanists. In practice, however, to attain and to preserve in specimens what the adjective *dried* denotes could have been no less than a "headache" for the field-workers who, wher-ever they went, had the elements as adversaries.[1]

The ideal specimen should consist of foliage, flowers, fruit and underground parts since important characters may exist in any one, perhaps in more than one, of these essentials. Few specimens did, or now do, attain perfection of content, with the result that tentative determinations of new plants often appear in the early treatises—the

8. Obviously, the greater a man's knowledge of plants the more valuable his records. Some kept were of course poor, others by distinction very fine.

Unfortunately, during the period of which I write, those who were studying and describing the material received were in the habit of disregarding the collector's data and of recording merely what they them-selves considered significant, with the result that since the importance of the *type*—the specimen which was described when the plant was named—has been recognized, this disregard of the collector's record has proved troublesome. I am told that this was a more or less universal failing of the period. Some say the British were worse offenders than the Americans!

9. When the acquisition of living material intended for propagation thousands of miles away was super-imposed, the collector's task was more than doubled. Even the collection of fertile seed (which neces-sitates being on the right spot at, actually, the right moment or before it falls) and the preservation of that seed in perfect condition until delivery to a distant destination, were no easy matters. I mention this form of collecting more than once in my chapters and shall not discuss it here.

1. Nonetheless these men acquired, prepared, preserved and got home much that was valuable.

It is to be regretted that, over the years, many collections did not receive the care now bestowed by appreciative custodians. One reads of valuable plants discarded when the papers which contained them were used for pipe spills (I am told that scores of types at Liverpool were destroyed in this manner), of bundles stowed away at the mercy of insects and of others lost, apparently forever. Perhaps not for-ever for some have come to light even in recent years.

While dried specimens may have little appeal to the lover of the living plant, they have been sufficient for the needs of the descriptive botanists and there are well-worn jokes intended to prove that plants had to be pressed before these men would attempt an identification. With the books in which plant descrip-tions old and new have appeared and with the content of herbaria available, a skilled taxonomist has all that he requires in order to pursue a happy and useful occupation!

botanists had to wait for additional (or what they were in the habit of calling "more ample") material before they could be sure that their classification was correct. We read complaints that collections were "scraps," "débris," and so on and in extenuation of such fault-finding should recognize that in describing a newly discovered plant the more complete the studied collection the more accurate and full can be the published description and the more readily can the proposed species be recognized and appraised by other botanists.

How the men carried the plants gathered during the course of a day's journey is left largely to the imagination; accounts of their daily procedure are extremely rare and disappointingly meagre. Drummond, on his Canadian journey, placed his plants in a container, the vasculum, now usually abandoned in favor of a portable press. Bradbury carried "a large inflexible port-folio"[2] and also utilized his hat. If a man had a horse, well and good; otherwise he shouldered his plants throughout the day—in addition to essentials. One man enumerated the paraphernalia of guns, ammunition, cooking utensils, blankets and much else that he suspended about his person. Douglas lists the equipment to be taken on his last journey and I noted with regret that he felt obliged to reduce his pocket handkerchiefs to one!

To work one's way thus encumbered through a pathless wilderness of swamps, undergrowth or fallen timber, up and down ravines, across creeks and rivers, in fair weather or in, veritably, foul or to traverse for days on end waterless deserts in horrible heat and permeating dust, was exhausting work, and the collector was not chosen because he qualified as a Paul Bunyan. Add the clouds of mosquitoes (orthography varied amazingly but not biting ability), of hoards of fleas which were to be expected after any contact with Indians (and it was the "thing to do" to shake hands with each inhabitant on arrival at and departure from an encampment or village), of "blow-flies" and others of their kind in the deserts and one can picture some of the minor distresses of a day's travel. Rats were active at night: the weary Douglas, taking a well-earned sleep, sprang to his gun in order to discover the specific characters of one such visitor departing with his inkwell.

Certainly we can be positive that a man was physically weary when he stopped for the day. It was then that the creature comforts of fuel (both for cooking and for warmth since even a hot desert can be cold at night), of water, of food and so on, had to be

2. Before leaving on my first collecting trip Professor Sargent presented me with a large and extremely heavy sole-leather portfolio which he had carried, hooked over a saddle-horn, on many journeys. The fact that it bore his name in full gave me a sense of adequacy! Mr. J. G. Jack, in charge of the party and a believer in lessons learned by experience, made no comment, then or later. Scrambling as we did—on foot —up and down mountain slopes, it became evident at once that I had not the strength to carry it and a young man of the party came to my assistance. Whenever new acquisitions had to be inserted the portfolio was laid open on the ground and the bulging content of papers and plants weighted down with stones because of the ever-present wind. It is still humiliating to remember that more than once at the end of a day my helpful assistant discovered that he had been transporting boulders as well as plants for some eight hours!

attended to. These essentials accomplished, the collector was able, and obliged, to attend to the plants gathered during the day. This involved not only arranging every specimen in proper form for pressing, but also writing up the records—the more numerous and the more specific all pertinent details of the plant's appearance and habitat, the more helpful for the man's sponsor who was of course in the dark about all such matters. The adjective *dried* may be said to have come into its own at this point. For, given that the day had been rainy or snowy, everything was wet when he made camp, unless by rare good fortune he had been able to protect the paper[3] in which he was to segregate each plant. One need only picture damp, wet, or snow-covered ground, wet wood or sometimes no wood, wet paper, wet plants, wet all things, to appreciate that to dry a specimen was not as simple as it sounds. Plants themselves are, of course, full of moisture; even under the best of conditions some may need to be transferred to dry paper every day for many days[4] before attaining the degree of crispness which proves to the knowing that further attention—except for keeping them crisp and free from mildew—is unessential. All of which indicates that the collector's night-task involved not only caring for what he had gathered on a particular day, but going over every specimen previously made until some—for no two dry at the same rate of speed—were in condition to separate from the still-moist as ready for shipment. But on most journeys they could not be shipped for weeks or months and, in some way or other, had to be transported during that time. Some bundles fell into streams when boats overturned or pack-animals lost their footing; some went over precipices when these creatures tried, as is their preference, to pass each other on the narrowest parts of trails; some were washed away in flash floods. They met with every imaginable form of mishap, even to shipwreck on the final lap of their homeward journey—one lot of records vanished between Dublin and London. These calamities meant tragedy for the collector and a disappointment at very least for the sponsor. As noted, the technique[5] of making specimens was not difficult to learn; what was difficult for the man in the field was to put each step of the proceeding into successful practice—and to get safely home in good condition what he had acquired.

3. One reads a good deal about paper and of how difficult it was to transport all that was required. One collector is reported to have shipped two tons for collecting purposes but how far he took it I have not been able to learn.

In the words of Gray writing to the missionary Spalding, collecting paper should be "soft, bibulous," and plants "should be dried between numerous thicknesses" thereof. Gray wrote from Cambridge. Spalding, in the wilds of Idaho, must have found the stipulation disheartening! Newspapers, now universally used for such purposes and available in every hamlet, were not printed in Idaho till the 1860's and it seems unlikely that Spalding possessed many from elsewhere.

4. Engelmann's specialties, all of the group called succulents (from the Latin *succus,* meaning "sappy"), are among the moistest. That very character ensures their survival in regions of aridity. The rainy season brings them into bloom so that a collector in the desert has both moist plants and moist weather—sometimes deluges—to make work arduous.

5. For plant collecting methods (from gathering the specimen to placing it, properly mounted and

Some collectors traveled in sailing ships along the Pacific coast, some up rivers in canoes or flatboats, some in steamboats on the Missouri and the Mississippi. Given that all went well, their transportation problem, in theory at least, was somewhat minimized. But every form of boat-travel had its peculiar disadvantages for the collector, the most serious deriving from the fact that he was supercargo, even a nuisance, to those with whom for reasons of safety, of convenience, or even duty, he was obliged to travel. Collectors of *anything* need time for their task and are famous for their unwillingness to hurry! And the primary object of expeditions whether by land or sea was to get somewhere, and as quickly as possible. Even now, when to keep going is rarely vital, how many motorists are eager to apply the brakes when a passenger expresses an interest in the vegetation? Menzies and von Langsdorff had their individual problems with ship-commanders, even with subordinates. Nuttall, with a man to pole his flatboat, had to assist his assistant if they were to move at all. Baldwin eventually decided that to attempt to acquire plants from a steamboat was the worst method a collector could choose, and even a titled visitor of scientific bent had to join in pushing and pulling when the vessel went aground, as all Missouri River steamboats in that day perpetually did. One reads of some objectionable feature or features characterizing every mode of travel.

But, *except for the fact that such difficulties were a handicap to good work,* we read not a complaint from these men. They were animated by the scientific spirit—than which none is stronger nor more persistent—and expected little material recompense, that is evident. Some very few planned to publish their collections and, of these very few, two at least had such hopes shattered by unethical men who knew better. The reward of the fortunate was to have, perhaps, a genus and/or a few species named in

labelled, in the steel file of a modern herbarium) one may turn to an article, "The preparation of botanical specimens for the herbarium," written by Ivan M. Johnston and published by the Arnold Arboretum in 1939. E. J. Alexander, writing on the same subject in the *Journal of the New York Botanical Garden* in 1946, comments upon a collector's night work:

"Drying of herbarium specimens is one of the most tedious of the field-collector's many tasks. Each evening after the day's collecting, every one of the specimens must be gone over, carefully arranged in the position in which it is desired to dry, all surplus parts removed, each sheet numbered, and all placed in presses between newspaper sheets for the final drying ... After all the day's specimens are put away to dry, which usually takes until midnight or later, it is necessary to write up the day's notes, journal and records ... Thus it is often two or three o'clock in the morning before one gets to bed, to be up again at daylight. Four or five hours of sleep were the most we usually had in the field ..."

This collecting was being done in Latin America, with artificial heat and other modern facilities; I do not feel sorry for Mr. Alexander, and add quickly that there is no reason to suppose that Mr. Alexander felt sorry for himself!

Plant collecting includes, in addition to hard work, much that is pleasurable, and not the least of the pleasures is the return home. One is full of enthusiasm for all that one has seen and looks forward to a welcome from old friends. Drawing into the city of Boston after several months away, with car and trailer loaded with strange-looking objects of all sorts, and with everything, passengers included, covered with dust, we drew up alongside a brilliant "super" something, driven by a gentleman whom I thought I knew pretty well! "Greetings," said I, "I'm back!" My smiling but sadly uninformed "friend" replied: "Why! I did not know you had been away. *Where,* in heaven's name, have you *been?*"

his honor, or to appear in the literature as having made fine specimens or as having acquired much that was valuable. But two collectors of my period seem to have been treated unjustly: Berlandier, who had his work and even his honesty discredited in the eyes of the botanical world, and Burke whose inability to ship home his collections with the promptitude expected in London appears to have been attributed to negligence. More commonly, if a man's work was unsatisfactory, his patron merely mentioned the fact in letters to his "cronies" and let it go at that. The standards of all sponsors were high but their vision of the problems confronting their envoys was sometimes myopic. Having sent a man into the field, they more than once forgot his existence,[6] or so it seems, failed to reply to his questions and let months, even years, go by without a word; out of touch with the world he had left, the collector craved news on arrival at a destination where receipt of mail was possible. But it was not a period of stenographers and typewriters and one can but marvel at some of the lengthy, detailed, longhand epistles that did get off to the worker. Unfortunately, the content was not always limited to praise and encouragement but was often interlarded with advice and scoldings which reawaken memories of letters from my homefront when school reports had not been of the best. Finally, I append the sad fact that more than one of the collectors met death while pursuing their task: Collignon in the Lapérouse disaster, Douglas in a cattle-pit, Baldwin and Gairdner from the tuberculosis which was certainly not curable by overexertion, Gambel from typhoid. Some, the reader may be glad to know, got home safely and presumably—if they did not start off immediately on more travels—"lived happily ever after."

6. Dr. E. D. Merrill vouches for the truth of one story about a forgotten collector. He was Richard C. McGregor, ornithologist in the Philippine services for many years. He had been selected for the post by Dean C. Worcester, member of the original Philippine Commission and Secretary of the Interior.

Worcester chose the start of the rainy season, season too of the typhoons, to make an inspection trip of the Batan and Babuyan Islands between Luzon and Formosa. During the period of the rains, which, roughly speaking, last from June to November, there is, because of the high seas, virtually no communication between these small islands and the larger Luzon, and only rarely do the small native sailing boats venture out. On this particular trip, in 1904, Worcester took McGregor along to collect birds. After visits of a day or two at various islands they reached Calayan. Conditions for collecting birds appeared excellent and it was decided that McGregor should stay there, Worcester agreeing to pick him up at the end of the month. It was an excellent arrangement, as planned, but Worcester forgot his end of the bargain and returned to Manila.

McGregor's wife was in California and fortunately he had notified her that he was about to start on the trip. Mail communication between the United States and Manila was slow at this period, a matter of about six weeks. But after months had gone by without word of any sort, Mrs. McGregor became desperate and wrote Worcester to find out what had become of her husband. Reminded, he sent out a relief expedition and brought the ornithologist back to Manila. I am told that the report on Calayan birds fails to mention the episode!

Dr. Merrill tells me that McGregor often expressed regret that he had not taken along a botanical press —he would then have had something to do during his sojourn of four or five months; as it was, his ammunition gave out at the end of four weeks and he could not even shoot birds. On later assignments he always took a press along, as a precautionary measure. For those who may wonder how he subsisted, the natives supplied him as best they could.

Let no one suppose that this enumeration of difficulties is exaggerated, nor that it was the lot of botanical workers alone. All the early explorers and pioneers labored under individual difficulties and universal hazards not the least of which was the omnipresent Indian—not even mentioned as yet. The plant collectors seem to have had little trouble with the savages—they were mistaken for medicine men and considered worthy of respect. Many a one was shown the Indian's medicinal herbs and had their uses explained. The Indian may have suspected a plant collector at first sight but, having seen what he was about, usually sat down with him over a fire and enjoyed a pipe of tobacco.

Samuel Wood Geiser, better than anyone else, has pictured the physical and psychological effects which frontier life wrought upon men of science and upon their work. Geiser's locale was Texas. But I have found his thesis repeated in the nearly universal lack of understanding which confronted the field-workers when their lot fell among those to whom mere living, even survival, was necessarily uppermost. To such the gathering of "weeds" was an enigma. On the other hand it is evident that, without the assistance received from the traders and trappers (individuals and companies), and from the settlers living from hand to mouth, the collectors could not have pursued their work at all—and this generously given help was forthcoming without expectation of any *quid pro quo*. As always, of course, a man's personality had a bearing upon his relationship to others.

Many of the collectors bore the title of "surgeon-botanist," for plants and their medicinal uses then had, as they still have, a close association. Others were designated as "naturalists," an appellation which, in this day of specialization, does not perhaps connote the distinction that it once did. Personally, I believe that those worthy to bear the title derive a special joy out of life. Their numbers are, I fear, diminishing.[7] Donald Culross Peattie, with a gift of expression, has noted that "Theirs are the eyes that understand what we all see."

7. I knew one "naturalist" although his acknowledged field was botany. I traveled with him more than once and his eyes missed nothing, animal, vegetable or mineral. He had the ability which all teachers should possess—but often lack—of pointing out the plant characters which, once *observed*, remain with the pupil forever. He had "an eye" for crosses and, strange as it may seem to those who avoid swampy thickets, loved to disappear into such uncomfortable surroundings—where only the waving of branches betrayed his whereabouts—in search of the complex hybrids of alder and of willow which may be found therein. He was John George Jack, whose heart and soul lay in the development of the Arnold Arboretum and of whom Professor Sargent once said that he never went on any trip without bringing back some valuable plant or plants. His knowledge of our native and introduced species and of the many interlocking factors upon which the successful propagation of any plant depends, is, I must believe, extremely rare. He was self-depreciating to a fault and avoided when he could putting his knowledge into print—the pages of *Garden and Forest* indicate that he should have written more than he did.

To demonstrate his eternal vigilance when in the field I tell one story. On a collecting trip to Glacier Park, Marcus H. Dall, son of the zoologist for whom one of the Rocky Mountain sheep was named, had been asked by an authority on ferns to bring back the *Polystichum Andersonii*, a rare species (or so we were given to understand) recorded from the Park. Mr. Dall, whose life had been spent on lines other

Were I to include an epilogue, it would constitute a tribute to four of the great descriptive botanists, mentioned time without number in these pages, who devoted their days—and their days were many and active virtually to the end—to a task to which they had veritably consecrated themselves from their youth. Instead I shall let two scriptural verses and some dates serve as an epitome of their lives:

Man goeth forth unto his work and to his labour until the evening.

Psalm civ: 23.

William Jackson Hooker. Born 1785. Died 1865.

John Torrey. Born 1796. Died 1873.

George Engelmann. Born 1809. Died 1884.

Asa Gray. Born 1810. Died 1888.

With the ancient is wisdom; and in length of days understanding.

Job xii: 12.

The subject matter of my chapters is by no means confined to the botanical work of the collectors although the fact that all participated therein explains their inclusion. Although the total of plant specimens gathered in the course of the sixty years must have been very large, the acquisitions of individual field workers varied greatly, but always understandably, both in quantity and in quality. On the other hand, as a participant in the opening of the trans-Mississippi west, every collector included here left an enduring legacy of another sort—his individual contribution to the overall picture of regions long since altered and of a way of life forever gone.

Often, like others before me, I have transcribed from the early published narratives. When, more recently, manuscript journals and records have been edited by persons familiar with a particular terrain and its plant life I have followed their transcripts which I did not consider myself, by geographical or by botanical knowledge, qualified to improve by reference to the originals. I have turned to unpublished sources when these contained matters of interest. For the most part I quote verbatim from the records—the old narratives certainly provide flavor. To have paraphrased would have

than botany, had a certain pride—based perhaps on a desire to uphold his father's scientific reputation—in accomplishing this commission, and he imbued our entire party with the belief that its kudos was at stake. We left no moist places unexplored—the books said it frequented damp spots—but we were a disappointed group as we rode out the trail on our last morning. At this crucial moment Mr. Jack, far in the rear, stopped and summoned us back. Pointing casually with his treasured botanical pick to a barren, arid hillside, he said: "Dall, you passed your Polystichum!" It grew in a "wrong" place and looked like no other *Polystichum* that any of us, exclusive of Mr. Jack, had ever seen. Although I am convinced that the species was exterminated in that particular locality, Mr. Dall's "specimens" lived for a time, strangely enough, in the heat of Washington, D. C.

meant fewer quotation marks but having sought, sometimes long, for the origin of much that has appeared on the printed page, I am convinced that the direct quote has merits.

The book is not a botany in the sense of a descriptive manual and, with exceptions, I have not included identifications of the innumerable plants mentioned by the collectors. Along their routes these men must have seen or come into intimate contact with a large proportion of the thousands of plants now known to be natives of the vast and diversified area west of the Mississippi River, species which have been distinguished, described and named in shelves of publications. Many of the plant names used by the collectors, and indeed by taxonomists of the period, are of the past and to enter the complex and still fluctuating realm of synonomy would serve no useful purpose here—anyone interested in a particular record need only turn to a botanical institution where, given that it is possible to do so, the identity of the species and its present appellation can be supplied. For these and other reasons I have not included plant names in the main index.

My reading has indicated that the now popular practice of relegating bibliographical matter and footnotes to an appendix, or even to several appendices, is so inconvenient that it becomes a matter of will power to seek out these important references. Instead, with a single exception, I have included the pertinent literature at the end of each chapter, believing that although some repetition was necessary, the convenience was compensatory. And I have inserted many footnotes[8] where I believed they "stood a chance" of being read. To these I have assigned individuals who, for one reason or another, did not deserve mention in the text but added interest to the overall picture; matters too which cast sidelights on my subject, botanical or other; comments as well which to some may appear trivial, for even botany in my belief need not be associated with unrelieved solemnity.

That this book constitutes research, in the revered sense of the term, I do not aver. It has involved *the act of searching*. For the included matters are widely scattered, botanical matters no exception—a man's collections may have been described years after he was in the field and in publications where one would little expect to find them. Even living as I do in the vicinity of great libraries, sources were not all readily available, nor did all offer much, if any, novelty once obtained; what on occasion may appear repetitious may also be viewed as labor-saving.

Except in a general way I do not pass judgment upon the accomplishments of the collectors but record, when such exists, the evaluation placed upon their work by those better qualified to assess its importance, the botanists of the period or their successors.

The sixty years is treated by decades, each ten year period opening with a short introduction which is followed by chapters upon the collectors who were in the field dur-

8. It would be difficult to better for convenience and readability some of the publications of the Francis P. Harper firm of New York dating back to the 1890's where footnotes, often long, appear on the appropriate page.

ing that time. Some men went on more than one journey and, when these warranted separate treatment, each expedition has been accorded a chapter. For example, Nuttall is assigned three (two in the decade of 1810–1820, one in the decade of 1830–1840). To place each collector in chronological relationship to others of the same decade was simple from 1790 to 1830, from 1830 to 1840 was more difficult, and from 1840 to 1850 became a matter of personal decision because of the numbers of individuals simultaneously in the field. Chronological rearrangements are possible but a man's placement within a given ten years does not seem to me important.

The maps with their captions explain themselves. In the text I have attempted to follow the collectors' journeys in terms of present day states and counties, omitting in most instances the qualifying words "what is now" or "the present" such and such. Because many of the expeditions crossed unexplored territory (often without identifiable landmarks as one participant explained) it is my belief that, even to those best informed upon a particular region's geography, certain portions of some routes must remain a matter of personal judgment, sometimes of guesswork. A number of the journeys have been traced by qualified persons and I have reported their findings although, even in such helpful instances, one may encounter gaps where one would welcome the complete or perfect record.

I am appreciative of the privilege, extended over the years by its successive administrators, of using the exceptional facilities of the Arnold Arboretum where, from various members of the staff, I have received unfailing help and far above the call of duty! In particular I am indebted to the late Professor Oakes Ames, Dr. Ivan M. Johnston and Dr. Clarence E. Kobuski for reading large parts of my manuscript and to Dr. Elmer D. Merrill for reading it from cover to cover. Mrs. Lazella Schwarten has assisted in compiling the literature citations, in procuring, not once but many times, the books and papers needed for study and—assistance which ranks high among numerous other forms of service—in encouraging the at-times-discouraged.

Mr. Walter Muir Whitehill, Director of the Boston Athenæum, has been a friend indeed. To him and to various members of his staff—Miss Marjorie L. Crandall and Miss Margaret Hackett in particular—I render thanks. From that remarkable library have been sent me (parcel post, uninsured) volume after volume which I considered it the part of wisdom to return in person.[9] I refer to some of its treasures in my chapters.

Dr. Erwin Raisz's geographical knowledge and ability as a cartographer needs no eulogy from me. But I pay tribute to his patience and to his skill in translating into graphic form the maps which I had envisioned.

Those who cross the threshold into any scientific field become immediately aware of their great obligation to the authors and learned societies who contribute without re-

9. Upon inquiry whether, by chance, the Athenæum possessed a copy of the first census of the United States, I was told that there were several: would I prefer to see "Thomas Jefferson's personal copy?" I should add that this was *not* entrusted to the mail, nor out of the building!

striction to the exchange of knowledge. Having quoted, often at length, from their papers, documents, manuscript records and the like, I am conscious that what these individuals and societies so generously supply goes far towards enhancing any usefulness which this book may possess.

To them and to the authors and publishers who have granted permission to quote from sources which they hold in copyright I express sincere appreciation. Should errors or omissions, either in requests or in acknowledgments, have occurred they are unintentional and I ask pardon.

American Board of Commissioners for Foreign Missions for permission to transcribe the Asa Gray–Henry Harmon Spalding letters on deposit at the Houghton Library of Harvard University.

American Philosophical Society for permission to quote from the *Original journals of the Lewis and Clark expedition 1804-1806,* edited by Reuben Gold Thwaites, published by Dodd, Mead and Company, 1904-1905.

Appalachian Mountain Club for excerpts from "Early American mountaineers" by Allen H. Bent, published in *Appalachia,* 1913.

Appleton-Century-Crofts, Inc. for passages from *The trans-Mississippi west (1803-1853)* by Cardinal Goodwin, copyright, 1922, by D. Appleton and Company; and from *Audubon the naturalist* by Francis Hobart Herrick, copyright, 1938, by D. Appleton-Century Company, Inc.

Stanley Clisby Arthur for permission to quote from *Audubon, an intimate life of the American woodsman,* published, 1937, by J. S. W. Harmanson, copyright by the author.

The Boston Athenæum for permission to transcribe from a journal kept by Isaac Sprague which is among the manuscripts in their collection.

Willard E. Ireland, Provincial Librarian and Archivist of the Provincial Archives, Victoria, British Columbia, for permission to quote from "Menzies' journal of Vancouver's voyage, April to October, 1792," edited by C. F. Newcombe and with botanical, ethnological and biographical notes by John Forsyth, *Memoir* No. V, *Archives of British Columbia,* 1923; from articles published in the *British Columbia Historical Quarterly* in 1937, 1938, 1940 and 1945; and to transcribe the record "Plants from Mr McLeod Localities &c" preserved among the manuscripts of William Fraser Tolmie.

California Historical Society for permission to quote from "The Russians in California" by E. O. Essig, Adele Ogden and Clarence John DuFour, *Special Publication* No. 7, 1933; and from articles published in the *Quarterly of the California Historical Society* in 1923, 1924, 1929, 1933, 1939 and 1945.

The California State Library for citations from the State Guide ("American Guide Series") *California* published by Hastings House Publishers, Inc., 1939, and copyright, 1939, by The California State Library. (See also: Hastings House Publishers, Inc.)

The Caxton Printers, Ltd. for excerpts from *Henry Harmon Spalding, pioneer of Old Oregon* by Clifford Merrill Drury, copyright, 1936, by the publisher; and from *Elkanah and Mary Walker* by the same author, copyright, 1940, by the publisher.

The Champlain Society for permission to quote from *The letters of John McLoughlin from Fort Vancouver to the Governor and Committee,* edited by E. E. Rich, with an introduction by W. Kaye Lamb, published, 1941, 1943, 1944, by the Society.

William Collins Sons & Co., Ltd. for selections from *The art of botanical illustration* by Wilfrid Blunt, copyright, 1950, by the publisher.

Edwin Corle for a brief quotation from *Listen, Bright Angel,* published by Duell, Sloan & Pearce, copyright, 1946, by the author.

The Public Library, City and County of Denver for passages from *The call of the Columbia. Iron men and saints take the Oregon Trail* ("Overland to the Pacific Series," vol. 4), edited by Arthur Butler Hulbert, published, 1934.

Bernard De Voto for permission to quote from *The year of decision 1846,* published by Little, Brown and Company, copyright, 1934, by the author; and from *Across the wide Missouri,* published and copyright, 1947, by Houghton Mifflin Company.

R. R. Donnelley & Sons Company for passages from three volumes of "The Lakeside Classics," edited by Milo Milton Quaife and copyright by the publisher: *Echoes of the past about California by General John Bidwell,* copyright, 1928; *Narrative of Zenas Leonard written by himself,* copyright, 1934; *Kit Carson's autobiography,* copyright, 1935.

E. P. Dutton & Co., Inc. for excerpts from *Edward Lear, landscape painter and nonsense poet (1812-1888)* by Angus Davidson, published, 1939 (American edition).

Joseph Ewan for permission to quote from his miscellaneous papers and from *Rocky Mountain naturalists,* published by the University of Denver, copyright, 1950, by the author.

Samuel Wood Geiser for permission to quote from his miscellaneous papers; from his *Naturalists of the frontier,* published serially in and copyright by *Southwest Review,* 1929, 1930, 1932, 1933 and 1937; and from the two editions of *Naturalists of the frontier,* published by Southern Methodist University, copyright, 1937, 1948, by the author.

Harvard University Press for brief quotations from *Douglas of the fir* by Athelstan George Harvey, copyright, 1947, by the President and Fellows of Harvard College.

Hastings House Publishers Inc. for excerpts from the State Guides ("American Guide Series"): *California,* copyright, 1939, by Mabel R. Gillis, California State Librarian (See also: The California State Library); *Colorado,* copyright, 1941, by The Colorado State Planning Commission; *Iowa,* copyright, 1938, by The State Historical Society of Iowa; *Kansas,* copyright, 1939, by The State of Kansas Department of Education; *Minnesota,* copyright, 1938, by Executive Council, State of Minnesota;

Montana, copyright, 1939, by Department of Agriculture, Labor and Industry, State of Montana; *Nebraska,* copyright, 1939, by The Nebraska State Historical Society; *Utah,* copyright, 1941, by the Utah State Institute of Fine Arts.

Houghton Mifflin Company for excerpts from *History of the American frontier 1763-1898* by Frederic L. Paxson, copyright, 1924, by the author; and for a short sentence from *Audubon's America* by Donald Culross Peattie, copyright, 1940, by the author.

Hudson's Bay Company for quotations from articles in *The Beaver,* Outfit 268 (September, 1937), Outfit 273 (September, 1942), Outfit 277 (September, 1946).

The State Historical Society of Iowa for permission to quote from *Marches of the Dragoons in the Mississippi Valley* by Louis Pelzer, published, 1917, by the Society.

W. Boulton Kelly, Executor of the estate of Howard Atwood Kelly, M.D., for short passages from *Some American medical botanists commemorated in our botanical nomenclature* by Howard Atwood Kelly, M.D., published by D. Appleton and Company, 1914 and 1929, copyright, 1914, by the publisher.

Sir Edward Salisbury, Director, Royal Botanic Gardens, Kew, for permission to transcribe letters written to William Jackson Hooker by Joseph Burke and by Alexander Gordon which are preserved in the manuscript collections of that institution.

J. B. Lippincott Company for excerpts from *Gardener's tribute* by Richardson Wright, copyright, 1949, by the author.

Longmans, Green and Company for citations from *The early far west. A narrative outline 1540-1850* by William James Ghent, copyright, 1931, by the publisher.

The Macmillan Company for an excerpt from *Economic beginnings of the far west* by Katherine Coman, copyright, 1912, by the publisher; and for passages from *The long ships passing* by Walter Havighurst, copyright, 1942, by the publisher.

August C. Mahr for permission to quote from "The visit of the 'Rurik' to San Francisco in 1816," *Stanford University Publications, University Series, History, Economics, and Political Science* vol. II, no. 2, 1932.

Violet Markham (Mrs. James Carruthers, C. H.) and to Hodder & Stoughton for permission to quote a short phrase from *Paxton and the bachelor Duke,* published, 1935.

Missouri Historical Society for quotations from *A journey to the Rocky Mountains in the year 1839 by F. A. Wislizenus, M.D.,* translated from the German with a sketch of the author's life by Frederick A. Wislizenus, published, 1912, by the Society.

University of Missouri for brief citations from "Lewis and Clark: linguistic pioneers" by Elijah Harvey Criswell, *University of Missouri Studies* no. 15, 1940.

Historical Society of Montana for quotations from "Affairs at Fort Benton from 1831 to 1869, from Lieutenant Bradley's journal. Period from 1831 to 1869," edited by Arthur J. Craven and published in the Society's *Contributions* vol. 3, 1900.

Oswald Mueller for permission to quote from *Texas with particular reference to*

German immigration and the physical appearance of the country described through personal observation by Dr. Ferdinand Roemer, translated from the German edition of 1849, published by Standard Printing Company, copyright, 1935, by the translator.

John Murray (Publishers) Limited for a short passage from *Life and letters of Sir Joseph Dalton Hooker* by Leonard Huxley, published, 1918.

Allan Nevins for permission to quote from *Frémont. Pathmarker of the west,* published and copyright, 1930, by D. Appleton Company.

Grace Lee Nute for excerpts from " 'Botanizing' Minnesota in 1838," published in *Conservation Volunteer,* 1945.

Oregon Historical Society for permission to quote from articles published in the *Quarterly of the Oregon Historical Society* in 1904 and 1905 and in the *Oregon Historical Quarterly* in 1933, 1939, 1940 and 1941; and from "The correspondence and journals of Captain Nathaniel J. Wyeth 1831-6," edited by F. G. Young and published in *Sources of the history of Oregon* vol. 1, 1899.

Oregon State Board of Control for excerpts from the State Guide ("American Guide Series") *Oregon,* published by Binfords & Mort, copyright, 1940, by Oregon State Board of Control.

Oxford University Press for excerpts from the State Guide ("American Guide Series") *Wyoming,* copyright, 1941, by Lester C. Hunt, Secretary of State; and from *A history of horticulture in America to 1860* by U. P. Hedrick, copyright, 1950, by the publisher.

The Pacific Northwest Quarterly for permission to quote from articles published in the *Washington Historical Quarterly* in 1906, 1912, 1930, 1931, 1932 and 1934.

Professor H. Humbert, Muséum National d'Histoire Naturelle, Laboratoire de Phanérogamie, Paris, for permission to transcribe from "Plants de l'Amérique Septentrionale" by A. Trécul, preserved in the manuscript collection of the Muséum.

The Historical Society of Pennsylvania for permission to transcribe letters of William Gambel and of Thomas Nuttall preserved among their manuscripts.

Princeton University Press for passages from "Norwest John and the voyage of the Juno," *Tales of an old seaport,* Pt. II, edited by Wilfred Harold Munro, copyright, 1917, by the publisher; and for passages from *John Torrey. A story of North American botany* by Andrew Denny Rodgers 3rd, copyright, 1942, by the publisher.

Elizabeth Putnam for permission to quote from privately owned letters written by her grandfather George Putnam and by Thomas Nuttall.

G. P. Putnam & Sons for permission to quote from *Frémont and '49* by Frederick Samuel Dellenbaugh, copyright, 1914, by the author.

The Ryerson Press (formerly William Briggs) for citations from *The remarkable history of the Hudson's Bay Company* by George Bryce, published, 1910 (3rd edition).

Charles Scribner's Sons for passages from *History of the State of Idaho* by Cornelius J. Brosnan, copyright, 1935, by the publisher.

Simon and Schuster for excerpts from *Green Laurels. The lives and achievements of the great naturalists* by Donald Culross Peattie, copyright, 1936, by the author.

The Texas State Historical Association for permission to quote from "A trip to Texas in 1828" by José María Sánchez, edited and translated from the Spanish by Carlos E. Castañeda and published in *Southwestern Historical Quarterly* in 1926.

Charles C. Thomas for excerpts from *Thomas Say. Early American naturalist* by Harry Bischoff Weiss and Grace M. Ziegler, copyright, 1931, by the publisher.

The Viking Press Inc. for a short sentence from *East of Eden* by John Steinbeck, copyright, 1952, by the author.

Washington State Historical Society for excerpts from the State Guide ("American Guide Series") *Washington*, published by Binfords & Mort, copyright, 1941, by the Society.

Wistar Institute of Anatomy and Biology for permission to quote from the *Autobiography of Isaac Jones Wistar 1827-1905*, published, 1914, by the Institute.

PROLOGUE

". . . I always travelled with my botanical collecting book and reams of paper to preserve my plants . . . I have been enabled to collect in 20 years . . . a most valuable Herbarium, rich in new species, rare plants, and complete Monographs . . .

"During so many years of active and arduous explorations, I have met of course all kinds of adventures, fares and treatment. I have been welcomed under the hospitable roof of friends of knowledge or enterprise, else laughed at as a mad Botanist by scornful ignorance . . .

"Such a life of travels and exertions has its pleasures and its pains, its sudden delights and deep joys, mixt with dangers, trials, difficulties, and troubles. No one could better paint them than myself, who has experienced them all . . .

"Let the practical Botanist who wishes like myself to be a pioneer of science, and to increase the knowledge of plants, be fully prepared to meet dangers of all sorts in the wild groves and mountains of America. The mere fatigue of a pedestrian journey is nothing compared to the gloom of solitary forests, where not a human being is met for many miles, and if met he may be mistrusted; when the food and collections must be carried in your pocket or knapsack from day to day; when the fare is not only scanty but sometimes worse; when you must live on corn bread and salt pork, be burnt and steamed by a hot sun at noon, or drenched by rain, even with an umbrella in hand, as I always had.

"Musquitoes and flies will often annoy you or suck your blood if you stop or leave a hurried step. Gnats dance before the eyes and often fall in unless you shut them; insects creep on you and into your ears. Ants crawl on you whenever you rest on the ground, wasps will assail you like furies if you touch their nests. But ticks the worst of all are unavoidable wherever you go among bushes, and stick to you in crowds . . . other obnoxious insects will often beset you, or sorely hurt you. Hateful snakes are met, and if poisonous are very dangerous, some do not warn you off like the Rattlesnakes.

"You meet rough or muddy roads to vex you, and blind paths to perplex you, rocks, mountains, and steep ascents. You may often loose your way, and must always have a compass with you as I had. You may be lamed in climbing rocks for plants or break your limbs by a fall. You must cross and wade through brooks, creeks, rivers and swamps. In deep fords or in swift streams you may lose your footing and be drowned. You may be overtaken by a storm, the trees fall around you, the thunder roars and strikes before you. The winds may annoy you, the fire of heaven or of men sets fire to the grass or forest, and you may be surrounded by it, unless you fly for your life.

"You may travel over an unhealthy region or in a sickly season, you may fall sick on the road and become helpless, unless you be very careful, abstenious [sic] and temperate.

"Such are some of the dangers and troubles of a botanical excursion in the mountains and forests of North America. The sedentary botanists or those who travel in carriages or by steamboats, know little of them; those who merely herborize near a city or town, do not appreciate the courage of those who brave such dangers to reap the botanical wealth of the land, nor sufficiently value the collections thus made . . .

". . . I never was healthier and happier than when I encountered those dangers . . . I like the free range of the woods and glades, I hate the sight of fences like the Indians! . . .

"The pleasures of a botanical exploration fully compensate for these miseries and dangers, else no one would be a traveling Botanist, nor spend his time and money in vain. Many fair-days and fair-roads are met with, a clear sky or a bracing breeze inspires delight and ease, you breathe the pure air of the country, every rill and brook offers a drink of limpid fluid . . . What sound sleep at night after a long day's walk, what soothing naps at noon under a shaded tree near a purling brook!

"Every step taken into the fields, groves, and hills, appears to afford new enjoyments, Landscapes and Plants jointly meet in your sight. Here is an old acquaintance seen again; there a novelty, a rare plant, perhaps a new one! greets your view: you hasten to pluck it, examine it, admire, and put it in your book. Then you walk on thinking what it might be, or may be made by you hereafter. You feel an exultation, you are a conqueror, you have made a conquest over Nature, you are going to add a new object, or a page to science. This peaceful conquest has cost no tears . . .

"Such are the delightful feelings of a real botanist, who travels not for lucre nor paltry pay . . .

". . . When nothing new nor rare appears, you commune with your mind and your God in lofty thoughts or dreams of happiness. Every pure botanist is a good man, a happy man, and a religious man . . .

"To these botanical pleasures may be added the anticipation of the future names, places, uses, history, &c. of the plants you discover. For the winter or season of rest, are reserved the sedentary pleasures of comparing, studying, naming, describing and publishing. A time may come, when if all plants are well known, little will be left to be done, except seeking rare plants or occasional deviations and varieties; but a long while will elapse before this may take place, since so few of our plants are completely known as yet . . ."

CONSTANTINE SAMUEL RAFINESQUE,
New Flora of North America (1836).

1786

INTRODUCTORY CHAPTER

*THE FIRST PLANT FROM WEST OF THE MISSISSIPPI
RIVER TO BE DESCRIBED IN A MANNER ACCEPTABLE
TO THE BOTANISTS WAS RAISED FROM SEED GATH-
ERED NEAR MONTEREY, CALIFORNIA, BY COLLIGNON,
GARDENER OF THE LAPÉROUSE EXPEDITION*

IN 1786, about four years before my story opens, a French expedition arrived on the west coast of North America and spent ten days, September 14-24, at Monterey, California. It was commanded by Jean-François de Galaup, Comte de Lapérouse, an eminent navigator, and had been sent out by Louis XVI during the period of comparative quiet preceding the French Revolution and the Napoleonic Wars. It was at Monterey five years before the Spaniard Malaspina, with whose arrival my story begins, visited the same port from September 13 to 23, 1791. The Lapérouse expedition ended in disaster.

In 1797 Louis Marie Antoine Distouff, Baron Milet-Mureau, edited and published in Paris the *Voyage de La Pérouse autour du monde,* a four-volume work. It contained an account of the origins and purposes of the expedition and included such records as had been sent back to France before the ships and those aboard disappeared in the South Pacific.

In 1937 the Institut Français of Washington, D. C., published, under the title *Le voyage de Lapérouse sur les côtes de l'Alaska et de la Californie (1786),* a reprint of such portions of the Milet-Mureau work as relate to the visit of the expedition along our west coast. It was edited by Gilbert Chinard and because of its generous format and the editor's enlightening footnotes, is a more readable volume than the earlier work. Both editions contain a frontispiece portrait of Lapérouse. The engraving in Milet-Mureau, taken from a miniature, suggests a younger man than the one included by Chinard, which was taken from a portrait, but both portray a cheerful face with twinkling eyes, humorous mouth, and cheeks and chin tending towards the stoutish.

Milet-Mureau spells the name "La Pérouse." But Chinard tells us in his introduction, that, although many variants have appeared, "Lapérouse" was the spelling acceptable to the commander himself and to his family, and I follow the usage adopted on such excellent authority. Unless otherwise stated my facts are taken from Milet-Mureau.

Lapérouse was born at "Guô près d'Albi," in Languedoc, France, not far from Toulouse, on August 23, 1741, and was put at nine years of age in the college of the

3

Jesuits at Albi. In 1756 he was made "garde de la marine," which seems to have meant entrance into the Royal Navy. Between 1778 and 1783 he fought against the British in the American Revolution. Soon after, he was appointed by the king to head the expedition in which we are interested. This voyage, perhaps inspired by the reports of Cook and other English navigators, was largely scientific in purpose although the "Mémoire du Roi" of June 26, 1785, indicates that information was desired as to the location, extent, condition, power and purposes of the Spanish possessions on the west coast of North America. The possibility of establishing French settlements and of participating in the fur trade also needed investigation.

The Academy of Sciences issued full directions. It was hoped that, in botany, research would be directed towards such useful objectives as a knowledge of the plants which the inhabitants of different countries used for food, medicine or the arts; collections were to be made in regions where the temperature did not differ appreciably from that of France and whose plants, naturalized in the French climate, might some day serve to ornament its plantings.

The Medical Society also proposed matters needing solution and information, the section dealing with "Matière médicale," in particular, itemizing aspects of the flora deserving study: the flavor and the fragrance of roots, woods, barks, leaves, flowers, fruits and seeds, of the saps flowing from trees, and so on. In addition to the herbarium for the botanists, one was to be made for the Medical Society, for botany and medicine went hand in hand in that day.[1]

There was also a "Mémoire pour diriger le jardinier dans les travaux de son voyage autour du monde; par M. Thouin, premier jardinier du jardin des plantes." [2] This covered not only the introduction of plant material into the countries visited, but also the acquisition of all forms of plant life, advice on how such might best be transported, sown, and so on. Chinard, writing of Thouin's instructions, comments in his introduction that the list of plants, seeds and young trees to be distributed in the course of the

1. In the "Mémoire physiologique et pathologique sur les Américains," written by "M. Rollin, Docteur en Médecine, Chirugien-major de la frégate la Boussole," and included in Milet-Mureau (4:36-61) is a long description of a plant used by the natives of California as a cure for venereal diseases and known to the Spaniards under the name *gouvernante*. To this description Chinard (p. 128, *fn.* 7) appends a footnote suggesting that this might well be a description of the creosote bush, which, though not indigenous about Monterey, might have been brought there by Indians of a neighboring tribe—perhaps from the San Joaquin valley where it does grow. Although Rollin's description is not altogether accurate, this may be true, for the name *gouvernante*—like the name *gobernadora*—is often applied to the *Larrea tridentata* (or *Covillea tridentata*), the creosote bush. If so, Rollin's description must be a very early one of the species in California.

2. According to Bretschneider: "The starting point of this celebrated Garden [the Royal Garden at Paris], now generally known under the name of 'Jardin des Plantes,' dates from the year 1626, when Hérouard, First Physician to Louis XIII, obtained from the King letters patent authorizing the establishment of a Botanical Garden in the suburb of St. Victor, of which Hérouard became Superintendent . . ." Also we are told that André Thouin became "Chief Gardener at the Royal Gardens" at the age of seventeen and held the post until his death in 1823.

voyage was curious and may perhaps explain the presence of European plants in regions where one would least expect to find them.

The scientific books entrusted to Lapérouse are enumerated. Those concerned with botany included Linnaeus' *Genera* et *Systema plantarum, Philosophia botanica,* and *Supplementum,* Forster's *Genera plantarum,* Plumier's *Plantarum genera,* and others.

There were two frigates, *La Boussole,* commanded by Lapérouse, and *L'Astrolabe,* commanded by De Langle. The scientific staff ("Ingénieurs, Savans et Artistes") in the first numbered ten; among them, and to us important, was "Collignon . . . Jardinier-botaniste." In the second were seven "Savans et Artistes," including a "Docteur en médecine, botaniste" De la Martinière, a naturalist Dufresne, and two artists—one a "Dessinateur pour la botanique, Prévost, oncle," the other his nephew[3] of the same name.

Sailing from Brest, France, on August 1, 1785, the ships went round the Horn and then north, by way of Chile and the Hawaiian Islands, to Alaska, reaching about 60 degrees of north latitude. From there they moved southward.

The fog lifting, they were able on June 23, 1786, to see Mount Saint Elias.

On July 3 they were in a bay to which Lapérouse gave the name "Port des Français." Here, ten days later, a tragedy occurred when six officers and fifteen men were drowned while attempting to pass through the entrance in canoes. Their comrades erected—on what they called "L'île du Cénotaphe"—a monument to their memory, burying at its base a bottle containing a short account of the disaster written by De Lamanon, physician of the *Boussole.* It began with the lines: "A l'entrée du port ont péri vingt-un braves marins; qui que vous soyer, mêlez vos larmes aux notres." It ended: "Émus par le malheur, et non décourages, nous partons le 30 juillet pour continuer notre voyage."

They were now on their way to California, to Monterey, where they were to replenish their supplies of wood and water before sailing across the Pacific.

Although points southward along the coast are mentioned, fog was nearly constant and weather overcast and they made no landings until, on September 14, 1786, they anchored in the Bay of Monterey. Lapérouse devotes two chapters to Monterey, and I give the substance of some of his remarks. He reported the shores along the bay to be low and sandy, the sea rolling to the base of the dunes with a noise that was to be heard more than a league distant; land to the north and south of the bay was elevated and covered with trees inhabited by the most charming birds; the soil was of an indescribable fertility and all manner of vegetables grew with perfect success; the crops of corn *(maïs),* barley, wheat and peas, could only be compared with those of Chile— European growers could not picture such fertility; fruit trees were still rare despite the fact that the climate was entirely suitable.

---

3. Guillaumin states that the nephew always refused to paint anything but plants and that, in a letter written from Macao on January 9, 1787, De la Martinière complains of this fact.

The expedition enriched the gardens of the governor and of the missions with different seeds brought from Paris. A few indigenous trees are mentioned:

"... Les arbres des forêts sont le pin à pignon, le cyprès,[4] le chêne vert, et le plantane d'occident: ils sont clair-semés, et une pelouse, sur laquelle il est très-agréable de marcher, couvre la terre de ces forêts ..."

From the time of arrival the party had been busy acquiring water and wood; of the latter they were permitted to cut all that the ships' boats would hold. The botanists occupied every moment enlarging their collections of plants. But the season was unfavorable, for the heat of summer had dried up everything and seeds had been scattered over the ground. Those plants that the gardener Collignon was able to recognize were:

"... la grande absinthe, l'absinthe maritime, l'aurone mâle, le thé du Mexique, la verge d'or du Canada, l'aster (œil de christ), la mille-feuille, la morelle à fruit noir, la perce-pierre (criste-marine), et la menthe aquatique[5]. .."

Collignon left mementos of the visit, giving the missions some potatoes from Chile, perfectly preserved—Lapérouse considered this to be not one of the least of their gifts. He believed that the root would succeed admirably in the light soil about Monterey. De Langle, too, having seen the native women grinding corn, a slow and laborious task, made a present of a mill[6] to the missionaries; it was believed that no greater service could have been rendered for it would make it possible for four women to do the work of one hundred, and time would be left for them to spin yarn from the wool of the sheep and to make rough cloth. Up to now the missionaries, more occupied with celestial than temporal matters, had greatly neglected the commonest "arts." De

4. Lapérouse here mentions the Monterey cypress (*Cupressus macrocarpa* Hartweg) which is such a conspicuous feature of the Monterey region. Sargent states in *The Silva:* "Although its seeds appear to have reached England in 1838, *Cupressus macrocarpa* was first made known to botanists in 1847 by Karl Theodor Hartweg, who had found it at Cypress Point [Monterey] the previous autumn."

5. Alice Eastwood, familiar with the flora of the Monterey region, made the following determinations of the plants which Collignon was said to have recognized:
   "L'absinthe maritime, *Artemisia pycnocephala* DC.
   L'armoise, *Artemisia heterophylla* Nutt.
   Le grand absinthe, probably *Artemisia californica* Less.
   L'aurone male, perhaps *Artemisia ludoviciana* Nutt. (This was the identification of the white-downy artemisia related to *A. heterophylla* of Asa Gray in the *Synoptical Flora of North America.*)
   La verge d'or du Canada, *Solidago californica* Nutt. Goldenrod.
   L'aster (oeil de christ), probably *Aster chilensis* Nees.
   La morelle a fruit noir, *Solanum Douglasii* Dunal. Nightshade.
   La perce-pierre (criste-marine), *Salicornia ambigua* Michx. Samphire.
   La menthe aquatique, probably *Mentha canadensis* L.
   Le thé du Mexique, *Micromeria Chamissonis* (Benth.) Greene. Yerba Buena.
   La millefeuille, *Achillea Millefolium* L. Milfoil or Yarrow."

6. According to Chinard, De Langle possessed two mills which he had constructed aboard the *Astrolabe.* Several later visitors to Monterey refer to the fact that, despite its laborsaving advantages, this extremely useful device lay unused.

Langle, writing the "Ministre de la Marine" from Monterey on September 22, 1786, reported that, when wind was light, two men could turn such a mill.

On September 24, after a stay of ten days, the expedition sailed westward, exploring the eastern shore of Asia from China to Kamtchatka, from which point a member of the scientific staff was sent overland to France bearing Lapérouse's journal and other records to date. From Botany Bay, New South Wales, Australia, in February of 1788, Lapérouse sent still more records, important journals, etc. Guillaumin states that Collignon wrote a last letter from that port on February 15, 1788—its content is mentioned later in this chapter. No more was heard of the ships after their departure from Australia. Donald Culross Peattie has stated that the famed zoologist Charles Alexander Lesueur, who left the expedition at Botany Bay, was the sole survivor.

Search was made when the disappearance of the expedition had become a matter of certainty. But it was not until some forty years later that Jules Sébastien César Dumont d'Urville and his party, sent to make further investigations, discovered the remains of one ship lying encrusted with coral off the island of Vanikoro, one of the Solomon group; his search had been directed to that point by a sea captain named Dillon, who had found relics of the expedition. Dumont d'Urville's party recovered an anchor, a cannon and other metal objects from the sea-floor, and learned from the natives of the sinking of two French ships. The leader was finally convinced that he had found the scene of the disaster.[7] The story is told in Dumont d'Urville's *Voyage pittoresque autour du monde* . . . , published in Paris in 1834 and 1835.

Guillaumin recounts that even before the Lapérouse expedition had met with annihilation De Langle and eleven of the party had been killed by savages on the island of "Maouna" on December 11, 1787.

The largest part of the scientific collections were, of course, lost with the ships. One plant, however, bears testimony to the California visit. I quote from Milet-Mureau:

"Le voyage de La Pérouse n'a pu . . . procurer un grand nombre de nouveaux végétaux; mais l'on doit distinguer parmi ceux qui ont été envoyés par le jardinier Collignon, une charmante plante herbacée qui a fleuri et fructifié au jardin des plantes, en 1789. Jussieu, qui l'observa le premier, reconnut qu'elle constituait un genre nouveau, appartenant à la famille des nyctages, et il lui donna le nom d'*abronia*, mot grec qui signifie en français, *beau, délicat.* (Voyez *Gen. Plant.* p. 448.) Lamarck en a donné une assez bonne figure dans ses *Illustrationes generum*, planche 150.[8] Les graines de cette plante avaient été récoltées en California."

Jepson repeats the statement in slightly different words:

"Two of the packets of seeds . . . sent were gathered at Monterey. The seeds of one packet were sown in the Jardin des Plantes in Paris, and produced a number of 'beautiful herbaceous plants.' This species was first observed by Jussieu who recognized it

7. In the Musée de la Marine, Musée du Louvre, Paris, some of the relics were once to be seen and, presumably, are still there.

8. Actually t. 105.

as belonging to his order Nyctagines, and made for it a new genus which he called *Abronia,* a full diagnosis being given in his Genera, published in Paris, in 1789. Two years later Lamarck in his Illustrationes gave to the new plant its specific name *umbellata. Abronia umbellata . . .* is the earliest described Californian plant."

Abronia umbellata is, I believe, not only the first plant from California but the first plant from the entire trans-Mississippi west to have been described in a manner acceptable to the botanists. This sand verbena with rosy or purplish flowers is common along our west coast from San Diego to the mouth of the Columbia River.

Jepson reports also: "The seeds of the other packet came with herbarium specimens of a pine collected at Monterey by Collignon." Of this pine Sargent wrote:

"Collignon . . . in 1787 sent to the Muséum d'Histoire Naturelle in Paris a Pine cone believed to have been gathered at Monterey, and said to resemble that of the Maritime Pine of Europe, but with the large seeds of *Pinus cembra.* Twelve plants were raised from these seeds, and were described about 1812 by Loiseleur de Longchamps as *Pinus Californiana.* Judging by the locality where Collignon is supposed to have obtained his cone, it might well belong to the Monterey Pine; but the large seeds suggest another species, while the description of the plants raised from them might apply as well to several other trees as to this. It is necessary, therefore, to pass over what is perhaps the earliest name of this tree as well as the specific name, *adunca,* published in 1816, and supposed to refer to cultivated plants raised from Collignon's seeds . . ."

The name now accepted for the Monterey pine is *Pinus radiata,* bestowed by David Don (*Trans. Linn. Soc. London* 17: 441) in 1836, and, according to Alice Eastwood, to a plant from the collection of Thomas Coulter.

A little more light has recently been shed upon our dim knowledge of the man who collected the *Abronia* by the publication of an article—"Collignon, jardinier du voyage de La Pérouse"—written by André Guillaumin, already quoted here.

His name was Jean-Nicolas Collignon. He was born at Metz on April 19, 1762, and was the son of a gardener, Pierre-Nicolas Collignon, and of "Barbe Simonin, veuve George . . ."

Guillaumin quotes Thouin to the effect that Collignon was young, active, intelligent, with some theoretical and practical knowledge of gardening, and writing sufficiently well to keep a journal of his observations; he had some knowledge, also, of plants. Letters were exchanged, before the departure of the Lapérouse expedition and en route (although none are from California) between Collignon and Thouin. The contents of all the letters are not given by Guillaumin but some are stated to have concerned the distribution of plant material, others to have been about seeds to be procured at Madeira and at the Canary Islands. One, dated January 27, 1787, and from Guillaumin's context presumably written at Macao,[9] records that Collignon was ship-

9. The small Portuguese colony a few miles from Hong Kong and, before Hong Kong was occupied by the British in 1841, a very important port. Hong Kong was ceded to the British in 1842 by the Treaty of Nanking.

ping, in three boxes of "fer blanc" enclosed in a wooden case, seeds gathered at various points, among others at Monterey. In this shipment, states Guillaumin in a footnote, was "... le type d'un genre nouveau de Nyctaginacées ..."—in other words the notable *Abronia.*

The last letter cited by Guillaumin was dated from Botany Bay, February 15, 1788, and in it Collignon notifies Thouin that he was continuing to revise his journal but that, in conformity to Lapérouse's orders, was not shipping it to Europe.

During the years following the disappearance of the ships Collignon's mother and sister, acccording to Guillaumin, became impoverished and were obliged to turn to Thouin for assistance. He also records that "Pirolle," writing in his *Horticulteur fran-çais,* mentions that C. de Tschudy gave Collignon's name to a particular graft which that celebrated native of Metz ("ce célèbre messin") had conceived ("avait imaginé").

According to Jepson, "The genus Collignonia, which includes some six species of herbs and undershrubs of the temperate region of the Andes from New Granada to Peru, was founded by Endlicher in his Genera Plantarum (1836-40). The type of the genus is *Abronia parviflora,* Kunth."

Lapérouse's scientists were interested in more fields than botany when in California. Stillman mentions that during the ten days at Monterey, Lapérouse made "... a good survey of the Bay of Monterey, which was published with his narrative, and also a rough sketch of San Francisco Bay as furnished him by the missionaries. This sketch of San Francisco Bay is the earliest printed, and the southern shore is the only part that is even approximately correct ..."

Alden and Ifft comment that ornithology was not neglected and that Lapérouse "... writes that many birds were seen and collected. Three are beautifully figured in the Atlas, two ... easily identified ... the California quail and the California thrasher." They remark also that De la Martinière, in his "Memoir concerning certain insects" (published in Milet-Mureau) was "... obviously out of his field, but is highly excited over polyps, siphonophores, nudibranch molluscs, and many other inhabitants of his bucket of sea water ..."

Collignon's sand verbena—a lowly plant and perhaps unimportant by comparison with the gigantic trees, spectacular shrubs and showy flowers which were later to be discovered—was, nonetheless, the first from the trans-Mississippi west to be considered worthy of a place in the floras still to be written by the botanists. As the first, it assumes an importance all its own and, if only as a memento of Lapérouse's tragic story, deserves remembrance.

ALDEN, ROLAND HERRICK & IFFT, JOHN DEMPSTER. 1943. Early naturalists in the far west. *Occas. Papers Cal. Acad. Sci.* 20: 9-12.

BANCROFT, HUBERT HOWE. 1886. The works of ... 18 (California I. 1542-1800): 428-438, *fn.* 5. Map ["La Pérouse's map." 434].

BRETSCHNEIDER, EMIL. 1898. History of European botanical discoveries in China. London.

DUMONT D'URVILLE, JULES SÉBASTIEN CÉSAR. 1834-1835. Voyage pittoresque autour du monde. Résumé général des voyages de découvertes de Magellan, Tasman, Dampier, Anson, Byron, Wallis, Carteret, Bougainville, Cook, Lapérouse, G. Bligh, Vancouver, d'Entrecasteaux, Wilson, Baudin, Flinders, Krusenstern, Porter, Kotzebue, Freycinet, Bellinghausen, Basil Hall, Duperrey, Paulding, Beechey, Dumont d'Urville, Lutke, Dillon, Laplace, B. Morrell, etc. Publié sous la direction de M. Dumont d'Urville, capitaine de vaisseau . . . 2 vols. Paris.

EASTWOOD, ALICE. 1939. Early botanical explorers on the Pacific coast and the trees they found there. *Quart. Cal. Hist. Soc.* 18: 335, 336, 344, *fn.* 6.

GUILLAUMIN, ANDRÉ. 1948. Collignon, jardinier du voyage de La Pérouse. *Bull. Mus. Hist. Nat.* sér. 2, 20: 96-100.

JEPSON, WILLIS LINN. 1893. Early scientific expeditions to California. I. *Erythea* 1: 185-190.

JUSSIEU, ANTOINE LAURENT DE. 1789. Genera plantarum . . . Paris. Appendix. 448, 449.

LAMARCK, JEAN BAPTISTE ANTOINE PIERRE DE MONNET DE & POIRET, JEAN LOUIS MARIE. 1791. Tableau encyclopédique. Botanique. I. 469. no. 2140. Abronia umbellata. t. 105.

LAPÉROUSE, JEAN-FRANÇOIS DE GALAUP, COMTE DE. 1797. Voyage de La Pérouse autour du monde, publié conformément au décret du 22 Avril, 1791, et rédigé par M. L. A. Milet-Mureau . . . 4 vols. Atlas. Paris.

―――― 1937. Le voyage de Lapérouse sur les côtes de l'Alaska et de la Californie (1786). Avec une introduction et des notes par Gilbert Chinard. Baltimore. (Institut Français de Washington. Historical Documents, Cahier 10.)

PEATTIE, DONALD CULROSS. 1936. Green laurels. The lives and achievements of the great naturalists. New York. 157.

SARGENT, CHARLES SPRAGUE. 1896, 1897. The Silva of North America. Boston. New York. 10: 104 (1896); 11: 105, *fn.* 1 (1897).

STILLMAN, JACOB DAVIS BABCOCK. 1869. Footprints of early California discoverers. *Overland Monthly* 2: 257.

1790-1800

INTRODUCTION TO DECADE OF

1790 – 1800

B Y 1790—the year of the first census of the United States—settlers, in a slow and contiguous advance, had made their difficult way westward across the Alleghenies and, in regions adjacent to the Ohio River, were erecting scattered homes and small colonies in the valley of the Mississippi River. This progress westward made it possible for the French botanist André Michaux to reach the eastern environs of the Mississippi River in 1795 and to collect plants from the mouth of the Ohio northward to opposite the thirty-year-old settlement of St. Louis. That he crossed to the far side of that river is not proven by the records.

The only botanical work accomplished in the trans-Mississippi west between 1790 and 1800 was on the Pacific coast.

Spain, then in possession of California and with the outpost of Nootka farther north, sent out in 1789 a great scientific expedition under the command of Alessandro Malaspina. In 1791 it reached California and for ten days in September its ships were anchored in the harbor of Monterey. Serving with Malaspina were two botanists, Thaddeus Haenke and Louis Née. Née had remained in Mexico when the expedition went north but, nonetheless, from branches brought him by members of the party, he was the first to describe two California trees in language acceptable to science. Haenke, during the Monterey visit, made a collection of dried plants—the first Californian one, it is said, to reach Europe. When the Malaspina expedition returned to Spain in 1794, Haenke stayed in Mexico and his collections were described by European, although not by Spanish, botanists.

From 1790 to 1795 the British navigator George Vancouver was making round-the-world surveys. In the years from 1792 to 1794, while off the west coast of North America, he conducted these surveys in the coastal waters of lands later to become the property of the United States—at various points between Juan de Fuca Straits and Puget Sound (to the north) and San Diego, California (to the south). Accompanying Vancouver as surgeon-naturalist was a Scottish gentleman, Archibald Menzies. Although Menzies could never penetrate far inland, all that he beheld in the plant-world was new, and it was because of the wonders which he reported on his return that British botanists and horticulturists in later years were to send a succession of collectors to our Pacific coast. Menzies was the first to gather plants along the shores of what is now the state of Washington and the second to do so on the California coast. He did not describe his own collections, and it has been said that some of his plants are still unstudied. A few were included by Frederick Pursh in his *Flora americae septentrionalis* (1814), the first such work to include any considerable number of plants from west of the Mississippi River.

13

Haenke and Menzies were the only botanists to collect plants west of the Mississippi River in the decade of 1790–1800.

Two events associated with this decade were to have an important influence upon the historical future of the west coast of North America and, although indirectly, upon the trend of botanical exploration:

First. In the year 1793 Alexander Mackenzie, traveling overland across Canada, reached the Pacific Ocean at the mouth of the Bella Coula River in present British Columbia. This was the first crossing of the North American continent north of what is now Mexico; and the feat pointed the way for the westward advance of the British fur trade to whose posts and officers innumerable explorers, including botanists and Americans as well as foreigners, were to be indebted for assistance over a long period of years.

Second. The Columbia River region was in later years to become a famous collecting-ground for plants—we shall hear again and again of men who worked along that waterway. The year 1792 was an important one in its history for, in the month of May, a Boston sea captain, Robert Gray, sailed his merchant vessel the *Columbia* through the river's perilous entrance. Gray's exploit, corroborated by his records, was to fortify the claims to priority of discovery[1] advanced by the United States at the time of the boundary settlement of 1846 with Great Britain.

1. In October and November of 1792 Vancouver delegated to his Lieutenant William Broughton, commanding the *Chatham*, the task of surveying this river, for long called "the Oregon," which he did in small boats, upwards for one hundred miles, or to a position which he named Point Vancouver. A number of place-names along the lower Columbia River still bear testimony to Broughton's visit.

The derivation of the name "Oregon" has been a matter of considerable discussion. H. H. Bancroft's *History of Oregon* devotes some eight pages to the subject. The name's first appearance in print has usually been credited to Jonathan Carver's *Travels through the interior parts of North America,* published in London in 1778, where it appears in the two spellings "Oregon" and "Oregan." But W. J. Ghent in *The road to Oregon* (1929) mentions that Major Robert Roberts cited the appellation in 1765, 1766 and 1772, in the two spellings "Ouragan" and "Ourigan." The name "Oregon" certainly derived its greatest publicity in William Cullen Bryant's "Thanatopsis," which first appeared in the *North American Review* in 1817, and, in somewhat altered form, in a small volume of Bryant's poems published in Cambridge, Massachusetts, in 1821:

> Where rolls the Oregon, and hears no sound,
> Save his own dashings.

Only in 1821, according to an unsigned article in the *North American Review* (n.s., 13:23, 1823), did congressional opinion in Washington concede that the "Oregon River" and the Columbia River were one and the same.

CHAPTER I

THE BOTANISTS HAENKE AND NÉE, ATTACHED TO THE MALASPINA EXPEDITION, MAKE NOTEWORTHY CONTRIBUTIONS TO THE KNOWLEDGE OF CALIFORNIAN PLANTS

SCIENTIFIC botany had its inception west of the Mississippi River when the Spanish expedition commanded by Alessandro, or Alejandro, Malaspina visited Monterey, California, in the year 1791. Two botanists[1] were of the party: Louis, or Luis, Née, who did not get to California but who, nonetheless, was able to describe two trees from that region; and Thaddeus Haenke, said to have made the first herbarium of Californian plants to reach Europe.

It was in 1885, or ninety-one years after the return of this expedition to Spain, before any of the journal kept by its commander was published.[2] The volume is rare in this country.

Although it is manifestly greatly abridged, this journal was issued as a quarto volume of some seven hundred pages, with plates and an excellent map showing the route of the expedition. The title-page reads: *Viaje político-científico alrededor del mundo por las corbetas Descubierta y Atrevida al mando de los capitanes de navío D. Alejandro Malaspina y Don José de Bustamente y Guerra desde 1789 á 1794. Publicado con una introducción por Don Pedro de Novo y Colson teniente de navío . . . Madrid*

1. In his paper, "Some fictitious botanists," John H. Barnhart mentions fourteen men whose biographies were included in *Appletons' Cyclopaedia of American biography:* One was Frederic August Lotter, who ". . . in 1789 was attached as botanist to the expedition that was sent by the Spanish government around the world under command of Capt. Malaspina . . ." The biography tells just where and when he was with the expedition and ends with a list of seven impressive titles, one of which was "Icones plantarum Americanarum rarium," and finally ". . . several less important works." States Barnhart: "Wholly fictitious. The botanists of the Malaspina expedition were Thaddäus Haenke and Luis Née."

The same cyclopaedia included an Isidore Charles Sigismond Née, pronounced, we are told, "(nay)," who traveled widely in South America and elsewhere and also published four books with resounding titles. Writes Barnhart: "The name is evidently in part compounded from those of Carl Sigismund Kunth . . . and Luis Née, both well-known botanists; but the rest of the sketch is wholly fictitious."

2. Safford states: "For a long time . . . [Malaspina's] history disappeared from view and investigations concerning it were made by the Società Geographica Italiana, the president of which, in his address of 1868 (Bolletino, 1868, pp. 73-74), announces its discovery in the archives of the hydrographic office at Madrid, and states that it is written in a great part in Malaspina's own hand. It is quite voluminous. A part of the narrative is said to have been published in the Anales Hidrograficos in 1871, but no such publication can be found in the official list . . ."

...*1885*.[3] The cover-title differs slightly in wording; both forms have been cited as the title. The authorship has sometimes been attributed to Novo y Colson, but it is evident that he merely wrote the introduction and was responsible for its publication.

Following the introduction is a general description of the voyage divided into three books. The first book (six chapters) takes the expedition as far as the Gulf of Panama. The second book (three chapters) covers the trip to its most northern point in Alaska and its return to Acapulco, Mexico; of these the third chapter (pp. 194-199) tells of the trip from Nootka off Vancouver Island southward along the coast to Monterey, California, where there was a halt from September 13 to 23, 1791, and of the departure southward. The third book (nine chapters) carries the expedition across the South Pacific to the Philippine Islands and elsewhere before its return to Spain. Included in the last chapter of this book (or perhaps merely as additions thereto) are a series of articles upon specific aspects of the voyage; of these the "Descripción física y costumbres de la California" (pp. 437-441) has some botanical interest.

References to the Malaspina expedition are found in the literature but the most valuable to our story are: Jepson's (1899) account of the California visit and (1929) his record of Haenke's discovery of the coast redwood; Safford's (1905) summary of matters of general interest; and Galbraith's (1924) overall picture of the voyage and her translation of such parts of the *Viaje* as relate to the California visit.

The voyage was planned by Carlos III of Spain. He died in 1788 but Carlos IV carried out the plans of his predecessor.

Galbraith tells that "Alexandro Malaspina was born in Italy, November 5, 1754, of noble lineage . . . His youth was spent in Italy; but foreign service then offering greater inducements to younger sons of noble families, he enlisted in the marine guard of Cadiz in 1774, being then twenty years of age, and henceforth called himself an adopted son of Spain. He rose rapidly in rank and went on many voyages in the Atlantic and China seas. In 1784 . . . he made a tour of the world . . . a good preparation for the scientific expedition to whose command he was soon to be assigned . . . Several years were spent in preparation for the voyage [of 1789-1794] . . . two vessels were built especially for the expedition, the finest instruments purchased, and everything provided that was conducive to the welfare of the officers and crews . . ."

We read of the fine impression made by the expedition in Archibald Menzies' journal under date of October 3, 1792. Menzies was then at Nootka and Malaspina had been there from August 13 to 28, 1791.

"Sr Malaspina who has been some time out on Discoveries with two Vessels under his Command they all agree is a very able Navigator & fitted out in the most ample manner for Discoveries with Astronomers Naturalists Draughtsmen &c.—He has al-

3. I have followed the Library of Congress copy. A copy in the Arnold Arboretum, also dated 1885, is a second edition ("Segunda Edición") according to the title-pages, but appears to be merely a second *printing*.

ready examined the Shores of South America & this coast & is now surveying the Philipine Islands.—He is to return to Peru & Chili round Cape Horn, to publish the result of his enquiries: So that the Spaniards mean to shake off now entirely that odium of indolence & secrecy with which they have been long accused."

The two ships were the corvettes *Descubierta* and *Atrevida*. The personnel of each is given in the *Viaje* (pp. 50, 51). Don Alessandro Malaspina commanded the first-named which carried 102 officers, scientists, crew and so on; of these Don Antonio Pineda was in charge of the work in natural history. Don José Bustamente commanded the *Atrevida* also carrying 102 persons. One of these was Louis (or Luis) Née, "botanico," who had been born a Frenchman but who had become a Spanish citizen; also José Robredo, a subaltern, and Manuel Ezquerra, a paymaster *(contador)*.

The ships left Cadiz,[4] Spain, on July 30, 1789 (*Viaje,* p. 53). They then skirted the west coast of North Africa, crossed the Atlantic to Montevideo, Uruguay, worked along the coast of Argentina and Patagonia, rounded the Horn and went north along the west coast of South America.

At Valparaiso, Chile, there was an important addition to the party in the person of the botanist "Tadeo Heenke" (Thaddeo, Thaddeus or Thaddaeus Haenke). His story appears in the *Viaje* (p. 86), also in the preface of Presl's *Reliquiae Haenkeanae,* written by Count de Sternberg. Haenke was born at Kreibitz, Bohemia, on October 5, 1761. He received the degree of Doctor of Philosophy at the University of Prague; he studied medicine but his chief interest, botany, was stimulated by residence with J. G. Mikan, professor in that field. He also studied under Jacquin in Vienna and it was on his recommendation and at the command of the Emperor Joseph II that he went as botanist on the Malaspina voyage. He was then twenty-three years of age. Arriving at Cadiz the day the ships sailed, but too late to get aboard, Haenke took the first sailing possible to Montevideo where, shipwrecked at the entrance to the Rio de la Plata, he is said to have lost all his possessions except his copy of Linnaeus and his collecting equipment. He had again missed the ships, but, undeterred, set out on foot and crossed "las Pampas ó llanuras de Buenos Aires y las cordilleras del Chile [*Viaje,* p. 86]," collecting on the way 1,400 plants, the greater part "nuevas ó no bien caraterizadas." On arrival at Santiago, Chile, he met by chance Malaspina and some of his officers who had gone to the capital city from Valparaiso, reported for duty and was assigned to the *Descubierta.*

The ships continued northward and on April 27, 1791, reached Acapulco, Mexico, where orders were received to proceed to Alaska and search for the northeast passage, reported by Maldonado and others. Although Malaspina doubted the existence

4. Bancroft records one interesting fact: "Malaspina seems entitled to the honor of having brought to California the first American who ever visited the country, and he came to remain, his burial being recorded on the mission register under date of September 13th, and name of John Groem, probably Graham, son of John and Catharine Groem, Presbyterians of Boston. He had shipped as gunner at Cádiz."

of such an interoceanic route, the search was carried out but no opening was discovered along the entire coast. The map of the *Viaje* indicates that the northernmost point reached was slightly south of the 60th parallel, in Prince William Sound, or "Ba[ja] Behring." From there they turned southward to Nootka, which their observations proved was situated on an island rather than on the mainland. Nootka was then under Spanish rule. They left this port in late August, 1791, bound for California and Bancroft reports that Malaspina ". . . had no observations of interest or importance until he reached California."

I shall quote from Galbraith's translation, following it in the *Viaje* and inserting the pages from which her citations are taken:

". . . On September 1 [*Viaje*, p. 194] they marked the entrance to the strait of Juan de Fuca . . . At the bay of Hecate, Malaspina notes that there were signs of a river but he continues onward . . . weather was mild and equable as far as Cape Mendocino . . . From Cape Mendocino to Pt. Reyes the passage was quickly made . . . unfortunately they encountered fogs . . . Still they sighted Pt. Reyes at intervals and located the entrance to San Francisco bay; but the fog thickening, they had to stand off shore as they made their way southward to Monterey bay. They missed the entrance because of the fog, and passed Pt. Pinos . . . they managed at eight o'clock on the morning of September 13 [*Viaje*, p. 195] to cast anchor in the bay of Monterey, half a mile from the presidio . . ."

They remained at Monterey ten days or until September 23.

Galbraith turns to the "Descripción física y costumbres de la California." I quote what is said of the vegetation of the Monterey region:

"Imagine what must have been the agreeable surprise to all of us at seeing in the month of September, some leagues around Monterey, common vegetation blossoming so fresh and abundant that the number of plants restored to Nature by this singular fertility was not less than one hundred. The fields were adorned with woods, now open, now dense, with pine, alder . . . oaks, and live oaks, and along with these on the higher summits the red pine, a tree much taller than the rest;[5] and various medicinal plants, some poisonous, and others useful or agreeable, making the number more than two hundred and fifty that Don Tadeo Haenke recognized. 'The soil consequently, having been fertilized with double strength, presents a land,' (says Haenke) 'black and rich, from one to two feet deep, made from myriads of decayed plants, and superimposed upon a shady, ash-colored clay, which is generally to be found in all the vicinity, except close to the sea . . .' [*Viaje*, pp. 437, 438]."

Two footnotes in the *Viaje* refer to the above passage: the first (p. 437) tells that Haenke had seen both ripe seed and flower buds on the laurel ("el laurel"), presumably the California laurel or *Umbellularia californica* Nuttall; the second (p. 438)

5. It has been suggested that the phrase ". . . the red pine, a tree much taller than the rest . . ." may have referred to the coast redwood.

distinguishes between the nonpoisonous and poisonous medicinal plants seen by Haenke:

"Entre les plantas medicinales cuenta este botánico, la malva, el *tropcolum-majus, la arthemisia-absintium, la arthemesia-dracumculus, la arthemesia-maritima; scorcornera-dentata, solidago-cricetorum, solidago-cinerea, gentiana-centaurium, salviafrutesceus, sambucus-racemosa, verónica anagallis, verbena carolina, rhannus maritimus, sichorium, virgetum, melissa prostrata, oxalis prostrata, tumarea achillea, millefolium;* etc.; y entre las venenosas *rhusradicans, rhustóxico, dendron, la cicutasiides; el hippomane discolor.*"

We are told of the agricultural work at the missions, and of the vegetation:

". . . it is clear that these continual fogs cannot be favorable to the cultivation of all kinds of seeds . . . many kinds of grains, and almost all cultivated fruits, need more or less of the sun's heat so that their seed can mature and ripen. This unfavorableness causes . . . a notable difference between the products of our missions on the shore and those situated in the interior of the country. Notwithstanding it is noticed that the maize . . . appears less sensitive, and particularly wheat, to the lack of the sun [*Viaje,* p. 438].

"We must . . . except from the above mentioned disadvantages, the missions fronting the Santa Barbara channel, where, maybe the islands that form it receive and check the fogs, or maybe its direction east and west, does not afford an opportunity for the northwesters to act with the same force as in other places. Certainly they obtain more natural and permanent heat, and with it, harvests more certain and abundant, as the mission fathers have repeatedly assured us [*Viaje,* p. 438].

"The kinds of grain principally sown and harvested in these missions of New California are wheat, barley, maize, beans, chick-peas, lentils, peas and vetch. Also many of them have fruits, those of San Buenventura and San Diego have grape vines, and in that of Santa Clara they raise particularly fine and abundant pears, peaches, and plums; the abundance of water and the beautiful clear and temperate climate of the country contributing to produce them. Likewise they have the fruit-stones or seeds of fruit trees, that Count de la Pérouse and Viscount de la Langle left when they were in Monterey in September, 1786, extending their generosity by leaving likewise different grains of the best quality which at the present time have greatly increased in the missions of San Carlos [*Viaje,* p. 438]."

". . . Monterey was very favorably situated, both for recreation and research in zoological[6] and botanical fields. 'Certainly,' says Malaspina, 'it is difficult to find another place better adapted to either purpose . . .'

"The vegetation at Monterey was a great surprise to the botanist, Haenke, especially along the borders of the Carmelo near its mouth, where there was found a great

6. Alden and Ifft comment: "The justness of these observations has since been strikingly affirmed by the establishment here of the Hopkins Marine Station for biological research."

variety of plants, whose seeds, he inferred, had been brought from the interior by the winter rains [*Viaje,* p. 196]."

What were the botanical results of the visit to California?

Louis Née remained in Mexico[7] when Malaspina left Acapulco in 1791 on the trip north. Although he did not see the living plants, Née named and wrote the first scientific descriptions of any California trees and, indeed of any trees from west of the Mississippi River. These were two oaks: *Quercus agrifolia,* the coast live oak, and *Quercus lobata,* the valley oak.

Of the first Née wrote: "I cannot give the height of this tree, of which I have only seen branches collected at Monterey and Nootka, by the marine officer, Don Joseph Robredo, and Don Manuel Esquerra, paymaster of the corvette Atrevida . . ." And of the second: "Of this species I have only seen branches brought from Monterey by Sres Robredo and Esquerra . . ." The descriptions appeared first in the *Anales de ciencias naturales* (Madrid) in 1801, and were reprinted in translation by Koenig and Sims in the *Annals of Botany* (London) in 1806.

Jepson (1899) pointed out that ". . . although Née had a decided taste for collecting, no California plants are ever attributed to him save Berberis pinnata, published in 1803 by Lagasca in 'Elenchus Plantarum' of the Royal Gardens at Madrid, but this was doubtless merely communicated to that author by Née."

Of Haenke's accomplishment in general, and at Monterey in particular, Jepson had this to say:

"The first botanist . . . to make an herbarium of Californian plants which reached Europe was Thaddeus Haenke . . . His collections from South America, from Nootka, Port Mulgrave, Monterey, San Blas, Acapulco, testify his scientific ardor . . .

"These, as most of his plants, were published by C. B. Presl, in the Reliquiae Haenkeanae, and consisted chiefly of grasses, rushes and sedges, and of the following exogens characteristic of the Californian summer and autumn hills and low plains: Datisca glomerata, Zauschneria Californica, and Frankenia grandiflora. Of exogens, either very few were collected or, as seems more likely, they shared the fate of certain South America bundles which were lost."

The question of who first, of our plant collectors, collected the coast redwood is of interest and it would seem to have been Haenke. Although the locality cited is "California," he might well have gotten to Santa Cruz from Monterey in 1791, where Menzies' collection, the second, was made in 1794. In an article, "Sequoia sempervirens in Granada," published in 1929, Jepson tells where he discovered the record:

"It has always been a thought with me that the Redwood must have been first collected (botanically speaking) by Thaddeus Haenke of the Malaspina Expedition . . .

---

7. Although Brewer, Sargent and others have stated that both Haenke and Née visited California this does not seem to have been the case. Bancroft published a "full list of the officers made at Monterey," and Née's name is marked with an asterisk indicating that he had stayed in Mexico.

This round the world expedition touched at Monterey in 1791 and Haenke botanized there in the dry season (September). Haenke's California material was published, after long delay, by C. B. Presl of Prague in the Reliquiae Haenkeanae, but no mention is made of any Redwood under any name, or of any conifer that could be construed as our species. A copy of Presl's great folio is in my library, but I do not possess his Epimeliae Botanicae, which lists some of Haenke's things and was published as late as 1849. But the last time I was at Kew I found on page 237 of the Epimeliae a record showing that Haenke collected the Redwood while in California. In all probability he collected seed as well as herbarium specimens. Since this was a Spanish expedition seed may have gotten back to Spain. Some of the ship's officers, it is known, collected seed of our native trees."

Since the "Epimeliae botanicae" may not, to some, be readily available, I quote the reference (p. 237, of reprint):

"Sequoia religiosa.—Taxodii spec. Douglas in Hook. compan. bot. mag. II. 150—Abies religiosa Hook. et Arn. in Beech. voy. 160 (nec Humb.).—Taxodium sempervirens Hook. et Arn. in Beech. voy. 392 (nec Lambert). Hook. ic. plant. t. 379.—Sequoia gigantea Endl. synops. conif. 198.—Habitat in California (Haenke).—Folia latiora et breviora quam in genuina Abiete religiosa habitu simillima. Spcimina [sic] haenkeana sterilia."

From Monterey the ships moved south on September 23, 1791, and were back at Acapulco in October. Here Pineda and Née rejoined the party, reporting favorably upon their own explorations. Haenke seized the opportunity, while the ships were being reconditioned, to make a trip to Mexico City and back! In December they started across the South Pacific, visited Guam in February, 1792, and reached the Philippines March 4, remaining until November. From there, by way of New Zealand, Australia (New South Wales), and other points they returned to the west coast of South America, reaching Peru the end of July, 1793. Here they learned of the rupture between Spain and the French Republic and next of Spain's declaration of war against France. "At this point the party broke up, most of the scientists proceeding overland to Buenos Ayres . . . the vessels sailed separately in October." At Montevideo the ships were conditioned for war, then left for Spain on June 21, 1794. They sighted "Cape St. Vincent," southernmost point of Portugal, on July 18 ". . . without having seen a sign of a hostile vessel [Viaje, p. 337]."

"Malaspina closes his narrative with the congratulatory remarks upon the good fortune attending their long, eventful voyage, '. . . during the space of five years and two months the corvets were so fortunate as to lose but ten persons on board or in the hospitals . . .' [Viaje, p. 338]."

The rest of the story is a sad one. Robert Greenhow wrote in 1845:

"The journals of Malaspina's expedition have never been published. A sketch of his voyage along the north-west coasts of North America is given in the Introduction to the Journal of Galiano and Valdes, in which the highest, and, in some places, the most extravagant praise is bestowed on the officers engaged in it. Yet—will it be believed?— the *name of Malaspina does not appear there or in any other part of the book.* The unfortunate commander, having given some offence to Godoy, better known as the Prince of the Peace, who then ruled Spain without restriction, was, on his return to Europe in 1794, confined in a dungeon at Corunna, and there kept a prisoner until 1802, when he was liberated, after the peace of Amiens, at the express desire of Napoleon. The name of one who had thus sinned could not be allowed to appear on the pages of a work published officially, by the Spanish government, for the purpose of vindicating the claims of its navigators."

Much the same story is told by Galbraith. She notes that Malaspina was freed on condition that he never return to Spain. He was taken to Genoa and from there ". . . retired to his old home in Lunigiana. His health was enfeebled by his long imprisonment, and he died four years later, April 9, 1809, at the age of fifty-five . . . the rancor of the government . . . was extended to all those who went on the expedition, and none of the reports or papers written by the officers and scientists were ever published by the government."

She quotes a letter, written in 1795 and included in the *Viaje,* which ". . . calculates that the contemplated history of the voyage 'will fill seven volumes . . . contain 70 maps, 70 drawings and cuts, and will cost two million reals.' "

When, on news of war, the Malaspina expedition was ordered home from Peru in July, 1793, Haenke remained in South America. He had intended, apparently, after crossing that continent to Buenos Ayres and collecting on the way, to meet the expedition at Montevideo. But the boats again left without him and, in this instance he was perhaps fortunate. I shall quote briefly what Safford has to tell of his later life and of what happened to his plant collections:

"In 1796 he established himself at Cochabamba . . . Bolivia. Here he established a botanical garden, gave medical assistance to his neighbors, and occupied himself with the study of natural science, making repeated excursions throughout the territory of what is now Chile, Peru, and Bolivia . . . Haenke looked forward to returning some day to Europe, but he was accidentally poisoned and died at Cochabamba in 1817. Only a small proportion of his herbarium reached Europe, the greatest part having been sent by the authorities to Lima, where it was lost. About 9,000 plants collected on the Malaspina expedition were sent, according to his wish, to the National Museum of Bohemia, at Prague. Others found their way to the Royal Garden at Madrid, with those of Née. Duplicates of these were sent to the University of Prague and the Musée Palatin at Vienna, and about 700 species to the Royal Herbarium at Munich. It was upon the collections at Prague and the notes accompanying them that the Reliquiae Haenkeanae of Presl was based."

Dr. E. D. Merrill tells me that, since Safford published the last part of his paper, nearly all of the Haenke and Née collections have been placed by other workers. A considerable number of Haenke specimens, several hundred species, are in the Bernhardi Herbarium[8] now at the Missouri Botanical Garden. Dr. Merrill also tells me that the Haenke collections were very badly mixed—certainly not by Haenke or Presl, but probably by some subordinate worker who misplaced the labels. In the *Reliquiae Haenkeanae* species of such tropical and subtropical genera as *Piper*[9] and *Begonia* are attributed to California where they could not possibly grow.

Safford wrote also of Née's later work and collections:

"Née left the Atrevida on the coast of Chile and proceeded overland, stopping at Talcahuano, Concepcion, and Santiago, and thence by way of the cordillera del Valle to Mendoza and over the pampas to Buenos Ayres. He rejoined the expedition May 10 [1794] . . . Née . . . took back with him 10,000 plants, nearly half of which were apparently new. His herbarium, together with descriptive notes and drawings, belong to the Royal Garden at Madrid. Many of his Guam plants were described by Cavanilles; among them are a number . . . that have not since been recognized, and no careful comparison has been made between the types in Madrid and material from the Pacific in England. Notes of both Née and Haenke are included in Malaspina's official narrative, lying in manuscript in the archives of the Madrid hydrographic office . . ."

Half a century later we find the topographer John Charles Frémont, one of our plant collectors, paying tribute to Malaspina's ability as a surveyor and quoting the great geographer Alexander von Humboldt's equally favorable opinion:

"The position adopted for Monterey and the adjacent coast, on the map now laid before the Senate, agrees nearly with that in which it had been placed by the observations of *Malaspina*, in 1791 . . . Vancouver removed the coast line as fixed by Malaspina, and the subsequent observations [of Beechey, Belcher, and Frémont] carry it back." And again:

"Of this skilful, intrepid, and unfortunate navigator, Humboldt (Essay on New Spain) says: 'The peculiar merit of his expedition consists not only in the number of astronomical observations, but principally in the judicious method which was employed to arrive at certain results. The longitude and latitude of four points on the coast (Cape San Lucas, Monterey, Nootka and Port Mulgrave) were fixed in an absolute manner.'"

8. *See:* Lamson-Scribner, F. 1899. "Notes on the grasses in the Bernhardi Herbarium, collected by Thaddeus Haenke, and described by J. S. Presl." *Report Missouri Bot. Gard.* 10: 35-59, 54 plates. And, for the acquisition of this Herbarium, *see: Ibid.,* 2: 25, *fn.* and 8: 19, *fn.*

9. "Piper Californicum Presl, *Hab. ad monte-Rey in California,*" serves as an example.

ALDEN, ROLAND HERRICK & IFFT, JOHN DEMPSTER. 1943. Early naturalists in the far west. *Occas. Papers Cal. Acad. Sci.* 20: 12-15.

24   BOTANICAL EXPLORATION OF THE TRANS-MISSISSIPPI WEST

BANCROFT, HUBERT HOWE. 1886. The works of . . . 18 (California I. 1542-1800): 490, 491.

——— 1884. *Ibid.* 27 (Northwest coast I. 1543-1800): 94, 249, 250.

——— 1886. *Ibid.* 33 (Alaska 1730-1885): 274, 275.

BARNHART, JOHN HENDLEY. 1919. Some fictitious botanists. *Jour. N. Y. Bot. Gard.* 20: 175, 176.

BREWER, WILLIAM HENRY. 1880. List of persons who have made botanical collections in California. *In:* Watson, Sereno. Geological survey of California. Botany of California 2: 553.

COLMEIRO Y PENIDO, MIGUEL. 1858. La botánica y los botánicos de la península Hispano-Lusitana. Estudios bibliográficos y biográficos. Madrid.

EASTWOOD, ALICE. 1939. Early botanical explorers on the Pacific coast and the trees they found there. *Quart. Cal. Hist. Soc.* 18: 336.

FRÉMONT, JOHN CHARLES. 1848. Geographical memoir upon upper California, in illustration of his map of Oregon and California, . . . *U. S. 30th Cong., 1st Sess. Misc. Doc.* No. 148. Map [by Charles Preuss].

GALBRAITH, EDITH C. *See:* Malaspina, A. 1924.

GREENHOW, ROBERT. 1845. The history of Oregon and California, and the other territories on the north-west coast of North America: accompanied by a geographical view and map of these countries, and a number of documents as proofs and illustrations of the history. ed. 2. Boston, 222, 223, & *fns.*

JEPSON, WILLIS LINN. 1899. Early scientific expeditions to California. II. *Erythea* 7: 129-134.

——— 1929. Sequoia sempervirens in Granada. *Madroño* 1: 242.

MALASPINA, ALEJANDRO. 1885. Viaje político-científico alrededor del mundo por las corbetas Descubierta y Atrevida al mando de los capitanes de navío D. Alejandro Malaspina y Don José de Bustamente y Guerra desde 1789 á 1794. Publicado con una introducción por Don Pedro de Novo y Colson teniente de navío académico correspondiente de la real de la historia. Madrid. [Cover-title reads: La vuelta al mundo por las corbetas Descubierta y Atrevida al mando del capitán de navío D. Alejandro Malaspina desde 1789 á 1794. Publicado con una introducción en 1885 por el teniente de navío Don Pedro de Novo y Colson.]

——— 1924. Malaspina's voyage round the world. [Edited by] Edith C. Galbraith. *Quart. Cal. Hist. Soc.* 3: 215-237. [Translation of that portion of Malaspina, A. 1885. Viaje . . . which relates to the visit to California.]

MENZIES, ARCHIBALD. 1923. Menzies' journal of Vancouver's voyage. April to October, 1792. Edited, with botanical and ethnological notes by C[harles] F[rederick] Newcombe, M.D. and a biographical note by J[ohn] Forsyth. *Memoir V. Archives British Columbia.* Victoria, B. C.

NÉE, LUIS. 1801. Descripcion de varias especies nueva de Encina (Quercus de Linneo). *Anales de ciencias naturales Madrid* 3: 260-278.

——— 1806. Description of several new species of oak from the Spanish of . . . *Annals of Botany* (Konig & Sims) 2: 98-111.

PRESL, KAREL BORIWOG. 1825-1835. Reliquiae Haenkeanae; seu descriptiones et icones plantarum, quas in America meridionali et boreali, in insulis Philippinis et Marianis collegit Thaddeaus Haenke . . . redegit et in ordinem digressit Carolus Bor. Presl . . . 2 vols. Prague. [The preface contains a biographical account of Thaddeus Haenke by Casparus Comes de Sternberg.]

—— 1851. Epimeliae botanicae. *Abh. Bohm. Ges. Wiss.* ser. 5, 6: 361-624. Read 1849. Reprinted, pp. 1-264, 1851. [For date of publication see Barnhart, J. H. 1905. *Bull. Torrey Bot. Club* 32: 590-591.]

SAFFORD, WILLIAM EDWIN. 1905. The useful plants of the island of Guam. *Contr. U. S. Nat. Herb.* 9: 1-416.

SARGENT, CHARLES SPRAGUE. 1895. The Silva of North America. Boston. New York. 8: 25, *fn.* 1.

CHAPTER II

MENZIES, NATURALIST OF THE VANCOUVER EXPEDI-

TION, INVESTIGATES THE FLORAL WEALTH OF THE

COASTAL REGIONS OF WASHINGTON AND CALIFORNIA

THE first of a series of British botanists to reach the Pacific coast of North America was Archibald Menzies.[1] According to John Forsyth, he was born at ". . . Stix or Styx, an old branch house of the Menzies of Culdares . . ." not far from Aberfeldy, Perthshire, Scotland; the Weem Kirk register entered his baptism as March 15, 1754. He was educated at Weem Parish School and received his first lessons in botany at Castle Menzies, an ancestral home dating back to 1057 ". . . where it was his privilege later to add many new varieties of trees which he had discovered during his travels." Forsyth remarks that ". . . nearly all the Menzies in the vicinity of Castle Menzies were either gardeners[2] or botanists.

"On leaving home, Menzies journeyed to Edinburgh and as a botanical student entered the Royal Botanic Garden . . . also at this time studying for the medical profession, and attended the Edinburgh University classes under Dr. John Hope, who is described as a genial and painstaking teacher and who took a deep interest in Menzies' education."

In 1778 Menzies made a trip to the Highlands and to the Hebrides; and for some years he served as assistant-surgeon in the British Navy.

It was Dr. Hope, curator of the Edinburgh Botanic Garden, who, by letter in 1786, recommended Menzies to the attention of the influential Sir Joseph Banks; the two corresponded and, more than once, Sir Joseph was of assistance in furthering his friend's botanical ambitions. It was he who, by interceding with Mr. Etches,[3] merchant owner of the ships *Prince of Wales* and *Princess Royal,* which were about to sail around

1. "The name Menzies is properly pronounced as if spelt Minges, but we in California have so long spoken the name phonetically that our pronunciation has come to have, in western America, the force of usage." (Jepson, 1929.)

"As is generally known, 'Z' in Scots is the equivalent of 'Y' and should be so pronounced . . ." (Balfour.)

2. Sir David Menzies reprints a page taken from an old garden book kept at Castle Menzies which tends to prove this. It shows a "Table of payments to garden staff . . . during the week ending December 24th, 1832." The wages were modest; no one was paid more than 10 pence for a day's work, for half a day three pence; the work was "collecting leaves and wheeling soil." Seven persons of the name of Menzies were employed that week: Duncan, Janet, Jean, Amelia, Catherine, the "Widow Menzies" who worked four days for a total of four shillings, and John.

3. Balfour states that it was the introduction of Menzies to Banks that ". . . led to his appointment as surgeon on the 'Prince of Wales' owned by the enterprising firm of John and Cadman Etches & Co."

the world in the interest of the British fur trade, had restrictions removed in order that Menzies might join the party and collect curiosities for Sir Joseph and his friends.

Menzies left London in September, 1786, and was back in England in July, 1789. Although he was off the northwest coast of North America in 1787 and 1788, he got no farther south than Nootka (the Nutka of the Spaniards) situated on the eastern side of an island which is separated from the western shore of Vancouver Island by Nootka Sound. When he returned to that region in 1792 he mentions occurrences of the earlier visit.[4]

"Having obtained some fame as a botanist," continues Forsyth, the British government appointed Menzies ". . . in 1790 as Naturalist to accompany Captain Vancouver . . . on a voyage round the world." This voyage lasted from 1791 to 1795; and it was when serving with Vancouver that, in 1792, 1793 and 1794, Menzies collected plants in what are now the states of Washington and California. So far as I know he did not do so in the state of Oregon.

Menzies kept a voluminous journal which is now the property of the British Museum. Two portions of this manuscript—both relating to periods in 1792 and 1793 when the expedition was along our Pacific coast—have been published: both were transcribed from a certified photostat[5] of the original.

The first of these published portions, covering the period from April 8 to October 13, 1792, was issued in 1923 as *Memoir V* of the Archives of British Columbia. Dr. C. F. Newcombe, who edited this portion of the journal, provided botanical and ethnological notes, an interesting preface, and much else, states that between these dates, ". . . Vancouver coasted along the mainland shore from latitude 35° 25′ north off what was then known as New Albion, but now included in Northern California, to latitude 52° 18′, where for that year the survey ended at Point Menzies, in Burke Channel."

The second published portion, covering the period from November 15, 1792, to December 9, 1793, was issued in the *Quarterly* of the California Historical Society for January, 1924, and was edited by Miss Alice Eastwood. It records visits along the

4. Visiting the "Village of *Tashees*" on September 5, 1792, he recognized that the wife of Chief Maquinna's brother was ". . . an old acquaintance the daughter of an elderly Chief . . . to whose friendship I owed much civility & kindness when I was here about five years ago. She & her Sisters were then very young, yet they frequently shewed so much solicitude for my safety, that they often warned me in the most earnest manner of the dangers to which my Botanical rambles in the Woods exposed me, & when they found me inattentive to their entreaties, they would then watch the avenue of the Forest where I enterd, to prevent my receiving any insult or ill usage from their Countrymen. But it was not till after I left them that I became sensible how much I owed to their disinterested zeal for my welfare by knowing more of the treacheries & stratagems of the Natives on other parts of the Coast . . ."

Menzies reveals his innate courtesy when he relates: "I emptied my pockets of all the little Trinkets they containd in her lap & begged her to come on board the Vessel with her Father . . . that I might have an opportunity of renewing our friendship by some gratifying present."

5. Now in the Archives of British Columbia at Victoria.

coast of California: two to San Francisco Bay, two to Monterey, and one to each of the following—Port Trinidad (not far north of present Eureka), Tomales Bay, Santa Barbara and San Diego. The editor wrote an introduction and identified many of the California plants mentioned by Menzies.

The manuscript journal in the British Museum stops on February 16, 1794. Vancouver was off the coast of California from November 3 until December 2 of that year. The ships put in at Monterey from November 6 to December 2 and, while there, a few excursions were made—one to Santa Cruz from November 27 to 29. But since Menzies' journal ends, as stated, on February 16 of that year, Vancouver's *Voyage* and one Menzies letter, containing only meagre facts, constitute the sources covering that visit.

I now turn to "Menzies' journal of Vancouver's voyage April to October, 1792," edited by Newcombe, quoting from the editor's preface and from such portions of his transcript of the journal as relate to Menzies' work along the northwest coast of what is now the United States. These portions are but a fraction of the whole. The modern place names and plant determinations which Newcombe includes marginally I have quoted between brackets and inserted in the text. The editor does not supply in his marginal plant-determinations the authors of the scientific names although some are recorded in his "Plants collected by A. Menzies on the north-west coast of America," [6] included as an appendix.

Menzies records that for "upwards of twelve months" he had been retained by the British government to go with Vancouver in the capacity of naturalist. Finding the long delay irksome, he asked to be sent as surgeon on the *Discovery,* ". . . promising at the same time that my vacant hours from my professional charge, should be chiefly employed in their service, in making such collections & observations as might tend to elucidate the natural history of the Voyage, without any further pecuniary agreement than what they might conceive me entitled to, on my return . . ."

After an undisclosed objection by Vancouver[7]—Menzies disclaims any desire to know what it was, ". . . being conscious of the rectitude of my own intentions . . ."—he was permitted to go in the original capacity of naturalist, Sir Joseph Banks having used

6. In his preface Newcombe expresses his indebtedness to Sir David Prain, Director of the Royal Gardens, Kew, "for photographs and notes of plants . . ." etc.; to Mr. James Britten "for information as to the Menzies' plants in the Banks Herbarium at the British Museum," and to Professor C. V. Piper, of the Department of Agriculture, Washington, D. C., "for varied assistance . . ."

7. Vancouver was a disciplinarian and, more than once, Menzies came into conflict with his superior officer on the voyage. A paragraph included in Forsyth's "Biographical note" suggests that such conflicts were anticipated before the voyage began:

"Sir Joseph Banks was apparently apprehensive as to the treatment he might receive, as witness his last letter to Menzies (August 10th, 1791) in which he says: 'How Captain Vancouver will behave to you is more than I can guess, unless I am to judge by his conduct toward me—which was not such as I am used to receive from persons in his situation . . . As it would be highly imprudent in him to throw any obstacle in the way of your duty, I trust he will have too much good sense to obstruct it.' "

his influence. Later, in September, 1792, when the surgeon of the *Discovery* was taken ill and returned to England, Menzies was asked to replace him which he did. Vancouver had urged it ". . . with a degree of earnestness that I could not well refuse, especially as he requested . . . that in case of my not accepting of it, to state my having refused it in writing, & as I did not know how far this might operate against my interest at the Navy Office, I with considerable hesitation accepted of the appointment, on Capt Vancouver's promising me that he would take care it should not interfere with my other pursuits . . ."

On the voyage Menzies' duties, in addition to collecting new and rare plants, included the distribution of seeds, of fruit trees, among them the orange, and of cattle. Newcombe points out that this was customary at the period, as evidenced by the mission of David Nelson when he sailed on the *Bounty* with Captain "Bread-fruit" Bligh.

At the request of Lord Grenville, Sir Joseph Banks furnished Menzies with formal instructions.[8] The substance of these is given by Forsyth:

"He was to investigate the whole of the natural history of the countries visited, pay attention to the nature of the soil, and in view of the prospect of sending out settlers from England, whether grains, fruits, etc., cultivated in England are likely to thrive. All trees, shrubs, plants, grasses, ferns, and mosses were to be enumerated by their scientific names as well as those used in the language of the natives. He was to dry specimens of all that were worthy of being brought home and all that could be procured, either living plants or seeds, so that their names and qualities could be ascertained at His Majesty's gardens at Kew. Any curious or valuable plants that could not be propagated from seeds were to be dug up and planted in the glass frame provided for the purpose . . ."

On his return Menzies' collections, with his journals, etc., were to be delivered to ". . . H. M. Secretary of State or to such person as he shall appoint to received them."

In transmitting these instructions to the Lords Commissioners of the Admiralty under date of February 23, 1791, Lord Grenville emphasized ". . . the necessity for impressing upon the commander of the ship that he was to afford every degree of as-

8. These were contained in a letter which F. R. S. Balfour published in 1945 in an appendix to an article about Menzies. Balfour, after the death of Menzies' great-grandnephew C. D. Geddes, in June, 1943, came into possession of the "papers and relics" of the botanist and distributed some of them: ". . . The diplomas, charters and relics I have presented to the Provincial Archives at Victoria, British Columbia, and certain of the papers to the Library of the State of Washington at Seattle." To the Linnean Society of London he gave a complete transcript, five volumes, of Menzies' journal. Further:

"Among the papers which I received and which Menzies had carefully preserved all his life, was a long letter from the great man, written 22 February 1791, from his house in Soho Square . . . Menzies' reply, preserved in the Banksian correspondence at the British Museum, South Kensington, is docile and subservient; indeed, to our ideas not at all the attitude that one man of science should adopt to another. . . It is hard for us in these days to quite realize the proudly domineering attitude of Banks in the world of science of his day . . ."

Banks's instructions were also published (from a holograph letter) in the British periodical *Forestry*, in an article by J. D. Sutherland (1931).

sistance to Mr. Menzies, as the service he has been directed to perform 'is materially connected with some of the most important objects of the expedition.' "

On the voyage the "sloop of war," or ship *Discovery,* was commanded by Vancouver, the "armed tender," or brig *Chatham,* by Lieutenant William Robert Broughton; the *Doedalus*[9] "store ship" joined the others from time to time. Unless exceptional circumstance arose, Menzies served only in the *Discovery.*

1792

From April 8 to 23 the *Discovery* was off the coast of California, sailing northward; from April 23 to 27 off the coast of Oregon; on the 27th off the mouth of the Columbia River—Menzies noted that the water ". . . appeard muddy like the over flowings of a considerable river"; on the 28th off the coast of Washington; on the 29th entering the "Juan de Fuca's Streights" and keeping, according to Newcombe's running title, along the "Washington Coast."

It was on the 29th that Vancouver's party ". . . spoke the *Columbia* of Boston commanded by Mr Gray . . ." Although Gray was generous in giving useful information and had considerable to tell of the "diabolic plots" and so on of the native tribes, he followed them into the Straits. Menzies commented on the 30th that the *Columbia* ". . . was seen again working out of the Streights . . . it would now seem as if the Commander of her did not put much confidence in what we told him of our pursuit, but had probably taken us for rivals in trade and followd us into the Streights to have his share in the gleanings of those Villages at the entrance, & this is conformable to the general practice among traders on this Coast, which is always to mislead competitors as far as they can even at the expence of truth."

It was about two weeks after this meeting, or on May 11, 1792, that Gray sailed into the river that was later to bear the name of his ship *Columbia.*[1] Vancouver, as we shall see, was in Admiralty Inlet on that date; but later in the year he sent Lieutenant Broughton and the *Chatham* to map the course of the river, which he did in small boats, and upward for one hundred miles. Gray's earlier discovery was confirmed in his records[2] and constituted one of the major claims advanced by the United States government at the time of the Oregon Settlement in 1846.

9. The name is so spelled throughout the Newcombe and Eastwood transcripts of Menzies' journal.

1. The name appears not infrequently as *Columbia Rediviva.* Cardinal Goodwin states that Gray's ship ". . . had won the distinction of being the first vessel to bear the American flag around the world."

2. Robert Greenhow's "Proofs and illustrations," forming an appendix of *The history of Oregon and California* . . . (1845), reprints extracts from ". . . the Second Volume of the Log-Book of the Ship Columbia, of Boston, commanded by Robert Gray . . ." covering the dates from May 7 to 21, 1792. This extract, Greenhow states in a footnote, ". . . was made in 1816, by Mr. Bulfinch, of Boston . . ." The log-book, he notes, ". . . was then in the possession of Captain Gray's heirs, but has since disappeared . . ."

On April 30 Menzies records that ". . . in the North East quarter a very solid ridge of Mountains was observd one of which was seen wholly coverd with Snow & with a lofty summit over topping all the others around it . . . This obtaind the name of *Mount Baker*[3] after the Gentleman [Lieutenant Joseph Baker] who first observd it."

From May 1 to 7 Vancouver was to explore the coast about Port Discovery.

It was on May 1 that Menzies made his first landing in a region that was later to become the property of the United States: this was on an island [*"I. de Carrasco* of Quimper, 1790"] in the Strait of Fuca, at the entrance to Port Discovery, Jefferson County, Washington. They named it Protection Island and Newcombe refers to the harbor as "Near the present Junction City, Wash." The "rural appearance" of the island ". . . strongly invited us to stretch our limbs after our long confined situation on board & the dreary sameness of a tedious voyage." Vegetation was well advanced, with ". . . a variety of wild flowers in full bloom, but what chiefly dazzled our eyes . . . was a small species of wild Valerian [*'Valerianella congesta* (Sea-blush)'] with reddish colord flowers growing behind the beach in large thick patches."

Menzies comments upon the fresh water, upon the "sloping bank coverd with green turf so even and regular as if it had been artificially formed," upon the "rich lawn beautified with nature's luxuriant bounties," and so on. They lit a fire and "regald" themselves "with some refreshment," returning to the ship about midnight "each well satisfied with the success & pleasure of this days excursion."

Back in the harbor of Port Discovery and well sheltered "by the favourite Island," Menzies on May 2 went ashore with Vancouver:

"Besides a variety of Pines we here saw the Sycamore Maple [*'Acer macrophyllum'*] —the American Aldar [*'Alnus orgona'* [4]]—a species of wild Crab [*'Pirus diversifolia'*] & the Oriental Strawberry Tree [*'Arbutus Menziesii'*], this last grows to a small Tree & was at this time a peculiar ornament to the Forest by its large clusters of whitish flowers & ever green leaves, but its peculiar smooth bark of a reddish brown colour will at all times attract the Notice of the most superficial observer.—We met with some other Plants which were new to me . . ."

While the ship was being overhauled on May 3 Menzies notes that ". . . my botanical pursuits kept me sufficiently engaged in arranging & examining the collections I had already made." On May 4 he made an excursion into the woods and ". . . met with

---

3. This was the first of a number of important peaks of the Cascade Range that were named on this voyage: Mount Rainier on May 7, for Vancouver's friend, a Rear Admiral; Mt. St. Helens on October 20, for Alleyne Fitzherbert, Baron St. Helens, British Ambassador at the court of Madrid; Mt. Hood on October 30, for Sir Samuel Hood.

A. H. Bent states that "In the first volume of his [Vancouver's] 'Voyage . . .' published in London in 1789, is an engraving of Mt. Rainier, for this, one of the last of the high mountains of the United States to be ascended, was the first to be pictured. The engraver was John Landseer, the father of Sir Edwin Landseer."

4. Actually *Alnus oregana* Nuttall.

vast abundance of that rare plant the *Cypropedium bulbosum*[5] ['*Calypso bulbosa* (False Lady-slipper)'] which was now in full bloom & grew about the roots of the Pine Trees in very spungy soil & dry situations . . . likewise . . . a beautiful shrub the *Rhododendron ponticum*[6] & a new species of *Arbutus* ['*Arctostaphylos tomentosum* (Manzanita)'] with glaucous[7] leaves that grew bushy & 8 or 10 feet high, besides a number of other plants which would be too tedious here to enumerate."

On May 5 everyone was busy, some cutting firewood, others "Brewing Beer[8] from a species of Spruce . . . the weather . . . favorable & vivifying . . ." May 6 ". . . being a day of relaxation parties were formd to take the recreation of the shore & strolling through the woods in various directions . . ."

From May 7 to 15 Vancouver, Menzies and some of the officers ". . . in three Boats manned & armed & provided with five days provision . . ." explored Admiralty Inlet and its bays.

On May 7, on the west side of the entrance to Admiralty Inlet ["Point Wilson"], Menzies ". . . found growing in the Crevices of a small rock . . . a new Species of *Claytonia,* & as I met with it no where else in my journeys, it must be considered as a rare plant in this country. I namd it *Claytonia furcata* & took a rough sketch of it which may be seen in my collections of Drawings." [9]

Ashore at ["Point Hudson"] he observed that ". . . the Soil though light & gravelly appeard capable of yielding in this temperate climate luxuriant Crops of the European Grains or of rearing herds of Cattle who might here wander at their ease over extensive fields of fine pasture . . . To the North east of us across Admiralty Inlet . . . we had . . . a most delightfull & extensive landscape, a large tract of flat country coverd with fine Verdure & here & there interspersd with irregular clumps of trees whose dark hue made a beautiful contrast . . . a rugged barrier of high mountains . . . terminated our prospect in lofty summits coverd with perpetual snow.

"After dinner we proceeded examining this southerly arm . . . Invited by the enchanting appearance of the Country & fine serene weather, I walkd with Capt Vancouver . . . along the Western shore for a considerable distance as it afforded me an opportunity of exploring for natural productions as I went along. After a long walk

5. Oakes Ames wrote me of this orchid: "Cypripedium bulbosum L. = *Calypso bulbosa* (L.) Oakes in Thompson, Hist. Vermont. The western plant is forma *occidentalis* Holzinger."

6. Identified by Newcombe as "*Rhododendron californicum* (Large-flowered Rhododendron)." Is probably *R. macrophyllum* G. Don.

7. Since Menzies specifies "glaucous" leaves Newcombe's tomentose plant is not suggested.

8. Menzies' advice was sometimes helpful to his companions; on July 31, in Simoon Sound, he wrote: "A party began to Water & another to brew Spruce Beer, but after erecting the Brewing Utensils on shore, they brought me word that there was none of that particular Spruce from which they used to Brew to be found near the landing place, on which I recommended another species (Pinus Canadensis) which answerd equally well & made very salubrious & palatable Beer."

Menzies' "Pinus Canadensis" was the western hemlock, *Tsuga heterophylla* (Raf.) Sargent.

9. Newcombe notes that "Neither the species nor the sketch are quoted by Hooker and other authors."

we met with a thick pine forest which obligd us to embark & the shore here ['Port Had-lock'] taking an Easterly direction we rowd along it & towards evening we found the arm . . . to . . . terminate in a small basin of shallow water being here divided only from the end of another arm by a flat muddy beach coverd with thick beds of marsh samphire ['*Salicornia ambigua* (Samphire or Glasswort)'] . . ."

On May 8, having finished the ". . . examination of the first small Arm which was namd Port Townsend . . .", the party walked along the shore of Admiralty Inlet to ["Basalt Point"] and later, by boat, reached "a bluff point" ["Foulweather Point"] from which they ". . . had a fine view of a very lofty round topped mountain coverd with Snow . . . which afterwards obtaind the Name of *Mount Rainier* . . ."

The next day, because of "thick rainy weather . . . we could not stir to any advan-tage.—As intervals of fair or clear weather permitted parties strolld along the Beach & met with some Oak Trees ['*Quercus Garryana* (Garry Oak)'] on which account our present situation was called Oak Cove, it stretches a little to the Westward & nearly meets the termination of Port Townsend . . ."

On May 10, proceeding along "the Southern Arm," and keeping to "the Starboard Shore," they breakfasted at Indian Cove ["Port Ludlow"], then crossed to ["Foul-weather Bluff"] and, after passing "a round clump of trees which had the appearance of an Island ['Hannon Point'] but which we found joind by a narrow beach to the West-ern Shore . . . ," they ". . . stopped for the night which was serene & pleasant on a snug Beach where we were very comfortable on the larb^d shore. The country on both sides of the arm . . . every where coverd with pine forests close down to the Beach & this afternoon I found on the western side a good number of hazle nut Trees ['*Corylus californica* (Hazel)'] for the first time on this side of America."

On May 11, following the same ". . . Arm . . . in a Southerly direction . . ." they came in the afternoon to a branch going off ". . . in a North West direction . . ." ["Dabop Bay"].

". . . At a place we landed on near the bottom of the Bay I saw vast abundance of a beautiful new species of Vaccinium with ever green leaves in full bloom ['*Vaccinium ovatum* (Evergreen Huckleberry)'], it grew bushy & was of a dark green colour like Myrtles which it much resembles in its general appearance. I had seen it before in several other places since we came into the Streights but no where in such perfection as here, I therefore employd this afternoon in making a delineation of it as we went along in the Boat."

May 12 they were ". . . close under that high ridge of Mountains with snowy sum-mits which support the Peaks of Mount Olympus & which now lay between us & the sea coast . . . their sides were every where coverd with one continued forest of Pinery." They were still following the ". . . Arm which still lead to the Southward . . ." This day Menzies mentions that he took a stroll in the woods, finding ". . . three different kinds of Maple & a *Rhamnus Arbutus* & *Ceanothus* ['*Rhamnus Purshianus* (Cascara Sagra-da) and *Ceanothus velutinus*'] that were new to me besides several others."

Having completed the exploration of Admiralty Inlet, the party started back on May 13 to rejoin the ships which they had left on the 7th. On May 14 they were again at ["Foulweather Bluff"] where Menzies met his second skunk.[1] May 15 they were back at Port Discovery,[2] ". . . wet, hungry & uncomfortable."

While preparation was being made for further explorations Menzies spent May 17 ". . . in getting on board some live plants which were new to me as I did not know that I should any where else meet with them, & in planting them in the frame on the Quarter Deck."

Starting on their way on May 18, Menzies was able, while Vancouver took a bearing off Protection Island, ". . . to have another stroll on that delightfull spot & among other Plants I collected I was not a little surprizd to meet with the *Cactus opuntia* ['Prickly Pear *(Opuntia polyacantha borealis)*'] thus far to the Northward, it grew plentifully but in a very dwarf state on the Eastern point of the Island which is low flat & dry sandy soil."

By way of Admiralty Inlet the ships passed Port Townsend, and anchored that day ". . . about 10 or 12 miles from the Entrance ['On west shore of Whidbey Id. ? Mutiny Bay']." On the 19th they passed "the bluff point," or Foulweather Bluff, then "a wide opening going off to the Northward ['Possession Sd.']," passed, on the right, "a pleasant point coverd with the richest verdure ['Restoration Point']," next "a round Island ['Blake Id.'] . . . coverd with wood," and anchored "close to the inner point of . . ." the opening or arm. They had taken, according to Newcombe, the "Colvos Passage."

Vancouver's party was now to work in the various arms of Puget Sound. Menzies traveled from May 20 to June 5 with a section commanded by "Lieutenant Puget in the Launch together with Mr Whidbey in the Cutter" and they got as far south as Totten, Eld and Budd inlets, farthest south, it would seem, in the last-named where they must have approached close to the site of what is now the city of Olympia.

On May 20 Menzies wrote that two boats were equipped to examine ". . . the Arm leading to the Southward, & though their mode of procedure in these surveying Cruizes was not very favorable for my pursuits as it afforded me so little time on shore at the different places we landed at, yet it was the most eligible I could at this time adopt for obtaining a general knowledge of the produce of the Country, I therefore embarkd next morning before day light . . . we enterd the Arm which led to the Southward . . . passed at noon a large opening . . . going off to the Eastward ['Dalco passage leading to Commencement Bay'] . . . we passed on our left hand in the forenoon an Island ['Vashon Id.'].

1. His first encounter had been on May 2, but the odor had been so objectionable that the creature had been abandoned. On the present occasion he was able to set his scientific spirit at rest. Although ". . . no one could remain any time within some distance of where it fell . . . ," Menzies "satisfied" himself "that it was the Skunk (*Viverra Putorius*)."

2. Menzies comments that they had afterwards found that Port Discovery, a year earlier, had been named "Port Quadra" by the Spaniards; he was in favor of retaining the earlier designation ". . . to prevent that confusion of names which are but too common in new discoverd countries."

"Up this Bay we had a most charming prospect of Mount Rainier which now appeard close to us though at least 10 or 12 leagues off, for the low land at the head of the Bay swelled out very gradually to form a most beautiful & majestic Mountain of great elevation . . . with a round obtuse summit coverd two thirds of its height down with perpetual Snow . . .

"We pursued our Southerly direction ['Through "The Narrows" '] . . . & . . . came to another arm leading off to the Westward ['Hale Passage'] which we enterd . . . we disembarkd on the Point to dine . . . Here I found some small trees of both the American & Mountain Ash ['Ash *(Fraxinus oregana),* Mountain Ash *(Pirus sitchensis)'*] neither of which I had before met with on this side of the Continent—The other Plants I saw in the course of this day were nearly the same as I had before examined in the other arm the former cruize."

On May 21 the party ran through a "narrow gut" ["Carr Inlet"], observed a "small Island" ["Herron Id."], and, after spending considerable time with some "Natives" whose intentions appeared for a time to be anything but friendly, they pitched their tents early ". . . on the western point of a narrow passage leading to the Southward ['South Head; Pitt Passage'] . . ." and opposite the "narrow gut" they had passed through in the morning. The next day they passed through ['Pitt Passage'], landed upon "a small Island ['Ketron Id.'] close to the Eastern Shore . . ." and camped "in a commodius place near the Point . . ."

On May 23, across from their camping place, they found ". . . a large Bay so flat & shallow ['Nisqually River and Flats'] that we could not approach near the shore . . . we did not see any appearance of a River . . ." After passing ["Anderson Id."], moving through ["Nisqually Reach"] and "a large opening going off to the Southward ['Dana Passage']," they landed on "an island at the further end of it ['Hartstene Id.']" where they remained for the night. On the 24th, they pursued the arm ["Case Inlet"], which then "trended more to the Northward," for some seven or eight miles until it ended "in shoal water & low marshy land"; they turned back and, through ["Pickering Passage"] reached "a large Island ['Hartstene Id.']" and camped. Menzies ". . . found the country exceedingly pleasant, & the Soil the richest I have seen in this Country— The Woods abound with luxuriant Ferns that grow over head."

On May 25, after passing ["Squaxin and Hope Ids."] they followed an arm "now about a Mile wide" which went "to the Southwestward ['Totten Inlet']," but, unable to reach within two miles of its termination because "of the shallowness of the water which was one continued flat," they returned whence they had come until at an opening they "struck off to the Eastward about two Miles & encamped on the point ['Cushman Point'] of another arm leading to the Southward."

Following an arm leading toward the south and coming to a fork they took, on May 26, "the Westernmost branch ['Eld Inlet']" and pursued it to its termination where they visited the encampment of some Indians of "peaceable disposition," and then turned back northward, keeping "the Continental shore," until they got back to their "... old ground by the large opening we had passed on the 23d so that we had now entirely finished this complicated Sound which afterwards obtaind the name of *Puget's Sound,* & after dining on the East point of the opening a favorable breeze sprung up . . . which we made use of to return to the Ship by the nearest route we could take." [3] This followed ["Nisqually Reach"]. They passed the island "on which we dind on the 22d ['Ketron Id.'] . . . When we came into the Main Arm . . . we continued under Sail all night & arrivd at the ship about 2 o'clock the next morning, but as they had removd her . . . towards the Point ['Restoration Point'] where the village was on we were obligd to fire off some Swivels which they answerd from the Ship & thereby discoverd to us her situation."

On arrival it was learned that the *Chatham* had returned two days earlier after "examining the North West side of the Gulph ['San Juan Archipelago'] . . ." and that Vancouver and Johnstone had left the day before "to examine the arm leading to the South Eastward ['East of Vashon Id.'] which we have already supposed to join with the one we were in." On May 28 Broughton and Whidbey, in order to save time, departed to ". . . follow back the opposite shore of this arm (which was presumd to be the Continent) . . ."

Menzies ". . . landed on the Point near the Ship ['Restoration Point'] where . . . a few families of Indians live in very Mean Huts or Sheds formd of slender Rafters & coverd with Mats. Several of the women were digging on the Point which excited my curiosity to know what they were digging for & found it to be a little bulbous root of a liliaceous plant which on searching about for the flower of it I discoverd to be a new *Genus* of the Triandria monogina ['One or more species of *Hookera (Brodiæa* of Smith)']. This root with the young shoots of Rasberries & a species of Barnacle . . . formd at this time the chief part of their wretched subsistance. Some of the women were employd in making Mats of the Bullrushes while the Men were lolling about in sluggish idleness . . .

"In the edge of the wood I saw a good deal of Ash & Canadian Poplar ['Ash *(Fraxinus oregana)* and *Populus trichocarpa* (Cottonwood)']."

On May 29 Vancouver's party was back having ". . . pursued the Arm they went to examine in a South East direction for about four leagues when they found it enter that extensive Bay running up almost to the bottom of Mount Rainier ['Commencement Bay and the vicinity of Tacoma'] which we have already described . . ." They had also covered southwestward much the same ground covered by Puget (and Menzies) although keeping to "the Larboard Shore" rather than "the Starboard." On May 30 the party left to rejoin the *Chatham* and sailed northward, anchoring near her about midnight.

On May 31 Menzies went ashore "on the point under which we lay ['Elliott Point']."
". . . Mr Broughton namd the point from the vast abundance of wild roses that

3. Newcombe notes marginally that "Budd Inlet and the site of Olympia, though not described, were included in this day's survey as shown by Vancouver's chart."

grew upon it *Rose Point*—A large Bay ['Everett Bay'] which went off to the North-ward was the most easterly situation which our Boats explord in this country . . . The land every where round us was still of a very moderate height & coverd with a thick forest of different kinds of Pine trees. In a marshy situation behind the Beach I found some Aquatic plants I had not before met with."

That evening the ships "came to again near an inland [*sic*] in mid-channel ['Gedney Id.'] for the night . . ."

On June 1, the northward arm dividing, the ships ". . . stood up the Eastermost which soon in the afternoon we found to terminate in a large Bay ['Port Susan'] with very shallow water & muddy bottom . . ." On the 2nd they returned by the arm they had come, keeping east of Camano Island, and, when they were ". . . a little below the point of division . . . anchord near the eastern shore abreast of a small Bay ['Tulalip Anchorage shown on Vancouver's chart'] . . ."

Here they remained for the 3rd and 4th. ". . . The latter being the King's Birth Day, Capt Vancouver landed about noon with some of the Officers on the South point of the small Bay ['South point of Tulalip Bay'] where he took posession of the Country with the usual forms in his Majesty's name & namd it *New Georgia* & on hoisting the English Colours on the spot each Vessel proclaimd it aloud with a Royal Salute in honor of the Day."

On June 5 they passed "the bluff point" ["Foulweather Bluff"] and anchored a few miles beyond on the west shore; they were again in Admiralty Inlet. The 6th they entered the Inlet and at midnight ". . . came to anchor on the outside near the North point of its entrance ['South of Marrowstone Point']."

". . . As Capt Vancouver & Mr Broughton were at this time going off in a Boat to . . . take bearing on a small Island ['Smith's Id.'] about 4 or 5 miles to the Northward . . . I accompanied them to examine it, at the same time for plants, but I found nothing different from what I had before met with in the Arms . . . Most part of the Island was faced with a sandy cliff & coverd with Pines densely copsed with Underwood."

At this point in the journal Menzies describes "The general appearance of the Country from this station . . ." I quote but a small portion:

". . . To the South West of us a high ridge of Mountains ran from the outer point of de Fuca's entrance in a South East direction,—gradualy increasing in height to form the rugged elevated peaks of Mount Olympus . . .

"To the South East of us down Admiralty Inlet was seen through a beautiful avenue formed by the Banks of the Inlet Mount Rainier at the distance of 26 Leagues . . . from it a compleat ridge of Mountains with rugged & picked summits covered here & there with patches of Snow & forming a solid & impassable barrier on the East Side of New Georgia, runing in a due North direction to join Mount Baker . . . to the North Eastward of us & from thence proceed in high broken Mountains to the North West-ward.

"Between us & the above Ridge & to the Southward of us between the two Mountains already mentioned a fine level Country intervened chiefly covard with pine forests abounding here & there with clear spots of considerable extent & intersected with the various winding branches of Admiralty Inlet . . . These clear spots or lawns are clothed with a rich carpet of Verdure & adornd with clumps of Trees & a surrounding verge of scatterd Pines . . . a beauty of prospect equal to the most admired Parks in England.

"A Traveller wandering over these unfrequented Plains is regaled with a salubrious & vivifying air impregnated with the balsamic fragrance of the surrounding Pinery . . .

"The Woods here were chiefly composed of the Silver Fir—White Spruce—Norway Spruce & Hemlock Spruce together with the American Abor Vitae & Common Yew; & besides these we saw a variety of hard wood scattered along the Banks of the Arms, such as Oak—the Sycamore or great Maple—Sugar Maple—Mountain Maple & Pensylvanian Maple—the Tacamahac & Canadian Poplars—the American Ash—common Hazel—American Alder—Common Willow & the Oriental Arbute, but none of their hard wood Trees were in great abundance or acquired sufficient size to be of any great utility, except the Oak in some particular places, as at *Port Gardner & Oak Cove.*[4] We also met here pretty frequent in the Wood with that beautiful Native of the Levant the purple Rododendron,[5] together with the great flowered Dog wood, Common Dog-wood & Canadian Dog-wood—the Caroline Rose & Dog Rose, but most part of the Shrubs & Underwood were new & undescribed, several of them I named, as *Arbutus glauca, Vaccinium lucidum, Vaccinium tetragonum, Lonicera Nootkagensis, Gaultheria fruticosa, Spiraca serrulata, Rubus Nootkagensis.* Others from particular circumstances were doubtful & could not be ascertained till they are hereafter compared with more extensive description &c. on my return to England.

"The wild fruits were Goosberries, Currants, two kinds of Rasberries, two kinds of Whattleberries, small fruited Crabs & a new species of Barberry . . ."

From June 7 to 24 Vancouver's ships were in the eastern end of the Juan de Fuca Strait along the northwestern coast of Washington, among the islands of the San Juan Archipelago,[6] and in the Strait of Georgia and its bays.

On June 7 the ships ". . . stood to the Northward near the Eastern side of the Gulph & having gone about 5 or 6 leagues . . . came to Anchor . . . near some Islands & broken land on the North side." The next day the *Discovery* anchored in ". . . Strawberry Bay ['Cypress Id.'] on the East side of the opening near the entrance . . ." Menzies tells that

4. Under date of June 2 Menzies had reported the return of two boats which had examined ". . . the termination of the western branch which was named Port Gardner ['Saratoga Passage. Gardner's name still in use for large inlet farther north.'] . . . They found Oak Timber more abundant in this arm than any we had yet explored . . ."

5. *Rhododendron macrophyllum* G. Don.

6. Because of faulty wording of the treaty of 1846, ownership of the San Juan Archipelago was long in dispute. In 1872 it was decided that Haro Strait to the east of Vancouver Island should constitute the boundary between United States and British possessions, giving the Archipelago to the United States.

"... Mr Broughton ... went off in the forenoon in a Boat to finish his Survey of the Islands that were to the Westward of us ['Orcas and smaller islands of the San Juan group'], on the North side of the Gulph, & as the rugged appearance of these seemed to offer a new field for my researches I accompanied him by a friendly invitation.

"... The Shores were almost every where steep rugged & cliffy which made Landing difficult of access from the rocky cliffs & chasms with which they abounded, but I was not at all displeased at the change & general ruggedness of the surface of the Country as it producd a pleasing variety in the objects of my pursuit & added Considerably to my Catalogue of Plants.

"I here found another species of that *new genus* I discoverd at Village Point in Admiralty Inlet, & a small well tasted wild onion ['Allium'] which grew in little Tufts in the crevices of the Rocks with a species of *Arenaria* both new. I also met with the *Lilium Canadense* & the *Lilium Camschatcense,* the roots of the latter is the *Sarana* ['The Saranne Lily, *Fritillaria camtschatcensis,* is not so common here as *F. lanceolata*. Both eaten by natives'] so much esteemed by the Kamtschadales as a favourite food . . ."

On June 11 they "... stood in for a large Bay ['Birch Bay'] . . ." and anchored about half a mile from the shore. While Vancouver and others were erecting an observatory, Menzies examined on the south shore of the bay "... the scite of a very large Village now overgrown with a thick crop of Nettles ['Nettle *(Utrica Lyallii)*. Much used formerly by Indians for making twine and nets'] & bushes ... along the Beach ... we found a delightful clear & level spot cropt with Grass & wild flowers . . ."

On June 12 Menzies "... landed at the place where the Tents were erected & walked from thence round the bottom of the Bay to examine the natural productions of the Country & found that besides the Pines already enumerated the Woods here abounded with the white & trembling Poplars together with black Birch ['Aspen Poplar *(Populus tremuloides* or *Vancouverensis)*']. In consequence of my discovery ... the place afterwards obtaind the name of *Birch Bay*. I also found some other plants unknown to me, two of which had bulbous roots & grew plentifully near the Tents, one of them was a new species of *Allium* from six to ten inches high & bore a beautiful number of pink colourd flowers, the other had a thick set spike of pale green colourd flowers & appeard to be a new species of *Melanthium* ['*Melanthium,* probably *Zygadenus venenosus* (Poison Camas)'] of which I made a rough drawing & collected roots of both to put in the plant frame as neither of them were at this time in Seed."

Vancouver and Puget left on June 12 for an eleven-day trip;[7] and Widbey and Broughton on the 13th took off in other directions. Menzies seems to have remained at Birch Bay. On June 16 he records that he "... landed on the opposite side of the Bay, where I enjoyd much pleasure in Botanical researches, in wandering over a fine rich meadow cropt with grass reaching up to my middle, & now & then penetrating

7. This took them to ["Semiahoo and Boundary Bays"].

the verge of the Forest as the prospect of easy access or the variety of plants seems to invite. Here I found in full bloom diffusing its sweetness that beautiful Shrub the *Philadelphus Coronarius* ['Syringa or Mock Orange *(Philadelphus Gordonianus)*'] which I had not met with before in any other part of this Country, & having collected a number of other Plants in this little excursion I returnd in the afternoon round the bottom of the Bay to have them examind & arrangd . . ."

June 17th was Sunday and "a day of recreation to all hands." On the 18th Menzies left with Johnstone and ". . . rowed across & landed upon the Eastermost of a group of small Islands ['Matia Id., probably . . .'] . . .

". . . Here the Shores were rocky, rugged & cliffy rising into hills of a moderate height composing a numerous group of Islands thinly coverd with stinted Pines . . .

"On a Point where we landed to dine we found growing some trees of Red Cedar [' "Red Cedar." Our Coast Juniper was formerly known under this name, which was originally applied to *Juniperus virginiana* . . .']; the Plants we met with in other respects did not differ much from the Plants I had collected a few days before on the Southermost of these Islands; a new species of the Genus *Epilobium* & another of the Genus *Polygonum* excepted. In the Cliffs of a small rocky Island I also found a species of *Saxifraga* I had not before met with . . ."

Returning from this excursion on June 19, they landed at a place where a few families of Indians lived ". . . in a few temporary huts formd in the slightest & most careless manner by fastening together some rough sticks & throwing over them some pieces of Mats of Bark of Trees ['Bark of *Thuja plicata* (Giant Cedar)'] so partially as to form but a very indifferent shelter from the inclemency of the weather."

They appeared to be "an harmless & inoffensive tribe." Before leaving the island Menzies walked ". . . for some distance along the sea side where we passd a low extensive Morass well cropd with Bullrushes of which large patches had been pluckd by the Natives & were now laid neatly out upon the Beach to season them for making their Mats, & it is probable that the conveniency of procuring a good supply of this Plant so necessary to their domestic comforts inducd these families to fix their temporary residence in the vicinity."

Newcombe notes that the party had probably been ". . . on the north shore of Orcas Id., where such a swamp exists opposite East Sound." The "Bullrushes" he identifies as "Tule *(Scirpus occidentalis)*."

Vancouver and Puget returned from their trip on June 23—Menzies gives a long account of where they went—and on June 24 the ships started on their way and ". . . soon passed Cape Roberts & stood up the great North West Arm . . ." They were in the Gulf of Georgia and, after crossing the 49th parallel, passed out of the region of this story.

Continuing northward through the Strait of Georgia and the northern waters separating the island of Vancouver from the mainland of British Columbia, the expedition proceeded through Queen Charlotte Strait, round the north end of Vancouver Island,

and arrived on August 28 at Nootka where the Sound of that name afforded a good natural harbor. Nootka[8] was then controlled by Spain and, before Vancouver left, he had conducted negotiations relative to its transfer to the British government. Vancouver's ships had been the first to circumnavigate the island of Vancouver.

Menzies' journal has much to tell of the visit to Nootka. He refers to the fine impression made by the Malaspina expedition which had been there in August, 1791. He also mentions another Spanish expedition:

"There were two Botanists attachd to the Spanish Squadron who visitd the Coast this Summer, one of them had been in the *Aranzaza* to the Northward & had made a considerable Collection of Plants from the different places they touched at, the other whose name was Don José Mozino[9] remaind at Nootka . . . together with an excellent draughtsman Sr Escheverea a Native of Mexico, who as a Natural History Painter has great merit. These told me that they were part of a Society of Naturalists who were employd of late years in examining Mexico & New Spain for the purpose of collecting Materials for a Flora Mexicana which they said would soon be publishd, & with the assistance of so good an Artist it must be a valuable acquisition."

Before leaving Nootka, Menzies was able to ship home on October 1 some of his

8. Balfour wrote in 1945: "Nootka is now a little-known place, but its name is familiar to us from the several well-known plants to which it is applied, all first seen by Menzies, e.g., *Cupressus nootkatensis, Rosa nutkanus,* and *Rubus nutkanus.*"

9. When enumerating those who had made plant collections in California, William H. Brewer wrote in 1880:

"Josef Mariano Moçiño was on the coast from California to Nootka in the year 1792, at the same time as Menzies . . . He afterwards botanized in Mexico, especially in its northern parts, along with Martin Sessé. The large collection of drawings which Moçiño brought to Europe after the death of Sessé, contains delineations of several Nootka species (such as *Rubus Nutkanus*), and apparently a few from California; but most of them were Mexican. This collection of twelve hundred drawings . . . was left by Moçiño in the hands of De Candolle, but after some years was suddenly reclaimed, upon which occasion copies of most of them were secured by the united labors of the principal ladies of Geneva. It is said that the herbarium made by Moçiño and Sessé went to Madrid . . ."

The phrases, "was on the coast from California to Nootka" and "apparently a few from California," suggested that the work of this expedition might belong in this story. Dr. Paul C. Standley of the Chicago Natural History Museum very kindly supplied the following information (*in litt.,* November 6, 1946):

"We do have here on loan from Madrid the Sessé and Moçiño Herbarium, and have had it since the beginning of the civil war in Spain. The material is supposed to contain all the collections made by the expedition, except for some duplicates. The greater part of it has been determined here, but not all of it, and some of the collections from the Pacific coast above Lower California remain without names. It is not certain to me at this time that any of the plants are from the coast of the present United States, although I have been under the impression that some of them were. There are a good many from the north Pacific coast, but they may all be from British Columbia.

"Colmeiro states that members of the expedition collected in the 'peninsula of California,' whatever that may mean. I do not believe any plants are from Lower California. I suggest you write Dr. H. W. Rickett . . . he is working on an account of the Sessé and Moçiño expedition . . . his information . . . is probably more accurate than mine . . ."

Dr. Rickett replied promptly to my query (*in litt.,* November 13, 1946):

"José Moçiño made a voyage from San Blas . . . to Nootka Sound. He left Mexico early in 1792, arriving at Nootka April 29, and returned to San Blas early in 1793. He was accompanied by the artists Echeverría

collections when the first lieutenant ("Mr. Mudge") of the *Discovery* sailed for England carrying dispatches.

". . . I sent home a collection of Seeds adressd to Sir Jos: Banks Bt for his Majesty's Garden & which I was afterwards happy to find that Mr. Mudge had taken great care in their preservation, by which some valuable Plants were added to the great collection at Kew through the uncommon skill & industry of Mr. Acton[1] in rearing them."

A few of Newcombe's comments upon Menzies' botanical work in general, and upon his work on the northwest coast of North America in particular, are included in his preface and are of interest:

"Sir William Hooker, himself an admirable botanical draughtsman, frequently speaks of the excellent drawings in his possession made by Menzies in the field. Some of these are reproduced[2] in the present volume; some seem to have been lost, as the Director of the Royal Gardens at Kew, writing in February, 1915, says that at that time after careful search no evidence has been found of the existence of a collection of plant drawings by Archibald Menzies." Again:

"Although . . . a generous donor of the great collections he made during his voyages, it was many years before his specimens were described and recorded. Amongst the earliest authorities to undertake this work were Sir J. E. Smith, founder and President of the Linnean Society of London (1791); R. A. Salisbury (1806); Esper (1800–1808); Turner, *c.* (1808–1819); Acharius (1810); Pursh (1814); and Lambert (1803).

and the anatomist Maldonado. He must have collected some plants, but I have no evidence of this. The artist made figures of 200 species. Nor do I have any evidence of any calls which they made along the way. Mociño wrote an account of his trip entitled Noticias de Nutka which was published . . . in 1813, with a biography of the author prepared by Alberto Carreño. Aside from this, there is no evidence that Sessé or Mociño, or any of their colleagues, entered what is now the United States, and much evidence that they did not. Much of my information is drawn from archives which I studied in Mexico in 1943. An account of the travels of the Expedition based upon these is now in press and will be published early next year . . . Much additional information is still buried in the archives and libraries of Mexico and of Guatemala . . ."

The story of the copying of the drawings mentioned by Brewer is told in the *Mémoires et souvenirs* of the elder De Candolle, edited by his son.

BREWER, WILLIAM HENRY. 1880. List of persons who have made botanical collections in California. *In:* Watson, Sereno. Geological survey of California. Botany of California 2: 553, 554.

CANDOLLE, AUGUSTIN-PYRAMUS DE. 1862. Mémoires et souvenirs de Augustin-Pyramus de Candolle . . . écrits par lui-même et publié par son fils [Alphonse de Candolle]. Genève. 283-291, *fn.* 291.

COLMEIRO Y PENIDO, MIGUEL. 1858. Martin Sessé. *In:* La botánica y los botánicos de la península Hispano-Lusitana. Estudios bibliográficos y biográficos. Madrid. 184, 185.

———— José Mariano Mociño. *Ibid.,* 185, 186.

MOCIÑO, JOSÉ MARIANO. 1813. Noticias de Nutka. Mexico. [Includes a biography of Mociño by Alberto Carreño.]

RICKETT, HAROLD WILLIAM. 1947. The royal botanical expedition to New Spain. *Chron. Bot.* 11: 1-86.

STANDLEY, PAUL CARPENTER. 1920. Trees and shrubs of Mexico. *Contr. U. S. Nat. Herb.* 23: 13-18.

1. William Aiton, Director of the Royal Gardens at Kew. It is evident here, as elsewhere, that Menzies revised his journal after his return to England.

2. One had been published in Hooker's *Flora boreali-americana* and two in his *Botanical miscellany*.

When Pursh was writing his Flora Americæ Septentrionalis he had in his charge a collection of plants made during the Lewis & Clark expedition of 1804–1806, and these had the first claim upon his attention.

"It will be noticed that many of the species from the coast, the types of which are attributed to these explorers, had already been mentioned or collected some years earlier by Menzies.

"Sir W. J. Hooker's work in connection with the description of Menzies' plants seems to have commenced in 1830 (Botanical Miscellany), and to have ended with his description of *Rhododendron californicum* in 1855. But his most frequent quotations of Menzies' collections, with descriptions of new species, are to be found in his Flora Boreali-Americana, 1829–1840.

"There can be little doubt that the voyage under consideration led to the journeys of Douglas and Scouler. The former, especially, was able to complete the work of his predecessor by sending home the seeds and living plants of trees and flowers of which Menzies, owing to the conditions in which he worked, could only make herbarium specimens. Both Douglas and Scouler acknowledged their indebtedness, and the latter uses the following words: 'While in London (1824) I received much important information from Dr. Richardson and Mr. Menzies with respect to the countries I was about to examine. The knowledge acquired from Mr. Menzies was peculiarly interesting, as he had already explored the very coast I had to visit, and cheerfully allowed me at all times to examine the plants he had collected on the North-west Coast, and to direct my attention to those which were most likely to be useful when cultivated in this country.' (The Edinburgh Journ. of Science, Vol. V, p. 196, foot-note.)"

As noted, the portion of Menzies' journal included in *Memoir V* which is concerned with his work on what is now the northwest coast of the United States ended on June 24, 1792. The transcript continues, however, to October 13 of that year when, setting sail from Nootka, Vancouver's ships shaped their course ". . . to the South East Ward . . ." bound for California.

I turn now to the second published portion of Menzies' journal, quoting from the transcript published in the *Quarterly* of the California Historical Society which covers the three periods (November 14, 1792–January 14, 1793; May 2–5, 1793; October 15–December 8, 1793) when Vancouver's party moved along the coast of California. The plant determinations which Miss Eastwood as editor supplies in footnotes I have italicized and quoted between brackets in the text.

1792

About one month after leaving Nootka (October 13, 1792), Vancouver on the *Discovery* sailed into the Bay of San Francisco, remaining until November 25. Bancroft places the anchorage "in front of Yerba Buena Cove." The *Quarterly* transcript begins with Menzies' entry of November 15 when he noted from the anchorage ". . . clear spots of Pasturage on which a number of Black Cattle were seen feeding in Herds; these inducd us to think favorably of the Country, which we should otherwise from general appearance be apt to pronounce naked & barren."

Menzies was ill and did not ". . . venture on shore till about mid-day . . ." when he took a short walk and found the country ". . . not so very barren as its appearance led us to suppose it." He could identify none of the bushes which were ". . . of a stinted appearance & not very numerous in variety, yet I was not able to ascertain their names, being so unfortunate as not to meet with a single Plant in flower in my whole excursion; I however observd that these Thickets were in a great measure composd of a species of Ever green Oak with Holly-like leaves which I took to be the *Quercus Cocciferus* ['*Quercus agrifolia* Née'] & which did not here grow above fifteen feet high. There was another Ever green nearly the same height but more ornamental & at this time plentifully cropped with red Berries which appeard to be a new species of *Cratoegus* ['*Heteromeles arbutifolia* (Lindl.) Roem.'] . . . also . . . near the sea side, what by its fruit seemd to be a dwarf species of Horse Chesnut ['*Æsculus californica* Nutt.'] & another bushy Plant which appeard to be a *Ceanothus* ['*Ceanothus thyrsiflorus* Esch.'] with some smaller shrubby Bushes of the Class *Syngnesia* ['Probably *Baccharis pilularis* DC. and *Ericameria ericoides* (Less.) Jepson']; two kinds of Willows ['*Salix lasiolepis* Benth. *Salix lasiandra* Benth.'] & the *Lonicera Nootkagensis* ['*Lonicera Ledebourii* Esch.']."

The party was received with every courtesy by the Spaniards, by the Commandant, and in particular by ". . . our good & kind friend the *Padri* who was our first visitor . . ." and who supplied all needs munificently. On November 16 the anchorage was moved into ". . . a Bay abreast of the Proesidio[3] about a quarter of a Mile off shore . . ." and Menzies comments that "What was pompously called by this name had but a mean appearance at a distance & near approach did not at all contribute to make its appearance more favorable."

On November 18 they made a trip of about ". . . four Miles through a hilly country . . ." to the Mission of San Francisco, established about fifteen years before. It had a ". . . small Garden fenced in containing about two Acres of Ground . . . pretty well filld with a variety of Potherbs & Culinary Vegetables . . . a number of Fruit Trees, such as Apples Peaches Figs & Vines, but none . . . very productive or had yet bore any good Fruit . . ."

Menzies was uncertain whether the plants were of poor quality or unsuited to the soil and climate. On the 20th he took another short walk but was again disappointed. ". . . I met with but few Plants in flower, but these were perfectly new to me, & made me more and more regret that it was not my lot to visit this Country at a more favorable Season for Botanical researches." He remained behind when Vancouver and others

3. For a time the name appears in this form in the Eastwood transcript.

made a visit (November 20-22) to the Santa Clara Mission.[4] Bancroft credits the party with "... being the first foreigners who had penetrated so far into the interior." [5]

On November 22 Lieutenant Broughton returned from his exploration of the Columbia River.

"On the afternoon of the 22d the Chatham arrivd in the Port & anchord along side of us. She left Columbia River[6] on the 10th of this Month after exploring it in the Boats upwards of a hundred Miles in a South East direction from its entrance, & in the greatest part of their way they found the Water perfectly fresh ... This examination was made by Mr. Broughton himself with two Boats manned & armed, who were absent from the Vessel ten days & were obligd to return for want of provision without being able to reach the source of the River nor the appearance of its termination[7] ..."

On November 25 the expedition left the Bay of San Francisco where it had been for ten days and proceeded southward, reaching Monterey on November 26 and remaining until January 14, 1793.

The *Doedalus* "store ship" had already arrived. As at San Francisco the party was cordially received by the Spaniards. Menzies was "so exceedingly weak" that he could not go ashore for some days and when he was able to do so was disappointed to find the country "... so exceedingly dry & parchd that there were but few Plants to be met

4. The party went on horseback, a method of travel about which they evidently knew little, for they set off at a "... full gallop & continued galloping & cantering on ... for upwards of twenty Miles before they halted ... By dismounting all were tired, some sore, but good humour prevaild in the highest degree ..." After lunching the same speed was resumed and on arrival at the mission all were "... cold stiff & fatigued ..." and two "... so ill that they were immediately obligd to go to bed ..." When they got back from the excursion they were "... much fatigued with their journey for they found the distance to be nearly eighteen leagues instead of eighteen miles ..."

They had found that the mission's gardens produced abundantly, the fruits doing better "... than at any other of their Northern Settlements in this Country, on account as was supposed of its inland situation." On the way the party had "... passed through Forests of fine Oaks ['*Quercus lobata* Née'] ... scatterd so far apart, that instead of incommoding or obstructing their way ... [they] contributed much to render it more delightfull ..."

5. Von Langsdorff in 1806 mentioned the Santa Clara Mission but did not state that he visited it.

6. When this name was first applied I do not know, but I should not have expected it to be in use in 1792. Perhaps this was one of Menzies' later insertions in his journal.

7. B. A. Thaxter states that Broughton "... went up the river ... to a point above Washougal ..." This is in Clark County, Washington, and not far from the present city of Vancouver.

H. W. Scott has written at some length about the early visitors to the Columbia River and has supplied the origin of many names along its lower reaches; certain of these names appear many times in this story. I mention but a few of those bestowed by Broughton:

"... To the ultimate point he reached he gave the name of Vancouver. All the way up and down he sprinkled names frequently. Walker's Island was named for one of his men. To Tongue Point he gave the name it bears to this day. Young's River and Bay he called for Sir George Young of the British Navy. To Gray's Bay he gave the name it bears as a compliment to the discoverer whose ship had lain in it some months before. When Broughton entered the river he found a small English vessel ... It was the bark *Jenny,* and her commander was Captain Baker. His name is perpetuated in Baker's Bay ... Baker, though disappearing then and there from history, has left his name to us forever." Mount Baker, as already noted, was named for one of Vancouver's lieutenants.

with in a state for investigation." On December 2 a trip was made to the "Mission of *Carmillo*," or San Carlos, four miles from the Presidio; they were pleasantly received —as Lapérouse had been six years earlier.

On December 5 Menzies went ashore with Broughton and Puget. "We strolld towards *Punta de Pinos* by a pleasant walk along the sea side, sometimes passing through Woods, the Trees of which were chiefly Pines & a species I had not met before on the Coast the *Pinus Toeda* ['*Pinus radiata* Don'] . . . Here & there we met clear spots of Pasture & Thickets of Brush-wood, consisting of various Shrubs many of which were new to me, & which I much lamented were not in a condition to be ascertained. Amongst them I observd the common Southern wood & several other species of the Genus *Artimesia* ['*Artemisia californica* Less. *Artemisia heterophylla* Nutt.'] These with a number of others diffusd in this dry Country an aromatic fragrance which was exceeding pleasant."

For December 6 and 7 Menzies ". . . remaind on board examining drawing & describing my little collection & such other objects of natural history as were brought me by the different parties who traversd the Country, & who were in general extremely liberal in presenting me with every thing rare or curious they met with."

On the 8th he joined "a sporting party" which lunched ". . . at a small House near a Garden . . . This we were told was the only Garden belonging to the Garrison, it was not well stockd with Vegetables . . . it was scarcely of a size to supply one fourth of the Inhabitants . . ." This caused him to reflect upon the ". . . indolence of the Spaniards . . . one would have supposd that a small Garden would afford to the Soldiers when off duty a most pleasant amusement & recreation, setting aside the advantages it would yield to his Family, but they live entirely on Garrison provision[8] . . ."

Back from the dunes along the shore were ". . . Clumps of Trees thinly scatterd of the Holly-leavd Oak *Quercus Coccifera* ['*Quercus agrifolia* Née'], & extensive fields of Pasture, but the greatest part of the Country here was coverd with stiff low Shrubs, many of them Evergreens & entirely new to me, but I was this day equally unfortunate in finding but very few plants in flower, many of these shrubby Plants appeard to be of the Class Lyngenesia[9] & were of a fragrant quality."

On December 13 Menzies ". . . traversd the woods & hilly ridge on the Western side of the Bay in Botanical researches & returnd . . . with several Plants & Birds . . . not before seen, which occupied my time in examining & describing for the two following days."

On the 18th he ". . . ranged the Country to the Eastward of the Proesidio for Plants. In the sandy Soil near the sea side I found a procumbent plant in flower which I considerd as a new genus of the Class Pentandria Monogynia & which on my return to

8. Not unlike comments which, one hundred years later, garden owners make to the effect that their gardeners prefer the canned to the home-grown vegetable!

9. Correctly *Syngenesia*.

England I found to be describd as such by *Jussieu* in his Genera of Plants under the name of ——— ['*Abronia arenaria* Menzies Hook. Exotic Fl. t. 193 (1827)=*Abronia latifolia* Esch. Mem. Acad. Petersb. X 281 (1825)'] I also found ... a new *Epilabium* ['*Epilobium franciscanum* Barbey'] a new shrubby species of *Solidago* ['*Solidago spathulata* DC.'] & another of *Polygonum* ['*Polygonum paronchia* C. & S.'] in flower; And in the Pastures the *Ranunculus repens* ['*Ranunculus californicus* Benth.'], a small species of *Oxalis* ['*Zanthoxalis californica* Abrams=*Oxalis*'] with a beautiful new shrubby species of *Mimulus* ['*Mimulus glutinosus* Wend.=*Diplacus glutinosus* Nutt.'] & a new *Verbena* ['*Verbena prostratus* R. Br.'] were in flower & pretty common."

On the 21st he went to the summit of a ridge west of the Presidio and found "... several of the more Northern plants, such as are commonly met with about Nootka & in New Georgia. Of these the following three beautiful Evergreens were here in abundance *Gualtheria fruticosa* ['*Gaultheria shallon* Pursh'], *Arbutus glauca*[1] ['*Arctostaphylos tomentosa* Pursh'], *Vaccinium lucidum* ['*Vaccinium ovatum* Pursh'], which are all new & peculiar as far as I know of to this side of America."

On December 26 the ships were replenished with vegetables, poultry, meat, etc., from the Santa Cruz Mission.

1793

On January 13 Menzies was again able to ship another collection to England when Broughton, commanding the *Chatham,* embarked on the *Activa* brig[2] on his way thither; he was taking to the British government dispatches relative to Nootka "... & as he was so good as promise to take charge of a box of Seeds for his Majesty's Garden ...", Menzies gave him as well "... a letter ... addressd to Sir Joseph Banks Bart."

Menzies comments upon "the great & sudden changes" in temperature experienced in California:

"... Mercury in the Thermometer exposd in a shaded place on shore rose sometimes at noon so high as 90 degrees of Fahrenheit's Scale, & at night fell so low again as 30 degrees ..."

1. Jepson, writing in 1929, notes that Menzies' journal shows "... that he covered exactly the stations where Arctostaphylos tomentosa Pursh grows and of which Menzies was the first collector. Mistakenly attributed to the 'Northwest Coast,' for a full century this species was thought to grow along the Oregon and Washington shores. Without doubt Menzies must have collected his specimens at Monterey, the shrub of the northern coast being a different one, namely Arctostaphylos columbiana Piper, long known, but not named until recent years."

2. The brig *Activa* was commanded by Don Quadra whom Vancouver's party had met at Nootka in 1792 and now met again at Monterey. Broughton was to accompany him as far as San Blas, Mexico, then cross the continent overland, a route which Menzies refers to as an "... arduous & interesting route to England."

These abrupt changes, he felt, affected ". . . every one more or less with rheumatic pains & catarrhous complaints . . ." Among those who were very ill with colds were the ". . . Sandwich Islanders . . . particularly the two Women[3] . . . [who] remain long in a doubtfull state of recovery . . ."

In the afternoon of January 14 Vancouver's ships ". . . weighed Anchor & made Sail out of the Bay of Monterrey." They were bound for the Hawaiian Islands.

From there they returned to California about fifteen weeks later and, from May 2 to 5 were at ". . . what the Spaniards called the *Port of Trinidad.*" Trinidad Head, Humboldt County, is a promontory at the entrance to the harbor. The only trips ashore mentioned by Menzies were made for the purpose of acquiring wood and water. However, Alice Eastwood states in her introduction that "Here Menzies collected *Ribes Menziesii*[4] and *Romanzoffia* and probably other species credited to 'Northwest America' but his notes are not so full as those of Captain Vancouver."

On May 5 the ships ". . . weighed Anchor & made sail out of the Bay . . ." of Port Trinidad; they were bound north and arrived at Nootka Sound on May 20.

The portion of Menzies' journal transcribed in the *Quarterly* resumes again on October 8 when, according to the Eastwood introduction, the expedition ". . . left Puget Sound[5] and sailed south."

They passed Cape Blanco, Curry County, Oregon, on October 14. Between October 15 and 20 Menzies served in the *Chatham* commanded by Puget who planned ". . . to examine an Inlet called by the Spaniards Port *Bodega* . . ." Menzies felt that

3. These natives of the Hawaiian Islands, traveling with Vancouver's party, seem to have been popular with all hands—and more than once they contribute a comic element to the story. They took to horseback riding like ducks to water and there is a story of how they shocked the Spanish matrons on a somewhat formal occasion.

4. Jepson, in 1929, states that ". . . according to his journal Menzies did not accompany the watering party ashore. While this is the type station for his new Ribes, R. Menziesii Pursh, it may be said that the journal shows that the various members of the ship's company . . . were in the habit of bringing objects of natural history to him."

And, writing of the coast redwood, he comments that although ". . . At no place in the journal . . . does Menzies mention seeing this most remarkable of all California coastal trees, either near Monterey or elsewhere, though indubitably he saw it at Trinidad, where it covers the high slopes near the shore with a stately forest." A footnote adds the deplorable fact that "The great trees of this forest have now been logged."

5. The phrase "left Puget Sound" suggested that, between May 20 and October 8, Vancouver's party might have worked along the coast of Washington, possibly of Oregon—outside the locale of the *Quarterly* transcript but in the region of this story. This, however, did not prove to have been the case.

Through the kindness of Drs. Alston and Crabbe of the British Museum photostats were obtained—some 125 pages of Menzies' beautifully written manuscript—covering this period. On October 5 Menzies was at Nootka Sound after four months of surveying northward along the coast of British Columbia. Between October 8 (day of departure from Nootka) and October 14 (when the ships were off Cape Blanco in southern Oregon) it would have been impossible for the expedition to have made a visit—of any importance certainly—to Puget Sound. Menzies wrote on October 5, at Nootka:

"As it was not intended to make much stay at this port but proceed to the Southward where we might

". . . this was likely to afford a good opportunity of landing on that part of California . . ." and ". . . to examine that part of the Coast for Plants . . ." On the 18th the ship ". . . got to within a few leagues of our intended Port . . ." but at the entrance ". . . stood off & on all night."

On October 20 Menzies landed on the west side of Tomales Bay; according to Eastwood, this was near Hog Island (their "Gibson's Island") in Marin County, California. Menzies observed that ". . . the grass & brush wood on this headland had been lately burned down so that I had little opportunity here to augment my botanical collection, the few plants I saw were not different from those I had before met with at San Francisco & Monterrey excepting a new species of Sisyrinchium ['*Sisyrinchium californicum* Ker'] with yellow flowers of which I brought on board live plants for the garden . . . The soil here in general was a loose sandy compost pretty deep & of a dark brown colour . . . We saw no fresh water & the arid aspect of the Country would indicate its being a scarce article if at all procurable . . ."

Setting sail again the ships passed what was ". . . supposd to be *Port Bodega* . . ." but did not enter because of the "exceeding thick" fog.

In the evening they passed *"Pta de los Reyes"* and on October 21 sailed into the port of San Francisco, anchoring close to the *Discovery,* which was "moord in her old birth." The expedition remained here until October 23.

On rejoining his "old ship mates" on the *Discovery* Menzies learned that, instead of meeting with the cordial reception enjoyed on their first visit to San Francisco (November 15 to 25, 1792), the activities of the entire party were now strictly regulated by the Spaniards. When Vancouver desired to take a short ride with three of his officers they were ". . . sufferd to ride about two Miles beyond the Garrison attended by a Soldier, but were positively refusd leave to go as far as the Mission." [6] Now Vancouver, in his turn, gave Menzies little latitude:

"As Capt Vancouver had already obtain leave for some of the Officers to go on shore on pleasure & even exceed the limits of the restrictions, I was in hopes he would be equally inclind to favor my pursuits, & therefore ask'd his leave to go on shore on the morning of the 22d if I should only have the scope of the parties who were daily

enjoy during the winter the mild climate & a supply of refreshments, a party was sent on shore on our arrival to fill our empty Casks with water which from the late rainy weather was now found very commodious in the Cove & another party was employd in cutting fire wood in order to compleat us with these necessary articles for consumption. The Carpenters too were laying in a stock of rough spars for various purposes & I employd my time in filling the frame on the Discovery's Quarter Deck with new live plants so that I might be able before our final departure from the Coast by carrying them with us to the Southward, to judge of those most likely to stand our long Voyage to England & select them accordingly."

6. Bancroft discusses at length the changed attitude of the Spaniards: the newly appointed governor, José Joaquin de Arrillaga, feared that in the past Vancouver and his party had been allowed too much freedom—he disapproved in particular of the earlier visit to the Santa Clara Mission where the visitors might have learned of the weakness of the Spanish settlements; he also seems to have feared for his own tenure of office and issued orders to the various Spanish posts that Vancouver's activities ashore should be curtailed.

landed & employd on the ship's duty, which he refus'd,[7] consequently I had no opportunity while we remaind here of collecting either plants or seeds for his Majesty's
Gardens, which I the more regretted as my state of health when here last year precluded me in a great measure from extending my excursions or examining the shores of
this Harbour with that minuteness I could wish."

The fathers of the San Francisco and Santa Clara missions visited the ships and regretted, although they could not explain, ". . . the singular restraints under which we
sufferd . . ." and which they considered unmerited: ". . . our general conduct at these
Settlements last year was such as deservd a more friendly reception on our return
again." The fathers said that ". . . their good wishes towards us still remain unal-
terable . . ."

On October 23 the party left San Francisco Bay, this time paying for all provisions;
for all except, perhaps, what the kind padres sent: ". . . a supply of Vegetables such
as Greens Radishes Pumpkins Water Melons & a parcel of hazle nuts, together with
a basket of pears & peaches[8] of the produce of Sta Clara . . ." Meeting the *Doedalus*
"store ship" the next day the party, which had been on short rations since leaving
Nootka, was able to resume ". . . the full allowance of provisions . . ." Hanson, commander of the *Doedalus,* brought news from Tahiti of Captain Bligh, then on his second voyage on the *Providence.*

On October 30 the *Discovery* reached Monterey, two days later than the *Chatham,*
which had sailed directly into the harbor despite ". . . dark & thick weather . . ." The
expedition remained at Monterey until November 6.

Orders about communication with the shore were as strict as at San Francisco;
Menzies gives as a possible reason that the new governor ". . . an intruder that came
from a neighbouring Province on the other side of the Gulph of California & took upon himself the command of this Province . . . acts at present as he himself says with-

7. Why Vancouver should have discriminated against Menzies is not clear. For the most part, certainly in
the period covered by this story, the two men seem to have gotten on well together; but Menzies observes
restraint as far as any criticisms of his superior are concerned.

Forsyth refers to Vancouver's health as being poor and mentions various occasions when his relationship with Menzies became unpleasant. Jepson (1929) comments at some length upon the same subject.

Vancouver was a strict disciplinarian—commanders of ships in those days seem to have been cut to
one pattern in this respect. Menzies comments very frankly, when the party was later at Monterey, upon
Vancouver's treatment of James Baily, seaman, who was missing for a week after a fireworks party and
had then ". . . claimd the protection of the Reverend Fathers . . ." Although the Padres interceded on the
man's behalf they could only obtain Vancouver's promise that he would not be punished while the ship
remained in Monterey; however, Baily ". . . remaind in irons till we went to sea & was the[n] punishd at
two different times with six dozen of lashes each time . . ." Menzies, who says very little about his own
difficulties, comments with astuteness, that ". . . a mitigation of this severe punishment at the instigation
of the Worthy Fathers would have been equally efficacious & a more creditable procedure, especially as
we did not find that the severity of the treatment in this case had prevented others from deserting before
our departure."

8. Menzies seems to have been looking "a gift horse in the mouth" when he comments that ". . . the latter were very indifferent fruit, but the Pears were pretty good."

out any instructions whatever . . ." This conduct ". . . was conceivd highly blameable even by his own Countrymen . . ."

On November 6 the ships started south and on the 10th the *Discovery* ". . . hauled into a Bay followd by the Chatham & Doedalus & anchord before the Presidio & Mission of Sta Barbara . . ." Here they were to remain until November 18. Their reception was more cordial at Santa Barbara. On November 11 parties landed ". . . on the duties of wooding & watering, the former was easily procurd at no great distance from the beach as there were some large trees of a kind of ever green Oak ['*Quercus agrifolia* Née'], which they were sufferd to cut down . . ."

On November 12, Menzies "Having previously obtain the Commandant's leave . . . set out pretty early . . . & ascended the hills to the eastward of the Presidio for the purpose of collecting Plants & examining the natural produce of the Country; the day was favorable for my pursuit, but the season of the year & the arid state of the Country was much against it, for though I was surrounded by new & rare objects in almost every step of my journey, yet finding very few of them either in flower or seed I was able to receive but little pleasure or advantage from my excursion; I went through beautiful groves of the Ever green oak ['*Quercus agrifolia* Née'] which here grew to pretty large trees, though at San Francisco & Monterrey the same plant seldom exceeded 15 feet high & grew in crabbed bushes, but here they had clear stems of nearly that height & no wise crouded but scatterd about to beautify the lawns & rich pastures with their shady & spreading branches, so that it was a delightful recreation to saunter through them; the thickets swarmd with squirrels & quails & a variety of other birds which afforded some amusement in shooting them as I went along."

The natives, chiefly old women, were gathering acorns for food.

On November 15 Menzies went westward of a bay which ". . . branch'd back into the Country among extensive salt Water Marshes on which grew vast quantities of Samphire (Salicornia herbacca) ['*Salicornia pacifica* Standley'] . . ."

The next day he devoted to "a solitary botanical excursion," and noted in the woody clumps west of the Presidio ". . . some Poplar ['*Populus trichocarpa* T. & G.'] & American Plane Trees ['*Platanus racemosa* Nutt.'] but they are mostly composd of the ever green oak . . . I was equally unfortunate this day in meeting with but few plants in flower, amongst these there was a beautifull new species of Mimulus ['*Mimulus longiflorus* Grant=*Diplacus longiflorus* Nutt.'] of which I preservd plants & seeds as I had not before seen it any where."

During Menzies' absence *"Padri Vincenti"* of the Buena Ventura Mission had visited the ship with nine mules loaded with various presents ". . . such as sweet & common Potatoes, Onions, Maize, Wheat, some Baskets of Figs & what are called prickly pears &c." He had feared that his mission would be passed by ". . . without his having the satisfaction of contributing to our comforts . . ." Menzies refers to his "benevolent disposition." He accepted a passage to Buena Ventura in the ships, sending word ahead to have a plentiful supply of refreshments ready for their arrival.

On November 18 Vancouver's party took leave of their "generous friends" at Santa Barbara and, with "our friend Padri Vincenti on board," set sail for San Diego. That evening they anchored within a league of the Mission of Buena Ventura and the next day attempted to land the padre but the seas were running too high; however they were able to do so on the 20th; on landing he ". . . receivd the caresses of his flock with tears in his eyes: Those who saw this meeting describd it as a very affecting one & they reported the Mission to be delightfully situated & in a higher state of improvement than any they had seen in the Country, particularly with respect to Horticulture . . ."

Again the ships were loaded with plentiful supplies of food. The next day was spent in recovering the anchor—the cable had "parted in the Clinch"—but they got off on the 22nd, sailed through Santa Barbara and San Pedro channels on the 24th and 25th, passed the island of Santa Catalina on the 26th, and reached a bay ". . . two leagues to the Northward of St Diego." On the 27th they passed the ". . . Promont of St. Diego which is called Point Limos . . . A reef runs near two miles out from this point covered with beds of Sea Weed mostly the Fucus pyriferus Lin. ['*Macrocystis pyrifera* (L.) Ag. Common long Kelp. W. A. Setchell']."

In the harbor the ships anchored opposite the San Diego Presidio. Here again the party met with a ". . . civil & favorable reception . . ." Nonetheless Menzies' activities were restricted by Vancouver:

". . . I might pursue my Botanical researches as far as the Presidio but not to go beyond it without particular leave for that purpose, for though I was not sufferd to go on shore myself this day, the Officers & Boats Crew . . . brought me off Plants & branches of the produce of the Shore that greatly excited my curiosity."

On November 29 he ascended a ridge from which there was an extensive view:

"The whole presented a naked dreary arrid prospect in which there was not a tree to be seen in any direction within our view. The soil on this ridge was sandy & exceeding dry & scorchd, yet it was mostly coverd with shrubbery & brushwood, amongst which I saw a vast variety of Plants that were entirely new to me, but to my no small mortification I met with only two plants in flower & very few in Seed during the whole excursion, there [? these] were a new species of *Euphorbia* ['*Euphorbia misera* Benth.'] & another of *Colutea* ['*Astragalus leucopsis* Torr.']—I also saw the Mesymbranthemum edulis ['*Mesembryanthemum æquilaterale* Haw.'] & five or six species of the Genus Cactus ['*Echinocactus viridescens* Nutt.; *Cereus Emoryi* Engelm.; *Mamillaria Goodridgii* Scheer.; *Opuntia Engelmannii littoralis* Engelm.; *Opuntia serpentina* Engelm.; *Opuntia prolifera* Engelm.'] . . ."

On November 30 "On the Sandy Beach near . . . the entrance of the Harbour I found a new plant in flower which I namd Morinda glauca ['*Heliotropum Curassavicum* L.'], its trailing branches with glaucous leaves contrasted with globular heads of light blue flowers were extremely ornamental to such a barren situation where a plant could hardly be expected to vegetate; I found nothing else in this excursion that I could ascertain what they were."

On November 31 "Different species of wormwood were met with here along shore & other aromatic shrubs with some Evergreens composd the thickets of brush-wood & impregnated the air with a refreshing fragrance."

On December 4 the party spent a pleasant day at the Presidio and on the 5th a large party of officials came for dinner on the *Discovery;* they ". . . seemd very partial to our little convivial parties . . ." and expressed regret at the restrictions to which Vancouver's party was subjected.

The padre of the San Diego Mission sent Menzies ". . . a branch in bloom of the Cassia ['*Parkinsonia aculeata* L.'] which I conceivd had been originally brought here from Mexico as I believe all the genus are tropical plants. He sent me also a quantity of fruit ['*Simmondsia californica* Nutt.'] in Kernels which he said were the natural produce of this Country, they were about the size of small kidney beans & in their taste somewhat like bitter Almonds; to these he ascribd many virtues . . . but what was most pleasant to me, he sent along with them some of the Plants that producd them, which were immediately planted in the frame on the quarter deck & I have the pleasure to add were brought alive to England & placd in his Majesty's Royal Garden at Kew, & as there were many other Plants growing on shore . . . which appeard new & ornamental, I employd two men this & the following day in digging them up & planting them in the same frame, till all the vacant space was filld up with such plants as were likely to be a valuable acquisition to the same royal collection."

On parting, December 7, with the "Padri Presidente," [9] Vancouver presented him with a small organ ". . . set to a miscellaneous collection of about thirty different tunes to the music of which he seemd very partial in his different visits on board last year . . ." The padre planned to take it back to the church at Monterey to ". . . be carefully preservd as a memento of our visit to this Country."

On December 8, 1793, Vancouver's ships left San Diego to cross the Pacific to the Sandwich Islands and the portion of Menzies' journal transcribed in the *Quarterly* ends at this point.

Since, according to his *Voyage,* Vancouver was along the California coast between October 17 and December 2, 1794, it seemed possible that there might exist an unpublished portion of Menzies' journal covering this period, but this does not seem to be the case. On my behalf Dr. E. D. Merrill obtained the following information from Dr. A. H. G. Alston of the British Museum (*in litt.,* March 22, 1945)—insertions between brackets are mine:

". . . There is a letter from Menzies to Banks dated Nootka Sound 1 Oct. 1794 (Banks Corresp. IX. f. 105) which relates that the expedition quitted the Sandwich Is. about the middle of March 1794 & made Trinity Island off Cadiak on the fourth of April. The coast of British Columbia was explored southward & eastward till 19 Aug.

---

9. Lasuen, according to Bancroft.

when they reached lat. 56°N. & returned to Nootka in Sept. Cook's Inlet, Prince William Sound and Port Mulgrave are mentioned.

"Another letter from Valparaiso April 1795 refers to a letter from *Monterey in California* dated Nov. 1794 & says that they left Monterey on Dec. 2 & *'made the coast of California in two or three places'* & then went to Marias Island, off San Blas [Nayarit Province, Mexico], Cape Corrientes ([Jalisco] Mexico), Cocos Isl. [off Costa Rica], passed Galapagos, then Masafuera [off the coast of Chile].

"As for the mss. [of Menzies' journal] it is in the manuscript department & will not be available until after the war as it has been evacuated. However I do not think that it would help as it ends 16 Feb. 1794.

"It is 'additional mss. 32641' & is catalogued on p. 169 of the 'Catalogue of additions to the manuscripts in the British Museum in the years MDCCCLXXXII-MDCCC-LXXXVII.' (1889).

"It was purchased at Sotheby's sale of 13 March 1886 & was probably lot 977 which is described as a journal of Banks.

"It would be interesting to know if there was a continuation & I suspect that the mss. sold at Sotheby's on 14 April 1886 as Vancouver's diary with 260 pp. may be a continuation, but I can't find out who purchased it . . ."

It is evident, therefore, that the only known source covering the California visit is *A voyage of discovery to the North Pacific Ocean, and round the world* . . . by George Vancouver. I follow the first edition of 1798.

1794

The *Discovery* left Nootka Sound on Friday, October 17, 1794. On November 3 it was off Mendocino, California. Vancouver relates on this date that he had intended, before proceeding to Monterey, ". . . to visit the bay of Sir Francis Drake, and from thence in our boats acquire a better knowledge than we had hitherto gained of port Bodega, our course after passing this promontory was directed along the coast to the south-east for that purpose."

But a bad storm arose and Vancouver decided that it was unsafe to attempt to enter ". . . a port, of which, we had so little knowledge . . . to keep the sea, was therefore our only prudent alternative . . ." On November 4 ". . . Monterrey was now lying s. 50 e. . . ." The storm had abated by November 6 ". . . and seeing the land at no great distance . . . we hauled off shore . . . when finding ourselves about 3 or 4 leagues from point Anno Nuevo, point Pinos in sight . . . and having a moderate breeze with fine pleasant weather, we steered for Monterrey, where about two in the afternoon we anchored, and moored nearly in our former situation."

The *Chatham* had arrived on November 2. The ships remained at Monterey for about three weeks—from November 6 to December 2.

From the above it is apparent that Vancouver did not put into any port between Nootka and Monterey.[1]

While the ships were at Monterey, excursions were made ashore. Vancouver wrote: "The weather, since the 8th [of November] had been delightfully pleasant . . . This . . . caused the water in the bay to be so very tranquil, that landing was easily effected on any of its shores, and rendered our intercourse with the country extremely pleasant. The same cause operated to invite the excursions of several parties into the country on foot and on horseback . . . I was . . . on Wednesday November 19 able to join in a party to the valley through which the Monterrey river flows . . . I had an opportunity of seeing what before I had been frequently given to understand; that the soil improved in richness and fertility, as we advanced from the ocean into the interior country. The situation . . . was an extensive valley between two ranges of lofty mountains, whose more elevated parts wore a steril and dreary aspect, whilst the sides and the intervening bosom seemed to be composed of a luxuriant soil. On the former some pine trees were produced of various sorts, though of no great size, and the latter generally speaking was a natural pasture, but the long continuance of the dry weather had robbed it of its verdure . . . yet the healthy growth of the oak, both of the English and the holly-leaved kind, the maple, poplar, willow, and stone pine, distributed over its surface as well in clumps as in single trees, with a number of different shrubs, plainly shewed the superior excellence of the soil and substratum in these situations to that which was found bordering on the sea shore."

Another three-day trip made during the sojourn has a greater interest. On November 27, while repairs were being made on the ship, Vancouver ". . . dispatched lieutenant Swaine on thursday morning with three boats over to the mission of Sta Cruz, in order to procure a supply of garden stuff, as the continuation of the dry weather, here, had made every species of esculent vegetable extremely scarce. Mr. Swaine returned on Saturday evening [the 29th] having been tolerably successful . . ."

Menzies' name does not appear as a member of Swaine's party and it is of course possible, as on other occasions, that plants were brought back to him by one of the men. But it seems probable that he went along and, at Santa Cruz, was the second botanist[2] to collect the coast redwood of California. Jepson wrote in 1923:

". . . It is inevitable that Menzies saw the Redwood at one or more points on the California coast in 1792 or 1793, but 1794[3] is the date of his collection. We often won-

---

1. Nor did he touch at any port in California after leaving Monterey; on December 2 the ships were sailing southeastward: "The coast of New Albion was still in sight . . . This was the last we saw of it . . ." On December 8 they passed the island of Guadalupe (lying off Baja California nearly opposite Sebastian Viscaino Bay) and were well down the coast of the peninsula. On the 14th they were off Cape San Lucas, the southernmost tip of Lower California and from there sailed for the Islas Tres Marias, and so on.

2. In 1923 Jepson was of the opinion that Menzies had been the first botanist to collect specimens of this tree, but by 1929 he had found a record proving that Thaddeus Haenke of the Malaspina expedition had been the first to do so. *See* pp. 20-21.

3. C. S. Sargent in 1896 attributed the discovery of the redwood to Menzies, but as of the year 1796,

dered where he collected this specimen. One day the writer was examining the Menzies specimen in the Herbarium of the British Museum . . . when he chanced to turn over the sheet and . . . written on the back, was the legend 'Santa Cruz, Menzies.' "

Jepson's story of the naming of this remarkable tree is of interest; it appears in his little volume *The trees of California* (ed. 2, 1923):

"Although carried to England in 1796 the Menzies specimen remained undescribed until published as new in 1823 by Aylmer Bourke Lambert . . . Of all genera of conifers known to botanists at that time this Pacific Coast tree is most like the genus Taxodium or Bald Cypress. Lambert consequently placed the new species in that genus and, since it was evergreen, gave it the name Taxodium sempervirens[4] to differentiate it from the previously known deciduous species, Taxodium distichum or Bald Cypress. The word sempervirens had no significance or connotation to Lambert other than as contrast to deciduous. Thus the species remained until Stephen Endlicher, a German botanist, studying the classification of the conifers, decided that this species represented a genus distinct from Taxodium and created the new genus Sequoia in 1847,[5] the Redwood thus becoming Sequoia sempervirens . . . The Redwood was therefore the first Sequoia to be made known, the character of the new genus having been published five years before the discovery of the second species, the Big Tree, Sequoia gigantea of the Sierras."

Since Jepson wrote, the big tree has been separated from the coast redwood and placed in the new genus *Sequoiadendron,* the specific name *giganteum* still retained. The transfer was made in 1943 by J. T. Buchholz of the University of Illinois on the basis of morphological differences. For the discovery of the big tree, see p. 921, *fn.* 9.

Jepson's comments upon the publication of Menzies' collections and his evaluation of the real importance of the man's journal are interesting:

"After the return of the expedition to England, the plant collection was not, unfortunately, worked up as a whole under one direction. While the first set went to the government and is now preserved in the National Herbarium at the Natural History Museum in London, special portions of the duplicates were presented to various botanists and publication of the new plants was done piecemeal in a scattering manner and usually after long delay. An unfortunate lack is the frequent absence of definite stations for the collections, which has resulted, in some cases, in no little confusion. . .

". . . it was at once obvious that one possessed a valuable historical document, but it was also apparent that it contained comparatively scanty records regarding the native vegetation . . . Menzies from time to time makes a few notes, in more or less general terms, of his botanical excursions ashore, but on account of the utter strange-

and "probably on the shores of the Bay of San Francisco." The date is incorrect since Vancouver's party "Arrived all well in the Thames the 20th of Oct. 1795."

4. In his *Desc. Pinus* 2: 24, t. 7, fig. 1 (1824).   5. In his *Syn. Conif.,* p. 198.

ness of the vegetation his comments are not, on the whole, of much significance. At that time the importance of a strictly scientific botanical journal with a numbering of specimens in sequence by stations and dated as collected was not appreciated and such careful methods in the way of field research work had not then been developed.

"On the other hand the journal is amply filled with other matters . . . one might suppose . . . that Menzies was the navigator or geographer of the expedition . . . in all probability, few things escaped his naturalist's eye. Faithfully and industriously he made every effort to carry out zealously the instructions of the Admiralty as prepared for that office by Sir Joseph Banks.

". . . the journal merits high praise in itself. Its statements are characterized by restraint and sobriety, and yet its descriptions are earnest and vivifying. The style, while occasionally a little pompous, is on the whole, nervous and forceful, showing an excellent command of the English vocabulary and exhibiting the writer as a man of deep thought and wide and careful observation, with a philosophical cast to his reflections. It is, by and large, an animated journal and difficult for any one interested in Pacific Coast history to lay down after once begun."

According to the records Menzies' collections are to be found in a number of repositories:

Asa Gray, writing in 1840 upon the content of European herbaria, mentions that the Linnean Society, London, ". . . possesses the proper herbarium of its founder and first president, Sir James E. Smith . . . In North American botany, the chief contributors are Menzies, for the plants of California and the Northwest coast . . ." And, in the British Museum, were "Two sets of plants collected by . . . Menzies in Vancouver's voyage . . . one incorporated with the Banksian herbarium, the other a separate collection. Those of this country are of the Northwest Coast, the mouth of the Oregon River, and from California. Many of Pursh's species were described from specimens preserved in this herbarium, especially the Oregon plants of Menzies . . ." Gray also mentioned that Lambert's herbarium contained a "considerable number" of Menzies' plants. This herbarium was sold at auction in 1842 and, according to Balfour, Menzies' specimens were purchased for the British Museum.

Alphonse de Candolle in 1880 mentions that Menzies' herbarium ". . . principalement de *Cryptogames* est au Jardin roy[al] de bot[anique] d'Édimbourg . . ." W. H. Brewer, in the same year, mentions "A set . . . at Kew . . ." C. V. Piper, in 1906, refers to "A very few . . . in the Gray Herbarium." The *Handbook and guide to the herbarium collections in the Public Museums Liverpool* (1935, 61, 63) states: ". . . The Liverpool Botanic Garden Collection contains . . . many plants from the herbaria . . . of Menzies . . ."

According to Balfour, writing in 1945: "Menzies was generous with his plants during his life and when he died bequeathed his large herbarium of cryptogams to the

Edinburgh Botanic Garden. Many of his plants at the time of his death, other than those given long before to Banks, remained with his family till about 1871, when they were presented to the New College in Edinburgh for the use of the Natural History classes.[6] These were acquired by exchange in 1866 by the British Museum from the Senatus of the University of Edinburgh. Some of his specimens from Oceania are reported to be in the Martius collection at Munich,[7] and his friend A. B. Lambert had others; these last are now in the British Museum . . ."

Balfour who, as I have noted, inherited a number of Menzies' possessions has this to tell of Menzies' later life:

"On retiring from the Navy he practised his profession in London, living in Chapel Place, Cavendish Square . . . At some time past middle life he married but had no child, and his wife predeceased him by six years[8] . . . From 1826 he lived at 2 Ladbrook Terrace, Notting Hill, and died there on 15 February, 1842, aged eighty-eight. It was at this house that Douglas visited him after his return from the NW. Coast in 1827, and previously Scouler had brought him a letter from Douglas . . . In Asa Gray's letters in 1839 there is frequent mention of his meeting Menzies in his own house and those of his friends . . ."

The references to Menzies in Gray's "letters"—they actually appear in his "Autobiography" and in the "Journal" of his first journey "in Europe" which are included in the *Letters of Asa Gray* edited by his wife—are not very numerous. Gray was then twenty-eight years of age, and Menzies "then over ninety" according to Gray, but actually eighty-five. They met in London; on the first occasion Gray was taken to see the old gentleman by N. B. Ward ". . . whose cultivation of plants in closed cases attracted much attention." The second time, after paying ". . . a visit . . . in company with Joe Hooker, to the Zoölogical Gardens in Regent's Park . . ." they ". . . called upon Lambert, Saturday being a kind of public day with him, and there met that Nestor of botanists, Mr. Menzies, whom I found a most pleasant and kind-hearted old man; he invited me very earnestly to come down and see him, which I will try to do some day. Meanwhile I expect to meet him on Tuesday at Mr. Ward's."

Most of the citations refer to Menzies as "kind-hearted," and are such as a courteous young man might express for one three times his own age. Whether they talked over Menzies' experiences along the Pacific coast we are not told—they had taken place almost fifty years earlier.

6. A disposition which, for an historical collection, seems scarcely suitable!

7. When Karl Friedrich Philipp von Martius died in 1868 his very large herbarium was purchased by the King of Belgium for the Jardin Botanique de l'État at Brussels.

8. Menzies and his wife were buried ". . . in Kensal Green Cemetery, London."

ALDEN, ROLAND HERRICK & IFFT, JOHN DEMPSTER. 1943. Early naturalists in the far west. *Occas. Papers Cal. Acad. Sci.* 20: 15-18.

BALFOUR, FREDERICK ROBERT STEPHEN. 1945. Archibald Menzies, 1754-1842, botanist, zoologist, medico and explorer. *Proc. Linn. Soc. London. 156th Session (1943-1944).* 170-183.

BANCROFT, HUBERT HOWE. 1886. The works of . . . 18 (California I. 1542-1800): 510-529. Map ["Map of 1792." 508].

BENT, ALLEN H. 1913. Early American mountaineers. *Appalachia* 13: 60, 61.

BREWER, WILLIAM HENRY. 1880. List of persons who have made botanical collections in California. *In:* Watson, Sereno. Geological survey of California. Botany of California 2: 553.

BUCHHOLZ, JOHN THEODORE. 1939. The generic separation of the sequoias. *Am. Jour. Bot.* 26: 535-538.

CAMERON, HECTOR CHARLES. 1952. Sir Joseph Banks, K. B., P. R. S. The autocrat of the philosophers. London.

CANDOLLE, ALPHONSE DE. 1880. La phytographie; ou l'art de décrire les végétaux considérés sous différents points de vue. Paris. 451.

EASTWOOD, ALICE. 1939. Early botanical explorers on the Pacific coast and the trees they found there. *Quart. Cal. Hist. Soc.* 18: 336, 337.

FORSYTH, JOHN. *See:* Menzies, Archibald. 1923.

GOODWIN, CARDINAL. 1922. The trans-Mississippi west (1803-1853). A history of its acquisition and settlement. New York. London. 211.

GRAY, ASA. 1840. Notices of European herbaria, particularly those most interesting to the North American botanist. *Am. Jour. Sci. Arts* 40: 1-18. *Reprinted:* Sargent, Charles Sprague, *compiler.* 1889. Scientific papers of Asa Gray. 2 vols. Boston. New York. 2: 1-21.

——— 1893. Letters of Asa Gray. Edited by Jane Loring Gray. 2 vols. Boston. New York. 1: 23, 121, 126, 141.

GREENHOW, ROBERT. 1845. The history of Oregon and California, and the other territories on the north-west coast of North America: accompanied by a geographical view and map of those countries, and a number of documents as proofs and illustrations of the history. ed. 2. Boston. 230-259.

JEPSON, WILLIS LINN. 1923. The trees of California. ed. 2. Berkeley. 22, 23.

——— 1929. The botanical explorers of California. VI. Archibald Menzies. *Madroño* 1: 262-266.

——— 1929. Sequoia sempervirens in Granada. *Madroño* 1: 242.

MEANY, EDMOND S. 1907. Vancouver's discovery of Puget Sound. Portraits and biographies of the men honored in the naming of geographic features of northwestern America. New York. 295-297, *fn.*

MENZIES, ARCHIBALD. 1923. Menzies' journal of Vancouver's voyage. April to October, 1792. Edited, with botanical and ethnological notes, by C[harles] F[rederick] Newcombe, M.D. and a biographical note by J[ohn] Forsyth. *Memoir V. Archives British Columbia.* Victoria, B. C.

——— 1924. Archibald Menzies' journal of the Vancouver expedition. Extracts covering the visit to California; with an introduction and notes by Alice Eastwood. *Quart. Cal. Hist. Soc.* 2: 265-340.

MENZIES, DAVID, BART. 1921. Dr. Archibald Menzies, R. N. *Gard. Chron.* ser. 3, 70: 320, 324.

PIPER, CHARLES VANCOUVER. 1906. Flora of the state of Washington. *Contr. U. S. Nat. Herb.* 11: 11.

SARGENT, CHARLES SPRAGUE. 1891. The Silva of North America. Boston. New York. 2: 90, *fn.* 2.

SCOTT, H. W. 1904. Beginnings of Oregon—Explorations and early settlement at the mouth of the Columbia River. *Quart. Oregon Hist. Soc.* 5: 101-110.

SHINN, CHARLES HOWARD. 1928. The Sequoias. *In:* Bailey, L. H. The Standard Cyclopedia of Horticulture 3: 3154-3156.

SUDWORTH, GEORGE BISHOP. 1918. Miscellaneous conifers of the Rocky Mountain region. *U. S. Dept. Agri. Bull.* 680: 21, 34.

SUTHERLAND, J. D. 1931. Archibald Menzies 1754-1842. *Forestry* (London) 5: 5-8.

THAXTER, B. A. 1933. Scientists in early Oregon. *Oregon Hist. Quart.* 34: 330.

VANCOUVER, GEORGE. 1798. A voyage of discovery to the north Pacific Ocean, and round the world; in which the coast of North-West America has been carefully examined and accurately surveyed . . . performed in the years 1790, 1791, 1792, 1793, 1794, and 1795, in the Discovery sloop of war, and armed tender Chatham . . . 3 vols. & Atlas. London.

1800-1810

INTRODUCTION TO DECADE OF

1800 – 1810

IN 1801—the year that Thomas Jefferson became third President of the United States—Spain relinquished to France by the secret Treaty of San Ildefonso her rights in "Louisiana," a vast region with ill-defined boundaries. It was, therefore, from France that the United States government by the Louisiana Purchase acquired the first great tract of land west of the Mississippi River. In the purchase, ratified on October 21, 1803, were included all, or portions of, thirteen future states: Louisiana,[1] Missouri, Arkansas, Iowa, Minnesota, Kansas, Nebraska, Colorado, North Dakota, South Dakota, Montana, Wyoming and Oklahoma.

On May 21, 1804, less than seven months after ratification of the treaty, Captains[2] Meriwether Lewis and William Clark left the town of St. Charles and started up the Missouri River on their way to the west coast. By September 23, 1806, they had returned to St. Louis from the mouth of the Columbia River and the long-cherished ambition of Jefferson to learn what lay between Mississippi River and Pacific Ocean had to some extent been realized. Not until the decade of 1840–1850, or nearly forty years later, did the United States government send other expeditions to the west coast: in 1841 the United States Exploring Expedition arrived there by sea and, in 1843, Frémont overland from the east.

Neither Lewis nor Clark possessed scientific training in any field. Yet, at the behest of Jefferson and with a small amount of instruction before they started on their journey, they made, despite many handicaps, a creditable showing in more than one scientific field. Such plant collections as got safely home and which the explorers had intended should be described by a botanist of their own choosing, were acquired by Frederick Pursh who, and apparently without authority, published the most noteworthy in his *Flora americae septentrionalis*. Of the one hundred and twenty-three species credited to Lewis and Clark by Pursh, it was stated in 1905 that ". . . fully one-half . . . stand as such to-day." No great number of plants perhaps, even as originally credited, but the first collected in lands acquired by the Louisiana Purchase and westward.

On the west coast Russia was in possession of Alaska. In 1806 Chancellor von Resanoff,[3] who had been sailing around the world with the famous von Krusenstern

---

1. Although more than half of the present state of Louisiana lies west of the Mississippi River, it has always been considered "southern" in its affiliations and, in this story, the "trans-Mississippi west" is considered, for its southernmost portion, to begin at that state's western boundary.

2. Although Clark was second lieutenant of artillery when he made his transcontinental journey, Lewis treated him as of equal rank with himself.

3. The name appears in the literature in many forms—Rezanov, Rezánof, etc., etc.

expedition, left the party and journeyed to the Russian colony at Sitka. Finding its inhabitants near starvation, he purchased the small ship *Juno* from an American sea captain and sailed south to the Spanish settlements in California; after a visit of six weeks at San Francisco he returned to Alaska having procured the much-needed food. With von Resanoff traveled as physician Georg Heinrich von Langsdorff, who had agreed to make the voyage when assured that his interest in botany and other lines of natural history would be furthered; but these promises of coöperation were not fulfilled and his scientific pursuits were, he claims, thwarted at every turn by superior and subordinates alike; with the result that his botanical achievements in California were of slight, if any, importance.

Von Resanoff died on his way back to the Russian capital, although he is said to have made a report on California before he left Alaska. Von Langsdorff reached St. Petersburg in March of 1808. It may have been that his favorable opinions of the agricultural possibilities and wealth of California were responsible, at least in part, for stimulating the Russian government to send expeditions in 1808 and 1811 to investigate the possibility of establishing a supply-base for her Alaskan colonies and for starting such an enterprise in 1812 at Fort Ross. If so, his contribution to botany, although indirect, may have been more important than he realized when he left California with his ambitions largely unfulfilled. Fort Ross was abandoned by the Russians on New Year's Day, 1842, but before then Russian men of science, one botanist at least among them, visited the colony; and Otto von Kotzebue in 1816 brought to California two men whose names are well known in botanical literature, von Chamisso and Dr. Eschscholtz.

Lewis and Clark and von Langsdorff were the only men to collect plants in the trans-Mississippi west during this decade. But, presaging more fruitful accomplishment to come, it was on the last day of December, 1809, that an Englishman, John Bradbury, reached St. Louis. He states that, from there, he made excursions into the wilderness during the ensuing spring and summer; he must therefore have been the first trained botanist to work from the Mississippi River westward.

It was in this decade that the American fur trade, the first great American enterprise west of the Mississippi River, had its beginnings. Chittenden tells its story in *The American fur trade of the far west*. For some forty years[4] practically every expedition into the far west, whether initiated by the government or by the private citizen, was to be in some measure, direct or indirect, indebted to the stalwart men of this profession: they opened routes through a wilderness, were well-informed about the Indian tribes and their movements, and were ever ready to advise the unknowing and, however

4. Chittenden defines the period thus: "The fur trade of the Missouri valley . . . did not assume large proportions until after the cession of Louisiana . . . and the exploring expeditions of Lewis and Clark and Pike. Its career thereafter continued practically unchecked until the tide of western emigration set in, about 1843. The true period of the trans-Mississippi fur trade therefore embraces the thirty-seven years from 1807 to 1843. In this trade the city of St. Louis was the principal, if not the only emporium . . ."

poor themselves, to share with those in need. Botanists, with the rest, owed much to their assistance.

It was in 1803 that the first flora of North America was published in Paris; this was the *Flora boreali-americana,* usually cited as the work of André Michaux although it is known that much of it was written by Louis Claude Marie Richard.[5] Since this flora included no plants which Michaux ". . . had not himself gathered or seen . . .", Asa Gray did not consider it ". . . an exhaustive summary of the country as then known . . ." I mention it as a landmark in the botanical literature concerned with North America. Gray, who cited it as the first of only two floras of North America ". . . ever published as completed works . . ."—that is before 1882, the date at which Gray wrote —estimated that "Leaving out the Cryptogams of lower rank than the Ferns . . . the Flora of Michaux . . . contains 1530 species, in 528 genera. No very formidable number . . ."

When, only eleven years later, Pursh published his *Flora americae septentrionalis,* that work recorded, as we shall see, a great increase in the number of recognized indigenous species, and many of them from west of the Mississippi River.

5. The work is cited as Richard's by some European, but never by American, botanists, or so I am told.

CHAPTER III

LEWIS AND CLARK COLLECT A CONSIDERABLE NUM-

BER OF NEW PLANTS ON THEIR TRANSCONTINENTAL

JOURNEY

BEFORE the Louisiana Purchase and before Meriwether Lewis and William Clark made their journey to the Pacific coast, Thomas Jefferson had been desirous of learning what lay west of the Mississippi River: he had considered sending George Rogers Clark, brother of William, into that region as early as 1783. In the last quarter of the eighteenth century he had arranged that the explorer John Ledyard should, from Kamtchatka, travel by Russian boat to Nootka Sound, then move south to the latitude of Missouri and cross the continent to the United States. It is said that this plan fell through when the Empress Catherine of Russia changed her mind and Ledyard was stopped about two hundred miles west of Kamtchatka. It is also said that Jefferson arranged to have Ledyard make the journey westward across North America but this came to nothing because of the explorer's death.

Jefferson's next plan included a botanist. In 1792 he proposed to the American Philosophical Society of Philadelphia that subscriptions be raised to send "some competent person," with a single companion, from east to west. Meriwether Lewis, anxious to go, was selected; also André Michaux, a French botanist who was already recognized as an authority on the plants of eastern North America. According to Michaux's journal,[1] the idea that he should participate in the exploration had been his rather than Jefferson's:

"Le 10 [Décembre, 1792] proposé à plusieurs membres de la Societé philosophique les avantages pour les Etats-Unis d'avoir des Informations Geographiques des Pays de l'Ouest du Mississipi et demandé qu'ils ayent à endosser mes traites pour la somme de 3600 lb., moyennant cette somme je suis disposé à voyager aux Sources du Missouri et même rechercher les rivièrres qui coulent ver l'Ocean Pacifique."

Michaux states that his proposition was accepted and that he named his terms to Jefferson, then Secretary of State: among others he was to retain all rights to discoveries in natural history and the Philosophical Society all information relating to geography. The *Proceedings* of the Society record that on April 19, 1793, it was decided to raise subscriptions; also that, on December 16, 1796, the "Michaux Committee" reported to the Society that the plan had failed. Various reasons for the failure have been offered: one was that Michaux had been instructed by Genêt, "Ministre Pleni-

---

1. Both Asa Gray (1842) and C. S. Sargent (1888) published transcripts of this portion of Michaux's journal. These vary in minor details. I quote from the Sargent transcript.

potentiaire de la Republique française," to continue his work elsewhere.[2] Lack of funds may have played a minor part for the same *Proceedings* record, on April 4, 1800, a list of subscriptions which, even in that day, could not have gone far towards meeting Michaux's requirements: Alexander Hamilton had contributed $12.50, George Washington $25.00, Jefferson $12.50, and so on—the total $128.50.

Unfortunately, when the expedition finally went into the field, it included no one with training in any scientific line.[3] Offsetting his lack of training were Lewis' natural endowments: Frederick Pursh, the first to publish upon the botanical collections of the expedition, mentions his "discerning eye" and Jefferson noted both his ". . . talent for observation, which had led him to an accurate knowledge of the plants and animals of his own country . . ." and his ". . . fidelity to truth so scrupulous that whatever he should report would be as certain as if seen by ourselves . . ." Moreover, aware that Lewis lacked ". . . familiarity with the technical language of the natural sciences . . ." Jefferson sent him to Philadelphia where members of the American Philosophical Society, among them Dr. Benjamin Rush, Dr. Caspar Wistar, and the botanist Dr. Benjamin Smith Barton, gave him some instruction in ". . . those objects on which it is most desirable he should bring us information." Barton, Lewis wrote Jefferson, had even considered accompanying him as far as the Illinois, but this plan was not put into effect.

Jefferson's instructions to Lewis were comprehensive: several copies of all notes were to be made in "leisure times," one copy upon the "paper of the birch" since this was less liable to be injured by dampness than "common paper." Observations were to be ". . . entered distinctly & intelligibly for others as well as yourself . . ." He was to note the ". . . soil & face of the country; its growth & vegetable productions; especially those not of the U. S. . . ."; the ". . . dates at which particular plants put forth or lose their flower or leaf . . ." and so on. The amazing thing is how many of the instructions seem to have been carried out.

Although a number of unauthorized accounts of the Lewis and Clark expedition came out earlier, no authentic record of the journey appeared until 1814. Lewis had

2. Elliott Coues considered that this explanation was presented ". . . with the reserve of the true diplomatist . . ." and that actually Jefferson had become aware that Michaux was a spy in the employ of the French government—this fact ". . . a matter of common tradition, if not of verifiable history."

3. In *A life of travels*, Rafinesque recounts that, in 1804, while considering whether to return to Italy or remain in America, he had had ". . . several offers of employment, not quite to my taste . . . I had once hesitated however when I was told that I might be admitted as Botanist in the expedition which Lewis & Clark were then preparing to survey the Missouri and cross the Oregon mountains. The dangers of this long journey would not have prevented me to join it; but the difficulty was to be admitted as Botanist or learned Surveyor: it appears that [Alexander] Wilson who wished to join the party as Ornithologist or Hunter, could not obtain the permission. The same might have happened to me; but I did not apply: this journey did not promise any reward, while I had the offer of a lucrative situation in Sicily, a country new to me."

We hear nothing of this from other sources and it may merely have been one of a number of suggestions made by Rafinesque's friends.

undertaken to publish it but, after his tragic and still unexplained death in 1809, this responsibility devolved upon Clark who, with the help of Nicholas Biddle and Paul Allen—the title-page states that Allen prepared it for the press—finally brought out the book some eight years after the return of the party. Publication had been unbelievably difficult from start to finish. Biddle had to seek for a publisher. Writes Elliott Coues, "The spectacle of a Biddle begging all Philadelphia to publish Lewis and Clark!" The net returns, before Allen's bill for five hundred dollars was received, were one hundred and fifty-four dollars and ten cents; Clark received a copyright to the copper plates and retained the right to bring out a second edition! More than two years after the book was issued he wrote Jefferson that he had ". . . not been so fortunate as to procure a single volume as yet." Reflecting, a century and a half later, upon the importance now placed upon the journey, the indifference long demonstrated, not only as to publication of the records but as to their safekeeping, is well-nigh unbelievable.

Several critical editions of the journals have been published and I found two of special value:

The first, in which the history of the expedition was "reprinted from the only authorized edition of 1814,[4] just mentioned, was published in 1893 and was the work of the ornithologist Elliott Coues. The editor's notes and commentaries are of great interest. Coues, who learned of the existence of the original manuscripts only when his book was nearing publication, has been criticized for the manner in which he incorporated some of that material and, very justifiably, for marring the originals with his own "verbal changes and comments," and in general treating "the material as though mere copy for the printer, which might be revised by him with impunity."

The second, in which the journals were "Printed from the original manuscripts," came out in 1904 and 1905 and was the work of the Reverend Reuben Gold Thwaites. In it the scientific data assembled by the explorers, and which they had planned to publish independently of the narrative, appeared for the first time, nearly one hundred years after the return of the expedition. Paxson has commented that it supplants the Coues edition "for matters of definite erudition."

These three editions, together with journals kept by other members of the party, have supplied the basic facts for much, if not most, of the miscellaneous literature concerned with the Lewis and Clark accomplishment.

I shall not repeat in detail the well-known journey but merely quote Coues's fine epitome of the explorers' achievements, mention a few dates and localities (ones which are referred to many times by later travelers), and then turn to the history of the botanical collections and botanical records. Wrote Coues:

". . . The History of this undertaking is the personal narrative and official report of the first white men who crossed the continent between the British and Spanish possessions. When these pioneers passed the Rocky mountains, none but Indians had

4. James K. Hosmer's history of the expedition was also transcribed from the Biddle edition of 1814.

ascended the Missouri river to the Yellowstone, and none had navigated the Columbia to the head of tide-water. The route was from Illinois through regions since mapped as Missouri, Kansas, Iowa, Nebraska, South Dakota, North Dakota, Montana, Idaho, Washington, and Oregon. The main water-ways on the Atlantic side of the mountains were the Missouri and Yellowstone; on the Pacific side, Lewis' river, the Kooskooskee, and the Columbia. The Continental Divide was surmounted in three different places, many miles apart. The actual travel by land and water, including various side-trips, amounted to about one-third the circumference of the globe. This cost but one life, and was done without any serious casualty, though often with great hardship, sometimes much suffering, and occasional imminent peril . . . The duration of the journey was from May, 1804, to September, 1806 . . . This is our national epic of exploration, conceived by Thomas Jefferson, wrought out by Lewis and Clark, and given to the world by Nicholas Biddle. Perhaps no traveler's tale has ever been told with greater fidelity and minuteness, or has more nearly achieved absolute accuracy . . ."

Seven months after the ratification of the Louisiana Purchase, or on May 21, 1804, Meriwether Lewis and William Clark with a party of less than fifty persons, started up the Missouri River. The Mississippi had been crossed on May 14.[5] Following the course of the Missouri they had, by November of that year, reached a point about fifty miles above present Bismarck, North Dakota. Here they erected quarters for the winter of 1804–1805, naming it Fort Mandan. The site was on the east bank of the Missouri River, in present McLean County, and, according to Chittenden, ". . . 7 or 8 miles below the mouth of Knife River and opposite, though a little above, the site where Fort Clark later stood." [6] It was, as the name suggests, in the territory of the Mandan Indians.

On April 7, 1805, the journey up the Missouri River was resumed. By late June they were in western Montana and spent eleven days portaging the Great Falls, a distance of some eighteen miles; caches were left at the beginning and end of the portage and I refer to these again. At the end of July they reached the Three Forks of the Missouri and followed the Jefferson and its tributary, the Beaverhead; crossing the Continental Divide through the Lemhi Pass they entered Idaho, turned northward and eventually crossed the Lolo Pass in the Bitterroot Mountains; turning southwest they reached the Clearwater (their Kooskooskee) and followed it to present Lewiston where they crossed into the state of Washington and followed the Snake River to its junction with the Columbia, descending that stream to its mouth. They had their first glimpse of the Pacific on October 7, 1805. For their winter quarters they first selected a site on the north or Washington side of the Columbia but changed to a less exposed

5. As late as 1916 a Lewis journal, covering the period from August 30, 1803, to May 14, 1804, was published for the first time, under the editorship of Milo M. Quaife (*Wisconsin Hist. Publs. Colls.* 22).

6. Fort Clark was in Oliver County, and Knife River enters the Missouri from the west, in Mercer County. Present Mandan, as distinguished from Fort Mandan, is in Morton County, near the mouth of Heart River which joins the Missouri from the west, and is about opposite Bismarck, Burleigh County.

one about three miles up the Lewis and Clark River (the "Netul") on the south or Oregon side, in present Clatsop County. They named their headquarters Fort Clatsop.

On March 23, 1806, the return journey was begun. The same route was followed until the party had again crossed the Lolo Pass. Here the two explorers separated for a time: Lewis turned northeastward, eventually reaching high up the Marias River and thence descending the Missouri River to its confluence with the Yellowstone; Clark followed a route across the Big Hole to the Three Forks where he crossed to the Yellowstone and followed that river to its mouth. Here the explorers were reunited. They then descended the Missouri River to St. Louis, arriving at their destination on September 23, 1806, having taken about six months to make the return journey.[7]

I now turn to the history of the botanical collections:

The first plants collected, between the starting point in Missouri and Fort Mandan, were sent down the Missouri River by barge on April 7, 1805, the day the expedition again started westward, and were consigned to the care of the United States Commandant at St. Louis, Captain Amos Stoddard, who was to forward them to the President. Clark noted on April 3, 1805, that ". . . we were all day engaged in packing up Sundery articles to be sent to the President of the U. S." The contents of various boxes are listed and Box 4 contained, among other things, "Specimens of plants numbered from 1. to 67. Specimens of Plants numbered from 1 to 60 . . . A specimon of a plant, and a parcel of its roots highly prized by the natives as an efficatious remidy[8] in cases of the bite of the rattle Snake or Mad Dog."

This first shipment of plants was entered in the Donation Book of the American Philosophical Society on November 16, 1805, and was then forwarded to Dr. Benjamin Smith Barton, some of the contents checked as having been received. Those that "came" were plants "numbered 1 to 60"; unchecked was the "efficatious remidy"; also specimens numbered "1. to 67."

Jefferson is said to have had some of the natural history collections sent from Fort Mandan at Monticello—although plants are not specified—and some went eventually to Charles Willson Peale's Museum in Philadelphia; finally, the plants of this shipment, although by a circuitous route, reached a safe haven in the Academy of Natural Sciences of the same city. I shall mention them again.

When the expedition arrived at the Great Falls of the Missouri they made at least two caches. A map from Clark's notebook, reproduced in Thwaites, shows one below the falls, or at the beginning of the portage; this was near Portage Creek—now known

7. H. M. Chittenden has stated: ". . . This celebrated performance stands as incomparably the most perfect achievement of its kind in the history of the world . . . The information gathered was so exhaustive and correct that *Lewis and Clark* continued to be the standard authority on the region traversed by the expedition for fully forty years thereafter."

8. Thwaites comments in his introduction to the journals that Clark, the ". . . less educated of the two, spelled phonetically, capitalized chaotically, and occasionally slipped in his grammar . . ."

as Belt Mountain, or Belt Creek.[9] Another is shown above the head of the falls or at the end of the portage; this was near "white bear" Islands which were situated above the confluence of the "Medicine," or Sun, and the Missouri rivers.

On June 28, 1805, Lewis recorded that ". . . my specimens of plants minerals &c. collected from fort Mandan to that place . . ." were deposited *"At the lower camp"* which, although slightly confusing, was apparently the cache at "white bear" Islands. For, back there on July 13, 1806, Lewis wrote:

"removed above to my old station opposite the upper point of the white bear island . . . had the cash opened found my bearskins entirely destroyed by the water, the river having risen so high that the water had penetrated. All my specimens of plants also lost . . ."

Numerous other caches are mentioned in the course of the narrative but, as far as plant content is concerned, this was undoubtedly the most important. Lewis' comments, just quoted, explain where and how the misfortune, alluded to by Frederick Pursh in the preface to his *Flora,* occurred:

". . . A much more extensive one, made on their slow ascent towards the Rocky mountains and the chain of the Northern Andes, had unfortunately been lost, by being deposited among other things at the foot of those mountains . . ."

Most of the plant collections which, when finally reassembled, were determined at the Gray Herbarium by Messrs. Robinson and Greenman in 1897, were dated either 1804 or 1806. Only fourteen of those listed (and dated) in Meehan's paper which supplies the determinations of these botanists, bear the date 1805 and, of these, all were collected later than June (the month of the portage of the Great Falls of the Missouri River).

Therefore, except for the Fort Mandan shipment, most of the existing specimens were gathered on the explorers' hasty return trip of only six months, the party leaving Fort Clatsop on March 23, 1806, and arriving at St. Louis on September 23. Pursh, who was certainly in a position to assess their value, refers to them as "A small but highly interesting collection of dried plants . . ."

This later collection, like the one from Fort Mandan, should, presumably, have gone to Benjamin Smith Barton, who is known to have received books and papers relating to the natural history of the expedition and who, from the first, seems to have been regarded as the person destined to publish upon this phase of the work. After Lewis' death Barton made a contract with Clark in which he agreed to undertake this task and complete it six months after the publication of the narrative of the expedition. But up to the time of his death in 1815 he is said to have done nothing towards fulfilling this obligation. F. W. Pennell, writing of Barton, has mentioned the fact that

9. One is reminded of the old saying: "If a donkey kicks you, don't never tell!" For Coues notes that Portage Creek ". . . is now known as Belt Mountains creek . . . It is called by mistake Bear creek on one map before me . . ." The erroneous name "Bear R." appears on the map of the Lewis and Clark route "Prepared by Elliott Coues" himself!

his health was poor and has commented that "... To us of a later time, Barton was greater in intentions than in practice, in the fields that he perceived should be cultivated than in ability to accomplish the needed tasks ..." While such personal characteristics may to some extent have explained the nonfulfillment of his contract with Clark, it is possible that there were other reasons—reasons involving Pursh and the plant specimens in particular.

Frederick Pursh whose birthplace and whose name have been variously cited,[1] was an emigrant to America, and had been employed by Barton to help him in the preparation of a flora, a work never brought to completion; this arrangement is said to have terminated in 1807. Pursh then went to New York; in 1811 (since in that city he "... found things in a situation very unfavourable to the publication of scientific works, the public mind being then in agitation about a war with Great Britain ...") he took his departure from America. In 1814 he published his *Flora americae septentrionalis* in London, the same year that the Biddle edition of the Lewis and Clark narrative was issued in Philadelphia.

A Frenchman is said to have defined opportunism as the preference of expediency to principle. And in that sense Pursh, on his own record, would seem to have been an "opportunist." The preface to his *Flora* is truly naïve in telling how, both in America and in England, he made a point of seeking out those who might be useful to his purpose. He extols the accomplishments and the knowledge of his various patrons, expresses regret that their labors have never been brought to completion, and then tells how valuable he himself has found their material, without even a plausible explanation of how he obtained authority to use it. He is known to have disregarded the rights of such botanists as John Bradbury and Thomas Nuttall; indeed, M. L. Fernald has characterized him as "One of the most active and apparently unscrupulous early Philadelphia botanists ..."

In the preface to his *Flora*, Pursh states that in 1806[2] he had returned from a collecting trip to the "Northern States," and, soon after, "... had the pleasure to form an acquaintance with Meriwether Lewis, Esq., then Governor of Upper Louisiana, who had lately returned from an expedition across the Continent of America to the Pacific Ocean, by the way of the Missouri and the great Columbia rivers, executed under the direction of the Government of the United States. A small but highly inter-

1. W. J. Hooker, writing "On the botany of America," recorded in 1825: "This celebrated Botanist, we believe, has been commonly, though erroneously, considered a native of Poland. While Professor Silliman was in Canada, in the autumn of 1819, he had a personal interview with Mr. Pursh, in the course of which the latter stated expressly that he was a *Tartar, born and educated in Siberia, near Toboltski.* 'Indeed,' says Professor Silliman, 'He possessed a physiognomy and manner different from that of Europeans, and highly characteristic of his native country.' "

On the other hand, Dr. H. W. Rickett, who does not mention the source of his information, wrote me on March 16, 1949: "Pursh was born in Saxony and originally named Friedrich Traugott Pursch. When he moved to England he called himself Frederick Pursh, and dropped the middle name."

2. According to Pennell the year was 1807.

esting collection of dried plants was put into my hands by this gentleman, *in order to describe and figure those I thought new, for the purpose of inserting them in the account of his Travels, which he was then engaged in preparing for the press* [italics mine]. This valuable work, by the unfortunate and untimely end of its author, has been interrupted in its publication; and although General Daniel [*sic*] Clark, the companion of Mr. Lewis, (to whom I transmitted all the drawings prepared for the work,) undertook the editorship after his death, it has not, to my knowledge, yet appeared before the public, notwithstanding the great forwardness the journals and material were in when I had the opportunity of perusing them."

Despite this frank statement that he had been handed the plants for a particular purpose, and despite the fact that the records were well on the way towards publication, he closes the subject with the comment that "The descriptions of those plants, as far as the specimens were perfect, I have inserted in the present work in their respective places, distinguishing them by the words '*v.s. in Herb. Lewis.*' "

In view of Pursh's reputation one wonders whether Lewis, if he turned the specimens over *to Pursh* at all, did not do so with the intent that they should be given to Barton whose assistant Pursh long was. For it seems unlikely that Lewis, and later Clark, should have made a similar arrangement with two persons. Unlikely also that Barton should have made his contract with Clark unless he believed himself in possession of the specimens which would be essential for his work. To find them gone—for Pursh took what he considered the most important to England—might well have had *some* bearing on Barton's failure to fulfill his obligation.

In any event, and justifiably or not, Pursh was the first to publish upon the Lewis and Clark specimens. He is said to have done the task well: Fernald refers to his ". . . amazingly brilliant 2-volume *Flora Americae Septentrionalis . . .*"

How many of the specimens Pursh took to England is not quite clear; in 1841 Asa Gray refers to "a few" in Lambert's herbarium in London. Not until 1898 did their whereabouts—indeed the fact that they were back in America—appear in print. The story is this:

In the Coues edition (1893) the plants mentioned in the journals were identified by Dr. Frank Hall Knowlton, a paleobotanist; he saw no specimens. In 1897 Charles Sprague Sargent, in two articles in *Garden and Forest,* reviewed these Knowlton determinations in so far as certain trees were concerned. The publication of these articles started a search for the specimens and in 1898 Thomas Meehan, associated with the Philadelphia Academy of Natural Sciences, recounted the outcome:

"What became of the complete collection has never been definitely ascertained up to this time . . . It was understood that Pursh took these plants to England, and that they were left by him with Mr. A. B. Lambert, Vice-President of the Linnæan Society . . . Lambert's herbarium was finally distributed and in some way not known to the writer, a number of Lewis' plants, forming Pursh's types, and marked 'from Lambert's Herbarium' became part of the Herbarium of the Academy of Natural Sciences of Philadelphia. Professor C. S. Sargent suggested . . . the possibility of some of the material being yet in the custody of the American Philosophical Society . . . After long and diligent search, packages of plants were found which could only be these . . ."

Meehan tells how the specimens were identified, describes their condition, which, despite the attacks of beetles, was in the main "fair," and tells of their placement in the Academy,[3] which was equipped to give them the requisite care. The collection, Meehan explains, was made "for the most part on the return trip. Many specimens . . . were collected and saved between the Rocky Mountains and Fort Clatsop, their winter quarters near the Columbia River." He continues:

"While in doubt as to the authorship of the labels attached to the specimens, note was made of an entry in the minutes of the American Philosophical Society under date of Nov. 15, 1805, that a box of plants was received from . . . Lewis. The seeds were sent to Mr. William Hamilton[4] and the 'Hortus Siccus referred to Dr. Barton to examine and report.' A full examination of the collection, revealed this package also. It contained the plants collected in 1804 between St. Louis and Fort Mandan. Pursh had evidently been over this . . . He evidently studied these collections before starting to Europe with them, leaving the duplicates, where there were any, and those which were too imperfect to be easily recognized . . ." !

Meehan goes on to tell that the rediscovered specimens were studied by Dr. B. L. Robinson and Mr. J. M. Greenman of the Gray Herbarium, Harvard University, and publishes their determinations—in parallel columns appear the accepted name of each plant and its treatment in Pursh's *Flora*. The Robinson-Greenman report was dated October, 1897.

"To the above detail by Dr. Robinson and Mr. Greenman, it may now be noted that this collection contains specimens of all but sixteen of Lewis's plants as described by Pursh in his *Flora*. Of these sixteen, seven . . . are represented already in the specimens from Lambert's herbarium, leaving but nine of the plants missing from the collection as described by Pursh. Only a few of these nine missing ones are of material

3. The name of the public-spirited individual who acquired these specimens seems not to be known. Pennell wrote in 1942:

"These specimens passed into Lambert's herbarium, for which they were mounted on uniform small sheets of paper and the source inscribed by Lambert on the reverse of an upper corner. After the botanist's death in 1842 his collections were sold at auction and widely scattered, but the North American portion, including those plants taken by Pursh from series that otherwise passed to the American Philosophical Society, were purchased and presented by some unknown donor to the Academy of Natural Sciences of Philadelphia."

4. Pursh's preface to his *Flora* tells that, near Philadelphia, he had visited ". . . the extensive gardens of William Hamilton, Esq., called the Woodlands, which I found not only rich in plants from all parts of the world, but particularly so in rare and new American species." I refer to Hamilton's part in growing some of the Lewis and Clark plants again, but note here that the late William H. Judd, of the Arnold Arboretum, made numerous attempts on my behalf to discover whether there still existed records of material sent to Hamilton or grown in his garden but could discover none. Mr. Judd had a wide acquaintance among horticulturists of Philadelphia.

importance. For all practical purposes, all the plants of Lewis and Clark's expedition are now deposited in the Academy."

The localities cited in this report were those recorded on the specimens. In 1898 Elliott Coues, familiar with the routes of the explorers, published a paper[5] in which he rendered these often vague localities as precise as the data permitted.

In the Thwaites edition of the journals and manuscripts, as stated, much of the scientific work of the explorers was published for the first time, some of botanical interest. Dr. William Trelease, then Director of the Missouri Botanical Garden at St. Louis, identified a number of the plants mentioned. That he examined the specimens is not stated.

Such, briefly, is the history of the plant material[6] collected on this famous journey.

Frederick Pursh, the first to estimate the scientific value of the plant collections, wrote in the preface to his *Flora:*

". . . the small collection communicated to me, consisting of about one hundred and fifty specimens, contained not above a dozen plants well known to me to be natives of North America, the rest being either entirely new or but little known, and among them at least six distinct and new genera."

Another report on the collections is found in the Thwaites edition. There Mr. Stewardson Brown, assistant to the Curators of the Herbarium of the Philadelphia Academy, after giving the story of the lost specimens—in much the same vein as Meehan—tells us that Pursh

". . . refers in his descriptions to one hundred and twenty-three (123) species . . . the majority of which he considered to be new to science . . . As might be expected, Pursh in his work of identification had fallen into some errors, as a subsequent critical study of the collections by Messrs. Robinson and Greenman . . . demonstrated; but nevertheless fully one-half of the one hundred and twenty-three species referred to by Pursh in the text of his *Flora,* stand as such to-day. The collection has also been found to contain, in addition to these, a number of species not recognized by Pursh as dis-

5. One may smile at Coues's opening paragraph: "Many years ago I prepared for publication in these *Proceedings* a paper on the plants of Fort Macon, N. C. It never appeared, because I submitted it to Professor Asa Gray, who told me it was a very good one, but asked me what was the use of printing it." !

6. In the introduction to his edition of the Lewis and Clark journals, etc., Thwaites tells how the many *documents* relating to the expedition were finally traced to a number of repositories until, ". . . seventy-five years after Jefferson's quest . . . [he had attempted to have the documents reassembled], there have at last been located presumably all of the literary records now extant of that notable enterprise in the cause of civilization . . ." Much of this summary was reprinted, with minor changes, on two occasions under the title "The story of Lewis and Clark's journals": first in 1904 (in the *Annual Report of the American Historical Association* of Washington, D. C.) and again in 1905 (in the *Quarterly of the Oregon Historical Society*).

Thwaites had wisely inserted *presumably,* for I read in the *New York Herald Tribune* of March 20, 1953, that records written by Clark, and with notations by Lewis, have recently been found in an old house in St. Paul, Minnesota, and turned over to the Minnesota Historical Society, and that Professor Ernest S. Osgood of the University of Minnesota considered them "priceless" and of much historical importance.

tinct, but which have proven to be the earliest collections of many of the species of subsequent authors. That some of these, although entirely new to Pursh, may have remained undescribed from what he considered a lack of sufficiently perfect material, seems probable . . . The collection, as preserved in the Herbarium of the Academy . . . to-day consists of specimens of one hundred and seventy-three recognizable species, mostly in fair condition; these include fifty-five of Pursh's types recognized as species at the present time, with thirty-eight additional ones, now for various reasons not considered tenable. The number lacks but fifteen of the species referred to by Pursh as contained in the Lewis Herbarium; of these, some may be included among the existing specimens, but, if so, are not recognizable from their descriptions. In Pursh's *Flora* he describes but five new genera; of this number four are based upon the plants received by him from Lewis, all of the specimens still existing in fairly good condition. Of these, three still retain the names conferred upon them by Pursh, two in honor of the heads of the expedition—represented each by a single species: *Lewisia rediviva* (Pursh Fl. 368), and *Clarkia pulchella* (Pursh Fl. 260, with an excellent figure); and the third, *Calochortus,* the genus of a handsome group of liliaceous plants confined to the Western United States and Mexico. The name selected by Pursh for the fourth new genus represented in the collection had, unfortunately, been proposed as early as 1775 by Aublet for another plant, which necessitated a new name. The discrepancy was noted by De Candolle, who in 1817 in the transactions of the Linnean Society, the original medium of publication of Pursh's new genera, renamed the genus in honor of Pursh. The species collected by Lewis, which is represented by an excellent specimen, now bears the name *Purshia tridentata* (Pursh) D. C."

The *Transactions* of the Linnean Society, mentioned by Stewardson Brown as the "original medium of publication of Pursh's new genera," fail to disclose any names published by Pursh,[7] although his *Flora* cited the genera *Calochortus, Lewisia,* and *"Clarckia"* as having been published in volume eleven of the *Transactions*—in the case of *"Clarckia"* Pursh indicated that he did not know the page ("v. 11, p. . . ."). The genus *Tigarea* was similarly cited.

Since papers appearing in the *Transactions* were read in the Society before publication, I made inquiry—for I confess to a certain scepticism regarding Pursh's statements —whether this preliminary step had been taken, and in this instance my suspicions were unfounded. Dr. S. Savage, in a letter to Dr. E. D. Merrill, dated May 24, 1945, supplied the information that four of Pursh's papers were read in orthodox manner although not necessarily by Pursh himself. These were: "Descriptions of Plants" (1812); "Four new genera of Plants" (1812); "North American Plants" (1812); "Hosackia" (1813). The letter reads:

"There is also a reference in the Council Minutes to a paper,—Catalogue of American Plants (1813). None of these papers were published by the Society. By good luck,

---

7. The only reference to Pursh which I have found in the *Transactions* was as donor of his *Flora* which is listed as no. 772 of the catalogue of the library of the Society.

the MS. of 'Four new genera of Plants,' 1812, has survived, and is before me now. I will give a short description of it . . . The paper is in the form of a letter addressed to A. B. Lambert, who no doubt 'read' it before the Society. 'Read Jany 21, 1812' Headed by Pursh—'London Jan. 20th 1812.' The general text gives some account of the collection made by 'Meriwether Lewis & Daniel [*sic*] Clark.' The scientific part gives generic and specific descriptions of

> Lewisia
> ————rediviva
>
> Clarkia
> ———— pulchella
>
> Calochordus [*sic*]
> ———— elegans
>
> Tigarea Aubl. guj.[8] 2.-p. 917.
> ———— tridentata

"Habitats are given and other notes. These texts are in two versions, although both are dated 20 Jan. 1812. One version has drawings of Lewisia rediviva and Calochordus elegans. Concerning Lewis & Clark's collections, Pursh states 'This valuable collection was put into my hands, by the late M. Lewis for examination;[9] I described all the plants & took drawings of many of them—but the unfortunate death of Mr. Lewis put a stop to the publication of his very interesting work. The drawings are now in the possession of one of the publishers in Philadelphia & the descriptions only are in my hands . . .' The MS. consists of 7 folios, of which 12 pages are written on."

I mention one more study of the Lewis and Clark plants which was apparently based, not on the specimens, but upon the descriptions found in the explorers' records. When Elliott Coues annotated geographically the localities cited on the specimens as recorded in the Robinson-Greenman report published by Meehan, he had this to say:

". . . there remains for some one the agreeable and useful task of reviewing Lewis and Clark's botanical *text* as distinguished from their *specimens*. For it is a curious fact, as I find on studying Mr. Meehan's paper, that the plants of which Lewis and Clark have most to say in their Journal, are not, as a rule, those of which specimens are now extant in their herbarium. Their botany, it may be said, runs in two parallel courses. One of these is represented by the specimens which they *collected,* and which became so many of Pursh's types; the other, by the herbs, shrubs and trees which they *observed,* and noted in their narrative, but did not actually collect. Oftentimes, to be sure, they describe what is in the herbarium, but I should imagine that fifty, if not a hundred species are to be found in the book, no specimens of which are known to be extant. This would appear to me to be a field of research at once alluring and stimulat-

8. The reference is to J. B. S. F. Aublet's *Histoire des plantes de la Guiane françoise* . . . 1775.

9. This, one notes, is a different statement from the one made in Pursh's *Flora*.

ing to some well equipped botanist, and I trust that the work may soon be done once and forever . . ."

A task of this very nature seems to have been undertaken by E. J. Criswell in his "Lewis and Clark: linguistic pioneers,"[1] published in 1940. This study contains innumerable references of one sort and another to plants, but the report upon the identifications of those appearing in the explorers' records are supplied in two lists: (1) supplying the "Usual Common Name" and "Lewis and Clark's Name" in parallel columns and arranged according to plant families (pp. lxxi-lxxviii) and (2) "The Botanical Index" which includes ". . . in alphabetical order complete lists not only of the new species discovered . . . also . . . all the species of . . . plants mentioned by the explorers . . . The nomenclature of the species . . . modernized as far as possible . . ." (pp. lxxix, xcviii-cxvii). These lists, involving an enormous amount of work, must of necessity include plants the identity of which remains a matter of individual interpretation of the plant descriptions. I surmise that, to botanists, they do not afford the same interest as the explorers' specimens. I comment upon one statement in Criswell's paper (p. lxxi):

". . . Much had been expected of the Osage orange, called by them the *Osage apple*,[2] for fabulous tales had been spread about it in the East. After they saw it, however, the explorers lost all interest in it; it did not in any way fulfill their expectations being useless for timber and producing no valuable fruit."

I refer elsewhere (p. 124, *fn.* 1) to what Sargent published upon the early history of this tree. It is of course possible that "fabulous tales" about it had been "spread" by word of mouth before Lewis and Clark started on their journey, but the first published record which Sargent was able to find was in the narrative of the journey of Dunbar and Hunter, made in 1804, the year Lewis and Clark started from St. Louis. From R. H. True's paper published in 1928 and to which I refer again, we know that Lewis saw the Osage orange growing in the garden of Pierre Chouteau at St. Louis— the same tree or trees that Bradbury saw in 1810 presumably. Lewis, as the content of True's paper proves, did not "lose all interest in it," but sent propagating material to Jefferson. The records do not indicate that they found other than cultivated plants on their travels[3] (and these young trees), so that it is not surprising that they were not impressed by its potentialities as "timber."

I might add that E. D. Merrill's *Index Rafinesquianus* (1949) cites a number of

1. The author expresses his indebtedness to Drs. Francis Drouet, H. W. Rickett and W. E. Maneval for assistance in various phases of the botanical work.

2. In Criswell's "The Botanical Index" (p. cvii) this is cited as "*Maclura pomifera* (Rafinesque) Nuttall, det. Thwaites (VII 296)," and the common name used by the explorers supplied as "Osage apple."

3. The distribution of the species, as given in Sargent's *Manual of the trees of North America (exclusive of Mexico)* (1922) is ". . . southern Arkansas to southern Oklahoma and southward in Texas to latitude 35° 36′; most abundant and of its largest size in the valley of the Red River of Oklahoma." Lewis and Clark did not get any great distance farther south than latitude 39° (when they were on the Missouri River between Kansas City and St. Louis).

new names which Rafinesque based on Lewis and Clark descriptions. See: *Pinus* (pp. 72, 73) and *Sambucus* (p. 222).

In 1928 R. H. True published a paper entitled "Some neglected botanical results of the Lewis and Clark expedition." It concerns the raising and distribution of plants grown from seeds and cuttings gathered by the explorers and presented by them to Thomas Jefferson. The author notes that the records concerned with such living material is "sparing and lacking in detail," but, nonetheless, he seems to have gathered together some interesting facts. The paper is based to a considerable extent upon letters[4] exchanged between Jefferson and Bernard McMahon, a famous nurseryman of Philadelphia.

Lewis wrote a letter to Jefferson from "St. Louis, March 26, 1804," stating that he was sending therewith ". . . some slips of the *Osage Plum,* and *Apple* . . ." He states that he ". . . obtained the cuttings . . . from the garden of Mr. Pierre Choteau[5] who resided the greater portion of his time for many years with the Osage nation. it is from this gentleman, that I obtained the information I possess with respect to these fruits . . . The *Osage Apple* is a native of the interior of the continent of North America, and is perhaps a nondiscript [that is, undescribed] production . . . Mr. Peter Coteau,[6] who first introduced this tree in the neighborhood of St. Louis, about five years since, informed me, that he obtained the young plant at the great Osage vilage from an Indian of that nation, who said he procured them about three hundred miles west of that place . . . the trees which are in the possession of Mr. Choteau have as yet produced neither flower nor fruit . . ." I have omitted a long description of this fruit, etc. True is of the opinion that "The herbarium material brought back by Lewis could certainly not have been collected at this time [that is, at St. Louis in March of 1804]. As seen in the herbarium of the Philadelphia Academy of Natural Sciences, the specimen seems to consist of a sterile branch from a young tree showing vigorous growth of wood and leaves of unusually large size. The mature condition of the leaves, the well-ripened wood and the fully developed winter buds suggest that the specimen was taken late in the season. No sign of flowers or fruit appears in the collection . . . this specimen, undated, was probably collected at St. Louis on the return of the Expedition in late September or October, 1806."

However, as True points out, Frederick Pursh, in the preface to his *Flora americae septentrionalis,* mentions that from a tree planted in a garden at St. Louis, "Perfect seeds . . . were given by Mr. Lewis to Mr. McMahon . . . who raised several fine plants from them, and in whose possession they were when I left Philadelphia." The story goes that Jefferson, learning from Lewis that he was to receive a variety of seeds, ar-

4. The paper is documented. Many of the letters were transcribed from *The papers of Thomas Jefferson* (ser. 2, vol. 5).

5. Correctly, Pierre Chouteau.

6. Whether "Peter Coteau" is merely a variant of the name Pierre Chouteau, I do not know; it seems probable.

ranged with McMahon and with William Hamilton, owner of the famous estate called "The Woodlands," in what is now Blockley Township, West Philadelphia,[7] to grow them and raised some himself at Monticello. At the request of Lewis, McMahon (from the correspondence) retained Frederick Pursh for a year "drawing and describing" these Lewis plants. But complications arose in regard to selling and describing these "new" plants. McMahon wrote Jefferson on December 24, 1809, in regard to this matter—the "priority" rights—after Lewis' death:

". . . In consequence of a hint . . . given me by Governor Lewis . . . I never parted with one of the plants raised from his seeds, nor with a single seed the produce of either of them, for fear they should make their way into the hands of any Botanist, either in America or Europe who might rob Mr. Lewis of the right he had to first describe and name his own discoveries in his intended publication, and indeed I had strong reasons to believe this opportunity was coveted by —— which made me still more careful of the plants. On Governor Lewis's departure from here . . . he requested me to employ Mr. Frederick Pursh . . . to describe and make drawings of such of his collection as would appear to be new plants . . . Pursh . . . took up his abode with me, began the work, progressed as far as he could without further explanation, in some cases, from Mr. Lewis . . . I thought it a folly to keep Pursh longer idle[8] and recommended him as Gardener to Doctor Hosack[9] of New York, with whom he has since lived. The original specimens are all in my hands, but Mr. Pursh, had taken his drawings and descriptions with him, and will, no doubt, on the delivery of them expect a reasonable compensation for his trouble. As it appears to me probable that you will interest yourself in having the discoveries of Mr. Lewis published, I think it a duty incumbent on me to give you . . . preceding information and to ask your advice as to the propriety of still keeping the living plants I have, from getting into other hands who would gladly describe and publish them without doing due honor to the memory and merit of the worthy discoverer."

The correspondence between Jefferson and McMahon kept up until 1815, according to True. He states that "Jefferson clearly gave up the idea of withholding knowledge of the Lewis plants until the publication of the records of the Expedition

7. A chapter is devoted to this interesting place and its inhabitants in Eberlein & Lippincott's *The colonial homes of Philadelphia and its neighbourhood*. As one travels on the Pennsylvania Railroad from Philadelphia to Baltimore one can still see the handsome old house, now standing in the midst of a cemetery. The authors refer to William Hamilton as a gardener born, and state that the Ginkgo tree and the Lombardy poplar as well as many other plants were introduced through his instrumentality.

8. According to the letter, it would seem that McMahon kept Pursh at the task from November [1807] until April, 1809.

9. This was David Hosack (1769-1835), a Doctor of Medicine, who founded the Elgin Botanic Garden in 1801; it was then situated about three or four miles from New York City, at what is now Rockefeller Center! In Dr. Howard Atwood Kelly's *Some American medical botanists commemorated in our botanical nomenclature* (New York. London. 97-103, 1929) a chapter is devoted to Hosack. *Hosackia bicolor* Douglas was named in his honor.

could take place . . ." As evidence of this he quotes a letter written by Jefferson to an aunt of General Lafayette, Madame de Tessé, an old friend. The date was December 8, 1813, Monticello. It concerned a plant which is now familiar to all gardeners and said to be naturalized in the eastern United States from Quebec to Virginia, Illinois and Missouri, although native to the Pacific slope.

". . . I have growing, which I destine for you, a very handsome little shrub of the size of a currant bush. Its beauty consists in a great produce of berries of the size of currants, and literally as white as snow, which remain on the bush through the winter, after its leaves have fallen, and make it an object as singular as it is beautiful. We call it the snow-berry bush, no botanical name being yet given to it, but I do not know why we might not call it Chionicoccus, or Kallicococcus. All Lewis's plants are growing in the garden of Mr. McMahon . . . to whom I consigned them and from whom I shall have great pleasure when peace is restored, in ordering for you any of these or of our other indigenous plants."

True notes that it is common knowledge that "several" of Lewis' plants eventually ". . . found their way into general cultivation and have long ago become widely distributed. Of the species recorded by Pursh and others as present in the herbarium material brought back by Lewis and represented in his seed collection . . ." he cites four: the Osage orange (*Toxylon pomiferum* Raf.[1]); the snowberry bush *(Symphoricarpos albus* Blake var. *laevigatus* Blake;[2] the golden, yellow-flowering or buffalo currant, also called the Missouri currant *(Ribes odoratum* Wendl.—*R. longiflorum* Nutt.); and the winter currant (*Ribes sanguineum* Pursh). True also notes that several other plants ". . . represented in the herbarium collection but not known to have been represented in the list of seeds are now of general value." He names the Oregon grape *(Mahonia Aquifolium* Nutt., *Berberis Aquifolium* Pursh); *Clarkia pulchella* Pursh, "a hardy annual plant . . ." and the bitter root (*Lewisia rediviva* Pursh). In terminating, True states that the list probably might be lengthened.

Although Lewis and Clark did not publish their own plants, the honor of having collected a goodly number of new species and a few new genera is nonetheless theirs. The species *rediviva* of the genus *Lewisia* (honoring the commander of the expedition) was collected by Lewis "On the banks of Clark's River," and is now the state flower of Montana. And the species *pulchella* of the genus *Clarkia* (named for the second in command) was collected by Lewis "On the Kooskooskee [Clearwater] and Clark's Rivers." The spelling "Clarckia," used by Pursh in describing this plant, seems to have been abandoned except by botanical "sticklers."

Lewis and Clark were the first to collect plants in the lands acquired by the Louisi-

---

1. Now *Maclura pomifera* (Raf.) Schneider.

2. In "A monograph of the genus Symphoricarpos," published in 1940, George Neville Jones cites this as a synonym of *Symphoricarpos rivularis* Suksdorf, *Werdenda* 1: 41, 1927—it has evidently been given many different names. He notes that "There is a colored plate of a fruiting branch of *S. rivularis* in Audubon's *Birds of America* 4: pl 375, 1835-38, and in the reprinted Macmillan edition of 1937." The bird of the plate 375 is the "Common redpoll *(Acanthis linaria)."*

ana Purchase and from the western margins of these lands westward to the Pacific. Not until Thomas Nuttall's journey of 1834–1836, just thirty years later, did another collector of plants cross overland from the Mississippi River to the west coast. Nuttall was a trained and discriminating botanist and his journey was rich in results. Lewis and Clark were to all intents and purposes untrained in any science and had as their major responsibility the safety of some fifty persons. That they were able to make any collections and to record as much as they did about the flora was a remarkable accomplishment. One may feel sure that the plant specimens which they gathered—although in a sense forgotten for the greater part of a century—are now numbered among the most cherished possessions of an old and respected institution.

BAKELESS, JOHN EDWIN. 1947. Lewis & Clark: partners in discovery. New York. [Fine map opp. p. 450.]

CHITTENDEN, HIRAM MARTIN. 1902. The American fur trade of the far west. A history of the pioneer trading posts and early fur companies of the Missouri Valley and the Rocky Mountains and of the overland commerce with Santa Fe. Map. 3 vols. New York.

COUES, ELLIOTT. 1898. Notes on Mr. Thomas Meehan's paper on the plants of Lewis and Clark's expedition across the continent, 1804-06. *Proc. Acad. Nat. Sci. Phila.* 1898: 291-315.

CRISWELL, ELIJAH HARRY. 1940. Lewis and Clark: linguistic pioneers. *Univ. Missouri Studies* 15: no. 2 .

EBERLEIN, HAROLD DONALDSON & LIPPINCOTT, HORACE MATHER. 1912. The colonial homes of Philadelphia and its neighbourhood. Philadelphia. London. 84-93, pl.

FERNALD, MERRITT LYNDON. 1942. Some early botanists of the American Philosophical Society. *In:* The early history of science and learning in America. *Proc. Am. Philos. Soc.* 86: 65, *fn.* 15.

GHENT, WILLIAM JAMES. 1931. The early far west. A narrative outline 1540-1850. New York. Toronto. 82-89, 91-100, 104-110.

GRAY, ASA. 1840. Notices of European herbaria, particularly those most interesting to the North American botanist. *Am. Jour. Sci. Arts* 40: 1-18. *Reprinted:* Sargent, Charles Sprague, *compiler.* 1889. Scientific papers of Asa Gray. 2 vols. Boston. New York. 2: 1-21.

——— 1842. Notes of a botanical excursion to the mountains of North Carolina, &c. with some remarks on the botany of the higher Alleghany Mountains, in a letter to Sir W. J. Hooker. *Am. Jour. Sci. Arts* 42: 1-49. *Reprinted: Ibid.* 2: 22-27.

HOOKER, WILLIAM JACKSON. 1825. On the botany of America. *Am. Jour. Sci. Arts* 9: 263-284.

JONES, GEORGE NEVILLE. 1940. A monograph of the genus symphoricarpos. *Jour. Arnold Arb.* 21: 209-214.

LEWIS, MERIWETHER & CLARK, WILLIAM. 1814. History of the expedition under the command of Captains Lewis and Clark, to the sources of the Missouri, thence across the Rocky Mountains and down the River Columbia to the Pacific Ocean, performed during the years 1804-5-6. By order of the government of the United States. Prepared for the press by Paul Allen, Esquire. In two volumes. Philadelphia. Published by Bradford and Inskeep; and Abm. H. Inskeep, New York. J. Maxwell, Printer. 1814 [Although not so stated on the title-page this was edited by Nicholas Biddle.]

────── 1893. History of the expedition under the command of Lewis and Clark, to the sources of the Missouri River, thence across the Rocky Mountains and down the Columbia River to the Pacific Ocean, performed during the years 1804-5-6, by order of the government of the United States. A new edition, faithfully reprinted from the only authorized edition of 1814, with copious critical commentary, prepared upon examination of unpublished official archives and many other sources of information, including a diligent study of the original manuscript journals and field notebooks of the explorers, together with a new biographical and bibliographical introduction, new maps and other illustrations, and a complete index, by Elliott Coues . . . 4 vols. New York.

────── 1903. History of the expedition of Captains Lewis and Clark 1804-5-6. Reprinted from the edition of 1814. With introduction and index by James K. Hosmer, LL.D. . . . ed. 2. 2 vols. Chicago.

────── 1904-1905. Original journals of the Lewis and Clark expedition 1804-1806. Printed from the original manuscripts in the library of the American Philosophical Society and by direction of its committee on historical documents together with manuscript material of Lewis and Clark from other sources including note-books, letters, maps, etc. . . . Now for the first time published in full and exactly as written. Edited, with introduction, notes, and index, by Reuben Gold Thwaites, LL.D. . . . 7 vols. & Atlas. New York. [For reprintings of portions of introduction see Thwaites, R. G. 1904. The story of . . .]

MEEHAN, THOMAS. 1898. The plants of Lewis and Clark's expedition across the continent, 1804-1806. *Proc. Acad. Nat. Sci. Phila.* 1898: 12-49.

MICHAUX, ANDRÉ. 1889. Portions of the journal of André Michaux, botanist, written during his travels in the United States and Canada, 1785 to 1796. With an introduction and explanatory notes, by C. S. Sargent. *Proc. Am. Philos. Soc.* 26: no. 129, 1-145.

PAXSON, FREDERICK LOGAN. 1924. History of the American frontier 1763-1893. Boston. New York. 137, *fn.* 2.

PENNELL, FRANCIS WHITTIER. 1942. Benjamin Smith Barton as naturalist. *In:* The early history of science and learning in America. *Proc. Am. Philos. Soc.* 86: 108-122.

PURSH, FREDERICK. 1814. Flora americae septentrionalis; or, a systematic arrangement and description of the plants of North America; containing, besides what have been described by preceding authors, many new and rare species, collected during twelve years travels and residence in that country. 2 vols. London.

SALISBURY, ALBERT & JANE. 1950. Two captains west. An historical tour of the Lewis and Clark trail. Seattle. [The maps are of interest.]

SARGENT, CHARLES SPRAGUE. 1897. The first account of some western trees. *Gard. & Forest* 10: 28-29; 38-40.

SCOTT, H. W. 1904. Beginnings of Oregon—Exploration and early settlement at the mouth of the Columbia River. *Quart. Oregon Hist. Soc.* 5: 101-119.

THAXTER, B. A. 1933. Scientists in early Oregon. *Oregon Hist. Quart.* 34: 330-344.

THWAITES, REUBEN GOLD. 1904. The story of Lewis and Clark's journals. *Ann. Rep. Am. Hist. Assoc. for the year 1903* [vol. 1] 105-129. *Reprinted: Quart. Oregon Hist. Soc.* 6:

26-53, 1905. [This is a reprint, with minor changes, of portions of Thwaites' introduction to *Original journals of the Lewis and Clark expedition* . . . 1: xxxiii-lvii, 1904.]

TRUE, RODNEY HOWARD. 1928. Some neglected botanical results of the Lewis and Clark expedition. *Proc. Am. Philos. Soc.* 67: 1-19. Read April 24, 1925.

WARREN, GOUVENEUR KEMBLE. 1859. Memoir to accompany the map of the territory of the United States from the Mississippi River to the Pacific Ocean, giving a brief account of each of the exploring expeditions since A.D. 1800 . . . *U. S. War Dept. Rept. expl. surv. RR Mississippi Pacific* 11: 17-19.

WHEELER, OLIN D. 1904. The trail of Lewis and Clark 1804-1904. A story of the great exploration across the continent in 1804-06; with a description of the old trail, based upon actual travel over it; and the changes found a century later. 2 vols. New York. London.

YOUNG, FREDERICK GEORGE. 1905. The higher significance in the Lewis and Clark exploration. *Quart. Oregon Hist. Soc.* 6: 1-25.

CHAPTER IV

VON LANGSDORFF, SURGEON-NATURALIST OF THE VON KRUSENSTERN EXPEDITION, ACCOMPANIES VON RESANOFF TO SAN FRANCISCO BAY IN THE SHIP JUNO

THE first Voyage of the Russians round the world," usually referred to as the Krusenstern Expedition, was sent at the direction of the Emperor Alexander the First and left Copenhagen[1] on September 8, 1803, or about six weeks before the ratification of the Louisiana Purchase.

There were two ships, the *Nadeschda,*[2] by some spelled "Nadezhda," and the *Neva.* Captain Adam Johann von Krusenstern was ". . . the proper chief of the expedition . . ." Another important member was the Chamberlain Nikolai Petrovich von Resanoff (Rezanov or Rezánof) ". . . going in the quality of Ambassador to Japan . . ." Still another participant was, according to Mahr, a lieutenant in the Imperial Russian Navy, Otto von Kotzebue,[3] who ". . . as a very young man had accompanied von Krusenstern on his own Arctic expedition . . ." But the individual of primary interest to our story was the naturalist-surgeon, Freiherr, or Baron, Georg Heinrich von Langsdorff, who was to visit "Saint Francisco," California, in April and May of the year 1806.

Von Krusenstern, organizer of the expedition and in command of the *Nadeschda,* wrote an account of the voyage which appeared in a number of Russian and German editions, with the first English translation in 1813. Captain Urey Lisiansky, commanding the *Neva,* also wrote of the voyage and his story was issued in English translation in 1814. But, since neither of these men visited California, I shall have no occasion to quote from their works.

A German edition of von Langsdorff's story was published at Frankfurt am Main in 1812. An English edition—*Voyages and travels in various parts of the world, during the years 1803, 1804, 1805, 1806, and 1807*—was issued in London in 1813 (volume one) and 1814 (volume two), and I shall take my account of the story therefrom. In 1927 T. C. Russell translated and revised that portion of the German edition concerned with the visit to California.

1. Bretschneider states: "The 'Nadeja' and the 'Neva,' with all the members of the expedition on board, took their departure from the roads of Kronstadt on Aug. 7th 1803 . . ." One botanist of the expedition, von Langsdorff, joined the ships at Copenhagen.

2. "*Nadeschda,* which was the name of our ship, signifies Hope." So states von Langsdorff.

3. Von Kotzebue was to command important expeditions at a later date, two of which took him to California in 1816 and 1824.

According to Bretschneider, von Langsdorff was born in "Hessen" [4] in 1774 and was, therefore, twenty-nine years of age when he went on the expedition.

His portrait appears as a frontispiece to his *Voyages* and, according to Bancroft (1885), shows ". . . a singularly unprepossessing face as portrayed . . ." There is, to be sure, a curious protuberance above the bridge of the nose! The book's dedication to the Tsar, "His Imperial Majesty Alexander the First," with its references to his "glorious reign," and "gracious condescension," and to the author's "unbounded gratitude and veneration," makes somewhat fulsome reading at the present time. With considerable satisfaction, or so it seems, von Langsdorff in his introduction enumerates his own qualifications for joining the expedition:

"To make travelling useful, a particular strength and turn of mind is requisite, which can only be acquired by beginning to travel early in life." This he had done "by several minor journies." He had ". . . obtained the degree of Doctor in Medicine and surgery, at Gottingen, in 1797 . . ." With Prince Christian of Waldeck he had gone, presumably as physician, to Lisbon, and in 1798 had accompanied him through several Portuguese provinces—a trip which the Prince ". . . alas! did not long survive . . ." Von Langsdorff remained in Portugal, not only because of the fine climate but because of the ". . . social circle of amiable and polished men . . . found at Lisbon . . ." Soon, unfortunately, he had ". . . acquired so extensive an acquaintance in many German, English, and Portuguese houses, and was honoured by them with so much confidence . . ." in his "professional capacity," that there was little time left for his favorite pursuit of natural history. Hoping for more leisure, he associated himself with an English auxiliary regiment for a time, then, after visiting London and Paris, returned to Germany in 1803, filled with a desire to go on some exploration. By now he had become a friend of "the first naturalists of France." This fact, together with ". . . the honours conferred . . . by the Imperial Academy of Sciences at St. Petersburgh, in naming [him] their correspondent . . ." so encouraged von Langsdorff that he decided to apply for membership in von Krusenstern's expedition. The ships, which had left Kronstadt a month earlier, were about to sail from Copenhagen "with the first fair wind," and another naturalist, Dr. Wilhelm Tilesius von Titenau (usually referred to as Tilesius), had been appointed. But von Langsdorff persisted, traveled by way of Lübeck and Travemunde to the Danish capital, where he found the party assembled, and ". . . entreated so earnestly of the Chamberlain Von Resanoff . . ." that his plea was granted. Von Krusenstern had also come to his support.

Two objectives of the expedition were to revive trade with Japan and China and to stimulate the Russian fur trade on the Pacific coast of North America.

After leaving Copenhagen the ships visited Falmouth, England, Teneriffe in the Canary Islands, crossed the Atlantic to Santa Catharina, Brazil, and went round Cape Horn. Thence, touching at various islands, they moved across the Pacific, eventually

4. The Grand Duchy of Hesse, once Hesse-Darmstadt.

reaching the "Sandwich" or Hawaiian Islands. From there they put into the bay of "Awatscha," in the northern part of which was situated "Petropaulowsk" (Petropavlovsk), on July 15, 1804. Leaving there September 7 and keeping to the east of Japan, they reached the harbor of "Nangasaki," or Nagasaki, on October 9, remaining until April 17, 1805. Here von Resanoff—in his ". . . quality of Ambassador to Japan . . ."—was completely frustrated, not only in his primary purpose of establishing trade relations, but in all other respects; while treated with baffling courtesy, he and his party were virtually prisoners for their entire stay; they were allowed ashore only in a small compound erected for their use. The story makes such interesting reading in the 1950's that I quote from the Emperor's ultimatum (as reported by von Langsdorff) which explains Japan's convictions about foreign relations:

" 'In former times, ships of all nations were allowed to come freely to Japan, and the Japanese were in the habit of visiting foreign countries with equal freedom. A hundred and fifty years ago, however, an emperor had strictly enjoined his successors never to let the Japanese quit the country . . . several foreign nations had, at various times, endeavoured to establish an intercourse of friendship and commerce with Japan; they were always . . . repulsed . . . because it was held dangerous to form ties of friendship with an unknown foreign power, which could not be founded on any basis of equality . . . Friendship . . . is like a chain, which, when destined to some particular end, must consist of a determined number of links. If one member . . . be particularly strong, and the others disproportionately weak, the latter must of necessity, by use, be soon broken. The chain of friendship can never, therefore, be otherwise than disadvantageous to the weak members included in it . . .' "

As to the presents sent from Russia, " 'If they were accepted, the Emperor of Japan must . . . send an ambassador with presents of equal value to the Emperor of Russia. But there is a strict prohibition against . . . quitting the country, and Japan is besides so poor, that it is impossible to return presents to any thing like an equivalent . . . Japan has no *great* wants, and has therefore little occasion for foreign productions: her few *real* wants . . . are richly supplied by the Dutch and Chinese, and luxuries are things she does not wish to see introduced . . .' "

This dignified expression of policy was written approximately fifty years before the arrival of Admiral Perry!

The Russian expedition finally sailed away carrying with it the rich presents which the Emperor of Japan had refused to accept. Von Resanoff, in late 1806, sent a ship to a Japanese-held island,[5] raided it—". . . asserting the prior right of the Russians . . ." —and removed as much loot as could be carried, as well as four Japanese prisoners. Writes von Langsdorff:

"In this manner, did the Japanese first learn, through the mortified *amour-propre* of the Ex-ambassador, to understand, in some degree, the extent of the Russian power

5. Sakhalin according to Langer (*An encyclopaedia of world history*, 898, 1940).

. . . The undertaking had, at least some colour of justice, and all possibility of M. Von Resanoff's ever being called to account for the measures he had pursued, was precluded by his death . . ."

On leaving Japan the expedition went north, by way of the Sea of Japan and the Sea of Okhotsk, to Kamtchatka. Following this route it was hoped ". . . to prosecute farther the discoveries made by the unfortunate and ever to be lamented La Perouse." [6]

Back at "Awatscha," the "harbour of St. Peter and St. Paul," (or Kamtchatka) on June 5, 1805, they obtained news from Europe: ". . . but of all that we received, none was so important and so unexpected as that Bonaparte was declared Emperor of France." [7]

Plans were changed. Von Krusenstern was to proceed to China and elsewhere before returning home by sea, and we part with him and the *Nadeschda* at this point, leaving him in process of ". . . clearing the ship of all superfluous objects, some of the principal of which were the presents intended for the Emperor of Japan, that he might proceed with the vessel lightened as much as possible . . ." We also part with Lisiansky and the *Neva*—von Langsdorff mentions that they went to Kodiak for the winter.

"The embassy to Japan having fulfilled its mission . . .", von Resanoff had planned to return to St. Petersburg to render a report thereon; but letters caused him to change his plans and he sent the report to the Russian capital by a courier, "one of the cavaliers of the embassy." Instead, he was to visit the Aleutian Islands and the northwest coast of North America, ". . . particularly the most distant possessions of the Russians in these parts, as plenipotentiary of the Russio-American Trading Company."

The second volume[8] opens with an introduction which describes the change of plans just mentioned and the reasons for von Langsdorff's decision to accompany von Resanoff:

"The Ex-ambassador von Resanoff, not judging it expedient to wander among the rugged, uncultivated, and inhospitable north-west coasts of America, without the attendance of a physician, made very advantageous proposals to me to accompany him . . . My choice . . . was made for accepting the ambassador's proposals; it seemed so much a debt due to science to undertake a journey to parts so little known, and which had received so little scientific examination, under auspices to all appearance particularly favourable, that I could scarcely consider myself as justified in declining it."

6. The *Voyages* contains several references to Lapérouse's records in this region and, in passing from Japan to the Okhotsk Sea, the ships went through La Pérouse Strait—on the map the "Canal de la Perouse."

7. He had crowned himself Emperor of the French on December 2, 1804.

8. Volume two of von Langsdorff's *Voyages* bears on its title-page the additional data: "Part II. containing the voyage to the Aleutian Islands and north-west coast of America, and return by land over the north-east parts of Asia, through Siberia, to Petersburg."

In a report to "The Aulic Counsellor, Professor Blumenbach[9] of Gottingen," dated from New Archangel, Norfolk Sound, in February, 1806, von Langsdorff later reiterates the reasons for his decision and expresses regret that his expectations were not fulfilled; among numerous difficulties he "... had never been in Russia previous to this expedition ..." and was handicapped by not understanding the Russian language.

"A blind zeal for Natural History, repeated promises both in writing and by word of mouth, of all possible support in the pursuit of my objects for the promotion of science ... induced me ... to accompany ... Von Resanoff to the north-west coast of America ... Since that time, buried in this remote part of the new world, my only consolation has been to sigh and long for the old."

Von Resanoff and von Langsdorff sailed from Kamtchatka June 14, "old stile," [1] 1805, in "... the galliot [or brig] Maria ... a heavy sailing two-masted vessel ... built at Ochotsk, and ... under the command of a lieutenant of the navy ... Andrew Wassilitsch Maschin; with him went as passengers the Chamberlain Von Resanoff, and his valet-de-chambre ..."—of whose reactions to the trip we, unfortunately, are told nothing. Also two lieutenants, one G. I. Davidoff, "a huntsman who was to serve as caterer," and who seems also to have served as von Langsdorff's interpreter. The crew was composed "... of adventurers, drunkards, bankrupt traders, and mechanics, or branded criminals in search of fortune." They were all poorly fed and clothed and the ship was overcrowded; illness soon became prevalent.

They reached "Oonalashka," or Unalaska, on July 16 and remained until the 21st. Here von Langsdorff remarks that "... though at Kamschatka large promises were made me, both in writing and orally, as to what should be done for the promotion of scientific undertakings, no alacrity had been shewn in fulfilling these promises, so that I began almost to repent having undertaken the voyage."

At Unalaska, at Kodiak and at Sitka there were establishments of the "Russio-American Company." They reached New Archangel, or Sitka, on July 26 where they found "... Von Baranoff[2] ... superintendent of all the Company's possessions in the

9. "Blumenbach's Lessons" had once "decided" von Langsdorff's attachment to natural history. Johann Friedrich Blumenbach (1752-1840) became professor of medicine and anatomy at Göttingen in 1778 and is said to have lectured for fifty years. Alexander von Humboldt was one of his pupils, as was Maximilian, Prince of Wied-Neuwied.

1. Von Langsdorff states on page 5 of volume two: "I shall ... here follow, without any exception, the universally received mode among the Russians, of reckoning according to the Julian computation of time, by which I hope the better to avoid errors."

According to *The World Almanac* ... for 1948 (p. 141, *fn.*): "... To change from the Julian calendar to the Gregorian calendar, add 10 days for the years 1582 to 1700; 11 days from 1700 to 1800; 12 days from 1800-1900; 13 days since 1900."

The computation of the difference between these calendars may perhaps explain why the dates cited by certain authors appear to differ slightly; I have used those supplied by von Langsdorff throughout his *Voyages*.

2. Alexander A. Baranoff had founded Sitka in 1804.

Aleutian Islands and America." Von Langsdorff was revolted by the exploitation and virtual enslavement of the Aleutian people. The Russian "Promüschleniks," although plundering, oppressing and tormenting the natives, were not themselves much better off; the greater part of them ". . . and the inferior officers of the different settlements are Siberian criminals, malefactors, and adventurers of various kinds . . ."

In his capacity of plenipotentiary, von Resanoff attempted certain much needed reforms and, since the inhabitants of Sitka were not only suffering from various forms of malnutrition but were in danger of starving during the coming winter, he decided, in order to obtain food supplies, to make a trip to the ". . . most northerly of the Spanish possessions in this part of the globe, St. Francisco, on the coast of New Albion . . . political reasons led to the choice . . ." He might instead have gone to the Hawaiian Islands.

There lay in the harbor of Sitka ". . . a vessel from the United States of America, the *Juno* [or *Unona*], of two hundred and fifty tons, from Bristol in Rhode-Island . . ." Its captain was "by name J. Dwolf," [3] whom von Langsdorff refers to as ". . . one of

3. "J. Dwolf" deserves identification—even in 1927 Russell refers to him as "John Wolf."

Leonard Bacon of Peace Dale, Rhode Island, referred me to the man's journal, "A voyage to the North Pacific and a journey through Siberia more than half a century ago," which was published as part two of Wilfred Harold Munro's *Tales of an old sea port* (1917). "J. Dwolf," owner of the *Juno*, was John DeWolf, or d'Wolf, called by his friends "Norwest John."

He had reached the Russian settlement on Norfolk Sound on August 14, 1805, a short time before the *Maria,* carrying von Resanoff and von Langsdorff, arrived there. DeWolf sold and delivered the *Juno* to the Russian American Company on October 5, making, it would seem, a good financial bargain. DeWolf wrote enthusiastically of von Langsdorff:

". . . by profession a doctor of medicine and surgery, and by taste a naturalist. He was a volunteer on the Russian American expedition, and was in pursuit of science . . . he was invited by Baron von Resanoff to accompany him to the Northwest coast of America, as his physician. He was particularly moved to accept the invitation by the opportunity which was thus offered for the collection of specimens of natural history. The Doctor, unlike Baron von Resanoff, spoke the English language fluently. As I lived under the same roof with him, we became almost inseparable, participating in each others' pleasures and troubles . . ."

Although invited to make the trip to California, DeWolf declined. He acquired the brig *Russisloff* and after the *Juno* had returned to Sitka from California, he set sail in his new ship for Okhotsk on June 30, 1806, von Langsdorff accompanying him. A year later they parted company at that place: "I took leave of my highly esteemed friend, Dr. Langsdorff, who intended to remain a week or so for the promotion of his favorite object . . ."

I shall not follow DeWolf on his homeward journey except to note that, shortly after he reached St. Petersburg, war was declared between Russia and England and he had difficulty getting out of Russia and in reaching Liverpool because of the blockade. He was back at Bristol, Rhode Island, on April 1, 1808, after ". . . an absence of three years and eight months."

Munro, editor of his journal, comments that "The German naturalist and the American sea captain were evidently 'two of a kind.' How highly the American esteemed his friend may be judged from the fact that he named his only son John Langsdorff."

Mark A. DeWolfe Howe wrote me on November 12, 1947:

"I am always glad to hear that some interest in 'Norwest John' DeWolf persists . . . he was a first cousin of my grandfather, John Howe . . . Your inquiry sent me to a DeWolf genealogy. There I find that there were two children of his marriage with Amelia Melville, an aunt, I believe, of Herman Melville . . .

the most compassionate and benevolent of men . . ." His boat, the *Juno,* was purchased, loaded with articles for trade and, ". . . the twenty-fifth of February old stile, or the eighth of March new . . .", 1806, the party set sail. "All . . . quitted with joyful hearts the miserable winter abode to which they had been doomed . . ." The crew was composed of riffraff and many were sick and incapable of work.

March 14: "On the fourteenth, at day-break, the horizon being particularly clear, we had the pleasure of discovering the long-wished for coast of New Albion. To the south were high chains of hills, to the north the land was low: directly in the east the landscape . . . terminated in the back-ground by a very high round peak, covered with snow . . . Captain Vancouver's Mount St. Helen's . . .

"Even without any astronomical observations, we might have presumed ourselves to be in the neighbourhood of a great river, as the sea had a dirty, troubled, and reddish appearance, and the water was mingled with a considerable quantity of clay. We soon discovered Cape Disappointment in latitude 46° 20′ . . . we expected soon to cast anchor . . . our hopes and schemes were frustrated by the wind shifting suddenly to the south-east . . . it was impossible to think . . . of running into an unknown harbour . . . we endeavoured . . . to pass the night without varying much from our present situation . . ."

March 15: ". . . towards noon on taking an observation, we found to our no small astonishment that we were in latitude 46° 58′; it seemed . . . incomprehensible how the current of the sea could have carried us so much to the north . . . our sleepy and dull Promüschleniks might in the night have steered in a wrong direction.

"In order . . . to ascertain our situation, in the afternoon a sailor and two Aleutians were dispatched in a three-seated baidarka to take a more accurate examination of the harbour: I entreated permission to join the expedition, and having obtained the consent of the principals, took my seat in the baidarka . . . it was agreed that we should return at six o'clock in the evening at the latest . . .

"We followed at a proper distance the foaming surf of the northern point of land, and soon perceived the opposite southern coast, which formed the other side of the entrance. Between them was a passage of about a sea-mile in breadth . . . perfectly navigable. We therefore rowed directly eastward . . . At length, not till a little before six, we reached the inner shore . . . I soon ascertained that . . . I was examining . . . Havre de Gray . . .

". . . In a north-westerly direction, and in the vicinity of a wood, I saw smoke rising

The older child . . . Nancy Melville, married one Samuel Downer . . . The other child was named John Langsdorff, who died in 1886. He too was married, but the genealogy gives no trace of his descendants . . ."

   The John DeWolf journal had been "Printed but not published" in Cambridge, Massachusetts, in 1861, according to a description of a copy with author's presentation inscription which was offered for sale at Goodspeed's in 1942, and now the property of Mr. Howe. See *The Month at Goodspeed's Book Shop* (14: no. 2, November, 1942, p. 37).

in several places, which plainly indicated that the country was inhabited.[4] I therefore ascended a sand-hill near the shore and fired my gun twice . . . perhaps the distance was too great for the guns to be heard, as my experiment was not productive of any consequences . . .

"It was already six o'clock in the evening, and the night was closing in very fast, so that I had no time to make a longer stay . . . At length, about half past ten . . . we arrived safe and well at the ship, to the great joy of our companions: they had given us up as . . . a prey either to the fury of the waves, or of the savages upon the shore.

"Immediately . . . the anchor was weighed . . . in seven hours we ran sixty-six miles and a half . . ."

March 16: ". . . the next morning, the sixteenth, rose to a violent storm which did not abate till evening . . ."

March 17, 18: ". . . the following days the same unfavourable . . . wind continued to blow . . ."

March 19: "On the nineteenth the weather was cloudy, rainy, and foggy, and it was not till evening that the . . . wind . . . began to abate . . . we could once more see the coast stretching to the east . . . to the north-east . . . rose a lofty peak covered with snow . . . undoubtedly the Mount Rainier which Vancouver observed . . . We proceeded slowly forwards . . ."

March 20: ". . . We . . . had, early in the morning of the twentieth of March, the pleasure of being presented with a clear view of the pleasing country along the shore . . . to the east lay the coast of Shoal-Water, to the south-south-east . . . Cape Look-out, and between them the northern promontory at the entrance of the River Columbia, called Cape Disappointment . . .

". . . we soon lost our favourable wind, and were once more driven to the north . . . Towards evening a calm came on . . . we . . . cast anchor . . . We lay at anchor all night . . ."

March 21: ". . . in the morning of the twentieth [?=twenty-first] directed our course towards Cape Disappointment . . .

"It was of no small importance to us to land if possible. Many of our sick, from the nature of their food on board the ship . . . daily grew more and more diseased . . . our commanders considered it as an absolute duty to put into the first harbour . . . Towards two o'clock we had doubled the high northern promontory, and towards five . . .

4. Katharine Coman states: "Oddly enough, on the very day (March 14, Old Style) that the captains [Lewis and Clark] broke camp, de Resanoff's ship, the *Neva* [= the *Juno*], attempted to run into the Columbia, but was prevented by the sudden shifting of the wind from northwest to southeast. Von Langsdorff entered Gray's Bay in a *bidarka* and saw the smoke of the Indian villages, but had no communication with the inhabitants."

plainly discerned . . . a violent surf . . . we had been driven . . . too near the southern cape called Cape Adams, having missed the passage. We . . . cast anchor . . . in the utmost danger . . . Lieutenants Schwostoff and Davidoff . . . represented to M. Von Resanoff that it was better . . . to abandon the idea of running into the River Columbia, and sail away directly for St. Francisco, where we should be among civilized people, and should no doubt find plenty of provisions and every thing necessary for the recovery of our sick."

On March 23 it was supposed that they saw the "high promontory" of Cape Mendocino; on March 27 they observed towards the south "Pinto de los Reys" and "Los Fallerones." They spent that night two or three miles off the entrance to the Bay of San Francisco and the next day entered the harbor. From a fort ". . . we were hailed by means of a speaking trumpet, and asked who we were, and whence we came. In consequence of our answer we were directed to cast anchor in the neighbourhood of the fort."

Shortly after arrival they received ". . . the news that England had declared war against Spain . . . when our vessel was first seen, it was supposed to be an English one, and an enemy . . ."

Von Langsdorff devotes four chapters to the San Francisco visit. The "proper commandant," Don Luis Arguello, was absent, but the Russians were cordially received and entertained by his wife and son, and in time the "Governor from Monterey," Don José Joaquin de Arrilaga himself arrived. Von Langsdorff was given every opportunity to observe at first hand the customs of the Spanish rulers, of the missions and padres, and of the Indians:

"During the time that the horses, oxen, and mules, of the several missions were employed in going backwards and forwards to bring us our cargo, the Governor, with his train, and the numerous family of the Commandant, did every thing in their power to make our stay agreeable to us. Almost every morning horses were upon the shore ready for us as soon as we chose to land, that we might take a ride about the neighbouring country: we had free permission to go every where except to the forts."

But it was not easy to make excursions. "Although the three missions of St. Francisco, Santa Clara, and St. Joseph, all lie near the south-eastern part of the Bay of St. Francisco, and a communication by water, from one to the other, would be of the utmost utility, it seems almost incredible, that, in not one of them, no, not even in the Presidency of St. Francisco, is there a vessel or boat of any kind. Perhaps the missionaries are afraid lest if there were boats, the escape of the Indians . . . might be facilitated . . ."

On March 29, the day after arrival, von Langsdorff made a short trip from the anchorage to the Mission of San Francisco, ". . . an ecclesiastical establishment, lying at the distance of a short German mile, eastward of the Presidency [Presidio]." On arrival "Our cicerone, Father Joseph Uria . . . understanding that I was a naturalist, took me by the hand when we were in the chapel, and made me take notice of a painting, which represented the *Agave Americana,* or large American aloe, from the midst of which, instead of a flower-stem, rose a holy virgin, by whom, he assured me, many extraordinary miracles were performed . . . I could not help thinking if the belief was really not assumed, this was the greatest miracle the Virgin could have wrought. From courtesy . . . I joined in his admiration . . . expressing . . . my extreme envy of the painter, who had seen so great a natural curiosity with his own eyes."

There is a short comment on the ". . . Mission of Santa Clara, which lies between St. Francisco and Monterey . . . one of the largest and richest." But we are not told that this was visited. Von Langsdorff notes that "In the province of New California, which extends from St. Francisco . . . to St. Diego . . . there are at present nineteen missions, each of which contains six hundred to a thousand converts."

Von Langsdorff did, however, get to the Mission of "St. Joseph," or San José, which lay "sixteen leagues" away from the ship's anchorage, across the Bay of San Francisco: ". . . they have no other intercourse but by land . . . are obliged to go round the bay, at least three times the distance." This trip was begun in three of the *Juno*'s boats but the first attempt, on April 14 and 15, was unsuccessful. On the 20th, von Langsdorff, in a baidarka, and walking a part of the way, reached his destination by sunset; the entire party were "exceedingly fatigued." He remained until the 23rd and it took two days to make the return trip. He wrote of the mission's gardens: ". . . the quantity of corn[5] in the granaries far exceeded my expectations. They contained at that time more than two thousand measures of wheat, and a proportionate quantity of maize, barley, pease, beans, and other grain. The kitchen-garden is extremely well laid out, and kept in very good order; the soil is every where rich and fertile, and yields ample returns. The fruit-trees are still very young, but their produce is as good as could be expected. A small rivulet runs through the garden, which preserves a constant moisture. Some vineyards have been planted within a few years, which yield excellent wine, sweet, and resembling Malaga."

Returning from the mission they spent one night in the open and the next morning landed on a shore[6] which ". . . appeared to be much higher and well wooded." Instead it proved to be ". . . a low boggy plain, overgrown with nothing but the salt weed, *salsola* . . ." This seems to have been the only native plant mentioned by von Langsdorff on the California visit; it is the *Suaeda californica* Watson, one of the so-called sea blites, and a member of the *Chenopodiaceae,* or saltbrush, family.

After another uncomfortable night in the open they got back to the ship. The excursion produced no collections:

"To my inexpressible concern, a number of objects of natural history which I had collected in my excursion, chiefly birds and plants, became a prey to the stormy sea, and I brought nothing home with me except the three sea-otters."

5. The word "corn" is evidently used in the sense of cereal or grain, since in the next sentence there is a reference to "maize" (*Zea Mays* Linnaeus), the Indian corn.

6. According to Stillman ". . . at the mouth of what is now known as Alameda creek . . ."

Von Langsdorff reports that, at this period, communication by land with Vera Cruz on the Gulf of Mexico had become a routine matter; indeed, "From St. Francisco any one may travel with the greatest safety even to Chili; there are stations all the way kept by soldiers." On the other hand information about the regions lying inward from the California coast was very slight—the Spaniards had no boats in which to ascend the ". . . four, or, as some say, five large rivers, which come from the east . . . Every year military expeditions are sent out to obtain a more exact knowledge of the interior of the country, with a view . . . of establishing, by degrees, a land communication between Santa Fé and the north-west coast of America."

One such military party returned during von Langsdorff's visit to the Mission of San José; it had penetrated 80 or 90 leagues eastward and ". . . had arrived in the neighbourhood of a high and widely extended chain of hills, covered with eternal snow . . . known to the Spaniards under the name of the Sierra Nevada, or Snowy Mountains." The party reported that dwellers in the Sierras were said to have seen, a few days' journey eastward, ". . . men with blue and red clothing, who entirely resembled the Spaniards of California: they were very probably soldiers of Santa Fé . . . sent . . . to examine the interior of the country westwards . . . a probability is . . . that, in time, a regular inland communication may be established between Santa Fé and St. Francisco."

The nonprogressive spirit of the padres was exemplified in the fate of the "little machine" for grinding corn which De Langle, of the Lapérouse expedition, had left with the mission fathers:

"The excellent and friendly La Perouse, with a view to lessening the labour, left a hand-mill here, but it was no longer in existence, nor had any use been made of it as a model from which to manufacture others. When we consider that there is no country in the world where windmills are more numerous than in Spain, it seems incomprehensible why these very useful machines have never been introduced here . . . the good fathers are actuated by political motives. As they have more men and women . . . than they could keep constantly employed the whole year, if labour were too much facilitated, they are afraid of making them idle by the introduction of mills." [7]

Von Langsdorff had this to suggest about a Russian colony in California:

"If Russia would engage in an advantageous commerce with these parts, and procure from them provisions for the supply of her northern settlements, the only means of doing it is by planting a colony of her own. In a country which is blessed with so mild a climate as California, where there is such plenty of wood and water, with so many other means for the support of life, and several excellent harbours, persons of enterprising spirits, might, in a few years, establish a very flourishing colony[8] . . . in

7. The Russians at Fort Ross appear to have been more progressive in the matter of mills than the Spaniards. Essig states that "A windmill for grinding flour was erected on the hill outside the stockade . . . and was still standing when John Bidwell took charge for Captain John Sutter in 1842."

8. Von Langsdorff was back in St. Petersburg on March 16, 1808. In that year the Russians sent emis-

time, Kamschatka and Eastern Asia would be amply supplied from hence with all kinds of vegetable and animal productions for the support of life. The Russio-American Company have already sufficient sources of wealth in their present possessions from the extensive fur-trade they yield, nor has any occasion to aim at increasing it by foreign dealings. Their settlements only want a better administration to rise with fresh vigour from their ruins; but to effect this, their strength must be concentrated, and they must abandon the mistaken policy of extending them to such a degree as to weaken every part."

Although von Langsdorff's *Voyages* occasionally contains lists of plants from specific regions visited on his long journey he includes none from California. In view of the promises made him before he left Sitka he appears to have had good reason for complaint:

"My researches in Natural History met with more obstacles in California than in any other part of our expedition. To detail all the petty circumstances which crossed me in this way would appear prolix . . . I shall only . . . mention some of the principal. Our regular habitation during our whole stay was on board the ship, and there we were constantly employed in loading and unloading goods. Several skins . . . that I had laid upon the deck to dry, were, I know not by what means, thrown overboard. The paper for drying plants disappeared one day when I was on shore, and I was informed was by mistake put under a quantity of goods which had in my absence been taken on board, so that it could not be got out without entirely unloading again, and this was impossible. Several live birds which I had purchased were, as soon as my back was turned, suffered to fly away . . . when I asked for a sailor to go upon a water excursion with me, I was told that these people had more important business to attend to, and that our expedition was not undertaken for the promotion of Natural History. One evening I brought home a number of ducks and other aquatic birds, intending to strip off the skins, and dry them, but the next morning I found them all with their heads cut off. By these, and numberless other occurrences of a similar kind, I was at length so entirely discouraged, that I relinquished all idea of attempting farther labours in the science, and resigned myself to the wishes of the Chamberlain Von Resanoff, that I should undertake the office of interpreter, and transact all our business with the missionaries relative to the purchase of corn and other articles."

After von Resanoff had fulfilled his mission of obtaining supplies,[9] the *Juno* sailed north, on the "tenth of May, old stile," 1806, and reached Sitka on June 8. The ship's

saries to investigate the possibilities of establishing a supply-base in California; they sent others in 1811; and in 1812 started their colony at Fort Ross. Whether von Langsdorff's opinions were taken into consideration in these matters I do not know, but if so, he made a contribution to our story. For before Fort Ross was abandoned by the Russians on January 1, 1842, they had sent to the colony a number of men interested in the natural sciences.

9. Von Resanoff, whom Robert Greenhow characterizes as "a singularly ridiculous and incompetent person," did not, apparently, employ "direct" methods only to attain his objective: a romance with Concepcion Maria Arguello, one of the fifteen children of the Commandant at San Francisco, is credited with

sailors, by contrast with the "living skeletons" who came to meet the boat, looked "plump, well-fed." The trip had consumed three months, the California visit almost six weeks.

From Sitka, von Resanoff left for home on the *Juno*. After leaving Okhotsk to travel overland to St. Petersburg, he contracted a fever but pressed on and, weakened by illness, fell from his horse and was killed. This was in March, 1807. Von Langsdorff visited his grave at Krasnojarsk on his way home; it was marked by ". . . a large stone, in the fashion of an altar, but without any inscription."

On June 19, 1806, von Langsdorff departed westward with "Captain Dwolf" on his small ship the *Rossüslaf,* or *Russisloff*. He "had been long enough at Sitka," and notes that ". . . I breathed more freely when I had completely lost sight of Mount Edgecumbe." By way of Kodiak and Unalaska he reach Kamtchatka (harbor of St. Peter and St. Paul) on September 13, ". . . old stile, or the twenty-fifth, new." On May 13,[1] 1807, he left for Okhotsk, where he received word of the death of von Resanoff. From there he traveled across Russia to Jakutsk, to Irkutsk capital of Siberia, to Tobolsk, Kasan, Moscow, eventually arriving at St. Petersburg March 16, 1808 ". . . after having traversed that vast empire through its whole extent from east to west." Here his narrative ends.

having played a part in his success. Von Langsdorff records that the young lady, "Donna Conception," ". . . was lively and animated, had sparkling love-inspiring eyes, beautiful teeth, pleasing and expressive features, a fine form, and a thousand other charms, yet her manners were perfectly simple and artless. Beauties of this kind are to be found, through not frequently, in Italy, Spain, and Portugal."

Her "bright eyes" had made a deep impression on von Resanoff:

". . . he conceived that a nuptial union with the daughter of the Commandant . . . would be a vast step gained towards promoting the political objects he had so much at heart. He had therefore nearly come to a resolution to sacrifice himself by this marriage to the welfare, as he hoped, of the two countries of Spain and Russia. The great difficulty . . . was the difference between the religion of the parties, but to a philosophic head like the Chamberlain's, this was by no means an insurmountable one . . ."

According to Bancroft, von Resanoff did not propose marriage "until all other expedients had failed," and he quotes a letter written by the Russian in which, as the historian truly states, "lies the evidence that rather unpleasantly merges the lover with the diplomat."

Before leaving California he had forwarded his romance to the point of betrothal; the story goes that he promised to return to St. Petersburg, obtain an appointment to Madrid ". . . as ambassador extraordinary from the Imperial Russian court . . ." and then return to San Francisco and claim his bride. However, he died on his way back to St. Petersburg. According to Bancroft, Donna Concepcion did not hear of his death for several years. She remained faithful to her lover—if such he sincerely was—and, although wooed by others, devoted the rest of her life to those in distress, and died at the age of eighty-seven in the Dominican convent of St. Catherine at Benicia.

1. On May 14, the day after the *Rossüslaf* left for Okhotsk, an episode occurred—a true "fish-story"— which is told in chapter 45 of Herman Melville's *Moby-Dick; or, the Whale*. After quoting von Langsdorff's story of the adventure, which tells how the ship under full sail had landed *on top of* a whale and had been raised "three feet at least out of the water," Melville continues:

"Now, the Captain D'Wolf here alluded to as commanding the ship in question, is a New Englander, who, after a long life of unusual adventures as a sea-captain, this day resides in the village of Dorchester near Boston. I have the honor of being a nephew of his. I have particularly questioned him concerning this passage in Langsdorff. He substantiates every word. The ship, however, was by no means a large one: a

Bretschneider, writing in 1898, states that von Langsdorff brought back from the expedition ("Admiral Krusenstern's circumnavigation of the globe") ". . . rich botanical collections from Japan, Kamtchatka, America and Siberia, which are now kept in the Museum of the Academy, St. Petersburg . . .

"As to the botanical collections formed by Langsdorff and Tilesius, during the circumnavigation, the former had given his plants to Fischer,[2] and Tilesius' plants are found in the herbarium of Ledebour . . . and in that of Fischer. These are now in the possession of the Botan. Garden, St. Petersb. to which, after the death of these botanists, their botanical collections devolved . . ."

Tilesius did not get to North America but traveled with von Krusenstern to China in 1805. Von Langsdorff refers to his artistic ability:

"Not less thanks are due to my friend and travelling companion, Counsellor Tilesius, who unites to the most extensive scientific knowledge exquisite taste in the fine arts, for the many sketches with which he has already favoured me, and for the many more promised by him to enrich and embellish the Second Part of my Travels . . . he has . . . in his possession a large collection of sketches of objects of Natural History, which he purposes by degrees to complete, and present to the public."

The plates in volume two of the *Voyages* depict only scenes in Alaska and California where, as stated, Tilesius did not travel. But in neither volume of von Langsdorff's work is the artist of the plates named, merely the engraver and publisher.

Bretschneider mentions that Fischer named two species in honor of von Langsdorff. Both were from Unalaska: *Viola Langsdorfii* and *Pedicularis Langsdorfii.* He mentions also that *"Viola japonica,* Langsdorff, in herb. Fischer, DC. Prod. I, 295, was discovered by Langsdorff near Nagasaki, in 1805." Still a third species, *Calamagrostis Langsdorfii,* from Siberia, was named for him by Trinius de Gram.

The genus *Langsdorffia,* from tropical America, was named in his honor by Martius.

The volume, *Plantes recuellies pendant le voyage des Russes autour du monde. Expédition dirigée par M. de Krusenstern. Publiées par G. Langsdorff et F. Fischer,* Tübingen, 1810, contained thirty beautiful plates of ferns. None, unfortunately, are from California.

Bretschneider supplies a few details of von Langsdorff's later life:

"In 1808 Langsdorff was elected Member of the Academy. Subsequently the Russian Government appointed him Consul General at Rio Janeiro, where he continued to collect plants for the Academy. L. retired from service 1831 and died in 1852, at Freiburg."

Russian craft built on the Siberian coast, and purchased by my uncle after bartering away the vessel in which he sailed from home."

Von Langsdorff, states Melville, refers to the creature as "An uncommon large whale, the body of which was larger than the ship itself . . ."

2. Dr. Friedrich Ernst Ludwig von Fischer, who is said by Bretschneider to have held many important positions in Russia among which was the Directorship of the Imperial Botanic Garden at St. Petersburg.

Brewer mentions that he is said to have visited California again "... in 1824, in connection with the second expedition of Kotzebue ..." but I have found no confirmation of this statement.

ALDEN, ROLAND HERRICK & IFFT, JOHN DEMPSTER. 1943. Early naturalists in the far west. *Occas. Papers Cal. Acad. Sci.* 20: 19-21.

BANCROFT, HUBERT HOWE. 1885. The works of ... 19 (California II. 1801-1824): 64-80, fn. 2-29.

——— 1888. *Ibid.* 34 (California pastoral 1769-1848): 332, 463, 464.

BRETSCHNEIDER, EMIL. 1898. History of European botanical discoveries in China. London. 313-316.

BREWER, WILLIAM HENRY. 1880. List of persons who have made botanical collections in California. *In:* Watson, Sereno. Geological survey of California. Botany of California 2: 554.

COMAN, KATHARINE. 1912. Economic beginnings of the far west. How we won the land beyond the Mississippi. 2 vols. New York. 1: 277.

DE WOLF, JOHN. 1917. Norwest John and the voyage of the Juno. A voyage to the north Pacific and a journey through Siberia more than half a century ago. *In:* Munro, Wilfred Harold. *Tales of an old sea port.* Part II. Princeton.

ESSIG, EDWARD OLIVER. 1933. The Russian settlement at Ross. *Quart. Cal. Hist. Soc.* 12: 191-209. *Reprinted: Cal. Hist. Soc. Spec. Publ.* 7: The Russians in California. 3-21. (1933).

GREENHOW, ROBERT. 1845. The history of Oregon and California, and the other territories on the north-west coast of North America: accompanied by a geographical view and map of those countries, and a number of documents as proofs and ilustrations of the history. ed. 2. Boston, 272-275.

LANGSDORFF, GEORG HEINRICH VON. 1812. Bemerkungen auf einer Reise um die Welt in den Jahren 1803-1807. 2 vols. Frankfurt am Main.

——— 1813-1814. Voyages and travels in various parts of the world, during the years 1803, 1804, 1805, 1806, and 1807. 2 vols. London.

——— 1927. Langsdorff's narrative of the Rezanov voyage to Nueva California in 1806 being that division of Doctor Georg H. von Langsdorff's Bemerkungen auf einer Reise um die Welt, when, as personal physician, he accompanied Rezanov to Nueva California from Sitka, Alaska, and back. An English translation, revised, with the teutonisms of the original hispaniolized, russianized, or anglicized, by Thomas C. Russell. San Francisco.

MAHR, AUGUST CARL. 1932. The visit of the "Rurik" to San Francisco in 1816. *Stanford Univ. Publ. Univ. Ser.; History, Economics and Political Science* 2: no. 2, 1-194.

MELVILLE, HERMAN. 1851. Moby-Dick; or, the Whale. ed. 1. New York. 231, 232. [First British edition published in London in same year under the title The Whale. 3 vols.]

STILLMAN, JACOB DAVIS BABCOCK. 1869. Footprints of early California explorers. *Overland Monthly.* 2: 258-260.

1810-1820

INTRODUCTION TO DECADE OF

1810–1820

BOTANICAL field-work made progress during these ten years, both from the Mississippi River westward and on the Pacific coast. Although plant collectors were few, what they accomplished was important.

Before the outbreak of the War of 1812 two British botanists, John Bradbury and Thomas Nuttall, made a journey up the Missouri River, availing themselves of the protection of a party of fur traders bound for the mouth of the Columbia River. The leader was Wilson Price Hunt and, since the party went at the behest of John Jacob Astor, its members were called the Astorians. The two botanists accompanied Hunt as far as the Arikara villages in northern South Dakota, thence traveling—Bradbury overland and Nuttall by boat—to the country of the Mandan Indians in North Dakota. From there they returned to St. Louis and eventually to New Orleans. Nuttall was fortunate enough to get to England before the War of 1812 made a crossing of the Atlantic impossible and in 1818 published his *Genera of North American plants*. The less fortunate Bradbury, who had returned to the United States from New Orleans, was unable to follow his plants to England until the war was over and, when he got there, found that Frederick Pursh had obtained access to his collections and had published the most important in his *Flora americae septentrionalis* in 1814. Bradbury, disheartened, never resumed plant collecting as a profession. However, in 1817, he published his famous *Travels in the interior of America, in the years 1809, 1810, 1811 . . .*, a work now considered one of the important sources of information for the region and period—perhaps a more generally useful book than any flora which he might have written!

After the war was over, Nuttall returned to the United States and pursued his botanical work in the east and, in 1819 and early 1820, made a remarkable journey up the Arkansas River to Fort Smith, from there entering eastern portions of what is now the state of Oklahoma, but then set apart for Indian occupancy. He was the first to collect plants in Arkansas and in Oklahoma. A year after his return, or in 1821, he published *A journal of travels into the Arkansa territory, during the year 1819*. Although not a financial success for the printer, the book, like Bradbury's *Travels,* is an illuminating picture of conditions then existing and is consulted by many whose interest in Nuttall's botanical publications is slight. Not until 1835–1836, or until some fifteen years after his return, did Nuttall find time to publish his "Collections towards a flora of the territory of Arkansas."

Another expedition—popularly known as the "Yellowstone Expedition"—traveled westward from St. Louis in this decade. This was sent into the field by the United States government and, until financial economies entered the Washington picture, was destined for the regions of the upper Missouri River. It consisted of two sections, one military, the other "scientific." The last was commanded by Stephen Harriman Long and, since it was the first government-sponsored expedition to include scientists, received some fame for that reason; not all deserved perhaps, since those who participated on the "scientific" basis paid for much of their own equipment, bought their own horses, and so on. William Baldwin, an American, served as botanist and physician for about six months of 1819, or from the time the expedition left St. Louis until it reached Franklin, Missouri. There Baldwin died; he had been suffering from tuberculosis when the journey began and neither his exertions nor his life in the steamboat *Western Engineer*—at that stage of steamboat development a most unsuitable form of travel from every point of view—had been conducive to curing his illness. When the journey was resumed in 1820, Edwin James had been appointed to take Baldwin's place, travel by steamboat had been stopped, and the party's destination had been altered and its activities abridged.

On the Pacific coast a Russian expedition, commanded by Otto von Kotzebue and traveling in the ship *Rurick,* reached California in 1816 and spent a month at San Francisco. In the party were two men, famous in the annals of California botany: Adelbert von Chamisso, botanist, poet, and author of *Peter Schlemihl,* a story of a man without a shadow; and Johann Friedrich Eschscholtz, by profession a physician and particularly interested in entomology, whose name will ever be remembered in the generic appellation of the California poppy. These two men collected many plants during their sojourn in the Bay of San Francisco and a goodly number of their discoveries were published in the decade of 1820–1830.

The portions of the trans-Mississippi west primarily affected by the War of 1812—which for a time halted the westward progress of the pioneers but which, by stimulating the American nationalistic spirit, eventually gave a new impetus to that progress —were the Mississippi River valley and the regions extending from the Great Lakes westward to the Pacific northwest where American interest was now in conflict with the better-established enterprise of Great Britain. The Treaty of Ghent (December 24, 1814) which technically terminated the conflict, guaranteed to the United States the possession of the "Old Northwest," a region comprising the present states of Michigan, Wisconsin, Ohio, Indiana, Illinois, and—important to our story—those northeastern portions of Minnesota which included the sources of the Mississippi and other great rivers.[1] The portion of Minnesota included in the settlement represented the second acquisition of territory by the United States in the trans-Mississippi west.

Two boundary questions were under consideration and were, at least temporarily, determined in this decade:

(1) In London, in 1818, the commissioners appointed to determine the boundary between the United States and Canada agreed that, from the Lake of the Woods west-

1. The more westerly portions of Minnesota had already been acquired by the Louisiana Purchase.

ward to the summit of the "Stony" or Rocky Mountains, the line should follow the 49th degree of north latitude; beyond that point to the Pacific Ocean there was to be joint occupation for ten years, after which further negotiations were to take place. The determination of this boundary as far west as the Continental Divide fixed the northern limits of the lands which the United States had acquired by the Louisiana Purchase.[2]

(2) The poorly defined boundaries of the lands acquired by the Louisiana Purchase had been a cause of controversy between Spain and the United States. In more than one region the resident Spaniards feared and resented the increasing encroachment of American traders and settlers. By the treaty of 1819 which consummated the Florida Purchase, lands in Texas west of the Sabine River[3] were allocated to Spain, in exchange for Florida which became part of the United States. And, by the same Florida Purchase, Spain surrendered to the United States her claims in the Oregon country. The provision of the treaty granting Spain control west of the Sabine River was resented by some Americans; and the growing interest of American citizens in the northwest coast was a matter of concern to Great Britain. Years of controversy and in some instances of bloody conflict were to pass before these territorial disputes were finally settled. Many of our plant collectors comment upon these matters and in some instances their work was affected thereby.

In London, in 1814, Frederick Pursh published his *Flora americae septentrionalis,* named by Asa Gray, writing in 1882, as the second of but two floras of North America ". . . ever published as completed works . . ." In this flora are to be found plants from west of the Mississippi River or from, writes Gray, ". . . the Great Plains, the Rocky Mountains, and the Pacific coast, although the collections were very scanty."

None of the plants described by Pursh from these western regions were of his own collecting. More than once he utilized the material of men whose stories are told here —some of Menzies' specimens, those of Lewis and Clark, the best of Bradbury's, a few of Nuttall's—perhaps there might have been more had not Nuttall been wary. But whatever Pursh's methods may have been, he is credited with having done a fine piece of work.

2. Precisely what points marked the 49th parallel, whether the sources of certain great rivers lay within American or British domain, what natural resources existed in this United States territory, and much else, still remained to be determined. And it was to acquire such knowledge that the United States government in two later decades (1820–1830 and 1830–1840) sent out exploratory expeditions which concern this story for the reason that they included men whose partial duty it was to collect plants and to report upon the vegetation.

3. In writing of the close of this decade, W. J. Ghent comments upon a state that had been admitted to the Union on April 8, 1812:

"Louisiana, though four-fifths of its territory lies west of the Mississippi, early passes from the category of a Western community; it becomes, in all respects, purely Southern, and is thus usually excluded from any treatment of the Western scene. For the next decade [Ghent's decade of 1821 through 1830] the Southern frontier stretches along the Sabine . . ."

After specifying that he was omitting "... all mention of more restricted works such as Nuttall's 'Genera of North American Plants,' which came out only four years after Pursh's Flora; also the 'Flora Boreali-Americana' of Sir William Hooker, which began in 1829, but was restricted to British America," Asa Gray estimated that Pursh's *Flora* "... contains 740 genera of Phænogamous and Filicoid plants, and some 3076 species,—just about double the number of species contained in Michaux's Flora of eleven years before."

CHAPTER V

AFTER COLLECTING IN THE ENVIRONS OF ST. LOUIS, BRADBURY TRAVELS WITH HUNT AND THE "ASTORIANS" UP THE MISSOURI RIVER TO THE ARIKARA VILLAGES AND FROM THERE MOVES NORTH TO THE COUNTRY OF THE MANDANS

ALTHOUGH Lewis and Clark made the first collection of plants from the Mississippi westward they had only had a slight indoctrination in such work before they started on their journey. The first trained botanist[1] to collect plants immediately west of the Mississippi River was the Englishman, John Bradbury. He had arrived at St. Louis, Missouri, on December 31, 1809, and his *Travels* tells that "During the ensuing spring and summer, I made frequent excursions alone into the wilderness, but not farther than eighty or a hundred miles into the interior."

The few facts of Bradbury's private life must be assembled from a multitude of sources. Most informative are: Harshberger (1899); True (1929), who quotes from

1. In "A biographical history of botany at St. Louis, Missouri," Perley Spaulding comments that ". . . For all practical purposes André Michaux may be said to have been the first botanist to work in the vicinity of St. Louis . . . the first botanical worker concerning whom published records have yet been found as having worked in the vicinity of St. Louis . . ."

This statement is true in the sense that Michaux was on the eastern side of the Mississippi directly across from St. Louis. But—interested in the identity of the first botanist to work in the eastern portion of the trans-Mississippi west—I cannot agree that Spaulding's further comment (that ". . . the evidence seems to indicate that he must have visited the west shore of the Mississippi . . .") is supported by the two records which he offers as proof.

The first of these records—that Michaux ". . . mentions St. Louis as being in a prosperous condition . . ." at the time of his visit to the Illinois Territory in 1795—is, I believe, a misinterpretation of the text of Michaux's journal. The American Philosophical Society very kindly sent me a microfilm enlargement of the portion of the journal covering the period from August 31 through October 5, 1795. This shows that C. S. Sargent's transcript thereof, although including an entry of September 20 which reads "St Louis florissant Pet. Cotes," omits two succeeding lines reading, respectively, "Vide Poche Ste Genevieve" and "Cap Girardeau N. Madrid Natchez." Taken together these three lines *of place-names* indicate a memorandum of some sort. The pages sent me include another very similar list mentioning towns which were never visited by Michaux in 1795. Moreover, the word "florissant," even if uncapitalized, is not an adjective descriptive of the "prosperous condition" of St. Louis, as Spaulding interprets the word, but is another name for the town of St. Ferdinand. R. G. Thwaites reproduced this error and added another in misinterpreting "Petites Côtes"—merely another name for the town of St. Charles—to mean prairies and hills. Bradbury's "Description of the Missouri Territory" (Appendix 4, p. 262 of *Travels*, Liverpool edition of 1817) and the Elliott Coues edition of *The expeditions of Zebulon Montgomery Pike* (1: 213, *fn.* 41, 1895) refer at some length to these and other places-names and their colloquial equivalents.

The second record—serving to Spaulding as possible evidence of Michaux's presence west of the Missisa rare and locally published English book, *Bygone Stalybridge* (1907), written by a Samuel Hill; and Harold William Rickett (1934).[2]

Britten and Boulger, and correctly, give the date of Bradbury's birth as August 20, 1768; True (citing Hill) as 1765, and at Souracre Fold or Far Souracre, Stalybridge, England. He had become interested in natural history at an early age—his father had given him the works of Linnaeus—and by the time he was twenty-two his writings and discoveries relating to plants and insects were known to the naturalists of London, among others to the great Sir Joseph Banks. Through his instrumentality Bradbury became a fellow of the Linnean Society about 1792. From sources other than Hill, True learned that in middle life Bradbury lived both at Liverpool and at Manchester. It was at Liverpool that he met William Roscoe, one of the founders of the Liverpool Botanic Garden, and William Bullock,[3] head of the Liverpool Museum. He became corresponding secretary of the Liverpool Philosophical Society.

In 1809 Bradbury was chosen to make an exploration in the United States. The

sippi River—is the citation in the *Flora boreali-americana* of a few plants from the Missouri River. Michaux was at Kaskaskia (Randolph County, Illinois); also at "Kaskia," now Cahokia (St. Clair County, Illinois), where, certainly, he was in "the vicinity of St. Louis." But the journal is fairly explicit as to itinerary and it seems fair to assume that, had he crossed the Mississippi, the fact would have been mentioned. More likely, as Spaulding thought possible, the plants were given Michaux by someone who had been on the Missouri.

MICHAUX, ANDRÉ. 1889. Portions of the journal of André Michaux, botanist, written during his travels in the United States and Canada, 1785 to 1796. With an introduction and explanatory notes, by C. S. Sargent. *Proc. Am. Philos. Soc.* 26: no. 129, 1-145. Cahier 9.

—— 1904. Journal of André Michaux, 1795–1796. *In:* Thwaites, Reuben Gold. Early western travels 1748–1846. 3: 25-104. [English translation of pp. 91-101, 114-140 of *Proc. Am. Philos. Soc.* 1889.]

SPAULDING, PERLEY. 1908. A biographical history of botany at St. Louis, Missouri. *Pop. Sci. Monthly* 73: 448-492.

2. Since this chapter was written Dr. Rickett has published letters written by or concerned with Bradbury. I refer to these and to the editor's notes at the end of this chapter and occasionally insert a footnote followed by the notation "Rickett (1950)."

3. The name of William Bullock appears more than once in this chapter. According to "Miss A. Macdonell," in the *Dictionary of national biography*, he had started a private museum in Liverpool, but in 1812 transferred it to London where, seven years later, it was sold at auction. This explains why his enterprise is sometimes referred to as the Liverpool, sometimes as the London, Museum.

Macdonell tells that Bullock made a trip to Mexico in 1822, bringing back many very miscellaneous "valuable curiosities." An account of this trip was published in 1824.

Five years later Bullock made a second trip to North America which was described in his *Sketch of a journey through the western states of North America, from New Orleans, by the Mississippi, Ohio, city of Cincinnati and falls of Niagara, to New York, in 1827* ... (London, 1827). Although the appendix includes a few pages headed "Botany" (taken from a work on Cincinnati by a "Mr. Drake"), it is apparent that Bullock's interest centered, not on natural history, but on the promotion of a real estate venture, "... a town of retirement, to be called Hygeia ... This will enable persons desirous of establishing themselves in this abundant and delightful country, to do so at a very moderate expense ..." As the name "Hygeia" suggests, Bullock was convinced of the healthful aspects of the Cincinnati region and stresses them to an amusing extent.

Mrs. Frances Trollope's *Domestic manners of the Americans* refers more than once to Bullock and his wife; she spent two years in Cincinnati and was entertained by them at their home "... about two miles

promoters of his trip, patrons as well of the Liverpool Philosophical Society, wished to learn what could be done to increase the supply of cotton used by the manufacturers. Bradbury refers to the Botanic Garden in Liverpool as his sponsor and that organization certainly supplied some of the funds for his undertaking. A letter from Roscoe to Jefferson, dated April 25, 1809, tells that "Among those who have encouraged his undertaking, in which he will be accompanied by his two sons, are the Proprietors of the Botanic Garden in Liverpool . . ." For this institution he was to collect seeds and plants. Whatever the main reason for Bradbury's visit, cotton growing[4] seems to have been neglected after his arrival in North America. And if "two sons" accompanied him, it is strange that they are never mentioned in his *Travels*. We do hear of one writing him from England.[5]

Bradbury begins the preface to his *Travels* thus:

"When I undertook to travel in Louisiana, it was intended that I should make New Orleans my principal place of residence, and also the place of deposit for the result of my researches. This intention I made known to Mr. Jefferson, during my stay at Monticello, when he immediately pointed out the want of judgment in forming that arrangement, as the whole of the country round New Orleans is alluvial soil, and therefore ill suited to such productions as were the objects of my pursuit."

With a letter of introduction from Roscoe, Bradbury had visited Jefferson for about three weeks. His host, we learn from True, wrote Meriwether Lewis, then a General, and as Governor of the Louisiana Territory resident at St. Louis, on August 16, 1809:

". . . Having kept him here about ten days, I have had an opportunity of knowing that besides being a botanist of the first order, he is a man of entire worth & correct conduct. as such I recommend him to your notice, advice & patronage, while within

below Cincinnati, on the Kentucky side of the river . . . a large estate, with a noble house upon it." I quote from the London edition (2 volumes, 1832):

". . . He and his amiable wife were devoting themselves to the embellishment of the house and grounds; and certainly there is more taste and art lavished on one of their beautiful saloons, than all Western America can shew elsewhere. It is impossible to help feeling that Mr. Bullock is rather out of his element in this remote spot, and the gems of art he has brought with him, shew as strangely there, as would a bower of roses in Siberia, or a Cincinnati fashionable at Almack's. The exquisite beauty of the spot, commanding one of the finest reaches of the Ohio, the extensive gardens, and the large and handsome mansion, have tempted Mr. Bullock to spend a large sum in the purchase of this place, and if any one who has passed his life in London could endure such a change, the active mind and sanguine spirit of Mr. Bullock might enable him to do it; but his frank, and truly English hospitality, and his enlightened and enquiring mind, seemed sadly wasted here. I have since heard with pleasure that Mr. Bullock has parted with this beautiful, but secluded mansion."

Donald Smalley's new edition (1949) of Mrs. Trollope's story contains numerous references to Bullock.

Stansfield (1951) quotes a Bradbury letter to William Roscoe which indicates that he had done some work in Bullock's museum.

4. ". . . Dr. [Joseph A.] Clubb [Curator of Museums at Liverpool] is inclined to regard Bradbury's relations with the cotton industry as more important than his connection with the Botanic Garden." (True)

5. According to the letters published by Rickett (1950), Bradbury's idea was to establish one son in New Orleans, there to grow and to ship to Liverpool the plants which his father collected. But this plan seems never to have been put through, doubtless for lack of funds.

your government or its confines. perhaps you can consult[6] no abler hand on your Western botanical observations."

Jefferson also wrote Benjamin Smith Barton on September 11, 1811, how, on his visit to Monticello, Bradbury passed ". . . every day in the woods from morning to night. He found, even on this mountain, many inedited articles . . ."

True tells us that, at the time of his visit to America, Bradbury ". . . was about 43 or 44 years of age and is described by an author as 'being in the prime of manhood, swarthy, broad-shouldered, and of medium height, amiable yet stubborn in disposition, temperate in his habits, and an excellent marksman. He was fond of music, active on his feet, and determined in his methods and opinions.' "

Of Bradbury's route from the east to St. Louis and of the precise time spent on that journey his *Travels* give no record. True comments: "It appears probable that in going from Monticello to St. Louis, Bradbury followed the usual route of those days down the Ohio River." [7] He was with Jefferson in August, 1809, and about four months later, or on December 31, 1809, arrived at his destination.

Bradbury tells how he spent what remained of the winter of 1809–1810 preparing for botanical excursions during the coming spring and summer—it was then that he made his trips into the wilderness about St. Louis. By the autumn of 1810 he was able to ship to Liverpool, by way of New Orleans, ". . . in seven packages, the result of my researches . . ." The boat to which these were entrusted was driven ashore some sixty miles below St. Louis; when Bradbury got to the scene he found that it had been able to continue on its way. But it was not until the end of November, 1811, that he learned of the receipt of these collections in Liverpool.

Bradbury's arrival in St. Louis antedated that of Thomas Nuttall, another British botanist, by some months; Nuttall reached that city from the east sometime during the autumn of 1810. In his "Travels and scientific collections of Thomas Nuttall" Francis W. Pennell states:

"During the late autumn of 1810 and the ensuing winter they likely made a number of excursions together. One such was to the Meramec River, as Nuttall speaks in 1821 about 'the beautiful white and friable sandstone which has been observed near a branch of the Merrimec by Mr. Bradbury and myself . . . in the winter of 1809 [= 1810].' Nuttall's 'Genera' gives a number of species as first seen by Bradbury on

6. Lewis' death occurred in October, 1809, and the two men could not have met.

7. The route is given in Letter 7 published by Rickett (1950). Bradbury had crossed ". . . that most Mountainous tract of country that lies betwixt Monticello and the falls of the Kenhawa (Kanawha) River a distance of about 240 Miles . . . I passed down the Kenhawa to its junction with the Ohio at point Pleasant 90 Miles from the falls here I bought a skiff . . . in which I rowed myself to the falls of the Ohio (Louisville) 325 Miles where I arrived the 27 Sepr. In a few days aterwards I fell sick of an Ague which . . . remained with me untill the 20 of November. On the 23 I quitted Louisville in a Boat down the Ohio in which I sailed to Shawaney Town (Shawneetown) about 10 Miles below the mouth of the *Wabash* and 350 from Louisville. here I quitted the River & got a Horse on which I travelled through an uninhabited country to Kaskaskia 145 Miles sleeping on the ground. At Kaskaskia I crossed the Mississippi into upper Louisiana and arrived here [St. Louis] Decr. 31."

this river . . . The relations between these two botanists . . . remained that of an incidental acquaintanceship which even their common interests failed to convert into a real comradeship."

After investigating what had happened to his collections in the shipwreck, Bradbury learned that some fur traders intended to follow the Lewis and Clark route to the mouth of the Columbia River. They were traveling in the interest of the Pacific Fur Company—subsidiary of the American Fur Company and, like the main organization, controlled by John Jacob Astor—and, as already mentioned, are generally known as the Astorians.[8] Their leader was Wilson Price Hunt[9] who, after he learned of Bradbury's objectives, ". . . in a very friendly and pressing manner . . ." invited him to accompany the party as far up the Missouri River ". . . as might be agreeable to my views." Although Bradbury had intended to spend the rest of the summer at ". . . Ozark, (or more properly Aux-arcs) on the Arkansas . . ." he changed his plans and accepted Hunt's invitation.

The main party of the Astorians, led by Hunt, left St. Charles, Missouri, on March 14, 1811, and arrived at their western destination on February 16, 1812. Returning, Robert Stuart led the main party eastward, leaving the post which had been built at the mouth of the Columbia River and named Astoria, on June 28, 1812, and arriving at St. Louis on April 30, 1813.[1]

From St. Louis Bradbury accompanied Hunt only as far as the Arikara villages in South Dakota, about 1,800 miles; from there he made a trip with Ramsay Crooks to the Mandan nation in North Dakota, another 200 miles farther north. Our interest in the Astorian expedition, therefore, covers only about four months and ends when Bradbury left the party to return to St. Louis. This was on July 17, 1811, the day before Hunt resumed his westward journey.

Of important books and articles concerned with the Missouri River journey I mention the following.

John Bradbury's *Travels in the interior of America, in the years 1809, 1810, and 1811*—the title is considerably longer—was originally published in Liverpool in 1817. A second edition, bearing the same title and including a map, was issued in London in 1819. The *Edinburgh Review* of December, 1818, discussing recent books

8. Washington Irving has told their story in *Astoria; or, incidents of an enterprise beyond the Rocky Mountains*, first published in Philadelphia in 1836.

9. W. J. Ghent refers to him as ". . . wholly unacquainted with the work to which he was assigned . . ."

1. The venture as a whole was unsuccessful. A supply ship, the *Tonquin*, sent round the Horn under command of Captain Jonathan Thorn, had arrived at the mouth of the Columbia River about a year before Hunt arrived there overland and had established the trading post Astoria. The *Tonquin* then moved north and, at Newtee Harbor, Vancouver Island, all but one of its company had been massacred by Indians. When the overland party, after various calamities, returned to St. Louis, it was only to learn that the War of 1812 had been in progress for some time.

Astoria, sold to the British in 1813 and renamed Fort George, was abandoned in 1823 by its new owners in favor of Fort Vancouver. The old post soon went to pieces and many later travelers refer to that regrettable fact.

about America, noted that "Mr. Bradbury is a botanist, who lived a good deal among the savages, but worth attending to."

In 1904 (in volume five of his *Early western travels*) R. G. Thwaites transcribed Bradbury's *Travels* from the London edition of 1819 and included a reproduction of the map.

Henry Marie Brackenridge[2]—although not traveling with Hunt's party but with the fur-trader Manuel Lisa—was on the Missouri River at the same time and, since he and Bradbury were friends, the two were together as much as possible. Brackenridge published two books which deal with events and personalities associated with the journey: the first was his *Views of Louisiana; together with a journal of a voyage up the Missouri river in 1811,* published in Pittsburgh in 1814; the second his *Journal of a voyage up the river Missouri; performed in eighteen hundred and eleven,* published in Baltimore in 1816 and referred to on the title page as a second edition *"Revised and Enlarged by the Author."*

In 1904 R. G. Thwaites transcribed Brackenridge's *Journal* (in volume six of his *Early western travels*) from the Baltimore edition of 1816. He omitted, as having no bearing on his series, Parts I, II and IV of Bradbury's Appendix.

Francis W· Pennell's "Travels and scientific collections of Thomas Nuttall" (1936) contains many references to Bradbury. Since Nuttall is not known to have kept a journal while with the Astorians, Pennell states that he used Bradbury's careful itinerary to check the dates and localities cited by Nuttall for his collections.

H. W. Rickett's "Specimens collected by Bradbury in Missouri Territory" (1934) gives a concise outline of the trip and a summary of some of Bradbury's interesting botanical discoveries.

Irving's *Astoria,* already mentioned, contains references to both Bradbury and Nuttall.

I shall transcribe from the Liverpool edition (1817) of Bradbury's *Travels* and from both the Pittsburgh edition (1814) and the Baltimore edition (1816) of the Brackenridge narrative.

The party traveled by boat and Bradbury seldom got far from the shores of the Missouri River. He did make one trip afoot (May 2-11) with Ramsay Crooks, but that was exceptional.

1811

From March 13 to April 27 the route lay through Missouri.

On March 13 Bradbury left St. Louis ". . . in company with a young Englishman of the name of Nuttall . . ." Hunt, who had left a day earlier, was overtaken at St. Charles, situated about twenty miles above the mouth of the Missouri, and the journey by boat began the next day. The season was too early for satisfactory plant collecting.

---

2. Chittenden characterizes Brackenridge as ". . . a young man of good education, very observing, and a promising young writer. His *View of Louisiana,* and his journal of his voyage up the Missouri, like Bradbury's *Travels,* are among our most reliable early authorities."

March 14: The first plant mentioned grew on Bonhomme Island where Bradbury ". . . observed in the broken banks . . . a number of tuberous roots, which the Canadians call *pommes de terre*. They are eaten by them, and also by the Indians, and have much the consistence and taste of the Jerusalem artichoke: they are the roots of *Glycine apios.*"

March 16: He noted ". . . abundance of *Equisetum hyemale,* called by the settlers *rushes* . . ." and esteemed as a winter food for cattle.

March 17: Before reaching "Charette" (La Charette, Warren County) he was impressed by the ". . . vast size to which the cotton wood tree ['Populus angulosa of Michaux, called by the French Liard (JB)'] grows. Many of these . . . observed this day exceeded seven feet in diameter, and continued with a thickness very little diminished to the height of 80 or 90 feet, where the limbs commenced."

It was on leaving La Charette that Hunt pointed out ". . . an old man standing on the bank . . . Daniel Boond, the discoverer of Kentucky." Having been given a letter to Boone, Bradbury went ashore for a talk: "He informed me that he was eighty-four years of age; that he . . . had lately returned from his spring hunt, with nearly sixty beaver skins." The next day John Colter, to whom Bradbury devotes a long footnote, joined the party for a few miles: "He seemed to have a strong inclination to join [and] accompany the expedition; but having been lately married, he reluctantly took leave of us."

March 18: Bradbury ". . . walked along the bluffs . . . beautifully adorned with *Anemone hepatica.*" Camp was ". . . near the lower end of Lutre (Otter) Island." From the north Loutre River empties into the Missouri in Montgomery County. The party was about opposite Hermann, Gasconade County.

March 20: Along the south side of the river were bluffs of whitish limestone: ". . . the tops are crowned with cedar, and the ledges and chinks are adorned with *Mespilus canadensis,* now in flower."

Camp was seven miles above the mouth of Gasconade River, joining the Missouri from the south in Gasconade County.

On March 21 we are told of an excursion with Nuttall. Bradbury ". . . went ashore . . . intending to walk along the bluffs, and was followed by Mr. Nuttall . . . we came to a creek or river, much swelled by the late rains; I was now surprised to find that Mr. Nuttall could not swim: as we had no tomahawk, nor any means of constructing a raft . . . we looked for no alternative but to cross the creek by fording it . . . I stripped, and attempted to wade . . . but found it impracticable. I then offered to take Nuttall on my back, and swim over with him; but he declined . . ."

Eventually they were able to cross on some driftwood and ". . . with some difficulty overtook the boat." This day they reached ". . . a French village, called Cote sans

Dessein, about two miles below the mouth of Osage river." This stream joins the Missouri from the south and forms the boundary between Osage County (east) and Cole County (west).

On March 27 they passed Manitou Rocks, near the southwestern corner of Boone County and camped a little above the mouth of Bonne Femme River which enters the Missouri from the north in Howard County. Here, notes Bradbury, ". . . the tract of land called Boond's Lick settlement, commences, supposed to be the best land in Western America for so great an area; it extends about 150 miles up the Missouri, and is near 50 miles in breadth."

On March 29 the party ". . . slept about a league above the settlements." And the next day were ". . . beyond all the settlements, except those of Fort Osage . . ."

On April 1 Bradbury comments upon the scratching quality of the prickly ash ["Zanthoxylon clava Hercules (JB)"] and upon the ". . . abundance of small prickly vines entwined among the bushes, of a species of *smilax.*" And, like all the early travelers, he describes his first encounter with a skunk!

April 2: "We passed the mouth of La Grande Riviere, near which I first observed the appearance of prairie . . ."

Grand River, boundary between Carroll County (west) and Chariton County (east), enters the Missouri a short distance above Brunswick. The prairie Bradbury defines as ". . . such tracts of land as are divested of timber . . ." and notes that it begins two or three hundred miles west of the Mississippi River and continues to the Rocky Mountains and, north and south, ". . . from the head waters of the Mississippi, to near the Gulf of Mexico; an extent of territory which probably equals in area the whole empire of China."

On April 8 they reached Fort Osage, near Sibley, Jackson County, and were met at the landing place by "Mr. Crooks"; and introduced to: ". . . Mr. Sibly, the Indian agent there, who is the son of Dr. Sibly[3] of Natchitoches." At the village of the "Petit Osage nation" Bradbury observed that ". . . the leaves of the flag, or *Typha palustris* . . ." were used for roof-mats and ". . . the pulp of the persimon ['Diospyros Virginiana (JB)'], mixed with pounded corn . . ." was made into cake ". . . in taste resembling gingerbread." This was called "staninca."

On leaving Fort Osage on April 10 Ramsay Crooks[4] (who had come down from "the wintering station" at the mouth of the Nodaway to meet Hunt) and twenty-six men were added to the party. From April 11 to 14 Bradbury notes that "We . . . employed our sail, wherefore I could not go ashore without danger of being left behind."

3. Dr. John Sibley had explored the Red River of the South from its mouth to some distance above Natchitoches in 1803–1804. The son was George C. Sibley.

4. Chittenden states that Crooks had entered the employ of the Pacific Fur Company in 1811. He eventually became president of the American Fur Company.

At a deserted settlement of some Kansas Indians he noted, on April 15, on "... the sides of the hills ... abundance of the hop plant (*Humulus lupulus*)."

On April 17 they arrived "... at the wintering houses, near the Naduet river ..." where some members of the expedition had been quartered since the previous autumn. Here they remained until April 21. The "Naduet river," or Nodaway, forms the boundary between Holt County (west) and Andrew County (east); the site of the "wintering houses" was about twenty miles above St. Joseph, Buchanan County. Chittenden locates it near "a small stream called the Nadowa."

While there Bradbury refers to the abundance of the pigeon ["Columbo migratorius (JB)"]—he shot 271 of them and then desisted. "The species of pigeon associated in prodigious flocks: one of these flocks, when on the ground, will cover an area of several acres in extent, and are so close to each other that the ground can scarcely be seen ..."

On April 21 the party embarked in four boats. On the 26th "The wind ... blew so strong, that we were obliged to stop during the whole day ... I resolved to avail myself of the opportunity to quit the valley of the Missouri, and examine the surrounding country ... I ascended the bluffs, and found the face of the country, soil, &c. were entirely changed. As far as the eye could reach, not a single tree or shrub was visible."

From April 27 until May 22 the boats seem to have followed the Nebraska, or western, side of the Missouri River.

On April 27 Bradbury comments: "This day I collected several new species of plants." On April 28 they reached the mouth of the Platte River, entering the Missouri from the west between Cass County (south) and Sarpy County (north); here, "As the ash discontinues to grow on the Missouri above this place, it was thought expedient to lay in a stock of oars and poles ..." This day Bradbury tells:

"... I availed myself of this opportunity to visit the bluffs four or five miles distant ... I found an extensive lake running along their base, across which I waded ... its surface was much covered with acquatic plants, amongst which were *Nelumbium luteum* and *Hydropeltis purpurea;* on the broad leaves of the former, a great number of water snakes were basking ..."

On April 29 he again visited the bluffs and found them "... to be of a nature similar to those on the north-east side." On April 30, with Ramsay Crooks, Bradbury walked to the "wintering house" where they were joined by the boats; here, according to Chittenden, near the present city of Omaha, Crooks and McLellan had their principal establishment. The plain they had crossed had been burned over the previous autumn and "... was now covered with the most beautiful verdure, intermixed with flowers."

On May 1 Bradbury was invited to accompany Crooks on a walk to the village of the Oto Indians (the "Ottoes") which, according to Chittenden, was some forty miles above the mouth of the Platte River; from there they were to walk to the village of the Omahas (the "Mahas") situated "... about 200 miles above us on the Missouri ..." There they were to rejoin the boats. Bradbury describes the essentials which he was to carry:

"Our equipments were, a blanket, a rifle, eighty bullets, a full powder horn, a knife, and tomahawk, for each. Besides these, I had a large inflexible port-folio, containing several quires of paper, for the purpose of laying down specimens of plants; we had also a small camp-kettle, and a little jerked buffaloe meat."

They started May 2 and ". . . travelled at a great rate, hoping to reach the Platte that night . . ." but stopped at the Corne du Cerf, or Elkhorn, River. This, Bradbury remarks, ". . . was very agreeable to me, as I was much exhausted . . . I was unable to eat at supper, and lay down immediately." They reached the Oto village on May 3 and remained until May 6. On May 7 they crossed the Elkhorn and on the 9th came to what they ". . . supposed to be Black Bird Creek, which falls into the Missouri . . ." This may have been Bell Creek which is shown on recent maps entering the Elkhorn in Washington County. Here Bradbury ". . . walked back to an eminence, to collect some interesting plants, having noticed them in passing." When a violent thunderstorm came up they moved out from under some trees—on Bradbury's advice—and lay down in their blankets on the open prairie. The botanist noted: "I put my plants under me." On May 11 they reached the Maha village. Bradbury's map shows the village on the Nebraska side of the Missouri, and Pennell adds ". . . some ten or twenty miles below Sioux City, Iowa." The expedition remained here until May 15.

The day of arrival Bradbury tells us of the interest which his plants aroused in the village.

". . . I saw an old Indian galloping towards me; he came up and shook hands with me, and pointing to the plants I had collected, said, 'Bon pour manger?' to which I replied, 'Ne pas bon;' he then said, 'Bon pour medicine?' I replied 'Oui.' He again shook hands, and rode away . . ."

On May 12, since "an old and respectable trader" by the name of James Aird was leaving for "the United States in a few days," and it was possible to forward letters, Bradbury ". . . was employed in writing until the 12th[5] at noon."

The expedition remained at the Maha village for May 13 and 14 and Bradbury comments that "The Maha's seem very friendly to the whites, and cultivate corn, beans, melons, squashes, and a small species of tobacco *(Nicotiana rustica)."*

On May 15 they ". . . embarked early and passed Floyd's Bluffs, so named from a

5. Among the letters published by Rickett (1950) is one (no. 18) from Bradbury to Roscoe, dated from the "Mouth of the River Naduet [Nodaway] 19 Ap 1811," with a postscript dated "Mahas Village 350 miles above St. Louis 12 May," which contains a notice of a shipment of plants not mentioned in the *Travels:*

". . . I send to St. Louis to be forwarded to you Sir for the Committee specimens of 25 Plants for which I can find no discription three of them may probably be valuable . . . Many of these have been collected in a Journey by Land which I undertook with an Indian Trader [Ramsay Crooks] . . . we quitted the Boats 200 miles below this place by water went to the Ottoes 40 miles & from them to this place 120 across the country . . ."

Bradbury refers to the same shipment of plants in another letter (no. 19), also included by Rickett, dated "Merion . . . Nr. St. Louis 16 Augt. 1811": ". . . I wrote to you Sir from the Mahas Nation & sent a small Packet of specimens . . ."

person of the name of Floyd,[6] (one of Messrs. Lewis and Clark's party) having been buried there." On May 17 and 18 the boats made such good progress that Bradbury had few opportunities for collecting—"I regretted this circumstance, as the bluffs had a very interesting appearance"—but on the 20th, a head wind stopping the boats, he notes: "I availed myself of this circumstance, and was very successful in my researches." On the 21st he ". . . walked the greatest part of the day, chiefly on the bluffs . . ." Bradbury mentions bluffs constantly. They doubtless afforded a better foothold for plants than the lowlands affected by the rise and fall of the river.

About May 22 they must have passed the mouth of the James River which enters the Missouri from the north, in Yankton County, South Dakota. Bradbury records:

"This day, for the first time, I was much annoyed by the abundance of the prickly pear. Against the thorns of this plant I found that mockasons are but a slight defence. I observed two species, *Cactus opuntia* and *Mamillaris*."

Having observed this day from the bluffs a bend in the Missouri—presumably the one between Bon Homme County, South Dakota (north) and Knox County, Nebraska (south)—Bradbury on the 23rd ". . . determined to travel across the neck. I therefore did not embark with the boats, but filled my shot pouch with parched corn, and set out, but not without being reminded by Mr. Hunt that we were now in an enemy's country."

The farther he got from the river the more the countryside improved. The hills and valleys were ". . . covered with the most beautiful verdure . . . I continued to travel through this charming country till near the middle of the afternoon, when I again came to the bluffs of the Missouri, where, amongst a number of new plants, I found a fine species of *Ribes,* or currant."

A somewhat alarming encounter with Indians terminated, fortunately for Bradbury, with the arrival of the boats. They passed ". . . the mouth of the river called L'Eau qui Court, or Rapid River." This was the Niobrara which empties into the Missouri from the south, in Knox County, Nebraska.

Between this river and White River, reached on May 26 or 27, the party had entered South Dakota and was to travel through that state until June 20. On the 27th "Little Cedar Island," which Bradbury locates as "1075 miles from the mouth of the Missouri," was reached; they were about opposite Chamberlain, Brule County. On the 28th Bradbury examined the island and also the bluffs:

". . . this delightful spot . . . is about three quarters of a mile in length, and 500 yards in width. The middle part is covered with the finest cedar, round which is a border from 60 to 80 yards in width, in which are innumerable clumps of rose and currant bushes, mixed with grape vines, all in flower, and all extremely fragrant. The currant is a new and elegant species, and is described by Pursh as *Ribes aureum.* Be-

6. Charles Floyd. The volume *Iowa* ("American Guide Series"), p. 312, states that the man died ". . . a short distance below Sioux City but his body was taken a little farther up the river and buried with military honors on this bluff." The site is in Woodbury County.

twixt the clumps and amongst the cedars the buffaloes, elks, and antelopes had made paths, which were covered with grass and flowers. I have never seen a place, however embellished by art, equal to this in beauty.

"Since our departure from L'Eau qui Court . . . the bluffs had gradually continued to change in appearance . . . I began to notice a number of places of a deep brown colour, apparently divested of vegetation . . . As we were now in an enemy's country, it was with reluctance Mr. Hunt suffered me to land . . . when I proceeded to examine one of these spots . . . nothing grew on it but a few scattered shrubs of a species of *Artemisia,* apparently a non-descript."

On May 29, availing himself of a short delay, Bradbury having ". . . hastened up the bluff . . .", observed ". . . a number of large white flowers on the ground, belonging to a new species of *œnothera,* having neither stem or scape, the flower sitting immediately on the root. On a signal being given from the boat, I was obliged to return . . ."

May 31 was devoted to a parley with some Sioux Indians who were persuaded to let the white men proceed only after Hunt explained "That we had come from the great salt lake in the east, on our way to see our brothers, for whom we had been *crying*[7] ever since they left us . . ."

On June 1 the party reached the great loop in the Missouri River which lies to the north of Lyman County and about midway between Chamberlain and Pierre.

"In the afternoon we entered upon the Great Bend, or, as the French call it, the Grand Detour, and encamped about five miles above the lower entrance. This bend is said to be twenty-one miles in circuit by the course of the river, and only 1900 yards across the neck."

The next day, after a visit from two Indians, Bradbury observed that ". . . in smoking the pipe they did not make use of tobacco, but the bark of *Cornus sanguinea,* or red dog wood, mixed with the leaves of *Rhus glabra,* or smooth sumach. This mixture they called kinnikineck."

On June 3 a boat arrived ". . . belonging to Manuel Lisa . . . we waited for it." Bradbury was ". . . much pleased . . . to find Mr. Henry Brackenridge was along with Mr. Lisa; I became acquainted with him at St. Louis, and found him a very amiable and interesting young man." Lisa's party had left St. Louis by boat on April 2 and had been trying to overtake Hunt, ostensibly for reasons of safety although the leader was suspected of other motives.[8] Brackenridge's *Journal* records the meeting as of June 2

---

7. In a footnote Bradbury attributes to Meriwether Lewis the statement that the Sioux ". . . are the vilest miscreants of the savage race, and must ever remain the pirates of the Missouri . . ."

At long distance to be sure, there was something Gilbert and Sullivanesque about the Indian reaction! When among the Osage Indians Bradbury had commented that this tribe, when stealing from the white men, were in the habit of weeping because ". . . they were sorry for the people whom they were going to rob."

8. Lisa was a partner in the St. Louis Missouri Fur Company, and was suspected of attempting to reach the Arikara Indians ahead of Hunt, representative of the the American Fur Company. Chittenden describes the race at some length. Hunt was inexperienced, whereas Lisa, who had made more than one trip to the upper Missouri and was familiar with the country, knew the tricks of the fur trade. Chittenden considers the suspicion unjustified.

and tells that "It was with real pleasure that I took my friend Bradbury by the hand." He had earlier explained:

"For my part I felt great solicitude to over take him [Hunt], for the sake of the society of Mr. Bradbury, a distinguished naturalist with whom I had formed an acquaintance at St. Louis . . . In the society of this gentleman, I had promised myself much pleasure, as well as instruction; and indeed, this constituted one of the principal motives for my voyage . . ."

About June 4 the party must have passed the mouth of "Teeton" River, actually the Bad River which enters the Missouri from the southwest in Stanley County and opposite Pierre. On the 5th Lisa had attempted to acquire the services of Hunt's interpreter, Pierre Dorion, and the episode nearly led to a fight which, Bradbury comments, "would certainly have been a bloody one." He notes on June 8:

"Since the affair of the 5th, our party has had no intercourse with Mr. Lisa, as he kept at a distance from us, and mostly on the opposite side of the river; this deprived me of the society of my friend Brackenridge. I regretted this circumstance . . . In the forenoon, we passed the mouth of Chayenne river . . . We encamped this night in a beautiful grove, ornamented with a number of rose and currant bushes, entwined with grape vines, now in bloom."

The Cheyenne River, which Bradbury sometimes refers to as the Chien or Chienne, enters the Missouri from the west, between Stanley County (south) and Armstrong County (north). On June 12 both parties of traders arrived at the Arikara villages, situated on the west bank of the Missouri, above Grand River which, from the west, enters the Missouri in Corson County, northern South Dakota.

The Arikara villages were to be Hunt's headquarters until he started for the Columbia River about a month later. His original plan had been to follow the Lewis and Clark route up the Missouri and the Yellowstone, but he had been persuaded by members of the company who had crossed the Rocky Mountains more than once to abandon the boats at the Arikara villages and start overland, thereby avoiding the territory inhabited by the dangerous Blackfeet Indians. This necessitated horses and, since a sufficient number could not be obtained from the Arikaras, it was arranged that Ramsay Crooks should proceed 150-200 miles farther up the Missouri to the Mandan villages where the Missouri Fur Company had a post (known as Lisa's post) and where Lisa agreed to help in procuring more. Bradbury decided to accompany Crooks who was to travel overland; Brackenridge and Nuttall were to accompany Lisa by boat.

While negotiations were going on with the Arikaras, Bradbury began a new form of collecting which must have added greatly to his labors. He notes that, on the 18th, he ". . . spent the remainder of the day examining the bluffs, to ascertain what new plants might be collected in the neighbourhood; having now, for the first time in the course of our voyage, an opportunity to preserve living specimens."

On the 13th he employed himself ". . . in forming a place for the reception of living specimens, a little distance below our camp, and near the river, for the convenience of water."

On June 14 he visited the upper Arikara village and was ". . . accosted by the *Medicine Man,* or doctor . . . He made me understand that he had seen me collecting plants, and that he knew me to be a *Medicine Man;* frequently shaking hands, he took down his medicine bag . . . and showed me its contents. As I supposed this bag contained the whole *materia medica* of the nation, I examined it with some attention. There was a considerable quantity of the down of reedmace, *(Typha palustris)* . . . used in cases of burns and scalds . . . also a quantity of a species of *Artemisia* . . . but that ingredient which was in the greatest abundance, was a species of wall-flower in character it agrees with *Cheiranthus erysimoides.* Beside . . . I found two new species of *Astragalus,* and some roots of *Rudbeckia purpurea* . . . I assured the doctor it was all very good, and we again shook hands . . ."

On the 15th Bradbury ". . . went out to collect, accompanied by Mr. Brackenridge, and proceeded farther into the interior . . ." than ever before and was ". . . rewarded by several new species of plants . . ." Indeed he was so occupied prior to the day of departure that he paid no attention to choosing his horse which, even before he had started, proved ". . . very bad . . . small, and apparently weak . . ." Bradbury would not delay the party while he selected another.

". . . I fixed my saddle, and we set out, having previously agreed with one of the men to take care of my plants during my absence . . ."

He and Crooks started on June 20, leaving the river for the interior. "Besides my rifle and other equipments, similar to those of the rest of the party, I had a portfolio for securing the specimens of plants. I had contrived already to collect some interesting specimens by frequently alighting to pluck them, and put them into my hat. For these opportunities, and to ease my horse, I ran many miles alongside of him. Notwithstanding this, by noon he seemed inclined to give up . . ."

This day they camped in the valley of "Cannon-ball river," which enters the Missouri from the west between Sioux County (south) and Morton County (north). They were in North Dakota. Bradbury notes:

"The alluvion of the river . . . is very beautiful, being prairie, interspersed with groves of trees, and ornamented with beautiful plants, now in flower. Amongst others which I did not observe before, I found a species of flax, resembling that which is cultivated: I think it is the species known as *Linum Perenne.* I rambled until it was quite dark, and found my way to the camp by observing the fire."

On June 21 they reached ". . . *Rivière de Cœur,* or Heart River, and encamped on its banks, or, more properly, lay down in our blankets." Heart River flows into the Missouri from the west, in Morton County, and they were opposite Bismarck.

Vast herds of buffalo were seen on June 22—their numbers were estimated to be upwards of 10,000. The horses ". . . began to appear much jaded, but mine [notes

Bradbury] in particular." Nonetheless they pushed on, passing Knife River, entering the Missouri from the west in Mercer County, and seven miles beyond ". . . the third village of the *Minetaree* or *Gros Ventres Indians* . . ." reached their destination, ". . . having travelled this day more than eighteen hours . . ." Although it was eleven o'clock at night, they were cordially received by ". . . Mr. Reuben Lewis, brother of Captain Lewis, who travelled to the Pacific Ocean. The mosquitoes were much less friendly to us . . . notwithstanding our excessive fatigue, it was next to impossible to sleep."

Fort Mandan, erected by Lewis and Clark as winter quarters in 1804–1805, was situated, according to Chittenden, ". . . on the left bank of the Missouri 7 or 8 miles below the mouth of Big Knife river and opposite, though a little above, the site where Fort Clark [Oliver County] later stood." [9] Since Bradbury records that they had proceeded more than seven miles beyond Knife River, they must have continued on to "Lisa's Fort" in Mercer County which, according to Chittenden, ". . . was situated on the right or south bank of the [Missouri] river, some ten or twelve miles above the mouth of the Big Knife . . ." Both forts are shown on Chittenden's map, but on opposite sides of the Missouri; both were in the territory of the Mandans and Bradbury and Nuttall of course visited the Indian villages. Since the horses were exhausted it was decided to remain at the fort for four or five days. Bradbury never rested. He noted on June 23 that "The bluffs here have a very romantic appearance and I was preparing to examine them . . . when some squaws came in . . . with a quantity of roots to sell. Being informed that they were dug on the prairie, my curiosity was excited, and on tasting I found them very palatable, even in a raw state. They were the shape of an egg . . . I found no vestige of the plant attached to them."

After finding a spot where the roots had been dug, Bradbury, to his surprise, ". . . discovered, from the tops broken off, that the plant was one I was well acquainted with, having found it even in the vicinity of St. Louis . . . and determined it to be a new species of *Psoralea,* which is now known as *Psoralea esculenta.*"

The country about the fort, bluffs especially, proved ". . . extremely interesting . . . The incumbent soil . . . of excellent quality, and was at this time covered with fine grass and a number of beautiful plants. The roots and specimens of these I collected with the greatest assiduity . . . I soon found the number to increase so much as I lengthened my excursions, that I resolved to remain at the fort . . ."

The fort, a square blockhouse, had ". . . a very pretty garden, in which were peas, beans, sallad, radishes, and other vegetables, under the care of a gardener, an Irishman, who shewed it . . . with much self-importance." Bradbury ". . . expressed . . .

9. Pennell evidently considers this to have been Bradbury's and Nuttall's headquarters, for he states that "Fort Mandan, near the villages of the Mandan Indians, was not at present Mandan [near the mouth of Heart River, Morton County], North Dakota, but was situated on the north side of the Missouri River in present McLean County, and almost opposite the later Fort Clark."

regret that he had no potatoes. 'Oh!' said he, 'that does not signify; we can soon have them; there is plenty just over the way.' " Reuben Lewis assured Bradbury ". . . that there really were potatoes at an English Fort on the river St. Peter's [the Minnesota], *only* from two to 300 miles distant." Bradbury relates on June 24:

"On the top of a hill, about four miles from the Fort, I had a fine view of a beautiful valley, caused by a rivulet . . . a branch of Knife River, the declivities of which abound in a new species of *Eleagnus,* intermixed with a singular procumbent species of cedar *(Juniperus.)* The branches are entirely prostrate on the ground, and never rise above the height of a few inches. The beautiful silvery hue of the first, contrasted with the dark green of the latter, had a most pleasing effect . . . the small alluvion of the rivulet was so plentifully covered with a species of lilly *(Lilium catesbæi),* as to make it resemble a scarlet stripe as far as the eye could trace it. I returned to the Fort much gratified . . ."

Having observed some corn planted by the Indians and now "nearly a yard high," Bradbury states in a footnote:

"This is about the full height to which the maize grows in the Upper Missouri, and when this circumstance is connected with the quickness with which it grows and is matured, it is a wonderful instance of the power given to some plants to accommodate themselves to climate. The latitude of this place is about 47 degrees geographically, but geologically many degrees colder, arising from its elevation, which must be admitted to be very considerable, when we consider that it is at a distance of more than 3000 miles from the ocean by the course of a rapid river. This plant is certainly the same species of *Zea* that is cultivated within the tropics where it usually requires four months to ripen, and rises to the height of twelve feet. Here ten weeks is sufficient, with a much less degree of heat. Whether or not this property is more peculiar to plants useful to men, and given for wise and benevolent purposes, I will not attempt to determine."

Bradbury, on June 25, ". . . had the pleasure of again meeting Mr. Brackenridge . . ." who had arrived by boat, and of learning that Lisa intended to remain a fortnight, for he ". . . was very glad to have so good an opportunity of examining this interesting country." Since Brackenridge had brought with him some of the botanist's small trading-articles, Bradbury was enabled ". . . to reward the gardener for his civility in offering me a place in the garden where I could deposit my living plants, and of this I availed myself during my stay."

When Crooks left on June 27 to return to the Arikara village Bradbury remained, planning to return with Lisa. On June 29 and 30 he "Continued to add to my stock . . ."

July 1 he notes: ". . . extended my researches up the river, along the foot of the bluffs . . ." and on July 2 mentions, as does Brackenridge in his *Journal,* some petrified trees. "Mr. Brackenridge and myself made an excursion into the interior from the river, and found nothing interesting . . . excepting some bodies of argillaceous schist, some parts of which had a columnar appearance. They were lying in an horizontal position, having something the appearance of the bodies of trees."

Bradbury, who seldom refers to Nuttall, mentions him on July 4: "This day being the anniversary of the independence of the United States, Mr. Lisa invited us to dine on board of his boat, and Messrs. Brackenridge, Lewis, Nuttall, and myself attended him . . ." Since Lisa announced his intention of departing in two days, Bradbury spent the 5th ". . . in packing up carefully my collection . . ."

They set out on July 6. "Our progress down the river was very rapid, as it was still in a high state. We did not land until evening, after making . . . more than 100 miles."

Back at the mouth of Cannon Ball River on July 7, Bradbury ". . . noticed and procured some additional specimens." That day he rejoined Hunt and learned that his party was soon to leave for the west. When invited to accompany them, Bradbury was at first inclined to accept, ". . . but finding that they could not assure me of a passage from thence to the United States by sea, or even to China, and recollecting also that . . . in passing over the Rocky Mountains, I should probably be unable to preserve or carry my specimens, I declined."

Instead, he arranged to return to St. Louis in one of Lisa's boats, but ". . . on consideration of being permitted to land at certain places which I pointed out, I offered to give him my boat as an equivalent. To this he readily agreed, and I continued to prepare for my departure." Before he had left for the Mandan country Hunt had presented Bradbury with the smallest of his boats, ". . . a barge built at Michillimakinac . . .", and three American hunters had agreed to help him in "navigating" it down the Missouri when he was ready to descend that stream. Since the Canadians, or "voyageurs," could not take their trunks, or "caisettes," overland, Bradbury now ". . . purchased from them seventeen, in which I purposed to arrange my living specimens, having now collected several thousands."

The plan for the return to St. Louis was the best the botanist could make; he would have preferred to descend the Missouri independently, and slowly, for he ". . . had noticed a great number of species of plants on the river, that, from the early state of the season, could not then be collected advantageously. These I had reserved for my descent . . ."

For some nine days, while waiting for departure, Bradbury has considerable to tell of miscellaneous matters, some concerned with the Arikara Indians. He describes their bows, the best of which were made from the horns of the mountain sheep; but "The next in value, and but little inferior, are made of a yellow wood, from a tree which grows on Red River, and perhaps on the Arkansas. This wood is called *bois jaune,* or *bois d'arc.* I do not think the tree has yet been described, unless it has been found lately in Mexico. I have seen two trees of this species in the garden of Pierre Chouteau, in St. Louis, and found that it belongs to the class *dioecia;* but both of the trees being females, I could not determine the genus. The fruit is as large as an apple, and is rough on the outside. It bleeds an acrid milky juice when wounded, and is called by the hunters the Osage orange.[1] The price of a bow made from this wood at the Aricaras is a horse and blanket . . ."

There is a further reference to the Indian medicine men:

"On account of my constant attention to plants, and being regularly employed in collecting, I was considered as the physician of the party by all the nations we saw; and generally the *medicine men* amongst them sought my acquaintance. This day, the doctor, whom Mr. Brackenridge and myself saw in the upper village, and who showed me his medicine bag [June 14], came to examine my plants. I found he understood a few French words, such as *bon, mal,* &c. . . . He showed me a quantity of a plant lately gathered, and by signs informed me that it cured the colic. It was a new species of *Amorpha.* I returned to the camp, accompanied by the doctor, who very politely carried the buffaloe robe for me."

Hunt was to leave with his party on July 18. On the 17th Bradbury wrote:

". . . I took leave of my worthy friends, Messrs. Hunt, Crooks, and M'Kenzie, whose kindness and attention to me has been such as to render the parting painful . . . I am happy in having this opportunity of testifying my gratitude and respect for them: throughout the whole voyage, every indulgence was given me, that was consistent with their duty, and the general safety."

He and Brackenridge, in the same boat, started downstream: ". . . we moved at the rate of about nine miles per hour." It was then that Bradbury learned ". . . that Mr. Lisa had instructed Mr. Brackenridge not, on any account, to stop in the day, but if possible to go night and day. As this measure would deprive me of all hopes of add-

1. Rafinesque described the Osage orange in 1817 (*American Monthly Magazine and Critical Review* 2: 118) under the name *Toxylon pomiferum.* Since this particular tree seems always to have aroused the interest of the early travelers, I quote what Sargent tells of its history in his *Silva;* insertions between brackets are mine:

"The earliest account of the Osage Orange appears in the narrative of [William] Dunbar and [George] Hunter's journey made in 1804 from St. Catherine's Landing on the Mississippi to the Washita River. It was first found by Mr. Dunbar at the post of the Washita, although traders with the Indians of the Red River [of the South] had doubtless been familiar with their Bois d'Arc before this, for in 1810 Bradbury found two trees growing in Pierre Chouteau's garden in St. Louis old enough to bear fruit. In the preface to Pursh's *Flora Americæ Septentrionalis,* published in 1814, allusion is made to its discovery by the expedition which crossed and recrossed the continent in 1804–1806 under the command of Captains Lewis and Clark, although there is no mention of the tree in their published journals. Early in this century seeds of the Osage Orange were received in Philadelphia by Bernard MacMahon and David Landreth, who raised plants from them; it was sent to England in 1818, and two years later was cultivated in the nurseries of Jacques Martin Cels in Paris."

Landreth and McMahon, or MacMahon (for whom the genus *Mahonia* was named by Nuttall), were well-known nurserymen and seedsmen of Philadelphia. Sargent tells that "In 1804 or 1805 David Landreth received from the Lewis & Clark Expedition seeds of the Osage Orange, which produced a number of plants . . ." He also notes that Dunbar ". . . praised the appearance of the Osage Orange, which he considered one of the most beautiful trees he had seen, suggested its probable value as a hedge-plant, and alluded to the dye obtained by the Indians from its roots."

The tree evidently naturalizes itself readily and, far from its native habitat or along roadsides in Essex County, New Jersey, I have noted many seedlings which must have originated from certain old hedges planted in the vicinity.

ing to my collection any of the plants lower down on the river, and was directly contrary to our agreement, I was greatly mortified and chagrined . . . Mr. Brackenridge felt sensibly for my disappointment . . . I . . . had the mortification[2] during the day, of passing a number of plants that may probably remain unknown for ages."

This ". . . breach of faith towards me by Mr. Lisa . . ." is mentioned in the preface to the *Travels* as a ". . . promise he neither did, nor intended to perform." Disappointed, Bradbury tried to console himself:

"Our descent was very rapid, and the day remarkably fine; we had an opportunity, therefore, of considering the river more in its *tout ensemble* than in our ascent . . ." !

By July 19 they had reached the upper end of the Great Bend and Bradbury was optimistic. Because of the steersman's ". . . timidity I had some hope of opportunities to collect." Sure enough, when a mere breeze "ruffled the surface of the river," the man put ashore and laid by for the rest of the day. They had descended ". . . about 280 miles in about two days and a half. I determined not to lose this opportunity to add a few species to my collection, and was accompanied . . . by Mr. Brackenridge, who employed himself in keeping a good look out for fear of a surprise by the Sioux, a precaution necessary to my safety, as the nature of my employment kept me for the most part in a stooping posture."

On the evening of July 20 a violent storm arose: "We stopped and fastened our boats to some shrubs, *(Amorpha fruticosa)* . . . Had our fastenings given way, we must inevitably have perished." After this experience Bradbury wrote:

"For myself I felt but little: two years in a great measure spent in the wilds, had inured me to hardships and inclemencies; but I felt much for my friend Brackenridge. *Poor young man,* his youth, and the delicacy of his frame, ill suited him for such hardships, which, nevertheless, he supported cheerfully."

Brackenridge described the same episode—Bradbury was in his early forties:

"For myself, I was accustomed to these things; but I felt for my friend Bradbury. Poor old man, the exposure was much greater than one of his years could well support. His amiable ardor in pursuit of knowledge, did not permit him for a moment to think of his advanced age; and wherever he may be, (for I have not heard from him for several years,) he carries with him the warmest wishes of my heart."

A renewal of the wind on July 21 gave Bradbury another opportunity, a little below the Niobrara River, ". . . to procure roots of the new species of currant, although with much pain and difficulty, having four miles at least to wade through water and mud . . ."

This day the plain was ". . . literally covered with buffaloes as far as we could see . . ." and both Bradbury and Brackenridge describe ferocious battle between the bulls. On July 25 they passed the Maha villages but did not stop; on the 27th they reached Fort Osage—where Bradbury had ". . . the pleasure to find Mr. Sibly had re-

2. The old narratives frequently use the word *mortification,* not in the sense of humiliation, but in the sense of disappointment or vexation.

turned . . . from his tour to the Arkansas . . ."—and on July 29 they arrived at St. Louis. The descent of the Missouri River from the Arikara nation had taken two weeks as against the three months required for its ascent.

Bradbury was welcomed into the house of his "worthy friend, Mr. Albert Gallatin." But another "worthy and respected friend, Mr. S. Bridge, from Manchester" arriving on the scene, he changed his plans and took up residence with him. The inducement probably lay in the fact that Bridge sent his "waggon" for the precious living plants and allocated to the botanist ". . . a piece of ground which, with much labour, I prepared in a few days, got it surrounded by a fence, and transplanted the whole of my collection."

In August, about ten days after arrival, Bradbury became ill with a fever which lasted until early December: "Its violence soon left me little hope for recovery." Besides which, on reaching St. Louis, he had visited the post office and had found ". . . letters from England, informing me of the welfare of my family. This pleasing intelligence was damped by a letter from my son, who informed me that those who had agreed to furnish me with the means of prosecuting my tour, to whom I had sent my former collection, had determined to withhold any further supply."

Nor was it until November that he ". . . received a remittance from those who had determined to withhold it, together with a letter from the person [a footnote states 'This man's name is Shepherd'] who managed the Botanic Garden at Liverpool, informing me that he had received my former collection, out of which he had secured in pots more than one thousand plants, and that the seeds were already vegetating in vast numbers."

In December a Mr. H. W. Drinker of Philadelphia, then living in St. Louis, gave Bradbury the opportunity of descending the Mississippi River to New Orleans in charge of a load of lead. On December 4 he took leave of his St. Louis friends—". . . several of whom from their polite attention to me, I have reason to hold in lasting remembrance . . ."—and on the 5th ". . . set off from St. Louis on the voyage to New Orleans, a distance of about 1350 miles." The dangers attendant upon navigation of of the Mississippi—". . . in particular to boats loaded with lead . . ."—are described. The greatest menaces were sunken trees of two sorts, "planters" and "sawyers." [3]

Perhaps Bradbury was dispirited by illness, perhaps he felt responsible for his cargo; but, whatever the reason, he makes few references to plants during this trip. He had much to tell, however, about a series of earthquakes, some violent, which began on December 15 and continued intermittently through the 21st; great damage was done to the town of New Madrid where Bradbury had spent the night of the 14th but which he had left on the morning of the 15th.

3. Mrs. Frances Trollope's *Domestic manners of the Americans* (chapter three) tells how the boat on which she was mounting the Mississippi River became involved with these obstructions. And Maximilian, Prince of Wied-Neuwied, has much to say about them on the Missouri River.

1812

On January 5, 1812, Bradbury reached "the port of Natchez." The next day he boarded a boat— ". . . she was a very handsome vessel . . . impelled by a very powerful steam engine . . ."—and on January 13 arrived at New Orleans where he consigned the lead to Drinker's agent. He had a final glimpse of his "friend Brackenridge" and, on January 20, set sail for New York. The narrative portion of the *Travels* ends at this point.

Pennell tells us that "Nuttall reached St. Louis again in time to have seen Bradbury . . . they might have journeyed down the Mississippi together, although Bradbury's silence is against the likelihood of this . . . Nuttall also descended to New Orleans 'in the latter end of the year 1811,' as Professor Barton phrases it . . . he . . . sailed directly for England . . . Nuttall was more fortunate than Bradbury in that, when the war came, he was on the home side of the ocean and able to prosecute diligently the naming of his plants . . ."

It seems strange that two men with such similar interests as Bradbury and Nuttall could have been thrown so closely together and yet have had little association; professional rivalry may have played a part but, more probably, the two were merely uncongenial. Bradbury mentions Nuttall but three or four times in his journal, and only casually. Brackenridge records delightful excusions with Bradbury but his references to Nuttall are trivial with one exception which has often been quoted:

"There is in company a gentleman of whom I have already spoken, Mr. Nuttal, engaged in similar pursuits, to which he appears singularly devoted, and which seems to engross every thought, to the total disregard of his own personal safety, and sometimes to the inconvenience of the party he accompanies. To the ignorant Canadian boatmen, who are unable to appreciate the science, he affords a subject of merriment; *le fou* is the name by which he is commonly known. When the boat touches the shore, he leaps out, and no sooner is his attention arrested by a plant or flower, than every thing else is forgotten. The inquiry is made *ou est le fou?* where is the fool? *il est apres ramasser des racines,* he is gathering roots. He is a young man of genius, and very considerable acquirements, but is too much devoted to his favorite pursuit, and seems to think that no other study deserves the attention of a man of sense. I hope, should this meet his eye, it will give no offence; for these things, often constituted a subject of merriment to us both."

An additional comment appears in Brackenridge's *Views of Louisiana* (p. 240):

"A characteristic anecdote of this gentleman was related to me . . . and shows to what an astonishing degree the pursuit of natural history had taken possession of his mind, to the exclusion of everything else . . ." When three hundred Arikara Indians thought to be "inimical" were rushing towards the boat and the party was on the point of firing, ". . . Nuttal, who appeared to have been examining them very attentively . . .

said . . . 'don't you think these Indians much fatter, and more robust than those [a party of Sioux] of yesterday.'"

Washington Irving's *Astoria* paints contrasting pictures of these two men:

"Among the various persons who were to proceed up the Missouri with Mr. Hunt, were two scientific gentlemen: one Mr. John Bradbury, a man of mature age, but great enterprise and personal activity, who had been sent out by the Linnean Society of Liverpool, to make a collection of American plants; the other, a Mr. Nuttall, likewise an Englishman, younger in years, who has since made himself known as the author of 'Travels in Arkansas,' and a work on the 'Genera of American Plants.' Mr. Hunt has offered them the protection and facilities of his party.

"The two naturalists . . . pursued their researches on all occasions. Mr. Nuttall seems to have been exclusively devoted to his scientific pursuits. He was a zealous botanist, and all his enthusiasm was awakened at beholding a new world . . . Whenever the boats landed at meal times, or for any temporary purpose, he would spring on shore, and set out on a hunt for new specimens . . . he went groping and stumbling along among a wilderness of sweets, forgetful of every thing but his immediate pursuit, and had often to be sought after when the boats were about to resume their course. At such times he would be found far off in the prairies, or up the course of some petty stream, laden with plants of all kinds. The Canadian voyageurs . . . were extremely puzzled by this passion for collecting what they considered were mere useless weeds. When they saw the worthy botanist coming back laden with his specimens, and treasuring them up as carefully as a miser would his hoard, they used to make merry among themselves at his expense, regarding him as some whimsical kind of madman. Mr. Bradbury was less exclusive in his tastes and habits, and combined the hunter and sportsman with the naturalist. He took his rifle or his fowling-piece with him in his geological researches, conformed to the hardy and rugged habits of the men around him, and of course gained favor in their eyes. He had a strong relish for incident and adventure, was curious in observing savage manners, and savage life, and ready to join any hunting or other excursion . . . he could not check his propensity to ramble . . ."

As already stated Bradbury's plans for returning to England from the Atlantic seaboard had to be put off because of the War of 1812. Although the Treaty of Ghent (December, 1814) technically ended the conflict, he does not seem to have left America even then, for the preface to the *Travels* tells that after the war he had ". . . made some arrangements[4] which caused a necessity for my stay some time longer."

By 1816 he was probably back in Liverpool, for in 1817 he published his *Travels;* True, quoting Hill, states that this taxed his slender means. Although this did not contain the botanical matter which Bradbury had hoped to include before Pursh made this

---

4. True explains that this further delay ". . . refers to business connections which he made somewhere in the East. This attempt to establish his financial situation involved a manufacturing enterprize in which the depression of the times assisted by the bad business morals of a partner brought Bradbury to grief . . ."

useless—I refer to this later—it was nevertheless an extremely valuable work. Chittenden thought highly of it:

"Bradbury's well-known book, *Travels in North America,* is one of the most useful works of this period, and one which the careful student of Louisiana history never fails to consult. It is the best existing authority on many points, and on some the only one . . ."

Bradbury's appendix, in its overall picture, is perhaps the most interesting portion of the *Travels.*

His "Description of the Missouri Territory" (part four of the appendix) makes clear that the author's concern was not with botany alone. By "Missouri Territory" he refers to the region acquired by the Louisiana Purchase, as yet but sparsely settled; he describes its resources in mineral wealth and its agricultural potentialities. He tells that, immediately along the Mississippi and extending westward for 100 to 250 miles, existed a thinly timbered country and beyond this, to the Rocky Mountains, was ". . . one vast prairie or meadow . . . [which] excepting on the alluvion of the rivers, and in a few instances, on the sides of the small hills, is entirely divested of trees or shrubs . . ." Its soil, he notes, ". . . is generally excellent, being for the most part black loam, and is tilled without much trouble."

Bradbury was more optimistic than his contemporaries: ". . . the belief in America is, that the prairie cannot be inhabited by the whites; even Mr. Brackenridge says it cannot be cultivated. My own opinion is, that it can be cultivated; and that, in process of time, it will not only be peopled and cultivated, but that it will be one of the most beautiful countries in the world."

He mentions the region's resources in timber and in fruits:

"The general character of this country is that of prairie, with scattered trees, and interspersed clumps. On the summits of the ridges, the timber is generally red cedar,[5] *(Juniperus virginiana)* on the prairie, post oak, *(Quercus obtusiloba)* black jack, *(Quercus nigra)* black walnut, *(Juglans nigra)* and shell bark hickory, *(Juglans squamosa.)* The alluvion of the rivers contain a greater variety, of which the principal are —cotton wood, *(Populus angulosa)* sycamore, *(Platanus occidentalis)* over-cup oak, *(Quercus macrocarpa)* nettle tree, or hackberry, *(Celtis crassifolia)* hoop ash, *(Celtis occidentalis)* honey locust, *(Gleditsia triacanthos)* black locust, *(Robinia pseudacacia)* coffee tree, *(Guilandina dioica)* peccan, *(Juglans olivæformis)* and many of the trees common in the states east of the Alleghanies . . ."

"The wild productions of the Missouri Territory, such as fruits, nuts, and berries, are numerous: of these the summer grape *(Vitis æstivalis)* appears to be the most valuable . . . The winter grape *(Vitis vulpinum)* is remarkable for the large size of its vine . . . the persimon, *(Diospyros virginiana)* . . . in appearance resembles a plum . . . The papaw *(Anona triloba)* is found in plenty on the alluvion of the rivers . . . Straw-

5. Bradbury places his commas before, rather than after, the parenthetical names.

berries are in vast abundance on the prairies . . . The pecan, or Illinois nut, is a kind of walnut . . . its shell . . . so thin as to be cracked between the teeth with the greatest ease . . . it has obtained the name of *Juglans olivæformis.*"

Bradbury depicts a land of abundance, as it doubtless was; a land where, under cultivation, are produced ". . . maize, wheat, oats, barley, beans, *(Phaseolus)* pumpkins, water and musk melons, and tobacco and cotton . . . Apples and peaches are very fine . . . They [the white settlers] pay great attention to gardening, and have a good assortment of roots and vegetables."

He was convinced that "The political and commercial advantages that will arise to the United States from the acquisition of Louisiana are incalculable, besides the vast revenue that will arise from the sale of lands." Bradbury was endowed with a prescience which many of the early explorers seem to have lacked and his predictions have certainly come to pass.

When commenting upon Part V of Bradbury's Appendix ("Remarks on the states of Ohio, Kentucky, and Indiana, with the Illinois and Western Territory . . ."), Thwaites suggests that Bradbury traveled through these regions after peace had been concluded.

But Bradbury does not tell when he went on these travels, merely prefacing the "Remarks" with the statement that "In a tour across the Alleghanies, and through the regions west of these mountains and east of the Mississippi river, I did not keep a regular journal, but contented myself with making general remarks, without any expectation that they would ever be submitted to public view. From these remarks I shall briefly extract such matter as may be useful to those who wish to visit the western country, or be read with interest by those who do."

Moreover, after his return to the east in 1812, we learn of business misfortunes and of attempts, in 1812 and 1816,[6] to obtain remunerative work—from the financial angle alone it would therefore have been strange had he been able to embark on a tour of such magnitude, and had he been financed in such an undertaking we should

6. True publishes two of Bradbury's letters which appeal to Jefferson for help in obtaining salaried positions and the ex-President did all that was possible to aid his friend:

In the first letter dated March 15, 1812, from New York—a reply was to be sent ". . . to the Postoffice Neward [*sic*] State of New Jersey"—Bradbury wrote that he had heard that a botanic garden was contemplated in the city of Washington and hoped that he might become its superintendent; the report was evidently unfounded, for Jefferson replied on March 21, 1812:

"there have been repeatedly applications by individuals, & one of them lately, for the use of some of the public grounds at Washington for the establishment of such a garden, and if the suspicion that it would be converted in to a mere kitchen-garden for the supply of the town market can be removed, it is in the power of the President and would probably be within his disposition so to dispose of it, but I do not believe the Government will or can do more."

In the second letter, dated from Wardsbridge, N. Y.—a place which True identifies as Montgomery, "about 12 miles west of Newberg [*sic*] on the Hudson"—on January 8, 1816, Bradbury had heard that a road was to be built from St. Louis to the northern boundary of "Louisiana" and hoped to obtain an appointment as commissioner; again he had been misinformed for no such road was in contemplation. He did not receive Jefferson's reply, having, True states, left for England.

learn of it from other sources. Bradbury's "Remarks" more probably were based upon observations made on his journey to St. Louis in 1809,[7] rather than on any subsequent trip.

Bradbury's "Catalogue of some of the more rare or valuable plants discovered in the neighbourhood of St. Louis and on the Missouri" (part six of his appendix) lists mainly herbaceous plants, about one hundred. No dates of collection are given and many of the localities are general ones. Pennell notes that ". . . these are nearly all included among the records in Nuttall's 'Genera.' " Most of Bradbury's plants were described and named by others and the reason is explained in the preface to the *Travels:*

"Immediately after my return to the United States [from the Louisiana Territory], and before I could make any arrangement, either for my return to England, or for the publication of the plants I collected, the war broke out with this country:—I waited for its termination, and made some arrangements which caused a necessity for my stay sometime longer . . . I had intended that this should have been accompanied by a description of the objects collected, that had not been before discovered; but on my return to England, I found that my design had been frustrated, by my collection having been submitted to the inspection of a person of the name of Pursh, who has published[8] the most interesting of my plants in an appendix to the *Flora Americæ Septentrionalis.*"

In the preface to his *Flora* Frederick Pursh tells how he acquired the Bradbury collection; he had gone to Liverpool and had seen the plants, recently arrived from America:

"I am also highly indebted to William Roscoe, Esq., who very obligingly communicated[9] to me Mr. Bradbury's Plants collected in Upper Louisiana. This valuable collection contains many rare and new species, having been collected in a tract of country never explored before: those which were entirely new I have described in the Supplement to the present work."

Pennell writes: "These specimens which were studied by Pursh seem to be all that were ever reported upon; they passed into the herbarium of A. B. Lambert, whose large collections were sold by auction piecemeal after his death in 1842. Asa Gray wrote in 1841 that Bradbury's collections, 'so far as they are extant,' were then in Lambert's herbarium. There seems to be no record of how they got there.[1] In a letter

7. For the route taken *see* p. 110, *fn.* 7.

8. In the *Travels,* under date of May 28, 1811, Bradbury in a footnote referable to the name Pursh, had commented:

"This man has been suffered to examine the collection of specimens which I sent to Liverpool, and to describe almost the whole, thereby depriving me both of the credit and profit of what was justly due to me."

9. Pursh's choice of word *communicated* is subject to two interpretations; it might mean that Roscoe showed him the specimens or gave them to him.

1. Among the letters published by Rickett (1950) is one (no. 23) from William Bullock to William Roscoe, dated London, November 4, 1813, which indicates that Roscoe had permitted Sir Joseph Banks ". . . to

to William Darlington, dated from St. Louis, June 11, 1819, William Baldwin wrote that Bradbury had requested him to describe certain plants ". . . and observed, that since Lambert had pirated from him his former collections, it was not his intention to publish independently . . ."

True notes that Bradbury does not seem to have resented Roscoe's action in turning over his collections to Pursh. Since Bradbury was employed by the Liverpool Society, Roscoe may have thought that they belonged to that institution to use as it thought fit. On the other hand he may have been quite as incensed as Bradbury, for Pursh's action disregarded the interests of the Society also.

As to the present whereabouts of Bradbury's collections, Pennell tells us that, after the sale of Lambert's herbarium in 1842, Bradbury's ". . . specimens may then have become widely dispersed; the largest series that I have yet discovered is at the Academy in Philadelphia. Like other specimens evidently from the same source and acquired at the same time, they had been mounted while in the Lambert Herbarium upon small sheets of a characteristic paper which bear on the reverse in the upper right-hand corner data in Lambert's handwriting. Eighteen species have been recently reported by H. W. Rickett as at the herbarium of Kew Gardens, and . . . he has given some notes and an itinerary of Bradbury's journey."

In this report—"Specimens collected by Bradbury in Missouri Territory" (1934)— Rickett states that, of the specimens which he enumerates as existing at Kew Gardens: ". . . fourteen represent species named by Pursh . . . from Bradbury's collections. It is improbable that they are the actual specimens seen by Pursh, who states . . . that he obtained Bradbury's plants from Roscoe . . . They may be regarded as isotypes. This is the more likely since one specimen is a species unknown to Pursh . . . The four remaining specimens are species named by Pursh before he saw Bradbury's plants (one from a collection by Lewis . . . the others probably from Nuttall's specimens)."

These Kew specimens, Rickett notes, ". . . came into W. J. Hooker's hands, through Dr. [Thomas] Taylor, before 1829."

Thomas Batt Hall, in "Some account of the Liverpool Botanic Garden" (1839), mentions that that institution ". . . sent out Mr. Bradbury, an excellent botanist, to collect for them in North-America; there are a great many of his dried specimens in the Herbarium, principally in [from ?] Louisiana." The greater part of these seem, for a

take a specimen of each of the duplicates of the Plants sent by Bradbury . . ." Bullock then, rather reluctantly it would seem, passes on a message from Lambert:

"Mr. Lambert has just called on me to request me to make a similar request to you on his part that I did for Sr. Josh. which I do only because I promised him not having the same reasons that I had in the other Case . . . you will please to determine for yourself in this case.—I would certainly select the duplicates myself were the case mine . . ."

Does not this suggest that, having seen Bradbury's collections in Liverpool, Pursh instigated Lambert to ask Bullock to make this request of Roscoe and that Roscoe complied? It certainly seems a plausible explanation of how they reached Lambert and became available to Pursh.

time, to have disappeared, for we read in True's sketch of Bradbury (p. 145, *fn.* 2) that "Dr. [Joseph A.] Clubb, Curator of Museums, Liverpool, under date of December 28, 1916, writes 'Some years ago we received from Liverpool Botanic Gardens a number of herbarium specimens and I find there are some 8 or 10 specimens recorded as being collected by Bradbury, but what has become of the mass of his collections, which must have been considerable, we have no records.' "

Inquiry has yielded some information about these lost specimens. On my behalf Dr. E. D. Merrill wrote Dr. Cotton at Kew who, in his turn, communicated with the Liverpool Garden; in reply he received a letter dated June 4, 1945, from Professor John McLean Thompson of the Hartley Botanical Laboratories of the University of Liverpool, which reads:

"First, I should tell you that these Bradbury plants have had a pretty sorrowful career. When I came to Liverpool and first saw the old house and potting sheds at Edge Hill (the site of the erstwhile Botanic garden) it was a custom of the garden hands to use these and other herbarium collections as a source of spills for their pipes. I've no idea what had actually been lost in this way . . . what could be saved was saved. And as the gardens really belonged to the Corporation, the remnants came mainly to belong to the Liverpool Museum . . . they have at last been put in some order and catalogued more or less as you will see in the short statement given on pages 42 and 43 of the accompanying little booklet [*Handbook and guide to the herbarium collections in the Public Museums Liverpool*].

"Just prior to the present War, Rickett wrote from Philadelphia asking for these plants to be returned . . . But the Corporation saw otherwise . . ."

The letter goes on to say that Dr. Helen Blackler had done some arranging and cataloguing of the specimens, that they had been sent for safe storage out of Liverpool, and were inaccessible for the time being. A few days later Professor Thompson forwarded a letter from "the official Keeper in Botany to the Liverpool Museums," Mr. Stansfield, dated June 6, 1945, which reads:

"We have about 250 plants by John Bradbury. Most of them have m/s notes by him giving locations, determinations, difficulties and so on. Many of them are signed by him. Nearly all the plants mentioned in 'Travels in the interior of America' 1817 are in our collection . . .

"During the early stages of the war, the plants were sent to safe storage in Wales. Our Museum in Liverpool was destroyed . . .

"In the Picton Reference Library (M/S), Liverpool, there is the correspondence between Bradbury and Roscoe . . . It is of biographical interest . . . A study of the [Stalybridge] parish records would be of interest to anyone wanting to write up John Bradbury."

As to Bradbury's later life, True, quoting from Hill, tells us that after the publication of the *Travels*, which had taxed his slender means, the man wished to leave England forever; he then had the good fortune to meet by chance on the streets of Liverpool an American sea captain and old friend, who offered him and his family a free passage to America. We are not told how many Bradburys[2] came to this country.

The preface to the second edition of the *Travels* (London, 1819) states that Bradbury returned to the United States after the issuance of the first edition and ". . . is now residing at St. Louis." This was probably in late 1817 or early 1818. By 1819, according to William Baldwin writing to Darlington on June 9 of that year, Bradbury was at St. Louis and about to erect a home on the adjacent prairie. True estimates that he did not stay there more than five years and quotes Hill to the effect that, ultimately, he ". . . became curator and superintendent of the Botanical Gardens at St. Louis[3] where he was not only placed beyond the fear of penury but was honored and respected by residents of that city." Perhaps so, but he did not remain long; for True quotes a letter to the editor of the *American Farmer,* dated Middletown, Kentucky, January 22, 1823, which refers to Bradbury as ". . . residing in this village." He also cites an obituary notice in the Missouri *Republican* of May 7, 1823:

"Died, at Middletown, Kentucky, on the 16th of March last, after a short illness, Mr. John Bradbury . . . known to the scientific world as among the first botanists and mineralogists . . . Never was there a better companion, nor a more sincere friend."

According to True, Hill's quoted account of Bradbury's death in the wilds of the west, and of his burial by the Indians are legendary. Mr. Stansfield's letter already quoted, states that he is buried in Stalybridge Parish churchyard.

As noted at the beginning of this chapter Dr. H. W. Rickett in 1950 published a paper entitled "John Bradbury's explorations in the Missouri Territory." The included letters—presumably those in the Picton Reference Library to which the Stansfield letter quoted above refers—were transcribed from photostats supplied him by the Liverpool authorities. Rickett, who had previously studied the Bradbury collections at Kew Gardens, states that he also examined those at Liverpool and at the Philadelphia Academy.

I have already commented, in footnotes, upon some of these letters when they offered facts not found elsewhere. While they paint a graphic picture of the hand-to-mouth existence of the entire Bradbury family, they do not, unfortunately, throw any great amount of new light upon Bradbury's life in America.

2. Rickett (1950) tells us that Bradbury had a wife and eight children. After reading some of the letters published by Rickett, especially those written by his wife Elizabeth and by his son John Leigh Bradbury, one realizes that the family was never far removed from penury but, notwithstanding, preserved their pride and high sense of integrity. Bradbury evidently sent his family a large proportion of the meagre amount which he received for his work; one hundred pounds *per annum* could not have gone far, however, to finance his own explorations and to support nine others as well.

3. The institution known as the Missouri Botanical Garden was not founded until 1889.

Rickett's analysis of the difficulties of Bradbury's task and his appraisal of the man's work are valuable; they follow his transcript of the letters.

"Bradbury was defeated in the great undertaking of his life chiefly by lack of appreciation (on his part and on that of his sponsors) of the difficulties of the task. To travel the American wilderness and bring back living plants from it required more money than had been foreseen or was forthcoming; though Bradbury himself soon realized this, he was unable to 'sell' the idea to his patrons in Liverpool. This was unfortunate; probably a few additional hundreds of pounds would have yielded results gratifying to all concerned. As for Bradbury's scientific endeavor, we can only speculate upon what he would have accomplished had not his collection been turned over to Lambert and Pursh. It may be suspected that he lacked the scientific acumen of either Pursh or Nuttall, and that the actual outcome, while distressing to Bradbury, was fortunate for posterity. On the other hand it is evident that he recognized the novelty of many of the species which he collected, and, given more means, could perhaps have described them properly. Whatever be the truth of this, we may share his just resentment at the circumstances which deprived him even of the chance to show what he could do. He seems never to have attempted botanical work again."

Rickett devotes some twelve pages to the history of the Bradbury specimens and to an analysis of the content of the collection.

". . . It is plain that (except for a few sent from St. Louis and from the Nodoway in January and April, 1811) most of the specimens were sent to John Leigh Bradbury in December, 1811. The latter made little or no attempt to follow his father's instructions to sell them, but sent them to Roscoe in Liverpool. At the Botanic Garden they were presumably sorted over and duplicates sent to various persons in exchange for their collections. It must have been in this way that Thomas Taylor obtained those which he later gave to Sir William Jackson Hooker . . . preserved in the Hooker Herbarium at Kew. Some of the duplicates reached Sir Joseph Banks . . . and at the instance of Bullock a set was sent to Alymer Bourke Lambert . . . Pursh had access to Lambert's herbarium, and on these specimens founded new species, besides determining such as represented species already known. At the sale of Lambert's herbarium in 1842, most of the North American specimens were purchased apparently by Dr. Thomas B. Wilson and by him presented to the Academy of Natural Sciences of Philadelphia [this on the authority of Dr. Pennell]. In that institution they are still preserved . . . The types, therefore of Pursh's species are the duplicates used for exchange by Roscoe.

"The specimens which remained in Liverpool—the 'originals' as contrasted with the duplicates—were seen in 1839 by Thomas Batt Hall. The collection of the Botanic Garden was transferred in 1909 to the Liverpool Public Museums . . ."

Rickett explains why the Bradbury collections at Liverpool (". . . about 115 unmounted specimens . . .") are "something of a miscellany" and supplies an annotated list of those at Kew, at Philadelphia and at Liverpool, ". . . in an effort to bring to-gether a complete list of his extant specimens."

Following this list, of 139 numbers, are two appendices. The first contains ". . . a list of 44 species described by Pursh in his *Supplementum* from specimens collected by Bradbury; of these 36 are accounted for by specimens enumerated in the fore-going list . . ."

The second appendix ". . . contains the names included in Bradbury's 'Catalogue of some of the more rare or valuable plants discovered in the neighbourbood of St. Louis and on the Missouri' . . ." Of the ninety-nine numbers listed, fifty are marked with an asterisk, indicating that they ". . . are represented by specimens mentioned in the present paper . . ."

Still another paper upon Bradbury has been published since this chapter was writ-ten: "Plant collecting in Missouri. A Liverpool expedition, 1809-11," by H. Stansfield (1951). I have already quoted from a Stansfield letter in this chapter.

The article concerns the ". . . 115 unmounted specimens—designated the John Bradbury Collection . . ." which are among the collections of foreign plants in the Herbarium of the City Museums of Liverpool. A map, prepared by Stansfield, depicts Bradbury's route up the Missouri River to Fort Mandan and designates seven locali-ties ("the places") where he collected plants. The author explains that ". . . Out of the many plants which he collected at each of these points only one or two are men-tioned in this account of his journey. These[4] and many more are in the Herbarium of the City Museums."

The article outlines Bradbury's journey and terminates with the following state-ments concerning his collections:

"Here in Liverpool very many of the plants collected . . . have reposed for 140 years, first at the Liverpool Botanic Garden, Olive Street; afterwards lost among the scattered effects of that institution when it was moved to Edge Hill in 1836 and later taken over by the Corporation in 1841. The plants were re-discovered in recent years among the foreign plants of the Herbarium in William Brown Street . . ."

After telling how Pursh obtained access to Bradbury's collections—". . . William Roscoe . . . loaned many duplicates to Pursh who was staying and working with Aly-mer Bourke Lambert at his estate at Boynton . . ." in England—and how he published the rarities in the supplement of his *Flora,* Stansfield continues:

"A number of Bradbury's plants described by Pursh were new species and there-fore TYPE specimens. Pursh, however, attributes the plants to Bradbury for in every in-

---

4. Stansfield supplies the now accepted names for seventeen species in the collection: three collected April 27 at the "Otto Village, La Riviere Platte"; six on April 28 on the bluffs and islands near the Platte River; two on May 11 near the "Maha Village"; three on June 5, at or near the "Big Bend" of the Missouri River; three on June 15, at or near the "Aricara Village."

stance he quotes, 'on the banks of the Missouri, Bradbury.-Herb. Brad.' The plants loaned to Pursh were never returned to Liverpool, but passed into the herbarium of A. B. Lambert whose large collections were sold piecemeal by auction, after his death in 1842. These plants were purchased by Thomas B. Wilson and presented by him to the Academy of Natural Sciences, Philadelphia, where they still repose. Although the Type specimens are in Philadelphia many of the exact localities of collection are missing from these plants, which can only be assigned to definite places by comparison with those in Liverpool. A few Bradbury plants found their way to the herbarium of Hooker and these are now at the Royal Botanic Gardens, Kew. The rest of the plants are at Liverpool. They are Isotypes (plants gathered at the same time and locality as the Type specimens).

"It is ironical that John Bradbury's claim to botanical fame rests chiefly upon entries in the work of a pirate botanist. The action of the authorities of the Liverpool Botanic Garden was not in their own best interest; it deprived Bradbury of the honour surely due to him, and the Liverpool Museum of very many historical plants . . ."

Anonymous. 1818. [Review of] Travels in the interior of America . . . By John Bradbury. 1817. *Edinburgh Review* 31: 133-150.

BRACKENRIDGE, HENRY MARIE. 1814. Views of Louisiana; together with a journal of a voyage up the Missouri river in 1811. Pittsburgh.

——— 1816. Journal of a voyage up the river Missouri; performed in eighteen hundred and eleven, by H. M. Brackenridge, Esq. Baltimore. [Is a second edition of Views of Louisiana . . .]

——— 1904. Journal of a voyage up the river Missouri; performed in eighteen hundred and eleven, by H. M. Brackenridge, Esq. *In:* Thwaites, Reuben Gold, *editor.* Early western travels 1748–1846. 6. [Transcribed from second edition, Baltimore, 1816. Omits Parts I, II, and IV of Appendix.]

BRADBURY, JOHN. 1817. Travels in the interior of America, in the years 1809, 1810, 1811; including a description of upper Louisiana, together with the states of Ohio, Kentucky, Indiana, and Tennessee, with the Illinois and western territories, and containing remarks and observations useful to persons emigrating to those countries. Liverpool.

——— 1819. [Title as in Liverpool edition of 1817.] Map ["Map of the United States of America, comprehending the western territory with the course of the Missouri. Engraved for Bradbury's Travels."]. London.

——— 1904. [Title as in editions of 1817 (Liverpool) and 1819 (London).] *In:* Thwaites, Reuben Gold, *editor.* Early western travels 1748–1846. 5. [Reprinted from second edition, London (1819); includes reproduction of map.]

BRITTEN, JAMES & BOULGER, EDWARD SIMONDS. 1931. A biographical index of deceased British and Irish botanists. ed. 2. London. 42.

CHITTENDEN, HIRAM MARTIN. 1902. The American fur trade of the far west. A history of the pioneer trading posts and early fur companies of the Missouri Valley and the

Rocky Mountains and of the overland commerce with Santa Fe. Map. 3 vols. New York.

DARLINGTON, WILLIAM. 1843. Reliquiae Baldwinianae: selections from the correspondence of the late William Baldwin, M.D., Surgeon in the U. S. Navy. Philadelphia.

GHENT, WILLIAM JAMES. 1931. The early far west. A narrative outline 1540–1850. New York. Toronto. 134-142.

GRAY, ASA. 1840. Notices of European herbaria, particularly those most interesting to the North American botanist. *Am. Jour. Sci. Arts* 40: 1-18. *Reprinted:* Sargent, Charles Sprague, *compiler.* 1889. Scientific papers of Asa Gray. 2 vols. Boston. New York. 2: 1-21.

HALL, THOMAS BATT. 1839. Some account of the Liverpool Botanic Garden. *Naturalist* 4: 395-400.

HARSHBERGER, JOHN WILLIAM. 1899. The botanists of Philadelphia and their work. Philadelphia. 153.

HILL, SAMUEL. 1907. Bygone Stalybridge. Stalybridge, England. *See* True, R. H. 1929. A sketch of . . .

IRVING, WASHINGTON. 1836. Astoria; or, anecdotes of an enterprise beyond the Rocky Mountains. 2 vols. Philadelphia.

MACDONELL, A. 1886. William Bullock. *Dict. Nat. Biog.* 7: 256.

PENNELL, FRANCIS WHITTIER. 1936. Travels and scientific collections of Thomas Nuttall. *Bartonia* 18: 1-51.

PURSH, FREDERICK. 1814. Flora americae septentrionalis; or, a systematic arrangement and description of the plants of North America; containing besides what have been described by preceding authors, many new and rare species, collected during twelve years travels and residence in that country. 2 vols. London.

RICKETT, HAROLD WILLIAM. 1934. Specimens collected by Bradbury in Missouri Territory. *Bull. Misc. Inf.* Kew. 49-61.

—— 1950. John Bradbury's explorations in Missouri Territory. *Proc. Am. Philos. Soc.* 94: 59-89.

SARGENT, CHARLES SPRAGUE. 1895. The Silva of North America. Boston. New York. 7: 86, *fn.* 5, 6, 9, 87, *fn.* 1.

SPAULDING, PERLEY. 1908. A biographical history of botany at St. Louis, Missouri. *Pop. Sci. Monthly* 73: 493-495.

STANSFIELD, H. 1933. Handbook and guide to the herbarium collections in the Public Museums Liverpool. 39, 42, 43.

—— 1951. Plant collecting in Missouri. A Liverpool expedition, 1809–11. *Liverpool Libraries, Museums & Arts Committee Bulletin* 1: no. 2, 17-31.

TRUE, RODNEY HOWARD. 1929. A sketch of the life of John Bradbury, including his unpublished correspondence with Thomas Jefferson. *Proc. Am. Philos. Soc.* 68: 133-150. [Includes extracts from Hill, Samuel. 1907. Bygone Stalybridge.]

CHAPTER VI

NUTTALL ACCOMPANIES HUNT AND THE "ASTORI-

ANS" UP THE MISSOURI RIVER TO THE ARIKARA VIL-

LAGES AND FROM THERE PROCEEDS BY BOAT TO THE

COUNTRY OF THE MANDANS

THOMAS Nuttall is an important figure in the botanical history of the United States. A mere glance at the maps which depict his journeys indicates the great amount of territory which he traversed, often on foot or on horseback, and both east and west of the Mississippi River. On all these travels he made plant collections and, to a major extent, was the one to publish upon them.

Asa Gray wrote in 1844—and praise from Gray, even with reservations, was praise indeed:

"Mr. Nuttall . . . first arrived in this country the very year [1807 or 1808] that the younger Michaux finally left it. And from that time to the present, no botanist has visited so large a portion of the United States, or made such an amount of observations in the field and forest. Probably few naturalists have ever excelled him in aptitude for such observations, in quickness of eye, tact in discrimination, and tenacity of memory. In some respects, perhaps, he may have been equalled by Rafinesque,—and there are obvious points of resemblance between the later writings of the two, which might tempt us to continue the parallel;—but in scientific knowledge and judgment he was always greatly superior to that eccentric individual . . ."

The most recent, comprehensive and best documented source of information regarding Nuttall's life and accomplishment is Dr. Francis W. Pennell's "Travels and scientific collections of Thomas Nuttall" (1936) and I shall cite from it often in my chapters on Nuttall. Pennell states:

"Besides incidental references and the internal evidence of his publications, our knowledge of Nuttall's life is nearly wholly derived from two obituary accounts, independently prepared in this city [Philadelphia] and both given to the world in early January, 1860, within four months of his death in England."

The first of these obituary accounts is the "Biographical sketch of the late Thomas Nuttall," which appeared in *The Gardener's Monthly* edited by Thomas Meehan. Pennell considers this ". . . the more trustworthy . . ." and believes ". . . must have been actually written, though with editorial additions . . ." by the botanist's intimate associate Dr. Charles Pickering. The second is the "Biographical notice of the late Thomas Nuttall," written by Elias Durand at the request of the American Philosophical Society, read before that body on January 6, 1860, and published in its *Proceed-*

ings for the same year. Pennell himself wrote "An English obituary account of Thomas Nuttall," which quotes fully from an article written by T. J. Booth, Nuttall's nephew, and is of particular interest insofar as the botanist's associations with England are concerned.

Nuttall was born according to this nephew "... of humble parentage in the year 1786, in the village of Long Preston, in Craven, in the West Riding of Yorkshire." Pennell gives the precise date as January 5, and tells that he first came to America, proceeding to Philadelphia, in "... the spring of 1807 or 1808, when he was twenty-one or twenty-two..." Nuttall wrote in the preface to *The North American Sylva:*

"Thirty four years ago, I left England to explore the natural history of the United States. In the ship Halcyon I arrived at the shores of the New World; and after a boisterous and dangerous passage, our dismasted vessel entered the Capes of the Delaware in the month of April... As we sailed up the Delaware my eyes were rivetted on the landscape with intense admiration. All was new!—and life, like the season, was full of hope and enthusiasm..."

Nuttall had been trained as a printer but, especially interested in all branches of natural history, soon turned to botany as his life work. Pennell has supplied, as fully as accuracy permits, a year-by-year record of his activities. In Philadelphia his associations—sometimes congenial, sometimes difficult[1]—were with the American Philosophical Society and the Academy of Natural Sciences. Whenever possible he returned to that city. Pennell, who separates Nuttall's sojourns in Cambridge, Massachusetts into two periods: *"Cambridge (1825–1829)"* and *"Cambridge (1830–1833),"* states:

"From 1825 Nuttall was under definite appointment at Harvard University. He was not made formal Professor, as had been his predecessor, William D. Peck, who had died in 1822; but, instead, he was appointed Lecturer on Natural History, also Curator of the Botanic Garden. This was the only formal post he ever held, and the accompanying salary was certainly his largest remuneration from science."

There seems to exist some question as to the precise year that Nuttall undertook his work in Cambridge.[2] But, having been refused leave of absence, we know that he

1. Pennell quotes a letter (in the Historical Society of Pennsylvania) written by Nuttall to John Torrey, October 27, 1838, which indicates that his work at the Philadelphia Academy was carried on under somewhat trying conditions: "... For any interest taken here in botany one might as well be amongst the Hottentots themselves. I am now very sorry, that they ever had a specimen of anything from me—but my hand is now effectually closed to them, nor shall my pen indite another line for their Journal ..." There seems to have been justifiable cause for complaint but Pennell notes "One suspects that a stove must have been installed ..." and records that Nuttall must have had a change of heart for "... he proved liberal in giving the Academy his western collections, although one sees why his later papers nearly all appeared in the Transactions of the American Philosophical Society."

2. In a letter dated February 3, 1954, from Kimball C. Elkins, Assistant in the Harvard University Archives, the year in which Nuttall's connection with Harvard College began is supplied, and it would seem conclusively. Writes Mr. Elkins:

resigned his appointment in 1833 in order to accompany Nathaniel Jarvis Wyeth on his transcontinental journey of 1834.

There are stories of how, while in Cambridge, Nuttall protected himself against interruption;[3] he no doubt had idiosyncrasies but these may have made possible his accomplishments. Setting aside its humorous aspects, Nuttall's absorption in his work has, for me, an appealing side.

Approximately thirty-three of Nuttall's seventy-three years were spent in North America. George Brown Goode considered, as have others, that he was ". . . so thoroughly identified with American natural history and so entirely unconnected with that of England that, although he returned to his native land to die, we may fairly claim him as one of our own worthies . . ." Actually, on inheriting a family estate, he returned to live in England about seventeen years before his death. This may have been due to financial necessity but nevertheless he was British by birth and evidently found it expedient to remain so.

The material side of life interested Nuttall but little. The anonymous author of the "Biographical sketch" wrote:

"He hated everything that savored of vanity or needless show, always aiming at the

"We have been able to find the following information in answer to your questions about Thomas Nuttall, Curator of the Botanic Garden.

"At a meeting of the Board of Visitors of the Massachusetts Professorship of Natural History held at the Athenæum, October 30, 1822, after the death of Professor William Dandridge Peck (Massachusetts Professor of Natural History, 1805–1822), it was voted that Thomas Nuttall be appointed curator of the Botanic Garden. This vote was submitted to the Corporation, and approved by them at their meeting of November 9, 1822. It was also voted by the Board and approved by the Corporation that 'if the Curator shall in the opinion of the Reverend and Honorable Corporation, be competent to deliver Lectures on Botany or any other branch of Natural History, this board would be gratified at his receiving their Authority and permission so to do, during the vacancy of the Professorship, and his being allowed such fees as they may think reasonable.' These votes appear in the minutes of the Visitors of the Professorship of Natural History, and in the College Records, both of which are catalogued and shelved here in the University Archives.

"As we told you on the telephone, the annual catalogue of Harvard University for 1823–24 is the first which lists Nuttall as Curator of the Botanic Garden. The *Historical Register of Harvard University* is apparently in error when it gives his dates as Curator of the Botanic Garden and Lecturer on Natural History, as 1825–1834. . . . "

Nuttall must have begun his work at the college promptly—sometime in the winter of 1822-1823. *See* pp. 185-187.

3. Asa Gray wrote Torrey from Cambridge, May 24, 1844, in regard to Nuttall's occupancy of the "Botanic Garden house":

". . . Mr. Nuttall . . . left some curious traces behind him. He was very shy of intercourse with his fellows, and having for his study the southeast room, and the one above for his bedroom, put in a trap-door in the floor of an upper connecting closet, and so by a ladder could pass between his rooms without the chance of being met in the passage or on the stairs. A flap hinged and buttoned in the door between the lower closet and the kitchen allowed his meals to be sent in on a tray without the chance of his being seen. A window he cut down into an outer door, and with a small gate in the board fence surrounding the garden, of which he alone had the key, he could pass in and out safe from encountering any human being . . ."

See also Fernald, M. L. (1942) for more about Nuttall's eccentricities.

real and substantial . . . He carried this habit of simplicity always with him. His dress . . . was chosen with a view to service . . . probably in no event of his life did pecuniary considerations influence him. His income was mainly derived from lectures . . . and the private sale of his collections and specimens . . ."

Another comment by the same writer appears significant:

"One great characteristic of the man was his readiness to listen to suggestions from any quarter respecting his favorite science, and much of his success was, doubtless, owing to this modesty of his nature."

Pennell attributes to Nuttall ". . . the faculty of making friends."

Nuttall made three journeys into the trans-Mississippi west. The first and the major and significant part of the second were made in the present decade (of 1810–1820).

On the first journey (1811–1812) he accompanied the expedition led by Wilson Price Hunt up the Missouri River from St. Louis to the Arikara villages in northern South Dakota and then, in company with the fur trader Manuel Lisa, proceeded to the country of the Mandan Indians in North Dakota. After his return to St. Louis, he descended the Mississippi River to New Orleans and from there sailed for England.

On the second journey (1818–1820) he ascended the Arkansas River, crossing the breadth of the state of that name and entering portions of eastern and northeastern Oklahoma.

On the third journey (made in the decade of 1830–1840) he accompanied Nathaniel Jarvis Wyeth from St. Louis to the Pacific coast, from there making two visits to the Hawaiian Islands and, before sailing from California for the eastern seaboard, collecting at a number of points between Monterey and San Diego.

This chapter concerns the first of these journeys.

Almost immediately after arrival in North America, Nuttall became acquainted with Professor Benjamin Smith Barton of Philadelphia, who had recently lost his assistant Frederick Pursh. Barton was working on a flora of North America and soon employed Nuttall as a helper, sending him on various field-trips. One, taking him northwestward, was carefully planned by his employer.

Among some recently discovered Barton papers published by Pennell (1936) with the permission of the owner Mrs. John R. Delafield, were the "Terms of Agreement, dated April 7, 1810, between Barton and Nuttall" and the "Directions for Mr. Thomas Nuttall."

According to the "Terms" Barton agreed to pay all expenses of the journey, Nuttall keeping ". . . an exact account of all expenditures." On the botanist's return to Philadelphia or ". . . within forty days . . ." thereof, he was to be recompensed ". . . at the rate of Eight dollars the month, lawful money of the United States." This contract Nuttall signed. He was to keep ". . . an exact Journal . . ." of his daily travels and of the observations made, these to be the exclusive property of Barton: ". . . no parts of them are to be communicated, without my consent, to any person." Barton promised ". . . to make a public acknowledgment . . ."—should his flora be published—that

Nuttall had made the journey. Nuttall was also to be entitled to ". . . a part of all the specimens of animals, vegetables, minerals, Indian curiosities, etc. . . ." which he collected, but could not dispose of any without Barton's consent: ". . . they might otherwise fall into the hands of persons who would use them to my [Barton's] disadvantage."

The "Directions" included an outline of the route; Pennell quotes this itinerary at some length. I repeat but a fraction. Nuttall was to go ". . . from Philadelphia to . . . Pittsburgh . . . to Sandusky . . . to Detroit . . . to Chicago . . . along the western shore of Lake Michigan . . . along Green Bay . . . to Keweenaw Point on Lake Superior . . . to Fond du Lac . . . to the Grand Portage . . . by the Lake of the Woods to Lake Winnipeg, which seems to have been a principal goal of the journey; thence on north-westward to the Beaver River at 55° N. in the northern part of the present Saskatchewan, which was to be the uppermost point of the journey. Somewhere in the far North-West . . . he was to winter. Returning he was to go southeastward to the Missouri River . . . descend it to about the mouth of the Illinois River . . ."

And so on, until he eventually got back to the starting point in Philadelphia.

"The method of transport was not specified, but the course indicated required more travel by land than by water . . . At several points it involved geographical exploration . . . It meant passing among many tribes of Indians, some like the Sioux already hostile to the Whites. It was hardly the undertaking for a lone adventurer, but from the distance of Philadelphia neither Barton nor Nuttall seems to have realized this."

The "Directions" had included a last and, in retrospect, a superfluous admonition to Nuttall:

". . . always remember, that, next to your personal safety, science, and not mere conveniency in travelling, is the great object of the journey. In pursuit of curious or important objects, it will often be necessary to court difficulties, in travelling."

The Barton papers also included a "List of articles belonging to me, which Mr. Th. Nuttall has in his possession: viz. . . ." Barton took no chances, apparently, and seven small articles are listed, among them "One thermometer, in a case . . . One steel-pen . . . Five blank Books . . . One dirk, in case . . ." And, before starting on his way, Nuttall evidently rendered a bill totalling four dollars and ninety-nine and one-half cents, and seems to have been "Paid 5 dollars, in full . . ." One wonders whether Nuttall returned the change! Barton acknowledged payment as of April 12.

After leaving Philadelphia in ". . . early or mid-April . . ." of 1810, the botanist, according to Pennell, eventually reached Prairie du Chien, Wisconsin, on the Mississippi River, and proceeded down that stream to St. Louis, arriving sometime in the autumn of 1810; he had departed from Barton's stipulated itinerary near the Detroit River. John Bradbury (whose story is told in the preceding chapter) had been at St. Louis for almost a year before Nuttall's arrival, or since December 31, 1809.

Independently of each other, these two men accepted an invitation to join the fur-trading expedition led by Wilson Price Hunt which was bound for the mouth of the Columbia River and which, since it went at the behest of John Jacob Astor, is usually known as the Astorian Expedition. In accepting Hunt's invitation, Nuttall—and probably wisely—departed still further from the terms of his contract with Barton. Pennell mentions possible reasons for this: "Above all, I fancy that he felt the hopelessness of carrying out his assigned project single-handed."

Despite Barton's stipulation that Nuttall should keep a journal, there is no evidence that he did so.[4] Bradbury kept one, however, and for the time the two men were together this supplies an itinerary. Pennell reports that he found both Bradbury's map and *Travels* helpful in checking the dates and localities of collection cited by Nuttall in the *Genera*.

In my chapter on John Bradbury I have followed in some detail the Missouri River route as he recorded it in his *Travels* and have referred as well to the journal of Henry Marie Brackenridge, and shall merely indicate, therefore, how the travels of Bradbury and Nuttall differed.

Both left St. Louis on March 13, 1811, and, after joining Hunt at St. Charles, followed the Missouri River until, on June 12, the expedition reached the Arikara villages, situated on the west bank of the Missouri just above the mouth of Grand River, in Corson County, northern South Dakota. Hunt remained there with most of his party until July 18 on which day he resumed his westward journey.

Bradbury, leaving the Hunt party at the Arikara villages on June 19, traveled overland with the fur trader Ramsay Crooks, to Fort Lisa, a Missouri Fur Company post, situated some ten miles above the mouth of Knife River, in Mercer County, North Dakota,[5] and arrived there on June 22. Nuttall, on the other hand, traveled up the Missouri with Lisa, who was bound for his post. Chittenden states that they arrived June 26 but Bradbury records meeting his friend Brackenridge (also traveling with Lisa) on June 25 so Chittenden may have been wrong. According to Bradbury's *Travels,* the three men dined together on Lisa's boat on the fourth of July.

Bradbury, Brackenridge and Lisa left the Mandan region on July 6 and were back at the Arikara village, Hunt's camp, on July 7. Bradbury and Brackenridge started by boat down the Missouri on July 17 and reached St. Louis on July 29 after an extremely rapid trip. Bradbury left St. Louis for New Orleans on December 4 and reached his destination on January 13, 1812; seven days later he sailed for the eastern states.

From Nuttall's notations in the *Genera,* Pennell assumes that, unlike Bradbury, Nuttall (although ". . . in a good geographical position to attempt to carry out Professor Barton's directions to go either to Lake Winnipeg or else to the upper Missouri River . . .") had remained in the Fort Mandan region "until frost" and then,

---

4. See the final paragraphs of this chapter for a recently published Nuttall "Diary" edited by Dr. Jeannette E. Graustein.

5. See p. 121, for the location of this post and of the first Fort Mandan.

probably with Manuel Lisa, returned to St. Louis sometime in the autumn of 1811.[6] From there he descended the Mississippi (stopping at the mouth of the Ohio and at Natchez) to New Orleans—nothing suggests that he made this trip with Bradbury—and from that port sailed for England early in 1812.

In sailing for England, Nuttall again departed from his Barton contract, which had stipulated his return to Philadelphia and the delivery of his collections to his employer. Pennell suggests various reasons for this change: ". . . the increasingly hostile sentiment towards England . . .", or the fear that his first change of itinerary ". . . was too flagrant for easy reconciliation . . .", or the need of ". . . access to larger herbaria and libraries . . ." than were to be found in America, and so on.

According to Pennell, Nuttall remained in England from 1812–1814, or until after the end of the War of 1812. Back in America he traveled into the southeastern states in 1815 and 1816, between and after these years working in Philadelphia and completing *The genera of North American plants, and a catalogue of the species to the year 1817* by the autumn of 1818. Pennell considers that, while in England, Nuttall may have compared species and written descriptions for the *Genera*.

Some of the seeds which he collected on the Missouri were certainly grown in the Fraser Nursery in London, for Nuttall prepared a catalogue[7] (issued by that firm in 1813) which, according to Pennell, listed ". . . 89 species, largely new to science . . ." from that region. And, "Between August, 1812 and November, 1816 at least 10 species, mostly from the Fraser establishment, were illustrated in the Botanical Magazine."

Some of Nuttall's plants were presented ". . . to the Botanic Garden of Liverpool, under the care of John Shepherd, to whom in 1818 he dedicated[8] the genus *Shepherdia* . . ."

Perley Spaulding notes that "Although Nuttall explored the Missouri country on two different occasions . . . he seems never to have published any considerable list of plants found by himself near St. Louis."

Writing of the *Genera,* Pennell comments that ". . . Nuttall's catalogue may quite properly be viewed as a summary of his own explorations up to the end of 1817." He reports that, unfortunately, Nuttall's labels accompanying his specimens bore ". . . but meagre data: just the name or abbreviation of the state, or other brief memorandum." And apparently only one letter (dated April 22 and sent from Pittsburgh) concerned with the trip of 1811–1812 has been found among the Barton papers.

6. The date is October 21, 1811, according to J. E. Graustein (pp. 12, 13, *fn.* 19).

7. For a reprint of this catalogue see Greene, Edward L., 1890, in *Pittonia.* Nuttall's part in its authorship is explained.

8. This dedication in the *Genera* reads: "In honor of Mr. John Shepherd curator of the Botanic garden of Liverpool, a scientific horticulturist, through whose exertions and the patronage of the celebrated Roscoe, that institution owes its present merit."

Throughout his discussion of Nuttall's Missouri River journey Pennell carefully cites, under the names used by Nuttall, a number of the species ("only a small portion" of those in the *Genera*) which represent Nuttall's own discoveries. Many were from Fort Mandan or beyond. He surmises that Nuttall must have ascended the Missouri beyond that point, reaching ". . . the northern extension of the Bad Lands . . ." but not the Rocky Mountains (or "Northern Andes") as he believed he had done.

Pursh's *Flora americae septentrionalis* came out on 1814 and Nuttall stated in the preface to his *Genera* (1818) that his own catalogue ". . . may be considered as supplementary . . ." to that earlier work, representing the ". . . result of personal collections and observations made from the year 1809 to the present time, throughout most of the states and territories composing the Union." In the phrase "result of personal collections and observations" Nuttall may have been contrasting his own methods with those of Pursh. Like other botanists of the period, Nuttall had reason to distrust Pursh and seems to have allowed him little access to his collections.

Pennell states that ". . . Pursh's 'Flora' contains only 15 species cited to Nuttall's herbarium . . . all of which are from the Missouri River. Only a part of these were new to science, and Nuttall's remark concerning *Bumelia serrata*[9] having been 'inadvertently described by Mr. Pursh' makes me suspect that he was unaware that any of these were to appear in Pursh's book. Nuttall's incommunicativeness may in good part have been due to obligation towards Barton, although he gives another reason in his comment on *Bartonia*."

Briefly, the *Bartonia* controversy was this: both Nuttall and Pursh desired to name a plant in honor of their patron. Nuttall had chosen a plant and name, *Bartonia superba*. He entrusted his notes, made from the living plant, personally observed on the Missouri, to Pursh and that botanist described it in *The Botanical Magazine* of 1812 (t. 1487) as *Bartonia decapetala*. The article ended with the statement (evidently by the editor John Sims): "For the above generic and specific characters, and indeed the whole communication, we are indebted to Mr. Frederick Pursh, author of a new Flora of North-American plants, now in the press." Nuttall claimed that, apart from "an imperfect capsule" collected by Meriwether Lewis on his descent of the Missouri at a time when no flowers were available, and except for Nuttall's own notes, Pursh had no knowledge of the plant.

"This unfortunate want of fidelity, prevented me from communicating to Mr. F. Pursh, many of the plants which now appear in this work [the *Genera*] . . . It was not surely honourable in Frederick Pursh, whom I still esteem as an able botanist, to snatch from me the little imaginary credit due to enthusiastic researches, made at the most imminent risk of personal safety!"

9. In the *Genera* Nuttall refers to *Bumelia* as "First noticed by Mr. Bradbury, near the lead mines of St. Louis on the Missisippi . . ." and adds that ". . . The *B. serrata,* inadvertently described by Mr. Pursh, was nothing more than a young branch of the *Prunus caroliniana* without flowers, which I had collected near the town of Natchez . . . I have thought it no less than my duty to the public to rectify this mistake, without, I hope, intending any personal reflection, as we are all equally liable to prevailing error."

Eventually Muhlenberg's earlier use of the name *Bartonia* for a "modest" plant of the *Gentianaceae* was accorded priority and the more striking *Bartonia* of the *Loasaceae*—whether named by Nuttall, Pursh or Sims, who is sometimes cited as author— is now transferred to the genus *Mentzelia* Linnaeus. But for some time this nomenclatorial controversy seems to have been a matter of general interest.[1]

Notwithstanding his desire to honor Barton, Nuttall failed to dedicate the *Genera* to him. Instead, Joseph Correa de Serra, Portuguese Minister to the United States, was so honored because of his active interest ". . . in the promotion of Natural Science, and . . . desire to elevate it to the rank of Philosophy . . ." Likely, De Serra had helped to finance publication; and Barton had died in 1815.

For the favorable reception[2] accorded Nuttall's *Genera* I quote from Pennell: "Elliott, perhaps the most discriminating of our early botanists, said of it in his Sketch: 'A work abounding in accurate information respecting the plants of this country.' Schweinitz wrote of it to Torrey: 'I think Mr. Nuttall's observations uncommonly excellent. His Genera has given me more light than any other book—it is so evident from all his remarks in that work, that they are the fruits of real personal acquaintance with the plants in nature.' And Torrey, in the preface to his 'Flora of the Northern and Middle Sections of the United States,' which he dedicated to Nuttall, remarked 'that it has contributed, more than any other work, to the advance of the accurate knowledge of the plants of this country.' "

Since this chapter was written Dr. J. E. Graustein has edited and published a "Diary" kept by Nuttall from April 12 through September 10, 1810—or to about the time of his arrival at St. Louis. The editor's preface contains matters of interest con-

1. William Baldwin wrote to William Darlington on March 21, 1819:

"It gives me great pleasure to inform you, that the views of our friend Collins coincide remarkably with our own, in relation to the great interests of Botany, in our Country . . . he pronounced most decidedly in favor of restoring the *Bartonia,* of Muhl. and rejecting, of course, the one about which Pursh and Nuttall have been contending."

The "friend" was Zaccheus Collins, for whom Nuttall named the genus *Collinsia.* There are many references to him in the correspondence of von Schweinitz and Torrey, most of them expressive of irritation at the man's slowness and hesitancy as far as determining plants was concerned. Von Schweinitz refers to him in a letter to Torrey, dated April 19, 1821:

". . . I sincerely deplore that his [Nuttall's] crypt. specimens have been swallowed by that retentive gulph, Mr. Collins, going *into* whose cave so many footsteps may be traced & none coming forth! I have among the rest written Z. Collins more than once, but have never been blessed with an answer . . ."

I may add that "gulph" according to the dictionaries is an obsolete form of "gulf" or "gulp"; not some strange animal as one might suppose from the context!

2. William Baldwin wrote Darlington, March 23, 1819 in less enthusiastic vein:

". . . The work of Nuttall is a convenient manual, and not without its merits: But it is . . . extremely faulty in numerous instances,—and the earlier it is superseded the better. The number of Nuttall's Genera requires to be abridged. He has placed too much reliance upon *habit,* in the formation of many of his new Genera: and this *method,* pursued, will inevitably end in *chaos,*—as it strikes at the root of everything like system. For example, the dogs of the frozen regions, clothed with a warm mantle of fur, must be made to constitute a distinct genus from those of the Tropics,—which are naked."!

cerning the discovery of this manuscript and her introduction gives a brief summary of the events in Nuttall's life. Included in three appendices are reprints of the "Terms of Agreement, Dated April 7, 1810, between Barton and Nuttall"; the " 'Directions for Mr. Thomas Nuttall' "; and the "List of articles belonging to me, which Mr. Th. Nuttall has in his possession: viz.," which were published in *Bartonia* by Dr. Pennell in 1936; also a "List of Animals Noted by Nuttall," and a "List of Plants Seen by Nuttall," or those animals and plants mentioned in the diary.

Among the illustrations which are of special interest are the "Catalogue of the Cambridge Garden (1818)," figures 6-9; "English Sale Catalog of Nuttall's Plants (1813)," plate 78, which is a partial reprint of the catalogue of plants offered for sale at Fraser's Nursery; and "Nuttall's List of Plants in the Cambridge Garden," plate 79. Also the map, figure 2, which first appeared in the second edition (London, 1819) of Bradbury's *Travels*.

Anonymous [CHARLES PICKERING ?]. 1860. Biographical sketch of the late Thomas Nuttall. *Gard. Month.* (Meehan) 2: 21-23.

BRACKENRIDGE, HENRY MARIE. For the published editions of his narrative *see* p. 137.

BRADBURY, JOHN. For the published editions of his *Travels . . . see* p. 137.

CHICHESTER, H. MANNERS. 1895. Thomas Nuttall. *Dict. Nat. Biog.* 41: 277, 278.

DARLINGTON, WILLIAM. 1843. Reliquiae Baldwinianae: selections from the correspondence of the late William Baldwin, M.D., Surgeon in the U. S. Navy. Philadelphia.

DURAND, ELIAS. 1860. Biographical notice of the late Thomas Nuttall. *Proc. Am. Philos. Soc.* 7: 297-315.

FERNALD, MERRITT LYNDON. 1942. Some early botanists of the American Philosophical Society. *In:* The early history of science and learning in America. *Proc. Am. Philos. Soc.* 86: 65-67.

GOODE, GEORGE BROWN. 1888. The beginnings of American science. The third century. *Proc. Biol. Soc. Washington* 4: 45.

GRAUSTEIN, JEANNETTE E. 1950–1951. *See* Nuttall, Thomas. 1950–1951.

GRAY, ASA. 1844. The longevity of trees. *North Am. Review* 59: 189-238. *Reprinted:* Sargent, Charles Sprague, *compiler.* 1889. Scientific papers of Asa Gray. 2 vols. Boston. New York. 2: 71-124.

——— 1893. Letters of Asa Gray. Edited by Jane Loring Gray. 2 vols. Boston. New York. 1: 326.

GREENE, EDWARD LEE. 1890. Reprint of Fraser's catalogue. *Pittonia* 2: 114-119.

HARSHBERGER, JOHN WILLIAM. 1899. The botanists of Philadelphia and their work. Philadelphia. 112, 151-159, 176.

NUTTALL, THOMAS. 1813. A catalogue of new and interesting plants, collected in Upper Louisiana, and principally on the river Missourie, North America. [By T. Nuttall] For sale at Messrs. Fraser's nursery for curious plants, Sloane Square, King's Road, Chelsea, London. *Pittonia* 2: 116-119, 1890.

——— 1818. The genera of North American plants, and a catalogue of the species to the year 1817. 2 vols. Philadelphia.

—— 1834. A description of some of the rarer or little known plants indigenous to the United States, from the dried specimens in the herbarium of the Academy of Natural Sciences of Philadelphia. *Jour. Acad. Nat. Sci. Phila.* 7: 61-115.

—— 1852. The North American Sylva; or a description of the forest trees of the United States, Canada, and Nova Scotia, not described in the work of F. Andrew Michaux, and containing all the forest trees discovered in the Rocky Mountains, the territory of Oregon, down to the shores of the Pacific, and into the confines of California, as well as in various parts of the United States, illustrated with 127 fine plates, By Thomas Nuttall, F.L.S., . . . &c. &c. &c. In three volumes.—Vol. I. being the fourth volume of Michaux and Nuttall's North American Sylva . . . Philadelphia . . . [Vols. II and III. being the fifth and sixth volumes of same.]

—— 1950–1951. Nuttall's travels into the Old Northwest. An unpublished 1810 Diary. Edited by Jeannette E. Graustein. *Chronica Botanica* 14: no. ½. 86 pp.

PENNELL, FRANCIS WHITTIER. 1936. Travels and scientific collections of Thomas Nuttall. *Bartonia* 18: 1-51.

—— 1938. An English obituary account of Thomas Nuttall. *Bartonia* 19: 50-53.

PURSH, FREDERICK. 1812. Bartonia decapetala. Ten-petal Bartonia. *Bot. Mag.* 36: t. 1487.

—— 1814. Flora americae septentrionalis; or a systematic arrangement and description of the plants of North America; containing, besides what have been described by preceding authors, many new and rare species, collected during twelve years travels and residence in that country. 2 vols. London.

RODGERS, ANDREW DENNY, 3RD. 1942. John Torrey. A story of North American botany. Princeton. [Contains numerous references to Nuttall and his various journeys.]

SARGENT, CHARLES SPRAGUE. 1891. The Silva of North America. Boston. New York. 2: 34, *fn.* 2.

SHEAR, CORNELIUS LOTT & STEVENS, NEIL EVERETT. 1921. The correspondence of Schweinitz and Torrey. *Mem. Torrey Bot. Club* 16: 142, 143.

SPAULDING, PERLEY. 1909. A biographical history of botany at St. Louis, Missouri. *Pop. Sci. Monthly* 74: 52-57.

STONE, WITMER. 1934. Thomas Nuttall. *Dict. Am. Biog.* 13: 596, 597.

CHAPTER VII

ESCHSCHOLTZ AND VON CHAMISSO, TRAVELING WITH

VON KOTZEBUE IN THE RURICK, *SPEND A MONTH AT*

THE BAY OF SAN FRANCISCO

THERE are a number of published accounts of the expedition considered in this chapter. I shall follow Otto von Kotzebue's own story, published in London in 1821, and translated from the German edition (Weimar, 1821) by Hannibal Evans Lloyd:[1] *A voyage of discovery into the South Sea and Beering's Straits, for the purpose of exploring a north-east passage, undertaken in the years 1815–1818 . . . in the ship Rurick, under the command of the Lieutenant in the Russian Navy, Otto von Kotzebue . . .*

Lloyd's preface explains that "The expedition was known to have originated in the enlarged views of that great patron of the sciences, His Highness Count Romanzoff,[2] Grand Chancellor of the Russian empire, and to have been fitted out with princely munificence at his sole expense."

The introduction was written by Adam Johann von Krusenstern, who explains that it had been Count Romanzoff's original plan to send two ships in search of a northern passage between the Atlantic and Pacific oceans, one from east to west, another from west to east, but this had been abandoned in favor of a single vessel and an exploration eastward; although von Krusenstern felt that there was little probability that such a passage would be found, ". . . its existence cannot be positively denied . . ." Other objectives were an examination of the coast of North America northward from Bering Strait and, by twice crossing the Pacific, the acquisition of greater knowledge of that ocean and its many islands: ". . . a rich harvest of objects of natural history was to be expected, as the Count had appointed, besides the ship's surgeon, an able naturalist to accompany the expedition." Von Krusenstern had superintended the building of the ship at Abo (then capital of Finland and some two days' march from St. Petersburg); its name, *Rurick*,[3] was chosen by Count Romanzoff.

In addition to "astronomical and physical instruments" and so on, the ship had three noteworthy innovations. One was a "life, or safety-boat," provided with "air-chests," the invention of "Mr. Fincham, a master ship-builder"; this proved to be so heavy that it was finally abandoned in Kamtchatka. A second innovation was "A dis-

---

1. Lloyd also translated the *Travels* of Maximilian, Prince of Wied-Neuwied.

2. Different spellings of the Count's name appear in the literature: Rumjanzoff (Mahr), Rumanzow (Alden & Ifft), Rumiantzof (Bancroft), etc.

3. Mahr spells the name "Rurik" but I use the spelling in von Kotzebue's *Voyage.*

covery lately made in England ... by Mr. Donkin, [which] consists in preserving
fresh meat, vegetables, soup, milk, in short eatables of every kind, for years together,
in a perfectly fresh state ..." This was not always satisfactory but must have been of
some value since, back from the voyage, von Kotzebue visited London and "... left
with the inventor several boxes of patent meat, as proof how well it had kept ..." The
third innovation was "... the fine discovery of the eminent natural philosopher, Mr.
Leslie, to produce ice by means of evaporation, even in the hottest room; by which it
is possible to have, even under the equator, the luxury of a cool beverage ..."

Two naturalists were of the party, also a physician and an artist. As first planned
these were to have been "Dr. Ledebour, professor of natural history in the university of
Dorpat ... he had proposed for his assistant Dr. Escholz,[4] who was to be also the
ship's physician ... Dr. Ledebour's health did not allow him to realise his wish, and
M. A. Von Chamisso,[5] of Berlin, accompanied the expedition as naturalist in his stead.
He was recommended ... by Professors Rudolph and Lichtenstein, as a thoroughly
well-informed man, passionately devoted to his department of science ..."

The painter was "A young man, of the name of Choris,[6] ... The richness of the
portfolio ... brought home ... and the praise ... bestowed upon him by the most
celebrated artists of St. Petersburg ... justify the choice of this young and deserving
artist."

Von Kotzebue's journal fills an approximate half of the three volumes of the Lloyd
translation of his *Voyage*.

The *Rurick* was launched at Abo on May 11, 1815, and on June 18 was anchored
"in the road of Cronstadt." By July 27 she was "furnished with provisions for two
years," and on July 30 left Cronstadt for Copenhagen. On August 27 she set sail,
crossed to Plymouth, England (where the "safety-boat" was taken aboard, "... our
whole crew being scarcely able to lift it into the Rurick ..."), and in late September left
for Teneriffe and the Canary Islands. From there she sailed to "St. Catherine's,"
Brazil, arriving in mid-December. Cape Horn was rounded on January 22, 1816, and
Avatscha Bay, the Harbor of St. Peter and St. Paul [or Petropavlosk], Kamtchatka,
reached in mid-July. Thence the expedition went to Kotzebue Sound[7] beyond Bering
Strait and then to Unalaska. From there, on September 14, the *Rurick* sailed for

4. "... Johann Friedrich (Ivan Ivanovitch) Eschscholtz ..." (Mahr).

5. "Louis Charles Adélaïde de Chamisso—'Adelbert' is a German adaptation, chosen by Chamisso himself ..." (Mahr).

   Some authors use the preposition *von,* others the preposition *de* before the family name Chamisso,
others merely refer to him as Chamisso. Although born a Frenchman, the man was an exile from France
and is most often referred to as a German. I use the form "von Chamisso" adopted in Kotzebue's *Voyage*.

6. "Louis (Login Andrevitch) Choris." (Mahr).

7. The names, Kotzebue Sound (north of the Seward Peninsula, in northwestern Alaska), Eschscholtz
Bay (an inlet of Kotzebue Sound in Bering Strait) and Chamisso Island (near the head of the Bay of Good
Hope in Kotzebue Sound), commemorate the visit of the expedition.

California—for us its important destination. It is not unusual to hear of botanists being late; on the departure for California Dr. Eschscholtz was not on hand when the party was ready to leave at daybreak: he had gone out the night before to botanize, had reached "an elevated spot," and had been afraid to descend in the dark!

1816

At midnight of October 1 they saw ". . . by moonlight the Cap de los Reyes . . ." and the next day ". . . at four o'clock in the afternoon dropped anchor in Port St. Francisco, opposite the Presidio. Our little Rurick seemed to throw the Presidio into no small alarm . . . we saw many soldiers on foot and on horseback, and in the fortress itself they were employed in loading the cannon . . . they enquired through a speaking trumpet, to what nation we belonged, the Russian Imperial flag not being known here."

The answer that ". . . we were Russians, and friends . . ." allayed fears and the party was hospitably welcomed by the Commandant, Don Luis Arguello, who had received orders from his government about the party and sent ". . . a courier to Monterey to acquaint the Governor of California of our arrival."

For one month the *Rurick* remained in the Bay of San Francisco for reconditioning and provisioning. While this was going on "Our naturalists were also employed, as there was much room for new discoveries in this country, so seldom visited by learned men. M. Choris was busily occupied in painting . . ."

The only trip inland from the bay mentioned by von Kotzebue was to the Mission of San Francisco on October 4:

"The weather was extremely fine, and an hour's ride brought us to our journey's end, though above half the road was sandy and mountainous. Only a few small shrubs here and there diversified the barren hills; and it was not till we arrived in the neighbourhood of the Mission, that we met with a pleasant country and recognized the luxuriant scenery of California."

Von Kotzebue thought unfavorably of the Spanish régime in California—the missions exploited the natives who, once baptized, were virtually slaves: "The rage for converting savage nations is now spreading over the whole South Sea, and causes much mischief, because the missionaries do not take pains to make men of them before they make them Christians . . ." The soldiers were in little better plight than the natives; they were destitute of clothing and had not been paid for seven years.

Nothing is said of any visit to the Russian settlement at Bodega which lay ". . . half a day's journey by water to the north of San Francisco, and is called by the Spaniards Port Bodega. The harbour is only for small ships."

On November 1, a month after arrival, "The Rurick was now again quite in order." The party ". . . had been abundantly supplied with provisions by the inhabitants . . ." and the ". . . crew were all in good health . . ." At nine A.M. the ship set sail and in an hour was out of the bay, bound for the Sandwich Islands; even when two miles out to sea they could still ". . . hear the loud howlings of the sea-lions, which were lying on the shore on the stones . . ."

The expedition remained in the Hawaiian Islands until mid-December.

1817

In 1817 it visited Radack in the Polynesian group and then returned to the Arctic regions, still searching for the northeast passage; in mid-July von Kotzebue gave up this quest because of ill-health and the *Rurick* started homeward, stopping at the Hawaiian and other islands of the Pacific and reaching the Philippines in mid-December.

1818

In late July, 1818, the *Rurick* resumed its way, doubled the Cape of Good Hope on February 30 and reached Portsmouth, England, in mid-June. From there it crossed to Copenhagen; "with an indiscribable emotion" von Kotzebue saw again on July 23 his native town of Reval left three years before, and on August 3 the ship ". . . cast anchor in the Newa [Neva], opposite to the palace of Count Romanzoff."

The itinerary is not difficult to follow in von Kotzebue's story. It is given in simplified form by August C. Mahr, to whose publication I refer later.

Most of the last half of volume two and the greater part of volume three of von Kotzebue's *Voyage* is devoted to the "Remarks and opinions of the naturalist of the expedition, Adelbert Von Chamisso," some fourteen pages of which describe the visit to California. The "Appendix by other authors" completes the third volume; its articles were for the most part written by Dr. Eschscholtz, although von Chamisso contributed a "Postscript and corrections and remarks." Neither the plates nor the "charts" included in the three volumes relate to California.

I quote from von Chamisso's "Remarks and opinions" (his chapter on California):

"A low ridge of mountains borders the coast of California, where we saw it, and intercepts the prospect into the interior of the country. It has not a volcanic appearance. The harbour of San Francisco ... enters through a narrow passage, receives some rivers from the interior, branches out behind the eminences, and forms into a peninsula, the country lying south of the entrance. The Presidio and Mission of San Francisco lie on this tongue of land, which, with its hills and downs, was the narrow field which lay immediately open to our researches ...

"The environs of San Francisco, in the northern hemisphere, are much poorer in natural productions than the coast of Chili, under the same latitude, in the southern. In the spring, when winter has afforded the earth some moisture, the hills and valleys are indeed adorned with brilliant iris and other flowers; but the drought soon destroys them.

"The fogs, which the prevailing sea-winds blow over the coast, dissolve in summer over a heated and parched soil, and the country exhibits in autumn only the prospect of bare scorched tracts, alternating with poor stunted bushes, and in places, with dazzling wastes of drift sand. Dark pine forests appear here and there on the ridge of the mountains, between the Punta de los Reyes and the harbour of San Francisco. The prickly-leaved oak, *Quercus agrifolia,* is the most common and largest tree. With crooked boughs and entangled branches, it lies, like the other bushes, bent towards the land; and the flattened tops, swept by the sea-wind, seem to have been clipped by the gardener's shears. The Flora of this country is poor, and is not adorned by one of those species of plants which are produced by a warmer sun. It however offers much novelty to the botanist. Well-known North American species[8] are found mixed with others belonging to the country;[9] and most of the kinds are yet undescribed. Only . . . Menzies, and Langsdorff, have made collections here; and the fruits of their industry are not yet made known to the world. The season was not very favourable[1] for us. We, however, gathered the seeds of several plants, and have reason to hope that we shall be able to enrich our gardens with them."

References to the flora of California end at this point.

Von Chamisso, like von Kotzebue, was not impressed regarding Spanish rule in California. His opinion of the missions was unfavorable: "The contempt which the missionaries have for the people, to whom they are sent, seems to us, considering their pious occupation, a very unfortunate circumstance." Like other visitors to the region he refers to a windmill, undoubtedly the one left by De Langle of the Lapérouse expedition: it ". . . creates astonishment, but does not find imitators." He found the political situation unstable.

"Spain has given way in the affair of Nootka [coveted by two nations]. England and the United States, without regarding its vain territorial possessions, are negotiating about the colony [the post of Astoria] at the mouth of the Columbia; and the Russian American Company have still a settlement [Fort Ross] a few leagues north of San Francisco."

Von Chamisso's *Tagebuch* (diary), included in his collected works, is more interesting reading than his "Remarks and opinions." The portion telling of the visit to San Francisco has been twice published in English translation: first by G. D. McElroy in 1873; again by A. C. Mahr in 1932, who reprints the McElroy translation with a few minor changes. I shall quote from the Mahr transcript and insertions between brackets are his.

At the time of the *Rurick's* visit, Spanish-Russian relations were none too good.

8. A footnote adds: *"Ceanotus, Mimulus, Oenothera, Solidago, Aster* [,] *Rhamnus, Salix, Aesculus?* &c. Species of wild grapes, which we did not see ourselves, are said to be very abundant in the interior, and to produce fruit of an agreeable taste."

9. *"Abronia, Eschscholzia, Cham.,* and others which are to be now described."

1. "October was a most unfavorable season to collect either botanical or entomological specimens in the small area accessible to them." (Essig)

Alexander von Baranoff had placed Ivan A. Kuskoff in charge of Fort Ross. The Spaniards resented the presence of the Russians.

"Bodega lying about thirty miles north of San Francisco, a half-day's journey, was considered by Spain, and not without some show of justice [justification], as her own soil and territory . . . Kuskoff, on Spanish soil and territory, had . . . built a handsome fort, defended by a dozen cannon, and there carried on farming . . . There he had his storehouses, for the smuggling trade with the Spanish forts [posts]; and there he captured by means of his Kodiakers, about 2,000 [several thousand] sea-otters . . . yearly, on the California coast . . .

"It did [does] not seem to me incomprehensible that the Governor of California should have been very angry when he received a late notice of this settlement. Various measures were [had been] taken to induce Herr Kuskoff to evacuate the place, but he referred all the Spanish emissaries to Herr Baranoff, who had ordered him here [there], and at whose command he would willingly leave the place, in case the order was given. So stood matters when we came to San Francisco . . ."

Spain's distrust of Russia was reflected in many small, sometimes comic, ways. When a question arose as to who should pay the first ceremonial visit, the Russian commander or the Spanish governor, von Kotzebue, purely by chance, happened to go ashore to visit the chronometer; ". . . as he stepped on shore, the Governor strode down the slope to meet him. The Captain, on his part, ascended the slope in order to receive the Governor; and so Spain and Russia, each going half-way, fell into one another's open arms!" But, whenever the question of Russian evacuation was broached, the Russians evaded the issue and ". . . so we were just as far advanced as before." [2]

Russia was not the only nation causing anxiety to Spain.

"The claims of Spain to the territories on this coast were not more highly respected by the Americans and English than by the Russians . . . [the mouth of the Columbia River, too, was regarded by Spain as her property]. The Spaniards and Mr. Elliot have given us very much the same accounts relative to the history of the colony there. The Americans had gone there, partly by land, partly by water, and founded a settlement[3] . . ."

Von Chamisso, botanist though he was, makes but two references to plants, both semijocose:

"The year was already old, and the country, which in the spring months (as Langsdorf has seen it) blooms like a flower-garden, presented now to the botanist only a dry, arid field. In a swamp, near by our tent, a water-plant had [is supposed to have] grown, which Eschscholtz asked me about after our departure. I had not observed it; he, however, had reckoned that a water-plant, my especial love, would not have es-

2. Discussions with Russia in 1816 were evidently much the same as those in the 1950's!

3. The reference is to W. P. Hunt's expedition and the establishment of Astoria. The sale of that post to the British during the War of 1812 is mentioned. It was in 1819, three years after the visit of the *Rurick* to San Francisco, that Spanish claims in the Oregon region were withdrawn.

caped me, and did not wish to get his feet wet. So much may one expect by relying on one's friends!"

"Don Pablo Vincenti, as he came to our tent from the Presidio . . . brought a present *á su amigo Don Alberto*—a flower which he had picked on the way, and which he gave me ceremoniously. It happened to be our wild tansy, or silver leaf *(Potentilla anserina),* and could not bloom more beautifully, even in Berlin [and it could not have bloomed more perfectly near Berlin.]"

Louis Choris' *Voyage pittoresque autour du monde* . . . , published in Paris in 1822, supplies some general information regarding the California visit but, although thirteen pages of text and ten colored plates are devoted to the region, no plants are mentioned save those growing at the Mission of San Francisco. The frontispiece is a portrait of ". . . le Comte N. Romanzoff, Chancelier de l'Empire, etc. etc." One plate includes two pleasing sketches: one of the "Cap de los Reyes à 6 miles de distance," the other "L'entrée du Port de Sn Francisco à 7 miles de distance."

The collections made in California by von Chamisso and Eschscholtz were published for the first time in three different periodicals:

(1) In *Horae physicae Berolinenses* . . . edited by C. G. Nees von Esenbeck. Here, in 1820, von Chamisso first described (in addition to the two genera *Romanzoffia* and *Euxenia*) the everywhere-admired California poppy, *Eschscholtzia californica;* here the generic name was spelled *Eschscholzia.*[4] The description was accompanied by a beautiful colored plate (t. XV)—"Prof. P. Guimpel ad nat. pinx. et sculp."—on which

4. Modern nomenclatorial rules call for the adoption of this original spelling, rather than the spelling *Eschscholtzia* now in common use among the botanists.

To give the reader a glimpse of the complexities of botanical nomenclature I mention the following: Lasègue (who refers to the California poppy correctly, as *"Eschscholzia californica"*) has this to say, and I translate:

"Mr. Lindley has nevertheless proposed to change the name *Eschscholzia* to *Chryseis* [meaning golden yellow] not only because the last seems to him more euphonious, but because the name *Elsholtzia,* so written by mistake, had been already given to another plant named in honor of Dr. Eschscholz's father, a double usage which is contrary to the rule established by the descriptive-botanists."

According to J. C. Willis, *Elsholtzia* Willd. (a genus of the *Labiatae*) still stood in 1931. This name, commemorating a different person and being published with a different spelling, is considered by botanists to be a totally different generic name from the one applied to the California poppy. Accordingly both *Elsholtzia* and *Eschscholzia* are properly used in modern botanical writing. And this despite the fact that, presumably, both father and son spelled their family name in the same way—*Eschscholtz*—which is not the spelling properly used to commemorate either!

Since, as recorded by Lasègue above, John Lindley attempted to change the name for a more "euphonious" one, I take pleasure in quoting his comments upon the naming of plants; they appear in his preface to *The Vegetable Kingdom* (xv-xviii, 1853):

"There is still another point in which the Author has endeavoured to effect some improvement, and that is the nomenclature. Since the days of Linnæus, who was the great reformer of this part of Natural History, a host of strange names, inharmonious, sesquipedalian, or barbarous, have found their way into Botany, and by the stern but almost indispensable laws of priority are retained there. If is full time, indeed, that some stop should be put to this torrent of savage sounds, when we find such words as Calucechinus,

the name was spelled *Eschholzia*. The genus was dedicated to von Chamisso's companion Dr. Eschscholtz[5]: "*Nomen* in honorem dixi consortis omnium laborum in itinere solertissimi, doctissimi, amicissimi Eschscholzii, Medicinae Doctoris, Botanicae aeque ac Etomologiae pertissimi." The plant's habitat was described as ". . . in arenis sterilibus siccis ad portum *Sancti Francisci Californiae.*"

In order to secure wider circulation of the description of *Eschscholtzia,* von Chamisso republished it, with minor changes, in *Linnaea* in 1826, stating: "Quam in Hor. phys. Berol. Bonnae 1820 editis divulgavimus descriptionem hic repetendam censuimus, quum illud opus paucis, nec inter botanicos satis notum appareat."

When first describing the California poppy in 1820 von Chamisso had noted that its seed had been planted: "Nunc, semine adlato, in hortis nostris, favente coelo, hospitabitur." In republishing the description in 1826 he reported that it had not grown satifactorily[6]: "Semen a nobis allatum spem fefellit, nec plantas produxit, quae gratum fuissent botanicis donum."

(2) In the *Mémoires de l'Académie Impériale des Sciences de St. Pétersbourg,* in 1826, Eschscholtz published his "Descriptiones plantarum novae californiae . . ."

Oresigenesa, Finaustrina, Kraschenninikovia, Gravenhorstia, Andrzejofskya, Mielichoferia, Monactineirma, Pleuroschismatypus, and hundreds of others like them, thrust into the records of Botany without even an apology. If such intolerable words are to be used, they should surely be reserved for plants as repulsive as themselves, and instead of libelling races as fair as flowers, or noble as trees, they ought to be confined to Slimes, Mildews, Blights, and Toadstools. The Author has been anxious to do something towards alleviating this grievous evil, which at least need not be permitted to eat into the healthy form of Botany clothed in the English language.

"No one who has had experience in the progress of Botany, as a science, can doubt that it has been more impeded in this country by the repulsive appearance of the names which it employs than by any other cause whatever . . ."

Lindley has considerably more to suggest on the subject and ends his comments with a plea for the use of "standard English names for Classes and Orders:"

". . . The author confidently believes that every intelligent reader will admit that such names as Urnmosses, Taccads, False Hemps, Pepperworts, Bristleworts, Chenopods, Hydrocharads, Scale-mosses, Birthworts, and Fringe-Myrtles are preferable to Bry-a-ce-æ, Tac-ca-ce-æ, Da-tis-ca-ce-æ, El-a-ti-na-ce-æ, Cheno-po-di-a-ce-æ, Des-vaux-i-a-ce-æ, Hy-dro-cha-ri-da-ce-æ, Jun-ger-man-ni-a-ce-æ, A-ris-to-lo-chi-a-ce-æ, Cha-mæ-lau-ci-a-ce-æ, and other sesquipedalian expressions."

To go back to the name "Kraschenninikovia" mentioned by Lindley above, the botanist Rafinesque offered at least five variants, possibly more, but I become bewildered: Kraschenikofia, Krascheninikofia, Kraschennikofia, Krasnikovia; *no;*—the fifth was Krascheninnikovia *Gueldensteeden,* not *Rafinesque!* See Merrill, Elmer D. 1949. *Index Rafinesquianus,* p. 271 (Index).

5. Essig notes that von Chamisso ". . . little thought of the honor he was bestowing upon his friend and himself in thus describing what afterwards became the state flower of the then sleeping Golden State."

6. Mahr comments upon this: ". . . As we see in the foregoing passage from *Linnaea* . . . the seeds did not produce satisfactory specimens. In the meantime, however, *Eschscholtzia californica* has been successfully cultivated in the gardens of Germany. I have seen it in wonderful specimens in the Vogelsberg Mountains (60 miles north of Frankfurt am Main), where it has adapted itself to the most unfavorable conditions, both of the climate and of the soil. The color was the same as the flower shows in California, while in the shops of German florists it is found in numerous varieties of color, from lemon yellow to a deep russet."

Following the title is the important notation: "Conventui exhibuit die 18. Junii 1823," indicating that this paper had been presented, or perhaps read, on June 18, 1823, or about one month before Eschscholtz started on his second voyage with von Kotzebue, a trip which took him to California for the second time in 1824. The plants described in his paper must, therefore, have been collected on Eschscholtz's first visit to California in 1816 although publication of his paper in the *Mémoires* was delayed until 1826, some three years after the presentation (or perhaps reading). When it appeared (ser. 5, vol. 10) it was along with numerous other papers on a variety of subjects such as cockroaches, bats, longitudes of Astrakan, etc., etc., which had been presented from two to forty-six years before they finally appeared in print. No reference has been found which indicates that Eschscholtz revised his original paper after he had been in California in 1824.

Eschscholtz's "Descriptiones plantarum . . ." was reprinted (in abstract) in the *Litteratur-Bericht zur Linnaea für des Jahr 1828* (*Linnaea* 3: 147-153, 1828). But in the reprint the important statement that the paper had been presented in 1823 was omitted. This omission, plus the fact, perhaps, that no dates of collection are cited by Eschscholtz, may be responsible for the conclusion, reached by one author at least, that the plants described were collected on Eschscholtz's second visit to California in 1824, which does not seem to have been the case.

(3) In *Linnaea. Ein Journal für die Botanik* . . . edited by Diedericus F. L. von Schlechtendal,[7] von Chamisso's California collections were published from 1826 to 1836 inclusive. Included were not only his new discoveries but, as well, rediscovered plants previously described from the same region or from elsewhere by other botanists. California plants are not segregated from those from other regions.

California plants are to be found in volumes 1 (1826), 2 (1827), 3 (1828), 4 (1829), 6 (1831) and 10 (1836) of *Linnaea*. The first article (volume 1) of the series in which they appeared was entitled "De plantis in expeditione speculatoria Romanzoffiana observatis rationem dicunt Adelbertus de Chamisso et Diedericus de Schlechtendal"; this title varies slightly in later volumes (in 2, 3, 4, 6), the words "rationem dicunt" changed to "dissere pergunt." In volume 10 the title is "De plantis in expeditione speculatoria Romanzoffiana et in herbariis regiis Beroliensibus observatis dicere pergit Adelbertus de Chamisso."

Von Chamisso turned over to other botanists certain families of plants which he had collected: we find plants of the *Labiatae* described by George Bentham (*Linnaea* 6, 1831), of the "Synantherae" or *Compositae* by Christian Friedrich Lessing (*Linnaea* 6, 1831) and of the *Leguminosae* by Theodor Vogel (*Linnaea* 10, 1835–1836; issued in 1836). Still other botanists—Schiede, De Gingins, etc.—are said by von Schlechtendal to have described plants collected by von Chamisso, but their publications do not appear to have included plants from California.

---

7. Diedrich Franz Leonhard von Schlechtendal.

When, in 1932, August C. Mahr published "The visit of the 'Rurik' to San Francis-co in 1816," he carefully segregated all the plants from California described in *Linnaea,* listing them in the order issued,[8] with volume and page references[9] and with various enlightening comments. Where new genera and species are described Mahr quotes the original descriptions in full; species collected in California but previously described by other botanists are also cited but without description. Included are ". . . the mod-ern names of all the plants mentioned, according to the nomenclature used by W. L. Jepson in his *Manual of the Flowering Plants of California* (Berkeley, California, 1923, 1925), or his descriptions."

Also—inserted by Mahr "in the places where Chamisso mentions them in his tax-onomy"—are ". . . Eschscholtz' descriptions of two species [out of his twelve novel-ties], published in *Mémoires* de l'Académie des Sciènces de St. Pétersbourg, Vol. X, pp. 283-284 . . ." These are *Lonicera Ledebourii* and *Ribes tubulosum* which had appeared in Eschscholtz's "Descriptiones plantarum," already mentioned under (2) above. Mahr states: "The descriptions of two new species of California plants and of a California butterfly, are the only contributions by J. F. Eschscholtz to this compila-tion of material concerning the visit of the 'Rurik' to San Francisco." A statement which can be true only if my supposition (that Eschscholtz's plants were all collected in 1816) is false.

In 1944 Miss Alice Eastwood published a short article entitled "The botanical col-lections of Chamisso and Eschscholtz in California." In this she lists the species col-lected by the two botanists, but omits the descriptions. The plants collected by von Chamisso and described in *Linnaea* are separated into two lists, the first including "new species," the second "Species also . . . collected, not considered as new . . ."

The author noted of von Chamisso's collections: "The locality cited was 'ad portum San Francisco' and all were collected, according to Chamisso, in the hills and downs about the Presidio. Very few if any are to be found there to-day." Further: "On the first expedition, during the month of October, sixty-nine species were collected. Among them were two new genera and thirty-three species. Three were synonyms having been previously described by other authors."

Further: "On the second expedition [1824], Eschscholtz named and described thirteen species, three among them previously described." As already noted, Esch-scholtz's "Descriptiones plantarum" . . . published in the *Mémoires* of the St. Peters-burg Academy had been presented in 1823 or before he started on his second expedi-tion—although this fact is not mentioned in the *Linnaea* abstract—and the plants must, therefore, have been collected on his visit of 1816. The error is understandable

8. It is stated in *Linnaea* (1: 35, *fn.*, 1826): "Sequimur in hac et sequentibus familis Candollii in Systemate vegetabilis propositum ordinem."

9. The pages in *Linnaea* cited by Mahr are not always in accord with those in the volumes of *Linnaea* examined by me where the pagination is sometimes peculiar. I am told that it was published only once.

since Miss Eastwood states that "Linnæa has been my source of information concerning these important collections."

Eastwood notes that "Eschscholtz does not give the exact place where his collections were made, but from the plants collected all but one could have come from San Francisco."

Asa Gray—writing in 1840 of the collections in the ". . . royal Prussian herbarium . . . deposited at Schöneberg (a little village in the environs of Berlin) opposite the royal botanic garden and in the garden of the Horticultural Society . . ."—reported that the general herbarium contained the ". . . botanical collections made by Chamisso . . . many . . . from the coast of Russian America and from California . . ."

In 1880 Alphonse de Candolle, in *La phytographie,* records the whereabouts of collections made on the voyage of the *Rurick* thus:

"Chamisso (de). Herb de 10 à 12,000 esp., à l'Académie imp. des sci. de Saint-Pétersbourg. Des doubles dans les herb. de Berlin[1] et de L'Univ. de Kiel."

"Choris. Herb. du Jardin imp. de Saint-Pétersbourg."

"Eschscholtz. Plantes de l'expéd. de Kotzebue, au jardin imp. de Saint-Pétersbourg. (1,300 esp.)."

After von Chamisso's death in 1837, his friend von Schlechtendal published a tribute to his memory in *Linnaea* (1839). I cite from the rather poor English translation published in the *London Journal of Botany* in 1843:

Von Chamisso—". . . very much a self-taught Botanist . . ."—first began to study plants at Copet, near Lake Geneva, Switzerland; this was near the home of Madame de Staël, whose son, Baron Auguste von Staël Holstein, first interested him in botany and was his companion on collecting excursions. Von Schlechtendal first became acquainted with von Chamisso in 1813. He describes their botanical trips and ". . . the many incidents of these herborizations . . . Chamisso was ever the foremost . . . An antique garb, once the state dress of a South Sea Chief, much worn, mended and stained, with a black cap of cloth or velvet, a large green box suspended by leather straps over his back, and a short pipe in his mouth, together with a rude tobacco pouch: such was the attire in which he sallied forth . . . when evening came . . . weary, travel-soiled, he did not make a very splendid appearance while bearing a pocket handkerchief crammed with plants, he met, on returning to Berlin, the *beau monde* . . . all in their Sunday attire . . ."

When on the *Rurick,* "The only individual who entered at all into his tastes, though he possessed not the same energy in collecting, was Eschscholtz. He too, gathered some plants and profited by the liberality of Chamisso, who exchanged duplicates and gave him specimens . . . Eschscholtz himself described only a few of his specimens . . . Chamisso was obliged to publish his collections at his own cost. Returning to Prussia . . . he presented the zoological and mineralogical portion to the University Museum at Berlin . . ."

1. Destroyed in World War II.

In 1819 von Chamisso was given the honorary degree of Doctor of Philosophy by the University of Berlin and was appointed "Assistant in the Berlin Botanical Institution," [2] and directed to ". . . pay particular attention to the Botanical Garden." After fire had destroyed his home at Neu Schöneberg he moved to Berlin where he and von Schlechtendal worked together in the "Royal Herbarium which contained Willdenow's collections." When the latter began his "exclusively *Botanical Journal, the Linnæa,*" von Chamisso started work on his own collections. Discussing his custom of sharing his collections with other botanists, the biographer comments:

"Happily this noble spirit of liberality is gaining ground among Botanists and superseding the narrow minded avarice with which naturalists were too apt to keep to themselves every thing but their opinions and dogmas on science . . ."

He states that ". . . only an inconspicuous looking plant among the *Amaranthaceæ,* described by his friend Künth, bears his name[3]. . ."

Mahr's introduction supplies an understanding picture of von Chamisso. He suggests that the romance, *Peter Schlemihl* [4] (which ". . . secured for its author a lasting place in German literature and became known all over the world in numerous translations . . .") was in a sense symbolic of von Chamisso's own "tragic condition of disconnectedness, of being a man without a country." Born of noble parentage in Champagne, France, in 1781, he was, when nine years of age, exiled with his family during the French Revolution; they finally made their home in Berlin. As a boy von Chamisso served as a page in the household of the royal family of Prussia and then entered the army. But he was never happy in army life; nor, when he visited his family after their return to France, did he ever feel at home in the land of his birth.

"When in 1813 the War of liberation broke out, Chamisso again became painfully conscious of his ambiguous national position; although he fully approved of Prussia's strife for liberty from French oppression, yet he found it impossible to bear arms against his native country."

After Napoleon's return from Elba, von Chamisso ". . . found himself in similar patriotic difficulty." It was then that he received his appointment as naturalist with von Kotzebue's expedition.

Aged fifty-six, he died on August 21, 1837, at his home in Berlin.

According to Mahr, "Johann Friedrich (Ivan Ivanovitch) Eschscholtz was born at Dorpat, in the Baltic Province of Russia, on November 12, 1793. He studied medicine and zoölogy and acquired the degree of *Medicinae Doctor*. He traveled on the 'Rurick' as ship's surgeon, and also accompanied Kotzebue on his second expedition, from 1823 to 1826 . . . He became professor of zoölogy at Dorpat, where he died in 1831."

2. Berlin's Royal Botanical Gardens, according to Mahr.

3. The genus *Chamissoa*, named by Humboldt, Bonpland and Kunth.

4. The rare first edition (1814) of this book is in the Boston Athenæum. The title-page reads: *Peter Schlemihl's wunderfame Geschichte mitgetheilt von Adelbert von Chamisso und herausgegeben von Friedrich Baron de la Motte Fouqué. Mit einem Kupfer Nürnberg bei Johann Leonhard Schrag 1814.*

Essig, referring to Eschscholtz's visit to California in 1816, notes that he turned ". . . his botanical collections over to Chamisso and was able to secure little of an entomological nature. What he did take was turned into [in to] the Imperial Museum at Moscow where it was later described by others." Eschscholtz's primary interest was entomology and Essig has more to tell of his work in that field when writing of von Kotzebue's visit to California in 1824.

ALDEN, ROLAND HERRICK & IFFT, JOHN DEMPSTER. 1943. Early naturalists in the far west. *Occas. Papers Cal. Acad. Sci.* 20: 21-28.

BANCROFT, HUBERT HOWE. 1885. The works of . . . 19 (California II. 1801–1824): 278-281.

BRETSCHNEIDER, EMIL. 1898. History of European botanical discoveries in China. London. 321.

BREWER, WILLIAM HENRY. 1880. List of persons who have made botanical collections in California. *In:* Watson, Sereno. Geological survey of California. Botany of California 2: 554.

CANDOLLE, ALPHONSE DE. 1880. La phytographie; ou l'art de décrire les végétaux considérés sous différents points de vue. Paris. 403, 410.

CHAMISSO, ADELBERT VON. 1820. Ex plantis in expeditione Romanzoffiana detectis genera tria nova offert Adalbertus de Chamisso, Ph.D. Soc. Nat. Cur. Ber. at Nat. Sci. Moscov. Sodal. Cet. Cum tabulis aeneis III. *In:* Esenbeck, Christian Gottfried Nees von, *editor.* Horae physicae Berolinenses collectae ex symbolis virorum doctorum . . . Bonnae. 73-76. 3 pl.

———— 1821. Remarks and opinions of the naturalist of the expedition, Adelbert von Chamisso. *In:* Kotzebue, Otto von. 1821. A voyage of discovery . . . 2: 351-433. 3: 1-318 [*California* 3: 38-51]. London.

———— 1842. Werke. 6 vols. ed. 2. Leipzig. [ed. 1, Leipzig, 1836.] [Contains: "Tagebuch von Adelbert von Chamisso von Unalaska nach Kalifornien, aufenthalt zu San Francisco," which was translated by McElroy, G. D. 1873. A visit to San Francisco . . . and reprinted (from McElroy as well as from the original German) in Mahr, A. C. 1932. The visit of the "Rurik" . . .]

———— 1873. A visit to San Francisco in 1816. From the German of Adelbert von Chamisso. [Translated by Georgie D. McElroy.] *Overland Monthly* 10: 201-208. *Reprinted:* Mahr, A. C. 1932. The visit of the "Rurik" to San Francisco . . . 29-51.

———— & SCHLECHTENDAL, DIEDRICH FRANZ LEONHARD VON. 1826–1836. De plantis in expeditione speclatorio Romanzoffiana observatis rationem dicunt . . . *Linnaea* 1-10. [For Californian plants *see* 1, 2, 3, 4, 6, 10. The title varies slightly in each volume.]

CHORIS, LOUIS. 1822. Voyage pittoresque autout du monde, avec des portraits de sauvages d'Amérique, d'Asie, d'Afrique, et des iles du grand ocean; des paysages, des vues maritimes, et plusieurs objects d'histoire naturelle; accompagné de descriptions par M. le Baron Cuvier, et M. A. Chamisso, et d'observations sur les crânes humains par M. le Docteur Gall, par M. Louis Choris, peintre. Paris.

EASTWOOD, ALICE. 1939. Early botanical explorers on the Pacific coast and the trees they found there. *Quart. Cal. Hist. Soc.* 18: 338.

—— 1944. The botanical collections of Chamisso and Eschscholtz in California. *Leafl. West. Bot.* 4: 17-21.

ESCHSCHOLTZ, JOHANN FRIEDRICH. 1826. Descriptiones plantarum novae californiae, adjectis florum exoticorum analysibus auctore J. Fr. Eschscholtz. Conventui exhibuit die 18. Juni 1823. *Mém. Acad. Imp. Sci. St. Pétersburg* 10: 281-289. *Reprinted* [in abstract and omitting the notation "Conventui exhibuit die 18. Juni 1823"]: *Linnaea,* Litteratur-Bericht 3: 147-153, 1828.

ESSIG, EDWARD OLIVER. 1933. The Russian settlement at Ross. *Quart. Cal. Hist. Soc.* 12: 207. *Reprinted: Cal. Hist. Soc. Spec. Publ.* 7: The Russians in California. (1933)

GRAY, ASA. 1840. Notices of European herbaria, particularly those most interesting to the North American botanist. *Am. Jour. Sci. Arts* 40: 1-18. *Reprinted:* Sargent, Charles Sprague, *compiler.* 1889. Scientific papers of Asa Gray. 2 vols. Boston. New York. 2: 1-21.

KOTZEBUE, OTTO VON. 1821. Entdeckungs-Reise in die Süd-See und nach der Berings-Strass zur Erforschung einer nordöstlichen Durchfahrt. Unternommen in den Jahren 1815, 1816, 1817 und 1818, auf Kosten Sr. Erlaucht des Herrn Reichs-Kanzlers Grafen Rumanzoff auf dem Schiffe Burick unter dem Befehle des lieutenants der Russisch-Kaiserlichen Marine Otto von Kotzebue. 3 vols. Weimar.

—— 1821. A voyage of discovery into the South Sea and Beering's Straits, for the purpose of exploring a north-east passage, undertaken in the years 1815–1818, at the expense of His Highness the Chancellor of the Empire, Count Romanzoff, in the ship Rurick, under the command of the Lieutenant in the Russian Imperial Navy, Otto von Kotzebue. 3 vols. London. [Translated by Hannibal Evans Lloyd from the Weimar edition. 1821.]

LASÈGUE, ANTOINE. 1845. Musée botanique de M. Benjamin Delessert. Notices sur les collections de plantes et la bibliothèque qui le composent; contenant en outre des documents sur les principaux herbiers d'Europe et l'exposé des voyages entrepris dans l'intérêt de la botanique. Paris. 212-213, 371-375.

MAHR, AUGUST CARL. 1932. The visit of the "Rurik" to San Francisco in 1816. *Stanford Univ. Publ. Univ. Ser.; History, Economics & Political Science* 2: no. 2. 194 pp.

SARGENT, CHARLES SPRAGUE. 1891. The Silva of North America. Boston. New York. 2: 39, *fn.* 4.

SCHLECHTENDAL, DIEDRICH FRANZ LEONHARD VON. 1839. Dem Andenken an Adelbert von Chamisso als Botaniker. *Linnaea* 13: 93-106. [English translation published in *London Jour. Bot.* 2: 483-491, 1843.]

WILLIS, JOHN CHRISTOPHER. 1931. A dictionary of the flowering plants and ferns. ed. 6. Cambridge, England. 240.

CHAPTER VIII

NUTTALL ASCENDS THE ARKANSAS RIVER TO FORT

SMITH AND FROM THERE MAKES TRIPS INTO EASTERN

OKLAHOMA, THEN SET APART FOR INDIAN OCCU-

PANCY

THOMAS Nuttall's second journey into regions west of the Mississippi River took place soon after publication of his *Genera of North American plants* in 1818. The trip terminated in February, 1820—of our next decade—but by January, 1820, the botanist was hastening to New Orleans and collecting little. The significant part of his trip, as the title of his journal indicates, belongs in the present decade and is all included here. Nuttall was the first to make plant collections in what are now the states of Arkansas and Oklahoma.

He described the journey in *A journal of travels into the Arkansa Territory during the year 1819. With occasional observations on the manners of the aborigines,* published in Philadelphia in 1821; included was "A map of the Arkansas River, intended to illustrate the Travels of Thos. Nuttall: constructed from his original manuscripts by H. S. Tanner."

R. G. Thwaites reprinted this journal in 1905 (volume 13 of his *Early western travels*), adding many interesting and helpful footnotes. F. W. Pennell's "Travels and scientific collections of Thomas Nuttall" (1936) gives the itinerary in concise form.

Nuttall made the trip on foot, by stage, by skiff and by flatboat, using any means of transportation available and, except for any assistants procurable at the moment, traveled alone. He left Philadelphia on October 2, 1818, and arrived at the confluence of the Ohio and Mississippi rivers on December 18, and our interest in his journey begins at this point. The Mississippi was filled with ice at this season and its descent was slow and difficult. Nor was it a suitable time for collecting plants; and, doubtless for this reason, there are few references to the flora although trees are mentioned occasionally. I cite from the Philadelphia edition (1821) of Nuttall's *Journal.*

1818

On December 20 Nuttall was near the "Iron-banks (called Mine au Fer by the French)," on the eastern side of the Mississippi and some twenty miles below the entrance of the Ohio River.

"The cyprus *(Cupressus disticha)* which continues some distance along the Ohio above its estuary, is here much more common, and always indicates the presence of annual inundation and consequent swamps and lagoons, but we do not meet with the long moss *(Tillandsia usneoides)*, a plant so characteristic of the prevalence of unhealthy humidity[1] in the atmosphere."

By evening of the 21st Nuttall was ten miles above the French settlement of New Madrid where the devastation caused by the earthquakes of 1811 was still noticeable.

"The river here appears truly magnificent, though generally bordered by the most gloomy solitudes, in which there are now no visible traces of the abode of man."

New Madrid, reached on the 22nd, was ". . . an insignificant French hamlet, containing . . . about 20 log houses and stores miserably supplied, the goods of which are retailed at exhorbitant prices . . ." At Point Pleasant, six miles below New Madrid and on the west side of the Mississippi, Nuttall, on December 23, saw ". . . the Catalpa *(Catalpa cordifolia)* in the forests, apparently indigenous, for the first time in my life, though still contiguous to habitations."

In a footnote he adds:

"In the immediate vicinity of the town I met with *Bœbera glandulosa, Erigeron (Cœnotus) divaricatum, Verbena stricta, V. Aubletia, Croton capitatum,* and *Helenium quadridentatum.* On the banks of the river *Oxydenia attenuata* and the *Capraria multifida* of Michaux."

Nuttall left Point Pleasant on Christmas Day and reached Little Prairie, on the west side of the Mississippi. All but one house had been swept away by the river which had risen after the earthquakes. Progress southward is recorded in names of islands and bluffs noted along the river.[2] On December 31, beneath coal beds at "the second Chicasaw Bluff," Nuttall observed in ". . . a friable bed of dark-coloured argillaceous and sandy earth . . . blackened impressions of leaves of an oak, like the red oak and the willow oak, with *Equisetum hiemale* or Shave-rush, and other vegetable remains . . ."

1819

On January 4 wind made it necessary for the boat to ". . . come to under fort Pickering . . ." near "the fourth Chicasaw Bluffs." Here a store bought articles from the Indians at a minimum and sold them to the settlers at an exhorbitant advance in price: ". . . whiskey, well watered . . ." was sold ". . . almost without restraint, in spite of the law, two dollars per gallon, and every thing else in the same proportion."

1. A similar statement appears in the *Journal* under date of January 28, 1820. Oakes Ames wrote me: "I doubt that Spanish moss indicates an unhealthy atmospheric humidity."

2. Nuttall was doubtless familiar with *The Navigator* published in Pittsburgh by the firm of Zadok Cramer, a guide book which went through a number of editions in the decade of 1810–1820, some of which are in the Boston Athenæum. Its maps reproduce sections of the meanderings of the Mississippi and other rivers with towns, physical features, etc., indicated thereon. The same library also possesses a rare little book published in 1810: F. Cuming's *Sketches of a tour to the western country . . . A voyage down the Ohio and Mississippi rivers . . .* —a work which evidently made a large contribution to the facts included in *The Navigator.* It was likewise published in Pittsburgh, by the firm of Cramer, Spear & Eichbaum.

"On the river lands I here first noticed the occurrence of *Brunichia, Quercus lyrata.* and *Carya aquatica (Juglans,* Mich.)"

On January 6 and 7 Nuttall was near the mouth of St. Francis River which empties into the Mississippi from the west, about nine miles above Helena in northeastern Arkansas. There were a few settlers.

"How many ages may yet elapse before these luxuriant wilds of the Mississippi can enumerate a population equal to the Tartarian deserts! At present all is irksome silence and gloomy solitude, such as to inspire the mind with horror."

A short distance above Big Prairie, consisting of four abandoned log cabins, Nuttall expresses his disappointment ". . . to meet with such a similarity in the vegetation, to that of the middle and northern states. . . . higher lands produce black ash, elm . . . hickory, walnut, maple, hackberry (*Celtis integrifolia,* no other species), honey-locust, coffee-bean, &c. On the river lands, as usual, grows plantanus or buttonwood, upon the seeds of which flocks of screaming parrots were greedily feeding,[3] also enormous cotton-wood trees *(Populus angulisans),* commonly called yellow poplar,[4] some of them more than six feet in diameter, and occasionally festooned with the largest vines which I had ever beheld. Here grew also the holly *(Ilex opaca), Aplectrum hiemale, (Ophrys hyemale,* Lin.), *Botrychium obliquum,* and *Fumaria aurea.* Nearly all the trees throughout this country possessing a smooth bark, are loaded with mistletoe *(Viscum verticillatum).*"

In the region of Big Prairie, on January 8:

"The scrub-grass or rushes, as they are called here *(Equisetum hiemale),* from about 50 to 60 miles above, to this place, appear along the banks in vast fields, and, together with the cane, which is evergreen, are considered the most important, and indeed, the only winter fodder for all kinds of cattle. The cane is unquestionably saccharine and nutritious, but the scrub-grass produces an unfavourable action on the stomach, and scours the cattle so as to debilitate and destroy them if its use be long continued."

On January 9 Nuttall saw one ". . . rude cabin . . . the only habitation . . . seen for 30 miles." On the 11th, nearing White River settlement, he ascended a bank and ". . . found the woods almost impenetrably laced with green briars *(Smilax),* supple-jacks *(Œnoplia volubilis),* and the *Brunichia,* and for the first time recognised the short podded honey-locust *(Gleditscia brachycarpa),* a distinct species, intermediate with the common kind *(G. triacanthos),* and the one-seeded locust *(G. monosperma),* differing from *G. triacanthos* in the persisting fasciculated legumes, as well as in their shortness and want of pulp."

On the 12th, at a "Mr. M'Lane's, a house of entertainment," Nuttall was advised to

---

3. A footnote adds that "Their most favourite food in the autumn is the seeds of the cuckold bur *(Xanthium strumarium).*"

4. The tulip tree, *Liriodendron tulipifera* Linn., is often called yellow poplar. Under date of January 21, Nuttall refers to the fact that this tree is not found on the banks of the Arkansas River.

proceed, by flatboat, ". . . to the port of Osark, on the Arkansa, by the bayou, which communicates between the White and Arkansas rivers . . ." It is clearly shown (Map VI, p. 109) in *The Navigator,* 6th edition, 1808. On the 13th, at the mouth of White River, some fourteen miles above the mouth of the Arkansas, Nuttall—obliged to await "the drunken whim" of a Yankee boatman whom he had hired out of necessity —endeavored to amuse himself "by a ramble through the adjoining cane-brake." Here he found abundance of the *Celtis integrifolia* and one-seeded honey-locust; ". . . also *Forrestiera acuminata* of Poiret (*Borya acuminata,* Wild.) . . . the *Senecio laciniata,* so common along the banks of the Mississippi, already showed signs of flowering."

For the next four months (January 14–May 15) Nuttall's travels were in Arkansas.

On January 14 he and the boatman ". . . proceeded up White river with considerable difficulty and hard labour . . ." The latter, before leaving, had increased his financial demands, and did not improve upon acquaintance: he would only steer, leaving his employer and another helper to struggle with the oar and the "cordelle" or towrope. They made no progress. In the woods Nuttall was ". . . shewn a scadent leguminous shrub, so extremely tenacious as to afford a good substitute for ropes, and commonly employed as a boat's cable. A knot can be tied of it with ease. On examination I found it to be the plant which I have called *Wisteria speciosa* (*Glycine frutescens.* Willd.) the Carolina kidney-bean tree."[5]

On the 15th they succeeded in passing through the bayou, eight or nine miles long, connecting the White and Arkansas rivers, and the next day began the ascent of the last. Water was very low and the unwieldy boat had to be dragged by the "cordelle," at a "very tedious and tiresome rate." By the evening of the 16th they were only six miles above the bayou. On the 18th Nuttall comments:

"No change, that I can remark, yet exists in the vegetation, and the scenery is almost destitute of every thing which is agreeable to human nature; nothing yet appears but one vast trackless wilderness of trees . . . All is rude nature as it sprang into existence . . ."

On January 19 they reached a small French settlement and took lodgings at a "Madame Gordon's." On the 20th Nuttall wrote:

". . . the weather . . . has felt to me like May . . . The birds had commenced their melodies; and on the high and open bank of the river near to Madame Gordon's, I had already the gratification of finding flowers of the same natural family as many of the early plants of Europe; the Cruciferæ; but to me they were doubly interesting, as the first fruits of a harvest never before reaped by any botanist."

Nuttall thought of taking passage with two ". . . elderly men out on a land speculation, who intended to ascend the river as far as the Cadron . . . 300 miles from hence by water, or to the Fort . . . 350 miles further . . ." But, since they "merely con-

5. He adds in a footnote: "The name of *Thyrsanthus,* given by Mr. Elliott, has been already employed for another genus."

descended" to let him accompany them and since he would have had to feed himself and work as a boatman, he decided against the plan. They ". . . appeared to be illiterate men, and of course . . . incapable of appreciating the value of science."

On January 21 he took a solitary ramble down the river's bank; and ". . . found along its shelving border, where the sun obtained free access, abundance of the *Mimosa glandulosa* of Michaux; also *Polypremum procumbens, Diodia virginica, Verbena nodiflora,* Lin. *Eclipta erecta,* Mich. *Poa stricta, Panicum capillaceum. Poa reptans* as usual in vast profusion, and *Capraria multifida.* The trees and shrubs are chiefly the Pecan, *(Carya olivæformis) C. aquatica;* the black walnut, *(Juglans nigra),* but very rare; *Fraxinus quadrangulata, Liquidamber* and *Platanus,* but rarely large or full grown; also *Celtis integrifolia;* the swamp oak *(Quercus aquatica),* nearly sempervirent, the red oak *(Q. rubra),* the scarlet oak *(Q. coccinea),* Spanish oak *(Q. falcata); Populus angulisans,* the cotton wood, of greater magnitude than any other tree in this country, with the wood yellowish, like that of the Tulip tree, answering the purpose of fence rails, and being tolerably durable. The smaller white poplar *(P. monilifera),* never so large as the preceding, commonly growing in groves like the willows, and presenting a bark which is white and even. Different kinds of honey locust, as the common species *Gleditscia triacanthos,* the one-seeded *G. monosperma,* and the short podded *G. brachyloba.* There is no sugar-maple, as I understand, nearer than the upper parts of the St. Francis and White river . . . The tulip tree *(Lyriodendron tulipifera)* . . . is not met with on the banks of the Arkansa."

Nuttall made a trip, on foot, from Madame Gordon's to ". . . the Post, now the town of Arkansas . . ." He left on January 22 and returned the next day. On the way he entered an oak swamp.

"The species are principally *Quercus lyrata, Q. macrocarpa* (the over-cup oak); *Q. phellos* (the willow oak); *Q. falcata* (the Spanish oak); and *Q. palustris* (the swamp oak); with some red and scarlet, as well as black and post oak on the knolls, or more elevated parts. In this swamp, I also observed the *Nyssa aquatica, N. pubescens* (Ogechee lime, the fruit being prepared as a conserve), as well as *N. biflora,* and *Gleditscia monosperma."*

After finally emerging from the "horrid morass" in which he had waded ankle-deep, Nuttall reached a "delightful tract" of high land where the fields of the settlers were a vivid green and the ground ". . . appeared perfectly whitened with the *Alyssum bidentatum.* The *Viola bicolor,* the *Myosurus minimus* of Europe, (probably introduced by the French settlers) and the *Houstonia serpyllifolia* of Michaux, (*H. patens* of Mr. Elliott) with bright blue flowers, were also already in bloom."

Nuttall was always conscious of the birds and notes that they ". . . were singing from every bush, more particularly the red bird . . . and the blue sparrow . . ."

Arkansas Post boasted some thirty or forty houses; it owed much to the enterprise of a "Monsieur Bougie." Ghent refers to him as Joseph Bogy. Thwaites cites his Christian name as Charles. Nuttall presented a letter of introduction.

"I soon found in him a gentleman, though disguised . . . in the garb of a Canadian boatman. He treated me with great politeness and respect, and, from the first interview, appeared to take a generous and active interest in my favour."

The Post was in a region of vast Spanish grants, with unsettled titles; since they comprised most of the best land, they were a deterrent to settlement; one, held by "Messrs. Winters, of Natchez," was no less than one million acres in extent. Nuttall felt that "Nature has here done so much, and man so little, that we are yet totally unable to appreciate the value and resources of the soil."

On January 23 he got back to Madame Gordon's after wading through "enswamped forests" and comments that "There are . . . alligators, though by no means numerous." He spent the 24th and 25th moving to "Monsieur Bougie's" in his own flatboat; he parted with it on arrival ". . . with a sort of regret . . . with all its difficulties . . . [it] had afforded . . . no inconsiderable degree of comfort and convenience." On the 26th he moved, in the pirogue of "Monsieur Bougie," to Arkansas Post which was to serve him as headquarters for about one month, or until February 25. The "insignificant" village was "destitute even of a hatter, a shoe-maker, and a taylor . . . containing about 20 houses . . ."

"In the meanest garb of a working boat-man, and unattended by a single slave, I was no doubt considered . . . one of the canaille, and I neither claimed nor expected attention; my thoughts centered upon other objects, and all pride of appearance I willingly sacrificed to promote with frugality and industry the objects of my mission."

At Arkansas Post Nuttall lodged with a "Dr. M'Kay" and ". . . found in him an intelligent and agreeable companion . . ." With his host he visited on January 28 a nearby lake.

"Here a vast prairie opens to view, like a shorn desert, but well covered with grass and herbaceous plants . . . Among other plants already in bloom in these natural meadows, we saw abundance of a new and fragrant species of *Allium* with greenish-white flowers, and destitute of the characteristic odour of the genus in common with *A. fragrans,* to which it is allied. The *Houstonia serpyllifolia* and *Claytonia caroliniana* were also in full bloom at this early season."

It was now the month of February and on the 3rd Nuttall walked six miles and found the prairie "almost one continued sheet of water."

"I observed springing up, the *Eryngium aquaticum,* occasionally employed as a medicine by the inhabitants . . . Crossing the prairie . . . we entered the alluvial forest, containing oak, box, elder [box elder] *(Acer negundo),* elm, &c. nearer the river cottonwood appears as usual. I saw here a prickly-ash *(Zanthoxylion Clava Herculis),* the size of an ordinary ash, but the same species as that of the southern states . . ."

Nuttall embarked on February 26 in a large skiff which was going up the river; he wished to get to the Cadron some 300 miles away by water. His ultimate objective was Fort Smith, not reached until April 24.

On March 1 he noted that ". . . the Red-bud *(Cercis canadensis)* was commonly in flower." Progress upstream is recorded by various farms along the river. On the 7th they reached "Mr. Morrison's" and remained through the "6th," or actually the 8th. They were in present Jefferson County. Here "The adjoining forest was already adorned with flowers, like the month of May in the middle states. The woods, which had been overrun by fire in autumn, were strewed in almost exclusive profusion with *Ranunculus marilandicus,* in full bloom . . ."

On March 9 he joined the party of a "Mr. Drope" who was traveling in a "large and commodious trading boat of 25 tons burthen . . ." For a time this gentleman figures in the *Journal* as "Mr. D."

On the 11th, "The forest was already decorated with the red-bud, and a variety of humble flowers. A species of *Vitis,* called the June grape, from its ripening at that early period, was also nearly in blossom . . . in leaf it somewhat resembles the *vigne des batures* (or *Vitis riparia* of Michaux), while the fruit, in the composition of its bunches, and inferior size, resembles the winter grape."

The next day Nuttall observed ". . . the first appearance of a hill in ascending the Arkansa. It is called the Bluff . . . a low ridge covered with pine . . ." This was near or at Pine Bluff, Jefferson County. On March 13 he ". . . walked along the beach with Mr. D., and found the lands generally dry and elevated, covered with cotton-wood *(Populus angulisans),* sycamore *(Platanus occidentale),* maple *(Acer dasycarpa),* elm *(Ulmus americana),* and ash *(Fraxinus sambucifolia* and *F. platicarpa).*"

On the 15th the land appeared still, for the most part, ". . . elevated above inundation. Some cypress clumps [in a footnote, *'Cupressus disticha'*] however, were observable . . . opposite we saw a cluster of Hollies *(Ilex opaca),* which were the first we had seen any way conspicuous along the bank of the river. The forests everywhere abound with wild turkeys . . ."

On March 20 they reached the home of a Mr. Daniels located only a few miles below Little Rock—in present Faulkner County. The region was evidently a thoroughfare, one road ("on the right") leading to St. Louis, one ("on the left") to Natchitoches on Red River about 250 miles away. From a Mr. Curran's the next day could be seen ". . . very considerable round-topped hills, one of them, called the Mamelle . . ."

"The cliffs bordering the river . . . were decorated with the red cedar *(Juniperus virginiana),* and clusters of ferns. After emerging . . . from so vast a tract of alluvial lands, as that through which I had now been travelling for more than three months, it is almost impossible to describe the pleasure which these romantic prospects again afforded me . . . Many of the plants common to every mountainous and hilly region in the United States, again attracted my attention, and though no way peculiarly interesting, serve to show the wide extension of the same species, under the favourable exposure of similar soil and peculiarity of surface. To me the most surprising feature in the vegetation of this country, existing under so low a latitude, was the total absence of all the usual evergreens, as well as of most of those plants belonging to the natural family of the heaths, the rhododendrons, and the magnolias; while on the other hand, we have an abundance of the arborescent *Leguminosæ,* or trees which bear pods, similar to the forests of the tropical regions. Here also the *Sapindus saponaria,* or soap-berry of the West Indies, attains the magnitude of a tree.

"On the banks of the river, near the precise limit of inundation, I met with a new species of *Sysimbrium,* besides the *S. amphibium,* so constant in its occurrence along the friable banks of all western rivers. This plant, which is creeping and perennial, possesses precisely the taste of the common cabbage *(Brassica oleracea)* . . ."

Since "Mr. D." and his boat were staying in this vicinity for several days, Nuttall on March 23 amused himself by sketching the hills.[6] Of one plant he wrote:

"Amidst these wild and romantic cliffs, and on the ledges of the rocks, where, moistened by springs, grew a cruciferous plant, very closely allied, if not absolutely the same, with the *Brassica napus* or the Rape-seed of Europe, and beyond all question indigenous."

They moved on and by the 25th the Cadron Hills were about six miles distant. On March 27 they reached the Cadron settlement, consisting of five or six families. It was situated some forty miles up the Arkansas River from Little Rock and near the mouth of Cadron Creek.[7] Here Mr. Drope's boat remained until March 31.

"Mr. M'Ilmery . . . is at present the only resident on the imaginary town plot . . . The Cadron was at this time in the hands of four proprietors, who last year commenced the sale of town-lots to the amount of 1300 dollars . . . What necessity there may be for projecting a town at this place, I will not take upon myself to decide, but a house of public entertainment, a tavern, has long been wanted, as the Cadron lies in another of the leading routes through this territory . . . To those southern gentlemen who pass the summer in quest of health and recreation, this route to the hot springs of the Washita[8] . . . would afford a delightful and rational amusement."

Nuttall found *his* amusement ". . . amongst the romantic cliffs . . . which occupy the vicinity of the Cadron . . .", finding ". . . vestiges of several new and curious plants, and among them an undescribed species of *Eriogonum,* with a considerable root, partly of the colour and taste of rhubarb. The *Petalostemons,* and several plants of the eastern states, which I had not seen below, here again make their appearance. The *Cactus ferox* of the Missouri, remarkably loaded with spines, appears to forebode the vicinity of the Mexican desert."

6. The copy of the *Journal* examined by me contains five "engravings," evidently reproductions of pencil drawings, but I have found no mention of the artist's name; it seems more than likely that they were made by Nuttall. Although they do not appear to demonstrate any artistic ability, they are quaint and, brown with age, blend pleasantly with the discolored pages of the book. They illustrate the following views: "Distant view of the Mamelle" ("P. 109."); "Mamelle" ("P. 110."); "Cadron Settlement" ("P. 114"); "Magazin Mountain" ("P. 130"); and "Cavaniol Mountain" ("P. 144"). The artist's idea of perspective is sometimes amusing.

7. Cadron Creek empties into the Arkansas River from the north between Conway County (west) and Faulkner County (east).

8. In October of 1820 Edwin James made a trip to the Hot Springs of the Washita and thence to Little Rock, presumably following the route mentioned by Nuttall.

Writing on March 28, Nuttall seems to have been depressed by conditions as he found them, but is encouraged for the future.

"It is to be regretted that the widely scattered state of the population in this territory, is but too favourable to the spread of ignorance and barbarism. The means of education are, at present, nearly proscribed, and the rising generation are growing up in mental darkness ... This barrier will, however, be effectually removed by the progressive accession of population, which, like a resistless tide, still continues to set towards the west."

The cost of commodities was high—sugar "at 25 cents the pound" among other items—and he felt that the settlers were not availing themselves of the country's native resources.

"There is a maple ... on the banks of White river, which has not come under my notice, called the sugar-tree (though not as they say, the *Acer saccharinum*), that would, no doubt, by a little attention afford sugar at a low rate; and the decoction of the wood of the sassafras and spice bush *(Laurus benzoin),* which abound in this country, are certainly very palatable substitutes for tea."

On March 31 they moved three miles above the Cadron: "In a small prairie adjoining ... a single tree of the bow-wood (or *Maclura*) existed, having a trunk of about 18 inches diameter." The next objective point was Dardanelle.

On April 2 they were nearly opposite ". . . the bayou or rivulet of point Remu, [Remove Creek of Conway County] from whence ... commences the Cherokee line ..." The ". . . highest grounds are thin and sandy, so much so, that occasionally the Cactus or prickly-pear makes its appearance." On the 3rd, "Still opposite point Remu ... there was a small sandy prairie, over which I found Cactus's and the *Plantago gnaphaloides* abundantly scattered."

The next day ". . . Over the vast plain ... appeared here and there belts of cypress, conspicuous by their brown tops and horizontal branches; they seem to occupy lagoons and swamps, at some remote period formed by the river ... On the shelvings ... of the mountain, I found a new species of *Anemone*."

On April 5 they ". . . passed the outlet of the Petit John." During the journey Nuttall had observed that "The insects which injure the morel cherry-trees so much in Pennsylvania ... here occasionally act in the same way upon the branches of the wild cherry, *(Prunus virginiana)*." By evening of the 6th they were still five miles below "the Dardanelle," and could see a ". . . magnificent empurpled mountain ... apparently not less than 1000 feet high, forming a long ridge or table, and abrupt at its southern extremity. From its peculiar form it had received the name of the Magazine or Barn ..."

This terminated in a headland called the Dardanelle, ". . . or as it is here more commonly called Derdanai ..." While "Mr. D." was busy with his trading at Dardanelle, Nuttall was occupied.

"I embraced this opportunity to make one of my usual rambles, and found an extraordinary difference in the progress of vegetation here, exposed to the south and sheltered from the north-western wind. Proceeding leisurely towards the summit of the hill, I was amused by the gentle murmurs of a rill of pellucid water . . . The acclivity, through a scanty thicket . . . was already adorned with violets, and occasional clusters of the parti-coloured Collinsia. The groves and thickets were whitened with the blossoms of the Dogwood (Cornus florida). The lugubrious vociferations of the whip-poor-will; the croaking frogs, chirping crickets . . . broke not disagreeably the silence of a calm and fine evening . . ."

On April 17 Nuttall notes that his ". . . rambles to day were rewarded with the discovery of a new genus, of the class Tetradynamia or Cruciferæ, allied to *Ricotia* and *Lunaria.*"

Nuttall parted from Mr. Drope and his trading boat on the 20th and started by pirogue with two French boatmen for Fort Smith, or "the garrison," about 120 miles away by water. On April 22 he camped on an island just below Mulberry Creek. Entering the Arkansas River from the north this forms, for a distance above the confluence, the boundary between Crawford County (west) and Franklin County (east). Nuttall mentions that "The beauty of the scenery was also enlivened by the melody of innumerable birds, and the gentle humming of the wild bees, feeding on the early blooming willows . . ." He identifies these as *"Salix caroliniana."*

Nuttall reached Fort Smith on April 24; the post was then in charge of Major William Bradford. The site had been selected by Stephen Harriman Long in 1817 and the fort, sometimes called Bellepoint or Belle Point, stood on a bluff at the junction of the Poteau and Arkansas rivers and was named in honor of General Thomas A. Smith. These streams meet in Sebastian County, Arkansas, directly on the present Arkansas-Oklahoma boundary. Nuttall made his headquarters here until mid-May in order to explore the region's natural history and to ". . . make it the depot . . ." of his collections.

During his stay he enjoyed the companionship of ". . . Dr. Russel, whose memory I have faintly endeavoured to commemorate in the specific name of a beautiful species of *Monarda* [in a footnote '*Monarda russeliana*']." Russell was physician at the post and died while Nuttall, in August of 1819, was making a trip from the Verdigris post to the Cimarron River.

On April 27 Nuttall walked five miles up the "Pottoe," or Poteau, River and, so doing, entered what is now the state of Oklahoma for the first time. The country resembled ". . . a cultivated park. The whole expanse of forest, hill, and dale, was now richly enamelled with a profusion of beautiful and curious flowers; among the most conspicuous was the charming Daisy of America ['*Bellis integrifolia*'], of a delicate lilac colour . . . intermingled, appears a species of *Collinsia,* a large-flowered *Tradescantia,* various species of *Phlox,* the *Verbena aubletia,* and the esculent *Scilla* . . ."

On the 28th he again ". . . walked over the hills bordering the Pottoe, about six miles, in order to see some trees of the yellow-wood *(Maclura)*, but they were scarcely yet in leaf, and showed no indications of producing bloom. Some of them were as much as 12 inches in diameter, with a crooked and spreading trunk, 50 or 60 feet high . . . From appearances, those few insulated trees of the Pottoe, are on the utmost limit of their northern range, and, though old and decayed, do not appear to be succeeded by others, or to produce any perfect fruit."

The next day he took ". . . an agreeable walk into the adjoining prairie . . . Like an immense meadow, the expanse was now covered with a luxuriant herbage, and beautifully decorated with flowers, amongst which I was pleased to see the Painted Cup of the eastern states ['*Euchroma coccinea* (*Bartsia coccinea*. Lin.)'], accompanied by occasional clusters of a white flowered *Dodecatheon* or American primrose."

On May 3 Nuttall went on horseback with "Dr. Russel" to ". . . Cedar prairie . . . about 10 miles south-east of the garrison . . ." Here he found ". . . a second species of that interesting plant, which my venerable friend, William Bartram, called *Ixia cælestina* ['*Nemastylis cælestina*']; the flowers of this species are also of a beautiful blue, and white at the base. The whole plain was, in places, enlivened with the *Sysirinchium anceps,* producing flowers of an uncommon magnitude; amidst this assemblage it was not easy to lose sight of the azure larkspur ['*Delphinium azureum*'], whose flowers are of the brightest ultramarine; in the depressions grew also the ochroleucous *Baptisia* ['*Baptisia leucophæa*'], loaded with papilionaceous flowers nearly as large as those of the garden pea."

Back at Cedar prairie on the 9th, Nuttall and the doctor "were gratified," on its wooded margin, ". . . by the discovery of a very elegant plant, which constitutes a new genus allied reciprocally to *Phacelia* and *Hydrophyllum.*" He appends the notation: "I have given it the trivial name of *Nemophila,* as, in this country, it now constituted the prevailing ornament of the shady woods."

From May 16 to June 21 Nuttall made a journey into what is now the state of Oklahoma, but was then the Indian Territory. He traveled with a party of soldiers led by Major Bradford, whose task it was to evacuate the whites who had settled in the territory of the Osage Indians—". . . the Kiamesha [Kiamichi] river being now chosen as the line of demarkation." The news of their eviction in favor of the Indians was announced to the white settlers as the party moved along:

"The people appeared but ill prepared for the unpleasant official intelligence of their ejectment. Some who had cleared considerable farms were thus unexpectedly thrust out into the inhospitable wilderness. I could not but sympathise with their complaints, notwithstanding the justice and propriety of the requisition. Would it had always been the liberal policy of the Europeans to act with becoming justice, and to reciprocate the law of nations with the unfortunate natives!"

From Fort Smith the party traveled mainly southwest. The objective point was the confluence of Kiamichi and Red rivers, in Choctaw County, on what is now the Texas-Oklahoma border. On May 16 they proceeded ". . . towards the banks of the Pottoe, and found the whole country a prairie, full of luxuriant grass about knee high, in which we surprised herds of fleeting deer, feeding as by stealth."

The next day, after twice crossing the Poteau, two picturesque mountains came into view, "the Cavaniol and Point Sucre," the last covered with thickets.

"These vast plains, beautiful almost as the fancied Elysium, were now enamelled with innumerable flowers, among the most splendid of which were the azure Larkspur, gilded Coreopsides, Rudbeckias, fragrant Phloxes, and the purple Psilotria."

On May 18, they ". . . passed the dividing ridge of the Pottoe and Kiamesha . . . very rocky . . ." where Nuttall ". . . was gratified by the discovery of a new shrubby plant allied to the genus Phyllanthus." Two days later, when about thirty miles from Red River, they were proceeding ". . . through a pathless thicket, equal in difficulty to any in the Alleghany mountains . . . The woods were now disgustingly infested with ticks, though free from musquetoes." On May 21 he notes that "For the last two days I was busily employed in collecting new and curious plants, which continually presented themselves." On the 22nd he observed: "Nothing could at this season exceed the beauty of these plains, enamelled with such an uncommon variety of flowers of vivid tints, possessing all the brilliancy of tropical productions." After crossing the Kiamichi they had a view of Red River, ". . . with the water very red and turbid . . . Here for the first time, I saw the *Maclura* (or Bow wood) in abundance, but almost a month past flowering, at least with the staminiferous plant."

On May 24 they reached Red River after passing through ". . . Horse-prairie, 15 miles above the mouth of the Kiamesha . . . This prairie derives its name from the herds of wild horses, which till lately frequented it, and of which we saw a small gang[9] on our return. It is very extensive, but flat, and in some places swampy. In these depressions we saw whole acres of the *Crinum americanum* of the West Indies, besides extensive glaucous fields of a large leaved and new species of *Rudbeckia*."

On May 26 Nuttall's difficulties began. Major Bradford started on the return trip to Fort Smith, but Nuttall, expecting to catch up with him later, ". . . delayed about two hours . . . for the purpose of collecting some of the new and curious plants interspersed over these enchanting prairies." Although he attempted to overtake Bradford for two days he never was able to do so and was fortunate, since he ". . . dared not to venture alone and unprepared through such a difficult and mountainous wilderness . . .", to be given refuge in the home of a "Mr. Styles." Nonetheless he was happy.

"My botanical acquisitions in the prairies, proved . . . so interesting as almost to make me forget my situation, cast away as I was amidst the refuse of society, without money and without acquaintance; for calculating upon nothing more certain than an immediate return, I was consequently unprovided with every means of subsistence."

On June 4 he was still ". . . at the house of Mr. Styles, without any very obvious prospect of regaining fort Smith." He kept at his collecting, walking over the prairie to a hill near Red River.

"The singular appearance of these vast meadows, now so profusely decorated with

9. Not slang! The word seems to have been used for a collection or herd of animals of one species as early as 1740 according to *A new English dictionary on historical principles*. (Oxford)

flowers . . . can scarcely be described. Several large circumscribed tracts were perfectly gilded with millions of the flowers of *Rudbeckia amplexicaulis,* bordered by other irregular snow-white fields of a new species of *Coriandrum.*"

After mentioning a number of the principal grasses observed, he continues:

"The common Milfoil, and sorrel *(Rumex acetocella),* are as prevalent, at least the former, as in Europe. In these plains there also grew a large species of *Centaurea,* scarcely distinct from *C. austriaca;* and along the margin of all the rivulets we met with abundance of the Bow-wood *(Maclura aurantiaca)* . . ."

On June 6 Nuttall walked ". . . five or six miles to collect specimens of the *Centaurea,* which, as being the only species of this numerous genus indigenous to America, had excited my curiosity . . . I now, for the first time in my life, notwithstanding my long residence and peregrinations in North America, hearkened to the inimitable notes of the mocking-bird . . . After amusing itself in ludicrous imitations of other birds . . . it at length broke forth into a strain of melody the most wild, varied, and pathetic, that ever I had heard from any thing less than human . . ."

Nuttall's interest, as his journal makes clear, was not confined to the flora! On June 8, on visiting Red River settlement, he learned of a party which was to leave for the Arkansas River in about a week. On the 14th this party called for him and he ". . . was obliged to make a hasty departure . . . my kind host and family . . . knowing from the first my destitute situation, separated from pecuniary resources, could scarcely be prevailed upon to accept the trifling pittance which I accidentally possessed."

But on the 15th the leader of the group became ". . . embarrassed by the accumulation of the mountains . . ." and the party got lost. This fact did not distract Nuttall from his pursuits and he records that, on the 16th, he saw in a lake ". . . the *Pontederia cordata, Nymphæa advena, Brassenia peltata,* and *Miriophyllum verticillatum,* all of them plants which I had not before seen in the territory . . ."

June 17 the party reached ". . . the three main branches of the river: Jack's creek to the south, Kiamesha to the east, and a third rivulet to the north . . ." Here, in a cove, Nuttall again found ". . . the *Ixia cælestina*[1] of my venerable friend Wm. Bartram."

Having missed, on the 19th, a gap in the mountains, the party with difficulty moved along the summit of a ridge ". . . through thickets of dwarf oaks *(Quercus chinquapin, Q. montana,* and *Q. alba),* none of them scarcely exceeding the height of a man . . ." This day they fortunately happened upon the trail of Major Bradford's party, and following it, came on the 20th to the Poteau and ". . . encamped in the valley of the third oak ridge that separated us from Cedar prairie." By noon of June 21 they reached Fort Smith where, Nuttall records, ". . . I had been long expected, and was very cordially welcomed by the Doctor and the Major." There is no suggestion that his prolonged absence of about three weeks had caused anyone any anxiety!

Nuttall notes that he remained at "the garrison," or Fort Smith, to "the end of the month," actually until July 6. He tells nothing of his activities during that time. On

1. Oakes Ames wrote me: "In Florida we call this Bartram's Ixia."

July 6 he started on a journey up the Arkansas River which again took him into Oklahoma and from which he did not return to Fort Smith until September 26 or 27. His
first objective was the trading post at the mouth of Verdigris River, about 130 miles
distant. He traveled in a boat belonging to a "Mr. Bougie, agent for Mr. Drope," the
"Mr. D." with whom he had traveled from March 9 to April 20. The first day he notes
that "Among three or four other new plants afforded me by examining the sand-
beaches, was a *Portulaca,* apparently the same with *P. pilosa* of the West Indies . . ."

July 8: "On Sambo island . . . we stopped to dine; and here on a bar of gravel I found
a new species of the Mexican genus *Stevia;* and never saw it afterwards in any other
locality. To the taste it was quite as bitter as many of the *Eupatoria.* This plant and the
Portulaca already mentioned, appear to have been recently, and almost accidentally,
disseminated from the interior."

July 10: ". . . on the second bar we arrived at . . . we found a few Chicasaw plumbs,
with natural orchards of which every beach abounds . . . for the first time, near the
Arkansa, we met with the hazel *(Corylus americana),* and the American raspberry
(Rubus occidentalis)."

On July 11 they passed ". . . the outlet of the Canadian . . ." This enters the Arkansas from the southwest, in Haskell County. Also the mouth of the Illinois River,
entering from the northeast, in Sequoyah County. Nuttall records:
"The variety of trees which commonly form the North American forest, here begin
very sensibly to diminish. We now scarcely see any other than the smooth-barked
cottonwood, the elm, box-elder *(Acer Negundo),* curled maple *(Acer dasycarpon),*
and ash, all of them reduced in stature. From hence the forest begins to disappear
before the pervading plain."
On July 14 they ". . . entered the Verdigris, where M. Bougie and Mr. Prior had
their trading houses." This was to be Nuttall's base until August 11, or even until
September 21 or 22 for he returned there from the Cimarron River. On the alluvial
lands between the Grand and the Verdigris rivers he saw ". . . larger trees than . . .
since leaving Fort Smith. Among them were lofty scarlet oaks, ash, and hackberry,
and whole acres of nettles *(Urtica divaricata),* with whose property of affording hemp,
the French hunters and settlers have been long acquainted . . . About eight miles from
the Arkansa, commences the great Osage prairie, more than 60 miles in length . . .
On entering the prairie I was greatly disappointed to find no change in the vegetation,
and indeed, rather a diminution of species. The *Amorpha canescens,* which I had not
heretofore seen, since leaving St. Louis and the Missouri, and a new species of *Helianthus* . . . instantly struck me as novel."
On July 17, with two companions, Nuttall started by canoe up "Grand," or Neosho,
River; they were to visit the Osage saltworks. The trip meant two days' travel each
way. On some cliffs, on the 18th, he ". . . recognised as new, a large shrub . . . a simple-
leaved *Rhus,* scarcely distinct from the *R. Cotinus* of the south of Europe and of our

gardens . . . The gravel bars were almost covered with *Amsonia salicifolia,* with which grew the *Sesbania macrocarpa* of Florida."

That evening, two miles below "the saline" and fifty miles from the Arkansas River, he notes that "In this elevated alluvion I still observed the Coffee-bean tree *(Gymnocladus canadensis),* the over-cup white oak *(Quercus macrocarpa),* the pecan *(Carya olivæformis),* the common hickory, ash, elm; and below, in places near the margin of the river, the poplar-leaved birch *(Betula populifolia)."*

Arriving at their destination on July 19, they found the salt mines nearly deserted. A murder had taken place and Nuttall had reason for self-congratulation for, about a month earlier, he had attempted to hire one of the "remorseless villains" as an assistant.

Intermittent fever was widespread at this time and on arrival Nuttall had his first attack.

"No medicines being at hand, as imprudently I had not calculated upon sickness, I took in the evening about a pint of a strong and very bitter decoction, of the *Eupatorium cuneifolium,* the *E. perfoliatum* or Bone-set, not being found in the neighbourhood. This dose . . . prevented the proximate return of the disease."

On July 20 he started back for the Verdigris post, using a compass and proceeding ". . . south by west, the distance being about 30 miles. Twenty miles of this route was without any path, and through grass three feet deep, often entangled with brambles, and particularly with the tenacious 'Saw brier,' *(Schrankia horridula)* . . . I lay down to sleep in the prairie, under the clear canopy of heaven;—but alone, and without the necessary comforts of either fire, food, or water . . . Every tender leaved plant, whether bitter or sweet, by thousands of acres, were now entirely devoured by the locust grashoppers, which arose before me almost in clouds. I slept, however, in comfort, and was scarcely at all molested by musquetoes. The next day, after spending considerable time in botanizing, I arrived at the trading houses."

The *Journal* for the remainder of Nuttall's stay at the Verdigris post, or until August 11, is largely devoted to an account of the Osage Indians. The man's illness had returned and botanical activities were undoubtedly curtailed.

On August 11, accompanied by a trapper named Lee, he started on horseback up the Arkansas River, bound for ". . . the Salt river, or first Red river of the Arkansa, called by Pike the Grand Saline, and about 80 or 90 miles distant . . ." The Cimarron, or "Grand Saline," enters the Arkansas in southeastern Pawnee County, Oklahoma. Nuttall did not get back to the Verdigris post until September 15. And he was fortunate to have survived the journey.

Starting out on August 11, he found in the prairie ". . . a second species of *Brachyris,* pungently aromatic to the taste, and glutinous to the touch; its aspect is that of *Chrysocoma."* The next day ". . . being oppressively hot and thirsty, I very imprudently drank some very nauseous and tepid water, which . . . produced such a sickness, that it was with difficulty I kept upon my horse . . ." Nonetheless they made twenty miles. On the 13th they came to a stream called ". . . the Little North Fork (or branch) of the Canadian . . ." This was Deep Fork which enters the Canadian from the north in McIntosh County. Here Nuttall was, as the chapter heading states, "Detained by sickness" from the 14th to the 16th. On the 17th Lee, and very wisely, attempted to persuade the sick man to return to the Verdigris before he was too weak to do so; but, states Nuttall: ". . . the idea of returning filled me with deep regret, and I felt strongly opposed to it whatever might be the consequences." On the 18th Lee again urged Nuttall's return ". . . in plainer terms than before." From August 19 to 23 the two men proceeded at the rate of a few miles a day. The heat was terrific, they were plagued by blowflies, and one of the horses having given out, the other had to do double duty. Indians were a constant menace; moreover Nuttall was desperately ill. He seems, nevertheless, to have retained his interest in plants for he wrote on the 24th:

"We passed by three or four enormous ponds grown up with aquatics, among which were thousands of acres of the great pond lily *(Cyamus luteus),* amidst which grew also the *Thalia dealbata,* now in flower, and, for the first time, I saw the *Zizania miliacea* of Michaux . . . across the desiccated corner of the pond . . . the *Ambrosias* or bitter weeds were higher than my head on horseback . . ."

Although Nuttall was still seriously ill they moved on, though very slowly, from August 26 to 31.

September 1: "We saw nothing far and wide but an endless scrubby forest of dwarfish oaks, chiefly the post, black, and red species."

September 2: "We found the small chinquapin oak by acres, running along the ground as in New Jersey. The *Portulaca* resembling *P. villosa,* which I had seen below in a similar locality, Mr. Lee picked for me to-day, growing in arid rocky places, where the soil had been nearly washed away."

On September 3 they reached ". . . the First Red Fork or Salt river . . . red and muddy . . . Its first view appeared beautifully contrasted with the broken and sterile country through which we had been travelling. The banks of cotton-wood *(Populus monilifera)* . . . resembled a verdant garden in panorama view . . ." Lee told Nuttall that, a few days' journey farther west, were ". . . extensive tracts of moving sand hills, accompanied by a degree of sterility little short of the African deserts."

On the 4th they went a few miles up the Cimarron, crossing the stream occasionally. But, since the Osages were evidently in the vicinity, it was decided on the 5th to return to the Verdigris. While Lee was examining some of his beaver traps his horse got inextricably mired. So it was decided to build a canoe but ". . . it was not even an easy matter to find a tree of sufficient size for this purpose. The largest timber was the cotton-wood *(Populus angulata)*." By September 8 Lee had built and loaded his canoe and the men started on their return journey, Nuttall keeping along the shore of the Cimarron River.

"Amongst several other new plants, I found a very curious *Gaura,* an undescribed species of *Donia*, of *Eriogonum*, of *Achyranthes, Arundo,* and *Gentian*. On the sandy beaches grew several plants, such as the *Uralepsis aristulata (Festuca procumbens,* Muhlenberg), an *Uniola*, scarcely distinct from *U. spicata* and *Sesuvium sessile* which I had never heretofore met with, except on the sands of the sea coast."

From September 9 to 13 they were descending the Arkansas River. But the water was very low. They were alarmed by the threatening behavior of the Osages—it was even dangerous to light a fire at night for the comfort of the sick Nuttall. To evade the savages they kept on at night ". . . amidst the horrors of a thunder storm, the most gloomy and disagreeable situation I ever experienced in my life." Quicksands were another peril; and Nuttall comments on ". . . the dreary howling of the wolves . . ."

Since it was impossible for the botanist, who was following the shore, to keep up with Lee in the boat, it was decided to part company, each returning as best he could to the post. The 14th was another miserable day for Nuttall but he finally recognized the Verdigris River; it was late and he had ". . . to lie down alone, in the rank weeds, amidst musquetoes, without fire, wood, or water . . ." By evening of September 15 he ". . . again arrived at the trading establishment of Mr. Bougie, an asylum, which probably, at that time, rescued me from death."

Nuttall remained at the Verdigris post until September 21 or 22, and then, in the boat of an *engagée,* descended the Arkansas River to Fort Smith—a trip of five days. He remained at Fort Smith until October 16, but his illness made botanizing impossible. Indeed, from the time he reached Fort Smith until he arrived at New Orleans the *Journal* has very little to say about the flora. For the remainder of 1819 Nuttall's narrative contains few dates.

He reached the Pecannerie settlement on November 3. He refers to the fact that this settlement, ". . . the most considerable . . . in the territory, except Arkansas, derived its name from the Pecan nut-trees *(Carya olivæformis),* with which its forests abound . . ." It boasted sixty families, ". . . living in a state of ignorance and mediocrity of fortune: many . . . were renegadoes from justice who had fled from honest society . . . indulging themselves in indolence, they became the pest of their more industrious and honest neighbours . . ."

The *Journal* at this point has something to tell of the "Hot-springs of the Washita" but, as Pennell states, "It is evident that Nuttall did not undertake to visit these, as is stated by Durand." I believe that Edwin James, of botanists, was the first to do so.

When Nuttall left the "Pecannerie" is not stated[2] in the *Journal,* but on December 18 he was at "Cadron" or Quadrant. This "imaginary town" consisted of four resident families and possessed a tavern crowded with travelers and emigrants but with only two "tenantable" rooms.

"Although I have been through life perfectly steeled against games of hazard, neither wishing to rob nor be robbed, I felt somewhat mortified to be thus left alone, because of my unconquerable aversion to enter this vortex of swindling and idleness."

---

2. Pennell gives the dates he was there as "Nov. 4–Dec. 17," and his arrival at Cadron as "Dec. 18."

1820

On January 4 Nuttall wrote: ". . . after waiting about a month for an opportunity of descending, I now embraced the favourable advantage of proceeding in the boat of Mr. Barber, a merchant of New Orleans, to whose friendship and civility I am indebted for many favours."

By evening of the 5th they were in view "of the pyramidal Mamelle," and on the 12th reached "Mr. Dardennes' " where Nuttall had been on March 9 of 1819; on the 15th came again to ". . . the post of Osark, or as it is now not very intelligibly called, Arkansas, a name by far too easily confounded with that of the river, while the name Osark . . . would have been perfectly intelligible and original."

The boat remained at Arkansas Post until January 19; on the 17th and 18th Nuttall revisited the prairie.

"The interesting plants and flowers which I had seen last year, at this time, were now so completely locked up in the bosom of winter, as to be no longer discernible, and nearly disappointed me in the hopes of collecting their roots, and transplanting them for the gratification of the curious.

"On the 19th, I bid farewell to Arkansas, and proceeded towards the Mississippi, in the barge of Mons. Notrebé, a merchant of this place, and the day following . . . arrived at the confluence of the Arkansa, a distance of about 60 miles. The bayou, through which I came in the spring, now ran with as much velocity towards White river, as it had done before into the Arkansa . . . We now found ourselves again upon the bosom of one of the most magnificent of rivers . . ."

On January 21 Nuttall ". . . now embarked for New Orleans in a flat boat, as the steam boats, for want of water, were not yet in operation." The descent of the Mississippi River took four weeks, or until February 18.

January 22: ". . . About 20 miles below the Arkansa, in the Cypress bend, we saw the first appearance of *Tillandsia* or Long moss."

January 24: ". . . we arrived at Point Chicot, which is included in the Arkansa territory . . . From the Chicasaw Bluffs downward, along the banks of the Mississippi, we perceive no more of the Tulip tree *(Liriodendron tulipifera),* and but little of the *Platanus,* greatly reduced in magnitude, compared with what it attains along the Ohio. The largest tree of the forest here is that which is of the quickest growth, the Cottonwood poplar *(Populus angulata).*"

January 28: ". . . we passed the settlement called the Walnut Hills . . ."

This was Vicksburg, Warren County, Mississippi. The same day, near "Warrington"—or Warrenton, Warren County—Nuttall again comments (as he had on December 20, 1819) upon the Spanish moss:

"The gloomy mantling of the forest communicated by the *Tillandsia usneoides* or long moss, which every where prevails, is a never-failing proof of an unhealthy humidity in the atmosphere."

On the 29th they passed ". . . the grand Gulf or eddy, near to which enters Big Black river." This stream enters the Mississippi from the northeast and the boat was keeping to the eastern side of the river. On a hill slope Nuttall observed ". . . the first trees of the *Magnolia grandiflora*. The small palmetto *(Sabal minor)* commences about Warrington . . ."

On January 31 they reached ". . . the well known and opulent town of Natchez . . ." We are told of the region's agricultural and horticultural products. The peach, fig, pear, quince and apple did extremely well; the cherry, gooseberry, and currant throve but were not fruiting, etc.; cotton was impoverishing the soil and little attention was given to its renovation—Nuttall had ideas about the best ways to do this.

Leaving Natchez on February 4, they reached Fort Adams, situated in the southwesternmost corner of Mississippi, the same day and on the 9th ". . . passed the thriving town of Baton-Rouge . . . Ever since leaving Natchez, we have had weather like summer, and vegetation already advances." From this point southward settlements were contiguous, or nearly so, and southward to the sea ". . . the whole country is alluvial and marshy."

February 18: "This morning we arrived at New Orleans, now said to contain about 45,000 inhabitants . . . In the neighbourhood of the city, and along the coast, the beautiful groves of orange trees, orchards of the fig, and other productions of the mildest climates, sensibly indicate our approach to the tropical regions, where the dreary reign of winter is for ever unknown. But little pains as yet have been taken to introduce into this country, though so thickly settled, the ornamental and useful plants which it is calculated to sustain. We yet neither see the olive, the date, nor the vineyard . . . That the date . . . would succeed, an accidental example in the city renders probable. This palm . . . has attained the height of more than 30 feet, with a trunk of near 18 inches in diameter, and has flowered annually for the space of several years . . . being only a staminiferous plant, it has not consequently produced any fruit."

Nuttall was appalled by the many graves; for, during the past summer, yellow fever had ". . . carried off probably 5 or 6000 individuals . . ." He was not impressed by the cultural aspects of the city: "Science and rational amusement is as yet but little cultivated in New Orleans."

The narrative portion of Nuttall's *Journal*[3] ends with his arrival at New Orleans. It

3. An Appendix contains four sections: (1) "An account of the ancient aboriginal population on the banks of the Mississippi, and the contiguous country"; (2) "The history of the Natchez"; (3) "Observations on the Chicasaws and Choctaws"; (4) "Thermometrical observations in the Arkansa territory, during the year 1819." Although correctly numbered "Sect. IV." in the "Contents," Nuttall's article itself is headed "Section III." Pennell comments that the appendices concerned with the Indian tribes ". . . would have delighted Professor Barton." As stated in my chapter VI, Barton had been instrumental in sending Nuttall on his first journey into western regions.

does not explain how, or when, the botanist left that city for the United States. Pennell states:

". . . it is evident that Nuttall returned directly to Philadelphia (or a neighboring seaport) by boat. No later work gives any suggestion of an intermediate stop . . . Nuttall would have reached our city about April, and the first allusion to him in the minutes of the Academy's meetings for 1820 is of May 2."

About one year after Nuttall's return to the east coast John Torrey wrote von Schweinitz from New York on March 22, 1821—the letter is published by Shear and Stevens:

". . . You enquire respecting Mr. Nuttall.—I mentioned just now that he had lately made a visit to this city, but he resides now in Philadelphia. Mr. N. returned last spring from another expedition up the Missouri,[4] & into the Arkansas Territory. He spent the year 1819 there & discovered a great number of new plants—probably about 300 species. He is now printing his *Journal,* but his botanical discoveries he is preparing to publish in the next volume of the American Philosophical Trans. of Phila. He found comparatively few cryptogamia, & all of them except the Ferns, he has given to Zaccheus Collins Esq., of Phil. This gentleman has undertaken to examine them, as Mr. Nuttall has not paid great attention to this department of Botany. I doubt very much however whether Mr. C. will consent to have his opinion of the specimens published, even if he should give any opinion. It is surprising how exceedingly cautious this gentleman is in this respect; for the (perhaps) hundreds of specimens which I have sent him, he has never returned me the name of one—You had better however write to him, as he may send you *specimens* if he will not give you *descriptions* & names of plants. Mr. Nuttall found on the Red and Arkansas Rivers, *Pilularia* & *Marsilea,* which have not before been observed in North America—I have duplicates for you . . ."

Nuttall's preface to *A journal of travels into the Arkansa territory* is dated from Philadelphia, November, 1821. In it he explains that ". . . so far from writing for emolument, I have sacrificed both time and fortune to it." Nuttall's dedication suggests that his trip was made possible by the "liberality of Joseph Correa de Serra, Zaccheus Collins, William Maclure and John Vaughan. The anonymous author of the "Biographical sketch of the late Thomas Nuttall" adds the name of Reuben Haines to the list of Nuttall's patrons; he comments that the *Journal,* although ". . . filled with highly interesting matter . . . [was] a poor speculation to his printer, through the absence of all anecdote and lightness, which, contrary to the strong advice of his friends, he would not admit . . . He hated everything that savored of vanity or needless show, always aiming at the real and substantial. He was . . . well aware that such a course did not please the public, and often deplored that 'he lived in an age that no longer tolerated the plain, unvarnished tale.' . . ."

Nuttall would have been gratified by a tribute paid his *Journal* by W. J. Ghent more than one hundred years later:

4. Nuttall had not been on the Missouri River since his trip with the Astorians in 1811–1812.

"The record of this journey . . . is one of the classics of the early frontier. Much of it, consisting of botanical descriptions, is of interest only to the scientist, but its descriptions of the settlements, of the customs and traditions of the Indians, its narration of travel incidents, its characterization of the outstanding personalities encountered, give it a permanent value not only to the historical student but to the general reader."

Although Nuttall's *Journal* was issued promptly, it was not until 1835–1836[5] that he published his important paper on the plant collections: "Collections towards a flora of the territory of Arkansas"; it appeared in the *Transactions* of the American Philosophical Society (vol. 5, new series) with the notation that it had been read before the Society on "April 4, 1834." Pennell attributes the delay in publication to more than lack of time: "In those early days there was both lack of sufficient specimens for comparison, and of a sufficient library for reference."

Pennell states that Nuttall ". . . presented descriptions from it before the Academy [of Natural Sciences of Philadelphia] in August and September, 1821, and in March, 1822, which were published in its Journal (2: 114-123, and 179-182)." The reference is to two articles describing plants ". . . recently introduced into the gardens of Philadelphia": in the first article, ". . . some new species . . . from Arkansa Territory . . ."; in the second, ". . . some rare plants . . ." from the same region.[6] But, referring again to

5. The publication date in volume 5 of the *Transactions* is 1837. But R. C. Foster has pointed out that Nuttall's paper, read April 4, 1834, came out earlier:

"The Transactions appear to have been published in parts, each containing twenty-six four-page signatures. Nuttall's paper began on the third page of V-2K, ended on the third page of V-3A, and was completely published before 1837 . . . it seems a justifiable assumption that pp. 139-184, at least had been issued by the end of 1835, and that the remainder, pp. 135-203, appeared in early 1836. It is beyond question that pp. 139-160 were issued by or soon after the middle of 1835. In any case the date, 1837, usually given in citing species described in this work, is certainly incorrect."

The publication date of Nuttall's paper should therefore be cited 1835-1836.

6. Some of Nuttall's plants from the Arkansas Territory were cultivated in "The Linnæan Botanic Garden," situated, according to its catalogues, at "Flushing, Long-Island, near New-York," which was owned by William Prince and by his son and namesake. J. H. Barnhart refers to this garden as ". . . a commercial enterprise, but one conducted with more regard for the advancement of American horticulture than for profit . . ."

The prefaces to the 22nd (1823) and 23rd (1825) editions of Prince's catalogues, on file in the Library of the Massachusetts Horticultural Society, mention Nuttall's plants—as well as those received from Edwin James—but, and possibly they did not survive in the Long Island climate, neither collection is mentioned in the 25th (1829) edition. The preface states:

". . . Thomas Nuttall, Esq. a gentleman celebrated as much for the liberality of his mind as for his great attainments in Botany, and other sciences, presented me [William Prince, the son] with seeds of all the species collected during his western tour—a very considerable number of which have flowered, and are now in a thriving state. During the recent Yellow Stone expedition [1820], under Col. Long, Dr. James also made a collection of eighty-four species on the Rocky Mountains, and elsewhere, which he very politely presented to this establishment, and among which it is expected will be found not only a large number of new species, but also some new genera."

These plants evidently aroused interest in botanical circles, von Schweinitz writing Torrey on November 24, 1822, that he had been told that Prince ". . . cultivates most of Mr. Nuttall's and other Missouri plants . . ." and, on April 2, 1823, enclosing ". . . a list of my *desiderata* in Am. Phaenogamy with a particular request to procure as many of them as possible either from collections, or from Prince's garden. I would go to some expence to get them."

the ". . . long pending report on the Arkansas plants of 1819 . . .", Pennell comments that ". . . probably, as shown by new species based on collections of that journey but which did not appear until Nuttall's later papers, this was not actually completed, but rather what was ready was drawn together for publication before setting out on the long journey now in prospect."

The trip "in prospect" was to be Nuttall's third journey west of the Mississippi River. On April 4, 1834 (the day when his "Collections towards a flora of the territory of Arkansas" was being read before the American Philosophical Society), Nuttall had already reached Loutre Lick (now Big Spring, Montgomery County, Missouri) and, in company with Nathaniel Jarvis Wyeth, was on his way to the mouth of the Columbia River.

As I have noted elsewhere (see p. 140, *fn.* 2) Nuttall's association with Harvard College began in 1822 with his appointment to the curatorship of the Botanic Garden. Since even Pennell, who has followed the botanist's yearly whereabouts with greater accuracy than anyone else, evidently knew but little about them for the period between 1820 and 1825 (when he states that his "definite appointment" at Harvard began) it would seem to be of interest to include a few records which have come to my attention since this book went to press. They are contained in two unpublished letters written by Nuttall from Cambridge, Massachusetts, on August 30 and September 16, 1823, and now the property of Miss Elizabeth Putnam of Boston, who has been kind enough to allow me to quote from them. They were written to her grandfather George Putnam, then a youth of sixteen and a student at Harvard College. They testify to Nuttall's interest in geology, supply a few additional records of localities visited in midsummer of 1823 (all in the vicinity of Worcester, Massachusetts), and refer to a hoped-for appointment at the college (over and above the curatorship of the Botanic Garden), which did not materialize. The letters are long and I quote but a fraction.

From Cambridge on August 3, 1823, young Putnam wrote his mother at the simple family home in Sterling:

". . .I recollect your charging me never to bring home any of my classmates, but as you never told me not to bring home a Proffessor, I shall introduce you to one when I go home. The case is this. I have become acquainted with Mr Nuttall, lecturer upon Natural history (particularly botany) and who will probably succeed the present professor of mineralogy who is soon to resign. He has travelled in almost all parts of the U. States, and now wants to scour the County of Worcester for new plants & minerals and as he has made me very fond of the latter interesting science, & given me more than 100 different specimens, I am to accompany him. You will probably see me with him on commencement day or the day before. I shall travel with him over our farm and the ones adjacent, in quest of plants. I shall learn their names, natures and uses as he is acquainted with every known plant in America & is the greatest botanist in the country. After this I shall walk with him to Worcester & elswhere as far as I am able.

You certainly would not have me neglect this very valuable opportunity of travelling with so good a natural Historian who can describe to me every natural object we meet with. For what shall we study in preference to nature? You need not be afraid of him for he is a plain, honest & social Englishman who in his travels among the western Indians has seen harder fare than he will find at your poor table . . . Tell Eliza to collect her herbs & flowers to have them explained . . . August 10th. I break open my letter to inform you that I shall more probably come home the Friday or Saturday before commencement. You need not make great parade as he is not used to it. Since I wrote my letter Mr Nuttal has been chosen Professor of Mineralogy & Geology."

The visit evidently took place. Nuttall's letter of August 30 describes his itinerary after parting from Putnam and details his findings at a number of mines, near Sterling, at Harvard, at "Boxborough (or Boxbury?)"—actually Boxboro—and at Worcester, from which point he took the coach and arrived ". . . in good time at Cambridge . . . [finding] a deserted house [doubtless the one belonging to the Botanic Garden] and locked doors, instead of my usual friendly reception . . ."

Nuttall's letter of September 16 begins with the statement: ". . . I am going on with my lectures to a large class who appear to be pleased with the subject—wh[ich] is always my highest reward . . ." Then:

". . . I thought it almost certain that I should have been the lecturer or professor of Mineralogy and Geology as you may remember from the seemingly qualified assurance of the President [then John Thornton Kirkland], but I find that there may be to me 'a slip betwixt the Cup and the Lip.' Who would have thought that my *seeming* friend Dr. Webster should have moved so far out of the honourable track as to endeavour to snatch from me this little additional employment and emolument, yet nothing is more certain than the fact of his endeavours to serve me, after all my confidence in him—this unexpected 'ill-turn.' To assist his pretensions, (like the introduction of the wooden horse into Troy,) he offers to deposite his cabinet[7] in the college, a great service no doubt to the College already encumbered by collections for wh[ich] room is wanting. I have, however, put in my claim, with what success time must show. If I am, however, debarred from this privilege, with the offer of wh[ich] I have, as you know, been tantalised, I intend never again to appear in any part of the college on any occasion. I shall look upon such treatment as a perfidy wh[ich] must be unpardonable and irreconcillable. As yet, I think it probable that the President is unacquainted with all this brewing, but I am prepared for the consequence, and if unfavourable I shall *never return* from England . . ."

7. The *Historical Register of Harvard University* lists but one Webster at this period: Dr. John White Webster, Lecturer in Chemistry, Mineralogy and Geology (1824–1826), Professor of Chemistry (1826–1827) and Irving Professor of Chemistry and Mineralogy (1827–1850), who is best known, I fear, as the murderer of Dr. George Parkman of Boston. Were this insufficient identification, the "cabinet" mentioned by Nuttall must have been the mineralogical cabinet from the sale of which the financially hard-pressed Webster attempted more than once to raise funds, as recorded at the trial.

Despite these determinations, and a number of gloomy predictions—one being the possibility that he might, like many others before him, be drowned on his way to England—Nuttall remained curator of the Garden for about ten years!! Indeed, Nuttall's frame of mind throughout this letter is so emotional that one wonders whether his supposition about Webster had any basis in fact. To end on a kind record, Miss Putnam tells me that her grandfather, in later life a well-beloved clergyman of Roxbury, visited the unfortunate Webster during his incarceration and tried to give him some spiritual help. By that time Putnam's old friend Nuttall was living permanently in England.

Anonymous [BIGELOW, JACOB ?]. 1823. [Review of] A journal of travels into Arkansa territory during the year 1819 . . . by Thomas Nuttall . . . Philadelphia. 1821. *North Am. Review* 16 (new ser. 7) : 59-76.

——— [PICKERING, CHARLES ?]. 1860. Biographical sketch of the late Thomas Nuttall. *Gard. Monthly* (Meehan) 2: 21-23.

BARNHART, JOHN HENDLEY. 1921. Biographical notices of persons mentioned in the Schweinitz-Torrey correspondence. *Mem. Torrey Bot. Club* 16: 290-300.

FOSTER, ROBERT CRICHTON. 1944. The publication-date of Nuttall's "Arkansas Flora." *Rhodora* 46: 156, 157.

GHENT, WILLIAM JAMES. 1931. The early far west. A narrative outline 1540–1850. New York. Toronto. 165.

NUTTALL, THOMAS. 1818. The genera of North American plants, and a catalogue of the species to the year 1817. 2 vols. Philadelphia.

——— [1821–1822.] A description of some new species of plants, recently introduced into the gardens of Philadelphia, from the Arkansa territory. *Jour. Acad. Nat. Sci. Phila.* 2: pt. 1, 114-123. "Read, August 7th, 1821." [Reprinted in Latin in *Linnaea,* Littratur-Bericht 4: 41, 1829.]

——— 1821. A journal of travels into the Arkansa territory, during the year 1819. With occasional observations on the manners of the aborigines. Philadelphia. Map ["A map of the Arkansas River, intended to illustrate the travels of Thos Nuttall: constructed from his original manuscripts by H. S. Tanner."].

——— 1835–1836. Collections towards a flora of the territory of Arkansas. *Trans. Am. Philos. Soc.* new ser. 5: 139-203. "Read April 4, 1834."

——— 1905. A journal of travels into the Arkansa territory, during the year 1819. With occasional observations on the manners of the aborigines . . . *In:* Thwaites, Reuben Gold, *editor.* Early western travels 1748–1846. 13. [Reprinted from the Philadelphia edition of 1821.]

PENNELL, FRANCIS WHITTIER. 1936. Travels and scientific collections of Thomas Nuttall. *Bartonia* 18: 1-51.

SHEAR, CORNELIUS LOTT & STEVENS, NEIL EVERETT. 1921. The correspondence of Schweinitz and Torrey. *Mem. Torrey Bot. Club* 16: 165, 175.

CHAPTER IX

BALDWIN, TRAVELING UP THE MISSOURI RIVER IN THE WESTERN ENGINEER *WITH LONG AND THE SCIENTIFIC SECTION OF THE "YELLOWSTONE EXPEDITION," GETS ONLY AS FAR AS FRANKLIN, MISSOURI*

STEPHEN Harriman Long, Major in the Corps of Topographical Engineers of the United States Army, was a graduate of Dartmouth College, class of 1809, and had taught mathematics at West Point. From the military academy he had been transferred to the Topographical Engineers in 1816 and is said to have been well-fitted for the tasks which he was called upon to perform. He commanded three important explorations[1] into the trans-Mississippi west and the second of these, made in 1819 and 1820, is considered in this chapter and the next. From the point of view of botanical work it is reasonable to separate the expedition's first year in the field from the second—which opens my next decade—for in 1819 William Baldwin, physician-botanist, served but a few months under Long, and when in 1820 the westward journey was resumed, Edwin James had been appointed to take Baldwin's place.

The history of the entire trip was compiled by Edwin James ". . . from notes of Major Long, Mr. T[homas] Say, and other gentlemen[2] of the exploring party . . ." and was first published in Philadelphia in 1823, under the title *Account of an expedition from Pittsburgh to the Rocky Mountains, performed in the years 1819 and '20 . . .* Another edition was issued in London in the same year. The title-pages of the American and British editions vary slightly.

R. G. Thwaites reprinted in 1905 (volumes 14 through 17 of his *Early western travels*) the narrative portion of James's *Account,* transcribing it from the London edition of 1823; the "Preliminary Notice," also included, he reprinted from the Philadelphia edition.

1. Long's first expedition, made in 1817–1818, crossed from the present state of Arkansas (following the river of that name) into eastern Oklahoma (then set apart for Indian occupancy), proceeded as far northwest as the mouth of the Canadian River, turned southeast across Oklahoma to Red River, and then northeast across Arkansas and Missouri to St. Louis. The journey had no botanical interest and Long's journal was never published. However, like the earlier explorations of Sibley, of Freeman, and of Dunbar and Hunter, it played a part in opening up the country lying west of the Mississippi River.

Long's third expedition of 1823 followed the Minnesota River and the Red River of the North to the 49th parallel and is described in my chapter XII since Thomas Say made some plant collections.

2. Although William Baldwin is not mentioned specifically in this phrase from the title of James's compilation, his notes are frequently quoted for the period when he served with the expedition.

In following the expedition's first year in the field I shall transcribe from the Philadelphia edition (1823) of James's *Account*.

The "Yellowstone Expedition" was divided into two sections; one was military, commanded by Colonel Henry Atkinson, and had to do with the establishment of military posts on the upper Missouri; its work bears no relation to the subject of this chapter. The second section, commanded by Long, was the first government-sponsored expedition to include scientists and for that reason, if for no other, may be considered memorable.

According to the orders of John C. Calhoun, Secretary of War, Long was to ". . . assume the command of the Expedition to explore the country between the Mississippi and the Rocky Mountains"; more particularly, to ". . . first explore the Missouri and its principal branches, and then, in succession, Red River, Arkansa and Mississippi, above the mouth of the Missouri. The object of the Expedition, is to acquire as thorough and accurate knowledge as may be practicable, of a portion of our country, which is daily becoming more interesting, but which is as yet imperfectly known . . . You will if practicable, ascertain some point on the 49th parallel of latitude, which separates our possessions from those of Great Britain . . . You will enter in your journal, every thing interesting in relation to soil, face of the country, water courses and productions, whether animal, vegetable or mineral. You will conciliate the Indians by kindness and presents, and will ascertain . . . the number and character of the various tribes, with the extent of the country claimed by each. The Instructions of Mr. Jefferson to Capt. Lewis, which are printed in his travels, will afford you many valuable suggestions . . ."

The "Preliminary Notice" (Philadelphia edition, 1823) in which these instructions appear, concludes: "It will be perceived that the travels and researches of the Expedition, have been far less extensive than those contemplated in the foregoing orders . . ." This change, which affected the expedition's second year of field work, was attributed to the state of the national finances during 1821—Long's party had returned in 1820—which had called ". . . for retrenchments in all expenditures of a public nature . . ."

Our interest in the expedition's first year in the field begins when William Baldwin and the rest of the scientific staff arrived at St. Louis on June 9, 1819, and ends with Baldwin's death at Franklin, Missouri, on September 1 of that year. Baldwin, born in Pennsylvania, was the first American with botanical training to collect plants westward from the Mississippi River.

On March 31, 1819, Long had issued instructions to members of his party assembled in Pittsburgh. Baldwin's task was no mean one.

"Dr. Baldwin will act as Botanist for the expedition. A description of all the products of vegetation, common or peculiar to the countries we may traverse, will be required of him, also the diseases prevailing among the inhabitants, whether civilized or savages, and their probable causes, will be subjects for his investigation; any variety in the anatomy of the human frame, or any other phenomena observable in our species, will be particularly noted by him. Dr. Baldwin will also officiate as Physician and surgeon for the expedition."

Thomas Say—who later earned fame as an entomologist and ornithologist—was zoölogist. Titian Ramsey[3] Peale was assistant naturalist with the duty of collecting specimens in the various branches of the scientific work; he is said by Sellers to have ". . . made many of the sketches used in illustrating the papers by members of the party." As son of the proprietor of Peale's Museum, he doubtless was instrumental in obtaining specimens for his father's institution.[4] There was also a geologist, Augustus Edward Jessup, and a painter of portraits, Samuel Seymour.[5] The expedition, in its membership, gave promise of great things.

Its work was not to be hampered by precedent: in geology, ". . . as also in Botany and Zoology, facts will be required without regard to the theories and hypotheses that may have been advanced on numerous occasions by men of science." And Baldwin wrote Darlington on April 5, 1819, that the volunteer journalist of the expedition, Major John Biddle of Philadelphia, ". . . is instructed 'not to interfere with the records to be kept by the Naturalists attached to the expedition.' " At the end of each trip all records and specimens were to be placed at the disposal of Long as agent for the United States government.

Darlington, in a letter dated December 23, 1818, had urged upon Calhoun Baldwin's appointment as botanist. The letter went without Baldwin's knowledge; he wrote Darlington on January 7, 1819, that he had heard from Long that "The

3. Peale's middle name occurs in the literature of the time in a variety of spellings, sometimes as "Ramey," but is usually abbreviated to "R." H. W. Sellers uses "Ramsey" and this is doubtless correct.

4. Joseph Ewan reports that "A large number of the zoological specimens taken on the expedition were deposited in Peale's Philadelphia Museum, founded in 1784 by Charles Wilson [Willson] Peale and containing as a nucleus the bones of a mammoth and a stuffed paddlefish . . . The expedition specimens were placed there due to the efforts of Titian R. Peale . . . and to Thomas Say . . . These zoological materials were all subsequently damaged or destroyed by dermestid infestations, with few exceptions . . ."

One finds many references to Peale's Museum in the literature of the period for it was customary, for foreigners in particular, to visit its wonders on arrival in Philadelphia. Charles Joseph Latrobe's *The rambler in North America* (ed. 2, London, 1836, p. 115) tells what that Australian visitor thought of its approach to scientific matters: when he asked why, ". . . in opposition to the opinion of the learned, the tusks of the enormous fossil Mastodon appeared . . . with their points turned down . . .", he was told by Peale—after he had given many reasons ". . . philosophical, physiological, zoological and osteological . . ."—that ". . . the ceiling of said museum was not lofty enough for him to place them with the points up."

Peale's enterprise was not the only one of its kind in America. William Bullock who, as noted elsewhere, had a museum of his own in Liverpool, visited a "Big Bone Museum" when he passed through New Orleans on his way to Cincinnati in 1827, and there carefully measured ". . . what are believed to be those of a stupendous crocodile . . ." recently found in a swamp near Fort Philip. He was of the opinion that the bones were those of a whale until otherwise convinced. Bullock was rapidly becoming "Americanized" for he ". . . offered a considerable sum for these immense remains, but the proprietor refused to part with them . . ."

5. Thwaites includes short biographies of Baldwin, Say, Jessup, Peale, etc., but notes that nothing is known of Samuel Seymour's life.

Botanic chair will be filled by yourself [Baldwin]; the Zoologic, by Dr. Say of Phila-delphia; and the Geologic by Dr. John Torrey[6] of New York."

1819

When Pittsburgh was left in early May, the party set off with "high expectations," everyone in good health except Baldwin. On arrival at Cincinnati on May 9 there was a delay on his account but by the 18th, although he had to be assisted on board the boat, he ". . . thought himself sufficiently recovered to proceed on the voyage . . ." By the 30th they had reached the mouth of the Ohio River and began the ascent of the Mississippi to St. Louis. The steamboat *Western Engineer* in which they traveled was in constant need of repairs and, although while these were being made opportunities for going ashore occurred, Baldwin was rarely able to join such excursions: "Plants were, however, collected and brought to him on board the boat, where he spent much of his time in the examination of such as were interesting or new." Baldwin's botani-cal notes, edited by James, are included in the text or in footnotes of the *Account*.

On June 9 the party was at St. Louis. Baldwin soon had the pleasure of meeting John Bradbury who had returned from England and was planning to make his resi-dence in the region which he had known ten years earlier.

"On the western borders of this prairie are some fine farms. It is here that Mr. John Bradbury, so long and so advantageously known as a botanist, and by his travels into the interior of America, is preparing to erect his habitation. This amiable gentleman lost no opportunity during our stay at St. Louis, to make our residence there agree-able to us."

On June 11 Baldwin refers to Bradbury—then fifty-four years of age—in a letter to Darlington:

"The venerable Mr. Bradbury called on me yesterday, and spent the day. His com-pany had a most exhilarating effect upon my health and spirits. In looking over my collection, I begged him to claim any thing he found, that might be his own. It turned out that a few, which I had marked for new, were known to him,—but he requested me to describe them: and observed, that since Lambert had pirated from him his former collections, it was not his intention to publish independently,—and that he would, with great pleasure, place in my hands all that he possessed, for publication:—

6. A letter from Torrey to Amos Eaton, dated February 16, 1819—it is quoted by Rodgers—describes the terms under which the scientists were to participate:

" 'The *terms* are,—that the naturalists will be provided with board, & receive protection—the papers, drawings &c are to be given up to [the] government, who are to have the entire disposal of them—the naturalists to furnish themselves at their own expence . . . the expedition is to traverse . . . about 30,000 miles! The time occupied . . . will probably be from three to five years . . . No compensation will be al-lowed the naturalists. I need hardly ask you, how you would have determined . . . in my situation . . .' "

These stipulations evidently did not offer any great inducement to Torrey—perhaps he could not afford to devote so many years of his life to a task which promised no financial recompense—for he did not join Long's party.

and that he should continue to pursue the Science for the intrinsic love he had for it,—and continue to furnish me with descriptions, and specimens, to be published as I might see fit, under his name and authority. As this was the first interview, many inquiries, which I had intended to make, were omitted."

The *Account* tells that "The grassy plains to the west of St. Louis, are ornamented with many beautifully flowering herbaceous plants. Among those collected there, Dr. Baldwin observed[7] the aristolochia Sipho, cypripedium spectabile ['C. parviflorum?'], lilium catesbeiana, bartsia coocinnea, triosteum perfoliatum, cistus canadensis, clematis viorna, and the tradescantia virginica. The borders of this plain begin to be overrun with a humble growth of black Jack ['quercus nigra'] and the witch hazel ['Hamamelis virginica'], it abounds in rivulets, and some excellent springs of water, near one of which was found a new and beautiful species of viburnum."

On June 21, after completing arrangements at St. Louis, the party ". . . left that place at noon . . ." and on the 22nd ". . . at ten o'clock . . . entered the mouth of the Missouri." Between this point and "Bellefontain," or Bellefontaine, four miles away, the boat grounded twice on one sand bar. At Bellefontaine "Dr. Baldwin found . . . a plant, which he considered as forming a new genus, approaching astragalus; also the new species of rose, pointed out by Mr. Bradbury, and by him called Rosa mutabilis. This last is a very beautiful species, rising sometimes to the height of eight or ten feet. The linden tree attains great magnitude in the low grounds of the Missouri; its flowers are now fully expanded."

A footnote, taken from *"Dr. Baldwin's MS. Notes,"* identifies the linden as "Tilia Americana" and adds: "The Podalyria alba, anemone virginiana, polygala incarnata (prairies) anagallis fluviatalis, carex multiflora, &c. were collected at Bellefontain."

The *Western Engineer* reached St. Charles, only about twenty miles from the starting point, on June 24. On the 25th Baldwin wrote Darlington:

" I think it was since I wrote last, that I found at St. Louis, I think, a new genus. Its habit is *Astragalus* . . . I have made out a description of it,[8]—and Mr. Bradbury has promised to attend to the flower, and communicate an account of it to me . . . It ought, perhaps, to be called Bradburya. Mr. Bradbury brought me, a few days after, another *outrè* plant—habit of *Astragalus* exactly . . . but the seed-vessel is too young for investigation . . . This Mr. B. is to describe and send to me."

At St. Charles some of the party—Say, Jessup, Peale and Seymour—decided to ascend the Missouri on foot to Loutre or Otter Island. "Dr. Baldwin, still confined by debility and lameness, was compelled to forego the pleasure of accompanying them." After their departure, while the boat lay at anchor one evening, ". . . Dr. Baldwin was able to walk a short distance on shore, but retur[n]ed much fatigued by his exertions."

7. The Philadelphia edition of the *Account* supplies the names of the plants mentioned without any attempt at consistency and neither capitalization nor italics are used with any uniformity. Some scientific names are supplied in footnotes and I have quoted these between brackets in the text.

8. The letter includes a somewhat lengthy description.

Nevertheless, he wrote of the region's plants and his comments are included in a foot-note of the *Account*:

"The vegetable productions of this place were, the populus deltoides, occupying the narrow margin of the river . . . the amorpha fruticosa,[9] and platanus occidentalis . . . The margin of the bluff produces the quercus rubra, juglans pubescens, carpinus Americana, (around the latter, we observed the celastrus scandens, entwined and in fruit,) and on higher grounds, the laurus sassafras and juniperus Virginianus. Of herbaceous plants, the only one in flower was the Rudbeckia fulgida. The higher parts of the hills were in many places thickly covered with species of Elymus and Andropo-gon . . ."

Under date of June 27 the *Account* continues:

". . . we crossed over to the right hand [north] side of the river . . . At evening we came to anchor half a mile below point Labidee, a high bluff . . . Here we were detained a day making . . . repairs . . . The shore . . . was lined with the common elder, (Sambu-cus canadensis) in full bloom, and the cleared fields were yellow with the flowers of the common mullein. This plant, supposed to have been originally introduced from Europe, follows closely the footsteps of the whites. The liatris pycnostachia here called 'pine of the Prairies,' . . . was now in full bloom . . . The Indian interpreter . . . showed me some branches of a shrub . . . much used among the natives, in the cure of Lues venerea . . . It is called 'blue wood' by the French, and is the Symphoria racemosa of Pursh . . . here rather taller, and the branches less flexuous than in the eastern states."

When the boat arrived at Loutre or Otter Island on July 2 the pedestrians were al-ready there: "They accomplished no more than they would have done on board the boat . . ." Moreover, they had had a miserable experience. Not in training for strenu-ous exercise, some of the men soon became "somewhat unwell," while all had suffered from thirst, exposure and exhaustion. Since they did not know the propensities of pack-horses, the only one they had, although "fettered," escaped twice and was last seen heading back to St. Charles. Baldwin wrote Darlington from Franklin on July 22 that after this loss the men ". . . were obliged to carry their equipage upon their backs . . . Young Peale has ever since been confined, with inflammation in his feet . . ."

Leaving Loutre Island on July 3, the boat came to anchor on the 5th ". . . above the village of Cote Sans Dessein." This town was about two miles below the main mouth of the Osage River and consisted of some forty families living in small log cabins. There was ". . . a tavern, a store, a blacksmith's shop, and a billiard table." The *Account* records that "The Cane ['Miegia macrosperma of Persoon'] is no where met with on the Missouri; but its place is in part supplied by the equisetum hiemale, which, remain-ing green through the winter, affords an indifferent pasturage for horned cattle and horses; to the latter, it often proves deleterious . . ."

9. A footnote to the footnote adds "This beautiful flowering shrub occupies the low lands of Georgia, on the sea coast, but is not confined to the margin of rivers, as appears to be the case on the Missouri."

Here we are told a good deal about *"milk sickness,"* the dissertation ending with the comment that "Dr. Baldwin was of the opinion that the *milk sickness* of the Missouri, did not originate from any deleterious vegetable substance eaten by the cows, but was a species of typhus, produced by putrid exhalations, and perhaps aggravated by an incautious use of a milk diet."

We hear of Baldwin's increasingly poor health:

"During the few days we remained at Cote Sans Dessein,[1] Dr. Baldwin, though suffering much from weakness, and yielding perceptibly to the progress of a fatal disease, was able to make several excursions on shore. His devotion to a fascinating pursuit, stimulated him to exertions for which the strength of his wasted frame seemed wholly inadequate; and it is not, perhaps, improbable that his efforts may have somewhat hastened the termination of his life."

On—possibly—July 13 they arrived at Franklin, Howard County, Missouri, across the river from "Boonsville," or Boonville, Cooper County. The town of Franklin, ". . . at present increasing more rapidly than any other on the Missouri, had been commenced but two years and a half before the time of our journey." In that short time it had already developed into an important settlement.

"Dr. Baldwin's health had so much declined that . . . he was induced to relinquish the intention of ascending farther with the party. He was removed on shore to the house of Dr. Lowry,[2] intending to remain there until he should recover so much strength as might enable him to return to his family. But the hopes of his friends, even for his partial recovery, were not to be realized. He lingered a few weeks after our departure, and expired on the thirty-first of August.[3] His diary, in which the latest date is the eighth of August, only a few days previous to his death, shows with what earnestness, even in the last stages of weakness . . . his mind was devoted to the pursuit, in which he had so nobly spent the most important part of his life."

James wrote: "To show the scope and accuracy of his method of observation . . . we subjoin a part of the observations registered in Dr. Baldwin's diary,[4] from July fifteenth, the time of our departure from Cote Sans Dessein, to its conclusion. From this the reader will be able to form a satisfactory idea of the vegetable physiognomy of the country on this portion of the banks of the Missouri."

1. The dates of arrival and departure from Cote Sans Dessein and of arrival at Franklin are confused in the *Account*. On p. 80 the date of arrival at the first is given as July 5; on p. 83 it is stated that they remained there a few days; on p. 85 the date of departure is given as July 6; on p. 95 as July 15. The arrival at Franklin is given on p. 88 as July 13.

2. John J. Lowry.

3. Lowry wrote Darlington that Baldwin died on September 1 and he was certainly in a position to know the facts.

4. Dr. H. W. Rickett informed me (*in litt.*, April 5, 1949): "Your information is correct that we [the New York Botanical Garden] have the manuscript of William Baldwin's diary. It is a collection of some 50 pages covered with very fine handwriting and apparently fragmentary. I should say that considerable work would be necessary to get the items in order and collated."

A long footnote is appended which I quote in part only:

"Above Cote Sans Dessein, we saw frequently the Juglans nigra, and J. pubescens, called white hickory, also a species of Cratægus which, though sometimes seen in Pennsylvania, appears to be hitherto undescribed. Its fruit is large, yellow when ripe, and of an agreeable flavour . . ."

"Franklin, July 15th. Portulacca sativa, Solanum nigrum, Urtica-pumila, Datura strammonium, and Phytolacca decandra, occur by the road side. Blackberries were now ripe, but not well flavoured. Campanula americana, the large Veronia mentioned at Cote Sans Dessein, now flowering.

"Some plants were brought in, among which we distinguished the Monarda fistulosa, Achillea millefolia, Cacalia atriplicifolia, called 'horse mint,' Queria canadensis Menispermum lyoni ? Verbena urticifolia. The Annona triloba is frequent about Franklin, also the Laurus benzoin, and the Symphoria now in flower, the Rhus glabrum, Cercis canadensis, Ampelousis quinquefolia, Eupatorium purpureum, in flower. Cucubalus stellatus, still flowering. The Prickly fruited Æsculus has nearly ripened its nut, Zanthoxylon clava herculis in fruit, a 'wild gourd' not in flower.

"July 26th. The Gleditschia is a small tree here, Geum album, Myosotis virginiana, Amarathus hybridus, Erigeron canadense, Solanum Carolinianum, very luxuriant and still flowering. The leaf of the Tilia glabra, I found to measure thirteen inches in length, and eleven in breadth. Bignonia radicans, Dioscorea villosa, a Helianthus with a leaf margined with spines, the narrow leaved Brachystemum, the Liatris pycnostachia, Rudbeckia purpurea, and various others in flower. Juglans porcina, and cinerea, Ostrya virginica, Rhus copallinum.

"August 4th. Dr. Lowry informed me he has seen Pyrus coronaris, forty feet in height in the forest about Franklin. He showed me a Rudbeckia about three feet high with a cone of dark purple flowers probably a new species.

"5th. Eupatorium hieracifolium beginning to flower, Menispermum canadense here called 'sarsaparilla,' its slender roots being substituted for that article.

"6th. A Mimulus is found here resembling M. ringens, but the leaves are not sessile; peduncle very short, flowers large, pink coloured, stem acutely quadrangular, Campanula Americana, three and a half feet high?"

James pays what, in this day and generation, one might call a somewhat "flowery" tribute to his predecessor William Baldwin. I quote but a fraction:

"He has left behind him a name which will long be honoured . . . His manuscripts were numerous, but his works were left unfinished. The remarks on the Rotbollia . . . are his only productions . . . hitherto before the public. His Herbarium, . . . has contributed to enrich the works of Pursh and Nuttall. He was the friend and correspondent of . . . Muhlenbergh, [Muhlenberg] and contributed materials for the copious catalogue of North American plants, published by that . . . botanist. In South America he met with Bonpland . . . and a friendly correspondence was established between them . . . He had travelled extensively . . . His notes and collections are extensive and valuable. During the short period of his connection with the exploring party . . . infirmities . . . could not overcome the activity of his mind, or divert his attention entirely from his favourite pursuit . . . in the course of the voyage from Pittsburgh to Franklin, [he] detected and described many new plants, and added many valuable observations relating to such as were before known . . ."

Leaving Baldwin in kind hands, Long and his party, on July 19, started on their way up the Missouri River. On September 17 the *Western Engineer* arrived at ". . . the trading establishment of the Missouri fur company, known as Fort Lisa . . ." This was on the west side of the Missouri River, about five miles below the promontory known as Council Bluff.[5] Two days after arrival, work was begun on "Engineer Cantonment," winter headquarters of the expedition. According to the *Account,* the site was ". . . on the west bank of the Missouri, about half a mile above Fort Lisa, five miles below Council Bluff, and three miles above the mouth of Boyer's river."—or in the southeastern corner of Washington County, Nebraska. On October 11 Long left for the east, the encampment having been completed early in the month. He got back to "Engineer Cantonment" on May 28, 1820, bringing with him Edwin James who had been appointed to take Baldwin's place. From there the westward journey was resumed on June 6. The story of the exploration of 1820 is told in my next chapter.

Baldwin's letters to his friend William Darlington give a franker picture of the difficulties experienced by the scientists on the trip up the Missouri than does the *Account* prepared by James for public consumption. According to Baldwin, the party's expectations had been high on leaving Pittsburgh on April 5, 1819: "Our boat is 65 feet keel, and draws but 19 inches of water. Our accommodations for books, clothes, &c. will be commodious and comfortable." Things looked less auspicious by May 1: "It was found, that in consequence of the weight of baggage, stores, &c. sinking the boat much deeper than was expected, our wheel . . . was also too deep . . ." The new machinery was stiff and, because fuel was bad in quality, the necessary steam could not be kept up—and so on *ad infinitum.* Baldwin tried to be optimistic: ". . . when we take into consideration the complicated structure of a steam boat . . . I do not see that any thing happened which might not reasonably have been expected . . ." He wrote from Shawaneetown on May 27: ". . . the field of botanizing is now becoming very rich; but I want a little more strength to make the most of it."

From St. Louis, on June 11: ". . . this boat,—hastily constructed, and built entirely of unseasoned timber,—is almost daily in want of repairs; and is so leaky and wet, that we have not a dry locker for our clothes. A great part of my stationary [presumably the collecting-paper] has been wet, and a portion of it entirely lost. It will be with the utmost difficulty that I shall save the specimens I may collect."

From St. Charles, on June 25, he wrote that the boat ". . . has done much better since we left St. Louis; and I hope I shall be able to save all my specimens."

But, finally, from Franklin, on July 22:

5. The present city of Council Bluffs is east of the Missouri, in Pottawatamie County, Iowa.

"I have at last the mortification to inform you, without hesitation, that a steam boat is not calculated for exploring . . . Slow as has been the progress of this boat, since our entrance into this river, little opportunity has been afforded to the Naturalists to do any thing. There has been no stopping, except to take in wood and water and to repair . . . Not one moment has been granted to the Naturalists, to explore, that could be avoided; and the most productive situations have all been passed by . . . No mode of travelling is so poorly calculated for Naturalists: and besides, it is the most expensive to government—the least expeditious, and safe, of any mode of travelling."

This was Baldwin's last letter and it ends thus: "The mail closes presently; and I feel myself too much indisposed to write, or to think much . . . I remain here until I recruit —if to recruit be my lot . . ."

Baldwin's story is a pathetic one,—a desperately sick man driven by enthusiasm for his task. He was only forty years old at the time of his death.

Dr. Lowry, in whose home Baldwin died, wrote Darlington on September 25, 1819, that the botanist had asked him ". . . to transmit his private papers to Mrs. Baldwin, and the botanical notes to yourself and Z. Collins, Esq., which will be done as soon as practicable." Darlington, who failed to receive the notes and who understood that Long had delivered them "to the late Mr. Collins," would have liked to possess these records but ". . . felt a delicacy in interfering." After they had come into Torrey's hands he was able to examine them: ". . . although replete with materials which might have been turned to good account by Dr. B. himself . . . they are, as he supposed, in too crude and imperfect a state to be used with much advantage, by other hands."

The plant collections made in the two years that Long's expedition was in the field[6] went eventually to Torrey, but not all at one time. Since they comprised not only Baldwin's specimens but those of James as well, I refer to them in my next chapter. Torrey wrote Darlington on August 26, 1826—the letter is quoted by Rodgers—that he had received the "remainder of the plants" comprising " '. . . all that Dr. Baldwin collected as well as Dr. James' plants. The former have Dr. B's labels attached to them. Some of the specimens, the labels of wh are written on rough draughts of letters, (I believe to his family) I will send to you. As you [were] a friend of the deceased botanist & know his memory these relics will I hope, be acceptable . . .' "

6. Darlington wrote of what must have been Baldwin's personal herbarium:
". . . although Mrs. Baldwin repeatedly informed me it was the Doctor's wish I should have such portion of it as I might desire,—I declined taking any part of it. The circumstances of the family required that the best disposition should be made of it, for their benefit . . . as I could not afford to pay as much for it as I thought it was worth, I disdained to take advantage of their generosity,—or in any degree to impair its value. Mr. Collins afterwards purchased it, with the view, as he told me, of placing it in the Philadelphia Academy of Natural Sciences. If I am not misinformed, however, his representatives *sold* it to the late Rev. Mr. Schweinitz,—who finally bequeathed it to the Philadelphia Academy."
Von Schweinitz's purchase was made in 1833 and Torrey wrote the buyer on November 2: "I rejoice in your acquisition of Dr. Baldwin's plants, though you may suspect that my joy is not without some selfish feeling,—for you generously offer me a share of your duplicates." One is amused, more than once, at the rapid dissemination of news in the botanical fraternity; also at the competition, usually friendly, which was indulged in by its more important members!

Only two botanical publications are attributed to Baldwin, and neither concerns this story. Harshberger comments:

"Fortunately his unpublished memoranda fell into the hands of Dr. Torrey, and though in a crude and fragmentary state . . . were used as their author would have wished, as contributions for Dr. Torrey's monograph of the *Cyperaceæ,* and for Dr. Gray's monograph of *Rhynchospora* . . . in Annals of the New York Lyceum of Natural History, vol. III. . . ."

Baldwin had wished to honor his friend Darlington and wrote him (January 21, 1819): "There can hardly be a doubt on the subject of my finding the *Darlingtonia.* It is a plant that I have been seeking for; and I shall cherish it as the choicest of my discoveries." To this the modest Darlington appends:

"I had jocosely suggested to Dr. B. that he might, perhaps, find the *Darlingtonia ignota,* somewhere along the margin of the Missouri: and it so happened, that his successor in the expedition, Dr. E. P. James, *did* find a species of the genus, afterwards named *Darlingtonia,* by Prof. De Candolle. *See Annals of New York Lyceum, Vol. 2, p. 101.*"

Baldwin himself was commemorated: in Nuttall's *Genera* is described the genus *Balduina,* "Dedicated as a just tribute of respect for the talents and industry of William Baldwin, M.D., late of Savannah, Georgia, a gentleman whose botanical zeal and knowledge has been rarely excelled in America." This herb belongs to the family of composites and the type species is *B. uniflora.* In Torrey and Gray's later *Flora of North America* the spelling was changed to *Baldwinia.* Baldwin wrote Darlington on March 21, 1819, of having seen in the collection of a German "of considerable botanical knowledge" a specimen labelled *Balduina uniflora* given him by Nuttall and ". . . was glad of this opportunity to tell him that the name would not be adopted in this country; and that it would shortly appear, named and described, from another quarter." Darlington explains in a footnote that Baldwin referred to a promise made by Elliott to publish upon the plant ". . . by another name; as Dr. B. was not pleased with the genus,—and was particularly dissatisfied with the orthography of *Balduina.* The genus, however, seems to be definitely established; though the orthography may be ultimately changed to *Baldwinia;* and such, I understand, is the intention of Drs. Torrey & Gray." [7]

In 1844, when reviewing Darlington's *Reliquiae Baldwinianae,* Asa Gray had a number of pleasant things to say of Baldwin and his work. But, to end on an anticlimax, the seventh edition of Gray's *Manual* refers to William Baldwin as a "Discriminating amateur botanist . . ."

The devotion existing between Darlington and Baldwin is evident in all their many letters. The renowned Dr. Howard A. Kelly, who wrote a short sketch of Baldwin, evidently felt the man's appeal and I confess to a similar reaction. While only a few

---

7. In the seventh edition of Gray's *New Manual of botany* the spelling *Balduina* is still retained.

months of the man's work have a direct bearing upon this story I shall tell a little of his life as recounted by Darlington in the "Biographical sketch" prefacing the collection of their correspondence:

Baldwin was born on March 29, 1779, of Quaker parents, in Newlin Township, Chester County, Pennsylvania, and received ". . . no other than the common school education afforded by the country schools . . . at that day . . . of very moderate pretensions." After teaching for a time, he turned to medicine as a profession, working with Dr. William A. Todd, a practitioner in Downingtown and attending medical lectures at the University of Pennsylvania; but, for financial reasons, these were not long continued. Darlington was a classmate and soon became an intimate friend. He tells how Baldwin, like a "ministering angel," nursed him through a severe illness. It was while at Downingtown that Baldwin became the intimate friend of Dr. Moses Marshall, nephew and heir of Humphry Marshall, owner of a famous garden at Marshalltown. The uncle, like the nephew, was "a respectable Botanist." This friendship, and the interest of Dr. Benjamin Smith Barton, stimulated Baldwin's "Botanical zeal."

In 1805, ". . . although, not yet an M.D. . . .", Baldwin sailed as surgeon on a merchant ship bound for Canton, by way of Antwerp. A fellow passenger reported that he embarked ". . . with only three shirts in his wardrobe!" Back from this trip in 1806, he got his medical degree in 1807 and began practice in Wilmington, Delaware, where he married Miss Hannah M. Webster ". . . a lady of superior intellectual endowments . . . whose education had received a classical finish quite unusual among American Females at that day." His leisure was devoted to plants and he began a correspondence with the Reverend Henry Muhlenberg which continued until the death of "that eminent and accomplished Botanist."

Because of his health—Darlington notes that "There was an hereditary predisposition to Pulmonary Consumption, which pervaded the whole family,—who were all finally swept away by that insidious destroyer . . ."—Baldwin moved to Georgia (to Savannah and to St. Mary's) where he pursued "Botanical research," traveling on foot with a knapsack on his back through Indian country, often entirely alone; but, ". . . such was his gentle, inoffensive demeanor . . ." that the savages always treated him well. Notes Darlington: ". . . there was . . . more of the genuine 'milk of human kindness' . . . in the composition of Doctor Baldwin, than in any man it was ever my happiness to know."

During the War of 1812 there was a period of medical service in the United States Navy with, finally, a commission and, for two years, an appointment at St. Mary's where ". . . for a considerable part of the time he had neither mate nor loblolly boy." [8] Harshberger tells that "For two years he ministered to the sick and distressed with no

8. Loblolly was a thick gruel or such, and the name was often used for a nautical dish or simple medicinal remedy. A loblolly boy assisted a ship's surgeon in his duties. See *A new English dictionary on historical principles* . . . (Oxford).

other aid than that of his wife." During a two-year period at Savannah he began a correspondence with Stephen Elliott, author of the *Sketch of the botany of South Carolina*. In 1816 there was a trip to eastern Florida—Darlington includes many letters written at this period. In 1817 he served on the frigate *Congress* with the "collateral duty" of studying the "vegetable productions" of the South American countries visited. According to Harshberger: "At all places [on this voyage] he made diligent use of his limited opportunities for collecting, and in the Philadelphia Academy are preserved many of the plants . . ." In 1818 Baldwin was back in Wilmington. Next came the appointment to the Long expedition with its sad termination.

Baldwin left a wife and four children. The ever-faithful Darlington tried—but in vain—to secure them a pension from the United States Navy Department.

Before leaving on the trip with Long, Baldwin's portrait had been painted by Charles Willson Peale whom Baldwin, and in this instance correctly for Peale had been born in 1741, refers to as "the old gentleman." It has been reproduced more than once and depicts a delicate, sensitive and kindly face. Baldwin wrote Darlington (March 6, 1819): "P. S. I have this morning (by request) sat for my portrait." Again (March 21): "My portrait is completely finished; and ought to be,—as I have sat little short of 12 hours. The old gentleman considered it one of his most finished performances; and spoke of sending it, (on this account—and not, I presume, on account of my beauty—) to the Academy of Fine Arts, as a specimen of his finished workmanship. We have all requested to remain nameless,—and to be deposited in a private apartment . . ."

Peale did portraits of others of the party. Baldwin considered the one of Long ". . . defective,—particularly about the eyes. The old gentleman complained of his never sitting well: the last time he sat, he was drowsy, from loss of sleep (in consequence of his wife's indisposition). This drowsiness is manifest in the picture."

The prospective "Yellowstone Expedition" had received much publicity and, even in 1819, reporters seem to have been active. Baldwin wrote Darlington, May 1, 1819:

"You may have seen a most erroneous and ridiculous account of us, in a New York paper of the 14th ult. which is most unfortunately ascribed to one of our young officers. It seems that some editors of newspapers are so fond of news, that they will publish extracts from letters, without seeing them. Certain it is . . . that we have not a young officer on board, ranking above a corporal, that would ever have written the extract . . . which is now going the rounds."

CHITTENDEN, HIRAM MARTIN. 1902. The American fur trade of the far west. A history of the pioneer trading posts and early fur companies of the Missouri Valley and the Rocky Mountains and of the overland commerce with Santa Fe. Map. 3 vols. New York. 2: 562-572.

DARLINGTON, WILLIAM. 1843. Reliquiae Baldwinianae: selections from the correspondence of the late William Baldwin, M.D., Surgeon in the U. S. Navy. Philadelphia.

EWAN, JOSEPH. 1942. Botanical explorers of Colorado—II. Edwin James. *Trail & Timberline* 282: 79-84.

———— 1950. Rocky Mountain naturalists. Denver.

GHENT, WILLIAM JAMES. 1931. The early far west. A narrative outline 1540–1850. New York. Toronto. 155-158.

GOODWIN, CARDINAL. 1922. The trans-Mississippi west (1803–1853). A history of its acquisition and settlement. New York. London. 47-54.

GRAY, ASA. 1844. [Review of] Reliquiae Baldwinianae . . . compiled by William Darlington . . . Philadelphia, 1834. *North Am. Review* 46: 192-195.

HARSHBERGER, JOHN WILLIAM. 1889. The botanists of Philadelphia and their work. Philadelphia. 119-125.

JAMES, EDWIN. 1823. Account of an expedition from Pittsburgh to the Rocky Mountains, performed in the years 1819 and '20, by order of the Hon. J. C. Calhoun, Sec'y of War; under command of Major Stephen H. Long. From the notes of Major Long, Mr. T. Say, and other gentlemen of the exploring party. Compiled by Edwin James, botanist and geologist for the expedition. 2 vols. Atlas. Philadelphia.

———— 1905. Account of an exploration from Pittsburgh to the Rocky Mountains, performed in the years 1819, 1820. By order of the Hon. J. C. Calhoun, Secretary of War, under the command of Maj. S. H. Long, of the U. S. Top. Engineers. Compiled from the notes of Major Long, Mr. T. Say, and other gentlemen of the party, by Edwin James, botanist and geologist of the expedition. *In:* Thwaites, Reuben Gold, *editor.* Early western travels 1748–1846. 14, 15, 16, 17. [The editor reprints the "Preliminary Notice" from the first volume of the Philadelphia edition of 1823. His text is reprinted from the first volume of the London edition of 1823.]

KELLY, HOWARD ATWOOD. 1929. Some American medical botanists commemorated in our botanical nomenclature. New York. Toronto. 104-112 [William Baldwin], 113-119 [William Darlington].

REDFIELD, JOHN HOWARD. 1883. Some North American botanists. VI. Dr. William Baldwin. *Bot. Gaz.* 8: 233-237.

ROBINSON, BENJAMIN LINCOLN & FERNALD, MERRITT LYNDON, *editors.* 1908. Gray's new manual of botany. ed. 7. New York. 842 [under *Balduina*].

RODGERS, ANDREW DENNY, 3RD. 1942. John Torrey. A story of North American botany. Princeton. 47, *fn.* 6, 52.

SELLERS, HORACE WELLS. 1934. Titian Ramsey Peale. *Dict. Am. Biog.* 14: 351.

SHEAR, CORNELIUS LOTT & STEVENS, NEIL EVERETT. 1921. The correspondence of Schweinitz and Torrey. *Mem. Torrey Bot. Club* 15: 279.

VIETS, HENRY R. 1928. William Baldwin. *Dict. Am. Biog.* 1: 547, 548.

WARREN, GOUVENEUR KEMBLE. 1859. Memoir to accompany the map of the territory of the United States from the Mississippi River to the Pacific Ocean, giving a brief account of each of the exploring expeditions since A.D. 1800 . . . *U. S. War Dept. Rept. expl. surv. RR Mississippi Pacific* 11: 23, 24.

Incidental references to William Baldwin are also found in the literature cited in my chapter X.

1820-1830

INTRODUCTION TO DECADE OF

1820 – 1830

I N the early years of this decade the United States government sent three expeditions into eastern portions of the trans-Mississippi west all of which brought back collections of plants. The first, and as far as plant collections are concerned the important one, was the continuation of an expedition which had started on its way in 1819; the last two followed upon a treaty defining a portion of the boundary between Canada and the United States. As to the first:

The scientific section of the "Yellowstone Expedition" commanded by Stephen Harriman Long which, in the last year of our earlier decade had gotten as far west as present Washington County, Nebraska, on its way to the upper Missouri River, resumed its journey in 1820 but in another direction, for it had been decided in Washington to curtail expenditures. Further travel up the Missouri in the steamboat *Western Engineer* was stopped and Long was ordered to take his party overland to the sources of the Platte River and, from there, by way of the Arkansas and Red rivers, to reach the Mississippi. William Baldwin having died in 1819, Edwin James had been appointed to take his place. James was an American and was the first botanist to cross the greater part of Nebraska from east to west, the first to skirt the eastern slopes of the central Rocky Mountains in Colorado and to reach the alpine flora[1] on a famous peak; the first to cross the extreme northeastern corner of New Mexico, to traverse the Texas "Panhandle," and to follow the Canadian River across Oklahoma as far as its confluence with the Arkansas River.[2] James's collection of plants was described by John Torrey and in a manner worthy of note. In the words of Asa Gray:

"As early as the year 1823, Dr. Torrey communicated to the Lyceum of Natural History descriptions of some new species of James' collection, and in 1826 an extended account of all the plants collected, arranged under their natural orders. This is the earliest treatise of this sort in this country, arranged upon the natural system;[3] and with it begins the history of the botany of the Rocky Mountains, if we except a few plants collected early in the century by Lewis and Clark, where they crossed them many degrees farther north, and which are recorded in Pursh's Flora . . ."

The boundaries of the great region known as "Louisiana" had been poorly defined at the time of its purchase from France, and their adjustment to the satisfaction of the

---

1. James's collection of the alpine plants of Colorado antedated the one made by David Douglas on the western side of the Rockies in the more northern regions of Oregon, Washington and Idaho; it was earlier too than the collection made by Thomas Drummond in the Canadian Rockies.

2. From the confluence of these rivers eastward to Fort Smith, Arkansas, James had been preceded by the botanist Thomas Nuttall, who had reached the meeting of these waters in 1819 when exploring westward from Fort Smith.

3. See p. 245, *fn.* 7.

nations concerned involved prolonged negotiations—in the north with Britain and in more southern regions with Spain and, later, Mexico.

After the termination of the War of 1812, Britain and the United States, in 1818, reached agreement upon a portion of the boundary between Canadian and American possessions: from the northwestern corner of the Lake of the Woods the line was to run directly to the 49th parallel of north latitude and follow that parallel westward to the summit of the Rocky Mountains.[4] Two of the three expeditions mentioned above were sent out by the United States government after this boundary settlement. They explored regions lying south of the line and adjacent to or included in what is now the state of Minnesota. Since virtually nothing was known of their natural resources, these were to be investigated; and information was to be obtained about the Indian tribes living therein—these peoples were to be conciliated.

The first expedition went into the field in 1820 and was commanded by Lewis Cass, Governor of the Michigan Territory, who had as assistant Henry Rowe Schoolcraft. Its primary purpose was to discover the source, then unknown, of the Mississippi River. The party reached a lake which they named "Cassina"—now Cass Lake—and returned under the impression that they had attained their objective. David Bates Douglass, an engineer, made a small collection of plants which was described promptly by John Torrey.

The second expedition was sent out by President James Monroe in 1823, and was led by Stephen Harriman Long, back from his more famous journey of 1819–1820. Its purpose was to explore St. Peter's, or Minnesota, River and the country which lay along the recently determined boundary and Lake Superior. After ascending the Minnesota River, Long's party crossed to the headwaters of Red River of the North, descended that stream for a time and, close to Pembina, North Dakota, raised the American flag on the 49th parallel. From there the expedition returned eastward, mainly through British possessions. The task of collecting plants had been assigned to Thomas Say who, in later years, was to attain greater fame as an entomologist and conchologist than as a botanist. Say's small collection of plants was described by Lewis D. von Schweinitz of Philadelphia.

The boundaries of the more southern portions of "Louisiana" were, as mentioned, matters of controversy between the United States and Spain and, after 1821, Mexico, in two regions in particular: in what is now New Mexico—centering about Santa Fe— and in Texas. Very naturally it would seem, many Spanish and Mexican settlers were bewildered by and resented the sale of "Louisiana" to France and its prompt resale to the United States, and were alarmed by the growing interest of Americans in lands which they themselves had long controlled. Aligned with Spaniards and Mexicans in their distrust of Americans were the Indians of northern New Mexico.

4. Since no satisfactory agreement could be reached in 1818 regarding the extension of this boundary westward from the Rocky Mountains to the Pacific, the commissioners agreed upon joint occupation by the citizens of the two countries concerned for a period of ten years, as well as for ten-year extensions until agreement could be reached.

Enterprising individuals from the United States, recognising in Santa Fe a good center for trade and a desirable base from which to extend such operations westward, made little headway in what were perilous attempts to gain access to that city until after the Mexican War. The United States government, like its citizens, was evidently aware of the importance of Santa Fe and, from 1825 to 1827, had surveys made for a road between Fort Osage (situated just east of the confluence of the Kansas and Missouri rivers) and Taos (not far north of Santa Fe); maps resulting from the survey were, for reasons unknown, filed away unpublished, together with what G. K. Warren describes as ". . . accurate and minute notes[5] and directions for the use of travelers . . ."

Interest in Santa Fe was also displayed by the botanists residing in our eastern states and by others living as far away as England and Europe. These men were all eager to send their collectors thither—they seem to have had no conception of the dangers of such a trip, nor were they aware, apparently, that the only feasible route to Santa Fe—and for many years—started from St. Louis, Missouri. One reads of their eagerness that a collector should turn his steps thither from Texas perhaps or even from California! But it was not until 1846 that an emissary of importance reached[6] the El Dorado of the botanists.

Americans had greater success in their attempts to penetrate Texas although it was long before they could live there with security.[7]

The southwestern boundary of the region included in the Louisiana Purchase, like its northern boundary, had never been clearly defined. After lengthy negotiations between Spain and the United States a treaty, resented by some Americans, was signed in 1819 by which, in exchange for Florida, the United States surrendered Texas. The line between Spanish and American possessions was to run from the Gulf of Mexico along the west side of the Sabine River to the 32nd parallel of north latitude, thence north to the Red River of the South and along its south side to the 100th meridian where it turned north to the Arkansas River—and so on, to the Pacific coast.

5. The records ". . . were published for the first time so far as known, in 1913, in the *Eighteenth Biennial Report* of the Kansas Historical Society." (Ghent, W. J. 1931. The early far west ... 1540–1850. New York. Toronto. 198, *fn.*.)

Although there were three commissioners, the survey is usually designated by the name of the topographer Joseph C. Brown.

6. Gambel was the first plant collector to pass through Santa Fe, on his way to California in 1841. Wislizenus was there, on his way to Mexico City, a short time before the city fell to the Americans in 1846. Fendler, the first botanist to do any important collecting in the region, reached Santa Fe not long after its conquest, but even then he did not dare venture far into the mountains.

7. In 1821 Stephen Austin obtained a grant from the Spanish government upon the condition that a certain number of families should be settled thereon; others received similar grants and it has been estimated that, by the beginning of the decade of 1830–1840, some twenty thousand Americans had availed themselves of such opportunities.

Mexico, having won its independence from Spain on February 23, 1821, soon became disturbed by the influx of foreigners into Texas and, in 1830, passed a law intended to halt further immigration, but the tide of settlers continued to flow westward.

When, in 1828, Mexico sent a commission to investigate the eastern portions of this boundary between American and Mexican republics, there traveled with the commission a botanist, Jean Louis Berlandier, who had been sent from Switzerland in 1826 to collect plants for the elder De Candolle and his friends. The young man, the first to do such work in Texas, pursued his task—regardless of storms, illness and difficulties unimaginable to his sponsor—but when his collections did not measure up to what the Genevese botanist expected, Berlandier was pictured to the world as delinquent and worse. In the light of records concerned with his accomplishment, Berlandier's rôle appears to have been more creditable than the one played by the renowned De Candolle.

Turning from the eastern portions of the trans-Mississippi west to those along the Pacific coast, one again enters regions where Spain (in California) had long been in power and where Great Britain (in the northwest) controlled the one important industry, the fur trade. In both these regions the interest of Americans—already demonstrated by the advent of Lewis and Clark and of Hunt and the Astorians—was known and, before many years had elapsed, was to be increasingly distrusted.

In California, where Mexico in 1822 had succeeded to the control once exercised by Spain, Spanish residents were not only resentful of changes inaugurated by the new authorities, but confused as well—the padres who had been granted rights by Spain still felt that their allegiance lay to that country. It was with the advent of Mexican rule that the decline of the missions began. There was a growing suspicion, largely instigated it would seem by the Mexican authorities, of any interest displayed by foreign nations—Russia's colony at Fort Ross, which had been established in 1812, was one cause for anxiety.

Consequently, when ships from foreign lands entered California ports they met with a cooler and less generous reception at presidios and missions than had awaited them not many years before. The British ship *Blossom,* commanded by Beechey, visited San Francisco and Monterery in 1826 and 1827 but found it impossible to replenish supplies and was obliged to turn to the Hawaiian Islands to fill its needs. Beechey's *Narrative* does not picture Mexican rule in California in any favorable light.

The plants collected in the course of the *Blossom*'s long voyage by the naturalist Lay and the surgeon Collie (who for many months acted as collector in the absence of Lay) were described in a famous treatise: *The botany of Captain Beechey's voyage,* a work which appeared in nine installments, fascicle 1 in 1830 and fascicle 9 (the last) in 1841. The authors, Hooker and Walker-Arnott, devoted a "Supplement" of nearly one hundred pages to the plants of California, but the collections described were mainly gathered in the decade of 1830–1840 by David Douglas, by "a friend of Mr Tolmie" and by others unconnected with Beechey's voyage; moreover, many of the included specimens were from regions far removed from what is now understood as "California."

A French ship, the *Héros* commanded by Duhaut-Cilly, also plied up and down the California coast during the years 1827 and 1828, in an attempt to establish trade with the ports and missions. Unlike Beechey and others who reached California at this period, Duhaut-Cilly had, in the words of Bancroft, ". . . nothing but kind words for all." He was not unobserving, however, and his *Voyage autour du monde* is replete with astute comments and interesting facts. In the *Héros* traveled a young man, Paolo Emilio Botta, who was making collections for the Museum of Natural History in Paris. In California, Botta made the collection of the road-runner from which the species was described by Lessing, and a *Godetia* from the state still bears his name. But the man's real claim to fame rests upon his discovery in later years of the ruins of Nineveh, notably of the palace of King Sargon.

When Spain in 1819 had exchanged Florida for Texas, she had also relinquished to the United States her rights in the Oregon country and in 1824 Russia took similar steps. Therefore, from the middle of the present decade onward, only British interests were in conflict with those of such enterprising Americans as might turn their steps towards the Pacific northwest—as many were to do in the next ten years.

British strength on the northwest coast was augmenting in the present decade. In 1821 the British North West Company merged with the older Hudson's Bay Company and, in the manner of the scientists, adopted the earlier name; this combine increased the power of the British fur trade from Hudson Bay to the Pacific. In 1824 John McLoughlin was appointed Chief Factor in the Columbia River district and was to hold that position for twenty-two years. All visitors to the region, whatever their nationality, were in some measure indebted to McLoughlin during that period although, very understandably, his responsibility lay in furthering British as against foreign enterprise.

No botanist from the United States reached the Pacific northwest—as none reached California—between 1820 and 1830. Two, however, arrived there from England, John Scouler and David Douglas. They arrived at the mouth of the Columbia River in April of 1825, in the Hudson's Bay Company ship, the *William and Ann*.

Scouler had taken the appointment of ship's surgeon in order to further his interest in the natural sciences in distant lands; his activities were determined by the coming and going of the ship and ended in October when she sailed for England. But during his visit of about six months Scouler made a considerable collection of plants along the Columbia River and at more northern coastal points.

Douglas was to remain in our northwest for about two years and then cross Canada to Hudson Bay on his way back to England. He had been sent out by the Horticultural Society of London[8] to collect living plant material which might be of value to the gardens of Britain, as well as the dried specimens needed for study by the technical botanists—in particular by Sir William Jackson Hooker, who was working on a flora of the northern regions of the North American continent. Douglas made a rich col-

8. The name was changed to Royal Horticultural Society in 1866.

lection of plants in what are now the states of Oregon, Washington and Idaho. He had an advantage over his predecessor Archibald Menzies—whose work was restricted to regions adjacent to the coast—for, thanks to the assistance of the Hudson's Bay Company, he was able to penetrate far into the interior. Douglas ranks high among plant collectors the world over, and probably highest among those who attempted similar work—much of it plant introduction—in this country.

Only a short seventeen years after Lewis and Clark had returned from their journey across what, to Americans, was an unknown region, the geography of the trans-Mississippi west was far better understood. In 1823 the *North American Review* had considerable to say on the matter; I quote a fraction:

". . . Any one who surveys the map of North America at the present day, and compares its features with those which it wore scarcely more than twenty years ago . . . cannot fail to be struck with the great changes it has undergone . . . The river Missouri, which has of late years been a fertile source of interest and wonder, was then only known as a tributary branch of the Mississippi of doubtful magnitude and extent. The Arkansa and other western streams were known little more than in name . . . the waters of the west, almost from the sources of the Mississippi and St Lawrence, were supposed to be gathered up by the fabulous Oregan or river of the west . . .

"We now see the Missouri stretching far to the north and west, not a tributary, but itself a principal and mighty river . . . beyond doubt the largest river of the known world. The Platte, the Arkansa, and other tributaries of this prodigious stream, would in the old continent be rivers of the first rate magnitude . . .

"We have spoken of the Missouri as the largest river known on the face of the globe . . . of course . . . including the part of the Mississippi which is below its confluence, and of which the Missouri is undoubtedly the true continuation . . . We know of no other river which draws from such an extent of country or connects together climates so remote and dissimilar . . ."

The article, which may well have been from the pen of Edward Everett, refers to the fact that, in the valley of the Mississippi, ". . . there repose more inhabitants than the United States contained at the beginning of the revolution . . ."

Two years later, or in 1825, Sir William Jackson Hooker[9] published in England an article entitled "On the botany of America"; G. Brown Goode, President of the Biological Society of Washington, when writing of the beginnings of science in America, accepted this as a tribute to what the botanists had been accomplishing:

". . . an event of importance . . . was . . . Dr. W. J. Hooker's essay on the botany of America, the first general treatise upon the American flora or fauna, by a master abroad, is pretty sure evidence that the work of home naturalists was beginning to tell."

9. In 1836 William IV conferred upon Hooker the title of Knight of the Hanoverian Order. Because I do not write as a Hooker contemporary, I use the title by which he is now universally designated.

CHAPTER X

JAMES, ACCOMPANYING LONG AND THE SCIENTIFIC SECTION OF THE "YELLOWSTONE EXPEDITION," REACHES HIGH ALTITUDES IN THE ROCKY MOUN- TAINS OF COLORADO AND CROSSES THE DESERTS OF NORTHERN TEXAS AND OKLAHOMA

IN my preceding chapter I have told how, until his death at Franklin, Missouri, William Baldwin served on the "Yellowstone Expedition" commanded by Stephen Harriman Long during its first year[1] in the field. With the resumption of the journey westward in 1820, Dr. Edwin James, twenty-three years of age and compiler of the records of the expedition, had been appointed to take Baldwin's place.

For the expedition's second year in the field—as for the first—I shall follow the *Account of an expedition from Pittsburgh to the Rocky Mountains, performed in the years 1819 and '20 . . .* which was compiled by Edwin James and published in Philadelphia in 1823. For the expedition's sojourn in Colorado I quote frequently from the writings of Joseph Ewan who is familiar with that region.

On April 24, 1820, Long was back in St. Louis, on his way to rejoin his party. With him was Captain John R. Bell, sent by order of the War Department, and Edwin James, ". . . appointed to serve as botanist and geologist, in consequence of recommendations from the Honourable Secretary of the Navy, from Dr. Torrey and Captain Le Conte." [2]

They traveled on horseback. On May 6 they passed Bon Homme settlements, St. Louis County, and Loutre Lick, Montgomery County, soon after entering ". . . that great woodless plain thirty miles in length, called the Grand Prairie . . ." On May 8 they reached Franklin, Howard County, remaining until the 14th on which day they came to Chariton, also in Howard County. On May 15, leaving the settlements, they entered "the wilderness," and crossed Grand River which enters the Missouri from the north between Chariton and Carroll counties. Long's map shows that they ascended that stream along its west bank. By the 16th they were approaching its

---

1. Long had remained with his party in 1819 until, in the autumn of that year, the winter headquarters, "Engineer Cantonment," had been built a few miles below the promontory known as Council Bluff (near the present town of Fort Calhoun, Washington County, Nebraska). Long had then returned to the east, descending the Missouri to St. Louis by canoe and from there proceeding to Philadelphia and Washington.

2. John Eaton Le Conte, of the Corps of Topographical Engineers; also known for work in botany and geology.

sources; on the 19th they turned northwest from the river and climbed ". . . to the level of the great woodless plain . . ." James comments:

". . . These vast plains, in which the eye finds no object to rest upon, are at first seen with surprise and pleasure, but their uniformity at length becomes tiresome . . . Nothing is more difficult than to estimate, by the eye, the distance of objects seen in these plains . . . A small animal, as a wolf or turkey . . . appears the magnitude of a horse . . ."

They had had several days of rain about this time and James had reported on the 18th that "Our encampment was completely inundated . . . The small portfolio, in which we had deposited such plants as we wished to preserve, had been placed for a pillow in the most sheltered part of the tent, and covered with a coat, but these precautions and all others . . . were unavailing, and the collection of plants we had then made was lost."

On May 25 they reached the wide valley ". . . and the yellow stream of the Missouri . . . at a point about six miles below the confluence of the great river Platte." This enters the Missouri just across from Mills County, Iowa. To reach this point from the mouth of Grand River, they had crossed northwestern Missouri and southwestern Iowa.

James noted that, on the precipitous hills bounding the Missouri ". . . we found the oxytropis lambertii, and the great flowering pentstemon; two plants of singular beauty. Here also we saw, for the first time, the leafless prenanthes, the yellow euchromia, and many other interesting plants. It would seem that several species of plants are distributed along the course of the Missouri, but do not extend far on either side. Probably the seeds of these have been brought down from their original localities, near the sources of the river . . . but the agency of rivers in this respect appears much less important, than . . . we might be inclined to imagine. In ascending the Missouri, the Arkansa, or any great river, every remove of forty or fifty miles brings the traveller to the locality of some plants, not to be seen before. This is perhaps less the case with rivers running from east to west, or from west to east, than with those whose course in a different direction, traverses several parallels of latitude."

On May 27 they camped ". . . near the mouth of the Boyer, about six miles from the wintering place of the party." This joins the Missouri from the northeast, in Pottawattamie County, Iowa. After crossing this stream and the Missouri about five miles below Council Bluff and three above the mouth of Boyer's River, next day, they reached "Engineer Cantonment" on the west (Nebraska) bank of the Missouri.

According to Chittenden, the operations of the entire "Yellowstone Expedition" [3] had been, up to this time, ". . . a huge fiasco . . . smothered in elaboration of method . . ." and the cost had been so great that Congress refused funds for further work. It ". . . was thus cut off before it was half completed, and as a half-hearted apology to the

3. A military section commanded by Colonel Henry Atkinson had set out in 1818, but its activities are of no interest to our story.

public for its failure, a small side show was organized for the season of 1820 in the form of an expedition to the Rocky mountains . . . under charge of Major Long . . ."

On arrival at "Engineer Cantonment" Long explained the change in plans. Further travel by boat up the Missouri was stopped. Instead ". . . an excursion, by land, to the source of the river Platte, and thence by way of the Arkansas and Red rivers to the Mississippi, is ordered." The work of the scientists was reallocated. "The duties assigned to Dr. Baldwin and Mr. Jessup, will be performed by Dr. E. James . . ."—with the exception of ". . . Comparative Anatomy, and the diseases, remedies, &c. known amongst the Indians . . ." which were to become the work of the "Zoologist, &c.", Thomas Say. James, as the list of personnel indicates, was "Botanist, Geologist, and Surgeon." Titian Ramsey Peale remained "assistant Naturalist."

Of the twenty-eight horses and mules serving twenty persons, sixteen were provided by Major Long and others, only six ". . . were the property of the United States . . ." The outfit consisted of bare essentials. James wrote on June 6:

"Several of the Indians about Council Bluff, to whom our proposed route had been explained . . . affected to laugh at our temerity, in attempting what they said we should never be able to accomplish. They represented some part of the country, through which we intended to travel, as so entirely destitute of water and grass, that neither ourselves nor our horses could be subsisted while passing it . . . Barony Vasquez, who accompanied Captain Pike in his expedition to the sources of the Arkansa, assured us there was no probability we could avoid the attacks of hostile Indians, who infested every part of the country . . . With these prospects, and with the very inadequate outfit . . . which was the utmost our united means enabled us to furnish, we departed from Engineer Cantonment, at 11 o'clock, on the 6th of June."

Leaving the Missouri River, the party was to cross nearly the breadth of Nebraska, following the Platte River and its South Fork, before entering Colorado on June 27.

The first objective was the Pawnee Loup Indian settlement. On the day of departure they crossed ". . . the Papillon, or Butterfly creek, a small stream discharging into the Missouri, three miles above the confluence of the Platte." On June 7 they reached and crossed ". . . the Elk-horn, a considerable river, tributary to the Platte."

They now entered the Platte River valley and James comments that "A species of onion, with a root about as large as an ounce ball, and bearing a conspicuous umbel of purple flowers, is very abundant about the streams, and furnished a valuable addition to our bill of fare."

On the march "the scientific gentlemen occupied any part of the line that best suited their convenience." On June 8 the route was along the north side of the valley ". . . which presented the view of an unvaried plain . . ." and about ". . . six miles above the place where it enters the valley of the Platte . . ." they crossed ". . . a small river, called La petite Coquille, or Muscleshell creek . . ." This stream, Shell Creek, enters the Platte from the northwest, in Colfax County. On June 9 they reached ". . . the valley of the Wolf river, or Loup fork of the Platte . . . called by the Indians the Little Missouri . . ." The confluence is in Platte County.

". . . In the fertile grounds, along the valley of the Loup fork, we observed several plants which we had not before seen . . . one belonging to the family of the *Malvaceæ*, with a large tuberous root which is soft and edible, being by no means ungrateful to the taste[4] . . . also the downy spike of the rabbit's-foot plaintain (Plantago *Lagopus,* Ph.) intermixed with the short grasses of the prairie. The long-flowered Puccoon (Batschia *longiflora,* N.) a larger and more beautiful plant than the B. *canescens* is here frequent. As we proceed westward, some changes are observed in the character of the soil and the aspect of vegetation. The Larkspurs and Lichnedias (species of Phlox and Delphinium,) so common and beautiful in all the country between St. Louis and Council Bluff, are succeeded by several species of Milk vetch, some Vicias, and the superb Sweet pea (Lathyrus *polymorphus*). Every step of our progress to the west brought us upon a less fertile soil."

To the above James adds in a footnote: "Astragalus *carnosus,* N.   A. *Missouriensis,* N.   A. *Laxmani,* Ph.   Gaura *Coccinnea,* N.   Troximon *marginatum,* Ph.   Hymenopappus *tenuifolius,* Ph.   Trichodium *laxiflorum,* Mx.   Atheropogon *oligostachyum* N.   Viola *palmata,* Ph. ? in fruit.   Hedeoma *hirta,* N.   Hordeum *jubatum,* Anemone *tenella,* Ph.   and other plants were among our collections of this day."

On June 10 they camped about eleven miles from the villages of the Grand Pawnees on Loup River, and the next day reached the first of these villages situated, according to Long's map, in the triangle formed by that river and "Willow," or Cedar, River, in Nance County. "The soil of this valley is deep and of inexhaustible fertility," states James. Before arrival a messenger had brought Long's party some "vaccine virus" for smallpox which had been sent to the War Department from Connecticut; it had been in a wreck on the way up the Missouri, ". . . thoroughly drenched, and the virus completely ruined."

Despite this fact, James relates: "We spent some time in attempting to explain to the chiefs, the nature and effects of the vaccine disease, and in endeavouring to persuade them to influence some of their people to submit to inoculation; but in this we were unsuccessful . . . We were, however, by no means confident, that they comprehended what we said on the subject of vaccination . . . We were not very solicitous to make the experiment . . . our virus, as before remarked, being unfit for use."

On June 12 they camped ". . . in front of the Pawnee Loup village." The next day, they crossed Loup River, suffering more than one "unexpected immersion" to the amusement of the Pawnees lining the shore. Thomas Say, who had already lost much of his equipment in a very similar experience elsewhere, ". . . was now, in a great measure, unencumbered with baggage." The crossing accomplished, they camped opposite the village "of the Pawnee Loups."

---

4. James comments in a footnote that ". . . It appears to be a congener to the two plants lately brought by Mr. Nuttall from Arkansa, and which have received the name of *Nuttallia*."

"The shore, opposite the Loup village, is covered with shrubs and other plants, growing among the loose sands. One of the most common is a large flowering rose, rising to about three feet high, and diffusing a most grateful fragrance. The Symphoria *glomerata* . . . is also a beautiful shrub . . . On the hills . . . we observed the Cactus fragilis . . . first detected on the Missouri by Lewis and Clark, [it] has been accurately described by Mr. Nuttall. The . . . joints of which it consists . . . separate from each other with great readiness, and adhere by means of the barbed spines . . . to whatever they may happen to touch."

On June 14 they camped on the Platte, 25 miles from the Pawnee villages. Near camp they ". . . observed a species of prickly pear (*Cactus ferox.* N.) to become very numerous. Our Indian horses . . . used the utmost care to avoid stepping near it. The flowers are of a sulphur yellow . . . A second species, the *C. mamillaris* N. occurs on the dry sandy ridges . . . The beautiful cristaria *coccinea. Ph.* (malva *coccinnea.* N.) is very frequent in the low plains along the Platte."

June 15 "Great Wood river," entering the Platte in Merrick County, was crossed and the Platte River ascended for 16 miles. On the 16th they saw ". . . in some small ponds near the Platte . . . the common species of pond weed (Potamogeton *natans* and P. *fluitans.* Ph.) also the Utricularia longirostris ? of Leconte, and an interesting species of Myrlophillum [*Myriophyllum*]."

To this a footnote adds: "Among other plants collected along the Platte on the 15th and 16th June are the Cheiranthus *asper. N.,* Helianthemum *canadense,* Athero-pogon *apludoides. N.,* Myosotis *scorpioides,* Pentstemon *gracile. N.* . . . along the river . . . a species of Plantago . . . manifestly allied to P. *eriophora* of Wallich . . . also P. *attenuata* of the same work . . . P. *attenuata,* Bradbury ?"

By June 18 they had ascended the Platte for some 200 miles above its confluence with the Missouri. It was a Sunday and a day of rest for both men and horses although ". . . some attention was given to the great objects of the Expedition . . . At Engineer Cantonment we had furnished ourselves with port folios of paper to receive specimens of such plants as we might collect, but we found that the precautions . . . to protect these from the weather had been insufficient, some of our collections being in part wet, and others having been made during the heavy rains . . . required much attention."

The Platte— ". . . called by the Otoes Ne-braska, (Flat river, or water,) . . ." —was shallow at this point, about three miles broad, with a rapid current and many islands ". . . covered with a scanty growth of cotton wood willows, the Amorpha fruticosa, and other shrubs." Near camp grew ". . . the wild liquorice, (glycyrhiza *lepidota,* N.) . . . The root is large and long, spreading horizontally to a great distance . . ." On June 19 they made about 30 miles, to a ". . . place where the hills on the north side close in, quite to the bed of the river." All were relieved to have left the prairie.

". . . The monotony of a vast unbroken plain, like that in which we had now travelled, nearly one hundred and fifty miles, is little less tiresome to the eye, and fatiguing to the spirit, than the dreary solitude of the ocean." Again James notes:

". . . some change is observed in the vegetable products of the soil. Here we first saw a new species of prickly poppy [in a footnote 'Argemony *alba*'], with a spreading white flower . . . On the summits of some of the dry sandy ridges, we saw a few of the plants called Adam's needles, (Yucca, *angustifolia*) thriving . . . in a soil which bids defiance to almost every other species of vegetation."

The yucca and its adaptation to environment is discussed at some length. And, in a footnote, James records:

"Other plants found here, were the great sunflower, Helianthus *giganteus,* Asclepias *obtusifolia,* Ph., A. *viridiflora,* Ph., A. *syriaca,* and A. *incarnata,* Amorpha *conescens,* N., Erigeron *pumilum,* N., A. *Veronica* approaching V. *beccabunga,* Scuttelaria *galericulata,* Rumex *venosus,* N., and several which are believed to be undescribed."

On June 22 they came to the confluence of the North and South forks of the Platte River, in Lincoln County, and estimated this to be 149 miles above the Pawnee Loup villages. They crossed the North Platte which they had been following, and on the 23rd the South Platte, camping on its south bank. They were to follow this river until July 11.

Here grew ". . . the beautiful white primrose (Œnothera pinnatifida. N.) with its long and slender corrolla reclining on the grass. The flower, which is near two inches long, constitutes about one half of the entire length of the plant."

A footnote records:

"Considerable additions were made, about the forks of the Platte, to our collections of plants. We found there, among others, the Pentstemon cristatum, N. Coronopus dydima, Ph. Evolvulus Nuttallianus, Roemer, and Shultz. Orobus dispar, Cleome tryphilla, Petalostemon candidum, Ph., and P. violaceum. Aristida pallens, N. two species of a genus approximating to Hoitzia, several species of Astragalus, and many others."

The route was now along the south side of the south fork of the Platte, the course ". . . inclining something more towards the southwest than heretofore." Many buffalo were seen and James notes that they were already being exterminated:

"It would be highly desirable, that some law for the preservation of game, might be extended to, and rigidly enforced in the country, where the bison is still met with: that the wanton destruction of these valuable animals, by the white hunters, might be checked or prevented . . . thousands are slaughtered yearly, of which no part is saved except the tongues."

They were encountering desert plants, cactus and sagebrush. "Prickly pears became more and more abundant as we ascended the river, and here they occurred in such extensive patches as considerably to retard our progress . . . The Cactus *ferox* is the most common, and, indeed, the only species which is of frequent occurrence." Again:

"Some extensive tracts of land along the Platte . . . are almost exclusively occupied by a scattered growth of several species of worm-wood, (Artemesia.) . . . The peculiar aromatic scent . . . is recognized in all . . ."

Six species are enumerated: "A. *Ludoviciana,* A. *longifolia,* A. *serrata,* A. *columbiensis,* A. *cernua,* A. *canadensis."* The fourth-named, James notes, was known to Lewis and Clark as "wild sage."

They were suffering from glare and heat, crossing "many tracts of naked sand."

"The Rocky Mountains may be considered as forming the shore of that sea of land, which is traversed by the Platte, and extends northward to the Missouri, above the great bend." They were never ". . . without some anxiety on the subject of Indian war-parties . . ."

On Sunday, June 25, they remained in camp. On the 26th James observed—what others have noted in regions where conditions of soil and moisture can support but a limited number of trees—that the cottonwoods ". . . from their low and branching figure, and their remoteness from each other . . . revived strongly in our minds the appearance and gratifications resulting from an apple orchard, for which from a little distance they might readily be mistaken, if seen in a cultivated region."

This day, June 26, the expedition had reached the border between Keith and Deuel counties, Nebraska, and was about to enter northeastern Colorado. It was to be in Colorado until July 29.

On June 27 the bottom lands along the river were "white with an effloresced salt."

". . . Among a considerable number of undescribed plants collected on the 27th, are three referrible to the family of the rough-leaved plants, (asperifoliæ) one of them belonging to a genus not heretofore known in the United States. It has a salverform corrolla, with a large, spreading, angular, plaited border. Another plant very conspicuously ornamental to these barren deserts, is a lactescent annual belonging to the family of the convolvulacæ, with a bright purple corrolla . . . also . . . the white stalked primrose, (Œnothera *albicaulis,* N.) a very small white flowered species of Talinum, and some others."

On June 28 three small creeks, ". . . discharging into the Platte from the northwest . . ." were passed. They were moving southwest across Logan County.

On June 29:[5] "The cactus ferox reigns as sole monarch, and sole possessor, of thousands of acres of this dreary plain . . . The rabbit's foot plantation, and a few brown and withered grasses, are sparingly scattered over the intervening spaces. In depressed and moist situations . . . the variegated spurge (Euphorbia *variegata,*) with its painted involucrum, and parti-coloured leaves is a conspicuous and beautiful ornament. The Lepidium virginicum . . . is here of such diminutive size that we were induced to search, though we sought in vain, for some character to distinguish it as a separate species."

On June 30, after traveling over the desert for about a week, the party was ". . . cheered by a distant view of the Rocky Mountains." This was at eight o'clock in the morning but not until noon was there ". . . a very distinct and satisfactory prospect of

5. According to Osterhout, ". . . in the vicinity of where Fort Morgan now is . . ."

them . . . Snow could be seen on every part of them which was visible above our horizon." By evening the view was better: three "conic summits" stood out, ". . . of nearly equal altitude. This we concluded to be the point designated by Pike as the Highest Peak." They were in error and had mistaken the mountain now designated as Long's Peak for the one named in honor of Pike.

By July 1 they were being subjected to extremes of temperature, great heat at midday and cool mornings and evenings.

"Many acres of this plain had not vegetation enough to communicate to the surface the least shade of green; a few dwarfish sunflowers and grasses, which had grown here in the early part of the summer, being now entirely withered and brown. In stagnant pools . . . we saw the common arrow head, (Saggittaria saggittifolia,) the alisma plantago, and the small lemna growing together . . ."

After traveling twenty-seven miles ". . . directly towards the base of the mountains . . . they appeared almost as distant in the evening, as they had done in the morning." They had been crossing Weld County and had now reached the point near present Greeley where the river makes an abrupt bend and where the Cache La Poudre River enters it from the northwest. July 2 was Sunday and a day of rest.

"A species of cone flower, (Rudbeckia *columnaris,* N.) was here beginning to expand. The showy R. *purpurea* . . . does not extend into the desolate regions. The common purslane (Portulacca *oleracea)* is one of the most frequent plants about the base of the Rocky Mountains . . ."

Wood was becoming more plentiful as they approached the mountains, which, on the evening of July 3 still looked twenty miles away. The route for several days, or since they left the bend in the river, had been inclining ". . . considerably to the south . . ." In the course of the day they ". . . passed the mouths of three large creeks, heading in the mountains, and entering the Platte from the northwest. One of these nearly opposite to where we encamped, is called Potera's creek . . ." This was, likely, St. Vrain's Creek which enters the South Platte nearly opposite Fort St. Vrain, Weld County.

On July 4 James notes: "We had hoped to celebrate our great national festival on the Rocky Mountains; but the day had arrived, and they were still at a distance . . . we did not devote the day to rest, as had been our intention . . ." Instead they searched for Long and others who had gotten lost. However,

"Several valuable plants were . . . collected, and among others, a large suffruticose species of Lupine.[6] The long leaved cotton-wood of Lewis and Clark [James adds in a footnote 'Populus *angustifolia,* J.'] . . . a species of populus, is here of very common occurrence . . . found intermixed with the cotton-wood, resembling it in size and general aspect. Its leaves are long and narrow . . . we also observed both species of the

6. According to Ewan (1942), "Along the South Platte in the vicinity of the mouth of St. Vrain Creek James collected what presumably is the 'large suffruticose species of lupine' mentioned by him, *Lupinus decumbens* . . ."

splendid and interesting Bartonia, the B. *nuda* in full flower, the *ornata* not yet expanded."

James gives quite a description of the genus *Bartonia* and adds in a footnote:

"Other plants were collected about this encampment . . . and interesting species of Ranunculus . . . Pentstemon *erianthera,* N., Poa *quinquefida,* Potentilla *anserina,* Scruphularia *lanceolata,* Myosotis *glomerata.* N.? &c., were also seen here."

On the 4th and 5th they were approaching the region of present Denver and, believing the mountains to be about five miles away, James and Peale on the 5th attempted to reach them by "the Cannon-ball creek," but after walking eight miles they were no nearer their objective. Among plants collected were ". . . the species of currant (Ribes *aureum* ?) so often mentioned by Lewis and Clark, the fruit of which formed an important article of the subsistence of their party while crossing the Rocky Mountains."

Also enumerated, in a footnote, were ". . . the common virgin's bower. Clematis *virginica,* Ph. Lycopus, *europeus?* Liatris *graminefolia,* Sium *latifolium* Œnothera *biennis,* and other plants, common in the east, with the more rare Linum *Lewisii,* Ph. and Eriogonum *sericeum,* &c."

Other streams tributary to the South Platte were passed. They were moving southward through Douglas County.

It was on July 6 that "At eleven o'clock we arrived at the boundary of that vast plain, across which we had journeyed for a distance of near one thousand miles; and encamped at the base of the mountains . . . At the foot of the first range, the party encamped at noon, and were soon scattered in various directions, being eager to commence the examination of that interesting region . . . From our camp, we had expected to be able to ascend the most distant summits then in sight, and return the same evening, but night overtook us and we found ourselves scarcely arrived at the base of the mountain."

On the sandstone ledges were collected ". . . a geranium[7] intermediate between the crane's-bill and herb robert, the beautiful calochortus (C. *elegans,* Ph.) and a few other valuable plants." They camped ". . . in front of the chasm, through which the Platte issues from the mountains." They had reached Platte Canyon.

On July 7 James and Peale crossed the river, planning to traverse ". . . the first range of the mountains and gain the valley of the Platte beyond but this they found themselves unable to accomplish." They did, however, climb over several ridges and collected a number of plants: they ". . . halted . . . several miles within the mountains, and elevated nearly to the limit of Phænogamus vegetation. The common hop, [H. *lupulus*][8] was growing in perfection, also the box elder, [Acer *negundo,* Ph.] the common sarsaparilla . . . [Aralia *nudicaulis*] the spikenard [A. *racemosa,*] and many other plants . . ."

7. "On July 6th the party encamped where the Platte emerges from the mountains; here were taken what became the type specimens of *Geranium caespitosum* James, and *Acer glabrum* Torr." (Ewan, 1942.)

8. The brackets in this and in some later paragraphs appear in James' *Account.*

In a footnote still others are mentioned:

"... we recognized the bear-berry Arbutus *uva-ursi L.* ... also the Dodecatheon *integrifolium, Ph.,* and a beautiful little plant, referrible to the genus Mentzelia of Plumier. On the higher parts of the mountain, an oak is common, approaching ... Quercus *banisteri, mx.,* also a small undescribed acer, the Juniperus *communis* and J. *virginiana.* In the ravines, the Rhus *toxico-dendron,* spiræa *opulifolia,* &c.; and at the base of the mountains, the Phrenanthes *rucinatum,* Saxifraga *nivalis, L.* a cerastium, &c."

For five days (July 8 to 12) the records in the text of the *Account* differ from those on its map. The party was still moving through Douglas County—there is a reference to "Castle rock" so they were evidently on the east branch of Plum Creek. They must soon have crossed into El Paso County. The *Account* states that they "... travelled nearly south, and crossing a small ridge dividing the waters of the Platte from those of the Arkansa, halted to dine on a tributary of the latter ..." From the region of Palmer Lake they had doubtless followed Monument Creek to Colorado Springs— passing near or through the famous Garden of the Gods.[9] It was on July 11 that the *Account* tells of one of James's notable acquisitions:

"In an excursion from this place we collected a large species of columbine, somewhat resembling the common one of the gardens, It is heretofore unknown to the Flora of the United States, to which it forms a splendid acquisition. If it should appear not to be described, it may receive the name of Aquilegia *cærulea.*" [1]

This columbine is now the state flower of Colorado.

James appends a short description and the plant still bears the name which he bestowed. Still other plants of the region are mentioned in a footnote:

"In passing from the head waters of the branch of the Platte called Defile [now Plum] creek, to those of one of the northern tributaries of the Arkansa, we noticed some change in the soil, and soon met with many plants we had not before seen. Several of these, as the common juniper, and the red cedar, (Juniperus *Virginiana, Ph.*) the black and hemlock spruce (Abies *nigra* and A. *canadensis*); the red maple (Acer *rubrum* Mx.) the hop hornbeam (Ostrya *virginica. L.*) the Populus *tremuloides* Mx. Pinus *resinosa* Pyrola *secunda.* Orchis *dilatata,*[2] &c. are common to mountainous districts in all the northern parts of the territory of the United States. A campanula,

9. Charles Elliott Perkins, for many years President of the Chicago, Burlington and Quincy Railroad, began buying this property (480 acres) about 1879. His daughter, Mrs. Edward Cunningham of Boston, tells me that although her father never carried out his original idea of building a summer home there, he held the property in order to protect it, and that after his death in 1907 his children gave it to the city of Colorado Springs for a park in fulfillment of his wish that it be preserved for the benefit of the public.

1. "Just after crossing the divide between the east slope's major drainage systems at Palmer Lake, on July 11th, the party ... collected the Blue Columbine, *Aquilegia coerulea,* in the scrub oak thickets of that region." (Ewan, 1942.)

2. Now *Habenaria dilatata* (Pursh) Hooker.

probably the C. *uniflora,* bearing a single flower about as large as that of the common hare-bell, occurs very frequently . . ."

Farther up the stream where they had halted to dine "Many fine plants were collected, several of which are hitherto undescribed." Among those named in a footnote with a lengthy description was ". . . a large and conspicuous plant of the natural family of the Cruciferæ, which may be referred to the new Genus *Stanleyea* [= *Stanleya*] of Nuttall, and distinguished as S. *integrifolia* . . . The whole plant, seen at a little distance, has a remote resemblance to Lysimachia *thyrsifolia.*"

James was now to devote several days to reaching, ascending and descending the mountain known as Pike's Peak—probably his most noteworthy excursion from a botanical point of view.

On July 12, having gone beyond the base of the mountain which they wished to ascend, they retraced their steps, finally camping in the timber on the bank of a creek.

"From this camp we had a distinct view of 'the Highest Peak.' It appeared about twenty miles distant, towards the northwest . . . it was determined to remain in our present camp for three days, which would afford an opportunity for some of the party to ascend the mountain."

Later in the *Account* James mentions that the ". . . creek on which the party encamped during the three days, occupied in making the excursion . . . is called Boiling-spring creek . . ." This had one of its principal sources in Boiling Spring, described as ". . . a large and beautiful fountain of water, cool and transparent, and highly aerated with carbonic acid . . ."

James's "Boiling-spring creek," now called Fountain Creek, has its origin at the Manitou Springs, lying to the west of Colorado Springs. It is from Manitou, according to the guide books, that a cog railway starts for the top of Pike's Peak. The spring was also known to the traders as the Fontaine qui Bouille, a name which was also applied to Fountain River, or Creek.

Several plants growing near the camp interested James. One, seen for the first time, was ". . . the great shrubby cactus, which forms so conspicuous a feature of the vegetable physiognomy of the plains of the Arkansa . . ." and which he tentatively identified as "Cactus *cylindricus* of Humboldt ?"; another ". . . highly interesting plant . . . is a cucurbitaceous vine . . ." named in a footnote as "Cucumis? *perennis,*[3] American

3. A reference to this plant is included in Pammel's biographical sketch of James which quotes a letter written to C. C. Parry by John U. Rauch from Burlington, Iowa, where James had taken up his residence in 1836:

". . . Mr. James . . . was in my office . . and in looking over the Plantæ Wrightianæ he was considerably amused to see that his opinion with regard to the *Cacurbita perennis* of Gray, he calling it *Cucumis perennis* was marked doubtful. He still thinks he is right, he told me Dr. Torrey first differed with him. He is as enthusiastic and ardent as ever, and remarked to me that he would walk one hundred miles to see a new plant, but would like to take the steam-boat back. You would have been delighted with him. He has his peculiarities, and the masses cannot appreciate him, he is at least two hundred years ahead of the time in many things . . ."

colycinth." James gives a long description and mentions that "Some plants of this interesting species are growing in the garden of the University at Philadelphia, also in that of Dr. Ewing, from seeds brought by Major Long, but they have not yet flowered." The camp was ". . . skirted with a narrow margin of cotton-wood and willow trees, and its banks produce a small growth of rushes on which the horses subsisted, while we lay encamped here. This plant, the common rush, (Equisetum *hiemale, Ph.*) . . . is eaten with avidity by horses . . ."

On July 13 "Dr. James being furnished with four men, two to be left at the foot of the mountain to take care of the horses, and two to accompany him in the proposed ascent to the summit of the Peak, set off . . ."

By eleven o'clock they were at the base of the mountain where they arranged the camp for the horses; then, on foot, they proceeded up the valley and ate their noon meal at the "Boiling spring." James and his two companions then began the climb. That night they ". . . could not . . . find a piece of ground large enough to lie down upon, and were under the necessity of securing ourselves from rolling into the brook . . . by means of a pole placed against two trees . . . we passed an uneasy night . . ."

At daylight of July 14[4] they started again. Some of the way was rugged and difficult but finally, after emerging from level country ". . . covered with aspen poplar, a few birches, and pines . . .", they saw ". . . almost the whole of the Peak, its lower half thinly clad with pines, junipers, and other evergreen trees; the upper half a naked conic pile of yellowish rocks, surmounted here and there with broad patches of snow; but the summit appeared so distant, and the ascent so steep, that we despaired of accomplishing the ascent, and returning on the same day." Again,

"In marshy places about this part of the mountain, we saw an undescribed white flowered species of caltha, some Spediculariæ, the shrubby cinquefoil (Potentilla *fructicosa, Ph.*) and many alpine plants."

Still higher, but before reaching ". . . the outskirts of the timber . . . the yellow flowered stone-crop (Sedum *stenopetalum, Ph.*), is almost the only herbaceous plant which occurs. The boundary of the region of forests, is a defined line encircling the peak . . . Above the timber the ascent is steeper, but less difficult than below . . . The red cedar, and the flexile pine [in a footnote 'Pinus *flexilis* J.'],[5] are the trees which appear at the greatest elevation . . . A few trees were seen above the commencement of snow . . . very small and entirely procumbent, being sheltered in the crevices and fissures of the rock . . .

"A little above the point where the timber disappears entirely, commences a region of astonishing beauty, and of great interest on account of its productions; the intervals of soil are sometimes extensive, and are covered with a carpet of low but brilliantly

4. According to Osterhout, this was ". . . undoubtedly the most notable day of the Expedition . . ."

5. According to Sudworth, "Dr. C. C. Parry is said to have been the first to introduce this pine into cultivation, plants having been raised in the Harvard Botanic Garden from seed he collected in Colorado in 1861 . . ."

flowering alpine plants. Most of these have either matted procumbent stems, or such as including the flower, rarely rise more than an inch in height. In many of them, the flower is the most conspicuous and the largest part of the plant, and in all, the colouring is astonishingly brilliant.

"A deep blue is the prevailing colour among these flowers, and the Pentstemon *erianthera,* the mountain Columbine (Aquilegia *cœrulea*) and other plants common to less elevated districts, were here much more intensely coloured, than in ordinary situations . . . May the deep cœrulean tint of the sky, be supposed to have an influence in producing the corresponding colour, so prevalent in the flowers of these plants?"

They rested about a mile above timber line, were somewhat refreshed, ". . . but much benumbed with cold." Realizing that it would be impossible to make the ascent and get back that day to the camp of the preceding night, they decided to keep on and spend the night where darkness overtook them.

"We met, as we proceeded, such numbers of unknown and interesting plants, as to occasion much delay in collecting, and were under the disagreeable necessity of passing by numbers which we saw in situations difficult of access."

Nearing the top, plants ". . . became less frequent, and at length ceased entirely." At the summit—a level plain of 10 or 15 acres—". . . scarce a lichen is to be seen." To the west was observed ". . . the narrow valley of the Arkansa . . ." and to the east the great plain. They had arrived at about four o'clock in the afternoon and only remained about thirty minutes, the mercury falling to 42 degrees. Strangely, ". . . the air in every direction [was] filled with clouds of grasshoppers, as partially to obscure the day . . ." They began the descent at about five p.m., reached timber a little before sunset, lost their way, and camped where they were.

At dawn of July 15 the thermometer stood at 38 degrees and, since they ". . . had few comforts to leave . . .", moved on with the first light. In about three hours they saw what they thought was the smoke of the encampment where they had left their equipment, but on arrival found that a fire ". . . was now raging over an extent of several acres." Curiously enough, although the small timber was ablaze, they seem to have felt no alarm about a forest fire; their concern was lest the blaze should be noted by Indians. Most of their cache was destroyed.

In order to avoid the "crumbled granite" which had troubled them in the ascent they now took a different route, only to find that the earlier handicap was ". . . nearly counterbalanced by the increased numbers of yuccas and prickly pears." By noon, however, they were back at the "Boiling spring," and soon thereafter reached the men who had been left with the horses. During the absence of the party the men left in camp had killed several deer. After dining, the united party moved back to the main encampment, arriving ". . . a little after dark, having completed our excursion within the time prescribed . . .

"Among the plants collected in this excursion, several appear to be undescribed. Many of them are strictly alpine, being confined to the higher parts of the mountain, above the commencement of snow. Most of the trees which occur on any part of the mountain are evergreen, consisting of several species of abies, among which may be enumerated the balsam fir, (Abies *balsamea, Ph.*) the hemlock, white, red, and black spruce, (A. *canadensis.* A. *alba.* A. *rubra* and A. *nigra,*) the red cedar, and common juniper, and a few pines. One of these, which appears to have been hitherto unnoticed in North America, has . . . five leaves in a fascicle . . . The branches . . . are also remarkably flexible, feeling in the hand somewhat like those of the Dirca *palustris.* From this circumstance, the specific name *flexilis,* has been proposed for this tree . . ."

Joseph Ewan wrote (1942) of James's accomplishments in Colorado:

"Naturally the account of James concerning the plants of the tundra interest us most for this was the first report upon the above-timberline floras at these latitudes . . . Upon that day [July 14] James collected what was described and named *Pinus flexilis* James, *Arenaria obtusa* Torr., *Tierella [Tiarella] bracteata* Torr., *Saxifraga Jamesii* Torr., most often referred to as *Boykinia Jamesii, Trifolium nanum* and *T. dasphyllum* both named by Torrey, *Primula angustifolia* Torr., *Androsace carinata* Torr., *Pulmonaria alpina* Torr., the present *Mertensia alpina, Pentstemon alpinus,* the 'name unfortunate, as the plant rarely reaches timber line and grows mostly upon the lower mountain slopes and foothills' (Pennell), *Chionophila Jamesii* Benth., the first described species of the genus, *Castilleia occidentalis* Torr., and *Actinea integrifolia* Torr., now placed in *Tetraneuris* as *T. brevifolia* Greene. It is difficult to identify all of James's citations in the narrative, and in fact many of the Alpine species are not mentioned by him in the textual account but appear in the 'Catalogue of plants' which he contributed to the American Philosophical Society."

Ewan mentions other collections made by James in the same region:

"We can recognize the 'orbicular lizards' about camp on Fountain Creek to have been 'horned toads' (*Phrynosoma douglasii hernandezi* Girard) and the 'crimson-necked finch', our familiar House Finch, or 'California Linnet' (*Carpodacus mexicanus frontalis* [Say]), first described to ornithological science from Say's collection taken in the vicinity of 'boiling spring,' present Manitou (cf. Osterhout in Oologist, 1920, 119)."

Of James's ascent of the peak which now bears Pike's name, Ewan has this to say:

"He describes the summit of Pikes Peak accurately when he says, the 'summit . . . is covered to a great depth with large splintery fragments of a rock, entirely similar to that found at the base of the Peak, except, perhaps, a little more compact in its structure' . . . It is interesting that James suggests this fragmentation to be due to lightning . . . The official altitude determined by Long's party for the peak was 11,500 feet; this was a notable shrinkage over Zebulon Pike's estimate. Long gave the name 'James Peak' to this summit but popular feeling overcame this baptism and effectively replaced Pikes Peak as the popular and official name, the name of James being transferred to a lesser summit to the south of Moffat Tunnel. John L. Jerome Hart has recorded these particulars, with full documentation . . . Grable's statement that James

and his party did not succeed in reaching the summit of Pikes Peak is, of course, per-
niciously erroneous . . . James himself retells in a letter of 1859 some details of his
ascent (letter of James to Parry, quoted by Pammel, 16). It should be noted that the
Long Expedition map indicates fairly accurately the 'Spanish Peaks', so that James's
reference in this letter to two peaks 'perhaps twenty miles' apart should not be construed as involving the historic peaks to the south."

And he summarizes James's accomplishments in Colorado thus:

"Edwin James was Colorado's first botanist. First to visit the state with intent to
discover, record, and collect the native plants. First to ascend any 14,000-foot peak
in North America![6] First to describe the treeless tundra of any 14,000-foot peak in
this country. First historian of Colorado's natural history . . ."

Long wrote on July 15, 1820, that, since ". . . no person either civilized or savage
. . ." had ever ascended to the summit of the mountain climbed by James he ". . .
thought proper to call the Peak after his name . . ." Long's proposed name—"James'
Peak"—never supplanted permanently what was perhaps the less-deserved appellation of "Pike's Peak."[7] Elliott Coues has stated that "The alternative names ran parallel for some years."

6. A. H. Bent notes that James ". . . seems to have been the first real climber in the Rocky Mountains.
The next was perhaps Captain Benjamin L. E. Bonneville . . ." who, in 1833, climbed what he thought was
the highest peak in North America, in the Wind River Mountains of Wyoming.

7. On his journey of 1806–1807, Zebulon Montgomery Pike had first seen the mountain on November
15, 1806:

"At two o'clock in the afternoon I thought I could distinguish a mountain to our right, which appeared like a small blue cloud . . . in half an hour they appeared in full view before us. When our small
party arrived on the hill they with one accord gave three cheers to the Mexican mountains."

Pike decided to reach "the high point of the blue mountain" and started off with three of his men on the
23rd.

By the 27th "The summit of the Grand Peak . . . now appeared at the distance of 15 or 16 miles from
us. It was as high again as what we had ascended, and it would have taken a whole day's march to arrive
at its base, when I believe no human being could have ascended to its pinical."

For this and other reasons Pike returned to camp, his climbing ambitions unfulfilled. On December 3
his party took the altitude "of the north mountain," the one under discussion. Coues reports the estimate to have been excessive, by more than 4,000 feet,—in part because the observation point was wrongly
determined.

*See: The expeditions of Zebulon Montgomery Pike, to headwaters of the Mississippi River, through
Louisiana Territory, and in New Spain, during the years 1805-6-7. A new edition, now first reprinted in
full from the original of 1810, with copious critical commentary, memoir of Pike, new map and other
illustrations, and complete index, by Elliott Coues . . . 1895. 3 vols. New York.*

This edition of Pike's travels makes extremely interesting reading, in large measure because of the editor's enlightening footnotes. Pike's name is associated with that of another botanist, in addition to James,
for it seems to have been an undisputed fact that he utilized without permission some of the maps compiled by the great Alexander von Humboldt.

G. Brown Goode stated in 1888 (*Proc. Biol. Soc. Washington* 4: 46):

"Jefferson has been heartily abused for not gratifying Alexander Wilson's request to be appointed
naturalist to Pike's expeditions . . ."

Probable it is that, had Wilson been the naturalist of Pike's party instead of Dr. John Hamilton Robin-

Forty-two years were to elapse before another botanist reached the summit of Pike's Peak. In a letter to John Torrey (reprinted in part in the *Transactions* of the Academy of Sciences of St. Louis) Charles Christopher Parry tells of his own ascent of that mountain on July 1, 1862. He comments that, since James first visited the summit on July 14, 1820, ". . . its peculiar vegetation has bloomed unheeded, and the meagre collection of plants made by Dr. James has not been duplicated in scientific herbaria . . ."

Ewan (1942) reports that "After proceeding down Fountain Creek to its confluence with the Arkansas River, where James records first seeing the wild gourd or calabazella, his 'American simblin,' or 'Arkansa simblin,' the party passed out of our territory[8] down the Arkansas . . ."

The departure from "Boiling-spring creek," or Fountain Creek, was on July 18 and the party started southwestward "to the Arkansa." For ten hours or for twenty-eight miles, and with the temperature at 90 to 100 degrees in the shade, they traversed a country with a loose, dusty soil, scattered over with a ". . . few dwarfish cedars and pines . . . and nearly destitute of grass or herbage of any kind. Our sufferings from thirst, heat, and fatigue were excessive, and were aggravated by the almost unlimited extent of the prospect . . . which promised nothing but a continuation of the same dreary and disgusting scenery."

By late afternoon, from the edge of a precipice, they looked down into the valley of the Arkansas; this was finally reached through a "rugged ravine." The valley proved to be ". . . a beautiful level plain, having some scattered cotton-wood and willow trees, and . . . good pasture for our horses." They camped near Turkey Creek. From here, on July 17 and 18, James, Bell and a couple of men made a trip up the Arkansas River. After thirty miles' travel they arrived ". . . at the spot where the Arkansa leaves the mountains." Here, in Fremont County, they had reached the eastern end of the Royal Gorge (or the great canyon of the Arkansas River) and took the bearings of "James' Peak" (Pike's Peak) which lay due north. They considered that the valley of the Arkansas, which, for the 30 miles traversed was ". . . arid and sterile, bearing only a few dwarfish cedars, . . . must forever remain desolate."

The expedition, with regret, left the mountains behind them on July 19.

son (who had "blooming cheeks," a "fine complexion and a genius speaking-eye" but whose chief contribution seems to have been in the hunting field), the natural history results of Pike's journeys might have been worth recording. Pike had received instructions to collect and to preserve ". . . specimens of everything curious in the mineral and botanical worlds, which can be preserved and are portable . . ."—a large assignment. His excuse for not fulfilling this task savors of the grandiloquent:

"With respect to the great acquisitions which might have been made to the sciences of botany and zoology . . . neither my education nor taste led me to the pursuit; and if they had, my mind was too much engrossed in making arrangements for our subsistence and safety to give time to scrutinize the productions of the countries over which we traveled, with the eye of a Linnaeus or Buffon . . ."

8. Long's party was to leave the state of Colorado on July 29.

"More than one thousand miles of dreary and monotonous plain lay between us and the enjoyments and indulgences of civilized countries. This we were to traverse in the heat of summer . . ."

The party's provisions were low and game was scarce. After another ten miles ". . . the barren cedar ridges, are succeeded by still more desolate plains, with scarce a green, or a living thing upon them, except here and there a tuft of grass . . . among the few stinted and withered grasses, we distinguished a small cæspitose species of Agrostis . . . Near the river and in spots of uncommon fertility, the unicorn plant, (Martynia *proboscidea, Ph.*) was growing in considerable perfection."

By afternoon they passed ". . . the mouth of the river St. Charles . . . which enters the Arkansa from the southwest . . .", in Pueblo County; they had made twenty-five miles.

On July 20 they passed the mouth of a creek ". . . called by the Spaniards Wharf creek . . ."—the name a corruption of Huerfano (Spanish for orphan). They made about the same distance as the day before. The country was ". . . almost destitute of vegetation of any kind."

On July 21 an Indian and his squaw, members of the Kaskaias tribe, called "by the French, Bad-hearts," were the first savages encountered in a long time for most were ". . . on a predatory excursion against the Indians of New Mexico."

"The low grounds, on the upper part of the Arkansa, have a sandy soil, and are thinly covered with cotton-wood, intermixed with the aspen poplar (P. *tremuloides. Mx.*) and a few willows. The undergrowth . . . consisting principally of the Amorpha *fruticosa* and a syngeneceous shrub, probably a vernonia. Along the base of the mountains and about this encampment, we had observed a small asclepias . . . rarely rising more than two or three inches from the ground . . . also the A. *longifolia* and A. *viridiflora* of Pursh. The scanty catalogue of grasses and herbaceous plants comprises two sunflowers, (H[elianthus] *gigantus* and H. *petiolaris.*) the great Bartonia, the white argemone, the Cactus *ferox,* the Andropogon *furcatum* and A. *ciliatum,* Cyperus *uncinatus,* Elymus *striatus,* and a few others."

They made camp ". . . about eighteen miles above the confluence of that tributary of the Arkansa, called in Pike's maps 'The First fork,' and, by our computation, near one hundred miles from the base of the mountain." Pike's "First Fork" was Purgatory River which enters the Arkansas from the south in Bent County. They were near the town of La Junta, Otero County.

At this point the expedition was divided: Long, James, Peale and others were to search, southward, for the sources of Red River. Bell, with Thomas Say and others were to descend the Arkansas by the most direct route to Fort Smith; the story of this contingent is told in six later chapters of James's *Account* and I mention their journey later. More important to our story is the section of which James was a member.

At five A.M. on July 24 Long's party crossed the Arkansas. Bell and his men were seen leaving camp and, as usual on such occasions, each party gave "three cheers." Long traveled a little east of south, intending to cross and ascend the "First Fork," or the Purgatory River; they made about twenty-seven miles. James noted, and described in a footnote, "A species of cone-flower (Rudbeckia *tagetes,*) with an elongated receptacle, and large red brown radial florets . . . about the margin of the stagnant pool, near which we halted. We also collected the Linum *rigidum?* and a semiprocumbent species of Sida which appears to be undescribed . . ."

The country was "sterile and sandy." On July 25 they crossed a region resembling that along the Arkansas River approaching the mountains. Ravines for the most part were ". . . destitute of timber, except a few cedars, attached . . . in the crevices of the rock. The larger vallies which contain streams of water, have a few cotton-woods and willow trees. The box elder, the common elder, (Sambucus *canadensis,*) and one or two species of Viburnum, are seen here."

The Purgatory, or Purgatoire, River enters the Arkansas a little east of Las Animas, Bent County; the party, having traveled "a little east of south," had probably reached the stream either in southern Otero County or northern Las Animas County; they had followed it, or its tributary Chacuaca, or Chaquaqua, Creek for a time.

"A beautiful Dalea, two or three Euphorbias, with several species of Eriogonum, are among the plants collected about this encampment. Notwithstanding the barrenness of the soil, and the aspect of desolation which so widely prevails, we are often surprised by the occurrence of splendid and interesting productions springing up under our feet . . ."

The tent was only large enough to shelter five men of the party of ten so that, when rain came, ". . . about the half of each man was sheltered under the tent, while the remainder was exposed to the weather. This was effected, by placing all our heads near together in the centre of the tent, and allowing our feet to project in all directions, like the radii of a circle."

On July 27, "with a feeling somewhat akin to that which attends the escape from a place of punishment," they took ". . . final leave of the 'Valley of the souls in Purgatory . . .'" and emerged upon an open plain.

"Here, the interminable expanse of the grassy desert burst suddenly upon our view . . . a boundless and varied landscape lay spread before us. The broad valley of the Arkansa, studded with little groves of timber . . . in the background . . . the snowy summit of James' Peak . . ."

"A large undescribed species of Gaura is common about the banks of all the creeks we had seen since leaving the Arkansa . . . We propose to call it Gaura *mollis.*"

They encamped on the 28th upon what James thought to be ". . . one of the most remote sources of the great northern tributary of the Canadian river."

The party had kept considerably east of the Raton Mountains and were about to cross the northeastern corner of New Mexico, entering that state in Colfax County or in Union County. The "tributary of the Canadian river" was, therefore, more probably a tributary of the Cimarron. They were to be in New Mexico until August 5. James seems to have been the first botanist to set foot in that state.

July 29 was marked by violent storms and "pelting hail," and the night was cheerless, especially for Peale who had an alarming attack of pain— ". . . somewhat relieved by the free use of opium and whiskey." The next day they left their "comfortless camp" early and descended to a dry stream in a ". . . deep and almost inaccessible valley . . ." and encamped. They believed the creek to be one of "the sources of the Red river of Louisiana . . ." They were not, however, on the sources of Red River, which lie in northwestern Texas, in the Staked Plains, but on a small tributary of the Canadian River which appears on some modern maps as Major Long's Creek.[9]

James tells that the country between the sources of the Purgatory and the present stream had been covered by a ". . . loose and scanty soil, in which sand, gravel, and rolled peebles are rarely seen . . . In traversing it we had collected many new and interesting plants . . . a large decumbent mentzelia, an unarmed rubus, with species of astragalus, pentstemon, myosotis,[1] helianthus, &c. Beside the common purslane, which is one of the most frequent plants about the mountains, we had observed on the Arkansa a smaller species, remarkably pilose about the axils of the leaves, which are also narrower than in P. *oleracea*. A very small cuscuta also occurs almost exclusively parasitic on the common purslane."

August 1 was a day of rest. They observed about camp ". . . a yellow flowering sensitive plant . . . Its leaves are twice pinnated, and manifestly irritable . . . also . . . two new species of Gaura, much smaller than G. *mollis,* also found here."

On August 2 the river valley widened and they once more entered ". . . a vast unvaried plain of sand. The bed of the creek had become much wider, but its water had disappeared." At night they had no fuel with which to cook, and the stagnant water had not been drinkable for an entire day. James includes a long description of ". . . a most interesting shrub or small tree, rising, sometimes to the height of twelve or fourteen feet. It has dioiceous flowers, and produces a leguminous fruit, making in several particulars a near approach to Gleditschia . . ." James writes of the mesquite.

On August 3 James complains that ". . . our sufferings from hunger and thirst, made our situation extremely unpleasant. We had travelled greater part of the day enveloped in a burning atmosphere . . ."

By night they found stagnant water, a little wood, and ate a badger ". . . with the addition of a young owl . . ." On the 4th the water of the river emerged to the surface, its suspended silt ". . . the colour of florid blood . . ." Its course to the southeast led them to the belief that they were on a tributary of Red River, and not until September 10, when they reached the confluence of the Red and Arkansas rivers, was the mistake recognized. Following Major Long's Creek they must, at about this time, have crossed

9. Ghent mentions that the party ". . . in the present Union County, New Mexico, reached a stream known as Major Long's Creek."

1. According to Osterhout, "The species of 'myosotis' may have been *Lappula* or *Oreocarpa*."

from New Mexico into extreme northwestern Texas and, after crossing Dallam and Hartley counties, have reached the Canadian slightly west of Tascosa, Oldham County. They were to follow that stream to its confluence with the Arkansas River.

"Two shrubby species of Cactus, smaller than the great cylindric prickly pear, noticed near the Rocky Mountains, occur in the sandy plains, we are now traversing. One . . . about four feet high, and very much branched, has long and solitary spines, a small yellow flower, and its fruit, which is about as large as the garden cherry, is very pleasant to the taste. The fruit of the C. *ferox* . . . was now ripe . . . nearly as large as an egg, and of a deep purple colour. The jatropha stimulosa . . . a cassia, an amorpha, and many new plants were here added to our collections."

The route lay through Texas until August 17.

On August 5 they were suffering from hunger and ate a wild horse, feeling, however, ". . . a little regret at killing so beautiful an animal, who had followed us several miles . . . and had lingered with a sort of confidence about our camp . . ." The next day, a Sunday, was spent in camp.

"The occurrence of the elm and diospyros indicated a soil at least approaching towards one adapted to the purposes of agriculture . . . we found here a gentiana, with a flower much larger than g. *crinita,* an orobanche, probably the o. *ludoviciana, N.,* a new croton, an ipomopsis, and many others."

Twenty miles further, on August 7, they camped near the river.

"In this region, the cenchrus tribuloides, a most annoying grass, supplies the place of the Cactus *ferox,* and the troublesome stipas of the Platte. The cenchrus bears its seed in small spikelets, which consist of a number of rigid radiating spines. These clusters of barbed thorns are detached at the slightest touch, falling into our mockasins, adhering to our blankets and clothing, and annoying us at every point. The clott-bur, (Xanthium *strumiarum*) which had occurred in every part of our route, began now to ripen and cast off its muricated fruit, adding one more to the sources of constant molestation."

One of the men brought in to camp an eight-inch centipede, ". . . biting at every thing . . . within its reach. Its bite is said to be venomous." On August 8 they crossed and recrossed the river several times—streams were dry—and advanced about twenty-six miles due east. James comments upon a "beautiful white flowered Gaura," undescribed, and before flowering resembling the "common flax." It was ". . . the fifth species of Gaura . . . met with west of the Mississippi . . ." Food was so scarce that they ". . . gave little attention to any object except hunting." On August 9 they were still eating the ". . . last of the horse-beef which, having been killed on the 5th . . . had suffered from long keeping." At noon the mercury registered 96 degrees.

On August 10 they ate a bison so "lean and ill-flavored" that nothing but "the most urgent necessity could have induced us to taste it." To add to their troubles they met a large band of ". . . Kaskaias, or Bad-hearts, as they are called by the French . . ."

They were led by a chief called Red Mouse. It was two days before they could get rid of these thieving companions and others who had joined them.

Although the heat was terrific (on August 13 it was 105 degrees in the shade), James begins to record the happy occurrence of edible fruits—grapes "ripe and delicious," and the "Osage plum" beginning to ripen.

On the 14th "The occurrence of the elm, the phytolacca, the cephalanthus, and other plants, not met with in a desert of sand, gave us the pleasing assurance of a change we had long been expecting in the aspect of the country."

However, it was still desert country for the most part: ". . . the yucca *angustifolia* and the shrubby cactus, the white argemone, and the night-flowering Bartonia, are the most conspicuous plants in these sandy wastes." On August 15 there was, near camp, a ". . . scattering grove of small-leaved elms . . . (the Ulmus *alata, N.*) . . ." and James describes it at some length. On the 16th they passed the mouth of a large creek entering the Canadian from the southwest. Drifting sand made traveling difficult and the heat became more intolerable; there was a violent hailstorm and James makes a gloomy prognostication:

"We have not spent sufficient time in the country near the eastern border of the Rocky Mountains, to enable us to speak with confidence of the character of its climate. It is, however, sufficiently manifest that in summer it must be extremely variable, as we have found it; the thermometer often indicating an increase of near fifty degrees . . . between sunrise and the middle of the day . . . If the wide plains of the Platte, the Upper Arkansa, and the Red river of Louisiana, should ever become the seat of a permanent civilized population, the diseases . . . will probably be fevers, attended with pulmonary and pleuritic inflammations, rheumatism, scrofula, and consumption." !!

It was probably on August 17 that the party crossed from Hemphill County, Texas, into Roger Mills County, Oklahoma, where, to the east of the state boundary, lie the Antelope Hills. They were to travel in Oklahoma until September 13.

"Our camp was on the southwest side of the river, under a low bluff, which separates the half wooded valley from the open and elevated plains. The small elms along this valley were bending under the weight of innumerable grape vines, now loaded with ripe fruit, the purple clusters crouded in such profusion as almost to give colouring to the landscape . . . The fruit . . . is incomparably finer than that of any other, either native or exotic, which we have met with in the United States . . . We indulged ourselves to excess . . . The great flowering hibiscus is here a conspicuous and highly ornamental plant, among the scattering trees in the low ground . . . the black walnut for the first time, since we left the Missouri, indicated a soil somewhat adapted to the purposes of agriculture."

James regretted the lack, in a country the geography of which was so little known, ". . . of ascertained and fixed points of reference." He suggests that, with greater knowledge, it might be possible to establish sectional divisions ". . . founded on the distribution of certain remarkable plants. The great cylindrical cactus, the American colycinth, (Cucumis *perennis,*) and the small-leaved elm, might be used in such an attempt, but . . . the advantages . . . would be for the most part imaginary."

They were beginning to feel some doubt as to the identity of the river which they had been following since August 4. Their direction for some days had averaged ". . . rather to the north, then south of east. This fact did not coincide with our previous ideas of the direction of Red River, and much less of the False Ouachitta or False Washita, which, being the largest of the upper branches of Red River from the north, we believed might be the stream we were descending."

On August 19 they left the river, which was becoming more and more serpentine, and kept east, thinking to rejoin it later. James was impressed by the elevated plains.

"The luxuriance and fineness of the grasses, as well as the astonishing number and good condition of the herbivorous animals of this region clearly indicate its value for purposes of pasturage."

He suggests that to burn off the native grasses and sow seeds of ". . . any of the more hardy cultivated graminæ . . ." might bring good results! They explored a ravine in search of plants.

"About the shelvings and crevices of the rocks the slender corrolla of the Œnothera *macrocarpa* and the purple blossoms of the Pentstemon *bradburii* lay withering together, while the falling leaves and the ripening fruit reminded us that the summer was drawing to a close."

Lack of water drove them back to the river on August 20 and they camped on its bank. In addition to the miseries inflicted by gnats and ticks, they were beset by swarms of "blowing flies" which made impossible the preservation of meat.

There was rain on August 21 and the next day there was running water in the river— they had seen none for two weeks or for 150 miles; on the 23rd the route was noticeably to the south.

On August 24 "The common post oak, the white oak, and several other species, with the gymnocladus, or coffee bean tree, the cercis, and the black walnut, which indicate a soil of very considerable fertility, now began to occur . . ."

The party's horses were exhausted and progress was slow. On August 25 James notes that "Our daily journies over desolate and uninhabited plains, could afford little to record, unless we were to set down the names of the trees we passed, and of the plants and animals . . .

"The country . . . has a soil of sufficient fertility to support a dense population, but the want of springs and streams of water, must long oppose a serious obstacle to its occupation by permanent residents."

Water had already subsided in the river. They moved southeast down its valley.

"Thickets of oak, elm, and nyssa, began to occur on the hills, and the fertile soil of the low plains to be covered with a dense growth of ambrosia, helianthus, and other heavy weeds."

Sunday, August 27, they rested in ". . . a delightful situation at the estuary of a small creek from the south . . . Herds of bisons, wild horses, elk, and deer, are seen quietly grazing . . ."

Unfortunately the party was beset by ". . . innumerable multitudes of minute, almost invisible seed ticks . . ." whose unpleasant qualities are described at length.

"Among many other plants . . . we observed the acalypha and the splendid lobelia *cardinalis,* also the cardiospermum *helicacabum* . . . a delicate climbing vine, conspicuous by its large inflated capsules. The acacia [robinia *pseudacacia*] the honey locust, and the Ohio æsculus are among the forest trees, but are confined to the low grounds. The common black haw [viburnum *lentago*] the persimmon or date plum, and a vitis unknown to us, occurred frequently and were loaded with unripe fruit. The misseltoe[2] . . . occurs here, parasitic on the branches of elms. In the sandy soils of the hills, the formidable jatropha *stimulosa* is sometimes so frequent, as to render the walking difficult . . . The cacti and the bartonias had now disappeared, as also the yucca, the argemone, and most of the plants . . . conspicuous . . . about the mountains. The phytolacca *decandra,* an almost certain indication of a fertile soil, the diodia *tetragona,* a monarda, and several new plants were collected in an excursion from our encampment."

On August 28, following the river, they reached an open plain where the grass was ". . . fine, thick and close fed." Here again were herds of bison, wild horses, antelope, deer, and also ". . . a large and uncommonly beautiful village of the prairie marmots . . . about a mile square . . ."

To avoid the winding river valley they climbed to open country on its north side, and had a view of an ". . . elevated country beautifully varied with gentle hills, broad vallies, fertile pastures, and extensive woodlands. The soil . . . of superior quality, the timber more abundant than in any region we have passed since we left the Missouri . . . Among the trees on the uplands are the black cherry, the linden, and the honey locust all indicating a fertile soil."

Since travel was laborious along the sandy bed of the river, they attempted, on August 30, to make their way through its wooded shore and ascend to the more open plains, but growth along the banks was ". . . so close and interlaced with scadent species of Smilax, Cissus and other climbing vines as greatly to retard our progress . . . we took a pleasure in observing the three American species of Cissus growing almost side by side. The C. *quinquefolia* [in a footnote, 'Ampelopsis quinquefolia of Michaux'], the common woodbine . . . grows here to an enormous size, and as well as the C. *hederacea,* seems to prefer climbing on elms. The . . . C. *bipinnata* ['C. *stans* of Persoon'] is a smaller plant, and . . . is rarely scadent. They all abound in ripe fruit . . . nauseous to the taste . . ."

2. The mistletoe, or *Phoradendron,* is now the state flower of Oklahoma.

On August 31 the party was at ". . . the western base of that interesting group of hills, to which we have attempted to give the name of the almost extinct tribe of the Oarks . . ." The name "Oarks," or Ozarks, is a corruption of the French *aux arcs*. Bradbury, some years before, had mentioned his plans, unfulfilled, to spend a summer at ". . . Ozark, (or more properly Aux-arcs) on the Arkansas . . ."

On September 1, following the bed of the river, the ". . . sycamore, the æsculus, the misseltoe, and the parroquet . . ." were conspicuous. After climbing some nearby hills they saw, to the east and south, ". . . a wild and mountainous region, covered with forests, where, among the bright verdure of the oak, the nyssa, and the castanea *pumila*, we distinguished the darker shade of the juniper, and others of the coniferæ."

The interpreter of Long's party, by name Adams, had been lost for five days. He was found on September 2. "When we discovered him, his appearance indicated the deepest despondency." They passed the mouth of a stream. Its considerable water ". . . was absorbed in the sands immediately at its junction with the larger stream." James noted:

"The Small leaved and White elm [in a footnote, 'Ulmus *americana,* and U. *alata*'], and the Nettle tree or Hackberry, the Cotton-wood, Mulberry, Black walnut, Pecan, Ash, Sycamore . . . are intermixed here to form the dense forests of the river valley; while in the more scattered woods of the highlands, the prevailing growth is Oak, with some species of Nyssa, the Dyospiros and a few other small trees."

On September 3 they traveled but a short distance, then searched for honey and collected plants. "Since leaving the open country, we had remarked a very great change in the vegetation." The dense shade produced fewer robust perennials and grasses than the arid, sandy deserts. "The sensitive cassia (C. *nictitans*.) the favorite food of the bees, some species of Hedysarum and a few other leguminæ are, however, common to both regions."

As on other days of attempted rest—Sundays—the ticks and "blowing flies" were in evidence; in addition, "An enormous black, hairy spider, resembling the mygale *avicularia* of South America, was often seen, and it was not without shuddering, that we sometimes perceived this formidable insect, looking out from his hole, within a few feet of the spot, on which we had thrown ourselves down to rest."

On September 4 James noted: ". . . we met with nothing interesting, except the appearance of running water in the bed of the river. Since the 13th of the preceding month, we had travelled constantly along the river, and in all the distance passed in that time, which could not have been less than five hundred miles, we had seen running water in the river, in one or two instances only; of those, one in it had evidently been occasioned by recent rains, and had extended but a mile or two, when it disappeared."

The country had become extremely fertile and on September 5 James saw many plants which he was unable to identify since both flowers and fruit had gone by. Recognized, however, was ". . . a very beautiful species of Bignonia, and the bow-wood or Osage orange." James, in a footnote identifies the last-named with "Maclura *aur-antiaca of Nuttall*." And, like all travelers of the period, he discusses the tree and its uses at some length. Whether the experiment proved really successful he does not say, but they attempted to use one of the tree's ingredients in a novel fashion:

". . . The bark, fruit, &c., when wounded, discharges a copious milky sap, which soon dries on exposure and is insoluble in water, containing probably, like the milky juices of many of the Urticaæ a large intermixture of Coatchouc or Gum Elastic. Observing this property in the milky juice of the fruit, we were tempted to apply it to our skin, where it formed a thin and flexible varnish, affording us, as we thought some protection from the ticks."

In transcribing from James's *Account,* I have omitted virtually all his references to other than plants; on this day a paragraph devoted to birds (and likely contributed by Thomas Say) seems worth quoting:

"We listened, as we rode forward, to the note of a bird, new to some of us, and bearing a singular resemblance to the noise of a child's toy trumpet. This we soon found to be the cry of the great ivory billed woodpecker, (Picus *principalis*) the largest of the North American species, and confined to the warmer parts. The P. *pileatus* we had seen on the 28th August, more than one hundred miles above, and this, with P. *ery-throcephalus,* were now common. Turkies were very numerous. The paroquet, chuck-wills-widow, wood robin, mocking bird, and many other small birds, filled the woods with life and music. The bald eagle, the turkey-buzzard and black vulture, raven, and crow, were seen swarming like the blowing flies, about any spot where a bison, an elk, or a deer had fallen a prey to the hunter. About the river were large flocks of pelicans, with numbers of snowy herons, and the beautiful ardea egretta."

On September 6 some rocky hills, running across the course of the river, had produced "an inconsiderable fall," which James believed would be an obstacle—should the river ". . . at any season of the year, contain water enough for the purposes of navigation . . ." He refers to it ". . . as the only spot, in a distance of six hundred miles, we can hope to identify by description . . ." and mentions that they called it ". . . the falls of the Canadian,[3] rather for the sake of a name than as considering it worthy to be thus designated . . ." James here gives his frank opinion of the country which the expedition had traversed after leaving the Rocky Mountains:

". . . In ascending, when the traveller arrives at this point, he has little to expect beyond but sandy wastes, and thirsty inhospitable steppes. The skirts of the hilly and wooded region extend to a distance of fifty or sixty miles above; but even this district is indifferently supplied with water. Beyond, commences the wide sandy desert stretching westward to the base of the Rocky Mountains. We have little apprehension of giving too unfavorable an account of this portion of the country. Though the soil is, in some places, fertile, the want of timber, of navigable streams, and of water for the

3. Until September 10 Long's party were under the impression that they were on Red River, so that "Canadian" suggests an appellation bestowed at a later date.

necessities of life, render it an unfit residence for any but a nomade population. The traveller who shall, at any time, have traversed its desolate sands will, we think, join us in the wish that this region may forever remain the unmolested haunt of the native hunter, the bison, the prairie wolf, and the marmot."

Less than ten miles below the falls, and entering from the north, the party came to ". . . the confluence of the Great North Fork, discharging at least three times as much water as we found above it . . ." The North Canadian River enters the Canadian, which they had been following since entering Oklahoma, in McIntosh County.

On September 8 travel was impossible except along the bed of the river where the water ". . . had now become so considerable, as to impede our descent . . .

". . . We . . . continued to make our way, though with great difficulty, and found our horses much incommoded, by being kept almost constantly in the water, as we were compelled to do, to cross from the point of a sand bar on one side of the river, to the next on the other. Quicksands also occurred, and in places where we least expected it, our horses and ourselves were made to *bite the dust,* without a moment's notice . . . the subsequent exertions necessary to extricate themselves, proved extremely harrassing to our jaded horses, and we had reason to fear, that they would fail us, in our utmost need."

Fearing that the collections would be "wetted" by these constant crossings they "made prize" of a canoe which they found along the bank and packed it with these valuables.

On September 10 they ". . . arrived at the confluence of our supposed Red river with another of a much greater size, which we at once perceived to be the Arkansa. Our disappointment and chagrin at discovering the mistake we had so long laboured under, was little alleviated by the consciousness that the season was so far advanced, our horses and our means so far exhausted, as to place it beyond our power to return and attempt the discovery of the sources of Red River . . .

"According to our estimate of distances on our courses, it is seven hundred ninety-six and half miles from the point where we first struck the Canadian to its confluence with the Arkansa. If we make a reasonable allowance for the meanders of the river, and for the extension of its upper branches some distance to the west of the place where we commenced our descent, the entire length of the Canadian will appear to be about one thousand miles. Our journey upon it had occupied a space of seven weeks, travelling with the utmost diligence the strength of our horses would permit."

They now crossed the Arkansas River, here forming the boundary between Muskogee County (west) and Sequoyah County (east). The day was spent in trying to work their way through "a dense and almost impenetrable cane-brake," where no vestige of a path could be found.

". . . Making our way, with excessive toil, among these gigantic gramina, our party might be said to resemble a company of rats traversing a sturdy field of grass. The cane stalks, after being trod to the earth, often inflicted, in virtue of their elasticity, blows as severe as they were unexpected . . . We received frequent blows and bruises on all parts of our bodies, had our sweaty faces and hands scratched by the rough leaves of the cane, and, oftentimes, as our attention was otherwheres directed, we caught with our feet, and had dragged across our shins the flexible and spiny stalks of the green briar . . . we laid ourselves down at dark, much exhausted by our day's journey.

"Our fatigue was sufficient to overcome the irritation of the ticks, and we slept soundly until about midnight, when we were awakened by the commencement of a heavy fall of rain, from which, as we had not been able to set up our tent, we had no shelter."

September 11, after more such effort, they reached the prairie and ". . . soon after discovered a large and frequented path, which we knew could be no other than that leading to Fort Smith." On the next day the valley of the Arkansas River lay to their right and beyond they could see the ". . . blue summits of the Point Sucre and Cavaniol mountains . . ." standing on opposite sides of the Poteau River—these appear on modern maps as Sugar Loaf Mountain and Cavanal Mountain, both in Le Flore County. Provisions were nearly exhausted but they found ". . . some pawpaw trees, with ripe fruit of an uncommon size and delicious flavor . . ." to allay their hunger. On camping, they ". . . collected . . . the beautiful vexillaria *virginica* of Eaton, which has the largest flower of any of the legumina of the United States, as is remarked by Mr. Nuttall. We saw also the menispermum lyoni, Hieracium *marianum,* Rhexia *virginica,* &c."

A party of men bound from Fort Smith to the Verdigris River brought news that Bell's party had arrived at Fort Smith "some days previous."

On September 13 they reached ". . . the little plantation opposite Belle Point . . . and found ourselves once more surrounded by the works of men. The plantation consisted of a single enclosure covered with a thick crop of maize intermixed with the gigantic stalks of the phytolacca *decandra,* and Ricinus *palma-christi,* forming a forest of annual plants, which seemed almost to vie with miegias and annonas, occupying the adjacent portions of the river bottom."

On arrival at the ferry they were ". . . not surprised to find our uncouth appearance a matter of astonishment both to dogs and men." Ferried across the river, they met with a cordial reception from Major Bradford commanding the fort, and greeted once more their former companions.

". . . Captain Bell, with Mr. Say, Mr. Seymour, and Lieutenant Swift, having experienced numerous casualties, and achieved various adventures, having suffered much from hunger, and more from the perfidy of some of their soldiers, had arrived on the 9th, and were all in good health. The loss most sensibly felt, was that of the manuscript notes of Mr. Say and Lieutenant Swift. Measures for the apprehension of the deserters, and the recovery of these important papers were taken immediately, and a reward of two hundred dollars offered. Mr. Glen[4] had kindly volunteered his

4. Hugh Glen, who was in command of the party met on September 11. He had a post on the Verdigris River.

assistance, and his influence, to engage the Osages in pursuit. But these efforts were unavailing."

At this point in the *Account* six chapters, ". . . from the MS. of Mr. Say . . .", recount the adventures of the Bell party. The theft had taken place the night of August 30. Three "infamous absentees" ". . . had deserted . . . robbing us of our best horses, and of our most important treasures. We endeavoured in vain to trace them . . . we returned from the fruitless search to number over our losses, with a feeling of disconsolateness, verging on despair.

"Our entire wardrobe, with the sole exception of the rude clothing on our persons, and our entire private stock of Indian presents, were included in the saddlebags. But their most important contents were all the manuscripts of Mr. Say and Lieutenant Swift completed during the extensive journey from Engineer Cantonment to this place. Those of the former consisted of five books, viz., one book of observations on the manners and habits of the mountain Indians, and their history . . . one book of notes on the manners and habits of animals, and descriptions of species; one book containing vocabularies of the languages of the mountain Indians; and those of the latter consisted of a topographical journal of the same portion of our expedition. All these being utterly useless to the wretches who now possessed them, were probably thrown away upon the ocean of prairie, and consequently the labour of months was consigned to oblivion, by these uneducated vandals."

James tells us that the site of Fort Smith was selected by ". . . Major Long,[5] in the fall of 1817, and called Belle Point in allusion to its peculiar beauty. It occupies an elevated point of land, immediately below the junction of the Arkansa and the Poteau, a small tributary from the southwest . . ."

On crossing the Arkansas River, Long's party had entered the state of that name, in northwestern Sebastian County. The gardens of the fort ". . . afforded green corn, melons, sweet potatoes, and other esculent vegetables, which to us had, for a long time, been untasted luxuries. It is probable we did not exercise sufficient caution, in recommencing the use of these articles, as we soon found our health beginning to become impaired . . ."

On September 19 Bell left for Fort Girardeau. The next day James and others departed ". . . to descend the Arkansa to the Cherokee agency, and to proceed thence to the Hot Springs of the Washita." On the 21st Long, Say, Seymour and Peale started for "Cape Girardeau," Missouri, about 140 miles below St. Louis, on the Mississippi.[6]

The "Cherokee agency" was reached September 23. This was at the mouth of Illinois Creek, in Pope County. From there James and others crossed the Arkansas River

5. According to Ghent, William Bradford and Long had chosen the site on their trip up the Arkansas and work began on the fort on December 25, 1817. At the time of James's arrival Bradford was in command of the fort.

6. Presumably Long and his companions had joined James, for the *Account* continues to use the pronoun "we" in narrating the journey.

and proceeded southeastward to the Hot Springs[7] of the Washita in Garland County, starting on the 25th. That day James observed that the hills south of the Arkansas were ". . . commonly covered with small and scattered trees. Several kinds of oak, and the Chinquapin (Castanea *pumila.* Ph.) attaining the dimensions of a tree . . . We distinguish here in the uplands two separate varieties of soil . . . sandstone . . . bearing forests of oak; and another resting upon a white petrosilicious rock . . . often covered with pine forests. The most common species, the yellow pine (P. *resinosa*) attains unusual magnitude; the P. *rigida* and some other species occur, but are not frequent. We also observed several species of vaccinium, the Mitchilla, the Kalmia *latifolia,* Hamamelis *virginica* ? Cunila *mariana,* and many other plants . . ."

At the Hot Springs, where they arrived on September 28, grew ". . . a number of interesting plants. The American holly, [Ilex *opaca*] is frequent in the narrow vallies within the mountains. The leaves of another species of Ilex, [I. *cassine,* the celebrated *Cassine Yaupon,*] which grows about the Springs, are there used as a substitute for tea.

"The Angelica tree, [Aralia *spinosa,* Ph.] is common along the banks of the creek, rising to the height of twelve to fifteen feet, and bending beneath its heavy clusters of purple fruit. The Pteris atropurpurea, Asplenium melanocaulon, A. ebenum, and other ferns are found adhering to the rocks. In the open pine woods the Gerrardia *pectinata,* considered as a variety of G. *pedicularia,* is one of the most conspicuous objects."

From Hot Springs the party traveled northeastward to Little Rock, Pulaski County, on the way crossing the Saline Fork on September 30. Little Rock was then ". . . a village of six or eight houses . . ." James seems to have been the first botanist to penetrate the country south of the Arkansas River. He notes at one point that "None of the tributaries to the Washita above the Hot Springs have hitherto been explored." He was referring to the country between the Cherokee towns and the Hot Springs, now allocated to Yell, Perry and Garland counties. The remainder of his journey, from Hot Springs to Little Rock, falls into the same category insofar, certainly, as botanical exploration is concerned. Nuttall in 1819 had spent some time at Dardanelle, Yell County, but had not penetrated far into the interior.

Leaving Little Rock on October 3, they traveled for four or five miles through ". . . the deep and gloomy forests of the Arkansa bottoms. Here we saw the Ricinus *palma christi* growing spontaneously by the road side, and rising to the height of twelve or fourteen feet . . . In the high and rocky country about White river, we fell in, with the route which had been pursued by Major Long and his party . . . following this we reached Cape Girardeau a few days after their arrival. The distance from Belle Point to Little Rock, by the way of the Hot Springs, is two hundred and ten miles, from Little Rock to Cape Girardeau three hundred, in the whole five hundred and ten miles."

7. This is still famous as a health resort—the springs considered to be beneficial for a number of chronic ailments such as rheumatism, gout, etc.

James has considerable to say of the bald cypress which filled extensive swamps in the region traversed and imparted ". . . a gloomy and unpromising aspect to the country." He refers to it as ". . . one of the largest trees in North America . . ." and comments upon the fact that ". . . The old error of Du Pratz,[8] with regard to the manner of the reproduction of the cypress, is still maintained with great obstinacy by numbers of people who never heard of his book.

" 'It renews itself,' he says, 'in a most extraordinary manner. A short time after it is cut down, a shoot is observed to grow from one of its roots, exactly in the form of a sugar loaf, and this sometimes rises ten feet high before any leaf appears; the branches at length rise from the head of this conical shoot.' p. 239.

"We have often been reminded of this account of Du Pratz by hearing the assertion among the settlers, that the cypress never grows from the seed. It would appear, however, that he could have been but little acquainted with the tree, or he would have been aware that the conic excrescences in question spring up and grow during the life of the tree, but never after it is cut down." Moving on,

"On the 12th October, the Exploring party were all assembled at Cape Girardeau." The steamboat *Western Engineer* in which Long's party had traveled up the Missouri in 1819 had recently arrived from St. Louis; it was to proceed to the "falls of the Ohio" when a rise in the river permitted. At this juncture most members of the party came down with "intermitting" fever. When, as soon as possible, the expedition broke up, James was still too ill to move. By November 22 he was sufficiently recovered to start, on horseback, for the falls of the Ohio. After a further delay because of "fever and ague" at Golconda, Pope County, Illinois, he started on his way to Philadelphia. The narrative portion of the *Account* ends at this point.

Chapters on specific phases of the scientific work follow. Also Long's contribution (chapter 18 of some 50 pages) entitled "General description of the country traversed by the Exploring Expedition . . ." which discusses the subject under sectional headings. I quote but a portion of his section V which describes ". . . the country situated between the Meridian of the Council Bluff, and the Rocky Mountains." Long notes:

". . . we do not hesitate in giving the opinion, that it is almost wholly unfit for cultivation, and of course uninhabitable by a people depending upon agriculture for their subsistence. Although tracts of fertile land, considerably extensive, are occasionally to be met with, yet the scarcity of wood and water, almost uniformly prevalent, will prove an insuperable obstacle in the way of settling the country . . . This region, however, viewed as a frontier, may prove of infinite importance to the United States, inasmuch as it is calculated to serve as a barrier to prevent too great an extension of our population westward, and secure us against the machinations or incursions of an enemy, that might otherwise be disposed to annoy us in that quarter . . ."

The *Account* was the subject of a long article in the *North American Review* for

8. Le Page du Pratz had emigrated to Louisiana in 1718. In 1758 he published his *Histoire de la Louisiane* in Paris. A chapter on trees is of historical interest only.

1823; although unsigned it was probably from the pen of Edward Everett, then the editor. The reviewer read "with shame and horror" one statement in the book:

"It will be perceived that the travels and researches of the expedition have been far less extensive, than those contemplated in the orders of the Secretary of War. The state of the national finances, during the year 1821, having called for retrenchments in all expenditures of a public nature, the means for the farther prosecution of the objects of the expedition were accordingly withheld.

"The state of our national finances! Some great calamity, perhaps, has befallen us . . . Detestable parsimony! The only country in the world, that has not been reduced to an avowed or virtual bankruptcy; the country, which has grown and is growing in wealth and prosperity beyond any other and all other nations, too poor to pay a few gentlemen and soldiers for exploring its mighty rivers, and taking possession of the empires, which Providence has called it to govern! One half of the wages of the members of Congress for the hours they have sagely devoted . . . to the nauseous petitions of Colonel Symmes[9] and his moon-stricken disciples, would have enabled this party of gallant officers intelligent and scientific travellers, to enlarge the known boundaries of all the kingdoms of nature. Poor, indeed, we are in spirit, if not in finance, if we will not afford to pay the expense of making an inventory of the glorious inheritance we are called to possess. England, staggering and sinking under her burdens, can fit out her noble expeditions . . . France has her intrepid naturalists . . . Russia, with her Krusensterns, and Kotzebues, and Lisianskis, is actually elbowing us out of the mouths of the Columbia . . . we cannot find a small party of discovery in powder and ball enough to hunt withal, or blankets and strouding enough to trade with the Indians . . . Honest poverty is no shame to the single man or the state. If we are poor, let us . . . truckle to the British, court the Russians, beg pardon of the Spaniards, and shake hands with the pirates. Let the President at Washington move into comfortable lodgings . . . and his white palace be leased out as a hotel. Put Congress back into the brick tenement, from which it lately escaped, and convert the capitol into a cotton factory . . ."

In terminating the *Account,* James wrote that the collections comprised sixty skins of "new or rare animals," "Several thousand insects," minerals, shells, fossils, sketches of animals, birds, insects, landscapes, and so on.

"Most of the collections made on this expedition have arrived in Philadelphia, and are in good preservation . . . The herbarium contains between four and five hundred species of plants new to the Flora of the United States, and many of them supposed to be undescribed."

I have cited but little from George E. Osterhout's paper, "Rocky Mountain botany and the Long Expedition of 1820" (1920). Ewan (1942) mentions it thus: "Interesting and carefully prepared running account with useful notes from a resident Colorado

9. Presumably John Cleves Symmes, who believed that the earth was hollow with open poles and a habitable interior. There is no indication that he acquired many disciples.

botanist who knew James's territory first hand. The list of plants which served as types of new species described by Torrey, Bentham, and others . . . is highly useful." Osterhout also wrote a paper "Concerning the ornithology of the Long Expedition of 1820" (*Oologist* 37: 118-120, 1920) which Ewan (*l.c.*) states is valuable ". . . for its notes upon the type localities of Say's collections."

Osterhout prefaces his list of James's plants, referred to above, with the comment that when plants are mentioned in the narrative ". . . it is an easy matter to determine the location; but only a few of them are so mentioned. In the published lists no dates are given, and the localities of the new species are often indefinite." Further:

". . . While the . . . list is probably quite incomplete, it is a remarkable list of new species, all or nearly all of them collected as the Expedition crossed the territory now comprised in the state of Colorado; and seemingly the first botanical collection which was made in this territory. The species are arranged according to the sequence in Rydberg's 'Flora of the Rocky Mountains and adjacent plains,' published in New York, 1917, the names follow this work throughout with the original synonyms cited where necessary."

The list enumerates seventy-three species and Osterhout states that "In general the native plants still grow and blossom as they did when Dr. James saw them in 1820; but a great change has been wrought in the country. Fruitful farms have replaced much of what seemed to be sterile soil, and towns and cities and a busy industry have come to their silent and uninhabited plains and hills."

John Torrey wrote three papers on James's collections, all published in the *Annals* of the Lyceum of Natural History of New York (now the New York Academy of Sciences):

(1) "Descriptions of some new or rare plants from the Rocky Mountains, collected in July, 1820, by Dr. Edwin James." This was read on September 22, 1823, and published in 1824. Ten species were described.

(2) "Descriptions of some new grasses collected by Dr. Edwin James, in the expedition of Major Long to the Rocky Mountains, in 1819–1820." This was read on May 17, 1824, and published in 1824. Eight species were described and one new genus, *Pleuraphis;* the type, *P. Jamesii,* was figured in Plate X.

(3) "Some account of a collection of plants made during a journey to and from the Rocky Mountains in the summer of 1820, by Edwin P. James, M.D. Assistant Surgeon U. S. Army." This was read on December 11, 1826, and published in 1826–1827.

Between the issuance of (2) and (3) above, or in 1825, James had published in the *Transactions* of the American Philosophical Society his "Catalogue[1] of plants col-

1. In the title and in the foreword to this "Catalogue" the author's name appears as "E. P. James." Torrey, as cited under (3) above, took up the middle initial "P." and we find it adopted by Darlington in *Reliquiae Baldwinianae* ("Dr. E. P. James did find . . ." on p. 299, and elsewhere) and by Sargent in *The Silva of North America* (3: 103, 1892). I have found no mention of the middle initial in Pammel's biographical account of James, nor has inquiry disclosed that he possessed one. It may have been an error of the printer.

lected during a journey to and from Rocky Mountains, during the summer of 1820."
This had been read in the summer of 1821. In a foreword James explains that his
collection contained "a considerable number of species not enumerated in this Catalogue" but that there had been no opportunity to make the requisite comparisons:
"Such only are given, with a few exceptions, as have already been ascertained to belong to the Flora of North America." James followed for the most part—in arrangement and nomenclature—Nuttall's *Genera*. The "Catalogue" was little more than a
classified list of names and but few localities of collection are cited.

Torrey wrote prefatory remarks to each of his papers and I quote from these in the
order enumerated above:

(1) "Among the many valuable discoveries in Natural History, made by the scientific gentlemen attached to the late expedition to the Rocky Mountains, commanded
by Major Long, those relating to the department of Botany are not the least interesting. Dr. Edwin James, who was the botanist in this hazardous journey, and whose zeal
in prosecuting his favourite science is so well known, having been called to accompany another expedition,[2] from which it is uncertain when he will return, has kindly
permitted me to commence the publication of the discoveries he made; particularly of
the plants from the summit of the Rocky Mountains, as well as the whole of the
Gramina. Owing to my having but part of Dr. James' collection in my possession, it
is impossible to preserve much order in the arrangement of the plants I shall describe;
they will therefore be published in occasional *Decades*, as I find leisure to determine
them . . ."

(2) "In a former part of this volume, I commenced an account of the alpine plants
collected in this expedition, by Dr. James, which I promised to continue, in occasional decades. Having been obliged, for the present, to defer the examination of the
remaining specimens, I beg leave to offer to the Lyceum descriptions of some new
grasses, collected by Dr. James in the same expedition. Their number might have been
considerably increased had not many of them been discovered a short time previous,
by Mr. Nuttall, in his Travels into Arkansas Territory. One species, of which there
were very perfect specimens, in the herbarium of Dr. James, is so peculiar that I have
proposed it as the type of a new genus, under the name of *Pleuraphis*."

(3) "Dr. James was the botanist appointed to succeed the unfortunate and lamented Dr. Baldwin, in the Expedition to the Rocky Mountains, commanded by
Major Stephen H. Long. He joined the Expedition in the spring of 1820. The region
explored on the west of the Mississippi, is included between 34° 40′ and 41° 30′ of
north latitude, and west to the Rocky Mountains. The expedition proceeded up the
Missouri as far as Council Bluffs, thence west to the Pawnee villages, up the Platte to
its sources, among the Rocky Mountains, south along the base of these mountains,
from the 40th to the 38th degree, when they descended the Arkansa about one hun-

2. The reference is to Long's expedition of 1823 to the St. Peter, or Minnesota, River. It is said that James
did not receive his orders on time and his place was taken by Thomas Say.

dred miles. The expedition here divided into two detachments, one of which continued to follow the Arkansa, while the other proceeded south to the sources of the Canadian, which they followed to its junction with the Arkansa, where they met,[3] and both detachments then continued their journey homeward to the Mississippi. As this extreme tract of country was traversed with great rapidity, and the party was exposed to great hardships and privations, little opportunity was afforded of making observations, or even of recording all the stations of the plants; and many of the specimens, owing to the same unfavourable circumstances, are injured or incomplete.

"Very few of the new species collected have yet been published. A considerable number were cursorily mentioned in the account of the Expedition compiled by Dr. James, but their characters are not given. In the summer of 1821, Dr. James communicated to the American Philosophical Society, a catalogue of all the plants collected on the journey, as far as he had then determined them. A large proportion of these were found east of the Mississippi, and are mostly well known species . . . The herbarium collected on the expedition, has recently been placed at my disposal, by my friend Dr. James. A few of the new or rare plants which it contained, I described several years since, in two papers . . . The following catalogue includes only the plants collected west of the Mississippi,[4] as it is chiefly these which present much interest to the botanist. In the examination of the specimens, I have been greatly assisted by my learned friend Thomas Nuttall, Esq. who has devoted more attention to the botany of this country than any other individual. This gentleman has not yet published the plants collected by him during his journey into the Arkansa country in 1819, though it is his intention to have his account of them succeed this in a short time.[5]

"Some of the species discovered by Dr. James, were found the year previous by Mr. Nuttall. These I have omitted to describe, as it would be improper to interfere with that gentleman's prior discoveries,[6] especially as he long since furnished me with almost a complete set of the specimens collected on his last journey.

"A considerable number of the new and rare plants enumerated below, are not contained in Dr. James' catalogue, as he had not the books or other necessary means

3. The two detachments were reunited at Fort Smith, Arkansas, as James stated.

4. Shear and Stevens include a letter written by Torrey to von Schweinitz on December 12, 1826:

" 'I am now busily employed in writing an account of the plants collected west of the Mississippi on Long's expedn. The whole will be arranged according to the Nat[ura]l Order. I have written much. The first part of the account is now printing in the Annals. You will be surprised to see what curious plants are in the collection—Many which were never before found north of Mexico.' "

5. Nuttall's "Collections towards a flora of the territory of Arkansas" was only published in 1835–1836.

6. Torrey's approach to the priority question was highly ethical and he reiterates it in a letter to William Darlington (quoted by Rodgers); after stating that a week earlier he had received from James the final consignment of plants collected on the Long expedition he notes:

" 'The . . . whole number of plants collected . . . by Dr. Baldwin and Dr. James, I think is about 700. Many of the latter are exceedingly interesting, & not a few are new species. They have never been ex-for examining them properly at his disposal. Among these are some which are peculiarly interesting to American botanists, such as *Pomaria glandulosa, Caltha sagittata,* &c which have not been hitherto found north of Mexico. Indeed the vegetation about the sources of the Arkansa and Canadian very much resembles that of Mexico, so that it was necessary to make a particular comparison of our plants with those described by Humboldt, Bonpland and Kunth."

In 1873 Asa Gray wrote of the Torrey publications:

"This is the earliest treatise of the sort in this country, arranged upon the natural system;[7] and with it begins the history of the botany of the Rocky Mountains, if we except a few plants collected early in the century by Lewis and Clark, when they crossed many degrees farther north, and which are recorded in Pursh's Flora . . ."

Edwin James was the first botanist to cross the greater part of Nebraska from east to west; the first to cross Colorado from north to south and, therein, the first to reach high altitudes in the central Rocky Mountains and to traverse the deserts of its southeastern portion; the first botanist to set foot in extreme northeastern New Mexico; and the first to cross, following the Canadian River, the Panhandle of Texas[8] and, except for its Panhandle section, the state of Oklahoma from west to east. In the northeastern portion of Oklahoma he had been antedated by Thomas Nuttall, who had reached the

amined by any botanist except myself. I am now making out an account of them which it will be hardly proper to publish till Mr. Nuttall publishes his prior discoveries made in a part of the country visited by Maj. Long. Mr. N. must not be long, however, or we shall both be anticipated by Europeans. There are now persons exploring towards the Rocky mountains & on the Missouri & their collections will be described by foreign botanists with the greatest dispatch . . .' "

The reference to European activity in the collecting and publishing fields would seem to refer to Berlandier whom the elder De Candolle had arranged to send to Mexico, as early as 1824 or 1825 according to Geiser. I know of no other "persons" exploring "towards" the Rocky Mountains at the date of Torrey's writing (1826). He had doubtless heard of De Candolle's *hopes* of sending Berlandier to the Santa Fe region.

7. The Natural System of plant classification shows the true relationship of plants as revealed by maximum agreement in all structures; in it the evolution of plants is taken into account. The Artificial (or Linnean) System on the other hand sorts, or forces, plants into an arbitrarily predetermined group of categories and generally brings unrelated plants together; it is an unnatural grouping in the sense that any alphabetical index is such.

8. For some time the question whether "Nicolas Antoine Monteil (1771–1835)" merited the distinction of having been the first botanist to work in Texas was under careful consideration, by one botanist at least. "Monteil's" life and works were first described (so far as is known) in *Appletons' cyclopaedia of American biography*. But "Monteil"—who was credited with having worked on the lower Trinity River—is now relegated to the considerable group of "fictitious" botanists whose biographies are found in that, for the most part, useful work.

Anonymous. 1888. Nicolas Antoine Monteil. *Appletons' Cyclop. Am. Biog.* 4: 385.

BARNHART, JOHN HENDLEY. 1919. Some fictitious botanists. *Jour. N. Y. Bot. Gard.* 20: 171-181.

GEISER, SAMUEL WOOD. 1930. Some frontier naturalists. *Bios* 5: 141.

—— 1934. That first Texas botanist. *Field & Lab.* 3: 11, 12.

—— 1937. Monteil, Nicolas Antoine (1771–1833). *In:* Naturalists of the frontier. Dallas. 329, 330.

confluence of the Arkansas and Canadian rivers on July 11, 1819. James had reached the same point on September 10, 1820.[9]

Nuttall had antedated James in Arkansas, when he ascended the river of that name in 1819. But when James, on his descent of the Arkansas, left that river near the Cherokee village and made an overland excursion to the Hot Springs of the Washita and thence traveled to Little Rock, he was, as already stated, the first botanist to penetrate any distance south of the Arkansas River.

While, as Torrey recorded, the plants gathered by Nuttall and by James were to some extent duplicated, it is evident that the two men did not travel, with minor exceptions, over the same territory.

In 1907, in the *Annals of Iowa,* Dr. L. H. Pammel published a biographical account of Edwin James which was based in part upon information supplied orally or in documentary form by relatives and friends of the botanist. My facts and quotations are derived therefrom.

James was born on August 27, 1797, at Weybridge, Addison County, Vermont,— ". . . the youngest of thirteen[1] children, ten of whom were sons . . ." The family home still stood in 1907 although much altered and ". . . has always remained the property of the family, having descended from father to son, in the older line." James ". . . prepared for college in the Addison County Grammar School, located at Middlebury, and graduated from Middlebury College in 1816, having walked five miles daily in his journeys to and from school." Pammel tells of a pamphlet containing a list of 551 plant species indigenous to Middlebury township which James prepared at about this time.

From Vermont James went to Albany where he studied medicine for two years. "In those days it was considered essential that a physician be well posted in natural history, especially in botany."

After his trip with Long in 1820 James married Clarissa Rogers of Gloucester, Massachusetts; they had one son, born in 1838 and named for his father. Soon after his marriage James was appointed surgeon in the United States Army and was stationed at a number of military posts. At Fort Crawford (now Prairie du Chien) he began a study of the Indian language, and at Mackinac ". . . became a great friend of the Chippewa Indians, into whose language he translated the New Testament from the Hebrew Chaldaic." Pammel reports that James made 500 converts in the course of his missionary work. It was at Fort Crawford that he became acquainted with John

9. Bell's section of Long's expedition (which, however, included no botanist) and Nuttall had covered a greater amount of similar territory: Bell had descended the Arkansas from near La Junta, Colorado, to Fort Smith, passing the mouth of the Cimarron River on the way; Nuttall had reached the mouth of the Cimarron, not far west of Tulsa, on September 3, 1819.

1. Ewan points out that James was the thirteenth child and George Engelmann the eldest of thirteen children.

Tanner, who, at the age of five, had been stolen from his home in Ohio by the Indians and reared among them, marrying a squaw and becoming very intemperate. James induced him to give up his bad habits and wrote the story of his life,[2] giving Tanner the proceeds of the sale of 1,000 copies. James eventually resigned from the army and lived in Albany where he edited a temperance journal. In 1836 he moved to Burlington, Iowa, where he lived for the rest of his life.

In a letter to Parry dated February 11, 1859, James recounts:

". . . I became a settler in Iowa twenty-two years ago and of course have seen great changes. The locomotive engine and the railroad car scour the plain in place of the wolf and the curlew. Mayweed and dog fennel, stink weed and mullein have taken the place of 'purple flox and the mocassin flower,' the Celt, the Dane, the Swede and the Dutchman are instead of Black Hawk and Wabashaw, Wabouse, Manny-Ozit and their bands . . ."

It was about the time of the move to Iowa, states Pammel, that some of James's ". . . peculiar traits . . . became more conspicuous. His mode of life, his opinions, and his views on moral and religious questions generally, were inclined to ultraism . . . he gradually assumed the habits of a recluse . . . With him to espouse a cause, was to carry it to the farthest possible extremes, often erroneous . . . at times positively wrong . . . it must be admitted that his errors were on the side of goodness . . . On the 25th of October, 1861, he fell from a load of wood, the team descending a small pitch of ground . . . both wheels passed over his chest . . . He lingered . . . until the morning of October 28th, when he expired at the age of 64 years . . ."

Dr. William Salter, friend of James, is quoted thus:

"The life of Edwin James is worthy your thorough study. He was a remarkable man in many respects—personal, scientific, historic, moral and religious—a unique character. Personally I only knew him as a mystic, a recluse, an abolitionist, a come-outer, an underground conductor for men 'guilty of a skin not colored like his own,' a non-resistant, in fact a 'John Brown' man, but never to the extent of taking up arms, more perhaps like a Tolstoi of to-day. I could never draw him out on his past life. He would not talk about himself . . ."

Pammel records: ". . . It is to be regretted that his manuscripts, collections and papers were destroyed. After the death of his wife, orders were given to his housekeeper, Mrs. Callahan, to burn them . . ."

The monotypic genus of the Saxifrage family, *Jamesia,* with type species *J. americana,* was named for Edwin James by Torrey and Gray. And many specific names do him honor: *Stellaria Jamesiana* Torr., *Parosela Jamesii* (Torr.) Vail, *Solanum Jamesii* Torr., *Saxifrage Jamesii* Torr.—now *Boykinia Jamesii* (Torr.) Engler—and *Chionophila Jamesii* Bentham.

2. *A narrative of the captivity and adventures of John Tanner, United States Interpreter at Saut de Ste. Marie, during thirty years residence among the Indians in the interior of North America. Prepared for the press by Edwin James, M.D.* New York. 1830. (*See* Pammel, 180, *fn.*)

Anonymous [? EDWARD EVERETT]. 1823. [Review of] Account of an expedition . . . Compiled by Edwin James . . . Philadelphia . . . 1823. *North Am. Review* 16 (new ser. 7): 242-269.

BENT, ALLEN H. 1913. Early American mountaineers. *Appalachia* 13: 56, 57.

CHITTENDEN, HIRAM MARTIN. 1902. The American fur trade of the far west. A history of the pioneer trading posts and early fur companies of the Missouri valley and the Rocky Mountains and of the overland commerce with Santa Fe. Map. 3 vols. New York. 2: 562-587.

DARLINGTON, WILLIAM. 1843. Reliquiae Baldwinianae: selections from the correspondence of the late William Baldwin, M.D., Surgeon in the U. S. Navy. Philadelphia.

EWAN, JOSEPH. 1942. Botanical explorers of Colorado—II. Edwin James. *Trail & Timberline* 282: 79-84.

———— 1950. Rocky Mountain naturalists. Denver.

GEISER, SAMUEL WOOD. 1937. Naturalists of the frontier. Dallas. 19, 325, 328.

GHENT, WILLIAM JAMES. 1931. The early far west. A narrative outline 1540–1850. New York. Toronto. 155-159.

GOODWIN, CARDINAL. 1922. The trans-Mississippi west (1803–1853). A history of its acquisition and settlement. New York. London. 47-54.

GRAY, ASA. 1873. John Torrey; a biographical notice. *Proc. Am. Acad. Sci. Arts* 9: 262-271. *Reprinted:* Sargent, Charles Sprague, *compiler.* 1889. Scientific papers of Asa Gray. 2 vols. Boston. New York. 2: 359-369.

JAMES, EDWIN. 1823. Account of an expedition from Pittsburgh to the Rocky Mountains, performed in the years 1819 and '20, by order of the Hon. J. C. Calhoun, Sec'y of War; under the command of Major Stephen H. Long. From the notes of Major Long, Mr. T. Say, and other gentlemen of the exploring party. Compiled by Edwin James, botanist and geologist for the expedition. 2 vols. Atlas. Philadelphia.

———— 1825. Catalogue of plants collected during a journey to and from the Rocky Mountains, during the summer of 1820. By E. P. James, attached to the expedition commanded by Major S. H. Long, of the United States Engineers. And by the Major communicated to the Society, with the permission of the Hon. J. C. Calhoun, Secretary of War. *Trans. Am. Philos. Soc.* new ser., 2: 172-190.

———— 1905. Account of an expedition from Pittsburgh to the Rocky Mountains, performed in the years 1819, 1820. By order of the Hon. J. C. Calhoun, Secretary of War, under the command of Maj. S. H. Long, of the U. S. Top. Engineers. Compiled from the notes of Major Long, Mr. T. Say, and other gentlemen of the party, by Edwin James, botanist and geologist of the expedition. *In:* Thwaites, Reuben Gold, *editor.* Early western travels 1748–1846. 14, 15, 16, 17. Map. [The "Preliminary notice" of the *Account* is stated to have been reprinted from volume 1 of the Philadelphia edition, 1823; the text from volume 1 of the London edition, 1823.]

OSTERHOUT, GEORGE E. 1920. Rocky Mountain botany and the Long Expedition of 1820. *Bull. Torrey Bot. Club* 47: 555-562.

PAMMEL, LOUIS HERMANN. 1907, 1908. Dr. Edwin James. *Annals of Iowa,* ser. 3, 8: 161-185; 277-295. *Reprinted: Contr. Bot. Dept. Iowa State Coll. Agri. Mech. Arts* No. 32.

PARRY, CHARLES CHRISTOPHER. 1862. Ascent of Pike's Peak, July 1st, 1862 by Dr. C. C.

Parry. From a letter addressed to Prof. Torrey, and communicated by him. *Trans. Acad. Sci. St. Louis* 2: 120-125.

———— 1862. Botanical necrology for 1861. Dr. Edwin James. *Am. Jour. Sci. Arts,* ser. 2, 33: 428-430.

RODGERS, ANDREW DENNY, 3rd. 1942. John Torrey. A story of North American botany. Princeton. 52.

ROSS, FRANK EDWARD. 1932. Edwin James. *Dict. Am. Biog.* 9: 576.

SARGENT, CHARLES SPRAGUE. 1891. 1892. The Silva of North America. Boston. New York. 2: 96, *fn.* 1 (1891); 3: 104 (1892).

SHEAR, CORNELIUS LOTT & STEVENS, NEIL EVERETT, 1921. The Correspondence of Schweinitz and Torrey. *Mem. Torrey Bot. Club* 16: 235.

SUDWORTH, GEORGE BISHOP. 1917. The pine trees of the Rocky Mountain region. *U. S. Dept. Agr. Bull.* 460: 7, *fn.* 1.

TORREY, JOHN. 1824. Descriptions of some new or rare plants from the Rocky Mountains, collected in July, 1820, by Dr. Edwin James. *Ann. Lyceum Nat. Hist.* 1: 30-36. Read Sept. 22, 1823.

———— 1824. Descriptions of some new grasses collected by Dr. E. James, in the expedition of Major Long to the Rocky Mountains, in 1819–1820. *Ann. Lyceum Nat. Hist.* 1: 148-156. Read May 17, 1824.

———— 1826–1827. Some account of a collection of plants made during a journey to and from the Rocky Mountains in the summer of 1820, by Dr. Edwin P. [*sic*] James, M.D. Assistant Surgeon U. S. Army. *Ann. Lyceum Nat. Hist.* 2: 161-164 (1826); 165-254 (1827). Read Dec. 11, 1826.

WARREN, GOUVENEUR KEMBLE. 1859. Memoir to accompany the map of the territory of of the United States from the Mississippi River to the Pacific Ocean, giving a brief account of each of the exploring expeditions since A.D. 1800 . . . *U. S. War Dept. Rept. expl. surv. RR Mississippi Pacific* 11: 19-22.

CHAPTER XI

DOUGLASS ACCOMPANIES CASS AND SCHOOLCRAFT

ON THEIR SEARCH FOR THE HEADWATERS OF THE

MISSISSIPPI RIVER

IN 1820 the United States government made its second attempt to reach the headwaters of the Mississippi River but this, like a similar attempt made by Pike seventeen years earlier,[1] was unsuccessful in attaining its objective. However, it merits a place in this story since it was the duty of one member of the party to collect plants.

The account of the expedition is more important as a picture of the United States government's relationship to the English, and to the Indian tribes living in the Great Lakes and Mississippi valley regions, than as a botanical record.

The story is told in Henry Rowe Schoolcraft's *Narrative journal of travels, through the northwestern regions of the United States extending from Detroit through the great chain of American lakes, to the sources of the Mississippi River. Performed as a member of the expedition under Governor Cass. In the year 1820.* It was published in Albany in 1821. The volume's dedication to the Honorable John C. Calhoun, "Secretary at War," stresses the contributions made in his administration ". . . to develope

---

1. In 1803 Zebulon Montgomery Pike had been selected by President Thomas Jefferson ". . . to trace the Mississippi to its source . . ." Pike ascended the river to what is now called Cass (at that time Upper Red Cedar) Lake, also examining Turtle River to its sources, Leech Lake, and Leech River to its confluence with the Mississippi. The trip consumed "eight months and 22 days," and Warren states that Pike's ". . . map of the river gives its general direction with considerable accuracy . . ."

Pike, in his preface to his account of this journey (it is quoted in the Philadelphia edition of his *Expeditions,* edited by Elliott Coues, published in 1895), refers to his nonaccomplishments along certain lines which the government had ordered investigated—among these, the natural resources of the country:

"In the execution of this voyage I had no gentleman to aid me, and I literally performed the duties (as far as my limited abilities permitted) of astronomer, surveyor, commanding officer, clerk, spy, guide, and hunter; frequently preceding the party for miles in order to reconnoiter, and returning in the evening, hungry and fatigued, to sit down in the open air, by firelight, to copy the notes and plot the courses of the day."

Although Pike had no "gentleman" to aid him, he did not travel alone, but with ". . . one sergeant, two corporals, and 17 privates . . ." He was merely drawing the distinction, universal in the early narratives, between those who ranked above noncommissioned officers (the "gentlemen") and those of less exalted station (the "men").

E. J. Criswell, writing of Lewis and Clark as "linguistic pioneers," makes this curious comment: "Lewis's greater sophistication is clearly revealed in his vocabulary. Some of the terms peculiar to him hark back to the more civilized social life he had led at home . . . He tells us that when they left St. Louis they set out 'under three cheers from the *gentlemen on the bank*'. Any other explorer would have said *men* instead . . ." !

250

the physical character and resources of all parts of our country,—to the patronage
... extended to the cause of science ... to the protection it has afforded to a very extensive line of frontier settlements, by stretching our cordon of military posts, through
the territories of the most remote and hostile tribes of savages,—and particularly, to
the notice it has bestowed upon one of the humblest cultivators of natural science."

The *Narrative journal,* throughout, reflects these objectives: it tells of quarries and
mines visited, of the geology, mineralogy, flora and fauna of the regions traversed, and
of arrangements made for garrisoning frontier posts—where direct opposition seems
to have come, not so much from the British, as from the Indian tribes. It gives, too,
a picture of the competition existing in the fur trade, an industry in which the British
had the upper hand, being well-equipped and holding advantageous positions for
carrying on such work. With the determination of the boundary between British and
United States territory, it was only necessary for them to cross, in many regions, from
one side of a river or lake to the other.

Schoolcraft's "Introductory remarks" outline "... the progress of discovery[2] in the
northwestern regions of the United States ..."

Nonetheless, "... amidst much sound and useful information, there has been
mingled no inconsiderable proportion, that is deceptive, hypothetical, or false ... A
new era has dawned in the moral history of our country ... its physical productions,
its antiquities, and the numerous other traits which it presents for scientific research,
already attract the attention of a great proportion of the reading community ... it is
eagerly enquired ... what are its indigenous plants, its zoology, its geology, its mineralogy, &c...."

"The specific objects of this journey, were to obtain a more correct knowledge of
... the Indian tribes—to survey the topography of the country, and collect the materials for an accurate map—to locate the site of a garrison at the foot of Lake Superior, and to purchase the ground—to investigate the subject of the northwestern
copper mines, lead mines, and gypsum quarries, and to purchase from the Indian
tribes such tracts as might be necessary to secure to the United States the ultimate advantages to be derived from them, &c...."

Lewis Cass, then Governor of the Michigan Territory, was the instigator of the expedition and its leader. On the title-page of the *Narrative journal* Schoolcraft is cited
as a member of four learned societies, and Calhoun appointed him mineralogist of
the expedition. He seems also to have been its artist.[3] Botanical interest centers in

2. We are told of Father Joseph Marquette the Jesuit, of Robert de la Salle, of the Franciscan Father
Louis Hennepin, of Baron La Hontan, of P. De Charlevoix, of the trader Alexander Henry — "... the first
English traveller of the region ..."—and of Jonathan Carver who, after "... having escaped ... the perils
of a long journey through the American wilderness ... perished for want, in the city of London, the seat
of literature and opulence." Samuel Hearne, Alexander Henry, "first English traveller of the region,"
and Pike are also mentioned.

3. As rarely happens, Schoolcraft's name appears on some of the plates. In one copy of the *Narrative*

"Capt. David B. Douglass,[4] Civil and Military Engineer," who made the plant collections. Unfortunately his main task was the taking of observations; and he is not even
mentioned by Schoolcraft in his rôle of botanist.

The expedition left New York on March 5, 1820, and, by way of Albany, Utica,
Geneva, Buffalo, etc., reached Detroit on May 8. There, supplies and means of transport—canoes modelled on those of the Indians—were assembled. With the ceremoni-
ous departure from Detroit on May 24, a day-by-day record of the journey begins. I
shall, however, tell the story briefly only up to the time, on July 17, when the canoes
entered the valley of the Mississippi River.

The party kept along the southern shore of Lake St. Clair, up the St. Clair River to
Fort Gratiot at the foot of Lake Huron, and along the south shore of that lake. On
June 6 it was at the island of Michilimackinac, or sometimes "Mackinac." From here
Schoolcraft, Douglass and others visited the gypsum quarries on St. Martin's Islands.

Fort Michilimackinac was situated just above the town. Fort Holmes, not then
garrisoned, occupied "the apex" of the island. It had been erected by the British ". . .
while they held possession of the island, during the late war, and by them named *Fort
George*. But after the surrender . . . the name was altered in compliment to the memory
of Major Holmes, who fell in the unfortunate attack upon the island . . ."

There was also a trading post ". . . of the American, or South West Fur Company,
under the direction of Messrs. Stuart and Crooks." [5] Schoolcraft includes a table listing the ". . . produce of the fur trade for one year, given by McKenzie . . .": among
the skins enumerated were 106,000 beaver, 32,000 martin, 1,800 mink, 1,650 fisher,
and so on. He comments: ". . . whether there is an increase or diminution of the total
amount, are the secrets of a business of which we are ignorant."

After provisions had been received from Detroit the party started on its way on
June 13. There were ". . . forty-two persons, embarked in four canoes," and twenty-
two soldiers in a 12-oared barge—this military escort was because of anticipated Indian hostility at "the Sault," or Sault de Ste. Marie. After passing through the "straits
of St. Mary," at the entrance to which was an English fortification on Drummond's
Island, they reached the foot of the rapids and, after passing two, came on June 15

journal examined there are six, rather than the original eight plates: nos. 2, 3, 5 and 6 are noted as "H. R.
Schoolcraft del." and nos. 4 and 8 as "H. Inman del." The "Doric Rock, Lake Superior," embellishing
the extremely ornate title-page (preceding a more business-like one) also bears Inman's name.

4. John H. Barnhart wrote a short sketch of the man: "Douglass, David Bates (1790–1849). United States
military engineer; graduate of Yale; professor at West Point throughout the period of Torrey's connection with the military academy; afterward professor at New York University, Kenyon College, and
Hobart College. He accompanied the Cass expedition to the upper Mississippi in 1820, and collected
plants in that region, then little known botanically. He was a son-in-law of Major Andrew Ellicott (1754–
1820), the famous surveyor."

5. Robert Stuart and Ramsay Crooks. It was with the latter that Bradbury had made his overland journey
from the Arikara villages to the Mandan region in 1811.

to the largest, the Sault de Ste. Marie, fifteen miles below the foot of Lake Superior.[6] On "the south or American shore" was the village. "On the north, or Canadian shore . . . the North-west Company's establishment . . ." which was well-equipped, possessing in addition to various buildings, a canal for drawing barges and canoes over the rapids, and a pier, forming an harbour, in which ". . . a schooner is generally lying to receive the goods destined for the Grand Portage, and the regions northwest of Lake Superior." This establishment was ". . . seated immediately at the foot of the Falls . . ." Schoolcraft relates that Sault de Ste. Marie (". . . on the outlet of Lake Superior, and at the head of ship navigation . . .") had been recognized by the French as an advantageous site and early occupied. "By this place all the fur trade of the northwest is compelled to pass, and it is the grand thoroughfare of Indian communication for the upper countries, as far as the arctic circle."

Traders, too, had always recognized that it occupied an important situation, although this had ". . . but lately been perceived by our government." Since one of the primary purposes of Cass's expedition was to prepare the way for the introduction of an American garrison, he summoned a council of the chiefs of the Chippeway tribe to explain to them the view of the government. The situation, as he described it, was this:

"By the treaty of Greeneville, of 1795, a saving clause had been inserted by Gen. Wayne, covering any gifts or grants of land in the Northwest Territories, which the Indians had formerly made to the French or English governments, and this clause had been renewed or confirmed by treaties with the same tribes since the conclusion of the late war."

The Indians, nonetheless, were opposed to the idea of an American garrison, disclaimed knowledge of the earlier grant, and then gave what ". . . amounted to a negative refusal." Cass then announced that, whether the Indians renewed the grant or not, the matter was already decided. The Indians hoisted the British flag over their encampment, Cass had it taken down, and so on. Serious trouble was averted and firmness seems to have won the day; the Indians "For this session of land . . . were paid on the spot, in blankets, knives, silver wares, broadcloths, and other Indian goods."

The Sault de Ste. Marie was left on June 17 and that night the party camped at ". . . Point aux Pins, on the Canadian side of the river . . . the only night during the whole expedition . . . passed in the Canadian territory." On the 18th the entrance to Lake Superior was ". . . in full view, presenting a scene of beauty and magnificence . . . rarely surpassed . . ." They followed the south shore of the Lake and on July 5 entered ". . . the mouth of the river St. Louis, which enters the lake at the head of the Fond du Lac." The passage of Lake Superior had taken eighteen days. Ascending the St. Louis River, they ". . . first saw in plenty the folle avoine, or wild rice[7] which . . .

---

6. Plate 3 of the *Narrative journal* includes a picture of the spot drawn by Schoolcraft ". . . to convey an idea of the unusual manner in which the maple, and the pine,—the elm, and the hemlock are intermingled . . ."

7. *Zizania aquatica* Linnaeus.

serves the Indians as a substitute for corn." The plant's nomenclature and method of propagation are discussed at some length.

At the Grand Portage some of the party took an overland route to Sandy Lake—in present Aitkin County, Minnesota; others followed the St. Louis River, made a portage and reached Sandy Lake by descending the Savannah River. This is clearly shown on the map.

At the solicitation of "the Sandy-lake Indians," of the Chippeway tribe, a council was held on July 16. Schoolcraft seems not to have thought highly of Indian "eloquence" but includes one speech from which I quote a comment:

". . . We can live a great while upon a little, but we cannot live upon nothing . . . our wild rice is all eaten up,—the buffaloes live in the land of our enemies, the Sioux . . . the President of the United States is a very great man . . . the Americans are a great people. Can it be possible they will allow us to suffer!"

Cass proposed to negotiate a peace between them and the Sioux and it was decided that ". . . some of their old men as embassadors [were] to accompany us to the Falls of St. Anthony, on our return from the sources of the Mississippi." According to agreement, this mission was duly undertaken on July 25.

1820

I now follow the daily progress of the expedition, through what is now north-central Minnesota and from there to Dubuque, Iowa, descending the Mississippi River.

On July 17 Cass, Schoolcraft, Douglass and two others of the command, with nineteen voyageurs and Indians, provisioned for twelve days, left the fort of the South West Fur Company in three canoes. The rest of the party were to await their return. They moved into the Mississippi valley by way of Sandy Lake River ". . . which discharges into the Mississippi, two miles below." Traveling upstream they had a strong current against them but made forty-six miles and passed six rapids. The water was ". . . reddish . . . a little turbid." We are told that the "alluvial banks" bore ". . . a forest of elm, maple, oak, poplar, pine, and ash. The elm predominates; maple and oak are common,—pine, ash, and poplar, sparing."

On July 18 they passed Swan River, rising in Itasca County and entering the Mississippi from the northeast, in Aitkin County. Also Trout and Prairies rivers.

"The timber has been much the same as yesterday,—elm and maple predominate . . . we passed several ridges of pine land . . . Tufts of willow, grass, and wild rice, skirt the water's edge."

The next day there was a portage around the "falls of Peckagama."

". . . Immediately at the head of the falls is the first island noticed in the river. It is small, rocky,—covered with spruce and cedar . . . In crossing this portage, I observed the small bush-whortleberry, *(vaccinium dumosum.)* A portion of the berries were already ripe. After passing the falls . . . a striking change is witnessed in the character of the country. We appear to have attained the summit level of waters. The forests of maple, elm, and oak, cease, and the river winds ... through an extensive prairie, covered with tall grass, wild rice, and rushes ... sitting in our canoes ... the rank growth of grass, rushes, &c. completely hid the adjoining forests from view, and it appeared as if we were lost in a boundless field of waving grass ... The monotony of the view can only be conceived by those who have been at sea ..."

After sixty miles they camped on the prairies, six miles above "Chevréuil, or Deer river," which enters the Mississippi in Itasca County. They found ". . . a very delicious species of wild raspberry, growing upon a small bush of the size of a strawberry vine ... as night approached, we first noticed the fire-fly, which has not before been seen upon the Mississippi."

On the 20th they passed the mouth of Leech River, flowing out of Leech Lake in Cass County. Thirty-five miles above they entered "Little lake Winnipec," having more the appearance of a marsh because of the ". . . rushes, spear grass, and wild rice . . ."

After following the Mississippi for ten miles they entered and crossed "Upper lake Winnepec"—Winibigoshish Lake on the boundary between Cass and Itasca counties—and camped on its north shore, at the mouth of "Turtle Portage river." "Its shores are ... covered, at the water's edge, with rushes, and wild oats. Upon its banks ... oak, maple, poplar, birch, and white pine." On the lake they met two Indian women in a canoe; they ". . . had come down the river for the purpose of observing the state of the wild rice, and at what places it could be advantageously gathered."

On July 21 they saw Pelican Island at the north end of the lake, with its ". . . myriads of water-fowl . . ." and soon after entered the mouth of the Mississippi inlet, ascending it ". . . fifty miles to its origin, in Upper Red Cedar or Cassina Lake . . ." This lies in Beltrami and Cass counties. Schoolcraft states in a footnote that he had suggested the name Cassina ". . . in order to prevent its being confounded with Red Cedar Lake, which is situated about 250 miles below. It is in allusion to Governor Cass."

About eight miles long by six broad, the lake ". . . presents to the eye a beautiful sheet of transparent water ... its banks are overshadowed by elm, maple and pine. Along its margin there are some fields of Indian rice, rushes and reeds ... It has an island towards its western extremity covered with trees, from which it derives its local name, but no red cedar is found around its shores.[8] This lake is supplied by two inlets called Turtle and La Beesh rivers, both tributary on the northwestern margin. The former originates in Turtle Lake ... La Beesh river is the outlet of Lake La Beesh,[9] which ... has no inlets."

8. The map of the *Narrative journal* includes a "perspective view" of Cassina Lake.

9. The map of the *Narrative journal* (1821) indicates that the route passed through Cassina Lake and ended at the mouth of the stream flowing out of Lake "Labeish." This lies further west than the stream flowing out of Turtle Lake. But the *Narrative of an expedition* ... (1834), describing Schoolcraft's second journey to the sources of the Mississippi River in 1832, refers to leaving ". . . the mouth of Turtle River

At a village of the Chippeways the party was presented with ". . . an abundance of the most delicious red raspberries . . ." We are told a story of "human misery," the tale of an unfortunate French trader who, in a helpless condition, had ". . . subsisted several months upon the pig weed which grew around his cabin . . . Governor Cass sent him a present of Indian goods, groceries, and ammunition, and engaged a person to convey him to the American Fur Company's Fort at Sandy Lake, where he could still receive the attention due to suffering humanity. These donations were swelled by every individual of the party . . ."

We are told that the distance from Sandy Lake to Cassina Lake, ". . . by the windings of the river is two hundred and seventy one miles, and from the Fond du Lac, at the head of Lake Superior, 429."

Writes Schoolcraft: "Cassina Lake, the source of the Mississippi, is situated seventeen degrees north of the Balize on the Gulph of Mexico, and two thousand nine hundred and seventy-eight miles, pursuing the course of the river. Estimating the distance to Lake La Beesh, its extreme northwestern inlet at sixty miles . . . we have a result of three thousand and thirty-eight miles, as the entire length of this wonderful river, which extends over the surface of the earth in a direct line, more than half the distance from the Arctic Circle to the Equator . . . its sources lie in a region of almost continual winter, while it enters the Ocean under the latitude of perpetual verdure; and at last, as if disdaining to terminate its career at the usual point of embouchure of other large rivers, has protruded its banks into the Gulf of Mexico, more than a hundred miles beyond any other part of the main. To have visited both the sources and the mouth of this celebrated stream, falls to the lot of few, *and I believe there is no person living, beside myself, of whom the remark can now be made.* On the 10th of July, 1819, I passed out of the mouth of the Mississippi, in a brig bound for New-York . . . on the 21st of July of the following year, I found myself seated in an Indian canoe, upon its source."

Schoolcraft at this point supplies an interesting analysis of the "physical character of the Mississippi" which, he states, ". . . may be advantageously considered under four natural divisions, as indicated by the permanent differences in the colour of its waters,—the geological character of its bed and banks,—its forest trees and other vegetable productions . . ." and so on. His "four natural divisions" were, briefly, these:

(1) The region extending from the sources of the Mississippi to the "falls of Peckagama," a distance of two hundred and thirty miles, where the river flows ". . . through a low prairie, covered with wild rice, rushes, sword grass, and other aquatic plants . . . This is the favourite resort of water-fowl, and amphibious quadrupeds."

(2) The region extending from the "falls of Peckagama" to the Falls of St. Anthony, a distance of six hundred and eighty-five miles:

". . . At the head of the falls of Peckagama, the prairies entirely cease; and below, a

(the spot of Gov. Cass' landing in 1820.)", and on the included map the mouth of Turtle River is designated as the "Ultimate point reached by Cass in 1820."

forest of elm, maple, birch, oak, and ash, overshadows the stream. The black walnut *(juglans nigra)* is first seen below Sandy Lake river, and the sycamore below the river De Corbeau . . . A few miles above the river De Corbeau, on the east side, we observe the first dry prairies, or natural meadows, and they continue to the falls of St. Anthony. These . . . are the great resort of the buffalo, elk, and deer . . . the only part of the banks of the Mississippi where the buffalo is now to be found . . ."

(3) The region extending from the Falls of St. Anthony to the mouth of the Missouri River, a distance of eight hundred and forty-three miles:

". . . The river prairies cease, and the rocky bluffs commence precisely at the falls of St. Anthony. Nine miles below it receives the St. Peter's from the west, and is successively swelled on that side by the Ocano, Iowa, Turkey, Desmoines, and Salt rivers, and on the east by the St. Croix, Chippeway, Black, Ousconsing, Rock, and Illinois. One hundred miles below the Falls of St. Anthony, the river expands into a lake, called Pepin, which is twenty four miles long and four in width. It is, on issuing from this lake, that the river first exhibits, in a striking manner, those extensive and moving sand bars, innumerable islands and channels, and drifts and snags, which continue to characterize it to the ocean . . ."

(4) The region extending from the mouth of the Missouri to the Gulf of Mexico, a distance of one thousand, two hundred and twenty miles:

"The fourth change in the physical aspect of this river is at the junction of the Missouri, and this is a total and complete one, the character of the Mississippi being entirely lost in that of the Missouri. The latter is . . . much the largest stream of the two, and carries its characteristic appearances to the ocean. It should also have carried the name, but its exploration took place too long after the course of the Mississippi had been perpetuated in the written geography of the country, to render an alteration in this respect, either practicable or expedient. The waters of the Mississippi at its confluence with the Missouri, are moderately clear, and of a greenish hue.—The Missouri is turbid and opake, of a greenish-white colour, and during its floods, which happen twice a year, communicates, almost instantaneously, to the combined stream its predominating qualities . . . it receives from the west, the Merrimack, St. Francis, White, Arkansas, and Red rivers; and from the east, the Kaskaskia, Great Muddy, Ohio, Wolf, and Yazoo. This part of the river is more particularly characterized by snags and sawyers,—falling-in banks and islands;—sand bars and mud banks;—and a channel which is shifting by every flood . . . The wild rice, *(zezania aquatica,)* is not found on the waters of the Mississippi south of the forty-first degree of north latitude, nor the Indian reed, or cane, north of the thirty-eighth. These two productions characterize the extremes of this river . . . The alligator is first seen below the junction of the Arkansas. The paroquet is found as far north as the mouth of the Illinois, and flocks have occasionally been seen as high as Chicago. The name of this river is derived from the Algonquin language . . . It is a compound of the word *Missi,* signifying *great,* and *Sepe,* a *river* . . ."

Schoolcraft reached the important conclusion that "There is no part of the Mississippi river which originates in the territories of British America. The northern boundary line of the United States will probably run a hundred miles north of its extreme source; but this is a point which still remains unsettled between the two governments.[1] . . ."

On July 22 the return journey from the supposed source of the Mississippi River was begun; it was more rapid than the ascent, for on the 24th they ". . . landed at the Southwest Company's Fort on Sandy Lake . . . having performed on our return, the same distance in three days, which we had occupied four and a half in ascending. We were rejoiced to find our friends in perfect health, and that no attempts had been made by the savages, during our absence, to molest them . . ."

There are six more chapters in the *Narrative journal:* Chapter X describes the "Journey, from Sandy Lake to the American garrison at St. Peter's"; XI, the "Journey, from St. Peter's to Prairie du Chien"; and XII, the "Visit, to the lead mines of Dubuque, on the upper Mississippi." The remaining chapters concern the return of the expedition to more eastern regions.

On July 25 the expedition left the fort on Sandy River (it was situated three miles from the Mississippi) and started for the mouth of St. Peter's River. It was traveling in three canoes and a barge and was accompanied by ". . . embassadors of peace from the Chippeway tribes to the Sioux of St. Peter's . . ." For Cass was to fulfill his promise, made on July 16, to attend a parley between these warring tribes.

On the 26th they passed the mouths of "the River au Solé (Alder river)" [2] and Pine River, and made one hundred miles. There is a comment about the universal pest of the traveler:

"He who is afflicted, without complaining, by an unexpected change of fortune, or the death of a friend, may be thrown into a fit of restless impatience by the stings of the mosquito . . . an enemy . . . too minute to be dreaded and too numerous to be destroyed."

The trees mentioned by Schoolcraft—oak, elm, maple, ash, etc., etc.—are always so similar as to suggest the repetitious; but he was crossing, for much of his journey, what was then one of the great hardwood belts of North America, and doubtless saw just what he said he did! At the camp of the 26th he, however, mentions two lesser plants:

". . . the wild rose *(rosa parviflora)* and a flower, resembling in some of its characters the ipomaea nil, with a short floriferous stem, and lance-oblong leaves: peduncle one-flowered, bell-shaped, white, downy. It appears to have escaped the notice of Pursh, in his botanical researches in the northwest."

1. Long's expedition of 1823 was to ascend the St. Peter's, or Minnesota, River, and to descend the Red River of the North to Pembina, North Dakota, near which it was to raise the United States flag on the 49th parallel.

2. Since *saule* is French for willow, rather than alder, this was probably Little Willow River, which empties into the Mississippi in Aitkin County, Minnesota.

On July 27 they reached the mouth of "the river De Corbeau," or Crow Wing River, where "... the Buffalo Plains commence, and continue downward, on both banks of the river, to the falls of St. Anthony ..." And, after making ninety miles, camped on the "left" or east bank of the Mississippi. The next day they spent some hours hunting buffalo below the mouth of Elk River.

"Naturalists have generally considered the American buffalo *(Bos Bubalus)* of the same species with the *Bison* and *Aurochs* of Europe and Asia ... The only part of the country *east* of this river, where the buffalo now remains, is included between the falls of St. Anthony and Sandy Lake, a range of about six hundred miles."

Schoolcraft's comments upon birds and animals are more interesting than those upon plants. Under date of July 29—the party camped that day about five miles below Crow River which enters the Mississippi from the southwest, between Wright and Hennepin counties, Minnesota—he records:

"In the course of the night a pack of wolves were heard on the opposite side of the river. There is something doleful as well as terrific in the howling of this animal, particularly when we start from a sound sleep during the stillness of night. It is, however, little to be dreaded ... another sound ... will frequently disturb the nightly rest of the traveller in the region of the Mississippi. It is the half-human cry of the Strix Nyctea, or great white owl ... seldom found south of the falls of St. Anthony."

On July 30 they reached the Falls of St. Anthony; they had "... a simplicity of character[3] which is very pleasing." In one view could be seen "... the copses of oak upon the prairies, and the cedars and pines which characterize the calcareous bluffs. Nothing can exceed the beauty of the prairies which skirt both banks of the river above the falls. They do not, however, consist of an unbroken plain, but are diversified with gentle ascents and small ravines covered with the most luxuriant growth of grass and heath-flowers, interspersed with groves of oak ..."

Schoolcraft's party portaged the falls and, nine miles below, came to the "... American garrison at St. Peter's ..." We are told that Pike, in September, 1805, had effected a treaty with the Sioux "... by which they cede to the United States the district of country from the junction of the St. Peter's with the Mississippi, to the falls of St. Anthony inclusive, and extending nine miles on each side of the river. The consideration for this grant was two thousand dollars. It could hardly have been anticipated at that time, when there were probably not more than a hundred American families in the extensive region now comprising the states of Indiana, Illinois, and Missouri, that in the short space of thirteen years the progress of our settlements would have de-

3. According to the volume *Minnesota* ("American Guide Series"), pp. 49, 93, 94, it was in 1847 that settlement began on the east side of the Falls and, two years later, on the west side. United by bridges, these settlements were to become the city of Minneapolis. The "simplicity of character" was doomed. But the Falls of St. Anthony were the decisive factor in making Minneapolis rather than St. Paul one of the most important milling centers of the world.

Schoolcraft records that "The European name of these falls is due to father Lewis Hennepin, a French missionary of the order of Recollects, who first visited them in 1680 ..."

manded the occupancy of a post in so remote a section of the union. Yet it was loudly called for even within that time, as a protection to the defenceless settlers on our north-western and southwestern frontiers—and as a check to the undue influence which the British traders have too long exercised over the Indian tribes inhabiting the territories of the United States. Yielding to this expression of the public voice, the government determined to establish a garrison[4] at St. Peter's . . .

"Since their arrival, the garrison have cleared and put under cultivation about ninety acres of the choicest bottom and prairie lands, which is chiefly planted with Indian corn and potatoes . . . Here we were first presented with green corn, pease, beans, cucumbers, beets, radishes, lettuce, &c . . . We found the wheat entirely ripe, and melons nearly so. These are the best commentaries that can be offered upon the soil and climate."

The *Narrative journal* tells what was then known of St. Peter's River—a stream which enters the Mississippi from the southwest, about midway between the present cities of Minneapolis and St. Paul and forms, for some distance above its mouth, the boundary between Hennepin County (north) and Dakota and Scott counties (south).

". . . It has never been explored except by voyageurs and traders . . . and remains to this moment, undescribed[5] in American geography. All questioned . . . both Indians and traders, agree in saying, that it is a long stream . . . of a great many tributaries, and flowing in its whole extent through a country of the most luxuriant fertility . . . Carver ascended it two hundred miles . . ."

Where St. Peter's River entered the Mississippi was ". . . a large island . . . covered with the most luxuriant growth of sugar maple, elm, ash, oak, and walnut . . . Among the forest trees upon its [the river's] banks we noticed the box-elder *(acer negundo)* or ash-leaved maple. The inner bark of this tree, boiled down with the common nettle into a strong decoction, is said to be used by the Indians as a remedy for lues venerea, and to be a sovereign cure for that disorder . . .

"Among the luxuriant herbage which characterizes the prairies of St. Peter's, is found a species of aromatic grass, upon which a high value is set by the aborigines. It throws off the most fragrant odour, and retains its sweetness, in a considerable degree, in the dried state. It is cut in a particular stage of its growth in the month of June, when it throws off its aroma most profusely, and continues to be gathered until it has run into seed, and is too dry to be plaited. The Indian women braid it up in a very in-genious manner . . . Whether this grass is the same with the *heracleum panaces* of Kamschatka, and of which the inhabitants distil an intoxicating liquor, similar in some

4. Colonel Henry Leavenworth had built a fort but a few months before Cass's arrival. The site was to be changed within a short time and a new fort erected under the direction of Colonel Josiah Snelling. It was named Fort Snelling in 1825.

5. This lack of knowledge was to be supplied in Keating's *Narrative of an expedition to the sources of St. Peter's River* . . . which compiled the information gathered on Long's expedition of 1823.

respects to brandy, I am unable to determine. It appears probable it may possess some properties in common with the holcus fragrans of Pursh."

On August 1, and with some rapidity it would seem, since Cass's party had only arrived the day before, a treaty of peace was ". . . concluded between the Sioux and Chippeways in the presence of Governor Cass, Colonel Leavenworth, Mr. Tallifierro, the Indian agent at St. Peter's, and a number of officers of the garrison. These two nations have been at war from the earliest times, and the original causes of it are entirely forgotten, but still the ancient enmity is carefully transmitted from father to son . . .

". . . Whether the peace will prove a permanent one, may be doubted. All their ancient prejudices will urge them to a violation of it, while past experience abundantly shews how difficult it has been to preserve a lasting peace between two powerful rival tribes of savages, whose predominant disposition is war . . . treaties of peace between Indian tribes, like those between civilized nations,[6] only amount to a momentary cessation of hostilities, unless the limits of their territories, and other subjects of dispute, are accurately defined, and satisfactorily settled."

To a reader who has become somewhat impregnated with Schoolcraft's account of savages, primitive conditions and so on, it comes as somewhat of a surprise to be told that, among some beautiful small lakes six miles west from St. Peter's cantonment, one (Calhoun Lake, ". . . stored with the most exquisite flavoured black bass . . .") ". . . has become a fashionable resort for the officers of the garrison. The intermediate country is a prairie, and is travelled in all directions on horseback . . . In the season of verdure, the waving heath-grass,—the profusion of wild flowers, and the sweet-scented Indian grass . . . fill the air with a refreshing fragrance, delight the eye with the richness and never-ending variety of their colours . . ."

Having fulfilled his promise to the Chippeways, Cass and his party started on August 2 for Prairie du Chien, a trip which consumed three days. On the 2nd they descended the Mississippi for thirty-eight miles, encamping at twilight ". . . upon the west shore, nine leagues below the village of La Petit Corbeau."

On August 3 they passed the mouth of the St. Croix, entering the Mississippi from the north and now constituting for a long distance the boundary between Minnesota (west) and Wisconsin (east)—"It is said to be the most practicable, easy, and expeditious water communication between the Mississippi river and Lake Superior." They passed the ". . . Sioux village of Talangamane, or the Red wing . . . handsomely situated on the west banks of the river, six miles above Lake Pepin . . ." and ascended ". . . an isolated mountain . . . called the Grange, from the summit of which you enjoy the most charming prospect. . . ." They entered Lake Pepin at one o'clock, and camped at six on its eastern shore. ". . . In the vicinity of our encampment, we observed the asparagus growing along the shore. The seeds had probably been dropped by some former traveller . . ."

6. Although the year was 1820, there is more than a little suggestion of present-day world problems.

Lake Pepin is described as a ". . . beautiful sheet of water . . . an expansion of the Mississippi river . . . twenty-four miles in length, with a width of from two to four miles . . ." On August 4 they finished the passage of Lake Pepin—". . . At the precise point of exit of the Mississippi river, from Lake Pepin, the Chippeway, or Sauteaux river, comes in from the east . . ."—and, sixty miles below, visited ". . . the Sioux village of Wabashaw . . . on the west bank of the Mississippi . . ." A few miles farther south they observed ". . . an isolated mountain, of singular appearance . . ." named by the voyageurs and rising out of the ". . . centre of the river, to a height of four or five hundred feet . . . The Mountain that sinks in the Water, *(La Montaigne qui Trompe dans l'Eau,)* an opinion being prevalent . . . that it annually sinks a few feet . . . It is further remarkable as being the only fast, or rocky island, in the whole course of this river, from the Falls of Peckagama, to the Mexican Gulf . . ."

After making seventy miles that day and ninety miles on August 5 they reached Prairie du Chien, situated on the eastern side of the Mississippi in present Crawford County, Wisconsin. Schoolcraft reports that the village had derived its name from a family of Fox Indians, once resident there, and distinguished ". . . by the appellation of *Dogs*. The present settlement was first begun in 1783, by Mr. Giard, Mr. Antaya, and Mr. Dubuque . . ." In 1820 it was ". . . the seat of justice for Crawford county, which has recently been erected in this part of the Michigan Territory, and a court of justice has already been established . . ."

Since Cass was to remain at Prairie du Chien for several days, Schoolcraft obtained permission to visit the lead mines of Dubuque, situated in the territory of the ". . . Fox, or Outagami Indians . . ." He left on August 6th and returned on the 9th. The mines were located seventy-five miles below Prairie du Chien, on the west side of the Mississippi. Julien Dubuque had obtained a grant from the Indians and had worked the mines until he died, much in debt, in 1810. His creditors were assigned his claim but this was disputed by the Indians; a settlement was still to be reached and in 1820 the Indians were working the mines themselves. The whole story, as told by Schoolcraft, is interesting. I quote from his description of "The valley of the Mississippi between Prairie du Chien, and the lead mines of Dubuque . . ." insofar as this relates to the flora and to the birds for which Schoolcraft had a "seeing" eye:

". . . a part . . . is prairie; and the remainder, covered with a heavy forest of elms, sugar tree, black walnut, ash, and cotton wood . . .

"Among the humbler growth, which adorns the borders of the forest, the cornus florida, the sarsaparilla, and the sumach, are frequently to be seen . . . the splendid foliage of the autumnal forest, is already visible in the rich hues of the fading maple, the heart-shaped aspen, and the populus angulata.

"The tall grass of the prairies . . . is . . . occasionally checquered with green copses of shrubby oaks, and beautified with the peculiar tribe of the heath-flowers . . . The channel of the river, is often expanded to an amazing width, and spotted with innumerable islands . . . some . . . crowned with a brushy growth, of young willows and slender cotton woods; others, present copses of the tallest trees . . . Perched upon these, we invariably find the heron, and king-fisher . . . The eagle, and the hawk, choose a more elevated seat to watch for their food, while the buzzard, with an easy wing, is continually sailing through the air . . . The white pelican . . . is always found upon the point of some naked sand bar . . . The duck and the goose . . . are always in motion . . . The pigeon, the snipe, the wild turkey, the raven, and the jay, are also common . . ."

After leaving Prairie du Chien on August 9 and descending the Mississippi River for three miles, Cass and his party began the ascent of the "Ousconsing," or Wisconsin, River on their way homeward.[7] The *Narrative journal* ends with Schoolcraft's arrival at Detroit, as his daily headings indicate, on the "CXXIII. Day.—(*September 23d*.)."

In 1822, in the *American Journal of Science and Arts,* appeared the "Notice of the plants collected by Professor D. B. Douglass, of West Point, in the expedition under Governour [*sic*] Cass, during the summer of 1820, around the Great Lakes and the upper waters of the Mississippi: the arrangement and description, with illustrative remarks, being furnished by Dr. John Torrey."

It opens with a letter from Torrey in New York, to Douglass, at the "West-Point Military Academy," dated August 4, 1821. Torrey evidently thought well of Douglass' collections:

"Inclosed I have the pleasure of sending you a catalogue of the plants from the North-West, which you forwarded me some time since for examination. Many of the species are very rare, others are from entirely new localities, and the whole are valuable in increasing our knowledge of botanical geography. To those species which are but little known or imperfectly described, I have added such remarks as I supposed would be useful."

Torrey also notes that "The Indian and popular names and localities are taken from your notes annexed to the specimens . . ." And, included, is an "Explanatory letter" from Prof. Douglass to the editor [of the *American Journal* . . .], dated "New-York Aug. 22, 1821":

"I must beg leave to observe, in the first place that the collection of plants was made by a person, who, besides not being a professed botanist, was almost constantly engaged with other objects of research. The formation of an Herbarium, requiring much leisure and frequent attention, could scarcely be expected, under such circumstances, and would not have been undertaken, except in the exigency of having no professed

7. On the return journey the party ascended the Wisconsin, portaged to Fox River and ascended that stream to Green Bay, northwestern arm of Lake Michigan, where they arrived on August 20. On the 22nd some of the party left for Michilimackinac. Cass and Schoolcraft proceeded along the western shore of Lake Michigan and arrived at Chicago on August 29. Thence Cass journeyed overland to Detroit, while Schoolcraft traveled to Michilimackinac along the eastern shore of Lake Michigan, arriving September 9. From there, by much the same route taken on the westward journey, he arrived at Detroit.

botanist attached to the Expedition. Secondly, the region of country traversed by the Expedition, particularly that bordering upon Lake Superior and the upper Mississippi, as well as a considerable portion of that on the Ouisconsin and Fox rivers and around Lake Michigan, is but indifferently rich in plants at best, and this collection is besides chiefly confined to such as flower in the course of the summer months. The deficiency I have endeavoured to supply as far as possible by notes, particularly on the forest growth, which I have interspersed in my journal; these however being at West Point, it is at present out of my power to communicate them.

"Finally, a part of the collection was injured by an accident on the Ouisconsin, in which my canoe was nearly filled with water before it could be got ashore. The consequence of which was that nearly all the plants in one case were completely spoiled before I was able to dry them. Such as the collection is, however, the catalogue is entirely at your service, and I am glad that so much interest has been given to it by Dr. Torrey. The *uvularia perfoliata* of this catalogue is the plant which I mentioned to you some time since, as efficacious in the cure of the Rattle-snake bite[8]—Of this I have been witness, but the efficacy of *Pedicularis Canadensis* for the same purpose, I can only state from report."

8. Under date of August 3, Schoolcraft had devoted nearly three pages to a discussion of various supposed cures for such bites. The expedition had climbed "the Grange."

"In ascending this mountain we first noticed the rattlesnake, *(crotalus horridus)* ... One of the most remarkable facts in the natural history of this dreadful animal, is, that its poison may be taken internally without any danger ... Charlevoix mentions a plant, which is an antidote ... called the rattle snake plant *(herbe a serpente a sonettes)* which grows abundantly throughout this country. 'This plant,' he remarks, 'is beautiful and easily known. Its stem is round and somewhat thicker than a goose-quill, rising to the height of three or four feet, and terminates in a yellow flower of the figure and size of a yellow daisy. This flower has a very sweet scent [and so on] ...' In another place, speaking of the *citron,* he remarks 'The root of this tree is ... a most sovereign antidote against the bite of serpents. It must be bruised and applied instantly on the wound: this remedy is immediate and infallible.' The plant alluded to in both instances, appears to be the common mandrake, or podophyllum peltatum of modern botany ...

"In our times the common plantain *(plantago major)* has been frequently mentioned as an infallible cure, both for the bite of the rattlesnake, and the tarantula ... but I cannot allude to any particular cases in which it has been successfully applied ...

"... Whether the Virginia snake-root, *(aristolochia serpentaria)* is applied as an antidote to the poison of serpents, I am unable to say. Ergotted rye, is also among the number of simples, which have been lately recommended in cases of the bite of the rattlesnake."

Although Fernald and Kinsey have little to say about plants as antidotes, their *Edible wild plants of eastern North America,* full of information and enriched by a goodly number of humorous comments, contains the gist of what one wants to know about the poisonous or edible properties of plants—some mentioned by Schoolcraft. Of the "Common plantain, *Plantago major*," the authors comment: "Only in emergency would most people use it, for the fibers are tough. Otherwise it would not be so common." ! Having once taken a field-trip with the late Dr. Fernald, I feel sure that he recommends no plant-diet which he himself had not sampled!

It pays to be careful and not, above everything, rely on common names. The story goes that a well-known author living in southeastern Maine cooked and ate a large mess of what was known in that region as "skunk cabbage." It was however, *Veratrum viride,* the Indian poke or white hellebore, known for its powerful action in reducing blood pressure. What treatments he received I do not know—there are antidotes—but I understand he still survives.

BARNHART, JOHN HENDLEY. 1921. Biographical notices of persons mentioned in the Schweinitz-Torrey correspondence. *Mem. Torrey Bot. Club* 16: 291, 292.

CRISWELL, ELIJAH HARRY. 1940. Lewis and Clark: linguistic pioneers. *Univ. Missouri Studies* 15: no. 2, xxi.

FERNALD, MERRITT LYNDON & KINSEY, ALFRED CHARLES. 1943. Edible wild plants of eastern North America. *Gray Herb. Special Publ.* 342.

PIKE, ZEBULON MONTGOMERY. 1895. The expeditions of Zebulon Montgomery Pike, to headwaters of the Mississippi River, through Louisiana Territory, and in New Spain, during the years 1805-6-7. A new edition, now first reprinted in full from the original of 1810, with copious critical commentary, memoir of Pike, new map and other illustrations, and complete index, by Elliott Coues . . . 3 vols. New York.

SCHOOLCRAFT, HENRY ROWE. 1821. Narrative journal of travels, through the northwestern regions of the United States extending from Detroit through the great chain of American lakes, to the sources of the Mississippi River. Performed as a member of the expedition under Governor Cass. In the year 1820. By Henry R. Schoolcraft, member of the New-York Historical Society . . . Embellished with a map and eight copper plate engravings. Albany.

SHAW, WILLIAM B. 1930. David Bates Douglass. *Dict. Am. Biog.* 5: 405, 406.

TORREY, JOHN. 1822. Notice of the plants collected by Professor D. B. Douglass, of West Point, in the expedition under Governour Cass, during the summer of 1820, around the Great Lakes and the upper waters of the Mississippi; the arrangement and description, with illustrative remarks, being furnished by Dr. John Torrey. *Am. Jour. Sci. Arts* 4: 56-69.

WARREN, GOUVENEUR KEMBLE. 1859. Memoir to accompany the map of the territory of the United States from the Mississippi River to the Pacific Ocean, giving a brief account of each of the exploring expeditions since A.D. 1800 . . . *U. S. War Dept. Rept. expl. surv. RR Mississippi Pacific* 11: 28, 29.

CHAPTER XII

SAY TRAVELS WITH LONG AND KEATING TO PEM-

BINA, NORTH DAKOTA, BY WAY OF THE "ST. PETER'S,"

OR MINNESOTA, RIVER AND THE RED RIVER OF THE

NORTH

IT was in 1823—or five years after an agreement had been reached between Great
Britain and the United States providing that the section of the boundary between "Louisiana" and Canada should follow the 49th parallel as far west as the
summit of the "Stony" or Rocky Mountains—that President James Monroe sent
Major Stephen Harriman Long, under orders of the War Department, to explore St.
Peter's River, now the Minnesota,[1] as well as the country along the recently determined
northern boundary of the United States lying between that river and Lake Superior.
After reaching Pembina, North Dakota, close to the 49th parallel, Long found it was
not practical to follow this parallel eastward and he traveled to Lake Superior by a less
direct route, much of it through British territory.

The story of the exploration, dedicated to President Monroe, is told in William
Hypolitus Keating's *Narrative of an expedition to the source of St. Peter's River, Lake
Winnepeek, Lake of the Woods, &c. &c. performed in the year 1823, by order of the
Hon. John C. Calhoun, Secretary of War, under the command of Stephen H. Long,
Major U. S. T. E. . . .* The title further explains that the *Narrative* was compiled by
Keating from the notes of Long, Say, Keating and Colhoun. Although the book was
issued in Philadelphia in 1824, or about one year after the return of the expedition,
Keating regretted this short delay: "Narratives of voyages of discoveries lose much of
their interest, if the publication be long deferred."

According to Keating's preface, ". . . the historical part of the narrative, together
with the topographical, and much of the descriptive matter . . ." was taken from Long's
notes; James C. Colhoun's manuscripts contributed to "the same departments," as well
as astronomical observations and "the greater part of the references to older writers."
Samuel Seymour, "Landscape Painter and Designer," contributed the paintings of the
country, the Indians, and so on. Keating, "Mineralogist and Geologist," was historiographer to the expedition; and Thomas Say,[2] "Zoologist and Antiquary," served
as botanist.

---

1. According to G. K. Warren, this ". . . was the first authentic exploration to the sources of the St.
Peter's, or Minnesota, River, although its lower portions had probably been visited by M. Le Sueur as
early as 1695."

2. Both Seymour and Say had served under Long when he commanded the "scientific section" of the

". . . From Mr. Say's notes, all that relates to the Zoology and botany of the country traversed has been obtained, as well as much of the matter relating to the Indians.

"The greater part of the appendix will be found to have been prepared by Mr. Say . . ."

Part I (Natural History) of this appendix was in two sections: the first (Zoology) was written by Say; the second (Botany), enumerating Say's collection of plants, was prepared by Lewis D. von Schweinitz.[3] As Say's life progressed, his major scientific interests were to become entomology and conchology. From the paucity of botanical information incorporated in the *Narrative* and from the fact that he took no part in the publication of his own collections but turned them over to von Schweinitz, it seems unlikely that he took any great interest in the botanical task assigned him.

The portion of the journey which crossed the region included in the locale of my story was only a small part of the whole—extending from the arrival of the party at the confluence of the "Wisconsan" and Mississippi rivers on June 19, 1823, to its departure on August 9, from Pembina, situated in the northeastern corner of present North Dakota.[4]

The United States government wished more light than it then possessed[5] upon ". . . the district of country bounded by the Missouri, the Mississippi, and the Northern Boundary of the United States. This triangular section includes about three hundred miles of longitude and seven hundred of latitude."

The War Department's orders to Long, dated April 25, 1823, outlined the route to be taken as well as the exploration's objectives. Long was to proceed from Philadelphia to Chicago, ". . . thence to Fort Armstrong or Dubuque's Lead Mines, thence up the Mississippi to Fort St. Anthony, thence to the source of the St. Peter's river, thence to the point of intersection between Red River and the forty-ninth degree of north latitude, thence along the northern boundary of the United States to Lake Superior, and thence homeward by the Lakes.

"The object of the expedition is to make a general survey of the country on the route pointed out, together with a topographical description of the same, to ascertain

"Yellowstone Expedition." It has been said that Edwin James, who had been botanist and surgeon on that early journey, was to have accompanied the present one in a similar capacity, but did not receive his instructions in time to do so.

3. Parts II, III and IV concerned astronomy, meteorology and Indian vocabularies.

4. From Pembina the expedition crossed into British territory—now the province of Manitoba, Canada. When it crossed the Lake of the Woods and followed Rainy River into Rainy Lake it may, for a short time, have been along the Manitoba-Minnesota boundary, for Paullin's *Atlas* (Plate 39B) indicates that this part of the return trip lay north, although in proximity to, the United States boundary when near the 95th and 96th degrees of longitude.

5. We are told that Jonathan Carver had visited St. Peter's River, but that the reliability of his story, published "about the year 1778," was considered questionable; Pike, on an eight-months' journey in 1805 and 1806, had lacked proper facilities to make accurate observations; and Governor Cass's journey in 1820 had thrown much light ". . . upon the history of the Upper Mississippi."

the latitude and longitude of all the remarkable points, to examine and describe its productions, animal, vegetable, and mineral; and to enquire into the character, customs, &c. of the Indian tribes inhabiting the same."

The expedition left Philadelphia on April 30, 1823, and, by way of Wheeling, West Virginia, and Fort Wayne, Indiana, reached "Fort Dearborn, (Chicago,)" on June 5.

"We were much disappointed at the appearance of Chicago and its vicinity . . . The village presents no cheering prospect . . . it consists of but few huts, inhabited by a miserable race of men, scarcely equal to the Indians from whom they are descended . . . Chicago is perhaps one of the oldest settlements in the Indian country; its name, derived from the Potawatomie language, signifies either a skunk, or a wild onion; and either of these significations has been occasionally given for it . . . As a place of business, it offers no inducement to the settler . . ."

The possibility that Chicago might become one of the points of communication between the northern lakes and the Mississippi is mentioned, but without optimism!

At Chicago Long had the option, as stated in the orders, ". . . of striking the Mississippi at Fort Armstrong, or at Dubuque's lead mines, and then ascending that river to Prairie du Chien. It appeared to him, however, that if the direct route to Prairie du Chien, across the prairies, was practicable, it would save several days; but . . . no person could be found who had ever travelled through, in that direction . . ."

Finally, ". . . an old French engagé . . . who had lived for upwards of thirty years with the Indians . . ." undertook the task, and, leaving Chicago on June 11, the party reached "the banks of the Wisconsan" on the 19th. Colonel Morgan, commander of the garrison, arrived with two boats.

"Although it was late . . . the party effected a crossing of the Wisconsan, and . . . the gentlemen proceeded under Colonel Morgan's guidance towards the Fort. It was . . . about eleven when they reached the Mississippi. This ride, at a late hour, was one of a most romantic character; the evening was fair and still . . . the moon shone in her full . . . All seemed to have their spirits excited by the sublimity of the scene. Even the Indian . . . appeared to have received an accession of spirits, and the loud whoops which he occasionally gave . . . enlivened the ride . . ."

The next morning with daylight they had their first view of the Mississippi River: "It is one of those grand natural objects, the sight of which forms an era in one's life."

Fort Crawford was situated above the mouth of the Wisconsin River, on the eastern side of the Mississippi, about one hundred and fifty yards from the bank: a "low and unpleasant" site and subject to inundations. The nearby village of Prairie du Chien, however, was located in a beautiful prairie which extended about ten miles down the river.

Between Chicago and Fort Crawford the party had seen "but one deer," "but a single wolf . . . of the kind called Prairie wolf," and "a badger."

"Among the birds observed, Mr. Say has recorded a single Red-headed Woodpecker, together with the Ferruginous Thrush, Towhee Bunting, Song Sparrow, Chipping Sparrow, Bartram's Sandpiper, Raven, Reedbird, and a Crow which was first heard near the Wisconsan.

"In the vegetable kingdom, the same gentleman observed that the Gerardria[6] was found, about the 15th, with its petals nearly of full length, but that afterwards they were found much shorter."

Little is recorded of the flora about Prairie du Chien and that little seems to have been taken from a manuscript journal kept by Long in 1817 when he traveled from Bellefontaine on the Missouri River to the Falls of St. Anthony and to the "Wisconsan" portage:

". . . The valleys are many of them broad, and appear well adapted to tillage and pasture; the highlands are also well calculated for the raising of grain. The country is generally prairie land, but the hills and valleys are in some places covered with a scattering growth of fine timber, consisting of white, red, and post oak, hickory, white walnut ['Juglans cinerea'], sugar tree, maple, white and blue ash, American box, &c."

Long's party remained at Prairie du Chien for five days while arrangements were being made for a military escort which was to accompany the party up St. Peter's River. This empties into the Mississippi from the southwest, about midway between the present cities of St. Paul and Minneapolis. Fort St. Anthony, to be renamed Fort Snelling in 1825, was situated on the bluff north of St. Peter's River and overlooked both that stream and the Mississippi. It had been occupied first in 1822 and was now under the command of Colonel Snelling. From Prairie du Chien the expedition traveled thither in two detachments: Long, Colhoun and four men went on horseback ". . . along the right[7] bank of the Mississippi . . ." Say and others ascended the river by barge. Dates are not always clear in the *Narrative,* but the start was apparently made on June 25. We are first told the experiences of Long's overland party.

They found travel difficult for, keeping back from the river, they were continually crossing tributaries and ravines—but farther inland would have meant fallen timber and no water according to their guide.

"The forests . . . consisted principally of oak, basswood, ash, elm, white walnut, sugar tree, maple, birch, aspen, with a thick undergrowth of hazel, hickory, &c.[8] In the bottoms the wild rice, horsetail, may-apple, &c. were found. The eye is charmed by the abundance of wild roses which are strewed over the country, and the palate is not less delighted with the excellence of the strawberry, which is remarkable for its fine fragrance, and which was, just at that time, in a state of perfect maturity."

6. Correctly, *Gerardia* Linn.

7. This was the west bank for Long was conveyed across the Mississippi before the boat party started and, at a later date, the *Narrative* mentions that the two parties separated ". . . on the west bank of the river . . ." It is often difficult in the old narratives to know what is meant by *left* and *right* banks—whether north, south, east or west—but as a rule the writers seem to have been looking downstream.

8. While the trees recorded in the *Narrative* are always much the same—as they also were in Schoolcraft's *Narrative journal* of 1821—one remembers that both parties were in one of the great hardwood belts of North America.

On June 28, when a valley widened, they came ". . . in sight of the majestic Mississippi . . . rolling its waters with an undiminished rapidity, in a bed checkered with islands . . . a spectacle, which, however often observed, always filled the mind with awe and delight."

On June 30 they visited an Indian village where, "As a compliment to the party, the United States' flag was hoisted . . .", over the cabin of the chief. The Indian custom to which all early travelers refer—shaking hands—is discussed at considerable length. The *Narrative* explains that it was, probably, ". . . received from the English . . ." Somewhat appropriately, the chief in this instance was named Shakea!

Say and his boating party had had a few interesting experiences. They traveled in an ". . . eight oar barge with a sail, or rather their *tent fly* . . . used as a substitute for one . . ." Unfortunately, while farewells were being exchanged with Long on the river bank, Say's "men" ". . . had broached the keg of liquor and helped themselves . . . so bountifully as to be soon affected by it . . . heated by the exercise of rowing, the effects of the whiskey became but too evident."

But Say was equal to the emergency; when one man exhibited "all the workings of a disordered imagination," he was lashed to the mast. "Mr. Say having administered to him the proper remedies, he gradually recovered, but finding it agreeable to abstain from work, feigned sickness . . . Mr. Say prescribed the use of an oar as a sudorific, by which he soon recovered the use of his lost senses."

They traveled at night to avoid "the torment of the mosquitoes." They ran into a gale and heavy rain and a thunderstorm; and rattlesnakes. In examining the head of one, Say had his thumb "punctured." Although this ". . . gave rise to much pain and numbness in the part . . ." it ". . . soon subsided, producing but little swelling." Despite such adversities they enjoyed the Mississippi:

". . . the magnificence of the scenery . . . so bold, so wild, so majestic . . . the rapidity of the stream delights us: it conveys such an idea of the extensive volume of water which this river ceaselessly rolls[9] towards the ocean . . ."

On June 28 they passed the mouth of Black River, entering the Mississippi from the northeast, between La Crosse County (south) and Trempealeau County (north). The *Narrative* reports this Wisconsin stream as ". . . one of the most important tributaries of the Mississippi . . . much resorted to for . . . obtaining timber . . . Not only does it supply the fort at Prairie du Chien, but even . . . much of the 'pine timber, used at St. Louis, is cut here.' "

On June 30 they reached the southern end of Lake Pepin. According to the guide, *"Le lac est petit, mais il est malin."* On the 31st they passed through the lake, really an enlargement of the Mississippi and about twenty-one miles long. Late on the 2nd

9. The "rollin' along" motion of the Mississippi seems to have been noted by more than one traveler—and more than one hundred years before Edna Ferber's dramatized *Show Boat,* Oscar Hammerstein 2nd's lyric, Jerome Kern's music, and Paul Robeson's voice, all combined to impress it on the minds and hearts of the American people.

of July they entered ". . . the St. Peter . . . the gentlemen spent the night on the south bank . . . Our boat made the trip in seven days and a half, which was considered the shortest that had been known of at the fort . . ."

On July 3 they crossed to Fort St. Anthony on the north bank, having traveled two hundred and eleven miles from Prairie du Chien. The *Narrative* had noted earlier:

"This may be considered as the first section of our journey; the whole distance from Philadelphia to this place, was near thirteen hundred miles, which were travelled in sixty-four days, stoppages included . . . an average of twenty miles per day."

The expedition remained at Fort St. Anthony until July 9.

"There were . . . about two hundred and ten acres[1] . . . under cultivation . . . one hundred were in wheat, sixty in maize, fifteen in oats, fourteen in potatoes, and twenty in gardens, which supply . . . an abundant supply of wholesome vegetables."

On July 6, while final preparations were being carried on at the fort, some of the party walked to the Falls of St. Anthony—nine miles away by the Mississippi and seven overland.

On the morning of July 9—after observing what must have been a display of northern lights—the expedition started up St. Peter's River;[2] there were thirty-three persons, including a ". . . new and more efficient escort . . .", one of which was Joseph Snelling, son of the fort's commander, who served as guide. Long, Keating, Seymour, etc., traveled in a skiff and canoes; Say, Colhoun and others by land. They were to "keep company together" as far as practicable and camp when possible at the same time. This division of the party lasted only a few days, or until July 14, for they did not travel at the same rate of speed.

On the 11th, at Little Prairie, the walking party was taken across the river—presumably to the west side thereof. On July 12th navigation of the boats was proving difficult—there were "snags" and sandbars; a canoe overturned and a keg of tobacco and considerable ammunition were lost; and a violent thunderstorm ". . . broke out with more violence than usually happens in our climates." They noticed a change in the vegetation.

"The forests, which had principally consisted of cotton-wood, birch, &c. were observed to become more luxuriant, and to be replaced by a heavy growth of oak and elm . . . The sandbars and small islands are covered with groves of willows."

On July 13 the *Narrative* records arrival at a point called ". . . the Crescent, from a beautiful bend which the river makes at this place. The two parties having united here, a day was spent in drying the baggage . . .

"As this is the highest spot on the St. Peter which we reached in canoes, it may be

1. In 1820 Schoolcraft had reported that ninety acres were in cultivation.

2. The *Narrative* comments: "The river is called in the Dacota language Watapan Menesota [I omit various accents], which means 'the river of turbid water.'. . . The name given to the St. Peter is derived from its turbid appearance, which distinguishes it from the Mississippi, whose waters are very clear at the confluence."

well to recapitulate the general characters of this stream, as we observed it from its mouth to the Crescent, a distance of one hundred and thirty miles by water."

After hearing what the boat party had observed, we learn that the land party had experienced great difficulties, "owing to the nature of the country."

". . . At times it was so marshy, that they could not proceed without much danger to themselves and their horses . . . The forests which they traversed, consisted chiefly of maple, white walnut, hickory, oak, elm, ash, linden, (Tilia Americana,) interspersed with grape-vines, &c. The absence of the black walnut on the St. Peter, and near Fort St. Anthony, was particularly observed. The rosin plant was not seen after leaving Prairie du Chien. The yellow raspberry was abundant in many places and ripe . . . The course of the party was generally in the valley of the St. Peter, not far from . . . the river . . . its principal defect is the want of objects to animate the scenes; no buffalo . . . no deer . . . no birds . . . the St. Peter rolls in silence its waters to the Mississippi; . . . where game is scarce, the Indian . . . finds no inducement to hunt, and hence the party frequently travelled for whole days, without seeing a living object of any kind."

It was decided to change the "mode of travelling." It had taken six days to get to the Crescent, water was becoming shallower all the time, provisions were insufficient for so large a party and game to date had only provided two ducks.

". . . Our guide further informed us, that if we continued . . . in canoes, we should . . . arrive on Red River after the buffalo had left it, and find it, probably, impossible to reach the head of Lake Superior before the winter season had commenced . . . we should be compelled to winter somewhere west of the lakes . . . this comported neither with Major Long's wishes, nor with the instructions . . . from the War Department . . . By proceeding all in one party on land, much time would . . . be saved, and the bends of the river need not be followed . . ."

On July 15, the party, reduced to twenty-four, left the Crescent.

". . . They were provided with twenty-one horses, two . . . disabled. Nine were allotted to the officers and gentlemen of the party; the remaining ten being required as pack-horses . . . the soldiers were all obliged to walk; which, however, as the country was fine prairie . . . was not considered a very hard duty . . ."

They camped upon ". . . one of a group of ponds, dignified with the appellation of the Swan Lakes, on account of the abundance of these birds said to exist in their neighbourhood . . . These lakes are more properly marshes . . ." On the 16th: ". . . The botany of the country was diversified by the reappearance of the Gerardria,[3] a plant which we had not seen since leaving Chicago . . ."

On July 18 the journey, crossing the prairies, was uninteresting.

". . . The monotony of a prairie country always impresses the traveller with a melancholy, which the sight of water, woods, &c. cannot fail to remove. During that day we enjoyed no view of the river, and the great scarcity of springs, and wood for cook-

3. This herb is so frequently mentioned that one sometimes wonders whether it was the only one which members of the party recognized!

ing, made the travelling uncomfortable; to these we must add a temperature of about 94° . . . but the greatest annoyance . . . was the mosquetoe, which arose in such swarms, as to prove a more serious evil than can be imagined by those who have not experienced it . . . The mosquetoes generally rose all of a sudden about the setting of the sun. Their appearance was so instantaneous . . . we had no time to prepare ourselves against them . . . Our horses fared even worse . . . during the day the big horsefly proved equally noxious . . .

"The party had frequent opportunities of remarking the difficulty which exists, to determine with accuracy the nature or size of objects seen at a distance. Sand-hill cranes, seen on the prairies, were by some of the company mistaken for elks . . .

". . . The herbarium was enriched by the addition of a beautiful specimen of the Lilium Philadelphicum, which was still seen flowering, though it had nearly ceased to bloom. Another great ornament of the prairies is the Lilium Superbum. The Gerardria was still occasionally seen. This plant is . . . considered by the Indians to be a specific against the bite of the rattlesnake; the root is scraped and the scrapings applied to the wound; it is said that if used upon a recent wound, a single application will suffice . . ."

On the 19th, five miles below their encampment, ". . . there is a place where the boats and their loads are carried for the distance of a mile . . . the place is called the Grand Portage. By this portage the canoes avoid thirteen rapids; these with twenty-six other rapids, constitute all the obstructions to the navigation of the river, from its source to its mouth . . ."

On the 20th or 21st they spent half a day at the *"Lac qui parle* . . . an expansion of the river about seven and a half miles long, and from one quarter to three quarters of a mile wide." On the 22nd they ". . . reached another, and the last expansion of the river . . ." or the Big Stone Lake.

". . . Our view to the west was . . . bounded by an extensive ridge or swell in the prairies, known by the name of the 'Coteau des Prairies.' It is distant from our course about twenty or thirty miles; its height above the level of the St. Peter is probably not short of one thousand feet . . .

"A second ridge or Coteau des Prairies is said to run in a direction nearly parallel to that . . . just described . . . In the valley between these two ridges, the Riviere de Jacques, or James River, runs and empties itself into the Missouri about the 43rd degree of latitude. Thus the Coteau des Prairies may probably be considered as changing the course of the Missouri, above the Mandan villages, from an easterly to a southerly direction . . .

"Its distance from our course prevented us from visiting the Coteau . . .

"The Coteau des Prairies may truly be considered as the dividing ridge between the tributaries of the Mississippi and those of the Missouri . . .

"After having left the Big Stone Lake, we crossed a brook which retains the name of the St. Peter, but which cannot be considered as part of that river; the St. Peter may, in fact, be said to commence in Big Stone Lake . . .

"By the route which we travelled, the distance, from the mouth of the St. Peter to the head of Big Stone Lake, is three hundred and twenty-five miles, of which we ascended one hundred and thirty by water . . ."

Three miles after leaving Big Stone Lake the party came to "Lake Travers," or Traverse, ". . . which discharges its waters by means of Swan or Sioux river into the Red river of Lake Winnepeek, whose waters . . . flow towards Hudson's Bay. The space between Lakes Travers and Big Stone, is but very little elevated above the level of both these lakes; and the water has been known, in times of flood, to rise and cover the intermediate ground, so as to unite the two lakes . . . both these bodies of water are in the same valley . . . Here we behold the waters of two mighty streams, one of which empties itself into Hudson's Bay at the 57th parallel of north latitude, and the other into the Gulf of Mexico, in latitude 29°, rising in the same valley within three miles of each other . . . We seek in vain for those dividing ridges which topographers and hydrographers are wont to represent upon their maps in such cases . . .

"The country which extends between the forty-fifth and forty-eighth parallels of latitude, and between the ninety-third and ninety-seventh of longitude, presents perhaps an example of the interlockage of the sources of rivers, which few, if any spot on the surface of the earth, can equal. Here, no high ridge extends to divide the sources of three of the largest streams that are known. The mighty Mississippi and many of its tributaries run from the same lakes or swamps, which supply the waters of Nelson's river and of the St. Lawrence . . . Carver . . . destroys all the value of his information, by placing in the same district, the sources of the Oregon, or Great River of the West."

Keating's *Narrative* is replete with geological information, but I have confined myself to his geographical interpretation of this region.[4] Virtually nothing is said about the vegetation. There is mention of a feast which the party attended—it was given by one of the tribes of "Dacota" Indians with lodges between Lac qui Parle and Lake Traverse—where one dish ". . . consisted of a white root, somewhat similar in appearance to a small turnip . . . called, by the Dacotas, tepsin, by the French, the 'Pomme blanche or Navet de Prairie [in a footnote "Psoralea Esculenta, Nuttall"].' It was boiled down into a sort of mush or hominy, and was very much relished by most of the party; had it been seasoned with salt or sugar, it would have been considered delicious. This was held, even by the guides, to be a great treat."

And, about halfway up Lake Traverse, they had observed at a post of the American Fur Company, ". . . lamb's quarter ['Chenopodium album'], which was more than seven feet high. This plant was, at this time, almost too old for use, but until then it had proved a very valuable addition, at our meals, to the extremely small ration of biscuit, which at that time was reduced to about one ounce per day for each man."

From July 23 to 25 the party remained at a post of the Columbia Fur Company which is shown on the map on the eastern shore of Lake Traverse. According to the

4. It was to be visited at a later date, in 1838 and 1839, by Nicollet who at that time was accompanied by the German botanist Karl Andreas Geyer.

Narrative, this British company had been created in 1822, those in charge having previously worked for either the Hudson's Bay Company or the "North-West" Company; after these had been consolidated,[5] they decided to establish themselves in United States territory and trade with the Indians south of the border—"under licences granted by the Indian agent at the mouth of the St. Peter." This company and the American Fur Company were the only ones then trading with the Indians "in that part of the United States." Volume two of the *Narrative* contains a frontispiece showing the view[6] from the fort.

With the addition of a guide and others from the fort, Long's party started on their way on July 26.

". . . Having ascended the St. Peter up to its head in Big Stone Lake, our next object was to proceed 'to the intersection between Red river and the 49th degree of north latitude;' and as we were informed that that stream runs nearly north and south, we determined to travel the usual route to Pembina and Fort Douglas, two of the posts of the Hudson's Bay Company, between which the 49th parallel was reported to strike the river . . .

"The first day of our journey was unpleasant . . . across dry prairies . . .

"The dullness of our morning ride was dissipated by the distant view of the buffalo grazing upon the prairie. We shall not attempt to depict the joy, which the first cry of 'buffaloes in sight,' created in the whole company; all were in activity. The practised hunters immediately gave chase . . . and before the sun set, three of these noble animals had been slain . . . We encamped early . . . The spot which we were obliged to select, was utterly destitute of wood, and the only fuel . . . was the buffalo dung . . . This made a fine warm fire, giving out no smell. The meat was cooked, and eaten with great delight . . .

"The spot of our encampment is called . . . Buffalo Lake; it is only an extension of Lake Traverse . . . immediately below this place . . . the lake assumes the characters of a stream, and receives the name of Sioux or Swan river . . .

"The next morning, as we proceeded, the buffaloes began to thicken before us; in every direction numbers . . . were seen. They generally collected in herds of thousands together . . ."

On July 27 the route lay along Sioux or, as it is called on the map Swan, River and along its eastern side. This was apparently the name applied to the upper waters of Red River, or Red River of the North, which now serves for a long distance as the boundary between North Dakota and Minnesota. The map portrays many tributary streams entering both Swan River and Red River and the *Narrative* explains the point

5. The consolidation took place in 1821.

6. This, like all the plates illustrating the narrative proper, are "Design'd by S. Seymour." He appears to have been a trained and skillful artist and his landscapes possess considerable charm although crudely reproduced. His drawings of the Indians show that he was an excellent draughtsman. It is not stated who made the two plates of shells included in Part I (Zoology), contributed by Say, but it was, likely, Say himself.

at which the river's name changed, in the parlance of the day, from the first- to the last-named:

". . . This stream [Swan or Sioux River] branches out, at about four miles above the place where we struck it; one of the branches rises . . . in Lake Travers . . . The other rises in Otter-tail Lake . . . By the Indians this branch is called Otter-tail River, and the stream continues, after the junction of the two, to be called by them Sioux or Swan river, until it receives the Red Fork that rises in Red Lake; they then apply to the stream the name of Red River; while the traders have bestowed this name to the branch that rises in Otter-tail Lake."

The journey was through the territory of the Dakota Indians and, more than once they seem to have manifested "hostile dispositions." Their proximity evidently quickened Say's imagination, for, as Sophocles noted, "To him who is in fear everything rustles." He was taking one of the night watches.

". . . It was, while watching on the night of the 29th, that Mr. Say's attention was suddenly directed to an object in the prairie. He saw it approaching with caution, and immediately the idea that it was probably an enemy, induced him to creep in the direction from which the object approached; it had the aspect of a wolf, but this he immediately conceived to be a stratagem of the wily Indian, who, to conceal his approach, had assumed a false garb. So intent was he upon this idea, that he scarcely considered it possible that it should in reality be but a wolf. He felt a strong temptation to fire upon it, but the fear of alarming the whole camp induced him to desist, and he was only satisfied of the true nature of the object . . . when the latter, alarmed at the rustling made by Mr. Say's creeping through the grass, scampered off on his four legs, with a rapidity and agility that satisfied him that this was its natural posture."

Long had been obliged to call a halt on ". . . the 'running of the buffalo,' . . . or the chasing of them on horse-back."

". . . Such a chace frequently extends over four or five miles, and the excitement which the horses themselves derive from it, is sometimes sufficient to impel them to run until their strength is completely exhausted . . . all who are not hunters, callous to the sight of a tortured animal, must regret the very indiscriminate slaughter which is usually made of the buffalo; yet it must be acknowledged that the sport has something dignified and highly interesting, and that it requires no small share of self-control to remain a passive observer of it. Notwithstanding the general orders issued to that effect, about fifteen buffaloes were killed in one day."

But the buffalo were diminishing as the party advanced and on July 30 the party killed one—it ". . . was the last that we saw."

Colhoun, the *Narrative* relates, ". . has endeavoured to trace the extent of country over which the buffalo is known to rove at present, or to have formerly inhabited . . . for its numbers have diminished so rapidly within a century, its rovings have been so restricted, that there is reason to apprehend that it will soon disappear from the face of the land . . ."

Some ten pages are devoted to Colhoun's opinions about the bison and they end with the curious observation that "... While in the vicinity of the buffalo we were entirely free from the torment of mosquitoes, from what reason we know not; we can scarcely believe that the animal attracts them all to itself."

By August 2 they were so far north that they were suffering from cold and variable temperatures. The thermometer which at noon of August 1 stood at 83° in the shade, at sunrise of the 2nd registered 43°. This variation was found "extremely unpleasant" and clothes because of heavy dews "... were as wet as if ... soaked in water." There were "... scarcely more than five hours of night ..." and, the moon being pretty full, they "... seldom experienced any darkness during the whole ... journey to Pembina."

On August 5 the party crossed Red River "... in a barge, opposite to the settlement called Pembina, where we remained four days.

"This completed a journey of two hundred and fifty-six miles, performed in eleven days, averaging therefore about twenty three miles per day ... The dull monotony of a journey upon prairie land never appeared to us so fatiguing. No trees were to be seen except those that fringed the water courses, these consisting principally of several varieties of oak, of the white, and some red elm, linden, gray ash, red-maple, cotton-wood, aspen, hackberry, ironwood, hop hornbeam, and white and red pine. On Red Lake we were told that the trees consist of fir, sugar-maple, and birch. The country is very flat, and remarkably deficient in water. There are no valleys, and but few brooks, streams, or even springs."

The settlement was in what is now the extreme northeastern corner of North Dakota, about one hundred and seventy miles above the mouth of Red River.

"Pembina constituted the upper settlement made on the tract of land granted to the late Lord Selkirk by the Hudson's Bay Company ...

"The Hudson's Bay Company had a fort here, until the spring of 1823, when observations, made by their own astronomers, led them to suspect that it was south of the boundary line, and they therefore abandoned it, removing all that could be sent down the river with advantage ...

"The main object of the party in visiting this place being the determination of the 49th degree of latitude, Mr. Colhoun lost no time in taking observations ... We ... pitched our camp a little further down on the bank of the river, and as near as we could judge to the boundary line. A large skin lodge ... lent to us, sheltered the gentlemen of the party ... our flies were pitched around it for the use of the soldiers. In honour of the President of the United States, this place received the name of Camp Monroe. A flag-staff was planted, which, after a series of observations, made during four days, was determined to be in latitude 48° 59′ 57⅓″ north ... the distance to the boundary line was measured off, and an oak post fixed on it, bearing on the north side the letters G.B. and on the south side those U.S. On the 8th of August, at noon, the flag was hoisted on the staff ... A national salute was fired at the time, and a proclamation made by Major Long, that 'by virtue of the authority vested in him by the President of the United States, the country situated upon Red river, about that point, was declared to be comprehended within the territory of the United States.' This declaration was made in the presence of all the inhabitants collected for that purpose. They appeared well satisfied on hearing that the whole of the settlement of Pembina, with the simple exception of a single log-house, standing near the left bank of the river, would be included in the territory of the United States. While fixing the posts, the colonists requested that they might be shown how the line would run; when this was done, the first observation they made was, that all the buffalo would be on our side of the line . . ."

Although much of interest is related about the region, botanical comments are as meagre as always in the *Narrative;* I quote them, together with Say's report upon the birds:

"Of the plants observed in this neighbourhood, besides [near] the Pembina, we can only mention the common hop; and the raspberry-bush, which yields fruit in great abundance and of a very superior quality; also a large kind of whortleberry, the fruit of which is double the size of ours, and more oval. The forest-trees are the same which we had previously seen on Red river. The zoology of the country is not very diversified. Among the birds seen by Mr. Say, during our stay at Pembina, were the turkey-buzzard, red-headed woodpecker, flicker, hemp-bird, king-bird, sparrow-hawk, house-wren, robin, chimney-bird, barn swallow, night-hawk, whip-poor-will, bald-eagle, hairy woodpecker, great heron, grakle, kildeer, blue-winged teal, ruddy duck, rose-breasted grosbeak, crow, raven, and pigeon, the last of which is very abundant in the woods."

Long had planned, in conformity to his orders from the War Department, to proceed eastward along the 49th parallel, or ". . . along the northern boundary of the United States to Lake Superior . . ." But he was told that the whole of the country was ". . . covered with small lagoons and marshes, which rendered it impenetrable for horses. The only practical mode was to follow the principal streams in bark canoes . . . Several routes were suggested; that by Lake Winnepeek appeared the best, and was adopted . . ."

On August 9 the expedition started on its way northward, crossing into what is now the Province of Manitoba, Canada.[7] Long, Say, Keating, Colhoun and Seymour reached Philadelphia on October 26, 1823.

7. According to a summary of the return journey supplied in Warren's "Memoir," the party ". . . continued down the valley of Red river . . . to Fort Gerry [Garry], at the mouth of Assiniboin river, and thence to Lake Winnipeg, and along its southern shore to Fort Alexander, at the mouth of Winnipeg river . . . ascended the Winnipeg river, to Lake of the Woods, across this to Rainy Lake river, up this to Rainy lake, across this lake, Sturgeon lake, and the chain of lakes, to Thousand lake, where they made the Portage du Prairie to the source of Dog river . . . down this stream to Fort William, on Lake Superior, making seventy-two portages after leaving Lake Winnipeg . . . From Fort William . . . along 'the dreary northern shore of Lake Superior' by water to the Sault Ste. Marie, which may be considered the terminus of the expedition . . ."

As stated earlier, the first part of the appendix to Keating's *Narrative* contained two subdivisions, Zoology and Botany: the first subdivision was by Thomas Say; the second, entitled "A catalogue of plants collected in the North Western territory by Mr. Thomas Say, in the year 1823," was compiled by Lewis D. von Schweinitz whose introduction states:

"Mr. Thomas Nuttall, who had taken upon himself the charge of examining this collection, and had begun to commit his remarks to paper, not having returned from Europe in time to complete his work, the plants collected by Mr. Say were entrusted to me, with a request to attempt their determination, and a description of such as appeared to be nondescripts. I have undertaken this task with great diffidence and sincere regret that it could not be executed by a gentleman, every way so exclusively competent as Mr. Nuttall is, both from his well known botanical talents in general, and his particular acquaintance with the western plants, my own knowledge of which is almost confined to what I owe to that gentleman's liberal and kind communications."

Nuttall's work ". . . unfortunately comprised only the five first plants of the present catalogue." Von Schweinitz is apologetic in regard to his new species: he was not only unfamiliar with the living material but the ". . . specimens . . . though generally well preserved, are mostly imperfect, rarely furnishing both flower and fruit together . . ."

The catalogue enumerated 130 species. Eleven were described by von Schweinitz as new: 4. Cyperus *alterniflorus* (p. 381); 13. Triticum *pauciflorum,* "Prairies of the St. Peter." (p. 383); 28. Gentiana *rubicaulis,* "Prairies of St. Peter's river." (p. 384); 43. Prunus *incana* (p. 387); 46. Crataegus *flexuosa* (p. 387); 50. Rosa *Sayi* (p. 388); 60. Stachys *velutina* (p. 390); 63. Melampyrum *brachiatum* (p. 391); 73. Vicia *tridentata* (p. 392); 81. Hieracium *scabriusculum* (p. 394); 83. Vernonia *corymbosa* (p. 394).

Of these new species only numbers 13 and 28 are cited from the region in which we are interested. No localities are given for numbers 50, 60, 63, 81 and 83. Nineteen numbers of plants cited from our region had already been described.

In his "Historical sketch of the science of botany in North America from 1635 to 1840," Frederick Brendel mentions that "As several boxes containing collections, and dispatched during the expedition, were lost, the botanical collection was very poor, only 130 species. As poor as the collection was, the description of the species called new by Schweinitz, are mostly riddles not yet solved. So the expedition, otherwise interesting, was unimportant as to botany."

Since this is the second expedition (of those described in this story) in which Thomas Say participated I shall tell something of his later life, a large portion of which was spent at the "ideal community" inaugurated by Robert Owen and situated on the Wabash River, at New Harmony, Indiana. Distinguished foreigners were wont to visit the colony and two at least mention Say.

After the expedition up St. Peter's River, Say for a time worked upon his collections in Philadelphia. According to the title-page of the first volume of his *American entomology,* published in 1824, he was then "Curator of the American Philosophical Society, and of the Academy of Natural Sciences of Philadelphia; Correspondent of the Philomatique Society of Paris; and Professor of Natural History of the University of Pennsylvania, and of Zoology in the Philadelphia Museum."

Of his scientific interests entomology and conchology[8] were the most important and scientists such as Nuttall seem to have been helpful in increasing his collections in both these fields.

Say joined the New Harmony group in 1826 and remained a member until his death, or for eight years. During this sojourn, the most influential member of the colony was William Maclure.[9]

Donald Culross Peattie's *Green laurels. The lives and achievements of the great naturalists* has considerable to tell about Say's life at New Harmony. He refers to Say as ". . . already the scientific hero of Major Long's expedition to the Colorado Rockies, which had failed so signally to locate the sources of the Red River of the South, to endear white Americans to red ones, or accomplish anything lasting save for Say's scientific soundings.[1] He had penetrated north to Pembina, too, and at that moment was the foremost authority upon the insects and molluscs of North America. It is said of him that he described and discovered more species of shells than any one else in the history of conchology, save only two very hasty and slipshod workers in the British Museum. As for our insects, there is no guessing at the number that go to Say's credit . . . Say's great *American Conchology* began to take shape at New Harmony . . . The exquisite hand of Lesueur[2] engraved the plates . . . This, the first book of its sort ever published in America, was set up, with infinite difficulties as to importing fonts of type and suitable paper stock, and keeping the printers sober. It was issued . . . in seven parts . . ."

Those interested in amusing "bits" about the naturalists at New Harmony should read Mr. Peattie's story; I have omitted, among many other items, his entertaining

8. Say's *American entomology* was published from 1824 to 1828, and his *American conchology* from 1830 to 1834.

9. It was for Maclure that Rafinesque named the genus *Maclura,* of which the Osage orange is the sole representative. Maclure was a geologist, a patron of the Philadelphia Academy and its president from 1817 to 1840. Although Owen's "social state" failed, Maclure remained for a long time at New Harmony, interested in promoting educational experiments.

1. Does Mr. Peattie belittle the accomplishments of Edwin James?

2. George Brown Goode—The beginnings of American science. The third century, *Proc. Biol. Soc. Washington* 4: 68, 69 (1888)—uses the same adjective: "No one ever drew such exquisite fishes as Lesueur . . . it is greatly to be regretted that he never completed his projected work upon North American Ichthyology . . . Twelve years of his life were wasted at New Harmony . . ." His full name was Charles Alexander Lesueur and he was born at Havre-de-Grâce, France, on January 1, 1778, and died at Havre on December 12, 1846, according to Goode.

account of Say's marriage, an elopement, to Miss Lucy Sistaire, one of the young ladies of the colony whose education was entrusted to Madame Marie Louise Duclos Fretageot, a highly esteemed member of the group. Peattie reports that Miss Lucy "found the utmost happiness in hand-painting" the shells of the *American conchology*.

Bernard, Duke of Saxe-Weimar Eisenach, was at New Harmony in 1826. Chapter 21 of his *Travels through North America*[3] is devoted to an account of the colony, its history, members, and so on. He had this to tell of Thomas Say and of the costume in which he looked so "comical":

"I renewed acquaintance here with Mr. Say, a distinguished naturalist from Philadelphia, whom I had been introduced to, at the Wistar Party there; unfortunately he had found himself embarrassed in his fortune, and was obliged to come here as a friend of Mr. M'Clure. This gentleman appeared quite comical in the costume of the society, before described, with his hands full of hard lumps and blisters, occasioned by the unusual labour he was obliged to undertake in the garden. . .

"There is a particular costume adopted for the Society. That for the men consists of wide pantaloons buttoned over a boy's jacket, made of light material, without a collar; that of the women of a coat reaching to the knee and pantaloons, such as little girls wear among us . . ."

3. The Duke's *Travels* is enjoyable! He landed in Boston on July 26, 1825. "It was ten o'clock . . . when I first placed my foot in America, upon a broad piece of granite! It was impossible to describe what I felt at that instant . . ." During his stay of two weeks he was considerably entertained and I quote his fascinating and somewhat Germanic approach to this hospitality: "The society, especially when ladies were not [sic] present, is uncommonly fine and lively—both sexes are very well educated and accomplished. So much care is bestowed upon the education of the female sex, that it would perhaps be considered in other countries as superfluous . . ." To this comment the translator might have added—as he did to the Duke's comments about an alligator—"[Nonsense.]—Trans."

While in Boston the Duke visited, among other places, Harvard College. "We were escorted through the botanical garden by Professor Nuttall, an Englishman, who had made several scientific journeys in the western parts of the United States. Among the green-house plants I observed a strelitzia, which had been raised from seed in this country, and also a blooming and handsome Inua gloriosa, and a Hedychium longifolium. The green-house and the garden are both small; in the latter I remarked no extraordinary shrubs or flowers, on the contrary, however, I saw many beetles . . . and many beautiful butterflies . . ."

Now garden and greenhouse—and presumably beetles and butterflies—are gone. Some noted their passing with regret. But, with the passage of time, the garden became of greater historical than botanical interest. Asa Gray's letters contain numerous references thereto, some recounting his attempts to acquire plants, and to maintain them in good condition once planted.

The Elgin Botanic Garden antedated the one in Cambridge; closed in 1814, Rockefeller Center in New York City now occupies its site. It was to this garden that Gray referred when, writing of *The Flora of North America*, he states that Pursh in 1807 ". . . took charge of the Botanic Garden which Dr. Hosack had formed at New York and afterward sold to the State, which soon made it over to Columbia University." Gray moralizes upon this disposition in a footnote: "Expecting, no doubt, that it would be kept up. But the Elgin Botanic Garden was soon discontinued. It occupied the block of ground now covered [Gray's paper was read in 1882] by the buildings of the College, and the surrounding tract—now so valuable—from which the College derives an ample revenue. *Noblesse oblige,* and it may be expected that the College, so enriched, will, before long, provide itself with a botanical professorship, and see to the careful preservation and maintenance of the precious Torrey Herbarium, which it possesses along with other subsidiary herbaria."

Another German visitor to New Harmony, in the winter of 1832–1833, was Maximilian, Prince of Wied-Neuwied. He was on his way to St. Louis where he was to embark on the journey up the Missouri River which is described in his *Travels in the interior of North America.* He describes Say's garden:

"Mr. Say's house was in a garden, where he cultivated many interesting plants of the interior of Western America. I there saw a large *Maclura aurantiaca* (Nuttall), the bow or yellow wood, or Osage orange, from the river Arkansas, of the wood of which many Indian tribes make their bows. It is a prickly tree, with very tough wood. There was one in St. Louis, in the garden of Mr. Pierre Chouteau, which did not, however, flourish. Dr. [Zina] Pitcher had the kindness to give me some seeds of this tree, which, however, have not succeeded. In Mr. Say's garden I likewise saw *Euphorbia marginata,* from Arkansas, several beautiful *phlox;* and the *Lonicera sempervirens,* was laden with its ripe fruit. The *Euphorbia marginata* flourishes exceedingly well at Bonn, where it was raised from seeds which I brought."

Weiss and Ziegler have written Say's biography. They describe him as ". . . tall and spare but apparently muscular and strong before his period of ill health, otherwise he could not have withstood the hardships of his western travels." But he suffered years of disability; Dr. B. H. Coates is quoted regarding ". . . Say's habit of going for long periods without food or very little of it . . ." and George Ord reported that Say and Maclure ". . . carried their habits of abstinence from food too far." Whether these dietary peculiarities were the cause of ill-health, or precautionary measures necessitated thereby, is not entirely clear! Weiss and Ziegler refer to Say's ". . . amiable disposition, his modesty, his sincerity, his conciliatory manner, his readiness to trust others, his willingness to help everybody whenever possible, his honesty and his love of truth . . ."—a large array of fine qualities certainly.

Say died at New Harmony on October 10, 1834. A monument was erected to his memory twelve years after his death[4] by William Maclure's brother Alexander.

4. The date of birth on the monument (the inscription is reproduced in Weiss and Ziegler, p. 223) is July 27, 1787, but his biographers cite it as June 27 (p. 29). It is not impossible that the monument's record is incorrect!

BERNHARD, KARL, Duke of Saxe-Weimar Eisenach. 1828. Travels through North America during the years 1825 and 1826. 2 vols. Philadelphia.

BRENDEL, FREDERICK. 1879. Historical sketch of the science of botany in North America from 1635 to 1840. *Am. Naturalist* 13: 76.

HOWARD, LELAND OSSIAN. 1935. Thomas Say. *Dict. Am. Biog.* 16: 401, 402.

KEATING, WILLIAM HYPOLITUS. 1824. Narrative of an expedition to the sources of St. Peter's River, Lake Winnepeek, Lake of the Woods, &c. &c. performed in the year 1823, by order of the Hon. John C. Calhoun, Secretary of War, under the command of Stephen H. Long, Major U. S. T. E. Compiled from the notes of Major Long, Messrs. Say, Keating, and Colhoun, by William H. Keating, A.M. Professor of mineralogy and chemistry as applied to the arts, in the University of Pennsylvania; geologist and historiographer to the expedition. Map ["Map of the country embracing the route of the expedition of 1823 commanded by Major S. H. Long."]. 2 vols. Philadelphia.

MAXIMILIAN ALEXANDER PHILIPP, Prince of Wied-Neuwied. 1843. Travels in the interior of North America. By Maximilian, Prince of Wied. With numerous engravings on wood, and a large map. Translated from the German by H. Evans Lloyd. To accompany the original series of eighty-one elaborately-coloured plates. Size, imperial folio. London.

PAULLIN, CHARLES OSCAR. 1932. Atlas of the historical geography of the United States. Edited by John K. Wright. *Carnegie Inst. Washington Publ.* 401: pl. 39B, p. 20.

PEATTIE, DONALD CULROSS. 1936. Green laurels. The lives and achievements of the great naturalists. New York. 254, 258.

SCHWEINITZ, LUDWIG DAVID VON. 1824. A catalogue of the plants collected in the north western territory by Mr. Thomas Say, in the year 1823. *In:* Keating, W. H. 1824. Narrative of . . . 2: 379-400.

WARREN, GOUVENEUR KEMBLE. 1859. Memoir to accompany the map of the territory of the United States from the Mississippi River to the Pacific Ocean, giving a brief account of each of the exploring expeditions since A.D. 1800 . . . *U. S. War Dept. Rept. expl. surv. RR Mississippi Pacific* 11: 24, 25.

WEISS, HARRY BISCHOFF & ZIEGLER, GRACE M. 1931. Thomas Say. Early American naturalist. Foreword by L. O. Howard. Springfield, Illinois. Baltimore.

CHAPTER XIII

SCOULER PAYS A VISIT OF SEVEN MONTHS TO THE

NORTHWEST COAST OF NORTH AMERICA

W HEN the renowned plant collector David Douglas sailed for the northwest
coast of North America in July, 1824, he considered himself fortunate ". . .
in having a companion in Dr. Scouler of Glasgow, a man skilled in several,
and devotedly attached to all branches of Natural History, a pupil of Dr. Hooker,[1] by
whom he was powerfully recommended to the H[udson's] B[ay] C[ompany] as surgeon to the vessel, in order that he might have an opportunity of prosecuting his favourite pursuit."

Two transcripts of a journal kept by Dr. John Scouler on this voyage have been
published:

The first appeared in three installments in *The Edinburgh Journal of Science,* conducted by David Brewster—one in 1826 (volume 5) and two in 1827 (volume 6)—
under the title "Account of a voyage to Madeira, Brazil, Juan Fernandez, and the Gallapagos Islands,[2] performed in 1824 and 1825, with a view of examining their natural history, &c. By Mr. Scouler. Communicated by the author." Although the editor
of the "Account" is not mentioned he may well have been Brewster. Whether the
manuscript from which the "Account" was transcribed is still in existence I do not
know.

The second transcript appeared in *The Quarterly of the Oregon Historical Society*
in three installments—in March, June and September of 1905 (volume 6)—under the
title "Dr. John Scouler's journal of a voyage to N. W. America," and was edited by
Frederick George Young. The March issue and a few pages of the June issue tell of
Scouler's travels before reaching our northwest coast and a few final pages of the September issue tell of his return voyage to England. The editor seems not to have known
of the earlier *Edinburgh Journal of Science* publication for he states:[3]

"The journal kept by Dr. Scouler during his explorations in the Pacific Northwest
has, I believe, never been published. Through the keen search for Oregon material,
conducted by Mr. Charles E. Ladd, of Portland, it was secured for the region to which
it mainly pertains, and it was generously turned over to the Oregon Historical Society."

1. William Jackson Hooker, then Professor of Botany at the University of Glasgow, and later Director
of the Royal Gardens at Kew.

2. Despite the fact that North America is not mentioned in the title, the two installments published in
1827 deal in the main with its northwest coast.

3. In a prefatory note to his transcript of W. J. Hooker's "Memoir" of David Douglas.

Scouler, like David Douglas, evidently made more than one draft of his journal. C. F. Newcombe comments[4] that "A comparison of the two renderings of the [Scouler] journal, shows that they only differ in minor details, such as the spelling of scientific names, but each contains information not common to both."

In my account of Scouler's activities and observations on this journey I shall quote from both of the transcripts cited above, distinguishing the sources by the notations (EJS)—*Edinburgh Journal of Science*—and (QOHS)—*Quarterly of the Oregon Historical Society*. When the subject matter is virtually the same in the two transcripts I have quoted what seems to me the more interesting version or, when they vary in substance, from both versions. Bracketed insertions in the *Quarterly* transcript are those of its editor.

After summarizing briefly the contributions made by Vancouver, by "Sir A. M'Kenzie," by Lewis and Clark, and by Archibald Menzies, and after paying tribute to the assistance rendered him by the Hudson's Bay Company, Scouler explains his appointment:

"The Hudson's Bay Company . . . were anxious to have a surgeon, (in their vessel about to undertake a voyage to the Columbia River,) who, in addition to his professional acquirements, was qualified to make collections in the various branches of natural history. Through the kind recommendations of Dr Hooker and Dr Richardson, I had the good fortune to meet with the company's approbation, and was appointed . . . As it is to the Hudson's Bay Company I am indebted for the means of making my collection, so, on my return, the objects I had procured would have been of very little use to the public, unless I had enjoyed the assistance of Dr Hooker, and the free use of his extensive library." (EJS)

A footnote tells of further assistance:

"While in London I received much important information from Dr Richardson and Mr Menzies with respect to the countries I was about to examine. The knowledge acquired from Mr Menzies[5] was peculiarly interesting, as he had already explored the very coast I had to visit, and cheerfully allowed me at all times to examine the plants he had collected on the North West Coast, and to direct my attention to those which were most likely to be useful when cultivated in this country. Through his advice I was induced to pay particular attention to the seeds of *Gualtheria Shallon*,[6] which have already produced young plants in the Botanic Garden at Glasgow. Dr Richardson also

---

4. In his bibliography of "Menzies' journal of Vancouver's voyage."

5. C. F. Newombe states in his preface to "Menzies' journal of Vancouver's voyage": "There can be little doubt that the voyage under consideration led to the journeys of Douglas and Scouler. The former, especially, was able to complete the work of his predecessor by sending home the seeds and living plants of trees and flowers of which Menzies, owing to the conditions in which he worked, could only make herbarium specimens. Both Douglas and Scouler acknowledged their indebtedness . . ."

6. *Gaultheria Shallon* Pursh. The foliage of this plant is now sold in great quantities by our eastern florists. "*Gualtheria*" was a not uncommonly used variant in the first part of the last century, and by a number of botanists.

gave me much instruction with regard to the best way of preserving animals, a subject on which his advice was of the utmost value; and I was farther indebted to him for specimens of many of those interesting plants he had collected while engaged in the Arctic expedition." (EJS)

1824

July 25: "On the 25th July 1824 we left Gravesend, with every thing necessary for the preservation of plants and animals. In a medical point of view we were also excellently supplied . . . during a long voyage of twenty-two months, we never once could detect a symptom of scurvy on any individual in the vessel. In the prospect of a long voyage, I esteemed myself fortunate in having for a companion Mr Douglas, a zealous botanist, who was engaged by the Horticultural Society of London to explore the vegetable treasurers of the North West Coast of America." (EJS)

We are told of visits to the island of Madeira and to Rio de Janeiro and are informed that ". . . the dangers of C. Horn have been greatly exaggerated & we invariably find that the most experienced sailor talks le[a]st about them." (QOHS) There were stops too at Juan Fernandez and at the Galapagos Islands. From the last it took more than seventy-three days to reach the entrance of the Columbia River. (QOHS)

1825

April 3: "This morning we saw Cape Dissapointment, a circumstance we had long & anxiously wished for . . . Our attempt to cross the bar, however, was unsuccessfull & we were under the mortifying necessity of putting again to sea." (QOHS)

April 7: ". . . we made another attempt & before the evening we were safely at anchor in Baker's bay. We lay about ¼ of a mile from the shore, & opposite two small rocky islands, which lay at the bottom of the bay. The land is very steap & uneven, but is covered completely by pine trees." (QOHS)

April 9: "In the afternoon in company with Mr. Douglass I made a short visit to the shore. The first we collected on North American continent was the charming *Gaultheria Shallon,* in an excellent condition. We then penetrated into those primeval forests never before explored by the curiosity of the botanist. Here the lover of musci & lichens enjoys ample opportunity of studying his favorite plants. The moisture of the climate is very favourable to the growth & variety of these plants & the trees & rocks are covered by them." (QOHS)

"On leaping from the boat, the first object which attracted our notice, was the *Gualtheria shallon* growing in abundance among the rocks, and covered with its beautiful roseate flowers. We then entered a forest of gigantic pine trees, among a brushwood of *Menziesia ferruginea* and different species of American currant, and the beautiful *Trilliums* and *Smilacinae* were beginning to expand their blossoms, and the *Mosses* and *Jungermanniae,* nourished by the winter rains, were covered with capsules. On our return we collected a few specimens of a small *Polypodium,* which is probably new to the American Flora.

"The appearance of vegetation differed considerably from that to which we had been previously accustomed. The whole country appeared one continued pine[7] forest; but on a closer examination, we found many places, which, from their marshy nature, refused support to the larger trees. These were covered by various grasses, and abounded in willows, and various kinds of currant. In more open places, as along the banks of the river, different kinds of brambles abound, many of them peculiar to this part of America, and equally distinguished for the beauty of their flowers, and the flavour of their fruit. But nothing is more worthy of notice than the verdure which is found throughout the year under the shade of the pine trees. This appearance arises from *Sallal (Gualtheria shallon)* whose evergreen leaves ornament these otherwise sterile situations, while they form the important article of support to the natives. At the time of our arrival there was no snow on the ground, and it is rarely seen even in winter. Vegetation at this time (April) was little more forward than in England at the same season, but it soon advanced with a rapidity unknown in England." (EJS)

April 10: "We landed again in Bakers bay, with the intention of going across Cape Dissapointment to the ocean. In this journey we met with many difficulties, not only from the steapness of the rocks, but from the deap pools of fresh water which were to[o] deap to pass. Our excursion was also obstructed by the immense profusion of *G. Shallon."* (QOHS)

What appears to have been the same trip is attributed to April 11 in *The Edinburgh Journal of Science:*

". . . Our excursion was not altogether useless, as we made a very considerable collection of acotyledonous plants, and obtained some curious species of land shells in the woods. From the great abundance of *Musci* and *Jungermanniæ* on the northwest coast, we had been led to expect a corresponding variety of new species, but in this we were disappointed, and most of the species we found were common to Europe; but, from the moisture and mildness of the climate, they acquired their full development; and we found many of them in a state of fructification, whose capsules are but rarely seen in Europe."

April 12: ". . . we landed (well craved) at Fort George, & were received in a very polite manner by Mr. McKenzie . . . He informed us that the other gentlemen were employed in building a new fort, about 80 miles further up the river, at Point Vancou-

---

7. At this period the genus *Pinus* was interpreted to include such genera as *Abies, Picea,* etc. Scouler, in this instance, may have been thinking of the so-called "Oregon pine," or Douglas fir, the *Pseudotsuga,* which made up much of the dense forest.

ver, & Ft. George had been ceded to the Americans by the treaty of Ghent, & they were expected to take possession of it very soon . . . We made a short excursion to the neighbouring woods, & collected a good number of *Musci* & *Jungermannias* besides Phænogamous plants." (QOHS)

April 13: "The rain detained us on board . . ." (QOHS)

April 14: "Since we have been in the Columbia River the rain has been incessant, & we have not had six [days] of uninterrupted dry weather since we anchored in Bakers bay . . . although it was very little better to-day we made an excursion to the shore."

April 15: "In this day's excursion, we met a number of Indians in the wood, chiefly women and children, who were employed in collecting vegetables, as the young shoots of different species of *Rubus* and *Rosa,* and, above all, the tender shoots of the horse-tail, *(Equisetum arvense,)* which attains a large size and is much esteemed by the Indians. We saw plenty of *Menziesia ferruginea,* but not yet in flower; we found various species of *Trillium* and *Smilacina;* but no plant we found gave us more pleasure than the *Hookeria lucens,* not only on account of its beauty, but as it brought to mind our distinguished botanical perceptor, to whose instructions we had been so much indebted." (EJS)

". . . I collected a considerable number of cryptogamous plants, & none of the plants I ever met with on the N. W. coast gave me greater pleasure than *Hookeria lanus.* I found beautifull specimens of the charming little plant, with its constant attendant, *Hypnum Splendens,* growing by the margins of a shady rivulet among a brush wood composed of *Menziesia ferruginea* . . ." (QOHS)

April 16: "As Mr Douglas was to set out next day for Fort Vancouver . . . we agreed to make an excursion to Tongue Point[8] (six miles from the fort) before we parted . . . We filled our boxes with various species of Claytonia and liliaceous plants; our pockets and handkerchiefs were filled with mosses and land shells . . ." (EJS)

The trip of April 16 is attributed to the 17th in the *Quarterly* and the distance to Tongue Point given as about five miles from Fort George.

April 22: "We have now so completely ransacked the neighborhood of Ft. George, that very few new plants now attract our notice, and our impatience is obliged to wait till the progress of spring lays open more plants to our curiosity. In this delemma I set out in quest of animals and was tolerabley successfull . . ." (QOHS)

April 29: "Since the 16th I have been employed in exploring the vegetable and animal productions of the country in every direction; but as the progress of spring did not keep pace with my wishes, I set out for Fort Vancouver, where the difference of soil would produce a corresponding variety of plants . . . when we landed in camp for the

8. Clatsop County, Oregon.

evening, we found ourselves at Cook Point, about thirty miles from Fort George."
(EJS)

April 30: "We were detained to gum two of our canoes, which gave me an opportunity of seeking for a few plants. The soil of Oak Point is marshy & alluvial, & the only vegetables that abound on it are Cyperaceæ, of which I amassed many species . . . Our progress to-day was very slow, as the wind was unfavourable & required that we should make frequent delays, which enabled me to obtain some excellent plants, viz., *Cornus Canadensis,* 2 sp. *Myosotis,* 1 sp. of *Fedice* [?], 1 of Valeriana, & some *Ranunculi.*"
(QOHS)

May 1: "We made very little progress on account of the strong wind that blew directly down the river." (QOHS)

May 2: ". . . at 11 o'clock we arrived at Ft. Vancouver . . . On rejoining my fellow traveller, Mr. Douglass, we made an excursion to Menzies island, where we found many interesting plants. These plants on this island belonged chiefly to the classes *Compositæ* & *Leguminosæ,* the *Phlox linearis, Collomia linearis,* Nuttall, & a beautiful *Myosotis,* which, as being the most beautifull nondescript plant we had yet seen, from the allusion contained in the Scotch name for the genus, we agreed to honour this plant with the name of *M. Hookeri.*

"Ft. Vancouver is built on the same plan as the other fort, but is not so large. Its situation is far more pleasant than that of Ft. George. It is situated in the middle of a beautifull prairie, containing about 300 acres of excellent land, on which potatoes & other vegetables are cultivated . . . The forests around the fort consists chiefly of *Pinue balsamea* & *P. canadensis,* while most *amentaceæ* are exceedingly rare. Within a short distance of the fort I found several interesting plants, as *Phalangium esculentum, Berberis nervosa, B. Aquifolium, Calypso borealis* & *Corallorhiza innata.*[9] The root of the *Phalangium esculentem* is much used by the natives as a substitute for bread. They grow abundantly in the moist prairies, the flower is usually blue, but sometimes white flowers are found. The bulbs are about the size of those of *Hyacinthus Menseriptus* [?], & are collected by women & children . . ." (QOHS)

". . . the tubers of a species of *Sagittaria,* which grows on the marshy banks of the river, affords an agreeable substitute for potatoes. In the neighbouring woods we found some of the choicest plants the N. W. coast can boast of.

"The *Linnæa borealis* . . . grew here in great profusion, and I afterwards found it equally common in the woods of Observatory Inlet, the northern limit of our voyage. The subjoined list[1] of the plants that were known to us may give the botanist some

9. Oakes Ames supplied the names now accepted for these two orchids: *Calypso borealis=Calypso bulbosa* (Linnaeus) Oakes: *Corallorhiza innata=Corallorhiza trifida* Chatelaine.

1. A footnote cites: "Calypso borealis, Corallorhiza innata, Berberis aquifolia, nervosa, Collinsia verna, Phlox linearis, Myosotis, *Nov. Sp.,* Sanicula, *Nov. Sp.*"

idea of one day's excursion, not above four miles from the fort . . . We usually spent the forenoon in botanising, and during the remainder of the day our time was fully occupied in arranging and drying the plants we had already obtained." (EJS)

May 10: "Since my arrival here my time has been entirely occupied in making excursions in every direction around the fort, & I had no reason to be dissatisfied with my collection." (QOHS)

On May 11 Scouler and Douglas set out for the coast: ". . . while stopped to prepare our supper I made a little excursion into the woods with Mr. Douglass & found *Dalibarda repens* [?], a large *Pyrola,* & a sp. of *Heuchera.*" (QOHS)
". . . we set out on our return to the coast, as the ship was to sail in a few days to visit some of the islands to the north of Nootka . . . on a fine dry beach, and while supper was preparing, we collected a few plants; *Dalibarda repens, Pyrola umbellata,* and a species of *Heuchera.*" (EJS)
During the night the canoes were allowed to drift downstream and they made fifteen miles.

May 12: ". . . At two o'clock we were of[f] Tongue point where the first plant that attracted my notice was a beautifull & a new species of *Mimulus,* & we collected specimens of it with the utmost enthusiasm. This little plant grows among *Musci* on the wet rocks & may be called *M. pusillus.* It is distinguished by the following characters & is the smallest of the genus. Leaf an inch in length, lower lip of the corolla spotted, leaves spatulate." (QOHS)
"The few minutes we spent here were not useless, for we had scarce leaped ashore, when a beautiful and new species of *Mimulus* attracted our attention, growing among mosses from the moist rock. This beautiful plant, perhaps the smallest of the genus, is not more than an inch and a half in height, and one solitary flower rises from the slender scape." (EJS)

Scouler's time after arrival at Fort George was evidently occupied in studying two Indians who had died of apoplexy. We are told that "Appoplexy is far from being a rare disease among the Cheenooks . . .", and the doctor's diagnosis seems to have been that its prevalence was due to ". . . the enormous quantity of fish & other kinds of animal food they eat, & their inordinate appetite for oil . . ." The daily journal-records (in both transcripts) do not begin again until thirteen days after arrival at Fort George:

May 25: "My time is now divided between making arrangements for our voyage to the Northward & completing as far as possible my collection of Columbia plants. We are about to leave the Columbia during the finest season of the year, but I anticipate a rich harvest at Nootka & Fuca straits." (QOHS)
"From this period till the first of June, when we embarked on our voyage to the more northern parts of this coast, our botanical labours suffered little intermission . . .

Thus the time passed away amid constant occupation till the vessel sailed . . . and Mr Douglas set out on a journey of several hundred miles to the interior." (EJS)

May 31: "We are now ready for our expedition . . . In the morning we landed at the Cheenook village to purchase salmon; before our departure I seized the opportunity of herborising & found some interesting plants as a fine sp. of *Triticum* & a sp. of *Spiræa*, 1 sp. of *Trifolium*." (QOHS)

June 1: ". . . we crossed for the second time the bar of the Columbia & stood to the north." (QOHS)

From June 1 to August 7 the *William and Ann* was in northern waters and such landings as were made were in what is now British territory. A. F. Hemenway made the statement that ". . . leaving Douglas, Dr. Scouler crossed the bar of the Columbia for the second time and sailed on a trading vessel along the coast of Washington up to Nootka. On July 7 the vessel started back and arrived at the Columbia September 7, 1825. During this trip Scouler visited almost every accessible bay or inlet which he passed."

C. F. Newcombe comments that "Mr. Hemenway's assertion that Scouler 'visited almost every accessible bay or inlet which he passed' is very far from being supported by the entries in that journal." It is misleading also in indicating that the ship merely went to Nootka and returned.[2]

Actually the *William and Ann* did not stop at Nootka on the way north but went directly to the Queen Charlotte Islands, passed through the "Skittigas" (Skidegate) channel and continued northeast to Portland Canal and Oservatory Inlet—at the western end of the present boundary between Alaska and British Columbia. The northernmost point of the voyage was some distance up Observatory Inlet and the period from July 7 to July 25 was spent between that point and the inlet's mouth. After passing through Skidegate channel, the ship proceeded south and reached Nootka on July 30, remaining until August 3. Under that date Scouler refers to matters already mentioned in this story—the tragedy of the Astorian supply ship *Tonquin,* and the one-time importance attaching to Nootka when Vancouver made his visit:

"Since visiting Nootka sound we have all been very curious to visit the village, & see what vestiges of the English & Spanish settlements remained. Although we received a very kind invitation from Moaquilla[3] to pay him a visit, the fate of the Tonquin which was cut of[f] a few miles to the S. had filled the minds of some on board with fearful apprehensions. Concerning the fate of the Tonquin the Indians were very reserved;

2. Possibly Piper followed Hemenway for he makes much the same statement: "From June until September Scouler spent on a trip to Nootka Sound and return, during which he is said to have visited nearly every harbor along that stretch of land."

3. The name is spelled 'Macuinna" in *The Edinburgh Journal of Science;* and in Menzies' "Journal" Maquinna. So far as I know the same individual is meant.

perhaps they had little to communicate . . . We know nothing authentic concerning the loss of this vessel . . . The Tonquin was the first ship the Americans sent with settlers to the Columbia; the captain after loosing two boat's crews on the bar of that river, whether by accident or on purpose, the stupid ferocity of the man renders it difficult to decide, he was sent on a trading voyage to the islands, where the loss of the ship I have not the smallest doubt was occasioned by his own negligence.

". . . Nootka, which excited so much contention between the courts of Madrid & London, is now completely neglected by every civilised power, & the state of poverty in which they are at present affords little inducement to the visits of mercantile adventurers . . ." (QOHS)

The *Quarterly* refers to ". . . the 3rd of August on which we left Nootka . . ." but under date of August 6 *The Edinburgh Journal of Science* records:

"On the 6th of August we left Nootka, and directed our course for the Straits of Juan de Fuca, concerning whose existence there has been so much discussion."

August 8: "This morning we entered the straits of Juan De Fuca . . ." (QOHS)

On the 9th, continuing down the straits, they anchored in the evening at Port Discovery, situated in what is now Jefferson County, Washington.

"The country, on both sides of the straits, had the most beautiful appearance of any part of the coast we had ever seen. It abounded in beautiful towns, interspersed with trees, and in some of the finest shrubs of American growth . . . we anchored in Port Discovery." (EJS) Surely "towns" is a typographical error—for *downs* perhaps?

Although the ship remained there from August 10 to 13 Scouler remarked that "As none of the boats ventured ashore I had no opportunity of examining the productions of the country." Leaving on the 14th they anchored at evening "in Strawberry Cove, Cypress Island." This lies to the east of Rosario Strait, off present Skagit County, Washington; it is one of the San Juan group of islands.

August 15: "We anchored in Strawberry cove of Vancouver, and, as the country here was uninhabited, employed the time in wandering through the woods and collecting plants . . . I hastened to a saline marsh . . . which afforded some interesting plants; and in a dry situation we found plenty of the yellow-flowered *myosotis* which grows so sparingly on the Columbia." (EJS)

"In the afternoon . . . we had an opportunity of visiting Strawberry Cove . . . We landed on a fine smooth sandy beach, which was bounded on all sides by low & marshy ground, covered with *Scirpi* & *Carices,* & abounding in dear trails. Along the beach we found abundance of *Berberis Aquifolia* & *B. Nervosa.*" (QOHS)

August 16: ". . . we left Cypress Island & before sunset we anchored opposite an Indian village. This tribe . . . is called the Lummie tribe . . ." (QOHS)

Having confidence in these Indians the party went ashore; on the 17th, ". . . As soon as I got ashore my attention was occupied with some interesting plants which grew on the beach; among these plants was a *Solidago* & a beautifull specimen of *Artemisia,* but what pleased me most was the vast profusion of *Myosotis Hookeri.* In an extensive saline marsh I found a sp. of *Salicornia* & a fine *Arenaria.* During my herborising the Indians watched my motions with considerable curiosity . . ." (QOHS)

August 18: "In the afternoon we proceeded farther up the Gulph of Georgia . . ." (QOHS)

On August 18 the *William and Ann* passed north of what is now the boundary between Washington and British Columbia and out of the region of this story. She got as far north as Point Roberts, British Columbia, and on August 26 was back at the location left on the 18th of that month.

August 27: "We went ashore again this afternoon to make a short botanical excursion . . . My botanising was very unfortunate & I was about to go on board without a single plant; on advancing a little farther into the woods, I had the good fortune to find a fine species of *Sanicula.* On examining this plant I found it to differ from any of the sp. of *Sanicula* I was acquainted with. From the down on the lower part of the stem & on the leaves it may be called *S. tomentosa.*" (QOHS)

On August 28 the ship was becalmed off "Tatooche," where they had last been on August 8. The island of Tatoosh lies off present Callam County, Washington, at the south side of the entrance to Juan de Fuca Strait. The Indians hesitated to come aboard for one of their number had been ". . . carried of[f] & sold as a slave by an American ship. We were already aware that such things had been done by an American vessel, but we had not seen any of his victims before. This villain, whose name is Ayres, once entered the Columbia & carried of[f] 12 men, seven . . . escaped . . ." (QOHS)

On September 1 the ship was off the Columbia River and on the 3rd ". . . crossed the bar of the Columbia, and again anchored in Baker's Bay." (EJS)

September 4: ". . . I set out for Fort Vancouver to join my old associate Mr Douglas . . . we passed Cheenook Point . . . the burying ground of Comcomly[4] . . . As it was sunset before we left the ship, we had to travel all night . . ." (EJS)

"This forenoon we left Baker's bay & proceeded to point Ellis . . . After dinner we left the ship & set out for Ft. Vancouver. We landed at Ft. George & found it entirely abandoned by the settlers & taken possession of by the Indians, who were rapidly reducing it to a state of ruin & filth. We left the Fort at 6 o'clock in an Indian canoe . . ." (QOHS)

September 5: ". . . next morning we breakfasted about 20 miles from Ft. George. The

4. According to the *Quarterly* transcript: ". . . here in the space of two years, the unfortunate old man had deposited the remains of 8 individuals of his family . . ."
    The old chief Concomly did not die until 1830. Nor did he then rest undisturbed, for in 1835 Dr. Meredith Gairdner was to steal his skull—in the cause of science!

place where we stopped was a low alluvial island, covered with willows & *Cyperaceæ*, but afforded some curious plants. I found *Solanum Nigrum, Sagittaria sagittifolia, Impatiens, Valeriana spiralis,* & fine nondescript species of *Sisyrhynchium,* with yellow flowers. This plant might be named in honor of Mr. Douglass, who has been so zealously employed in collecting the vegetable productions of the N. W. Coast." (QOHS)

". . . we breakfasted on a small alluvial island, about twenty miles from the ship. At this place I had a few minutes to botanize, and, although the season was far advanced, we found some curious plants. The wapito or Indian potato grew in abundance. *Valisneria* grew in the mud of the river, and I also found a beautiful *Sisyrinchium* with yellow flowers." (EJS)

The date of arrival at Fort Vancouver is not stated in either transcript of the journal. The *Quarterly* notes that ". . . Mr. Douglass had gone up to the cascades, but was expected every day." *The Edinburgh Journal of Science* notes that ". . . Mr Douglas was just returned from a short excursion he had made into the interior." According to Douglas he was back on September 13, finding Scouler there. Not until September 20 does Scouler again date his entries. Earlier we are told:

". . . My stay at Ft. Vancouver was principally employed in making excursions along with Mr. Douglass & in examining our specimens. I, however, collected very few plants, as the weather had been exceedingly dry and most of the summer flowers had dissapeared & the autumnal ones were by no means numerous." (QOHS)

"On arriving at Fort Vancouver we were kindly received . . . While I had botanized along the coast, from the Columbia in lat. 45 N. to Queen Charlotte's island in 52, Mr Douglas had taken a most extensive range in a different direction, and through a very different country, so that our respective herbariums contained entirely different sets of plants. His first excursion[5] had been up the Multuoma or Wilhamut river, which takes a southerly course from the Columbia. He followed this river for about fifty miles in a country abounding in salt springs, and where the Indians cultivate the *Nicotiana rustica.* His next journey[6] was to the falls of the Columbia, where he found a rich country for the botanist . . . During my stay at the fort, our time past pleasantly away in making excursions, arranging our collections, and procuring specimens of the animals of the country." (EJS)

September 20: "I left Fort Vancouver to return to the ship, which was soon to sail . . ." (EJS)

". . . I took my final leave of Fort Vancouver . . . Our voyage down the river was very uncomfortable as it rained almost incessantly & the wind was very unfavourable. At Mount Coffin I availed myself of the opportunity of examining the mode of in-

5. Douglas' own records show that he had made his trip to the Multnomah, or Willamette, River from August 19 to 30, 1825.

6. Douglas' records for this journey were September 1 to 13, 1825. It was doubtless the trip "to the cascades" mentioned in the *Quarterly*.

terment, & to procure a specimen of their [the Indians'] compressed skulls ..."
(QOHS)

September 22: "... we breakfasted at the Kowlitch village ..." (QOHS)

Under date of November 10 the *Quarterly* transcript mentions "... leaving the
Columbia on the 25[7] October ..."

The Scouler journal in *The Edinburgh Journal of Science* ends on September 20
with the sentence:

"The stormy weather ... continued without interruption till we left the Columbia,
where we exchanged its dense fogs and constant showers for a brighter sky and a more
pleasing climate."

In 1827 the same *Journal* (6: 251-253) published an article "On the temperature
of the north west coast of America. By Mr Scouler. Communicated by the author."
But since the tables of temperature refer to years antedating Scouler's visit the data
must have been collected by someone else.

The *Quarterly* transcript contains entries relating to the return journey to England
and ends on March 30, 1826:

"To-day we spoke a small French vessel bound for the Great Bank of Newfoundland. We visited them & had the satisfaction of hearing that all was well at home, &
obtained a supply of potatoes."

Appended are a few comments on the climate of the Columbia River district and a
table recording the rainfall on the northwest coast for 1822 and 1823, 1823 and 1824,
and 1824 and 1825.

In quoting from the transcripts of the Scouler journal I have confined my citations
mainly to his routes and to the plants which he observed. In a "Notice respecting Mr
Scouler's and Mr Douglas's recent voyage to the north-west coast of America," published in *The Edinburgh Journal of Science* (1826), the anonymous author comments
that Scouler's "... attention was not wholly occupied by the botany and zoology of the
country; he lost no opportunity of acquiring as complete a knowledge, as the nature of
the circumstances would allow, of the manners and customs of the Indians; add to
which he collected many articles of curiosity, such as the dresses, arms, domestic utensils, skulls of the natives, and a well-preserved mummy."

As an example of Scouler's interest in zoology I shall quote a few of his comments
upon the albatross. His work was not confined to the land; even at sea he collected fish,
sea birds and much else. He seems to have had no superstition regarding the bird's
association with good weather.[8] After rounding Cape Horn, Scouler wrote on November 8, 1824: "... we succeeded in catching 20 of them ... all of dusky black
colour, & belonged to the species *D[iomedea] fulginosa* ... They were very large and

7. The Douglas records state that he had hastened down the Columbia River to see his old shipmates, but
had arrived (October 24) just one hour after the boat had sailed.

8. The "Rhime of the Ancient Mariner" was published in 1837.

one of them measured seven [feet] from wing to wing. The physiognomy of these animals is very curious. Their flat forehead & large eyes & very convex *corneæ,* make them resemble an owl, & renders it probable that they seek their prey by night . . ." (QOHS)

Their anatomy, the content of their stomachs, etc., etc., are described. On December 5 a different species was obtained:

"Mr. Douglass caught three albatross . . . very different from those we had procured further south. Their plumage was more light . . . & their bill was of a milk white colour. These birds were very large, one of them weighed 18 pounds, & measured 12 feet between the extremities of the wings. They appeared to be *D[iomedea] exulans* . . ." (QOHS)

On February 17, 1825, he acquired four more of the same species and reached conclusions at variance with those of Cuvier:

". . . It may be proper to notice some mistakes into which Cuvier has fallen . . . with respect to this bird. He says: Ils habitent tous les mers Australes, vivent de froi de poissou de mollusques . . . He also mentions the *D. exulans* as being a great enemy to the flying fish. The first of these mistakes, that the *Diomedea* is entirely an antar[c]tic bird, we have now had abundant means of rectifying & saw the bird in equal abundance in 40 degrees north latitude as we did of[f] Cape Horn . . . not a single individual has ever been found in the North Atlantic. M. Cuvier also represents the *D. exulans* as living very much on flying fish. During all our voyage we never saw an albatross within many degrees of the region of flying fish. I have noticed these errors because they are the only ones I ever could detect in the Régne Animal of this distinguished naturalist, & who is undoubtedly better acquainted with the structure of organized bodies than any man in Europe." (QOHS)

When, in 1828, in his *Botanical Miscellany,* W. J. Hooker published the new genus *Scouleria*—with type species *S. aquatica*—he based it upon a collection made by Scouler as well as upon specimens gathered by Thomas Drummond in the Rocky Mountains and on the Columbia River:

"It gives me great pleasure . . . to find a plant belonging to his favourite tribe, the *Musci,* which, constituting a new genus, I . . . dedicate to him. During three years that Mr. Scouler attended the College course of botanical lectures, I witnessed . . . his increasing love for natural history; and although the anatomy and physiology of animals be his most favourite pursuit, and the one by which it is to be expected that he will . . . rise to much fame,—yet botany has occupied a great share of his attention, and his herbarium includes much of novelty from the countries which, like a second Menzies, he has visited, and as a naturalist successfully investigated."

Scouler had found the plant on July 1, 1826, while at or near Salmon Cove in Observatory Inlet, present British Columbia, and had recorded in his journal:

"The little valley around Salmon Cove has a beautiful verdant appearance, and a small brook supplies it with abundance of excellent water. The channel of this stream is everywhere covered with aquatic mosses, particularly *Fontinalis antipyretica* and *F. squamosa,* and among them I found one of the rarest and most beautiful of the musci of America, which, from the remarkable structure of its capsule and operculum, will doubtless form a new genus." (EJS)

A footnote adds: "*Scouleria,* Hooker's MSS."

Among Pacific coast species named for Scouler, Piper's "Flora of the state of Washington" cites a St. Johnswort (*Hypericum Scouleri* Hooker), a mountain harebell (*Campanula Scouleri* Hooker), a "wild foxglove" (*Pentstemon Scouleri* Lindley) and a catchfly (*Silene Scouleri* Hooker); and there are others.

It is said that a mineral, *Scoulerite,* was named in his honor.

Asa Gray noted in 1841 that, in Hooker's herbarium in Glasgow—which ". . . comprises the richest collection of North American plants in Europe . . ."—are to be found "nearly complete sets" of the plants collected by Franklin, Drummond, David Douglas, "as well as those of Professor Scouler, Mr. Tolmie, Dr. Gairdner, and numerous officers of the Hudson's Bay Company . . ."; that Sir William Jackson Hooker took some of Scouler's collections to Kew when he became director of that institution is probable. Britten and Boulger (1931) refer to Scouler's plants ". . . at Kew and Dublin . . ." Piper (1906) states that "The best set is in the British Museum."

According to the *Dictionary of national biography*—upon which most references to Scouler's personal life seem to have been based—he was born at Glasgow on December 31, 1804, and received his rudimentary education at Kilbarchan,[9] in County Renfrew.

After studying medicine for a time at the University of Glasgow, he went to Paris where he studied at the Jardin des Plantes. He then went on the trip to North America which has been recorded in this chapter; subsequently on another to India, and so on.

Returning to Glasgow, he received his medical degree in 1827 and afterwards practised as a physician. In 1829 he was appointed ". . . 'professor of geology and natural history and mineralogy' in the Andersonian University . . . In 1834 [he was] appointed professor of mineralogy, and subsequently of geology, zoology, and botany, to the Royal Dublin Society, a post he held till his retirement on a pension in 1854, when he returned to Glasgow."

Scouler died in Dublin on November 13, 1871, and was buried at Kilbarchan. We are told that "He bequeathed his books, which included many of great rarity, to Stirling's Library, Glasgow." He was joint-editor of a medical and of a scientific journal and published "upwards of twenty papers . . . between 1826 and 1852."

9. Kilbarchan, like Paisley about five miles away, was a manufacturing center for the famous Paisley shawls.

Anonymous. 1826. Notice concerning Mr Scouler's and Mr Douglas's recent voyage to the north-west coast of America. *Edinburgh Jour. Sci.* 5: 378-380.

BRITTEN, JAMES & BOULGER, GEORGE SIMONDS. 1931. A biographical index of deceased British and Irish botanists. ed. 2. London. 271.

GRAY, ASA. 1840. Notices of European herbaria, particularly those most interesting to the North American botanist. *Am. Jour. Sci. Arts* 40: 1-18. *Reprinted:* Sargent, Charles Sprague, *compiler.* 1889. Scientific papers of Asa Gray. 2 vols. Boston. New York. 2: 1-21.

HEMENWAY, ANSEL FRANCIS. 1904. Botanists of the Oregon country. *Quart. Oregon Hist. Soc.* 5: 207-214.

HOOKER, WILLIAM JACKSON. 1828. Scouleria aquatica. *Bot. Miscel.* 1: 33-35, pl. 18.

MENZIES, ARCHIBALD. 1923. Menzies' journal of Vancouver's voyage. April to October, 1792. Edited, with botanical and ethnological notes, by C[harles] F[rederick] Newcombe, M.D. and a biographical note by J[ohn] Forsyth. *Memoir V. Archives British Columbia.* Victoria, B. C.

PIPER, CHARLES VANCOUVER. 1906. Flora of the state of Washington. *Contr. U. S. Nat. Herb.* 11: 13.

SARGENT, CHARLES SPRAGUE. 1896. The Silva of North America. Boston. New York. 9: 66, *fn. 5.*

SCOULER, JOHN. 1826, 1827. Account of a voyage to Madeira, Brazil, Juan Fernandez, and Gallapagos Islands, performed in 1824 and 1825, with a view of examining their natural history, &c. By Mr Scouler. Communicated by the author. *Edinburgh Jour. Sci.* 5: 195-214; 6: 51-73, 228-236.

―――― 1905. Dr. John Scouler's journal of a voyage to N. W. America. [Edited by Frederick George Young.] *Quart. Oregon Hist. Soc.* 6: 54-75, 159-205, 276-287.

WOODWARD, B. B. 1897. John Scouler. *Dict. Nat. Biog.* 51: 122, 123.

YOUNG, FREDERICK GEORGE. 1904. Literary remains of David Douglas, botanist of the Oregon country. *Quart. Oregon Hist. Soc.* 5: 215-222. [Editorial comments prefacing the reprint of W. J. Hooker's A brief memoir of the life of Mr. David Douglas . . .]

CHAPTER XIV

DOUGLAS MAKES HIS FIRST VISIT TO THE NORTH-
WEST COAST OF NORTH AMERICA AND THEN SPENDS
TWO YEARS IN ENGLAND

DESPITE the fact that, as a collector and introducer of plants, David Douglas has had few rivals, a biography of the man has yet to appear.[1] The basic facts of the Douglas story are to be found in his own writings, his journals and his letters, while the tangible evidence of his success exists in the many important plants which through his discrimination and extraordinary effort, are to be found in living collections. His tragic death is probably the most widely publicized event of the man's short but fruitful career.

I shall not dwell long on Douglas' life before his collecting days began[2]—the oft-repeated stories of his fondness for everything related to natural history, while doubtless true, have acquired in the retelling a tinge of *de post facto*. But undoubtedly, from the interest taken in his advancement by those who came in contact with the boy and young man, he must have demonstrated special abilities where plants were concerned as well as a desire to learn. There is a reference to his excellent memory.

David Douglas was born at Scone,[3] near Perth, Scotland, the son of a stone mason John Douglas and of, writes Hooker, "Jean Drummond, his wife." The date of birth is usually cited as 1799.[4] He attended the village and parish schools and, at the age of ten or eleven, was employed in the nursery-gardens of the Earl of Mansfield at Scone. He was then promoted to the "ornamental department" or flower-gardens, where "Mr. M'Gillvray, a young man who had received a tolerable education," took an interest in his instruction. After a further period in the "forcing and kitchen garden," he moved to the estate of Sir Robert Preston near Culross, "a place then celebrated for a very select collection of plants." Here he was permitted access to the owner's botanical li-

1. Since this chapter was written, A. G. Harvey published in 1947 his *Douglas of the fir. A biography of David Douglas botanist*. I include references to this in footnotes, citing Harvey (1947). The same author's earlier short article on Douglas I distinguish as Harvey (1940).

2. Harvey (1947) devotes his first chapter to Douglas' early life in Scotland, tells of his family connections, his education, and his interests. The reader will find it more enlightening than other publications covering the same period.

3. "Scone, pronounced by the natives Skoon, was the ancient capital of Pictavia, now Scotland, as early as 710, and the coronation place of Scottish kings from 1153 till 1488 . . ." (Wilson). It was the theft of the coronation "Stone of Scone" that in the 1950's caused such excitement in Great Britain.

4. Loudon and others cite 1798. Harvey (1947) states that 1799 is correct.

brary. Next he was admitted to the Botanical Garden at Glasgow whose curator, Stewart Murray, took an interest in the lad's education and he was enabled to attend botanical lectures given by Dr. William Jackson Hooker at the University, and to make botanical excursions into the "Highlands and islands of Scotland." At this point Hooker's influence upon Douglas' life seems to have become a direct one for he writes:

"It was our privilege, and that of Mr. Murray, to recommend Mr. Douglas to Joseph Sabine, Esq., then Honorary Secretary of the Horticultural Society, as a Botanical Collector; and to London he directed his course accordingly, in the spring of 1823 . . . he was despatched, in the latter end of May, to the United States . . ."

The miscellaneous literature concerned with Douglas' journeys and accomplishments is abundant but also repetitious, in the sense that—except for occasional publications which supply corroborative or explanatory facts—it turns for its substance to two published memoirs which, between them, contain transcripts of all Douglas journals known to be extant, as well as transcripts of most of the letters which he is known to have written from the field; and Douglas' journals and letters are, and promise ever to be, the foundation of his story.

The first of these memoirs, published promptly after Douglas' death, represented the work of William Jackson Hooker; the second, which appeared almost eighty years later, was published by the Royal Horticultural Society of London.

Neither record—according to the statements of its editor—furnishes a verbatim transcript of the Douglas manuscripts: Hooker made corrections and deletions and added matter to the originals; and the Royal Horticultural Society volume included changes and additions when the editor considered these to be advisable. Douglas is known to have made copies of his journals and of his letters and, since these vary in wording and to some extent in content, one must turn from one version to another of his published writings[5] in order to learn, let us say, his whereabouts at a particular time, what he found on a particular occasion, and much else.

Since these memoirs represent, as noted, the major source of information upon Douglas, extracts both long and short have been transcribed therefrom.

If one is to follow Douglas' life with any degree of understanding, I believe that it is essential to know what is contained in, as well as absent from, these two important records and I shall, therefore, discuss them at length before turning to the man's travels. First, the invaluable Hooker "Memoir."

A little over two years after David Douglas' death, which occurred on July 12, 1834, Sir William Jackson Hooker, in the *Companion to the Botanical Magazine,* began the publication of a paper which included one, and extracts from a second, Douglas journal, as well as a number of letters which he wrote from the field. Entitled "A

5. Both journals and letters would seem to warrant publication in parallel columns. Such a presentation I believe would be valuable from the point of view of botany, of horticulture and of regional history alike.

brief memoir of the life of David Douglas, with extracts from his letters," [6] publication began in October, 1836, and continued through January, 1837, despite the fact that volume II of the periodical in which the entire paper was printed is dated 1836.[7] Whether the "Memoir" still exists, or ever existed, in manuscript form I do not know.

The "Memoir" opens with a biographical sketch of Douglas up to the year 1824 (pp. 79-82); this is introductory and bears no specific heading. It was largely based— insofar as it antedates Douglas' connection with the Horticultural Society of London— upon information which Hooker had received from William Beattie Booth,[8] ". . . townsman and intimate friend . . ." of Douglas:

"It is to Mr. Booth . . . that I am indebted for almost all that relates to the subject of this memoir, previous to his entering the service of the Horticultural Society, and for the copies of some letters, as well as several particulars relative to his future career."

Next follows "A sketch of a journey to the north-western parts of the continent of North America, during the years 1824, 5, 6, and 7. By David Douglas, F.L.S." (pp. 82-140). This journal which describes Douglas' first journey to the Pacific northwest —but his second journey to North America—was the first to be published and is prefaced by Hooker thus:

". . . we should have known little or nothing of his adventures, were it not for a Journal which he kept with great care . . . and which has been deposited in the library of the Horticultural Society of London. From that Journal is here selected whatever is likely to prove interesting to our readers . . ."

Interspersed through the journal are ". . . some occasional observations and extracts from the few letters that were received by his friends during the continuance of this mission . . ."

The late F. R. S. Balfour[9] offers a possible explanation for the fact that this journal

6. That Hooker considered this the title of his entire paper may be deduced from the important type in which it is printed; all other sectional headings are in type of lesser importance.

7. T. A. Sprague, writing in the *Bulletin of miscellaneous information* of the Royal Botanic Gardens, Kew (362-364, 1933), makes clear that the content of the *Companion to the Botanical Magazine* was issued in a series of numbers, those in which Hooker's paper appeared being: no. 15, 79-96, October 1, 1836; no. 16, 97-128 (with portrait of Douglas), November 1, 1836; no. 17, 129-160, December 1, 1836; and no. 18, 161-192, January 1, 1837.

8. A short biographical account of Douglas written by Booth was published in Loudon's *Gardener's Magazine* for May, 1835, but is said to have appeared earlier in the *West Briton and Cornwall Advertiser*. Harvey (1947) cites a Booth "Memorandum on David Douglas" among manuscripts at the Royal Botanic Gardens, Kew.

9. Frederick R. S. Balfour, who wrote a short biographical account of Douglas and who eventually fell heir to some of his possessions, manuscripts, etc., is mentioned more than once in this chapter as well as in my chapter on Menzies. Although he lived in Scotland and in England, some of his early years were spent in America and it is my belief that he was more familiar with the trees of this country than are most native Americans.

Before his death in 1945, he was owner of the beautiful estate of "Dawyck" at Stobo, Peebleshire, Scotland, which, from "the 13th to the end of the 17th century," had been the property of the Veitch family

differs in wording and to some extent in content from two other Douglas journals published at a later date but covering the same period:

"It seems to me probable that the one published by Hooker must be a revision made by Douglas after his return home and was possibly to be the foundation of a book of his travels which John Murray had invited him with a 'handsome offer' to write after his return in 1827. He began preparing this for the press but it was never completed . . . That Douglas made copies of his Journals is certain . . ." [1]

Following the journal is more biographical material, mainly concerned with the years (1827–1829) which Douglas spent in England[2] between his first and second visits to the west coast of North America (pp. 140-146). Incorporated in Hooker's biographical account is a "List of plants, introduced by Mr. Douglas, in 1826-7" (pp. 140-142); it is a mere list of names, one hundred and sixty-four in all, and—needless to say—the botanists have made revisions since 1836.

Next the "Memoir" publishes the "Account of Mr. Douglas' second visit to the Columbia; his excursions in California; and his visit to Mouna Roa in the Sandwich Islands; with particulars respecting his death" (pp. 146-182). This is a covering title for Douglas' activities from 1829 to 1834.

It opens with some introductory biographical facts (p. 146).

Next follow a number of Douglas letters (pp. 146-161). They are important since, with a few others published elsewhere, they contain the only firsthand records of Douglas' work from the time of his departure from England in October, 1829, to his departure from the Columbia River for the Hawaiian Islands in October, 1833.

Next, under the heading "Mr. Douglas' voyage from the Columbia to the Sandwich

of horticultural fame. There Balfour grew many trees, shrubs and herbaceous plants from foreign lands; many were from the northwest coast of North America where he had lived for a number of years because of business interests and where he made later visits. W. Balfour Gurley has written an account of his cousin's work in arboriculture and in horticulture as well as in other branches of natural history (*Proc. Linn. Soc. London*, 1944–1945, 63-65) as well as the story of "Dawyck" (*Jour. Royal Hort. Soc. London* 72: 5-9, figs. 5-13, 1947); one figure shows a fine specimen of *Picea Breweriana* S. Watson, a species which Rehder's *Manual of cultivated trees and shrubs* (1940) refers to as ". . . rare in cult[ivation]."

Alastair Norman Balfour inherited "Dawyck" and, I am assured, his father's interest in its plants.

1. The basis for considering that the journal published by Hooker was based upon a "revision" of Douglas' field-journal is not clear: it evidently discounts Hooker's statement quoted above that the ". . . Journal which he [Douglas] kept with great care . . ." had constituted his source.

Harvey (1947) also refers to the journal included by Hooker as composed of ". . . extensive selections from what is believed to be a revision of his journal which Douglas prepared for Murray . . ."

Hooker comments upon Douglas' attempts to publish his own journal:

"As some further compensation for his meritorious services, the Council of the Horticultural Society agreed to grant him the profits from what might accrue from the publication of the Journal of his Travels, in the preparation of which for the press, he was offered the assistance of Mr. Sabine and Dr. Lindley; and Mr. Murray of Albermarle-street was consulted on the subject. But this proffered kindness was rejected by Mr. Douglas, and he had thoughts of preparing the Journal entirely himself. He was, however, but little suited for the undertaking, and accordingly, although he laboured at it during the time he remained in England, we regret to say, he never completed it."

2. These years in England are discussed at the end of this chapter.

Islands, and the ascent of Mouna Roa," we are given "extracts" from a Douglas journal (pp. 161-177) covering the short period between his departure from the Columbia on October 18, 1833, and January 30, 1834 (the last cited date). It is prefaced by Hooker thus:

"As already mentioned, the only Journal of Mr. Douglas' Second Expedition, which has reached this country, is that commencing with his departure from the Columbia, including the voyage to the Sandwich Islands, and the ascent of Mouna Roa. From this, with the loan of which we have been favoured by its possessor, Mr. John Douglas, we make the following extracts . . ."

We are then told how news of Douglas' death reached England (pp. 177-178). And the details of his last days are supplied in two letters (pp. 178-182): the first is a "Copy of a letter from the missionaries of Hawaii[3] to Richard Charlton, Esq., His Brittanic Majesty's Consul at the Sandwich Islands," and the second a "Copy of a letter from Mr. Charlton to James Bandinel Esq. (Inclosing the above.)." Hooker ends the "Memoir" with a reference to the fate—it would seem to have been a happy one—of Douglas' little Scottie and mentions certain shipments of birds, seeds and roots (p. 182).

From the above it is evident that, for Douglas' last journey to North America, the only journal published by Hooker—and still the only journal of the period ever published—is a fragment covering the last few months of the botanist's life. There seem to have been two reasons for the absence of such records—the first an accident in which he lost most of his possessions, the second his break with the Horticultural Society.

As to the first: Douglas had kept a journal from the time he left England in the autumn of 1829 until June 13, 1833. His last letter to Hooker (dated "Woahoo," May 6, 1834) relates that, on the morning of June 13, ". . . at the Stony Islands of Fraser's River . . . my canoe was dashed to atoms,[4] when I lost every article in my possession, saving an astronomical journal, book of rough notes, charts, and barometrical observations, with my instruments. My botanical notes are gone, and, what gives me most concern, my journal of occurrences also,[5] as this can never be replaced, even by myself. All the articles needful for pursuing my journey were destroyed, so that my voyage for this season was frustrated. I cannot detail to you the labour and anxiety this occasioned me, both in body and mind, to say nothing of the hardships and sufferings I endured . . . The collection of plants consisted of about four hundred species—two

3. Joseph Goodrich and John Diell.

4. An account of this accident, as well as a few dates relating to Douglas' itinerary, were supplied to Hooker in a letter written him by "Archibald M'Donald" from "Edinburgh, 20th January, 1833." This important letter is included by Hooker in the "Memoir" (pp. 159, 160, *fn.*).

5. Since Douglas is said to have merely numbered his specimens and to have kept in his field-books under similar numbers the records of locality and so on, this was indeed a serious loss. In earlier years he had made copies of such records and sent them back to the Horticultural Society from time to time, but on this journey he seems not to have taken this precaution.

hundred and fifty of these were mosses, and a few of them new. This disastrous occurrence has much broken my strength and spirits . . ."

Despite Douglas' reference to his broken health and spirit, his account of the episode ends with the comment: "After this misfortune, in June, I endeavoured, as far as possible, to repair my losses and set to work again . . ."

Nevertheless there is no evidence that he resumed the keeping of a journal during the four-month period dating from the accident to his departure from the Columbia River for the Hawaiian Islands on October 18, 1833, when the journal "extracts" published by Hooker begin.

The second reason for the absence of a journal—or for the fact that Douglas did not send copies of the records which he kept before the accident back to England on this journey—may perhaps be explained by the fact that, ten months before the episode, or in September, 1832, he had severed his connection with the Horticultural Society. He was in the Sandwich Islands when he did so.[6] The Hooker "Memoir" has this to say of Douglas' resignation:

"The expenses of this mission were, in great part, to have been defrayed by the Horticultural Society of London of which Mr. Sabine was still Secretary; but when those changes took place in the Institution[7] . . . in consequence of which Mr. Bentham became the Honorary Secretary in the room of Mr. Sabine, Mr. Douglas wrote from the Columbia, resigning his appointment of Collector to the Society, and he withdrew altogether from its service, sending it, however, at the same time, all the collections he had made up to that period, but declaring his intention, nevertheless, to transmit all seeds and living plants he might procure, as a present to the Garden. This determination, which arose from some misunderstanding, is deeply to be regretted, not only because we know, from our acquaintance with Mr. Bentham's character and feelings on the subject, that this gentleman would have exerted himself to the uttermost to further Mr. Douglas' success: but because to this circumstance may perhaps be attributed the loss of nearly the whole of his Journals. To that Society, during the former expedition, they were from time to time carefully despatched; but now there was no one to whom he was bound to communicate the result of his investigations and labours: and with the remnant of his collections, sent home after his death, no Journal has appeared, save that of his *Voyage from the Columbia to the Sandwich Islands* and the *Ascent of Mouna Roa*." (HM, 146)

6. The break is recorded by George Bentham in the Society's *Transactions* (ser. 2, 2: 375, 1842). Bentham had, by then, replaced Joseph Sabine as secretary of the Society.

". . . on his arrival at the Sandwich Islands on his way back to the Columbia, he received intelligence that his friend, Mr. Sabine, had resigned the office of Secretary of the Horticultural Society in consequence of which he was induced by some misconception to resign his own appointment of collector . . ."

7. "A Report to the Council" on May 1, 1840 (*Trans. Hort. Soc. London* ser. 2, 2: 375, 1842), states: ". . . a material change was made in the management of the Horticultural Society, after the Report of the Committee appointed on the 2nd of February, 1830 . . ."

Hooker does not enlighten his readers upon the reason for the change in secre-
taries, nor as to the basis of the "misunderstanding" motivating Douglas' resignation.[8]
The Horticultural Society was in dire financial difficulties at the time and there was
considerable dissension between the staff and the members as to its management.[9]
Douglas' activities must certainly have been financially curtailed by his resignation although it would be impossible to suspect him of having anything mercenary in his
makeup—when in the wilds he may be said to have lived "on a shoestring." But the
break must have left him with inadequate means to carry on[1] his work indefinitely.
Whatever the whys and wherefores of the episode, the consequences of the break were
regrettable.

So much for the first publication upon David Douglas. In so far certainly as
Hooker's "Memoir"[2] is concerned, the *Companion to the Botanical Magazine* is inade-
quately indexed.

As to the second memoir concerned with Douglas and his work:

In 1914, or eighty years after Douglas' death, the Royal Horticultural Society of
London published three of his hitherto unpublished journals and reprinted the "extracts" from the Hawaiian journal included in the Hooker "Memoir" with the exception of the opening lines of its first paragraph which are important since they tell, in the
first person, of Douglas' departure from the Columbia and supply a brief record of
his second stay in California.

The preface to the volume, written by the Society's secretary, W. Wilks, explains
the long delay in publication: the handwriting was "nowhere easy" to read, in places
"most difficult," occasionally "quite impossible"; the ink had faded, which added to
the difficulties, and ". . . only occasional spare time could be given to a work which

8. Harvey (1947) seems to have interpreted it largely on a loyalty-to-Sabine basis.

9. *The Gardener's Magazine* for 1830 has considerable to tell of the situation and some of it, in retrospect,
has a humorous side. The Society was, of course, reorganized and put on a better financial basis. Harvey
(1947) reports at some length on the subject.

1. Bancroft's *History of the northwest coast* (2: 509, *fn.* 3) supplies a record showing that the Horticultural Society took no financial risks:
    "Had not Douglas been recalled, or his supplies cut off, it is doubtful if ever he would have left his
fascinating forests. A letter from Alexander Seaton, Esq., treasurer of the Horticultural Society, to William Smith, Esq., informs us that 'David Douglas has ceased to be in the service of the society, and that the
society will not repay any further advances made to him.' *Douglas' Private Papers*, MS, 1st ser. 75."

2. In 1904 and 1905 *The Quarterly of the Oregon Historical Society* reprinted the Hooker "Memoir" in its
entirety, with interesting commentaries. The editor, F. G. Young, mentions in his prefatory chapter:
"Probably not a copy of the work containing the Douglas narrative is to be found in Oregon."
    Portions of the "Memoir" had, however, been published earlier—notably in the *Hawaiian Spectator*
(1838) and in the account of Douglas written by Barnston (1835).
    Later, as we shall see in discussing the volume issued in 1914 by the Royal Horticultural Society, this
reprinted practically all the Hawaiian journal which had appeared in the "Memoir" as well as the letters
written by the Hawaiian missionaries and by Charlton describing Douglas' death.

had already waited nearly one hundred years for publication . . ." Further, when one journal had been prepared for the press, a second had come to light.

The content of the volume is described thus on its title-page: *Journal kept by David Douglas during his travels in North America 1823–1827 Together with a particular description of thirty-three species of American oaks and eighteen species of pinus With appendices containing a list of the plants introduced by Douglas and an account of his death in 1834.*

Following a short preface and the table of contents is the "Journal kept by David Douglas during his travels in North America on behalf of the Royal Horticultural Society,[3] 1823 and following years" (pp. 1-30). This describes Douglas' first trip to North America which extended from June 3, 1823 (departure from London for Liverpool), to January 10, 1824 (return to London)—he had arrived in New York on August 3, 1823, and had sailed for England on December 12 of the same year. Since the trip did not take Douglas west of the Mississippi, it has no bearing on my story. His travels were mainly confined to our Middle Atlantic States although, by boat, he coasted the southern shore of Lake Erie as far west as the Detroit River, from there visiting Amherstburg in southern Ontario, Canada, whence he made a trip into the state of Michigan. He met many persons of importance in botanical and horticultural circles and those who are interested in the region and period should find this journal of value.[4]

Next follows "Some account of the American oaks, particularly of such species as were met with during a journey from   *   –   *   , in the year 1823, by David Douglas, in a letter to the Secretary, Joseph Sabine, F.R.S." (pp. 31-49). This is prefaced by a note: "A number of side-notes, chiefly from Pursh or Michaux, have been added to the MS. in Douglas' handwriting. These have been incorporated in the text. —Ed." Most of the paper relates to species indigenous to the eastern United States. However, *Quercus agrifolia* Née ("Found upon the north-west coast of America, about Nootka Sound . . .") and *Q. Garryana* Hooker ("Common . . . on the low banks of the Columbia . . .") must certainly have been later additions to the paper—indeed the editor comments to that effect in a footnote to the last-named.

Next follow two journals, both of which describe Douglas' first trip to the Pacific northwest in 1824–1827.

The first of these journals (pp. 51-76) covers the same period and bears the same title ("Sketch of a journey . . .") as the journal published by Hooker. But, although the opening paragraph in both is the same, the contents are not identical. The editor, Wilks, explains:

"The following pages . . . consist of a condensed account in Douglas' own hand-

3. Only in 1866 did the Horticultural Society change its name to Royal Horticultural Society.

4. Harvey (1947) describes the journey in his chapter 3 (pp. 22-36).

writing of the journeys which are afterwards expanded and recounted in detail. It has been thought better to reproduce it exactly as it stands, as it is in many respects easier reading than the expanded account, and also because it contains several not uninteresting little additions . . ."

This is, apparently, the "Diary" which is referred to in the Wilks preface:

"After the Diary of his journey to North Western America had been prepared for the press and set up in type, a second manuscript was discovered which was at first sight taken to be a duplicate, but which on closer examination was found to contain a great deal of additional information. It had therefore to be compared word for word with the Diary and the additions inserted in their proper places."

The second of the journals (pp. 77-293)—the "expanded account" mentioned by Wilks—is entitled "Journal of an expedition to North-West America; being the second journey undertaken by David Douglas, on behalf of the Horticultural Society." Since, as the editor mentions above, additions had been inserted in their proper places, it is not a verbatim copy of the original.

There have been published, therefore, three journals (the one in the "Memoir" and the two in the volume under discussion) each different, but all describing the journey of 1824–1827.[5]

Next follow eight Appendices (pp. 295-349).

Appendix I, by "W. W[ilks]," is a "Memoir of David Douglas" (pp. 295-297). It supplies little that is not found in the Hooker "Memoir." It does, however, mention some Douglas manuscripts:

"In the 'Proceedings of the Royal Society' under date April 27, 1837, it is recorded that Mr. Sabine[6] received from Douglas several volumes of lunar, chronometrical, magnetical, meteorological and geographical observations,[7] together with a volume of field sketches.[8] The geographical observations of latitude and longitude refer to two distinct tracts of country: first, the Columbia River, and its tributaries, and the district to the westward of them; and secondly, California. Douglas very judiciously selected

5. Chapters 4 through 12 (pp. 37-143) of Harvey (1947) relate to this journey.

6. Edward Sabine, brother of Joseph Sabine.

7. These were undoubtedly the ". . . astronomical journal . . . charts, and barometrical observations . . ." which were saved in the Fraser River accident as mentioned in Douglas' letter to Hooker of May 6, 1834, as well as in another to W. E. Hartnell of November 11, 1833, which was only published in 1923.

Some of these records may form part of the manuscript mentioned by Harvey (1947) as in the Royal Society, London: "Sabine, Edward. 'Observations taken on the Western Coast of North America by the late Mr. David Douglas, with a report on his Papers.' 1837."

8. The "volume of field sketches" may be the one which Harvey (1940) mentions thus:

"Douglas also undertook the making of field-sketches along the route and did a continuous series of the country travelled from Fort Okanagan to the Quesnel River, showing the natural configuration, with notes on trees, vegetation, and soil. I succeeded in finding Douglas' note-book containing these sketches in England in 1938, and it has since been deposited in the British Columbia Archives."

the junctions of rivers, and other well-characterised natural points, as stations for geographical determination . . .

"What may have become of these interesting volumes I have not been able to trace, the only books in the Royal Horticultural Society's possession being the journals reproduced in the present volume. . ."

Appendix II(pp. 298-317) reprints from the *Companion to the Botanical Magazine,* under the title "Extracts from David Douglas's journal of a subsequent expedition in 1833-4. Which journal was sent to his brother Mr. John Douglas," the "extracts" from the Hawaiian journal published by Hooker in the "Memoir," but omitting, as I have stated, the first and important portion of the opening paragraph.

Appendix III (pp. 318-323) reprints the letters from the missionaries and from Bandinel which are published in the "Memoir."

Appendix IV (p. 324) is merely an extract from a California newspaper concerning a monument to the memory of David Douglas.

Appendix V (p. 325)—"List of papers written by David Douglas"—enumerates eight of his publications, of which numbers 1, 4, 5 and 7 relate to plants; number 8 to the volcanoes in the Sandwich Islands; number 2 concerns a vulture and number 3 two "undescribed" species of mammals.

Appendix VI (pp. 326-336)—"Plants introduced by David Douglas during the years 1826-34"[9]—represents the work of Dr. H. R. Hutchinson. Wilks, referring to the plants mentioned in the Douglas journals, notes that "All the botanical names have been very carefully looked up by Mr. H. R. Hutchinson and the name given to the plant in *Index Kewensis* or by other later authority is quoted at the bottom of the page with a reference to the author responsible for it." He also expresses his indebtedness to Hutchinson for ". . . the identification of the plants mentioned, which latter work adds enormously to the value of the publication and to the ease with which it may be consulted by Botanists and Horticulturists." Hutchinson's careful (and undoubtedly laborious and time-consuming) paper has, like the list in Hooker's "Memoir," undergone some revisions[1] since 1914.

9. Writing a year after the publication of the Royal Horticultural Society volume, the British forester William Somerville, comments upon the list of plants supplied in Appendix VI:

". . . it contains the names of many forest trees of the first rank. First and foremost . . . *Pseudotsuga Douglasii* . . . In the list we also find *Abies amabilis, A. grandis, A. nobilis, Pinus insignis, P. Lambertiana, P. Coulteri, P. monticola, P. ponderosa, P. Sabiniana,* and *Picea sitchensis.* Amongst wood-leaved [*sic;* ? broad-leaved] trees we have *Acer circinatum* and *Acer macrop[h]yllum,* while of shrubs and herbaceous plants there are 20 species and varieties of *Ribes,* 25 of *Lupinus,* 18 of *Pentstemon,* 13 of *Œnothera."*

1. Oakes Ames, who more than once urged that I should bring up to date the plant names mentioned in the early narratives, wrote me (July 20, 1947) that he had changed his opinion; referring to the Royal Horticultural Society volume, he notes that ". . . H. R. Hutchinson attempted to run down to their nomenclatorial counterparts, all of the plant names used by Douglas in the years 1823 to 1827. You will remember that I urged you to modernize the nomenclature . . . It would be a hazardous task! . . . About two hundred and twelve years ago, Dillenius in a letter to Linnaeus, wrote these pregnant words: 'We all know the nomenclature of Botany to be an Augean stable, which C. Hoffman, and even Gesner, were not able to cleanse.' . . . So—even two hundred and twelve years ago 'the nomenclature of Botany,' as it does

Appendix VII, "Ice Lettuce," tells of seeds of this plant which were brought to England by Douglas in 1823 (p. 337).

Appendix VIII (pp. 338-348), "Some American pines," treats of seventeen species, only six of which are now referable to *Pinus*. The history of the paper is supplied by Wilks in a foreword:

"At the time when this volume had already been pronounced ready for the press, one of the Society's officers, engaged in turning out a very old packing-case, which had apparently not been opened since the Society moved from South Kensington in 1887, came upon two manuscripts,[2] which . . . proved to be in David Douglas' handwriting, and, as far as can be judged, are of about the same date as the other manuscripts from which the rest of this volume has been printed . . .

". . . there are two distinct manuscripts[3] covering precisely the same ground, but in which one will from time to time contain a few words or a sentence which does not occur in the other. These Manuscripts have been most carefully compared and collated with the following result . . . the original expressions have been scrupulously retained even at the cost, in places, of grammatical accuracy and clearness of meaning."

The conifers included in this paper are ones which Douglas must have observed on his journey of 1824–1827.

The Royal Horticultural Society volume (sometimes cited hereafter as RHS) has an excellent index (pp. 349-364).

to this hour, suggested the stable of Augeas. . . . I think you have followed a wise course in avoiding the attempt to modernize the botanical names used by your heroes. Even Hutchinson failed in many cases to anticipate the vagaries of modern 'New Deal Botanists.' . . ."

2. J. M. Murray, British forester, wrote in 1831 that many years after Douglas' death:

". . . there were unearthed in the offices of the Horticultural Society notes in the handwriting of Douglas. These on examination, proved to be descriptions of conifers which he had found. Certain sentences in these are of particular interest to foresters. *Pseudotsuga taxifolia* he says has a timber resembling larch, but 'If we judge from the quantity of charcoal produced it will not prove so durable as the Larch.' He says *Picea sitchensis* resembles *P. Douglasii*, and although not so large as that species 'it may nevertheless become of equal if not greater importance, as it possesses one great advantage over that one by growing to a very large size on the Northern declivities of the mountain [in] apparently poor, thin, damp soils; and even in rocky places, where there is scarcely a sufficiency of earth to cover the horizontal wide-spreading roots, their growth is so far from being retarded that they exceed one hundred feet high and eight feet in circumference. This unquestionably has great claims on our consideration as it would thrive in such places in Britain where even *P. sylvestris* finds no shelter. It would become a useful and large tree.' *Pinus contorta* was found by him only on the coast and he dismisses it with 'Little can be said in favour of this tree either for ornament or as a useful wood.' "

3. Balfour probably refers to these, although he mentions no dates, when he notes that

"The R. H. S. in their archives have a thick bundle of sheets in Douglas' writing: on each he describes a plant of his collecting. They have never been published."

Balfour, himself, then publishes two of Douglas' descriptions: one of the canoe cedar, *Thuja plicata* D. Don, and the other of the golden-leaved chestnut of the Coast Ranges, *Castanopsis chrysophylla* (Hooker) A.DC.

The last-named tree was of great interest to the British botanists, and Hooker, as we shall note on occasion, urged the collection of propagating material by Hudson's Bay Company officials and by collectors. In recent years the tree has been placed in a new genus, *Chrysolepis,* by Hjelmquist.

I turn now to David Douglas' first journey to the northwest coast, made in 1824–1827. He was to be vastly indebted throughout these years to the assistance of the Hudson's Bay Company. *The Champlain Society Publications* record the instructions issued by that organization's officials:

". . . Governor and Committee to Chief Factors in charge of the Columbia District, July 22, 1824, 'Mr. Douglas is a Passenger in the *William & Ann,* and is sent by the Horticultural Society for the express purpose of collecting Plants and other objects of natural history, he will remain with you till next Season, and we desire you will afford every assistance in promoting the object of his Mission' . . ."

Chief Factor John McLoughlin's reply was dated "Fort Vancouver *6th Octr.* 1825":

". . . In compliance with your directions we have given every assistance to Mr. Douglas which our means afforded and I am only Sorry Situated as we are it has been out of our power to make him as comfortable as we would wish. He expressed a desire of going across the Continent in the Spring. But I informed him this would depend on the Instructions we would receive from York this fall as we might be so situated as not to have it in our power to accomodate him with a passage."

My purpose in recounting the story of this journey is to supply as accurate an itinerary as possible (based on data found in the three journals describing the trip or in any other reliable source), to note some of the important plants which Douglas mentions as he moved about, and to recount a few of his more noteworthy experiences and reflections.

On this particular journey Douglas' travels in what is now the United States were confined to the states of Washington, Oregon and Idaho. To return to England he crossed Canada—from present British Columbia to Hudson Bay.

My story is primarily concerned with the period between April 7, 1825 (when Douglas reached Baker's Bay at the mouth of the Columbia River) and April 19, 1827 (when, starting homeward, he crossed the 49th parallel[4] at the point where the Columbia River crosses the line between Washington and British Columbia).

The foundation of the story rests upon the journal in the Hooker "Memoir"(pp. 82-140) and upon the two journals in the Royal Horticultural Society volume (pp. 51-76 and 77-293). The Hooker "Memoir" includes three letters written by Douglas (pp. 105, 106, *fn.*; 107, 108, *fn.*; pp. 113, 114). The Royal Horticultural Society volume includes two (pp. 197, 198 and 247). Other sources are listed in the literature cited at the end of this chapter.

1824

On July 26 Douglas sailed from Gravesend, England in the Hudson's Bay Company's ship *William and Ann.* Dr. John Scouler was serving as ship's surgeon.

---

4. The precise date on which the crossing of the present boundary line between the United States and Canada occurred, was kindly supplied by Harvey (*in litt.,* March 22, 1947).

1825

On February 10 they were in sight of a river, the Columbia, in longitude 134° W. Because of weather conditions, the ship was unable to cross the sand bar at the river's entrance for six weeks. On April 7 it anchored in Baker's Bay, on the north side of the river's mouth. Even before landing, Douglas comments upon three conifers which he observed from the ship:

"The ground on the south side of the river is low, covered thickly with wood, chiefly *Pinus canadensis, P. balsamea,* and a species which may prove to be *P. taxifolia.*[5] The north (Cape Disappointment) is . . . covered with wood of the same kinds[6]. . ."

Accompanied by Scouler, Douglas made his first landing on April 9 on Cape Disappointment. The two men had very similar interests and worked together whenever possible. Indeed, Scouler's journal sometimes accounts for certain short periods[7] not recorded by Douglas; occasionally their journals differ slightly as to dates but the discrepancies are not important ones.

Between April 11 and 18 the ship moved up the Columbia, coming to anchor at Point Ellis, opposite Fort George—earlier known as Fort Astoria. With the Chief Factor, John McLoughlin, Douglas traveled by boat, April 19 and 20, to Fort Vancouver.[8] This new post of the Hudson's Bay Company was situated some ninety miles from the sea and—in anticipation of the possibility that the Columbia River might eventually constitute the boundary between the United States and Canada—had been located on the north side of the river; the site was some twelve miles below Point Vancouver, or about opposite present Portland, Oregon. Douglas notes that Point Vancouver was the spot where the officers of Vancouver's expedition had ". . . terminated the survey of the river in 1792." Douglas takes the occasion (April 19) to insert a tribute to McLoughlin.[9]

5. *Leisure Hour* (1883), mentions "some magnificent specimens" of this tree growing in England, among them ". . . one in the grounds of the Palace at Scone . . . grown from seed sent home by Douglas in 1827 . . . it is now seventy-five feet high and seven feet in girth . . ."

6. J. M. Murray, in the British periodical *Forestry,* wrote: "Among the first plants mentioned by Douglas were *Pseudotsuga taxifolia* and *Tsuga Albertiana.* The size of the former was always to him a matter of wonder. His earlier experiences on the eastern coast led Douglas into blunders with species of the west. He thus mistakes *Tsuga Albertiana* for *Tsuga canadensis, Larix occidentalis* for *Larix europea,* and *Thuja gigantea* for *Thuja occidentalis.* This last error he corrected at a later date and named the western species *Thuja Menziesii,* although the name was afterwards not found to be valid."

7. Excursions which they made on April 10-12 and April 14-17 are mentioned in my chapter on Scouler.

8. Scouler did not reach Fort Vancouver until May 2.

9. This was one of the insertions, obviously added at a later date, which are found quite frequently in Douglas' journals; he notes that McLoughlin had received him ". . . with much kindness . . . In the most frank and handsome manner he assured me that everything in his power would be done to promote the views of the Society. Since I have all along experienced every attention in his power, horses, canoes and people when they could be spared to accompany me on my journeys . . ."

From April 21 to May 9 Douglas worked in the vicinity of Fort Vancouver. He mentions a visit to a plain seven miles below the Fort on the north side of the Columbia (May 1) and another to Menzies Island (May 2); Scouler, who had reached the Fort on May 2, went along.

The longer of the journals supplied by the Royal Horticultural Society is interrupted at intervals by lists of plants: Douglas' no. 11 was "*Alnus* sp.; a tree 50 to 70 feet high . . ."; no. 38 was "*Acer macrophyllum,* of Pursh; one of the largest and most beautiful trees on the Columbia River . . . 6 to 16 feet in circumference; 60 to 90 feet high . . ."; no. 55 was "*Taxus* sp.;[1] a tall tree, 20 to 60 feet high . . ."

From May 10 to 16 Douglas made a trip down the Columbia River with Scouler; he returned without his companion who was about to depart on the *William and Ann* for a trip northward. It was on this journey that Douglas found the beautiful madroña, *Arbutus Menziesii* Pursh: ". . . among other curious plants a noble species of *Arbutus, A. procera.*" In a list of plants dated May 10, his no. 184 is "*Arbutus* sp. . . . This very ornamental tree . . . attains the height of 60 feet and sometimes 2 in diameter . . . The leaves of the young shoots where there are no flowers answer the description of *A. laurifolia.* I have no doubt it is that noble species; the fruit is not yet ripe . . . Do not fail[2] to put up a treble supply of its seeds; being evergreen it is the more desirable."

From May 17 to 30 short trips were made about Fort Vancouver—one of ten to thirty miles on May 20—and he took time to arrange his collections. From May 31 to June 3 he made a trip towards the Grand Rapids, following the north bank of the Columbia.

From June 20 to July 6 Douglas traveled up the Columbia with Company canoes going to the interior posts. He went as far as a few miles above the Great Falls, or about two hundred miles above the sea. According to F. G. Young, ". . . to a point a few miles above Celilo Falls." He was back at Fort Vancouver on the 6th and, from the 7th to the 18th, worked on his collections and made short trips about the fort. From July 19 to August 5 he made a trip by canoe to the mouth of the Columbia and back—paying a visit to an Indian chief on the day of his return. This trip had been ". . . principally for the purpose of searching for and inquiring after the tuberous-rooted *Cyperus* mentioned by Pursh in his preface, the root of which is said to afford the natives food something like potatoes when boiled. After a laborious route of twelve days along the shore north of Cape Disappointment, I was obliged reluctantly to return without being fortunate enough to meet it . . ."

From August 6 to 18 he worked upon his collections.

From August 19 to 30, with a party of hunters and traveling in canoes, Douglas made a trip to and up the "Multnomah," or Willamette, River ascending it for about

1. J. M. Murray notes that the alder ". . . may have been *Alnus oregona* . . . *Acer macrophyllum* . . . was introduced to this country [England] by Douglas who sent seed in 1825 . . . he records . . . a species of *Taxus* (*Taxus brevifolia* Nuttall), although no seed was sent by him . . ."

2. One of Douglas' numerous reminders to himself.

twenty-five miles above the Falls, situated at present Oregon City, Clackamas County, Oregon. For ten days he remained in the vicinity of the "Calapoore"—Calapooie—Indians. He was back at Fort Vancouver on the 30th and spent the 31st arranging his collections. It was on this expedition that he learned for the first time of the existence of the sugar pine upon which he was later to bestow the name *Pinus Lambertiana*.

From September 1 to 3 Douglas made a trip by canoe to the Grand Rapids of the Columbia, climbing (September 3-5) to the summit of a mountain on the north side of the river; he refers to this as ". . . one of the most laborious undertakings I ever experienced the way was so rough . . . very little paper could be carried . . ." He also ascended (September 7-9) to the summit of some hills on the south, or Oregon side, which he found ". . . easier of ascent . . ." and where he ". . . had the good fortune to find two new species of pine, *Pinus nobilis* and *P. amabilis*,[3] two of the noblest species of the tribe . . ." He also mentioned, among other plants, one of the bear grasses, "*Helonias tenax*"—the generic name is now *Xerophyllum*—and "In rocky places of the mountain *Arbutus tomentosa* was not a stranger." This last is now placed in the genus *Arctostaphylos*. On getting back to Fort Vancouver on the 13th, Douglas found Scouler, returned from his voyage to the north.

Since the *William and Ann* was soon to leave for England, Douglas spent the days from September 14 to October 3 in writing letters, transcribing his journal and arranging his collections for shipment to his sponsors. The short RHS journal records of the last days of September:

"The remainder of this month was devoted to packing up my gleanings of dried plants, consisting of sixteen large bundles of American and eight from other places,[4] a large chest of seeds, one of birds and quadrupeds, and one of various articles of dress &c. A large portion of each of the varieties of seeds was reserved for the purpose of sending across the continent in the ensuing spring."

Scouler's journal records that he left Fort Vancouver to return to the ship on September 20, so that Douglas lost his congenial companion.

From October 4 to 21 Douglas was incapacitated by an injury to his knee, but, feeling better on the 22nd, and hearing that the departure of the ship had been delayed, he left for the coast where he hoped to see his old shipmates. But he found on arrival (October 24) that the *William and Ann* had sailed an hour earlier. He mentions "*Helonias tenax*," seen on the journey, as "a very desirable plant for cultivation."

From the coast, and although the infected knee was still troubling him, Douglas started on a trip which was to last twenty-five days and take him into southwestern Washington—into what are now Pacific, Lewis and Cowlitz counties.

3. Writes Balfour: "The latter was not seen again till Sargent, Parry and Engelmann rediscovered it in the autumn of 1880 on Silver Mountain on the Fraser River. Sargent soon afterwards found it where Douglas first saw it, at Grand Rapids on the Columbia." These were the firs: *Abies nobilis* Lindley, the red fir, and *Abies amabilis* (Loudon) Forbes, the white fir.

4. Collected on the journey out from England.

After he had been ferried across Baker's Bay, he borrowed a canoe and portaged Cape Disappointment (October 25, 26), kept along the coast to "Cape Foulweather" whence he sent the canoe back to the Columbia and, on foot, went overland to the mouth of the "Cheecheeler," or Chehalis, River or to where it empties into "Whitbey," or Gray's, Harbor (October 30); he started by canoe up Chehalis River (November 7) and had made sixty miles by November 11; he then abandoned the canoe and, on foot with a pack-horse, traveled across country until he reached the "Cow-a-lidsk," —Cowlidsk or Cowlitz—River (November 13); there he procured a boat and descended the Cowlitz to its entry into the Columbia (November 14) and, ascending the Columbia, got back to Fort Vancouver (November 15).

It had been a miserable as well as a foolhardy journey for a sick man and Douglas notes: "I had suffered so much . . . that little hope was left for me being able to do much good for this season, at least in botany." From November 16 to December 31 he remained at Fort Vancouver—the rainy season, moreover, made excursions impossible.

1826

From January 1 to March 19 he was still at Fort Vancouver, arranging his collections and preparing for a trip up the Columbia River into the interior. He left on this trip on March 20 and did not get back to Fort Vancouver until August 29. He traveled with John McLeod, James Ermatinger, and others of the Company. It took them from March 20 to April 21 to reach Fort Colvile,[5] situated not far from Kettle Falls, present Stevens County, in northeastern Washington. They traveled up the Columbia to the Dalles and the Great Falls (March 23-25); to "Wallawallah" (March 26-28); one day was spent at Fort Walla Walla (March 29); they camped at Priest Rapids, present Yakima County (April 1); from there they moved on to the Hudson's Bay Company's post at the confluence of the Okanogan[6] and Columbia rivers, in present Okanogan County (April 2-8); thence to the mouth of the Spokane River (April 11)— this stream enters the Columbia between Lincoln County (south) and Stevens County (north); here Douglas (April 11-18) made up his records, etc. Kettle Falls was reached April 21.

From April 22 to May 8 Douglas worked in the vicinity of Kettle Falls. It was towards the end of his stay that he found *Larix occidentalis* of Nuttall, the tamarack or western larch; he confused it with the eastern species, calling it *"P[inus] Larix."*

"P. Larix is found in abundance in the mountain valleys, much larger than any I

---

5. The first fort of this name was British and was named Colvile, in honor of Andrew Colvile, one of the Hudson's Bay Company's officials; the name of the later American fort, as well as that of the still later town, was spelled Colville.

6. "The name is spelt *Okanogan* in the United States; *Okanagan* in Canada. It has the distinction of having been spelt in no less than forty-five different ways . . ." (Harvey, 1940)

have seen on the other side of the continent or even read of. I measured one some 30 feet in circumference, and several . . . blown down . . . 144 feet long."

Having broken his gun, Douglas was now obliged to make a trip overland to the abandoned Hudson's Bay Company post on the Spokane River where lived the only man within hundreds of miles who could repair it. It was a hard trip; it took him three days to get there (May 9-11) and two to return (May 13, 14).

From May 15 to June 5 he worked in the vicinity of Kettle Falls. From June 6 to 9 he was on his way back to the Company's post at "Wallawallah," or Walla Walla.[7] From June 10 to 16 he made trips in the vicinty of Walla Walla.

From June 17 to 30 Douglas was to make two trips into the Blue Mountains[8] of northeastern Oregon. The first was from June 17 to 24, the second from June 26 to 30.

He reached the base of the Blue Mountains by following the Walla Walla River (June 18), climbed to the summit of the "highest peak" (June 20) and then returned to Walla Walla (June 24)—his guide had dissuaded him from crossing into the valley of the Grande Ronde River to the southeast of the Blue Mountains.

After spending a day (June 25) at Walla Walla, he set out again, this time by the north bank and north fork of the Walla Walla River; he reached snow (June 28) and returned (June 30) to the post.

It was on the first of these short trips into the Blue Mountains that Douglas collected our native peony, a plant of which he seems to have been particularly proud. ". . . In these untrodden regions on the verge of eternal snow were *Paeonia Brownii,* the first ever found[9] in America . . ." He wrote of it on June 25, when he was spending a day at the post:

7. Barnston refers to what must have been this post as "Wallander, an establishment just below Lewis and Clarke's Fork, then considered the key to the navigation of the upper Columbia."

Dr. A. E. Porsild, Chief Botanist, National Museum of Canada, Ottawa, sent me the following information:

"I can find no reference to a Hudson's Bay Post named Wallender on the Columbia, but from the context of Barnston's sketch and from the following quotation from MS notes in the National Museum Library 'Historic Forts and Trading Posts of the French Regime and the English Fur Trading Companies' compiled by E. Voorhis, 1. c. 175, says: 'Fort Walla Walla originally a North West Co. fort on left bank of Columbia river, at mouth of Walla Walla river, 5 miles below mouth of Lewis or Snake river. Built 1818. The Hudson's Bay Co. succeeded to this fort in 1821 and rebuilt it in 1841 . . .' "

8. The next important botanist to reach these mountains seems to have been Thomas Nuttall who, on his transcontinental journey with N. J. Wyeth, crossed the range on August 31 and September 1, 1834.

9. *Paeonia Brownii* Douglas was described in 1829 in Hooker's *Flora boreali-americana* (1: 27). The type locality was "near the confines of perpetual snow on the subalpine range of Mt. Hood." The collector was Douglas and the year 1826. Since Mount Hood is in Hood River County, Oregon, the plant was not described from the first of Douglas' collections, in the Blue Mountains. *P. Brownii* was for a long time considered to be the sole representative of the genus on the American continent but another species, *P. californica* Nuttall, is now recognized. Nuttall's manuscript description was quoted by Torrey and Gray in their *Flora of North America* 1: 41, 1838. See Stebbins, G. Ledyard, Jr., 1938. The western American species of Paeonia. *Madroño* 4: 252-260.

"In the early part of the day I placed in dry paper some of the plants, and under the presser the more recently gathered ones, and then put up some seeds which I had hung up to dry ten days since. I come now to take from my note-book the following collected on my journey: (149) *Paeonia* sp. . . . flowers small petals same length as the stamens, centre and the outside deep purple, on the edge and inside bright yellow; a low plant, six inches to a foot high; in great abundance . . . flowering in perfection on the confines of perpetual snow; lower down it is seen in feeble enervated plants, and in the more temperate regions completely disappears; this valuable addition will I trust be an acquisition to the garden; if in my power, seeds of it must be had."

From July 1 to 8 Douglas remained at Walla Walla and from July 9 to 15 descended the Columbia to the Dalles—twelve miles below the Falls. He had planned to go to Fort Vancouver in the hope of finding mail from home, but meeting a "brigade" of the Company (July 10) who were on their way upstream and who had brought his letters, he turned back with them. They reached Walla Walla on the 15th and Douglas spent the 16th in gathering and packing seed. The brigade was moving north and he determined to accompany them. They left on July 17 and Douglas was not to get back to Walla Walla until August 25. The first major objective was the ". . . fork of Lewis and Clarke's River,[1] about 150 miles from the Columbia." They started at daylight and walked till ten in the morning, making an additional fifteen or twenty miles each evening, and reached the fork of the Clearwater on July 24.

It was from here that Douglas made a short trip into the mountains lying to the southeast. He was under the impression that these were the Blue Mountains, writing: ". . . I was desirous of making a trip to the mountains, distant about sixty miles, the same ridge I visited last month further to the south-east[2] . . ." It is generally accepted that his trip took him into the Craig Mountains of northwestern Idaho, a range which seems not to appear by name on many modern maps. For the Blue Mountains of northeastern Oregon terminate to the north near latitude 46° N. and are cut off from the Craig Mountains[3] by the canyon of the Snake River. Before it was considered safe to make such a trip a "conference" was held with the Indians of the region and it was not

1. Barnston refers to this as the "Fourche de l'Eau clair," or the fork of the Clearwater. The Clearwater joins the Snake at present Lewiston, Nez Perce County, Idaho.

2. As noted, he had visited the Blue Mountains from June 17-24 and from June 26-30. The abridged RHS journal under date of July 25 reads: "From this point [the fork] I again visited the Blue Mountains, the same ridge I had been on two weeks ago." And the running-title in the amplified RHS journal is "Blue Mountains."

3. Dr. F. Marion Ownbey, curator of the Herbarium of the State College of Washington, identified them thus (*in litt.* April 7, 1947): "The Craig Mountains referred to in Piper's Flora are a low range at the west end of the Salmon-Clearwater divide. For some reason they do not appear on most maps although they are a rather conspicuous feature. They are about the same height as the Blue Mountains and appear as a continuation of that range across the deep canyon of Snake River . . ."

Piper's "Flora," mentioned above, states that "Douglas collected about . . . Lewiston and in the adjacent Craig Mountains . . ."

until July 28th that Douglas—accompanied by a companion supplied by the commander of the brigade—started on his way. The trip lasted only until the 29th.

They set off ". . . in a southeasterly[4] direction, the country undulating and very barren," and reached the mountains at nine a.m. (July 28), breakfasting at the spot ". . . pointed out to me by the Indians where Lewis and Clarke built their canoes,[5] on their way to the ocean, twenty-one years ago."

Alone, Douglas records that he "Reached the highest peak of the first range at 2 p.m., on the top of which is a very remarkable spring, a circle 11 feet in diameter . . . I could find no bottom . . . at a depth of 60 feet . . . I have called the spring Munro's Fountain . . . Found a few seeds of *Paeonia,* but not so ripe as I could have wished . . ."

By six p.m. Douglas had rejoined his companion and, traveling all night, reached the camp at the fork at sunrise of July 29.

It should be of interest to quote from a letter dated August 14, 1947, which Dr. Ownbey—already mentioned here—wrote me about Douglas' trip into the Craig Mountains just described:

". . . In the time allotted he would have reached the Craig Mountains in the vicinity of Lake Waha . . . As you indicate, the spring described as 'Munro's Fountain' should be critical in determining just where he did go, and indeed it has proved to be . . . The difficulty is that such a spring no longer exists . . . it was necessary to dig into the geological literature of the region to find mention of what I take it to be . . . I think that we can rule out the possibility that Douglas visited the Canoe Camp of Lewis and Clark as exceedingly unlikely. At first, I was inclined to believe this possible, so much so that about ten days ago I took advantage of the opportunity to investigate the south wall of the Clearwater Canyon opposite Ahsahka and Orofino . . . Four considerations . . . are critical in eliminating the possibility that Douglas visited the Canoe Camp . . . These are: (1) the probability of accomplishing this in the time allotted; (2) the absence of a spring even remotely resembling his description of 'Munro's Fountain'; (3) the fact that this locality is east, not southeast, of Lewiston . . . (4) the negative evidence for the occurrence of *Paeonia* in this area. The arguments against his visiting the east end of the Blue Mountains at this time are similar, but not quite so convincing. This would mean that he would have gone southwest instead of southeast from the junction of the rivers . . . If Douglas went southeast of the fork of the river,

4. The *Blue Mountains* lay to the southwest rather than to the southeast.

5. Douglas' informants must have been in error for the spot where the Lewis and Clark party halted (September 27–October 7, 1805) is described by C. J. Brosnan thus:

"Building the Canoes.—On September 24 the travellers followed down the present Jim Ford Creek and two days later reached the confluence of the North Fork of the Clearwater with the main stream. The site selected for their headquarters . . . was at 'Canoe Camp,' . . . situated on the south side of the Clearwater and not far from the present site of Ahsaka."

Ahsaka, or Ahsahka of some maps, is in Clearwater County, Idaho, and certainly thirty miles *east to northeast* of present Lewiston. Douglas notes that he traveled *southeast* from that point.

as he states, he would have reached the foot of the Craig Mountains at the end of a short day's journey . . . It is only about eighteen miles from Lewiston to the mountains and he could have covered much more distance easily . . . He would have reached the mountains in the immediate vicinity of Lake Waha since there is a deep canyon both to the right and to the left. At this point on the face of the mountain, nearer the bottom than the top, there are three small lakes, from top to bottom and largest to smallest, Waha, Blue (Middle), and Mud, formed by a landslide damming the valley of Waha Creek . . ."

Dr. Ownbey then describes these lakes in detail, records local opinion on the subject of springs, and refers to a paper which treats of their nonexistence. Following this he adds:

"There is considerable difficulty in reconciling the situation as I have reconstructed it with the description of 'Munro's Fountain.' There is enough agreement, though, that I think it likely that it was this place that Douglas was describing, particularly since I know of no other which will satisfy the requirements as well. The spring, issuing from the mountainside (at least 1000 feet below the top) would have formed a stream of sufficient size, the fall is about right, and the marsh at the foot of the mountain would have been Mud Lake. *Paeonia Brownii* and *Thuja plicata* are apparently absent, but both may have been there in Douglas' time, and may still be . . . What he was doing from nine to two o'clock is a mystery. He could have climbed the mountain in an hour . . . and what he called the foot of the mountain if he considered the spring on top, is a further complication. Of course springs simply do not come out on top of the highest peak in a range. The water must have a source higher than its outlet. For this reason, I think we may discount the accuracy of the statement that the spring was on top of the mountain. The one at Lake Waha was well up on the side, but . . . at least 1000 feet below the top. In spite of these discrepancies and contradictions, it seems most likely that it was here that Douglas went. All of his statements simply cannot be true, but if this solution should be the correct one, there are no great inaccuracies in his account which cannot be explained as typographical errors or slight misstatements of fact . . ."

I believe that all must agree that, as an example of the coöperative spirit, Dr. Ownbey's letter, cannot be surpassed.[6]

On July 30 the brigade moved camp to the ". . . northern shore of the north branch . . ." of the Clearwater River, presumably. On August 1 they left to descend the Snake River. Douglas, with three companions, started overland for Kettle Falls, and reached the Company's old establishment on the south side of Spokane River on August 3 where he saw again "old Mr. [Jacques Raphael] Finlay," the gunsmith. From there, crossing the Barrière and Cedar rivers—in the last-named losing his collection of

6. Harvey (1947) includes some references to the identification of Douglas' "Munro's Fountain" with one of the outlets of Lake Waha (p. 85, *fn.* 12).

seeds, his notebooks, etc.—he reached Kettle Falls (August 4) by the route which he had taken once before (May 13, 14, 1826). He notes that he had been away for "two months"—he had left Kettle Falls last on June 6.

He learned that a vessel—the *Dryad*—was to sail for England on September 1 and, since this gave him an opportunity to ship some of his collections home, he spent August 5-18 collecting seed and packing his specimens, having been told that ". . . this may be the last ship for some years going direct to England . . ."

On August 19 Douglas left Kettle Falls for Fort Vancouver.

His first objective point was Fort Okanogan ". . . distant 250 miles north-west [actually southwest] of this place." Keeping to the south side of the Columbia River, he got to within ten miles of the Spokane River (August 19), crossed the Spokane and made fifty miles to the Grand Rapids of the Columbia (August 20), passed the Grand Coulee (the 21st) and arrived at the "establishment" at Okanogan (the 22nd).

He then descended the Columbia, camping on the "Piskahoas" River (the 23rd), passed the Stony Islands and Priest Rapids (August 24), and reached Fort Walla Walla (the 25th); he traveled all night (the 26th), reached the Great Falls and the Dalles (the 27th), camped fifteen miles above the Grand Rapids (the 28th) and reached Fort Vancouver[7] at midday, August 29, "after traversing nearly eight hundred miles of the Columbia Valley in twelve[8] days . . ."

From August 30 to September 19 Douglas wrote letters, packed his collections for shipment to England (under date of September 1 he notes: ". . . saw my chests placed in one of the boats . . . going with the cargo to the ship."), and prepared for his trip to the "Multnomah," or Willamette, River. This trip, which was to become one of the most famous of his journeys, lasted from September 20 until November 19. Douglas' purpose in making it was to obtain seeds of the sugar pine of which he had heard on his trip of about a year before (August 19-30, 1825) when on the lower Willamette. He traveled with some of the Company trappers, led by A. R. McLeod, who were bound for the very place he wished to visit. Their objective point was the "Umpqua River" in southwestern Oregon.[9]

7. *The Champlain Society Publications* include a letter from John McLoughlin to the Company officials, dated "Fort Vancouver *1st Sept.* 1826" which states:

". . . Mr. Douglas who came Out in the *William and Ann* Arrived four days ago from the Interior and now Intends to pass some time in the Willamette . . ."

8. Douglas' computation of time and distance is difficult to reconstruct from his records.

9. The heading in the Hooker "Memoir" to this portion of the Douglas journey is: "Excursion to North California and the Umptqua or Arguilar River . . ."

F. G. Young, editor of the reprint of the "Memoir," comments: "The text shows that Mr. Douglas did not penetrate to the boundary of what is now 'North California.' "

As the crow flies, Looking Glass Valley near present Roseburg, which represents the southernmost point of Douglas' journey, is about eighty to one hundred miles north of the present boundary between Oregon and California.

Douglas left Fort Vancouver on September 20 by boat and joined McLeod's party which had taken their horses overland, at a point some sixty-five miles up the Willamette (September 22); they remained there a few days (September 23-26), then moved southward by slow marches, camping on the "Yamhill" River (on the 28th[1]), and passing Mount Jefferson, lying 25 to 30 miles to the east (October 5).

On October 9 Douglas wrote: "On the summit only low shrubs and small oaks and a species of *Castanea*. This handsome tree I saw at the foot of the hill, but very low, not more than 4 feet high . . . but was agreeably disappointed to find my little shrub of the valley change on the mountains to a tree 60 to 100 feet high, 3 to 5 feet in diameter *(Castanea chrysophylla[2])* . . . I cannot say whether it is new or not, but am inclined to think it rare; at least I had a laborious search to find it in fruit, which I did on the very highest peak and only one tree . . ."

On October 16 the party reached the Umpqua River. Rising in the Cascade Range, its north and south forks unite about nine miles northwest of present Roseburg, Douglas County, Oregon; it empties into the Pacific some twenty-two miles north of present Empire City. The day of his arrival he found the handsome California laurel, *Umbellularia californica* (Hooker & Arnott) Nuttall:

"In the deep dark valleys of this stream, Red Deer River, I had the pleasure to find *Laurus regia,* a beautiful evergreen tree, a decoction made from the bark of which was used by the hunters as a beverage. So exceedingly powerful is the fragrant scent which it emits by the rustling of its leaves that it produces sneezing; the smell is precisely like that of the well-known *Myrtus Pimenta.*"

From October 17 to 19 Douglas made his first attempt to reach the mountains southeast of the Umpqua River but was unwell and had to return to camp. His second attempt (October 20-29) was successful.

Following the left (presumably south) bank of the Umpqua he traveled west towards the sea (October 20), parted from McLeod, who was to continue west (October 23), and turned due east for seventeen miles (October 24). At great risk of his life he succeeded in finding[3] the sugar pine and in securing a few cones (October 26), and, after various difficulties, and ill, he finally got back to the company camp on the 29th.

1. It was on this day, September 28, 1826, that Douglas rediscovered the madroña, *Arbutus Menziesii* Pursh, which he had found first on May 9, 1825, on the Columbia River.

"On the low hills on gravelly soil *Arbutus procera* attains a greater size than any on the Columbia; they are frequently 15 inches to 2 feet in diameter near the root, and 30 to 45 feet high."

2. *Castanopsis chrysophylla* (Hook.) A. DC.

Balfour tells us that, in Britain, "*Castanopsis chrysophylla,* despite Douglas' hopes for it, is still a rare tree . . . and seldom bears fertile fruit. Though discovered by Douglas seeds did not reach this country till 1845."

3. The volume *Oregon* ("American Guide Series"), p. 321, states that Douglas found the sugar pine west of present Roseburg, on Sugar Pine Mountain. The date mentioned is October, 1828, but 1826 is correct.

Balfour mentions that "The place of his discovery . . . is likely to be permanently commemorated in Looking Glass Valley, 18 miles from the modern Roseburg, where it is proposed that an area of 2,000 acres of Sugar Pine Forest is to bear his name."

From October 30 to November 6 Douglas remained in camp, making one twelve-mile trip only—to a point "below" the camp (November 3). Young states that it was to the lower Umpqua.

Since he had obtained what he most desired, since the rainy season was setting in, and since some of McLeod's men were to take out a despatch, Douglas decided to accompany them back to Fort Vancouver; moreover, he had only four months before he was to leave—on March 1, 1827—to cross the continent on his way to England. It took him, returning by the same route he had come, twelve days (November 7-19) to make the trip. He had been gone two months. One mishap occurred (November 17) when his collections became saturated in crossing the "Sandiam," or Santiam, River.

Back at Fort Vancouver, he spent about ten days (November 20-31) in arranging his collections; he was, moreover, fatigued and suffering from the effects of some eighteen months of fighting the elements. Nonetheless, he did not waste his last opportunities: from December 1 to 8 he made a collection of the various timber trees in the forests near the fort and, from December 9 to 25, made a trip down to the sea ". . . in quest of Fuci, shells, or anything that might present itself . . ."

He traveled by canoe, reaching Fort George (December 11), crossing to the north side of the Columbia (December 12), and, proceeding along Baker's Bay, portaged Cape Disappointment to the bays "near Cape Shoalwater of Vancouver" (December 14). On the 15th he reached the home of one of his Indian friends, Cockqua. But he was ill and decided to return to Fort Vancouver; he reached Oak Point (December 18), passed the mouth of the Willamette River (December 20) and, at noon of Christmas Day, was back at the post.

The botanical results of this trip, like the one a year earlier at much the same season (October 22 to November 25), Douglas considered unsatisfactory.

From December 26 to 31 he went out occasionally, gathering woods and mosses.

1827

From January 1 to March 1 rain and snow made collecting impossible.

From March 2 to 9 he again traveled down to the sea, but the botanical results were again poor. After his return to Fort Vancouver, he spent his time until March 19 preparing for his journey to Hudson Bay.

The transcontinental journey began on March 20. According to Young, Douglas traveled ". . . with the Annual Express for England . . ." In the company were Edward Ermatinger, bound for Hudson Bay, and John McLoughlin on his way into the interior. Starting from Fort Vancouver, they followed the Columbia River.

They reached Point Vancouver and the Cascades (March 21), reached a point midway between Grand Rapids and the Dalles (March 22) and were four miles below the Falls (March 23), seven miles below John Day's River (March 24), nine miles below "the big island" (March 25), at the upper end of the same island (March 26) and seven miles below Walla Walla post (March 27). They camped three miles below Lewis and Clark's River, or the Snake (March 28), and noted the Blue Mountains "enwrapped in snow to their base." On the 29th they camped above the "commencement of the clayey hills," were fifteen miles below Priest Rapids on the 30th, halfway up the Rapids on the 31st.

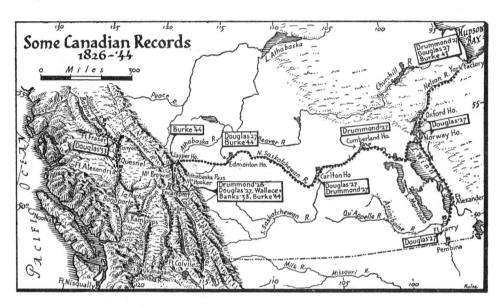

On April 1 they camped on the south side of the Columbia seventeen miles above the Rapids and on April 5 reached Okanogan post.

Douglas comments that, having walked eleven days, his feet were painful, and for the next day and a half he traveled in the boat—around the "big bend."

They camped fourteen miles above Okanogan post (April 6), passed the Dalles and spent the night at "the big stone . . . more properly a natural column . . ." (April 7), passed the junction of Spokane River and Kettle Falls—"covered with snow"—and reached "Fort Coville on the Kettle Falls" (April 12). From April 13 to 17 Douglas busied himself at Fort Covile in arranging his collections and writing letters. They passed the Dalles (April 18) after making twenty-nine miles, and passed "Flathead River" after another thirty miles on April 19.

It was on this day, April 19, 1827, that Douglas crossed the 49th parallel,[4] later to form the boundary between the United States and Canada in this region, and passed beyond the locale of this story.

It took four months to cross Canada. The express stopped at the Hudson's Bay post at Edmonton; again at Norway House at the northern end of Lake Winnipeg. From there Douglas traveled south by boat to another company post at Fort Garry, at the

4. For the precise date I am indebted to Harvey (*in litt.*, March 22, 1947).

junction of the Assiniboine and Red rivers, a trip which consumed one month (July 11–August 7). From Fort Garry he made trips west, up the Assiniboine.[5] From Norway House the brigade traveled northeastward, to Oxford House on Oxford Lake and thence, on August 28, 1827, reached York Factory situated on the peninsula lying between the mouths of Hayes and Nelson rivers. Douglas notes his pleasure at ". . . seeing in the bay riding at anchor the company's ship from England . . ." by name, the *Prince of Wales*.

To now I have confined myself mainly to Douglas' itinerary. It seems fitting at this point to tell a little of what his work in the field had entailed. This is made abundantly clear in his journals, although in a matter-of-fact, all-in-the-day's-work manner of telling. His experiences were not, of course, unique—each of our collectors had his individual difficulties—but the Douglas story supplies an excellent picture of what such a task, if well done, represented in unremitting effort and in disregard of all personal considerations such as congenial companionship, comfort, safety, and the like.

The journals and letters describing Douglas' journey of 1824–1827 contain many references to persons holding positions of importance in the Hudson's Bay Company. Traveling and living under difficult conditions, often for weeks on end, with some of these men—a test of adaptability if there ever was one—Douglas parted from them on terms of friendship.[6] His journals contain many expressions of appreciation for kindnesses rendered: one such refers to Alexander Roderick McLeod with whom he was traveling when in search of the sugar pine. Hooker took note of it under date of November 4, 1826: "A memorandum in poor Douglas' hand-writing in the margins of his journal at this place is, 'Remember, on arriving in London, to get him a good rifle-gun as a present.' "

Field-workers in science met few persons of similar interests; Douglas' contacts with botanists—although more numerous perhaps than those of his predecessors— were rare. It was for the most part a lonely life.

On the voyage to the west coast he had had the companionship of Scouler.

5. Fort Garry was not far north of Pembina, North Dakota, where, in August, 1823, Stephen Harriman Long had raised the American flag on the 49th parallel which, five years before, had been determined as the boundary between British and American possessions in that region.

6. There was one exception necessary to prove the rule, and this took place on Douglas' later journey of 1829–1834. The story has been told more than once, by Bancroft among others. Douglas, while enjoying the hospitality of Samuel Black in charge of Fort Kamloops, is said to have made the remark: " 'The Hudson's Bay Company is simply a mercenary corporation; there is not an officer in it with a soul above a beaver-skin.' " As Bancroft remarks: "For one who had received from the . . . Company nothing but kindness, David Douglas was somewhat free with his comments." A duel failed to materialize when Douglas, summoned to the fray, "declined the invitation."

Bancroft describes Black as ". . . an educated man of no small attainments . . . he managed to command the respect of his associates, if not by his learning, then by his enormous stature . . ." I have found no reference to Douglas' stature, but it seems doubtful that fear motivated his withdrawal. The man was exhausted at the time and fatigue can explain much that is out-of-character—as his comment seems to have been.

"At Gravesend I met Dr. John Scouler of Glasgow, who was going on the same voyage to officiate in the capacity of surgeon. This was to me news of the most welcome kind, being previously acquainted with each other and on the strictest terms of friendship." And again:

". . . I held myself fortunate in having a companion in John Scouler, M.D., skilled in several, and devotedly attached to all, branches of natural history, who undertook the voyage . . . that he might have an opportunity of prosecuting his favourite pursuit."

At Madeira the two men ". . . visited the summit of one of the highest mountains . . . [and] collected for our respective herbaria several interesting though not new plants." We are told of a rare self-indulgence when they shared in the purchase of ". . . 1/16 of a pipe of wine (about 6 and ½ dozen) for which we paid £7." They made their first trip ashore, on Cape Disappointment, together. On returning to Fort Vancouver (October 12, 1825), Douglas found Scouler, just returned from a trip to Nootka and other northern points, in "possession" of his house and was ". . . delighted to hear of his success. We sat and talked over our several journeys, unconscious of time, until the sun from behind the majestic hills warned us that a new day had come. We therefore retired for a few hours."

When Scouler left, with the departure of the *William and Ann,* Douglas confesses—in a letter to Hooker dated March 24, 1826—that he ". . . felt very lonely during the first weeks after Dr. Scouler had sailed." While making "the portage of nine miles" around Priest Rapids (April 3, 1826) he wrote his friend:

". . . Since you left me, there has been no one to join me in my walks, and for several weeks I felt very uncomfortable, being especially grieved at not having seen you before your departure . . . How glad I shall be to join you in our usual trip of [to] Ben Lomond, where we shall have more time and a keener relish for talking over our journeys in North-West America . . . I have neither time nor convenience for writing . . . this is penned upon the top of my specimen board, under which are some exceedingly interesting things . . ."

On his way home (August 15, 1827) Douglas was still reminiscent about their past experiences:

"Found and laid by specimens of *Linnaea borealis* (in fruit). This is the first time I have ever seen this plant in this state. It is rare. Mr. Scouler informed me that he had found it in 1825, in perfect fruit, in the shady forests of Nootka Sound . . ."

Thomas Drummond—serving as botanist with Captain John Franklin's second expedition (1825–1827)—had arrived at York Factory the day before Douglas' arrival in 1827. Douglas had heard of Drummond's presence in Canada while at Fort Vancouver (November 16–December 31, 1825) when McLeod told of having met the Franklin party on Cumberland Lake:

"I learned there was a Mr. Drummond attached to them as botanist; he accompanied Mr. McLeod as far on his route as the foot of the Rocky Mountains and is to pass the summer in the country towards Peace and Smoky Rivers. This I take to be Drummond of Forfar, from the description given of him."

He heard more of Drummond on the Athabasca River (May 3, 1827):

"Informed that Dr. Richardson had in February arrived at Cumberland House; that Captain Franklin had met a ship in the North Sea; and that Mr. Drummond, who spent last summer [1826] in this neighborhood, had in November gone to Fort Edmonton on the Saskatchewan River."

A week later (May 19), in his capacity of botanical-collector, Douglas was slightly apprehensive:

"Learned that Mr. McDonald, the person who had in charge my box of seeds addressed to be left at Fort Edmonton . . . had endured much misery descending the Athebasca . . . in company with Mr. Drummond. Hope my box is safe (do not relish botanist coming in contact with another's gleanings)."

But his fears were soon allayed and he wrote (May 22, 23) at Fort Edmonton:

"I found Mr. F. McDonald here, who took charge of my box last year . . . it had sustained injury, it having been broken . . . I must mention the particular kindness of Mr. Rowland . . . Thinking that they would be better to have the paper changed, he had them *in his presence* [italics mine] examined by Mr. Drummond, who was there at that time . . . This was kind . . ."

"Only eighteen papers had suffered, amongst which . . . is *Paeonia* . . . one of the finest plants in the collection. It often happens that the best goes first."

Finally, at Carlton House (June 1) the two botanists met:

"Here I found Mr. Thos. Drummond had come down to meet Dr. Richardson in spring . . . In the evening had an account of his travels and progress . . . He appears to have done well. I must state he liberally showed me a few of the plants in his possession . . . in the most unreserved manner."

Three days later they went on an excursion together and Douglas was able to record by June 10:

"Morning and part of the forenoon spent looking over Dr. Drummond's collection from the Rocky Mountains. Many fine Alpine plants . . . Considering the opportunities, he had many fine plants all arranged together."

It is fortunate that the men did not meet their death together at this point. Douglas had reached York Factory on July 28, Drummond the day before and he tells of the episode. With Kendall and others the two men had visited the *Prince of Wales* anchored about five miles from the fort. Returning in the evening, a wind of hurricane velocity suddenly arose and, before the ship's boat could come to the rescue, their small craft was driven some sixty to seventy miles out to sea. Not until noon the next day when the wind abated were they able to make the shore by rowing, and assisted, fortunately, by a landward tide; their lives had been despaired of. It is always "the other fellow" who suffers from seasickness and Drummond claimed that he had "endured little inconvenience," but that Douglas had been "dreadfully ill" [7] and that

7. At one point on his journey of 1826 Douglas comments that, although he feared shooting rapids, he was a good sailor!

"Although I am no coward in the water and have stood unmoved, indeed with pleasure, at the agitation

neither he nor Kendall had recovered ". . . the full use of their limbs until their landing in England." It was shortly after this boating experience, or on September 15, that they set sail for England in the *Prince of Wales,* and, according to Drummond, ". . . arrived in London on the 15th of October, 1827 . . ."

Dr. John Richardson led, with E. N. Kendall, a secondary Franklin party which explored the northern coast of Canada between the mouths of the Mackenzie and Coppermine rivers. Douglas had just missed Richardson at Edmonton House but caught up with him at Cumberland House (June 9, 1827). He refers to his collections as "princely":

"Here I was greeted by Dr. Richardson, safe from his second hazardous journey from the shores of the Polar Sea . . . The Doctor has a splendid herbarium and superior collection in almost every department of natural history."

Douglas met the famous John Franklin himself at Norway House. When that naval officer left (July 9, 1827) with Richardson "in their canoe for Canada," Douglas wrote one of his interesting memoranda: "Feel obliged to Captain Franklin (good man); will see Mr. Sabine."

Hooker—not so much concerned with the companionship of kindred souls as with botanical accomplishment—was satisfied with the results attained by these men. He mentions Douglas first:

"It was fortunate that he directed his scientific researches chiefly to the western side of the Rocky Mountains; for, during the very time that he was carrying on his investigations there, his countrymen, Dr. Richardson and Mr. Drummond, were exploring the territories to the eastward of that vast stretch of the Cordillera: the former chiefly in the Arctic regions, the latter in nearly the same parallels of latitude with Mr. Douglas: and the result of their combined exertions has been a mass of collections that have thrown a new light on the Natural History of those interesting regions and have supplied the material for Dr. Richardson's works . . . and of our *Flora Boreali-Americana . . .*"

Like every traveler in a strange land, Douglas always longed for letters from home. The arrival of any mail was a rarity. He wrote, November 18, 1825:

". . . the express . . . arrived from Hudson's Bay which they left on the 21st of July . . . there is only an annual post . . . I hastened to the landing-place, congratulating myself on the news from England . . . there were no letters, parcel, or any article for me. I was given to understand they left Hudson's Bay before the arrival of the ship which left London the May before, so that if Mr. Sabine wrote to me, the letter will remain on the other side of the continent till next November. I was exceedingly disappointed."

When he finally received some letters in June and July of 1826 he confessed:

"I am not ashamed to say (although it might be thought weakness by some) I rose

of the ocean raging in the greatest pitch, yet to descend such a place I can never do unless necessity calls for it."

from my mat four different times during the night to read my letters; in fact, before morning I might say I had them by heart—my eyes never closed."

Replying to one from Joseph Sabine, he wrote: "To say I am happy would only convey but a faint idea of the rapture I enjoy in hearing from you."

For his part, Douglas tried to keep in touch with friends, botanical and other, in England. From Fort Vancouver he wrote Archibald Menzies (October 2, 1825[8]), who had given helpful advice when the trip was in prospect; he corresponded with Hooker regularly; occasionally with Joseph Sabine. Only occasionally does he indicate that letter writing was arduous: "I can only think of Mr. Turner, Lindley, and Munro. It is impossible I can write them a single line."

But if there was little time for writing, he could name plants, or mountains, for his friends; for those of whom he could "only think": a lupine, "I have placed Mr. Turner's name behind it"; for John Lindley, *Oenothera Lindleyana;* also, ". . . this splendid plant . . . I cannot refer to any in the American flora . . . For the present I call this *Munroa speciosa,* after Mr. Munro." For Alymer Bourke Lambert, patron of botanists, he named *Pinus Lambertiana,* a "truly grand tree" and the one of all others which impressed Douglas most on this trip. He wrote Scouler (April 3, 1826):

"This is unquestionably the most splendid specimen of American vegetation—what would Dr. Hooker give to dine under its shade? As for Mr. Lambert, I hardly think he could eat at all, if he saw it."

For Sabine he named, from the Blue Mountains, ". . . *Lupinus*[9] *Sabinii* (whose beautiful golden blossoms gave a tint to the country . . .);" also a phlox: ". . . this exceedingly beautiful species I name *P. Sabinii,* in honour of Jos. Sabine, Esq., for the zeal he has taken in illustrating this beautiful genus of plants, when described or figured I beg it will be adopted . . . I hope it will ere long decorate the garden at Chiswick. S[eed]." For Richardson, *Pentstemon*[1] *Richardsonii.* For Robert Brown, in addition to a mountain,[2] Douglas named the rare western peony, *Paeonia Brownii*: "This valuable addition will I trust be an acquisition to the garden; if in my power, seeds of it must be had." For his special friend, a *Mimulus:*

"This interesting species I call *M. Scouleri,* after John Scouler, who has been the agreeable companion of my long voyage from England and walks on the solitary Co-

8. This letter, so far as I know, has never been published but should be interesting.

9. Piper's "Flora of the state of Washington" includes twelve species of lupine which bear names bestowed by Douglas.

1. Douglas seems to have bestowed names upon twelve of the twenty-four species of *Pentstemon* cited in Piper's "Flora" as indigenous to the state of Washington.

2. On the trip across Canada Douglas had gone through Athabasca Pass in the neighborhood of which he climbed (May 1, 1827) a peak the altitude of which is now estimated at 9,150 feet. Douglas wrote in the abridged RHS journal:

"This peak, the highest yet known in the northern continent of America, I felt a sincere pleasure in naming Mount Brown, in honour of R. Brown, Esq., the illustrious botanist . . . A little to the south is one

lumbia, who first noticed it when in company with me[3] . . . a fine plant for cultivation."
There were other tributes—far too many to enumerate here.

If a phrase symbolizing David Douglas were to be taken from his writings—it appears in a letter to Mrs. Richard Charlton—it is this: "Labor I will . . ."

He was young, only twenty-seven, when he landed on Cape Disappointment in April of 1825, and only thirty-five when he met his death. A sense of responsibility to the Society which sent him on his travels is evident—he was reared in the school where what was worth doing was worth doing well. But, above all else, devotion to his work was the impelling force of his life: it drove him to excesses of endeavor which left him physically, and to some extent mentally, exhausted. Young refers to his work as carried on with ". . . one might almost say desperate, zeal during the major part of the time from 1825 to 1833." The difference between Douglas and many of our plant collectors seems to have lain in the pressure which he exerted upon himself: his work was of a seasonal character, the productions of a vast territory called for acquisition, time went all too quickly.

Plant introduction, Douglas' major ambition, involved the collection of properly matured seed: a week too early, if have it he must, meant another visit a year later; even a day too late and the seed was dispersed. The journals are full of reminders to get, or to get more of, some particular kind. Once obtained, seed needed expert care, proper drying, subsequent protection against moisture, and proper packing for shipment—the neglect of any such essentials meant the difference between success and failure. The shipment of living plants was another and perhaps even more difficult task; they had to be kept moist—but not too moist; to care for them until shipped meant eternal vigilance and to pack them properly for a journey of thousands of miles was an art in itself. Douglas often included cultural directions, even samples of soil. For *Gaultheria Shallon*—the first plant he took in his hands after landing on Cape Disappoinment, "So pleased was I that I could scarcely see anything but it"—he advised:

"It might be worth mentioning to Mr. Munro to try it in rich decomposed vegetable soil, being its natural way of propagating. I am sorry that I have it not in my power to send specimens of the fruit in spirits . . . that put up . . . was by some evil disposed person stolen for the sake of the spirits they were in . . ."

nearly of the same height . . . which I named Mount Hooker, in honour of my early patron . . . Dr. Hooker, to whose kindness I, in a great measure, owe my success hitherto in life . . . I was not on this mountain."

Douglas had had but little training in surveying, etc., before his journey and has been criticized for attributing excessive altitudes to these peaks.

Chapter 11 of Harvey (1947), "Putting mountains on the map," is most entertaining reading; the author absolves Douglas of any desire for self-glorification in his overestimates of the heights of the mountains Brown and Hooker.

A. H. Bent refers to Douglas as ". . . the first to do any actual climbing in Canada."

3. Scouler mentions the discovery in his journal, under date of May 12, 1825.

Roots of a plant which he named for Munro were ". . . sent in a jar of dry sand, I hope they will keep and vegetate." Seed of the *Mimulus* named for Scouler was ". . . sent in the earth it grew in." *Ribes aureum*—"perhaps it might be well to try it in a *very dry poor soil with a little lime.*" Such suggestions abound in the journals and may well have played a part in the success with which Douglas' introductions were raised; but such observations and their recording took infinite time. Long nights, days when collecting was impossible, were devoted to the care of his acquisitions.

Soon after arrival on the Columbia River Douglas described his system of records: he numbered his plants in sequence and recorded at least the generic name in his journal:

"Of those marked with an S. seeds were sent home and a small portion of each kept where they could be divided . . . those kept are, as I am instructed, to be either taken or sent home across the continent. A dry specimen of each is also kept for reference to the collecting of seeds, and will be sent or taken home by sea, being too bulky to cross the continent, though a few of the most interesting may."

For the year 1825 his journal records a plant series, numbers 1-510,[4] made between his arrival on the west coast and November 16. For 1826 there was a series, numbers 1-227, collected between March 23 and August 7. From that time the journal includes no list until June 17, 1827, when a series begins with collections made between Fort Edmonton and Norway House (numbers 1-94), the series for the entire year amounting to 291 numbers.

Douglas had to show discrimination: "The difficulty of carrying as many of the different things collected as would appear desirable is very great: many times on my journeys I am under the necessity of restricting myself to a small number, to give place for a smaller proportion of all."

In a land of plentiful rain the making of his herbarium (dried reference specimens) necessitated much drying and changing of paper, not once but many times—preferably daily—until the plant was in proper condition for shipment. Newspapers, available to modern collectors, did not exist[5] in the Pacific northwest in Douglas' day, and great quantities of paper had to be brought out from England and a certain amount carried wherever the collector went. On Douglas's trip into the interior in March, 1826, he records that "By the kindness of Mr. McLoughlin I was enabled to pack up thirty quires of paper weighing 102 lb., which, with the whole of my other articles, is far more than I could expect when the difficulty and labour of transportation is taken into consideration."

4. Of these many must have been shipped home in late November, 1825, on the ship which took Scouler back to England.

5. The volume *Oregon* ("American Guide Series"), p. 135, records that *The Oregon Spectator,* first issued at Oregon City on February 5, 1846, was the first newspaper to be published west of the Rocky Mountains. And the volume *Washington* ("American Guide Series"), p. 114, tells that the *Columbian,* printed on an old Ramage press, was first issued on September 11, 1852, at Olympia. A yearly subscription cost five dollars and it came out every Saturday.

The drying of these collecting papers, and the plants they contained, once accomplished, a storm or a river-crossing might make it necessary to repeat the operation.

"I laboured under very great disadvantage by the almost continual rain; many of my specimens I lost, and although I had several oilcloths, I was unable to keep my plants and my blanket dry or to preserve a single bird . . . Only two nights were dry . . . before I could lie down to sleep my blanket drying generally occupied an hour . . ." Again:

"Although my plants were covered with a double oilcloth, I found it inadequate to keep them dry, and, lest any should be injured, such as were wet I put in dry paper, and placed under some pieces of bark near the fire for the night."

The ice-cold Barrière River in northern Washington was crossed by swimming.

"I made two trips on my back, one with my paper and pen, the other with my blanket and clothes—holding my property above water in my hands . . . being nearly half-an-hour in the water I was so benumbed with cold that I was under the necessity of kindling a fire."

After recrossing it in the same way three days later, Douglas notes that it continued ". . . very rainy during the whole night . . . I felt cold, my blanket and clothing being wet. As I could not sleep I rose at two o'clock and with some difficulty dried my blanket and a spare shirt, in which I placed my paper containing the few plants collected . . ."

For the next two days he was ill, but ". . . regretted it the less as the weather was so rainy and boisterous, with thunder, that I could have done but little good although in good health." So it went, from day to day, week to week and year to year.

The elements were not the only things with which Douglas had to contend; there was also the fauna!

"Last night I was much annoyed by a herd of rats, which devoured every particle of seed I had collected, cut a bundle of dry plants almost right through, carried off my razor and soap-brush. One . . . was in the act of depriving me of my inkstand, which I had been using before I lay down . . . I lifted my gun . . . and gave him the contents. I found it a very strange species, body 10 inches long, tail 7 . . ."

Follows immediately a description of both male and female. Almost invariably, after describing some misfortune, the journal, in the same breath so to speak, turns to more important matters. Lest he be thought to complain unduly, "When my people in England are made acquainted with my travels, they may perhaps think I have told them nothing but my miseries. That may be very correct, but I now know that such objects as I am in quest of are not obtained without a share of labour, anxiety of mind, and sometimes risk of personal safety."

On April 19, 1825, or just after Douglas had arrived on the Columbia River, he had expressed a certain amount of enjoyment in the outdoor life:

"I have been only three nights in a house since my arrival, the first three on shore . . . I have a tent when it can be carried, which rarely can be done . . . In England people shudder at the idea of sleeping with a window open; here, each individual takes his blanket and with . . . complacency . . . throws himself on the sand or under a bush . . . Now I am well accustomed to it, so much so that comfort seems superfluity."

But by June 20 fatigue was entering the picture; Douglas' solace at such times was always the British cup of tea:

"The luxury of a night's sleep on a bed of pine branches can only be appreciated by those who have experienced a route over a barren plain, scorched by the sun, or fatigued by groping through a thick forest, crossing gullies, dead wood, lakes, stones, &c. Indeed so much worn was I three times by fatigue and hunger that twice I crawled, for I could hardly walk, to a small abandoned hut.[6] I killed two partridges . . . which I placed in my little kettle to boil for supper . . . I awoke at daybreak and beheld my supper burned to ashes and three holes in the bottom of my kettle . . . I had to make a little tea . . . This I did by scouring out the lid of my tinder-box and boiling water in it!"

And the same day: "I had to cross a plain nineteen miles without a drop of water, of pure white sand, thermometer in the shade 97°. I suffered much from the heat and reflection of the sun's rays; and scarcely can I tell the state of my feet in the evening . . ."

Many a night's rest was destroyed by insects of one sort or another. Douglas does not refer as often as most travelers of the period to mosquitoes, but occasionally, as when, "Having no tent, was dreadfully annoyed by mosquitoes." Worse, if possible, were the fleas, sure to be present when there was any contact with Indians or Indian dwellings.

". . . I laid myself down early on the floor of the Indian Hall, but was very shortly afterwards roused from my slumber by an indescribable herd of fleas, and had to sleep out among the bushes; the annoyance of two species of ants, one very large, black, ¾ of an inch long, and a small red one, rendered it worse, so this night I did not sleep and gladly hailed the returning day."

Douglas, moreover, had certain disabilities to contend with. One was an injury to his knee, caused ". . . by falling on a rusty nail when employed packing the last of my boxes . . ." Of necessity he indulged in a rest—from October 4-22, 1825—yet, before the infection was cured, he was off to the Chehalis River, a trip which involved portages, lack of food, cold and rain, and sitting up for nights on end in order to keep warm. He wrote: "I arrived at Fort Vancouver . . . being absent twenty-five days, during which I experienced more fatigue and misery, and gleaned less than in any trip I have taken in this country." The knee was still troubling him in late March of 1826. But his eyes were his most serious affliction. He wrote Scouler on April 3, 1826:

". . . my sight . . . which was always weak, is much impaired during the last months; without pain or inflammation, a dimness has come on which is a great loss to me, especially in the use of the gun, which, as you know, I could handle to some advantage . . ."

By June 15 of the same year he was more and more handicapped: "My eyes began

6. In August, 1826, Douglas comments: "I have a tent, but generally am so much fatigued that the labour of pitching it is too great. Here it could not be done for want of wood, and tent-poles cannot be carried."

to trouble me much, the wind blowing the sand, and the sun's reflection from it is of great detriment to me . . . My eyes so inflamed and painful that I can scarcely see distinctly an object ten yards away." Snow-climbing in the Blue Mountains only made matters worse. He wrote on June 28 (he had walked from noon of the previous day): ". . . I . . . gained the Columbia . . . at nine in the morning, much worn down and suffering great pain from violent inflammation of the eyes. To read or write I cannot . . ."

Douglas is said to have suffered from *ophthalmia,* an affliction which is now recognized to take many forms. Certainly his eyes got worse rather than better and on his last visit to the Columbia and elsewhere they were in an alarming condition. In 1826 his journal opens at Fort Vancouver, with gloomy predictions:

". . . I am now here, and God knows where I may be the next. In all probability, if a change does not take place, I will shortly be consigned to the tomb. I can die satisfied with myself. I never have given cause for remonstrance or pain to an individual on earth. I am in my twenty-seventh year."

In March, 1826, Douglas, on his own responsibility, made the decision to remain a year longer than his instructions had called for.

"From what I have seen in the country, and what I have been enabled to do, there is still much to be done . . . I have resolved not to leave for another year to come . . . I cannot in justice to the Society's interest do otherwise . . . If the motive which induces me to make this arrangement should not be approved of, I beg it may at least be pardoned[7] . . . two considerations presented themselves: first, I am incurring very little expense; second being laid up an invalid last autumn during my seed harvest, I lost doubtless many interesting things which I would have otherwise had. Lest the former should be made objection to, most cheerfully will I labour for this year without any remuneration, if I get only where with to purchase a little clothing."

Of "clothing" Douglas seems to have had but the minimum. He wrote Hooker, March 24, 1825:

"My store of clothes is very low, nearly reduced to what I have on my back,—one pair of shoes, no stockings, two shirts, two handkerchiefs,[8] my blanket and cloak: thus I adapt my costume to that of the country, as I could not carry more, without reducing myself to an inadequate supply of paper and such articles of Natural History."

There are frequent allusions to an inadequacy of garments, also to the fact that at times the officials of the Hudson's Bay Company replenished his wardrobe.

For the most part Douglas appears to have been on friendly terms with the Indians, certainly with those whom he came to know and who came to know him. He was interested in their use of the native tobacco:

"I have seen only one plant before, in the hand of an Indian two months since at the

7. In the preface of the *Transactions* of the Horticultural Society, dated January 20, 1827, appears this interesting comment; the italics are mine:

"Mr. Douglas was expected to have returned home in the last year, but finding the objects of his research still unexhausted, he *had been induced to remain another season* . . ."

8. On Douglas' next and better-equipped journey, he lists but one!

Great Falls of the Columbia, and although I offered him 2 ozs. of manufactured tobacco he would on no consideration part with it. The natives cultivate it here, and although I made diligent search for it, it never came under my notice until now . . . Fortunately I met with one of the little plantations and supplied myself with seeds and
specimens without delay. On my way home I met the owner, who, seeing it under my
arm, appeared to be much displeased; but by presenting him with two finger-lengths of
tobacco from Europe his wrath was much appeased and we became good friends. He
then gave me the above description for cultivating it. He told me that wood ashes
made it grow very large. I was much pleased with the idea of using wood ashes. Thus
we see that even the savages on the Columbia know the good effects produced on the
vegetation by the use of carbon. His knowledge of plants and their uses gained him
another finger-length. When we smoked we were all in all . . ."

Many Indians served him in the capacity of guide though not always satisfactorily
by any means. He knew how best to impress them:

"A large species of eagle . . . was perched on a dead stump close to the village; I
charged my gun with swan shot, walked up to within forty-five yards of the bird, threw
a stone to raise him, and when flying brought him down. This had the desired effect:
many of them placed their right hands on their mouths—the token of astonishment or
dread . . . I find it to be of the utmost value to bring down a bird flying when going
near the lodges, at the same time taking care to make it appear as a little thing and as
if you were not observed."

Since, as all collectors know, cones prefer the topmost branches, there were times
when Douglas could not reach them: "All the trees were too large to be cut down with
my small hatchet, and as to climbing, I have already learned the propriety of leaving no
property at the bottom of a tree."

Yet he wrote: ". . . it is worthy of notice that there has hardly ever been an instance
of dishonesty known when trust was placed in them by depositing property in their
hands." The Indians were often mystified by his habits.

"They think there are good and bad spirits, and that I belong to the latter class, in
consequence of drinking *boiling* water, lighting my tobacco pipe with my lens and the
sun . . . above all, to place a pair of spectacles on the nose is beyond their comprehen-
sion: they immediately place the hand tight on the mouth, a gesture of dread and astonishment."

On parting with one guide who had served him well, Douglas relates:

". . . he requested I would shave him, as he had pretensions to civilisation and aped
with nicety European manners. I . . . did so, and invited him to come at the New Year
to see me, when I would give him a dram, a smoke, and shave him again. He told me
. . . to let all King George's chiefs know of him when I spoke to them with paper."

But there were times when Douglas' life was in grave danger from the savages. One
such occasion was when, after months of eager anticipation, he finally beheld the
sugar pine which he had traveled many miles to see. He had learned of the tree's existence in August of 1825:

"The first thing that gave me any knowledge of it, was the very large seeds and scales of the cones which I saw in an Indian's shot-pouch; after treating him to a smoke, which must be done before any questions are put, I enquired and found it grew a little south on the mountains . . ."

But it was not until more than a year later—or on October 26, 1826—that he was able to reach a spot where the tree was growing. Douglas was alone at the time except for his ever-present little dog:

"The large trees are destitute of branches, generally for two-thirds the length of the tree; branches pendulous, and the cones hanging from their points like small[9] sugar-loaves in a grocer's shop, it being only on the very largest trees that cones are seen, and the putting myself in possession of three cones (all I could) nearly brought my life to an end. Being unable to climb or hew down any, I took my gun and was busy clipping them from the branches with ball when eight Indians came at the report of my gun. They . . . seemed to me anything but friendly . . . To save myself I could not do by flight, and without hesitation I went backwards six paces and cocked my gun . . . I was determined to fight for life . . . I stood eight or ten minutes looking at them and they at me without a word passing, till one at last . . . made a sign for tobacco, which I said they should get on condition of going and fetching me some cones. They went,[1] and as soon as out of sight I picked up my three cones and a few twigs, and made a quick retreat to my camp, which I gained at dusk . . ."

The journal then supplies a description of the cones: ". . . all containing fine seed." [2] It had been, presumably, before the arrival of the Indians that Douglas had taken the dimensions of the largest tree he ". . . could find that was blown down by the wind: Three feet from the ground, 57 feet 9 inches in circumference; 134 feet from the ground, 17 feet 5 inches; extreme length, 215 feet." [3]

Under date of April 26, 1827, Douglas included an itemized record of his mileage for the years 1825, 1826 and 1827: "I shall just state as near as possible their extent . . . My notes will show by what means it was gained." In 1825 he had traveled 2,105 miles, in 1826, 3,932 miles; and—to date of writing in 1827—995 miles. A grand total of 7,032 miles.

1827–1829

Douglas was back in London on October 15, 1827, and remained in England until late October of 1829 when he set out again for our northwest coast. Neither the Hooker

9. The height of the tree makes them appear "small." Sargent's *Manual of the trees of North America* (edition 2) describes the length of the cones as 11-18 inches, or rarely 21 inches, long.

1. To me the most remarkable part of the story!

2. Balfour comments: "It is to be regretted that there are so few trees of it in Great Britain to-day. There are growing in the valley of the Tay close to Douglas' home the finest specimens in Europe of Douglas Fir, Sitka Spruce, Thuya and the great Firs which Douglas named and introduced, but alas! the Sugar Pine has disappeared from Scotland."

3. The tallest cited by Sargent in his *Manual of the trees of North America* is 220 feet.

"Memoir" (pp. 140-146) nor the short Wilks "Memoir" in the Royal Horticultural Society volume (pp. 295, 296) gives more than a meagre picture[4] of these years. Such Douglas letters as were included in Hooker's "Memoir" relate largely to plans for his next journey.

On Douglas' return to England he was promptly elected a Fellow in the Linnean, the Zoological and the Geographical Societies of London, "free of all expense" !

Hooker quotes a letter written by William Beattie Booth:

" '. . . His company was now courted, and unfortunately for his peace of mind he could not withstand the temptation . . . of appearing as one of the *Lions* among the learned and scientific men in London . . . Flattered by their attention, and by the notoriety of his botanical discoveries . . . he seemed for a time as if he had attained the summit of his ambition . . . when the novelty of his situation had subsided, he began to perceive that he had been pursuing a shadow instead of a reality!' " [5]

Hooker himself comments that Douglas' ". . . temper became more sensitive than ever, and himself restless and dissatisfied; so that his best friends could not but wish, as he himself did, that he were again occupied in the honourable task of exploring North-west America. The Hudson's Bay Company, as upon the former occasion, made a most liberal offer of assistance, and it was resolved that he should go again to the Columbia River, partly at the expence of the Horticultural Society, and partly with the assistance of the Colonial Office[6] . . ."

Even allowing for the difference between what was considered adequate compensation in Douglas' day and now, no one can accuse the man's sponsors of over-generosity. The Horticultural Society was in dire financial difficulties at the time which explains, even if it does not justify, a somewhat penurious approach in its disbursements to Douglas.[7]

4. Chapters 13 and 14 (pp. 144-165) of Harvey (1947) are far more enlightening; Douglas is pictured, in the records of those who knew him at the time, as critical and in some ways difficult.

In extenuation of such faults there can be little doubt that the return to civilization must have required considerable readjustment, both physical and psychological.

5. In a letter written from Monterey dated November 20, 1831, Douglas indicates that he could then look back on his London years in better perspective:

"Some of my London fashionable friends . . . were glad to get rid of me . . . should I ever return I shall make but little exertion to renew my acquaintance with them, however much I may lose by it."

6. Douglas, writing on October 27, 1829, mentions that the Colonial Office ". . . pays the principal part of my expenses, and will give me compensation for my charts and for the information I may bring home on my return."

Of course he never did return. Harvey (1947) mentions that this "compensation" was not forthcoming after Douglas' death.

7. Harvey (1947) refers to the great amount of material received from Douglas on his journey of 1824–1827 with this astute comment:

"Perhaps the enthusiasm was greater from the fact that the entire cost of the expedition, including Douglas' remuneration, was under £40. According to George Bentham, 'His whole expenses for food, etc., whilst among the Indians, three years, amounted to £66 including a wager of £5 he lost to an Indian chief . . .' . . . One species of shrub alone, it was said, justified the entire expense of the expedition—

Bentham, in his Report to the Council (*Trans. Hort. Soc. London* ser. 2, 373) dated May 1, 1840, discloses a little of the Society's financial situation ten years earlier:

"It is well known, that at . . . the commencement of the year 1830, the affairs of the Society had fallen into a state of irregularity and confusion. The large amount of debt, then first made public, and the check given to its credit . . . were much increased by the loss of a considerable source of income, in consequence of the resignation of a large number of Fellows.[8] The new Council . . . found it necessary . . . to restore the Society's credit to a healthy state. Every subject of expenditure not of the most urgent necessity was unsparingly cut off . . ."

Douglas can scarcely have frittered away his entire sojourn in London. He must have spent a considerable amount of time upon his specimens—certainly they usually represent the first task undertaken by a collector back from the field. And we know that he must have written some papers—the Royal Horticultural Society volume lists eight and its Appendix VII still a ninth. Wilson reports that "In the Royal Society's catalogue, Douglas is credited with fourteen papers, which are in the transactions and journals of the Royal, Linnean, Geographical, Zoological and Horticultural Societies . . ." And when else did he have time to prepare his papers on pines and his papers on oaks to which I have already referred? And Hooker has mentioned the time spent in attempting, even if unsuccessfully, to revise his journal for publication by Murray.

Moreover, he took lessons in surveying and so on, thanks to the interest of Captain—later Sir Edward—Sabine. Wilks observes that "During the interval between his return and starting on his new journey Douglas is related to have worked no less than eighteen hours a day perfecting himself in various scientific and technical ways in which he felt himself to be deficient." [9]

Hooker considered that Douglas' last journey ". . . was undertaken under far more favourable auspices than the previous one." Perhaps he was thinking of Douglas' value to science—enhanced by his past experience and the additional knowledge ac-

the beautiful red-flowering currant *(Ribes sanguineum)*, which soon embellished the gardens of Europe."

The quotation from Bentham was from a manuscript "Autobiography" at Kew. The reference to the income from *Ribes sanguineum* is taken from John Lindley, writing in the *Botanical Register* (16: t. 1349).

8. The "considerable loss of income" occasioned by "resignations" must have been largely *on paper! The Gardener's Magazine* (6: 114, 234-252, 1830), edited by J. C. Loudon, devotes considerable space to the "Horticultural Society and Garden," to its inefficient management and extravagance. In regard to the dues a footnote (p. 236) supplies the following tidbit:

"A considerable number of Fellows are in arrear with their subscriptions, and upwards of twenty have not yet paid their admission fees. The list includes one king (George IV.), for his subscription to the garden, 500 guineas; one duke, one marquess, thirteen earls or lords, seventeen honourables or sons of lords, and twelve clergymen. The amount of unpaid admission fees amounts to upwards of 100£., and they are almost entirely due by honourables. In the list of commoners there is not a name of an obscure individual, of a nurseryman, or serving gardener; these were made to pay regularly."

9. Douglas wrote Hooker in 1832: ". . . Capt. Sabine goes so far as to say, that he can suggest to me no improvement in the manner of taking my astronomical or other observations, or in the way of recording them . . . Capt. Sabine feels . . . too true a regard for my welfare not to point out my faults, and as this letter adverts to none, I may take it for granted, I trust, that he is well pleased with me."

quired from Sabine which would make his work more significant along geographical lines—or perhaps he was merely considering the man's equipment. Douglas, himself, seems to have been pleased with this, writing Hooker that he now possessed ". . . a beautiful assortment of chronometers, etc. . ." etc.:

"Nothing pleases me so much as the addition of £20 which has been given me by the Colonial Office; I asked for £60 to provide books, tables and charts, and they sent me £80, as also some instruments, which, though previously used by other persons, are in perfectly good order. I ought to think myself a very lucky fellow, for indeed every person seems to take more interest than another in assisting me." [1]

From the "Interior of the River Columbia," in April of 1833, Douglas wrote of what he was taking into the field:

"My outfit is five pounds of tea, and the same quantity of coffee, twenty-five pounds of sugar, fifteen pounds of rice, and fifty pounds of biscuit: a gallon of wine, ten pounds of powder and as much of balls, a little shot, a small silk fishing-net, and some angling tackle, a tent, two blankets, two cotton and two flannel shirts, a handkerchief, vest, coat, and a pair of deer-skin trousers . . . two pairs of shoes, one of stockings, twelve pairs of mocassins, and a straw hat. These constitute the whole of my personal effects; also a ream and a half of paper, and instruments of various kinds; my faithful servants, several Indians, ten or twelve horses, and my old terrier . . . who has guarded me throughout all my journies, and whom . . . I mean certainly to pension off, on four pennyworth of cat's meat per day!"

Although it is impossible to imagine Douglas with (like von Resanoff) a "valet-de-chambre," he did have a servant on his last visit to America. Archibald McDonald, writing Hooker of the accident on the Fraser River, mentions his "man, Johnson." [2] As far as I have been able to discover, Douglas never once refers to him in his journals. One hopes that he proved helpful!

Douglas, still in London, outlined his future plans in a letter to Hooker dated August 6, 1829:

". . . my principal objects are to make known the vegetable treasure of the Interior of California, from the northern boundaries of Mexico, near the head of the Gulf . . . I am not quite certain, but that when I have completed my expedition on the Continent of America, I may cross to the opposite shore, and return in a southerly line, near the Russian frontier with China[3] . . . People tell me that Siberia is like a rat-trap,

1. At one point of the "Memoir" Hooker pays tribute to ". . . the generosity of Mr. Douglas' heart, and his grateful disposition whenever any act of kindness was shown him."

2. Henry F. Reed (*Oregon Hist. Quart.* 34: 314-323, 1933)—although he does not refer to the man's services with Douglas—mentions him thus:

"In the light of the historical evidence now available, William Johnson, a British subject and retired servant of the Hudson's Bay Company, is entitled to the honor of being the first white man to settle permanently on the site of Portland, or for that matter within the boundaries of Multnomah County!"

3. Later, September 14, 1829, he indicates that he was making such plans as he could for this trip:

". . . I spent ten days with Captain Lutke . . . he gave me letters to Baron Wrangel, Governor of the

which there is no difficulty in entering, but from which it is not so easy to find egress. I mean at least to put this saying to the test . . . trifles do not stop me . . ."

Just before departure from England Douglas wrote Hooker (October 27, 1829):

"I cannot tell you how pleased I am to have seen the first part of your *Flora Boreali-Americana* before sailing, and that I am enabled to take it with me to America. The map is good . . . The plates are truly beautiful; but I see you have not given a picture of *Pæonia (P. Brownii)* . . ."

Douglas was then ". . . hourly expecting the summons to sail, and . . . not aware that we shall touch at any place, except the Sandwich Islands, where it is intended to make a short stay."

Douglas' third and last journey to North America took place in the decade of 1830–1840 and is described in my chapter XVIII.

Russian Colonies in America, and of the Aleutian Islands, as also circulars to Siberia. The Baron is a man of vast information, and joins heart and hand with all those who have scientific views."

As we shall see, the prospective trip never materialized.

ALDEN, ROLAND HERRICK & IFFT, JOHN DEMPSTER. 1943. Early naturalists in the far west. *Occas. Papers Cal. Acad. Sci.* 20: 32-37.

ANDERSON, RUFUS. 1864. The Hawaiian Islands: their progress and condition under missionary labors. Boston.

Anonymous. 1862. The giant-pine discoverer and his fate. *Leisure Hour* 11: 454, 455.

—— 1883. David Douglas. The story of a botanical collector. *Leisure Hour* 32: 206-210.

BALFOUR, FREDERICK ROBERT STEPHEN. 1942. David Douglas. *Jour. Royal Hort. Soc.* 67: 121-128; 153-162. Map. [This map, fig. 42, is a reproduction of the one in Hooker's *Flora boreali-americana*, pt. 1, 1829, which shows Douglas' travels in North America in 1825–1827, etc.]

BANCROFT, HUBERT HOWE. 1885. The works of . . . 20 (California III. 1825–1840): 403-405, & *fn.* 37-41; 777, 778.

—— 1884. The works of . . . 28 (Northwest coast II. 1880–1846): 507-511, & *fn.* 1-4.

BARNSTON, GEORGE. 1860. Abridged sketch of the life of Mr. David Douglas, botanist, with a few details of his travels and discoveries. *Canad. Nat.* 5: 120-132, 200-208, 267-278, 329-349.

BENT, ALLEN H. 1913. Early American mountaineers. *Appalachia* 13: 60.

BENTHAM, GEORGE. 1835. Report on some of the more remarkable hardy ornamental plants raised in the Horticultural Society's garden from seeds received from Mr. David Douglas, in the years 1831, 1832, 1833. *Trans. Hort. Soc. London* ser. 2, 1: 403-414, 476-481.

—— 1842. Report on the progress of the Horticultural Society of London, from May 1, 1830, to April 30, 1840. *Trans. Hort. Soc. London* ser. 2, 2: 373-377.

BOOTH, WILLIAM BEATTIE. 1835. Some account of the late Mr. Douglas, the botanist. *Gard. Mag.* (Loudon) new ser. 1: 271, 272. [Stated to have been reprinted from *West Briton and Cornwall Advertiser*.]

BOULGER, GEORGE SIMONDS. 1888. David Douglas. *Dict. Nat. Biog.* 15: 291.

BREWER, WILLIAM HENRY. 1880. List of persons who have made botanical collections in California. *In:* Watson, Sereno. Geological survey of California. Botany of California 2: 554, 555.

BRITTEN, JAMES & BOULGER, GEORGE SIMONDS. 1931. A biographical index of deceased British and Irish botanists. ed. 2. London. 94.

BROSNAN, CORNELIUS JAMES. 1935. History of the state of Idaho. rev. ed. New York. 45.

CANDOLLE, ALPHONSE DE. 1880. La phytographie; ou l'art de décrire les végétaux considérés sous différents points de vue. Paris. 406.

DOUGLAS, DAVID. 1827. An account of a new species of Pinus [*P. Lambertiana*], native of California: in a letter to Joseph Sabine, Esq., F.R. and L.S., Secretary of the Horticultural Society. By David Douglas, A.L.S. Communicated by Mr. Sabine . . . *Trans. Linn. Soc. London* 15: 497-500. Read November 6, 1827.

———— 1833 [1829–1833]. Description of a new species of the genus Pinus [*P. Sabiniana*]. By David Douglas, F.L.S. Communicated by the Horticultural Society. *Trans. Linn. Soc. London* 16: 747-749. Read April 3, 1832.

———— 1834. Extract from a private letter addressed to Captain Sabine, R.A., F.R.S., by Mr. David Douglas, F.L.S. [Dated: "Woahoo (Sandwich Islands) 3rd of May, 1834."] *Jour. Royal Geog. Soc. London* 4: 333-345.

———— 1838. The great crater of Mouna Loa, Hawaii. Extracts from the unpublished correspondence of the late David Douglas, Esq. *Hawaiian Spectator* 1: 98-106.

———— 1923. A David Douglas letter. [Dated: "Monterey, Upper California, November 20, 1831."] *Quart. Cal. Hist. Soc.* 2: 223-227.

———— For journals and letters of David Douglas *see also:* Hooker, W. J. 1827; 1836–1837; Parry, C. C. 1883; Royal Horticultural Society (London) 1914; Wilson, W. J. 1919; Jepson, W. L. 1933.

DRUMMOND, THOMAS. 1830. Sketch of a journey to the Rocky Mountains and to the Columbia River in North America. *Bot. Miscel.* 1: 178-219.

DU FOUR, CLARENCE JOHN. 1933. The Russian withdrawal from California. *Quart. Cal. Hist. Soc.* 12: 240-249. *Reprinted: Cal. Hist. Soc. Publ.* 7: The Russians in California. 1933.

EASTWOOD, ALICE. 1939. Early botanical explorers on the Pacific coast and the trees they found there. *Quart. Cal. Hist. Soc.* 18: 341.

GOLDIE, JOHN [3rd]. 1938. In memory of David Douglas. *Brit. Col. Hist. Quart.* 2: 89-94.

GRAY, ASA. 1840. Notices of European herbaria, particularly those most interesting to the North American botanist. *Am. Jour. Sci. Arts* 40: 1-18. *Reprinted:* Sargent, Charles Sprague, *compiler.* 1889. Scientific papers of Asa Gray. 2 vols. Boston. New York. 2: 1-21.

———— 1844. The longevity of trees. *North Am. Review* 59: 189-230. *Reprinted: Ibid.* 2: 71-124.

———— 1866. William Jackson Hooker. *Am. Jour. Sci. Arts* ser. 2, 41: 1-10. *Reprinted: Ibid.* 2: 321-332.

———— 1873. John Torrey; a biographical notice. *Proc. Am. Acad. Arts Sci.* 9: 262-271. *Reprinted: Ibid.* 2: 359-369.

———— 1893. Letters of Asa Gray. Edited by Jane Loring Gray. 2 vols. Boston. New York. 1: 58, 123.

HANSEN, CARL. 1892. Pinetum Danicum. Conifers collected and observed by Professor Carl Hansen . . . Copenhagen . . . Notes sent to the Conifer Conference held at Chiswick, October, 1891. *Jour. Royal Hort. Soc.* 14: 257-480.

HARVEY, ATHELSTAN GEORGE. 1940. David Douglas in British Columbia. *Brit. Col. Hist. Quart.* 4: 221-243.

———— 1947. Douglas of the fir. A biography of David Douglas botanist. Cambridge, Mass.

HEMENWAY, ANSEL FRANCIS. 1904. Botanists of the Oregon country. *Quart. Oregon Hist. Soc.* 5: 207-214.

HOOKER, WILLIAM JACKSON. 1827. Account of an expedition under Captain Franklin, and of the vegetation of North America, in extracts of letters from Dr. Richardson, Mr. Drummond, and Mr. Douglas. Communicated by Dr. Hooker. *Edinburgh Jour. Sci.* 6: 107-117.

———— 1829-1840. Flora boreali-americana; or, the botany of the northern parts of British America; compiled principally from the plants collected by Dr Richardson & Mr Drummond on the late northern expeditions, under command of Captain Sir John Franklin, R.N. to which are added . . . those of Mr Douglas, from north-west America, and of other naturalists. Map. 2 vols. London.

———— 1836-1837. A brief memoir of the life of Mr. David Douglas, with extracts from his letters. *Comp. Bot. Mag.* 2: 79-192. [Issued: 79-96, Oct. 1, 1836; 97-128, Nov. 1, 1836; 129-160, Dec. 1, 1836; 161-192, Jan. 1, 1837.] *Reprinted: Quart. Oregon Hist. Soc.* 5: 215-222 [Literary remains of David Douglas, botanist of the Oregon country . . . Editorial prefatory notes.], 223-229, 325-369 (1904); 6: 76-97, 206-227, 286-309, 417-449 (1905). Edited by Frederick George Young. [For the content of Hooker's A brief memoir . . . see my pp. 300-305.]

———— & WALKER-ARNOTT, GEORGE ARNOTT. 1830-1841. The botany of Captain Beechey's voyage; comprising an account of the plants collected by Messrs Lay and Collie, and other officers of the expedition, during the voyage to the Pacific and Bering's Strait, performed in His Majesty's ship Blossom, under the command of Captain F. W. Beechey, R.N. in the years 1825, 26, 27, and 28. London. ["California.—Supplement." 316-409. Issued from 1840-1841.]

HORTICULTURAL SOCIETY OF LONDON. 1825-1827. 1827-1831. *Transactions* 6: iv, v (1826); 7: ii, iii (1830).

HOWELL, JOHN THOMAS. 1937-1939. A collection of Douglas' western American plants. *Leafl. West. Bot.* 2: 59-62 (1937); 74-77, 94-98, 116-119, 139-144 (1938); 170-174, 189-192 (1939).

———— 1949. Concerning David Douglas. *Leafl. West. Bot.* 3: 160-162.

JACKSON, BENJAMIN DAYDON. 1905. George Bentham. London.

JEPSON, WILLIS LINN. 1899. Epitaph of David Douglas. *Erythea* 7: 174.

———— 1933. David Douglas in California. *Madroño* 2: 97-100.

KUYKENDALL, RALPH SIMPSON & DAY, ARTHUR GROVE. 1848. Hawaii. A history. From Polynesian kingdom to American commonwealth. New York.

LAMBERT, AYLMER BOURKE. 1832. A description of the genus Pinus, with directions relative to the cultivation, and remarks on the uses of the several species: also descriptions of many other new species of the family of Coniferae. 2 vols. London. ed. 3. 1: 57, 58, t. 34 [*Pinus Lambertiana*, Gigantic Pine].

LASÈGUE, ANTOINE. 1845. Musée botanique de M. Benjamin Delessert. Notices sur les collections de plantes et la bibliothèque qui le composent; contenant en outre des documents sur les principaux herbiers d'Europe et l'exposé des voyages entrepris dans l'intérêt de la botanique. Paris. 207, 268.

LOUDON, JOHN CLAUDIUS. 1836. Biographical notice of the late Mr. David Douglas, F.L.S., the traveller and botanist; with a proposal to erect a monument to his memory; and the list of plants which he introduced. Drawn up from various communications by the conductor of The Gardener's Magazine. *Gard. Mag.* (Loudon) new ser. 2: 602-609.

MCLOUGHLIN, JOHN. 1941. The letters of John McLoughlin from Fort Vancouver to the Governor and Committee First Series, 1825–38. Edited by E. E. Rich . . . with an introduction by W. Kaye Lamb . . . Toronto. *Publications of the Champlain Society* 4: 15, 37, 338. Appendix A.

MURRAY, J. M. 1931. David Douglas. *Forestry* (London) 5: 144-148.

PARRY, CHARLES CHRISTOPHER. 1883. Early botanical explorers of the Pacific coast. *Overland Monthly* ser. 2, 2: 409-416.

PIPER, CHARLES VANCOUVER. 1906. Flora of the state of Washington. *Contr. U. S. Nat. Herb.* 11: 12, 13.

ROYAL HORTICULTURAL SOCIETY (London), Published under the direction of. 1914. Journal kept by David Douglas during his travels in North America, 1825–1827. Together with a particular description of thirty-three species of American oaks and eighteen species of pinus. With appendices containing a list of the plants introduced by Douglas and an account of his death in 1834. London. [For the content of this volume see my pp. 305-309.]

SARGENT, CHARLES SPRAGUE. 1891. The Silva of North America. Boston. New York. 2: 94, *fn.* 1. [Throughout this work (14 vols.) are many references to trees discovered by David Douglas.]

SCOULER, JOHN. For his journals *see* p. 298.

SOMERVILLE, THOMAS. 1871. An early hero of the Pacific. Douglas, the botanist. *Overland Monthly* 7: 105-113.

SOMERVILLE, WILLIAM. 1915. David Douglas. *Quart. Jour. Forestry* (London) 9: 151-157.

SPRAGUE, THOMAS ARCHIBALD. 1933. The dates of Hooker's "Companion to the Botanical Magazine." *Bull. Misc. Inf.* Kew: 362-364.

STILLMAN, JACOB DAVIS BABCOCK. 1869. Footprints of early California discoverers. *Overland Monthly* 2: 256-263.

WAGNER, HENRY RAUP. 1921. The plains and the Rockies. A bibliography of original narratives of travel and adventure 1800–1865. San Francisco. Nos. 31 (p. 25), 50 (p. 34).

WILSON, WILLIAM FREDERICK, *compiler.* 1919. David Douglas, botanist at Hawaii. Honolulu.

WYETH, NATHANIEL JARVIS. 1899. The correspondence and journals of Captain Nathaniel J. Wyeth 1831—6. A record of two expeditions for the occupation of the Oregon country; with maps, introduction and index. Edited by F. G. Young. *In: Sources of the history of Oregon* 1: parts 3-6. Eugene.

YOUNG, FREDERICK GEORGE, *editor. See:* Hooker, W. J., 1836–1837, and Wyeth, N. J., 1899.

CHAPTER XV

COLLIE AND LAY VISIT CALIFORNIA IN THE BLOS-

SOM COMMANDED BY BEECHEY

FREDERICK William Beechey, English navigator and son of the portrait-painter Sir William Beechey, was born in 1796 and died in 1856. In 1818 he accompanied Captain John Franklin on his first voyage to the Arctic and a year later served with Edward William Parry when that navigator was searching for a northwest passage. His next trip, the subject of this chapter, was described in two volumes published in London in 1831: *Narrative of a voyage to the Pacific and Beering's Strait, to co-operate with the polar expeditions: performed in his Majesty's ship Blossom, under the command of Captain F. W. Beechey, R.N. . . . in the years 1825, 26, 27, 28 . . .* Beechey's *Narrative* is easy to follow, for year and month appear marginally on every page so that daily records are clear. There is also a good map. The book was dedicated "To the King." [1]

The introduction explains the reason for the expedition. In 1824 Captain Parry had been directed to search for the northwest passage by way of Prince Regent's Inlet. Captain Franklin was ". . . to connect his brilliant discoveries at the mouth of the Coppermine River with the furthest known point on the western side of America, by descending the Mackenzie River, and, with the assistance of his intrepid associate, Dr. Richardson, by coasting the northern shore in opposite directions towards the two previously discovered points . . ."

Should these two expeditions reach the west coast it was anticipated that their resources would be exhausted; Franklin, moreover, would have no ship in which to return to England. Beechey's orders were to meet the two explorers at Chamisso Island in Kotzebue Sound and take Franklin back to England; Parry, his supplies replenished, was apparently to return in his own ship.

On leaving England, Beechey was to cross the Atlantic to Rio de Janeiro, go round Cape Horn, visit certain islands—mainly of the Polynesian group—in the South Pacific, proceed to the Hawaiian Islands, and from there to Bering's Strait; he was to be at Kotzebue Sound not later than July 10, 1826, and there await the arrival of the explorers until the end of October or as late as the season permitted. Should the explorers fail to arrive by that time he was to seek a more auspicious wintering-place and be back at Chamisso Island by August 1, 1827. The entire trip and its various contingencies are carefully outlined.

In his introduction to the *Narrative* Beechey describes his ship, a sloop:

---

1. Presumably to the British monarch William IV, brother and successor of George IV, who had died on June 26, 1830, for the *Narrative* was published in 1831.

"The vessel selected for this service was his Majesty's ship Blossom, of twenty-six guns, but on this occasion mounting only sixteen; and on the 12th of January, 1825, I had the honour of being appointed to the command of her."

Among the officers—"most of them men distinguished for their abilities"—was a naturalist, George Tradescant Lay, and a surgeon, Alexander Collie.[2] Edward Belcher, who was later to visit western North America in the *Sulphur,* was assistant-surgeon. The ship sailed with a complement of one hundred. The cargo included presents for the kings of the Society and Sandwich Islands. "The College of Surgeons sent bottles of spirits for the preservation of specimens, and the Horticultural Society enhanced our extra stores with a box of seeds[3] properly prepared for keeping."

Instructions from the Lords of the Admiralty included the following:

"As we have appointed Mr. Tradescant Lay as naturalist on the voyage, and some of your officers are acquainted with certain branches of natural history, it is expected that your visits to the numerous islands of the Pacific will afford the means of collecting rare and curious specimens in the several departments of this branch of science. You are to cause it to be understood that two specimens, *at least,* of each article are to be reserved for the public museums; after that the naturalist and officers will be at liberty to collect for themselves.[4] You will pay every attention in your power to the preservation of the various specimens of natural history, and on your arrival in England transmit them to this office; and if, on your arrival at any place in the course of your voyage, you should meet with a safe conveyance to England, you are to avail yourself of it to send home any despatches you may have, accompanied by journals, charts, drawings, &c., and such specimens of natural history as may have been collected. And you will, on each of your visits to Owhyhee, deliver to his Majesty's consul at that place duplicates of all your previous collections and documents, to be transmitted by him, by the first opportunity, to England."

In the event of war breaking out while the ships were away from England Beechey was instructed to see that no hostile act was committed, ". . . the vessel you command being sent out only for the purpose of discovery and science, and it being the practice of all civilized nations to consider vessels so employed as excluded from the operations of war . . ."

Finally, as to the publication of the records:

"The collections of botanical and other specimens of natural history have been reserved for separate volumes, being far too numerous to form part of an appendix to the present narrative. His Majesty's government having liberally appropriated a sum

2. Beechey pays tribute to Collie ". . . for his unremitting attention to natural history, meteorology, and geology . . ."

3. These were intended for distribution. This practice began very early. I am told that the navigator, Pedro Fernandez de Queiros, recorded planting maize on the Marquesas Islands on his voyage of 1605. The custom explains the presence of many plants in regions far removed from their native land.

4. A more liberal provision than that offered by the United States government when, for example, the scientists of the "Yellowstone Expedition" went into the field. But Britain had had more experience in such matters!

of money to their publication, I hope, with the assistance of several eminent gentlemen, who have kindly and generously offered to describe them, shortly to be able to present them to the public, illustrated by engravings by the first artists. The botany, of which the first number has already been published, is in the hands of Dr. Hooker, professor of Botany, at Glasgow, who in addition to having devoted the whole of his time to our collection, has borne with the numerous difficulties and disappointments which have attended the progress of the publication of this branch of natural history . . . The department which he has so kindly undertaken will extend to ten numbers 4to., making, in the whole, about 500 pages, and 100 plates of plants, wholly new, or such as have been hitherto imperfectly described . . . notwithstanding the liberal assistance of his Majesty's government, there is so little encouragement for works of the above-mentioned description, that they could not be published unless the contributions were gratuitously offered to the publishers." [5]

I now follow the *Narrative*. The year was 1825.

The *Blossom* sailed from Spithead, Hampshire, England, on May 19, left Rio de Janeiro on August 13, and "made" Cape Horn on September 16; after touching at Conception and at Valparaiso, Chile, she reached Easter Island on November 16. From December 5 to 21 she visited Pitcairn Island—approximately seventy pages are devoted to an account of the mutiny on the *Bounty* and to the subsequent adventures of the mutineers, their condition and mode of life at the time of the *Blossom's* visit. [6]

1826

On March 18, 1826, the *Blossom* reached "Otaheite," [7] in the Society Islands; and, on May 20, anchored off ". . . Honoruro, the principal port of the Sandwich Islands . . ."

Some of those on the ship had been "afflicted with dysentery" after leaving "Otaheite" and ". . . on the 7th, great apprehensions were entertained for Mr. Lay, the naturalist . . ." On leaving "Woahoo" on May 31, Lay was left behind. [8]

"At the time of our departure the health of Mr. Lay was by no means restored, and

5. The edition of *The botany of Captain Beechey's voyage* must have been limited to a few hundred copies. It now commands $100 per copy and is almost never offered by dealers.

6. Of the original nine white male settlers there remained but one, John Adams (known as Alexander Smith when on the *Bounty*), who recounted and signed the story transcribed in the *Narrative*. Beechey's party found the inhabitants (then sixty-six in number) hospitable, law-abiding, and satisfied with their extremely simple living conditions. There was little illness on Pitcairn. They were fearful lest the *Blossom* might have brought disease in some form and asserted ". . . that the Briton left them headaches and flies . . ."

7. Here, at Tahiti, they received a visit from ". . . Captain Charlton, his Majesty's consul for the Society and Sandwich Islands . . ." This was undoubtedly the Charlton who, after David Douglas' death in the Sandwich Islands in 1834, forwarded his effects, together with the letters of the missionaries describing what was known of the tragedy.

8. He rejoined the expedition only when it returned to the Hawaiian Islands on January 25, 1827.

as it appeared to me that his time during the absence of the ship could be more profit-
ably employed among the islands of the Sandwich groupe than on the frozen shores of
the north, he was left behind . . . Mr. Collie took upon himself the charge of naturalist,
and acquitted himself in a highly creditable manner."

From the Sandwich Islands the *Blossom* moved on to "Kamschatka," and reached
Petropaulski[9] on June 27. There were dispatches telling of Captain Parry's return to
England so that, henceforth, Beechey had only to consider the needs of Captain
Franklin. When Kamtchatka was left on July 1, the *Blossom* had but fourteen days
to reach Kotzebue Sound and she only got to Chamisso Island on July 25—about ten
days later than the date specified in Beechey's orders.

While waiting for Franklin's arrival, no time was wasted. A survey (August 17-
September 9) was made to the northward and a detachment reached a "tongue of
land" which Beechey named Point Barrow. Caches were made and instructions left at
various points as arranged with Franklin. When the explorer had not appeared by
October 14, Beechey, for the safety of his ship and his men, started south, passed
through the Aleutian Chain and on November 5 the party was in sight of "New Albion." Our interest in the voyage begins at this point.

November 5: ". . . we made the high land of New Albion about Bodega, and soon
afterwards saw Punta de los Reyes, a remarkable promontory, from which the general line of coast turns abruptly to the eastward, and leads to the port of San Francisco. We stood to the southward during the night . . . The weather was very misty,
and a long swell rolled towards the reefs . . ."

November 6: "The next evening we passed Púnta de los Reyes, and awaited the return of day off some white cliffs . . . in all probability those which induced Sir Francis
Drake to bestow upon this country the name of New Albion . . ."

The second volume of the *Narrative* begins at this point, with some eighty pages devoted to the visit to California:

November 7: "When the day broke, we found ourselves about four miles from the
land. It was a beautiful morning, with just sufficient freshness in the air to exhilarate
without chilling. The tops of the mountains, the only part of the land visible, formed
two ranges, between which our port was situated . . . We bore up for the opening between the ranges . . . At length two promontories, the southern one distinguished by a
fort and a Mexican flag, marked the narrow entrance of the port."

Although everyone was anxious to land, the *Blossom* was becalmed at this juncture but eventually it ". . . dropped . . . anchor in the very spot where Vancouver had
moored his ship thirty-three years before." As they passed the fort ". . . a soldier protruded a speaking-trumpet through one of the embrasures, and hailed us with a sten-

9. Here they found Baron Wrangel and his ship *Modeste*. David Douglas, who was in touch with the
Baron, thought highly of him.

torian voice, but we could not distinguish what was said. This custom of hailing vessels has arisen from there being no boat belonging to the garrison, and the inconvenience [must have been] felt by the governor, in having to wait for a report of arrivals, until the masters of the vessels could send their boats on shore."

Once past the fort the ". . . port of San Francisco . . . breaks upon the view, and forcibly impresses the spectator with the magnificence of the harbour. He then beholds a broad sheet of water, sufficiently extensive to contain all the British navy, with convenient coves, anchorage in every part, and, around, a country diversified with hill and dale, partly wooded, and partly disposed in pasture lands of the richest kind, abounding in herds of cattle . . . So poorly did the place appear to be peopled that a sickly column of smoke rising from within some dilapidated walls, misnamed the presidio or protection, was the only indication we had of the country being inhabited.

"The harbour stretches to the S.E. to the distance of thirty miles, and affords a water communication between the missions of Sán José, Sánta Clára, and the presidio . . . A creek on the N.W. side of this basin . . . leads up to the new mission of Sán Francisco Solano . . . a strait to the eastward, named Estrécho de Karquines, communicates with another basin into which three rivers discharge themselves . . . named Jesus Maria, El Saraménto, and Sán Joachin . . . we noticed seven American whalers at anchor at Sausalito, not one of which showed their colours; we passed them and anchored off a small bay named Yerba Buena, from the luxuriance of its vegetation . . . I immediately went on shore to pay my respects to Don Ignacio Martinez, a lieutenant in the Mexican army, acting governor in the absence of Don Louis . . ."

They asked at once about the possibility of obtaining supplies; but ". . . it seemed that with the exception of flour, fresh beef, vegetables, and salt, which might be procured through the missions, we should have to depend upon the American vessels for whatever else we might want, or upon what might chance to be in store at Monterey, a port of more importance than Sán Francisco, and from being the residence of a branch of a respectable firm in Lima, better supplied with the means of refitting vessels after a long sea voyage."

Most visitors to Californian ports had been critical of Spanish rule. Now, with Mexico in control since 1821, there had been a further deterioration in morale and in living conditions. Beechey discusses the situation at great length and from many angles. The governor, the soldiers, the mission fathers, had much to contend with. The salary of the acting-governor ". . . was then eleven years in arrears . . ." The Mexican government had attempted to pay its dependents in cigars instead of in dollars, even sending a boatload of this commodity but ". . . cigars would not satisfy the families of the soldiers, and the compromise was refused . . . Fortunately for Martinez and other veterans in this country, both vegetable and animal food are uncommonly cheap, and there are no fashions to create any expense of dress . . . The neglect of the government to its establishments could not be more thoroughly evinced than in the dilapidated condition of the building in question [the governor's abode]; and such was the dissatisfaction of the people that there was no inclination to improve their situation, or even to remedy many of the evils which they appeared to have the power to remove.

"The plain upon which the presidio stands is well adapted to cultivation; but it is scarcely ever touched by the plough, and the garrison is entirely beholden to the missions for its resources."

The soldiers, unpaid for many years, found living costly. Duties, forty-two and one half per cent, were imposed not only on imported foreign goods but on those from Mexican territory: ". . . under the old government, two ships were annually sent from Acapulco with goods, which were sold duty free, and at their original cost . . ." Certain privileges related to land ownership had been rescinded and tenure of property was uncertain—Mexico was considering sending settlers to California and wished to retain its hold on desirable property in the event that it might be needed.

Discontent pervaded the missions also. The salary of the padres, four hundred dollars *per annum,* had been discontinued. They were obliged to swear allegiance to the reigning authorities and regarded this as a violation of their earlier oath to the king of Spain:

". . . until he renounced his sovereignty over the country, they could not conscientiously take it . . . so much did they regard this pledge, that they were prepared to leave the country, and to seek an asylum in any other[1] that would afford it them."

Another grievance of the church fathers was the order to free all converted Indians[2] of good character. While theoretically an excellent idea it did not work when put into practice, for not only did it deprive the missions of their workers, but the Indians possessed ". . . neither the will, the steadiness, nor the patience to provide for themselves." It was a vicious circle: ". . . while the missions furnish the means of subsistence to the presidios, the body of men they contain keeps the wild Indians in check, and prevents their making incursions on the settlers."

All in all Beechey was not optimistic. "Husbandry is still in a very backward state, and it is fortunate that the soil is fertile, and that there are abundance of labourers to perform the work, or I verily believe the people would be contented to live upon acorns. Their ploughs appear to have descended from the patriarchial ages . . ."

There were but few settlers unconnected with either the garrisons or the presidios.

"In this part of California, besides the missions, there are several pueblos, or villages, occupied by Spaniards and their families, who have availed themselves of the privileges granted by the old government, and have relinquished the sword for the ploughshare. There are also a few settlers who are farmers, but, with these exceptions, the country is almost uninhabited."

At this point the *Narrative* gives an account of an overland trip from San Francisco to Monterey where it was hoped to replenish the ship's supplies. Beechey did not make the trip but compiled his account from the journals of those who did. The story con-

1. "Manilla" was one of the places considered.

2. Beechey has more to tell of this order when he returned to California in 1827.

tains nothing of botanical interest but does give a good picture of what the author calls ". . . the deserted state of the country . . ." It seems to have been the first overland trip in California made to date by any of the expeditions with which my story is concerned—other visitors had, apparently, kept close to port. Collie, acting-naturalist in the absence of Lay, went along; also the purser, named Marsh, and as interpreter a man named Evans. They had an escort and the padre of the mission loaned horses[3]—and free of charge! On November 9 the party started, led by a "dragoon of California" gaudily dressed, with a "tremendous pair of iron spurs, secured by a metal chain," and mounted on a saddle—evidently the Mexican variety—which ". . . retained him in his seat by a high pummel in front and a corresponding rise behind." This individual carried the passport. Like all mariners once in the saddle, they set off "at a round trot."

The first day they got as far as the Mission of Santa Clara, about forty miles from San Francisco. They had gone by way of the Mission of San Francisco, followed along the "Sierre de Sán Bruno"—noting on the summit of one of its branches stretching into an arm of the sea called "Estrecho de Sán José" a thick wood ". . . called Palos Colorados from its consisting principally of red cedar pine . . ."—and had nooned at ". . . a small cottage called Burri Burri, about twelve miles from Sán Francisco." They had then descended upon a plain named Las Salinas—probably because it was "overflowed occasionally by the sea." Passing hills known as "the Sierra del Sur," the road ". . . opened out upon a wide country of meadow land, with clusters of fine oak free from underwood. It strongly resembled a nobleman's park: herds of cattle and horses were grazing upon the rich pasture, and numerous fallow-deer . . . The resemblance . . . could be traced no further. Instead of a noble mansion . . . a miserable mud dwelling . . . a number of half-naked Indians . . . basking in the sun . . . This spot is called Sán Mateo, and belongs to the mission of Sán Francisco.

". . . they arrived at a farm-house about half way between Sán Francisco and Santa Clara, called Las Pulgas (fleas); a name which afforded much mirth to our travellers, in which they were heartily joined by the inmates of the dwelling, who were very well aware that the name had not been bestowed without cause."

The plain continued, but now ". . . the noble clusters of oak were . . . varied with shrubberies, which afforded a retreat to . . . Californian partridges, of which handsome species of game the first specimen was brought to England by the Blossom, and is now living in the gardens of the Zoological Society . . ."

On arrival at Santa Clara mission the whole party ". . . soon fell asleep—thanks to excessive weariness—and slept as soundly as *las pulgas* would let them." Santa Clara "On the whole . . . is one of the best regulated and most cleanly missions in the country. Its herds of cattle amount to 10,000 in number, and of horses there are about 3000."

3. The procurement of these animals was always a serious problem, for there were many tricks in the trade; a favorite one seems to have been to sell the horses one day and set them free at night—making it necessary for the purchaser to rent other steeds the next day, or walk.

It possessed ". . . some excellent orchards, producing an abundance of apples and pears. Olives and grapes are also plentiful, and the padres are enabled to make from the latter about twenty barrels of wine annually. They besides grew a great quantity of wheat, beans, peas, and other vegetables."

Leaving the Mission of Santa Clara, they reached, the second day, the Mission of San Juan, fifty-four miles distant. They had first passed through a ". . . beautiful avenue of trees, nearly three miles in length . . ." leading from the mission to the pueblo of San José, ". . . the largest settlement of the kind in Upper California . . ." Its five hundred inhabitants—retired soldiers and their families—lived in mud houses ". . . miserably provided in every respect . . ." After eighteen miles they had reached ". . . the banks of a limpid stream, the first . . . seen in their ride." Passing hills near the coast called *"El ójo del cóche,"* they had then descended into the fertile plain of *"Las Llágas"* near the end of which were ". . . some cottages, named *Ranchas de las Animas* . . ." Here they were fed "jerk beef" served from silver dishes, ate with silver spoons and drank from silver cups. While the setting seemed incongruous, it was the Spanish custom. When they arrived at the Mission of San Juan, they were welcomed ". . . by padre Arroyo, who in hospitality and good humor endeavoured to exceed even the good father of Santa Clara."

Since the following day was a holiday the vaquéros had gone off with the saddles and the party was obliged to remain at the mission. They were taken to see some Indian dwellings:

". . . thirty huts belong to some newly converted Indians of the tribe of Tooleerayos *(bulrushes)* . . . The exterior . . . of these wretched wigwams greatly resembles a bee-hive . . . They seemed to have lost all the dignity of their nature; even the black-birds . . . had ceased to regard them as human beings, and were feeding in flocks among the wigwams . . ."

The padre fed his visitors well—". . . various exquisite dishes . . . successively smoked on the board of the generous priest . . ."—and they were surprised to be told, in such a remote spot, that a Spaniard had effected the Northwest Passage. Their informant was merely describing "the fictitious voyage of Maldonado." [4] Padre Arroyo was a simple[5] man; he was also a "grammarian," and had written a vocabulary and grammar of the Indian languages. He gave his visitors a ". . . disquisition on the etymology of the name of the Peninsula of California." His opinion on the derivation of the name is given at some length in the *Narrative*.

Leaving the "plain of Sán Juan" on the fourth day, the party ascended some steep hills ". . . commanding a view of the spacious bay of Monterey. Then winding among

4. Both Quadra and Vancouver, states Beechey, had ". . . satisfactorily proved the voyage in question to have been a fabrication."

5. Simple indeed, as Beechey points out: after a youthful Indian couple had fallen in love and eloped, they had been brought back to the mission where the padre, ". . . to punish this misbehavior, incarcerated them together, and kept them thus confined until they had expiated their crime."

valleys, one of which was well wooded and watered, they entered an extensive plain called 'Llano del Rey,' which, until their arrival, was in the quiet possession of numerous herds of deer and jackals. This tract of land is bounded on the north, east and south-east, by mountains which extend with a semicircular sweep from the sea at Santa Cruz, and unite with the coast line again at Point Pinos. It is covered with a rank grass, and has very few shrubs. In traversing this plain, before they could arrive at some ranchos, named Las Salinas . . . the party had to wade through several deep ditches and the Rio del Rey, both of which were covered with wild ducks. The cottages called Las Salinas are on the farm of an old Scotchman, to whom the land was granted in consequence of some services which he had rendered to the missions. They rested here . . .

"The remainder of the plain over which they passed toward Monterey was sandy, and covered with fragrant southernwood, broken here and there by dwarf oaks, and shrubs of the syngenesious class of plants. As they approached the town, pasture lands . . . succeeded this wild scenery . . . [then] trees of luxuriant growth, houses scattered over the plain, the fort, and the shipping in the bay, announced the speedy termination of their journey. At five o'clock in the evening they alighted in the square at Monterey, and met with a kind reception from Mr. Hartnell,[6] a merchant belonging to the firm of Begg and Co. in Lima, who was residing there, and who pressed them to accept the use of his house while they remained in the town—an offer of which they thankfully availed themselves."

The governor was Gonzales, ". . . an officer who had been raised by his own merit from the ranks to be captain of artillery and governor of Monterey: his family were residing with him, and having been educated in Mexico, complained bitterly of their banishment to this outlandish part of the world, where the population, the ladies in particular, were extremely ignorant, and wretched companions for the *Mexicanus instruidas* . . . there are no balls or bull-fights in Monterery; and for all the news they heard of their own country, they might as well have been in Kamschatka . . ."

As far as the needed supplies and medicines were concerned ". . . there were no medicines to be had . . . some stores . . . essential to the ship could nowhere be procured. The exchange on bills was favourable, but there was no specie . . ." Marsh, the purser, bought what he could from the inhabitants and from the "shipping in the roads," and ". . . arranged with a person who had come out from Ireland for the purpose of salting meat for the Lima market, to cure a quantity for the use of the ship, and to have it ready on her arrival in Monterey."

The horse-problem delayed departure from Monterey and also made it necessary to return to San Francisco by the same route; nor when the party again reached the Mission of San Juan did they find the padre as helpful as he had been a few days before. Seventeen shillings per horse was the purchase price demanded ". . . in California . . . considered so exhorbitant that our shipmates did not think proper to suffer the imposition, and awaited the horses belonging to the mission. After a day's delay,

6. This presumably was the W. E. P. Hartnell with whom David Douglas was to reside while at Monterey in 1831.

during which they again heard many invectives against the new government of Mexi-co . . . they resumed their journey, and arrived at Sán Francisco on the 17th of No-vember."

The entire trip had taken eight days.

"In this route it will be seen that, with the exception of the missions and pueblos, the country is almost uninhabited; yet the productive nature of the soil, when it has been turned up by the missions, and the immense plains of meadow land over which our travellers passed, show with how little trouble it might be brought into high culti-vation by any farmers who could be induced to settle there."

Beechey was obliged to give up his intention of completing certain surveys left un-finished by Vancouver and, instead, had to seek supplies. The consul at the Hawaiian Islands had been asked to assist in this and Beechey decided to proceed in that direc-tion rather than to Lima or to Canton. By Christmas Day of 1826, he notes:

". . . we had all remained sufficiently long in the harbour to contemplate our depar-ture without regret . . . to use the expression of Donna Gonzales, California appeared to be as much out of the world as Kamschatka . . . On the 28th we took leave of our hospitable and affable friends, Martinez and Padre Tomaso . . . weighed anchor, and bade adieu to the Port of Sán Francisco . . ."

The next chapter of the *Narrative* is devoted to observations upon California, its trade, climate, natural resources, and so on. Beechey was impressed with the coun-try's importance: it ". . . possessed every requisite to render it a valuable appendage to Mexico . . ." Nevertheless, and he quotes Vancouver, the territory which had been subjugated and colonized through great expenditure of time and labor, had been " '. . . turned to no account whatever . . .' "

"Indeed it struck us as lamentable to see such an extent of habitable country lying almost desolate and useless to mankind, whilst other nations are groaning under the burthen of their population. It is evident . . . that this indifference cannot continue; for either it must disappear under the present authorities, or the country will fall into other hands, as from its situation with regard to other powers upon the new continent, and to the commerce of the Pacific, it is of too much importance to be permitted to re-main long in its present neglected state. Already have the Russians encroached upon the territory by possessing themselves of the Farallones, and some islands off Sánta Barbara; and their new settlement at Rossi [Fort Ross], a few miles to the northward of Bodega, is so near the boundary as to be the cause of much jealous feeling;[7]—not without reason it would appear, as I am informed it is well fortified . . .

"The tract situated between California and the eastern side of the continent of North America, having been only partially explored, has hitherto presented a formid-able barrier to encroachment from that quarter; but settlements are already advancing far into the heart of the country, and parties of hunters have lately traversed the in-

7. Beechey's *Narrative* was published ten years before the withdrawal of the Russians from California, and seventeen years before the Mexican Cession.

terior, and even penetrated to the shores of the Pacific . . . with ease, compared with the labour and difficulty experienced by Lewis and Clarke, who had not the benefit which more recent travellers have derived from the establishment of inland depôts by the American fur companies. One of these depôts we were informed by a gentle-man[8] belonging to the establishment, whom we met at Monterey in 1827, is situated on the western side of the rocky mountains on a fork of the Columbia called Lewis River, near the source of a stream supposed to be the Colorado." [9]

The chapter has little to tell of the flora of California:

"The forests of this part of California furnish principally large trees of the pinus genus, of which the *p. rigida* and the red cedar are most abundant, and are of sufficient growth for the masts of vessels. Two kinds of oak arrive at large growth, but near the coast they do not appear to be very numerous. There is here a low tree with a smooth reddish-brown bark, bearing red berries, which from the hardness of its wood, would serve the purpose of lignum vitae; there are also some birch and plane trees; but there are very few trees bearing fruit which are indigenous; the cherry tree and the goose-berry bush, however, appear to be so.

"The shrubs covering the sand hills and moors are principally syngenesious,[1] or of the order rhamnus, while those which prefer the more fertile and humid soils are a gaudy-flowered currant bush, and a species of honeysuckle; but the most remarkable shrub in this country is the yedra,[2] a poisonous plant affecting only particular con-stitutions of the human body, by producing tumors and violent inflammation upon any part with which it comes in contact; and indeed even the exhalation from it borne upon the wind, is said to have an effect upon some people. It is a slender shrub, preferring cool and shady places to others, and bears a trefoil crenated leaf. Among other useful roots in this country there are two which are used by the natives for soap, *amole* and *jamate*."

8. Probably Jedediah S. Smith. Wagner writes: "Curiously enough, Capt. Beechey seems to have met Smith while he was in Monterey. He refers to the captain of a band of American beaver trappers as very intelligent, stating that he had received from him considerable valuable information in regard to the char-acter of the country beyond the Tulares."

9. Perhaps Beechey refers to the second of the rendezvous which was held, according to Carl P. Russell, on "Weber River," in 1826. W. J. Ghent states that William H. Ashley had sold to Jedediah Smith and others in July of 1826 his rights in the organization which was to become known in 1830 as the Rocky Mountain Fur Company; Smith had attended the rendezvous on Weber River in Ogden Valley and from there, starting in late August, had eventually reached California ". . . by way of Cajon Pass and San Bernardino, and arrived at San Gabriel Mission on November 27. No white American, so far as known, had ever before crossed the desert country to California." The San Gabriel Mission was near San Diego and he may well have gone from there to Monterey, where the *Blossom* stopped from January 1 to 5, 1827.

1. The term seems to have been a favorite of Beechey's for he uses it more than once. *Syngenesia* is a Linnean class—the *Compositae*.

2. The name "yedra" is usually applied by the Spaniards to the ivy (*Hedera helix* Linn.) but Beechey cer-tainly had the poison oak (the sumach, *Rhus diversiloba* Torrey & Gray) in mind. I am told that the Spaniards were in the habit of misapplying Spanish names to species (whether in South America, Cali-fornia or the Philippines) which merely bore some rough resemblance to the original Spanish plant.

Bound for the Sandwich Islands, the *Blossom* left San Francisco and took two days to get to Monterey, slightly more than a hundred miles by sea.

"On the last day of the year we passed Punto año nuevo, which with Punto Pinos forms the bay of Monterey . . . a spacious sandy bay about twenty miles across . . . The mission of Santa Cruz is situated at the north extremity of the Bay near Punto año nuevo . . ."

1827

"We dropped our anchor in Monterey Bay on the first of January, and with the permission of the governor, D. Miguel Gonzales, immediately commenced cutting the spars we required; for each of which we paid a small sum. Through the assistance of Mr. Hartnell, we procured several things[3] from the missions . . ."

The Mission of San Carlos was about a league from Monterey and the party paid it a visit.

". . . The ride from the Presidio to Sán Carlos on a fine day is most agreeable . . . The road leads principally through fine pasture lands, occasionally wooded with tall pine, oak and birch trees; but without any underwood to give it a wildness, or to rob it of its park-like aspect."

On January 5, 1827, after less than a week at Monterey, the *Blossom* sailed for the Sandwich Islands. On January 25 they saw the island of "Owyhee" and on the 26th anchored in the ". . . harbour of Honoruru, the capital of the Sandwich Islands . . . we had . . . the pleasure to find Mr. Lay the naturalist ready to resume his occupations. During our absence, he had unfortunately been prevented pursuing his researches among the islands by a severe illness."

On March 4 they left the islands, sought in vain for Wake Island shown on "Arrowsmith's chart," passed Assumption of the Ladrone group, and reached the Bashee Islands on April 7; on the 8th they were off Formosa and on the 10th anchored "in the Typa," or, I presume, at Taipei situated at the northern end of the island. Here they received a most ungracious welcome from the Portuguese governor—". . . the Portugese at Macao . . . claim the Typa as their own, under the emperor's original grant of Macao to them for their service to China." Various letters were exchanged with the authorities. One, from the Hong Kong merchants, asked to have an order enforced ". . . on the said Peitchie's cruiser to go away and return home. She is not allowed to linger about . . . We . . . send our compliments, wishing you well in every respect."

Despite this order, Beechey took his time about leaving and departed on April 30

3. One treasure in the San Carlos Mission church Beechey tried, in vain, to acquire: ". . . a drawing of the reception of La Perouse at the mission, executed on board the Astrolabe, by one of the officers of the squadron."

The Gilbert Chinard edition of Lapérouse's voyage contains a plate which is cited in the list of illustrations as "Réception de Lapérouse à la mission de Carmel (Musée naval de Madrid)." The name of the artist is not mentioned but it might well have been the elder Prévost. Whether this was the painting coveted by Beechey I do not know—it might, in the course of time, have gotten to Madrid. It is, as reproduced, a charming little painting and Beechey seems to have had good taste.

only after he had procured the necessary supplies from "the Company's establishment." On July 3, 1827, he was again at Petropaulski and from there, by a slightly more western route than his earlier one, reached Bering Strait and on August 5 was at Chamisso Island, filled with ". . . curiosity and interest in the fate of our countrymen . . ." Nothing left the year before had been seriously molested by the Esquimaux. Unfortunately, the ". . . swarms of mosquitos that infested the shore at this time greatly lessened our desire to land." Franklin did not arrive at the meeting-place before the *Blossom* departed a second time for the south on October 6. Beechey left additional provisions; also warnings about some of the natives who had proven hostile.

"Having taken our final leave of Beering's Strait, all hope of the attainment of the principal object of the expedition in the Polar Sea was at an end; and the fate of the expedition under Captain Franklin, which was then unknown to us, was a subject of intense interest . . . The enterprising voyage of Captain Franklin down the Mackenzie, and along the northern shores of the continent of America, is now familiar to us all, and, considering that the distance between the extremities of our discoveries was less than fifty leagues, and that giving him ten days to perform it in, he would have arrived at Point Barrow at the precise period with our boat, we must ever regret that he could not have been made acquainted with our advanced situation, as in that case he would have been justified in incurring a risk which would have been unwarrantable under any other circumstances . . ."

The *Blossom* again reached Monterey on October 29, 1827: ". . . we were apprised of our approach to the coast of California by some large white pelicans, which were fishing a few miles to the westward of Point Pinos." When it was learned from some old English newspapers of the death of "His Royal Highness the Duke of York," the ship carried the flag at half-mast during the time she remained in port.

In the ten months of their absence from Monterey the Mexican government had come to realize more fully the importance of the California missions and the padres had had their salaries restored, with a promise of recompense for what was in arrears. Orders concerned with liberating the Indians had been modified for, once given their liberty, they had not known how to use it and had been a source of annoyance or worse to the settlers.

On November 17 the ship left for San Francisco where good water could be had; it was scarce at Monterey and ". . . very unwholesome, being brackish and mingled with the soapsuds of all the washerwomen in the place, and with streams from the bathing places of the Indians . . ."

1828

The return journey to England included stops on the west coast of Mexico. In January of 1828 the *Blossom* was at San Blas. British residents there and at Acapulco were alarmed by the political unrest in Mexico and it devolved upon Beechey to take back certain moneys which would be safer in England. This meant a delay of about five months.

"On the 3d June all the specie was embarked, and we put to sea on our way to Brazil; passed the meridian of Cape Horn on the 30th . . . and . . . arrived at Rio Janeiro on the 21st July . . . after a passage of forty-nine days arrived at Spithead, and on the 12th October paid the ship off at Woolwich . . .

"In this voyage, which occupied three years and a half, we sailed seventy-three thousand miles . . ."

Although Beechey's *Narrative* contained little about the flora of California, the voyage resulted in a volume of great interest to botanists. Its introduction had stated that the plant collections made on the voyage warranted special treatment and were to be published separately. William Jackson Hooker and George Arnott Walker-Arnott undertook the task. Entitled *The botany of Captain Beechey's voyage; comprising an account of the plants collected by Messrs Lay and Collie, and other officers of the expedition . . . performed in His Majesty's ship Blossom, under the command of Captain F. W. Beechey . . . in the years 1825, 26, 27, and 28,* it appeared in separate parts over the years 1830–1841. The title page bears only the latter date.

It was dedicated "To Captain F. W. Beechey, R.N., F.R., & A.S., commander of her[4] M.S. Blossom, and subsequently of her M.S. Sulphur,[5] under whose auspices, and by whose zealous encouragement, the plants described in the present volume were chiefly collected . . ." This is followed by a "List of places visited by the expedition . . ." and, curiously, the only reference to any California visit was "Monterey, November," and as of the year 1827, after Lay had rejoined the expedition. Although Beechey pays tribute to Collie's zeal[6] when substituting for Lay, the fact that no mention is made of his visit to San Francisco (November 7–December 28, 1826) and to Monterey (January 1-5, 1827) suggests that Collie did not accomplish anything worth mentioning when the real naturalist of the party was absent.

Although twenty-one pages (134-165) of *The botany* are devoted to "California. (Collected at San Francisco, and a few at Monterey Bay.)", it is apparent that Hooker and Walker-Arnott were basing their descriptions—certainly for the most part—upon collections made by Menzies, Douglas and others, rather than upon material gathered in that region by persons in the *Blossom*.

4. Queen Victoria had begun her reign in 1837.

5. Beechey had been appointed to the command of the *Sulphur* when that ship started around the world in late December, 1835; but, on arrival at Valparaiso, Chile, in February of 1836, ill-health necessitated his return to England and, after various changes in officers, Captain Edward Belcher was given command. As far as that particular voyage is concerned certainly Belcher, rather than Beechey, is associated with the *Sulphur*.

6. Appended to the list is a note: "The Botanical Collections were made by Mr. Lay the Naturalist, and by the officers of the ship generally; but in particular by Mr. Collie, who, during the absence of Mr. Lay, zealously undertook the care of the department with which that gentleman was entrusted, and whose notes, as well as those of Mr. Lay, have been of much service in drawing up the following account. The collections of the different countries will be described nearly in the order of the places given in the above list . . ."

Pages 316-409 (issued 1840–1841) are headed "California.—Supplement." And a note states: "Where not otherwise mentioned, it is understood that the following species are from the collection of Mr Douglas. They were presented by the Horticultural Society of London, in whose service Mr Douglas was at the time he gathered them."

A foreword to this Supplement refers to the fact that it had been intended to terminate *The botany* of Beechey's voyage with a description of the Rio de Janeiro plants, the first and the last station at which collections were made. But the collection was so small and ". . . the specimens in such wretched condition,[7] and those plants which can be determined so well known, that we think it unnecessary to enumerate them.

"We believe we shall further the cause of science much more by occupying the remaining pages of our work with a Supplement to the *Californian Collection,* which we are enabled to do from that made, chiefly at Monterry and San Francisco, (at no great distance from the coast,) by the unfortunate Douglas . . . and from another, very recently sent us by Mr Tolmie, from the *'Snake Country,'* in the interior of California . . . The specimens, in beautiful preservation, were gathered, in the summer of 1837, by a friend of Mr Tolmie,[8] who conducted a party from Fort Vancouver, on the Columbia, to the rendezvous of the American Trappers, in the interior of California . . ."

From this statement it is apparent that the California Supplement bears little relation to the work of Lay and Collie; little, too, to what is now understood as California. For the "rendezvous of the American Trappers" in 1837 was held on Green River, in western Wyoming.

Alden and Ifft mention Lay and Collie in a flattering light—I quote what they say about their work on zoology:

"It is unfortunate that very little can be learned of either Collie or Lay, but a glance at the report on the zoology and botany of Captain Beechey's voyage will show that they were exceedingly observant and careful collectors. Collie was obviously an able man, astute in his observations and careful in his dissections, checking any peculiarity on a second specimen. Lay is responsible for some half a dozen miscellaneous papers appearing between 1829 and 1842. Their material is extensively reported on by Rich-

---

7. Leonard Huxley's *Life and letters of Sir Joseph Dalton Hooker* (1: 106, 1918. London) suggests—perhaps only in jest—a reason for the poor quality of the collections. Sir Joseph was averse to social activities but, while in Tasmania in 1840, he visited the "somewhat imperious" Lady Franklin, wife of the Governor Sir John Franklin, and was evidently much entertained, to the detriment of his botanical collecting.

"His own disinclination to spend his time in meaningless amusements can be gathered from letters of the period. Herein he was fortified by a letter from his Glasgow friend Arnott, who warns him to collect, not to dance or amuse himself: 'H[er] M[ajesty] does not pay for this.' He quotes the example of Lacy [sic] and Collie, who were not employed to play the fiddle on Beechey's voyage, yet that seemed the principal part of their occupation!"

8. See my chapter XXX where this "friend of Mr Tolmie" is supplied with a name.

ardson, Vigors, Owen, Hooker, and others. The notes of Collie and Lay were carefully written and illustrated—so well (it is perhaps to be regretted) that, as was the custom at the time, many new species were set up from descriptions alone. Methods of preservation were so inadequate that much of the material was too poor to describe accurately, and this led to further errors. San Francisco Bay in November, 1826, and Monterey Bay on January 1, 1827, received the *Blossom,* and the California jay, pygmy nut-hatch, California towhee, and redshafted flicker represent new species collected from these regions. As usual, the names of the officers were commemorated in bird, beast, and fish . . ."

In 1845 Lasègue reported that some of the plants collected on Beechey's voyage were in the Delessert herbarium—even "quelques-uns des échantillons uniques" described in *The botany.* This herbarium had been built up in Paris but when De Candolle published *La phytographie* in 1880, it had been transferred to Geneva.[9]

In 1840 Asa Gray mentioned that "The herbarium of Dr. Arnott of Arlary . . . is . . . interesting to the North American botanist, as well for the plants of the 'Botany of Captain Beechey's Voyage,' etc., published by Hooker and himself, as the collections of Drummond and others, all of which had been studied by this sagacious botanist."

Britten and Boulger (1931) refer to Lay's plants as at Kew. Since W. J. Hooker must have possessed a series of the specimens made on the voyage of the *Blossom,* he doubtless took them to Kew when he transferred there from Glasgow.

The name of George Tradescant Lay is perpetuated in the attractive Californian herb known as *Layia,* an officially conserved generic name.

William Jackson Hooker himself drew some of the fine plates included in *The botany.*[1]

9. De Candolle states, and I translate: "His [Delessert's] herbarium, described by Lasègue . . . , after having been owned by his brother François, was given to his nieces at Geneva, where it was placed in a building of the botanical garden and arranged according to the natural system . . ."(p. 407)

1. Wilfrid Blunt, in *The art of botanical illustration* (London, 1950), has much to say about William Jackson Hooker's botanical drawings, referring to him as ". . . no mean artist . . . his great botanical knowledge more than compensated for any lack of training or technical skill . . ." (p. 223)

Referring to Curtis' *Botanical Magazine,* he notes that, in 1826, ". . . Hooker took upon himself the whole burden of its illustration, remaining for nearly ten years its only draughtsman. Further, most of the drawings made for the additional volumes in the new edition of Curtis's *Flora Londinensis* (1817-28) were his work. He also made the admirable drawings for his *Flora Boreali-Americana* (1829-38). That such distinguished botanists as Hooker, Herbert and Lindley were prepared to 'turn artist' is also evidence of the advantage that they realised was to be gained from the keenness of observation acquired by such work, especially when it involved the drawing of dissected flowers enlarged." (pp. 223, 224)

ALDEN, ROLAND HERRICK & IFFT, JOHN DEMPSTER. 1943. Early naturalists in the far west. *Occas. Papers Cal. Acad. Sci.* 20: 30, 31.

BANCROFT, HUBERT HOWE. 1885. The works of . . . 20 (California III. 1825–1840): 120-125.

BEECHEY, FREDERICK WILLIAM. 1831. Narrative of a voyage to the Pacific and Beering's Strait, to co-operate with the polar expeditions: performed in his Majesty's ship Blossom, under the command of Captain F. W. Beechey, R.N., F.R.S. &c. in the years 1825, 26, 27, 28. 2 vols. London.

BREWER, WILLIAM HENRY. 1880. List of persons who have made botanical collections in California. *In:* Watson, Sereno. Geological survey of California. Botany of California 2: 554.

BRITTEN, JAMES & BOULGER, GEORGE SIMONDS. 1931. A biographical index of deceased British and Irish botanists. ed. 2. London. 182, 183.

CANDOLLE, ALPHONSE DE. 1880. La phytographie; ou l'art de décrire les végétaux considérés sous différents points de vue. Paris.

GHENT, WILLIAM JAMES. 1931. The early far west. A narrative outline 1540–1850. New York. Toronto. 216.

GRAY, ASA. 1840. Notices of European herbaria, particularly those most interesting to the North American botanist. *Am. Jour. Sci. Arts* 40: 1-18. *Reprinted:* Sargent, Charles Sprague, *compiler.* 1889. Scientific papers of Asa Gray. 2 vols. Boston. New York. 2: 1-21.

———— 1869. George A. Walker-Arnott. *Am. Jour. Sci. Arts,* ser. 2, 47: 140-141. *Reprinted: Ibid.* 2: 347-348.

HOOKER, WILLIAM JACKSON & WALKER-ARNOTT, GEORGE ARNOTT. 1830–1841. The botany of Captain Beechey's voyage; comprising an account of the plants collected by Messrs Lay and Collie, and other officers of the expedition, during the voyage to the Pacific and Bering's Strait, performed in His Majesty's ship Blossom, under the command of Captain F. W. Beechey, R.N. in the years 1825, 26, 27, and 28. London. ["California.— Supplement." 316-409. Issued from 1840–1841.]

JACKSON, BENJAMIN DAYDON. 1893. Biographical notes. Botany of Beechey's Voyage and Flora of North America. *Jour. Bot.* 32: 297-299.

LAPÉROUSE, JEAN FRANÇOIS DE GALAUP, COMTE DE. 1931. Le voyage de Lapérouse sur les côtes de l'Alaska et de la California (1788) Avec une introduction et des notes par Gilbert Chinard. Baltimore. (Institut Français de Washington. Historical Documents, Cahier 10.)

LASÈGUE, ANTOINE. 1845. Musée botanique de M. Benjamin Delessert. Notices sur les collections de plantes et la bibliothèque qui le composent; contenant en outre des documents sur les principaux herbiers d'Europe et l'exposé des voyages entrepris dans l'intérêt de la botanique. Paris. 85.

LAUGHTON, J. K. 1885. Frederick William Beechey. *Dict. Nat. Biog.* 4: 121, 122.

RUSSELL, CARL PARCHER. 1941. Wilderness rendezvous period of the American fur trade. *Oregon Hist. Quart.* 42: 5-11.

WAGNER, HENRY RAUP. 1921. The plains and the Rockies A bibliography of original narratives of travel and adventure 1800–1865. San Francisco. No. 29 (p. 24).

CHAPTER XVI

BOTTA, ACCOMPANYING DUHAUT-CILLY IN THE HÉROS, STOPS AT SOME OF THE PORTS OF CALIFORNIA

IN his enumeration of those who made plant collections in California W. H. Brewer mentions that "Several Californian species were described, by or after 1834, as having been collected by Mr. P. E. Botta, but I have no more definite information as to when or where he collected here. He was a travelling naturalist, collecting for the Museum of Natural History of Paris."

The man was Paolo Emilio Botta[1] who traveled in the *Héros* commanded by August Bernard Duhaut-Cilly.[2] The story of the voyage in which he participated was first published in Paris in 1834: *Voyage autour du monde, principalement a la Californie et aux iles Sandwich pendant les années 1826, 1827, 1828, et 1829.* In 1841 this was translated into Italian by Botta's father.[3] In 1929 Charles Franklin Carter published in the *Quarterly of the California Historical Society* a translation of the portion relating to California: "Duhaut-Cilly's account of California in the years 1826–1829."

The voyage was made at the behest of certain French bankers and industrialists who, according to Duhaut-Cilly's introduction, hoped to make business contacts and to open up trade relations in the Hawaiian Islands, in California and elsewhere. The idea had been suggested by a man who figures only as "M[onsieur] R——" in the *Voyage;* Carter supplies the name Rives.[4]

---

1. Botta was born in Italy but in later life became a consul of France and his name sometimes appears in the Italian, sometimes in the French form Paul Émile Botta.

2. According to C. F. Carter, who translated a portion of Duhaut-Cilly's *Voyage:*

"Duhaut-Cilly's works are catalogued under *Bernard* by the Library of Congress, and information from that institution, as well as from the Bibliothèque Nationale in Paris, discloses that the author's patronimic was indeed originally 'Bernard Duhaut-Cilly'(*Revue des Provinces de l'ouest,* Nantes, 1853) or, perhaps more correctly, 'Bernard du Haut Cilly' (*Repertoire générale de bio-bibliographie bretonne,* Rennes, 1889). It became the custom, however, to call the members of the family Duhaut-Cilly only, as we see our author wrote it on his title page, while keeping Bernard as a middle name."

3. Carlo Guiseppe Guglielmo Botta, an historian, whose most important work is said to have been the *Storia d'Italia dal 1789 al 1814.* He also wrote a history of the American Revolution.

4. Manley Hopkins' *Hawaii* (ed. 2. New York. 1869) states that Rives had, it seems, accompanied Liholiho, or Rio-Rio, the king of the Hawaiian Islands and his queen on their visit to England. Both royal visitors died there—of measles apparently—and were taken back to the islands; but Rives remained behind in England. He appears to have been an unreliable character and when Duhaut-Cilly left California for the last time he sailed without him. Bancroft explains that "The Signor R—— had in the meantime run away to Mexico."

The *Héros* spent its time moving from one California port to another from January 26, 1827 (when it first arrived in California) to April 30 when (from San Diego) it left for Mazatlán, Mexico. Back at San Diego on June 10, it continued its visits along the California coast until, on October 20, it sailed for Lima, Peru. Back again at Monterey on May 3, 1828, the ship proceeded to the Russian colony of "Ross," near Bodega Bay where it stayed three days.[5] After further visits to coastal ports the *Héros* sailed away for the last time on August 27, 1828, having spent, writes Duhaut-Cilly, almost two years in California. Most frequently visited of California ports had been San Francisco, Santa Cruz, Monterey, Santa Barbara, San Pedro and San Diego. While the ship lay at anchor in these harbors the commander and others made visits to adjacent missions, for it was through these establishments—and through the ports —that commerce was mainly carried on. Duhaut-Cilly met with a cordial reception everywhere. His story is very readable.

Bancroft reports in considerable detail upon the voyage of the *Héros* and evidently thought well of its commander:

"From the preceding outline of the French trader's movements, it is seen that his opportunities for observation were more extensive than those of any foreign visitor who had preceded him. No other navigator had visited so many of the Californian establishments. His narrative fills about three hundred pages devoted to California, and is one of the most interesting ever written on the subject. Duhaut-Cilly was an educated man, a close observer, and a good writer. Few things respecting the country or its people or its institutions escaped his notice. His relations with the Californians, and especially the friars, were always friendly, and he has nothing but kind words for all. The treachery of his supercargo[6] caused his commercial venture to be less profitable than the prospects had seemed to warrant . . ."

The *Voyage* mentions Botta more than once and he and Duhaut-Cilly collected together, birds mainly, it would seem. I have found no reference to plant collecting. However, I translate two interesting comments upon plants, one on the redwood. When at Fort Ross, Duhaut-Cilly was taken to see the lumbering being done; many firs ("sapins") were being cut but, for the most part, it was the tree ". . . called *palo colorado* (red wood, or 'bois rouge'). This fir has no other quality than that of being very regular and of splitting with the greatest ease; moreover, it contains little resin and is quite brittle. It is the largest tree that I ever saw. Mr. Chelicoff called my attention to the trunk of one which had been recently felled: it was twenty feet in diameter, the dimension taken two feet from the ground . . . the entire trunk was more than thirteen feet in breadth: I measured two hundred and thirty feet from the base to the lowest branches ['la naissance de la cime'] or to where the crown had been separated from

---

5. Duhaut-Cilly was enthusiastic about the appearance and management of the colony although, because of the diminishing number of sea otters and seals, he doubted its value to the Russian company. Included in the *Voyage* is a drawing of the settlement, dated 1828, which Bancroft mentions as "probably the only one of early times extant." It is reproduced in "The Russians in California" (*Cal. Hist. Soc. Spec. Publ.* 7).

6. Rives.

the trunk. One can figure the enormous quantity of boards that could be produced from a tree of such girth. . . All the palos coloradoes are not of such prodigious size; but it is very common to see ones which three men would have difficulty in encircling . . ."

Also of interest are Duhaut-Cilly's references to an introduced mustard.[7] On one occasion, at San Pedro, he had gone hunting with one of his officers but this sport had been frustrated by the difficulty of making their way through a veritable forest of that plant—an additional drawback had been the rattlesnakes with which the mustard was infested. The same plant is mentioned again in the environs of the Pueblo de los Angeles. Botta was along on this occasion:

". . . Leaving the prairies we met with only a forest of mustard, with stems higher than a rider's head,[8] and forming what seemed like dense walls on both sides of the way. In the last few years this plant has become a formidable scourge in a part of California. It has invaded the fine pastures and threatens to extend over the entire country. The inhabitants might have fought this enemy at the outset by exterminating in their entirety the first troublesome plants; but, neglecting to do so, the pest has grown in a manner scarcely to be corrected by such a small population. Even fire has proved ineffectual. When the stalk is dry enough to burn it has already sown a great part of its seed, and fire only makes the soil more suitable for the reproduction of the plant which it was intended to destroy."

Although the *Voyage* fails to mention that Botta collected plants, he evidently did so, for Edouard Spach mentions a few of Botta's specimens in his papers. One was

7. Which mustard is referred to I do not know. Jepson, in *A manual of the flowering plants of California*, lists two which attain a considerable height: *Brassica campestris* Linn., "('Common Yellow Mustard')," growing to six feet, and *Brassica nigra* (Linn.) Koch, "Black Mustard," sometimes "even 12 ft. high."

Very likely it was the last-named, which Asa Gray, in "The pertinacity and predominance of weeds," mentions among the "prevalent weeds" of California and "perhaps introduced by way of western South America."

Gray's paper should be read by the jokers who enjoy asking how they should distinguish a "weed" from a "flower"! For Gray cites various definitions of a "weed" and supplies several excellent ones himself:

"We may . . . consider weeds to be plants which tend to take prevalent possession of soil used for man's purposes, irrespective of his will; and, in accordance with usage, we may restrict the term to herbs. This excludes predominant indigenous plants occupying ground in a state of nature. Such become weeds when they conspicuously intrude into cultivated fields, meadows, pastures, or the ground around dwellings. Many are unattractive, but not a few are ornamental; many are injurious, but some are truly useful . . . In the United States, and perhaps in most parts of the world, a large majority of the weeds are introduced plants, brought into the country directly or indirectly by man . . .

". . . any herb whatever when successfully aggressive becomes a weed . . ."

Engelmann, by 1843, thought it worth while to report upon the numbers of introduced plants which were naturalizing themselves about the Missouri River settlements. He was then making a catalogue of some of K. A. Geyer's collections. See p. 770, *fn.* 1.

8. I was interested in a sentence in John Steinbeck's *East of Eden* (p. 4, 1952); the author was describing the plants of the Salinas Valley, Monterey County, California: "When my grandfather came into the valley the yellow mustard was so tall that a man on horseback showed only his head above the yellow flowers."

Godetia Bottae, described by Spach in his "Synopsis monographiae onagrearum" published in 1835 in the *Nouvelles Annales des Sciences Naturelles.* In "A synopsis of the North American Godetias" (1907) Jepson notes that the ". . . type is preserved at the Jardin des Plantes, Paris. One specimen is labeled by Spach as follows: 'Godetia Bottae, nob. (Spach, 1839), California, M. Botta.' The other he labeled 'Godetia Bottae Spach,' and in another hand on the label is the legend, 'Monterey, M.P.E. Botta, 1829.' . . ." "Holostigma Bottae," described by Spach in his monograph is now *Oenothera bistorta* Nuttall.

C. S. Sargent mentions that "*Æsculus Californiana* was first noticed by Dr. P. E. Botta . . ." Spach's name for this buckeye, supplied in his "Generum et specierum Hippocastanearum revisio" (*Ann. Soc. Nat.,* ser. 2, 2: 62, 1845), was *Calothyrsus californicum,* but *Aesculus californica* (Spach) Nuttall is now the accepted name.

Jepson refers to Botta as ". . . mainly a collector of reptiles." And H. M. D. de Blainville's paper "Descriptions de quelques espèces de reptiles de la Californie" (*Nouv. Ann. Mus. Hist. Nat.,* ser. 3, 4: 233, 1835) records that ". . . La Faculté des sciences fit l'acquisition du plus grand nombres des objets rapportés, et autres furent cédés du Muséum d'histoire naturelle ou à M. le prince Masséna, sous la condition expresse qu'ils ne pourraient etre publiés que par M. Botta ou avec son agrément . . ."

I mentioned that Botta and Duhaut-Cilly, who seem to have been congenial, more than once collected birds together. I shall translate from the French edition of the *Voyage* the description of a species of humming-bird ("une jolie espèce d'oiseaumouche") which frequented the shrubby tracts about San Francisco. It had a flamingred head and throat and was, perhaps, the smallest in existence.

"When this charming little creature lighted for brief instants on a dry branch, one might have thought it a ruby sphere or a little ball of glowing iron throwing off rays of sparks. When several were on the same branch an Arabian, admirer of marvels, might have mistaken it for a branch loaded with the jewels of which he dreams when reading the Thousand-and-one-nights."

There is also a good description of a road runner; it was observed in the region of San Diego, in April, 1827:

"One finds, about the anchorage, that running bird which I have already mentioned by the name of Churay, to which is attributed the ability to kill snakes for food. The Churay is slightly larger than a magpie; as to form it is very similar to that native of our country. Like it, it has a long tail which it raises at times to an almost perpendicular position. Its color is tawny with green feathers and irridescences. It seldom flies and only for short distances; but it runs almost as fast as a horse. It is said that, when it finds a sleeping snake, it erects a rampart made of the spiny branches of a cactus about it and, when this is accomplished, wakes the snake suddenly, by crying; the snake, expecting to flee, impales itself on the long barbs which decorate its prison, and the bird finishes it off with blows of its beak."

It is possible that this was by Botta who wrote an article about this bird— "Description du *Saurothera Californiana (Coucou Churea)*" (*Nouv. Ann. Mus. Hist. Nat.* 4: 1835). He stated that it had already been described by Lesson[9] in the "Suppléments à Buffon (tome VI, p. 420)." Botta evidently collected the type specimen. And his article included a plate. The title cited appears in the table of contents (p. 434).

Botta was later to earn lasting fame as the first to discover the Assyrian ruins of Nineveh; specifically, a part of the palace of King Sargon. He was then ". . . consular agent in Mosul,[1] on the upper Tigris," and in the service of the French government. *Larousse du XXe siècle* (1: 791) states that the reliefs which he sent to the Louvre in Paris constituted the beginning of that Museum's Assyrian collection. He wrote a book entitled *Monuments de Ninive découverts et décrits par Botta, mesurés et dessinés par E. Flandin,* which was published from 1847 to 1850 (Paris. 2 vols.).

Bancroft, having only "a fragment" of the French edition of Duhaut-Cilly's *Voyage* in his library, based his references to the trip on the Italian edition, published in Turin in 1841. He notes:

". . . The portion added to this translation, *Botta, Osservazioni sugli abitanti delle isole Sandwich e della California de Paolo Emilio Botta. Fatte nel suo viaggio intorno al globo col Capitano Duhaut-Cilly,* occupies p. 339-92 of vol. ii; that part relating to Cal. is found on p. 367-78. These notes had originally appeared as *Botta, Observations sur les habitans de la Californie,* in *Nouv. Annals des Voyages,* lii, 156-66 . . ."

Unfortunately, these "Observations sur les habitans de la Californie"—which form part of Botta's paper on the inhabitants of the Sandwich Islands—contain no references to plants.

---

9. It is now *Geocóccyx californianus* (Lesson) according to F. M. Bailey's *Birds of New Mexico.* 1928.

1. A chapter is devoted to Botta's archaeological persistence and accomplishments in that very fascinating book *Gods, graves and scholars. The story of archaeology* by C. W. Ceram, first published in German but recently translated by E. B. Garside. (New York. 1951.) Botta's work was continued by Austen Layard whose fame was greater perhaps than that of his predecessor.

ALDEN, ROLAND HERRICK & IFFT, JOHN DEMPSTER. 1943. Early naturalists in the far west. *Occas. Papers Cal. Acad. Sci.* 20: 31, 32.

BANCROFT, HUBERT HOWE. 1885. The works of . . . 19 (California II. 1801–1821): 650; 20 (California III. 1825–1840): 128-131, *fn.* 31.

BOTTA, PAOLO EMILIO. 1831. Observations sur les habitans de la Californie. *In:* Nouvelles annales des voyages 52: 156-161.

—— 1835. Description du Saurothera Californiana (Coucou Churea). *Nouv. Ann. Mus. Hist. Nat.* (Paris) 4: 121-124. pl. 9; 434.

BREWER, WILLIAM HENRY. 1880. List of persons who have made botanical collections in California. *In:* Watson, Sereno. Geological survey of California. Botany of California 2: 555.

DUHAUT-CILLY, AUGUST BERNARD. 1834–1835. Voyage autour du monde, principalement a la Californie et aux iles Sandwich, pendant les années 1826, 1827, 1828, et 1829. 2 vols. Paris.

—— 1929. Duhaut-Cilly's account of California in the years 1827–1828. Translated from the French by Charles Franklin Carter. *Quart. Cal. Hist. Soc.* 8: 131-166, 214-250, 306-336.

GRAY, ASA. 1879. The pertinacity and predominance of weeds. *Am. Jour. Sci. Arts,* ser. 3, 18: 161-167. *Reprinted:* Sargent, Charles Sprague, *compiler.* 1889. Scientific papers of Asa Gray. 2 vols. Boston. New York. 2: 234-242.

JEPSON, WILLIS LINN. 1907. A synopsis of the North American Godetias. *Univ. Cal. Publ. Botany* 2: 319-352.

SARGENT, CHARLES SPRAGUE. 1891. The Silva of North America. Boston. New York. 2: 62.

SPACH, EDOUARD. 1834. Generum et specierum Hippocastanearum revisio. *Ann. Sci. Nat.* (Paris), ser. 2, 2: 62 *(Calothyrsus Californica).*

—— 1855. Synopsis monographiae onagrearum. *Nouv. Ann. Mus. Hist. Nat.* (Paris), ser. 3, 4: 336 *(Holostigma Bottae),* 393 *(Godetia Bottae).*

CHAPTER XVII

BERLANDIER IS ABLE TO ENTER SOUTHEASTERN
TEXAS WITH A MEXICAN BOUNDARY COMMISSION

I SHALL preface the story of Jean Louis (or Luis) Berlandier by translating from the *Mémoires et souvenirs de Augustin-Pyramus de Candolle* which, as the title page explains, was written by the elder De Candolle and published (1862) by his son Alphonse. For it was the criticism, written by one of the most respected of European botanists and emanating from Geneva, Switzerland—a great center of European culture at the time—which for many years was responsible for blackening the reputation of Berlandier in the eyes of the world. S. W. Geiser, mentioning some of the handicaps—in climate merely—which Berlandier faced, comments: "I doubt whether DeCandolle ever experienced such difficulties in his botanical travels." While, perhaps, the story might more properly be placed at the end rather than at the beginning of this chapter its early inclusion may serve to stress the difficulties of a task which, from Geneva, looked so simple.

De Candolle was citing some of the foreigners who had taken his botanical courses and studied his collections:

"Berlandier was a young man born into an extremely poor family near Fort-de-l'Écluse, France. He began as a pharmaceutical apprentice. He showed energy, enthusiasm for natural history, and had acquired by his own efforts a certain amount of classical education. Touched by these efforts, I arranged to include him among my students and allowed him to work among my collections. He did a piece of work on the Grossulariæ which was inserted in the *Memoirs of the Society of Natural History of Geneva*, and from which he subsequently extracted the article of the *Prodromus*. This piece of work, while not a distinguished one, had certain merits for a beginner. I saw to it that the author was selected by the Museum to go to Marseilles to receive a living ostrich[1] which had been sent us. I admitted him also as a student in my botanical field-classes. The idea then came to us—for we would not let ourselves be discouraged by Wyndler's[2] lack of success—to send a botanical collector to Mexico, and

1. Evidently an honor although, from the little that I know about ostriches, not one to covet.

2. According to Alphonse de Candolle, his father and "Philippe Dunant, Moricand, [and] Mercier" had defrayed Berlandier's expenses and, from the context, those of Wyndler as well. "Wyndler" had been sent to Porto Rico where, almost at once, he had contracted yellow fever and consequently, according to Alphonse de Candolle, could send back only "very inferior" collections; perhaps he too needs a champion! Lasègue, who at least credits the man with the initial "H," states that he sent back "... une collection assez nombreuse de plants ..." Alphonse de Candolle's *La phytographie* refers to Wyndler's numbered collections as mainly in the herbaria of "... de Candolle, Delessert, et Webb ... aussi au Musée d'hist. nat. de Nancy."

365

we chose Berlandier; but his senselessly ambitious, restless, vain and independent makeup did not brook certain teasings which were perpetrated by those of us who were making the arrangements for the trip and he left already ill-disposed. We had thought of Mexico because its native wealth was then little known, and because I had an affiliation with Mr. Alaman, Minister of the Interior, who promised me protection for my employee. Indeed, this protection he never lacked, and among other favors [Alaman] saw to it that he was made a member of a great expedition to the Mexican government which had as its purpose the marking of the northern boundaries; but Berlandier profited but little from his opportunities. He sent only a small number of dried plants, and these poorly selected and poorly prepared; he neglected entirely to send animals, seeds, and notes upon the country. After a certain length of time he even neglected to write us, so that we were long in doubt whether he was alive or dead. With the result that we found ourselves in the position of having spent about 16,000 francs for dried plants which were scarcely worth one quarter that amount! This outcome, taken in conjunction with our preceding experience, left us completely disgusted with expeditions of this sort."

To these comments of his father's Alphonse de Candolle added for good measure:

"Berlandier, ashamed of his conduct, pretended to be dead. I found, in Paris, that he had written a letter to the Museum offering his services, which was dated December 20, 1838, or twelve years after his departure. I later learned that he had, on his own authority, taken up the practice of medicine, that he had been employed by a Mexican general on boundary matters, that he had settled at Matamoras, where he had, honorably and disinterestedly enough, practised medicine, and that in addition, he had been sent by Arista to General Taylor to ask that he should not cross the Colorado River; and, finally, that he had died (this time actually) while crossing the San Fernando River, in the summer of 1851. He left manuscripts on the geography and natural history of the country which were purchased at Matamoras by an officer of the United States, Lieutenant Couch, who presented them to the Smithsonian Institution, and who even had the generosity to send some dried plants to us as compensation for our expenditures in sending Berlandier to America."

For long Berlandier's reputation rested upon this prejudiced and, for him, injurious story. It was passed on in an even more unpleasant form by Asa Gray who recorded—in his obituary notice of Charles Wilkins Short—that some of Berlandier's collections were sent in particular to two Swiss botanists ". . . who had contributed to send out Berlandier to Mexico as a collector, but from whom (apparently through Berlandier's dishonesty) they had failed to receive any adequate return."

The opinions of the two De Candolles had been published in 1862, Asa Gray's in 1863. But, so far as I know, it was not until 1933—or some seventy years later—that anyone published an exoneration of Berlandier. When Geiser did so, he left Berlandier's slate clean. That a vindication—Geiser refers to a ". . . subsequent conspiracy of silence entered into by obsequious botanists . . ."—was not published sooner, by someone, is unexplainable, for Berlandier's defender has this to tell of the amount of material which reached Geneva:

". . . some qualification should be made of DeCandolle's estimate. Among the archives in the library of the United States National Museum is a little volume in Berlandier's handwriting labeled 'Expédition.' It is a list of shipments of plants, seeds, and animals sent to DeCandolle and to Moricand in Geneva. The list gives a full invoice of all items included; and from it I learn that between April 25, 1827, and November 15, 1830 (the approximate date when the Commission was dissolved), Berlandier sent in all 188 packets of dried plants totaling some 55,077 specimens; 188 packets of plant seeds; 935 insects; 72 birds; 55 jars and bottles of material in alcohol; and more than seven hundred specimens of land and fresh-water mollusks, mostly from Texas. These are but the chief collections sent.

"It may be thought that for some reason DeCandolle did not receive all the items dispatched to him. I know only that Berlandier's manuscript lists 2,320 'numbers'; and in a catalogue sent by Alphonse DeCandolle to Asa Gray, giving the names of plants collected by Berlandier, received by his sponsors in Geneva, and by them distributed, there are 2,351 numbers. The manuscript catalogue[3] in the Gray Herbarium library and the covering letter from Alphonse DeCandolle, dated April 24, 1855, are all the evidence needed to show that the shipments entered by Berlandier in his private book, for his eye only, reached their destination."

Why, when the size of the collection was known to Alphonse de Candolle in 1855, he still saw fit, in 1862, to publish his father's and his own criticisms is difficult to understand. Berlandier had died in 1851 but it was not too late to clear his memory. Geiser comments that "Berlandier soon became aware of DeCandolle's outspoken dissatisfaction with his work in Texas . . . What defense could this twenty-two-year-old youth make that would be satisfying to the great European scientist?"

As mentioned by De Candolle, Berlandier had been permitted by the Mexican government to join an expedition which ". . . had for its purpose the marking of the northern boundaries . . ." By the treaty with Spain which was signed on February 22, 1819, the United States government had been willing to relinquish "Texas" for Florida. But, states W. J. Ghent:

"Upon one point De Onis had been unyielding. No part of what was regarded as Texas would be ceded. Monroe and Adams therefore consented to the dividing line of the Sabine River . . ."

Geiser describes in considerable detail the complexities of this boundary dispute. And H. H. Bancroft's *History of Mexico* has this to say:

". . . President Victoria's administration appointed a commission in 1827, with

3. This "manuscript catalogue" bears the inscription: "Plantes de Berlandier Liste de noms que Mr Moricand avait receuillis d'après ses propres déterminations et d'après les ouvrages ou les indications de divers botanistes, accrue de quelques noms qui se trouvaient chez Mr de Candolle. Communiqués par ce dernier en 1855." It lists 2,351 numbers.

General Manuel Mier y Teran as its chief, to ascertain the boundary line between the Mexican and American republics under the treaty of 1819. Teran then had an opportunity of observing the giant springing up in that portion of his country. He went as far as Laredo and San Antonio de Béjar, and examined nearly the whole of that region. Austin's colonists were almost all Americans, and of the same nationality were those who in 1826 settled the western line of the Colorado and Nueces. For this reason the American government and people became the more anxious for the acquisition of Texas, which was made manifest in various ways; namely, by throwing obstacles in the way of the treaty of limits to delay its becoming effective, and by attempting to extend the boundary of Louisiana beyond the Sabine River."

There are two published sources dealing with the work and the routes of the commission headed by Don Manuel de Mier y Teran:

(1) The *Diario de viage de la Comisión de Limites* . . . which was published in Mexico in 1850. It was the joint work of "D[on] Luis Berlandier," botanist and zoologist of the commission, and "D[on] Rafael Chovel," mineralogist. Much of it deals with periods when Berlandier was in what is now Mexico proper rather than the United States. It is divided into chapters, some out of chronology; the first chapter, "Diario de viage," most nearly concerns our story, and as to pages 92-119.

(2) A diary written by José María Sanchez, translated from the Spanish by Professor Carlos E. Castaneda: "A trip to Texas in 1828," which was published in *The Southwestern Historical Quarterly* for 1926. It covers the period from November 10, 1827, when the commission was about to start from Mexico, until its arrival at Nacogdoches, on June 3, 1828. Berlandier, who was then seriously ill, had parted from the commission on May 29 to return to San Antonio, or "Bejar." Sanchez served in the capacity of draughtsman or topographer.

In addition to the above is Samuel Wood Geiser's paper "In defense of Jean Louis Berlandier," which first appeared with critical commentaries and documentation in the *Southwest Review* for 1933, and later, with revisions, formed a chapter in the first (1937) and second (1948) editions of his *Naturalists of the frontier*. While, as far as the routes taken by the commission are concerned, Geiser seems to have followed the *Diario de viage* and the Sanchez diary, he also cites from a manuscript journal of Berlandier's for periods not included in the other publications. His paper first supplies the routes taken by the commission during the period when Berlandier was a member of the party and follows this with comments upon the man's botanical accomplishments at certain points in the journey. I shall, as anyone writing of Berlandier must do, depend much upon Geiser, and I cite his thrice-published and twice-revised paper under the date of issue. Insertions between brackets are mine.

Geiser tells that arrangements for sending Berlandier ". . . to Mexico were . . . made, some time in 1824 or '25, with Lucas Alamán, a former student in Geneva of De Candolle. Alamán, now Minister of Foreign Affairs of the newly established Republic of Mexico, had decided to survey and establish the boundary between the Mexican

Republic and the United States . . . At some time in 1826 . . . Berlandier left Europe for Mexico . . ." (1933, pp. 436, 437)

"Berlandier's manuscript account of his journey to Mexico indicates that he left Le Havre on the American brig *Hannah Elizabeth* . . . on October 14, 1826, and on December 15 landed at Panuco, near Tampico, on the Mexican coast." (1948, p. 34)

Collecting on the way, he finally reached Mexico City, remaining there until November 10, 1827, when the commission started on its way, and ". . . reached Laredo exactly thirteen weeks after his departure from the capital." (1948, p. 37) Sanchez wrote in his diary, under date of February 2, 1828:

". . . we discerned in the distance the peaceful waters of the Rio Bravo del Norte whose treeless banks displayed the water lying like a silver thread upon the immense plain . . . It was decided to cross the river. This was accomplished with little difficulty, and at last we entered the *Presidio de Laredo,* situated on the opposite bank."

He states that the commission remained at Laredo until February 20, 1828, when it started northward. I shall follow the Sanchez diary until May 29.

February 20: ". . . we left Laredo and traveled along a level road lined with shrubs. After traveling a league we noticed that the vegetation is more plentiful and the grass or pasturage that covers the ground of better quality. At nightfall we stopped on the banks of a small pond . . . This place we called *el Chacon.*"

February 21: ". . . we resumed our march over ground that was as level as that of the previous day but more pleasing to the eye . . . We found a great number of tree trunks petrified . . . Large pieces of these were taken for Messrs. Chovell[4] and Berlandier to analyze . . . we were obliged to halt . . . on the edge of a small lake . . . called *El Pato* . . ."

February 22: ". . . We stopped at last in a place called *La Parida* . . . There we heard for the first time the frog whose astounding call resembles the bellowing of a bull. This frog is rarely seen and is as large as the head of a man . . ."

While the commission had been stationed at Laredo the *Diario de viage* had commented upon a few plants: on February 1 upon ". . . una *Didynancia*[5] *(Zenillda* de los indígenas [)], la *Mimosa oechinoides,* una *Rutacea* y una hermosa *Mimosa arborescente* con flores amarillas." At La Parida, on February 22, upon an *"Eryngium,"* a small *"Plataginea,"* a spiney *"Leguminosa"* whose yellow flowers resemble those of a *"Rutacea,"* and so on.

February 23: "At a distance of six leagues from *La Parida* lies the Nueces River . . . Travelers often have to wait eight or ten days to try to ford it . . ."

On the 24th Sanchez remarks that the vegetation ". . . seems to increase as one ap-

---

4. The name is always spelled in this form by Sanchez.

5. Correctly, *Didynamia,* one of the Linnean orders.

proaches the interior of Texas . . ." They halted that day at ". . . a beautiful watering place formed by a ravine among the hills . . . The place is known as Cañada Verde." Geiser identifies it as ". . . Green Branch, in present McMullen County . . ." (1948, p. 43) The next day they were still in McMullen County:

February 25: ". . . we crossed another ravine called Guadalupe, and stopped on the banks of Rio Frio . . ."

On February 26 they reached San Miguel. The diary records that ". . . increased vegetation and the numerous small trees scattered everywhere made the view delightful . . ."

February 27: ". . . We . . . finally stopped on the bank of a small but beautiful creek called *La Parrita* . . . The evergreen oaks and other trees . . . make it beautiful . . ."

On the 28th, after making "six leagues," the party stopped at "Rancherías." Sanchez notes that "Near this spot there is a hill known as the Hill of San Cristobal that contains much iron ore . . ." On the 29th they reached the Medina River and, after crossing, camped on the north bank:
"The crystal waters and the large trees along the banks make the river extremely beautiful. Its peaceful stream joins that of the San Antonio and Guadalupe rivers, and empties into the gulf of Mexico."

March 1: ". . . We crossed the Cíbolo, a small creek (and at a short distance saw the mission of *La Espada* . . . Bejar is at a distance of about two leagues . . ."

They had this day arrived at the ". . . city of San Fernando de Bejar . . . capital of the beautiful Department of Texas . . ." It was situated near the "Mission of San Antonio de Bejar." [6] Sanchez, writing of the year 1828, observed that "The Americans from the north have taken possession of practically all the eastern part of Texas, in most cases without the permission of the authorities. They immigrate constantly, finding no one to prevent them, and take possession of the *sitio* . . . that best suits them without either asking leave or going through any formality other than that of building their homes . . ."
Geiser explains that the route taken by the commission ". . . from Laredo to Bexar [or San Antonio] followed the old Bexar Road (not the Presidio Road, which crossed the Rio Grande farther up), through present Webb, McMullen, Atascosa, and Bexar Counties . . ." (1948, p. 37)
The *Diario de viage* had commented upon the plants observed along the route: between the Nueces and the Cañada Verde, February 24, upon the spiney *"Leguminosas"* and the "Gramineas," and upon the lesser stature of the *Cacti* or *Opuntia;* between

6. See the volume *Texas* ("American Guide Series"), pp. 327, 328, for a short but interesting account of the early establishment—missions, a fort, a villa, etc.—which were eventually consolidated into San Antonio de Bexar.

the Cañada Verde and the Rio Frio, February 25, upon a beautiful *"Anémona,"* and an occasional *"Physalis";* about the camp at San Miguel, February 26, grew *"Draba,"* *"Corydalis"* and *"Sisymbrium";* at *"la Parrita,"* February 27, the same herbaceous plants and a *Celtis;* at "la Ranchería," February 28, *"Leguminosas"* and many annuals, a lesser number of *"Mimosas"* and *"Anémonas,"* and most abundant were the oaks and walnuts, also *"Carex, Linaria, Draba,* y el *Alyssum;"* at the river Medina, February 29, there were again walnut trees, *"Juglans."* At "Bejar," March 1, there is a reference to a peach, "El duranzo *(Percica Vulgaris),"* to the willow and to a white poplar or "álamo" which had flowers and leaves until the end of February.

Berlandier and Chovel's *Diario de viage* ends with the commission's arrival at San Antonio de Bejar on March 1 and does not begin again until July 14, 1828, when Berlandier left Bexar for Laredo. But the itinerary is supplied in the Sanchez diary and I follow that source until May 28. After remaining a month and a half at San Antonio the commission started for San Felipe de Austin on the Brazos River.

April 13: "We left San Antonio de Bejar for Nacogdoches on Sunday . . . taking an eastward course along verdant hills covered with spring flowers, we arrived at Salado Creek at five and camped on the western bank of the said creek, whose small stream lies between two hills, bordered on either side with live oaks, walnuts, plum trees, and elms . . ."

April 14: "We continued along rolling hills, woods, and small valleys bedecked with beautiful flowers where numerous butterflies flitted about . . . we halted on the Cíbolo Creek . . . a permanent stream, though small . . ."

They traveled for seven hours on the 15th and on the 16th ". . . reached the Guadalupe River and crossed on a ferry boat . . . On the eastern bank . . . are built six wooden cabins inhabited by three North American men, two women and two girls[7] of the same nationality, and a Mexican, all of whom form the village of Gonzales . . ."

Geiser explains that the commission was following ". . . the so-called Middle Road, which led from present San Antonio to Gonzales [Gonzales County] . . . on the Guadalupe. To the eastward of Gonzales the road, after crossing the La-Bahía road, continued to the Colorado. Here it joined the Atascosito road leading to San Felipe [Austin County] . . ." (1948, p. 38)

". . . Berlandier found the Guadalupe country near Gonzales attractive botanically . . . although Berlandier had but a short stay . . . he made excellent use of his time. Six years later, in June of 1834, Thomas Drummond was to explore extensively in the region . . ." (1948, p. 44)

April 17: "Various tasks of Señor Berlandier caused us to start our march at ten o'clock . . . and to halt at about three or four o'clock in the afternoon at Tejocotes

---

7. One of the girls is described as a "flower of the desert." Sanchez writes in a somewhat ornate style, embellished with references to the perfume of the flowers, the singing of the birds, and so on.

Creek . . . at about two in the morning a furious storm broke out and lasted almost two hours . . .”

April 18: “. . . the light rain that remained after the storm ceased, and we had only the cold wind of the northwest to contend with. We crossed the creek, not without trouble . . . We halted at a place which, because it had no name, the general called *Loma Grande* . . .”

April 19: “. . . we crossed the creek called La Baca, and at about three leagues from this place the wheel of one of the instrument wagons broke . . . making it necessary to stay in a beautiful meadow. We . . . gave this place the name of *Campo de la Rueda* . . . We remained at this place all the following day [the 20th] to repair the wheel.”

On the 21st they “. . . crossed *La Navidad* and *El Metate* creeks . . .”

April 22: “. . . we pitched camp near a small creek. Not knowing what they call this stopping place, and seeing how unattractive it was I suggested that they name it after me, and from then on it was, for us, Sanchez’ Camp. In the afternoon it rained considerably.”

April 23: “. . . we succeeded in reaching the Colorado after about five and a half hours. A very good house, belonging to Mr. Wis, an American from the United States of the North, is built on this spot. We were given excellent lodging there during our entire stay . . .”

April 26: “. . . We crossed two or three muddy creeks, halted, and made our camp on one . . . known as S[an] Bernardo.”

April 27: “. . . we arrived at a distance of four or five leagues from the settlement of San Felipe de Austin, where we were met by Mr. Samuel Williams, secretary of the empresario, Mr. Stephen Austin; and we were given lodging in a house that had been prepared for the purpose.

“This village has been settled by Mr. Stephen Austin, a native of the United States of the North. It consists . . . of forty or fifty wooden houses on the western bank of the large river known as the *Rio de los Brazos de Dios* . . . In my judgment, the spark that will start the conflagration that will deprive us of Texas, will start from this colony. All because the government does not take vigorous measures to prevent it. Perhaps it does not realize the value of what it is about to lose.”

The wagons needed repairs, and the party was obliged to remain at San Felipe from April 28 to May 9.

“. . . with much regret we noticed the river begin to rise as is customary at this time of year . . . the stream began to bring down enormous tree trunks . . . on the 30th . . . the most terrible storm I have ever seen was raging. The rain was so heavy that it seemed as if the entire sky, converted to rain, were falling on our heads. The woods were afire with the vivid flashes of lightning . . . The shock of the shrill howling winds was horrible and it continued until eight the next morning . . . I gave thanks to the Almighty for having come out unharmed from such a furious storm. We remained in the village, the flood preventing our crossing the river . . ."

Finally it was decided to attempt a crossing and on May 10, with everything and everybody aboard a ferryboat, they started: "A drunk American held the rudder and three intoxicated negroes rowed, singing continuously . . ." But the boat was carried two leagues before it was possible to make a landing. When they did so, it was only to learn that the wagons which had been sent ahead the day before were stuck in the mud. Since they could not reach the party until the next day, the general slept in the open:

"To the unbearable heat were added the continuous croakings of frogs, the discordant singing of the drunken negroes, and the numberless legion of mosquitoes that bit us everywhere, all of which kept us from sleeping a wink. When the longed-for dawn broke we saw the terrible onslaught these cursed insects had made upon us, leaving us full of swollen spots, especially on the face of the general, which was so raw it seemed as if it had been flayed."

May 11 was spent in getting the wagons across a bayou. On the 12th, since the horses were suffering for lack of forage, Sanchez and Chovel went to the house of a Mr. Groce, an American, where they bought corn.

". . . Later, they asked us into the house for the sole purpose of showing us the wealth of Mr. Groce and to introduce us to three dogs called Ferdinand VII, Napoleon, and Bolivar. The indignation at seeing the name of the Columbian Liberator thus debased, caused Mr. Chovell to utter a violent oath . . ."

Mr. Groce, from what Sanchez had to tell, seems not to have been a very desirable character. As they were unable to procure food on the night of the 12th and went without breakfast on the 13th, they got on their way and, after a short march, stopped at a place which they called ". . . the Camp of the Virgin, because the general that night observed the pivot star of that constellation on the meridian." On May 14 they stopped on a hill ". . . in front of Mr. Groce's second house." The 15th was spent pulling their wagons out of mudholes; the heat was extreme and that afternoon ". . . a furious rain came down that lasted until midnight . . . we were in the most pitiable condition imaginable."

May 16: "In spite of the rain we continued our painful march through the flooded woods and after seven hours of fatigue, during which we advanced but one league because of the mud holes, we camped near the house of a North American who lives in this vast solitude alone with his young wife."

May 17: ". . . Mr. Berlandier and John, the cook, were sick with fever . . . we advanced about a quarter of a league in order to reach the house of a North American called Nolland. We carried the sick men in the carriage, and at the house were provided with milk and chickens to feed them."

It took until midday of May 18 to get the wagons across a mudhole where they had passed the night—everything had to be unloaded and packed on mules.

"... Our patients continued to grow worse, and a corporal ... was ... stricken with the same illness ... it was decided to make a bed in the carriage for Mr. Berlandier and another under it, in a hammock, for the cook, John. Mr. Chovell took charge of the sick ... Batres and I took charge of the kitchen about which neither he nor I understood a thing ... we started to perform wonders impelled by necessity. In the afternoon, the general fell ill with the same fever ... a furious storm broke out which lasted until dawn ... the general ordered that a buffalo skin be thrown over his bed to protect him ... he perspired so freely that the following day he had no fever."

May 19 & 20: "We remained in the same place and the sick men continued to grow worse ..."

It took three hours to cross "Jaranames Creek" on the 21st. Water was high. And the instrument wagon kept breaking down. Indeed they spent from the 22nd through the 24th crossing creeks and mending wagons. On May 25 they reached the Trinity River—there were two houses, "... each ... a miserable log cabin ..." Daily, from the 26th to the 28th, more men were becoming ill.

"The mosquitoes, the lack of food, and the inability to cross the river, all added greatly to this calamity. May the Lord have pity on us!"

It was decided at this point to send Batres, Berlandier and Chovel back to San Antonio. Sanchez was to proceed with the general to Nacogdoches.

Geiser clarifies the route which had been followed between Gonzales, reached April 17, and the Trinity River where, on May 29, Berlandier was to leave the commission for a time at least; insertions between brackets are mine:

"The route ... carried them over Peach Creek and through the high country of the Lavaca, where, near present Schulenberg [Fayette County] at the Loma Grande [reached April 18], they enjoyed the magnificent prospect ... Thence they passed on to ... present Columbus [Colorado County] on the Colorado River [where the party halted from April 23-25]. Here, at Beeson's Ferry, the cavalcade halted [because of high water] ..." (1948, p. 44)

After being held up at the Brazos by high water they finally got across on May 9— at what Geiser calls "... the Atascocito Crossing at San Felipe ..." (1948, p. 47) They then started northeastward, "... up the east-of-the-Brazos road towards Jared Groce's plantation, 'Bernardo,' near present Hempstead in Waller County ... made camp [May 13] ... not far from present Hempstead. At this place the road they had been following joined the Magdalena road and led them approximately up the route of the present Hempstead-Navasota highway ... they encamped [May 14] not far from Groce's 'Second House,' near present Courtney [Grimes County] ... [On May 15 they were near] ... present Navasota [Grimes County] ... they reached [May 16] Holland's Place, near present Anderson [Grimes County] ..." (1948, p. 38)

It was here that Berlandier became ill and, states Geiser, ". . . continued seriously ill until [May 28] the party reached the Sertuche Crossing,[8] where the road from Bexar to Nacogdoches (the old Camino Real or Upper Road) crossed the Trinity." (1948, p. 38)

That Berlandier's specimens were not the first described from the San Felipe, or Austin, region was not the man's fault, as Geiser makes clear:

". . . Berlandier employed his time in making botanical collections. This locality was later to be very carefully explored for plants by Thomas Drummond (1833-4) and Ferdinand Lindheimer (1839 and 1844). Of the collections made here on this journey in 1828 by Berlandier, relatively few, apparently, ever reached DeCandolle and the other Genevese botanists. Most of the species that were collected by Drummond and Lindheimer were described as new by the British botanist, Sir William Hooker, and by Asa Gray. The loss of the specimens destined for DeCandolle is doubtless to be ascribed to the conditions under which the Commission worked, to Berlandier's serious illness, and to the inclement weather that prevailed at the time, which must have jeopardized very seriously the collections he had made . . ." (1948, p. 45)

When General Mier y Teran left for Nacogdoches, he took Sanchez with him and directed the scientific staff, including Berlandier, to return to Bexar and meet him again at Matamoras "at the end of summer." Geiser reports that Berlandier, still ill, ". . . could collect hardly any plants on the return trip to Bexar over the Upper Road; and the great botanical expedition that DeCandolle had set so much store by was over . . . Disasters at the Brazos and the Colorado on the way back still further damaged such specimens as Berlandier had preserved." (1948, pp. 51, 52)

He states that ". . . the scientific staff . . . left the Trinity River on May 30 . . . camped during incessant rains on the Brazos, June 3-6; passed the Colorado on the 12th, narrowly escaping drowning in a sudden rise of water after a cloudburst in that region; crossed the San Marcos River on the fifteenth, and reached Bexar on the eighteenth of June. Here they remained about a month, delayed in their departure for Matamoras, as Berlandier said, 'by much and continued rain.' . . ." (1948, p. 39)

A section of the *Diario de viage,* headed "Diario de San Fernando de Bejar a Matamoras" (pp. 120-137), describes the trip from Bexar to Laredo (July 14-25). Leaving Bexar on July 14, they reached the Medina River on the 15th and crossed it on the 16th; on the 18th they were at "la Parrita," and on the 19th at the "rio de San Miguel." Berlandier comments upon the *"Leguminosas"* known as mesquites, upon the *"Graminea"* with spiney fruit known as *"Cadillo,"* and upon a species of "Syngenesia"—or composite—with white flowers.

Reaching the river Frio on July 20, they crossed it the next day, spent the 21st and 22nd repairing a wagon; the soldiers, we are told, ate the red fruit of the *"Raqueta,"*

8. Presumably identical with "Robbin's Crossing" of Geiser's map (1948, p. 37), in southern Houston County, or near the point where the counties of Houston, Madison, Walker and Trinity adjoin. It would seem to have been the most northern point reached by Berlandier.

one of the Opuntias. July 23 they were at the "Cañada de Aqua Verde," and on the 24th at the "rio de las Nueces, en el paso de Sta Barbacoa." Here there were trees such as ash, oak, elm and a species of "Leguminosas" with yellow flowers known as "retama." The "Ricin con flores rojizas" was, perhaps, a *Jatropa.* On July 25 they reached the "presidio de Laredo," and evidently remained in the region until August 11 when they crossed the Rio Grande into what is now Mexico, and from there reached Matamoras August 20, 1828.

Under date of July 27, and while at Laredo, Berlandier supplies (pp. 133, 134) a description of one of the figworts which he calls ". . . le Cenicilla *(Terania)* . . . un paqueño arbusto de la familia *Anthirrineas,* notable por su aspecto. Se eleva á la altura de seis á siete piés: sus hojas, de un blanco agrisado, son pequeñas y numerosas: sus flores son grandes y violadas. Los indígenas de la tribu de los carrizos, ponen sus ramas en infusión, y se sirven de ésta como de un febrífugo." [9]

On the same date Berlandier also mentions the creosote bush: "La Gobernadora *(Zigophyllum resinosum)* comienza à presentarse . . ." Laredo should be near the eastern limit of the range of what is now called *Larrea tridentata.*

On July 28 there is a reference to the persimmon, "zapote," and, among plants with edible fruits, to a species of *Opuntia,* "la *Raqueta* de Téjas," and to "las *Capsulas carundas* de una especie de *Yuca,* de la que no hemos podido procurarnos sino los granos."

Having, as noted, reached Matamoras on August 20, 1828, they awaited the arrival of the general. When, unexpectedly delayed, he did not appear, Berlandier after a few weeks returned to Bexar.

The *Diario de viage* (pp. 249-282) contains a chapter, "Gaza del oso y cibolo on el N.O. de Tejas," describing a hunting trip (November 19 to December 18) for bear and buffalo which Berlandier made with Lieutenant Francisco Ruiz and some Comanche Indians.

The party seems to have gone northwestward (across northern Bexar and Kendall counties into Gillespie County), then southwestward (across Kerr, ? Bandera, and Uvalde counties to near present Knippa, in the last-named), and then eastward (across Medina County) to the starting point in Bexar County. Geiser mentions that, at Kerr-

9. P. C. Standley, in his "Trees and shrubs of Mexico" (p. 1306) writes of *Leucophyllum texanum* Bentham:
"The plant is employed locally as a remedy for fever and ague. It is probably this species which was mentioned by Berlandier [*Diario de viage,* p. 276] as occurring in Texas where it is known as 'cenicilla.' Berlandier states that an infusion of the leaves was used by the Indians as a febrifuge. He proposed a new genus for the plant and called it *Terania frutescens,* in honor of General Mier y Teran, the director of the expedition, but no formal description of the genus was published."
The plant is mentioned more than once by Berlandier and—on p. 276 cited by Standley (it was under date of December 7, 1828, when Berlandier was on a hunting trip)—he wrote: "*La Cenicilla Teránea Frutescens,* vegeta sobre algunos peñascos, pero es muy rara."
As Ivan M. Johnston points out, Berlandier *did* publish the plant's description, in both Latin and Spanish, in what was probably 1832. I mention this later in this chapter.

ville, ". . . Ruiz shot a buffalo." And, more important, records that "On this trip Berlandier seems to have made few or no botanical collections, and but few botanical observations." However, it was on this trip that he did bestow a name upon one of the common trees of western Texas; on December 7 the party traveled through the beautiful Uvalde Canyon—"El Cañon de D. Juan de Ugalde." Berlandier wrote:

"A la orilla de los torrentes, y sobre todo, en la del arroyo principal, se encuentram nogales de una especie natural, cuyos frutos muy pequeños, parecidos á una grande avellana [hazel nut], tienen un Endocarpo muy duro, y por esto se ha descrito bajo el nombre de *Juglans Microcarpa*." *(Diario de viage,* p. 276)

Juglans microcarpa is the accepted name for the small-fruited walnut of western Texas.

When the year 1829 began, Berlandier seems still to have been at Bexar. From a Berlandier manuscript Geiser supplies an account of two trips which he made from that point. The first was with "the commander of the presidio of Bexar," Lieutenant Colonel Antonio Elosua, to Goliad to ". . . quell an uprising against the Commandante at the presido there." And Geiser comments that "Berlandier's delightful account of the trip (filling some twenty pages of manuscript) gives a good view of the country." The excursion lasted from February 3 until February 14. (1948, p. 41)

The second trip began February 25 when Berlandier ". . . left Bexar for Aransas Bay with a party of Mexicans." They had hardly started when their horses were stolen but they set off again on the 28th and in five days reached Goliad:

". . . Berlandier met there the captain of the galette *Pomona* and decided to accompany him to New Orleans. On March 7 they set out for the port of Aransas Bay [on the Gulf of Mexico] . . . On March 12 they embarked . . . They arrived at the Belize[1] on April 1.

"The notes describing Berlandier's stay in New Orleans are missing. The manuscripts contain only his meteorological observations made on board the *Pomona* in the port of New Orleans, April 25–May 5 . . .

"The naturalist left New Orleans for Texas on the eighth of May . . . reached Texas on the thirteenth, and the next day was . . . in Aransas Bay. On the seventeenth he reached Goliad . . . In less than three days he rejoined his companions on the Commission in Bexar." (Geiser, 1948, pp. 41-42)

Not having met General Mier y Teran at Matamoras on August of 1828 as directed, they apparently set off to do so in 1829, leaving Bexar on July 14, or the same day as the year before. The itinerary, as given by Geiser, is nearly identical; they were again at the Rio Frio on the 20th, passed the Nueces on the 25th and on the 28th reached the presidio at Laredo; when they

". . . arrived at Matamoras, sometime about the thirtieth of August, Berlandier's work with the Commission was virtually over; the Commission itself seems to have

1. A United States pilot station on the Mississippi River, a few miles above the head of the delta.

been dissolved in November[2] . . . Berlandier determined at this time to take up residence at Matamoras, and he was to live there until his death in 1851.

"Although in later years Berlandier made other excusions for botanical collecting, notably a journey to Goliad and Bexar in the spring of 1834, his place among the Naturalists of the Frontier depends primarily upon the fact that his work with the Boundary Commission was the first extensive collecting done in Texas, antedating by five or six years Drummond's important work in the vicinity of San Felipe and Gonzales . . ." (1948, p. 42)

Alphonse de Candolle mentioned that it was through the generosity of a Lieutenant Couch that, after Berlandier's death, the botanists at Geneva came into possession of some of Berlandier's plant collections. The *Ninth Annual Report* of the Smithsonian Institution tells something of Couch and of the wealth of Berlandier manuscripts which reached Washington. The "Report of the Secretary," Joseph Henry, to the Board of Regents (1855) relates:

". . . About the beginning of the year 1853, Lieutenant D. N. Couch, U. S. A., communicated to the Smithsonian Institution a proposition to make at his own expense a scientific exploration in the States of Mexico, adjoining the Lower Rio Grande . . . it was commended by me in a letter to the Secretary of War,[3] and a request made that Lieutenant Couch might have leave of absence for the purpose of carrying out his design . . . this young officer soon after embarked on his expedition . . . his attention was especially directed to the existence in Mexico of a valuable collection of manuscripts and specimens in natural history, of which information had been communicated to us. He was requested to examine and report as to its character. He found the manuscripts to contain a large amount of historical and geographical information, chiefly pertaining to the States of the old republic which lay between the Sabine and Sierra Madre,[4] and a series of maps and results of the topographical and meteorological observations. The collections in natural history consisted of specimens in botany, zoology, mineralogy, &c.

"These collections were made by Luis Berlandier, a native of Switzerland, and a member of the Academy of Geneva. He came to Mexico in 1826, for the purpose of making a scientific examination of that country. Soon after his arrival he was appointed one of the Boundary Commission organized by the then new republic, with the object of defining the boundaries, extent, resources, &c., &c., of the northern or frontier states. This position gave him unusual facilities for observation and investigation relative to the character of the country, and for making collections of its natural history.

2. Ivan M. Johnston states: "According to Alcocer, Naturaleza ser. 2, iii. 556-557 (1901), the commission operated as late as 1830 or 1831."

3. Then Jefferson Davis.

4. The Sierra Madre Oriental as against the Sierra Madre Occidental, which, in a rough sense, is a continuation of the Rocky Mountains.

He, however, never returned to his native country, but married and settled in Mexico, and continued his researches until the period of his death in 1851.[5] Lieutenant Couch purchased the whole collection from the widow of the deceased, and transmitted it immediately to the Institution, which bore the expense of transportation. It contains matter which would be valuable to the general government, and which it is hoped will be purchased, and a sufficient sum paid to reimburse the cost of procuring it. In the appendix will be found a catalogue of the manuscripts.

"Lieutenant Couch himself collected a large number of specimens in natural history, which were presented to the Institution, and have already been examined and described . . ."

In S. F. Baird's "Report on American explorations in the years 1853 and 1854," included in the appendix of the *Ninth Annual Report* of the Smithsonian, there is a notice of Couch's work and of his purchase of the Berlandier collections which represented ". . . the result of many years of labor in the province of Tamaulipas, and proved to be of extraordinary value." Also included in the appendix (pp. 396-398) is a "Catalogue of Berlandier's historical and geographical MSS.," which opens with this estimate of the accomplishments of the man repudiated by the two De Candolles:

"The results of his extensive labors are in manuscripts . . . comprising an amount of information of the country west of the Sabine of the highest importance."

Then follows "a brief catalogue of the MSS." A few of the listed items which relate, in part at least, to Texas, are:

"Travels in Mexico and Texas, 1826 to 1834, inclusive. Containing notes upon the statistics, early settlements, and Indian tribes between the Sabine and Pacific, &c., &c., &c.—7 vols. . . .

"Diary of the Commission of Limits in Northern Mexico, 1830.—3 vols. . . .

"Com. Limits, Geography and stat. Physiques, &c., &c., by various authors, with notes by Berlandier.—Notes.

"Astronomical observations made in Northern Mexico, Texas, &c.—6 volumes . . .

"Maps in MSS. of journeys, routes, valleys, defiles, suburbs of towns, villages, &c., &c., of the country lying between the Sabine and Rio Grande and valley of Mexico . . ."

The maps are divided into three groups, one of which is described thus:

"Topography in detail of the route from San Antonio to Laredo . . . &c., &c. . . . Various routes of ancient Texas . . . Different plans of the State, sections, ancient maps, &c., &c. 25 maps in MSS . . ."

The United States government evidently did not see fit to buy the botanical collections—perhaps they acquired the manuscripts, "&c., &c."—for their purchaser was Charles Wilkins Short,[6] doctor of medicine who had a great interest in botany and

---

5. As Alphonse de Candolle stated, he had died—by drowning—when crossing the San Fernando River, a stream which Geiser (1948, p. 54) locates as south of Matamoras.

6. Short had a number of plant species named in his honor but is immortalized in the generic name of

amassed a fine herbarium, sharing with many individuals and institutions the collections which he was able to acquire. Asa Gray thought highly of Short and received from him a fine share of Berlandier's plants; he wrote of Short's transaction:

". . . He purchased, at a liberal price, the important botanical collections of Texas and northern Mexico, left by Berlandier, which Lieutenant (now General) Couch acquired of his widow and sent on to Washington; and, retaining one set for his own herbarium, he caused the rest to be distributed among the botanists to whom they would be most useful . . ."

In 1924 Ivan M. Johnston published a short article entitled "A neglected paper by Jean Louis Berlandier." The crudely printed little document is in the Gray Herbarium library. As Johnston states, "The first numbered page bears at its head: Memorias de la Comision de limites. Historia Natural Botanica. por El General Teran y L. Berlandier." Johnston fixes its date as 1832 and notes of the content:

"In substance the paper consists of dual Latin and Spanish descriptions of eleven newly proposed species and four new genera. The descriptions are carefully prepared and are evidently the work of Berlandier. The identifications of the proposed genera and species, which appear to have been completely neglected and which are not listed in the Kew Index, has been greatly facilitated by the specimens from the Berlandier herbarium . . . in the Gray Herbarium. These in a number of cases are labeled in Berlandier's handwriting with the names published in the brochure. Further help has been derived from the volume of Berlandier's unpublished plates . . . in the Gray Herbarium . . . It seems probable that the brochures and the Berlandier manuscripts all came to the Gray Herbarium through the gift of Dr. Short . . ."

Of the included plants described by Berlandier as new, two—*Leucophyllum frutescens*[7] and *Ehretia anacum*—bear the specific names which he bestowed and are re-

---

Shortia galacifolia Gray, of romantic history. First described by Gray from an unidentified specimen which, in 1839, he saw in Paris in the herbarium of the elder Michaux, the plant was accidentally found a second time in 1877 by a young boy, G. M. Hymans, near the town of Marion, McDowell County, North Carolina. In 1886 C. S. Sargent found it again in the high mountains of the same state and presumably near the very spot where Michaux had first gathered it nearly a century earlier. The plant's story having been publicized, Sargent (*Garden & Forest* 1: 506, 507, 1888) comments that Hymans ". . . reaped a rich harvest during a year or two by selling plants (and, it is to be feared, by exterminating them) for herbarium specimens, at extravagant prices . . ."

The thrill of rediscovery—and fame in botanical circles—still awaits the individual fortunate enough to happen upon the evasive "Gordonia" or "Franklinia"—to the botanists *Franklinia alatamaha* Marshall—which was found by the two Bartrams on the Altamaha River of Georgia in 1765, but which at the present day is known only in cultivation.

And, while my thoughts wander far from Berlandier, I take this opportunity to suggest that plant collecting can still have its excitements, although not, perhaps, the financial remuneration of the discovery of oil! Read "The rediscovery of a lost orchid" written by Oakes Ames and published in the *American Orchid Society Bulletin* for September, 1939. As the author ends his paper, ". . . sooner or later, when chance knocks loudly, somebody says, with what we moderns call bromidic commonplaceness: 'Truth is sometimes stranger than fiction.' "

7. See p. 376, *fn.* 9.

corded from Texan as well as from Mexican localities, the first-named from ". . . prope San Antonio de Bejar in Texas . . .", the latter from ". . . Texas prope la Bahia del Espiritu Santo . . ."

Although, after his service on the boundary commission, Berlandier lived at Mata-moras and did most of his botanical collecting south of the Rio Grande, he gathered in that river's lower valley numbers of plants from which many of the common species of Texas and of Mexico were described. In the history of Texas botany, he will be re-membered as the first of the profession to have investigated, in the years 1828 and 1829, the flora of the Gulf Coastal Plain, the Balcones Escarpment, and the Edwards Plateau, all famous collecting grounds for plants.

His name will be remembered in the genus *Berlandiera,* described by De Candolle in the *Prodromus* (5: 17, 1836) and illustrated in the *Icones selectae plantarum* (t. 26, 1839). It is a member of the family *Compositae.*

Geiser (1933, p. 432, *fn.* 2) states that:

"Berlandier manuscript materials are to be found in the following American li-braries: Yale University; Library of Congress; Library of the Smithsonian Institution; Library of the Gray Herbarium of Harvard University; and Library of the University of Texas."

Of Berlandier's plants the best collection is said to be in the Gray Herbarium. Gray gave a set to John Torrey which is in the New York Botanical Garden.

Gray in 1840 mentioned ". . . a set of Berlandier's Texan and Mexican plants . . ." in ". . . the Royal Museum at the Jardin des Plantes or Jardin du Roi . . ." in Paris.

I translate from Lasègue's *Musée botanique de M. Benjamin Delessert* (1845):

"Between 1827 and 1850 Mr. Berlandier sent several shipments of plants from Mexico, chiefly gathered as follows: first in the state of 'Coahuila-et-Texas,' in the vicinity of Austin, of the Brazos River, of the Trinidad River, of Saltillo [Coahuila, Mexico], of 'San-Antonio-de-Bejar,' and the colony of 'San-Felipe de Austin,' in the lagoons of 'San-Nicolas' near the bay of 'Aransasua' [Aransas] . . ."

Also from *La phytographie* of Alphonse de Candolle (1880):

"His numbered collections from Mexico are many; the most complete in the her-baria of Moricand, de Candolle and Delessert, at Geneva. Others in the herbaria of Webb at Florence, of Harvard University, of Cardinal de Haynald, of the Universities of Kiel and Leipzig, of the British Museum . . . and of the Palatin Museum at Vienna . . ."

Winkler (1915) records that ". . . A set of Berlandier's Texas plants was secured by the Smithsonian Institution in 1855. The Smith. Inst. 1846-1896, p. 707." [8]

8. The reference is to *The Smithsonian Institution, 1846–1896. The history of its first half century,* by George Brown Goode, published in 1897.

BAIRD, SPENCER FULLERTON. 1855. Report on American explorations in the years 1853 and 1854. *Smithsonian Inst. Ann. Rept.* 9: 86, 87.

BANCROFT, HUBERT HOWE. 1885. The works of . . . 14 (Mexico V. 1824–1861): 154, 155.

BERLANDIER, LUIS & CHOVEL, RAFAEL. 1850. Diario de viage de la comision de limites que puso el gobierno de la republica, bajo la direccion del Exmo. Sr. general de division D. Manuel de Mier y Teran. Lo escribieron por su órden los individuos de la misma comision D. Luis Berlandier y D. Rafael Chovel. Mexico.

CANDOLLE, ALPHONSE DE. 1880. La phytographie; ou l'art de décrire les végétaux considérés sous différents points de vue. Paris. 396. 460.

CANDOLLE, AUGUSTIN-PYRAMUS DE. 1862. Mémoires et souvenirs de Augustin-Pyramus de Candolle . . . écrits par lui-même et publiée par son fils [Alphonse de Candolle]. Genève. 336, 337 and *fn.* 2.

GEISER, SAMUEL WOOD. 1933. Naturalists of the frontier XI. In defense of Jean Louis Berlandier. *Southwest Review* 18: 431-459.

———— 1937. Naturalists of the frontier. Dallas. 38-72, 311, 318.

———— 1948. *Ibid.* ed. 2. 30-54, 265, 266, 271.

GHENT, WILLIAM JAMES. 1931. The early far west. A narrative outline 1540–1850. New York. Toronto. 167, 168.

GRAY, ASA. 1840. Notices of European herbaria, particularly those most interesting to the North American botanist. *Am. Jour. Sci. Arts* 40: 1-18. *Reprinted:* Sargent, Charles Sprague, *compiler.* 1889. Scientific papers of Asa Gray. 2 vols. Boston. New York. 2: 1-21.

———— 1863. Charles Wilkins Short. *Am. Jour. Sci. Arts,* ser. 2, 35: 451-453. *Reprinted: Ibid.* 2: 312-314.

HENRY, JOSEPH. 1855. Catalogue of Berlandier's historical and geographical MSS. *Smithsonian Inst. Ann. Rept.* 9: 396-398.

———— 1855. Report of the Secretary [to the Board of Regents] *Smithsonian Inst. Ann. Rept.* 9: 15.

JOHNSTON, IVAN MURRAY. 1924. A neglected paper by Jean Louis Berlandier. *Contr. Gray Herb.* new ser. 70: 87-90.

LASÈGUE, ANTOINE. 1845. Musée botanique de M. Benjamin Delessert. Notices sur les collections de plantes et la bibliothèque qui le composent; contenant en outre des documents sur les principaux herbiers d'Europe et l'exposé des voyages entrepris dans l'intérêt de la botanique. Paris. 207, 268.

SÁNCHEZ, JOSÉ MARÍA. 1926. A trip to Texas in 1828. *Southwestern Hist. Quart.* 29: 249-288. Translated by Carlos E. Castañeda. ("From the *Archivo General de Guerra y Marina, Fracción 1, Legajo núm.* 7. Mexico City.")

SARGENT, CHARLES SPRAGUE. 1890. The Silva of North America. Boston. New York. 1: 82, *fn.* 1.

STANDLEY, PAUL CARPENTER. 1920. Trees and shrubs of Mexico. *Contr. U. S. Nat. Herb.* 23. 1306.

WINKLER, CHARLES HERMANN. 1915. The botany of Texas. An account of botanical investigations in Texas and adjoining territory. *Univ. Texas Bull.* 18: 5.

1830-1840

INTRODUCTION TO DECADE OF

1830–1840

THESE ten years show a considerable increase in the numbers of plant collectors working in lands west of the Mississippi River. Their activities were confined, roughly speaking, to the outer margins of that great territory but this peripheral belt widens and shows fewer breaks. Starting in the southeastern portion of this belt (in southeastern and central Texas) and advancing north and northwest, then west, and finally south (through California) I shall indicate the fields investigated by these men. The southern portion of the belt—or the regions extending from southern California to the starting point in Texas—still remained unexamined by plant collectors in 1840.

By 1836 the first phase of the conflict between American and Mexican factions in Texas had been brought to a close with the recognition of the Texas republic. It was during the turbulent years preceding this event that Thomas Drummond, by way of St. Louis and New Orleans, reached Texas and worked in the southeastern portion of what is now that state from 1833 to the autumn of 1835. Although he was there during a period of social unrest and was handicapped by great floods and a cholera epidemic, he nonetheless sent back a fine collection of plants to his sponsor, William Jackson Hooker. Drummond never returned to Britain but died in Havana on his way thither.

North of what was then Mexican-held Texas, and separated therefrom by the Red River of the South, lay regions which the United States government had set apart for Indian occupation; now included in eastern Oklahoma, these lands were then designated as the "Indian Territory," and I use the appellation in this restricted sense rather than in the sense of Indian lands in general. Despite the allocation of this region, an adventurous and not always desirable element of white settlers was penetrating the Indian Territory and a number of army posts were established for its protection.

The first of these frontier posts was Fort Smith, established in 1817; it had been used as headquarters by Thomas Nuttall after he had ascended the Arkansas River in 1819 and was situated at the confluence of the Arkansas and Coteau rivers, in present Sebastian County, Arkansas—or just east of the boundary of the Indian Territory. Fort Smith was abandoned in 1824 and Fort Gibson, higher up the Arkansas River and within the eastern portion of the Indian Territory, was established in the same year. Also established in 1824 was Fort Towson, located on the eastern side of Kiamichi River and on the north side of Red River, or at the junction of these streams. There were other army posts: one at the confluence of the "False Washita," or Washita, River and referred to in the early narratives as Camp Washita. Fort Washita, established in 1842, was situated higher up the river.

Zina Pitcher, surgeon in the United States Army, was stationed at Fort Gibson from 1831 to 1834 and, when his busy life permitted, collected plants for John Torrey. Melines Conkling Leavenworth, also in the medical service of the army, was stationed more than once at Fort Towson and spent one summer at Camp Washita, from there making a trip westward to the Cross Timbers, a belt of hardwoods which extended, roughly speaking, north and south across the Indian Territory and beyond. Leavenworth was assigned to a number of other army posts and it was from Fort Jesup and Camp Sabine, both near the western boundary of Louisiana, that he was able to travel west of the Sabine River, or into Texas. His plants went to John Torrey. Availing himself of army protection, the German botanist Heinrich Karl Beyrich traveled from St. Louis to Fort Gibson and from there in 1834 accompanied the United States Dragoons to Camp Washita and thence to an undetermined point in the Cross Timbers. The expedition had as its purpose negotiations with certain Indian tribes and was first commanded by General Henry Leavenworth and later by Colonel Henry Dodge. Like many other unfortunates in the party, Beyrich died in the course of his return journey to Fort Gibson. The artist George Catlin was present and described the man's death. Among distinguished visitors to Fort Gibson were Washington Irving (serving on a commission having to do with the resettlement of certain Indian tribes) and Charles Joseph Latrobe (later to become the first governor of Victoria, and the founder of the Melbourne Botanic Garden) with his young Swiss protégé, Count Albert Pourtales; although Latrobe has been credited with being the third in "a cavalcade of botanists" to enter Oklahoma, neither he nor any member of the Irving-Latrobe party displayed much interest in plants on their highly enjoyable hunting trip of one month into the Indian Territory.

Northward along my imaginary periphery, but still within eastern portions of the trans-Mississippi west, plant collectors entered what is now Minnesota and adjacent regions to the east, west and south. One such collector, best known as a geologist, was Douglass Houghton who, in 1832, accompanied Henry Rowe Schoolcraft in his search for the sources of the Mississippi River; they reached Lake Itasca and returned to the Mississippi by a meandering journey which brought them to the mouth of Crow Wing, or "De Corbeau," River. Another to enter this general region was a German botanist, Karl Andreas Geyer, who was employed to collect plants by Joseph Nicolas Nicollet when that famous topographer was mapping the "Hydrographical Basin of the Upper Mississippi River" in 1838 and 1839. On his own initiative Geyer had made an earlier journey up the Missouri River in 1835 which had produced few, if any returns, and had collected plants about St. Louis. Geyer's most important work was to be accomplished in the decade of 1840–1850.

More than one individual had gathered plants in the Missouri River valley before the present decade: Lewis and Clark had sent back a shipment of plants from Fort Mandan, North Dakota, in April of 1805; John Bradbury and Thomas Nuttall had collected along the Missouri between St. Louis and the country of the Mandan Indians in

North Dakota in 1811; William Baldwin in 1819 had collected between St. Louis and Franklin, Missouri, where he died; Edwin James, Baldwin's successor, had continued up the Missouri and overland to the mouth of the Platte (where he turned west) in 1820. In the present decade Maximilian, Prince of Wied-Neuwied, a German scientist traveling for his own delectation and enlightenment, ascended the Missouri by steamboat to Fort Union, Montana, and from there continued up the Missouri to Fort McKenzie, situated near the mouth of Marias River in the same state. After Maximilian's trip of 1833 the country lying within easy reach of the Missouri River had been visited by qualified collectors to about the 48th parallel of north latitude and to about the 111th of longitude.

Other collectors of plants crossed the trans-Mississippi west to the Pacific northwest in this decade: in 1833 Nathaniel Jarvis Wyeth traveled to the Columbia River and back, on the way east acquiring a few specimens ("from the falls of the Columbia to the first navigable waters of the Missouri") for his friend Thomas Nuttall; on a second overland journey to the Columbia in 1834 Wyeth was accompanied by Nuttall. After leaving the Missouri River this party soon entered Nebraska and, after crossing that state as well as Wyoming, Idaho and northeastern Oregon, it reached Fort Walla Walla (or Fort Nez Percés), Washington, and finally Fort Vancouver. Because of Nuttall's presence this was an important botanical journey.[1] Nuttall was an Englishman and the first trained botanist to cross North America south of the Canadian possessions.

Following the outer margins of the lands west of the Mississippi, I have reached the Pacific northwest. American interest in the Oregon country was increasing and the Hudson's Bay Company, although helpful to new arrivals of whatever nationality, was aware of what this interest portended. Spain in 1819 had relinquished its rights in the Oregon region to the United States, and Russia in 1824 had, by agreement with the United States, withdrawn its claim south of latitude fifty-four forty; so that, in the decade of 1830–1840, only the United States and its citizens offered a serious threat to the interests of the British company. American interest in the northwest had been demonstrated as early as the 1820's when Hall Jackson Kelley of Massachusetts had organized a society (incorporated in 1831) for the colonization of the region, and Wyeth had had ambitious projects to do so in 1833 and 1834. Missionaries, too, were showing concern for the spiritual and material well-being of the Indians, and Jason and Daniel Lee and their followers (who had traveled west with Wyeth) had established

---

1. For Wyeth the expedition was a failure. He had had plans for competing in the fur trade, but three American companies were already operating in the Rocky Mountains and the Hudson's Bay Company controlled the trade in the northwest, and, moreover, because of the ruthless destruction of the fur-bearing animals he was entering the field too late—the best days of the American fur trade terminated, according to Chittenden, about 1843. Nor did Wyeth's plans to establish a salmon industry and to start agricultural projects in the Willamette River valley materialize. Nonetheless, in the words of H. H. Bancroft, Wyeth's "... influence on Oregon occupation and settlement was second to none ... He ... more directly than any other man, marked the way for the ox-teams which were so shortly to bring the Americanized civilization of Europe across the roadless continent."

themselves in the Willamette River valley. In 1836 Marcus Whitman and his wife Narcissa opened their station in the vicinity of Fort Walla Walla, with plans for colonization[2] as well as for redemption of the savages—Edward Belcher, commanding the British ship *Sulphur*, commented in 1839 that the Hudson's Bay Company was making a mistake not to select its own countrymen to render "spiritual assistance." Persons of scientific bent were arriving with the rest and the influx constituted a problem, if nothing more, to those at Fort Vancouver—as is evidenced in a letter written by Peter Skene Ogden[3] on February 25, 1837. Because of the date of this letter and because of the enthusiasm for collecting birds displayed by Nuttall and his ornithological friend John Kirk Townsend, I feel sure that they must have been among the troublesome arrivals:

". . . by the Snake country we had an assortment of Am. Missionarys the Rev. Mr. Spaulding & Lady two Mr. Lees & Mr. Shephard surely clergymen enough when the Indian population is now so reduced but this not all there are also two more Gent. as follows 2 in quest of Flowers 2 killing all the Birds in the Columbia & of the U. States and you know it would not be good policy not to treat them politely they are a perfect nuisance . . ."

Except for Wyeth and Nuttall all other plant collectors to visit the Pacific northwest between 1830 and 1840 arrived by sea. David Douglas, continuing the work begun in the 1820's, spent six months in 1830 and from October, 1832, to October, 1833, in that region. During his last months in the northwest Douglas suffered the loss of most of his invaluable records in the Fraser River of British Columbia and this disaster had an injurious effect upon his health and spirit, as well as upon his accomplishments. During the last year of Douglas' visit to the northwest, two medical men with botanical predilections arrived at Fort Vancouver: William Fraser Tolmie and Meredith Gairdner. Tolmie was to make the northwest his home for the remainder of his life, but his most notable botanical accomplishment (in the period of my story certainly) seems to have been an excursion to the lower slopes of Mount Rainier in 1833. Gairdner was ill almost from the time of his arrival on the Columbia and, after about thirty months, moved to the Sandwich Islands where he died. Although interested in more than one scientific line, Gairdner's medical work and petty duties as a clerk gave him little time for such pursuits and during his sojourn in the northwest he felt unhappy and frustrated, or so the story goes. The British ship *Sulphur,* commanded by Edward Belcher, stayed at the Columbia River for six weeks in 1839. With Belcher traveled Richard Brinsley Hinds who was later to collaborate with George Bentham in

2. In the next decade (1840–1850) the Henry Harmon Spaldings were to settle among the Nez Percés at Lapwai, Idaho, and Elijah White, in 1842, was to lead the second wagon train of emigrants into the Oregon country.

3. The letter is included in the "Journals and correspondence of John McLeod, Senior, Chief Trader, Hudson's Bay Company, who was one of the earliest pioneers in the Oregon Territory, from 1812 to 1844. Copied from the originals in the Dominion Government Archives by R. E. Gosnold." This document is in the Library of Congress.

writing *The botany of the voyage of H.M.S. Sulphur,* a work in which Hinds supplied general remarks upon the flora, the climate and so on, of the Pacific northwest.

Except for David Douglas, and to a lesser extent Tolmie, the collectors mentioned did not get far into the interior. However, in 1837, some plant specimens were collected between Fort Vancouver and the rendezvous of the trappers in northwestern Wyoming by a "friend of Mr Tolmie" who proves to have been John McLeod, one of the traders of the Hudson's Bay Company. McLeod's specimens, "in beautiful preservation," and said to have come from ". . . the *'Snake Country,'* in the interior of California," were later described by Hooker and Walker-Arnott in *The botany of Captain Beechey's voyage . . .*

We now turn southward to California, which since 1822 had been controlled by Mexico. The period of Mexican rule, lasting until 1846, was marked by the gradual destruction ("decline" or "secularization") of the missions. In the decade of 1830–1840, as in the previous one, visitors to Californian ports found the once-generous padres less openhanded; not, probably, because their spirit had changed, but because they had received orders to that effect and because they were becoming more and more impoverished. Nonetheless, David Douglas, who stayed at many of the Californian missions, had a better word for the padres than some of his compatriots who knew them less well. The population of California was heterogeneous but, however much divided along racial, religious, social and other lines, it seems to have been in accord in its growing resentment of Mexican domination. Some Americans, reaching California by sea from time to time, had remained as settlers or traders; soon, in the early years of the decade of 1840–1850, they were to arrive overland with their families.

Four important botanists reached California between 1830 and 1840, and all arrived by sea. David Douglas was there for nineteen months (December, 1830–August, 1832) and again for one month in November of 1833. He records that he gathered no plants on this last visit. On the first he made a notable collection which he shipped back to England from the Hawaiian Islands. His travels extended from a short distance north of San Francisco southward to the region of Santa Barbara, possibly to Ventura, and were confined to the coastal ranges and to the valleys west of the Sierra Nevada. Thomas Coulter, who was sending his collections to the elder De Candolle in Geneva, arrived in 1831 from Mexico. He is reported to have remained in California for a number of years (three or even four) but of that time only a brief four months are accounted for in his record of a two-way trip from Monterey across the Colorado Desert to the confluence of the Gila and Colorado rivers. At this point, in the vicinity of present Yuma, Coulter must have been the first botanist to enter Arizona. In March of 1836, Thomas Nuttall reached California from the Sandwich Islands and moved down the coast from Monterey to San Diego in a sailing vessel, collecting plants, birds and shells when he could get ashore. He returned to the United States in a ship made famous in Dana's *Two years before the mast*. The fourth botanist to reach California was Richard Brinsley Hinds, voyaging in the *Sulphur* commanded by Edward Belcher. In 1837, and again in 1839, the ship lay off San Francisco, and on the last visit Hinds was able to travel some distance up the Sacramento River; he was also able to get ashore at Monterey, Santa Barbara, San Diego and other points. Basing his opinions upon field-observations, Hinds wrote at length about the vegetation of the regions which he visited, comparing the plants of one section of the country with those of other sections and discussing the effects of climate and environment; his was the first such report to be published about the vegetation of any part of the trans-Mississippi west. The collections made in California were described in *The botany of the voyage of H.M.S. Sulphur,* the joint work of Hinds and George Bentham.

In his introductory words to the volume Hinds noted that ". . . Upper California has already been tolerably examined . . ." He referred, presumably, to the region's plants as of the time of his visit, which was approximately fifty years after the Lapérouse expedition had stopped at Monterey. The comment could, however, have been applicable only to a small portion of what is now the state of California—to ports and areas along the coast, to the more accessible slopes of the coast ranges and to some of the valleys lying east of these mountains, regions which lay, generally speaking, from the environs of the Bay of San Francisco southward. No botanist or plant collector by 1840 had worked in the Sierra Nevada nor made the dangerous overland journey from the Columbia River to and through northern California. Douglas, when searching for the sugar pine in 1826, had, from the north, approached closest to the present Oregon-California boundary. Moreover, the vast regions lying east of California were virtually unknown to plant collectors. What are now the adjacent states of Nevada and Arizona (except for Coulter's very brief sojourn at the confluence of the Colorado and Gila rivers) had not been entered by such men, nor Utah, nor New Mexico (except for Edwin James's short passage of its extreme northeastern corner). Colorado had been traversed from north to south by James, who kept to the eastern side of the Rocky Mountains—he seems to have made the nearest approach to California from directly east. The plant collectors who reached the Pacific coast overland before 1840 had kept much farther north, through Wyoming, Montana and Idaho, and were all bound for the Columbia River.

The botanists who were describing the plants sent home by their emissaries felt that, although much remained to be accomplished, their work was beginning to bear fruit. In his "Advertisement [preface] to the American edition" of *An introduction to the natural system of botany* by John Lindley which was published in New York in 1831 —or at the beginning of the present decade—John Torrey wrote:

"The catalogue which I have prepared, embraces a considerable number of genera and species which are not described in the latest general Floras, but it is by no means asserted to be complete. There are extensive districts in North America which have never been visited by a Botanist, and even in the United States there are large spaces which are but little known or very imperfectly explored. There are also many plants collected by Douglass [*sic*], Richardson, Drummond, Scouler, Nuttall, and others, which have not yet been published, so that it is probable that North America, excluding the Mexican states, contains not less than 5000 phenogamous plants."

Torrey's estimate would doubtless have been larger had it been made at the end rather than at the beginning of this decade—or after Drummond's Texan collections, Douglas', Coulter's and Nuttall's Californian collections, as well as others from the Indian Territory and elsewhere had become known to him. On the other hand, Torrey's conclusions regarding the existence of "extensive districts in North America" (if we consider only the trans-Mississippi west) which "had never been visited by a Botanist," would have held true in 1840.

CHAPTER XVIII

DOUGLAS VISITS NORTHWESTERN NORTH AMERICA

FOR A SECOND TIME AND SPENDS TWENTY MONTHS

IN CALIFORNIA

UNDER the last decade I devoted a chapter to the first two trips which David Douglas made to North America, the second of which in 1824–1827 had taken him to the Pacific northwest, and to the years from 1827–1829 which he spent in England on his return.

The present chapter concerns Douglas' third and last visit to North America and would be incomplete without an account of his trip into New Caledonia, now British Columbia, in the course of which he lost his invaluable records, and without reference to his last visit to the Hawaiian Islands where he met his death. Both regions, however, lie outside the locale of my story.

In my earlier chapter I discussed the content of the two important publications (the Hooker "Memoir" and the Royal Horticultural Society volume) which, between them, contain the important records of Douglas' entire life. For the man's first journey to the northwest in 1824–1827 these two volumes included, in addition to letters, three journals describing his day-by-day activities. For the present journey of 1829–1834 there exists no journal describing his years in the northwest and in California and knowledge of where he went and what he accomplished during those years is mainly derived from the letters which he wrote William Jackson Hooker from the field, and from "extracts" from a journal describing his last months in the Hawaiian Islands, all of which were published by Hooker shortly after Douglas' death. That no journal describing his work in the northwest and in California exists is explained by two circumstances, one factual (the loss of his records) and the other contributory (his resignation from the service of the Horticultural Society of London to which, on the earlier journey he had sent copies of his records as a precautionary measure). In the absence of any journal, Douglas' letters supply, therefore, the only firsthand, descriptive account of what was a highly important journey.

Although the Hooker publication ("A brief memoir of the life of Mr. David Douglas, with extracts from his letters," commonly referred to as the Hooker "Memoir") supplies the foundation of the present story it does not offer a complete, nor at times a clear, day-by-day account of Douglas' whereabouts, for records such as dates are scattered through the letters and the picture must be pieced together. However, in the miscellaneous literature concerned with Douglas, one finds corroborative, explanatory and other facts which help to supplement the Hooker records. Of this additional material by far the most important, in connection with Douglas' visits to California, is a list of localities known to have been visited by him on specified dates—twenty-three in 1831, one in 1832, one in 1833. This list was compiled by A. G. Harvey and published with his approval by J. T. Howell in 1942.[1] The greater part of the list cites stations where Douglas took latitudes and longitudes at the California missions (as recorded in "a manuscript at the Royal Society, London") although a few are taken from miscellaneous sources. Bancroft's *History of California* (3: 403, *fn.* 39) supplies a somewhat similar list,[2] but since no dates are cited, it is valueless as an itinerary; it does, however, mention a few stations not included in the Harvey list. Of the supplementary material concerned with Douglas' sojourn on the northwest coast the most helpful is a short biographical sketch of Douglas written by George Barnston;[3] the two men seem to have had similar interests and made one trip at least together. Douglas occasionally wrote Barnston after their work took them in different directions.

I shall attempt to reconstruct Douglas' journey of 1829–1834, taking my facts from the above sources or from wherever they may be found, and citing the authority in every instance. The title of the work in which the record appears is given in full in my bibliography of Douglas (pp. 338-341) but, for convenience, I refer to the authors most frequently quoted as follows:

B (Barnston); Ba (Balfour); Bt (Bancroft, *Northwest Coast*); Bt2 (Bancroft, *California*); H (Harvey, 1940); HH (Harvey in Howell, 1942); HM (Hooker, "Memoir"); HS (*Hawaiian Spectator*); J (Jepson, 1933); RHS (Royal Horticultural Society, London); RGS (Royal Geographical Society, London); W (Wilson).

Harvey's *Douglas of the fir* (1947) was published after this chapter was written and I refer to it (largely in footnotes) as Harvey (1947); seven of its chapters cover the Douglas journey of 1829–1834. The Harvey biography is interesting not only as a picture of Douglas but of the period in which he worked.

My earlier chapter left Douglas awaiting in England the sailing of the Hudson's Bay Company's ship *Eagle*. Even before he left he had learned that conditions in northwestern North America were to make his work more difficult than it had been on his last visit—the Indians were less peaceable for one thing. On March 11, 1829, the *William and Ann* had gone down at the entrance of the Columbia River and all forty-six

1. It was reprinted, slightly rearranged (and as to dates of 1831 only) in Harvey's *Douglas of the fir* (p. 176, *fn.* 6, 1947).

2. This was a "table of geographical positions" (with latitudes and longitudes) which Douglas supplied to the Mexican governor of California, José Figueroa, on November 25, 1833 and now preserved in the Bancroft collection.

3. According to *The dictionary of Canadian biography*, George Barnston was a fur trader and naturalist, born in Edinburgh, Scotland, in 1800. He entered the service of the Hudson's Bay Company in 1820, crossed the Rockies in 1825 and established the first trading post on the Fraser River. He retired from the service of the Company in 1867 with the rank of chief factor.

Bancroft's *Northwest Coast* (2: 515, *fn.* 11) mentions that "Up to the spring of 1831, for some time, Fort Walla Walla had been in charge of George Barnston ..." We are told that "... he was a man of good intellectual attainments, and was universally respected ..."

persons on board had been lost. Douglas wrote Hooker from Greenwich on September 14, 1829:

". . . It was the vessel in which Mr. Scouler and I went out in 1824 . . . It is stated that those who escaped from the wreck were destroyed by my old friends, the Chenooks. This may be true, though I confess I entertain some doubts, for I have lived among those people unmolested for weeks and months. The temptation, however, of obtaining the wreck, may have overcome their better (if they possess any) feelings. Though this is far from agreeable news, and though the name of my new Captain (Grave) may sound ominous, I shall yet venture among these tribes once again. I doubt not I can do as much as most people, and perhaps more than some who make a parade about it . . ." (HM, 144)

He had heard more of the Indians when he wrote from London on October 27, 1829:

". . . I am sensible of the great advantage I derive from my former experience of travelling in the country, of hunting, collecting, &c.; and certainly if I find the Indian tribes as quiet as when I left them, much good may be effected. Of this, however, I feel considerably afraid, in consequence of the destruction of the Hudson's Bay Company's Ship's crew, and the murder of some parties of Americans, by which I am warned to walk with great caution, and more reservedly than before. If I find the natives hostile to the 'Man of Grass' . . . I must shift my quarters to some other part of the country . . ." (HM, 145)

On arrival at the Columbia he learned of another shipwreck, of more Indian murders, and of still a worse peril—intermittent fever. He mentions them all in a letter to Hooker dated October 11, 1830:

"The ship which sailed along with us was totally wrecked[4] on entering the Columbia River; I am happy to say, however, no lives were lost. To this vessel I had first been appointed, and then changed to the one in which I came. But for this fortunate alteration I should have lost my all . . ." (HM, 148)

The murders had occurred in the region of the Willamette River:

". . . a party of hunters were all killed, save two, who returned to tell the melancholy fate of their companions; and again a second party has nearly shared the same fate. You may judge of my situation, when I say to you that my rifle is in my hand day and night; it lies by my side under my blanket when I sleep, and my faithful little Scotch terrier, the companion of all my journies, takes his place at my feet. To be obliged thus to accoutre myself, is truly terrible. However, I fail not to do my best, and if unsuccessful in my operations, can make my mind easy with the reflection that I used my utmost endeavours . . .

---

4. This was doubtless the wreck mentioned by Bancroft:

"The ship *Isabel*, Captain Ryan, was wrecked on Sand Island while entering the Columbia on the 23d of May 1830. Immediately she struck, the crew deserted her. Had they remained at their post, they might have saved the ship, as there was little difficulty in saving the cargo." (Bt, 2, 515)

"A dreadfully fatal intermittent fever broke out in the lower parts of this river about eleven weeks ago, which has depopulated the country. Villages, which had afforded from one to two hundred effective warriors, are totally gone; not a soul remains! The houses are empty, and flocks of famished dogs are howling about, while the dead bodies lie strewed in every direction on the sands of the river. I am one of the very few persons among the Hudson Bay Company's people who have stood it, and sometimes I think, even I have got a *great shake,* and can hardly consider myself out of danger, as the weather is yet very hot." (HM, 148)

This epidemic seems to have persisted for the duration of Douglas' visit. He wrote Hooker as late as April 9, 1833, from the interior:

"Fever still clings to the native tribes with great obstinacy, and not a few of the Hudson's Bay Company have suffered very severely from it. Only three individuals out of one hundred and forty altogether escaped it, and I was one of that small number[5] . . . I never was in better health." (HM, 158)

For convenient reference I preface my more detailed account of Douglas' journey with a brief outline:

Douglas left England in October, 1829, in the ship *Eagle.* The voyage took about seven months and included a stop at the Hawaiian Islands. He arrived at the mouth of the Columbia on June 3, 1830. He worked in what are now the states of Washington and Oregon (June 3 to December 10, 1830); spent approximately nineteen months in California (December 22, 1830 to August 18, 1832); returned, by way of the Hawaiian Islands (where there was a stop for less than a week in September[6]), to the Columbia River region where he remained for about one year (October 14, 1832 to October 18, 1833). During that period he made a journey from Fort Vancouver into the northern interior of British Columbia (then called New Caledonia) in the summer of 1833, losing his possessions on the Fraser River on June 13. He sailed from the Columbia (October 18, 1833) for the Hawaiian Islands, spending on the way about one month in California (November 4 to 29); he reached the Islands (December 23, 1833) and there met his death (July 12, 1834).[7]

5. Barnston tells us that ". . . small pox could not have made a more destructive sweep. It remains a question for physicians to solve, how this intermittent, until then quiescent in that quarter, should have broken out with such violence without any apparent reason . . . The disease has taken permanent root . . . Douglas was ill like the others but being something of a leech, had an early recovery, and recruited perfectly by following up his wonted perambulations."

Barnston appears to have based his statement upon a letter from Douglas, dated November 29, 1830, which comments: "I was ten days in that state between hope and fear, but never laid down." Perhaps he considered the attack, which he mentioned to Hooker on October 11, 1830, so slight as not to be worth mentioning or perhaps he meant that he had been *in dread of* the disease but never ill. Certainly he wrote Hooker in 1833 that he had "altogether escaped" the scourge.

6. Harvey (1947) states that Douglas was ". . . at Honolulu from September 7th to 12th."

7. The following dates in the above paragraph are supplied in Harvey (1947): December 10, 1830; August 18, 1832; October 14, 1832; November 29, 1833 (cited by others as November 28 and November 30).

1829

In the *Transactions* of the Horticultural Society of London (7: iii, 1830) it is recorded that "Mr. Douglas re-embarked on the 26th of October, 1829 . . ." And Douglas, in London, mentions that on the 27th he was ". . . hourly expecting the summons to sail[8] . . ." (HM, 145)

1830

From the entrance of the Columbia River on October 11, 1830, Douglas wrote Hooker of the journey which included a stop at the Hawaiian Islands:

"I must now pass from London to Oahu, in the Sandwich Islands, all in one line! The ship touched no where on the eastern shore of South America, which to me was a great loss and disappointment, for I had anticipated much advantage from researches made on that Continent and the Islands of the South Seas. It was not my fortune to climb the snowy Peak of Mouna Kaah . . . nor could I get to Mouna Roa, which at this instant is dreadfully agitated by volcanic fires . . . but I did what was of more service to Botany, in scaling the lofty and rugged peaks of Mouna Parrii . . . The season was unfavourable, very rainy, and being just the conclusion of winter, I could only obtain Mosses and Ferns. I hope yet to visit this place again, under more favourable circumstances . . ." (HM, 147)

The same letter mentions that Douglas had reached the entrance of the Columbia River ". . . on the 3rd of June, in eight months, from London . . . The lateness of my arrival, for it was the 1st of July before I could leave the coast for the Interior, has been a serious drawback; the season proving unusually early, all the vernal plants, which are by far the most numerous, beautiful, and curious here, were withered and decayed. It took me twenty-four days of hard labour to reach the very lofty chain of mountains on which I was in July, 1826;[9] I again found my *Pæonia* . . . and including all my labours, there are, I should think, full one hundred new species, and perhaps some new genera, though I have yet only determined one, which is akin to *Œnothera*." (HM, 147)

Barnston had accompanied Douglas from the coast on this journey "of twenty-four days" into the interior, which took them as far as Fort Walla Walla and lasted from July 1 to July 23:

"Lewis and Clarke's Fork being a place of some note as the point of confluence of

8. Harvey (1947) reports that the ship left on October 31.

9. From *July* 27 to 28, 1826, Douglas had been in the Craig Mountains of Idaho and he did not get so far north on the present occasion. To him the Craig and the Blue Mountains were one and the same range. He had been in the Blue Mountains from *June* 17 to 24 and from *June* 26 to 30, 1826. In both ranges he had found the peony of which he was so proud.

the two great branches of the Columbia[1] . . . Mr. D. desired to adopt it as one of his principal stations for astronomical and magnetical observations. I therefore had the pleasure of his company up to Wallawalla, to the charge of which post I had been appointed. On the route, whenever an opportunity offered we were on shore together, and I was much surprised to remark the quickness of sight he displayed in the discovery of any small object or plant on the ground over which we passed. When in the boats . . . he would frequently spring up abruptly in a very excited manner, and with extended arms keep his finger pointed at a particular spot on the beach or . . . rocks, where some new or desirable plant had attracted his notice. This was the signal to put on shore . . . at Wallawalla he was immediately busied in taking observations . . . To enable him again to visit the Blue Mountains . . . I furnished him with five horses, and our interpreter with a sturdy boy. After a few days absence on this excursion he returned . . . regretting that the wild disposition of the Shoshonees . . . prevented his penetrating further to the southward . . . To have attempted that would have been attended with great risk and danger. At this time we adopted a very successful method of catching lizards, single horse-hair lassos, tied to the end of a wand . . ." (B, 268, 269)

Douglas, having fulfilled his desire to re-visit the Blue Mountains, now descended the Columbia to Fort Vancouver, leaving Walla Walla on July 23 and arriving at his destination on the 25th. Barnston states:

"On the 23rd July Mr. Douglas left Wallawalla and I felt his absence as a sad blank . . . His thoughts were turned towards California, and he availed himself . . . of a return boat to Fort Vancouver to return to the coast." (B, 269)

Douglas wrote Barnston that the first night of the journey had been spent below "Day's" (or John Day's) River, the second at the Cascades, from where, the next day, he "proceeded" to the Fort. (B, 270)

He was still at Fort Vancouver eleven days later for Barnston received a letter from there dated August 4th. (B, 270)

Soon after he must have started on a trip up the Willamette River, or into present Oregon. He mentions it in a letter to Barnston dated from Fort George (the old Fort Astoria) on November 29:

"I had an extensive journey to the Willamette with Mr. McLeod[2] . . . The highest land on which I was on the Willamette Ridge is 1043 feet elevated above the apparent base . . . I find 11,320 feet for Mount Jefferson . . . I would have had good reason to rejoice, had I not had a misfortune on my return. On one of the rapid tributaries of the Mattnomah [Multnomah, or Willamette], I lost my zoological collections, a dreadful

---

1. The junction of the Snake ("Lewis and Clarke's Fork") with the Columbia lies between what are now Walla Walla County (south) and Franklin County (north) and is not far north of Fort Walla Walla which was situated near present Wallula, Walla Walla County.

2. Alexander Roderick McLeod.

loss . . . It is curious; on the 17th November 1826, I lost *everything I had at the same place,*[3] when returning from my southern journey! . . . Since that time I have made my intended trip to the cascades . . . I accomplished all I wished . . . Your friends will have told you of . . . a fatal intermittent fever . . . in the lower parts of the river . . . I am thus far on my way to California . . ." (B, 271, 272)

On October 11, from the entrance of the Columbia River, Douglas wrote Hooker about his trip to the Willamette—for he could have meant nothing else by the phrase "after sixty days of severe fatigue." It must have consumed most of August, all of September and perhaps a few days in October:

"I have now just saved the sailing of the ship, and, after sixty days of severe fatigue, have undergone, as I can assure you, one of still more trying labour, in packing up three chests of seeds, and writing to Mr. Sabine and his brother. The Captain only waits for this letter, after which the ship bears away for Old England; I am truly sorry to see her go without my dried plants, but this is unavoidable, as I have not a bit of well-seasoned wood in which to place them, and should, moreover, be unwilling to risk the whole collection in one vessel; and the sails are already unfurled, so that it would be impossible to attempt dividing them. I however transmit one bundle of six species, exceedingly beautiful, of the genus *Pinus.* Among these, *P. nobilis* [*Abies nobilis*] is by far the finest. I spent three weeks in a forest composed of this tree, and day by day could not cease to admire it . . . I have added one new species during this journey, *P. grandis* [*Abies grandis*], a noble tree . . . growing from one hundred and seventy to two hundred feet high. In the collection of seeds, I have sent an amazing quantity of all kinds. Your specimens are in every way perfect. I have a few Mosses and a considerable number of *Fuci:* this is a department in which I fear the Flora will be deficient; but as I am to spend this winter entirely on the coast, you may expect to receive all that are found within the parallels of the British possessions on the Pacific side of this continent . . ." (HM, 147, 148)

Douglas was considering California but was uncertain how to get there:

"On the direction of my next year's route, I am not yet decided; but my desire is to prosecute my journey in North California, in the Valley of Bonaventura, through which a stream of considerable magnitude flows, and mingles its streams with the ocean in the Bay of Montérey. If I can venture thither in safety by land, I will do so;[4] if not, I shall go by sea to Montérey . . ." (HM, 148)

Douglas arrived at Monterey on December 22, 1830, and remained in California until August 18, 1833. According to Bancroft ". . . he came down from the Columbia on the *Dryad* to investigate the flora . . . He brought letters from Captain Beechey to

3. The earlier accident had taken place on the Santiam (or "Sandiam") River, after Douglas had made his successful journey in search of the sugar pine.

4. The first botanist to travel overland from the Columbia River into California was William Dunlop Brackenridge who made the journey in September and October of 1841.

Hartnell,[5] with whose family he became very intimate . . ." (Bt, 2, 403) Douglas refers to him as ". . . an English gentleman, in whose house I lived at Montérey . . ." (HM, 154)

On November 23, 1831, Douglas wrote Hooker from Monterey, describing conditions when he arrived, and—although very sketchily—where his travels had taken him up to date of writing. He was then hoping to leave California for the northwest, or, should this not be possible, for the Hawaiian Islands. Actually, he was obliged to remain in California until August 18, 1832. I shall italicize the phrases which prove helpful in reconstructing his itinerary and shall occasionally alter the paragraphing.

"On the 22nd of December last (1830), I arrived here by sea, from the Columbia, and obtained leave of the Territorial Government[6] to remain for the space of six months, which has been nearly extended to twelve, as the first three months were occupied in negociating this affair . . . I shall now endeavour to give you a brief sketch of my walks in California . . .

"Early as was my arrival on this coast, spring had already commenced; the first plant I took in my hand was *Ribes speciosum,* Pursh . . . remarkable for the length and crimson splendour of its stamens; a flower not surpassed in beauty by the finest *Fuchsia;* and for the original discovery of which we are indebted to the good Mr. Archibald Menzies, in 1779. The same day I added to my list *Nemophila insignis* [*Nemophila Menziesii* Hooker & Arnott, commonly known as 'baby blue-eyes'] . . . a humble but lovely plant, the harbinger of Californian spring, which forms, as it were, a carpet of the tenderest azure hue. What a relief does this charming flower afford to the eye from the effect of the sun's reflection on the micaceous sand where it grows. These, with other discoveries of less importance gave me hope . . .

"From time to time, *I contrived to make excursions in this neighborhood, until the end of April, when I undertook a journey southward, and reached Santa Barbara . . . in the middle of May, where I made a short stay, and returned* [to Monterey] *late in June, by the same route, occasionally penetrating the mountain-valleys which skirt the coast.*

"*Shortly afterwards I started for San Francisco, and proceeded to the North of that port.* My principal object was to reach the spot whence I returned in 1826, which I re-

5. Bancroft's "Pioneer register and index" (*California* 3: 777, 778) contains a long account of William Edward Petty Hartnell (1798–1854) who first went to California as a member of a mercantile firm serving as agents for a firm in Lima, Peru, and another in Liverpool and Edinburgh. After financial reverses Hartnell retired and ". . . undertook the life of a ranchero at Alisal . . ." He evidently held many positions of responsibility in California and was much respected. He was a fine linguist ". . . a master of the Spanish, French, and German languages besides his own."

Du Four mentions that "In the course of a visit to Fort Ross in 1833, Baron Wrangell established William E. P. Hartnell as agent for the Russian American Company at Yerba Buena . . ."

Many visitors to Monterey mention the man.

6. Bancroft reports that, on Douglas' present visit, ". . . he explored Alta California by permission of the Mexican government, under the promise to make no sketches of what they called their military defences . . ." (Bt, 508)

gret to say, could not be accomplished. *My last observation was at 38° 45', which leaves an intervening blank of sixty-five miles.* Small as this distance may appear to you, it was too much for me!

"My whole collection of this year in California, may amount to five hundred species, a little more or less. This is vexatiously small, I am aware; but when I inform you that the season for botanizing does not last longer than three months, your surprize will cease. Such is the rapidity with which spring advances . . . the plants bloom here only for a day. The intense heats set in about June, when every bit of herbage is dried to a cinder. The facilities for travelling are not great,[7] whereby much time is lost . . . It would require at least three years to do any thing like justice to the Botany of California, and the expense is not the least of the drawbacks . . ." (HM, 149, 150)

By filling in the crude itinerary supplied by Douglas (and italicized by me in paragraphs three and four of the above quotation) with the specific localities supplied in Harvey's list and elsewhere, it is possible to follow Douglas' whereabouts with some degree of accuracy. He had, as noted, arrived at Monterey on December 22, 1830.

1831

From the date of his arrival "until the end of April," Douglas made "excursions in the neighbourhood" of Monterey. He was in that city on January 29. (HH, 161)

On February 4 Douglas was at the mission of San Juan Bautista, for his description of *Pinus Sabiniana* (*Trans. Linn. Soc. London* 16: 747-749, 1833) is dated from the "Mission of St. John's, Upper California, February 4, 1831." He was there ten days later, February 14. (HH, 161)

This mission faced the plaza[8] in the town of San Juan, situated "at the base of the Gabilan Range." (Ca, 384) It lay northeast of Monterey, in present San Benito County.

The Bancroft list of stations cites "Cerro de Gavilan (top.) . . ."

The fact that Douglas sent his description of *Pinus Sabiniana* from the San Juan mission is fairly good proof that he obtained the plant in the Gavilan, or Gabilan, Mountains where, according to Sudworth's *Forest trees of the Pacific slope* (p. 55), it is common. But Douglas undoubtedly saw the tree more than once in California, for the species has a wide distribution. I mention this pine again in connection with Douglas' plant discoveries in California.

On February 20 Douglas was at the mission of Santa Clara. (HH, 161) This was situated considerably north of Monterey, in what is now northern Santa Clara County.

---

7. Harvey (1947) states: "Travel facilities were very poor. It was a case of riding horse-back or walking, for the El Camino Real in some places was only a trail. Most of Douglas's journeys were made on foot . . . There were no roads or trails through the interior . . ."

8. For some of the names, locations and facts about the California missions mentioned in this and succeeding paragraphs, I am indebted to the volume *California* ("American Guide Series") which I cite as Ca.

The mission—"Santa Clara de Asís," or Assisi—which Douglas visited must have been the third, "dedicated in 1822," for two earlier ones had been destroyed by flood and by earthquake. (Ca, 380, 381)

In "February" he was at the mission of Santa Cruz. (HH, 161)

This had been founded under the name "Misión la Exaltación de la Santa Cruz (the elevation of the holy cross)" and was located in the town of Santa Cruz (Ca, 334-336), which is situated on the northern shore of Monterey Bay, in southern Santa Cruz County.

It was, likely, while at Santa Cruz that Douglas saw the coast redwood which he refers to as "a species of *Taxodium,* which gives the mountains a most peculiar, I was almost going to say awful, appearance." (HM, 150) However, he might also have found it in the Santa Lucia Mountains for he is known to have visited and climbed in that range. Douglas had more than one opportunity to enter the Santa Lucia Mountains but he only records one visit, in March of 1832. Sudworth's *Forest trees . . .* (p. 147) mentions that the redwood extends southward, from the San Francisco region, through the Santa Cruz Mountains, stops "for a few miles around Monterey Bay," but occurs again "in Santa Lucia Mountains (Monterey County) . . . chiefly on seaward side of range . . ."

Douglas was at Monterey on March 2 and March 27. (HH, 161) He was there on April 20, for Bancroft tells that he obtained ". . . in April a *carta de seguridad* to prosecute his researches for six months . . ." and notes that this *"Carta"* is "dated April 20, 1831 . . ." (Bt 2, 403 & *fn.* 38) He was at Monterey on April 21. (HH, 161)

In his letter to Hooker Douglas mentions that he remained at Monterey "until the end of April," and then "undertook a journey southward," reaching "Santa Barbara . . . in the middle of May" and making "a short stay." (HM, 149)

The first mission visited on the southward journey was at Soledad, on April 25. (HH, 161)

This was the "Misión de Nuestra Senora de la Soledad (Our Lady of Solitude)." (Ca, 388) It was situated in the valley of Salinas River and across that stream from the present town of Soledad, in Monterey County.

Douglas next stopped at San Antonio, April 27. (HH, 161)

This was the "Misión San Antonio de Padua." The site was about six miles from present Jolon, Monterey County. (Ca, 388)

He was at San Miguel on May 1. (HH, 161)

This was the "Misión San Miguel Arcangel," dedicated to St. Michael the Archangel. The town of San Miguel, in what is now San Luis Obispo County, "clusters near" the mission. (Ca, 389)

He was at San Luis Obispo on May 3. (HH, 161)

"In a bowl-shaped valley at the southern base of the Santa Lucia Mountains is San Luis Obispo . . . seat of San Luis Obispo County. The town grew up around Misión San Luis Obispo de Tolosa (St. Louis, Bishop of Toulouse) . . . which is said to have

been so named because two of the pyramidal volcanic peaks in the neighborbood suggested a bishop's mitre." (Ca, 391)

Here he had, of course, an opportunity to enter the Santa Lucia Mountains, but it
seems doubtful that he did so for he seems to have been moving about with considerable rapidity.

On May 5 he was at Purisima. (HH, 161)

The "Misión La Purisima Concepcion" (Ca, 347) was situated to the north of the
Santa Ynez River, not far northeast of the present town of Lompoc, Santa Barbara
County.

On May 6 Douglas was at Santa Ynez. (HH, 161)

The mission of Santa Ynez was near the river of the same name, southeast of present Buellton, Santa Barbara County and ". . . on the outskirts of the hamlet of
Solvang." (Ca, 392)

Douglas reached Santa Barbara "in the middle of May" and made a "short stay."
(HM, 149) Probably he lived at the Santa Barbara mission which was situated in the
town. While there it is possible that he went southward, about thirty miles, to the mission of San Buenaventura, which was in the town of Ventura, Ventura County. For
one of the stations in the Bancroft list is "Cerro de Buenaventura (top.) . . ."

On leaving Santa Barbara, Douglas returned to Monterey, arriving there "late in
June." He had traveled "the same route, occasionally penetrating the mountain-valleys which skirt the coast." (HM, 149) He was at Monterey on July 15. (HH, 161)

It was "shortly" after his return that Douglas started northward, bound for "San
Francisco." (HM, 149)

The first recorded date was July 27, at San Rafael. (HH, 161)

The present San Rafael ". . . seat of Marin County . . . grew up around the Mission
San Rafael Arcangel . . ." (Ca, 365)

He was, two days later, at San Francisco Solano, on July 29. (HH, 161)

The "Misión San Francisco Solano" was named "for the saint known as the Apostle
to the Indies." (Ca, 362) It is situated on the plaza of the town of Sonoma, present
Sonoma County.

On August 5 and August 9 Douglas was at San José. (HH, 161)

The "Mission San José . . . clusters around La Misión del Gloriosisimo Patriarcha
Señor San José (the mission of the most glorious patriarch, St. Joseph) . . ." (Ca, 598)
It was near Alameda River (between the present towns of Irvington and Warm
Springs, in Alameda County) and about fifteen miles north of San José, Santa Clara
County which, according to Bancroft, was ". . . the first [pueblo] in California, [and]
was founded . . . under the name San José de Guadalupe, that is San José on the River
Guadalupe." (*California* 1: 312)

It was at this point that Douglas made his trip to the north of San Francisco
with the desire ". . . to reach the spot whence I returned in 1826, which I regret to say
could not be accomplished. My last observation was at 38° 45′, which leaves an in-

tervening blank of sixty-five miles . . ." Since the latitude cited is essentially that of Sonoma, Napa and Sacramento counties, Douglas did not get within some two hundred and fifty miles of the present northern border of California, and Roseburg (near the spot in Douglas County, Oregon, where Douglas found the sugar pine) lies nearly one hundred miles north of that border. These mileages are only rough estimates. Just what places he did visit before reaching the latitude named he does not disclose.

Douglas undoubtedly was at San Francisco but he writes nothing of it to Hooker.

Eastwood felt sure that he ". . . must have been on Mount Diablo [in Contra Costa County, to the east of San Francisco Bay], because he collected *Calochortus pulchellus* which has never been found elsewhere . . ."

But this opinion is not in line with Jepson's of a few years earlier—that the ". . . form of Calochortus pulchellus that he obtained might have been collected in the Sonoma region . . ." His visit to that mountain rests, therefore, upon the question of which botanical determination is considered correct! The collector himself never mentions Mount Diablo.

Douglas had returned to Monterey by August 25. (HH, 161)

He informed Hooker (in a letter from the "River Columbia, Oct. 23, 1832.") that he had written him ". . . from Montérey, in Upper California, in October 1831, and sent it by way of Mexico, under care of our Consul at the Port of San Blas; there I detailed to you the extent of my travels in that territory, and the progress of my collections, as well as gave you a brief notice of the country . . ." (H, 151, 152)

There have been a number of references in the literature to this so-called "lost" letter written from Monterey in October, 1831. It is my opinion that it never existed, and that Douglas, writing from the field about a year later, merely made the very natural slip of stating that he had written Hooker in *October* when he meant *November*. One finds very similar slips in Douglas' writings, as in those of others of our field-workers, but, since most do not concern a potentially valuable record,[9] they have passed unnoticed.

Douglas was obliged to remain in California longer than he intended or desired.

9. Two letters written by Douglas from Monterey in *November*, 1831, have been published:

(1) is dated "Monterey, Upper California, Nov. 20, 1831," and was published in the *Quarterly of the California Historical Society* (2: 223-227; 1923).

(2) is dated "Montérey, Upper California, Nov. 23rd, 1831." and was published in the Hooker "Memoir" (149-151). This was reprinted in large part in *The Overland Monthly* for 1883 (411-413) in an article written by C. C. Parry (409-416); to the extent of its extracts this is a verbatim copy of the one published in the "Memoir."

In publishing (1) above, the *Quarterly* states:

"That which we print is from a copy in the Douglas correspondence in the Provincial Library at Victoria, B.C., kindly furnished us by the librarian, Mr. John Forsyth. Although without address, it was most probably written to the celebrated botanist, Sir William Hooker. Douglas wrote another letter from Monterey, to which he refers in his published letter [from the "River Columbia, Oct. 23, 1832."], which he says was descriptive of the country. All efforts to locate that letter up to the present time have proved

Indeed he did not get away until August of 1832. Barnston explains that "So little intercourse was there in those days with San Francisco that he was detained all winter and spring there[1] without finding an opportunity of shipping himself off, but his time was comfortably spent at the presidias, where the hospitality of the padres was extended to him in so kind a manner that he ever warmly remembered it." (B, 273)

Douglas refers to his prolonged sojourn in a letter to Hooker, dated from the "River Columbia, Oct. 23, 1832":

"The Hudson Bay Company's vessel did not arrive on the coast of California in November [1831], as had been expected, which, in some measure, frustrated my projects. No opportunity having offered for proceeding, either to the Columbia or the Sandwich Islands in the winter [of 1831–1832] or spring [of 1832] of last year, I continued to consider California *as still new to me,* and set to work a second time, finding new plants, and drying specimens of those which I formerly possessed . . . I might have effected more; but being in constant dread of a vessel arriving, and sailing without me, I could not venture to be absent more than fifteen or twenty days at a time from the coast; however, as I did my best, I try to feel content." (HM, 152)

1832

Bancroft states that Douglas' ". . . name appears on the rolls of the compañia extrangera in January 1832 . . ." (Bt 2, 403) Such documents were issued at Monterey.

Douglas was in the Santa Lucia Mountains in March of 1832. He wrote Hooker:

"I will now mention another new *Pinus* to you *(P. venusta),* which I discovered last March, on the high mountains of California (you will think that I manufacture Pines at my pleasure). As my notes are not at hand, I must describe from memory. . . . This tree attains a great size and height, and is, on the whole, a most beautiful object. It is

unavailing . . . Some extracts from this letter [(1) above] were published in the *Overland Monthly,* Volume II, 1883, page 409."

But the letter (2) published in *The Overland Monthly,* as noted above, merely reprints extracts from the one published earlier by Hooker; whereas letter (1) varies not only as to date but considerably as to content. Douglas made copies of his letters and journals and I believe that letter (1)—because of its earlier date, because of the fact that it is not addressed to anyone, and because of its content—was merely an earlier draft of letter (2) which omits matters contained in letter (1) and is carefully reworded.

H. R. Wagner also mentions the "not published" Douglas letter in *The plains and Rockies* (p. 34), stating that Douglas ". . . did, however, write two letters from Monterey to Hooker, one in October and the other dated November 23, 1831. The second one, which is almost entirely devoted to botany, was printed by Hooker, but the first, which Douglas states in a subsequent letter, details his travels in California and gave a brief notice of the country, was not published."

But letter (2) *does* contain an account of Douglas' travels in California, an account of his collections there, and "a brief notice" certainly, of that region. I can see no reason why, within a month, Douglas should have written Hooker on precisely the same matters.

1. Not at San Francisco, but in California, and probably most of the time at Monterey.

never seen at a lower elevation than six thousand feet above the level of the sea, in latitude 36°, where it is not uncommon.[2]

"I saw for a second time, and in a new habitat, *Pinus Lambertiana,* more southerly on the mountains of Santa Lucia, in Upper California. Its cones were in fine condition, though perhaps a little too young and somewhat longer than those I had discovered further to the North in 1826 . . ." (HM, 152)

He was at Monterey in August for Bancroft mentions that in the list of stations which Douglas furnished Figueroa ". . . the variation of the compass at Monterey is dated August 1832." (Bt 2, 403) Both Bancroft (Bt 2, 403) and Barnston (B, 273) mention that he sailed in August; it was on an American ship which was bound for Honolulu.

Before considering Douglas' later travels it is of interest to know what he felt about his botanical accomplishment in California. He wrote Hooker from Monterey on November 23, 1831:

"Of new genera I am certain there are nineteen or twenty, at least, and I hope you will find many more. Most of them are highly curious. As to species, about three hundred and forty may be new. I have added a most interesting species to the genus *Pinus,* *P. Sabinii,*[3] one which I had first discovered in 1826, and lost, together with the rough notes, in crossing a rapid stream on my return northward. When compared with many

2. Douglas was writing of *Abies venusta* (Douglas) K. Koch, the Santa Lucia fir or the bristlecone fir. Although he merely mentions that he collected it "on the high mountains of California" he undoubtedly found it in the Santa Lucia Mountains for the tree's habitat is limited to that range. Sudworth's *Forest trees of the Pacific slope* (p. 122) refers to its distribution as ". . . mainly in Monterey County. Scattered in patches of several or a few hundred trees in heads of canyons on both slopes of seaward part of Santa Lucia Mountains (Monterey National Forest) . . ." Jepson's *Manual* (p. 54) gives its habitat as "Rocky mountain peaks and deep cañons, Santa Lucia Mts."

According to J. M. Murray, ". . . the discovery . . . is not usually credited to Douglas . . . Dr. Coulter, who met Douglas in Monterey, is frequently given as the discoverer of this tree. But neither Douglas nor Coulter collected seed, and it was not till 1835 that Lobb introduced it to Britain." William Lobb collected plants in western North America for the Veitch firm of Exeter, England.

For a time the tree bore the name *Abies bracteata* bestowed by D. Don.

3. Hooker refers the reader to a footnote (p. 106 of the "Memoir") but this seems to concern *Pinus Lambertiana,* the sugar pine. Douglas' "*P. Sabinii,*" or *P. Sabiniana,* is the Digger or gray pine, a Californian species which he found in February, 1831, when visiting the San Juan Bautista mission. In 1826 he was in southern Oregon and the accident mentioned was on the Santiam River. The occurrence of *P. Sabiniana* in Oregon is certainly questionable.

Balfour has commented:

"After Douglas' death Sir Edward Sabine claimed that the name was in compliment to himself; till now the name has been thought to be in honour of his brother, Joseph Sabine, as indeed it probably was." (Ba, 156)

This was certainly a curious misunderstanding in view of the dedication appended to Douglas' description of this tree (*Trans. Linn. Soc. London* 16: 747-750, 1833):

"The active and enlightened zeal which Joseph Sabine, Esq. has ever taken . . . for the introduction of new, choice and useful plants . . . induces me to affix his name to one of the most beautiful objects in nature . . ."

individuals of the genus inhabiting the western parts of this continent, its size is in-considerable, from 110 to 140 feet high, and 3 to 12 feet in diameter . . . But the great beauty of Californian vegetation is a species of *Taxodium*,[4] which gives the mountains a most peculiar, I was almost going to say awful, appearance . . . I have repeatedly measured specimens of this tree 270 feet long and 32 feet round at three feet above the ground. Some few I saw, upwards of 300 feet high; but none in which the thickness was greater than those I have instanced. I possess fine specimens and seeds also. I have doubled the genus *Calochortus*;[5] *C. luteus* . . . is especially deserving of attention, as the finest of all. To *Mimulus*[6] I have also added several, among them the magnificent *M. cardinalis* . . . an annual, three or four feet high, handsomer than *M. luteus; Clarckia elegans* . . . is a pretty species, but hardly equal to *C. pulchella;* it grows to four or six feet, and has entire petals. It is to *Gilia, Collomia, Phlox,* and *Heuchera,* that the great-est additions have been made: indeed they are too numerous to mention. Something is also done among the *Onagrarieæ.* Besides the new genus (*Zauschneria* of Presl) . . . I possess another new genus, and a multitude of *Œnotheras.* Also four unde-scribed kinds of *Pentstemon,* two of which far exceed any of the known species, and are shrubs; and among the *Papaveraceæ,* two, if not three, new genera. One is frutes-cent, with a bifoliate calyx and four petals, it has the stamens of *Papaver* and the fruit of *Eschscholtzia,* with *entire leaves!* This is my *Bichenovia*[7] . . . By far the most singu-lar and highly interesting plant here belongs to a genus, in some respects akin to *Salvia;* it is annual, and I have called it *Wellsia*[8] . . . This, with many others, I trust you may yet have the pleasure of describing from living specimens, as I have sent to London up-wards of one hundred and fifty nondescript plants, which I hope will bloom next sea-son . . ." (HM, 150, 151)

Balfour comments: "It is surprising that though the coast of Monterey alone is the native home of *Cupressus macrocarpa,* now the most widely planted of all Cypress-es, it was never mentioned by Douglas." It is indeed curious. For we know that he

4. "This is the first mention of the coast redwood by a botanist in the field. It had, however, been de-scribed and published in 1824 by Lambert from material brought home by Menzies in 1795. Lambert's name *Taxodium sempervirens* stood till the genus Sequoia was created in 1847, and its introduction took place about the same time." (Ba, 156)

   As noted elsewhere, the botanist Haenke seems to have been the first to collect the coast redwood in 1791. Menzies' specimen is dated from Santa Cruz in 1794. But, as Balfour states, Douglas is the first to mention it *while in the field.*

5. In *A manual of the flowering plants of California* Jepson lists six species of *Calochortus* which were named by Douglas : *C. luteus* (yellow mariposa), *C. venustus* (white mariposa), *C. splendens* (lilac mari-posa), *C. macrocarpus, C. pulchellus* (golden lantern) and *C. albus* (white globe lily).

6. Jepson, *ibid.,* cites four *Mimulus* named by Douglas: *M. moschatus, M. floribundus, M. alsinoides,* and *M. cardinalis.*

7. This is the so-called bush poppy, the *Dendromecon,* described by Bentham (*Trans. Hort. Soc.* ser. 2, 1: 407, 1834).

8. According to Hooker, *Audibertia incana* Bentham, one of the sages.

must have seen these trees since one station cited in the Bancroft list was "Monterey (Cipres Pt) . . ." Cypress Point commands Carmel Bay and between there and Point Lobos grow some of the most famous of these Monterey cypresses. Lapérouse was, I believe, the first of our botanists to mention "le cyprès" and merely in a list of California's forest trees ("arbres des forêts . . ."). Sudworth's *Forest trees of the Pacific slope* (p. 159) estimates that some of the larger of the cypresses "are doubtless over 200 years old." Sargent states in *The Silva* (10: 104, 1906) that "Although its seeds appear to have reached England in 1838, *Cupressus macrocarpa* was first made known to botanists in 1847 by Karl Theodor Hartweg, who had found it at Cypress Point the previous autumn. It is now the most universally cultivated coniferous tree in the Pacific states, where it has proved hardy from Vancouver's Island to Lower California . . ."

The collections made in California were shipped by Douglas after he reached the Hawaiian Islands.

"From the Sandwich Islands, I shipped on board the Sarah and Elizabeth, a South-Seaman of London, and bound for that port, nineteen large bundles of dried plants, in two chests, together with seeds, specimens of timber, &c. The Captain, a worthy little man, placed these articles in his own cabin, which gives great relief to my mind as to their safety. I have written to the Horticultural Society of London (should such exist), requesting the Council to permit four of the bundles of dried plants, destined 'for Dr. Hooker of Glasgow,' to be despatched without delay, and further 'begging that they will permit me to transfer the publication of each and all these plants, saving those which the Society may consider as coming within their plans,[9] to that gentleman, either for an Appendix to his *Flora Boreali-Americana,* or in any other works in which he may be engaged. No one is more able or willing to do the Society justice, while such a proceeding would be peculiarly gratifying to me.' I have still at Fort Vancouver a good bundle of plants, perhaps about seventy species which I shall try to send, through Mr. Garry, overland this spring, for publication, with *Mosses* and *Sea-weeds,* so that your *Flora* may be as complete as possible." (HM, 152, 153)

Douglas had Hooker's needs ever in mind. California had provided seaweeds, but he had done less well on mosses. He wrote from the Columbia about them:

"What a blank we have in the department of sea-weeds! You must still look to Mr. Menzies as the main stay, though you will find some fine species in my collection from California. Fearing I may not have it in my power to visit the numerous groupes of islands, so particularly rich in this class of vegetables on the North-West parts of the continent, I have written to all friends, American as well as English, residing there, and requested them to collect every thing in the shape of sea-weed, and that I may put them to as little trouble as possible, I have told them simply to dry them in the sun. They can, like Mosses, be revived and put in order afterwards. Scarcely a Moss exists

9. While at the Hawaiian Islands Douglas had resigned his position of collector—hence his stipulations.

in California! But when we consider the excessive dryness of the climate, our surprize may cease. Perhaps no where else in the world is such drought felt, if we except the deserts of Arabia, Egypt, and the plains of Ispahan . . ." (HM, 154)

Douglas, who must have known the mission fathers of California extremely well, evidently esteemed them more highly than many other visitors to California. He wrote to Barnston:

"I spent 19 months in California and amassed a collection, of such an extent, as can only be equalled by its novelty and beauty . . . I lived almost exclusively with the fathers who without an exception, afforded me the most essential assistance, hospitality to excess, with a thousand little courtesies which we feel and cannot express. I had no bickerings about superstition, no attempt at conversion or the like . . . When there I was under no restaint . . . feast day and fast all the same, the good men of God gave me always a good bed, and plenty to eat and drink . . . A more upright and highly honorable class of men I never knew. They are well educated; I had no difficulty from the beginning with them, for saving one or two, exceptions, they all talk Latin fluently, and though there be a great difference in the pronunciation between one from Auld Reekie, and Madrid, yet it gave us but little trouble. They know and love the sciences too well to think it curious to see one go so far in quest of grass. The Mexican territorial government as applied to California is abominable, and that is the mildest word I can use. The secular part of the community is . . . some good and many bad . . ." (B, 274, 275)

He wrote Hooker on the same subject:

"I have received a copy of Capt. Beechey's book. I entertain a great respect for that gentleman, but I think he has been too severe on the Catholic Missionaries in California. Any man who can make himself well understood by them, either in Castilian or Latin, will discover very shortly that they are people who know something more than their mass-book, and who practise many benevolent acts, which are not a little to their credit, and ought to soften the judgment of the stranger, who has probably had more opportunity of seeing men and things than the poor priests of California. Their errors are the errors of their profession, and not of their hearts . . ." (HM, 154)

As stated, Douglas left Monterey in August.[1] He went first to the Hawaiian Islands and he wrote Hooker of the voyage:

". . . Anxious that no time should be lost, I sailed from Montérey for those islands in an American vessel of forty-six tons burden! and had a passage of only nineteen days. What would have been thought, forty years ago, of passing over more than half of the great Pacific in such a craft? If steam-boats and rail-roads are not in our way, we, poor wanderers, must take what offers, sometimes good and sometimes bad." (HM, 153)

On arrival at the Islands, Douglas heard of the resignation of his friend Joseph

1. August 18. (Harvey, 1947, x)

Sabine from the secretaryship of the Horticultural Society and sent in his own resignation as collector.[2] It is mentioned in George Bentham's "Report on the progress of the Horticultural Society of London . . ." read May 1, 1840, where some two pages are devoted to Douglas (*Trans. Hort. Soc. London,* ser. 2, 2: 375-377, 1840).

The *Publications of the Champlain Society* (The letters of John McLoughlin . . . First series 4: 338) include a letter written by Duncan Finlayson, dated "British Consulate Woahoo 10*th* Septr. 1832," which mentions that ". . . Mr. David Douglas the Botanist arrived here from California 3 days since . . ." which places his arrival[3] on September 7.

Douglas wrote Hooker of the visit and of his hopes to return there in the future:

". . . At the Sandwich Islands a violent rheumatic fever prevented me from venturing at all to the hills during my short stay, and I sat and fretted about it! I have indeed had some hard work since I quitted England, of which I occasionally feel the effects, particularly in cold weather . . ." (HM, 153)

"I have a great desire to become better acquainted with the vegetation of the Sandwich Islands, as I am sure much remains to be done there, and before quitting that country I made conditional arrangements with Capt. Charlton, our Consul, to aid me, should I return. This I shall earnestly endeavour to do. The Consul is a most amiable and excellent person. In *Ferns* alone, I think there must be five hundred species." (HM, 155)

One piece of good fortune occurred as the ship approached the Columbia River:

". . . On my way to this river, and not far from its entrance, I had the pleasure to meet my old ship the Eagle, and old friend Lieutenant Grave, R.N., who handed me a parcel from Soho Square, containing the second and third parts of the *Flora Boreali-Americana!* Singular indeed it was that I should receive this, just in the nick of time, for had it not been for a *kind unfavourable wind,* which obliged my vessel to go considerably out of her way, I should have missed her, and of course lost the pleasure of a sight of the *Flora.* I cannot really express how much I am obliged to you for writing to me. If it were not for your letters, and the information they convey, I should be utterly without news, for nobody has sent me any." (HM, 153)

According to Barnston, Douglas reached the Columbia ". . . towards the end of October."[4] (B, 273) He wrote two letters to Hooker from the River Columbia, one on October 23, one on October 24, and mentions them as sent ". . . from the entrance of the Columbia river, by the last vessel which sailed for England, commanded by my excellent friend, J. E. Grave, Lieut. R.N. . . ." (HM, 157) Since winter was setting in, he occupied himself with "astronomical observations" for a time.

2. I refer to this in my earlier chapter on Douglas.

3. Harvey (1947, x) gives the date of departure from the Islands as September 12.

4. Harvey (1947, x) cites the date of arrival as October 14.

". . . The Columbia was closed with ice for four weeks at Menzies' Island,[5] where it rather exceeds a mile in breadth, the thermometer indicating 22° of Fahrenheit, which is bitterly cold for the shores of the Pacific, in the parallel of 45. This gave me an excellent opportunity of multiplying my astronomical observations, on the angular distance between the moon's limb and the sun; the planets Venus, Mercury, Saturn, and Mars, and the fixed stars; not less than eight thousand observations in about six hundred sets, separately computed, for the purpose of ascertaining the absolute longitude of Fort Vancouver." (HM, 157)

1833

At the opening of the new year he was still at such work.

". . . I observed the beautiful eclipse of the moon on the night of January the 5th . . . with many of the eclipses of Jupiter's satellites. Indeed my whole skill was exerted on these operations . . . I merely mention these things that you may not tax me with idleness, a character with which I am charged by the Londoners, and perhaps more deservedly in that great metropolis than elsewhere. I hope you have not finished the fine Order *Coniferæ* in the *Flora Boreali-Americana,* that you may include the *Pines* discovered in my late journeys, viz. *Pinus venusta, Sabini,* and *grandis.*" [6] (HM, 157)

Barnston reports that Douglas, in ". . . March of 1833 . . . made a short tour by the Cowlidsk river to Puget Sound,[7] where he took a rapid survey of the bays and headlands, determining their latitude and longitude, and obtaining the altitudes, bearings and distances of the snowy peaks . . . A number of mosses and algae were collected in this quarter . . ." (B, 273)

Douglas wrote Hooker of the trip:

". . . I have made a journey . . . North of the Columbia, to New Georgia, and a most laborious one it was. My object was to determine the position of the Head Lands on the coast, and the culminating points of the many prodigiously high snowy peaks of the Interior,[8] their altitudes, &c., and as I was favoured with exceedingly fine clear weather, this was effected much to my satisfaction. On this excursion I secured about two hundred species of *Mosses;* but as I am rather ignorant of this tribe, there may be

---

5. Opposite Fort Vancouver.

6. As noted, Douglas had discovered both *Abies venusta* and *Pinus Sabiniana* in southern California in February, 1831, and *Abies grandis* on his trip to the Willamette River in late 1830.

7. Balfour comments that Douglas was ". . . doubtless in the Olympic Mountains." Possibly, but it seems more likely (from the length of the trip—he was away from Fort Vancouver less than three weeks—and from the fact that he approached Puget Sound from the south by the Cowlidsk River) that the journey merely took him to the Nisqually region where, in 1832, the Hudson's Bay Company was erecting a post or fort. The Olympic Mountains are chiefly in Clallam and Jefferson counties.

8. These were peaks in the Cascade Range for Douglas refers on April 19 to "Mount Jefferson of Lewis and Clarke," to Mount Hood and to "Mount St. Helens of Vancouver."

a few more or less: certain it is, however, that there are many fine kinds that are totally unknown to me; and perhaps even you may find some of them new. I have also some interesting *Fuci* from Puget's Sound, collected on the same journey . . . I have bespoken the services of all the Captains on the North-West coast to bring me all sorts of sea-weeds, simply coiled up, dried, and put in a bag . . ." (HM, 157)

He seems to have been back at Fort Vancouver on March 17, for he wrote Barnston from there on that date.

Douglas was now to embark on his journey into New Caledonia, or the northern interior of British Columbia.[9] Although Barnston (B, 277) and McDonald (in HM, 159, *fn.* 1) mention the date of departure as March 20, Douglas gives it as the 19th. After he had been on his way almost a month he wrote Hooker his last letter from the northwest (dated "Interior of the River Columbia, lat. 48°, 5′ N., long. 119°, 23′ W. April 9, 1833."):

"I quitted the ocean on the 19th of March, and followed the course of the river to this spot, picking up a few of the early-flowering plants, and better specimens of others which I had already possessed: among them are some novel species of *Platyspermum, Thysanocarpus,* and *Ranunculus:* a new *Phlox,* and a few Mosses. The disparity of climate between this point and the coast is very striking, though the difference of latitude be only 3°, and of longitude 6°. There, in the middle of March, many plants were in bloom; while here last night we had a new fall of snow of some depth, and the ground is still sprinkled with old snow . . .

"As soon as the season permits, which I trust will be in a few days, I shall leave this spot for the northward, travelling sometimes in canoes, or on horseback, but far more generally on foot. The country is mountainous and very rugged, the rivers numerous, and there are not a few lakes of considerable extent. Perhaps I shall cross Mackenzie's track, at Fraser's River . . . and proceed northward, among the mountains, as far as I can do so with safety, and with the prospect of effecting a return. The country is certainly frightful; nothing but prodigious mountains to be seen: not a deer comes, say the Indians, save once in a hundred years—the poor natives subsist on a few roots . . ." (HM, 157, 158)

When Archibald McDonald (one of the traders of the Hudson's Bay Company) was in Scotland in 1835 Hooker evidently asked him to supply particulars of Douglas' disastrous voyage into New Caledonia, and the McDonald report is included in the Hooker "Memoir" (159, *fn.* 1). I shall quote the portion relating to the journey,[1] most of which lay beyond the region considered in this story, for Douglas crossed the 49th parallel on his way north sometime between April 17 and 25, and, on his way south, sometime between June 23 and July 11. McDonald tells that, early in March, Douglas

9. According to Wilson: ". . . The name British Columbia was given in 1858 at the direction of the late Queen Victoria in order to satisfy the objections made in France to the name of New Caledonia as being borne by an island [in the South Seas] then claimed by the French." (W, 77, *fn.* 41)

1. Harvey (1947) devotes his chapter 16 to the journey, supplying dates and considerable detail.

met him ". . . at Puget's Sound and we returned together to Fort Vancouver, on the 20th of the same month, when he embarked with our people, who were crossing over to Hudson's Bay. He landed at Okanagan, whence he proceeded with the cattle party to Thompson's River, Alexandria, and Upper Caledonia. At Stuart's Lake he found one of the Company's officers preparing to set out on an exploring expedition, down Simpson's River, which falls into the Pacific, two or three degrees North of M'Kenzie's small river, and was much disposed to accompany him: but fearing they could not reach the sea, or any of our settlements on the coast, and would in that case lose time, and be disappointed in other projects he had in view, he did not join the party. With his man, Johnson, he shipped himself in a small bark canoe down to Fort George[2] . . . remained a day or two with Mr. Linton, and, on the second day after he commenced descending the stream . . . experienced the disaster . . . communicated in a letter to yourself. From Alexandria Mr. Douglas got back to Thompson's River, and Oakanagan, by the same route that he went, and with the same means that he had from our people in Spring. At Oakanagan he took two Indian canoes, and when halfway down to Walla-wallah, on the 14th of July, met Mr. Conolly, of New Caledonia, and myself, on our way up the river, with supplies for the Interior. He continued some days at Walla-wallah, with Mr. Pambrun, making occasional journies to the Blue Mountains, and finally attempted the ascent of Mount Hood. In the month of September, 1833, I received a letter from him, stating that he was on the eve of sailing again for the Sandwich Islands."

Douglas wrote Hooker of the accident—in his last letter, dated from "Woahoo, Sandwich Islands, May 6th, 1834." I have quoted much of it in my earlier chapter on Douglas and shall merely repeat that it occurred on the "13th of June," and that the unfortunate man lost "every article" he possessed "saving an astronomical journal, book of rough notes, charts, and barometrical observations, with my instruments. My botanical notes are gone, and, what gives me most concern, my journal of occurrences also, as this is what can never be replaced, even by myself . . . I passed over the cataract and gained the shore in a whirlpool below, not however by swimming, for I was rendered helpless, and the waves washed me on the rocks."

Douglas has more to tell of his experience in a letter to his old friend W. E. P. Hartnell which was published by Jepson (1940):

". . . I have been in the Snowy Mountains as high as the 60° over a dreary unhospitable country, where I suffered extreme hardship—from hunger, indeed nearly utter starvation. Intense cold in the mountains then scorching weathering heat—and to compleat my misfortunes I lost the *whole* of my collection at a dreadful cataract of a river nearly as large as the Columbia in the mountains. The only articles saved were my *Instruments and astronomical* Journal not even a morsel of food of any kind, bedding, etc. When I tell you that I was an hour and 40 minutes in the water in the rapids myself and after escaping had 300 miles over a barbarous country without food or

---

2. According to Harvey (1947), now called Prince George. It was situated on the Fraser River.

shelter you will form an idea of my condition, I am thankful to God that no lives were lost though some from exhaustion suffered greatly while others could not endure the privation of the want of food." (J, 99)

Douglas, despite his troubles, had had his friend Hooker in mind, and wrote him to that effect:

"You will probably enquire why I did not address you by the despatch of the ship to Europe last year. I reached the sea-coast greatly broken down, having suffered no ordinary toil, and, on my arrival, was soon prostrated with fever. My last letter to you was written from the interior of the Columbia, and bore date about the middle of April [April 9], 1833 . . . just before starting on my northern journey. Therein I mentioned my intention of writing a few lines to you daily, which I did, up to the 13th of June, a most disastrous day for me . . . (HM, 159)

"After this misfortune . . . I endeavoured, as far as possible, to repair my losses, and set to work again; and I hope some good new species were obtained for the *Flora Boreali-Americana,* which I am very anxious should reach you without delay. It is more than probable that I may have the pleasure of presenting these to you myself, say in March next, as it is my intention to return to England by the very first opportunity; and I hope this small collection may give you some satisfaction, as it is all I can now offer you from North-West America . . ." (HM, 160)

Harvey mentions the particular site of the disaster: ". . . Two years ago, after considerable inquiry, I succeeded in finding the exact spot . . . His 'Stony Islands' are small rocky islets in what now is known as Fort George Canyon, between Quesnel and Prince George . . ." (H, 232, 233) The "Mr. Linton" with whom Douglas stayed at Fort George, was George Linton, and, strangely enough, he, four members of his family and three others were to lose their lives not long afterwards in much the same— if not the very same—spot where the Douglas accident occurred.[3]

Of the New Caledonia region Douglas wrote Hooker:

"The country over which I passed was all mountainous, but most so towards the Western Ocean:—I have written to Mr. Hay, [British] Under Secretary of State, respecting the boundary line on the Columbia, as the American government is anxious to obtain a footing there." (HM, 159)

Having reached Fort Walla Walla on July 15, Douglas remained there until the 25th (H, 243). According to the MacDonald report, just quoted, he made while there ". . . occasional journies to the Blue Mountains, and finally attempted the ascent of Mount Hood." [4] (HM, 159, *fn.* 1) From Fort Walla Walla Douglas returned to Fort Vancouver and evidently spent the remainder of his time in the northwest in that region.

Douglas wrote that "On Friday, the 18th of October, 1833, we quitted Cape Dis-

3. An account of this Linton accident is told in *The history of the northern interior of British Columbia (formerly New Caledonia)* by A. G. Morice (ed. 3, 193, 1905).

4. Harvey (1947, 208) states that the attempt was unsuccessful.

appointment, in the Columbia River . . ." (HM, 161, 162) He was again bound for the Hawaiian Islands, by way of California. It was a slow trip; and in his letter to Hartnell, Douglas refers to the fact that the vessel ". . . after a tempestuous passage of 21 days has landed here [in the Bay of San Francisco] nearly a wreck. The Governor of the Columbia, Mr. Finlayson, is on board . . ." He wrote Hooker that ". . . after encountering much variety of weather, and many baffling gales [we] anchored off Point de Los Reyes on the 4th of November, and remained there till the 28th of the same month, our attempts to beat out of the Harbour of Sir Francis Drake having proved, several times, ineffectual." (HM, 162)

The letter to Hartnell bears the heading: "At my tent on the Hill of Yerba Buena[5] Novm 11, 1833." (J, 98, 99) One of the stations cited in the Bancroft list was "S. Francisco (Yerba Buena) . . ." and Bancroft states that the list was furnished to Figueroa on November 25, 1833. (Bt 2, 3: 402, *fn.* 39) Douglas, who comments that he did "nothing in the way of Botany" while at San Francisco (HM, 160), wrote briefly of his stay in California and of his departure for the Islands:

"On the 29th, I accompanied Mr. Finlayson in a small boat to Whaler's Harbour, near the neck of the bay, which leads up to the hill of San Rafaele, the highest peak[6] in the immediate vicinity of the port. We landed at Mr. Read's farm-house, placed on the scite of an old Indian camp, where small mounds of marine shells bespeak the former existence of numerous aboriginal tribes. A fine small rivulet of good water falls into the bay at this point. Returning the same afternoon, we cleared the Punta de los Reyes, on the 30th, and, descrying the mountains of St. Lucia, South of Montérey, at a distance of forty or fifty miles, steered southward for the Sandwich Islands." (HM, 162)

Douglas had left California for the last time and in so doing passed out of the region considered in this story.

What he had to tell of the period between his departure from Cape Disappointment on October 18 and his departure from California on the 29th or 30th of November, 1833, was published by Hooker in the "Memoir" and taken from ". . . the only Journal of Mr. Douglas' Second Expedition . . . with the loan of which we [Hooker] have been favoured by its possessor, Mr. John Douglas . . ." (HM, 161)

Douglas arrived at Oahu ("Woahu") on December 23 and spent Christmas Day ". . . with two English ladies, the wife of our Consul, Mr. Charlton, and her sister . . ." (HM, 160) From there, on the 27th, he left for Hawaii, stopping on the way at

5. Jepson comments:

" 'The Hill of Yerba Buena' was probably Telegraph Hill, then known as Loma Alta, which name was sometimes applied to the Cove of Yerba Buena. Douglas arrived in the 'Dryad' . . . Apparently while in port Douglas set up a tent ashore for his observations and collection activities. The landing was at Clark's Point, about Broadway and Battery streets, and presumably his tent was not far distant.

"There was no settlement at Yerba Buena at this time."

6. The reference is to Mount Tamalpais.

"Rahaina Roads" (Lahaina, island of Maui) "for the purpose of procuring more ballast." Here the "American Missionary, Mr. Spaulding[7] having come on board, I accompanied him on shore . . ." (HM, 162)

1834

He reached Byron's, or Hilo, Bay on the eastern side of Hawaii on January 2 and notes that he took up his abode ". . . with the Rev. James Goodrich, an American missionary, from whom I have received great kindness. I have since made successive journeys to the summits of the mountains and volcanoes—my first being Mowna Kaah [Mauna Kea], my second to Kiraueah [Kilauea], and my third Mowna Roa [Mauna Loa] . . ." (RGS, 333)

He wrote Hooker that the journey to the summit of ". . . *Mouna Kaah* . . . occupied fourteen days . . ." and that he ". . . amassed a most splendid collection of plants, principally Ferns and Mosses: many . . . truly beautiful . . . Of Ferns alone I have fully two hundred species, and half as many Mosses; of other plants comparatively few, as the season is not yet good for them, nor will be so until after the rains . . ." (HM, 160, 161)

The Hawaiian journal contains a long account of this journey, which seems to have lasted from January 7 to 15. Scattered through the account of the trip there are numerous references to the plants which he saw and collected:

". . . Large timber trees were covered with creepers and species of *Tillandsia,* while the *Tree Ferns* gave a peculiar character to the whole country . . . Above ['the Saw Mill'] . . . the *Banana* no longer grows, but I observed a species of *Rubus* among the rocks . . . The number of *Filices* is very great, and towards the upper end of the wood, the timber trees, sixty or seventy feet high, and three to ten inches in circumference, are matted with Mosses, which, together with the *Tillandsias* and *Ferns,* betoken an exceedingly humid atmosphere. The wood terminates abruptly . . . I delayed ascertaining the exact altitude of the spot where the woody region ends, (a point of no small interest to the Botanist,) until my return, and sate down to rest . . . where the ground was thickly carpeted with species of *Fragaria,* some of which were in blossom, and a few . . . in fruit . . . The last plant that I saw upon the mountain was a gigantic species of the *Compositæ*[8] . . . with a column of imbricated, sharp-pointed leaves,

---

7. Henry Harmon Spalding.

8. Hooker inserts the name: "*Argyrophyton Douglasii,* Hook. Ic. Plant. t. 75." The plant has since been placed in the genus *Argyroxiphium,* with the unlovely specific name *Sandwicense.*

It was this plant, doubtless, which gave Asa Gray some "food for thought." He wrote Hooker from New York, on April 7, 1836: ". . . The parcel contained a specimen of a Composita (from Mouna Kea) which puzzled me extremely, and I was unable to ascertain its genus by Lessing. The anthers are free, or slightly coherent, in all the flowers I examined . . ."

I have never seen the living plant but in photographs it suggests—in form of inflorescence and in leaf-rosette—*Yucca Whipplei* of California. Wilson calls it the silver sword plant of Hawaii and includes an excellent photograph (plate 12).

densely covered with a silky clothing. I gathered a few seeds of the plants which I met with, among them a remarkable *Ranunculus,* which grows as high up as there is any soil . . . the species of *Compositæ* mentioned before . . . together with a small *Juncus,* grows higher up the mountain than any other plant. The great difference produced on vegetation by the agitated and volcanic state of this mountain is very distinctly marked. Here there is no line between the Phenogamous and Cryptogamous Plants, but the limits of vegetation itself are defined with the greatest exactness, and the species do not gradually diminish in number and stature, as is generally the case on such high elevations." (HM, 162-164)

Referring to what he calls the "Woody Country" which, from the context, seems to have been at or near timber line, Douglas comments that "Three kinds of timber, of small growth, are scattered over the low knolls, with one species of *Rubus* and *Vaccinium,* the genus *Fragaria,* and a few *Gramineæ, Filices,* and some alpine species. . . ." (HM, 164)

He mentions that he had obtained a fine collection on the 15th:

". . . I gathered a truly splendid collection of Ferns, of nearly fifty species, with a few other plants, and some seeds, which were tied up in small bundles, to prevent fermentation, and then protected by fresh Coa bark. Several beautiful species of Mosses and Lichens were also collected . . . I repeatedly sate down, in the course of the day, under some huge spreading Tree-Fern, which more resembled an individual of the Pine than the Fern tribe . . . On the higher part of the mountain, I gathered a Fern identical with the *Asplevium viride* of my own native country . . . we quitted the saw-mill for the bay and arrived there in the afternoon, the arrangement and preservation of my plants affording me occupation for two or three days. It was no easy matter to dry specimens and papers during such incessantly rainy weather . . ." (HM, 166)

Douglas, according to his Hawaiian journal, next made a trip to "the volcano and to Mauna Roa," leaving on January 22; he had a guide and nine men. (HM, 166) The "volcano" was Kilauea, on the eastern slope of Mauna Loa, and one gazetteer gives its altitude as 4,400 feet, that of Mouna Loa as 13,675 feet. Both are active volcanoes.

". . . On leaving the bay, we passed through a fertile spot, consisting of Taro patches in ponds . . . Here were fine groves of *Bread-fruit* . . . little plantations of vegetables and of *Morus papyrifera,* of which there are two kinds, one much whiter than the other. The most striking feature in the vegetation consists in the Tree-Fern, some smaller species of the same tribe, and a curious kind of *Compositæ,* like an *Eupatorium* . . ." (HM, 167)

Douglas reached the crater of Kilauea on the 23rd and spent the night in a ". . . tent . . . pitched twenty yards back from the perpendicular wall of the crater . . ." He spent the 24th there and on the 25th started for "Kapupala," or Kapapala, where he made preparations, hired guides and so on, and on the 28th started for Mauna Loa. Not much is told of the vegetation along the way to the crater, but in starting out he refers to "a strong kind of Raspberry-bush, destitute of leaves," the fruit of which he was told was white, and to collecting some small stems of a heath-like plant, and dried stalks of the composite he had found on Mauna Kea. (HM, 173) He reached the crater on the 29th and, on the 30th, ". . . arrived at Kapupala at four p.m. . . . the sky clear." (HM, 177) This, states Hooker, ". . . is the closing sentence of Mr. Douglas' Journal; penned, indeed, by the date, some months previous to the *letter* . . . (May 6th, 1834) . . . which was certainly among the last, if it were not the very last, that he addressed to any friend in Europe . . . Of the events which happened between that period and the melancholy accident which occasioned his death, a space of little more than two months, there is, unfortunately, no information . . ." (HM, 177)

I have omitted Douglas' many, long, and very interesting descriptions of the volcanoes; his trips to reach them were arduous—he suffered now from heat, now from severe cold, and his eyes troubled him greatly. Following the ascent of Mouna Loa he wrote Mrs. Richard Charlton from "Byron's Bay,[9] Hawaii, Feb. 7, 1834." The letter was published in the *Hawaiian Spectator* and I quote but a fraction:

". . . I have just about an hour ago arrived from Mauna Loa . . . I sit down to tell you the story of a traveller. A sight of the volcano fills the mind with awe . . . I have exhausted both body and mind, examining, measuring, and performing various experiments, and now I learn that I, know nothing . . . I assure you Madam, that these islands offer reward to the naturalist, over all others . . . I must return to the volcano, if it is only to look—to look and admire . . . One day there . . . is worth one year of common existence . . . With thankfulness and joy, the beautiful constellation of Orion being my guide, I rose to descend to a climate more congenial to my nature, and the habitations of men, the land of flowers and the melody of birds." (HS, 102, 103)

After his various ascents Douglas returned to Oahu for a time, for he wrote Edward Sabine from there under date of "Woahoo (Sandwich Islands), 3d of May, 1834." Portions of this letter were published in the Royal Geographic Society's *Journal* in 1834; therein Douglas states that he was intending ". . . to sail for England by the first opportunity; but as this is not likely to occur till August or September, I shall continue to labour at these islands to the best of my ability." (RGS, 334) In similar vein he wrote Hooker from "Woahoo . . . May 6th, 1834,"—his last letter: "I go immediately to Hawaii to work on these mountains. May God grant me a safe return to England." (HM, 161)

---

9. More than once Douglas mentions the visit of "Lord Byron" to the Islands. The man was a cousin of the poet, and the reader is referred to Kuykendall and Day (1848) and Rufus Anderson (1864) who tell the story. In brief it was this:

When, during the reign of George IV, the king of the Hawaiian Islands, Liholiho, his queen and members of his court visited England, the party was stricken with measles and both members of the royal family died. Their bodies were taken back to the Islands in the 46-gun British frigate *Blonde*, commanded by Captain the Right Honorable (George Anson) Lord Byron, R.N. The ship reached Honolulu in May 1825. Kuykendall and Day state that, "After leaving Oahu, the *Blonde* sailed to Hilo, where the harbor was surveyed and came to be known as 'Byron's Bay.' . . ."

In a letter written by Richard Charlton to James Bandinel on August 4 we are informed that Douglas, after recovering from an attack of rheumatism, had left Oahu for Hawaii, "on the 3rd ultimo," which must have been July 3. He must have landed at Kohala ("Rohala") Point with the intention of traveling from there to Hilo by way of Mauna Kea.[1]

For some reason, as stated by the missionaries, Douglas dismissed his guide and on July 12th started across the mountain. When his body was found the next day it lay at the bottom of one of the pits which, covered with branches, were used to trap bullocks. Douglas had been warned, apparently, to be careful about these traps. Whether the bullock fell in after, or before, Douglas, or whether Douglas had been murdered and thrown in, was all subjected to careful investigation at the time, but no positive solution seems to have been arrived at and the matter promises to be of perennial interest. The person who last saw Douglas alive was an ex-convict, Edward, or "Ned," Gurney, and his past record was not forgotten by those who advanced the murder theory. Douglas' faithful little dog was found a short distance away, guarding his master's effects. For those who are interested in his fate, he was given to James Bandinel and, let us hope, led a happy life.

No one seems to have suggested that Douglas' failing eyesight may have been the direct cause of the accident—the fact might well be that he did not *see* the pit, if he approached it from his blind side. His sight had been poor when he left the Columbia as early as 1826. On April 9, 1833, he wrote Hooker:

". . . while my left eye is become infinitely more delicate and clear in its power of vision, the sight of my right eye is utterly gone; and, under every circumstance, it is to me as dark as midnight. If I look through a telescope or microscope, I see objects pretty well at a distance, but the least fatigue brings on a *doubling* of the object, with a surrounding vapory haze, that soon conceals every thing. These results were owing to an attack of ophthalmia, in 1826, followed by snow-blindness, and rendered irretrievable by the scorching heat of California. I use purple goggles to diminish the glare of the snow, though most reluctantly, as every object, plants and all, is thus rendered of the same colour." (HM, 158)

One of the last entries in the Hawaiian journal refers to the painful condition of his eyes:

". . . I found myself instantly seized with violent pain and inflammation in my eyes, which had been rather painful on the mountain, from the effect of the sun's rays shin-

---

1. Curiously, the letter of the missionaries Joseph Goodrich and John Diell (as it is quoted in the Hooker "Memoir") mentions that ". . . Mr. D. sailed from Lahaina to Rohala [Kohala] Point . . ." (HM, 179) and that ". . . Mr. D. left Rohala Point . . . and proceeded to cross *Mouna Roa* [italics mine] on the north side . . ." (HM, 178) It would certainly seem that Mauna Kea must have been intended—Mauna Loa was not on the route to Hilo.

Harvey (1947, 232) gives the scene of the accident as Mauna Kea.

ing on the snow; a slight discharge of blood from both eyes followed, which gave me some relief, and which proved that the attack was as much attributable to violent fatigue as any other cause[2] . . ." (HM, 177)

Douglas' badly mutilated body was taken down from the mountain to Hilo on July 13 and from there removed to Oahu on August 3 and buried the next day in the Kawaiahao Church Cemetery. Wilson, writing in 1919, states that "The exact spot . . . is not now known . . . In 1856, Mr. Julius L. Brenchley . . . author of 'The cruise of H.M.S. Curacoa among the South Sea Islands,' . . . forwarded from England a tombstone of white marble . . . the tombstone has been placed on the face of the south-west wall of Kawaiahoa Native Church . . . in a few more years from this date, the inscription will become illegible."

How Julius Lucius Brenchley—who had "no personal acquaintance" with Douglas—took steps to see that "the resting-place of an eminent scientist should neither be neglected nor forgotten" is told by John Goldie[3] in the *British Columbia Historical Quarterly* for 1938.

According to Balfour, ". . . in 1930 the R. H. S. . . . had this marble tablet removed to the inside of the church and they sent out . . . two bronze tablets; on one is the original inscription and the English translation, on the other a record of their action. The Society thus made amends for their earlier neglect to set up a monument to their distinguished servant." (Ba, 160, 161)

The inscriptions have been published more than once; by Jepson in 1899, by Wilson in 1919, who included an illustration of the stone, and by Balfour in 1942.

Just one hundred years after Douglas' death, or on July 12, 1934, a stone pyramid with a bronze tablet was unveiled at the scene of the accident. (*American Forests* 48:

2. Douglas, of necessity, was his own diagnostician. Double vision may have many causes I am told, but it seems unlikely that the exertions to which he subjected himself would now be recommended as a cure for any form of that affliction.

3. The author of this paper, I was interested to learn, was the grandson and namesake of the John Goldie (1793–1886) whose name appears more than once in the botanical literature of the period of my story and who wrote a paper, "Description of some new and rare plants discovered in Canada, in the year 1819," which came out in the *Edinburgh Philosophical Review* for 1822. The paper was "communicated" by Hooker and he named one of the plants (*Aspidium Goldianum*) after the collector. This elder Goldie seems to have held the "all-time-high" for loss of collections by sea: two lots he never heard of again after they had been put aboard ships bound for Scotland, and a third perished on a vessel which sank in the St. Lawrence River. He took the precaution to *carry* his fourth lot—all that remained after "three years spent in botanical researches."

The paper contains a reference to Frederick Pursh whom Goldie met in Montreal and who advised him to turn "towards the north-west country . . . and promised to procure . . . permission to accompany the traders leaving Montreal." When the time came Pursh was away but Goldie learned that ". . . even if present, his interest would scarce have been sufficiently strong to have obtained the assistance and protection which I desired."

Goldie the elder seems never to have collected plants in lands west of the Mississippi, merely in Canada where he eventually settled. He had become a friend of David Douglas when the two men were studying under William Jackson Hooker at Glasgow University.

1942) Another monument to his memory was erected at his birthplace, Scone; but its architecture, states Balfour, ". . . is not of the kind that appeals to us to-day."

Douglas, I believe, would have preferred commemoration in the plants which he loved. Wilson cites four Hawaiian species which, for a time at least, bore Douglas' name.

He would have been proud, too, of some of the tributes to his accomplishment. I quote but a few, written by American botanists:

"He collected nearly five hundred species in the State [of California] . . . he added more to the knowledge of the botany of that region than all the botanists who had gone before." W. H. Brewer (1880)

"Douglas is said to have introduced two hundred and seventeen species of plants into English gardens, the list including many valuable and beautiful trees . . . No other collector has ever reaped such a harvest in America, or associated his name with so many useful plants. By an unfortunate hazard of fate the noble Douglas Fir, the most important timber tree introduced by Douglas, and one of the most valuable trees in the world, does not . . . perpetuate his name in the language of science, and it is a humble primrose-like herb,[4] which commemorates this explorer of forests and discoverer of mighty trees." C. S. Sargent (1891)

"When David Douglas began the exploration of northwestern America in 1825, the scientific world began to receive its first accurate knowledge of the vast coniferous forests of this region, and of the great number of peculiar species of trees composing them. During the following seven years, Douglas discovered over twenty species; the remaining forty-six species, being either too far inland or too local and secluded, escaped this sharp-eyed collector . . ." J. G. Lemmon (1894)

"The extent and amount of this man's collections during the three seasons he spent in the Northwest almost surpass belief . . ." C. V. Piper (1906)

"He was the first botanical collector in California in residence for any extended period and during this time he traveled through the Coast Ranges from Monterey north to the Mission San Francisco Solano (Sonoma) and south to the Mission of Santa Barbara. He was not only the first traveler to collect the extensively rich and varied spring flora of the Coast Ranges, nearly all the species of which were new to botanical science, but also the first to leave some written description of it. Hundreds of new species, our most familiar plants, were based on the Douglas collections, study of which is not exhausted even at this day . . . While most of his field work was plainly done in the Coast Range valleys, yet on account of the fact that he collected certain plants it would seem an inevitable inference that he penetrated eastward as far as the inner Coast Range. For example he collected Thelypodium flavescens Hook., which develops abundant colonies in the inner range but has never been found near Monterey. At Kew one of the collections of Thelypodium flavescens made by Douglas carries on the sheet a waif of another species, an extremely tiny plant of Strepthantus

4. *Douglasia* Lindley.

hispidus Gray, which, it is not likely, would have been gathered anywhere but in the inner South Coast Range country where it is a restricted endemic . . ." W. L. Jepson (1933)

Of Douglas' work on other lines than botany the *Hawaiian Spectator* (1838) comments that his name is ". . . identified with the progress of discovery in the various departments of natural science in . . . the Hawaiian group. No traveler has accomplished more in developing the inexhaustible 'vegetable treasure' every where spread through the kingdom of nature, in these sections of the Pacific." Alden & Ifft (1945) refer to his "contributions to zoology" and call him "a skilled natural scientist."

Balfour states that Douglas' ". . . collections of dried plants are divided between the Hookerian and Bentham collections at Kew,[5] the Lindley Herbarium at Cambridge and the British Museum[6] at South Kensington." These represented, of course, the bulk and the most important of his specimens. But duplicates were distributed: and A. de Candolle in *La phytographie* (1880) referred to some such ". . . dans l'herb. imp. de Saint-Pétersburg, et dans l'herb. de Candolle (environs 300)." C. V. Piper mentions a "few of his duplicates . . . in the Gray Herbarium." [7]

The duplicates at St. Petersburg to which De Candolle referred above were sent back to this country in the 1930's for determination and eventually reached California. Mr. J. T. Howell of the California Academy of Sciences published a paper upon them, "A collection of Douglas' western American plants," which appeared in seven installments in *Leaflets of western botany*. I quote what he has to tell of the collection and then his comments upon the publication of Douglas' plants:

"Of Douglas' plants there were 90 specimens, 17 from northwestern America and 75 from California. The former carried the printed label: 'Herb. Hort. Soc. Lond. America Boreali-occidentalis. D. Douglas'; and the latter were labelled: 'Herb. Soc. Lond. Nova California. Douglas, 1833.' In spite of the fact that many of the specimens are duplicates of the type collections of certain species, almost none carried a determination and only a few had the generic name written on the label. This would seem to indicate that the specimens were distributed by the Royal Horticultural Society very early and that the set sent to St. Petersburg was sent in the same condition as those sent to the English herbaria, to Asa Gray, to DeCandolle, and to others. In the Rus-

---

5. Bentham presented his herbarium to Kew and this, with the Hooker Herbarium, formed the great nucleus of the now vast herbarium at that institution.

6. Harvey (1947, 240, *fn.* 11) mentions that the Horticultural Society ". . . in 1856 . . . sold all his herbaria, 1460 specimens, to the British Museum . . ." This sale, presumably, included his collections from the Hawaiian Islands which, when the man's estate was settled, went to the Horticultural Society "for an unknown sum" according to the same reference.

7. Possibly these are the ones mentioned by Asa Gray in the journal of his first visit to Europe in 1838–1839. The reference appears in volume one of Gray's *Letters* (pp. 123, 124):

"London, January 24, 1830.—I have . . . had one of two botanical sittings with Bentham . . . He immediately had all the remaining parcels of Douglas's Californian and Oregon plants sent down to his house, and has supplied me as well as he could; and a valuable parcel I shall have of them . . ."

sian set nearly all the specimens are numbered but the numbers were undoubtedly assigned in a taxonomic sequence for the distribution and not according to any order the specimens might have had when they were originally collected." (pp. 60, 61)

"The botanical importance of Douglas' Californian collections is due not to any intrinsic value in the specimens themselves, but rather to the botanical opinions that have been expressed about them, opinions which were expressed in the form of new species and varieties and which constitute most of the ground work of our western botanical taxonomy. Hooker and Arnott's Californian supplement to the Botany of Capt. Beechey's Voyage (1841) is the most important single work in which Douglas' Californian collections were listed and described, but many were also included by Torrey and Gray in the Flora of North America (1838–1843) and by DeCandolle and his collaborators in his Prodromus. Many species were named from plants grown from seeds of Douglas' collecting by Lindley in the Botanical Register and by Bentham in the Botanical Magazine and in the Transactions of the London Horticultural Society. However, most of the species which Bentham named from Douglas' collections were based directly on specimens collected in California, and were described in the great monographic works of that illustrious botanist." (p. 95)

That one peak in the Province of Manitoba was named for David Douglas has already been mentioned. Balfour refers to another:

". . . the state of Oregon has given his name to a peak in sight of the Eugene-Klamath Falls of the Columbia. One hundred years elapsed before the first scientific explorer was thus commemorated."

The locality cited was so peculiar that I turned to Dr. F. Marion Ownbey for enlightenment and he came to my assistance (*in litt.*, August 19, 1947):

". . . concerning Balfour's statement about David Douglas Peak . . . I had no more success in finding it than you had until someone suggested that I write the Oregon Historical Society. The reply of Howard M. Corning, Research Director of that organization, cleared it up immediately. It is a conspicuous peak on the north side of Salt Creek Valley opposite the Cascade line of the Southern Pacific (railroad) Company in southeastern Lane County. It is shown on the map of the Willamette National Forest (south half), published in 1935, in S. 24, T. 22 S., R. 5 E., and the elevation is given as 6253 feet. A detailed reference to the peak and its naming is found in 'Oregon Geographical Names,' by Lewis A. McArthur, published in 1944. David Douglas Peak, then, is in sight of the railroad between the towns of Eugene and Klamath Falls, and the allusion to a falls of the Columbia (River) bearing the hyphenated name of Eugene-Klamath is inaccurate and misleading. That this peak can be seen from any point on the Columbia River is exceedingly unlikely, so that I am inclined to dismiss the whole matter as a . . . typographical error . . ."

Before bidding farewell to Douglas I should like to record some of the persons—most of whom are mentioned elsewhere in this story—whom he met on his last journey to North America. For the most part it was a lonely life as far as congenial companionship was concerned.

On October 11, 1830, he wrote Hooker how pleased he was at the prospect of having someone to work with him, but who the companion was to have been we are not told, and he seems never to have materialized:

"... I should indeed be delighted to have such a companion as the gentleman whom you describe, and whom I have hitherto only known by report. More than ten times as much could be effected by the united exertions of two." (HM, 147)

To Barnston he wrote in March of 1833:

"To console myself for the want of friends of a kindred feeling in this distant land, for an exchange of sympathy or advice, I vary my amusements; by day it is a barren place that does not afford me a blade of grass ... during the stillness of a cloudless night ... localities are determined, altitudes measured, the climate ... analyzed. Thank God my heart feels gladness in these operations; without such to pass away an hour, my time would be blank." (B, 277)

Douglas kept Hooker informed of those whom he did meet. On the Columbia River, in the autumn of 1833, he had seen Dr. Meredith Gairdner, but had missed Mr. William Fraser Tolmie.[8]

"My meeting with Dr. Gairdner afforded me heart-felt satisfaction ... he is a most accomplished and amiable gentleman, devotedly attached to Natural History ... he told me of your health, and that of your family: the additions to your Herbarium, &c. I endeavoured to show him the attentions to which every friend of yours is justly entitled at my hands, and only regret that our time together was so short; for he is a person whom I highly respect. Mr. Tolmie had quitted the Columbia for the North-West coast before I arrived, and thus deprived me of the pleasure of seeing an old student of yours. I wrote to him twice, indicating those parts of the country which promised to yield the best harvest to the Naturalist, and particularly requesting his attention to the sea-weeds; but have not heard from him since, nor indeed at any time. I much regret not having seen this gentleman, for I could have told him many things useful for a young man entering this country as a Botanist or traveller to know. However, I have explained them all to Dr. Gairdner ... I have given Dr. Gairdner my notes on some more species of *Pinus*. This gentleman, and Mr. Tolmie, will have a good deal to contend with. Science has few friends among those who visit the coast of North-West America, solely with a view to gain. Still with such a person as Mr. M'Loughlin on the Columbia, they may do a great deal of service to Natural History." (HM, 158, 159, 160)

Douglas' plan to visit the Russian settlements in Alaska and to proceed thence by way of Russia to Europe, and which he had taken steps to further before leaving Eng-

8. Writes Hooker: "This accomplished gentleman [Gairdner], together with Mr. Tolmie, one of my most zealous Botanical students, I had the pleasure of recommending to medical appointments in the Hudson Bay Company's possessions on the North-West coast of America ..." (HM, 158, *fn.* 2)

land, never materialized. He may have been planning to get to Alaska when he made his trip into New Caledonia in 1833 (March to July). But he does not say so, merely that "... Perhaps I shall cross Mackenzie's track, at Fraser's River ... and proceed northward, among the mountains, as far as I can do so with safety, and with the prospect of effecting a return ..." But he had kept in touch with Baron Wrangel, thought highly of him, and the Baron had been helpful.

He wrote Hooker on November 23, 1831; and again October 24, 1832:

"I have met the Russian Authorities twice ... and have received the utmost kindness from them. Two days ago I received a letter from Baron Wrangel, Governor of the Russian Possessions in America and the Aleutian Isles, full of compliments, and offering me all manner of assistance, backed by Imperial favour from the Court. This nobleman is ... the Captain Parry of Russia, keenly alive to the interests of Science, and anxious to assist, in every way, those who labour in this field." (HM, 151)

"I have had two most kind letters from Baron Wrangel ... to whom I was made known through the Russian Minister at the Court of London. In his first he writes thus ...:—'J'ai appris avec une vive joie vôtre intention de faire une tournée dans nos environs. Soyez sûr, Monsieur, que jamais visite ne m'a été plus agréable, et que des bras ouverts vous attendent à Sitka. Si vous avez l'intention de retourner en Europe, par la Sibérie, je puis vous assurer qu'au mois de Mai de l'année prochaíne, vous pourrez commodement aller sur un de nos navires à Okotsk, où, d'après des nouvelles que je viens d'apprendre, on vous a dejà preparé un gracieux accueil.' This is more than kind, and the facilities offered for May, 1832, of course hold good for ensuing years. This letter was accompanied by a copy of a volume ... containing some very interesting accounts of the Russian expedition to Mount Ararat; also an outline of Mertens' labours with Capt. Lutke's Pendulum and Experiments ... The Baron wrote me a second letter, and being fearful that I might not have received his first, took care to give me the same information, backed with additional assurances of his good will. I have had the advantage of seeing Cyrill Klebinkoff, Chief Director of the Russian-American Fur Company, an excellent man, who has great claims on my gratitude, as well as several Officers of the Imperial Navy. Indeed they seem to be a set of people whose whole aim is to make you happy ..." (HM, 155)

Perhaps dangerous conditions, perhaps the difficulty of travel in New Caledonia, or perhaps the desire to revisit the Hawaiian Islands, altered Douglas' original plans. At all events we hear no more of the proposed trip.

Douglas does not mention Nathaniel Jarvis Wyeth, but his lost journals may have done so. Wyeth, however, refers to Douglas three times, twice in January and once in June of 1833; they had evidently had some difference of opinion: "... as I had some dispute with Mr. David Douglass about the grouse of this country I subjoin a description ..." It was a lengthy one—Wyeth intended to prove his point.

At Monterey, California, Douglas met Thomas Coulter who had reached there from Mexico. I refer to Douglas' comments in my next chapter for the date of the letter to Hooker in which they are contained—November 23, 1831—fixes, as nearly as it is possible to do so, the date of Coulter's arrival from Mexico.

He wrote in 1832, from the Columbia:

"I heard of M. Klotzsch from Mr. Ferdinand Deppe[9] of Berlin, whom I had the pleasure to meet in California. Formerly M. Deppe devoted his time wholly to Natural History, Zoology in particular; but now he is partly engaged in mercantile pursuits . . ." (HM, 154)

To Douglas' comment upon Deppe—and the man seems to have been a backslider as far as botany was concerned—Hooker added the information:

"Many of the new plants discovered by this gentleman in Mexico were published by Schlechtendahl and Chamisso in the volumes of the 'Linnæa'; the same work also contains some interesting accounts of his excursions."

Douglas sent messages of a jesting sort to his old associate Scouler; writing Hooker on October 24, 1832:

"I will trouble you to offer my kind regards to . . . Dr. Scouler, and say . . . that I have a tolerable collection of bones for him, but as I thought he would himself enjoy the task of cleansing them, I have only cut away the more fleshy parts, by which means, too, they hang better together. They consist of a *Sea Otter,* entire; *Wolves, Foxes, Deer,* a *Panther's Head,* &c. I shall send them by the earliest opportunity. You may also tell him that human heads are now plentiful[1] in the Columbia, a dreadful intermittent fever having depopulated the neighbourhood of the river . . ." (HM, 155)

Later collectors were to enter the very regions explored by Douglas and, as conditions made such journeys feasible, were to enter new regions where he had been unable to penetrate; these men found, of course, plants overlooked or never encountered by him. Conditions for field-work did not, however, improve by leaps and bounds and for many years his successors were to undergo, each in his particular way,

---

9. W. H. Brewer states that Deppe ". . . visited California in 1831 or 1832. But his name is seldom met with in connection with California botany."

An unsigned article entitled "Ferdinand Deppe's Reisen in Kalifornien" was published in 1847 (*Zeitschrift für Erdkunde* 17: 383-390, Johann Gottfried Lüdde, editor); it tells that for the last six years of his stay in America (the years are not given) Deppe made business trips between Acapulco, Mexico, and Monterey, California, and on one occasion sailed from Mazatlan to La Paz and thence to San Loreto (both in Lower California), from there traveling overland with a pack train to Monterey, a distance of 290 leagues, covered in 35 days. The author specifies that only birds and a few mammals were collected and names what they were.

Deppe evidently kept up his visits to California for some years after Douglas met him there in 1832, for Bancroft (*California* 4: 105 *fn.* 54) mentions that he was a passenger on the *Rasselas* when that vessel put in at Monterey on her way from Sitka to Honolulu in October, 1836.

I. M. Johnston tells me that some specimens of plants given as collected by Deppe in California exist in European herbaria; but, in at least two instances, he believes them mislabelled and very probably from Hawaii. (See *Contrib. Gray Herb.* 68: 87, 88, 1923.)

1. A reference to Scouler's acquisition, at some risk to his life, of a compressed Indian skull, just as he was leaving the Columbia River in 1825.

dangers and hardships comparable to, or possibly worse, than Douglas had met with. He was not the only one to encounter death while pursuing his task. However, of the plant collectors of my region and period, David Douglas stands out as the romantic figure. His was an enthusiasm unquenchable by vicissitudes and even by calamity, and there emanates from his story an aura of the adventurous and the chivalrous which is lacking in the stories of those who accomplished as much, perhaps, but in a more matter-of-fact, less spectacular way. The same aura, for me, emanates from the story of John James Audubon.

For the literature referred to in this chapter see pp. 338-341.

CHAPTER XIX

COULTER TRAVERSES THE COLORADO DESERT OF

CALIFORNIA AND SETS FOOT IN ARIZONA AT THE CON-

FLUENCE OF THE GILA AND COLORADO RIVERS

I KNOW of only two published articles which throw any important light on Thomas Coulter's visit to California:

The first is Coulter's "Notes on Upper California," published in 1835 in the *Journal* of the Royal Geographical Society of London. This paper ("Communicated by Dr. Thomas Coulter. Read 9th March, 1835.") is mentioned in the author's text as a "letter." An excellent map shows the routes which he took between Monterey and the junction of the Colorado and Gila rivers in 1832.[1] Coulter had intended to supply a fuller account of his California sojourn, for his "Notes" end with the statement:

"I am sorry to be obliged to content myself with offering the Society so desultory and imperfect a sketch as this; but I have many claims on my time, the most urgent of which is the preparation of a work in detail on the entire subject of California. Whatever is here defective will there, I hope, be found supplied."

The second published article is F. V. Coville's "The botanical exploration of Thomas Coulter in California" (1895). It includes Coulter's valuable map and, from the point of view of plant collecting, is more enlightening than Coulter's "Notes."

In 1943 Rogers McVaugh published "The travels of Thomas Coulter, 1824–1827." The author states: "This account has been drawn up from Coulter's own note-book, lent by his nephew, Joseph A. Coulter, to Dr. F. V. Coville, and now on deposit in the files of the Division of Plant Exploration and Introduction, Bureau of Plant Industry." For those who are interested in Coulter's work in Mexico this is a valuable document but, since the ". . . final entry is dated October 11, 1827 . . .", it ends several years before Coulter is known to have reached California. Coulter's later papers are reported to have been lost between London and Dublin. McVaugh's article does, however, contain some miscellaneous information about the collector which is of interest to this story.

The approximate date of Coulter's arrival in California is known, for in a letter to William Jackson Hooker, dated from Monterey, November 23, 1831, David Douglas wrote of his joy at meeting another botanist:

"Since I began this letter, Dr. Coulter, from the Central States of the Republic of

1. This map is reproduced in Bancroft's *History of California* (3: 407) but for some reason the portion south of San Felipe (a region which Coulter was the first botanist to explore) is not included.

428

Mexico, has arrived here, with the intention of taking all he can find to De Candolle at Geneva. He is a man eminently calculated to work, full of zeal, very amiable, and I hope may do much good to Science. As a salmon-fisher he is superior even to Walter Campbell, of Islay, Esq., the Izaak Walton of Scotland; besides being a beautiful shot with a rifle, nearly as successful as myself! And I do assure you, from my heart, it is a *terrible pleasure* to me thus to meet[2] a really good man, and one with whom I can talk of plants."

Bancroft mentions that the date of Coulter's arrival in California appears on the list of the Compañia Extranjera:

"The second name on the list was that of Thomas Coulter . . . an English scientist, who after extensive travels in Mexico had arrived in California in November, 1831, by what route or conveyance I have been unable to learn, but probably by sea." [3]

Coulter's one recorded trip in California is described in his "Notes," to which his map is an invaluable addition. I shall follow this narrative but rearrange it somewhat in order to correlate Coulter's dates with his descriptions of the country.

He makes a few general statements:

"It will not be necessary to enter, at present, into much detail of my journeys in the country, of which the principal was that from Monterey to the junction of the Rios Colorado and Gila." (p. 60)

"I shall not at present go into any examination of the vegetation of California, though this, as well as its Fauna, is well worthy of the most attentive consideration." (p. 68)

"The rainy season of 1832 ended late in February . . . and I started so soon as the country was passable, which it is not at all during the rains, nor for some time afterwards." (p. 64)

While on this trip Coulter took geographical positions with the aid of a chronometer; he had but one, and ". . . trusted to it solely." For the longitude of Monterey, point of departure and return, he used Beechey's observations:

". . . doubts have been expressed of the possibility of using a chronometer on shore, from the difficulty of transporting it safely, particularly on horseback. I am satisfied . . . that this difficulty is not so great as has been imagined. All that appears to be necessary, is to carry the chronometer belted tight against the abdomen, and wear it so day and night." (p. 61)

---

2. According to Coville, "These two botanists evidently worked together, to the great delight of Douglas, during the winter of 1831-2 and the following spring, Douglas finally sailing from Monterey to the Sandwich Islands and Coulter setting out on his trip to Arizona."

I have found no record, other than Coville's (and his statement is based upon the Hooker "Memoir" of Douglas), which indicates that they "worked together." Of course it would have been natural for them to have done so. Coville's reference to Coulter's "trip to Arizona" must refer to the man's sojourn at the confluence of the Gila and Colorado rivers.

3. C. C. Parry refers to the route as "probably by way of San Blas [Jalisco] . . ."

Adopting this method, all went satisfactorily except for a slight change of rate during Coulter's stay at the Colorado River where the heat stood frequently at 140° Fahrenheit. He then made what seems like a reasonable concession to the heat (of present Yuma):

"Perhaps this degree of heat ought not to affect the chronometer; but I found it so intolerable, that I was obliged to leave off the belt in which I carried it, and to allow it to lie horizontally . . . which may . . . have contributed to produce the disturbance." (p. 62)

Coulter's list of "Observations in the order of their dates" is a helpful supplement to his map. There is, however, no way of knowing positively whether the observations were taken on the day of departure from a given station but, for reckoning purposes, I assume this to have been the case. The map, which Coulter refers to as "very rude, and in many respects certainly not very correct," shows alternate routes between stations in some instances and a few side-trips. It is not always possible to determine which of these alternate routes was followed on Coulter's southern, which on his northern, journey nor when the side-trips were made although the length of time consumed between stations offers some indication.

In outlining Coulter's routes I have inserted when possible present-day counties. Some of his place-names are difficult to identify.

1832

Coulter took observations at Monterey on January 22 and on February 22, as well as a final one before he started southward:

March 20. Monterey, Monterey County. Thence to the next station (Santa Barbara) consumed sixteen days. Or three days longer than Coulter's return journey.

Leaving Monterey the map indicates that Coulter kept to the west of the "Rio San Buenaventura," which he calls the "Monterey River" [presumably the Salinas River], then turned still farther west to reach San Luis Obispo. Although he does not specify missions they, rather than towns, were doubtless his objectives.

Between San Luis Obispo, San Luis Obispo County, and Santa Inez, Santa Barbara County, two routes are indicated on the map:

One, farther west than the other, went directly to La Purisima, Santa Barbara County. Coville mentions that the mission La Purisima Concepción was then situated on the "old wagon road between San Luis Obispo and Santa Inez." This was the route taken by Coulter on his way south, as he himself mentions. It crossed the "Guadalupe River" near its mouth.

The other route lay inland and was evidently more direct. South of the "Guadalupe" (identified by Coville as the Santa Maria River "between San Luis Obispo and

Santa Barbara Counties") the map shows a break in the route but this may have been a cartographical error. It crossed the "Guadalupe" higher up; this crossing would presumably have been easier to accomplish in July (on the return journey) than in March.

From Santa Inez the route, which touched the coast at one point, led directly to Santa Barbara.

Of this portion of his journey Coulter wrote:

"The rivers, which in the dry season are mere beds of sand, are quite impassable when swollen; and even for some weeks after they have fallen low, the danger and difficulty of crossing some of them, on horseback, are very considerable . . . the sand comes down mixed with a vast quantity of mud which settles together with it; so that even when the stream becomes so low that a small animal can walk across, a horse or man cannot. It is not until the mud is gradually washed out of the surface of the deposit that this becomes possible. We have then a bed of hard sand resting upon one of semi-fluid mud and sand; and it is very difficult to say when and where it is safe to make the passage. On this occasion I had to pass the Guadalupe, in this state, between San Luis Obispo and La Purissima; and it was only after long search that I found a place where a bear had passed, and trusting to his sagacity I followed his steps. The stream was broad, very shallow, and the bed of clear sand on the surface of the deposit must have been very thin, for it swagged under foot like the surface of a quagmire . . .

"From Monterey southward the road runs through a series of narrow ravines, as far as San Luis Obispo; but about Santa Inez, south of San Luis, and again in the neighbourhood of Santa Barbara, it runs on, or close by the beach . . ." (p. 64)

April 4. Santa Barbara, Santa Barbara County. Thence to the next station (San Gabriel) seventeen days. Or three days less than the return journey.

The route ran along the coast to San Buenaventura, Ventura County. From there to San Fernando, Los Angeles County, alternate routes are indicated on the map:

One crossed what must have been the Santa Clara River near the coast and later turned nearly due east to San Fernando.

The other kept up the Santa Clara River for a time, crossed it and joined the other route to San Fernando. From San Fernando the route was southeast to San Gabriel.

Coulter records that the road, from Santa Barbara southward, ". . . keeps chiefly along the west foot of the mountains, separated from the sea by low sand-hills, in some places of considerable breadth, as at San Gabriel, where they are almost twelve leagues broad." (p. 64)

April 23. San Gabriel, Los Angeles County. Thence to the next station (Pala) six days. Or thirteen days less than the return journey.

The map shows a direct route to Pala: it is indicated as considerably west of San

Bernardino and would appear to have run between the Santa Ana Mountains (west) and the San Bernardino Mountains (east) and across southern Los Angeles and western San Bernardino Counties into northern San Diego County.

It also shows three side-trips from San Gabriel: (1) almost due west to Pueblo, perhaps merely *a pueblo,* Los Angeles County; (2) to San Pedro, Los Angeles County; (3) southward along the coast to San Luis Rey and thence to San Diego, both in San Diego County.

Since the overall number of days between Pala and Santa Barbara on the northward return journey was thirty-nine as against twenty-three for the southern journey, it seems very probable that the side-trips from San Gabriel were made on the way north, or between May 27 and July 5.

Of this section of the route Coulter wrote:

"The best way to the Colorado, in the dry season, is to follow the coast road as far as San Luis Rey, and thence ascend the Pala stream [or the San Luis Rey River], which runs in a very narrow ravine behind the maritime ridge, crossing the summit level between its head and that of the small stream of San Felipe, which runs south-eastward till it reaches the border of the sand plain at Carizal, where it sinks . . ." (p. 64)

April 30. Pala, or La Pala, northwestern San Diego County. Thence to the next station ("Ford" on the Colorado River) eight days. Or two days less than the return journey.

According to the volume *California* ("American Guide Series"), p. 525, "Pala grew up around La Asistencia de San Antonio de Pala . . . deserted by its padres in 1829 . . ." It was situated on San Luis Rey River.

From there Coulter's route crossed to the "small stream of San Felipe" and then left it for "Carinal," or Carizal. According to Coville's interpretation of the journey, San Felipe was the stream "heading in Warner Pass, San Diego County." Of Carizal Coville states: "According to Coulter's narrative, the Carinal is the sink of San Felipe creek. According to his map, however, it is on Carrizo Creek, the next stream to the south, thus corresponding to the place known as Carrizo [San Diego County] in the Pacific railroad reports and on recent maps. *Carinal* on Coulter's map is a misprint."

Coulter's "Notes" mention that, although subterranean, San Felipe's ". . . course across the plain, when swollen, which it rarely is, is marked by a dry channel, in many points of which a little water, usually very bad, is to be had by deep digging." (p. 64)

Continuing his description of the route to the Colorado River which, beyond Carrizo, was in Imperial County, Coulter relates:

"There is not much difficulty in any part of the journey up to this point,—the Carizal [Carrizo Creek]; but from hence across the plain, which is here about one hundred miles broad, and totally destitute of pasture, cattle suffer extremely. It is always possible to carry water enough for a party of men; but horses and mules must pass the first two days absolutely without water or food,—and even then get only brine at the point called the Aqua Sola, from its being the only pond on the plain. When I passed, the water I found at this place was so strong that it purged both men and cattle. There is here some rush and reed which mules will eat, though horses usually refuse them.

"From hence there is still another day's journey to the Rio Colorado." (pp. 64, 65)

Coville identifies "Aqua Sola" as "probably the 'lagoons' at Alamo Mocho of the Pacific railroad reports and recent maps." I have, unfortunately, been unable to locate "Alamo Mocho" on "recent maps." But the Alamo River (like San Felipe Creek and Carrizo Creek) lay across Coulter's route and I believe that the reference may be to some pond of surface water along what was, for the most part, a dry stream bed. All these streams lie south of and empty into the Salton Sea according to the map of the State of California issued by the Department of the Interior, U. S. Geological Survey, in 1929. The Geological Map of the State of California, issued by the State Mining Bureau in 1916 (and reprinted in 1929) shows a "Mesquite Lake" which is evidently an enlargement of the Alamo River, lying about midway between present Brawley and El Centro, and, it would seem, situated directly on Coulter's route.

May 8. "Ford" on the Colorado River, ". . . six miles below its junction with the Gila . . ." In this vicinity Coulter remained for nine days, or until May 17, when he started on his return journey.

Coville places the ford "at or near the site of the present town of Yuma." If Coulter crossed the Colorado, as he doubtless did, he was the first botanist to enter the present state of Arizona—Yuma being now the most important town in southwestern Pima County. He wrote of his sojourn at the confluence of the Gila and Colorado:

". . . a few remarks upon the climate of Mexico. In an early part of my letter I stated that the thermometer had frequently stood at 140° Fahr. This . . . was the temperature of the atmosphere a few feet above a plain excessively dry and heated by the sun; and something may also be due to the circumstance of the thermometer, although carefully screened from the sun, being exposed to the radiation of the soil, which was very great and frequently oppressive. This very high temperature is not . . . of very frequent occurrence, but always owing to some local and temporary causes; one of which, and indeed a necessary condition for its attaining its greatest height, being that there should be no wind. Such was the case during the latter three days of my stay at the ford on Red [Colorado] River. The wind, after having blown for many days from the S.W., suddenly lulled three days before my departure on my return, and continued dead calm for the first day of my journey, that is, until reaching Agua Sola. This was one of the most painful days I have spent, notwithstanding that the excessive dryness of the atmosphere necessarily exempted me from that oppression felt in damp situations even at very inferior temperatures . . ." (p. 68, 69)

Coulter refers to travel in Sonora, Mexico, but there is no reason to suppose that he journeyed to the regions which he mentions. "Alta" is doubtless the Altar of recent maps, situated on the river of the same name which is a branch of the Rio de la Concepción.

"After passing the river the same difficulties [lack of water] continue for seven days farther, on the Sonora road, as far as Alta; but this part of the journey, from its greater length, it is extremely imprudent to attempt without a proper guide. The only water to be had is found in the ravines, frequently at some distance from the road, in excavations called Tinajas, made by the Indians, who were formerly much more numerous in this neighbourhood than they are at present." (p. 65)

On the return journey to Monterey the dates and stations were as follows:

May 17. "Ford" on the Colorado River. Thence to the next station (Pala) ten days.

May 27. Pala. Thence to the next station (San Gabriel) nineteen days.

June 15. San Gabriel. Thence to the next station (Santa Barbara) twenty days.

July 5. Santa Barbara. Thence to the next station (Monterey) fourteen days.

Coulter considered his journey as ". . . the most interesting, the longest, and by far the most laborious of those I made in California." (p. 64) He was back at Monterey on July 19, 1832. The trip had consumed approximately seventeen weeks and what he did and where he went during the remainder of his stay—estimated at from two to three years—we do not know. As stated, he had arrived in California in November, 1831. Douglas wrote Hooker from the Columbia River on October 23, 1832:

"I left in California my friend Dr. Coulter, who will not, I trust, quit the country till he has accomplished every thing, for he is zealous and very talented. To De Candolle, who is his old tutor, he sends all his collections; and who can wonder at his giving him the preference? Dr. Coulter expects to be in England in the autumn of 1833; I have given him a letter of introduction to you."

Coville, on the other hand, refers to a later departure: "After further collecting in California, apparently always in the vicinity of Monterey, Coulter returned to Europe by way of Mexico, in the year 1834 . . ." David Don mentions that he spent "nearly three years" in California, and ". . . visited regions unexplored by Mr. Douglas."

Coulter had, of course, been the first botanist to cross the Colorado Desert and to set foot—we must believe—in Arizona. He and Douglas had made somewhat similar journeys between Monterey and Santa Barbara, Coulter in 1832, Douglas in 1831. If we can trust the routes shown on his map, Coulter was the first botanist to travel overland between Santa Barbara and San Diego, much of the way following the coast. At San Diego he had been antedated by Menzies, who had spent twelve days there in 1793 (November 27 to December 8) and about three weeks in 1794 (November 6 to December 2). It was not until seven years after Coulter's visit that Hinds reached San Diego.

Of some sixty miscellaneous species named for Coulter which are listed by Coville, twenty-two are from California, and, so far as I know, none were from regions which would have been inaccessible to him on the journey just described.

In the column "Botanical Information" Hooker wrote in the *Journal of Botany* in 1834:

"In the south-west of N. America . . . Dr. Coulter, a most indefatigable and accomplished Botanist, has spent many months. He joined Mr. Douglas at Monterrey, and afterwards proceeded south to the Rio Colorado, at the head of the Gulf of California, in lat. 34° N.; and when our informant left California, Dr. Coulter was preparing to ascend that river. If he should proceed to Santa Fè, in New Mexico, as was probable, it would not surprise us were he to meet Mr. Drummond there."

Although only a small proportion of Coulter's stay in western North America has been accounted for, it seems certain that, had he ascended, or even attempted to ascend, the Colorado River that fact would have become known. By Coulter's day only two white men are known to have made the journey.[4] Or, had Coulter reached Santa Fe by any route, that fact, too, would have become known. Drummond found it impossible to get there from Texas without returning to St. Louis and abandoned the idea. This was doubtless wishful thinking on Hooker's part: the botanists, in the security of their offices thousands of miles away, seem to have had no conception of what it was possible for their emissaries to accomplish in the way of travel.

Coulter's paper—as its presentation to the Royal Geographical Society suggests— is largely concerned with the physical aspects of California. From that angle it served to correct two misconceptions:

One concerned Tulare Lake, at that period usually referred to as Tule Lake or Lakes. As Stillman points out, Coulter's map ". . . represents the lakes as discharging into the Bay of San Francisco at San Jose!" This may have been one of the aspects of his map which Coulter considered "not very correct." Bancroft notes that Coulter ". . . corrects the 'great popular error' respecting the Tule Lakes which has 'raised these comparatively insignificant ponds to the rank of a vast inland sea.' . . ." The "Notes" state clearly:

"The Tule Lakes . . . discharge . . . very little, if any, water into San Francisco . . . Such at least is the account given by the American hunters. A severe accident prevented my crossing this ground myself . . . I hope further observation, whilst it must correct, will confirm the general view I have taken of the country." (p. 60)

Another misconception concerned the position of ". . . the junction of the Rios

4. Edwin Corle's delightful book, *Listen, Bright Angel* (1946), tells how the two Patties, James Ohio and Sylvester, father and son, in the 1820's, ". . . reached the lower Colorado in the vicinity of what is today Yuma and worked upstream, following the general course of the river, all the way through Utah and into Wyoming. Most of the time they didn't know where they were except 'out west.' . . . But one thing was certain: the Patties did not like it [the Grand Canyon region] and said so. In this they were not alone. The Patties left the only record of the country between the Spanish priests in 1776 and Lieutenant [Joseph C.] Ives in 1858. . ." Corle's book was, he notes (*in litt.*, February 27, 1954), ". . . reissued as The Story of the Grand Canyon with some additional material in 1951 . . ."

Colorado and Gila . . ." This Coulter laid down ". . . nearly forty miles farther north than Lieutenant Hardy[5] has done . . . This point, which was the site of the two missions of San Pedro and San Pablo [both shown on Coulter's map], has long attracted a good deal of attention . . ." It was ". . . usually the only practicable ford on the river below the junction . . ." and important in the attempt to establish ". . . a communication over-land between Sonora and California . . ." These missions, San Pedro on the north and San Pablo on the south side of the river, were abandoned by the time of Coulter's visit. He notes that "The junction of the two rivers is not a mile above this point . . ."

Coulter's opinions upon California as a place of settlement are interesting:

"The only settled portion of Upper California lies along the coast; the missions being nearly all within one day's journey from it. The only point where a mission has any settlement farther inland is at San Gabriel, where the Rancho of San Bernardino is at the head of the valley, some thirty leagues from the port of San Pedro. This is indeed the only point of either Californias, south of San Francisco, capable of sustaining a large population. The valley . . . is in many places very fertile, and wheat, where it can be irrigated, yields better here than in any other part of the Mexican territories that I have seen. The vine also thrives better, and is beginning to be extensively cultivated.

". . . the general government is now making considerable efforts to colonise Upper California from Mexico, under the apprehension that, if not done, the North Americans will get in in too great numbers. This apprehension appears to be hardly rational . . . Any efforts made for the purpose of colonising Upper California should be directed towards the portion of the country north and east of San Francisco and east of the Tule lakes, which is fertile, well wooded and watered, and of sufficient extent to make its colonisation worth while as a speculation; the rest of the country south of San Francisco and west of the Tule lakes, possessing, with the exception of San Gabriel, too little cultivable ground, and of this a very small portion irrigatable—the soil . . . where it is arable, being usually rich.

"The number of the white inhabitants has . . . increased very rapidly, and I believe is now not under six thousand, though I cannot state their numbers very exactly until I have examined the statistical materials . . . collected.

"The reverse . . . is the case with the aboriginal inhabitants[6] . . . one would suppose they ought at least not to have lost ground, not having been driven from their homes,

---

5. G. K. Warren's "Memoir" . . . mentions that Lieutenant Hardy, R.N., ". . . visited the whole coast of the Gulf from Mazatlan around by the mouth of the Colorado to Loredo [Loreto], in search of pearl fisheries. He did not determine any positions by astronomical observations, and his map has not been used by me." The years in which Hardy examined the Gulf of California are given as "1825 to 1828."

6. Coulter notes that because of the ". . . much smaller number of [Indian] women living than of men . . . in Upper California, in almost all the missions, a great many of the men cannot find wives. The mission of San Luis Rey is the only remarkable exception . . . the destruction of the missions, will enable the white inhabitants to acquire possession of the great bulk of the mission lands . . ."

as in the United States, nor having had ardent spirits at all within their reach until lately." (pp. 65, 66)

The fullest account of Thomas Coulter's life was supplied by the Reverend Dr. Thomas Romney Robinson, ". . . who, from long acquaintance and intimate friendship with the deceased, was enabled to speak minutely of his personal career; and whose extensive and profound acquaintance with almost every department of knowledge . . . enabled him to judge correctly of the aspect in which the labours of Doctor Coulter should be placed."

His paper was read at the meeting of the Royal Irish Academy on March 16, 1844 and (with eulogies of other Academy members who had died in 1843) was published in its *Proceedings:*

"Thomas Coulter was born in 1793, near Dundalk [capital of the county of Louth, and some forty-five miles from Dublin] . . . From an early age he was devoted to field sports, which he followed with a minute attention to the habits of his game, that belonged more to the naturalist than to the sportsman. Bees were another favourite object, and he possessed that remarkable power of handling these irritable insects with impunity . . . He had in after-life the same privilege as to serpents; of which some members may recollect an amusing exhibition in this room . . .

"He was prepared for college by Dr. Neilson, the author of a well-known Irish grammar . . . In the University . . . under . . . the late Dr. Lloyd . . . the prevailing bias of his mind prevented him from equalling in mathematical attainments some of his fellow-pupils . . . But in practical mechanics, in Chemistry, Physiology, and above all, in Entomology and Botany, he far outstripped his college contemporaries, and while yet an undergraduate, his collections of Irish insects and mosses were such as might have been owned with credit by a veteran . . . his success made him only the more conscious of his deficiencies, and determined him to seek abroad the means of supplying them. Having spent one or two summers in Paris, where he made very extensive dried collections of the plants in the Jardin des Plantes, he established himself in 18—, at Geneva, where, under the auspices of De Candolle, he found all that he could desire. How well the three or four years which he spent there were employed, appears from the memoir on the Dipsaceæ, which he then published, and still more from his Herbarium, of which the European part was then formed . . . a work, which . . . as the result of almost unaided individual exertion, may well be called gigantic." [7]

We are told that Coulter had made plans ". . . on his return from the Continent, in 1824 . . . to explore a considerable portion of America," but, instead, accepted employment ". . . as medical attendant[8] to . . . the Real del Montè Mining Company for

7. According to a Robinson footnote, Coulter's herbarium, including Mexican and Californian plants, contained about 150,000 specimens.

8. Robinson does not mention that Coulter had any medical training but, cited among Life Members of the Royal Irish Academy (as of March 16, 1834), is "Coulter, Thomas, M.D. . . ."

three years, during which time he hoped to complete the Mexican Flora . . ." Next he took charge of the company's mines, the Veta Grande, and ". . . under his management the concern became productive." Next ". . . he passed to California, where, and in Sonora, he spent four years, always actively employed on his primary objects, and in spirit-raising adventure . . .

". . . the herbarium which he collected under such circumstances contains upwards of 50,000 specimens of 10,000 to 11,000 species,[9] the far greater proportion of which was collected and preserved by himself . . . in connexion with the herbarium he had gathered specimens about the size of a 16mo. book of nearly 1,000 descriptions of woods, most, if not all . . . accompanied by dried specimens of the foliage and inflorescence of the trees from which they were taken; the whole gathered by himself, and being, perhaps, the largest collection of this particular kind ever made by any unaided individual.

". . . he returned to Europe with an immense increase to his collection, but with a constitution irreparably injured by the hardships which he had encountered . . . at home he was destined to meet a severe loss. In the transport from London to Dublin, a case containing his botanical manuscripts, and the materials of a personal narrative, disappeared, and could never be traced . . . of the latter, nothing remains except a brief account of Upper California . . . and the former are totally lost, except some communications to De Candolle and Lambert. After this his chief anxiety was to secure the herbarium, which has cost him so much, from dispersion and neglect; and in this . . . he was not disappointed. It has become the property of our University, and the task of arranging it was the employment of his few remaining years . . . It was completed for the European part, and about 8,000 of the American specimens; but the remaining packages are well furnished with memoranda, so that for them also the arrangement is practicable . . ."

Coville's opinion of Coulter's work was laudatory:

"Coulter was the first botanist who penetrated the Colorado Desert, remarkable for the aridity of its climate and the peculiarities of its flora. His collections were very large, and their enumeration, had it been published as a single report, would have formed probably the most valuable contribution to North American botany ever issued . . . this effort to record an outline of his work will serve to show to some extent the importance of his scientific explorations to the advancement of botany in America."

Of the publication of his collections he has this to say:

"The collections of Cactaceae sent to De Candolle seem to have contained the only ones of Coulter's plants that reached an avenue of publication previous to his death, with the exception of five new Californian pines described in the paper by David Don, and the *Cupressus coulteri* of Forbes. But Professor W. H. Harvey, upon his ap-

9. Coville estimates the collection at ". . . over 50,000 specimens, probably representing between 1,500 and 2,000 species . . ."

pointment as Coulter's successor,[1] in 1844, proceeded with the arrangement of the Californian and Mexican plants,[2] issued three short papers on them, and, apparently in the years 1846 and 1848, distributed the greater part of the duplicates, the first set going to Kew and others to Dr. Asa Gray and Dr. John Torrey in America. The specimens were sent out under more than 1,700 numbers, unfortunately arranged systematically instead of chronologically."

After listing "the titles of special papers based on Coulter's collections,[3] together with two of Coulter's own publications[4] ... chronologically arranged ...", Coville continues:

"The descriptions of most of Coulter's new species have come into the literature of botany irregularly and incidentally in various reports and monographs prepared at the great herbaria which received his collections, so that the complete collation of them would be a matter of great difficulty and probably not worth the effort. As indicative, however, of some of the avenues of publication, I append a list, chronologically arranged, of most of the species that bear his name, together with a few critical notes on other species regarding which mistakes appear to have been made."

This is followed by a list of fifty-seven species, twenty-two of which appear to have been collected in California.

Coville ends his paper with a plea, and with a suggestion—which seems never to have been fulfilled—that he might write more on Coulter:

"I earnestly request that any known facts regarding Thomas Coulter not included in this paper be communicated to me. These together with other matter, including the letters of Coulter to Augustin Pyramus and Alphonse DeCandolle, to which through

1. Brewer refers to Harvey as "curator of the herbarium of Trinity College, Dublin ..." And Sargent refers to Coulter's appointment as "curator of the herbarium in the Botanic Garden at Dublin ..." Although the men would seem to have held different positions, I am told that the School of Botany is a part of the Botanic Garden, which is a part of Trinity College, which is a part of Dublin University.

2. Harvey's *Memoir* contains two short references to this task. One is in a letter dated February 16, 1844: "I ... have been working on Coulter's Mexican and Californian plants ..." and the other is dated October 24, 1844: "I have been very busy ... sorting Mexican plants every night."

3. The reference is to two papers by A. P. de Candolle: (1) "Revue de la famille des Cactées" (*Mém. Mus. Hist. Nat. Paris* 17: 107-119, 1828); (2) "Mémoire sur quelques nouvelles espèces de Cactées, et principalement sur celles envoyées du Mexique par le docteur Coulter" (27 pp., 12 plates, published Paris, 1834). I have cited its half title.

According to Coville, both these papers describe a collection (fifty-seven species of living plants) sent De Candolle in 1828. In paper (2) "Almost all this new matter is based on cultivated specimens from Coulter's first consignment in 1828."

Robinson records that "... seventy species and varieties of Cacti ... [went] to the late Provost, the Rev. Dr. Lloyd ... to be presented to the College for their botanic garden ... At the same time, a similar collection [went] to ... De Candolle, for the Geneva Botanic Garden ... One [plant] ... Cereus Coulteri ... may now be seen ... in the College Botanic Garden." W. J. Hooker mentions these collections as "particularly interesting."

4. Coulter's "Notes on Upper California." And his "Mémoire sur les Dipsacées" (written while a student at Geneva) and published in 1823, and reprinted in 1824 (*Mém. Soc. Phys. de Genève* 22: 13-60).

the courtesy of Dr. Casimir DeCandolle I have recently had access, may hereafter be incorporated into a more detailed account of Coulter's labors."

Just when the collections made by Coulter were transferred from Geneva to Dublin is not recorded. Brewer mentions that it must have been either when Harvey became curator of the herbarium at Trinity College, "or previously." The answer is probably the simple one that Coulter, deciding to work upon his own collections, had them sent from Geneva to Dublin which had become the seat of his activities after his return from North America. His "botanical manuscripts" and other papers were, as his friend Robinson noted, lost in transit between these cities. Just why he himself went to Dublin instead of to Geneva is not explained. The elder De Candolle in 1821 had begun work on the *Prodromus,* and although his son Alphonse once commented that his father's motto was "Il faut entreprendre quatre fois plus qu'on ne peut faire," he probably realized that the additional task of helping with Coulter's large collection would be impossible. Augustin-Pyramus de Candolle died in September,[5] 1841, about six or seven years after Coulter's return from America.

In his *Mémoires et souvenirs* De Candolle refers to Coulter as an "Irlandais, fort instruit." His opinions of the man were far more favorable than those expressed for his other unfortunate student Berlandier but his comments, like those of Asa Gray, were on the accurate as against the overenthusiastic side. And his humor was slightly ponderous; I translate:

"Dr. Coulter . . . worked intelligently and while with me prepared a monograph on the 'Dipsacées' which had merit . . . He is a well-informed man, a trustworthy and agreeable conversationalist, one for whom I have retained a friendship." Again:

"He had developed a sort of passion for studying the habits of reptiles and lived surrounded by snakes and lizards. He usually had several living ones in his pockets; he would place them on his hand and make them stay motionless by whistling them tunes. He lamented the fact that there were no snakes in Ireland, and on his departure thought that it might be a patriotic act to ship to that country a crate of living snakes. However, he did not carry this form of herpetological patriotism to the point of sending them."

Harvey described the beautiful Matilija poppy of California with the comment:

"Had there been no genus *Coulteria,* it is to this that I should have affixed the name of Dr. Coulter; but De Candolle having in this matter long anticipated me, I desire, as the next greatest respect that I can pay to Dr. Coulter's memory, to bestow upon this fine plant of his discovery, the name of his most distinguished and one of his most intimate friends. I . . . inscribe it to the Rev. T. Romney Robinson, the Astronomer of

---

5. Strange as it may seem, his son records two different dates, sixteen days apart, for his father's death; in the opening line of his preface to the *Mémoires et souvenirs:* "Mon père est mort le 9 septembre 1841." In a footnote (p. 489): ". . . il est mort le 25 septembre 1841 . . ." J. H. Barnhart (*Jour. N. Y. Bot. Gard.* 22: 130, *fn.* 3, 1921) cites the first.

Armagh ... I regret that an elder *Robinsonia*[6] prevents me from making use of Dr. Robinson's family name ... in calling the present genus *Romneya,* I follow a sufficiently established precedent."

The plant remains *Romneya Coulteri* Harvey to this day!

Harvey based the genus *Dithyrea* and, with W. J. Hooker, the genus *Lyrocarpa* (with type species *L. Coulteri*) upon collections made by Coulter in California—both belong to the mustard family. Also the genus *Whitlavia* (now reduced to *Phacelia*), with two species *grandiflora* and *minor,* noting that "Among Dr. Coulter's Californian plants are two remarkably handsome species of *Hydrophyllaceæ* with larger flowers than any others of the order ..." He believed that these plants, once introduced to gardens, "... will probably become as universal favourites as the *Nemophilæ* and *Giliæ,* which they rival or perhaps excel in beauty ..."

Among pines described by David Don in 1837 from Coulter's Californian collections were *Pinus muricata,* the prickle cone or bishop pine, and *Pinus Coulteri,* the big cone or Coulter pine; the first he found near San Luis Obispo, the last in the Santa Lucia Mountains.

Pinus Coulteri, according to Stillman, bears "... the heaviest cone of all the pine-trees known ..." When in southern California one Christmas season I had the fun of gathering many of them on the Palms to Pines Highway in the San Jacinto Mountains and shipping them to friends in the east. Full of resin, they throw off, when lit, an astounding amount of heat. I have since classified the recipients into two groups: those who burned them and those who preserved them as objects of beauty! Even after fifteen years some may still be seen on a number of Boston hearths. It is by this pine, rather even than by the beautiful *Romneya,* that I remember Thomas Coulter.

---

6. The origins of botanical names are, at times and to say the least, surprising: this was named in "... honour of Robinson Crusoe ..."

BANCROFT, HUBERT HOWE. 1885. The works of ... 20 (California III. 1825–1840): 406, *fn.* 46. Map [partial reproduction of "Coulter's Map." 407].

BREWER, WILLIAM HENRY. 1880. List of persons who have made botanical collections in California. *In:* Watson, Sereno. Geological survey of California. Botany of California 2: 555.

CANDOLLE, AUGUSTIN-PYRAMUS DE. 1862. Mémoires et souvenirs de Augustin-Pyramus de Candolle ... écrits par lui-même et publiée par son fils [Alphonse de Candolle]. Genève. 328, 329, 332.

CORLE, EDWIN. 1946. Listen, Bright Angel. New York. 75, 76.

COULTER, THOMAS. 1835. Notes on Upper California. *Jour. Royal Geog. Soc. London* 5: 59-70. Map ["Upper California, to illustrate the Paper by Dr. Coulter."].

COVILLE, FREDERICK VERNON. 1895. The botanical explorations of Thomas Coulter in Mexico and California. *Bot. Gaz.* 20: 519-531. pl. 34 [reproduction of Coulter's map of Upper California].

Don, David. [1834–1837]. Descriptions of five new species of the genus Pinus; discovered by Dr. Coulter in California. By Mr. David Don, Libr. L.S. Read June 2nd, 1835. *Trans. Linn. Soc. London* 17: 439-444.

Eastwood, Alice. 1939. Early botanical explorers on the Pacific coast and the trees they found there. *Quart. Cal. Hist. Soc.* 18: 341.

Harvey, William Henry. 1845. Description of a new genus of Papaveraceae, detected by the late Dr. Coulter, in California. *London Jour. Bot.* 4: 73-76. pl. 3 [*Romneya Coulteri*].

—— 1845. Characters of two new genera of Cruciferae, discovered by the late Dr. Coulter, in California. *London Jour. Bot.* 4: 76-78. pl. 4 [*Lyrocarpa Coulteri* Hook. & Harv.], pl. 5 [*Dithyrea californica* Harv.].

—— 1846. Description of a new genus of Hydrophyllaceae, from California. *London Jour. Bot.* 5: 311, 312. pl. 11, 12 [*Whitlavia* Harv.].

—— 1869. Memoir of W. H. Harvey, M.D., F.R.S., etc., etc. Late Professor of Botany, Trinity College, Dublin. With selections from his journal and correspondence. London. 147, 156.

Hooker, William Jackson. 1834. Botanical information. *London Jour. Bot.* 1: 175.

—— 1836–1837. A brief memoir of the life of Mr. David Douglas, with extracts from his letters. *Comp. Bot. Mag.* 2: 151.

McVaugh, Rogers. 1943. The travels of Thomas Coulter, 1824–1827. *Jour. Wash. Acad. Sci.* 33: 65-70. Map ["The routes followed by Coulter in Mexico, 1825–1827."].

Parry, Charles Christopher. 1883. Early botanical explorers of the Pacific coast. *Overland Monthly* ser. 2, 2: 414.

Robinson, Thomas Romney. 1844. List of members of the Royal Irish Academy deceased since the 16th of March, 1843. *Proc. Royal Irish Acad.* 2: 551, 553-557, v.

Sargent, Charles Sprague. 1892. 1897. The Silva of North America. Boston. New York. 3: 84, *fn.* 2 (1892); 11: 99, 139 (1897).

Stillman, Jacob Davis Babcock. 1869. Footprints of early California discoverers. *Overland Monthly* 2: 263.

Veitch, James & Sons. 1881. A Manual of the Coniferae, containing a general review of the order; a synopsis of the hardy kinds cultivated in Great Britain; their place and use in horticulture, etc., etc. London. 326.

Warren, Gouveneur Kemble. 1859. Memoir to accompany the map of the territory of the United States from the Mississippi River to the Pacific Ocean, giving a brief account of each of the exploring expeditions since A.D. 1800 ... *U. S. War Dept. Rept. expl. surv. RR Mississippi Pacific* 11: 27.

CHAPTER XX

PITCHER, WHILE STATIONED AT FORT GIBSON IN THE

"INDIAN TERRITORY," SPENDS HIS FREE TIME COL-

LECTING PLANTS

ZINA Pitcher was born in New York State in 1797 and received a medical degree at Middlebury College, Vermont. Information about the man's life is mainly to be found in medical journals, obituary notices and the like, and their substance is repetitious; nonetheless the little that is known of Pitcher's interest in natural history—botany in particular—is derived therefrom. A short chapter in Dr. Kelly's delightful volume, *Some American medical botanists,* is devoted to Pitcher but, except in its manner of presentation, even this offers little that is new.

Pitcher served for some fifteen years as surgeon in the United States Army, much of the time at frontier posts, and later practiced his profession in Detroit, Michigan. As a member from 1837 to 1852 of the Board of Regents of the University of Michigan, he played an important rôle in organizing its Medical School. He held many positions of responsibility but since these were mainly connected with his medical career I do not enumerate them here.

The fullest biography of Pitcher is contained in an address delivered on the University of Michigan's Founders Day, 1908, by Frederick J. Novy and published the same year in *The physician and surgeon.* In the sense that this tells where Pitcher was stationed during his service in the army it has a direct bearing on his botanical activities.

Immediately after acquiring his "diploma" in medicine Pitcher ". . . entered the army medical service as an assistant surgeon with the rank of first lieutenant." The next eight years were spent ". . . at various frontier posts in the then, as yet unbroken wilderness of the Territory of Michigan." Further:

". . . The records of the war department seem to be incomplete as to his details prior to 1831, but from occasional mention of his contemporaries, and from statements made by himself, we know that he was at Saginaw in 1823, Detroit in 1824, at Fort Brady (Sault Sainte Marie) in 1828, and at Fort Gratiot in 1829. The sojourn at various times in Detroit as well as the station at Fort Gratiot brought him into intimate association with the prominent men of that city, such as General Cass, Henry R. Schoolcraft . . . and other men who were called upon to mold the future state of Michigan."

We are then told of his assignment to regions west of the Mississippi River.

"The first records on file in the war department show that he was stationed from November, 1831, to July, 1834, at Fort Gibson[1] among the Creeks, Cherokees and

---

1. H. I. Featherly mentions Charles Joseph Latrobe as the third in a "cavalcade of botanists" to visit

other tribes of the Indian Territory. At this time, while his cherished friend General Cass was secretary of war under Andrew Jackson, he received his commission as major and surgeon . . ."

Fort Gibson, established by General Arbuckle in 1824, was located on the eastern side of the Neosho River, entering the Arkansas River from the north, and in what is now northeastern Muskogee County, Oklahoma; it lay northwest of Fort Smith—Nut-

Oklahoma, his predecessors having been Thomas Nuttall and Edwin James. Latrobe's trip through what is now eastern Oklahoma extended from October 10 to November 9, of 1832; while the records indicate that Zina Pitcher was stationed at Fort Gibson (in what is now Muskogee County, Oklahoma) from "November, 1831, to July, 1834." The matter of precedence is unimportant. Pitcher, however, did some plant collecting, while the indications are that Latrobe spent his time in hunting in a region where the amount of game, incidentally, was astounding.

Latrobe's visit to the southwest is such a famous one that—despite the fact that the "botanist" seems to have displayed little interest in plants at the time—I shall refer to it briefly.

Latrobe had come to America in the capacity of "guide, philosopher and friend" to a very young Swiss gentleman, Count Albert Pourtales. While on his way across the Atlantic he became a close friend of Washington Irving and, after travels in the eastern United States, the two visited Niagara Falls and had plans to proceed from there to the Canadian provinces. But when Irving was asked to serve on a government commission concerned with the transfer of certain Indian tribes to lands set apart for their use, plans were changed and Latrobe and the Count accepted an invitation to accompany him to the headquarters of the commission at Fort Gibson. From the east they followed the usual route down the Ohio and up the Mississippi to St. Louis and from here ascended the Missouri to Independence; thence they crossed the Missouri frontier into the country of the Osage Indians and then traveled south to their destination.

They left Fort Gibson on their hunting trip on October 10, 1832. Roughly speaking, their route took the form of a loop through the eastern portion of central Oklahoma: first northwest (to the region of present Tulsa), southwest through what are now Payne, Logan and Oklahoma counties (to the region of Norman, Cleveland County), thence—homeward—southeast and northeast to Fort Gibson, reached on November 9. They made two passages of what must have been the northern portion of the timber belt known as the Cross Timbers. The volume *Oklahoma* ("American Guide Series") mentions that some of their camping places are now recorded by markers.

Irving told the story in *A tour of the prairies*, Latrobe in *The rambler in North America: 1832–33*, and both accounts are remarkably readable. The last, full of historical interest, would seem to hold its own with some of the early classics of the frontier such as Bradbury's *Travels*, Nuttall's account of his journey into the "Arkansa" territory, Schoolcraft's *Narrative journal* and Keating's *Narrative of an expedition*. Latrobe visited many of the regions mentioned in these earlier stories and was an observant traveler; his opinions upon the government's Indian policy were perspicacious: he deplored the exploitation of the tribes by unscrupulous traders and agents and believed that the government placed too much power in the hands of such men. Chittenden seems to have held much the same beliefs.

Unfortunately, Latrobe has virtually nothing to say about the plants of the Indian Territory; visiting the region in the autumn he saw only such late-blooming ones as the "*Heliotrope* and *Solidago* species," "*Astres* [sic]," "the red and white *Eupatorium*," the "*Liatris*, or rattlesnakes'-master," the "*Gentiana* with their deep blue," and so on. Like all foreign visitors, he was impressed by "the glories of the autumnal Flora."

Latrobe returned to the eastern United States after his journey but was back at St. Louis in early October of 1833. From there he traveled, overland and by boat, to the Falls of St. Anthony. In a month he was back at St. Louis where he remained ten days and then descended the Mississippi to New Orleans, arriving about "the end of the year." From there he sailed on a vessel bound "to the port of Tampico in Mexico" in the "second week in January, 1834."

Maiden's *Records of Victorian botanists* cites from a speech delivered by A. J. Ewart which mentions that Latrobe, first Governor of Victoria,

tall's headquarters after he had ascended the Arkansas River in 1819—which was just across the eastern boundary of the Indian Territory,[2] in present Sebastian County, Arkansas. Still a third post established for the protection of the Indian Territory was Fort Towson at the junction of the Kiamichi River with Red River which, until 1848, formed the boundary between Mexican-Texas and the Indian country; the site was in present Choctaw County, Oklahoma. There can be no doubt that Pitcher's duties from 1831 to 1834 took him to all these posts.[3]

Novy includes a list of plants (prepared by C. A. Davis) all of which, for a time at least, bore the specific name *Pitcheri;* none of the localities of collections are more than general ones, but four were collected while he was stationed in the Indian Territory.

"*Carduus Pitcheri* (Torrey) Porter. Pitcher's Thistle. This plant is only found growing in the barren sands of beaches and sand dunes on the shores of Lakes Michigan, Huron and Superior. 'First found by Doctor Zina Pitcher on the great sand banks of Lake Superior.' Torrey in 'Eaton's Manual,' Fifth Edition, 1829.

"*Gaura Pitcheri* (Torrey and Gray), Small. Pitcher's Gaura. *Gaura biennis* var. *Pitcheri* (Torrey and Gray in 'Flora of North America'). 'Arkansas; Doctor Pitcher.'

". . . founded the Melbourne Botanic Gardens, selecting the site and educating public opinion on the subject. He took the warmest interest in the early development of the Garden, visiting it frequently . . . He appointed the first three Curators—Arthur, Dallachy, and Mueller. His interest was not merely of the official kind; he had a real knowledge of plants, exotic and native; some who knew him personally agree on this . . . He is commemorated by *Glycine Latrobeana* Benth."

The same speech records "a charmingly egotistic remark"—and certainly a backhanded compliment to Latrobe—made by a Mr. John G. Robertson in a letter to Sir William Hooker:

" 'With the exception of the late Mr. Robert Lawrence, Mr. Ronald Gunn, and our much-respected ex-Governor, Mr. Latrobe, I never met any individual resident who knew anything more about Australian plants than myself.' "

ANONYMOUS. 1892. Charles Joseph Latrobe, C. B. *Dict. Australasian Biog.* 269.

BRITTEN, JAMES & BOULGER, GEORGE SIMONDS. 1931. A biographical index of deceased British and Irish botanists. ed. 2. London. 180, 181.

CHITTENDEN, HIRAM MARTIN. 1902. The American fur trade of the far west. A history of the pioneer trading posts and the early fur companies of the Missouri valley and the Rocky Mountains and of the overland commerce with Santa Fe. 3 vols. Map. New York. 2: 803 [location of the Cross Timbers described].

FEATHERLY, HENRY IRA. 1943. The cavalcade of botanists in Oklahoma. *Proc. Oklahoma Acad. Sci.* 23: 10.

GHENT, WILLIAM JAMES. 1931. The early far west. A narrative outline 1540–1850. New York. Toronto, 229.

IRVING, WASHINGTON. 1883. A tour on the prairies. *In:* The works of . . . 4 (The Crayon Miscellany 1): 17-198. Author's rev. ed. New York.

LANE-POOLE, STANLEY. 1892. Charles Joseph Latrobe. *Dict. Nat. Biog.* 32: 182, 183.

LATROBE, CHARLES JOSEPH. 1836. The rambler in North America: 1832-33. ed. 2. 2 vols. London.

MAIDEN, JOSEPH HENRY. 1908. Records of Victorian botanists. *Victorian Naturalist* 25: 109, 110.

2. A region set apart by the United States government in the early 1830's for the use of certain Indian tribes formerly living east of the Mississippi River.

3. The locations of these and other army posts are shown on the map of "Indian Territory *and* the southern plains 1817–1860" in James Truslow Adams' *Atlas of American History,* p. 104 (1945)

"*Falcata Pitcheri* (Torrey and Gray), Kuntze. Pitcher's Hog Peanut, 'Red River, Arkansas;' Doctor Pitcher. Torrey and Gray in 'Flora of North America,' I, page 292.

"*Clematis Pitcheri,* Torrey and Gray, Pitcher's Clematis. 'On the Red River, Arkansas;' Doctor Pitcher. Torrey and Gray in 'Flora of North America,' I. page 10. This plant, by a revision of names,[4] is now known as *Clematis Simsii,* Sweet.

"*Arenaria patula,* Michx., Pitcher's Sandwort, (*Arenaria Pitcheri,* Nuttall): 'Plains of Arkansas;' Doctor Pitcher. Torrey and Gray in 'Flora of North America,' I, page 180."

Kelly comments that "It would seem almost reasonable to expect plants named for Zina Pitcher to be our familiar 'pitcher' plants, but a thistle, a peanut, a clematis and a sandwort commemorate the botanist."

According to Geiser, Pitcher "Collected fossils and plants in Texas along the Red River, opposite Fort Towson (1833). The fossils were sent to Dr. S. C. Morton of Philadelphia, who described them in his synopsis of Cretaceous fossils of the United States (1834) . . ."

And Novy mentions that "A fossil oyster, collected by him in Texas was named *Gryphæa Pitcheri.*"

Nuttall, from Fort Smith, had been the first botanist to enter what is now the state of Oklahoma, and when he journeyed to the Cimarron River he must have passed close to the later site of Fort Gibson. Although he had made his famous trip up the Arkansas River in 1819, it was fifteen years before he was able to undertake the task of describing the plants he then collected. According to Pennell, he had returned to Philadelphia from his post in Cambridge "about the middle of January," of 1834, and he was planning to leave on his trip to the west coast with Wyeth—which began in March. Pennell notes that he wrote, among various papers, one ". . . on the new and interesting specimens in the Academy's collections." Among these acquisitions were plants ". . . from Arkansas by Dr. Pitcher . . ."

Some of Pitcher's collections went to John Torrey at Columbia University; after Torrey's death his herbarium was deposited at the New York Botanical Garden when that institution was established in the 1890's. Dr. Rickett (*in litt.,* March 14, 1951) tells me that "Plants collected by Zina Pitcher . . . are in the Torrey Herbarium here." They may, in part at least, have been those mentioned by Torrey in a letter to his friend von Schweinitz, dated October 22, 1832, which is transcribed by Shear and Stevens:

"A few days ago I received a large collection of plants from the Arkansas country,—embracing about 300 species. Many of them are exceedingly curious and interesting—& not a few of them quite new. I have a few duplicates of the rarer species, which I will send to you . . . Some of the plants have been collected by Nuttall, in 1819 & 1820. These have mostly been described & sent to De Candolle for publication . . ."

Novy has this to say of Pitcher's interest in natural history:

4. Because of the priority rule.

"His leisure hours at the outposts were by no means wasted. Though an eager student of history and medicine, he found the opportunity to study nature as well. His writings show an extensive knowledge of the plants found along the entire frontier from Lake Superior to the Gulf. He was in touch with the leading men in botany and geology at that time, and the closeness of his observations may be gathered from the fact that several species of plants and fossils which he discovered received his name. He invariably cultivated the acquaintance of the Indian medicine men . . . and . . . acquired an intimate knowledge of Indian life, their medical customs, and learned to know the plants employed by the aborigines. The information thus acquired served him . . . in the preparation of an extended monograph on 'Indian Medicine,' which appeared in . . . Schoolcraft's great work." [5]

Life at the frontier posts had not been without privation and hardship but Novy states that it developed in Pitcher ". . . that sturdy self-reliance and that independence tempered with due regard for the authority of his superiors which characterized the man in later years." He quotes from the narrative of Dr. J. L. Whiting, friend of Pitcher, who in 1823 had been sent to Saginaw to attend a sick garrison:

" 'I found the whole garrison sick . . . and Doctor Zina Pitcher, the surgeon in charge, was the sickest of the lot. He was completely broken up. He had some one hundred twenty souls, old and young—sixty enlisted men, with officers, laundresses, and children—under charge, and all of them sick but one, with one of the most abominably distressing fevers imaginable. He was all alone, one hundred miles from anywhere, with an appalling amount of work on hand, and no wonder he broke down. When I reached Saginaw he was being carried all over the garrison on a mattress by men well enough as yet to move about or lift anything, giving opinions and advice, and a dreadful sight he presented, I can assure you.' "

Two years after Pitcher left Fort Gibson, or in 1836, he resigned from the army service ". . . although but three numbers removed from the head of the list of surgeons which would have soon brought him to the rank of surgeon-general . . ." He took up the practice of medicine in Detroit, ". . . then a city of about six thousand inhabitants and the capital of the new state of Michigan . . ." He played an important rôle ". . . in the history of the free schools of Detroit, in the history of the University, of the medical department and of the medical profession of the city and state." As a regent of the University he ". . . showed the greatest interest in getting together and personally arranging the collections in geology, zoology and botany." Novy records that, ". . . at one of the first meetings of the Board of Regents (1838) . . . Doctor Pitcher secured the appropriation of $970.00, an enormous sum at that time, for the

5. According to the "List of papers and addresses by Zina Pitcher" which is included in Novy's article, this was:

"1854.—Medicine; or Some Account of the Remedies Used by the American Indians in the Cure of Diseases, and the Treatment of Injuries, to which they are Liable, and Their Methods of Administering and Applying Them. Part IV, pages 502-519 of Schoolcraft's 'Information Respecting the History, Condition and Prospects of the Indian Tribes of the United States.['] Philadelphia: 1854."

purchase of one of the most prized works in our library today, namely Audubon's magnificent work on 'American Ornithology.' "

Novy lists thirty-nine papers published by Pitcher between 1832 and 1869; they are mainly on medical subjects.

Dr. E. D. Merrill tells me that he "would make a guess" that Pitcher had much to do with the selection of Asa Gray as professor of natural history at the newly chartered University of Michigan. In his "Autobiography. 1810–1843," which is included in the *Letters of Asa Gray,* Gray states that he had been appointed botanist of the United States Exploring Expedition in 1836, but resigned in favor of William Rich, ". . . having been about that time appointed professor of natural history in . . . the University of Michigan . . ." Gray left Michigan University when, "Some time in April [of 1842] I received a letter from President Quincy, telling me that the Corporation of the university [Harvard] would elect me Fisher professor of natural history if I would beforehand signify my acceptance . . . I came in July, in the midst of vacation, before Commencement, which was then in September . . ."

Although Zina Pitcher's contributions to botanical knowledge were undoubtedly of lesser importance than those which he made in the line of his profession, they were not without usefulness; for it was from a plant here and another there that the substance of the floras was compiled. Pitcher was but one of the group of extremely busy men who spent what free time they possessed upon the task of plant collecting.

BARNARD, FREDERICK AUGUSTUS PORTER. 1878. Representative men of Michigan. Cincinnati. 112, 113.

GEISER, SAMUEL WOOD. 1937. Naturalists of the frontier. Dallas. 331.

GRAY, ASA. 1893. Letters of Asa Gray. Edited by Jane Loring Gray. 2 vols. Boston. New York. 1: 21, 27.

HINSDALE, BURKE AARON. 1906. History of the University of Michigan . . . with biographical sketches of regents and members of the university senate from 1837 to 1906. Edited by Isaac N. Demmon. Ann Arbor. 30, 91-93, 174.

JACKSON, JOHN DAVIES. 1872. Report on American medical necrology. *Trans. Am. Med. Assoc.* 23: 591.

KELLY, HOWARD ATWOOD. 1929. Some American medical botanists commemorated in our botanical nomenclature. New York. London. 145-150.

NOVY, FREDERICK G. 1908. Zina Pitcher. *Physician & Surgeon* 30: 49-64.

PENNELL, FRANCIS WHITTIER. 1936. Travels and scientific collections of Thomas Nuttall. *Bartonia* 18: 34, 35, fn. 76.

PHALEN, JAMES M. 1934. Zina Pitcher. *Dict. Am. Biog.* 14: 636, 637.

SHEAR, CORNELIUS LOTT & STEVENS, NEIL EVERETT. 1921. The correspondence of Schweinitz and Torrey. *Mem. Torrey Bot. Club* 16: 277.

STUART, MORSE & BRODIE, WILLIAM. 1874. Memorial. *Trans. Mich. State Med. Soc.* ser. 2, 6: 241-248.

T., D. H. 1869. Analysis of the life of Dr. Zina Pitcher. *Richmond & Louisville Med. Jour.* 7: 685-689. *Reprinted: Mich. Univ. Med. Jour.* 3: 99-110, 1872-1873.

CHAPTER XXI

HOUGHTON ACCOMPANIES SCHOOLCRAFT TO THE
SOURCES OF THE MISSISSIPPI RIVER

I N an earlier chapter I have told of the expedition commanded by Lewis Cass, governor of the Michigan Territory, which attempted to reach the sources of the Mississippi River in 1821, and which had returned under the mistaken belief that Cass Lake (which they called "Cassina") and its two northern tributaries, La Biche (or "La Beish") and Turtle rivers, constituted its sources. Henry Rowe Schoolcraft had served as topographer and historian on that expedition and after his return had been appointed Indian Agent and had been stationed at Sault Sainte Marie. David Bates Douglass had made a small collection of plants.

In 1831, or ten years later, the United States government ordered another expedition to proceed, under command of Schoolcraft, ". . . into the Chippewa country, north-west of Lake Superior, in the execution of duties connected with Indian affairs." Because instructions arrived too late, the party did not get started until June of 1831. "In the mean time, means for more extensive observation were provided, a physician and botanist engaged . . ." and a detachment of troops was assigned to accompany the party. The "physician and botanist" was Douglass Houghton.

After the expedition had started another change of plans was necessary:

". . . the low state of the water on the Upper Mississippi, would render it difficult, if not impracticable, to reach the [Indian] bands at its sources, during the drought of summer. Public reasons were . . . urgent for visiting the interior bands, located between the groupe of islands at the head of Lake Superior, and the Mississippi[1]— where a useless and harassing conflict was kept up between the Sioux and Chippewa nations."

The country crossed by this expedition in 1831 lay east of the Mississippi River

1. Houghton evidently participated in a Michigan State survey concerned with botany. The reference is found in H. T. Darlington's "Taxonomic and ecological work on the higher plants of Michigan. Part I. A brief account of the Michigan flora and botanical survey." (*Michigan State College and Agric. Exper. Station. Tech. Bull.* 201, 6, 1945.) In a section on "Earlier botanical exploration in Michigan" Darlington notes:

"Apparently the first official report on the plant life of the state came out in 1839. This was the result of a botanical survey of the two southern tiers of counties. The expedition, organized by Douglass Houghton, was in immediate charge of John Wright, another physician and botanist . . . Wright had trouble caring for his specimens, and his progress across the forest wilderness was very slow . . . Another pioneer explorer of the period was William Burt, who often accompanied Houghton. Burt, though not a trained botanist, collected plants at various times during his work as a land surveyor. A list of these, identified by [Dennis] Cooley, was published in 1846 in Jackson's *Lake Superior* . . ."

Wright published upon the plants: Wright, John, M.D. 1839. "Catalogue of the phaenogamous and filicoid plants collected in the Geological Survey of Michigan." *Legislative Report* 27: 17-44. Detroit.

and is of interest to my story only because the plants collected by Douglass Houghton in 1831 were included with those which he acquired in 1832—or on the expedition to be discussed in this chapter. Whether a full account of the 1831 expedition was ever published I have not been able to discover: perhaps it was intended that Houghton should write one, for Schoolcraft states:

"A narrative of this expedition, embracing its principal incidents, and observations on the productions of the country, is in preparation for publication by one of the gentlemen of the party. In the mean time, the official report transmitted to the Government, and submitted to Congress by the War Department . . . are subjoined in the appendix to this volume."

The "volume" just mentioned was published in New York in 1834, under the title: *Narrative of an expedition through the upper Mississippi to Itasca Lake, the actual source of this river; embracing an exploratory trip through the St. Croix and Burntwood (or Broule) Rivers: in 1832. Under the direction of Henry R. Schoolcraft.* The included map bears the caption "Sketch of the sources of the Mississippi River, drawn from Lieut. Allen's observations in 1832, to illustrate Schoolcraft's inland journey to Itasca Lake."

I shall transcribe the journey from the Schoolcraft *Narrative*. After leaving Leech Lake the route is somewhat difficult to follow at times.

1832

"Early in 1832, the plan of visiting the source of the Mississippi was resumed." On June 7 the party left "St. Mary's." From that point to Cass Lake Schoolcraft merely gives a "brief sketch" of the route, having described it fully in his *Narrative journal* of 1821; after leaving Lake Superior it lay through the present state of Minnesota.

Although one objective of the journey was to find the headwaters of the Mississippi, another objective, and perhaps a more pressing one, was to make peace between the Sioux and the Chippewas. The *Narrative* contains stories of Indian unrest—increased it would seem by the existing rivalry between British and American traders. The Indians were confused. They had first traded with the French. Later, "The War of 1812, found all the northern tribes confederated with the English . . ." Still later, ". . . In 1816, a law was passed by Congress to exclude foreigners from the trade." This was followed by the establishment of American frontier posts: in 1819 at St. Peter's, in 1822 at Sault Sainte Marie, and so on. The North West Company[2] had transferred to the Astor Company, after the determination of the boundary between Canada and the

2. The North West Company, an amalgamation of Canadian fur-trade interests, had, over the years, become a powerful competitor of the older Hudson's Bay Company and Schoolcraft comments that it ". . . waged one of the most spirited and hard contested oppositions against the Hudson's Bay Company, which has ever characterized a commercial rivalry . . ." In 1821 the two companies were merged under the name of the older company.

United States, all their posts "south and west of the lines of demarkation" but they maintained an active trade "along the lines." One of the best means of procuring the allegiance of the Indians was to sell them what Schoolcraft always speaks of as "ardent spirits." The Indian Office was determined to stop this practice and in this had the fullest coöperation possible from Schoolcraft.

At Cass Lake a "select party" was organized to explore the sources of the Mississippi; Indians were engaged as guides, as well as seven "engagés and a cook." Schoolcraft was in command and other important members were

"Lieut. J. Allen . . . Doct. Douglass Houghton, the Reverend Wm. Boutwell [sometimes spelled Boutuell], Mr. George Johnston . . . sixteen persons. These, with their travelling beds, were distributed among five canoes, with provisions for ten days . . . The detachment of infantry was left in our encampment."

Cited as part of the equipment was "an herbarium"!

Johnston was attached to the Indian Office and was familiar with Indian languages. We find him translating Boutwell's sermons to the savages. James Allen played an important rôle. According to G. K. Warren, writing in 1859, "To Lieutenant Allen we are indebted for the first topographical and hydrographical delineation of the source of the Mississippi; and this, somewhat improved by Mr. Nicollet, is our authority to the present day for the Mississippi above the mouth of Swan river . . ."

Warren quotes from Allen's report to Congress (a body which is not pictured in a very generous light):

"I was not furnished with . . . any instruments by which to fix, from astronomical observations, the true geographical positions of points necessary to be known for the construction of an accurate map . . . a compass, the only instrument I had, was placed in my canoe, where it was constantly under my eye; and, as the canoe proceeded in the line of a river, I carried my observations from the compass to a field-book at every bend or change of direction, thus delineating . . . all the bends of the river precisely as they occurred; and by establishing a scale of proportions in the lengths of the reaches, I was . . . enabled to lay down and preserve the general curve of a river with surprising accuracy, as was tested afterwards . . . On the portion of the Mississippi above Cass Lake . . . I bestowed on the tracing and computing of distances the most unremitted attention."

Allen had attempted to procure the proper instruments at Fort Brady before starting, and doubtless at his own expense, but in vain.

The *Narrative,* unfortunately, tells nothing of Douglass Houghton in his capacity of botanist. One of his important duties during the trip was to vaccinate the Indians against smallpox. His report upon the subject is included in the appendix of the *Narrative.* He is, of course, best known as a geologist.

From the time of the party's arrival at Cass Lake—bisected by the present boundary between Beltrami County (north) and Cass County (south)—on July 10, 1832, Schoolcraft's story assumes "a more personal character."

On July 10, leaving the mouth of Turtle River[3] on the "right [north]," and proceeding across Cass Lake, the party landed on "Grand Island, or Colcaspi," which on
the map occupies a large part of the southwestern end of the lake. Here the "select
party" already mentioned was organized. On the 11th it started westward, through a
strait between Grand Island and the lake shore (passing "Garden" and "Elm" Islands) and started up the channel of the Mississippi which, near its entrance into Cass
Lake, broadened into two small lakes. Beyond the second, "Tascodiac," were "Hills
of sand, covered with yellow pines . . . and the river exhibits for several miles above,
either a sand bank, or a savannah border. Time is the only measure of distance, which
we had the means of referring to . . . fifteen miles above Cass Lake, the meadow lands
cease." After passing some ten rapids in the next twenty to twenty-five miles, they
reached ". . . the most northern point of the Mississippi, which is marked by the fine
expanse of . . . Lac Traverse . . ." This appears to have been Lake Bemidji of recent
maps, in southern Beltrami County. A landing was made. "The soil . . . appeared to
be rather rich, bearing a growth of elm, soft maple and white ash."

The party now moved southward, passing out of the lake by a short channel into a
small bay or lake, called on the map "Lake Irving." Here ". . . the Mississippi, which
has now been pursued to its utmost northing, is ascended directly south. About four
miles above this bay, the Mississippi has its ultimate forks, being formed of an east
and west branch . . . the west branch is decidedly the largest, and considerably the longest."

The guide advised following first the east branch (designated on the map as the
"Plantagenian or south fork") which the party did, to its source, and from there portaged to the west fork (the "Itascan or west fork of the Mississippi") which they descended to the forks—making a loop as it were. The *Narrative* notes that none of the
earlier maps ". . . indicate the ultimate separation of the Mississippi, above Cass Lake,
into two forks." The expedition now started up the east fork.

They crossed two lakes or river-expansions ("Marquette" and "La Salle" on the
map). Leaving the last of these, the stream was ". . . strikingly diminished in volume,
with a limited depth, and a vegetation of a more decidedly alpine character." Four
miles above was a larger lake called "Kubbakunna, The Rest in the Path," which on
the Schoolcraft map, and on some recent ones is called Lake Plantagenet and is situated in Hubbard County. It presented ". . . a pleasing aspect, after the sombre vegetation, we had passed below. Rushes . . . were abundant towards its head . . . the ground
too low and wet for camping." Higher up they put ashore.

"The soil . . . appeared to be of the most frigid character. A carpet of moss covered
it, which the foot sank deep into, at every step. The growth was exclusively small grey
pine, with numerous dead branches below, and strikingly festooned with flowing moss.

3. The map of Schoolcraft's *Narrative* indicates that the mouth of Turtle River was the "Ultimate point
reached by Govr. Cass in 1820." Schoolcraft explains that in 1820 water had been too low to ascend either
Turtle or "La Beish" rivers.

Nearer the margin of the river, alder, tamarack, and willow, occupied the soil . . . Notices of the natural history of the country, during this day's journey are meagre. The principal growth of forest trees, out of the immediate valley, is pine. The plants appear to present little variety, and consist of species peculiar to moist, cold, or elevated situations."

On July 12, above Lake Plantagenet, the river resembled ". . . a thread wound across a savannah valley. A species of coarse marsh-land grass, covers the valley. Clumps of willow fringe this stream. Rushes and Indian reed are gathered in spots most favorable to their growth. The eye searches in vain, for much novelty in the vegetation. Wherever the stream touches the solid land, grey pine, and tamarack are conspicuous, and clumps of alder here take the place of willow. Moss attaches itself to almost every thing . . . there is a degree of dampness and obscurity in the forest, which is almost peculiar to the region."

They then passed a number of rapids and a tributary stream called the "Naiwa" (on some recent maps spelled Niawa), entering from the southeast. On July 13 the channel above this tributary was ". . . diminished to a clever[4] brook, more decidedly marshy in the character of its shores, but not presenting in its plants or trees, any thing . . . to distinguish it from the contiguous lower parts of the stream . . . It presents some small areas of wild rice."

At the foot of "Ossowa" (on the map "Usawa") Lake, the party halted briefly. "It exhibits a broad border of aquatic plants, with somewhat blackish waters . . . It is the recipient of two brooks, and may be regarded as the source of this fork of the Mississippi . . . We entered one of the brooks, the most southerly . . . It possessed no current and was filled with broad leaved plants, and a kind of yellow pond-lily." Portaging along this brook, they reached its source: ". . . the prevailing growth . . . is spruce, white cedar, tamarack, and grey pine . . ."

Having reached the source of the east fork of the Mississippi, the party portaged for six miles to the west fork, or to Itasca Lake; in so doing it had crossed from Hubbard County into southeastern Clearwater County, where Lake Itasca is situated, or ". . . from the east to the west branch of the river . . . Beginning in a marsh, it soon rises into a little elevation of white cedar wood, then plunges into the intricacies of a swamp matted with fallen trees, obscured with moss. From this the path emerges upon dry ground . . . soon ascends an elevation of oceanic sand . . . bearing pines . . . another descent . . . another elevation. In short . . . a series of deluvial sand ridges, which form the height of land between the Mississippi Valley and Red River . . . locally denominated Hauteur des Terres . . . It is . . . the table land between the waters of Hudson's Bay and the Mexican Gulf . . . probably, the highest in Northwestern America, in this longitude."

Thirteen times during the course of the portage the party stopped to rest.

"Ripe straw berries were brought to me, by the men, at one of the resting places. I

4. Probably in the old sense of active or nimble.

observed a very diminutive species of the raspberry, with fruit, on the moist grounds. Botanists would probably deem the plants few, and destitute of much interest. Parasitic moss is very common on the forest trees, and it communicates a peculiar aspect to the grey pine, which is the prevailing growth on all the elevations.

"Every step we made . . . seemed to increase the ardor with which we were carried forward . . . What had been long sought, at last appeared suddenly. On turning out of a thicket . . . the cheering sight of a transparent body of water burst upon our view. It was Itasca Lake—the source of the Mississippi."

The length of the Mississippi River, writes Schoolcraft, ". . . assuming former data as the basis, and computing it, through the Itascan, or west fork, may be placed at 3160 miles,[5] one hundred and eighty-two of which, comprises an estimate of its length above Cass Lake." We are told that "Itasca[6] Lake, the *Lac La Biche* of the French, is . . . a beautiful sheet of water, seven or eight miles in extent, lying among hills . . . surmounted with pines . . . The waters are transparent and bright, and reflect . . . the elm, lynn, maple, and cherry . . . with other species more abundant in northern latitudes."

On the lake's single island, states Schoolcraft, grew ". . . the forest trees above named . . . promiscuously with the betula and spruce."

"Having gratified our curiosity in Itasca Lake, we prepared to leave the island, but did not feel inclined to quit the scene without leaving some memorial, however frail, of our visit. The men were directed to fell a few trees at the head of the island . . . creating an area, for the purpose of erecting a flag staff. This was braced by forked stakes, and a small flag hoisted to its place. Taking specimens of the forest growth of the island, of a size suitable for walking canes, and adding a few species to our collections of plants and conchology, we embarked on our descent. The flag which we had erected continued to be in sight for a time, and was finally shut out from our view by a curve of the lake." [7]

On July 13 Schoolcraft and his party left Lake Itasca and started on the return trip down the west fork which, as it left the lake, was only about ten feet wide, the channel winding, shallow and full of rapids and small falls. After some twelve miles it ". . . displays itself . . . in a savannah valley, where the channel is wider and deeper . . . circuitous, and bordered with sedge and aquatic plants. This forms the first plateau. It ex-

5. The distance from Itasca Lake to the mouth of the Mississippi is now estimated to be about 2,550 miles, or slightly less.

6. George Bryce reports the origin of the name to have been this:
". . . Schoolcraft . . . asked Boutwell the Greek and Latin names for the headwaters or true source of a river. Mr. Boutwell could not recall the Greek, but gave the two Latin words,—*veritas* (truth) and *caput* (head). These were written on a slip of paper, and Mr. Schoolcraft struck out the first and last three letters, and announced to Boutwell that 'Itasca shall be the name' . . . Boutwell . . . always maintained his story of the name, and this is supported by the fact that the word was never heard in the Ojibeway mythology."

7. According to the volume *Minnesota* ("American Guide Series"), p. 309, Itasca State Park includes within its boundaries ". . . Lakes Itasca, Elk, and Hernando de Soto, each bordered by magnificent virgin pines. Established in 1891, it has since been enlarged to nearly 32,000 acres . . ."

tends eight or nine miles." After traveling through an "almost continued series of rapids," they made camp. "Wearied with the continued exertion, the frequent wettings, and the constant anxiety, sleep soon overshadowed the whole party . . ."

There were more rapids on July 14 and in one, Allen's canoe was ". . . reversed, with all its contents . . . The canoe-compass was irrevocably lost. He fetched up his fowling piece himself. Other articles went over the falls.

"The character of the stream, made this part of our route a most rapid one. Willing or unwilling we were hurried on . . . Less time was given to the examination of objects than might otherwise have been devoted. Yet I am not sure that any important object was neglected . . . who ever has devoted his time in going thus far up the Mississippi, will have made himself so familiar with its plants, soil and productions, that 'he who runs may read.' The pine, in its varieties, is the prevailing tree . . . out of the narrow alluvions of the valley, arenaceous plains appear. Among the plants that border the river, the wild rose . . . is very often seen. The salix . . . presents itself on the first plateau, and is afterwards one of the constant shrubs on the savannahs. The Indian reed . . . is here associated with wild rice."

The river continued "to descend in steps." About one hundred and four miles below Lake Itasca the "Piniddiwin" entered it from "the left." Although it had the appearance of a marsh, its name was "Lac la Folle." On July 15, after traveling all night, the party passed "Travers," or Bemidji, Lake at daylight and by nine a.m. reached the permanent encampment on Cass Lake, having been in the canoes more than eleven hours.

"The day being the Sabbath, the Reverend Mr. Boutwell, devoted a part of it, as he had done on the previous Sabbaths of our route, in giving religious instruction.

"Mr. Johnston's readiness in scripture translation, put it in the power of Mr. B. to address them [the Indians] on the leading doctrines of the gospel. With what effects . . . cannot be fully stated. Strict attention appeared to be paid by the Indians . . . Incredulity and bold cavilings were more observable, I think, at the most remote points of our route; and most interest manifested . . . in the villages situated nearest the frontier posts . . . it is to be hoped that no circumstances will prevent Mr. B. from communicating his observations to the christian public at an early period."

After a friendly council with the Chippewas, the expedition moved on on July 16 to "Pike's Bay" at the southern end of Cass Lake and then started for Leech Lake, situated in northwestern Cass County. The trip began with a portage, ". . . on the edge of an open pine forest, interspersed with shrub oak. The path is deeply worn, and looks as if it might have been used by the Indians, for centuries. It lies across a plain presenting the usual aridity of similar formations, and exhibiting the usual growth of underbrush and shrubbery. I observed the alum root, harebell and sweet fern, scattered through the more prevalent growth of wortle berry, L.[8] latifolia, &c. . . ."

8. No genus is cited. The common name "wortle," or whortle, berry is often applied to the *Vaccinium* (blueberry) or *Gaylussacia* (huckleberry), but "L." stands for neither.

In crossing the small "Moss Lake," the men ". . . brought up on their paddles, a species of moss of a coarse fibrous character . . . quite a characteristic trait of the lake." From this lake to "Lake Shiba" [9] was "an unvaried sand plain . . . sometimes brushy . . ." From the south end of Lake "Shiba" they portaged to an arm of Leech Lake, crossed it and camped on its eastern shore and near its southern outlet which was ". . . so tangled and wound about in a shaking savannah, covered with sedge, that every point of the compass seemed to be alternately pursued."

According to Schoolcraft, "Leech Lake is one of the most irregular shaped bodies of water that can be conceived of." A subjoined map, made from "the notes of Lieut. Allen," bears out this statement. The party did not explore the lake and Allen's data was obtained from the Indians. July 17 was spent at the lake, conferring with the Chippewa tribe who were, as always, having trouble with the Sioux. Schoolcraft comments upon the tribes of this region:

"There is a remarkable conformity in the external habits of all our northern Indians. The necessity of changing their camps often, to procure game or fish, the want of domesticated animals, the general dependance on wild rice, and the custom of journeying in canoes, has produced a general uniformity of life . . . emphatically a life of want and vicissitude. There is a perpetual change between action and inanity, in the mind, which is a striking peculiarity of the savage state. And there is such a general want of forecast, that most of their misfortunes and hardships, in war and peace, come unexpectedly . . . The only marked alteration which their state of society has undergone, appears to be referable to the era of the introduction of the fur trade, when they were made acquainted with, and adopted the use of, iron, gunpowder, and woollens . . . But it brought with it the onorous evil of intemperance, and it left the mental habits essentially unchanged . . ."

One exception to the above was ". . . the Guelle Plat . . . ruler of the Pillager band . . . the sentiments he uttered on the Sioux war, the fur trade, and the location of trading posts and agencies, were such as would occur to a mind which had possessed itself of facts, and was capable of reasoning from them." There was a feast, with speeches. "Many applications were made for the extraction of . . . teeth . . . for blood letting . . . a favorite remedy among the northern Indians." Much of Douglass Houghton's time was ". . . employed in the application of the vaccine virus, which constituted one of the primary objects of the visit . . . Little difficulty was found in getting them to submit to the process. The ravages of the small pox in this quarter, about the year 1782, were remembered . . ." The account of the Indian council ends with a description of the Guelle, or Guele, Plat's appearance as he came down to the shore to see the expedition depart. He was ". . . dressed in a blue military frock coat, with red collars and cuffs, with white underclothes, a linen ruffled shirt, shoes and stockings, and a neat citi-

9. The name, we are told, was "composed of the initials of the names of the gentlemen of the party."

zen's hat . . . this foreign costume . . . he knew we would consider . . . a mark of respect[1] . . ."

The party moved slightly farther down the inlet to Leech Lake where they could get fire wood and make an early start the next day. Allen's command and Johnston remained behind some hours longer—Houghton too, who was still busy giving vaccinations.

Having completed his investigation of the sources of the Mississippi River and having held a council with the Indians of the region, Schoolcraft was now to start his return journey, from Leech Lake crossing overland to the sources of the "De Corbeau," or Crow Wing River, and descending that stream to its entry into the Mississippi. Crow Wing River drains parts of Wadena and Cass counties and enters the Mississippi between Cass County (north) and Morrison County (south).

Leaving Leech Lake on July 18, there were numerous portages between a series of small lakes: "This day's journey was a hard and fatiguing one, to the men." [2] Camp was ". . . on the source of the Kagági, or De Corbeau river."

"We had now reached the *fourth* source of the primary rivers of the Mississippi, and all heading on the elevation of the Hauteur des Terres, within a circle of perhaps seventy miles. These sources are Itasca Lake, its primary, Ossawa, east fork, Shiba Lake and river, source of Leech Lake, and the primary source, The Long Water, being the source of the De Corbeau, or Crow-wing river."

They started at sunrise[3] of July 19 and passed through "the Kagi Nogumaug," source of the De Corbeau, ". . . a handsome sheet of pure water presenting a succession of sylvan scenery. Its outlet is a narrow brook overhung with alders. It may average a width of six feet . . ." Next through another lake, the "Little Vermilion," its banks covered with ". . . birch and aspen, with pines in the distance . . . Tamarack is a frequent tree on the shores, and the pond lilly, flag and Indian reed, appear in the stream." They came next to "Birch Lake," then to "Lac Ple," with part of its shores ". . . prairie, interspersed with small pines."

The map shows a chain of small lakes joined by the De Corbeau River. I shall not name them all. Many do not appear on modern maps, certainly not by the appellations

1. The chief had received his first medal from the British at Fort William, on Lake Superior, but it had been taken from him by Pike in 1806. Schoolcraft ". . . renewed it, by the largest class of solid silver medals, July 19th, 1828."

2. Nor could the night have been restful for Schoolcraft! They had no sooner encamped before they were overtaken by the Guelle Plat who expressed surprise that they had progressed so far. He had brought with him ". . . his companion and pipelighter . . . a tall, gaunt, and savage looking warrior . . . the murderer of Gov. Semple." After describing the horrible murder, Schoolcraft comments that although the man appeared to be ". . . quiet and passive . . . I could not divest my mind of the recollection . . ."

3. Before departure Schoolcraft presented the Guelle Plat with a lancet in order that he might continue the vaccinations himself!

supplied in the *Narrative*. Some, perhaps, do not even exist at the present day for, although described as lakes, they were doubtless of a temporary nature[4] and, in the course of more than one hundred years, may have altered in appearance and character. To what extent this is true could be ascertained only by a careful, on-the-spot survey of Schoolcraft's route. I name a few which, to one familiar with the region, may still be identifiable. Some, as was customary on all exploring expeditions, were named for members of the party. One (in the chain of lakes depicted on the map) was "Allen's Lake," after the topographer who had visited a tributary stream at the lake's northern end; another, "Johnston's Lake"—Johnston had landed on its shore "to fire at a deer"! It had ". . . handsomely elevated hard wood and pine shores . . ." An hour beyond they came to the entrance of the eleventh, ". . . and last lake of the series called Kaichibo Sagitowa, or the Lake which the River passes through one End of, or Lake Leelina. Not many miles below this point, the river forms its first forks, by the junction of Shell river, a considerable stream . . ." This is shown on some maps as an upper tributary of Crow Wing River. The expedition camped about fourteen miles below the forks. There had been rain and everything was wet; Schoolcraft paid tribute to the cheer and efficiency of his crew of voyageurs who, after ". . . submitting to severe labor, both of the night and day, on land and water . . . are not only ready for further efforts, but will make them under the enlivening influence of a song."

July 20: "The ensuing portion of our voyage down this stream, occupied a day and a half, during which we probably descended a hundred and twenty miles. Its general course, from the forks, is south-east. It is swelled by two principal tributaries from the west, called Leaf [entering Crow Wing River in Wadena County] and Long Prairie [entering the same near the city of Motley, Morrison County] rivers. . . . On the other [or northern] bank, it is joined by the Kioshk, or Gull river . . . of inferior size."

Gull River enters Crow Wing River not far west of its confluence with the Mississippi. Schoolcraft describes Crow Wing River as ". . . a stream of noble size . . . There is no part of it, which can be called still water; much of it is rapid. For about seventy miles below the junction of Shell River, there is a regular series of distinct rapids . . . Below the junction of Leaf River, this characteristic becomes less noticeable, and it disappears entirely, below the entrance of the Long Leaf branch. Its banks are elevated, presenting to the eye, a succession of pine forests, on the one hand, and an alluvial bend, bearing elms and soft maple, on the other. There is a small willow island . . . and several small elm islands . . . but nothing at all comparable, in size, soil and

4. The volume *Minnesota* ("American Guide Series"), p. 19, writing of the state's "Flora," comments that ". . . Most lakes and ponds have heavy growths of aquatic and subaquatic plants . . . All of these plants tend to fill the lake and finally, when a drought comes and the water level falls, another lake has been obliterated. So gradual is this transition that only the most observant realize that it is the fate of all these lakes to disappear as have those of the older glacial-drift area of the southern part of the State. During its transition a lake may survive as a swamp filled with tamarack, or become a sedge-covered meadow, and eventually a wet prairie."

timber, with the large and noted island, called *Isle de Corbeau* [Crow Wing Island], which marks its junction with the Mississippi . . .

"The forest of this fork of the Mississippi, abounds in almost every variety of the pine family. We observed the sugar maple less frequently on our whole route, than would be inferred from the knowledge, that this tree is spread over the sources of the Mississippi, and flourishes, even in its most northern latitudes . . . perhaps . . . river alluvions, and low grounds generally, are unfavorable to its growth. Its true position is the uplands . . . other species of the maple, frequently exhibited their soft foliage, over the stream, together with the elm, and the ash, and some varieties of the oak. Pine is, however, by far the most abundant and valuable timber tree, disclosed along the immediate banks of this river . . ."

It was on July 21 that the party had again reached the Mississippi River: "On issuing out of the upper channel [that is, above the Island of "De Corbeau," or Crow Wing], and entering the broad current of the majestic Mississippi, we beheld the opposite shore lined with Indian lodges, with the American flag conspicuously displayed."

The map of the *Narrative* pictures an Indian tepee surmounted by the flag, with the caption "Council 21st July." At this council, held the day of the expedition's arrival, the spokesman for the Chippewas (the same Guelle Plat) seems to have shown considerable common sense:

"The Grosse Guele, observed that, as the line was a question between the Chippewas and Sioux, a firm peace could never exist, until the line was surveyed and marked, so that each party could see where it ran . . . The Sioux, were in the habit of trespassing . . . when their own [the Chippewas'] hunters went out . . . they did not like to stop short of the game, and they saw no marked line to stop them . . . it had been promised at the treaty at Prairie du Chien . . . he wished . . . [Schoolcraft] to convey his words on the subject to the President . . ."

The speech ended with the hope that ". . . the same advice given to the Chippewas, would be given to the Sioux. If the Sioux would not *come* over the lines, they . . . would not *go* over them . . ."

After the council the expedition proceeded eighteen miles down the Mississippi to ". . . the trading house of a Mr. Baker . . ." which was situated on the eastern shore of the river at "Prairie Piercee," or, on the map, "Pierced Prairie." The *Narrative* ends at this point, Schoolcraft explaining that at the mouth of the De Corbeau River his route ". . . intersects that of the expedition to the sources of the Mississippi, under the direction of the present Secretary of War, Gov. Cass, in 1820."

The expeditions commanded by Schoolcraft seem always to have been accomplished smoothly and successfully. He had two hard and fast rules: no alcohol and a day of rest on the Sabbath. The expedition just described, he notes, ". . . has been accomplished, from beginning to end, without the use of so much as a drop of ardent spirits, of any kind, either by the men upon whom the fatigues of the labor fell, or by the gentlemen who composed the exploring party . . . No Sabbath day was employed in travelling. It was laid down as a principle, to rest on that day . . ." Rather than increasing the time ". . . employed on a public expedition in a very remote region . . . the result was far otherwise . . . an equal space had been gone over, in less time, than it had ever been known to be performed, by loaded canoes, or . . . by light canoes, before."

From De Corbeau River the expedition descended the Mississippi as far as the St. Croix River which, a short distance below the Falls of St. Anthony, enters from the northeast. After ascending this stream to its source in Upper Lake St. Croix, a portage was made to the "Bois Brulé River" (entering Lake Superior to the east of Fond du Lac Bay). This trip—all east of the Mississippi—is described in Schoolcraft's "Exploratory trip through the St. Croix and Burntwood (or Brulé) Rivers," which forms part of the same volume as his *Narrative*.

Part three of the appendix of Schoolcraft's *Narrative* is entitled "Localities of plants collected in the northwestern expeditions of 1831 and 1832. By Douglass Houghton, M.D., Surgeon to the expedition." A foreword tells that "The localities of the following plants are transcribed from a catalogue kept during the progress of the expedition, and embrace many plants common to our country, which were collected barely for the purpose of comparison. A more detailed account will be published at some future day."

Letters quoted by Rodgers in his biography of John Torrey (pp. 101-104) indicate that Torrey had been in touch with Schoolcraft before the expedition of 1832 got under way and that the explorer had offered to send him ". . . 'the botanical rarities' . . ." collected on the journey and to let him ". . . 'examine the collections made by Dr. Houghton . . .' " during the last expedition—evidently the one of 1831. Houghton wrote Torrey of what he had acquired on that occasion and of his expectation that the journey of 1832 would produce more valuable plants; I quote a portion of the letter (dated March 20, 1832) as transcribed by Rodgers:

" 'In the expedition referred to I acted as naturalist and have now the entire collection of plants, as well as parts of the other collections in my possession. You are undoubtedly well aware of the numerous difficulties which are presented in preserving and securing plants during a long and tedious canoe trip. With the utmost care, I was unable to preserve many of my duplicate specimens, & others were entirely lost, or much injured . . .

" 'I am attached to an expedition for the ensuing summer, in the double capacity of Surgeon and Naturalist, which bids fair to be of far greater interest in a botanical point of view than that of last summer . . .' "

When the expedition had returned, Torrey wrote Schoolcraft (October 5, 1832): " 'Dr. H[oughton] had not arranged his collections, but he expected to have them in order before many days, & he promised me a sight of them as well as a portion of the duplicates . . . You must send all your queer & doubtful things to him . . .' "

This would certainly suggest that Houghton did not send his entire collection to Torrey for determination. And the fact that, in publishing his plants—a mere list of names and localities—Houghton expressed no indebtedness to Torrey also suggests that he made the determinations himself. However, Rodgers states that ". . . While the list was published under Houghton's name, obviously Torrey did much of the work of examining the 1831 and 1832 plants."

Rodgers ends his comments upon the collections made on the 1831 and 1832 trips by stating that Torrey wrote Schoolcraft (the date is not cited) that

"Dr. Houghton had sent 'the rarer plants collected in the last Exped[ition],' and he had furnished Torrey with his remarks on them; but Schoolcraft's boxes and specimens had not arrived. 'You[r] boxes & packages of specimens must have been detained on the way by the closing of the Canal,' Torrey explained. Botany sent from the West came to the East in those days by the Erie Canal which was closed in winter, when frozen."

Brendel states that Houghton ". . . collected about 250 species of plants, only a few of which were new." I am informed by Dr. Rickett (*in litt.,* March 14, 1951) that "Plants collected by . . . Douglass Houghton . . . are in the Torrey Herbarium here [the New York Botanical Garden]."

I find no record that the ". . . more detailed account . . ." of the collections, which Houghton's foreword said would be published at some later time, ever appeared.

Biographies of both Schoolcraft and Houghton are to be found in the *Dictionary of American biography,* and the Michigan Historical Society published in 1904 an article upon Houghton which was written by I. C. Russell. Facts about both men are, therefore, readily available. They become living personalities, however, in Walter Havighurst's story of the Great Lakes, *The long ships passing.* I quote a small part of his pictures of both men.

Schoolcraft lived at Sault Sainte Marie and ". . . from 1822-41 held the semi-diplomatic office of Indian Agent on the Northwest Frontiers." He was an explorer, a student and a scientist—famed both as geologist[5] and ethnologist.

". . . His repute in scholarly circles was so established that a diploma of membership in the Royal Geographic Society of London was addressed to 'Henry Rowe Schoolcraft, United States of America.' The diploma was missent to St. Mary's, Georgia, and only after months did it find its way to the cold St. Mary's fronting the Canadian hills . . .

"He spent years issuing rations of bacon and beans and tobacco, and making gifts of axes, traps and kettles . . . They [the Indians] came to him . . . begging, wheedling, gossiping, quarreling, in grief and anger and need, asking him to satisfy their hunger and settle their disputes. Finally he refused to see them on the Sabbath, and refused

5. Goode accords Schoolcraft the distinction of having printed, in 1829, ". . . his 'Transallegania,' a mineralogical poem, probably the last as well as the first of its kind written in America . . ."

to see any intoxicated Indian, explaining that 'the President has sent me to speak to *sober* men only.' With patience and ardor he came to know them and their country . . .

". . . He worked for years on a lexicon of their [the Chippewa] language, and on his *Algic Researches . . .*

". . . A Cambridge poet found the riches of the *Algic Researches* and made a song that all Americans could cherish."

Havighurst tells of the names left in the Lake Superior country by the French Jesuits, by the voyageurs ("the . . . picturesque and living beside the staid names from English nobility . . .") and by the Indians:

"Schoolcraft loved those names . . . He served on a commission that gave Indian names to Michigan's northern counties. But Charles O'Malley of the state legislature, having a quarrel with Schoolcraft, attacked him by changing Indian names to Irish, and so a tier of counties bear the names of Clare, Roscommon, Emmet, and Antrim. It was a blow to Schoolcraft, and a loss to the map of Michigan . . ."

In 1833 Schoolcraft moved to Mackinac Island, and left the region of the Great Lakes permanently in 1841.

Turning to Havighurst's picture of Douglass Houghton: he ". . . stood just five feet four inches in his explorer's boots, but he cast a shadow over all of Lake Superior. His strength was meagre and his health was frail throughout the thirty-four years of his life. But he cruised five times the wild southern shore of Lake Superior in a birch canoe and he mapped the mineral resources of a region that was to have an epic history. A scholar more than an empire builder, he foresaw the empire that was coming. Despite exposure, illness, rheumatism, he led an arduous life and kept an ardent spirit . . .

"He was twenty-one when he came to Detroit in 1830 as a teacher of science . . . Ten years later he was . . . mayor and had declined the presidency of the University of Michigan . . .

"Following [the journey of 1832 with Schoolcraft] . . . he persuaded the legislature of the young state to appropriate funds for a geological survey. In fact he was appointed State Geologist, and in 1837 began extensive surveys of both the upper and lower peninsulas of Michigan . . . What he found drew men to that land like a magnet . . . So the first towns appeared on Lake Superior . . . copper towns . . ."

After mentioning the fortunes made in such ventures as the Calumet Mining Company, Havighurst describes the perils endured by men working in the field. Douglass, returning from an exploring expedition, was caught in a blizzard as he coasted the shore of the lake in a mackinaw boat.

". . . 172 inches of snow have been recorded in that country. The average over many years was 114 inches . . .

"The copper country was opened, but not without struggle and hardship. It took the life of the man who uncovered its treasure.

"There is a grim saying that Lake Superior never gives up its dead. But in the spring of 1846, while Little Silver River ran brimful of snow water and the moccasin flowers were bright in the woods, on the murmuring shores of Keweenaw was found the body of Douglass Houghton."

BRENDEL, FREDERICK. 1879. Historical sketch of the science of botany in North America from 1635 to 1840. *Am. Naturalist* 13: 768.

BRYCE, GEORGE. 1910. The remarkable history of the Hudson's Bay Company including that of the French traders of north-western Canada and of the North-West, XY, and Astor Fur Companies. ed. 3. Toronto.

GOODE, GEORGE BROWN. 1888. The beginnings of American science. The third century. *Proc. Biol. Soc. Washington* 4: 53.

HAVIGHURST, WALTER. 1942. The long ships passing. The story of the Great Lakes. New York. 43-52, 167-169.

HOUGH, WALTER. 1935. Henry Rowe Schoolcraft. *Dict. Am. Biog.* 16: 456, 457.

HOUGHTON, DOUGLASS. 1834. Localities of plants collected in the northwestern expeditions of 1831 and 1832. By Douglass Houghton, M.D. Surgeon to the expedition. *In:* Schoolcraft, H. R. 1834. Narrative of an expedition ... Appendix 1, No. 3, 160-165.

MERRILL, GEORGE P. 1932. Douglass Houghton. *Dict. Am. Biog.* 9: 254-255.

RODGERS, ANDREW DENNY, 3rd. 1942. John Torrey. A story of North American botany. Princeton. 102, 103, 104.

RUSSELL, I. C. 1904. Douglass Houghton. *Fourth Rep. Mich. Acad. Sci.* 160-162.

SCHOOLCRAFT, HENRY ROWE. 1821. Narrative journal of travels, through the northwestern regions of the United States extending from Detroit through the great chain of American lakes to the sources of the Mississippi River. Performed as a member of the expedition under Governor Cass. In the year 1820. By Henry Rowe Schoolcraft ... Map. Albany.

—— 1834. Narrative of an expedition through the upper Mississippi to Itasca Lake, the actual source of this river; embracing an exploratory trip through the St. Croix and Burntwood (or Broule) Rivers: in 1832. Under the direction of Henry R. Schoolcraft. Map ["Sketch of the sources of the Mississippi River, drawn from Lieut. Allen's observations in 1832, to illustrate Schoolcraft's inland journey to Itasca Lake."]. New York.

WARREN, GOUVENEUR KEMBLE. 1859. Memoir to accompany the map of the territory of the United States from the Mississippi River to the Pacific Ocean, giving a brief account of each of the exploring expeditions since A.D. 1800 ... *U. S. War Dept. Rept. expl. surv. RR Mississippi Pacific* 11: 27-29.

CHAPTER XXII

TOLMIE CLIMBS THE LOWER SLOPES OF MOUNT RAINIER

WHEN writing of the last visit of David Douglas to North America, I mentioned the epidemic of "intermittent fever" then prevalent along the lower Columbia River. In his *History of Oregon,* H. H. Bancroft relates that "Previous to 1833 there had been no physician at Fort Vancouver, except Dr. McLoughlin, who, through the epidemic of 1830 and the several seasons of fever that followed, suffered much fatigue from care of the sick, and much annoyance from the interruption of his business. In 1833 two young surgeons came out from Scotland, Gairdner and Tolmie. They had for their patron Sir William Hooker. Gairdner . . . was surgeon at Fort Vancouver from 1833 to 1835, but being troubled with hemorrhage of the lungs, went to the Hawaiian Islands in the autumn of the latter year, where he died . . . William Frazer Tolmie, his associate, was from the University of Glasgow, and made botany a study. He had been at Fort Vancouver but a few months when he was assigned to the post on Millbank Sound. Returning to Fort Vancouver in 1836, he served in the medical department for several years."

William Jackson Hooker—less concerned with health problems on the northwest coast than with knowledge of that region's flora—inserted a notice, supplemented by a footnote, in the column "Botanical information" in the *Journal of Botany* for 1834:

"The district of the Columbia too and others of the vast possessions of the Hudson's Bay Company, may be expected to be more thoroughly known, from the liberality of the Company just mentioned, who have appointed two medical gentlemen, Dr. Meredith Gairdner and Dr. Tolmie, well versed in Natural History, to reside in that country. They embarked in August, 1832, and the news of their arrival is anxiously expected by their friends." According to the footnote:

"While this sheet is in the press we have the pleasure of being able to say, that by letters, now received (March, 1834) from Dr. Gairdner, we learn that they had a safe but long passage of eight months' duration, to Fort Vancouver on the Columbia. Mr. Tolmie was stationed at Nisqually House, a new station of the Hudson's Bay Company, at the head of Puget Sound, a spot that has scarcely been visited by any Botanist since the voyage of Captain Vancouver."

According to Bancroft's *Northwest coast,* the men had come out in "the bark Ganymede." Douglas wrote Hooker that on arrival it had afforded him "heart-felt satisfaction" to meet Gairdner, but that Tolmie had "quitted the Columbia for the North-West coast" and that he could only leave him letters and suggestions. This chapter concerns the activities of William Fraser Tolmie.

Among "Documents" published in *The Washington Historical Quarterly* are transcripts of three portions of a journal kept by Tolmie after his arrival at the Columbia

River in 1833. I shall cite these according to chronology of subject matter, not by date of publication. The editor in every instance appears to have been Edmund S. Meany.

(1) The first transcript, published in 1912, was entitled "Journal of William Fraser Tolmie–1833." It covers a twelve-day period, beginning April 30 and continuing through May 11 of 1833. The editor explains that the original ". . . was deposited by the family in the Provincial Library of British Columbia at Victoria. There several copies were made, from one of which the following portion of the journal is printed. A visit was made to the above named library to check carefully the copy with the original, but it was found that Dr. Tolmie's son had withdrawn the original for his own studies. Comparison was then made with copies held by Clarence B. Bagley of Seattle and by George H. Himes of the Oregon Historical Society. It was then ascertained that the original copy had been blurred in some spots by moisture. It is believed that the following record is as accurate as it is possible to obtain."

(2) The second transcript, published in 1932, was entitled "Diary of Dr. W. F. Tolmie." It covers the same period as (1) above but continues through June 5, 1833. According to Mr. Meany's introduction, "The . . . document involves the writing of two outstanding pioneers of the Puget Sound region—Doctor William Fraser Tolmie and Edward Huggins. Mr. Huggins is responsible for the document in its present form . . . his prefatory note . . . tells how he got access to the original diary . . . and why he chose that portion giving the first detailed description, in 1833, of the region about Fort Nisqually. Internal evidence shows that Mr. Huggins wrote the document in 1892."

After describing the roundabout way in which the document reached the *Quarterly,* the editor explains that the original of the Tolmie journal ". . . has been carefully cherished by the family at Cloverdale, near Victoria, British Columbia. Portions of it has been copied and extracts have been published. It is believed that the portion here copied by Mr. Huggins has never before been put in type."

After pointing out that Tolmie and Huggins had married sisters, "the daughters of Chief Factor John Work," the introduction ends thus:

"The document is . . . reproduced as Mr. Huggins wrote it. He headed it 'Leaves from the diary of the late Dr. William Fraser Tolmie, Hudson's Bay Company service.' . . . he made a few eliminations and . . . occasionally he interpolated an explanatory remark, sometimes signing his initial . . . he had an unquestioned right to do such editing if he chose. His relative by marriage had a place in his highest esteem."

The portion of the Huggins transcript which covers the same period as (1) above— an approximate half of the 1932 transcript—certainly suggests rewriting. It eliminates a number of rather delightful passages indicative of Tolmie's predilections; there are few, if any, factual changes. The preface, written by Huggins, begins:

"I have been kindly permitted by Mr. John Tolmie, Cloverdale, Victoria, B.C. to make a few extracts from the private Journal of . . . his father, and have selected those pertaining to his first arrival in this country in 1833 and descriptive of his trip made that year across the portage to Nisqually House, afterwards Fort Nisqually, situated near the mouth of the Squally river, in Pierce County, and now the property of the writer . . .

". . . the writer has copied the Doctor's journal almost verbatim, and has endeavored to confine himself to that which refers to the portion of the country about Fort's Vancouver and Nisqually . . ."

(3) The third transcript from the Tolmie journal, published in 1906, was entitled "First attempt to ascend Mount Rainier." It covers entries for a ten-day period, August 27 through September 5, 1833.

A foreword relates that Miss Jennie W. Tolmie, daughter of W. F. Tolmie, copied the entries from her father's diary and sent them to Clarence B. Bagley.[1]

Although interspersed with comments upon the natural beauties of the countryside, its flowers and trees, Tolmie's journal—by contrast with those kept by David Douglas —lacks for the most part any botanical interest. The portions seen contain, however, interesting sidelights on the region and period. I quote, omitting much, from the transcript of 1912—my (1) above—between the dates of April 30 and May 11, 1833.

April 30: "Off Cape Disappointment . . . At 9 . . . the land was distinctly seen on either side . . . a series of low undulating hills alternating with flats . . . the whole supported a luxuriant growth of tall trees . . .

"Came on deck at 1. Cape Disappointment had just been recognized . . . It is a bluff, wooded promonitory, and the contiguous land of same character (that of a rolling country) stretches away to N. or N. by W. . . . the land from N. to N.N.E. has an insulated appearance . . . Chenooke Point . . . was distinguished by a triangular yellow patch on an adjoining hill which the gloomy aspect of the surrounding forest made conspicuous . . . Point Adam was seen . . . low, flat, and clad with trees . . . At 2 the C. thought it prudent to stand out to sea . . .

"In entering, the chief danger consisted in passing between the Cape and the South Spit, a narrow point which runs off from the Middle Grounds . . . that part of the bar above water . . . In passing between Middle Ground and Chenooke Point you are between Scylla and Charybdis . . ."

May 1: Fort George. ". . . found that the Gannymede was making for the Cape with a favorable breeze . . . we stood in and passed within 150 yards of the Cape . . . at 9 were sailing across Baker's Bay in safety . . . The Cape is a steep, precipitous crag about 200 feet high, its sides grassy and shrubby and summit crested with pines . . .

". . . off Point Ellice . . . Mr. Fisk . . . in charge of Fort George, arrived bringing the intelligence . . . that Dr. McLoughlin was at Ft. Vancouver with but very few assistants, that Mr. Douglas had accompanied the hunting party to New Caledonia, and

---

1. Transcript (3) above was published in 1905 as chapter 64 of Ezra Meeker's *Pioneer reminiscences of Puget Sound*. The chapter was written by Clarence B. Bagley and was entitled "In the beginning." It seems also to have been issued as a pamphlet in 1905.

is expected to return . . . in June. Fort George seen from Point Ellice where the Gany-mede lay at anchor did not much resemble its namesake in Scotland,—a few cottages perched on a green knoll close to beach . . . all around it a trackless forest. Set out in . . . canoe for Fort George, distant six miles. Rather rough passage, got wetted . . .

". . . At 7 our canoe was ready and we embarked . . ."

As noted under May 2, the party slept at ". . . Tongue Point, 6 miles above Fort George . . ."

May 2: ". . . I thank my stars that I had got a cloak instead of a heavy gray coat, for G[airdner] was put to many shifts to keep . . . comfortable with a great coat cloak of gray plaid . . . The scenery along the banks has been of a monotonous character, a dense unbroken forest of pines covers them and the surrounding hills, the only inter-ruption to this is where low sandy points project . . . clothed with stunted willows and bushes . . . Arrived at Kahelamit village . . . Breakfasted at a sunny little cove under right bank . . . Banks becoming more elevated, still invested with dark pines . . . on our right . . . low and flat for some distance up and enlivened by a bright green foliage of willow and aspens. It terminates . . . in a wooded knoll termed Oak Point . . . Began to read Cowper's Table Talk. Metricle errors occur in almost every line, but the ideas are fine and seriously expressed . . . Have been paddling along in the merry moonlight and since it became too dark for reading have been rousing the echoes with Auld Lang Syne, &c. . . . Now encamping on a small wooded islet . . ."

May 3: ". . . Have been coasting left bank and now arrived at Tawallish, a small [In-dian] lodge . . . The mouth of Tawallish river . . . appears a little above huts. Canoes going to Fraser's river ascend it . . . Have come along right bank rugged and jutting out into bluffs adorned with saxifragas and sedums in flower . . . On a high bluff . . . the habitations of the dead are very numerous . . . For several miles the edge of river is bordered with willows, aspens, birch, etc. The scenery assumes a softer character . . . Have been coasting along Deer Island . . . Slender elegant-looking trees ornament its surface . . . Have been paddling for an hour reading Cowper's 'Progress of Error.' With the arrows of polished but cutting satire he attacks the modish follies of the day and rises to higher themes toward conclusion, addressing Lord Chesterfield, or rather his shade, under the name of Petronius, he condemns his epistles with just severity . . . Have distributed brandy among the Indians and are now going to court 'Nature's sweet restorer' in the bottom of canoe."

May 4: Fort Vancouver. ". . . reached our destined port after nearly an eight months' pilgrimage. Knocked at the gate which, after some delay, was opened by the gardener . . . a Celt . . . Governor McLoughlin . . . appeared in shirt and trousers on the stair-case of the common hall and welcomed us with a cordial shake of the hand . . . Our fare was excellent . . . superb salmon, fresh butter and bread, tea, with . . . milk and mealy potatoes . . . about 7 . . . visited garden. Young apples are in rich blossom and extensive beds sowed with culinary vegetables are layed out in nice order, and under a long range of frames melons are sown. Afterwards visited patients, which are pretty numerous, and have been divided between us . . . engaged in putting apothecary's Hall in some degree of order, visited and prescribed for my patients . . . In the evening putting apoth. hall, which is to be our temporary domicile, to rights, and am now, 10½, going to turn in. From what I have seen of Gov. like him and think my first pro-possessions will be confirmed by a longer acquaintance."

May 5: ". . . Having conversed frankly with G[airdner] last night, proposed to him that we should reside permanently together in the present domicile . . . G. stated his wish that we should be separated, and from that we talked on our former differences[2] and finally became reconciled, which am glad of, as it will add materially to our mutual comfort and happiness . . . Read Bogatsky[3] before breakfast. Afterwards visited patients and attended Episcopalian morning service read by Gov. in dining hall . . ."

The text of the journal is garbled at this point but apparently Tolmie took a walk over a plain:

"Its surface is diversified with ckumps of trees and lakes of water, and profusely bedecked with beautiful flowers, among which I noticed particularly a large species of lupin, a blue orchidous looking plant called kames and the root of which is baked under ground and eaten by the Indians. A great variety of others did not attract so much attention . . . In retraversing the pine wood, the Gov. pointed out to me a tall slender tree having a profusion of *syngenesius* flowers called here devil's wood . . . Sugar maple also grows in this wood . . . looked over introduction to 1st No. of the Canton Miscellany begun in 1831. It is well written on the whole, though diffused and prosy. Rode out with Gov. . . . to see the farm which extends along the banks of river to east of Fort. There are several large fields of wheat and pease, and one of barley, with extensive meadows . . . Have agreed with G. to have alternate days of taking patients under charge . . ."

May 6: The entries for this day include a description of the layout of the quarters which Tolmie and Gairdner were to occupy, a description of the medicines and medical equipment available, and an account of a burial service. Tolmie had evidently wanted to perform an autopsy ("that body be inspected") but McLoughlin considered

---

2. In an article upon Meredith Gairdner (1945), A. G. Harvey mentions that on the trip from England the relationship between the two men had become strained:

"... Gairdner's brilliance was accompanied by conceit. Cool and reserved in manner, he left it to Tolmie to chat with the ladies and to attend them and the crew professionally, although sometimes he was called in for consultation. Tolmie became much annoyed by what he called Gairdner's 'haughty airs of superiority,' and towards the end of the voyage they were barely on speaking terms. He says that the two women (Mrs. Charlton, wife of the British Consul at Honolulu, and her sister, Mrs. Taylor) ... summed up Gairdner as 'proud & vain of his attainments & aiming at singularity in his manner,' an estimate in which he concurred ..."

3. Presumably *The Golden Treasury* by the German theologian K. H. Bogatzky.

this inadvisable "as from the force of Canadian prejudices such a thing had never been done." Tolmie comments: "Must endeavor to overcome these prejudices when I become better acquainted with their nature and extent . . .

"In the evening had some conversation with the Gov. on farming. Wheat here yields a return of 15-fold; barley from 40 to 50; maize requires the richest soil, barley, hay, then wheat, and lastly oats or peas."

May 7: ". . . wrote log and conversed with G. until past 12 . . . By our labours we have brought apartment somewhat near to state of order and tidiness but there is still much to be done. Borrowed from Gov. first and second vol. of Humboldt's Personal Narrative of Travels in So. America. Sowed Dahlia seeds[4] in garden under a frame, visited a store; it seemed in a state of confusion . . . Now, 9½, going to begin Humboldt."

May 8: "Began Cowper's poem on tenth but soon laid it aside and accompanied G. along river's bank for a short distance upward. There is a nice pebbly beach . . . edged with verdant trees and brushwood, and elegant wild flowers of various species . . . After breakfast resumed labours in dispensing and busy till 5 p. m. . . . At 6 G. and I set out to walk along the farm with guns and I having vasculum . . . we struck up toward the wood and then walked along an upland plateau . . . Its breadth is about ¼ mile and it presents a rounded bluff face to northward, beautified with elegant columbines, luxurious lupines and other plants equally attractive but unknown . . . G. and I. walked along the plateau by the border of . . . wood, now admiring the rich groves of lupin amidst the trees mixed with handsome columbines, sun flowers, and a great variety of herbacious plants in flower . . . there were some enchanting spots and my heart bounded with delight and enthusiasm as I surveyed them . . . the glimpses of the magnificent Columbia obtained through interruptions to the belt of wood which skirts its northern shore showed it to flow placidly and musically along. On its southern shore, great trees extended in a narrow strip along lowlands, but behind, a range of undulating hills perhaps 500 feet high stretched east and west and in the background the colossal Mt. Hood . . . reared his lofty summit above the clouds. The tout ensemble was the finest combination of beauty and grandeur I have ever beheld . . ."

At this point Tolmie had his first encounter with a skunk and notes that he was ". . . giving G. the polecate as I shall not have time to examine . . . its skin. Collected a specimen of the Devil's tree used as a purge . . . and tried it in a few instances."

May 9: "Up at 6. Examined plants procured last night. It is I think *Cornus Florida* which, in the U. S. is sometimes substituted for cinchona in a doze of . . or . . powdered bark; its composition is *Cinchonanie quinine* and gum. After breakfast visited patients

4. Dorothy O. Johansen, in an article about Tolmie, reports that he brought the dahlia seed from the Hawaiian Islands (where the *Ganymede* had stopped on the way out) and that "He introduced, besides the dahlia, the domestic strawberry, and imported quail from California . . ."

who are all in an improving way, and was, on my return, informed by Mr. McL. that I am to be despatched to northward in the St[eame]r Vancouver which is to set out on a trading voyage in a few days along the coast. Shall probably be left with Mr. Finlayson at the new fort on Millbank Sound which is to supplant Fort Simpson. The situation of the settlement is . . . on an island which forms the south bank of north branch of Salmon river, at the entrance to Sound about latitude 51 30 N. Long. 127 W. The projected establishment to N. is in latitude 57°, Long. 132°. . . . It will not interfere with the Russians as they have no posts to south of Norfolk Sound. I would have preferred remaining here but *il n'import* . . . as we are to coast a great part of the way and touch at several stations in Puget Sound and the Gulf of Georgia, the voyage I anticipate will be agreeable.

"In the north must be constantly armed to the teeth as the Indians are dangerous. Busy during the day in acquiring information regarding medicines necessary to be taken, etc. In the evening walked out with G. along Vancouver Plain . . . Below Fort for some way it is covered with gigantic relics of the primeval forest . . . the magnificence and grandeur of its colossal tenants was very impressive and the ground was beautifully carpeted with wild flowers and low creeping evergreen shrubs . . . What an excellent cricket field this part of the plain would make. The site of it would throw Wilkinson into estacies."

We are told of the danger from Indians not only on the Willamette—as reported by "Mr. McKay," and "Mr. McDonald," recently returned—but of their experiences in the "Snake country." Also something about the hunting:

"This country last season produced 10,000 beaver skins which generally weigh about 1 pound each and are sold in London at 25 sh. per pound. There are no buffaloes and few deer in this country and fish is the support of the hunters . . . McKay has had many encounters with the bear and the best way he says when a wounded bear rushes at you is to stand and reload and when he comes near, if your gun is unloaded, look at him steadily and he will not attack but raised on his hind legs will continue to return your gaze until tired of his position, when he betakes himself quietly off."

On May 10 Tolmie was busy preparing for the trip—as to clothing, a gun and so on. "My disbursements at the store amount to 14.11-2." On May 11 he was "Up at 7. After breakfast showed rifle to McDonald who got Depote a noted marksman to try it; but he, after three shots, declared the barrel poor and the sight improperly constructed and it is now in the hands of the carpenter who understands the thing and is to make the necessary alterations. Busy all day getting up Vancouver's chest and invited in the evening to have some ball practice with G. but was requested by Gov. to copy a correspondence between Company and Russia for Compy, with which have been employed since 6 p.m."

Transcript (1) ends at this point and I now follow the Huggins transcript—my (2) above.

May 12: Mail from York Factory arrived unexpectedly, ". . . being five or six months earlier than the usual period that the letters are received. Some parties had been dispatched from York to the Saskatchawan during winter, and the opportunity of forwarding the Columbia letters by them embraced. It would be very gratifying to those on this side if the same thing could be done annually, but it has only occurred once before. Between Fort Colville and the Boat Encampment, the Posts nearest the Mountains, on either side, and 300 miles distance from each other, travelling for one or two individuals is almost impracticable, and the Society of Indians in that country makes it precarious and difficult to obtain an escort . . . During our ride the Doctor [McLoughlin] unfolded to G. and me his views regarding the breeding of Cattle here. He thinks that when the trade of Furs is dying out, which at no very distant day must happen, the servants of company may turn their attention to the rearing of cattle, for the sake of the hides and tallow, in which he says, business could be carried on to a greater amount than that of the Furs collected West of the Rocky Mountains. Furs are already becoming scarce, and the present supply is obtained by an almost exterminating system of hunting. In 1792 the North West Company sent more furs from a comparatively small space of country than is now sent to Britain from all the H.B.Co's country and the Government posts in Canada . . ."

May 17: "Macdonald has proposed that I should accompany him by land to Nisqually on Pugets Sound where he is to take charge of a Fort the Company is establishing there. The 'Vancouver' is to go round with goods, and I join her there, and proceed in her to the Northward. I cordially agreed to accompany Mac, and having obtained the Doctors sanction, set about making the necessary arrangements. We are to ascend the Cowlitz River to its source, proceed thence on horseback to the bottom on Pugets Sound, and holding afterwards a Northern course to arrive by land at Nisqually. Our journey will probably occupy eight or nine days. I shall take fowling piece, pistols and 'Skenedher' (dagger on knife)."

May 18: ". . . Put arms in order. Gave Dr. McLoughlin the Acacia seeds[5] got at Oahu. At 11.30 bade adieu to all in Fort and followed Mr. MacD. to strand, where our canoe was lying . . . Had some fine glimpses of Mount St. Helens, its summit conical, and sides more rounded than those of Mt. Hood. It is invested with a pure sheet of snow . . ."

May 19: ". . . Paddled across to right bank and entered Jolifie river on its lower bank, and just opposite Coffin Island is the site of an Indian village . . . only its superior verdure distinguished the spot from the surrounding country. Intermittent Fever . . . here

5. According to Harvey (1940, 238): "Tolmie . . . had brought with him some acacia seeds which he planted at Fort Vancouver. Some of the trees are still standing. Later, Dr. Tolmie took some of their seeds to Victoria, on Vancouver Island, and from them grew the beautiful trees now at the old Tolmie homestead."

committed its fullest ravages and nearly exterminated the villagers, the few survivors deserting a spot where the pestilence seemed most terribly to wreck its vengeance . . ."

May 22: "Entered the Cowlitz . . . At 5:30 Arrived at Coal bed and set the men to work with pickaxe and shovel. The Coal did not seem rich . . . laid aside a quantity . . . to be carried to Fort Vancouver by returning Canoe. The river now is a continuous rapid . . ."

May 24: ". . . reached the second Coal mine . . . Having landed explored bank for some extent around Coal . . .

"At 5. arrived at a small inhabited village (Cowlitz landing) where the portage commences . . . Mac set out . . . to a village . . . to engage horses for our journey . . ."

May 25: ". . . Mac has decided on taking the canoe-men on to Nisqually, if the cattle party do not return tonight . . . Set out in a canoe to search for coal along the banks upwards . . . after being once nearly upset disembarked, and climbing the high and rugged bank arrived at a beautiful prarie (Cowlitz Prarie) extending . . . at least 4 or 5 miles, nearly a mile broad and very level . . . Bearing East the pyramidal St. Helens appeared in immaculate whiteness . . . The soil of prairie seemed very fertile. It was covered with a luxuriant, but not rank grass, and adorned with a much greater variety of flowers, than either the Cattlepootle or Jolifie plains, and much fewer trees . . . Found ripe strawberries on a sunny brae with a eastern exposure . . ."

May 26: ". . . arrived shortly before sunset . . . on the S. Western border of Prairie . . . The Prairie now seemed encircled with trees . . . St. Helens . . . towered high . . . and the other mountains (Rainier or 'Tacoma') called by the Indians, 'Pus-ke-youse' bore E.N.E. at summit, divided into two rounded eminences, with a narrow intervening hollow . . . The ascent seems most practicable from S.E. . . . We are now four or five days journey from Nisqually . . ."

May 27: ". . . About 4 arrived at first stage of portage about 20 miles from Cowlitz and encamped in a long narrow prairie . . ."

May 28: "Encamped on Grand Prairie . . . started about 8 on foot . . . Crossed two lower ridges, occasionally meeting with new plants . . ."

May 29: ". . . walked in the Grand Prairie, along its N.E. border for some miles . . . Crossed a steep hill and much incommoded by the heat and glare of the Sun. Prairies dry and sandy or gravelly, but as Mac observed admirably adapted for Sheep . . ."

May 30: Arrived at Nisqually . . . Forded the Nisqually about 3 miles from its mouth . . . arrived at the proposed site of Nisqually Fort on a low flat about 50 paces broad, on the shores of Puget's Sound. The most conspicuous object was a store half finished, next a rude hut of cedar boards, lastly a number of Indian lodges constructed of mats hung on poles . . . Bathed in the sound, which was as smooth as crystal . . . Went up to prairie with Mac and saw the proposed site of Fort and Farm . . ."

We are told of Macdonald's various plans for the settlement, in buildings and so on.

". . . A small garden about 40 yards square had been formed five weeks ago and sowed with Onions, Carrots, Turnips and Cabbage, which all appeared above surface, but seemed to suffer from drought. Some rows of potatoes planted at the same period, looked well . . ."

June 2: ". . . had an excellent view of a long range of Snow speckled Mountains, in the penninsular opposite . . . to the highest summit is given the classical name of Mount Olympus . . . Had a solitary walk in the prairie . . . and came to a beautiful lake, nearly circular . . . the broad leafed Nymphea floated on its unruffled bosom, and Flora adorned its margin with a profusion of yellow ranunculi and others unknown . . ."

June 3 seems to have been devoted to writing and to talking with Macdonald on "the affairs of the company." On June 4 the two men explored the prairie, passing a small stream "denominated the Coe." Tolmie comments upon the abundance of oak. Back at camp in the evening he gave his companion "a brief lecture on the circulation of the blood."

On June 5 they canoed along the banks of Nisqually River and in the evening visited the farm where the ". . . potatoes were greatly improved by the rains of Monday and Tuesday, and also the Carrots and Turnips . . ."
The Huggins transcript of Tolmie's journal ends this day.

Instead of moving on to his new post on Millbank Sound, Tolmie was detained at Nisqually until November because of an accident to one of the men of the party. It was in late August, of 1833[6] that he visited Mount Rainier. The introductory words to the transcript of Tolmie's journal describing the ascent state that ". . . The trip was a botanizing expedition, and as such was a success, while the attempt to reach the summit of the great mountain was a failure. The diary is also remarkable in that it speaks of glaciers . . .

"It is . . . seen from this diary that Doctor Tolmie discovered the Rainier glaciers twenty-four years before the trip made by Lieutenant Kautz[7] . . ."

6. Piper's date for the climb is "1837," but, unless the ascent was attempted a second time, this would seem to be incorrect.

7. According to A. H. Bent, the summit of Mount Rainier was not reached until 1870 ". . . when General Hazard Stevens and Philemon B. Van Trump accomplished it."

However, "In 1857 Lieutenant . . . August V. Kautz, U.S.A., led a party to within a thousand feet of its top." His attempt to proceed farther had been halted by a glacier.

Tolmie's feat was commemorated one hundred years later. His son, the Honorable Simon Fraser Tolmie, who in 1928 became Premier of the Province of British Columbia, had this to say:

"On September 2, 1933, the National Park Service of the United States and the Rainier National Park Advisory Board invited a number of people from the state of Washington and from British Columbia to attend a centennial celebration and the dedication of a Mowich, or Tolmie, Entrance to Mount Rainier National Park . . . On the gateway is a bronze plaque recording the significance of the celebration, to serve as a permanent record of this first exploration of the mountain by Dr. Tolmie in 1833. Several descendants of Dr. Tolmie were present, and I was very proud to be asked to take part in the ceremony.

"The trip to Mount Rainier by Dr. Tolmie was fruitful in more ways than one. He discovered and recorded many new plants which bear his name. He also secured the skins of many birds, including that of the Macgillivray's Warbler *(Oporornis tolmiei),* which was named after him by Dr. J. K. Townsend." [8]

I quote from the transcript of the Tolmie journal, my (3) above, again omitting much:

August 27: "Obtained Mr. Herron's consent to making a botanizing excursion to Mt. Rainier, for which he has allowed 10 days. Have engaged two horses . . . and Lachalet is to be my guide. Told the Indians I am going to Mt. Rainier to gather herbs of which to make medicine, part of which is to be sent to Britain and part retained in case intermittent fever should visit us when I will prescribe for the Indians."

August 28: ". . . No horses have appeared. Understand that the mountain is four days' journey distant—the first of which can only be performed on horseback. If they do not appear tomorrow I shall start with Lachalet on foot."

August 29: "Prairie 8 miles N. of home. Sunset. Busy making arrangements for journey, and while thus occupied the guide arrived with 3 horses. Started about 3, mounted on a strong iron grey, my companions disposing of themselves on the other two horses, except one, who walked. We were 6 in number. I have engaged Lachalet for a blanket, and his nephew, Lashima, for ammunition to accompany me and Nuckalkut and Poyalip (whom I took for a native of Mt. Rainier) with 2 horses to be guide on the mountain after leaving the horse track, and Quilliliaish, his relative, a very active, strong fellow, has volunteered to accompany me. The Indians are all in great hopes of killing elk and chevriel (deer), and Lachalet has already been selling and promising the grease he is to get. It is in a great measure the expectation of finding game that urges them to undertake the journey. Cantered slowly along the prairie and are now at the residence of Nuckalkut's father, under the shade of a lofty pine, in a grassy amphi-

8. B. A. Thaxter notes that ". . . during the month of May our hillside thickets resound with the song of the Tolmie warbler, a bird more often called in Oregon the MacGillivray warbler."

theatre, beautifully interspersed and surrounded with oaks, and through the gaps in the circle we see the broad plain extending southward to Nisqually. In a hollow immediately behind is a small lake whose surface is almost one sheet of water lilies about to flower. Have supped on sallals . . ."

August 30: "Sandy beach of Poyallipa River . . . as I dozed in the morning was aroused by a stroke across the thigh from a large decayed branch which fell from the pine overshadowing us . . . Got up about dawn, and finding thigh stiff and painful thought a stop put to the journey, but after moving about it felt easier. Started about sunrise . . . Made a northeasterly course through prairie . . . The points of wood now became broader, and the intervening plain degenerated into prairions . . . Ascended and descended at different times several steep banks and passed through dense and tangled thickets, occasionally coming on a prairion. The soil throughout was of the same nature as that of Nusqually. After descending a very steep bank came to the Poyallipa . . . Rode to the opposite side through a rich alluvial soil plain, 3 or 4 miles in length and ¾ to 1 in breadth. It is covered with fern about 8 feet high in some parts. Passed through woods and crossed river several times . . . encamped on the dry part of the river bed, along which our course lies tomorrow. The poyallipa flows rapidly and is about 10 or 12 yards broad. Its banks are high and covered with lofty cedars and pines . . . Lachalet has tonight been trying to persuade me from going to the snow on the mountains."

August 31: ". . . Have traveled nearly the whole day through a wood of cedar and pine, surface very uneven, and after ascending the bed of river a couple of miles are now encamped about ten yards from its margin in the wood. Find myself very inferior to my companions in the power of enduring fatigue. Their pace is a smart trot which soon obliges me to rest . . ."

September 1: "Bank of Poyallipa river. It has rained all night and is now, 6 A.M., pouring down . . . The prospect is very discouraging. Our provisions will be expended and Lachalet said he thought the river would be too high to be fordable in either direction . . . Our course lay up the river, which we crossed frequently . . . Have been flanked on both sides with high, pineclad hills for some miles. A short distance above encampment snow can be seen . . . Have supped on berries . . . Propose tomorrow to ascend one of the snowy peaks above."

September 2: "Summit of snowy peak immediately under Rainier.[9] Passed a very uncomfortable night . . . Ascended the river for 3 miles to where it was shut in by amphitheatre of mountains and could be seen bounding over a lofty precipice above. Ascended that which showed most snow. Our track lay at first through a dense wood of pine, but we afterwards emerged into an exuberantly verdant gully, closed on each side

9. S. F. Tolmie records that "On September 2, and again the following day, he reached the summit of a peak now known as Tolmie's Peak."

by lofty precipices. Followed gully to near the summit and found excellent berries in abundance. It contained very few Alpine plants . . . After tea I set out with Lachalet and Nuckalkut for the summit, which was ankle deep with snow for ¼ mile downwards. The summit terminated in abrupt precipice northwards and bearing N.E. from Mt. Rainier, the adjoining peak. . . . Collected a vasculum of plants at the snow, and having examined and packed them shall turn in . . ."

September 3: "Woody islet on Poyallipa. It rained heavily during the night, but about dawn . . . frost set in. Lay shivering all night . . . At sunrise, accompanied by Quillili-aish, went to the summit . . . The snow was spangled . . . It was crisp and only yielded a couple of inches to the pressure of foot in walking. Mt. Rainier appeared surpassingly splendid and magnificent; it bore, from the peak on which I stood, S.S.E., and was separated from it only by a narrow glen, whose sides, however, were formed by inaccessible precipices. Got all my bearings correctly to-day, the atmosphere being clear and every object distinctly perceived . . . the S. Western aspect of Rainier seemed the most accessible . . . Its eastern side is steep on its northern aspect. A few small glaciers were seen on its conical portion . . ."

Tolmie has considerable to tell of Mount Rainier's appearance; and to those familiar with the mountain it might be of interest. On September 4 they were near "the commencement of prairie." And on the evening of the 5th were back at "Nusqually," where "all is well."

Bancroft's *History of the northwest coast* includes an account of Tolmie's life; but probably the most complete is the one supplied by his son, Simon Fraser Tolmie; and I take the following facts therefrom:

William Fraser Tolmie was born at Inverness, Scotland, on February 3, 1812. He received his education at private schools in Edinburgh, studied medicine at Glasgow University and received his medical degree in 1832.

He entered the service of the Hudson's Bay Company as physician and surgeon on September 12, 1832, and two days later sailed for the northwest coast of North America, reaching Fort George on May 1, 1833.

Soon after arrival he was transferred from Fort Vancouver to Fort Nisqually situated at the southern end of Puget Sound. The trip, overland, consumed twelve days. He arrived May 30 and remained until mid-December. It was in August and September that he made his trip to Mount Rainier.

From there, by boat, he went to Fort McLoughlin on Milbank Sound; he arrived on December 23 and remained five months.

On May 30, 1834, he sailed in the *Dryad* with Peter Skene Ogden who was bound for the Stikine River; on June 15 they reached Fort Simpson, "then situated on the Nass River." The reason for this trip is explained by Tolmie's biographer and I found it interesting. In 1833 Roderick Finlayson had been sent on ". . . a scouting trip, to find a site for a trading-post which would intercept the furs then reaching the Russian establishments in Alaska, from what is now the Northern Interior of British Columbia. The Stikine was found to be the chief trading route of the district, and in 1834 . . . Ogden was sent to erect a fort on that river; but he found the Russians blocking the entrance to the territory, and the Indians of the area allied with them . . . unable to come to any understanding with the Russians, Ogden and his party had to withdraw. On the return voyage southward Ogden moved Fort Simpson from the site . . . on the Nass River, to the present site . . . the old post . . . was abandoned to the natives on August 30."

After remaining for a time at Fort Simpson, Tolmie returned to Fort McLoughlin, arriving November 3, 1834.

In the spring of 1836 he returned to Fort Vancouver, remaining until 1841 when, on leave, he made the overland trip[1] with the Company express to York Factory on Hudson Bay and from there sailed for England.

Back at Fort Vancouver in 1843, Tolmie was again sent to Fort Nisqually where he was made "superintendent in charge of the Nisqually farms of the Puget Sound Agricultural Company, which had taken over the farming and kindred operations of the Hudson's Bay Company." Tolmie retained this post for some sixteen years, or until 1859.

"In 1846 the Oregon boundary award gave the Americans title to what is now Oregon and Washington. As a result, the headquarters of the Hudson's Bay Company was removed from Fort Vancouver, Washington, to Victoria, which thereafter became the chief trading-post. In July, 1859, Dr. Tolmie . . . moved to Victoria, where he took over the management of the Puget Sound Company's farms on Vancouver Island, and also became one of the three members of the Board of Management of the Hudson's Bay Company. He retired from both services in 1870."

Tolmie had purchased a tract of land at Cloverdale near Victoria and this was his home until his death on December 8, 1886. His son devotes several pages to his interests in later life; one was Cloverdale.

"Among the striking features in the grounds are the original acacia trees imported from Fort Vancouver, one of the largest oaks on Vancouver Island, estimated to be 800 years old, some very excellent Oregon ash, and a fine specimen of California redwood, planted about the time the house was built . . . Among . . . interesting things that my father did was to import quail from California . . . The descendants of these quail are still to be seen every day in the grounds of Cloverdale."

Of the many years spent by Tolmie within regions now included in the United States, the published records from his journals cover only a brief forty-six days and these

1. A member of the party was George T. Allen, whose diary—"Journal of a voyage from Fort Vancouver, Columbia River, to York Factory, Hudson's Bay, 1841" (*Trans. Oregon Pioneer Assoc.* 9: 38-55, 1881)—is said to describe the journey.

refer only to time spent within what is now the state of Washington. But he evidently made one trip at least into Oregon, for Bancroft's *Northwest coast* (2: 615, *fn.* 18) mentions that "Obtaining in 1840 a year's respite from medical duties, he spent his time travelling over the Willamette plains and elsewhere . . ." And Piper reports that some of his specimens are labelled "Multnomah River," another name for the Willamette. The Bancroft Library may contain documents telling of this excursion. Mrs. Alice B. Maloney tells me (*in litt.,* February 3, 1948) that "The Huntington Library, San Marino, California, in the Soliday Collection, has the Nisqually Journals kept by Dr. Tolmie from 1833 to, I believe 1850 when Edward Huggins took charge . . ."

While some of the narratives written by visitors to the Columbia region are certainly more entertaining than Tolmie's I did not find these as devoid of interest as did Bancroft:

"It is well nigh heart-rending to see the fires of struggling genius smothered by the very vastness of the surrounding vacuum, to see ideas dissipated, melting into nothingness by reason of the rarity and illimitableness of their mental atmosphere. Tolmie's *Journal,* kept at Nisqually House in 1833, at Fort McLoughlin in 1834-5, and at Fort Vancouver in 1836, is an example. Educated only through the medium of books, the mind cut and trimmed by the conventionalities of old societies, when thrown upon its own resources and left alone with nature it had nothing to think of, nothing to say. Hence this shrewd young Scotch medical man, instead of telling us something of himself, the strange country he is in, the people, white and copper skinned, their aims, failures, and destinies, sighs over what he did this day a year ago in Scotland. Then he goes on with scores of pages of nothings, covering months of non-existence, until the reader wonders afresh how it were possible for so wise a man to write so much and say so little. No small portion of the writer's time was now spent reading books such as Paley, Dwight, and Guthrie, upon whom he piously discourses, and with much learning for so young a man . . . But none like Tolmie at Nisqually can be found in all the noble army of north-west traffickers . . ."

Is it possible that Bancroft's opinion reflects some prejudice, based upon Tolmie's unwillingness to disclose to the persistent historian[2] all the information which he desired to obtain?

"To the literature of the coast, and to my library, Tolmie has contributed two manuscript volumes; one a copy of the journal kept at Nisqually House, Fort McLoughlin, and Fort Vancouver in 1833-6, and the other a *History of Puget Sound and the Northwest Coast.* The first contains comparatively little valuable information, though composed of many words; the other is in answer to direct questions, written for the most part by Mrs Bancroft and myself during our visit to Victoria in 1878. We found Tolmie rather a difficult subject. He could have told more than he did, and would have done

2. John Walton Caughey's *Hubert Howe Bancroft, Historian of the west,* published by the University of California in 1946, gives an amazing picture of the acquisition of the records now in the Bancroft Library.

so but for his diplomatic instincts, and dislike to full, free, straightforward statements. Nevertheless, for what he did give us, which is most valuable, let us be duly thankful."

In his volume *British Columbia 1792–1867* (p. 63), Bancroft records what must have been a major contribution on Tolmie's part to life in the northwest:

"W. F. Tolmie states that . . . at Milbank Sound . . . in connection with Chief Trader Donald Manson, he 'conceived the idea of establishing a circulating library among the officers of the company . . . It was readily taken up by Dr McLoughlin and Mr Douglas. A subscription library was formed which did much good for about ten years, soon after which time it was broken up. The officers subscribed, sent the order for books and periodicals to the company's agent in London; the books were sent out, and as everybody had subscribed, they were sent to all the forts throughout the length and breadth of the land. The library was kept at Fort Vancouver, subscribers sending for such books as they wanted, and returning them when read . . . The Hudson's Bay Company, by their ships, sent out the *Times* and other leading papers for circulation. This was the first circulating library on the Pacific Slope, extending from 1833 to 1843.' "

According to Britten and Boulger, Tolmie's plants are in the Kew Herbarium.

Piper mentions that "Tolmie's specimens are mostly labeled 'Fort Vancouver,' 'Multnomah River,' and 'N.W. Coast.' Many specimens collected in the 'Snake country' of south Idaho and described in the Botany of Beechey's Voyage, are usually accredited to Tolmie, though he expressly states that they were gathered for him by a friend." [3]

Torrey and Gray gave Tolmie's name to two plants native of British Columbia, and both of the saxifrage family: one was the genus *Tolmiea,* the other the species *Saxifraga Tolmiei.*

3. For the identity of this "friend" see my chapter on John McLeod.

BAGLEY, CLARENCE B. 1905. In the beginning. *In:* Meeker, Ezra. Pioneer reminiscences of Puget Sound. Seattle. 467-554. [This includes extracts (474-479) from Tolmie's journal describing his visit to Mount Rainier in 1833. *See* Tolmie, W. F. 1906. Documents . . .]

BANCROFT, HUBERT HOWE. 1884. The works of . . . 28 (Northwest coast II. 1800–1846): 525, *fn.* 2, 615, *fn.* 18, 616, 617, 628, *fn.* 10.

—— 1886. The works of . . . 29 (Oregon I. 1834–1848): 34, 35.

—— 1887. The works of . . . 32 (British Columbia 1792–1887): 63.

BENT, ALLEN H. 1913. Early American mountaineers. *Appalachia* 13: 62, 63.

BRITTEN, JAMES & BOULGER, GEORGE SIMONDS. 1931. A biographical index of deceased British and Irish botanists. ed. 2. London. 302.

GRAY, ASA. 1840. Notices of European herbaria, particularly those most interesting to the North American botanist. *Am. Jour. Sci. Arts* 40: 1-18. *Reprinted:* Sargent, Charles Sprague, *compiler.* 1889. Scientific papers of Asa Gray. 2 vols. Boston. New York. 2: 1-21.

HARVEY, ATHELSTAN GEORGE. 1940. David Douglas in British Columbia. *Brit. Col. Hist. Quart.* 4: 237-239 & *fn.* 44.

———— 1945. Meredith Gairdner: Doctor of Medicine. *Brit. Col. Hist. Quart.* 9: 89-111.

HOOKER, WILLIAM JACKSON. 1834. Botanical information. *Jour. Bot.* 1: 175.

————1836–1837. A brief memoir of the life of Mr. David Douglas, with extracts from his letters. *Comp. Bot. Mag.* 2: 158, 159.

———— & WALKER-ARNOTT, GEORGE ARNOTT. 1830–1841. The botany of Captain Beechey's voyage; comprising an account of the plants collected by Messrs. Lay and Collie, and other officers of the expedition, during the voyage to the Pacific and Bering's Strait, performed in His Majesty's ship Blossom, under the command of Captain F. W. Beechey, R.N. in the years 1825, 26, 27, and 28. London. ["California.—Supplement." 316-409. Issued from 1840–1841.]

JOHANSEN, DOROTHY O. 1937. William Fraser Tolmie of the Hudson's Bay Company 1833–1870. *Beaver, Outfit* 268: 29-32.

PIPER, CHARLES VANCOUVER. 1906. Flora of the state of Washington. *Contr. U. S. Nat. Herb.* 11: 14.

THAXTER, B. A. 1933. Scientists in early Oregon. *Oregon Hist. Quart.* 34: 339.

TOLMIE, SIMON FRASER. 1937. My father: William Fraser Tolmie, 1812–1886. *Brit. Col. Hist. Quart.* 1: 227-240.

TOLMIE, WILLIAM FRASER. 1906. Documents. First attempt to ascend Mount Rainier. [Edited by Edmond S. Meany.] *Wash. Hist. Quart.* 1: 77-81. [*See* Bagley, C. B. 1905. In the beginning . . .]

———— 1912. Documents. Journal of William Fraser Tolmie—1833. [Edited by Edmond S. Meany.] *Wash. Hist. Quart.* 3: 229-241.

———— 1932. Documents. Diary of Doctor W. F. Tolmie. [Copied ". . . almost verbatim . . ." by Edward Huggins.] [Edited by Edmond S. Meany.] *Wash. Hist. Quart.* 23: 205-277.

CHAPTER XXIII

GAIRDNER SPENDS TWO AND ONE-HALF YEARS IN THE COLUMBIA RIVER REGION BUT IS UNABLE TO DEVOTE MUCH TIME OR STRENGTH TO NATURAL HISTORY

IN two earlier chapters I have referred to the fact that Meredith Gairdner and William Fraser Tolmie, both doctors of medicine, were sent out together from England to the Columbia River region when a virulent epidemic of "intermittent fever" was wiping out the Indian population and making serious inroads as well among the whites. Chief Factor John McLoughlin of the Hudson's Bay Company, who had been obliged to tend the sick in addition to performing multifarious other duties, sent a plea for assistance to the London office of the Company. In writing of Tolmie I mentioned William Jackson Hooker's part in the selection of these two men.

More has been published upon Tolmie than upon Gairdner; his life after 1833 was spent, except for short periods, in the Pacific northwest, while Gairdner's sojourn was brief—only some twenty-nine months, for he reached the Columbia River on the last day of April, 1833, and left for the Hawaiian Islands in September of 1835, dying there about a year and a half later.

A. G. Harvey's "Meredith Gairdner: doctor of medicine," published in the *British Columbia Historical Quarterly* in 1945, is the only source known to me which provides an over-all picture of the man considered in this chapter. The article, of some twenty pages, is carefully documented. Many of its facts are taken from manuscript material at Kew and in the Provincial Archives at Victoria. It deals primarily with Gairdner's medical work in the northwest not only because this was the author's main thesis, as his title suggests, but also because Gairdner was unable to pursue to any extent his hoped-for work in the natural sciences while there. For the Company, when medical tasks abated, kept him busy at what seems to have been little more than selling goods over the counter. Although specified in the contract[1] Gairdner had other plans in mind when he accepted the position and resented his inability to turn to intellectual pursuits and make his contribution to science. He was a sick man, almost from the time of arrival, and the story of his thwarted ambitions is rather pathetic.

Harvey tells us that the spelling of the name Gairdner ". . . is a variation of Garde-

---

1. *The Champlain Society Publications* include a short sketch of Gairdner which mentions that he ". . . was engaged on a contract for five years at £100 per annum for his services as doctor, plus an additional salary for his duties as a clerk. He was to be eligible for promotion in the service."

ner, which accords with the Scottish pronunciation ...." Meredith's father was a Scotsman and practised medicine in Edinburgh. The son was born in London, on November 27, 1809, and by the time he was twenty had acquired a medical degree at the University of Edinburgh. He then went abroad and worked under C. G. Ehrenberg— a German scientist famous for microscopic work along many lines—and by the time he was twenty-two had published a number of papers and ". . . a comprehensive work on mineral and thermal springs." Having by then returned to Scotland, Gairdner became acquainted with Hooker who was Professor of Botany at the University of Glasgow. Harvey states that, after the plans had been made with the Hudson's Bay Company, the two recruits, Gairdner and Tolmie, visited ". . . the gardens of the Horticultural Society of London, to see the wonderful plants which had been collected in Northwest America by Hooker's pupil, David Douglas."

The voyage out in the Hudson's Bay Company's bark *Ganymede* (which left Gravesend on September 15, 1832, went round the Horn and stopped at the Hawaiian Islands) consumed seven and one-half months and Gairdner reached Cape Disappointment on April 30, 1833.

Tolmie's journal, as noted in my last chapter, mentions that after arrival at Fort Vancouver, he had ". . . conversed frankly with G[airdner] . . . proposed to him that we should reside permanently together in the present domicile [the 'apothecary's Hall'] . . . we talked on our former differences and finally became reconciled . . ." Harvey gives a side light on Gairdner's character which, as demonstrated on the voyage, appears to have been on the arrogant side. In any event after the conversation and for the short time the men were together, they seem to have spent what available time they had in walks through the countryside and in amicable conversations about their common interests; Tolmie's journal refers more than once to "G."

Whether Gairdner had contracted tuberculosis before he reached the Columbia or whether he developed it after arrival we do not know, but the fact that he suffered from that disease soon became apparent. As Harvey points out ". . . not for many years after did the germ origin[2] of the disease become known . . ." and in Gairdner's day "bleeding" and climate were considered to be the important factors in a cure. But, even had the need of complete rest and proper nourishment been understood, Gairdner was not in a position to observe such requisites.

Harvey paints the picture thus:

"Gairdner had found his position at Fort Vancouver quite unlike what he had expected. When he had engaged with the Company he understood that his duties would not take up all his time, and that in his leisure he could travel about the country and indulge his bent for botany, geology, and other natural history subjects . . . things turned out to be very different. His work was onerous, owing chiefly to the recurrent

2. The bacillus was discovered about fifty years later, or in 1882, by the German physician Robert Koch.

outbreaks of malaria, and when there was any let-up, instead of being allowed to take nature-study jaunts, he was expected to assist at the Indian Hall in trade with the natives. He had little time for hobbies.

" 'Opportunities of visiting even the environs of the Fort are few and far between,' he wrote Hooker a few months after his arrival. 'My collections of plants in N. W. America are as yet but small having made but one small journey into the country of the Walamet river, ground already traversed by Douglas.' He envied that botanist's freedom, continuing: 'The true method of examining this country is to follow the plan of Douglas, whether with the view of investigating the geognostic, botanical, or zoological riches of the country.' More than a year later he again complained to Hooker of the close confinement to duty: '. . . scientific researches are quite out of the question *in* their [the Company's] service however liberal they may be in encouraging them in persons unconnected with them.' He also got his father to write Richardson, asking him to intercede with the Company . . .

"The only natural history trip he was able to make was the one to the Willamette Valley referred to above. A proposal to climb Mount St. Helens . . . had to be abandoned. A study of the salmon of the Columbia and their habits did not take him from Fort Vancouver. Daily meteorological observations that he made at the Fort were dull routine. Regarded as . . . a scientist of wide knowledge . . . he nevertheless was denied the opportunities for further knowledge, and required to serve behind a counter and weigh out sugar, and measure tobacco for the natives.

"This was the disappointment that he blamed for his illness."

In May of 1835 Gairdner went for a change of climate to Fort Walla Walla but since the importance of rest was not understood,

". . . he could not forego the opportunity of exploring the country, an opportunity he had been awaiting for two years. Accordingly he made a horseback trip over the Blue Mountains and down into the Grande Ronde Valley[3] and back, a journey of 200 or 300 miles and lasting several days. Notwithstanding this, he returned to Fort Vancouver after an absence of about ten weeks feeling improved in health. However, the malaria had broken out once more, and the pressure of his professional and other duties brought on a relapse. Then, seeing no prospect of his work easing, he decided to leave his position for the time being and make an extended sojourn in the Sandwich Islands."

Gairdner left Fort Vancouver "late in September, 1835," and the remainder of his

3. Harvey mentions the date of Gairdner's visit to Fort Walla Walla as "in the latter part of May, 1835." He must, therefore, have been the second of the botanical persuasion to enter the Grande Ronde Valley, Thomas Nuttall having been there on his way west with Wyeth in late August and early September of 1834. Tolmie's "friend"—by name John McLeod, a trader associated with the Hudson's Bay Company and the man whose plants are cited in the California Supplement of *The botany of Captain Beechey's voyage*—passed through the Grande Ronde in 1837.

short life was spent in the Islands. He died ". . . on March 26, 1837, in his 28th year, at the home of Mr. Sullivan, next door to that of the Rev. John Diell[4] . . . at Honolulu, and two days later was buried in the graveyard of the mission."

Before leaving the Columbia, Gairdner made one last effort on behalf of science. Harvey again supplies the story in an article entitled "Chief Concomly's skull," published in the *Oregon Historical Quarterly* in 1939. Concomly, the author states, ". . . was the principal chief of the confederacy of all the Chinook tribes along the Columbia between the Cascade Range and the sea except the Clatsops . . ."

He had been a victim of the "intermittent fever" epidemic of 1830, had first been buried in a canoe and later interred. Gairdner knew the spot and, anxious to do ". . . something in the interest of science . . . something out of the ordinary . . .", he risked his life, went alone to the burial place, dug down to the coffin, decapitated the body, and got safely away with the head. This he packed, took to Honolulu, and shipped to his friend Dr. John Richardson. According to Harvey (writing in 1939), the skull is preserved in the Museum of the Royal Naval Hospital at Haslar, near Portsmouth. In 1938, "screwed up and tucked away inside the skull," was found a copy of a letter from Gairdner to Richardson, dated November 21, 1835, which stated among other matters, that the writer intended if he returned to the Columbia, "to procure you the whole skeleton." As Harvey comments, ". . . strange things have been and still are done in the cause of science."

Of tributes to Gairdner, Harvey records the following:

"As is fitting, Gairdner's name has been given to four species in the realm of nature that he loved so much. Two of them are wild flowers. *Carum Gairdneri* Gray is a caraway found in dry open places and in thickets in various parts of the Pacific Northwest. Its sweet, nutty roots were an Indian food. *Pentstemon Gairdneri* Hook. is a beardtongue found in the dry regions of Washington, Oregon, Idaho, and Nevada. Next is *Dryobates pubescens gairdneri,* or Gairdner Woodpecker, a bird of the West Coast, resembling, but smaller than, the common Harris Woodpecker. Last is *Salmo gairdneri* (so named by Richardson) the well-known steelhead salmon or salmon trout, found in the sea and coastwise streams from California to Alaska."

Gairdner's plant collections are at Kew. And Britten and Boulger report his wood collections in the herbarium of the British Museum.

---

4. This was the missionary who, with another of his profession, wrote what was known of the death of David Douglas. The Hawaiian fern *Diellia* was dedicated to him by Brackenridge (*U. S. Expl. Exped.* 16: 271 [1854]).

BANCROFT, HUBERT HOWE. 1886. The works of . . . 28 (Northwest coast II. 1800–1846): 525, 526.

——— 1886. The works of . . . 29 (Oregon I. 1834–1848): 34, 35.

BRITTEN, JAMES & BOULGER, GEORGE SIMONDS. 1931. A biographical index of deceased British and Irish botanists. ed. 2. London. 118.

GRAY, ASA. 1840. Notices of European herbaria, particularly those most interesting to the North American botanist. *Am. Jour. Sci. Arts* 40: 1-18. *Reprinted:* Sargent, Charles Sprague, *compiler.* 1889. Scientific papers of Asa Gray. 2 vols. Boston. New York. 2: 1-21.

HARVEY, ATHELSTAN GEORGE. 1939. Chief Concomly's skull. *Oregon Hist. Quart.* 40: 161-167.

—— 1940. David Douglas in British Columbia. *Brit. Col. Hist. Quart.* 4: 237-239.

—— 1945. Meredith Gairdner: Doctor of Medicine. *Brit. Col. Hist. Quart.* 9: 89-111.

—— 1947. Douglas of the fir. A biography of David Douglas botanist. Cambridge, Mass. [Contains a number of references to Gairdner.]

HOOKER, WILLIAM JACKSON. 1836–1837. A brief memoir of the life of Mr. David Douglas, with extracts from his letters. *Comp. Bot. Mag.* 2: 160.

MCLOUGHLIN, JOHN. 1941. The letters of John McLoughlin from Fort Vancouver to the Governor and Committee, First Series, 1825–38. Edited by E. E. Rich . . . with an introduction by W. Kaye Lamb . . . Toronto. Publications of the Champlain Society 4: 344. Appendix B.

PIPER, CHARLES VANCOUVER. 1906. Flora of the state of Washington. *Contr. U. S. Nat. Herb.* 11: 14.

THAXTER, B. A. 1933. Scientists in early Oregon. *Oregon Hist. Quart.* 34: 338, 339.

CHAPTER XXIV

DRUMMOND'S VISIT TO TEXAS COINCIDES WITH A CHOLERA EPIDEMIC, THE "GREAT OVERFLOW" AND CONDITIONS OF WIDESPREAD LAWLESSNESS

THOMAS Drummond made two visits to North America. On the first his travels were confined to Canada. On the second he made a journey, famous in the annals of botany, into eastern portions of Texas—the subject of this chapter. As an introduction thereto I shall mention Drummond's Canadian journey since it gives an insight into the character of the man and a picture of his earlier accomplishment.

On the ". . . recommendation of Professor Hooker and other eminent scientific men . . ." Drummond was appointed assistant naturalist on Captain John Franklin's second expedition of 1825 to 1827 to the polar regions of North America. With George Back and others, Franklin descended the Mackenzie River to the Arctic sea and sailed west along the coast to within some one hundred and fifty miles of Point Barrow. In another chapter I have told of two unsuccessful attempts by F. W. Beechey commanding the *Blossom* to make rendezvous with Franklin in Kotzebue Sound. A secondary party led by Dr. John Richardson (chief naturalist and surgeon of the expedition) and E. N. Kendall (assistant surveyor) explored the territory between Mackenzie and Coppermine rivers.

At Cumberland House, Saskatchewan, Drummond had been detached from the expedition and assigned the task of exploring westward into the Canadian Rockies; this assignment began on June 28, 1825, and after it had been completed he sailed from Hudson Bay for England in mid-September, 1827, in the same ship as David Douglas. The journey had taken Drummond into remote regions of Saskatchewan, Alberta, southeastern British Columbia[1] and, returning eastward, Manitoba.

Drummond's "Sketch of a journey to the Rocky Mountains and to the Columbia

1. Drummond got no farther into British Columbia than "Boat Encampment," westernmost end of the portage across the Rocky Mountains at Athabasca Pass, and it was there that he had his only glimpse of the Columbia River although the title of his "Sketch" suggests that he saw more. He states that he reached ". . . the Boat Encampment on the Columbia, the 17th of October [1826]. On the following day, the brigade [of the Hudson's Bay Company] pursued their voyage, and I began to prepare for re-crossing the Rocky Mountains . . ."

David Douglas had been at "Boat Encampment" on April 27, 1827, and Franklin mentions that Douglas ". . . crossed the Rocky Mountains . . . by the same portage-road that Mr. Drummond had previously travelled . . ."

Hooker was interested in a complete coverage of these northwestern regions and in this instance his emissary-collectors seem to have made extraordinarily good connections.

River in North America," published in 1830 in Hooker's *Botanical Miscellany*, is the most important source of information for his trip, but Richardson's *Quadrupeds* contains an abridged account as does Preble's publication upon the Athabasca-Mackenzie region.

In the "Sketch" Drummond described his daily schedule during one portion of the journey:

"The plan I pursued for collecting was as follows. When the boats stopped for breakfast, I immediately went on shore with my vasculum, proceeding along the banks of the river, and making short excursions into the interior, taking care, however, to join the boats, if possible, at their encampment for the night. After supper, I commenced laying down the plants gathered in the day's excursion, changed and dried the papers of those collected previously; which occupation generally occupied me till daybreak, when the boats started. I then went on board and slept till the breakfast hour, when I landed and proceeded as before. This I continued daily until we reached Edmonton House, a distance of about 400 miles . . ."

Drummond's months in the field included periods of solitude, hardship and occasional danger, but these he never stresses; nor did he allow them to interfere with his objectives. After describing an alarming encounter with a grizzly ("Grisly") bear —a creature universally respected by travelers in the far west—he adds casually: "My adventure . . . did not, however, prevent my accomplishing the collection of the *Jungermannia*. It is No. 17 of the 'American Mosses.' "

After his trip Drummond became curator of the Botanical Garden at Belfast. His mosses were never published except in the form of an *exsiccata* (dried specimens with tickets or labels), never included in library catalogues. It was to an *exsiccata* that Hooker referred in the *Journal of Botany* (1: 53, *fn.*, 1834), citing the title thus: " '*Musci Americani,* or dried specimens collected in British North America, and chiefly among the Rocky Mountains during the Second Land Arctic Expedition, under the Command of Captain Sir John Franklin, R.N., by Thomas Drummond, Assistant-Naturalist to the Expedition,' In 2 volumes, quarto." The Farlow Reference Library of Harvard University contains an *exsiccata*.[2] Hooker estimated that the number of species exceeded two hundred and forty and, with varieties, two hundred and eighty-six. ". . . the whole of the continent of North America has not been known to possess so many Mosses as Mr. Drummond has detected in this single journey." Geiser comments that, in addition to mosses, many new species of flowering plants were added ". . . to the known flora of America, some . . . so rare as to have escaped the ken of naturalists since Drummond's day."

Richardson, Drummond's immediate superior, paid tribute more than once to his

2. Mrs. Constance Ashenden, Librarian of the Farlow Reference Library, wrote on June 11, 1951, that their copy bore the heading: "Musci Americani; or specimens of the Mosses collected in British North America, and chiefly among the Rocky Mts, during the Second land Arctic expedition under the command of Captain Franklin, R.N. By Thomas Drummond, assistant Naturalist to the expedition. Glascow. 1828," and contained fifty-four pages, bound in two volumes.

subordinate, and Franklin, in a letter written to Hooker from Great Bear Lake on November 10, 1825, refers to Drummond as ". . . my main stay in the botanical and entomological departments . . ." Most important of all, Hooker was satisfied with the work of Douglas, Drummond and Richardson (whom he had also recommended): ". . . the result of their combined exertions has been a mass of collections that have thrown a new light on the Natural History of these interesting regions and have supplied the material for Dr. Richardson's works . . . and our *Flora Boreali-Americana* . . ."

I now turn to Drummond's second visit to North America. On April 26, 1831, John Torrey wrote to his friend Lewis D. von Schweinitz:

"After so much about unsuccessful collectors you will not perhaps wish to hear of a new proposition—but I will venture to mention it to you. Mr. Drummond, the celebrated collector and muscologist . . . has just arrived here from Scotland, bringing me letters from Drs. Hooker & Greville. Mr. D. is about proceeding on a journey to the West of the Mississippi for the express purpose of collecting specimens in all the branches of Natl. History *for sale to any one who chose to purchase* them. He expects to spend several years in this country, & to explore all those parts which have hitherto been little or not at all examined. Many gentlemen in England & Scotland have engaged to take full sets of all that he collects & Dr. Hooker has fixed the price for the plants—which tho' rather high is not extravagant for rare new ones—viz. £2 per hundred. He will allow American botanists to make selections of such plants as they need. You may calculate to what an extent Mr. D. expects to collect, when he has sent out to New Orleans, two tons of paper.[3] Mr. D. asks nothing in advance but he would like to form some estimate what number of specimens would probably be taken in America.—He will leave here in a few days & if you would like to engage two or three hundred specimens please let me know. Dr. Hooker has kindly sent me a set of Mr. Drummond's mosses, collected in Franklin's 2nd journey—about 280 species—many quite interesting." [4]

3. Under "Scientific explorers in America and Africa," Rafinesque records in his *Atlantic journal and friend of knowledge* (1832) the same bit of botanical gossip, though slightly garbled as to dates and localities:

"Mr. Drummond, the botanist, has been exploring the Oregon Mountains for two years past, chiefly for plants and seeds. He was sent by some English botanists and gardeners. It is said that he took to St. Louis two tons of paper for preserving plants."

4. Von Schweinitz was receptive to Hooker's suggestion, replying with promptitude:

". . . The proposals of Dr. Drummond are . . . so tempting, that notwithstanding impoverished circumstances I cannot help requesting you to secure for me the right of getting two hundred species from him on the conditions proposed, begging you to undertake their selection for me."

By August Torrey had the matter arranged:

"Your request . . . shall be faithfully attended to. I shall order duplicates of the very species which I desire for my own Herbm. . . . it seems quite out of the question for me to go beyond 300 species unless I can tempt him with some of our New England Mosses in the way of exchange. . . ." (Shear and Stevens)

Drummond had, therefore, started on the journey that was eventually to take him into Texas.

W. J. Hooker published an account of Drummond's second visit to America in the years 1831 to 1835. Entitled "Notice concerning Mr. Drummond's collections, made in the southern and western parts of the United States," the paper is largely made up of "extracts" from letters which Drummond wrote from the field with, interspersed, Hooker's comments and a catalogue of the plants collected before the man reached Texas. It came out in seven installments, issued as Drummond's letters and specimens were received: two appeared in Hooker's *Journal of Botany* for 1834, five in his *Companion to the Botanical Magazine* (three in 1835, the last two in 1836). Three slight variations in the title occur. Because of the explicitness of Drummond's letters, the "Notice" provides a fairly clear picture of the man's itinerary and experiences.

Hooker's "Notice" seems to have supplied much of the subject matter, certainly as to routes and botanical collections, for Samuel Wood Geiser's "Thomas Drummond"; but, in addition, the author includes from little-known sources much that is interesting about conditions in Texas at the time of the botanist's visit. Geiser's article has been published three times—first in a series of documented papers upon naturalists of the frontier which came out in the *Southwest Review* (1930), and again in the first (1937) and second (1948) editions of the volume *Naturalists of the frontier* where these papers were assembled. Here the chapters lack documentation but include short bibliographies. The chapter on Drummond varies slightly in each of the three issues. The volume pleads the cause of the man of science when confronted with primitive conditions and deprived for the most part of intellectual stimulus and companionship. I have found Geiser's theme repeated, with variations, in the lives of many of the men considered in my story. But Geiser, who has only praise for Drummond and his achievements on the Franklin expedition, does not think so highly of the man's reaction to his months in Texas.

"As Drummond is revealed in his letters to Hooker, he does not wear the habiliments of heroism. We demand a hero with the strength of a Hercules, the will of a Loyola, and the impetuousness and zeal of a Vesalius. In the Texas episode Drummond seems almost entirely lacking in these qualities. His bitter complaints against country and people left as ill an opinion of him in Texas as he had formed of his surroundings. His letters, published after being edited by Hooker, evoked from Mary Austin Holley a rejoinder which, as the only contemporary record of Drummond in Texas, I quote in its entirety:

" 'Mr. Thomas Drummond of Glasgow has done more than any other man towards exploring the botany of Texas. He sent home many plants and seeds which have been successfully cultivated there, and drawings of them have been given in late numbers of Curtis's Botanical Magazine. He had made arrangements to settle his family in Texas, where he could have devoted himself with ardor to his favorite science, and where *with his land* and *his cows,* to use his own language, he could *have been more independent in a few years than he could ever have hoped to be in Great Britain.* Unfortunately for science, as for himself, Mr. Drummond took the year of flood and cholera, 1833, to make his first, and only visit, to his adopted land; and in common with every body else, suffered much inconvenience and consequent sickness. Hence his views of the country are partial and drawn from present personal experience. He saw through jaundiced eyes—and not with the eyes of a philosopher. Notwithstanding he liked nothing, and nobody, he sent home seven hundred new specimens[5] of plants; and a hundred and fifty preparations of birds, obtained in a very few excursions; and resolved there to live and die; no poor compliment, surely, to any place, however we may, for the time being, abuse it.' "

I did not derive, in reading Drummond's letters to Hooker, a sense that the man lacked "heroism" or made "bitter complaints"; he did not, I agree, have the "strength of a Hercules." And the fact that he considered bringing his family to Texas—"if sufficient funds can be obtained"—can be taken as a compliment without perverse implication. Drummond's sojourn (from March, 1833 until mid-December, 1834) coincided, as Geiser states, with "the cholera epidemic" and "the Great Overflow," as well as with "the growing unrest over the encroachments of the Mexican government in Texan affairs and the increasing social strain," which would seem to cover the miscellaneous difficulties—over and above illness and floods—that Drummond experienced. Geiser himself refers to the fact that the ". . . frequent revolutions in Mexico and the resulting administrative changes in Texas induced a condition of anarchy which gave all good men grave concern." There can be no doubt that Drummond chose a poor time for his visit but there is no reason to suppose that he exaggerated. As I interpreted the letters, Drummond's complaints—if such they may be called— had their basis in concern for the quality of his work, not for himself: moisture, if no more, makes the preservation of plants extremely difficult, floods hamper travel and increase the peril of shipment, never too safe in those days as many a collector's loss testifies. Moreover, of the months spent in Texas no small portion was, for Drummond, a period of illness. Immediately on arrival he contracted cholera and he had other afflictions; in distinction to Job he complained very little although he does mention that it affected his work—and, nowadays, boils and felons are considered worthy of respect. At all events read the letters and note whether they are open to two interpretations.

I shall follow Drummond's journey in the Hooker "Notice" and include some of Geiser's enlightening comments. To differentiate subject matter I break down some of Hooker's dense paragraphs, abridge somewhat and, because it is clear when Hooker is citing from Drummond's letters, use but one set of quotation marks.

Hooker opens his paper with praise of Drummond's earlier accomplishments:

5. Geiser inserts "species."

". . . His collections both in Zoology and Botany have been admired by all who have seen them, for the manner in which the specimens are preserved, as well as for the judgment with which they have been selected, and they reflect the highest credit on his zeal and assiduity."

The idea of further investigations (in ". . . some of the less known parts of the *Southern* and *Western United States* of North America, and if practicable, in visiting those interesting and hitherto unexplored and mountainous regions of Mexico and California bordering upon the United States . . .") had occurred to Hooker and [Robert] Graham, and funds were raised ". . . by the liberality of our natural history friends . . ." Drummond was given letters to all who might be of assistance. There was ". . . a powerful letter . . . to Mr. Aster, the head of the United States Fur Company, whose influence extends from the Mississippi to the Pacific Ocean . . ." and so on. There is a long list of those who contributed, pecuniarily or otherwise. Drummond's letters do not indicate that he had any money to waste for he is often concerned with costs. Extracts from the letters now begin.

1831

On April 28 Drummond wrote Hooker from New York:

"I arrived here on the 25th instant, after an excellent passage from Liverpool, and immediately delivered your letters of introduction to Drs. Torrey and Hosack . . . I am likely to meet with little difficulty in prosecuting my journey to Santa Fé at the proper season . . . Vegetation is not yet sufficiently advanced to induce me to make any delay . . . I . . . intend setting off to-morrow for Philadelphia . . . I expect to reach St. Louis by the end of May . . . I am in excellent health, which I hope will continue, and enable me to fulfil the expectations of those friends who have so kindly assisted me in the present undertaking."

On May 7 he wrote from Philadelphia:

"Having been delayed here for several days . . . procuring introductory letters to the interior . . . I have put up such plants as I observed in flower. There is . . . nothing interesting[6] among them, but they may serve as a *memento* of my having been in Philadelphia . . . I shall proceed immediately to Baltimore and Washington . . . thence to Wheeling, walking across the Alleghanies . . . I have now divested myself of all my luggage, except what is absolutely necessary, and still it amounts to a considerable weight . . . I am provided . . . with maps of all the country I am likely to visit, as far as Santa Fé; but I find the purchase of them very expensive."

6. To this Hooker appends a footnote: "It is but justice to this most disinterested man to observe, that he invariably speaks of his collections as of less value than they really prove to possess. In regard to the parcel in question, it contained, independently of some very rare and some new *Mosses,* an excellent set of the spring plants of Pennsylvania, especially of the *Vaccinium* tribe . . . &c. and the little known *Floerkia* . . ."

A letter from St. Louis, July 19, was written, notes Hooker, "under considerable depression of spirits, in consequence of severe indisposition." The next, from New Orleans, December 6, recapitulated "nearly the whole of the information" which it had contained and Hooker confines himself to "extracts." It tells that, at Fredericks-town, Drummond began his walk across the Alleghenies; a wagon carried his equipment. He considered these mountains "mere *ridges*" and as he was "a month too early" the young specimens turned black in drying. On reaching Wheeling seventeen days later, he had to await the arrival of his heavy luggage for ten days. He then started for Louisville, by steamboat rather than by a small boat as he would have preferred; three days after arrival he contracted "fever" and was laid up for ten days. As soon as possible he started for St. Louis (a trip of five days) where he ". . . had a relapse of the fever, with confirmed ague, and had immediately recourse to medical advice, but without deriving any advantage. Thus I lost a considerable number of specimens . . . to the drying . . . I was totally unable to attend . . ."

In two weeks he was better but had another relapse, with further unpleasant complications, which reduced him to "little else but skin and bone." Recovery was slow but before leaving he was able to do some collecting. ". . . I flatter myself that my collection will yet be such as to give a tolerable idea of the general nature of the vegetation round St. Louis." [7] Since his collecting paper did not arrive he notes that he was ". . . under the necessity of purchasing paper at a very high price, and pasteboard was still more difficult to be procured . . .

"I do not consider the amount of species I have yet collected can exceed 500. They, and the other objects of Natural History, shall be despatched by the first vessel that goes direct to the Clyde. My health is now tolerable, and I trust I may consider myself acclimated."

Drummond was unable to proceed to Santa Fe as intended:

"Unfortunately, owing to the lateness of my arrival at St. Louis, it was impossible for me to proceed up the Missouri, the Fur traders, whom I wished to have accompanied, generally leaving their head-quarters on the first of May, or even sooner."

However, he was hopeful about getting there at another time. He had had ". . . most liberal offers of assistance . . . *The American Fur Company* . . . generally go out by the route of Santa Fé, assembling at a small village about a hundred miles from that place called Toas [Taos], from whence they proceed to the mountains. The 2d company was under the direction of General Ashley, and he still retains a considerable interest in it. Their hunting-ground is near the source of the Missouri. In short there will be no difficulty in getting to the mountains."

---

7. Travelers bound for the west at this period usually started from St. Louis and the roster of those who collected plants thereabouts is given in "A biographical history of botany at St. Louis, Missouri" by Perley Spaulding. Bradbury recorded that he had gathered plants in that region in 1809 and Engelmann was later to publish upon a collection which Geyer made in 1842, in Missouri and in Illinois. But I have seen but little reference in the literature to collecting about St. Louis until Fendler, Lindheimer and Engelmann arrived on the scene.

Hooker notes that "This letter was soon followed by the arrival of a collection of roots, chiefly from St. Louis, *via* Liverpool . . ." Drummond's next letter, from New Orleans and dated December 14, described the content.

"Among them you will find a gigantic *Grass*, which I hope may arrive alive. I should never have considered it to be a *Grass*, had I not seen the flowers . . . It is No. 27 of the Catalogue . . . my No. 7 . . . is probably the *Silphium gummiferum* of Elliott . . . No. 54 is a singular water-plant, floating on the surface, after the manner of *Lemna*, but I do not know to what Order it belongs[8] . . . In the box you will find a basket containing some *Shells* . . . I intended to have sent the *Seeds* . . . with this parcel . . . I shall defer them . . . as the damp arising from the living plants might injure them. I have between two or three hundred kinds, such as they are.

". . . I consider myself fortunate in the route I have pursued; there being very little variation in the plants of the Mississippi about St. Louis, from those of the more northern territories. I flattered myself all along, that, when I reached that place, I should be in the Prairie country; but there is nothing of the kind: the woods consisting of stunted *Oak*, with very few timber-trees. On the Illinois side of the river is something like a Prairie, which is called the 'American Bottom.' The *Silphium (gummiferum)* . . . was the most interesting plant I found there: but the country is so unhealthy, that there are few settlers in it, although it is of great extent, and the richest land I have seen. Fever and ague are universal about St. Louis, not one out of fifty escaping, either among natives or strangers.

"The first appearance of a change of vegetation, at least in the forest-trees, takes place about the mouths of the Ohio; *Cupressus (Taxodium,* Rich.) *disticha* here making its first appearance . . . as you descend the river, this tree becomes covered with *Tillandsia usneoides,* and the American *Misseltoe (Viscum flavescens,* Pursh.) In travelling . . . by the steam-boats, you have very few opportunities for collecting. The only time is when they stop to take in wood, which, being usually kept in flat boats in readiness, is very short indeed.

"The country around New Orleans is swampy . . . at the present season of the year entirely covered with water . . . You will be surprised, when you receive the specimens, to find almost a total absence of *Ferns* and *Orchideous* plants. The most abundant genus is *Verbena,* and I believe there may be some species not described by Pursh."

1832

Hooker records that "On the 18th May . . . three chests arrived in excellent condition. Their contents are best described by Mr. Drummond himself, in his letter, dated New Orleans, January 3d, 1832."

"No. 1 contains an assorted and complete collection of nearly 700 species, (exclusive of *Cryptogamiæ*): the specimens are numbered, and I keep a list under such

8. Hooker inserts: "(this was the curious *Azolla Caroliniana.*)"

names as I can again recognize them by; so that I can at once give any information about any species that may be required. It had been my intention to number all the specimens in the various collections . . . this . . . occupies more time than I can spare from more important avocations. The same box likewise contains a large quantity of duplicates. In box No. 2, will be found several sets for those friends who were so kind as to assist in my outfit, with collections of *Seeds* and several species of *Acorns* and *Pine-cones,* which should be distributed with the respective specimens; and Reptiles in spirits for Dr. Scouler. No. 3 contains the *Mosses* and *Hepaticæ,* gathered during the journey, two boxes of *Shells,* a box of *Coleopterous Insects* . . . and collections of *Seeds* . . . &c. . . ."

Hooker was obliged to keep subscribers to Drummond's sets informed about what was being received and was evidently anxious to get more purchasers in Europe.

"During the spring and summer, Mr. Drummond explored the neighbourhood of New Orleans, with his accustomed zeal, and thrice visited the opposite shore of the Lake Pontchartraine; and during these excursions formed another ample collection of nearly 300 species of plants (exclusive of *Cryptogamiæ*), and many *Insects* and *Shells,* which were received in Glasgow in two very large chests, in August of the present year, (1832.) These will be distributed to the respective subscribers as soon as uniform numbers can be put to all the species . . . It will be my object, in an early No. of the *Miscellany,* to give a list of names, corresponding with these Nos.; by which means the value of the species will be considerably increased to the subscribers.

"By the last accounts . . . received from Mr. Drummond . . . it was his intention to set out for Natchatoches on the Red River, whence he hoped to despatch a collection, *via* New Orleans, and proceed to Texas; but his exact route, or the length of his visit there, must depend upon a variety of circumstances, and upon the success attending the disposal of his collections in Europe. It is the expense alone attending the transport of his baggage in so unfrequented a country, that has prevented this enthusiastic traveller from being among the mountains of Mexico . . ."

Although Drummond sent a letter from Covington, St. Tammany County, Louisiana on September 2, Hooker quotes only a short paragraph about shells; he notes, however, that Drummond sent also ". . . an ample collection of plants, among which are many interesting ones, besides a considerable number of mosses and above 100 kinds of seeds, together with roots for cultivation. Among the seeds, are those of a most beautiful *Nuttallia* (*N. Papaver* . . .) and among the latter, plenty of the little known *Sarracenia psittacina,* Mich."

Hooker confesses to having distributed some of these collections ". . . as from 'Alabama;' but . . . afterwards ascertained that the Covington here alluded to is in Louisiana."

Next was received a box of plants from Jacksonville, Florida, ". . . sent early in the present year, 1833 . . . together with one again from the vicinity of New Orleans, which arrived in July last." These shipments, Hooker notes, ". . . complete the Louisiana collection, amounting, Mr. Drummond reckons, in all (exclusive . . . of Cryptogamia) to 1000 species. As the selecting and distribution of these has devolved entirely upon myself, I have found it impossible to put numbers to the whole, as I had intended . . . The first portion of the *New Orleans collection,* is alone distributed with Numbers."

At this point in the "Notice" Hooker acknowledges the help rendered him by British and American botanists in determining the collections received to date. And, after explaining how these were to be distributed to subscribers, begins a catalogue of the plants collected before Drummond's trip into Texas:

"Besides the services I trust this catalogue may render to those who possess any portion of Mr. Drummond's plants, it may be considered useful as showing the geographical range of the species it embraces."

The catalogue begins in *The Journal of Botany* (1:185-202, 1834) and is continued in the *Companion to the Botanical Magazine* for 1835 (1:21-26, 46-49, 95-101) and 1836 (1:170-177; 2: 60-64).[9]

Under "Botanical Information" in the *Companion to the Botanical Magazine* (1835) Hooker refers to Drummond's trip to Texas,

". . . a country recently claimed by the United States from the dominion of Mexico, but hitherto almost untrodden by the foot of a Botanist . . . it had attractions for Mr. Drummond . . . perhaps, increased by the circumstance of a small collection of plants falling into his hands, which were gathered in that country by Mr. Berlandier . . . The particulars of his stay in Texas, will be given in the introductory notice to the remarks we shall have to offer on the plants themselves: suffice to say . . . that he has sent at three separate periods several chests of dried plants, of which the last, and by far the most interesting arrival, still remains to be distributed; and that he has . . . enriched our gardens with seeds and roots of several new, or little known plants: among them are five species of *Cactus,* some handsome species of *Phlox,* a most remarkable new Cruciferous plant allied to the beautiful *Streptanthus,* (*Bot. Mag. t.* 3317,) and two kinds of *Pentstemon,* which, I think, may be reckoned, by very far, the handsomest of this very handsome genus: of these, one had been previously discovered by Mr. Nuttall, on the Red River, and called by that gentleman, on account of the great size and general appearance of the flower, *P. Cobœa;* the other and more beautiful one appears to be quite new . . ."

But, before the "particulars of his stay in Texas" could be printed, Hooker had learned of Drummond's death and he prefaces the extracts to the man's letters (*Comp. Bot. Mag.* 1: 39-49, 183) with a statement to that effect:

"Little did I foresee that . . . the painful duty would devolve upon me of recording

9. The years of issue are supplied by T. A. Sprague in the Kew *Bulletin of Miscellaneous Information* (362-364, 1933).

his death, which took place at Havanna, in Cuba, in the month of March of the present year [1835]."

Hooker then prints the letters and, in a final paragraph, records the little that he then knew about Drummond's death; if more information was sought and obtained, it was not published as far as I have been able to learn. Hooker's statement may well be included at this point and the letters read with what ultimately happened to Drummond in mind: is it not possible that the infected thumb, the ulcers and similar afflictions from which the collector was suffering during, certainly, his last two months in Texas, culminated in a general and fatal infection? Hooker related:

"Of the nature of the illness . . . we are not informed. Some fears for his safety, I confess, came across my mind when, in the end of June of this year (1835) I received from Cuba, *viâ* Hamburgh, three boxes, which instead of being filled with plants . . . only contained his little personal property, clothes, bedding, &c.; together with a very few ill-dried plants and insects . . . my worst fears [were] realized, by the arrival of a letter which H. B. M.'s Consul at Havanna, C. D. Tolmie, Esq., had the kindness to write to me, dated 11th March, 1835, enclosing a certificate of Mr. Drummond's death . . . and referring me for particulars to another letter . . . dispatched by an earlier packet, but which has, unfortunately, not yet reached its place of destination. But it is time to leave this painful subject and to proceed to the more agreeable task of continuing the list of Mr. Drummond's discoveries in the United States previous to his visit to Texas . . ."

1833

Drummond's first letter from Texas was from the "Town of Velasco, mouth of Rio Brazos . . ." and Geiser supplies the date of May 14, 1833.

"We had a favourable passage from New Orleans to this place, and on our arrival found the river so high that it occasioned a delay of a week before we could reach the town of Brazosia . . . only twenty miles up the river. The country, in general, is low and swampy, and ever since we came here, it has been flooded by the river: it consists almost entirely of prairies, except that the water-courses are bordered by woods, consisting chiefly of Live Oak and Poplar, with an under-growth of Carolina Cherry.

"I remained a few days at Brazosia, and having an opportunity of sending by a vessel to New Orleans, I despatched the specimens which I collected without delay.

"Never having seen any part of the sea-coast in this neighbourhood, I determined on returning to the mouth of the Rio Brazos, and commencing my operations there. I accordingly came back to this place, which nearly proved fatal to me, for when I had been here about ten days, and completed a collection of the few plants then in flower . . . I was suddenly seized with cholera . . . I fortunately took what was proper for me, and in a few hours the violent cramps in my legs gave way to the opium with which I dosed myself . . . the Captain and his sister . . . died, and seven other persons . . . a

large number for this small place, where there are only four houses . . . All cases terminated fatally, except mine . . . The weather was particularly cold and disagreeable for
more than a week before the cholera appeared; indeed the air here is constantly saturated with moisture, so as to render the proper preservation of specimens a work of
absolute impossibility. I am almost afraid that the accompanying collection, which I
have taken the utmost pains to dry sufficiently, may not reach you in good order.

"My recovery from cholera was very slow. When my appetite returned, I was nearly starved for want of food, the few individuals who remained alive being too much
exhausted with anxiety and fatigue to offer to procure me any thing. I am now, thank
God, nearly well again, though my face and legs continue much swollen . . .

"As far as possible, I am endeavouring to replace the specimens which were
spoiled during my illness, and have just packed up the whole, consisting of about an
hundred species of plants, and as many specimens of birds . . . about sixty species,
some snakes, and several land-shells . . . Among the plants are several which I would
particularly recommend as deserving of notice for their beauty: two are species of
Coreopsis, one with flowers twice as large as those of *C. tinctoria,* and extremely handsome . . . also a syngenious plant, allied to *Rudbeckia* . . . the blossoms are copper-
coloured, and the whole rises to about a foot high . . . I may safely say, that I have seen
more than a hundred flowers, open on it at the same time. Also a fine procumbent
Œnothera, much like *Œ. macrocarpa*[1] . . . and a charming *Ixia,* of which I send roots
. . . The want of my tent and the chief part of my ammunition, which I was obliged to
leave at St. Louis, proves a serious inconvenience to me.

"To-morrow I intend making an attempt to reach Brazosia again, but the greater
part of the journey is waist-deep in mud and water; thence I shall go to San Felipe,
whither my baggage is already sent, sixty miles beyond Brazosia. Above the latter
place, the river is navigable for boats, so that my luggage must go in waggons. I feel
anxious about my collections, which I leave here, to await a vessel going to New Orleans; but there is no help for it, and from the interior . . . it is still more difficult to obtain conveyances, the charge for freight being so enormous as to exceed the value of
the collections. The cost from Brazosia to New Orleans is forty cents. per foot, and
the amount of my passage and luggage hither was fifty dollars. Boarding averages six
dollars a-week, and that of the roughest kind. It is, however, so long since my hope of
being able to realize any thing more than will cover my expenses has been dispelled,
that I am not disappointed, and my only desire is to remunerate those who have contributed to my outfit, and by the collections of Natural History specimens which I
shall send home, to give a good general idea of the productions of this part of the
world. Of the genera *Pentstemon* and *Sabbatia* (?), which are beautiful and numerous,
I send many specimens and seeds; also of a lovely *Rudbeckia,* which is a great ornament to the prairies here. I could ask a thousand questions about my plants, for I am

1. Hooker identifies this as "*Œ. Drummondii,* Hooker in *Bot. Mag.* t. 3661."

shut out from all information; though Pursh's American Flora is among my luggage, I can hardly get a sight of it.

"You may form an idea of the difficulties I have to encounter in this miserable country (more miserable, however, as to its inhabitants than in any other respect) when I tell you, that all the bird-skins I sent you were removed with a common old penknife, not worth two cents, and that even this shabby article I could not have kept had the natives seen any thing to covet in it . . . I am obliged to leave behind my blanket and the few clothes that I have brought, because of the difficulty of carrying them, though I feel pretty sure that I shall never see them again. These trifles I only mention to give you some idea of my present situation; they do not affect me much, except as preventing me from pursuing the objects of my journey with the success that I could wish.[2]

"I have not yet positively fixed my future plans, but I wish to go westward from San Felipe, and crossing the Rio Colorado [of Texas], to trace it to its sources, if it be practicable."

Geiser tells us that Velasco ". . . at the time of Drummond's arrival, was but a small village, having been laid out the year before . . . Stephen F. Austin places the population at about twenty . . . Brazoria [Drummond's 'Brazosia'] was fifteen miles distant by land from Velasco, and thirty miles if one followed the meanders of the Brazos . . . In 1833 . . . it was the most important shipping point in Texas . . . San Felipe, the capital of Austin's colony, had been laid out in 1824 . . . at a distance from Velasco of eighty miles by land or one hundred and eighty miles by the Brazos River. In 1832 it was a settlement of about thirty families, with several stores and two taverns where travelers, such as Drummond, might stay as guests, living on the very simple fare to which Texans were accustomed . . ."

On August 3 Drummond wrote from San Felipe de Austin. He had been successful in reaching "Brazosia" and in

". . . proceeding higher up the country. This plan I accomplished, though in an unexpected manner, for the river had risen to a height so unprecedented, that a boat brought me across the prairies, which were flooded to a depth of from nine to fifteen feet!

". . . at Brazosia, I found the whole town overflowed, and the boarding-house floor was covered with water a foot deep. I determined . . . that my stay should be as short

---

2. It may have been the above paragraph which particularly aroused the resentment of Mrs. Holley. Yet Geiser quotes from Stephen F. Austin's ". . . Address of the Central Committee to the Convention of April 1, 1833 . . ." (delivered only forty-four days, therefore, before Drummond wrote his comment):

" '. . . a total disregard of the laws has become so prevalent, both amongst the officers of justice, and the people at large . . . that the protection of our property our persons and lives is circumscribed almost exclusively to the moral honesty or virtue of our Neighbor.' "

Drummond, who evidently had not picked a good neighbor, refers only to his own experiences, and as "trifles." Austin, who was certainly in a position to know whereof he spoke, presented a more blanketing indictment.

as possible, and took the first opportunity of a boat to Bello,[3] where I was so happy as to see some dry land; a commencement of the prairie country, which extends uninterruptedly to the West.

"I had been very uneasy about my luggage, which had preceded me . . . but all was safe, and after remaining a few days at Bello, to recruit my strength for the journey, I commenced my walk to this place, collecting what plants I could find by the way . . .

"The collection which I left at the mouth of the river, amounted to one hundred species, and my list now contains three hundred and twenty, which are packed in excellent order; also, seeds, roots, and bulbs, with some bottles of reptiles. I hope these may reach Europe safely; but I am not without fears, on that score, as the cholera is raging in this neighbourhood, and has nearly depopulated Brazosia.

"My health continues to be good, since I recovered from that disease, although I am necessarily much exposed from the nature of my pursuits; the weather, too, is extremely hot, probably nearly 100° of Farenheit . . ."

". . . I intend to proceed immediately to a distance of about forty miles, near the source of the Brazos, when I shall be nearly half way to the Colorado river; but I have no prospect except of carrying the requisite stock of my paper myself, together with a change or two of linen, which this warm climate renders absolutely necessary . . . I trust you will give me your advice as to my movements. If you think that the risk will be adequately repaid, I am most willing to proceed, nay, I am anxious to do so, that I may be able to communicate a good general idea of the Botany of Mexico.

"About one-third of the plants collected on my route, were destroyed by the overflowing of the river. Vegetation is now recommencing, but I never witnessed such devastation; it has extended even two hundred miles higher up the river than this place, it is impossible for me to collect any thing like a given number of species in a certain time, though vegetation scarcely receives any check, even during this winter, in this climate."

He wrote again from San Felipe de Austin on October 28:

"I have this day forwarded a box of specimens, together with some growing plants, and several bottles, containing the fruit of a shrub, and some curious lizards and snakes. Amongst the roots is one, apparently an *Amaryllis,* from which I anticipate a curious inflorescence; and in the packets of seeds, are several very choice plants, not excelled in beauty by any species now in cultivation.

"The intention of pursuing my way westwardly . . . was carried into effect, and I returned here about ten days ago. The journey has produced about one hundred and fifty species of plants, bringing up my Texas list to nearly five hundred; and I have sent numerous samples of almost every kind. This collection might give you some idea of what might be expected, if I could reach the mountains; my prospect of effecting this

3. Bells Landing according to Geiser.

would be . . . very precarious . . . as the Indians have been very troublesome on the frontiers, and have killed several Americans on the Colorado river this autumn.

"During the approaching winter, I think of visiting the sea-coast: probably Harrisburg, near Galveston Bay, whence I may forward such things as I can collect, to New Orleans. I do not expect to make a very great addition to my number of plants, but rather anticipate that they will be of a different class; for instance, the *Cacti,* of which I have got but three, are said to be numerous.

"After spending next summer in Texas, I should wish, before returning to Scotland, to visit the extreme western parts of Florida . . .

"Since commencing this letter, two or three nights of frost have destroyed every vestige of vegetation.

"There are a great many *Gramineæ* in this collection . . . I have not marked any as distinct except I am perfectly convinced that they are so."

1834

After a visit to the coast Drummond wrote again from San Felipe de Austin on April 24:

". . . I mentioned my plan of wintering on the sea-coast, which I accordingly did, in Galveston Bay; but . . . my principal object has, to a great degree, been defeated. I was in hopes of being able to collect a goodly number of birds there; but, for some unknown cause, there were scarcely any birds in the bay during the past winter. I spent the month of January in Galveston Island, said to be the greatest resort of sea-fowl on the whole coast, and with difficulty could procure enough to eat—the island being uninhabited, and the weather so bad that it rained incessantly for three months . . . After remaining in the bay till the last instant . . . I returned hither with one hundred and eighty specimens; fifty of them had not been seen before . . .

"It is my desire this summer, to advance as far into the interior as possible; but several difficulties lie in the way. The Indians are becoming very dangerous . . . The necessity of having all *the luggage carried,* is another great hindrance to my movements; I may state that I had to navigate an old canoe from Galveston Bay to Harrisburg . . . from eighty to one hundred miles, all by myself, and with hardly any provision; for, owing to the failure of last year's crops, famine is threatening the inhabitants of that district . . . arrived there, I was obliged to hire a cart and oxen to come to this place, for which I paid sixteen dollars . . . there is one blessing . . . I enjoy excellent health . . . it has been tried with such fatigue as would have broken down thousands.

"I have added a few plants, lately . . . some . . . very handsome; especially four or five species of *Phacelia,* and two of *Coreopsis,* with a bulbous-rooted plant, like an *Ixia,* but hexandrous. I am glad to find that you have figured the species of *Nuttallia*[4] . . . it is a very fine flower. I have also seen another, apparently quite new . . .

4. Hooker identifies it as "Nuttallia *Papaver: see Bot. Mag.* t. 2387."

"This is the worst country for insects I ever saw,[5] the custom of burning the prairies probably accounts for it . . ."

He next wrote from San Felipe de Austin on September 26:

". . . I stated my intention of proceeding to the Upper Colony,[6] as soon as possible. This I did, and had reached the Garrison,[7] one hundred miles above this place; and made arrangements for joining a band of friendly Indians, who were going to hunt near the sources of Little River, one of the tributaries of the Rio Brazos, when the news that a packet of letters was here, which might contain instructions for my movements, reached me, and I returned hither to take them up, and, consequently, lost the chance of accompanying the Indians.

"I am sorry that it is perfectly impracticable to accomplish your plan for my reaching either the mountains or Santa Fè. This settlement does not extend to within one hundred miles of the former, and the intervening country is full of hostile Indians . . . a dozen men are requisite to protect any traveller who should venture among them. As to Santa Fè, it is at an immense distance from this place, and there is no intercourse. From the towns of the Interior, there is communication sometimes with it; but the best way of going thither is from St. Louis, or from Tampico, or Matamozos [Matamoras], which are frequently visited from New Orleans.

"The name of *Linum Plotzii* . . . had better, perhaps, be changed to *Berlandieri*, who was the person who discovered it. *Psoralea arenosa* is, in this collection, in fruit; with the two *Coreopsides*.

"I am sorry to say that I have found no insects, as they are very scarce in these and all the prairie countries, owing to the frequent burning of those lands.

"The whole country, from the Rio Colorado [of Texas] to the Guadaloup, a distance of eighty or ninety miles, is as destitute of verdure as the streets of Glasgow, except some small patches along the creeks.

"After returning to San Felipe, for my letters . . . I joined a waggon . . . bound for Gonzales, in Guadaloup, one hundred miles distant; but having exposed myself to the burning sun, in the middle of several days, I was seized with bilious fever, which was nigh proving fatal, and has been followed by violent boils and a disease, here called *Felon,* in my thumb. The latter rendered my hand useless for two months, and I caused the place to be opened, and several bits of bone to be removed; and some other pieces have since worked out, so that I have been threatened with the loss of my thumb; but I hope to escape this disaster.

"Were it practicable for me to reach the mountains, I could easily double the seven hundred species, which is the number I have collected in Texas."

According to Hooker: "This is the last letter that was received from Texas, and the

5. Drummond, let Texans note, was writing as an entomologist (as his next letter shows) and was complaining of a dearth, not of a superabundance, of insects as I at first supposed!

6. Geiser adds, "of Austin."

7. Or "Tenoxtitlan" according to Geiser.

Collections made there, mentioned in the two following letters, were all dispatched from New Orleans and proved exceedingly rich and valuable, both in what concerns the number and the rarity of the species, no less than the excellency of their preservation."

On December 20 another letter was written from New Orleans:

"I arrived here yesterday, from Texas, bringing all the specimens I had collected last season, and a box which had been omitted to be forwarded, containing some which had been gathered during the preceding year . . .

"My last opportunity of writing you was from San Felipe, in October,[8] and it is needless to recapitulate what I then said; my Texas collection of plants now amounts to seven hundred species.

"If practicable I shall proceed immediately to Florida, going northward, as the season advances. Perhaps I may reach Baltimore, whence I can take shipping for Europe . . .

"I am sorry to say that I have had a violent attack of diarrhœa, accompanied with such a breaking out of ulcers, that I am almost like Job, smitten with boils from head to foot, and have been unable to lie down for several nights: but, as I am a little better, I hope to be well in a short time."

On Christmas Day Drummond was at New Orleans and acknowledges a "kind letter" from Hooker, enabling him to arrange his "plans for next summer." Hooker had evidently reiterated his desire to have the collector go directly to Santa Fe but Drummond stood firm—the trip would be long and Indians were dangerous; however, it might be possible to get there from St. Louis.

"This plan would occupy at least two years . . . I am becoming very anxious to see my family, and must, in consequence, endeavour to be in Scotland by this time next year . . . The question naturally arises as to what I shall do at home, and as I do not think it would be advisable for me to remain there, I have determined, if sufficient funds can be obtained, to return with my family to Texas, where I can buy a league of land for one hundred and fifty dollars, and if I can add the purchase of a dozen cows and calves, which cost ten dollars each (that is, the cow and calf), a few years would soon make me more independent than I can ever hope to be in Britain."

As a further possibility for financial independence the region might be a good one from which to pursue his collecting activities:

8. According to Geiser: "Evidently a letter written in October, 1834, miscarried, for although Drummond refers to it in his next letter to Hooker, it is not to be found among the letters published in Hooker's account . . ."

Is it not possible that Drummond had in mind his just-quoted letter from San Felipe, dated September 26? Letters from the field were not always completed on the day begun; and, two months later, it might have been difficult to remember whether one had written in late September or in early October. The letter of December 20 cites the same number (700) of species as the one of September 26.

"I should then have an opportunity of exploring the country from Texas to the city of Mexico, and west to the Pacific, which would occupy me seven years at least . . . I have been given to understand that the Mexican Government wishes particularly to have the Natural History of its territories examined, and would liberally reward the person who did it. Now I am not vain enough to expect much remuneration for what I could do, still, with your assistance, I think I might, in the course of two or three years, publish a tolerably complete catalogue of the plants of that country, and, were proper application made, a grant of land would certainly be given me. These plans I mention, that you may kindly consider them at your leisure.

"In the collection now sent is a box, containing several species of *Cacti,* some very interesting. Three are allied to *Mammillaria,* one to *Melo-cactus,* and several to *Opuntia.* They are all from Gonzales . . .

"By the 1st of January I expect to leave this place, but am not decided on the exact route . . .

"I find it would be absolutely necessary for me to return to Britain, in order to purchase a stock of necessaries, clothing, instruments for collecting insects, &c. Upon such articles as knives and forceps, a person who could afford to lay out two or three hundred dollars, would make cent. per cent. here, and a thousand per cent. on many things, so that the journey would cost nothing.

"Pray write to me at Charleston . . ."

Added, under date of December 28, is the comment that, since writing the above, Drummond had been able to secure passage on a ship sailing for "Apalachicola, in Florida, in two days from this time." He mentions:

"My health is better, though one of my thumbs is still unhealed, so that I have only the use of one hand."

1835

There was still another letter, from Apalachicola, dated February 9:

". . . There is no means of getting from this place by land to the extreme south of Florida, which I chiefly wish to visit, therefore I shall probably go to Havanna, whence there are always vessels for Key West . . .

"My health is tolerable, though I am much pained by a severe ulcer on one leg, for which the Saw Palmetto is but an indifferent doctor . . .

"I sail this evening for Havanna."

I have already quoted what Hooker had to tell of Drummond's death in Cuba. The few and meagre biographical references which I have found in the literature of the time not only confuse him with his brother James Drummond, known for his travels in Australia, but also contain what, in view of the content of his letters, may be described as misconceptions. One such notice appeared with the description of *Phlox Drummondii* Hooker (*Bot. Mag.* t. 3441, 1835) and seems to have supplied the substance for the short obituary notice in *The Gardener's Magazine* (Loudon) (new ser. 1: 608) which I quote, italicizing what seems questionable at least:

"Mr. *James Drummond* . . . has fallen a *victim to the climate of Cuba* . . . He *had made arrangements* for a grant of land in the interior of Texas, so that *his prospects* for the maintenance of his family *were brighter than ever . . .*"

Although species, such as the beautiful phlox just mentioned, pay honor to Thomas Drummond, it is his brother James who is commemorated in the genus *Drummondita*[9] Harvey.

David Douglas died at the age of thirty-five. Since the date of Thomas Drummond's birth appears not to be known[1] we can only surmise that he still had useful years in prospect. Hooker wrote:

"Thus have perished, while engaged in the cause of science with a zeal of which history presents few examples and nearly at the same time, two men in the prime of life, of about the same age . . ."

In his *Fauna boreali-americana* John Richardson also pays tribute to these two men. I quote first from his volume on quadrupeds and then from his volume on fish:

". . . a zone of at least two degrees of latitude in width, and reaching entirely across the continent, from the mouth of the Columbia to that of the Nelson River of Hudson's Bay, has been explored by two of the ablest and most zealous collectors that England has ever sent forth . . ."

". . . I may also notice here, the deaths of two naturalists to whom the former volumes of this work are much indebted. Their walk in science was indeed far beneath the lofty platform which Cuvier constructed, but they were unrivalled in the paths they chose for themselves . . . to Mr. David Douglas . . . we owe many of the most beautiful hardy flowers which ornament our gardens. He perished miserably . . . Thomas Drummond, of Forfar . . . was my friend and associate on Sir John Franklin's second expedition. An enthusiastic admirer of animals and plants, he was eminently qualified for a collector of objects of natural history, by an extreme quickness and acuteness of vision, and a wonderful tact in detecting a new species. His favourite pursuits were

---

9. I believe that this generic name is still accepted. Persistence had won the day! I quote from Jackson's biographical notice of James Drummond:

". . . The genus *Drummondia* was created by De Candolle to commemorate his botanic services, but that genus is now merged in *Mitellopsis*. *Drummondia* of Hooker [which, since a moss, may have been intended to honor Thomas Drummond] has not been accepted by bryologists, the species being referred to *Anodontium* of Bridel, but finally *Drummondita*, a genus of Diosmeae, was founded by Dr. Harvey in 1855."

1. Spaulding places it "about 1780," as "supplied by Mr. J. R. Drummond, grandson of Thomas." Geiser mentions it as "probably . . . about the year 1790." Jackson omits the date of birth altogether, referring to him as the younger brother of James who was born "? 1784." He was probably in the early fifties.

carried on under circumstances of domestic discomfort and difficulties, that would have quelled a meaner spirit . . . In his company, and by his aid, most of the birds described in the second volume of this Fauna were procured: by his unremitting industry and strenuous exertions a very great proportion of the plants included in Dr. Hooker's *Flora Boreali-Americana* were obtained; and the *Musci Americani* . . . will be a lasting monument of his activity and penetration as a cryptogamic botanist. After making very large collections of plants in various parts of the United States, and in the province of Texas, he died in the Island of Cuba, where he had landed on his way to the Florida Keys."

But I like best Dr. Geiser's comments upon Drummond's work in Texas—the final paragraphs of his chapter, where Drummond's "succession of jeremiads" seems to have been forgotten:

". . . Had he made his permanent home here, the botanical history of Texas would have been written very differently. There would have been no Lindheimer, no Wright, no Reverchon, no S. B. Buckley, no Lincecum, collecting plants for Asa Gray and Elias Durand and George Engelmann. Before their day the flora of Texas would have been described by Hooker, Bentham, Lindley, David Don, and other British botanists. By the time that Charles Wright and John James Audubon came to Texas, the botany of all that part of Texas which had been wrested from the Indians would have been open to the world. And the work in Mexico, begun by Berlandier, would have been greatly advanced. For where Berlandier was weak, there Drummond was strong.

"A man of tremendous physical energy, of persistence, of unsuspected idealism, of complete devotion to science: forgetful of self, pursuing his unreasoning love of botany without any reckoning or calculating of the end—such was Thomas Drummond. It seems an unnecessarily cruel fate that kept him from bringing to completion his work in Texas."

After Hooker ended the publication of the Drummond letters (*Comp. Bot. Mag.* 1: 39-49, 1835) his "Notice" contained three more installments enumerating the collector's plants, the last ending with the statement that these were "To be continued." No more appeared, however, and Hooker never published an enumeration of Drummond's Texan collections.

In 1847 Francis Boott published his *Descriptions of six new North American Carices*. These were based on Thomas Drummond's 1833 collections—two from New Orleans, the other four from Texas.

BOOTT, FRANCIS. 1847. Descriptions of six new North American Carices. *Boston Jour. Nat. Hist.* 5: 112-116.

DRUMMOND, THOMAS. 1828. [Abridged account, "in his own words," of Drummond's journey from Cumberland House to the Rocky Mountains.] *In:* Franklin, John. 1828. Narrative of . . . 308-313.

—— 1828. "*Musci Americani,* or dried specimens collected in British North America, and chiefly among the Rocky Mountains during the Second Land Arctic Expedition, under the command of Captain Sir John Franklin, R.N., by Thomas Drummond, Assistant-Naturalist to the expedition. In 2 volumes, quarto." [This title is cited from Hooker, W. J. 1834. *Jour. Bot.* 1: 53, *fn.* It is not a published volume but an *exsiccata. See* p. 487, and *fn.* 2.]

—— 1830. Sketch of a journey to the Rocky Mountains and to the Columbia River in North America. *Bot. Miscel.* 1: 178-221.

FRANKLIN, JOHN. 1828. Narrative of a second expedition to the shores of the polar sea in the years 1825, 1826 and 1827, by John Franklin, Captain R.N., F.R.S., &c. and commander of the expedition . . . London. xi, xxi, 308-313.

GEISER, SAMUEL WOOD. 1930. Thomas Drummond. *Southwest Review* 15: 478-512.

—— 1937. Naturalists of the frontier. Dallas. 73-105.

—— 1948. *Ibid.* ed. 2. Dallas. 53-78.

HOOKER, WILLIAM JACKSON. 1827. Account of the expedition under Captain Franklin, and of the vegetation of North America, in extracts of letters from Dr Richardson, Mr Drummond, and Mr Douglas. Communicated by Dr Hooker. *Edinburgh Jour. Sci.* 6: 107-117. [Contains a Drummond letter "*Dated* Rocky Mountains, April 26, 1836."]

—— 1830. Flora of the British possessions in North America announced. *Bot. Miscel.* 1: 92-94.

—— 1834. Notice concerning Mr. Drummond's collections, made in the southern and western parts of the United States, (to be cont'd.) *Jour. Bot.* (Hooker) 1: 50-60, 183-202 [1st installments of Drummond's letters to Hooker].

—— 1835. Phlox Drummondii. Mr. Drummond's Phlox. *Bot. Mag.* 62: pl. 3441. [An extract from this was reprinted in *Gard. Mag.* (Loudon), new ser. 1: 608 (1835) under the heading "Obituary. Mr. James Drummond."]

—— 1835. Botanical information. *Comp. Bot. Mag.* 1: 16.

—— 1835. 1836. Notice concerning Mr. Drummond's collections made chiefly in the southern and western parts of the United States. (Cont'd from *Jour. Bot.* 1: 202.) *Comp. Bot. Mag.* 1: 21-26, 39-49, 95-101 (1835); 170-177 (1836); 2: 60-64 (1836). [Although noted as "To be continued," no more was published.]

—— 1829–1840. Flora boreali-americana; or, the botany of the northern parts of British America: compiled principally from the plants collected by Dr Richardson & Mr Drummond on the late northern expeditions, under command of Captain Sir John Franklin . . . to which are added . . . those of Mr Douglas, from north-west America, and of other naturalists . . . Map. 2 vols. London. [First issued in installments from 1829–1840.]

JACKSON, BENJAMIN DAYDON. 1888. James Drummond. *Dict. Nat. Biog.* 16: 33.

—— 1888. Thomas Drummond. *Ibid.* 16: 41.

PREBLE, EDWARD ALEXANDER. 1908. A biological investigation of the Athabasca-Mackenzie region. *U. S. Dept. Agric., North American Fauna* 27: 58-61.

RAFINESQUE, CONSTANTINE SAMUEL. 1832. Scientific explorers in America and Africa. *Atlantic Journal and Friend of Knowledge* 1: 26.

RICHARDSON, JOHN. 1829. 1831. 1836. Fauna boreali-americana; or the zoology of the northern parts of British America: containing descriptions of the objects of natural history collected on the late northern land expeditions, under the command of Sir John Franklin, R.N. By John Richardson, M.D., F.R.S., F.L.S. . . . surgeon and naturalist to the expeditions . . . Part first, containing the quadrupeds . . . xiv-xviii, 27, 28 (1829). Part second, The birds . . . xv (1831) [William Swainson is cited as co-author.] *Ibid.* Part third, The fish . . . xiv, xv (1836). London.

SHEAR, CORNELIUS LOTT & STEVENS, NEIL EVERETT. 1921. The correspondence of Schweinitz and Torrey. *Mem. Torrey Bot. Club* 16: 248, 251, 252, 225.

SPAULDING, PERLEY. 1909. A biographical history of botany at St. Louis, Missouri. *Pop. Sci. Monthly* 74: 48-50.

SPRAGUE, T. A. 1933. The dates of Hooker's "Companion to the Botanical Magazine." *Bull. Misc. Inf. Kew.* 362-364.

CHAPTER XXV

WYETH COLLECTS A FEW PLANTS WHEN RETURNING

EASTWARD FROM HIS FIRST JOURNEY TO THE COLUM-

BIA RIVER

IN the present decade of 1830–1840 Nathaniel Jarvis Wyeth made two overland journeys from Boston, Massachusetts, to the Columbia River and both have a botanical interest, for Thomas Nuttall published the small collection of plants made by Wyeth when returning from the west coast on the first journey (1833) and himself accompanied Wyeth from St. Louis, Missouri, to the Pacific on the second in 1834.

The present chapter concerns the first expedition of 1832–1833 and my information thereon is derived from three sources:

(1) An account written by John B. Wyeth, a young man with a desire for adventure, who accompanied his "kinsman" as far west as Pierre's Hole in eastern Idaho where, with other dissatisfied persons, he deserted his cousin and returned east—the hardships of the journey were not to his taste. On his return, and with the assistance of Dr. Benjamin Waterhouse,[1] who wished to discourage similar undertakings, he wrote a book entitled *Oregon; or a short history of a long journey from the Atlantic Ocean to the region of the Pacific by land; drawn up from the notes and oral information of John B. Wyeth, one of the party who left Mr. Nathaniel J. Wyeth, July 28th, 1832, four days' march beyond the ridge of the Rocky Mountains, and the only one who has returned to New England.* It was "Printed for John B. Wyeth" in Cambridge, in 1833.[2] One of N. J. Wyeth's letters characterizes the book as "one of *little lies* told for gain."

Reuben Gold Thwaites edited this journal and published it in 1905 as volume twenty-one of his *Early western travels 1748–1846.*

The remaining sources for the journey are (2) N. J. Wyeth's correspondence (taken from his "letter-book"); and (3) his journals, published together in one volume in 1899. These records, privately owned, were edited by Frederick George Young, secretary of the Oregon Historical Society, and came out as volume one of the *Sources of the history of Oregon,* under the title "The correspondence and journals of Captain

---

1. "Instruction in Botany at Harvard began in the first course of organized lectures in Natural History offered in the United States, a course given from 1788 to 1809 by Dr. Benjamin Waterhouse of the Medical staff." (Bailey, Irving Widmer. June, 1945. "Botany and its application at Harvard. A report to the Dean of the Faculty of Arts and Sciences.")

2. The copy of this rare little book which I examined is inscribed "For the library of the Boston Athenaeum from Benjn Waterhouse Cambridge 1833."

Nathaniel J. Wyeth 1831–61. A record of two expeditions for the occupation of the Oregon country with maps, introduction and index." [3] For the route between Pierre's Hole and the west coast as well as for the return journey, Wyeth's journal is my main source of information although the headings of his letters and sometimes their content occasionally elucidate his diary.

Wyeth was born in Cambridge, Massachusetts, in 1802 and was, therefore, about thirty years of age when he started on his first journey. He is said to have founded the ice industry of New England. When in frozen regions of the far west his journal harks back to this earlier occupation, but it is apparent that he longed for more exciting ventures! He seems to have had a faculty for making friends. H. H. Bancroft comments: "They liked him in Boston, and they liked him at Fort Vancouver; they believed in him everywhere."

In Wyeth's day American pioneers had advanced into western Iowa, Missouri and Arkansas; but beyond lay the vast prairies, the deserts and the Rocky Mountains which, until the arrival of the railroads, offered no inducement to settlement and which had to be crossed, unless the adventurous chose the route round Cape Horn, before the northwest coast could be attained. The next decade was to witness the hardships involved in any overland crossing. In his introduction to the Wyeth records, Young points out that the British, crossing Canada, could avail themselves of the "... wonderfully convenient natural facilities of reticulated water courses making easy water transits."

Nonetheless, Wyeth was not deterred by the difficulties, had a strong faith in the opportunities awaiting the courageous in the Oregon region, and made many plans: he hoped to participate in the fur trade, to engage in agricultural pursuits (among such the growing of tobacco), to start a salmon industry, and more. But three American fur companies were already active in the Rocky Mountains and the Hudson's Bay Company had a virtual monopoly in furs on the west coast. Moreover, he was entering that enterprise during the last years of the hunting-bonanza which, because of the ruthless destruction of the fur-bearing animals, was to be short-lived. On the west coast the British Company, while helpful to new arrivals, had its own interests to consider and these, very naturally, did not include the successful establishment of American competitors in proximity to their own enterprises. Wyeth lacked sufficient financial backing to carry out his plans in such a remote and—to most Americans in the east—still untested region and, moreover met with a series of misadventures, one the loss of a trading ship (the *Sultana,* Captain Lambert) which some of his Boston supporters had sent round the Horn. After the failure of all his hopes he finally returned to the east in 1836 a disappointed man. He once wrote his wife: ". . . the success of what I have undertaken is life itself to me . . ."

3. In 1934 A. B. Hulbert, in the fourth volume of the series "Overland to the Pacific" *(The call of the Columbia . . .)* reprinted a portion of Wyeth's journal (June 6 through October 29, 1832) with editorial comments, as well as letters relating to the journey.

Nevertheless, historians of the period in which Wyeth was active regard him as an important figure and devote considerable space to his endeavors. H. H. Bancroft paid him tribute in these words:

"Though to himself, and pecuniarily, Wyeth's Oregon adventures were a failure, his influence on Oregon occupation and settlement was second to none . . . He it was who, more directly than any other man, marked the way for the ox-teams which were so shortly to bring the Americanized civilization of Europe across the roadless continent. Thus may we easily trace the direct influence of Boston, far greater than that of New York with its Astor, upon the American possession. . . ."

But we are primarily interested here in Wyeth's plant collection and in where it was made. He evidently acquired some specimens on his way west in 1832 and after his arrival on the west coast in 1833, and shipped these to his friend Thomas Nuttall, but they were never received. Nuttall states that the collection which he described ". . . was made wholly on the returning route of Mr. W. from the Falls of the Columbia to the first navigable waters of the Missouri; when pursuing the remainder of his route down the rapid current of the Missouri, scarcely any additional opportunity of adding to the fasciculus occurred."

This being the case the only portion of Wyeth's eastward journey in 1833 which has a botanical interest is included between the dates of February 10 (when he passed the Falls of the Columbia) and August 27 (when he started his rapid descent of the Missouri from Fort Union). Of his trip west I give, therefore, but a brief outline.

Wyeth reached St. Louis from the east on April 18, 1832, traveled by boat up the Missouri River and at Independence joined forces with a party[4] led by the experienced trapper William L. Sublette who was bound for the "American Alps, the Rocky Mountains." They left Independence on May 12, kept south of the Kansas River to near present Topeka where the river was crossed, went up the Big and the Little Blue

4. Also traveling west with Wyeth was John Ball (1794–1884) of Lansingburg, New York, whom A. B. Hulbert refers to as an "amateur geologist of ability and information." Although Ball wrote much that was interesting about the journey I have been unable to find that he collected plants; indeed, in a letter which he wrote from "Head Waters Lewis River July 15, 1832" (quoted by Hulbert, p. 169), he states: "The botany of my journey has been much neglected." In citing from Wyeth's journal under date of June 16, where Wyeth mentions that in spite of the desolation of the region ". . . the earth is decorated with a variety of beautiful flowers and all unknown to me . . ." and comments that "hard travelling disenables our botanist to examine them," Hulbert identifies "our botanist" with John Ball (p. 115, fn. 72).

Ewan's *Rocky Mountain naturalists* (Denver, 1950, 156) mentions that Ball's writings ". . . are notable for general notes on vegetation, wild life, etc., made by an unschooled voyageur." This is true, but in going through what has been published by and about Ball I find nothing to indicate that he made any addition to the botany of the trip. Certainly Nuttall, in his paper upon the plants of Wyeth's first transcontinental expedition, does not mention any Ball contributions.

BALL, JOHN. 1834. Geology and meteorology west of the Rocky Mountains. (Communicated by Prof. Amos Eaton, of the Rensselaer School.) *Am. Jour. Sci. Arts* 25: 351-353.

―― 1835. Remarks upon the geology, and physical features of the country west of the Rocky Mountains, with miscellaneous facts; by John Ball of Troy, N. Y. *Am. Jour. Sci. Arts* 28: 1-16.

―― 1902. Across the continent seventy years ago. Extracts from the journal of John Ball of his trip across

rivers, then overland to the Platte, reached near Grand Island; following the main Platte, its South Fork, Sweetwater "Creek," which led into South Pass, and Lewis' Fork of the Snake River, they arrived on July 8 at the rendezvous[5] of the trappers in Pierre's Hole, not far west of the Wyoming-Idaho boundary. Here Wyeth was deserted by seven of his party. With eleven who remained faithful and in company with Milton G. Sublette (brother of William) he started on August 29 on his way westward. Hardly had they left Pierre's Hole before they were engaged in a battle with some Blackfeet Indians.[6] After this experience the party proceeded to the Columbia River, arriving at Fort Walla Walla on October 14 and at Fort Vancouver two weeks later. The trip had taken in all between five and six months.

Wyeth remained on the west coast only a few months, starting east as soon as travel was possible, for he was eager to get home and organize his second expedition.

I shall follow Wyeth's return journey (as it is given in Young's transcript of the man's journal) as nearly as possible in terms of a modern map. Not, however, in any detail, for there are weeks on end when the route meanders through unnamed mountain ranges, and along nameless creeks, and when we are supplied with merely the daily mileage and a point-of-the-compass direction. The map[7] included ("First Expedition —Routes in the farther West. 1832–3") is crude but helpful. No one seems to have published his route in any detail and Chittenden, whose outline of the journey is the fullest I have seen, merely covers a great portion of it with the adjective "tortuous." Wyeth ignores punctuation and capitalization, but probably did well to keep any journal under the circumstances.

1833

On January 11 Wyeth records that the Columbia River was closed with ice and that he was detained at Fort Vancouver "until it opens." On February 2 he began the ascent of the river.

the Rocky Mountains, and his life in Oregon compiled by his daughter [Kate N. B. Powers]. *Quart. Oregon Hist. Soc.* 3: 82-106.

——— 1925. Autobiography of John Ball. Compiled by his daughters Kate Ball Powers, Flora Ball Hopkins, Lucy Ball. Grand Rapids, Michigan.

HULBERT, ARCHER BUTLER. 1934. Overland to the Pacific. Volume 4. The call of the Columbia. I. Iron men and saints take the Oregon trail. Edited, with bibliographical résumé, 1830–1835, by Archer Butler Hulbert. With maps and illustrations. The Stewart Commission of Colorado College and the Denver Public Library.

5. According to C. P. Russell, this was "The Eighth Rendezvous (1832)—Pierre's Hole."

6. This is described by Washington Irving in *The adventures of Captain Bonneville, U. S. A.* (Author's revised edition, New York, 73-80, 1849); also by C. P. Russell and by De Voto. The latter describes the site (p. 49) thus:

"Pierre's Hole, called the Teton Basin nowadays, is in Teton County, Idaho. The south fork of Teton River flows out of it to the north ... A range of lesser mountains called the Big Holes forms its western and southwestern walls ..."

7. Possibly the work of the editor F. G. Young, but this is not stated.

". . . we started for Wallah Walla   I had with me two men and am in company with Mr. Ermatinger of the H.B.Co. who has in charge 3 boats with 120 pieces of goods and 21 men. I parted with feelings of sorrow from the gentlemen of Fort Vancouver . . . Doct McGlaucland the Gov. of the place is a man distinguished as much for his kindness and humanity as his good sense and information and to whom I am so much indebted that he will never be forgotton by me   this day we came to the Prairie du Li[s] . . ."

They reached the "foot of the Cascades" on February 5, on the 8th the Dalles and, on the 9th, passed the "Great Dalls" and the "little dalles" and came to the "Shutes or falls of the Columbia." These were passed on the 10th. It must have been near or at this point that Wyeth began his collection of plants.

On February 12 they ". . . made the mouth of a considerable stream coming from the S. called John Days River from a hunter of that name . . .", passed the Umatilla ("River Ottilah") on the 14th and by afternoon arrived at "Wallah Walla," remaining through the 19th. Fort Walla Walla was situated near present Wallula, Washington, north of the river of the same name and on the east side of the Columbia which at that point runs nearly north and south. Wyeth mentions:

". . . gras[s] just getting green . . . the river W[alla] W[alla] . . . bottoms good but not extensive and no wood   the corn for this post 150 bushells last year was raised at least 3 miles from the fort   none was stolen by the Indians a good test of their honesty as they are all most always starving."

Leaving Fort Walla Walla on February 20, Wyeth and his party kept up the Columbia to the mouth of Snake River which enters the Columbia from the northeast: ". . . our course was . . . over a country which would be considered light sandy land with but little sage grass good and in tufts . . ."

They were bound for Spokane River and, instead of following the longer route by the Columbia, they turned up the Snake and, after following it, and one or more of its northern tributaries for a time, turned northeast, reaching Spokane River on February 25. The route had taken them through miles of fallen timber ". . . one third of the trees prostrated last year by southernly gale   their trunks much obstructed the path before us . . . the scene reminds me of my Ice men at work by torch light . . ."

They must have reached Spokane River near the eastern end of the southern boundary of Stevens County, Washington, for the site of Spokane House was not far northwest of present Dartford; "Spokan River" at that point was ". . . a stream now about half as large as the Snake River . . . at this place are the remains of the old Spokan House one Bastion of which only is now standing which is left by the Indians from respect to the dead one clerk of the Co. being buried in it . . . the country on approaching this river from the South resembles the pine plains of N. Hampshire near Concord . . ."

The day of arrival they evidently got most of their baggage across the river and camped in "a large plain devoid of wood a deep valley running N. . . ."

They had been told by some Indians that the "road to Colville" was "impassable for snow" but nonetheless Wyeth must have made the journey thither. The map shows the route—northwestward. There are only fragments of letters—one dated "Fort Colville March 12th 1833"—and the journal stops on February 25 and begins again on the "27th March" when Wyeth notes that they arrived at camp—presumably the one near Spokane River—and found "all well and in better order." There can be no doubt that the journey was made, for Wyeth mentions that "the compass in one place would not Traverse   this happened while going to Coville from Spokan and coming from there back." Young estimates that the trip ". . . consumed exactly a calendar month." It had taken them into northern Stevens County.

The journey was resumed from Spokane River on March 27 and the direction was now northeastward and, until April 2, across Idaho. The map indicates that they kept south and southeast of Lake Pend Oreille, an expansion of Clark's River, and on April 1 came to its eastern end or to ". . . the entrance into it of the River Fete Plate." [8] They must have crossed northern Kootenai County and southeastern Bonner County and have reached the Montana boundary.

Travel had been excessively bad, Wyeth commenting on March 31: ". . . we had no human comfort except meat enough to eat and good." He mentions that Lake Pend Oreille[9] ". . . is a large and fine sheet of water . . . surrounded by lofty and snowy Mts . . . their summits are timbered . . ." and the "difficult" swampy land through which they traveled ". . . had the largest cedars apparently the same as those of the N. E. that I have ever seen   I measured one at my height from the ground of 31 feet circumferance and I presume some were larger . . . camped on acc. of my horses having had no feed lately . . . my Indian brought me in some onions and two kinds of trout   some of the trout I have bought of the Indians as large as 10 lbs.   they are plenty and taken with the hook   there are plenty of ducks and ge[e]se   the Ducks are the [same] as the tame ducks of N. E."

From April 3 to May 29 the route was through Montana: ascending "Clark's Fork R. or Flathead" and, according to the map and despite journal-records of many crossings, following its western side, until on May 15, they crossed it and camped on a creek tributary to the upper waters of the Missouri River—probably the Bitterroot River encountered near present Missoula.

On the above portion of the journey they had reached, by April 7, ". . . the Flathead post[1] kept by Mr. Rivi and one man . . . through thick wood not very good trail

8. Wyeth's "Fete Plate" was of course, his French for Tête Platte, or Flathead. Clark's River has gone by a multitude of names: Clark's, or Clarke's, River, Flathead River, Clark Fork of the Columbia River and so on. But on modern maps the name Flathead seems to be applied to the river which runs out of the southern end of Flathead Lake, Montana, and enters Clark's River near present Paradise, in Sanders County—in other words to an upper tributary of Clark's River.

9. Wyeth does not refer to this lake by name until April 7 when he calls it "Flathead or Ponderay lake."

1. Chittenden's map places Flathead Post some fifteen or twenty miles almost due south of Flathead Lake and a little northeast of the point where Flathead River turns rather sharply northwest to enter, eventually, Clark's Fork. See p. 514, fn. 2.

and a snow storm which loaded the pines in such a manner as to bend them down to the ground . . . trees loaded down in this way and frozen . . . constitute much of the difficulty of the route from Flathead or Ponderay lake to this place . . . this place is scituated on a fine prairie   2 mil[e]s long 1 wide and some pleasant after coming through thick woods and mountains . . ."

Wyeth's party remained at Flathead Post[2] from April 11 to 21 and, since fifteen of their forty-seven horses had gotten lost, men and Indians went on search parties. Wyeth himself took the opportunity (April 11-17) ". . . to see if there were many beaver in the country . . . found few . . ." Ermatinger and his men, who had been traveling by boat, arrived at the post on April 12. Wyeth found the country beautiful.

"This valley is the most romantic place imaginable   a level plain of two miles long by 1 wide   on the N a range of rocky and snow clad Mts.   on the S. the Flathead river a rapid current and plenty of good fishing running at the immediate base of another lofty Snowy and Rocky range of Mts. Above and below the vall[e]y the mountains of each range close upon the river so as apparently to afford no outlet either way . . . it is really a scene for a poet . . ."

They left the Flathead Post on April 22 and the next day Wyeth mentions that they were ". . . now fairly in the dangerous Country . . ."—or in the vicinity of the Black-feet Indians. On the 24th: ". . . the salts here whiten in the ground and the animals are almost crazy after it which makes them bad to drive   the morning was sult[r]y and I travelled without my coat but in the afternoon we had a fine [s]hower with some thunder of good quality   the vall[e]y we left today abounds with the finest Kamas I have yet seen   as provisions are scarce in camp the women dug much of it"

Wyeth, except presumably for Indians, was traveling with only two "men." On the 28th he demonstrated the disciplinary qualities essential in a leader:

". . . quarreled and parted with my man Woodman   he appeared to think that as I had but two he might take libertys under such circumstances   I will never yield an inch   I paid him half as I conceive he had gone half the route with me . . ."

On the 29th he inserts an omission:

"Forgot to mention in proper place that I saw Plumb trees at the place we left W. branch of the Flathead river   these are said to be good   about [one] inch through

2. In reply to my inquiry regarding the location of this post in Wyeth's day Dr. Paul C. Phillips supplied the following information (*in litt.*, October 22, 1953): "The old Saleesh House, built by David Thompson in 1809 probably was located about one mile west of the present Eddy [Sanders County] and about 3½ miles east of the Thompson River . . . Some years later the post was moved about four miles west to the bank of the Thompson River. It gradually acquired the name of Flathead Post or Flathead Fort and this is shown on the Ferris map. The 'Old House' on this map was the original Saleesh House . . ." The map referred to is dated 1836. *See* Ferriss, W[arren] A[ngus]. 1940. *Life in the Rocky Mountains A diary of wanderings on the sources of the rivers Missouri, Columbia, and Colorado from February, 1830, to November, 1835.* Paul C. Phillips, *editor*. Denver. Other references to these posts are to be found in Cleland, Robert Glass. 1950. *This reckless breed of men. The trappers and fur traders of the southwest.* New York. And in: *Quart. Oregon Hist. Soc.* 14: 366-388 (1913); *Pacific Northwest Quart.* 33: 251-263 (1942).

ripe in Sept. and found nowhere else but at this place   I tried hard to get some stones but could not . . ."

It was this day that they reached a large settlement of Flathead Indians,

". . . 110 Lodges Containing upward of 1000 souls with all of which I had to shake hands   the Custom in meeting these indians is for the Coming party to fire their arms then the other does the same   then dismount and form single file both sides and passing each other shake hands with men women and children a tedious job . . . there is much kamas in this region   we find little meat in the Indian Camp and are therefore much short[e]ned for food"

On the 30th Wyeth mentions that he "went out to collect some flowers for friend Nuttall, afterwards to see the Camp . . ." His comments on the Flatheads are interesting and highly favorable:

". . . Theft is a thing almost unknown among them and is punished by flogging as I am told but have never known an instance of theft among them   the least thing even to a bead or pin is brought you if found and things that we throw away   this is sometimes troublesome   I have never seen an Indian get in anger with each other or strangers. I think you would find among 20 whites as many scoundrels as among 1000 of these Indians   they have a mild playful laughing disposition and their qualities are strongly portrayed in their countenances. They are polite and unobtrusive and however poor never beg except as pay for services . . . they are moderate and faithful but not industrious.   they are very brave and fight the blackfeet who continually steal their horses and kill their straglers with great success beating hollow equal numbers . . . the more peaceable dispositions of the Indians than the whites is plainly seen in the children   I have never heard an angry word among them nor any quarreling altho there are here at least 500 of them together and at play the whole time at foot ball bandy and the like sports which give occasion to so many quarrells among white children"

He continues on May 1:

"Same camp   the day reminds me of home and its customs . . . I find an Indian Camp a place of much novelty   the Indians appear to enjoy their amusements with more zest than the whites   altho they are simple   they are great gamblers in proportion to their means bolder than the whites"

Moving on, Wyeth's party traveled slowly, spending as much if not more time in camp than on the trail. On May 11 Wyeth went hunting. ". . . We are short of provisions . . . the river . . . is now jammed in between the hills . . ." On the 13th he notes ". . . we approach the head of this river [Bitterroot] fast . . ."

It was on May 15 that Wyeth started up a creek the main branch of which appeared to "run about E[ast] from the plain" and led into the mountains. "This mountain divides us from the heads of the Missouri." On the 18th, after following a pass [? Gibbons Pass] for eight hours, they ". . . struck a little thread of water which during 28 mil[es] increased gradually to a little river and S. E. to another coming from the S. and both go off together N.   this is one of the heads of the Missouri   we crossed it and camped here   we found both Bulls and cows which make all merry   this pass is good going when there is no snow   now there was about one foot . . . there is a visible change in the appearance   vegetation is not so forward the trees appear stinted and small the land poorer and covered with Sedge. . . ."

They had probably encountered the Big Hole River, headwater of Jefferson River. They seem not to have moved far or fast at this time, perhaps because of the elements, of which they had a good example on the 24th:

"A double portion of the usual weather viz. rain Hail snow wind rain and Thunder into the bargain   we are so near where they make weather that they send it as if [it] cost nothing."

The journal records that they were moving east and southeast before reaching the "Salmon river." On May 29 they ". . . struck a small creek [? Lemhi River] going into Salmon river . . ." When they reached the main stream Wyeth noted that it was ". . . here a small creek running through a fine open plain valley about 6 miles wide and extending each way as far as the eye could reach   the river runs here W. by N. On the S. side is a high range of snow mountains perhaps not covered the whole year   this range is parrallel with the river . . ."

To get from the upper waters of Salmon River to Wyeth's destination, the rendezvous at Pierre's Hole, was to take from the end of May until July 9 and the route lay through (?) Lemhi Pass into Idaho. The trip was one of short marches and long periods in camp, sometimes as long as one week. On June 2 the journal records:

". . . through an open plain nearly level   finished the streams of Salmon river and struck one called Little Godin   it terminates near the three butes in a little lake   here goes S. E. through the valley the mts. appear terminating on both sides . . ."

They kept along this valley on the 3rd and 4th and on the 5th ". . . Saw the three Butes come in sight one by one and then the Trois Tetons   the Butes S. E. by S. 20 mils distant about   so far this river rapid and little brush and no beaver grass worse and worse."

Apparently, having descended the valley of Lost River—as the Little Godin or Godin River is now usually called—Wyeth and his party turned northeastward and traveled across the northeastern end of the Snake River plain, camping on "Kamas River" (June 20-23) and arriving (July 7) at Henry's Fork of Snake River in Fremont County. From there they moved (July 9) southeastward to Pierre's Hole in what is now called the Teton Basin, and were just west of the Idaho-Wyoming border. From there they moved, mainly southeast, to Bonneville's Fort (reached July 15), then ten miles southeast, and finally camped (July 17) with the "Rocky Mountain Fur Co.," remaining until the 24th. They had crossed (perhaps by Teton Pass, July 11 and 12) to a point west of Jackson's Hole in Teton County, Wyoming, and from there

(July 14) had moved into Sublette County. C. P. Russell states that "The small mining town of Daniel, Wyoming, now occupies the site [of the rendezvous]."

The portion of Wyeth's journey traced above, much of it across arid regions, had presented its own peculiar difficulties. On June 7 there had been ". . . a gale from the S. W. which blew down the lodges accompanied with a little rain and enough dust to suffocate one . . ." On the 8th ". . . a fine rain & Hail and Thunder . . ." On the 18th ". . . Severe hail & snow yesterday afternoon and rain most of last night and until noon today."

There are but few references to the vegetation. On the 19th there was ". . . snow in patches in shaded places but the country green with herbage and mostly in blossom . . ." The next day, on "Kamas River," we are told that the root of the plant, for which the creek was named, was ". . . so abundant as to exclude other vegetation." By early July "cotton wood" and "narrow leafed cotton wood" are mentioned and there was "some fine luxuriant clover."

On June 7 the journal makes an interesting comment about the appearance of what was presumably a wild horse; it was ". . . extremely fat   it is surprising how fat a horse gets by being left to himself   no grooming that I have ever seen will make a horse appear as beautiful as to be left to his own resources . . ."

On June 26 Wyeth killed a grouse and includes a long description of the bird—evidently intending to prove his side of an argument which he had had with David Douglas—they must have met on the Columbia in 1832: ". . . as I had some dispute with Mr. David Douglass about the grouse of this country I subjoin a description . . ."

Bonneville was in the vicinity. Wyeth had an "express" from him on June 22 and on July 1 some of the Captain's men had arrived in Wyeth's camp; on the 3rd they camped ". . . on same creek with Mr. Bonneville with about 40 men bound for Green River . . ." and were there four days. It was from this camp ("Heads of Lewis River July 4th 1833") that Wyeth penned a number of letters, one to his friend Nuttall in Cambridge, mentioning a package of plants which, however, never reached its destination:

"I have sent through my brother Leond of N. York a package of plants collected in the interior and on the western coast of America somewhere about Latt. 46 deg. I am afraid they will be of little value to you. The rain has been so constant where I have been gathering them that they have lost their colors in some cases, and they will be liable to further accident on their route home . . . I have several times attempted to preserve birds to send to you but have failed from the moisture and warmth . . . P.S. By the notes on the paper my journal will show the place from which the plant comes if kept in its proper sheet until I get home."

Moving on from this camp on the "Heads of Lewis River" on July 7, Wyeth arrived at Bonneville's Fort on the 15th, stayed two days, and then moved southeast down Green River (July 17) and camped, as noted, with the "Rocky Mountain Fur Co."

On the way he had passed (July 10) ". . . about 2 miles from the battle ground of last year [July 6, 1832] with the Gros Ventres . . ." and (July 12) the point where ". . . 9 men under Capt. Stevens were attacked by about 30 Blackft. a little later than this time last year and several of them killed.[3] Mr. Bonneville informs me that when he passed last year in August their bones were lying about the valley."

Russell lists this as "The Ninth Rendezvous (1833)—Fort Bonneville and Horse Creek." He mentions the presence of Sir William Drummond Stewart[4] and his artist companion Alfred J. Miller whose names appear elsewhere in my story. This was, according to Russell, ". . . Stewart's first appearance among the trappers." The rendezvous broke up on July 24.

Leaving the rendezvous, Wyeth's party moved southeastward across Sublette County reaching tributaries of Green River and of the Big and Little Sandy (July 24, 25, 26) and a branch of the Sweetwater (July 27). Still moving southeastward and following the Sweetwater (encountered on July 28) the party entered Fremont County, went through South Pass—though the journal does not mention the crossing—and soon turned north or northeast to the ". . . Po[r]poise in a small rapid thread running through sandstone banks . . ." (July 30). On July 31 they moved on ". . . to the junction of Great Po[r]poise river . . . then . . . to the junction of Wind river . . . Altogether they form a large and muddy river but fordable now which is after a heavy rain."

Wyeth's "Great Po[r]poise" was the Popo Agie River which, uniting with Wind River, forms the Big Horn. The map indicates that from July 31, when he first reached the Big Horn, until August 12 when he got back to that stream, he had traveled considerably west of that river, encountering the upper waters of its many tributaries; the journal mentions some by name—one, Grey Bull River. Following the Big Horn they must have crossed Hot Springs, Washakie and Big Horn counties until, on August 12 or thereabouts they entered southern Montana between Carbon County (west) and Big Horn County (east); on August 17 they reached the confluence of the Big Horn and the Yellowstone rivers, in Treasure County.

Until August 12 Wyeth had been traveling overland; from that date onward to St. Louis he traveled by boat, for most of the way in one which he himself built, his famous bull boat. The journal records on the 12th:

". . . this day went out to get Bull Hydes for boat   got enough and employed the rest of the day in making a Boat . . ."

On August 13 and 14[5] he "Remained at same camp   made a bull boat . . ." On the 15th he tried it out; and again on the 16th "Made a start in our Bull Boat   found it to answer the purpose well   large enough   runs well   leaks a little   made 3 miles

3. According to Chittenden, this battle had taken place in the vicinity of Jackson's Hole and had involved some of the party, led by Alfred K. Stevens, which had deserted Wyeth in July, 1832.

4. Stewart is the "unifying force" in Bernard De Voto's *Across the wide Missouri* (1947).

5. According to Chittenden, it was on the 14th that Wyeth ". . . entered into a contract . . . with Milton Sublette, on behalf of the Rocky Mountain Fur Company, giving bond for its faithful performance, to

N. E. stream   rapid shoals at places . . . Too much liquor to proceed therefore stopped." On the 16th:

"Made a start in our boat   found travelling quite pleasant but requires much caution on account of some snaggs and bars . . . We grounded about 6 times this forenoon it is surprising how hard a thump these bull Boats will stand . . . All feel badly today from a severe bout of drinking last night . . . The mosquitoes have anoyed me much today they affect me almost as bad as a rattle snake . . ."

On August 17 they reached the Yellowstone; it was ". . . of clear water and did not mix with the waters of the Big Horn which was at this time dirty for some miles   about 3 miles below the mouth of the Big Horn we found Fort Cass[6] one of the Am. F. Co. . . . we were treated with little or no ceremony by Mr. Tullock . . . we therefore pushed on next morning."

On the 19th they passed "the mouth of Rose Bud a river coming from S.S.W." and on the 21st the mouth of Powder River. On the 24th they reached the confluence of the Yellowstone and Missouri rivers—near present Buford, Williams County, and just over the Montana-North Dakota border. By the 22nd the bull boat had been "getting quite rotten" ! Instead of continuing down the Missouri from this point Wyeth turned up that river, westward for about five miles, to Fort Union, another post of the American Fur Company, ". . . pleasantly scituated on the N. bank of the Missouri 6 miles above the junction of Yellow stone . . ." Here he was ". . . met with all possible hospitality by Mr. McKensie the Am. F. Co. agent in this country . . ."

Whether Nuttall considered the Missouri River, or the point at which Wyeth began to travel down the Big Horn in his bull boat, as the "first navigable waters of the Missouri" and the point where Wyeth ceased collecting plants, I do not know but, from either point navigation must have kept him fully occupied. The rest of his journey down the Missouri River lasted from August 27 to October 9. I merely mention his progress downstream. Milton Sublette severed his connection with the party at Fort Union and Wyeth's journal records his own courage:

". . . we are . . . left without any one who has descended the Missouri but I can go down stream."

By September 2 he had reached Fort Mandan, North Dakota; by the 8th ". . . Fort Pier[r]e pleasantly scituated on the right bank rather low but withall romantic . . ." and in South Dakota, and by the 11th, after six hours of "pulling," had passed the Great Bend of the Missouri. On September 21 the party camped at "the old post of Council Bluffs . . . now grown up with high weeds a memento of much money spent to little purpose . . ." and were then in Nebraska. Below Council Bluffs Wyeth found

bring out next year before a stated time three thousands dollars' worth of merchandise required by the company for the prosecution of their trade the following year."

Wyeth's journal does not refer to this agreement but when, on June 18, 1834, he reached the rendezvous on his second westward journey, the Rocky Mountain Fur Company repudiated their contract.

6. Fort Cass was situated "on the E. bank of the Yellow stone river" according to Wyeth.

growing ". . . the Hic[k]ory Shagbark Sicamore and Coffee Bean trees not seen above also Night Shade Brier." On September 23 they were near the mouth of the Nodaway River, in Missouri, at ". . . a trading house of the Am. F. Co. called Rubideau Fort at the Black Snake Hills and on the N. bank of the river . . ." where he saw the "Black Locust" for the first time. On the 27th they were at "Cantonment Leavenworth" situated on the west side of the Missouri and north of present Leavenworth, Kansas; on the 28th at Liberty, Clay County, Missouri, where, two days later, Wyeth made plans to travel on a boat bound for the garrison at St. Louis. His journal ends on September 30. Wyeth's letter XCII records his arrival at St. Louis on October 9 and Chittenden reports that he was back in Cambridge on November 7.

As to the plants which Wyeth had collected: Pennell, writing of Thomas Nuttall's sojourn in Philadelphia in 1834, states that among other tasks which he had in prospect was ". . . the study of the small collection of plants which Capt. Wyeth had gathered on his adventurous eastbound course in 1833 . . . These could not have reached Nuttall before early November, 1833, but his study was sufficiently finished to be read before the Academy, February 18, 1834."

This paper, entitled "A catalogue of a collection of plants made chiefly in the valleys of the Rocky Mountains or Northern Andes, towards the sources of the Columbia River, by Mr. Nathaniel B.[7] Wyeth, and described by T. Nuttall," was published in the *Journal* of the Academy of Natural Sciences of Philadelphia in 1834.

Nuttall's introductory paragraph, the first lines of which I have already quoted, reads:

"This collection was made wholly on the returning route of Mr. W. from the falls of the Columbia to the first navigable waters of the Missouri; when pursuing the remainder of his route down the rapid current of the Missouri, scarcely any additional opportunity of adding to the fasciculus occurred. The number of species and their interest to the botanist will therefore be duly appreciated, and particularly when it is known that this was the first essay of the kind ever made by Mr. W.; and yet I can safely say, that besides their number, there being many duplicates, they are the finest specimens probably, that ever were brought from the distant and perilous regions of the west by any *American* collector." [8]

The species collected numbered one hundred and thirteen.

7. Wyeth's middle name was Jarvis and the initial "B." was obviously a slip. His cousin's middle initial was "B." but I have been unable to learn what it stood for.

8. Had Nuttall known nothing of the collections made by Lewis and Clark and by Edwin James, or did he consider them inferior to those of Wyeth?

BANCROFT, HUBERT HOWE. 1884. The works of . . . 28 (Northwest coast II. 1800–1846): 557-575.

CHITTENDEN, HIRAM MARTIN. 1902. The American fur trade of the far west. A history of the pioneer trading posts and early fur companies of the Missouri valley and the Rocky Mountains and the overland commerce with Santa Fe. Map. 3 vols. New York. 1: 434-448.

DE VOTO, BERNARD. 1947. Across the wide Missouri. Illustrated with paintings by Alfred Jacob Miller, Charles Bodmer and George Catlin. Boston.

GHENT, WILLIAM JAMES. 1931. The early far west. A narrative outline 1540–1850. New York. Toronto. 252-256.

GREENHOW, ROBERT. 1845. The history of Oregon and California, and the other territories on the north-west coast of North America: accompanied by a geographical view and map of those countries, and a number of documents as proofs and illustrations of the history. ed. 2. Boston. 359, 360 & fns.

HULBERT, ARCHER BUTLER. 1934. Overland to the Pacific. Volume 4. The call of the Columbia. I. Iron men and saints take the Oregon Trail. Edited, with bibliographical résumé, 1830–1835, by Archer Butler Hulbert. With maps and illustrations. The Stewart Commission of Colorado College and the Denver Public Library. 105-154.

NUTTALL, THOMAS. 1834. A catalogue of a collection of plants made chiefly in the valleys of the Rocky Mountains or Northern Andes, towards the sources of the Columbia River, by Mr. Nathaniel B. Wyeth, and described by T. Nuttall. Jour. Acad. Nat. Sci. Phila. 7: 1-60.

PENNELL, FRANCIS WHITTIER. 1936. Travels and scientific collections of Thomas Nuttall. Bartonia 18: 1-51.

PIPER, CHARLES VANCOUVER. 1906. Flora of the state of Washington. Contr. U. S. Nat. Herb. 11: 14.

RUSSELL, CARL PARCHER. 1941. Wilderness rendezvous period of the American fur trade. Oregon Hist. Quart. 42: 24-27.

SCHAFER, JOSEPH. 1936. Nathaniel Jarvis Wyeth. Dict. Am. Biog. 20: 576, 577.

WYETH, JOHN B. 1833. Oregon; or a short history of a long journey from the Atlantic Ocean to the region of the Pacific, by land; drawn up from the notes and oral information of John B. Wyeth, one of the party who left Mr. Nathaniel J. Wyeth, July 28th, 1832, four days' march beyond the ridge of the Rocky Mountains, and the only one who has returned to New England. Cambridge. Printed for John B. Wyeth, 1833.

—— 1905. Wyeth's Oregon; or a short history of a long journey. In: Thwaites, Reuben Gold, editor. Early western travels 1748–1846 21: 1-106. [Reprint of the Cambridge edition of 1833.]

WYETH, NATHANIEL JARVIS. 1899. The correspondence and journals of Captain Nathaniel J. Wyeth 1831–6. A record of two expeditions for the occupation of the Oregon country; with maps, introduction and index . . . Edited by F. G. Young. In: Sources of the history of Oregon 1: parts 3-6. Eugene.

CHAPTER XXVI

MAXIMILIAN, PRINCE OF WIED-NEUWIED, ASCENDS

THE MISSOURI RIVER TO FORT MCKENZIE AT THE

MOUTH OF MARIAS RIVER

MAXIMILIAN Alexander Philipp, Prince of Wied-Neuwied, is best known—as far as botanical accomplishment is concerned—for his work in Brazil. And it is doubtless true that, although he collected plants with assiduity on his journey up the valley of the Missouri River from St. Louis to Fort McKenzie, Montana, which is described in this chapter, the expedition's enduring importance derives largely from the Prince's detailed information about the Indians of the Great Plains; for, traveling with officials of the American Fur Company and staying at their trading posts, Maximilian had many opportunities—and missed none—to study the tribes at first hand. Moreover, he had brought with him a young Swiss artist, Charles, or Karl, Bodmer,[1] whose task it was to paint all that he saw, Indians and landscapes in particular. His accomplishment supplements, if it does not surpass, that of his patron.

Maximilian's Missouri River journey began with his departure from St. Louis on April 10, 1833, and he was back in that city on May 27, 1834.

The account of the Prince's visit to North America was first published in German under the title *Reise in das innere Nord-America in den Jahren 1832 bis 1834* (Coblenz, 1839–1841); it was accompanied by an *Atlas* in imperial folio containing Bodmer's plates and an elaborate map. A French edition in three volumes was published in Paris in 1840–1843, with an Atlas.

Under the title *Travels in the interior of North America* Maximilian's work was translated into English by Hannibal Evans Lloyd[2] and published in London in 1843. Included is "a large map." In the Atlases (associated with the English edition) which I examined, the plate captions are in German, French and English. The "Translator's preface" to the *Travels* states:

"The prospectus of the German original announced that the work would consist of two large quarto volumes, accompanied by a portfolio of above eighty beautifully col-

1. Bodmer was born in Zurich, Switzerland, on February 12, 1809. He became a painter and engraver. After his trip to North America with Maximilian (1838–1841) he returned to France and lived at Barbizon, the famous artist colony near Fontainebleau. He received medals at the Expositions of 1851, 1855 and 1863 and in 1876 was awarded the cross of the Legion of Honor. *See: La grande encyclopédie* (7: 26).

2. Lloyd (1771–1847) was born in London and became known as a writer, translator and philologist. I have referred to his translation, from the Weimar edition, of Otto von Kotzebue's narrative—*A voyage of discovery into the South Sea and Beering's Strait*—in my chapter VII.

·oured copper-plates, executed by eminent artists at Paris, from the original drawings. Some specimens of the plates having been brought to London, were so much admired by many competent judges, that Messrs. Ackermann were induced to agree with the Paris publisher for a limited number of copies of the plates; and as it might justly be presumed that the English purchasers would be desirous of having the narrative of the travels, it was resolved to publish a translation compressed into a single volume . . ."

In 1905 Reuben Gold Thwaites published a critical edition of Maximilian's *Travels* (volumes 22-24 of his *Early western travels 1748–1846*) and an Atlas (volume 25) in which Bodmer's plates are reproduced in black and white and much reduced; while useful for reference they are in no way comparable to those in the folio Atlas. Thwaites reprinted the Lloyd translation and, in an appendix, included a number of Maximili-an's papers (on Indian vocabularies, on birds, on meteorology, etc.) which do not ap-pear in the Lloyd edition.

None of this additional matter concerns Maximilian's work in botany. Two papers on plants had appeared in the Lloyd translation, and are reprinted by Thwaites, but only one—and I refer to it later—concerns plants found on the Missouri River jour-ney. The Prince reports in the "Author's preface" that "Several men, of great eminence in the learned world," had contributed to his *Travels* and that "President Nees von Esenbeck has undertaken the determination and description of the plants which I brought home . . ."

I might add that Maximilian and Bodmer are mentioned many times in De Voto's *Across the wide Missouri* (1947). The author's sense of humor goes far, at times, to-wards leavening the somewhat heavy loaf of the Prince's scientific approach!

I shall follow the narrative in the Lloyd translation, omitting, with regret, many matters which, though interesting, have little bearing on my story.

Maximilian's purpose in visiting North America was scientific. As announced in his preface it was to present ". . . a clear and faithful description of the country . . ." rather than "statistical information," since the work was intended ". . . for foreign, rather than for American readers . . ." He was to describe ". . . those cheerless, deso-late prairies, the western boundary of which is formed by the snow-covered chain of the Rocky Mountains, or the Oregon, where many tribes of the aborigines still enjoy a peaceful abode . . .

"The vast tracts of the interior of North-western America are, in general, but little known, and the government of the United States may be justly reproached for not hav-ing done more to explore them. Some few scientific expeditions, among which the two under Major Long produced the most satisfactory results for natural history, though on a limited scale, were set on foot by the government; and it is only under its protec-tion that a thorough investigation of those extensive wildernesses, especially in the Rocky Mountains, can be undertaken. Even Major Long's expeditions are but poorly furnished with respect to natural history . . ."

There is an apology for the fact that, although the entries in the journal are numerous, the variety was "very trifling." The reason is explained—although it seems doubtful that much thought was given to insurance when the fire raged:

"I need . . . indulgence with respect to many observations on natural history . . . for this the loss of the greater part of my collections will be a sufficient excuse. The cases containing them were delivered to the Company, to be put on board the steamer for St. Louis, but not insured; and when the steamer caught fire, the people thought rather of saving the goods than my cases, the contents of which were, probably, not considered to be of much value, and so they were all burnt. This may be a warning to future travellers not to neglect to insure such collections."

Nor, because of "unfavorable circumstances"—Indians—was Maximilian able to reach the Rocky Mountains.

1832

The *Travels* begins with the embarkation from Hellevoetsluis, or "Helvoetsluys," a port not far from Rotterdam in South Holland, ". . . on board an American ship, on the 17th of May,[3] in the evening . . .", and on May 24 Maximilian saw ". . . Land's End, Cornwall, vanish in the misty distance, and bade farewell to Europe."

On July 3 ". . . Cape Cod lay to the south . . ." On July 4 they ". . . cast anchor at India Wharf, Boston, on the forty-eighth day of our voyage . . ." Boston, including some outlying towns, was ". . . an extensive city, with above 80,000 inhabitants . . ." The Prince attended the Fourth of July celebration on "the Commons," which was conducted with ". . . no impropriety or unseemly noise." He visited "Bunker's Hill," and "Cambridge College"—he felt that this might have been very interesting if he ". . . had known that Mr. Nuttal, one of the most active naturalists and travellers in North America, held an office there."

Although the journey through the eastern states is worth reading I move on to "New Harmony, on the Wabash," where Maximilian went to "visit the naturalists," remaining from October 19 to March 16, 1833.

1833

He found that ". . . Harmony had fallen into decay, and the people whom Mr. Maclure had settled there, were in part dispersed." However, Thomas Say (". . . well known as having accompanied Major Long in his two journeys into the interior . . .") was still there, and Charles Alexander Lesueur. Say had ". . . a fine library of books on Natural History . . . constantly enriched with the most valuable new works published in Europe . . . a printing press, a copper-plate press, and an engraver . . . Mr. Say . . . lives in a very retired manner, devoted to the study of natural history, and to literary pursuits."

3. According to an entry of May 7, 1833, Germany had been left on May 7, 1832.

Say's ". . . zoological collection was confined to insects and shells." Maximilian reports that his ". . . entomological collection was continually damaged by the rapacious insects, which are much more dangerous and destructive here than in Europe." I have told more of the visit to New Harmony in my chapter on Thomas Say.

Before leaving the colony Maximilian took the precaution of finding out that there were ". . . satisfactory accounts of the sanitary state of the southern and western parts of the United States." Cholera (which had been prevalent in the eastern states when he had been there and which had spread over the country) had abated at Cincinnati and "on the Ohio . . . had generally ceased." Learning that St. Louis was "perfectly healthy," the Prince made preparations ". . . for proceeding . . . as soon as our collections were packed up and sent off." By way of the Wabash and "Tenesee" rivers he finally arrived at the mouth of the Ohio and then started (March 20) by steamboat up the Mississippi, ". . . keeping to the left or eastern bank." On March 24 ("to our great joy") St. Louis was beheld: in 1816 there had been only about 2,000 persons living there, but now it was ". . . a rapidly increasing town, with 6,000 or 8,000 inhabitants . . ."

Because it was Maximilian's intention ". . . to travel through the interior of the western part of North America, and, if possible, the Rocky Mountains, St. Louis was unquestionably the most proper basis for such an enterprise. The question was, whether it was more advisable to go by the caravans by land to Santa Fé, or to proceed by water up the Missouri . . . the plan of following the course of the Missouri seemed to be the most suitable for my purposes . . ."

We are given a short account of the fur trade and told that ". . . foreign travellers cannot expect to succeed in their enterprises without the assistance of . . ." the American Fur Company which was now in control south of "British North America." Maximilian was advised to obtain passage with "Mr. McKenzie,"[4] on the "Yellow Stone steamer," bound on company business to ". . . Fort Union, at the mouth of the Yellow Stone River . . . the 10th of April . . . was the day fixed for our departure."

Maximilian's route, from St. Louis as far as Fort Clark in the heart of the Mandan settlements of North Dakota, had already been followed by plant collectors: by Lewis and Clark both westward (1804–1805) and eastward (1806)—indeed their first botanical specimens had been shipped east from Fort Mandan; by Bradbury and Nuttall (1811), who had made the trip from St. Louis to Fort Mandan when traveling with the Astorians; by Baldwin (1819), who had accompanied Long as far as Franklin, Missouri, and by his successor James (1820), who had reached "Engineer Cantonment," winter headquarters of the Long expedition, below the promontory of Council Bluff in eastern Nebraska. Therefore, by the time of Maximilian's visit many of the plants indigenous to the Missouri River valley between St. Louis and southern North

4. This was Kenneth McKenzie whom Chittenden refers to as the ". . . ablest trader that the American Fur Company ever possessed . . . The noted post of Fort Union . . . the best built and the most commodiously-equipped post west of the Mississippi was McKenzie's creation . . ." (1: 384-387)

Dakota must have been collected more than once and the *Travels* throws little new light, if any, upon the region's flora. Maximilian was familiar with the writings of those who had preceded him, for he refers frequently to the journals of Lewis and Clark (always spelled "Clarke" in the *Travels*), of Brackenridge, of Bradbury, and of Long (and his collaborator Say). Some of these references appear in the text, some in footnotes; there are even comments on Mrs. Trollope's *Domestic manners of the Americans* (". . . the authoress is probably right in many points . . .") which, if he saw a copy only after his return to Germany, suggests that he kept pace with the literature.

Although changes had taken place since the earlier journeys—settlements were advancing up the Missouri, some towns had increased in size, others had been abandoned—there is little to suggest that steamboat navigation had improved, for the *Yellowstone,* like the *Western Engineer* of Long's day, had continual trouble with "snags" [5] and "sawyers," with low water—to unload and load again was a frequent necessity—and with much else. But delays had advantages: they gave the German scientist, intensely interested in everything that was to be seen, an opportunity to go ashore with frequency, and the artist Bodmer was able to paint and sketch and to collect plants and much else. At Fort Pierre, on June 5, Maximilian and the other passengers were transferred to the *Assiniboine* which took them to Fort Union.

The entire trip is described in great detail by Maximilian but I shall merely mention a few landmarks along the way, tell of the Prince's experiences, quote some of his comments about the countryside, and so on.

On April 10, ". . . at eleven o'clock . . . the Yellow Stone left St. Louis . . . There were about 100 persons on board . . . most . . . were those called *engagés,* or *voyageurs* . . . lowest class of servants of the Fur Company . . ."

He comments this day that ". . . The bushes of wild plums were covered with snow-white blossoms, and those of the *Cercis Canadensis,* with their red flowers; and I could not help remarking that, in this country, most of the trees and bushes have their flowers before their leaves . . . Towards evening the lofty plane trees, with their white branches, were beautifully tinged with the setting sun . . ."

On April 12 "The hills were covered with forests, where many trees were putting forth leaves, especially the very delicate green foliage of the sugar maple . . . Near some habitations the European peach trees were in blossom . . ."

Already ". . . The Yellow Stone had several times struck against submerged trunks of trees, but it was purposely built very strong, for such dangerous voyages. This was its third voyage up the Missouri. The Fur Company possess another steamer called the Assiniboin which had left St. Louis to go up the Missouri before us . . ."

April 13: ". . . Near the Gasconade, where we took in wood, many interesting plants were in blossom . . . The hills near it are frequently covered with the white and the

5. Bodmer's Plate VI of "Snags (sunken trees) on the Missouri," portrays the steamer in the midst of these obstructions to navigation. As depicted, they are slightly suggestive of diminutive icebergs.

yellow pine, which supply St. Louis with boards and timber for building . . . we came to Côte-Sans-Dessein, an old French settlement of six or eight houses . . . It must have been formerly much more considerable, since Brackenridge calls it a beautiful place. The river has destroyed it . . . While Mr. Bennett, the master of our vessel, landed to visit his family, who lived here, we botanized on the opposite bank, where oaks of many kinds were in blossom, and where the Monocotyledonous plant is found, which is called here Adam and Eve. Its roots consist of two bulbs joined together, of which it is said that, when thrown into the water, one swims and the other sinks. It is held to be good for wounds. The flower was just beginning to appear.

". . . you soon come to Jefferson City . . . the capital, as it is called of the State of Missouri, where the governor resides. It is at present only a village, with a couple of short streets, and some detached buildings on the bank of the river . . ."

April 15: ". . . Near the mouth of Chariton River, there are several islands, covered with willows, poplar, and hard timber. The river here makes a considerable bend; the numerous sand banks did not permit us to proceed in a direct line . . . Some parts of the banks were rent in a remarkable manner by the rapid stream, when the water was high. In many places, large masses, fifteen or eighteen feet in height, had sunk down, with poplars thirty or forty feet high, as well as entire fields of maize, and piles of timber, which form together a wild scene of devastation . . .

"The drift wood on the sand bank, consisting of large timber trees, forms a scene characteristic of the North American rivers; at least I saw nothing like it in Brazil, where most of the rivers rise in the primeval mountains, or flow through more solid ground . . ."

April 16: ". . . At ten o'clock we came to some excessively dangerous parts, where our vessel frequently struck, and we were obliged to stop the engine, and to push by poles. The vessel stuck fast in the sand, and it was necessary to fasten it to the trees on the bank till it could be got afloat again. At this point the great forests begin to be interrupted by open places, or prairies, and we were at the part called Fox Prairie . . .

"To-day we saw, for the first time . . . the prairies of the Lower Missouri covered with luxuriant young grass . . ."

For most of the 17th, 18th and 19th the *Yellowstone* was in difficulties of one sort and another; a flatboat was procured "to lighten our vessel, by landing a part of the cargo, which was piled up in the wood, on the bank, and covered with cloths. Mr. Bodmer made a faithful sketch of this scene. (Plate IV.)"

At Liberty, Clay County, on the 20th, Maximilian mentions that ". . . the beautiful flowers of the red bud tree, bright green moss, and a thick carpet of verdure, chiefly consisting of the leaves of the May-apple *(Podophyllum,)* everywhere covered the mountains. The papaw trees were just opening their buds. This is about the northern limit for the growth of this tree . . . Before evening we became acquainted with the quicksands of the Missouri . . ."

April 21: ". . . we reached the mouth of the river Konza, or Konzas, called by the French, Rivière des Cans . . . At the point of land between the Konzas and the Missouri, is the boundary which separates the United States from the territory of the free Indians . . .

"We are now in the free Indian territory, and felt much interested in looking at the forests, because we might expect to meet with some of their savage inhabitants. We examined the country with a telescope, and had the satisfaction of seeing the first Indian, on a sand bank, wrapped in his blanket . . .

"The underwood of the forest consisted chiefly of *Laurus benzoin* and *Cercis Canadensis;* the ground was covered with *Equisetum hyemale,* from one and a half to two feet high . . ."

On April 22 they came to ". . . the landing-place of the cantonment Leavenworth, a military post, where four companies . . . of infantry . . . were stationed to protect the Indian boundary . . .

"We were stopped . . . and our vessel searched for brandy, the importation of which is prohibited; they would scarcely permit us to take a small portion to preserve our specimens of natural history . . ."

April 23: ". . . Early in the morning a large branch of a tree, lying in the water, forced its way into the cabin, carried away part of the door case, and was left on the floor . . . one might have been crushed in bed . . . Our navigation was attended with many difficulties to Independence River . . ."

On the 24th they came to the Blacksnake Hills. Near the bank was the trading house occupied by an agent of the Fur Company named "Roubedoux." His house " had a very neat appearance, and Mr. Bodmer sketched this pretty landscape . . ." This artist, I may say, seems to have earned his salt, for at this point he also ". . . brought some beautiful plants from the prairie, among which were the fine orange-coloured flowers of the *Batschia canescens,* which we here saw for the first time."

On April 25 the steamer passed the mouth of the "Nadaway," or Nodaway, River; on the 26th the mouth of the "Grand Nemahaw"; and on the 27th of the "Little Nemahaw." At the last " . . . it appeared to be in imminent peril." For it was driven into a sandbank by a great storm from the southwest.

". . . One of our chimneys was thrown down, and the fore-deck was considered in danger; the large coops, which contained a number of fowls, were blown overboard, and nearly all of them drowned . . . our cables . . . happily, held fast . . . Some of our hunters and Mr. Bodmer appeared on the banks, and wanted to be taken on board, but the boat could not be sent, and they were obliged to seek shelter from the storm in the neighbouring forests . . . persons acquainted with the Missouri, assured us they had never encountered so violent a storm in these parts . . .

"On the following day we were obliged to lighten the ship before we could proceed, by landing the wood which we had taken in the previous day, and many other articles. Our vessel, however, soon ran aground again, and as we could not proceed, we made the vessel go backwards to the right bank, where we passed the night . . ."

At the "Narrows of the Nishnebottoneh," or Nishnabatona, on April 29, Maximilian was returning to the ship when ". . . the pilot called out that there was a rattlesnake very near me, the rattle of which he heard; I looked, and immediately found the animal, and having stunned it with some slight blows, I put it into a vessel in which there was already a live heterodon and a black snake, where it soon recovered. The three agreed very well together, but were afterwards put into a cask of brandy to go to Europe. This rattlesnake was of the species *Crotalus tergeminus,* first described by Say, which is very common on the Missouri . . ."

May 1: ". . . during the night we had observed some fireflies . . . About noon a white cat-fish was caught by one of the lines which we had thrown out; a second broke the strong line as we were drawing it in. The first we had caught weighed sixty pounds, and we soon took another weighing sixty-five pounds, and a third weighing 100 lbs, in the jaws of which was the hook of the line that had been broken . . . It is only on the Upper Missouri that this fish attains so large a size."

May 2: ". . . we reached Five Barrel Islands . . . The forest was picturesque but not very lofty; the bird cherry was in flower, but the blossoms of the red bud had lost their bright colour. Vines twined round the trunks of the trees, and the numerous blossoms of the Phlox formed blue spots amongst the rocks. Towards night we met a canoe, with two persons on board, one of whom was M. Fontenelle, clerk to the Fur Company, who resided near at hand at Belle Vue. He was a man who had much experience in trade with the Indians, and had often visited the Rocky Mountains. As he was shortly to undertake an expedition to the mountains, with a body of armed men, he turned back with us."

On May 3 they passed the two mouths of "La Platte River," and then ". . . Papilion Creek, and saw . . . the buildings of Belle Vue, the agency of Major Dougherty . . . At two in the afternoon . . . M. Fontenelle's dwelling . . ." On May 4 they came to ". . . the white buildings of Mr. Cabanné's trading post . . ." Here were gathered many Omaha and Oto Indians. Bodmer sketched an Omaha boy whom ". . . the father first daubed with red paint. He took vermilion in the palm of his hand, spat upon it, and then rubbed it in the boy's face. The head of this boy was shaved quite smooth, excepting a tuft of hair in front, and another at the back. (See the wood-cut.) . . ."

Maximilian refers to the fact that the "Omahas, or, as some erroneously call them, Mahas," had once been a numerous tribe but had been reduced by wars with their neighbors and by the dreadful ravages of the smallpox,[6] till now there remained "but few vigorous young men among them."

6. The "Translator's preface" (pp. viii-x of the *Travels*) contains a transcription of a letter, dated "New

May 5: ". . . we passed the mouth of Boyer's Creek on the east bank, where the Missouri makes a bend, and saw the ruins of the former cantonment, or fort, at Council Bluffs. This military post was established, in the year 1819 . . . In the year 1827 . . . troops were withdrawn and stationed at Leavenworth . . . At present there were only the stone chimneys, and . . . a brick storehouse under roof . . . everything of value had been carried away by the Indians . . ."

They were passing regions where the ". . . Omaha Indians hunt on both banks . . . of the river; they are said to be the most indolent, dull, unintellectual, and cowardly of the Missouri Indians."

May 7 was ". . . the anniversary of our departure from Germany . . ."

May 8: ". . . we came to Floyd's Grave, where the sergeant of that name was buried by Lewis and Clarke . . . A little further up is Floyd's River . . . About half a league beyond . . . is the mouth of the Big Sioux River, interesting from the circumstance of its being the boundary of the territory of the Dacota, or Sioux nation . . ."

There had been a violent thunderstorm that day and the *Yellowstone* had been obliged to take shelter under a bank, but the journey was resumed on the 9th.

". . . We passed along wild, desolate banks, then a green prairie, by a chain of steep hills, partly bare, partly covered with forests, or with isolated fir trees and picturesque ravines, with dark shadows, into which the close thicket scarcely allowed the eye to penetrate. We here saw, for the first time, a plant which now became more and more common; namely, the buffalo-berry-bush (*Sheperdia argentea,* Nutt.), with pale, bluish green, narrow leaves . . ."

May 10: ". . . we had been exactly four weeks since we left St. Louis. At the spot where we now were, it is said that large herds of buffaloes are seen in the winter, but we have not yet met with one of these animals. The character of the country was much changed; it is, for the most part, naked, and without woods. The trees which are found here are no longer lofty and vigorous, as on the Lower Missouri; yet the wild vines are still seen climbing on the bushes, though this too, entirely ceases further up the river [.] Near the mouth of Vermilion Creek, the green hills of the prairie approach very near the river . . . At the mouth . . . we saw wild ducks and geese . . . the latter, with six young ones, anxiously endeavoured to escape us. The female remained faithfully with the young ones, while the male flew away."

On May 11 they reached the mouth of "Jacques," or James, River. They caught up with the *Assiniboine* steamer, which had been held up by low water, and went aboard. ". . . In this steamer there were two cabins, much lighter and more pleasant than

Orleans, *June 5, 1838*," which as H. Evans Lloyd notes, is "very affecting." It describes the horrible epidemic which broke out in June of 1837 among the Mandans. Whether it was written by Maximilian is not stated. The letter ends: "According to the most recent accounts, the number of the Indians who have been swept away by the small-pox, on the western frontier of the United States, amounts to more than 60,000."

those in the Yellow Stone; the stern cabin had ten berths, and the fore cabin twenty-four, and between decks was the large apartment distinct for the *engagés* . . . we took advantage of the fine weather to make an excursion into the prairie.

". . . In the prairie itself there were many pools of water, and we found several interesting plants, among which were some with long roots like carrots, especially the yellow flowering *Batschia longiflora* (Pursh.), and the *Oxitropis Lamberti B.* The great yellow-breasted lark (*Sturnella,* Vieill.), was everywhere seen in pairs, and its short, coy call, and its pleasing, whistling note, were heard from every side . . . we saw the prairie hen, and the long-billed curlews *(Numenius longirostris)* . . . Skeletons of buffaloes were scattered in the plain . . ."

They were among the "Punca Indians" and Bodmer painted one of their chiefs— "(See the portrait, Plate VII, Fig. 3, which is a good resemblance.)" On May 12 both steamers were having trouble with low water but eventually they got started and ran for a time downstream, with such rapidity that some Indians aboard ". . . became giddy, and sat down on the floor." After landing them at their village, Maximilian witnessed ". . . a great prairie fire, on the left bank. The flames rose from the forest to the height of 100 feet—fiery smoke filled the air: it was a splendid sight! A whirlwind had formed a remarkable towering column of smoke, which rose, in a most singular manner, in graceful undulations, to the zenith . . ."

May 13: ". . . the Yellow Stone passed the mouth of the Running-water River (l'eau qui court)[7] . . . The Assiniboin was before us. We reached the mouth of Punca Creek . . . In this neighbourhood are many villages of the prairie dogs (*Arctomys ludoviciana,* Ord.), in the abandoned burrows of which, rattlesnakes abound . . . On a sudden, three Punca Indians appeared and hailed us . . . Though they made signs to us to take them on board,[8] we did not stop, but renounced the pleasure of more closely observing these interesting people. The trees on the edge of the prairie, by which we passed, were old, thick, and low, with their summits depressed and cramped. They were the resort of the Carolina pigeon . . . The red cedars, in particular, were stunted and crippled, often thicker than a man's body in the trunk, and very frequently wholly withered . . . We proceeded along the hills of the left bank . . . where red cedar abounded, and we stopped to fell a number of these trees. A wild lateral ravine here opened to the Missouri, up the steep sides of which our wood-cutters climbed, and cut down the cedars, which were loaded with their black berries. The wood of this tree emits a very aromatic scent and is much used by the steam-boats for fuel, because it supplies a great deal of steam, and the berries, as we were told, are eaten by the Indians for certain medicinal purposes. At the bottom of the narrow ravine, there was a thicket of elm, cedar, bird-

7. In a footnote the Prince mentions that "It likewise bears the name of Rapid River. Bradbury gives the names of some plants which he gathered on its banks." It is now commonly called the Niobrara.

8. These are the first "thumbers" encountered in my reading! One was smart enough to keep up with the boat, which moved slowly, and eventually got aboard. "Major Bean gave him some tobacco, powder, lead, and ball . . . he returned, well satisfied, to his comrades."

cherry, clematis, celtis, celastrus, vine, and other shrubs; and the neighbouring of lofty verdant hills of the prairie produced many beautiful plants, among which was *Stanleya pinnatifida,* with its splendid long bunches of yellow flowers . . .”

On May 14 they “. . . had a very difficult navigation, and were even obliged to put back, so that the Assiniboin overtook and passed us . . .” The steamers kept close to each other, first one ahead, then the other—one, certainly, always in difficulty!

“. . . In the preceding year, the whole prairie was seen from the steamer to be covered with herds of buffaloes, but now there were no living creatures, except a few wild geese and ducks, which had likewise become scarce, since the termination of the great forest below La Platte River . . .”

May 15: “. . . we reached the place where we had stopped the preceding night[9] . . . we again overtook the Assiniboin . . . At this place there was a narrow deep ravine of a small stream, now dry . . . We climbed from the bottom of the ravine up the singular eminences of the prairie, and collected some interesting plants, particularly the wild turnip. Two species of cactus were not yet in blossom; they are, probably, not sufficiently known to botanists. One of them has been taken for the *Cactus opuntia;* and Captain Back, too, says, that it is found on an island in the Lake of the Woods; but this is certainly not the above-named plant . . .”

May 16: “. . . we reached . . . the Cedar Island, which is said to be 1,075 miles from the mouth of the Missouri[1] . . .”

On May 17 they hunted for “the pomme blanche, which was very common . . .” On the 18th they saw their first buffalo, but the hunters got none, only a large buck antelope. Maximilian found “. . . the ground, on the long-extended ridge, covered with the blue flowers of the *Oxitropis Lamberti* (Pursh.), which grew in tufts about a foot high . . . In the daytime we suffered great heat in these excursions, while there was also a high wind, and the ground was hard and dry; the soles of our shoes became so polished on this ground and the hard dry grass, that it was difficult and fatiguing to walk on the slopes. We were forced to remain here many days, because the water was very shallow . . .”

Actually the steamer was floated on the 22nd, but at a poor moment for Maximilian:

“. . . We happened to be on the hills when the bell summoned us on board, and hastened as quickly as possible to the bank, but came too late, and were compelled to follow the vessel for a couple of hours, clambering over fragments of stone, pieces of

9. Indicating that they even went backwards at times!

1. In his *Travels* (p. 78, *fn.*) Bradbury mentions “Little Cedar Island” as “1075 miles from the mouth of the Missouri.” The Missouri River boasts a number of “Cedar” Islands. Maximilian was now between Charles Mix and Gregory counties, South Dakota. Bradbury's island was further upstream and was the one on which Fort Recovery was located, and which was situated, according to Chittenden, “. . . a mile below the modern city of Chamberlain [Brule County, South Dakota] . . .” (3: 952, 953)

rock, to creep through thickets full of thorns and burs, or to wade through morasses; and not until eleven o'clock did we get on board . . ."

The *Yellowstone* went aground again on the 24th, and Maximilian and McKenzie explored "the neighbouring eminences," finding "many rare plants." On May 25 they reached the "Sioux Agency, or, as it is now usually called, Fort Lookout[2] . . ." The boat stopped above it but the Prince "returned, in a heavy rain, through the bushes and high grass," in order to learn something about the Indians whom he calls the ". . . Dacotas . . . or the Sioux of the French . . ." Bodmer painted a portrait of one of "the most considerable men among them," called "Wahktageli," or the Big Soldier, and Maximilian considered it "a capital likeness"; the Indian had "remained the whole day in the position required." On May 27 the *Yellowstone* moved on with some Indians aboard:

". . . as the vessel proceeded very quickly, our Indians laid down their heads as a sign that they were giddy, but they were soon relieved, as the water became shallow. We lay to not far above the stream which Lewis and Clarke call the Three Rivers.[3] Here we again had leisure to make an excursion in the wood, where the ground was covered with the pea vine *(Apios tuberosa),* and a plant resembling convallaria . . ."

May 28: ". . . We . . . reached the Big Bend which the Missouri takes round a flat point of land; following the course of the river, it is twenty-five miles round, while the isthmus is only one mile and a half across . . . The Little Soldier sat by the fireside, smoking his pipe, in doing which, like all the Indians, he inhaled the smoke, a custom which is, doubtless, the cause of many pectoral diseases. The tobacco, which the Indians of this part of the country smoke, is called kini-kenick, and consists of the inner green bark of the red willow, dried, and powdered, and mixed with the tobacco of the American traders. According to Say, they also smoke the leaves of the arrow-wood *(Viburnum),* when they have none of the bark."

May 29: ". . . we were nearly at the end of the Big Bend, and stopped . . . to cut down cedars . . . To the south, we saw the tops of the Medicine Hills . . . we suddenly espied a canoe, with four men in it . . . a boat was put out, and brought back two of the strangers, who proved to be Mr. Lamont, a member of the Fur Company, and Major Mitchell,[4] one of their officers, and Director of Fort McKenzie, which is situated near the falls of the Missouri. They came last from Fort Pièrre, and were on their way to St. Louis, but we persuaded them to return with us . . ."[5]

2. It was situated about ten miles above present Chamberlain, South Dakota.

3. These rivers or creeks were in the region of the present Crow-Creek Indian Reservation of Buffalo and Lyman counties, South Dakota.

4. Chittenden identifies him as David D. Mitchell, first ". . . a clerk and then . . . a partner in the Upper Missouri Outfit. He was the builder of Fort McKenzie in 1832. . ." (1: 388, 389)

5. The plans of executives in the 1840's seem to have been delightfully flexible! Mr. Fontenelle on May 2 had also found it quite possible to adapt his activities to the wishes of those whom he met.

May 30: ". . . Before six, in the evening, we reached the mouth of Teton River, or the Little Missouri, which the Sioux call the Bad River . . . Fort Pièrre (so called after Mr. Pièrre Chouteau) was erected higher up, on the west bank, opposite an island.[6]

"The steamer had proceeded a little further, when we came in sight of the Fort, to the great joy of all on board; the colours were hoisted, both on the steamer and on the fort . . . we . . . landed at Fort Pièrre, on the fifty-first day of our voyage from St. Louis. A great crowd came to welcome us . . . There were many Indians among them . . . There seemed to be no end of shaking hands . . .

"The Sioux, who live on Teton River, near Fort Pièrre, are mostly of the branch of the Tetons; though there are some Yanktons here. The former are divided into five branches, and the latter into three . . .

"During our stay here, on board the vessel, we were continually besieged by Indians, who did not move from the spot. Our time was, therefore, divided between these visitors and our excursions into the prairie. On the 2nd of June, 7,000 buffalo skins and other furs were put on board the Yellow Stone, with which it was to return to St. Louis. We took this opportunity of sending letters to Europe: the Assiniboin was assigned us for the continuation of the voyage . . . I made, on the 4th of June, an interesting excursion into the prairie . . ."

The *Assiniboine* left Fort Pierre on June 5; according to Maximilian, it was "perfectly equipped for the voyage up the river." On the 6th he notes that "The addition to our Flora was very considerable," for the steamer had again run aground and they stayed in one spot till June 7.

". . . This delay gave us time to make an excursion. In company with Mr. Bodmer, I ascended the slippery, very steep eminences along the river . . . The lower part of these eminences is generally covered with plants, particularly grasses, while the upper is bare . . . The climbing up these high, slippery ascents in the heat of the day was rather fatiguing. When we came into the clefts between the pyramids, we found the ground, in general, slimy, and so adhesive that we were almost compelled to leave our shoes behind. In such places, some old red cedars, groups of the bird cherry, ashes, roses, &c., were nourished by the moisture. Near the hills, and in the plain, a cactus, with roundish, flat joints, grew in abundance. It was not yet in blossom, and I cannot say whether it is the plant taken by Nuttall, for *Cactus opuntia;* probably it is *Cactus ferox* . . ."

June 8: ". . . Frequently taking soundings, we proceeded but slowly in the shallow Missouri, and, early in the afternoon, reached the place where the timber for building Fort Pièrre had been felled. From this place it is fifteen miles to the mouth of the

6. Fort Pierre replaced Fort Tecumseh. Chittenden (3: 955, 956) locates it ". . . 3 miles above the mouth of the Teton and about a quarter of a mile from the Missouri . . ." Work on the fort was begun in 1831 and it had been christened a year, nearly to the day, before Maximilian's visit.

Chayenne River . . . we could not reach the mouth of the Chayenne till . . . the following morning . . ."

On June 10th and 11th game seems to have been more abundant than hitherto. Near the mouth of the Little Cheyenne River they saw many elk, Mr. Bodmer shot a "very large male antelope," and there were many traces of beaver. At the mouth of "Moreau's River," they had reached, we are told, ". . . the southern boundary of the territory of the Arikkaras, though they often make excursions far beyond it."

June 12: ". . . our cannon, muskets and rifles were loaded with ball, because we were approaching the villages of the hostile Arikkaras. We came to Grand River . . .

"The Arikkaras . . . are a branch of the Pawnees, from whom they have long since separated . . . Manoel Lisa, a well-known fur trader, had formerly built a trading house in this country, of which nothing now remains; though the place is still called Manoel Lisa's Fort . . ."

On June 14 they reached "Cannon-ball River."

". . . This river has its name from the singular regular sand-stone balls which are found in its banks, and in those of the Missouri in its vicinity. They are of various sizes, from that of a musket-ball to that of a large bomb . . . Such sand-stone balls are met with in many places on the Upper Missouri; and former travellers have spoken of them . . . A mile above the mouth of the Cannon-ball River, I saw no more of them. The Missouri has risen considerably; and during the night, our people were obliged to keep off, with long poles, the trunks of trees that came floating down the river, without being able to prevent our receiving shocks which made the whole vessel tremble."

June 15: ". . . the river has risen nine inches, and brought down much wood and foam, which was expected, for it is reckoned that, in the month of June, the Missouri is twice much swollen from the melting of the snow in the Rocky Mountains . . . we saw, on the eastern bank, a chain of table hills, quite flat at the top, which extends for a considerable distance. The river turns, to the westward, towards this interesting chain, which is called the Mountains of the Old Mandan Village . . . We came to the site of the old Mandan village, which was situated, at the foot of the hills, in a fine meadow near the river; some poles, that were still standing, were the only remains of it; there was no village here at the time of Lewis and Clarke's journey . . . We are now in the territory of the Indian tribe of the Mandans . . ."

June 16: ". . . We soon reached the mouth of Heart River . . ."

June 17 was spent in a long parley with some Sioux, "of the branch of the Yanktonans," and McKenzie seems to have handled the situation well; Maximilian mentions that "The Yanktonans are represented as the most perfidious and dangerous of all the Sioux, and are stated frequently to have killed white men, especially Englishmen, in these parts."

June 18: ". . . We left at an early hour the place of meeting, from which it was twelve miles to Fort Clarke . . . we passed a roundish island covered with willows, and reached the wood on the western bank, in which the winter dwellings of the Mandan Indians are situated . . . As we drew nearer the huts of this village, Fort Clarke,[7] lying before it . . . came in sight, with the gay American banner waving from the flag-staff . . . The Assiniboin soon lay to before the fort . . . 600 Indians were waiting for us . . ."

In charge of the fort was James Kipp. Maximilian devotes only a few pages to Fort Clark and the Indians at this point for he was to visit the post on his return journey. He refers, however, to the Indian custom of shaking hands, and to the necessity of letting ". . . them examine us on all sides." The savages were no respecters of princely personages!

". . . This was sometimes very troublesome. Thus, for example, a young warrior took hold of my pocket compass which I wore suspended by a ribbon, and attempted to take it by force, to hang as an ornament round his neck. I refused his request, but the more I insisted in my refusal, the more importunate he became. He offered me a handsome horse for my compass, and then all his handsome clothes and arms into the bargain, and as I still refused, he became angry, and it was only by the assistance of the old Charbonneau,[8] that I escaped a disagreeable and, perhaps, violent scene . . ."

On June 19 the steamer got under way and reached "the second Mandan village, on the south bank . . ." The country on that side ". . . appeared to us to have some resemblance with many parts on the banks of the Rhine . . ." The population of the region flocked down to the banks:

". . . The appearance of this vessel of the Company, which comes up, once in two years,[9] to the Yellow Stone River, is an event of the greatest importance to the Indians; they then come from considerable distances to see this hissing machine, which they look upon as one of the most wonderful medicines (charms) of the white men. . . ."

Maximilian considered that the spectacle of the Indians grouped on the bank was "The most attractive sight which we had yet met with upon this voyage . . ."

". . . The steam-boat lay to close to the willow thicket, and we saw, immediately before us, the numerous, motley, gaily painted, and variously ornamented crowd of the most elegant Indians on the whole course of the Missouri . . .

7. The fort was named for William Clark. Chittenden (3: 957) gives the site as ". . . on a bluff in an angle of the river and on its right [west] bank, 55 miles above the N[orthern] P[acific] R. R. bridge at Bismarck, N. D. . . ." It was about ten miles below the confluence of Big Knife River, Oliver County, North Dakota, and on the opposite side of the Missouri from Fort Mandan, winter quarters of Lewis and Clark. It was undoubtedly one of the most important posts on the river.

8. The interpreter.

9. The *Yellowstone* had been in service since April of 1831. The *Assiniboine* was on its maiden voyage. Chittenden (1: 338-341) has much to tell of the interest which steamboat service on the Missouri aroused, not only in the United States but in Europe.

"The expression of their remarkable countenances, as they gazed at us, was very various; in some, it was cold and disdainful; in others, intense curiosity; in others, again, good-nature and simplicity . . ."

Although the Prince expresses regret that it was impossible, "by any description, to give the reader a distinct idea of such a scene," he described what he saw very graphically. Bodmer, unfortunately, had not "sufficient time" to make a drawing of the assemblage.

By June 21 they reached the ". . . mouth of the Little Missouri, which is reckoned to be 1670 miles from the mouth of the Great Missouri . . . Before noon we reached the territory of the Assiniboins . . ." Maximilian found two of these Indians ". . . to be very quiet, obliging men. Thus, for instance, they never returned from an excursion on shore, without bringing me some handfulls of plants, often, it is true, only common grass, but they had observed that we always brought plants home with us."

On the 22nd the *Assiniboine* caught fire and, although it was soon put out, the danger ". . . was not inconsiderable, as we had many barrels of powder on board." Next the current of the river was so powerful that the ship was driven ashore and "the lower deck gallery was broken to pieces." Next the paddle box was broken and carried away by the current—it seems to have been one of the worst days of the trip.

On June 23 they had their first meeting with some Assiniboine Indians, "a branch of the Sioux." They were given presents of food which "seemed to please them much. They said that since they came to these parts in the spring, they had suffered much from want of food, buffaloes being scarce . . .

"After dinner, we proceeded along the side of a prairie, where we heard the note of the great curlew. The valley of the river was bounded on both sides by very remarkable whitish-grey, obliquely stratified ridges, with singular spots of red clay, and bushes in the ravines; at their feet was the prairie, covered with pale green artemisia; and on the tongues of land, at the windings of the Missouri, there were fine poplar groves, with an undergrowth of roses in full bloom, buffalo-berry bushes, and many species of plants . . .

". . . we saw, for the first time, the animal known by the name of the bighorn, or the Rocky Mountain sheep, the *Ovis montana* of the zoologists. A ram and two sheep of this species stood on the summit of the highest hill, and, after looking at our steamer, slowly retired. These animals are not frequent hereabouts, but we afterwards met with them in great numbers."

June 24: ". . . we came to the mouth of Muddy River (the White Earth River of Lewis and Clarke), which issues from a thicket on the north bank . . . Following the numerous windings of the Missouri . . . we reached, at seven o'clock in the evening, the mouth of the Yellow Stone, a fine river, hardly inferior in breadth to the Missouri at this part . . . The two rivers unite in an obtuse angle; and there is a sudden turn of the Missouri to the north-west . . . A little further on lay Fort Union, on a verdant plain, with the handsome American flag, gilded by the last rays of evening, floating in the azure sky . . .

"As the steamer approached, the cannon of Fort Union fired a salute, with a running fire of musketry, to bid us welcome, which was answered in a similar manner by our vessel. When we reached the fort, we were received by Mr. Hamilton, an Englishman, who, during the absence of Mr. McKenzie, had performed the functions of director, as well as by several clerks of the Company, and a number of their servants (*engagés or voyageurs*), of many different nations, Americans, Englishmen, Germans, Frenchmen, Russians, Spaniards, and Italians . . . with many Indians, and half-breed women and children. It was the seventy-fifth day since our departure from St. Louis, when the Assiniboin cast anchor at Fort Union."

From the time Maximilian left Fort Clark in North Dakota (June 19) until his return there (November 8) he was in country where very little, if any, plant collecting had been done; he was, as we shall see, to ascend the Missouri from Fort Union to Fort McKenzie at the mouth of Marias River.

Lewis and Clark had shipped east from Fort Mandan plants collected before reaching that point, but others gathered between there and the Rocky Mountains were lost by the flooding of a cache. Lewis, returning east, had reached high up the Marias River and from there had descended the Missouri to the mouth of the Yellowstone—Maximilian's route in reverse—but his trip had been rapid and even the combined collections of the two explorers on their entire eastward journey was said by Pursh to have been "small."

Wyeth, returning from the Pacific coast in 1833, reached Fort Union by way of the Yellowstone but while in that neighborhood[1] was occupied in building his famous bullboat. Once embarked in that craft his plant collecting ceased—at "the first navigable waters of the Missouri."

Maximilian was, therefore, virtually the first in what should have been a fruitful botanical field. Unfortunately he was there at a poor season for plants and, moreover, it was not a region where white men could then wander about in safety.

A chapter of the *Travels* is devoted to "Fort Union and its neighbourhood." The author tells that the fort had been begun by McKenzie in 1829 and, by 1833, was one of the most important posts of the American Fur Company, having ". . . superintendence of the whole of the trade in the interior, and in the vicinity of the mountains." There were two posts "higher up"—Fort Cass, 200 miles up the Yellowstone River, and "Fort Piekann [Piegan], or as it is now called, Fort McKenzie[2] . . . 850 miles up the Missouri . . ." The last traded with ". . . the three tribes of the Blackfoot Indians."

Maximilian describes the size of the fur enterprise backed by Astor and his part-

1. Wyeth's sojourn at Fort Union extended from August 24-27. Maximilian was there both before (June 24 –July 8) and after (September 9–October 30) Wyeth.

2. Chittenden's map shows Fort Piegan to the north of the mouth of Marias River, Fort McKenzie to the south; both were on the west side of the Missouri River and both in what is now Choute County, Montana.

ners, involving many agents at the different posts, great numbers of trappers in the field, and so on. At Fort Union alone ". . . artisans of almost every description are to be met with, such as smiths, masons, carpenters, joiners, coopers, tailors, shoemakers, hatters, &c." But the company was making hay while the sun shone, rather than looking to the future, for the Prince notes that ". . . Wild beasts and other animals, whose skins are valuable in the fur trade, have already diminished greatly along this river, and it is said that, in another ten years,[3] the fur trade will be very inconsiderable. As the supplies along the banks of the Missouri decreased, the Company gradually extended the circle of their trading posts, as well as enterprises, and thus increased their income . . ."

Maximilian's presentation of the subject is well worth reading, but I shall merely quote from an included statement which summarizes an average year's trade in pelts:

"I. Beavers: about 25,000 . . . 2. Otters: 200-300. . . 3. Buffalo cow . . . 40,000 to 50,000 . . . 4. Canadian weasel . . . 500 to 600. 5. Martin . . . about the same quantity. 6. Lynx, the northern lynx . . . 1,000 to 2,000. 7. Lynx; the southern or wild cat . . . ditto. 8. Red Foxes . . . 2,000. 9. Cross foxes . . . 200 to 3,000. 10. Silver foxes: twenty to thirty. Sixty dollars are often paid for a single skin. 11. Minks . . . 2,000. 12. Musk-rats[4] . . . from 1,000 to 100,000. 13. Deer . . . from 20,000 to 30,000 . . . The elk . . . is not properly comprehended in the trade, as its skin is too thick and heavy, and is, therefore, used for home consumption . . ."

More is said of the buffalo:

". . . It is difficult to obtain an exact estimate of the consumption of this animal, which is yearly decreasing and driven further inland. In a recent year, the Fur Company sent 42,000 of these hides down the river, which were sold, in the United States, at four dollars a-piece. Fort Union alone consumes about 600 to 800 buffaloes annually, and the other forts in proportion. The numerous Indian tribes subsist almost entirely on these animals, sell their skins after retaining a sufficient supply for their clothing, tents, &c., and the agents of the Company recklessly shoot down these noble animals for their own pleasure, often not making the least use of them, except taking out the tongue. Whole herds of them are often drowned in the Missouri; nay, I have been assured that, in some rivers, 1,800 and more of their dead bodies were found in one place. Complete dams are formed of the bodies of these animals in some of the morasses of the rivers; from this we may form some idea of the decrease of the buffa-

3. This was apparently an excellent estimate for Chittenden records that "The true period of the trans-Mississippi fur trade . . . embraces the thirty-seven years from 1807 to 1843." (1: 2)

In 1845 Robert Greenhow reported that the fur trade was ". . . declining every where." Among the papers appended to Greenhow's history is one entitled "Furs and the Fur Trade" (pp. 411-413) which contains much that is interesting about the uses made of the furs and about the important fur markets of the world.

4. He adds in a footnote: "At Rock River which falls into the Mississippi, the Indians caught, in 1825, about 130,000 musk-rats; in the following year, about half the number; and, in about two years after, these animals were scarcely to be met with . . ."

loes, which are now found on the other side of the Rocky Mountains, where they were not originally met with, but whither they have been driven."

Of the vegetation about Fort Union Maximilian wrote:

". . . The hills are partly bare, and very few flowers were in blossom; the whole country was covered with short, dry grass, among which there were numerous round spots with tufts of *Cactus ferox,* which was only partly in flower. Another *cactus,* resembling *mammillaris,* with dark red flowers, yellow on the inner side, was likewise abundant. Of the first kind it seems that two exactly similar varieties, probably species, are found everywhere here; both have fine, large, bright yellow flowers, sometimes a greenish-yellow, and, on their first expanding, are often whitish, and the outerside of the petals, with a reddish tinge; but in one species, the staminae are bright yellow, like the flower itself, and, in the other, of a brownish blood red, with yellow anthers. The true flowering time of these plants begins at the end of June.

". . . Between the hills, there are, sometimes, in the ravines, little thickets of oak, ash, negundo maple, elm, bird-cherry, and some others, in which many kinds of birds, particularly the starling, blackbird, &c., build their nests . . .

"Not far above and below the fort there are woods on the banks of the Missouri, consisting of poplars, willows, ash, elm, negundo maple, &c., with a thick underwood of hazel, roses, which are now in flower, and dog-berry, rendered almost impassable by blackberry bushes and the burdock *(Xanthium strumarium),* the thorny fruit of which stuck to the clothes. In these thickets, where we collected many plants, the mosquitos were extremely troublesome. In such places we frequently heard the deep base note of the frogs; and in those places which were not damp, there were patches of two kinds of solidago; likewise *Gaura coccinea* (Pursh.), and *Cristaria coccinea,* two extremely beautiful plants; and, on the banks of the river, the white-flowering *Bartonia ornata* (Pursh.), and the *Helianthus petiolaris* (Nutt.), which were everywhere in flower, &c., &c."

Maximilian remained at Fort Union from June 24 until July 6—on July 4 it had been just one year since he had landed in Boston—and devoted most of his time to studying the Indians. Bodmer was kept extremely busy with paintings and vignettes. The Prince's private servant, Dreidoppel by name,[5] does not appear in the *Travels*— or so I believe—until after Fort Union had been reached; but he too was kept busy, hunting and collecting; one of his important discoveries was a tree ". . . on which the corpses of several Assiniboins were deposited . . ." and where he collected a skull that had fallen to the ground and ". . . in which a mouse had made its nest for its young; and Mr. Bodmer made an accurate drawing of the tree, under which there was a close thicket of roses in full blossom, the fragrant flowers of which seemed destined to veil this melancholy scene of human frailty and folly. (See Plate XXX.)"

Preparations were made for a trip to Fort McKenzie, in the Company's keelboat

5. Dreidoppel is reported to have served with Maximilian in Brazil.

Flora. The number aboard, wrote Maximilian, ". . . including us travellers, amounted to fifty-two persons . . ." McKenzie went only a little way; Mitchell was in charge and Culbertson's presence is mentioned in the course of the journey. After fireworks, salutes, and so on, they got started on July 6. The boat had to be towed and rowed, and by the 8th pushed as well; on rare occasions sails were used; there were "many difficulties" and they did not reach their destination until August 9. Maximilian devotes two chapters to the journey (XVII, XVIII).

July 8: ". . . The forest through which our men passed, had, in these parts, a very thick underwood of rose and buffalo-berries . . . we came to the mouth of a stream . . . called, by the Canadians, La Rivière aux Trembles, and by Lewis and Clarke, Martha's River[6] . . ."

July 10: ". . . The bushes of dog-wood, symphoria, and roses were so full of mosquitoes, that when we had discharged our pieces, it was difficult to reload them . . . In the neighbouring prairie we found the cactus plant, which we have before mentioned, covered with the most beautiful flowers, which attracted vast numbers of insects . . ."

July 11: ". . . We had now passed a place called L'Isle au Coupé (the cut-off) . . . *Helianthus petiolaris,* in full size and beauty, as well as the two species of willows *(Salix longifolia* and *lucida)* already mentioned, grew on the banks of the river . . . The prairie extended, without interruption, as far as the eye could reach; it is called Prairie à la Corne de Cerf, because the wandering Indians have here erected a pyramid of elks' horns . . ."

July 12: ". . . Several interesting plants were gathered, among which were the *Asclepias speciosa,* with large fragrant flowers, and a new species of *lactuca* or *prenanthes* . . ."

Juy 13: ". . . The wood here was so thickly matted with willows, roses, dog-berry, and many burrs and other troublesome plants, and likewise so full of dry broken wood and rubbish, lying on the ground, that it was excessively difficult to penetrate. I followed, alternately, the paths trodden by buffaloes, elks, bears, and deer, and at length got into such an intricate thicket, that it was not till after many hours of painful and fatiguing exertion, that I was so fortunate as to find our vessel; but all my clothes were completely torn to rags . . ."

July 14: ". . . We . . . saw, for the first time, a beautiful plant, which is frequent from hence further up the river, the *Rudbeckia columnaris* (Pursh), the petals of which are half orange-colour and half brown . . ."

On July 15, after considerable exertion, Mitchell and Culbertson took a swim and

6. The stream—now known as Big Muddy River—enters the Missouri from the north in Roosevelt County.

Maximilian, with opinions upon every subject, including when it was wise to bathe, comments that ". . . my American friends, heated as they were, threw themselves into the water to refresh themselves." On the 16th, after an unsuccessful and "fatiguing excursion of six hours" in pursuit of buffalo, Dreidoppel was discovered roasting an antelope:

". . . while he was busy in preparing the skin . . . for my zoological collection, he suddenly perceived two large white wolves standing about ten paces from him . . . He might have shot them both, had not the ramrod of his rifle been broken . . ."

I am still uncertain whether Dreidoppel was an expert with a gun or the reverse! This day "The wood . . . was full of gooseberries, of a pleasant acid taste . . . The shrub which bears these black berries is thickly set with reddish thorns, almost like *Robinia hispida*."

On the 18th Maximilian ". . . could not help making comparisons with my journeys on the Brazilian rivers. There, where nature is so infinitely rich and grand, I heard, from the lofty, thick, primeval forests on the banks of the rivers, the varied voices of the parrots, the macaws, and many other birds, as well as of the monkeys, and other creatures; while here, the silence of the bare, dead, lonely wilderness is but seldom interrupted by the howling of the wolves, the bellowing of the buffaloes, or the screaming of the crows. The vast prairie scarcely offers a living creature, except, now and then, herds of buffaloes and antelopes, or a few deer and wolves. These plains, which are dry in summer, and frozen in winter, have certainly much resemblance, in many of their features, with the African deserts . . ."

At this period the *Travels* has much to tell of hunting excursions for grizzlies, buffalo, and so on; and much to tell of birds also. Because the boat moved slowly the passengers had ample opportunity to go ashore. There are few references to landmarks along the way, and doubtless few existed. On July 19 they passed "Milk River, on the north bank," on the 21st they were at the mouth of "Big Dry River, which joins the Missouri on the south side." This stream, as shown on modern maps, flows into a southern arm of Fort Peck Reservoir. As they approached the Bad Lands, or "Mauvaises Terres," on the 25th Maximilian comments that ". . . vegetation was for the most part withered: the *Allium reticulatum,* with its white flowers, quite dried up; *Cactus ferox,* poor and shrivelled, and the bones of the buffaloes, bleached by exposure to the air, bore testimony, even in this solitude, to the uncertainty of life . . ."

On July 26 the prairie was covered with grasshoppers—"the whole surface of the ground seemed to be alive . . ." On the 27th they could see Musselshell River which enters the Missouri between Garfield County (east) and Petroleum County (west):

"Muscleshell River, the Coquille of the Canadians, joins the Missouri on the southwest side . . . It is reckoned that its mouth is half way between Fort Union and Fort McKenzie . . . the navigation of the Missouri, from the mouth of the Muscleshell upwards, is more easy than before . . ."

On July 28 the Prince had ". . . occasion to reflect on the rude manners of our crew . . .":

"For some time past we had made a numerous and interesting collection of natural history, many articles of which we were obliged, for want of room, to leave on deck. The skins, skulls of animals, and the like, some of which it had cost us much trouble to procure, were generally thrown into the river during the night, though Mr. Mitchell had set a penalty of five dollars on such irregularities. In this manner I lost many highly interesting specimens; and on board our keel-boat, with the most favourable opportunities, it was hardly possible to make a collection of natural history, if I except the herbarium, which we kept in the cabin, under our eyes, so that we brought but a small part of what we had collected to Fort McKenzie."

The same day Bodmer and Dreidoppel were ". . . employed in collecting most interesting impressions of shells, and very beautiful baculites, of the latter of which there were large, very fine, opalescent specimens." A footnote reports that "The fine collection of all these impressions and petrifactions made on this occasion, has, unfortunately, not reached Europe . . ."

". . . The prairie now alternated with woods of tall poplars, and these trees, probably, do not form, in any part of the globe, such fine and lofty forests as they do here."

Having learned on July 30 that a large bear had been shot on the bank of the Missouri and abandoned—". . . a piece of news which was very agreeable to me, and of which I resolved to take advantage . . ."—Maximilian, in company with Mitchell, Bodmer, Dreidoppel and others, set out on the 31st to find it. They discovered the remains, untouched, took the skin and left the bones tied up in a tree, ". . . intending to take them on our return, after they had been a little more cleaned by the birds of prey and insects." [7]

". . . Laden with our booty, and with the plants we had collected, we proceeded . . . and were just in sight of the keel-boat when, taking advantage of a favourable wind, it hoisted its sail, and left us no alternative but to follow it . . . for three or four hours. Our fatiguing way led through a rough prairie, covered with hard grasses, with the epinette de prairie, helianthus, and prickly cactus, through thick skirts of the forest, with a thorny undergrowth of roses, gooseberries, and burs . . ."

August 1: ". . . we are now approaching the most interesting part of the Mauvaises Terres . . . We here collected several plants, and Mr. Bodmer made a sketch of the mountain tops. (Plate XXXV., Fig. 28)."

August 2: ". . . We saw several islands, among which was doubtless Lewis and Clarke's Good Punch Island, a name which is unworthy of being transmitted to posterity[8] . . ."

7. On September 19 nothing was to be found but bone fragments and the tree bore marks of a bear's claws.

8. One wonders whether Maximilian was an "abstainer" !

By August 3 bighorn sheep were plentiful and some were killed. When cooked Maximilian thought that the meat had ". . . an unpleasant peculiar taste . . . I cannot agree with Ross Cox, who calls it delicious meat; probably because he could find nothing better in many parts of the interior of North America."

We are told that ". . . Mr. McKenzie had promised a hunter to give him a horse if he would bring a young bighorn alive; but, up to this time, he had not been able to procure one. The names of bighorn and grosse corne, given to this animal by the English and French, are properly taken from the large thick horns of the ram, which often weigh forty pounds the two, and make the animal's head appear quite small . . ."

On August 5 the party had many anxious hours when some five hundred Gros Ventres Indians swarmed into and about the keelboat. Mitchell seems to have dealt with them fearlessly and skillfully and nothing calamitous happened but Maximilian refers to the situation as ". . . everything but agreeable."

August 6: ". . . The part of the country called The Stone Walls, which now opened before us, has nothing like it on the whole course of the Missouri; and we did not leave the deck for a single moment the whole forenoon . . . They are the continuation of the white sand-stone which occurs in such singular forms at the Blackhills . . ."

Bodmer of course sketched "very accurately" a number of the "fantastic forms . . . (Plate XXXIV., Figs. 6, 7, 8, 9, and Plate XXXV., Figs. 22, 23, 24, 27, 29)." And many others also. As the boat moved along Maximilian noted that ". . . Extremely stunted and often strangely contorted cedars *(juniperus)* grew among these rocks; but the pines *(Pinus flexilis)* were well grown and flourishing, though not above forty feet high . . . In the prairie beyond the Stone Walls, *Cactus ferox* grew, and at their foot, the beautiful *Bartonia ornata,* with its large snow-white flowers."

On August 7 they passed through ". . . what is called the Gate of the Stone Walls . . ." and on the 8th ". . . again saw before us the summits of the Rocky Mountains . . .

". . . Turning round a point of land, we saw before us a long table-formed range of hills, behind which is Fort Mc Kenzie, which we might have reached by land in half an hour. In the front of these hills, on the north bank, is the mouth of Maria River, called, by the Canadians, Marayon; after we had passed it, we saw . . . on the same bank, the ruins of the first fort, or trading post, which Mr. Kipp, clerk of the American Fur Company, had built, in the year 1831, in the territory of the Blackfeet. This fort was abandoned in 1832, and the present Fort Mc Kenzie built in its stead, and this, too, is soon to be abandoned. In this manner the Fur Company continues to advance, and firmly establishes itself among nations that are but little known, where the fur trade is still profitable . . . we saw two Indians on horseback, who galloped off as soon as they perceived us, doubtless to carry the news of our arrival to the fort . . ."

Mitchell, at this time, was apprehensive as to ". . . the safety of the fort and our expedition . . . we had . . . reason to proceed with much caution." There was a night watch put on, and Mitchell with four men went on a reconnoitering expedition; they returned without having obtained any information ". . . of the state of the fort . . . we had nothing to do but patiently to wait for daylight." The next morning, the 9th, they were welcomed by a party from the fort:

"We passed the last winding of the river, and a most interesting scene presented itself. A prairie extends along the north bank, at a point of which, projecting towards the river, we saw Fort Mc Kenzie, on which the American flag was displayed. A great number of Indian tents was erected in the plain, which was covered with the red population in various groups, all of whom hastened to the bank. Near to the fort, the men (about 800 Blackfeet) were drawn up in a close body, like a well-ordered battalion . . . The palisades and the roof of the fort, as well as the neighbouring trees, were occupied by Indian women and their children . . . the whole prairie was covered with them . . .

"We approached the landing-place, and at length set foot on shore, amidst a cloud of smoke caused by the firing of the Indians and of the *engagés* of the fort . . . we were received by the whole population, with the Indian chiefs at their head, with whom we all shook hands . . . When we arrived at the fort there was no end of the shaking of hands; after which we longed for repose, and distributed our baggage in the rooms. We had happily accomplished the voyage from Fort Union in thirty-four days, had lost none of our people, and subsisted during the whole time by the produce of the chase."

Maximilian made Fort McKenzie his headquarters until September 14. On August 16 he did make a trip as far as the mouth of a stream ". . . called by Lewis and Clarke, Snow River[9] . . . the most extreme point of my journey on the Upper Missouri, though at that time I still hoped to reach the principal sources of that river, the Jefferson, the Madison, and the Gallatin." But at the time of the Prince's visit conditions were far from peaceful because of Indian uprisings and he could not, obviously, venture far from the surroundings of the fort. Before turning to his own account of his sojourn, I shall quote what was written about his visit to Forts Union and McKenzie by someone who saw him "in action" so to speak. This "someone" is generally credited with having been Alexander Culbertson,[1] who was undoubtedly there at the time.

"In this year an interesting character in the person of Prince Maximilian, from Coblentz on the Rhine, made his first appearance in the upper Missouri. The Prince

9. This is the present Shonkin Creek which enters the Missouri River near Fort Benton, Chouteau County.

1. Chittenden (1: 388) includes a biographical sketch of Alexander Culbertson. He was an important man in the American Fur Company which he joined in 1829. For a long time he was in charge of Fort Union.

The account of Maximilian's visit to Fort McKenzie is told in an article entitled "Affairs at Fort Benton from 1831 to 1869. From Lieut. Bradley's Journal." It appeared in the *Contributions of the Montana Historical Society* (3: 206-210, 1900) and, although not directly credited to Culbertson, a note of the editor, A[rthur] J. C[raven] (p. 287) mentions the ". . . invaluable assistance of Major Culbertson . . ." Bernard De Voto, in *Across the wide Missouri* (p. 137) refers to Culbertson as author of the quotation.

Under date of August 28, 1834, J. J. Audubon quotes a description of ". . . a skirmish which took place at Fort MacKenzie . . ." which he says (see *Audubon and his journals*) he copied from Culbertson's manuscript. It does not mention Maximilian, however, and bears no resemblance in wording to the transcription in the *Contributions* mentioned above.

was at that time nearly seventy[2] years of age, but well preserved, and able to endure considerable fatigue. He was a man of medium height, rather slender, sans teeth, passionately fond of his pipe, unostentatious, and speaking very broken English. His favorite dress was a white slouch hat, a black velvet coat, rather rusty from long service, and probably the greasiest pair of trousers that ever encased princely legs. The Prince was a bachelor and a man of science, and it was in this latter capacity that he had roamed so far from his ancestral home on the Rhine. He was accompanied by an artist named Boadman and a servant whose name was, as nearly as the author has been able to ascertain its spelling, Tritripel, both of whom seemed gifted to a high degree with the faculty of putting their princely employer into a frequent passion, till there is hardly a bluff or a valley on the whole upper Missouri that has not repeated in an angry tone, and with a strong Teutonic accent, the names of Boadman and Tritripel.[3]

"The Prince had ascended the Missouri from St. Louis to Fort Union in the steamer Assinaboine, ranging the shore at every opportunity in quest of new objects to add to his collection of small quadrupeds, birds, botanical specimens and fossils; keeping his artist as busy as his easy nature allowed in making sketches of the scenery on the route. Arrived at Fort Union, he requested permission to accompany Mitchell's keel-boat to Fort McKenzie and was allowed to do so. During the voyage he improved the opportunities it afforded, and made constant additions to his collections. He remained at Fort McKenzie about a month, when he was furnished with a small mackinac boat, in which with his party he descended to the Mandan Village, leaving a hearty invitation to Mitchell and Culbertson to visit him in Europe and the promise to send the former the present of a double barreled rifle and the latter a fine meerschaum. He remained at the Mandan Village the following winter, when he had a severe attack of the scurvy, but aided by the restorative qualities of wild onions was enabled to recover and return home to write an account of his travels, which was published in German, with illustrations, and afterwards translated into English. McKenzie subsequently visited him in his palace at Coblentz, where he lived in a style befitting a prince, and was received with great cordiality and entertained with lavish hospitality. He inquired whether the double barreled gun and the meerschaum had reached their destination, as he had remembered his promise and forwarded them soon after his return to Europe. They had not, and never were received, for it subsequently appeared that the vessel in which they were shipped was lost, so they are probably now among the ill-gotten hoards of the Atlantic.

"During the Prince's stay at Fort McKenzie he had an opportunity to witness an Indian battle. In the latter part of August, while a trading party of thirty lodges of Piegans . . . were encamped under the walls of the fort, they were suddenly charged . . . by about fifteen hundred Assinaboine warriors . . . it was supposed that they were

2. Maximilian was born in 1782 and was therefore, it would seem, in the early fifties.

3. "Boadman" was, of course, Bodmer, and "Tritripel" was Dreidoppel.

attacking the fort, and its entire garrison . . . rushed to arms and opened fire upon the advancing swarms. The Prince, too, seized his gun and manned one of the port holes of the upper bastion. His gun was already loaded, but overlooking it in the excitement of the moment, he rammed down another charge of a size proportioned to the extreme gravity of the situation. Then discerning, through his port hole, an Assinaboine warrior within range of his weapon he levelled his gun, covered the person of the miserable Assinaboine with a careful aim, pulled the trigger and proceeded to revolve with great rapidity across the bastion till he came in severe contact with the opposite wall and fell stunned to the floor.[4] The garrison had by this time discovered the real object of the attack, and under the orders of Mitchell had ceased to fire, after inflicting upon the Assinaboines the loss of one man, possibly, though not probably, the Prince's intended victim[5] . . ."

More is told of the battle but not of Maximilian's participation therein.

The *Travels* contains little of interest about the flora of the Fort McKenzie region. There are, however, occasional references to the uses made of certain plants by the Blackfeet Indians; I quote a few scattered examples:

". . . Very often they adorn themselves with a braided necklace, composed of a sweet-smelling grass, probably *anthoxanthum* . . ."

". . . Their tobacco consists of the small, roundish, dried leaves of the sakakomi plant *(Arbutus uva ursi)* . . . The Blackfeet, like most of the tribes of the Upper Missouri, sow the seeds of the *Nicotiana quadrivalvis,* having first burnt the place[6] where they intend it to grow . . ."

". . . From the vegetable kingdom they obtain many roots; the pomme blanche, or white turnip, is very common in their prairies. The women and children dig them up with a particular kind of wooden instrument, and bring them in strings to the Whites for sale . . . Another turnip-like root, called by the Canadians, *racine à tabac,* is buried in the earth with hot stones, and is said to become black, like tobacco, as soon as it is fit to eat; it has a sweetish taste, like parsnips . . ."

"Their country does not produce any wood suitable for bows; and they endeavour to obtain, by barter, the bow wood, or yellow wood *(Maclura aurantiaca),* from the River Arkansas."

---

4. If Bodmer made a picture of his patron at this moment, it is not reproduced in the *Atlas* nor in the *Travels.*

5. Maximilian allowed nothing grewsome to interfere with his scientific pursuits; he wrote: "The Indian who was killed near the fort especially interested me, because I wished to obtain his skull. The scalp had already been taken off, and several Blackfeet were engaged in venting their rage on the dead body . . . Before I could obtain my wish, not a trace of the head was to be seen."

Bodmer's Plate XLII ("Fort McKenzie, with the combat of 28th August 1833") is full of action and indicates that the artist was a trained anatomical draughtsman. It is, possibly, *the* already-scalped Assiniboine who lies in the foreground.

6. David Douglas had been pleased to note that the Indians along the Columbia River had appreciated "the good effects produced on the vegetation by the use of carbon."

"The arrows of the Mandans and Manitaries are neatly made; the best wood is said to be that of the service berry *(Amelanchier sanguinea)*."

States Maximilian:

"It was my intention to pass the winter in the Rocky Mountains, and I had the execution of this project much at heart ... A great number of the most dangerous Indians surrounded us on all sides, and had in particular occupied the country towards the Falls of the Missouri, which was precisely the direction we should have to take."

He was advised against making the trip and, changing his plans, asked for a boat in which to descend the Missouri. Those at Fort McKenzie were fearful of an attack by the Assiniboines and Mitchell "willingly co-operated." Possibly he did not enjoy being responsible for a foreigner and a prince to boot! After planks had been cut for a new "Mackinaw boat," it was built by a shipwright in the courtyard of the fort and, on September 11, was taken down to the Missouri for loading:

"... the necessary arrangements for the voyage were made; large cages were made for my two live bears; and kitchen utensils and beds were procured. The cases, containing my collections, filled a large part of the boat, which, unfortunately, proved too small. I had received from the company Henry Morrin as steersman, and ... three young, inexperienced Canadians ... who were ill qualified for such a voyage, and did not even possess serviceable fire-arms. Thus there were only seven persons in the boat, but the time was most valuable, and I fixed my departure for the 14th of September."

After taking leave of Mitchell and of Culbertson the boat got off on the 14th as planned. Maximilian was evidently in command and one derives the impression that he did not particularly enjoy the journey!

"By noon all our effects were put on board the new boat, and it became more and more evident that we had not sufficient room ... The great cages, with the live bears, were placed upon the cargo in the centre, and prevented us from passing from one end of the boat to the other ... there was not room for us to sleep on board ... a most unfavourable circumstance, because it obliged us always to lie to for the night ...

"Towards four o'clock a thunder-storm came on ... As we had reason to be on our guard against the Indians we regretted that my two bears were unusually dissatisfied with their confinement, and manifested their feelings by moaning and growling, which might very easily have attracted some hostile visitors ... we fastened the boat to some trunks of trees, and passed a very uncomfortable night, lying on our deck, while a heavy rain prevented us from sleeping."

September 15: "... we were in a lamentable plight ... all of us, more or less, wet and benumbed, as the boat had no deck, and we found, to our great dismay, that this new vessel was very leaky, so that the greater part of our luggage was wet through ... as soon ... as we had baled out the greater part of the water, we hastened to proceed ... When the sun had risen a little higher, we landed ... and made a large fire ... Our drenched buffalo robes and blankets were brought on shore to dry, and I discovered, to my great regret, that the pretty striped squirrel (*Tamias quadrivittatus,* Say), which I had hoped to bring alive to Europe, was drowned in his cage . . ."

Maximilian notes that at any other time he would have been highly interested to examine "the remarkable features" of the Stone Walls, "but now I was extremely impatient to know the extent of our loss." So another stop was made:

"As the sun now shone with considerable power, we hastened completely to unload the boat, to open and unpack all the chests and trunks, one by one. How grieved were we to find all our clothes, books, collections, some mathematical instruments, in a word, all our effects, entirely wet and soaked. The chests were, for the most part, open in all the joints, and quite useless; but what afflicted me the most, was my fine botanical collection of the Upper Missouri, made with labour and expense of time, which I could not now put into dry paper, and which therefore, was, for the most part, lost, as well as the Indian leather dresses, which became mouldy. We had now no resource but to remain where we were till most of our things were dried; a most distressing necessity. A large spot of the prairie was covered with our scattered effects, and a wind arising caused some disorder among our goods, and we were obliged to take care that nothing might be lost. My extensive herbarium had to be laid, on account of the wind, under the shelter of the eminences of a small lateral ravine, which took me the whole day, and yet all the plants became black and mouldy . . . after . . . all the chests [were] put on board again, we continued our voyage . . ."

The Prince's obvious desire on this trip was to avoid Indians, not to learn more of their customs! The party moved on as rapidly, and as quietly, as possible; more than one zoological specimen was allowed to escape because it was thought "prudent to avoid all unnecessary noise," and nothing is said about plants or plant collecting. Whether due to good fortune or to wise precautions, not a single human being was encountered for the entire trip of fifteen days.

Maximilian remained at Fort Union from September 29 to October 30. It was autumn and the country ". . . had much changed . . . The whole prairie was naked, dry, and withered; the plants were in seed, which were then covered with flowers; the woods had put on their yellow tint . . ."

As far as botanical work was concerned he evidently relaxed, and by October 27 there was a heavy fall of snow. After exchanging the mackinaw boat for a larger one he and his party left for Fort Clark on October 30. There is a mention of his bears, etc.:

". . . My live animals, which would not eat pork, were half famished, and the bears especially made an incessant growling, which was in every respect highly disagreeable . . . a few grapes were still hanging on the branches, but they were very small and indifferent, and did not suit the taste of even my little fox."

On November 3 Maximilian collected some rocks. ". . . I increased my collections with the most interesting series of the rocks of the Upper Missouri, which, I regret to say, have not reached Europe, as they were irrevocably lost."

And on the 8th some specimens of a petrified tree: ". . . the tree . . . is supposed to be part of an old cedar *(juniperus)*; it is the lower part of a hollow trunk, with a portion of the roots; and though this mass still perfectly shows the formation of the wood, it is now converted into a sounding stone. As the whole of this interesting specimen was much too ponderous to be removed, I carried off a good many fragments, without, however, disfiguring the tree,[7] which will, doubtless, some day, find a place in some museum in the United States. This kind of petrified wood is not, by any means, unfrequent on the Missouri. Of the many interesting specimens of this kind which I had collected, very few have found their way to Europe."

On November 8 Maximilian and his party reached Fort Clark, remaining there for the winter. The Prince spent his time studying the Indian tribes of the Upper Missouri.

Two chapters of the *Travels* (XXIV and XXV) are devoted to the Mandans, one (XXVI) to the "tribe of the Manitaries, or Gros Ventres," and one (XXVII) to "a few words respecting the Arikkaras." In his preface Thwaites points out that Maximilian's opinions about these tribes, written while they were still powerful and before their numbers were reduced by the smallpox epidemic of a few years later, are especially valuable.

I shall quote only a few of the Prince's comments upon the uses made of certain plants by the Mandans:

". . . They smoke the leaves of the tobacco plant, which is cultivated by them; the bark of the red willow *(Cornus sericea),* which they obtain from the traders, is sometimes mixed with the tobacco, or the latter with the leaves of the bearberry *(Arbutus uva ursi).* The tobacco of the Whites, unmixed, is too strong for the Indians, because they draw the smoke into their lungs; hence they do not willingly smoke cigars.

". . . The plants which they cultivate are maize, beans, French beans, gourds, sunflowers, and tobacco *(Nicotiana quadrivalvis),* of which I brought home some seeds, which have flowered in several botanic gardens.

"Of maize there are several varieties of colour, to which they give different names. The several varieties are: —1. White maize. 2. Yellow maize. 3. Red maize. 4. Spotted maize. 5. Black Maize. 6. Sweet maize. 7. Very hard yellow maize. 8. White, or red-striped maize. 9. Very tender yellow maize." [8]

"The beans are likewise of various sorts—small white beans, black, red, and spotted beans. The gourds are—yellow, black, striped, blue, long, and thick-shelled gourds.

"The sunflower is a large helianthus, which seems perfectly to resemble that culti-

7. Maximilian would doubtless have removed the entire tree had it been possible! Nevertheless he is the first and only collector met with in my reading who displays an interest in sparing a plant—petrified or other.

8. To this is appended a footnote: "I brought to Europe specimens of the several kinds of maize grown among the Mandans; these have been sown . . . The heads have by no means attained the same size, on the Rhine, as in their native country . . ."

vated in our gardens . . . There are two or three varieties, with red, and black, and one with smaller seeds. Very nice cakes are made of these seeds . . .

"The wild plants of the prairie are used by the Mandans, and other people of the Upper Missouri; and to those before mentioned, I can only add the feverolles *(Faba minor equina),* a fruit resembling the bean, which is said to grow in the ground, but which I did not see . . ."

Maximilian describes the manner of cultivating and of cooking these plants, what they tasted like, and much else.

Maximilian's visit at Fort Clark lasted until April 18, 1834. Two chapters (XXVIII and XXIX) describe the visit.

Soon after his arrival, word came that cholera had broken out again at St. Louis; although it had not reached the Upper Missouri, McKenzie prepared for its approach by taking a young doctor with him to Fort Union.

For a time Maximilian was able to do a little plant collecting:

"Our first employment was to go on hunting excursions into the prairies round the fort, which afforded us an opportunity of collecting the seeds of the dried plants of the prairie."

By November 22 he and his party were given fine quarters in a new building: there was "space to carry on our labours," windows "afforded a good light for drawing," tables, benches, and shelves were provided, the floor was boarded over, "the door was furnished with bolts on the inside," firewood, "covered with frozen snow, was piled up close to the chimney." Soon after they moved in everybody "felt indisposed, and . . . obliged to have recourse to medicine, but this was, probably, to be ascribed principally to the way of living and the state of the weather . . ." Maximilian ". . . examined all the medical stock of the fort, and found neither peppermint nor other herbs, which would have been serviceable at this time; only a handful of elder flowers, and rather more of American camomile, which has a different taste to the European . . . unfortunately we were without a medical man."

Soon there were heavy snowstorms, the weather became very cold, and the Prince found it hard to get about; nevertheless, on foot and when opportunity offered, he visited the various Indian villages. Dreidoppel hunted with assiduity; one of his remarkable feats (or so it seems to me) was the killing of ". . . a couple of wolves, which he allured by imitating the voice of a hare . . ." He is described dragging ". . . the entrails of a hare about the prairie . . ." and then concealing himself: ". . . he soon saw six wolves follow the scent and approach him; but it was so cold that he could not wait for them."

1834

By January the roof of the fine new quarters was leaking, ". . . the water dripped on our books and papers from the loft which was covered with snow." And, evidently despite the inside bolts, Indians "unceremoniously . . . came to dry themselves before our fire; this was not very agreeable . . ." for their robes and hair were wet and evidently dripped over everything. Besides, they were inclined to steal and took, among other things, Maximilian's thermometer. By mid-March the inhabitants of the fort were attacked by what Culbertson diagnosed as scurvy. Maximilian describes his symptoms and notes that ". . . our provisions were very bad and scanty."

". . . During the tedium of my confinement to bed, I was enlivened by the frequent visits of the Indians, and I never neglected to continue my journal, which, from fever and consequent weakness, was often very fatiguing. Mr. Kipp kindly sent me some new-laid eggs every day, as well as rice which he had reserved for me, and from which I derived great benefit. The inmates of the fort had nothing to eat but doughy maize bread and maize boiled in water; but Mr. Kipp, who did not like the latter, was obliged to fast."

Bodmer painted assiduously but had his own difficulties with freezing paints, and his Indian models—like every "sitter"—had their own ideas about how they should be dressed, etc., etc. Nor should Bodmer's collecting activities go unmentioned. Aged twenty-eight, he enjoyed adventure probably. On one occasion with a companion and "well armed," he visited some abandoned huts of the Arrikara Indians ". . . in order to procure . . . some skulls and prairie bulbs . . . They brought back two well pre-served male skulls which," states Maximilian, "I added to my collection; one of these is now in the anatomical museum of the University of Bonn, and the other in the col-lection of Mr. Blumenbach at Göttingen." Bodmer's painting of Fort Clark, captioned "Fort Clark[9] on the Missouri (February 1834)" is very fine.

Maximilian, in his chapter XXIV, has a little to tell about the plants of the Fort Clark region:

"The extensive prairies, and their hills, certainly produce a great variety of plants, of which a part only have been described. Bradbury collected many plants about the Mandan villages, which were described by Pursh; and Nuttall's works likewise con-tain many; but there is, undoubtedly, much remaining to be done, especially in the chain of the Black Hills. The country, about the Missouri, has its peculiar botanical characters. The tongues of land at the bends of the river are generally covered with wood; other parts of the banks more rarely so; the species of trees and shrubs which occur here have already been mentioned. There are no pines in the vicinity of Fort Clarke; but they are found higher up the river; nor are there any birch trees; indeed, I did not meet with one on the whole course of the Missouri . . . In the prairies on the Missouri, near Fort Clarke, the same species of cactus are found as near Fort Union; the grasses are not of so many species as might be supposed; *Chondrosium oligosta-chyum* (Nees), which grows to the height of ten or twelve feet, and *Bryzophyrum spicatum,* are, however, found there. As I had no opportunity of botanizing here in

9. The name Clark is correctly spelled in the *Atlas.* In the copy of the *Atlas* which I examined this particu-lar plate bears no number nor does it appear in the list of plates.

the summer time, my list of plants of this part of the country is very incomplete; but Bradbury and Nuttall were more fortunate. Many officinal plants grow here, but there are no physicians to direct the use of them.

"In the forests about Fort Clarke, only a very small quantity of useful timber is found. The poplar burns quickly, and emits much heat, and the bark serves for the winter food of the horses . . ."

Spring was beginning. ". . . The grass began to sprout, and some young plants appeared in the prairie, even a pulsatilla, with purple blossoms, apparently the same as the *P. vulgaris* of Europe; the Indians call this plant the red calf-flower . . ."

The entry above was of April 9. Maximilian left Fort Clark on April 18 and reached St. Louis on May 27. I shall not give the journey in detail. On April 21 the party passed the mouth of Cannonball River where "not a vestige of verdure was to be seen." At Fort Pierre (April 28, 29) "the bushes were just beginning to bud." At White River (May 2) they made an excursion "to the ravines and hills. Cactus and yucca grew here in abundance, and some plants of the prairie were already in flower." On May 12, at Boyer's Creek, they "saw the first plane tree on the Missouri . . . this species of tree becomes more and more common as you descend the river . . ." By May 18 they reached Cantonment Leavenworth, where Maximilian was annoyed by the treatment accorded his party. The next day they were at Williams Ferry (the present site of Missouri City); the bears, I was delighted to learn for the first time, had survived the winter. "My live bears attracted all the inhabitants in the neighbourhood; nay, the people here were more eager to see the much-dreaded grizzly bears than even in Europe."

The Mormons are mentioned and Maximilian had opinions upon them as upon all subjects:

"Among our inquisitive visitors there were several men belonging to a religious sect known here by the name of Mormons. They complained bitterly of the unjust treatment which they had lately received. They had lived on the other side of the Missouri, and, as they asserted, had been expelled, on account of their doctrines, by the neighbouring planters, their dwellings demolished and burnt, their plantations destroyed, and some of them killed, on which they settled on the north bank of the river. I was not able to learn whether all this was true, or why, after an interval of one or two years, they have not obtained redress from the government. So much, however, is certain, that, if these people spoke the truth, it would be a great disgrace to the administration of justice in this country, which calls itself the only free country in the world[1] . . ."

On May 21 the party reached Fire Prairie where they ". . . again saw magnificent forests, the trees of which were so lofty that our guns were unable to reach the birds perched on the upper branches.[2] The ground was covered with flowers, among which

1. An interesting comment in the year 1834.

2. Since the *Travels* is not given to exaggeration, one wonders whether the gun or the height of the tree— or perhaps Dreidoppel's aim—was responsible.

was a beautiful sky-blue iris, and with an undergrowth of papaw trees, above which rose the tall forest trees, such as *Gleditschia triacanthos,* sassafras, tulip trees, &c., entwined with the *Vitis hederacea* . . ."

Franklin and Boonville were passed (May 23) and Côte sans Dessein (May 25). From St. Charles (May 27) Maximilian went overland to St. Louis, hiring a wagon ". . . in order to reach the Mississippi as speedily as possible . . . Dreidoppel was to go by water, with my collections, &c. . . . At noon we alighted at the Union Hotel in St. Louis, after an absence of above a year.

"St. Louis was now healthy, and not suffering from the cholera as we expected . . . No change of consequence had taken place since our last visit. At the factory of the American Fur Company I found very agreeable letters from Europe . . . I greatly regretted the absence of General Clarke . . ."

Maximilian seems to have seen everything there was to see in St. Louis, among other things an exhibition of George Catlin's paintings. Unfortunately he does not disclose what his "opinion" was.

"Here we saw a collection of Indian portraits and scenery by Mr. Catlin, a painter from New York, of which we were able to form an opinion after our recent travels in the country . . ." [3]

Maximilian regretted that he was unable to see the ". . . tame buffaloes, which Mr.

3. Chittenden's estimate of the relative importance of the work of Catlin and of Maximilian is of interest:

"In 1832 George Catlin, the well-known Indian painter, ascended the Missouri to Fort Union on the first steamer that ever reached that point. He descended by skiff in the fall, having made a large number of sketches of scenes in the upper country and many portraits of the Indians and traders. Subsequently Catlin visited other portions of Western America . . . He undoubtedly did a great work in preserving in pictorial form a condition of life that no longer exists except in history . . . His works, like those of Maximilian, will always be resorted to by students of the native races and early conditions of the Missouri valley. The contrast between the methods of these two men, however, was most marked. Maximilian was a scientist—discriminating and accurate in his observations and careful and conservative in recording his impresssions. Catlin was a visionary enthusiast upon a single theme, the American Indian. He saw everything pertaining to the natives through highly colored glasses, and . . . he recklessly exaggerated his impressions when he attempted to record them with his pen and pencil. He was distrusted by those who knew him in the West and was more than once taken to task by his contemporaries. Audubon . . . flatly insinuates that he was dishonest. Parkman characterizes him as a 'garrulous and windy writer.' It is regrettable that one who did so much work of real worth should have marred it by a characteristic which throws doubt upon the accuracy of it all.

"In the year 1833 Maximilian, Prince of Wied, ascended the river, visiting Forts Union, McKenzie and Clark, and wintering at the latter post. He traveled at this time under the name of the Baron of Brausenburg, and is always so referred to in the correspondence . . . He was accompanied by a small party among whom was a skilful artist by the name of Bodmer. The Prince, upon his return to Europe, published the results of his labors in the most complete and elaborate work ever prepared upon this region. It is much to be regretted that it is so inaccessible to the public . . . The work covers a great variety of subjects—narrative, anecdote, history, ethnology, geology, natural history—and omitting a few minor errors, it is exceedingly accurate, discriminating and judicious in all its scope. It ranks with *Astoria, Bonneville,* and the *Commerce of the Prairies,* as a descriptive work, while it is far ahead of the others in the elaborate character of its illustrations. In this respect it has no competitor, or only a feeble one in Catlin." (2: 637, 638)

In the preface to his volume on the fur trade Chittenden had referred to Maximilian as ". . . the most

Pièrre Chouteau kept on his estate near the town, though I should have been very glad, for many reasons, to have seen these animals in a domestic state."

After quite a dissertation upon various disputed points relative to the crossing of the wild buffalo with "the tame species," he makes an assertion—and he certainly was in a position to know the facts:

"There is another point on which I differ from Mr. Gallatin,[4] namely, his denial of the great decrease in the number of buffaloes in general. For when we consider how far these animals have been driven up the country, and that, in these very parts, they are even less numerous than formerly, we have a fact that at once proves a great decrease, of which nobody in the interior of the country can entertain a doubt."

After about one week in St. Louis Maximilian and his party started down the Mississippi (June 3) by steamboat, and the next day reached the mouth of the Ohio. Thence, after a visit with his friends at New Harmony, the Prince returned to the east by a somewhat different route from the one taken westward. He sailed from New York on the "packet-boat Havre" (July 16), and (August 8) reached Havre-de-Grâce at the Seine estuary; ". . . at half-past eleven o'clock the Havre cast anchor in Europe."

It seems doubtful that any traveler up the Missouri, before or since Maximilian, made better use of his opportunities; and he had made a point of seeing that Charles Bodmer also did so.

Appendix VII (pages 511-520) of the *Travels* bears the title "Systematic view of the plants brought back from my tour on the Missouri, drawn up by President Nees von Essenbeck [sic], at Breslau." Unfortunately, no dates and extremely few localities of collection are supplied, so that it is impossible to associate the plants, except for a possible half dozen, with regions described in the *Travels*.

Brendel (1879) comments that Maximilian ". . . brought back to Germany a small collection of about two hundred species, which were published by Nees v. Esenbeck . . . There was nothing new except the genus Sarcobatus, proposed by Nees and afterwards described again by Torrey under the name of Fremontia." By the 1830's so many plants were being collected and described from west of the Mississippi River that those who were classifying and publishing this material had to be alert—not only lest descriptions which came out in obscure publications pass unnoticed, but to make sure that what was described as new was properly classified. We have an example of this need for vigilance in "*Sarcobatus Maximiliani,* (Pulpy Thorn, Lewis and Clarke

reliable published authority upon the early history of the American Fur Company on the upper Missouri," and had mentioned his own good fortune in being able to study a copy of the *Travels:*

". . . The extensive library of Americana belonging to the Hon. Peter Koch of Bozeman, Montana, possesses the very unusual treasure of a copy of Maximilian's book. The loan of this work during several months made it possible to draw from the distinguished author much information, which, in an ordinary perusal, would have been overlooked." (1: xiv)

4. Is this the "Albert Gallatin" who welcomed John Bradbury on his return to St. Louis in 1811?

Iter.)," which Nees von Esenbeck described in 1839 from a specimen collected by Maximilian "In regione Mississippi fluvii superiori tractus latos investit hic frutex." Fruit was lacking and the author referred it to the nettle family, the "Urticeæ," or *Urticaceae*. Nine years after Maximilian collected his specimen, Frémont, on his expedition of 1842, found the same species; and, on the basis of his specimen, Torrey (in 1843, in his "Catalogue of plants" appended to the Frémont report) described the new genus *Fremontia,* with type species *F. vermicularis;* Torrey placed the plant correctly in the family *Chenopodiaceae.* In 1845 W. J. Hooker (*London Jour. Bot.* 4: 480, 481, *fn.*) after mentioning the fact that the Frémont report where the Torrey description had appeared had a limited circulation, reprinted the description with the comment that *Fremontia* ". . . is the *Sarcobatus Maximiliani* of Nees described by Dr. Seubert in the 'Botanische Zeitung,' for Nov. 1, 1844; but was previously noticed under the same name, in a work as little likely to fall into the hands of Botanists as the 'Report' above mentioned, namely 'Prince Maximilian v. Wied Reise ins innere Nordamerika,' I. p. 510, and II. p. 477 . . . Some observations on this Genus will be found at p. 1, of the present volume of our 'Journal,' from the pen of Dr. Lindley, who was, however, unacquainted with the character of *Fremontia* . . ."

Since Hooker, on behalf of Maximilian's translator Lloyd, had revised Nees von Esenbeck's paper,[5] we see how the identity of the Maximilian and the Frémont specimens came to his notice. Hooker, noting that Seubert was unaware of this identity, thought best to clarify the matter for the sake of other botanists. I mention Torrey's *Fremontia* again (p. 929, *fn.* 6).

The discovery of one new plant was certainly no remarkable botanical accomplishment[6] for such an extended journey into a region which, from the country of the Mandans in North Dakota—visited by Bradbury and Nuttall in 1811—westward to Fort McKenzie near the mouth of Marias River in Montana, had been untouched by plant collectors if we except the rapid return passage of the Lewis and Clark party. However, I have told Maximilian's story at some length because it is one of the best, if not *the* best, picture of the Missouri River valley, and of travel therein at this period, that I have read. Bodmer's plates enhance the value of this picture immeasurably, and I believe accurately.

Maximilian had been born in 1792 at Wied-Neuwied, situated not far from Coblenz, in the Rhine province of Prussia. After service in the German army he retired in 1815 with the rank of major general and devoted the rest of his life to science. According to

---

5. H. Evans Lloyd states, in his "Translator's Preface," that "The papers in the Appendix, giving an account of the plants collected, are . . . inserted entire, and have been kindly revised by my friend Sir William Hooker." Only one paper in the appendices of the *Travels* relates to the Missouri River journey.

6. Sargent augmented Maximilian's plant discoveries by one, stating in 1891 that *"Salix Missouriensis* was first collected at Fort Osage on the Missouri River by the German naturalist, Maximilian, Prinz von Neuwied, and was first described by Nils Johan Andersson, the Swedish salicologist [as a variety of *Salix cordata*]." "Fort Osage" was another name for Fort Clark.

Sargent, "From 1815 to 1817 he traveled in the interior of Brazil with the naturalists Neircriss and Sellow, the scientific results of his journey appearing in a number of memoirs." Maximilian died in 1867.

Spaulding, writing of botanists associated with St. Louis, includes a short biographical account of Maximilian. He explains that the Prince's expedition to Brazil was undertaken ". . . in order to satisfy a keen desire to add to the world's knowledge, imparted to him by the celebrated Professor Johann Friedrich Blumenbach,[7] of whom he was a favorite pupil . . . His resulting publications gave him a high rank among the scientists of the period, and his 'Reise nach Brasilien in den Jahren 1815 bis 1817' was soon translated into the French, English and Dutch languages . . ." Further:

"After his return to his native city Prince Maximilian worked over his collections with the aid of a number of experts, and published several papers upon his results . . . His collections are preserved in the museum of his native city, where he died in 1867. Martius honored him by naming a genus of Brazilian and West Indian palms, *Maximiliana,* thus very appropriately connecting him with the botany of that country, of which he was one of the pioneer explorers."

De Candolle in *La phytographie* (1880)—and without stating how much of the herbarium consisted of Brazilian as against North American plants—refers thus to the whereabouts of the Maximilian collections: "Wied-Neuwied (Prince de). Avait donné son herb. à Martius. D'autres pacquets, contenant des esp. décrites par Schrader, Nees, etc., sont chez son neveu le comte de Solms." The herbarium of Karl Friedrich Philipp von Martius, particularly rich in South American types, is now at the Jardin Botanique de l'État in Brussels, King Leopold having purchased it for the Jardin Botanique after Martius' death.

We find under "Notes" in the periodical *Nature* (10: 150, 1874) the following item: "On the 3rd inst. [June, 1874] the corner stone of the American Museum of Natural History in New York was laid by the President of the United States.[8] The ground belonging to the Museum measures about eighteen acres, and the building when completed according to plan will be larger than the British Museum[9] . . . The Trustees have purchased the collection of Prince Maximilian, of Neuwied, on the Rhine . . ."

In reply to my query regarding the content of the above collection Dr. W. M. Faunce, Assistant Director of the Museum, wrote me on November 7, 1952: "The collection of Prince Maximilian of Wied (or Neuwied) consisted of 4,000 mounted birds, 600 mounted mammals and 2,000 fishes and reptiles mounted and in alcohol. No plants were included, according to our records. These specimens are still in existence but not as one collection. They have been dispersed to the various departments concerned."

7. It was Blumenbach, a professor at Göttingen, who had also inspired in his student von Langsdorff a love of natural history.

8. Then Ulysses S. Grant.

9. Whether this forecast, with its somewhat "American" ring, has come to pass I do not know!

BRENDEL, FREDERICK. 1879. Historical sketch of the science of botany in North America from 1635 to 1840. *Am. Naturalist* 13: 770.

CANDOLLE, ALPHONSE DE. 1880. La phytographie; ou l'art de décrire les végétaux considérés sous différents points de vue. Paris. 459.

CHITTENDEN, HIRAM MARTIN. 1902. The American fur trade of the far west. A history of the pioneer trading posts and early fur companies of the Missouri valley and the Rocky Mountains and the overland commerce with Santa Fe. Map. 3 vols. New York. 1: xiv, 2: 338-341, 384-389; 2: 637, 638; 3: 952, 953, 955, 956, 957.

CRAVEN, ARTHUR J., *editor*. 1900. Affairs at Fort Benton from 1831 to 1869. From Lieut. Bradley's journal. Period 1831 to 1839. *Contr. Hist. Soc. Montana* 3: 206-210.

DE VOTO, BERNARD. 1947. Across the wide Missouri. Illustrated with paintings by Alfred Jacob Miller, Charles Bodmer and George Catlin. Boston.

GREENHOW, ROBERT. 1845. The history of Oregon and California, and the other territories on the north-west coast of North America: accompanied by a geographical view and map of those countries, and a number of documents as proofs and illustrations of the history. ed. 2. Boston. 411-413.

HOOKER, WILLIAM JACKSON. 1845. [Note on the genus *Fremontia.*] *London Jour. Bot.* 4: 480, 481, *fn.*

MAXIMILIAN ALEXANDER PHILIPP, Prince of Wied-Neuwied. 1839-1841. Reise in das innere Nord-America in den Jahren 1832 bis 1834. 2 vols. & Atlas. Coblenz.

———— 1843. Travels in the interior of North America. By Maximilian, Prince of Wied. with numerous engravings on wood, and a large map. Translated from the German, by H. Evans Lloyd. To accompany the original series of eighty-one elaborately-coloured plates. Size, imperial folio. London.

———— 1905. Travels in the interior of North America. *In:* Thwaites, Reuben Gold. *editor.* Early western travels 1748–1846. 22, 23, 24, 25. Atlas. [The text is transcribed from the London edition of 1843, translated by H. E. Lloyd. The plates of the Atlas are in black and white and much reduced.]

SARGENT, CHARLES SPRAGUE. 1891. The Silva of North America. Boston. New York. 9: 138, *fn.* 2 & 3.

SPAULDING, PERLEY. 1909. A biographical history of botany at St. Louis, Missouri. *Pop. Sci. Monthly* 74: 50-52.

TROLLOPE, FRANCES. 1832. Domestic manners of the Americans. 2 vols. London. Printed for Whittaker, Treacher & Co. Ave-Maria-Lane.

WEISS, HARRY BISCHOFF & ZIEGLER, GRACE M. 1931. Thomas Say. Early American naturalist. Foreword by L. O. Howard. Springfield, Illinois. Baltimore. 153, 154.

CHAPTER XXVII

LEAVENWORTH COLLECTS A FEW PLANTS WHEN STA-TIONED AT A NUMBER OF ARMY POSTS IN THE "INDIAN TERRITORY"

ALTHOUGH Asa Gray, writing the obituary notices of the man whose story is told in this chapter, expressed the hope that ". . . some one acquainted with them would put on record the incidents of his life . . .", it was seventy-five years before anyone attempted to do so. This was in 1947 when Rogers McVaugh, in *Field & Laboratory,* published "The travels and botanical collections of Dr. Melines Conkling[1] Leavenworth." This paper, for many of the facts concerned with Leavenworth's ancestry, family connections, medical training and early interest in botany, drew its information from a genealogy[2] of the family, and McVaugh states that he was able to "summarize" Leavenworth's "active life" ". . . quite adequately from his own letters to Torrey, now preserved at the New York Botanical Garden, and from his reports and other records in the War Department Archives in Washington, D. C."

Included in the paper is a photograph-reproduction of the doctor dressed in army uniform and evidently well along in years. The face is one to remember—by no means handsome but kindly, keen and quizzical and—although his "insight" in matters scientific may not have been "great" as Gray carefully stated—I feel sure that he was extremely perspicacious in "sizing up" members of the human race.

Before discussing McVaugh's paper I shall refer to earlier publications dealing with Leavenworth.

His death in 1862 was noted briefly in the *American Journal of Science and Arts* in 1863 and, in the same volume under "Botanical Necrology, 1862," Asa Gray amplified what was undoubtedly his earlier notice. I quote both in turn:

"Melines C. Leavenworth.—Dr. Leavenworth died near New Orleans, La., in December,[3] while acting as Surgeon in the 12th Connecticut regiment. He was among the oldest of American botanists, his first paper, 'on four new plants from Alabama,' having appeared in vol. vii of the first series of this Journal, in 1834. Dr. Leavenworth has

---

1. Leavenworth's given names had appeared in a variety of spellings: "Mellins C.," "Melines Conkling," "Melinas C.," "Melines Conklin," or most often, and for obvious reasons perhaps, abbreviated to "M. C." Leavenworth's middle name, as McVaugh points out, was *Conkling,* and his father's elder brother bore the name *Melines*—facts which seem to indicate how the doctor's names should appear.

2. *A Genealogy of the Leavenworth Family in the United States, with Historical Introduction, Etc.,* by Elias Warner Leavenworth L.L.D., pp. 376. illus. Syracuse, N. Y., 1873.

3. A number of United States Army registers state that he died ". . . 16 Nov., 1862." McVaugh is in accord with this.

resided, since his retirement from the medical service of the United States army, in Waterbury, Conn., until at the call for volunteers, well advanced in years and by no means firm in health, he went cheerfully to offer his life a sacrifice for his country."

"Dr. Melines C. Leavenworth, as already announced in this Journal, died in the vicinity of New Orleans, in December last, while acting as Surgeon to the 12th Connecticut regiment, at the age probably above three score years. It is to be desired that some one acquainted with them would put on record the incidents of his life. He was formerly and for many years a surgeon in the United States Army; from which, however, he retired about twenty years ago. While in the army, at frontier posts in Arkansas, Louisiana, and Florida, he indulged his strong botanical tastes, and did useful service, by observing and collecting the plants within his reach, which he communicated to Dr. Torrey along with copious notes. These were the more important as his dried specimens were seldom neatly preserved. The pages of the *Flora of North America,* upon which his name so often occurs, testify to his zeal and success as a botanical explorer and pioneer. His ardent love of botany—fostered, we believe, by the late Dr. Tully[4]—must have early developed; for as much as forty years ago he discovered 'four new plants from Alabama,' which he described in the seventh volume of this Journal, in 1824. Among the rare plants which he detected, a very peculiar one— the *Amphianthus pusillus* of Torrey—which he found in the upper part of Georgia, is so very scarce and local that it has never been met with since. A pretty and strikingly marked Cruciferous genus, one species of which (if indeed distinct from the other) was discovered by Dr. Leavenworth, dedicated to him by Dr. Torrey,[5] commemorates his botanical services:—which services, indeed, were continued to the last. For no sooner had he landed with his regiment upon our southern coast than he zealously began to collect the plants he met with, and to note their peculiarities. *Although his scientific acquirements and insight were not great* [italics mine],[6] his zeal and devotion to botany were thorough and genuine."

A number of army registers which I have consulted cite the year of Leavenworth's

4. William Tully of the Yale Medical Faculty, according to McVaugh.

5. *Leavenworthia* Torrey.

6. Even in obituary notices Gray apparently found it difficult to bestow more praise than strict accuracy warranted! When reading his papers I have smiled more than once at his tart comments, as when, in a review of a "popular exposition" by M. A. Curtis on the woody plants of North Carolina, he refutes the author's statement—that "Botanists will of course find fault with it ..."—with the remark that ".... we quite enjoy a glimpse of Flora en deshabille and slip-shod ..."!

Gray could wield a sharp pen when he chose and when he sensed criticism of botanical science he really spoke his mind. One notable example was his review of John Ruskin's *Proserpina: Studies of wayside flowers while the air was yet pure among the Alps and in the Scotland and England which my father knew* first published in *The Nation* in 1875. Ruskin satirized the botanical approach to plants, made erroneous statements about them, but loved them (or certain of them) nonetheless. I found the little book enjoyable and had some laughs. Not so Gray, who "tore it to ribbons" so to speak, from his opening phrase "Mr. Ruskin, 'having been privileged to found the School of Art in the University of Oxford,' now proposes to found a new school of botany"—to his final paragraph concerned with Ruskin's intimation ".... that even Dr. Darwin must be ranked among 'the men of semi-faculty or semi-education who are more or less in-birth as 1796 and his birthplace as Connecticut. To the first McVaugh adds January 15; to the last, Waterbury. They indicate by an asterisk that he was *not* a graduate of the Military Academy and state that he received the appointment of "Assistant Surgeon" on September 1, 1833, and resigned from the service on September 30, 1840. They fail to mention that Leavenworth served as "Acting Assistant Surgeon" from October, 1831, to May, 1833, which made his first overall military duty more nearly ten than seven years—as, again, McVaugh points out. Heitman's *Historical register . . .* mentions that when the doctor re-entered the service on February 12, 1862—or some twenty-two years later—it was again as Assistant Surgeon with the 12th Connecticut Infantry.

The army registers are uninformative regarding the stations where Leavenworth saw service. Gray merely mentioned ". . . frontier posts in Arkansas, Louisiana, and Florida . . ." and Sargent's short sketch of Leavenworth in *The Silva* was obviously taken from Gray.

The "Arkansas" referred to by Gray was the region lying north of Red River (of the South) and along the courses of the Arkansas, the Canadian and the Washita rivers and now comprised in eastern Oklahoma. Set apart for occupation by the "Five Civilized Tribes" transplanted from east of the Mississippi River, it was called the "Indian Territory." Leavenworth was stationed at two posts in this region:

One was Fort Towson, established in 1824 on the Kiamichi River near its junction with Red River. The site was not far from present Fort Towson, Choctaw County, Oklahoma; it was directly across from Red River County, Texas. Here, as recorded in his only paper (of four) mentioning plants from the region of my story,[7] Leavenworth gathered or observed a number of plant species, such as:

"Ulmus crassifolia, *Nuttall* . . . Found on the Red River prairies in the vicinity of Fort Towson . . ."

"Sophora affinis . . . Abundant on a rocky ridge about one mile from Fort Towson, near the road leading to the landing on Red River."

"Sapindus . . . frequently occurs on rocky eminences near Fort Towson . . ."

capable of so much as seeing, much less thinking about color,' etc. . . ." Curiously, the one "botanist" for whom Ruskin had something even slightly favorable to say was Gray himself.

Wilfrid Blunt, in *The art of botanical illustration* (1950), includes a chapter entitled "John Ruskin: a digression," which, although it includes no mention of Gray's review, is understanding of Ruskin's purpose; and the many persons who love flowers, but prefer them unanalyzed and undissected, may agree more nearly with Blunt's comments than with Gray's. The chapter opens thus:

"Ruskin's *Proserpina* (1874–86), in spite of its untidy diffuseness, its irritatingly didactic tone and its ill-digested science, remains the most stimulating book ever written about flowers. 'It may be described, in one aspect of it,' writes his biographer, Cook, 'as a series of drawing lessons in flowers.' . . .

"Ruskin loved flowers full-heartedly; but his approach to Nature, searching and inquiring though it was, remained that of the poet and the painter . . ."

7. This paper, entitled "On several new plants," was published in 1845 in the *American Journal of Science and Arts*. The author's purpose was threefold: "1st. The description of new and unknown plants. 2nd. Additional memoranda relating to those that are rare and little known. 3rd. To enlarge the knowledge of the range of certain plants."

"Sanguinaria Canadensis . . . It is found at Fort Towson, on the rocky bluff near Gales' Creek[8] . . ."

"Philadelphus hirsutus, *Nuttall* . . . Abundant on the bluff at Fort Towson . . ."

"Nemostyles gemmiflora, *Nuttall* . . . Found by myself in the vicinity of Fort Towson . . ."

The second post in "Arkansas" was one which Leavenworth, in a letter to Torrey dated November, 1836, and quoted in Rodgers' biography of Torrey (p. 176), refers to as a place where he "spent one Summer," and which was situated ". . . in the vicinity of the mouth of the False Washita,[9] one of the large tributaries of Red River . . ." *Fort Washita,* situated on the eastern side of Washita River and some twenty-five (?) miles above its confluence with the Red River, was not established until 1842. The site referred to by Leavenworth must have been identical with, or near, the spot where the regiment of dragoons, first commanded by General Henry Leavenworth and later by Colonel Henry Dodge, encamped in the summer of 1834, which was directly at the confluence of the "False Washita" with Red River. With the dragoons at the time were the artist George Catlin and the botanist Heinrich Karl Beyrich. In his paper "On several new plants" Leavenworth mentions that *Sophora affinis* was "Found by myself and Mr. Beyrich in the Red River prairies . . ." which suggests that these men collected together, although it does not offer proof positive.[1] Writing of *Ulmus crassifolia* Nuttall in the same paper, Leavenworth refers to it as "Found on the Red River prairies in the vicinity of Fort Towson; from thence westward to the Cross Timbers, about thirty five miles beyond the mouth of the False Washita, the last large tributary of the Red River from the north. The Cross Timbers are said by the hunters to be a line of timbers extending from the Red River to the Missouri . . ."

8. According to McVaugh, "Gates' Creek." He notes: "This species was re-collected at Leavenworth's locality in 1947, at what appears to be its southwestern limit of range."

9. Although the name "False Washita" appears frequently in the early narratives, I have not found in the literature—although it doubtless exists—any explanation of how the appellation arose. On one map— "Map of the United States and their territories between the Mississippi and the Pacific Ocean; and a part of Mexico. Compiled in the Bureau of the Corps of Topogl Engs under a resolution of the U. S. Senate, from the best authorities which could be obtained. 1850."—the name seems to designate the lower reaches of the Washita River. Possibly the adjective "False" was used to distinguish the Oklahoma stream from the Washita, or Ouachita, River which from Arkansas crosses into Louisiana and enters Red River not far from its confluence with the Mississippi. Bradbury's *Travels* (242, 1817) notes that "The most southerly of the salt rivers that rise in the region . . . is a branch of Red River, called by . . . the French Fouxoacheta . . ."

1. The story of the dragoon expedition is told in my chapter XXVIII.

The journal of the trip, kept by Lieutenant Thompson B. Wheelock, records that from July 1 to 5 contingents of dragoons, from an encampment on the Washita, were crossing that river. On July 4 Wheelock's troops "encamped about four miles west from camp Washita," and on this day the botanist "Beyrick" and the artist Catlin joined them for the journey westward.

Catlin's Letter 39 (written at the "Mouth of False Washita, *Red River*") describes the position of this camp as ". . . on the banks of the Red River, having Texas under our eye on the opposite bank. Our encampment is on the point of land between the Red and Washita rivers, at their junction . . ."

As I shall mention later, it must have been from Camp Washita that Leavenworth made a westward journey to the Cross Timbers and, presumably, observed the range of this elm.

Turning now from Leavenworth's frontier posts in "Arkansas" to the two others mentioned by Gray, "Florida" is, of course, outside the region considered in my story. "Louisiana" also,[2] except for the fact that it was undoubtedly from posts near the western boundary of that state (where, according to McVaugh, Leavenworth was stationed at Fort Jesup and at Camp Sabine) that the doctor collected or observed plants growing within the state of Texas. Some such are recorded in his paper "On several new plants"; viz.,

"Stillingia ligustrina, *Michaux* . . . Found by me in eastern Texas, near the residence of Dr. Veatch[3] about forty miles west of the Sabine River."

"Ulmus crassifolia, *Nuttall* . . . I have also seen it in Texas skirting a small prairie two miles from the Sabine, near the road to Natchitoches, La., to Nagadoches, Texas . . ."

"Nemostyles coelestina, (*Ixia coelestina,* Bartram.) . . . Frequent in western Louisiana, near the Sabine River; still more frequent on the Texas side of that river; occasionally also in the vicinity of Fort Jessup, La. . . ."

"Nemostyles gemmiflora, *Nuttall.* . . . Found by myself in . . . western Louisiana, eastern Texas . . ."

I turn now to McVaugh's account of Leavenworth's "travels and botanical collections," in particular to those associated with the trans-Mississippi west.

The author tells of the doctor's years in our southeastern states (in Alabama, Georgia, etc.) and of his application for a ". . . commission as Assistant Surgeon, United States Army . . ." in 1831, when he was thirty-five years of age. With the approval of this application ". . . Melines began his Army career September 22, 1831, probably at Jefferson Barracks. He accompanied a detachment of troops down the Mississippi and to Camp Jesup, Louisana, where he began a period of almost exactly nine years spent in one frontier post after another . . .

"From October, 1831, to May, 1833, he was stationed at Camp Jesup as Acting Assistant Surgeon . . .

"In June, 1833, Leavenworth met the Medical Examining Board at Jefferson Barracks, and finally received his appointment as Assistant Surgeon . . ."

After spending "most of the rest of the summer" at Opelousas, Louisiana, there was a period at Fort Jesup (situated we are told ". . . in Sabine Parish . . . southwest of Natchitoches on the road to Gaines Ferry on the Sabine River, and to Nacogdoches, Texas . . .") "until the end of April, 1834." It must have been during this residence at

2. For my exclusion of the state of Louisiana, the major part of which lies west of the Mississippi River, see pp. 63, *fn.* 1; 105, *fn.* 3.

3. John Allen Veatch.

Fort Jesup, in 1834, that Leavenworth made what McVaugh records at a later point in his story as "Possibly . . . Leavenworth's first visit to Texas, although of course he may have made others during his stays at Fort Jesup from 1831 to 1833 . . ."

"In a letter to Torrey, May 3, 1837, the collector included a list of 'Plants found in Texas in 1834—during a trip from Fort Jesup to Nagadoches in April'; this list comprises 9 species. The most interesting plant recorded is what the collector called *Ribes rotundifolia,* from 'Rock St. Augustine,' Texas. This is one of the very few reports of any native currant or gooseberry from Texas; the specimen was cited by Torrey and Gray in the *Flora* as *Ribes gracile,* and presumably the reference to Texas in the *North American Flora* under *Grossularia curvata* (N. Amer. Fl. 22: 222. 1908) is taken from the same specimen."

From Fort Jesup Leavenworth ". . . was transferred to duty at Ft. Towson, Arkansas. We know from a letter of May 7 that he left Ft. Jesup on April 29 and reached Ft. Towson May 6 . . . Apparently he left Ft. Towson the very same day, May 7, 'on detached service' at a camp near the mouth of the False Washita . . . In a report some years later . . . Leavenworth wrote as follows: 'Proceeded to Fort Towson under orders & on my arrival there remained one night & immediately releived Dr. Hogan . . . & was on field duty from 15th May 1834 to 15th Oct 1834 most of which time was spent either at the Cross Timbers or . . . near the mouth of the False Washita. In Oct 1834 I returned to Fort Towson where I remained until releived by Dr. Wells in Novr 1835.' "

At this point McVaugh refers, as I have already done, to the presence of the regiment of dragoons at the camp at the mouth of the "False Washita" and states that, according to the Leavenworth *Genealogy,* ". . . Melines was with his uncle [General Henry Leavenworth] at the time of the latter's death at the Cross Timbers. The exact locality[4] is unknown to me . . ." For proof that the botanist Beyrich and M. C. Leavenworth collected together—the latter, as I have stated, had referred to the fact that he and Beyrich had both found *Sophora affinis* [Torrey & Gray] in the Red River prairies —McVaugh offers the following:

". . . More convincing evidence is found in the *Flora of North America* where under *Psoralea linearifolia* (v. 1, p. 300) appears the following citation: "Arkansas, *Beyrich.* Communicated by Dr. Leavenworth." Evidently the two men did meet; presumably they exchanged interesting specimens, and some of those given to Leavenworth found their way eventually to Torrey."

---

4. McVaugh cites a number of records as to the exact spot where the General died. Its location is of interest as fixing the westernmost point reached by the doctor who was with his uncle at the time. States McVaugh: "Presumably the farthest point reached by Leavenworth . . . was somewhere in present Love or Carter County, Oklahoma."

According to Ghent (*The early far west* . . . p. 309), when General Leavenworth had become incapacitated he had been ". . . left at a temporary halting place, Camp Smith . . . Dodge, returning with his command to Camp Smith, learned that Leavenworth had passed away on July 21." The Wheelock journal, under date of August 5, 1834, records that the general ". . . died at his camp, near 'Cross Timber' on 21st of July."

Continuing the recital of Leavenworth's activities, McVaugh tells that "After his summer on the False Washita, Dr. Leavenworth returned to Fort Towson. . . . and spent the next thirteen months there. Apparently he was able to find a considerable amount of time for botany, for there were no very active Army movements during this period . . . On December 13, 1835, Short[5] says he has received some good specimens from Leavenworth, and it may be supposed that these were collected in 1835 in the vicinity of Ft. Towson." [6]

McVaugh relates that Leavenworth moved from Fort Towson to a number of army posts in Florida (October, 1835–September 17, 1836), then to Fort Jesup, Louisiana (October, 1836–January, 1837), and next to Camp Sabine, Louisiana—situated ". . . in what is now Sabine Parish, roughly northeast of Sabine Town, Texas, and east of Gaines Ferry, where the Nacogdoches road crossed the Sabine River . . . According to Leavenworth it was . . . 21 miles from Fort Jesup . . ." He was there about sixteen months (January, 1837–mid-May, 1838) when he was again moved to a number of posts in Florida. His service "in the South and Southwest" ended with a furlough which evidently began with his arrival (by boat and accompanying a detachment of sick men) in New York on June 20, 1839. Although he saw service in Michigan in 1839–1840, McVaugh comments that his plant-collecting days had ceased and that ". . . the *Flora of North America* contains not a single reference to any plant collected at Fort Gratiot, Michigan . . ."

It was while at the Louisiana posts mentioned above that Leavenworth ". . . found time to make at least three botanical collecting trips into Texas during the summer of 1837 . . . Camp Sabine was not far from the road between Natchitoches and Nacogdoces, and it was along this road[7] that he seems to have made some of his excursions in Texas. . . .

"In May, 1837, according to his letters, he 'spent sometime in Texas.' He penetrated into the state at least as far as Nacogdoches, where he collected a number of speci-

5. Charles Wilkins Short. I am told that his herbarium is at the Academy of Natural Sciences in Philadelphia.

6. One pleasant little anecdote supplied by McVaugh may have related to these 1835 collections:

"In a letter to the Surgeon General, March 25, 1836, Leavenworth requested that he be allowed to return from his post (then in Florida) to Fort Towson, where he had been obliged to leave behind, 'from want of transportation[,] one of the most valuable collections of the Plants of the fa[lse] Wash[ita] that has ever been made by any individual, containing many manuscript species & genera.' Apparently the request was not granted; we can imagine the feelings of his superiors in the Army at his request to be relieved from a post in a theater which was then an active seat of war . . . to go to the rescue of some bundles of dried plants! In spite of this, Torrey received and cited enough 'Arkansas' plants from Leavenworth to make it clear that at least a good part of the collection eventually came through safely. About 75 species from 'Arkansas' [*i.e.* Oklahoma] are cited in the *Flora of North America* on the basis of Leavenworth's collections, or half again as many as from any other state or region."

7. Here he had observed the *Ulmus crassifolia* Nuttall mentioned in his paper "On several new plants" from which I have quoted; and the "Nemostyles cœlestina, (*Ixia cœlestina*, Bartram.)" had been "more frequent on the Texas side of" the Sabine River than on the Louisiana side.

mens, including what he thought was an interesting composite, a *Tetragonotheca*. On August 3 he writes: 'I have lately made the acquaintance of Dr. Veatch living near Zavala[8] Texas—who has a taste for botany,' and adds: 'During the past month I have travelled 4 days in Texas—embracing a circuit of 120 miles, ¾ths of this distance in the region of the long leafed pine.'

"Probably this 'circuit of 120 miles' was a trip from Camp Sabine to Jasper and Zavala, and return possibly by some slightly different route. Assuming that Leavenworth crossed the river at Sabine Town, as seems likely, he would have had to travel almost exactly 120 miles to make this trip (judging from modern highway distance), and his allusion to the 'long leafed pine' [*Pinus palustris*] suggests that most of the journey was indeed in the area between San Augustine and Zavala, where this species abounds."

The second trip into Texas was in September of 1837: "On October 4 Leavenworth wrote again to Torrey, mentioning another 5-day trip to Texas, to [unspecified] places not previously visited. He sent for Torrey's inspection what must have been a species of *Cooperia* collected 'early in September' near San Augustine, Texas. I suspect that he visited Dr. Veatch again at this time . . ."

McVaugh refers to what must have been the third trip into Texas thus:

"The last reference to Texas in Leavenworth's letters was on May 22, 1838, when he was about to sail from New Orleans to Florida; he was sending Torrey 'a small collection' from Texas, including, as he said, specimens of *Cacalia tuberosa*."

As to Leavenworth's accomplishment while there: "The *Flora of North America* includes about 30 references to Texan collections made by Leavenworth, and about 12 species and varieties described for the first time in this work are based wholly or partly on these collections. Now when so much more is known about the flora of Texas, it is hard to realize that these 30 collections were relatively important ones in 1838; only Drummond, of all the early field workers, had collected more Texan material that was available to Torrey and Gray!"

With the termination of Leavenworth's work in Texas and in Oklahoma our specific interest in his activities ends. One may be sure that his medical services had been no easy one and resent, on his behalf, the tragedy which McVaugh recounts "struck suddenly."

"The middle-aged medical man, counting heavily upon his long practical experience, arranged to take an examination for promotion, before an Army Medical Board. The examination was held in Philadelphia, and Leavenworth returned to Fort Gratiot about June 20 [1840], only to learn soon after his arrival that during his nine years on the frontier, Army medical standards had been raised to such a level that his own knowledge was not even enough to qualify him, in 1840, for the post to which he had

8. See p. 1056, *fn*. 1. McVaugh states in a footnote that "Zavala" is "(not to be confused with the modern Zavalla, Angelina County)." The one to which Leavenworth refers was "in Jasper County."

been certified in 1831 and again in 1833. His resignation from the Army was accepted as of September 30, 1840, and he retired to his home in Waterbury to live until called forth by the Civil War.

"In the first year of that struggle our retired practitioner, at the age of 66, volunteered for service and was accepted in this emergency by the Army that had rejected him twenty years before . . ."

After devoting a paragraph to the man's Civil War service, McVaugh's paper ends thus:

"We who follow botanical paths today must pause a moment to salute such men as this Dr. Leavenworth, not only as in this case when they die as heroes, but for their hard and dangerous pursuits of their avocations in botany, on the frontiers of America. These travelers and surgeons and lieutenants, who must have loved the lives they lived, fed the genius of the Torreys and Grays who wrote the history and science of botany; without them and their devotion to botany we should have fared poorly before travel became the easy thing it is now."

Gray had mentioned the cruciferous genus—*Leavenworthia*—established by Torrey in honor of Dr. Leavenworth. McVaugh states that ". . . Students of the American flora are familiar with *Carex Leavenworthii, Vicia Leavenworthii, Eryngium Leavenworthii, Coreopsis Leavenworthii,* and *Solidago Leavenworthii* . . ."

McVaugh's paper, which I have quoted at great length, is an additional tribute—as well as an understanding and well-deserved one—to a man whose efforts on behalf of botany had gone largely unnoticed hitherto.

Anonymous. 1863. Obituary [of Leavenworth]. *Am. Jour. Sci. Arts,* ser. 2, 35: 306.

BLUNT, WILFRID. 1950. The art of botanical illustration. London. 230-235.

CATLIN, GEORGE. 1841. Letters and notes on the manners, customs, and condition of the North American Indians . . . Written during eight years' travel amongst the wildest tribes of Indians in North America, in 1832, 33, 34, 35, 36, 37, 38, and 39. 2 vols. New York. 2: 36-86. [With 400 illustrations, "carefully engraved from his original paintings."]

GEISER, SAMUEL WOOD. 1948. Naturalists of the frontier. ed. 2. Dallas. 277.

GHENT, WILLIAM JAMES. 1931. The early far west. A narrative outline 1540–1850. New York. Toronto.

GRAY, ASA. 1863. Botanical necrology, 1862. *Am. Jour. Sci. Arts,* ser. 2, 35: 451.

———— 1860. The trees of North Carolina. *Am. Jour. Sci. Arts,* ser. 2, 30: 275, 276. *Reprinted:* Sargent, Charles Sprague, *compiler.* Scientific papers of Asa Gray. 2 vols. Boston. New York. 1: 115-117.

———— 1875. Ruskin's Proserpina. *The Nation,* No. 528, August 12, 1875. *Reprinted: Ibid.* 1: 199-204.

HEITMAN, FRANCIS BARNARD. 1903. Historical register and dictionary of the United States Army, from its origination, September 29, 1789, to March 2, 1903. 1: 622.

KELLY, HOWARD ATWOOD. 1929. Some American medical botanists commemorated in our botanical nomenclature. New York. London. 216.

LEAVENWORTH, MELINES CONKLING. 1824. Notice of four new species collected in Alabama. *Am. Jour. Sci. Arts* 7: 61-63.

——— 1825. List of the rarer plants found in Alabama . . . (Communicated in a letter to Charles Hooker, M.D.). *Am. Jour. Sci. Arts* 9: 74.

——— 1831. Description and history of a new plant, Tullia Pycnanthemoïdes. *Am. Jour. Sci. Arts* 20: 343-347.

——— 1845. On several new plants . . . in a letter to the Junior Editor [Benjamin Silliman, Jr.]. *Am. Jour. Sci. Arts* 49: 126-131.

McVAUGH, ROGERS. 1947. The travels and botanical collections of Dr. Melines Conkling Leavenworth. *Field & Laboratory* 15: 57-70.

POWELL, WILLIAM HENRY, *compiler.* 1900. List of officers of the Army of the United States from 1779 to 1900 . . . 428.

RODGERS, ANDREW DENNY, 3RD. 1942. John Torrey. A story of North American botany. Princeton. [Contains a few references to Leavenworth.]

RUSKIN, JOHN. 1879. Proserpina: studies of wayside flowers while the air was yet pure among the Alps and in the Scotland and England which my father knew. New York. J. Wiley & Sons. vol. 1.

SARGENT, CHARLES SPRAGUE. 1892. The Silva of North America. Boston. New York. 3: 66, *fn.* 1.

WHEELOCK, THOMPSON B. 1834. Journal of the expedition of the regiment of dragoons into the Indian country in 1834. *U. S. 23rd Cong., 2nd Sess., Sen. Doc.* 1: No. 1, 73-93.

CHAPTER XXVIII

BEYRICH ACCOMPANIES DODGE AND THE UNITED

STATES DRAGOONS INTO COUNTRY LYING BETWEEN

THE CANADIAN RIVER AND THE RED RIVER OF THE

SOUTH

URING the administration of Andrew Jackson and when Lewis Cass was Secretary of War, the United States Dragoons were sent, in mid-summer of 1834, into the country lying between the Canadian and Red rivers which is now included in eastern Oklahoma; it had been set apart for occupation by Creek, Seminole, Cherokee, Choctaw and Chickasaw Indians—the "Five Civilized Tribes"— and was officially known as the Indian Territory. General Henry Leavenworth was in command of the expedition until he became incapacitated, when Colonel Henry Dodge was appointed to take his place. Negotiations were to be carried on with certain tribes who were endangering the safety of Americans and warring among themselves.

The story of the expedition, from Fort Gibson and back, is told by one of Dodge's lieutenants, Thompson B. Wheelock—"Journal of the expedition of the dragoons into the Indian country in 1834"—which was published as a Senate Document in 1834. The volume contains messages by Jackson and Cass, mentioning calamities casually, to say the least. For conciseness Wheelock's "Journal" would be hard to surpass. Army training may have taught him to confine his statements to ungarnished fact —he mentions death as incidentally as the daily mileage.

Grant Foreman's *Pioneer days in the early southwest,* published in 1926, devotes some fifty pages to the expedition, its purposes, its horrible journey, and the conference held at Fort Gibson with the heads of various Indian tribes after its return. The volume includes interesting footnotes upon persons and place-names and a useful map.

Louis Pelzer's *Marches of the dragoons in the Mississippi valley . . . ,* published in 1917, contains information regarding the route taken from St. Louis to Fort Gibson.

Traveling with the Dragoons from St. Louis to Fort Gibson and beyond was a German botanist, Heinrich Karl Beyrich, usually mentioned in the literature as Charles Beyrich or Beyrick. How far west he got from Fort Gibson (it was somewhere in the Cross Timbers) is uncertain. He perished, like many others, on the journey. The artist Catlin was with him at the time and described his death.

Dodge was evidently successful, temporarily at least, in his dealings with the Indians. But from all other aspects the trip was disastrous and largely because of the ignorance of those who inaugurated the plans. For the Dragoons were sent into one

569

of the hottest sections of the southwest during the hottest season of the year, dressed in clothes which would have been uncomfortable even in Arctic regions, marched during the hottest hours of every day and were soon out of food and water, drinking from buffalo-wallows when other liquid was unobtainable.

Foreman paints such a picture of what can only be called the "get-up" of the Dragoons, which he based upon facts supplied in the *Army and Navy Chronicle,* that I turned to that publication for further enlightenment.[1] Under the heading "Dress of the Army," the regulation apparel for various branches of the service is itemized. While the Dragoons had both a dress and an undress uniform, neither betokened comfort and I do not distinguish the differences here—either would have relegated Gilbert and Sullivan's British Dragoon to second place. I cite a few of its most distressing and inappropriate features, ignoring the paragraphing of the original:

"Coat—dark blue cloth, double breasted, two rows of buttons, ten in each row, at equal distances . . . stand up collar, to meet and hook in front, and no higher than the chin . . . buttons . . . gilt, the lace gold, the collar, cuffs, and turnbacks, yellow, the skirt to be ornamented with a star . . . the length of the skirt to be what is called *three-quarters* . . . Trousers—blue grey mixture . . . with two stripes of yellow cloth three fourths of an inch wide, up each outward seam . . . Cap—. . . to be ornamented with a gilt star, silver eagle, and gold cord; the star to be worn in front, with a drooping white horse-hair pompon . . . Boots—ankle. Spurs—yellow metal. Sabre—steel scabbard, half basket hilt, gilt with two fluted bars . . . the outside, fish skin gripe, bound with silver wire . . . Knot—gold cord with acorn end. Sash—silk net, deep orange color . . . to be tied on the right hip . . . Waist-belt—black patent leather, one and a half inch wide, with slings, hooks, and plate . . . Stock—black silk. Gloves—white . . ."

And so on, ad infinitum, including a "Great coat," with ". . . stand up collar, cape to meet, and button all the way in front, and reach down to the upper edge of the cuff of the coat . . ." Wheelock's terse description of the appearance of his men after some ten weeks in the desert is sufficient commentary: they were ". . . a sorry figure; but one that looks like service."

In the *Army and Navy Chronicle* for June 30, 1834 (1: 3, 1835) is a contemporaneous picture, probably from the pen of its publisher and editor Benjamin Homans, of the expedition's purposes. It expresses the hope that its members might have time to return "before cold weather sets in."! I quote therefrom:

"The expedition, it is understood, will be accompanied by several gentlemen of science, who go at their own expense. The object of the expedition is to give the wild Indians some idea of our power, and to endeavor, under such an imposing force, to enter into conferences with them, to warn those Indians who have been in the habit of robbing and murdering our people who trade among them, of the dangers to which they will be exposed in case they continue their depredations and massacres.

1. Despite the extremely fine print, closely set type and somewhat grim overall appearance, its pages—like those of Senate and House Documents—are full of interesting facts.

"Several delegations of the newly emigrated Indians, now settled beyond the limits of the States and Territories to the westward of the Mississippi, as well as of the Osages and other tribes near them, will accompany the expedition, in the hopes of making treaties of friendship with the wild tribes, and thus prevent, for the future, the recurrence of those wars which are so common among the Indians.

"The expedition, it is hoped, will result in much good; it will afford protection to the civilized Indians, to our frontiers, to our trade with the natives, and cover the Santa Fé caravans trading with Mexico; and, perhaps, enlighten the Indians generally as to the humane policy of the United States towards them, and also as to their own true interests . . ."

I now follow the journey, transcribing from Wheelock's "Journal":

1834

June 15: "The nine companies distined for the campaign . . . began their movement from Camp Jackson[2] on the 15th of June, and . . . encamped on the west bank of the Arkansas three miles from Fort Gibson; thence moved 18 miles westwardly[3] to camp 'Rendezvous;'—strength of the regiment, five hundred."

The "officers for the campaign," then commanded by Leavenworth, included "Colonel, Henry Dodge; Lieutenant Colonel S. W. Kearney[4] . . ." On the "Staff," among others, was "1st Lieutenant, T. B. Wheelock," author of the journal and Philip St. George Cooke to whom I refer again. One company was left at Camp Jackson, while "eight were assembled at 'Camp Rendezvous' on the evening of the 20th of June."

June 21: "(S. W., 20 miles.)—Twenty-three men, pronounced by the surgeon unfit for the campaign, sent back to Fort Gibson. The regiment took up its march for the Washita, upon the new road made by General Leavenworth . . . crossed the north fork of the Canadian,[5] and encamped one mile thence; difficulty with wagons ascending the bank of the stream . . ."

Here, according to arrangement, the party was joined by four bands of Indians, interpreters, hunters and guides of the Osage, Cherokee, Seneca and Delaware tribes, who were to assist in dealings with the Pawnees. And, accompanying the party, were two Indian girls taken prisoners some years earlier, one by the Osages, one by the Pawnees; by restoring them to their respective nations it was hoped to conciliate the Indians and pave the way for treaties.

2. Situated about a mile from Fort Gibson.

3. Their destination was the Washita River, reached on June 28.

4. Stephen Watts Kearny.

5. They were in central Hughes County. Foreman places them near what is now Holdenville.

Wheelock's journal continues under the heading "Camp 'Cass.' "

June 22: "(W. 15 miles) . . . company ('I,') left in rear, on account of breaking down of company wagon; *wagons great drawback to military expeditions;* route to-day chiefly through timber; here and there small prairies; water scarce . . ."

June 23: "(W. by S. 17 miles)—Marched from Camp 'Cass' . . . alternate prairie and timber; water less scarce than before, but warm, of a milky color and *in pools.*"

June 24: "(W. by S. 21 miles) . . . road to-day chiefly through timber . . ."

June 25: "(W. 13 miles.) Colonel Dodge and staff reached Camp Canadian, on the west bank of the Canadian, 15 miles from last camp . . . reported to General Leavenworth, whom we found in camp . . . road to-day through open level prairie, well watered; crossed the Canadian half a mile below the mouth of Little River . . ."

June 26: "(W. 32 miles.) . . . Colonel Dodge . . . found General Leavenworth at Camp Osage, 5 miles south of Cave creek . . . the regiment under command of . . . Kearney . . . left 27 sick men at Camp Canadian . . . Lieutenant Cooke was left sick . . ."

June 27: "(W. 23 miles.) . . . Left Camp Osage . . . crossed Blue river, 10 miles from Camp Osage . . . met with and killed the first buffaloe . . ."

June 28: "(W. 25 miles.) . . . encamped at Bon [Bois] d'Arc creek; passed a herd of buffaloe . . . Indians with us killed six . . . road to-day chiefly over bushy prairie and through timber . . . in general, small, post oak and black jack, some trees of bond [bois] d'arc a wood valuable to Indians for bows, a yellow elastic wood of great tenacity; entered the Washita Bottom, 8 miles on the day's route; elm trees, sycamore and ash; health of our party good."

June 29: "(W. by S. 15 miles) . . . reached Captain Dean's camp . . . a mile or two from the Washita[6] . . . encamped near . . . road to-day timber and brushy prairie . . ."

June 30: "General Leavenworth declares his intention of commanding in person the expedition to the Pawnee country . . . some companies of infantry were to accompany us . . . remained in camp."

July 1: "The regiment, under Col. Kearney, arrived . . . and encamped near the Washita . . . 45 men and three officers sick . . . The surgeon attributes the sickness to exposure in the heat of the day; seventy five horses and mules disabled; rapid marching in the day, and poor grazing at night, are supposed to have been the causes; remained in camp."

July 2: "Remained in camp."

6. Totaling Wheelock's mileage from June 21 through June 29, the expedition had moved west approximately one hundred and eighty miles from Fort Gibson.

July 3: "Preparations for crossing the Washita . . . whole day occupied in the passage of the left wing; horse and mule lost in crossing . . . great disappointment in not receiving . . . horse shoe nails, sent blacksmith to Fort Towson to make nails. Lieut. Edwards arrived with 23 men who were left sick at the Canadian on the 26th ult.; men chiefly recovered; Lieut. Cooke had gone back to Fort Gibson on surgeon's certificate of ill health."

July 4: "(W. 4 miles.)—The right wing of the regiment crossed the Washita and joined the left wing; encamped about four miles west from camp Washita; four horses drowned . . .

"The Washita is a narrow stream about 45 yards in width . . ."

This day the botanist Beyrich and the artist Catlin are mentioned by Wheelock for the first time.

". . . Monsieur Beyrick, botanist, &c., joined us to day with the view of accompanying the regiment to the prairie; Mr. Catlin, portrait painter, is also with us. Gen. Leavenworth declares his intention of sending Col. Dodge with 250 men to the Pawnee Villages; Gen. L. changes his determination to command in person . . ."

The party remained in camp on July 5 and, when they moved on the next day, more sick men were left behind. On the 6th they moved ". . . westwardly eight miles to camp 'Leavenworth'[7] . . ."

July 7: "(W. 5 miles.)—. . . killed several buffaloes. Gen. Leavenworth joined us a short time previous to setting out from camp Leavenworth; left him there. Regiment . . . reorganized; left 109 men for duty, and 86 sick . . .

"New arrangement of officers. Field and Staff, Col. Dodge, Lieut. Col. Kearney . . . and Wheelock . . ."

July 8: "(Remained in camp.)—Waiting for lost horses. A stupid sentinel, last night mistook a horse for a hostile Indian, fired and killed him; alarmed the camp and sent off in a *stampedo* the rest of the horses; recovered all save ten. The men of the regiment are excellent material, but unused to the woods; they every day discover deficiencies in this kind of service . . ."

July 9: "(N. W. 14 miles.) . . . Colonel Dodge received . . . instructions from General Leavenworth to send back a field officer to command at Camp Leavenworth . . . Kearney was ordered to report . . . for that duty; ten men, whose horses were lost on the 7th instant, were sent back to Camp Leavenworth . . ."

On July 10 they marched west sixteen miles and evidently reached the "Cross timbers." The country was ". . . rough and broken, with but little water . . . not much water at camp."

7. Totaling the mileage of July 4-6, the site of camp "Leavenworth" must have been about twelve miles west of the Washita River.

July 11: "(W. 20 miles.) . . . Country to-day, small rolling prairies, bushy, ravines, scrubby oak ridges; want of good water on road; bad water at camp to night . . ."

July 12: "(W. 12 miles.) . . . "

July 13: "(W. by N. 23 miles.)—Passed through the last of the cross timbers, and entered upon the grand prairies. Marched . . . from Camp 'Choctaw,' west by north 23 miles, and encamped on a creek; highly beautiful country; tolerably well watered. Command impeded to-day by sick men in litters . . . Rear guard did not come up until 10 [p.m.]; kept back by the sick men falling in the rear."

July 14: "(W. 17 miles.)—. . . number of sick decreased . . . Country to-day beautiful; open prairie . . . provision threatens to be scarce . . . uncertainty of reaching the Pawnees much lessened."

This day they met with a band of "Camanchees." They told Dodge ". . . that they were a very numerous people. Colonel D. answered, that *we* were a very numerous people; that more troops were coming behind, with large guns." On the whole the meeting was amicable and Wheelock notes that the appearance ". . . of a Camanche fully equipped, on horseback . . . [was] beautifully classic."

July 15: "(N. W. 24 miles.) . . . one [Comanche] . . . remained as guide; he assures us that we shall reach the camp of the Camanches to-morrow . . ."

July 16: "(N. by W. 12 miles.) . . . An accident . . . last night; Serjeant Cross was shot, by a dragoon, in the hip . . ."

On reaching the camp of the "Camanches" that day ". . . they invited us to cross the creek and encamp with them; Colonel Dodge however, preferred leaving the creek between us and our red friends . . . Our camp, 'Camanche,' an admirable position . . . Orders issued that no man should visit the Camanche camp; nor officer without special permission. The Camanches have hoisted an American flag over their camp . . ."

We are told that the expedition was ". . . now in sight of a chain of peaks, so called; mountains, bearing south and west; behind these are the Toyash[8] villages . . . Number of sick twenty-nine, in litters four . . ."

July 17: "(Remained in camp.) Camanche chief still absent . . . Dodge hopes to be able to induce him to accompany us to the Toyash villages . . ."

July 18: "(W. 7 miles.)—The chief has not arrived . . . The Camanches visit our camp and trade with us. Monsieur Beyrick, the botanist, left us on the 7th instant . . ."

Since, according to the Wheelock journal, Beyrich had joined the party on July 4 he would only have been with the Dodge contingent for three days had he left on the "7th." This date may have been an error for *17th*. Or Beyrich may have changed from

8. The name is usually spelled Tawehash.

one contingent of the troops to another. Catlin states that Beyrich traveled from Fort Gibson ". . . to the False Washita [where Wheeler records his presence on July 4], and the Cross Timbers and back again [to Fort Gibson]." Wheelock had first mentioned the Cross Timbers[9] on July 10 and had mentioned leaving them on July 13. I have been unable to figure precisely how far west Beyrich traveled.

On July 18 Wheelock records: "Number of sick to-day thirty-three; three officers sick . . ." and that the ". . . command [was] delayed two hours, waiting for the litters to come up; six litters including Mr. Catlin's . . . Our sick incumber us so much that Colonel Dodge resolves to leave them."

July 19: "(S. W. 23 miles.)—Marched . . . for the Toyash villages; command reduced to 183 men; left in sick camp, covered by a breastwork of felled timber, seventy-five men, thirty-nine of these sick . . . left our jaded horses . . . two miles from camp began to ascend hills, apparently a ridge of mountains running south by east . . . road rough, leading over rocky ravines and close passes in the mountains. Our guide seems to have chosen the most uneven and circuitous route; height of these mountains from 200 to 1,500 feet. Wagons nor artillery could possibly pass these hills . . . No buffaloe. Our unshod horses suffered very much to-day. Wild horses in abundance and bears . . . Scanty allowance of provisions for our men. We march too fast to be able to hunt much on the road . . . baggage reduced to three pack-horses to each company."

July 20: "(W. 37 miles.) . . . Road literally of granite road for miles; after a few miles, struck high prairies, thinly scattered with bushes; then ravines and difficult passes . . . Many horses gave out to-day . . . We encamped five miles from the Toyash village[1] which River is situated on a branch of Red river . . . Water to-day at our camp, salt; width of the branch of Red river about 500 feet from bank to bank; water low . . . the Cherokee guide, very ill; Kiowa girl ill also."

July 21[2]: "(Northwest five miles) . . . marched . . . for the Toyash village . . ."

9. Chittenden identifies the "forest section known as the Cross Timbers" thus: ". . . being mainly in Texas and the Indian Territory, and . . . extended from the Brazos river northwesterly to the Canadian river. South of the Canadian a branch of the Cross Timbers extends westwardly and then northerly across the North Fork of the Canadian, where it finally disappears. This hilly, wooded section is the line between the lower well-watered prairies and the high, arid plains. It is covered with a thick growth and a wide variety of trees mostly of dwarfish, stunted size. The underwood is so overgrown with vines and shrubbery as to be in places almost impassable." (2: 803)

There are many references to the Cross Timbers in the narratives of those who traveled in the Indian Territory.

1. The "Toyash village" was the Tawehash or Wichita village. The branch of Red River [or North Fork of Red River] to which Wheelock refers was Elm Fork which enters the larger stream from the north or northwest in Greer County, or not far north of the boundary between Greer and Jackson counties. The Toyash village was situated to the west of the Wichita Mountains.

2. According to Ghent, ". . . the farthest point reached, [was] in present Jackson County, Oklahoma . . ." The party was, therefore, in the southwestern corner of that state.

A band of Indians came out to meet the Dragoons and reported that the chief was away in the "Pawnee Mohawks country." Wheelock comments that the fields of the Indians ". . . are well cultivated, neatly enclosed, and very extensive, reaching in some instances several miles. We saw . . . melons of different kinds, squashes, &c."

The Indians were alarmed by the approach of the troops and begged that they should not be fired on. Dodge promised them safety; ". . . the uncle of the Pawnee girl . . . embraced his relation and shed tears of joy on meeting her." The Dragoons camped about a mile from the village which was situated near the river and consisted of about two hundred "grass lodges." They were ". . . hospitably entertained; our own provisions were almost entirely exhausted; we . . . found most excellent fare in the dishes of corn beans . . . they . . . for desert, gave us water-melons and wild plumbs . . ."

Indeed, during their stay of three days, the party was dependent upon what the Indians provided in dried meat and green corn.

The Wheelock journal devotes several pages to Dodge's talks with the assembled Indians.[3] Dodge asked about a white man—"His name was Abby[4]"—and the answer was that he had been killed. The murderers, it was said, were "the Oways, who live south . . ." A white boy, "Matthew Wright Martin," seven years of age, was with the Pawnees and he and the Pawnee girl were exchanged. The Indians had with them a Negro and evidently included him for good measure, the chief stating "You can take him, and do as you please with him." It was this Negro who told the Americans of Martin's presence when the Indians denied it. On July 23 and 24 there were further talks and the ". . . following interesting ceremony took place." It is explained:

"The boy whom we recovered yesterday, is the son of the late Judge Martin, of Arkansas, who was killed, by a party of Indians, some weeks since. The son was with the father on a hunting excursion, and being parted from him, (his death, however, he did not witness, and is now in ignorance of it,) the boy relates, that after being parted from his father, the Indians who had taken him were disposed, save one, to kill him; this one shielded him, and took care of him in sickness. Colonel Dodge, as a reward for this noble kindness, gave him a rifle, and, at the same time, caused the *little boy* to present him, with his own hand, a pistol . . ."

After two more days of speechmaking it was decided that further conferences should be held at Fort Gibson. The expedition now turned eastward.

"It is on all accounts desirable to remove from here; our provisions prove unhealthy for our men; consisting entirely of green corn, and dried horse and buffalo meat; the weather has been excessively hot and dry; our men, many of them sick, are without a physician or medicines; two or three officers are, and have been for several days, ill of fevers.

3. According to Ghent, there were ". . . Comanches, Kiowas, Tawehashes, and some other Indians." (p. 309)

4. The name appears in various spellings. Catlin uses Abbé.

"The Camanche squaws are very troublesome; they steal every thing that they can secrete; the Toyash woman [sic] are infinitely more respectable . . ."

July 25: "(E. 6 miles.) . . . The command, with the Indians [for guides], the white boy, and the negro in company, marched at 3 o'clock, halted at 5 o'clock, and encamped on a creek . . ."

July 26: "(E. 21 miles.) . . . our guide . . . left us; he was no loss, for he had led us over an uselessly long route, over rocks and hills, through deep ravines, all of which our guide to day, a Toyash, has avoided . . . we have passed through a beautiful valley . . . over an open level prairie, leaving the granite roads on our right and left in the mountains . . . water scarce . . ."

July 27: "(E. 23 miles.) . . . reached the sick camp . . . Mr. Catlin very ill; 29 sick men in both camps . . . Wheelock's servant left sick on the 19th instant; died in our absence . . . reached 'Roaring river,' a short stream . . . empties in the Red river . . . Col. Dodge and all the officers unable to account for not hearing from Gen. Leavenworth . . ."

July 28: "(E. by N. 12 miles.)—Broke up the sick camp, and marched . . . with the whole command . . . Excessive hot weather; 43 sick; 7 in litter . . . The heat to-day has been overpowering, both to men and horses; water tolerable; coursed north from our trace going out; camp to-night about six miles from former trace. Col. Dodge sent an express in search of Gen. Leavenworth . . ."

July 29: "(E. by N. 15 miles.) provisions very short . . . road to day over rolling prairies, between two forks of the Washita . . ."

July 30: "(N. E. 14 miles.) . . . weather excessively hot . . . course interrupted by frequent deep gullies, totally impassable for wagons; 9 miles from camp passed the Washta; good water to-day; encamped on a fine stream; large fishes visible from the bank; timbered creeks; black jack, elm, and mulberry trees . . ."

July 31: "(N. by E. 10 miles.) . . . encamped on a branch of the Canadian . . . no news from the express; anxiously looked for . . ."

August 1: "(N. by E. 15 miles) . . . crossed the Canadian[5] . . ."

August 2: "(Remained in camp)—Rest! Welcome rest for men and horses; occupied in killing and drying buffalo meat for the anticipated march to Fort Leavenworth; probable distance thither 400 miles. Our men not infrequently lost in hunting; in several instances, absent from the camp all night. Our men find an excellent substitute for tea and coffee in a wild sage plant . . ."

5. From the Toyash villages the expedition had moved east and then northeast and, on arrival at the Canadian, was no great distance south of present Oklahoma City.

August 3: "(1 mile.)—Moved a mile . . . for change of grazing and police; our horses are in bad order . . . they may not be equal to a march to Fort Leavenworth; may . . . be compelled to move to Fort Gibson, to recruit and shoe them. Little Martin flourishes, and is a great favorite in the command; he is an uncommonly fine boy."

August 4: "(S. 8 miles.) . . . along the Canadian[6] . . . Grass very much dried; scarce affording subsistence for our horses; Col. Dodge has decided for marching for Fort Gibson. The prairie took *fire to-day near our camp* . . . with difficulty extinguished."

August 5: "(Remained in camp.)—Rested for the day . . . The express to Gen. Leavenworth returned; intelligence from Capt. Dean, 3d infantry; announces the death of Gen. Leavenworth;[7] he died at his camp, near 'Cross Timber' on 21st of July. Lieutenant McClure of the regiment, died at the Washita on the 20th July. Bilious fevers; 150 men sick at the Washita."

August 6: "(S. E. 23 miles.) Marched . . . for the Fort at the mouth of Little River . . . road through 'Cross Timbers.' This is a timbered thicket, small black jack saplings so close as to frequently require the axe to make a road for a horseman. Five litters in our train; men in them extremely ill. Col. Dodge sent an express to Col. Kearney, who is at Camp Smith, near the mouth of the Washita, directing him to move his command to Fort Gibson . . . Encamped in timber, in the bottom of a branch of Little River; found excellent grazing in the pea vines. Litters came up several hours after the command."

August 7: "(S. by E. 18 miles.) . . . still in the 'Cross Timbers,' which promise to continue till we strike the road to Fort Gibson . . . Our route to-day . . . on the dividing ridge, between the Canadian and Little river . . ."

August 8: "(E. by S. 18 miles.) . . . exceedingly warm day; stubborn thickets. Crossed and encamped in the bottom of Little river . . . No water from morning till the halt for the night. Passed many creeks, the beds . . . entirely dry; our horses looked up and down their parched surfaces, and the men gazed in vain at the willows ahead, which proved to mark only where water had been. The timber is larger here. Black walnut and sycamore . . . The woods abound to-day in plums, and a variety of finely flavored grapes . . . Sick report numbers thirty men and three officers."

August 9: "(N. E. 20 miles.) . . . Cross Timbers, but more open than for the last three days . . . encamped . . . in open timber, near where we struck the road from Fort Gibson to the Washita, which was three miles from the post at the mouth of Little river."

6. Moving along the Canadian southward for nine miles (August 3 and 4) the party could not have been far from present Norman, the county seat of Cleveland County.

7. After parting from Colonel Dodge and his contingent on July 6, General Leavenworth had attempted to move westward but had gotten no great distance beyond the Washita River. The precise site of "Camp Smith" I do not know. Leavenworth was buried for a time at Fort Towson but was eventually interred at Delhi, New York. *See: Army and Navy Chronicle* 1: 374, 1835.

August 10: "(Remained in camp.)—Dragoon camp, 'Canadian:' We drew from Lt. Holmes, commander of the infantry, camp 'Canadian,' at the mouth of Little river,[8] provisions for four days . . . vast many sick; on *our sick list*, thirty."

August 11: "(22 miles.) . . . left our sick, whom we brought in litters, at the infantry camp . . . our men happy with pork and flour."

August 12: ". . . express returned from Camp Smith . . . Kearney reports many sick; 71 for duty, 41 sick; 8 for duty at Camp Washita, and 70 sick; many of our horses disabled led by men in rear of the columns; tolerable water, wholly in pools. It is worthy of remark that the mules of the command look better than when we started on the campaign, while it would be difficult to select *ten horses* in good order. The command ordered to walk and ride one hour alternately; this relieves the horses."

August 13: "(17 miles.) . . . reached the Creek settlements, at the north fork of the Canadian . . . informed here that the mother of little Martin has recently offered two thousand dollars for his recovery; she will soon be happy by his restoration without ransom or reward."

August 14: "(20 miles) . . . to our former camp 'Rendezvous,' from whence the regiment started on the 21st June. Our horses are exceedingly worn . . . our men present a sorry figure; but one that looks like service; many of them literally half naked; sick list reduced to *nineteen*."

August 15: "(14 miles.) . . . an officer was sent in advance to purchase corn . . . encamped 3 miles from the west bank of the Arkansas; Col. Dodg[e], and staff, together with the Indians, crossed the river late in the evening and reached Fort Gibson."

August 16: "(Fort Gibson.) . . ."

August 24: "Colonel Kearney's command arrived yesterday; great number of sick men, and worn down horses . . . Lieutenant Swords, (sick,) Lieutenant Van Deever, (sick,), Lieutenants Bowman, Ury, and Kingsbury, assistant surgeon Hailes, (very sick.) . . . little Martin is still with Colonel Dodge, and the negro we brought from the Toyash village has been delivered to his master."

So ends the Wheelock journal.

Some twenty-three years later Philip St. George Cooke, in his *Scenes and adventures in the Army: or the romance of military life,* expressed his opinion of the government's action in sending untrained troops on such a mission. After recounting the hardships which these inexperienced men had already undergone during the winter before they got started from Fort Gibson, Cooke relates—and I quote but a fraction:

". . . the regiment marched full six weeks too late, when it is considered that we

8. The party was back at Fort Canadian, or Fort Holmes, where they had been on June 25.

580 BOTANICAL EXPLORATION OF THE TRANS-MISSISSIPPI WEST

were to traverse the burning plains of the South: and the thermometer having previously risen to 105° in the shade, there was every prospect of a summer of unexampled heat.

"It is painful to dwell on this subject . . . On, on they marched, over the parched plains whence all moisture had shrunk, as from the touch of fire; their martial pomp and show dwindled to a dusky speck . . . disease and death struck them as they moved; with the false mirage ever in view, with glassy eyes, and parched tongues, they seemed upon a sea of fire. They marched on, leaving three-fourths of their number stretched by disease in many sick camps; there, not only destitute of every comfort, but exposed with burning fevers to the horrors of the unnatural heat — it was the death of hope. The horses too were lost by scores. In one sick camp, they were in danger of massacre[9] by a horde of Camanche Indians . . . near by . . .

"General Leavenworth and his aid stopped. They both lost their lives. Colonel Dodge, with 150 of the hardiest constitutions, persevered and overcame every obstacle; they reached the Tow-e-ash village . . . such he discovered to be the name of a numerous tribe, who together with Camanches, Kiawas, and Arrapahoes had hitherto been confounded under the name of Pawnees.

"There, perhaps within the boundary of Mexico [Texas], was made this first though feeble demonstration of the power and ubiquity of the white man. Some breath was expanded in an effort to mediate peace between these wandering savage robbers and their red neighbors of our border; as availing as it would be to attempt to establish a truce between the howling wolf of the prairies and his prey . . .

"The shattered and half famished remnants of the regiment were gathered together at Fort Gibson,[1] in August. The thermometer had risen in the shade to 114°. There, in tents and neglected, many more suffered and died . . ."

The letters of the artist George Catlin who, as stated, traveled with the Dragoons from Fort Gibson westward and back, some of the time in a litter, are also frank in their criticism. I cite from his *Letters and notes* . . . published in New York in 1841, from those numbered 37 through 45.[2] These often elaborate the events so tersely described by Wheelock.

Letter 39 was written at the "Mouth of False Washita, *Red River*." In it the writer explains that "We are, at this place, on the banks of the Red River, having Texas under our eye on the opposite bank. Our encampment is on the point of land between the Red and False Washita rivers, at their junction . . . We are distant from

9. Cooke tells that the ". . . judgment and determination of . . . the lamented Izard . . ." probably saved these men.

1. The conference with the Indian tribes was held at Fort Gibson in September, 1834. Despite the fact that the agreements reached were not always adhered to the results were beneficial in a number of ways. *See:* Foreman (p. 154)

2. In the edition studied these letters bear no dates and these are also omitted in the Catlin text.

Fort Gibson about 200 miles, which distance we accomplished in ten days." He pictures the flora:

"Scarcely a day has passed, in which we have not crossed oak ridges, of several miles in breadth, with a sandy soil and scattering timber; where the ground was almost literally covered with vines, producing the greatest profusion of delicious grapes, of five-eighths of an inch in diameter, and hanging in such endless clusters, as justly to entitle this . . . wilderness to the style of a vineyard (and ready for the vintage) . . .

"The next hour we would be trailing through broad and verdant valleys of green prairies . . . our progress completely arrested by hundred of acres of small plum-trees, of four or six feet in height . . . every bush . . . was so loaded with the weight of its delicious wild fruit, that they were in many instances literally without leaves on their branches, and bent quite to the ground, Amongst these, and in patches, were intervening beds of wild roses, wild currants, and gooseberries. And underneath and about . . . and . . . interlocked . . . huge masses of the prickly pears, and beautiful and tempting wild flowers . . ."

Letter 40, dated from the "Mouth of False Washita," tells of being detained because of ". . . the extraordinary sickness[3] which is afflicting the regiment, and actually threatening to arrest its progress."

Letter 44, "Camp Canadian, *Texas*," describes what the party was enduring:

". . . From day to day we have dragged along exposed to the hot and burning rays of the sun, without a cloud to relieve its intensity, or a bush to shade us, or anything to cast a shadow, except the bodies of our horses. The grass for a great part of the way, was very much dried up, scarcely affording a bite for our horses; and sometimes for . . . many miles, the only water we could find, was in stagnant pools . . . in which the buffaloes have been lying and wallowing like hogs in a mud-puddle. We frequently came to these dirty lavers, from which we drove the herds of wallowing buffaloes, and into which our poor and almost dying horses, irresistibly ran and plunged their noses, sucking up the dirty and poisonous draught . . . in some instances, they fell dead in their tracks—the men also (and oftentimes . . . the writer of these lines) . . . laded up and drank to almost fatal excess, the disgusting and tepid draught, and with it filled their canteens . . . sucking the bilious contents during the day . . ."

Of the many miseries described by Catlin the above should suffice. At the time of writing he himself was ill: ". . . I plainly see that I shall be upon a litter, unless my horrid fever leaves me, which is daily taking away my strength, and almost, at times, my senses . . ." That his surmise was correct is indicated in his next letter, number 45, written from "Fort Gibson, *Arkansas*." He had been taken back to this fort on a litter in company with others,—those who had died were buried on the spot.

---

3. How the "sickness" should be diagnosed I do not know. Lasègue records it as cholera. Catlin describes it as ". . . a slow and distressing bilious fever, which seems to terminate in a most frightful and fatal affection of the liver." Chittenden mentions that ". . . a deadly malaria prevailed all the time."

It is in this letter 45 that Catlin writes of the botanist Beyrich and describes his death:[4]

"On the same day was buried also the Prussian Botanist, a most excellent and scientific gentleman, who had obtained an order from the Secretary at War to accompany the expedition for scientific purposes. He had at St. Louis, purchased a very comfortable dearborn waggon, and a snug span of little horses to convey himself and his servant with his collection of plants, over the prairies. In this he travelled in company with the regiment from St. Louis to Fort Gibson, some five or six hundred miles, and from that to the False Washita, and the Cross Timbers and back again. In this Tour he had made an immense, and no doubt, very valuable collection of plants, and at this place had been for some weeks indefatigably engaged in changing and drying them, and at last, fell a victim to the disease of the country, which seemed to have made an easy conquest of him, from the very feeble and enervated state he was evidently in, that of pulmonary consumption. This fine, gentlemanly and urbane, excellent man, to whom I became very much attached, was lodged in a room adjoining to mine, where he died, as he had lived, peaceably and smiling, and that when nobody knew that his life was in immediate danger. The surgeon who was attending me, (Dr. Wright,) was sitting on my bed-side in his morning-call at my room, when a negro boy, who alone had been left in the room with him, came into my apartment and said Mr. Beyrich was dying—we instantly stepped into his room and found him, not in the *agonies* of death, but quietly breathing his last, without a word or a struggle, as he had laid himself upon his bed with his clothes and his boots on. In this way perished this worthy man, who had no one here of kindred friends to drop tears for him; and on the day previous to his misfortune, died also, and much in the same way, his devoted and faithful servant, a young man, a native of Germany. Their bodies were buried by the side of each other, and a general feeling of deep grief was manifested by the officers and citizens of the post, in the respect that was paid to their remains in the appropriate and decent committal of them to the grave."

This same letter describes Catlin's own efforts on behalf of science—he had attempted to collect fossils and minerals. Although heavier than plants, they should have required less work, once gathered; however, the task was too much for Catlin:

"On our way, and while we were in good heart, my friend Joe and I had picked up many minerals and fossils of an interesting nature, which we put in our portmanteaux and carried for weeks, with much pains, and some *pain* also, until the time when our ardour cooled and our spirits lagged, and then we discharged and threw them away; and sometimes we came across specimens again, still more wonderful, which we put in their place, and lugged along till we were tired of *them,* and their weight, and we discharged them as before; so that from our eager desire to procure, we lugged many pounds weight of stones, shells, &c. nearly the whole way, and were glad that their

4. The *Beitrage zur botanischen Zeitung* (2: 222, 223, 1844), in a notice of Catlin's *Letters and notes . . .* (London, 1841), quotes some of the artist's remarks about Beyrich.

mother Earth should receive them again at our hands, which was done long before we got back."

As noted in Catlin's just-quoted letter, Beyrich had traveled with the regiment of Dragoons ". . . from St. Louis to Fort Gibson, some five or six hundred miles, and from that to the False Washita, and the Cross Timbers and back again . . ."

Lasègue's *Musée botanique de M. Benjamin Delessert* (p. 466) includes a short account of Beyrich's life which confirms Catlin's statement regarding the trip from St. Louis and, supplies some additional facts about the man's work in North America. I translate the statement:

"*Charles Beyrich.*—Charles Beyrich, a Prussian gentleman, visited western America under the auspices of his government, and devoted the greater part of his time to botany. He spent the summer of 1833 mainly in the two Carolinas and in Georgia where, as well as in some of the neighboring states, he amassed a collection of 1500 species of plants during one season. Visiting the city of Washington in the following winter, he learned that a military expedition into the Indian territory west of the Mississippi was being planned for the following spring and obtained permission to accompany it. He joined the detachment at St. Louis, in the spring, and repaired to different stations on the frontier. On the return of this expedition, and laden with collections found in a new and unknown region, Beyrich was attacked by cholera and died at Fort Gibson in September 1834."

I have already mentioned Pelzer's *Marches of the dragoons in the Mississippi valley*. According to the author's preface, three regiments of Dragoons appear "upon the military rosters of the United States" but the book deals mainly with the "First Regiment of Dragoons." Pelzer does not mention Beyrich but, since he doubtless collected plants on his way to Fort Gibson, it should be of interest to mention briefly what is told of the route from Jefferson Barracks southward:

The march began on November 20, 1833, and they made "three or four miles," and stopped at "Camp Burbees." The second day (the 21st) they made twenty miles, and the third (the 22nd) twenty-three.

"Sunday, the 25th of November, 1833, was spent in marching seventeen miles . . . On the 8th of December the soldiers came upon the silent ruins of an old town of the Delawares . . . The boundary line between Arkansas Territory and Missouri was crossed, and the little town of Fayetteville next noted the passage of the regiment. Soon heavy forests of oak, elm, and pecan were left behind, and the marches then averaged twenty-five miles per day. On December 15th Illinois Creek was forded and that night the dragoons camped upon its banks. Fort Gibson having been passed on the afternoon of December 17, 1833, a temporary encampment named Camp Sandy was made nearby upon a sandbar which projected half-way across the Grand River . . ."

The journey had taken some twenty-seven days and we are told that "Camp Jackson, laid out in a little bit of woods one and a quarter miles from Fort Gibson, became the winter quarters for the regiment."

Pelzer's account carries the Dragoons to the Pawnee Pict and Comanche villages and back, but this later part of the journey has already been described in the words of participants. Pelzer's story is full of interesting comments and quotations.[5]

How many of the plants collected by Beyrich on his journey west of the Mississippi were, and are now, preserved I do not know. Perhaps none, perhaps all. The army may have shipped his effects to Germany after his death.

Asa Gray, writing in 1840 of European herbaria and in particular of "The royal Prussian herbarium . . . deposited at Schöneberg[6] (a little village in the environs of Berlin), opposite the royal botanic garden, and in the garden of the Horticultural Society . . .", mentions that "The principal contributions of the plants of this country to the general herbarium, garden specimens excepted, consist of the collections of the late Mr. Beyrich, who died in western Arkansas,[7] while accompanying Colonel Dodge's dragoon expedition, and a collection of the plants of Missouri and Arkansas by Dr. Engelmann, now of St. Louis . . ."

In *La phytographie* Alphonse de Candolle mentions that Beyrich collections from Brazil and the United States are in the herbaria at Kiel and in the Imperial Garden at "Saint-Pétersbourg (600 esp[èces]) . . ."

And Lasègue, already quoted, mentions that, in the herbarium of [Carl Sigismund] Kunth[8] in Berlin, are plants collected by Beyrich in America; and others in [Carl Friedrich von] Ledebour's herbarium in Russia, but from what country these came is not stated. The man had collected in Brazil at one time.

Von Chamisso and von Schlechtendal named the genus *Beyrichia* in his honor; this is a Brazilian genus of the Figwort Family *(Scrophulariaceae).*

Other plant collectors had been in what is now the state of Oklahoma before Beyrich: Nuttall, from his headquarters at Fort Smith, Arkansas, had entered some of its eastern portions and James had crossed it following the Canadian and Arkansas rivers on his way to Fort Smith; Pitcher and M. C. Leavenworth had made collections while stationed at army posts in, or adjacent to, Oklahoma.

But, except for the younger Abert, who made a trip down the Canadian River in 1845 of our next decade, botanical collectors did no further work in the state until after 1850.

---

5. The expedition had few humorous aspects. One welcomes, therefore, Pelzer's quotation from the journal of a Dragoon who accompanied one of the regiments between Jefferson Barracks and Fort Gibson in May and June of 1834:

"'This country', concludes the journal, 'is remarkable for insects such as snakes, Ticks, & Cattipillars.'"

6. The Royal Botanic Garden was moved to Dahlem-Berlin in 1907 (*Beibl. Bot. Jahr.* 40: no. 90, 62, May, 1907). It was destroyed by fire in World War II.

7. Arkansas Territory.

8. The Kunth herbarium was destroyed by fire in World War II.

ARMY AND NAVY CHRONICLE. Benjamin Homans, *editor*. 1835. Washington, D. C. 1: 3, 383, 384, 391, 392, 399, 400.

CANDOLLE, ALPHONSE DE. 1880. La phytographie; ou l'art de décrire les végétaux considérés sous différents points de vue. Paris. 396.

CATLIN, GEORGE. 1841. Letters and notes on the manners, customs, and condition of the North American Indians ... Written during eight years' travel amongst the wildest tribes of Indians in North America, in 1832, 33, 34, 35, 36, 37, 38, and 39. 2 vols. New York. 2: 36-86. [With 400 illustrations, "carefully engraved from his original paintings."]

CHITTENDEN, HIRAM MARTIN. 1902. The American fur trade of the far west. A history of the pioneer trading posts and early fur companies of the Missouri valley and the Rocky Mountains and of the overland commerce with Santa Fe. 3 vols. New York. 2: 631-633, 803.

COOKE, PHILIP ST. GEORGE. 1857. Scenes and adventures in the army: or romance of military life. Philadelphia. 215-227.

FOREMAN, GRANT. 1926. Pioneer days in the early southwest. Cleveland. Map. 103-156.

GHENT, WILLIAM JAMES. 1931. The early far west. A narrative outline 1540–1850. New York. Toronto. 808, 809.

GRAY, ASA. 1840. Notices of European herbaria, particularly those most interesting to the North American botanist. *Am. Jour. Sci. Arts* 40: 1-18. *Reprinted:* Sargent, Charles Sprague, *compiler*. 1889. Scientific papers of Asa Gray. 2 vols. Boston. New York. 2: 1-21.

JACOBS, EMIL. 1875. [Biographical notice of Heinrich Karl Beyrich.] *Allgem. Deutsche Biographie* 2: 605.

LASÈGUE, ANTOINE. Musée botanique de M. Benjamin Delessert. Notices sur les collections de plantes et la bibliothèque qui le composent; contenant en outre des documents sur les principaux herbiers d'Europe et l'exposé des voyages entrepris dans l'intérêt de la botanique. Paris. 334, 340, 466, 482, 505.

PELZER, LOUIS. 1917. Marches of the dragoons in the Mississippi valley. An account of marches and activities of the first regiment United States dragoons in the Mississippi valley between the years 1833 and 1850. Iowa City. 1-48.

WHEELOCK, THOMPSON B. 1834. Journal of the expedition of the regiment of dragoons into the Indian country in 1834.[9] *U. S. 23rd Cong., 2nd Sess., Sen. Doc.* 1: No. 1, 73-93. [References to the same expedition are found in "Message from the President of the United States ..." (p. 16) and in "Report of the Secretary of War ..." (pp. 25, 26).]

9. The title of Wheelock's journal is cited as it appears in the index of the volume in which the journal was published, p. viii.

CHAPTER XXIX

NUTTALL TRAVELS TO THE COLUMBIA RIVER WITH

WYETH'S SECOND TRANSCONTINENTAL EXPEDITION

AND LATER VISITS SOME COASTAL PORTS OF CALI-

FORNIA

IN an earlier chapter I told something of Nathaniel Jarvis Wyeth and his importance in the development of the far west and discussed his first transcontinental expedition of 1832–1833, concentrating upon that portion of the return journey when he made a small collection of plants.

On Wyeth's second expedition to the Columbia River in 1834, now described, he
was accompanied by the botanist Thomas Nuttall. This was Nuttall's third journey
west of the Mississippi River and one of his most important field trips. If he kept a
journal, as he did when ascending the Arkansas River, all trace of it has been lost.

Another member of Wyeth's party who was interested in scientific pursuits was
John Kirk Townsend of Philadelphia, a physician and ornithologist. The most valuable record of the trip is his *Narrative of a journey across the Rocky Mountains, to the
Columbia River, and a visit to the Sandwich Islands, Chili, &c. with a scientific appendix,* which was published in Philadelphia in 1839 and, under a different title, in
London in 1840.[1] An "Advertisement" written by Townsend and prefacing the Philadelphia edition states:

"The Columbia River Fishing and Trading Company was formed in 1834, by
several individuals in New York and Boston. Capt. Wyeth, having an interest in the
enterprise, collected a party of men to cross the continent to the Pacific, with the purpose chiefly of establishing trading posts beyond the Rocky Mountains and on the
coast.

"The idea of making one of Capt. Wyeth's party was suggested to the author by the
eminent botanist, Mr. Nuttall, who had himself determined to join the expedition
across the North American wilderness. Being fond of Natural History, particularly the
science of Ornithology, the temptation to visit a country hitherto unexplored by naturalists was irresistible; and the following pages, originally penned for the family-
circle, and without the slightest thought of publication, will furnish some account of
his travels."

1. In 1934 A. B. Hulbert, in the fourth volume of the series "Overland to the Pacific" (*The call of the
Columbia . . .*), printed "Extracts from a Private Journal, Kept by Mr. John Townsend, During a Journey
across the Rocky Mountains, in 1834." Hulbert states that it ". . . furnishes no major addition to the well-
known *Narrative* story, and is, in good part similar to that document . . ."

In 1905 Reuben Gold Thwaites reprinted the Townsend *Narrative* from the Philadelphia edition of 1839 (in volume 21 of his *Early western travels 1748—1846*) with an introduction and notes. He omitted the scientific appendix as well as Townsend's account of the Sandwich Islands, Chile and other regions unrelated to the locale of his series.

In his *Narrative* Townsend refers constantly to "Mr. N. and myself." He and Nuttall were much together but, although we acquire a picture of the botanist's assiduity and absorption in his task, we are told virtually nothing of his collecting-methods and specific accomplishments. The youthful author was full of vigor and spirit and his story is very readable; as the journey progresses, however, his sense of humor and exuberance appear to diminish somewhat!

Wyeth also kept a journal and this, like the one kept on his first expedition, was edited by Frederick G. Young; it begins May 5, 1834, at the crossing of the Kansas River, and ends April 13, 1835; but it is not a complete nor always a clear record and the major portion concerns the author's activities after arrival at Fort Vancouver on September 14, 1834. Published with the journal are a number of letters.

After he had reached the west coast Nuttall made two trips to the Sandwich or Hawaiian Islands (December, 1834–April, 1835, and October, 1835–March, 1836) and Townsend accompanied him on the first of these. After they had returned to the Columbia River in April, 1835, they went different ways for the most part although there is no indication of estrangement. Townsend took charge of the hospital at Fort Vancouver for about one year after Dr. Meredith Gairdner's ill health had made it necessary for him to give up that position. Gairdner and Nuttall sailed for the Sandwich Islands on the same ship in October, 1835.

After his second winter in the Islands Nuttall visited California arriving there in March of 1836. He went down the coast in the brig *Pilgrim,* going ashore when the vessel put into port. Pennell has suggested that there was a stop at San Francisco but the records are mainly from Monterey to San Diego. On May 8, 1836, he sailed from San Diego in the *Alert,* a ship made famous in Dana's *Two years before the mast—* and is said to have reached Boston on September 20, 1836.

Brewer (1880), Coville (1899), Jepson (1934), Pennell (1936), Eastwood (1939) and others have written briefly about Nuttall's visit to California. *The North American Sylva* and Nuttall's botanical papers published in the *Transactions* of the American Philosophical Society and elsewhere contain records of the collections made in California, and on his journey westward.

Wyeth had been in correspondence with Nuttall about the proposed journey while he was on his way east in 1833, writing him on July 4 from "Heads of Lewis River":

"I shall remain here one more year. You if in Camb[ridge] may expect to see me in about a year from the time you receive this. I shall then ask you if you will follow another expedition to this country in pursuit of your science. The cost would be less than living at home."

By January 15, 1834, plans had been made and the meeting place was being discussed. By February 4 Wyeth had evidently heard that Townsend wished to join the party:

". . . I am pleased that there will be one more added to our society. As he will probably have no servant I would not recommend to him to take many goods. His stock of clothing would I should think be all that he will require, I should advise him to take three mules but three horses or two mules would do. What little clothing he may want in the country he shall have on such terms that he will have no reason to complain. I do not see that he need provide anything before reaching St. Louis more than he has unless he carrys *implements of science*.

". . . Can you get some cherry, peach, apple, pear, apricot, plumb, and nectarine stones to take with us? . . ."

Nuttall must have asked where they were to meet, for Wyeth wrote him from St. Louis on March 4:

". . . I can not tell you exactly what time I shall leave St. Louis, but of this you may be certain that you will have no time to spare after you receive this and when you arrive at St. Louis call on Mess Von Phull & McGill who will inform you if I have gone up the river in which case follow as fast as you can. At Liberty or Independence you will hear if I have started and how long. If I have not been gone more than three or four days with a good horse you will easily overtake me before you come to any dangerous country following the trail of my horses . . ."

We are told by the anonymous author of the "Biographical sketch of the late Thomas Nuttall" that the botanist had returned to Philadelphia in 1833, ". . . determined as he said, to resign his professorship [at Harvard], as the College authorities would not grant him leave of absence, and he then made arrangements for his great journey to the Pacific Coast. He wrote to the Governor of the Hudsons Bay Company for protection and hospitality in case he should visit any of their posts, but received a very unsatisfactory reply, which Nuttall said was not much more than he expected, as the subordinates, in such cases, he had always found to sympathize with his objects more than the officials."

I shall follow Wyeth's westward journey in the Philadelphia edition of Townsend's *Narrative*, supplementing this with comments by Chittenden, Bancroft, Pennell, Russell and other writers when they offer enlightenment on various aspects of the journey.

1834

Nuttall and Townsend reached St. Louis on March 24 in "the steamboat Boston from Pittsburgh." Wyeth was already there and supervised a shopping party on the 25th:

". . . He accompanied us to a store in the town, and selected a number of articles for us, among which were several pairs of leathern pantaloons, enormous overcoats, made of green blankets, and white wool hats, with round crowns, fitting tightly to the head, brims five inches wide, and almost hard enough to resist a rifle ball." [2]

March 28: "Mr. N. and myself propose starting to morrow on foot towards the upper settlements, a distance of about three hundred miles. We intend to pursue our journey leisurely, as we have plenty of time before us, and if we become tired, we can enter the stage which will probably overtake us."

March 29: ". . . Mr. N. and myself started on our pedestrian tour . . . I was glad to get clear of St. Louis, as I felt uncomfortable in many respects while there, and the bustle and restraint of a town was any thing but agreeable to me . . . at Florisant . . . we spent the night . . ."

May 30: ". . . At about noon, we crossed the river on a boat worked by horses, and stopped at a little town called St. Charles.

"We find it necessary . . . to travel very slowly, as our feet are already becoming tender, and that we may have an opportunity of observing the country, and collecting interesting specimens. Unfortunately for the pursuits of my companion, the plants (of which he finds a number that are rare and curious) are not yet in flower, and therefore of little use to him. The birds are in considerable numbers, among the principal of which is the large pileated woodpecker, *(Picus pileatus.)*

"Mr. N. and myself are both in high spirits. We travel slowly, and without much fatigue, and when we arrive at a house, stop and rest, take a drink of milk, and chat with those we see. We have been uniformly well treated; the living is good, and very cheap . . . the inhabitants . . . live comfortably, and without much labor . . ."

From March 31 through April 3 they visited in the home of a chance acquaintance, "superior to those we had been accustomed to meet." When he learned that Townsend and Nuttall were "naturalists, and were travelling in that capacity," he took "considerable interest." While Townsend enjoyed himself with the man's three "fine looking" daughters, "Mr. N. was monopolized by the father, who took a great interest in plants."

April 4: "I rose . . . at daybreak, and left Mr. N. dreaming of weeds, in a little house at which we stopped last night, and in company with a long lanky boy . . . set to moulding bullets in an old iron spoon, and preparing for deer hunting . . .

"We . . . turned our attention to the turkies . . . rather numerous in the thicket . . . we killed four, and returned to the house, where, as I expected Mr. N. was in a fever at my absence, and after a late, and very good breakfast, proceeded on our journey.

"We find . . . less timber . . . when a small belt appears, it is a great relief, as the monotony of a bare prairie becomes tiresome.

2. This headgear is referred to by De Voto as "the progenitor of the cowboy's Stetson." !

"Towards evening we arrived at Loutre Lick.[3] Here there is a place called a *Hotel*
. . . a pig-stye would be a more appropriate name . . ."

April 6: "We were overtaken by a stage . . . going to Fulton, seven miles distant,
and as the roads were somewhat heavy, we concluded to make use of this conveyance.
The only passengers were three young men . . . travelling homeward . . . we had jokes
without number. Some of them were not very refined . . .

". . . at Fulton . . . saw the villagers . . . parading along to church. The bell . . . gave
rise to many reflections. It might be long ere I should hear the sound of the 'church-
going bell' again . . .

". . .We took a light lunch at the tavern . . . I shouldered my gun, Mr. N. his stick
and bundle, and off we trudged again, westward ho! . . ."

On the 7th they were at Boonville — "the prettiest town I have seen in Missouri" —
where Townsend observed ". . . vast numbers of the beautiful parrot of this country,
(the *Psittacus carolinensis.*) They flew around us in flocks . . . They seem entirely
unsuspicious of danger . . . It is a most inglorious sort of shooting; down right, cold-
blooded murder."

On April 9 a steamboat arrived with Wyeth aboard.

"We embarked immediately, and soon after were puffing along the Missouri, at
the rate of seven miles an hour. When we stopped . . . we were gratified by a sight of
one of the enormous catfish of this river and the Mississippi, weighing full sixty
pounds . . . they are sometimes caught of at least double this weight. They are excellent eating . . ."

The "pedestrian tour" had consumed eleven days. On April 14 they reached
"Independence landing." The town was six miles away from the river on a well-selected
site, but, as a town, was "very indifferent." They remained in a house "on the landing"
for ten days to acquire what was needed for the journey westward and to assemble
the members of the party.[4]

". . . We were very much disappointed in not being able to purchase any mules here,
all the saleable ones having been bought by the Santa Fee traders, several weeks
since. Horses, also, are rather scarce, and are sold at higher prices than we had been
taught to expect, the demand for them at this time being greater than usual. Mr. N.
and myself have, however, been so fortunate as to find five excellent animals amongst
the hundreds of wretched ones offered for sale, and have also engaged a man to attend
to packing our loads, and to perform the various duties of our camp.

"The men of the party, to the number of about fifty, are encamped on the bank of
the river, and their tents whiten the plain for the distance of half a mile . . .

"We have amongst our men, a great variety of dispositions . . . Some have evidently

3. Now Big Spring, Montgomery County, Missouri.

4. Wyeth wrote in a letter dated Independence, April 25, 1834: ". . . It is like keeping a bag of fleas together to keep the men in this whisky country."

been reared in the shade, and not accustomed to hardships, but the majority are strong, able-bodied men, and many are almost as rough as the grizzly bears . . .

"During the day the captain keeps all his men employed in arranging and packing a vast variety of goods for carriage . . . The bales are usually made to weigh about eighty pounds, of which a horse carries two.

". . . Captain W. . . . appears admirably calculated to gain the good will, and ensure the obedience of such a company, and adopts the best possible mode of accomplishing his end. They are men who have been accustomed to act independently; they possess a strong and indomitable spirit which will never succumb to authority, and will only be conciliated by kindliness and familiarity . . . Captain W. may frequently be seen sitting on the ground, surrounded by a knot of his independents, consulting them as to his present arrangements and future movements, and paying the utmost deference to the opinion of the least among them.

"We are joined here by Mr. Milton Sublette, a trader and trapper of some ten or twelve years' standing. It is his intention to travel with us to the mountains, and we are glad of his company, both on account of his intimate acquaintance with the country, and the accession to our band of about twenty trained hunters . . . his men are enthusiastically attached to him.

"Five missionaries, who intend to travel under our escort, have also just arrived. The principal of these is Mr. Jason Lee . . . his nephew, Mr. Daniel Lee, and three younger men . . . who have arrayed themselves under the missionary banner, chiefly for the gratification of seeing a new country, and participating in strange adventures.

"My favorites, the birds, are very numerous in this vicinity, and I am therefore in my element . . .

"The little town of Independence has within a few weeks been the scene of a brawl . . . happily settled without bloodshed. It had been for a considerable time the stronghold of a set of fanatics, called Mormons, or Mormonites, who . . . showed an inclination to lord it over the less assuming inhabitants of the town . . . the whole town rose, *en masse,* and the poor followers of the prophet were forcibly ejected from the community. They took refuge in the little town of Liberty, on the opposite side of the river . . ."

On April 28 the caravan started on its way and, leaving the Missouri River, entered Kansas, traveling through that state until May 17 or thereabouts. The party included ". . . seventy men, and two hundred and fifty horses . . . Captain Wyeth and Milton Sublette took the lead, Mr. N. and myself rode beside them; then the men in double file, each leading, with a line, two horses heavily laden, and Captain Thing (Captain W.'s assistant) brought up the rear. The band of missionaries, with their horned cattle, rode along the flanks.

"I frequently sallied out from my station to look at and admire the appearance of the cavalcade . . . it was altogether so exciting that I could scarcely contain myself . . . We were . . . a most merry and happy company . . .

"Our road lay over a vast rolling prairie, with occasional small spots of timber at the distance of several miles apart, and this will no doubt be the complexion of the track for some weeks.

"In the afternoon we crossed the *Big Blue* river at a shallow ford . . ."

On April 29 there was a heavy rain followed "by a tremendous hail storm." Townsend tells of the panic created among the horses by the hailstones, "as large as musket balls."

"Camping out to-night is not so agreeable as it might be . . . the ground being very wet and muddy, and our blankets (our only bedding) thoroughly soaked; but we expect to encounter greater difficulties than these ere long, and we do not murmur."

May 1: "We encamped this evening on a small branch of the Kansas river . . . we were joined by a band of Kansas Indians, (commonly called *Kaw* Indians.) . . ."

May 3: "We arrived at the Kansas river, a branch of the Missouri . . . a broad and not very deep stream, with the water dark and turbid, like that of the former . . ."

Until the 5th[5] the party had been in the settlements of the Kansas Indians, trading and so on, but this day they ". . . rode out of the Kaw settlement, leaving the river immediately, and making a N. W. by W. course . . . We encamped in the evening [of the 6th] on a small stream called Little Vermilion creek . . ."

On May 8 Sublette ". . . left . . . to return to the settlements . . . His departure has thrown a gloom over the whole camp . . . I had become so much attached to him, that I feel quite melancholy about his leaving us[6] . . ." On the 9th the party was in the vicinity of the Pawnees, of ". . . thieving propensities, and quarrelsome disposition . . ."

By the 15th they were ". . . travelling along the banks of the Blue river,—a small fork of the Kansas. The grass is very luxuriant and good, and we have excellent and beautiful camps every night."

On May 18 they ". . . arrived at the Platte river . . . from one and a half to two miles in width, very shoal . . ." They were now, certainly, in Nebraska and were to be crossing that state until June 1.

"The prairie is here as level as a race course, not the slightest undulation appearing throughout the whole extent of vision, in a north and westerly direction; but to the eastward of the river . . . is seen a range of high bluffs or sandbanks, stretching away to the south-east until they are lost in the far distance.

"The ground here is in many places encrusted with an impure salt . . .

"We are now within about three days' journey of the usual haunts of the buffalo . . ."

5. Townsend's dates are sometimes confused; he refers to the 5th as the "20th."

6. In a footnote we are told that the man's affected leg, the cause of his departure, was twice amputated and that "the disease" caused his death a year later.

On May 20 they saw "many thousands" of these animals and spent the next three days in hunting them and Townsend seems to have been torn between enjoyment of the chase and revulsion at the slaughter ". . . merely for the tongues, or for practice with the rifle . . ." Their computed distance from the Missouri settlements was "about 360 miles."

May 24: ". . . we forded the Platte river, or rather its south fork, along which we have been travelling during the previous week . . . Instead of the extensive and apparently interminable green plains . . . here was a great sandy waste, without a single green thing to vary and enliven the dreary scene. It was a change . . . and we were therefore enjoying it . . . when we were suddenly assailed by vast swarms of most ferocious little black gnats . . . so exceedingly minute that, singly, they were scarcely visible . . . their sting caused such excessive pain, that for the rest of the day our men and horses were rendered almost frantic . . . a day of most peculiar misery.
". . . poor Captain W. was totally blind for two days afterwards."

May 25: "We made a noon camp to-day on the north branch or fork of the river, and in the afternoon travelled along the bank of the stream . . .
"In the evening we arrived upon the plain again; it was thickly covered with ragged and gnarled bushes of a species of wormwood, *(Artemesia,)* which perfumed the air, and at first was rather agreeable. The soil was poor and sandy, and the straggling blades of grass . . . brown and withered . . .
"The birds thus far have been very abundant. There is a considerable variety, and many of them have not before been seen by naturalists. As to the plants, there seems to be no end to them, and Mr. N. is finding dozens of new species daily . . ."

May 26: ". . . the whole camp was suddenly aroused by the falling of all the tents. A tremendous blast swept as from a funnel over the sandy plain . . . The men crawled out from under the ruins . . . muttering imprecations against the country and all that therein was . . .
"During the whole day a most terrific gale was blowing in our faces, clouds of sand were driving and hurtling by us, often with such violence as nearly to halt our progress . . ."

On May 28 they reached Scotts Bluff, in the county of the same name, and were approaching the Wyoming border:
". . . we diverged from the usual course, leaving the bank of the river, and entered a large and deep ravine between the enormous bluffs.
"The road was very uneven and difficult, winding from amongst innumerable mounds six to eight feet in height, the space between them frequently so narrow as scarcely to admit our horses . . . These mounds were of hard yellow clay, without a particle of rock of any kind, and along their bases, and in the narrow passages, flowers of every hue were growing. It was a most enchanting sight; even the men noticed it, and

more than one of our matter-of-fact people exclaimed, *beautiful, beautiful!* Mr. N. was here in his glory. He rode on ahead of the company, and cleared the passages with a trembling and eager hand, looking anxiously back at the approaching party, as though he feared it would come ere he had finished, and tread his lovely prizes under foot.

"The distance through the ravine is about three miles. We then crossed several beautiful grassy knolls, and descending to the plain, struck the Platte again, and travelled along its bank. Here one of our men caught a young antelope, which he brought to the camp upon his saddle. It was a beautiful and most delicate little creature, and in a few days became so tame as to remain with the camp without being tied, and to drink, from a tin cup, the milk which our good missionaries spared from their own scanty meals. The men christened it '*Zip Coon*,' and it soon became familiar with its name, running to them when called, and exhibiting many evidences of affection and attachment. It became a great favorite with every one. A little pannier of willows was made for it, which was packed on the back of a mule, and when the camp moved in the mornings, little *Zip* ran to his station beside his long-eared hack, bleating with impatience until some one came to assist him in mounting."

May 31: ". . . We came to green trees and bushes again . . . We encamped in the evening in a beautiful grove of cottonwood trees, along the edge of which ran the Platte, dotted as usual with innumerable islands.

"In the morning, Mr. N. and myself were up before the dawn, strolling through the umbrageous forest . . . I think I never before saw so great a variety of birds within the same space. All were beautiful, and many of them quite new to me . . .

"None but a naturalist can appreciate a naturalist's feelings—his delight amounting to ecstacy—when a specimen such as he has never before seen, meets his eye, and the sorrow and grief which he feels when he is compelled to tear himself from a spot abounding with all that he has anxiously and unremittingly sought for.

". . . We had long been travelling over a sterile and barren tract, where the lovely denizens of the forest could not exist, and I had been daily scanning the great extent of the desert, for some little *oasis* such as I had now found . . . yet the caravan would not halt for me; I must turn my back upon the *El Dorado* of my fond anticipations . . .

"What valuable and highly interesting accessions to science might not be made by a party, composed exclusively of naturalists, on a journey through this rich and unexplored region! The botanist, the geologist, the mamalogist, the ornithologist, and the entomologist, would find a rich and almost inexhaustible field for the prosecution of their inquiries . . ."

June 1: ". . . we arrived at Laramie's fork of the Platte, and crossed it without much difficulty . . ."

Wyeth's party was now in Wyoming, where the Laramie River enters the Platte in western Goshen County. They were to travel through that state until July 6 when they entered southeastern Idaho.

It was at the crossing of the Laramie that they found some of William Sublette's men engaged in building a fort.[7]

On June 2 they crossed the Black Hills[8] (where the ". . . general aspect . . . was dreary and forbidding . . .") and came to the plain and the Platte's North Fork. By June 7 the country was more level but ". . . the prairie is barren and inhospitable looking to the last degree. The twisted, aromatic wormwood covers and extracts the strength from the burnt and arid soil. The grass is dry and brown, and our horses are suffering extremely for want of food."

After passing the "Red Butes" on June 7, they left the Platte River on the 8th, and camped on the "Sweet-water" by the "Rock Independence" on the 9th. This landmark on the Oregon Trail is in Natrona County, on the north side of the Sweetwater and no great distance west of the confluence of that stream and the North Fork of the Platte; since passing travelers inscribed their names upon it, it became known as the "Register of the Desert."

On June 10 the Wind River Mountains of Fremont County were visible, "about ninety miles to the west." They kept along the Sweetwater and must have gone through South Pass, although this passage is not mentioned[9] by Townsend. On the 14th they ". . . left the Sweet-water, and proceeded in a south-westerly direction to Sandy river, a branch of the Colorado of the west . . . a hard and most toilsome march for both man and beast. We found no water on the route, and not a single blade of grass for our horses . . ." On the 16th they saw ". . . large herds of buffalo on the plains of Sandy river, grazing in every direction on the short and dry grass. Domestic cattle would certainly starve here, and yet the bison exists and even becomes fat . . ."

June 17: "Our course was still down the Sandy river, and we are now looking forward with no little pleasure to a rest of two or more weeks at the mountain rendezvous on the Colorado. Here we expect to meet all the mountain companies who left the States last spring, and also the trappers who come in from various parts, with the furs collected by them during the previous year . . ."

June 19: "We arrived to-day on the Green river, Siskadee, or Colorado of the west,—

7. According to Chittenden (3: 967), Fort William, named for William L. Sublette, was begun on June 1, 1834, and was situated at the confluence of the North Platte and Laramie rivers, on the left bank of the last and "about a mile above its mouth." It was later sold and rechristened Fort John, for John B. Sarpy. "Before 1846 another post was built about a mile farther up stream and to this the name *Fort Laramie* was given . . . About 1849 the American Fur Company sold out to the government and moved some distance down the river. The famous military post of Fort Laramie then began its career . . ."

8. "Early name for the Laramie Range, encountered just west of Laramie River, which was crossed at the future site of Fort Laramie, Wyoming." Hulbert, p. 171, *fn.* 114.

9. Pennell's map indicates this route and Chittenden states that they crossed the pass.

a beautiful, clear, deep, and rapid stream, which receives the waters of Sandy[1]—and encamped upon its eastern bank . . ."

Townsend went out looking for birds and on his return discovered that the camp had been moved, he knew not where. But his horse was sufficient to the emergency and, as these creatures always do when left behind by a pack train, started in pursuit at a speed which Townsend "found some difficulty in restraining." On arrival at camp he discovered that, in crossing a stream, he had lost the coat which he had attached "carelessly" to his saddle. The garment had ". . . contained the second volume of my journal, a pocket compass, and other articles of essential value to me . . . although I returned to the river, and searched assiduously until night, and offered large rewards to the men, it could not be found.

"The journal commenced with our arrival at the Black Hills, and contained some observations upon the natural productions of the country . . . as well as descriptions of several new species of birds, and notes regarding their habits, &c., which cannot be replaced.

"I would advise all tourists, who journey by land, never to carry their itineraries upon their persons; or if they do, let them be attached by a cord to the neck, and worn under the clothing . . .

"In consequence of remaining several hours in wet clothes, after being heated by exercise, I rose the next morning with so much pain, and stiffness of the joints, that I could scarcely move. But . . . I was compelled to mount my horse . . . and . . . ride steadily . . . for eight hours. I suffered intensely during the ride . . .

"When we halted, I was so completely exhausted, as to require assistance in dismounting, and shortly after, sank into a state of insensibility . . . Then a violent fever commenced . . . I think I never was more unwell in my life; and if I had been at home, lying on a feather bed instead of the cold ground, I should probably have fancied myself an invalid for weeks."

In a footnote Townsend comments: "I am indebted to the kindness of my companion and friend, Professor Nuttall, for supplying, in a great measure, the deficiency occasioned by the loss of my journal." The lost records must have covered about two weeks.

Wyeth's party had reached the rendezvous on June 18. And it was on this day that the Rocky Mountain Fur Company refused to honor the contract which it had made

1. Green River rises in the Wind River Mountains of western Wyoming and, flowing southward, enters northeastern Utah; there it turns nearly due east into the northwestern corner of Colorado; from there it turns southwest and again enters Utah and continues on a southernly course through the eastern portion of that state until it joins the Colorado in the southeastern corner of Wayne County.

The Big Sandy River rises not far from South Pass, flows southwest and enters Green River in northwestern Sweetwater County. The Little Sandy rises in eastern Sublette County and joins the Big Sandy not far south of present Farson, in northwestern Sweetwater County. All these streams are mentioned by travelers over the trail to Oregon by way of South Pass, whether bound east or west, and by those who visited the various rendezvous in the adjacent mountains.

with him on August 14, 1833,—to purchase three thousand dollars worth of merchandise. According to agreement, Wyeth had brought this with him from the east. William Sublette seems to be held responsible for the fur company's breach of faith.[2]

Townsend resumes his narrative on June 22, and reports upon the assemblage at the rendezvous. It was a "heterogeneous" one—Indians of various tribes, "French-Canadians, half-breeds, &c.," and camp was "a perfect bedlam."

". . . A more unpleasant situation for an invalid could scarcely be conceived. I am . . . compelled all day to listen to the hiccoughing jargon of drunken traders, the *sacré* and *foutre* of Frenchmen run wild, and the swearing and screaming of our own men . . . scarcely less savage than the rest, being heated by the detestable liquor which circulates freely among them.

". . . The principal liquor in use here is alcohol diluted with water. It is sold to the men at *three dollars* the pint! . . ."

All purchases were paid for in beaver skins, buffalo robes, and so on.

By June 30 Townsend was able to move about again:

"Our camp here is a most lovely one in every respect . . . Our tents are pitched in a pretty little valley or indentation in the plain, surrounded on all sides by low bluffs of yellow clay. Near us flows the clear deep water of the Siskadee . . . The river . . . contains a great number of large trout, some grayling, and a small narrow-mouthed white fish, resembling a herring. They are all frequently taken with the hook, and, the trout particularly, afford excellent sport to the lovers of angling. Old Izaak Walton would be in his glory here, and the precautionary measures which he so strongly recommends in approaching a trout stream, he would not need to practise, as the fish is not shy, and bites quickly and eagerly at a grasshopper or a minnow . . ."

According to Russell the tenth rendezvous—of 1834—had been held at "Ham's Fork of the Green River."[3] Some of Wyeth's letters are dated "Ham's Fork." He wrote a number from there and two mention Nuttall. One to his wife comments that "Mr. Nuttall is well and cursing the tittle tattle of Cambridge in high style." Another to E. W. Metcalf: "Mr. Nuttall is with me and well and has made an immense collection of new plants preserved also there is a Mr. Townsend who has found a good variety of new birds and preserved them."

On July 2 Wyeth's party left the rendezvous with horses and men "very much recruited" by their rest. Many of Wyeth's men had left the party but about thirty Indians had joined him: "Flat-heads, Nez Percés, &c., with their wives, children and

2. Townsend says nothing of the episode, but it is told by both Chittenden and Russell.

3. Chittenden mentions that "There is no little confusion as to the precise locality . . . but from Wyeth's journal it seems to have been about 12 miles above the mouth of the Big Sandy in the valley of Green River. The parties, however, did not remain here but moved over to Ham's Fork, a short day's march, on the 18th . . ."

Ham's Fork is a small stream which joins the Black Fork of Green River not far from present Granger, in western Sweetwater County, Wyoming.

dogs." They sought protection while traveling through the country of "their enemies, the Black-feet." Townsend reports too that they were joined ". . . at the rendezvous by a Captain Stewart,[4] an English gentleman of noble family, who is travelling for amusement, and in search of adventure. He has already been a year in the mountains, and is now desirous of visiting the lower country, from which he may probably take passage to England by sea. Another Englishman, a young man, named Ashworth,[5] also attached himself to our party, for the same purpose."

Wyeth's party continued up Ham's Fork for two days and left it ("now diminished to a little purling brook") on July 4. They then ". . . passed across the hills in a north-westerly direction for about twenty miles, when we struck Muddy creek[6] . . . a branch of Bear river, which empties into the Salt lake, or 'lake Bonneville,' as it has been lately named, for what reason I know not . . . This being a memorable day, the liquor kegs were opened, and the men allowed an abundance. We, therefore, soon had a renewal of the coarse and brutal scenes of the rendezvous. Some of the bacchanals called for a volley in honor of the day . . . we who were not 'happy' had to lie flat upon the ground to avoid the bullets which were careering through the camp.

"In this little stream, the trout are more abundant than we have yet seen them . . . These fish would probably average fifteen or sixteen inches in length, and weigh three-quarters of a pound . . ."

July 5: ". . . early on the afternoon arrived on Bear river, and encamped. This is a fine stream of about one hundred and fifty feet in width . . . At the distance of about one hundred miles from this point, the Bear river enters the Salt lake . . ."

Although Wyeth's party was now near the northeastern corner of Utah it did not enter that state but crossed from Wyoming into Bear Lake County, Idaho, following Bear River; it was bound for the Snake River. Until August 24 the route was in Idaho.

July 6: ". . . we crossed the river, which we immediately left, to avoid a great bend, and passed over some lofty ranges of hills and through the rugged and stony valleys between them; the wind was blowing a gale right ahead, and clouds of dust were flying in our faces . . . The march . . . has been a most laborious and fatiguing one both for man and beast . . . at length we struck Bear river again and encamped . . ."

July 7: ". . . we travelled but twelve miles, it being impossible to urge our worn-out horses farther. Near our camp . . . we found some large gooseberries and currants, and made a hearty meal upon them. They were to us peculiarly delicious. We have lately

4. Sir William Drummond Stewart. He had been at the rendezvous of 1833 also.

5. Ashworth is mentioned several times in Townsend's *Narrative*. He got to Fort Walla Walla and probably to Fort Vancouver, as did Stewart. Ashworth's Christian name is not disclosed and I have found no reference to him in any other book dealing with the period.

6. Thwaites mentions that the stream no longer bears this name but he supplies no other appellation.

been living entirely upon dried buffalo, without vegetables or bread . . . Game is very scarce . . .

"The alluvial plain here presents many unequivocal evidences of volcanic action, being thickly covered with masses of lava, and high walls and regular columns of basalt appear in many places . . . the hills are thickly covered with a growth of small cedars, but on the plain, nothing flourishes but the everlasting wormwood, or *sage* as it is here called."

July 8: "Our encampment . . . was near what are called the 'white-clay pits,' still on Bear river. The soil is soft chalk, white and tenacious . . . in the vicinity are several springs of strong supercarbonated water . . . The taste was very agreeable and refreshing, resembling Saratoga water, but not so saline . . ."

They were in the vicinity of Soda Springs, in southwest Caribou County, or just north of a great bend made by Bear River as it changes its course from north to south. They spent the 9th in camp to refresh the horses, and were joined by ". . . Mr. Thomas McKay, an Indian trader of some note in the mountains . . . a step-son of McLaughlin, the chief factor at Fort Vancouver." He had been hunting with a party of Canadians and Indians.

". . . This party is at present in our rear, and Mr. McKay has come ahead in order to join us, and keep us company until we reach Portneuf river, where we intend to build a fort."

They moved on July 10 and in the afternoon came upon ". . . Captain Bonneville's company resting after the fatigues of a long march. Mr. Wyeth and Captain Stewart visited the lodge of the 'bald chief,' and our party proceeded on its march.

". . . Wyeth gave us a rather amusing account of his visit to the worthy captain. He and Captain Stewart were received very kindly . . . and every delicacy that the lodge afforded was brought forth to do them honor. Among the rest, was some *methiglen* [metheglin] or diluted alchohol sweetened with honey, which the good host had concocted . . . Cup after cup was drained, until the hollow sound of the keg indicated that its contents were nearly exhausted, when the company rose, and . . . were bowed out of the tent with all the polite formality which the accomplished captain knows so well how to assume."

That night they reached ". . . Blackfoot river, a small sluggish, stagnant stream, heading with the waters of a rapid rivulet passed yesterday, which empties into Bear river . . ." They were now in, or near, Bingham County. Here they had an encounter with a grizzly bear (weighing "at the least, six hundred pounds . . .") and—to the regret of all—the pet antelope "Zip Koon" had to be killed after it had been rolled on by a horse. It had been with the party since late May. They had tried other wild pets, young grizzly bears and young buffalo calves, but the first, even when no larger than puppies, were "cross and snappish," and the horses kicked the buffalo whenever they had the opportunity.

July 11: ". . . We are now in a part of the country which is almost constantly infested with the Blackfeet . . . we know we are narrowly watched . . .

"Our encampment . . . is on one of the head branches of the Blackfoot river, from which we can see the three remarkable conic summits known by the name of the 'Three Butes' or 'Tetons.'[7] Near these flows the Portneuf, or south branch of Snake or Lewis' river. Here is to be another place of rest . . ."

July 12: "In the afternoon we made a camp on Ross' creek, a small branch of Snake river. The pasture is better than we have had for two weeks, and the stream contains an abundance of excellent trout. Some . . . are enormous, and very fine eating. They bite eagerly at a grasshopper or minnow, but the largest fish are shy, and the sportsman requires to be carefully concealed in order to take them. We have here none of the fine tackle . . . of the accomplished city sportsman . . . only a piece of common cord, and a hook seized on with half-hitches, with a willow rod cut on the banks of the stream; but with this . . . we take as many trout as we wish, and who could do more, even with all the curious contrivances of old Izaac Walton or Christopher North?"

July 14: ". . . we travelled but about six miles . . . and we pitched our tents upon the banks of the noble Shoshoné or Snake river. It seems now, as though we were really nearing the western extremity of our vast continent. We are now on a stream which pours its waters directly into the Columbia, and we can form some idea of the great Oregon river by the beauty and magnitude of its tributary. Soon after we stopped, Captain W., Richardson, and two others left us to seek for a suitable spot for building a fort . . ."

Wyeth planned to establish a permanent trading post, and to leave there the merchandise which the Rocky Mountain Fur Company left on his hands at the rendezvous. A suitable location was found and they moved there on the 15th.

". . . This is a fine large plain on the south side of the Portneuf, with an abundance of excellent grass and rich soil. The opposite side of the river is thickly covered with large timber of cottonwood and willow, with a dense undergrowth of the same, intermixed with service-berry and currant bushes.

"Most of the men were immediately put to work, felling trees, making horse-pens, and preparing the various requisite materials for the building, while others were ordered . . . to make a hunt, and procure meat for the camp. To this party I have attached myself . . .

"Our number will be twelve . . . no rum or cards are allowed."

7. The reference must have been to the Buttes lying north of the Snake River and not to the Tetons of northwestern Wyoming.

The hunting party left on July 16 and got back on the 26th, after adventures of various kinds.

". . . My companion, Mr. N., had become so exceedingly thin that I should scarcely have known him; and upon my expressing surprise at the great change in his appearance, he . . . remarked that I 'would have been as thin as he if I had lived on old *Ephraim* for two weeks, and short allowance of that.' I found . . . that the whole camp had been subsisting . . . on little else but two or three grizzly bears which had been killed in the neighborhood . . . I wished my *poor* friend better luck for the future . . .

"At the fort, affairs look prosperous: the stockade is finished; two bastions have been erected, and the work is singularly good . . . The house will now soon be habitable, and the structure can then be completed at leisure by men who will be left here in charge, while the party travels on to . . . the Columbia."

Townsend, Nuttall and Wyeth had supper with Thomas McKay who was camped not far away. Townsend admired ". . . the order, decorum, and strict subordination which exists among his men, so different from what I have been accustomed to see in parties composed of Americans . . ."

July 27: ". . . being the Sabbath, our good missionary, Mr. Jason Lee, was requested to hold a meeting, with which he obligingly complied . . . the greater part of our men . . . Mr. McKay's company, including the Indians, attended . . .

"A meeting for worship in the Rocky mountains is almost as unusual as the appearance of a herd of buffalo in the settlements. A sermon was perhaps never preached here before.[8] . . .

"Mr. Lee is a great favorite with the men, deservedly so . . ."

On July 30 McKay left for Fort Vancouver, accompanied by Stewart and the missionaries. They wished to travel slowly because of the horned cattle which they were driving "to the lower country . . . we hope to meet them all again at Walla-Walla, the upper fort on the Columbia."

August 5: "At sunrise . . . the 'star-spangled banner' was raised on the flag-staff at the fort, and a salute fired by the men . . . All in camp were then allowed the free and uncontrolled use of liquor . . . Such scenes I hope never to witness again . . ."

Chittenden (3: 974) describes the site of Fort Hall as ". . . on the left bank of the Snake river, a little above the mouth of the Portneuf . . . Its history as a trading post is almost entirely associated with the Hudson Bay Company, to whom Wyeth sold it in 1836. It was an exceedingly important point during the emigration period, and later became a military post of considerable note."

8. The Idaho historian C. J. Brosnan wrote: ". . . an event occurred on Sunday afternoon, July 27, 1834, that should be honored in our history . . . Oregon's first Methodist missionary . . . Jason Lee . . . conducted the first religious service, not only in Idaho, but in the Pacific Northwest . . . Jason Lee delivered the first sermon ever preached in Idaho . . ."

Bancroft mentions its potential importance at the time it was built and, in another passage, refers to its strategic position at the time of the migrations:

"The building, at this time, of a substantial fort midway between Leavenworth and the mouth of the Columbia by Americans, though the establishment afterwards fell into the hands of the Hudson's Bay Company, for a time was a very important affair . . . it signified occupation, domination. All this region was still debatable ground, and every move of this kind had its influence in subsequently fixing the dividing line between British and United States domain."

"The post became famous and performed good service during the several great overland emigrations . . . it was central, and valuable in scores of ways. From this point in time radiated roads in every direction: to Missouri, to California, to Utah, to Oregon, and to British Columbia . . ."

August 6: ". . . we . . . bade adieu to 'Fort Hall.' . . . We crossed the main Snake, or Shoshoné river, at a point about three miles from the fort. It is here as wide as the Missouri at Independence, but, beyond comparison, clearer and more beautiful . . .

"We shall now, for about ten days, be travelling through the most dangerous country west of the mountains, the regular hunting ground of the Blackfeet Indians . . ."

When travelers over the Oregon Trail reached the confluence of the Portneuf and Snake rivers they usually turned down the Snake, followed it for a time (or to present Glen's Ferry), then turned northwestward to the Boisé, following that river to its entrance into the Snake. Wyeth, however, turned northwest from Fort Hall, crossed the Snake River Plain, and entered the mountains lying between the Lost River Range (east) and the Sawtooth Range (west); after trying, unsuccessfully, to cross one pass he made his way through another; finally, turning southwest, he came to Camas Prairie and the upper waters of Boisé, or Big Wood, River; following that stream he reached its confluence with the Snake on August 23. Townsend has much to tell of this two-weeks' journey and its hardships:

August 7: "We were moving . . . with the dawn, and travelled steadily the whole day, over one of the most arid plains we have seen, covered thickly with jagged masses of lava, and twisted wormwood bushes. Both horses and men were jaded to the last degree . . . We saw not a drop of water during the day, and our only food was the dried meat . . . which we . . . chewed like biscuits as we travelled . . . The air feels like the breath of a sirocco . . . We were a sad and most forlorn looking company . . . One of our men . . . cast himself resolutely . . . to the ground, and declared that he would lie there till he died . . . to infuse a little courage . . . proved of no avail . . . so we left him to his fate[9] . . .

"Soon after night-fall . . . the foremost of the party found a delightful little cold spring; but they soon exhausted it, and then commenced, with axes and knives, to dig

9. "Poor Jim," a mulatto, was later dragged into camp and, I judge, recovered.

it out and enlarge it. By the time Mr. W., and myself arrived, they had excavated a large space which was filled to overflowing with muddy water. We . . . drank until we were ready to burst . . ."

On August 8 they reached ". . . Goddin's creek, so called from a Canadian . . . who was killed[1] in this vicinity by the Blackfeet." Godin's Creek is the Lost River of present maps. They left this stream on the 9th and ". . . travelled for ten miles over a plain, covered as usual with wormwood bushes and lava." They had the good fortune to kill a stray buffalo.

Leaving the plain ". . . we struck into a defile between some of the highest mountains we have yet seen . . . late in the afternoon . . . we reached a plain . . . covered with excellent grass, and a delightful cool stream . . . Here we encamped, having travelled twenty-seven miles.

". . . In the mountain passes, we found an abundance of large, yellow currants, rather acid, but exceedingly palatable to men long living on animal food exclusively . . . some of our people became so much attached to the bushes, that we had considerable difficulty to induce them to travel again."

They were again on Godin's Creek on August 10 and spent the 11th in hunting since food was very scarce: ". . . we have now abandoned the 'wasty ways' which so disgraced us when game was abundant; the despised leg bone . . . is now polished of every tendon . . ."

On August 12 they moved on and ". . . entered a defile between the mountains . . . covered, like the surrounding country, with pines: and, as we proceeded, the timber grew so closely added to a thick undergrowth of bushes, that it appeared almost impossible to proceed with our horses . . . obstacles of various kinds impeded our progress;—fallen trees . . . large rocks and deep ravines, holes in the ground . . . an hundred other difficulties, which beggar description . . ."

Having found a clear piece of ground, water and grass, they stopped and Wyeth and Richardson went to search for a practicable pass. They returned on the 13th having had no success, so there was nothing to do but return by the way they had come.

". . . We have named this rugged valley, 'Thornburg's *pass*,' after one of our men of this name, (a tailor,) whom we have to thank for leading us into all these troubles. Thornburg[2] crossed this mountain two years ago, and might therefore be expected to know something of the route, and as he was the only man in the company who had been here, Captain W. acted by his advice, in opposition to his own judgment, which

1. The horrible murder is described later by Townsend. It took place at Fort Hall.

2. Townsend tells us more about Thornburg at a later date (July 4, 1835); he came to a sad end—killed by a gunsmith on Wapato Island. The affair was investigated and the verdict was "justifiable homicide" and the killer was given "a properly attested certificate" to that effect. Townsend relates that Thornburg had once imbibed the content of a two-gallon bottle of alcohol in which the naturalist had preserved some lizards and snakes: ". . . I did not discover the theft until too late to save my specimens . . ."

had suggested the other valley as affording a more probable chance of success. As we are probably the only white men who have ever penetrated into this most vile and abominable region,[3] we conclude that the name we have given it must stand, from priority."

Back on the plain they evidently turned up "the other valley" mentioned above.

". . . The entrance was very similar in appearance to that of Thornburg's pass, and it is not therefore very surprising that our guide should have been deceived. We travelled rapidly along the level land at the base of the mountain . . . then began to ascend . . . The whole journey . . . from . . . the heights, until we had crossed the mountain, has been a most fearful one . . ."

Townsend evidently did not enjoy precipices! By shutting his eyes and letting his horse have his head, he got through safely. Late in the afternoon they entered ". . . a fine rich valley or plain, of about half a mile in width, between two ranges of the mountain. It was profusely covered with willow, and through the middle of it, ran a rapid and turbulent mountain torrent, called Mallade river. It contains a great abundance of beaver . . . in the night, when all was quiet, we could hear the playful animals at their gambols, diving from the shore into the water, and striking the surface with their broad tails . . ."

On August 14 and 15 the party continued down the Malade River, following an Indian trail and crossing hills and valleys, one of which was "covered profusely with a splendid blue Lupine," and finally camped on a branch of "Mallade river."[4] They were still crossing branches of that "tortuous" stream on the 16th, and on the 17th came to ". . . 'Kamas prairie,' so called from a vast abundance of this esculent root which it produces, (the *Kamassa esculenta*, of Nuttall.) . . ."

". . . The plain is a beautiful level one . . . in spring, the pretty blue flowers of the Kamas are said to give it a peculiar, and very pleasing appearance. At this season, the flowers do not appear, the vegetable being indicated only by little dry stems which protrude all over the ground among the grass.

"We encamped here, near a small branch of Mallade river; and soon after, all hands took their kettles and scattered over the prairie to dig a mess of Kamas . . . We were . . . furnished thereby with an excellent wholesome meal . . . The Indian mode of preparing it, is . . . the best . . . fermenting it in pits under ground, into which hot stones have been placed. It is suffered to remain in these pits for several days; and when removed, is of a dark brown color, about the consistence of softened glue, and sweet, like molasses. It is then often made into large cakes, by being mashed and pressed together, and slightly baked in the sun. There are several other kinds of bulbous and tuburous roots, growing in these plains, which are eaten by the Indians, after

3. Thwaites was of the opinion that the party had been in a group of mountains which bore the local name of Devil's Bedstead.

4. This was a creek which entered the Malade River near Ketchum, Blaine County. Northeast of Ketchum is Sun Valley Lodge, of present-day skiing fame!

undergoing a certain process of fermentation or baking. Among these, that which is most esteemed, is the white or biscuit root, the *Racine blanc* of the Canadians,— (*Eulophus ambiguus, of Nuttall.) . . .*"

After more rugged country on August 18 they came on the evening of the 19th ". . . to a fine large plain, and struck Boisée, or Big Wood river, on the borders of which we encamped. This is a beautiful stream, about one hundred yards in width, clear as crystal, and, in some parts, probably twenty feet deep. It is literally *crowded* with salmon, which are springing from the water almost constantly . . . we are not provided with suitable implements for taking any, and must therefore depend for a supply on the [Snake] Indians, whom we hope to meet.

"We found, in the mountain passes, to-day, a considerable quantity of a small fruit called the choke-berry, a species of prunus, growing on low bushes. When ripe, they are tolerable eating, somewhat astringent . . . producing upon the mouth the same effect, though in a less degree, as the unripe persimmon. They are now generally green . . . We have seen, also, large patches of service bushes, but no fruit. It seems to have failed this year . . ."

The party had encountered the South Fork of Boisé River, in Elmore County, and followed it.

August 20: ". . . some distant hills in the south-west . . . we suppose indicate the vicinity of some part of Snake river.

"We have all been disappointed in the distance to this river, and the length of time required to reach it. Not a man in our camp has ever travelled this route before, and all we have known about it has been the general course."

August 21: "The timber along the river banks is plentiful, and often attains a large size. It is chiefly of the species called balsam poplar, *(Populus balsamifera.)*"

They were in the country of the Snake Indians and the young Englishman, Ashworth, who had joined Wyeth's party at the rendezvous had an unpleasant experience with one of their bands. Townsend refers to him as a ". . . calm, intrepid, and almost fool-hardy young man . . ." On August 22 Townsend visited one of the Snake camps but did not remain long:

". . . although I had been a considerable time estranged from the abodes of luxury, and had become somewhat accustomed to . . . a partial assimilation to a state of nature, yet I was not prepared for what I saw here. I never had fancied any thing so utterly abominable, and was glad to escape to a purer and more wholesome atmosphere.

"When I returned to our camp . . . trading was going on as briskly as yesterday. A large number of Indians were assembled . . . all . . . had bundles of fish . . . The price of a dried salmon is a straight awl, and a small fish hook, value about one cent; ten fish are given for a common butcher knife that costs eight cents . . . A beaver skin can be had for a variety of little matters, which cost about twelve and a half cents; value, in Boston, from eight to ten dollars!"

August 23: "Towards evening, we arrived on Snake river, crossed it at a ford, and encamped near a number of lodges along the shore. . . . Captain W., with three men, visited the Indians . . . to trade for fish . . . They treated Captain W. with the same insolence and contempt which was so irritating from those of the other village.

"This kind of conduct is said to be unusual among this tribe [the 'Bannecks'], but it is probably now occasioned by their having recently purchased a supply of small articles from Captain Bonneville, who . . . has visited them within a few days.

"Being desirous to escape from the immediate vicinity of the village, we moved our camp about four miles further . . ."

W. J. Ghent states that, on his arrival on the Boisé River, Wyeth ". . . found to his chagrin, that the Hudson's Bay Company, alarmed at the building of Fort Hall, had just erected a post of its own, Fort Boisé." C. J. Brosnan mentions this as the British company's ". . . first Idaho trading post . . . built by Chief Trader Thomas McKay on the Reed or Boise River about ten miles from its mouth. In 1838 the site was changed to the east bank of the Snake River and a short distance north of the mouth of the Boise River." Neither Townsend nor Wyeth mentions the new fort—the latter does refer to the "very impudent Pawnacks."

On crossing the Snake River, Wyeth's party had entered Malheur County, Oregon, and was to travel through that state until, after crossing the Blue Mountains, it entered Washington in early September.

August 24: "We passed . . . over a flat country, very similar to that along the Platte, abounding in wormwood bushes, the pulpy-leaved thorn, and others, and deep with sand, and at noon stopped on a small stream called *Malheur's creek* . . .

". . . We followed the course of the creek during the afternoon, and in the evening encamped on Snake river, into which the Malheur empties . . ."

August 25: "Early in the day the country assumed a more hilly aspect. The rich plains are gone. Instead of a dense growth of willow and the balsam poplar, low bushes of wormwood, &c., predominated, intermixed with the tall, rank prairie grass . . .

"In the afternoon, we deviated a little from our general course, to cut off a bend in the river,[5] and crossed . . . a part of an extensive range which we had seen for two days ahead, and which we suppose to be in the vicinity of Powder river, and in the evening encamped in a narrow valley, on the borders of the Shoshoné."

August 26: "We commenced our march early, travelling up a broad, rich valley, in which we encamped last night, and at the head of it, on a creek called Brulé, we found one family . . . of Snake Indians . . .

5. Presumably the bend made by the Snake along the extreme northeastern boundary of Malheur County.

"We bought of this family, a considerable quantity of dried choke-cherries . . . This fruit they prepare by pounding it with stones, and drying it in masses in the sun. It is then good tasted, and somewhat nutritive, and it loses, by the process, the whole of the astringency which is so disagreeable in the recent fruit.

"Leaving the valley, we proceeded over some high and stony hills, keeping pretty nearly the course of the creek . . ."

On August 27 they were still following Brulé, or Burnt, River, but on the 28th turned ". . . due north and . . . struck Powder river, a narrow and shallow stream, plentifully fringed with willows. We passed down this river for about five miles and encamped." The Oregon Trail left Snake River at the mouth of Burnt River and ascended that stream; but the trail had been lost and although they searched for it the next day it was not until the 30th that they beheld "a broad, open trail stretching away to the S. W."

August 31: "Our route . . . was over a country generally level and free from rocks; we crossed . . . one short, and very steep mountain range, thickly covered with tall and heavy pine trees, and came to a large and beautiful prairie, called the *Grand ronde.*[6] Here we found Captain Bonneville's company, which has been lying here for several days . . .

"After making a hasty meal, and bidding adieu to the captain . . . we . . . rode off. About half an hour's brisk trotting brought us to the foot of a steep and high mountain, called the *Blue.* This is said to be the most extensive chain west of the dividing ridge, and, with one exception, perhaps the most difficult of passage. The whole mountain is densely covered with tall pine trees, with an undergrowth of service bushes and other shrubs, and the path is strewed . . . with volcanic rocks . . . We travelled . . . until after dark . . ."

On September 1 they ascended and descended one ridge, ascended another and then again started downward; at nine o'clock at night they came to the first water seen all day. Townsend calls it "the most toilsome march I ever made." On the morning of the 2nd Wyeth hastened ahead to Fort Walla Walla whence he planned to send back provisions. The rest of the party, after four miles, reached *"Utalla river,"* or the Umatilla River.

"As we were approaching so near the abode of those in whose eyes we wished to appear like fellow Christians . . . Mr. N.'s razor was fished out from its hiding place in the bottom of his trunk, and . . . our encumbered chins lost their long-cherished ornaments; we performed our ablutions in the river, arrayed ourselves in clean linen, trimmed our long hair, and then arranged our toilet before a mirror . . . I admired my own appearance considerably . . .

"Having nothing prepared for dinner to-day, I strolled along the stream . . . and

6. The Grande Ronde Valley is in Union County, and is traversed by the river of the same name.

made a meal on rose buds . . . on returning, I was surprised to find Mr. N. and Captain T[hing] picking the last bones of a bird which they had cooked . . . the subject was an unfortunate owl which I had killed in the morning, and had intended to preserve, as a specimen. The temptation was too great to be resisted by the hungry Captain and naturalist, and the bird of wisdom lost the immortality which he might otherwise have acquired . . ."

On September 2 the party "struck the Walla-walla river," took a rest, and then traveled until nearly dark when, from a hill, they had a view of the Columbia River:

". . . I gazed upon the magnificent river, flowing silently and majestically on, and reflected that I had actually crossed the vast American continent, and now stood upon a stream that poured its water directly into the Pacific. This, then, was the great Oregon . . ."

Fort Walla Walla, also known as Fort Nez Percé, was in sight:

". . . on the borders of the little Walla-walla, we recognised the white tent of our long lost missionaries. These we soon joined . . . Mr. N. and myself were invited to sup with them . . . they had travelled comfortably from Fort Hall, without any unusual fatigue . . . Their route, although somewhat longer, was a much less toilsome and difficult one, and they suffered but little for food, being well provided with dried buffalo meat, which had been prepared near Fort Hall. . . .

". . . they had engaged a large barge to convey themselves and baggage to Fort Vancouver, and . . . Captain Stewart and Mr. Ashworth were to be of the party. Mr. N. and myself were very anxious to take a seat with them, but . . . were told that the boat would scarcely accomodate those already engaged. We had therefore to relinquish it, and prepare for a journey on horseback to the *Dalles,* about eighty miles below, to which place Captain W. would precede us in the barge, and engage canoes to convey us to the lower fort . . ."

September 3: " . . . the next morning, we visited Walla-walla Fort, and were introduced . . . to . . . Pierre S. Pambrun, the superintendent . . .

"The fort . . . stands about a hundred yards from the river, on the south bank, in a bleak and unprotected situation, surrounded on every side by a great, sandy plain, which supports little vegetation, except the wormwood and thorn-bushes. On the banks of the little river . . . there are narrow strips of rich soil, and here Mr. Pambrun raises the few garden vegetables necessary for the support of his family . . . Indian corn produces eighty bushels to the acre."

The site of Fort Walla Walla was at present Wallula, on the east side of the Columbia and less than a mile above the Walla Walla River. After the missionaries in their barge had started down the Columbia, Wyeth's company crossed the Columbia and camped there until September 5 when they started for Fort Vancouver. On the 10th they reached the Dalles.

". . . The entire water of the river here flows through channels of about fifteen feet in width, and between high, perpendicular rocks; there are several of these channels . . . and the water foams and boils through them like an enormous cauldron . . ."

It was decided to change from horses to canoes: ". . . we should all much prefer the water conveyance, as we have become very tired of riding." But after they had started on the 11th the wind ". . . rose to a heavy gale, and the waves ran to a prodigious height . . ." and they were obliged to put ashore.

September 12: "The gale continues with the same violence . . . we do not . . . think it expedient to leave our camp. Mr. N.'s large and beautiful collection of new and rare plants was considerably injured by the wetting it received; he has been constantly engaged since we landed yesterday, in opening and drying them. In this task he exhibits a degree of patience and perseverance which is truly astonishing; sitting on the ground, and steaming over the enormous fire, for hours together, drying the papers, and re-arranging the whole collection, specimen by specimen, while the great drops of perspiration roll unheeded from his brow. Throughout the whole of our long journey, I have had constantly to admire the ardor and perfect indefatigability with which he has devoted himself to the grand object of his tour. No difficulty, no danger, no fatigue has ever daunted him, and he finds his rich reward in the addition of nearly *a thousand* new species of American plants, which he has been enabled to make to the already teeming flora of our vast continent. My bale of birds, which was equally exposed to the action of the water, escaped without any material injury."

Wyeth, fearing that his business at Fort Vancouver would "suffer by delay," set off in his canoe despite the weather. "Mr. N. remarked that he would rather lose all his plants than venture his life in that canoe." When the wind changed, they started again on the 13th and the next day ". . . arrived at the 'cascades,' and came to a halt above them . . . These cascades, or cataracts . . . extend, for perhaps half a mile . . . the water dashes and foams most furiously . . .

". . . It is . . . necessary to make a portage, either by carrying the canoes over land . . . or by wading in the water near the shore . . . and dragging the unloaded boat . . . by a cable. Our people chose the latter method . . . In the meantime, Mr. N., and myself were sent ahead to take the best care of ourselves that . . . circumstances permitted . . ."

Proceeding along a trail, they came upon ". . . our old friend Captain Stewart, with the good missionaries, and all the rest who had left us at Fort Walla-walla on the 4th . . . they were most abundantly soaked and bedraggled . . ." There were other misadventures and much hard labor before the canoes and their contents were finally assembled at the lower end of the cascades. To add to the difficulties it had rained heavily. Townsend, after recounting his exhaustion, adds an amusing footnote:

". . . It was by far the most fatiguing, cheerless, and uncomfortable business in which I was ever engaged, and truly glad was I to lie down at night on the cold, wet ground, wrapped in my blankets, out of which I had just wrung the water, and I think I never slept more soundly or comfortably than that night.

"I could not but recollect at that time, the last injunction of my dear old grand-mother, not to sleep in damp beds!!"

On September 16 they reached Fort Vancouver, and ". . . stepped on shore at the *end of our journey*.

"It is now three days over six months since I left my beloved home. I, as well as the rest, have been in some situations of danger, of trial, and of difficulty, but I have passed through them all unharmed, with a constitution strengthened, and invigorated by healthful exercise . . .

"On the beach in front of the fort, we were met by Mr. Lee, the missionary, and Dr. John McLoughlin, the chief factor, and the Governor of the Hudson's Bay posts in this vicinity . . . The missionary introduced Mr. N. and myself in due form, and we were greeted and received with a frank and unassuming politeness . . . He requested us to consider his house our home, and provided a separate room for our use, a servant to wait upon us, and furnished us with every convenience . . . I shall never cease to feel grateful . . . for his disinterested kindness to . . . travel-worn strangers."

After describing Fort Vancouver ("situated on the north bank of the Columbia on a large level plain, about a quarter of a mile from the shore"), Townsend describes the farm. Because the states of Washington and Oregon have since developed some of the greatest fruit-growing centers in the United States, I thought his comments upon the apples of special interest:

"Mr. N. and myself walked over the farm with the doctor . . . He has . . . several hundred acres fenced in, and under cultivation . . . it produces abundant crops, partic-ularly of grain, without requiring any manure. Wheat thrives astonishingly; I never saw better in any country, and the various culinary vegetables, potatoes, carrots, parsnips, &c., are in great profusion, and of the first quality. Indian corn does not flourish so well as at Walla-walla, the soil not being so well adapted to it; melons are well flavored, but small; the greatest curiosity, however, is the apples, which grow on small trees, the branches of which would be broken without the support of props. So profuse is the quantity of fruit that the limbs are covered with it, and it is actually *packed* together precisely in the same manner that onions are attached to ropes when they are exposed for sale in our markets . . ."

The missionaries, having selected a site for their establishment on the Willamette River,[7] were planning to leave on September 26. Wyeth himself wished to erect a fort and by the 25th had learned that ". . . the brig from Boston, which was sent out by the

7. They first settled at French Prairie, in the angle formed by the junction of Pudding River with the Willamette; in 1840 Lee moved to the vicinity of present Salem.

Bancroft's *History of Oregon* (1: 71) includes an extremely helpful map entitled "French Prairie" which shows many of the localities adjacent to the Willamette River which are not shown on more recent maps.

company to which Wyeth is attached, had entered the river, and was anchored about twenty miles below, at a spot called Warrior's point, near the western entrance of the Wallammet.

"Captain W. mentioned his intention to visit the Wallammet country, and seek out a suitable location . . . and Mr. N. and myself accepted an invitation to accompany him . . .

"On the 29th, Captain Wyeth, Mr. N., and myself, embarked in the ship's boat for our exploring excursion . . .

"At about five miles below the fort, we entered the upper mouth of the Wallammett. This river is here about half the width of the Columbia . . . and navigable for large vessels to the distance of twenty-five miles. It is covered with numerous islands, the largest of which is that called *Wappatoo Island,* about twenty miles in length . . ."

On September 30 they ascended the Willamette to the falls, situated at present Oregon City, Clackamas County, and then turned back down the river to the Columbia.

A full month is unaccounted for in the *Narrative* which begins again on November 1. By then, evidently, Wyeth had decided to erect his post, "Fort William," on Wapato (now Sauvie or Sauvé) Island and Townsend and Nuttall accompanied him there:

November 1: ". . . we arrived at the brig. She was moored, head and stern, to a large rock near the lower mouth of the Wallammet. Captain Lambert with his ship's company, and our own mountain men, were all actively engaged at various employments; carpenters, smiths, coopers, and other artisans were busy in their several vocations; domestic animals, pigs, sheep, goats, poultry, &c., were roaming about as if perfectly at home . . . An excellent temporary storehouse . . . has been erected . . . It is intended . . . to build a large and permanent dwelling . . . which will . . . include the store and trading establishment, and form the groundwork for an *American fort* on the river Columbia."

November 5: "Mr. N. and myself are now residing on board the brig, and pursuing with considerable success our scientific researches through the neighborhood. I have shot and prepared here several new species of birds . . . My companion is of course in his element; the forest, the plain, the rocky hill, and the mossy bank yield him a rich and most abundant supply . . .

"Mr. N. and myself have been anxious to escape the wet and disagreeable winter of this region . . . we concluded to take passage in the brig, which will sail in a few weeks for the Sandwich Islands. We shall remain there about three months, and return to the river in time to commence our peregrinations in the spring."

Townsend and Nuttall were at Fort Vancouver on November 23 and on December 2 started down the Columbia and boarded the brig which set sail on the 3rd. On the 4th they were off *"Mount Coffin,"* an Indian burial "sanctuary." "Mr. N., and myself visited the tombs." On December 8 they ". . . anchored off *Fort George,* as it is called

... it scarcely deserves the name of a fort ... There is but one white man residing here, the superintendent of the fort ... The establishment is, however, of importance, independent of its utility as a trading post, as it is situated within view of the dangerous cape, and intelligence of the arrival of vessels can be communicated to the authorities at Vancouver in time for them to render assistance to such vessels by supplying them with pilots, &c. This is the spot where once stood the fort established by the direction of our honored countryman, John Jacob Astor. One of the chimneys of old Fort Astoria is still standing, a melancholy monument of American enterprise and domestic misrule ..."

The ship reached Baker's Bay on the 9th and anchored "within gunshot of the cape." Here a "... wide bar of sand extends from Cape Disappointment to the opposite shore,—called Point Adams ..." and conditions were such that they were unable to make the passage. While waiting, on the 10th, "Mr. N. and myself visited the sea beach, outside the cape, in the hope of finding peculiar marine shells, but although we searched assiduously ... had but little success ..."

On December 11 conditions were improved: "In about twenty minutes we had escaped all the danger, and found ourselves riding easily in a beautiful placid sea. We set the sails ... and the gallant vessel ... rushed ahead ..."

Townsend's *Narrative* now devotes some one hundred and thirty pages to the trip to the Sandwich Islands, to the sojourn there (where "Mr. N., and myself" were entertained by missionaries, royalty and everyone else), and to the return voyage to the Columbia River.

1835

On April 16 the ship *May Dacre* was off Cape Disappointment, "... crossed the dangerous bar safely, and ran direct for the river." She anchored that day off Oak Point (near present Quincy, Columbia County) and on April 17 reached Wapato Island and stopped at Warrior Point.

The rainy season was over by May 12 when Townsend resumes his narrative. On May 20 he started with Wyeth for Fort William and for the falls of the Willamette River; Nuttall's presence in the party is not mentioned. How long this trip lasted is not stated, but Townsend was certainly at Fort Vancouver on July 2, for on the 4th a letter came from Fort William stating that the tailor Thornburg,[8] whom Townsend "... had seen but two days previously, full of health and vigor, was now a lifeless corpse."

Nuttall is mentioned in the *Narrative* only twice after he and Townsend had returned from the islands. The first occasion was on the "11*th*"[9]; "Mr. Nuttall ... has just returned from the dalles, where he has been spending some weeks ..."

8. See under August 12, 1834.

9. No month is cited; the preceding date had been *"July 4th,"* and the succeeding one was *"August 19th."*

The second occasion is under date of October 1 when Townsend mentions the departure of Dr. Meredith Gairdner for the Sandwich Islands:

". . . My companion, Mr. Nuttall, was also a passenger in the same vessel. From the islands, he will probably visit California, and either return to the Columbia by the next ship, and take the route across the mountains, or double Cape Horn to reach his home."

These are but scanty records for Nuttall's five and one-half months in the Columbia River region in 1835. As Townsend mentioned under date of September 29, 1834, the upper mouth of "the Wallammett" was "about five miles below" Fort Vancouver so that, from the collecting-angle, it is unimportant whether the botanist's base of operations was at Fort Vancouver or at Fort William.

Piper states that he made his ". . . headquarters during 1835 . . . on Sauvie Island, at the mouth of the Willamette River, then called Wappatoo Island. Nuttall made but few and short excursions from his base, apparently finding enough to occupy his energies there. He did, however collect about the Willamette Falls, Fort Vancouver, and the mouth of the Columbia . . ."

Dr. F. W. Pennell evidently went into the matter of the botanist's whereabouts with great care and tells us that "More can be gained by bringing together the dated observations scattered through the second edition of his 'Land Birds,' and into the outline so formed the undated ones from the 'Sylva' and the papers in the Transactions of the American Philosophical Society for 1841 and 1843 all fit. But there are puzzling discrepancies."

Some of these "discrepancies" he points out:

"The first ornithological date, April 16, is assigned to the mouth of the 'Wahlamet river,' on the day before their supposed arrival at that point. 'Banks of the Columbia' and 'of the Wahlamet' rivers seem likely enough for May, and probably the first month was busily spent in this neighborhood . . . On May 29 and in June are bird observations from near 'the outlet of the Wahlahmet.' On June 21 is a record from Fort Vancouver . . . but it is likely that this base was touched repeatedly in the season. On July 4 are two records from Point Chinhook (Wash.), far down on the estuary toward Cape Disappointment. All these would suggest that Nuttall had not traveled far from the lower course of the Columbia, and they leave one unprepared for Townsend's entry supposedly of July 11 concerning 'Mr. Nuttall, who has just returned from the Dalles, where he was spending some weeks.' If the last record be actually August 11 the situation would become clear, except that toward the 'end of July' is a record from 'prairies of the Wahlahmet,' suggesting that Nuttall had gone instead up the Willamette valley. Such is the meagre itinerary for a season's busy work."

Pennell adds a footnote to the above:

"In answer to the question whether some of these dated records, all of which concern birds, could not have been supplied by Townsend, is the fact that on July 4 he was at Wappatoo Island and could not possibly have supplied the two records from Chinhook. Townsend's entry just mentioned, is merely '11th' and occurs between that for 'July 4th' and 'August 19th,' so that the above suggestion is quite plausible."

As I interpret Nuttall's itinerary discussed above, "his first ornithological date, April 16," was obviously no more than a slip—he was at the Willamette River within twenty-four hours or less of that date. He may well have been at Wapato Island on "May 29 and in June," and by "June 21" have been at Fort Vancouver (as one record mentions) on his way to the Dalles; Townsend states that he had spent "some weeks" there which indicates (given that the "11*th*" was *July* 11) that he was there in late June and early July. He might well have been back again in the Willamette prairies by the "end of July." As far as the "two records" from Point Chinook are concerned it seems strange, if Nuttall went so far down the Columbia, that he should have returned with so little to show for it. There would have been ample time for him to visit the mouth of the Columbia between his return from the Dalles (were that visit in July *or* August) and his departure for the Sandwich Islands in September of 1835, and he may have done so. It is also possible that someone might have given him the Chinook specimens, but *not* Townsend, who was at Fort Vancouver on July 4. The solution, given existing records, would seem to be "anyone's guess" !

Before considering Nuttall's visit to California, what of Wyeth's ambitions in Oregon, and what of Townsend's remaining months on the Columbia?

Wyeth wrote his uncle, Leonard Jarvis, on September 20, 1835:

". . . I am surrounded with difficulties beyond any former period of my life and without the health and spirit requisite to support them. In this scituation you can judge if memory brings to me the warnings of those (wiser and older) who advised a course which must at least have resulted in quietness. Yes memory lends its powers for torment . . ."

Historians have treated Wyeth with respect, despite the failure of all his enterprises. He was not a visionary. Chittenden's estimate of why his efforts were frustrated seems a fair one. After speaking of the man's "indomitable pluck" which "won for him the admiration of every one and particularly of . . . the venerable Dr. John McLoughlin," Chittenden continues:

"But in spite of Wyeth's herculean efforts his business did not prosper. His British friends, though kind and hospitable . . . could not, of course, aid him in his schemes of commercial rivalry, and gradually, yet surely, he saw the fabric of his hopes crumble to ruin. The salmon business did not develop as was expected . . . Townsend . . . in his journal of July 11, 1835, says: 'It really seems that the "Columbia River Fishing and Trading Company" is devoted to destruction; disasters meet them at every turn, and as yet none of their schemes has prospered. This was not for want of energy or exertion. Captain Wyeth has pursued the plans which to him seem best adapted to success, with the most indefatigable perseverance and industry, and has endured hardships without murmuring which would have prostrated many a more robust man. Nevertheless he has not succeeded.'. . .

"Having now given up all hope of successfully continuing the business Wyeth turned his thoughts to the matter of closing it out as advantageously as possible. He went to Fort Hall in the fall of 1835[1] and remained there during the winter. Returning to Fort William in the spring he placed it in charge of a Mr. C. M. Walker with directions to lease it for fifteen years if he could, and then went back to Fort Hall where he arrived on the 18th of June. A week later he left for the east . . . and reached home early in the autumn of 1836. In the following year Fort Hall and all its appurtenances were sold to the Hudson Bay Company. Fort William was not disposed of, and Wyeth's possession of the island became the foundation of a claim for it at a later date under United States laws.

"Thus ended in failure, but not in dishonor, one of the celebrated enterprises of the American Fur Trade. The causes of its failure are plainly apparent, and may all be summed up in want of adequate preparation, and in the apathy of the United States government to its true interests on the Pacific . . . It was no dishonor to Wyeth that he failed. His plans were in themselves well conceived, no less so than those of Astor, but success in both cases unfortunately depended upon questions of an international character over which private individuals had no control."

In addition to letters and journals Wyeth wrote a very interesting short memoir upon the northwest which appeared in 1839 in Cushing's "Territory of Oregon (Supplementary Report")". Of this Greenhow, writing in 1845, had this to say:

"Captain Wyeth's expeditions, though unprofitable to himself, have been rendered advantageous to the world at large; for his short memoir on the regions which he visited, printed with the report of the committee of the House of Representatives on the Oregon territory, in February, 1839, affords more exact and useful information, as to their general geography, climate, soil, and agricultural and commercial capabilities, than any other work yet published."

Townsend remained on the Columbia after Dr. Meredith Gairdner, ". . . the surgeon at Fort Vancouver, took passage . . . to the Sandwich Islands . . . In his absence, the charge of the hospital will devolve on me, and my time will thus be employed through the coming winter . . ." So Townsend wrote on October 1, 1835.

According to Bancroft's *Northwest Coast,* Townsend finally set sail ". . . in the Hudson's Bay Company's bark *Columbia,* Captain Royal, the 21st of November 1836 . . . again visited the Hawaiian Islands, and thence proceeded to Valparaiso, where on the 22nd of August 1837 he reëmbarked on board the brig *B. Mezick,* Captain Martin, for Philadelphia, where he landed the 17th of November."

1. Fort Hall was far removed from any base of supplies as well as from protection. Indians threatened, attacked his emissaries, stole his goods and horses; his employees were unreliable and deserted when they felt like it. It was, at the period when it was built, impossible to maintain. At the period of the overland migrations, as Bancroft stated, it "became famous and performed good service . . ." But Wyeth was a decade too early to profit from this turn in events.

Townsend's *Narrative* ends, as far as his sojourn on the northwest coast is concerned,[2] on November 30, 1836—a different date from the one given by Bancroft above—when the bar was crossed. "We are now out, and so good bye to Cape Disappointment and the Columbia, and now for *home,* dear home again!"

1836

Nuttall spent a few weeks in California after his second sojourn in the Sandwich Islands. According to his preface to *The North American Sylva:*

"... I now arrived on the shores of California, at Monterey. The early spring (March) had already spread out its varied carpet of flowers; all of them had to me the charm of novelty, and many were adorned with the most brilliant and varied hues. The forest trees were new to my view. A magpie, almost like that of Europe, (but with a yellow bill,) chattered from the branches of an Oak, with leaves like those of the Holly, *(Quercus agrifolia.)* A thorny Gooseberry, forming a small tree, appeared clad with pendulous flowers as brilliant as those of a Fuchsia. A new Plane tree spread its wide arms over the dried up rivulets. A Ceanothus, attaining the magnitude of a small tree, loaded with sky-blue withered flowers, lay on the rude wood-pile, consigned to the menial office of affording fuel. Already the cheerful mocking-bird sent forth his varied melody ... The scenery was mountainous and varied, one vast wilderness, neglected and uncultivated; the very cattle appeared as wild as the bison of the prairies ... In this region the Olive and the Vine throve with luxuriance and teemed with fruit; the Prickly Pears *(Cactus)* became small trees, and the rare blooming Aloe *(Agave americana)* appeared consigned without care to the hedge row of the garden."

Pennell, who devotes a paragraph to the California visit, states that "The fullest locality records are to be found in his botanical papers in Trans. Amer. Philos. Soc. N. S. 7: 283-453 and 8: 251-277." He suspects "... that the vessel from the Hawaiian Islands had already touched at San Francisco as *Bahia artemisiaefolia* was obtained there." After mentioning two plants collected at Monterey in March, he continues:

"Soon the brig *Pilgrim*[3] arrived, and Nuttall took passage in it southward along the coast. Much of April must have been spent at Santa Barbara, for there he made his most extensive collections [three species are cited] ... In the 'Land Birds' are records from Santa Barbara, credited to April 13 and to late April. At San Pedro, still in California, the stay must have been brief [two plants are cited] ... But at San Diego, where

2. I do not follow Townsend's activities in the northwest after Nuttall's departure. On March 16, 1836, he mentions that Dr. Tolmie relieved him at the hospital and that he had the chance of "... peregrinating again in pursuit of specimens ..."

3. C. C. Parry: "... from Dana's narrative we learn that he [Nuttall] shipped down the coast on the hide ship Pilgrim, stopping to take in hides at the ports of Santa Barbara and San Pedro, and reaching San Diego April 16 ..."

the Pilgrim is said to have arrived on April 15,[4] there is again a long list of species . . . [one] was credited to April, and . . . [two] to May."

Two other papers which deal with the California visit should be mentioned: one by F. V. Coville, "The botanical explorations of Thomas Nuttall in California" (1899), and one by W. L. Jepson, "The overland journey of Thomas Nuttall" (1934).

Coville points out: "It may seem strange . . . that Nuttall, wishing to go to California from the Columbia, did not make the journey overland, or at least take a vessel down the coast. The fact is he did not do this simply because he could not. Up to that time there was no land route from the Willamette to the Sacramento across the mountains of the Umpqua and the Rogue rivers and the terrible Siskiyous. As for a coast-wise vessel from the Columbia to a California port, that was a rare occurrence. The trade of the Columbia was exclusively a fur trade, and, while the trading vessels went frequently to the Hawaiian Islands to get provisions or sometimes to take on a cargo of sandal-wood for delivery at some eastern Asiatic port, they seldom had occasion to stop in California as they sailed to or from Cape Horn . . .

"In the absence of any direct account of Nuttall's movements in California, it seems best to collate the type localities of the new species of plants described by him as collected in that State, and with this in view a search has been made through the works in which most of these California collections were published, namely the seventh and eighth volumes of the Transactions of the American Philosophical Society, new series, 1840 to 1843, and in Torrey and Gray's Flora of North America, 1838 to 1843. As a result, it appears that Nuttall's California collections were made at Santa Barbara, San Pedro (the port of Los Angeles), and San Diego, in March, April, and May, 1836. He did not visit the California coast north of Monterey."

Coville's paper includes a "List of principal new species based on Nuttall's California collections" and a prefatory paragraph states:

"It is important that the new species based on Nuttall's California collections be critically identified, and since to many Californian botanists both the type specimens and the original descriptions are not readily accessible, the following list of species has been prepared. The list, arranged by type localities, includes the species described in Torrey and Gray . . . and in . . . the Transactions . . . After the original name is given the current equivalent, if different from the original, and any additional information suggested by the first description, such as habitat, precise locality, date of collecting or flowering, probable misidentification, or incorrect use of name. No attempt has been made to identify the species critically . . ."

The list includes seventeen species from Monterey, sixty-five from Santa Barbara, two from San Pedro and forty-four from San Diego. W. L. Jepson limits Nuttall's work to ". . . the vicinity of the seaports. While in California he did not see Quercus douglasii, Acer macrophyllum, Acer negundo var. californicum, Arbutus menziesii,

4. Parry reports the length of the visit as "barely twenty-four days."

Pseudotsuga taxifolia, Pinus sabiniana, Pinus lambertiana, Abies bracteata or Fraxinus oregana. All of these grow in Monterey county except possibly the last, while Quercus douglasii is exceedingly common on the old trail from Monterey to Mission San Antonio and Mission San Miguel."

Nuttall sailed from California for the east coast in the *Alert*, made famous in Richard Henry Dana's *Two years before the mast. A personal narrative of life at sea*, first published anonymously in New York in 1840. It contains numerous references to Nuttall.

Dana records on Friday, April 15, that at San Diego "Arrived, brig Pilgrim, from the windward." When the *Alert* was "one week out" to sea, on Sunday, May 15, he wrote:

"This passenger—the first and only one we had had, except to go from port to port, on the coast, was no one else than a gentleman whom I had known in my better days; and the last person I should have expected to have seen on the coast of California— Professor N————, of Cambridge. I had left him quietly seated in the chair of Botany and Ornithology, in Harvard University; and the next I saw of him, was strolling about San Diego beach, in a sailor's pea-jacket, with a wide straw hat, and barefooted, with his trowsers rolled up to his knees, picking up stones and shells. He had travelled over land to the North-west Coast, and come down in a small vessel to Monterey. There he learned that there was a ship at the leeward, about to sail for Boston; and, taking passage in the Pilgrim, which was then at Monterey, he came slowly down, visiting the intermediate ports, and examining the trees, plants, earths, birds, &c., and joined us at San Diego shortly[5] before we sailed. The second mate of the Pilgrim told me that they had got an old gentleman on board who knew me, and came from the college that I had been in. He could not recollect his name, but said he was a 'sort of an oldish man,' with white hair, and spent all his time in the bush, and along the beach, picking up flowers and shells, and such truck, and had a dozen boxes and barrels, full of them. I thought over everybody who would be likely to be there, but could fix upon no one; when, the next day, just as we were about to shove off from the beach, he came down to the boat, in the rig I have described, with his shoes in his hand, and his pockets full of specimens. I knew him at once, though I should not have been more surprised to have seen the Old South steeple shoot up from the hide-house. He probably had no less difficulty in recognising me. As we left home about the same time, we had nothing to tell one another; and owing to our different situations on board, I saw but little of him on the passage home. Sometimes, when I was at the wheel of a calm night, and the steering required no attention, and the officer of the watch was forward, he would come aft and hold a short yarn with me; but this was against the rules of

---

5. Pennell considers that "All this accords well with Nuttall's records from the California ports, but not with Dana's previous statement . . . that the Pilgrim arrived as early as April 15. Perhaps she went back to Santa Barbara, or else some other coasting vessel brought Nuttall the latter part of the way."

The *Alert* must have sailed May 7 or 8 for Dana reports her as "one week out" to sea on May 15. Dana may have used the word "shortly" in a general sense—three weeks in port, in retrospect, is not a very long time.

the ship, as is, in fact, all intercourse between passengers and the crew. I was often amused to see the sailors puzzled to know what to make of him, and to hear their conjectures about him and his business. They were as puzzled as our old sailmaker was with the captain's instruments in the cabin . . . The Pilgrim's crew christened Mr. N. 'Old Curious,' from his zeal for curiosities, and some of them said that he was crazy, and that his friends let him go about and amuse himself in this way. Why else a rich man (sailors call every man rich who does not work with his hands, and wears a long coat and cravat) should leave a Christian country, and come to such a place as California, to pick up shells and stones they could not understand . . . however, an old salt, who had seen something more of the world ashore, set all to rights, as he thought, —'Oh, 'vast there!—You don't know anything about them craft. I've seen them colleges, and know the ropes. They keep all such things for cur'osities, and study 'em, and have men a' purpose to go and get 'em. The old chap knows what he's about. He a'n't the child you take him for. He'll carry all these things to the college, and if they are better than any that they have had before, he'll be head of the college. Then, by-and-by, somebody else will go after some more, and if they beat him, he'll have to go again, or else give up his berth. That's the way they do it. This old covey knows the ropes. He has worked a traverse over 'em, and come 'way out here, where nobody's ever been afore, and where they'll never think of coming.' This explanation satisfied Jack: and as it raised Mr. N.'s credit for capacity, and was near enough to the truth for common purposes, I did not disturb it."

The *Alert,* after dangerous days and nights of battling with ice and gales, came in sight of Cape Horn, and Dana refers again to Nuttall:

". . . even Mr. N., the passenger, who had kept in his shell for nearly a month, and hardly been seen by anybody, and who we had almost forgotten was on aboard, came out like a butterfly, and was hopping round as bright as a bird.

"The land was the island of Staten Land, just to the eastward of Cape Horn; and a more desolate-looking spot I never wish to set eyes upon . . . Yet, dismal as it was, it was a pleasant sight to us . . .

"In the general joy, Mr. N. said he should like to go ashore upon the island and examine a spot which probably no human being had ever set foot upon; but the captain intimated that he would see the island—specimens and all,—in—another place, before he would be out a boat or delay the ship one moment for him."

The ship reached Boston on September 20, 1836, according to Parry. The passage had consumed one hundred and thirty-five days. The anonymous author of the "Biographical sketch of the late Thomas Nuttall" mentions one tribute to the botanist:

"The owners of the vessel, that brought him home from the Pacific, Messrs. Sturgis and Bryant, of Boston, to their honor be it said, would not take a cent of passage money from him, 'as,' they said, 'you travel for the benefit of mankind.' "

Nuttall wrote in the preface to *The Sylva:*

"After a perilous passage around Cape Horn, the dreary extremity of South America, amidst mountains of ice which opposed our progress in unusual array, we arrived again at the shores of the Atlantic. Once more I hailed those delightful scenes of nature with which I had been so long associated . . . But the 'oft told tale' approaches to its close, and I must bid a long adieu to the 'new world,' its sylvan scenes, its mountains, wilds and plains, and henceforth, in the evening of my career, I return, almost an exile, to the land of my nativity!"

This was presumably written in 1841, for at the end of that year Nuttall left the United States to take up his abode in England.

For Nuttall's years after his return from the west coast one turns again to Pennell's valuable paper:

He worked (1836–1841) at the Academy in Philadelphia upon his various collections, ". . . mineralogical and zoological, as well as botanical . . ." His "new marine Shells from Upper California" and the "Crustacea" gathered in conjunction with Townsend, were reported upon by T. A. Conrad and J. W. Randall respectively. He, himself, brought his *Land Birds* up to date in a second edition.

His descriptions of many new species of Polypetalous plants ". . . were incorporated into Torrey and Gray's 'Flora of North America,' . . ." and he ". . . commenced the preparation of a series of studies, almost monographic in character, as reports upon the plants of the great journey. Two were issued in the Transactions of the American Philosophical Society: 'Descriptions of New Species and Genera of Plants in the Natural Order Compositae, collected on a Tour across the Continent to the Pacific, a Residence in Oregon, and a Visit to the Sandwich Islands and Upper California, during the Years 1834 and 1835,' read October 2 and December 18, 1840; and 'Descriptions and Notices of new or rare Plants, Lobeliaceae, Campanulaceae, Vacciniaceae, Ericaceae, collected in a Journey over the Continent of North America, and during a Visit to the Sandwich Islands and Upper California,' read December 3, 1841 . . .

". . . The series of western monographs never passed beyond the three that have been mentioned. The project of extending Michaux's 'Sylva' by the addition of three more volumes was put through in haste, and the result left to others to see through the press . . . it was filled with original taxonomic work . . ."

We are told also that ". . . Nuttall's study of American plants abruptly ended with the return to England, except for his paper 'On Simmondsia, a new Genus of Plants from California,' in 1844, and what he accomplished on his last visit to America."

Nuttall's residence in England lasted from 1842 until his death in 1859, but was broken by a visit to America, from late 1847 into early 1848. Pennell reports that during this time he worked at the Philadelphia Academy, publishing in 1848 his paper upon William Gambel's collections made in the Rocky Mountains and in California.[6]

Nuttall's return to England had, as its basis, the acceptance of an inheritance, Nut-Grove Hall in Lancashire, ". . . near Prescot, a village near St. Helens on the road

6. This paper described as well a considerable number of Nuttall's own western collections.

from Liverpool." The terms of the inheritance apparently stipulated that he should be in residence for the greater part of every year and it was only by combining periods of permissible absence that he was able to make the visit to the United States. It seems doubtful that much money went with the property but the rumor that it *did* had evidently given Nuttall's botanical friends some concern, Lewis von Schweinitz writing Torrey (July 6, 1824), "It is rather a dangerous experiment to get a large fortune as I hear he has—it is even more so than Mineralogy." [7]

As a British landowner, Nuttall became interested in the culture of rhododendrons and, if the cause of his death was as reported, it was in character: in his eagerness to open a shipment believed to contain some of these plants, he overstrained himself. He died at Nut-Grove Hall on September 10, 1859, and was buried at Christ Church, Eccleston.

According to Pennell, his passing ". . . seems to have been scarcely noticed in English journals of the time . . . On our side of the Atlantic he has claimed more attention . . ."

Pennell refers to Nuttall as ". . . not only a pathfinder but also among the keenest students of our plants." He then quotes Asa Gray (as I did in my first chapter on Nuttall) to the effect that ". . . no botanist has visited so large a portion of the United States, or made such an amount of observations in the field and forest. Probably few naturalists have ever excelled him in aptitude for such observations, in quickness of eye, tact in discrimination, and tenacity of memory."

Gray's next sentence, comparing Nuttall's later writings with those of Rafinesque, seemed to Pennell ". . . unjust, and I think that it betrays the fact that in 1844 Gray had not grasped what a remarkable proportion of endemism characterizes every portion of western North America. There is also the fact that Nuttall was inclined to the formation of small concrete genera and Gray to retain large comprehensive ones."

According to G. N. Jones, Nuttall made some contributions to the science of geology:

"The fame of Thomas Nuttall rests chiefly on his monumental work on North American plants and his valuable contributions to American ornithology.[8] Of his achievements little has been written, probably because their value was not perceived by students in other branches of science. Yet he was the first to lay the foundations for the solving of problems of geological correlation and of identifying geological formations. On his Mississippi expedition [doubtless his 'Arkansa journey'] he made extensive collections of fossils by means of which he correlated the limestones of the Mississippi River with those of the mountainous regions of England. On the basis of

7. The fact that both Nuttall and Torrey were interested in geology was another cause of anxiety to the botanists and the Torrey-von Schweinitz correspondence contains a number of rather ponderous jokes on that subject.

8. The first ornithological club in America was named in honor of Nuttall. See: "An account of the Nuttall Ornithological Club 1873 to 1919" by Charles Foster Batchelder which appeared in the *Memoirs* of that Society in December, 1937.

fossils collected near the Mandan village [on his trip with the Astorians] he made possible the first definite recognition of beds of Cretaceous age in North America."

G. Brown Goode, addressing the Biological Society of Washington of which he was president, mentions the growing interest in geology and mineralogy at this period. His comment upon Nuttall's ability is terse and blanketing: "Nuttall was not great as a botanist, as a geologist, or as a zoölogist, but was a man useful, beloved and respected."

Pennell supplies much information about Nuttall's plant collections, tells where they are located and appraises their importance. I quote what he has to say of those in the Philadelphia Academy, in other herbaria of the United States, and finally in Britain. The following seven paragraphs are quoted from his pages 42, *fn.* 96, 44-45, *fn.* 107.

"The only important series of Nuttall's specimens in America is at the Academy of Natural Sciences of Philadelphia. Here was his American headquarters for the study of our flora. Before 1830 the contemporary 'Handbook to the Museum of the Academy of Natural Sciences of Philadelphia' credited him with having given the institution over 4,000 species of plants from the Missouri and Arkansas Rivers, the Great Lakes, Carolina, Georgia, Louisiana, etc. Before 1837 he had followed this by the gift of his large exotic herbarium, which had presumably been built up by exchanges . . . In 1837 were added an extensive series from his recent transcontinental expedition, and these concern us now.

"Comparing these specimens at Philadelphia with those retained by Nuttall in his own herbarium, one notes that they are merely single or a few plants from the more ample material that was retained. The data are slightly more abbreviated, but in this regard there is little distinction. Especially, one is impressed by the fact that the same names uniformly occur in both series. This would show that both reflect Nuttall's determinations as made before the end of 1841, and clearly indicate how extensively he had already studied his plants. For many western species and genera, based subsequently upon later collections or nearly simultaneously upon David Douglas' specimens, Nuttall had selected his own definite names. It is indeed tragic that the great work of describing these plants was interrupted . . ."

"Some collections passed to Dr. John Torrey of New York, especially of the far western collections reported in Torrey and Gray's 'North American Flora,' and these are now at the New York Botanical Garden; but they do not amount to a 'nearly complete suite' of the plants of the transcontinental journey as announced in the preface to this 'Flora.' Also some of these and of the Arkansas plants are at Harvard University, acquired directly or from Durand . . . they are relatively few . . . Nuttall's collections other than botanical seem to have been all given to the Philadelphia Academy, though only those of minerals are mentioned in the Academy's Handbooks."

"Nuttall's own herbarium passed on his death to the British Museum in London, where it is now in excellent custody. It would appear that up to the publication of the

'Genera of North American Plants' in 1818 he kept few or no specimens for himself, presenting a complete series of his plants to the Academy of . . . Philadelphia so that there repose his early types . . . Beginning with the Arkansas collections, more ample material is to be found at the British Museum than at Philadelphia, and so we may consider that his later types are in London, with isotypes at Philadelphia.

"Nuttall's specimens are unusually small, even for the period, and it would seem that they were intentionally so in order to facilitate transportation. At the British Museum there seem to be more specimens, rather than ones of larger size than at Philadelphia. Doubtless, many that he would have considered as suitable duplicates for distribution as exchanges, have been retained and now mounted together so as to supply the full sheets at London. The data are unfortunately always meagre, even though in publication Nuttall had much to say about his plants. Did his interesting remarks come from his memory or from actual records?

". . . I think the latter . . . Usually Nuttall's localities prove to be accurate . . . There should have been once an extensive and invaluable series of manuscript books to tell both of his travels and collections. I made inquiry for such at the British Museum in 1930, but was informed that no papers had been received . . . with his herbarium. Yet Dr. Jepson[9] . . . tells us that Dixon Nuttall assured him that 'all such material [was] taken to London shortly after his death.' Perhaps . . . a collection of them exists somewhere, their nature wholly unsuspected. Or perhaps . . . these precious records were long ago destroyed, and we shall never know what they would have revealed to us."

"There are also many specimens from Nuttall in the Liverpool Botanic Garden Collection as announced in the 'Handbook and Guide to the Herbarium Collections in the Public Museums, Liverpool,' page 55, 1935 . . ."

Writing of European herbaria in 1840, and mentioning the collections in the "Royal Museum at the Jardin des Plantes or Jardin du Roi" in Paris, Asa Gray stated that in the herbarium of Mr. Webb[1] ("only second to that of Baron Delessert") was ". . . a North American collection, sent by Nuttall to the late Mr. Mercier of Geneva . . ." Also ". . . a set of Nuttall's collections on the Missouri . . ." in the Lambert Herbarium; a year after Gray wrote, Lambert's Herbarium was sold at auction and scattered. W. H. Brewer (1880) mentions that Nuttall collections are ". . . also at Kew . . ."

9. Jepson had visited Nut-Grove Hall. "It was my hope to turn up, perhaps, some of Nuttall's journals, but all such material, taken to London shortly after his death, disappeared."

1. Philip Barker Webb. His entire herbarium is at the Instituto Botanico, Florence, Italy.

Anonymous [PICKERING, CHARLES ?]. 1860. Biographical sketch of the late Thomas Nuttall. *Gard. Month.* (Meehan) 2: 21-23.

BANCROFT, HUBERT HOWE. 1884. The works of . . . 28 (Northwest coast II. 1800–1846): 576-599.

——— 1886. The works of . . . 29 (Oregon I. 1834–1848): 14, 15, 59-73, 85, *fn.* 11. Map ("French Prairie," 71).

BREWER, WILLIAM HENRY. 1880. List of persons who have made botanical collections in California. *In:* Watson, Sereno. Geological survey of California. Botany of California 2: 555.

BROSNAN, CORNELIUS JAMES. 1935. History of the state of Idaho, rev. ed. New York. 70, 93.

CHITTENDEN, HIRAM MARTIN. 1902. The American fur trade of the far west. A history of the pioneer trading posts and early fur companies of the Missouri valley and the Rocky Mountains and the overland commerce with Santa Fe. Map. 3 vols. New York. 1: 448-456; 3: 967, 974.

COVILLE, FREDERICK VERNON. 1899. The botanical explorations of Thomas Nuttall in California. *Proc. Biol. Soc. Washington* 13: 109-121.

[DANA, RICHARD HENRY]. 1840. Two years before the mast. A personal narrative of life at sea. ed. 1. New York. 332, 359-361, 412, 413.

DE VOTO, BERNARD. 1947. Across the wide Missouri. Illustrated with paintings by Alfred Jacob Miller, Charles Bodmer and George Catlin. Boston. 184.

DURAND, ELIAS. 1860. Biographical notice of the late Thomas Nuttall. *Proc. Am. Philos. Soc.* 7: 297-315.

EASTWOOD, ALICE. Early botanical explorers on the Pacific coast and the trees they found there. *Quart. Cal. Hist. Soc.* 18: 341, 342.

GHENT, WILLIAM JAMES. 1931. The early far west. A narrative outline 1540–1850. New York. Toronto. 257-260.

GOODE, GEORGE BROWN. 1888. The beginnings of American science. The third century. *Proc. Biol. Soc. Washington* 4: 71.

GRAY, ASA. 1840. Notices of European herbaria, particularly those most interesting to the North American botanist. *Am. Jour. Sci. Arts* 40: 1-18. *Reprinted:* Sargent, Charles Sprague, *compiler.* 1889. Scientific papers of Asa Gray. 2 vols. Boston. New York. 2: 1-21.

GREENHOW, ROBERT. 1845. The history of Oregon and California, and the other territories on the north-west coast of North America: accompanied by a geographical view and map of those countries, and a number of documents as proofs and illustrations of the history. ed. 2. Boston. 359, 360, *fn.,* 376-404.

HULBERT, ARCHER BUTLER. 1934. Overland to the Pacific. Volume 4. The call of the Columbia. Iron men and saints take the Oregon Trail. Edited, with bibliographical résumé, 1830–1835, by Archer Butler Hulbert. With maps and illustrations. The Stewart Commission of Colorado College and Denver Public Library. 184-226.

JACKSON, BENJAMIN DAYDON. 1922. Thomas Nuttall (1786–1859). *Jour. Bot.* 60: 57.

JEPSON, WILLIS LINN. 1934. The overland journey of Thomas Nuttall. *Madroño* 2: 143-147.

JONES, GEORGE NEVILLE. 1937. Thomas Nuttall. Botanist, ornithologist, geologist. *Little Gardens* (Seattle) 8: 6-25.

NUTTALL, THOMAS. 1841. Descriptions of new species and genera of plants in the natural order Compositae, collected on a tour across the continent to the Pacific, a residence in Oregon, and a visit to the Sandwich Islands and upper California, during the years 1834 and 1835. *Trans. Am. Philos. Soc.* new ser., 7: 283-453.

—— 1843. Description and notices of new or rare plants in the natural orders Lobeliaceae, Campanulaceae, Vacciniaceae, Ericaceae, collected in a journey over the continent of North America, and during a visit to the Sandwich Islands, and upper California. *Trans. Am. Philos. Soc.* new ser. 8: 251-272.

—— 1848. Descriptions of plants collected by William Gambel, M.D., in the Rocky Mountains and upper California. *Jour. Acad. Nat. Sci. Phila.* new ser. 1: 149-189. 3 plates.

—— 1850. Descriptions of plants collected by Mr. William Gambel in the Rocky Mountains and upper California. *Proc. Acad. Nat. Sci. Phila.* 4: 7-26. pl. 22, 23, 24. Read in February, 1848. [Preliminary diagnosis.]

—— 1852. The North American Sylva; or a description of the forest trees of the United States, Canada, and Nova Scotia, not described in the work of F. Andrew Michaux, and containing all the forest trees discovered in the Rocky Mountains, the territory of Oregon, down to the shores of the Pacific, and into the confines of California, as well as in various parts of the United States. Illustrated with 127 fine plates. By Thomas Nuttall, F.L.S., . . . &c. &c. &c. In three volumes.—Vol. I. being the fourth volume of Michaux and Nuttall's North American Sylva . . . Philadelphia . . . [vols. II. and III. being the fifth and sixth volumes of same.]

PARRY, CHARLES CHRISTOPHER. 1883. Early botanical explorers of the Pacific coast. *Overland Monthly* ser. 2, 2: 414, 415.

PENNELL, FRANCIS WHITTIER. 1936. Travels and scientific collections of Thomas Nuttall. *Bartonia* 18: 1-51.

PIPER, CHARLES VANCOUVER. 1906. Flora of the state of Washington. *Contr. U. S. Nat. Herb.* 11: 14.

RUSSELL, CARL PARCHER. 1941. Wilderness rendezvous period of the American fur trade. *Oregon Hist. Quart.* 42: 27-30.

SHEAR, CORNELIUS LOTT & STEVENS, NEIL EVERETT. 1921. The correspondence of Schweinitz and Torrey. *Mem. Torrey Bot. Club* 16: 119-181.

STANSFIELD, H. 1935. Handbook and guide to the herbarium collections in the Public Museums Liverpool. 35, 41, 43, 55. [For Stansfield's authorship see p. 9.]

TORREY, JOHN & GRAY, ASA. 1838-[1843]. A Flora of North America, containing abridged descriptions of all the known indigenous and naturalized plants growing north of Mexico, arranged according to the natural system. 2 vols. in 1. New York, etc.

TOWNSEND, JOHN KIRK. 1839. Narrative of a journey across the Rocky Mountains, to the Columbia River, and a visit to the Sandwich Islands, Chili, &c., with a scientific appendix. Philadelphia.

—— 1905. Townsend's narrative of a journey across the Rocky Mountains, to the Columbia River. *In:* Thwaites, Reuben Gold, *editor.* Early western travels, 1748–1846 21: 105-369. [A reprint of pp. 1-186, 218-264 of Philadelphia edition of 1839. Townsend's account of the visit to the Sandwich Islands, to Chile, etc., as well as his scientific appendix are not included in the reprint.]

—— 1934. Townsend's original journal. *In:* Hulbert, A.B. 1934. Overland to the Pacific. Volume 4. The call of the Columbia . . . 184-226.

WYETH, NATHANIEL JARVIS. 1839. Mr. Wyeth's memoir. Washington. *U. S. 25th Cong., 3rd Sess., House Doc.* 3: No. 101, 6-22. *In:* Cushing, C. Territory of Oregon (Supplementary Report).

——— 1899. The correspondence and journals of Captain Nathaniel J. Wyeth 1831–6. A record of two expeditions for the occupation of the Oregon country, with maps, introduction and index . . . Edited by F. G. Young. *In:* Sources of the history of Oregon 1: parts 3-6. Eugene.

CHAPTER XXX

THE "FRIEND OF MR TOLMIE" WHO MADE COLLEC-

TIONS OF A QUALITY PLEASING TO HOOKER AND

WALKER-ARNOTT PROVES TO BE THE FUR-TRADER

JOHN McLEOD

IN the foreword to the "California Supplement" of *The botany of Captain Beech-ey's voyage* it is stated that many of the plants described were gathered by David Douglas rather than by the botanists attached to the ship *Blossom*. And special reference is made to another included collection received by the authors, William Jackson Hooker and George Arnott Walker-Arnott. This was

". . . very recently sent to us by Mr. Tolmie, from the '*Snake Country*,' in the interior of California. This is a name given to the vast extent of Prairie through which Lewis' branch, or the *Snake River*, holds its course. Fort Hall is situated at the confluence of Blackfoot with Snake River . . . Snake Fort is built at the junction of Reed's River[1] with the Snake . . . The specimens, in beautiful preservation, were gathered, in the summer of 1837, by a friend of Mr Tolmie, who conducted a party from Fort Vancouver, on the Columbia, to the rendezvous of the American trappers, in the interior of California. Some few of the specimens are from the '*Green River*;' for the meeting of the Beaver Trappers . . . was held in that year in the valley of the '*Green River*,' a stream which is considered to be probably the main branch of the Rio Colorado, and which empties itself into the Gulf of California. There is not, perhaps, in the whole of North America, a district more interesting to the Botanist than that from which these plants are derived; situated near the western foot of the Rocky Mountains, at an immense distance from the coast, and at a great elevation, as may be inferred from the fact of its being near the sources of two great rivers, the one having its course to the north (into the Columbia), the other to the south (into the Gulf of California); and whose respective windings seem to circumscribe the whole of New California, except that portion of it which is washed by the Pacific Ocean. If other gentlemen attached to the hunting expeditions of the Hudson's Bay and American Companies would thus occupy a portion of their leisure time, we should soon be as well acquainted with the vegetation of the interior of this vast continent as we now are with that of its coasts."

Believing that it would be interesting to learn the identity of William Fraser Tolmie's "friend" whose collections were in such "beautiful preservation" and whose name has

---

1. Reed's River is Boisé River and "Snake Fort" is undoubtedly Fort Boisé.

627

gone unrecorded for more than one hundred years, I determined to find out, if possible, who the man was.

Michel LaFramboise of the Hudson's Bay Company was suggested as a possibility, but since he was hunting about the Bay of San Francisco in 1837 and was planning to visit the mouth of the Colorado River, he was not journeying in the right direction. It was suggested that Mrs. Alice Bay Maloney might be able to help and she lived up to her reputation as "the authority" on the Hudson's Bay Company in California, writing me (February 3, 1948) as follows:

"From what evidence I can find the collector of the botanical specimens you mention was John McLeod.

"In Volume IV of the Hudson's Bay Record Society Publications appears a letter of Chief Factor John McLoughlin to the Governor, Deputy Governor and Committee, Honble, H. Bay Compy., dated Fort Vancouver 31st Octr., '37. Paragraph 18, page 208 reads:

" 'As mentioned in the 15th Paragraph of mine of 20th March last to the Governor and Council at York, Mr. McLeod left this on 18th April with an Outfit, and proceeded to the American Rendezvouse which he reached on 28th June, on Green River, a Branch of the Rio Colorado, about 200 Miles S. E. of Salt Lake.[2] On the 18th July the Americans arrived from St. Louis . . .'

"W. J. Ghent in Early Far West (1936 edition) p. 262 records the presence of Thomas McKay and John McLeod at the Green River Rendezvous of 1837.

"According to my notes Michel LaFramboise was in the Sacramento and San Joaquin valleys of California in 1837. Neither he nor Tom McKay possessed qualifications for making careful botanical collections; nor does it seem at all likely that Dr. Tolmie would refer to LaFramboise as a personal friend. Caste lines were drawn at Fort Vancouver rather strictly. John McLeod was a 'gentlemen of the Company.'

"If you are interested in further research on John McLeod a letter to the Archivist of British Columbia at Victoria would no doubt produce dates for the numerous letters of John McLeod in their collection, with possible information on his 1837 journey . . ."

But there proved to have been three, possibly four, men named John McLeod who served in the North West and Hudson's Bay Companies! Two were readily eliminated as possibilities. And, after reading the records of a third which are in the Library of Congress,[3] it was apparent that he was not working in the Columbia River district in

---

2. Mrs. Maloney (*in litt.* February 15, 1948) notes: "Dr. McLoughlin had never attended a rendezvous (at least I have no record that he did so), and consequently his geography in the letter I quoted, was probably very inaccurate."

Certainly the rendezvous on Green River was more nearly northeast than southeast of Great Salt Lake.

3. "Journals and correspondence of John McLeod, Senior, Chief Trader, Hudson's Bay Company, who was one of the earliest pioneers in the Oregon Territory, from 1812 to 1844." "Copied from the originals in the Dominion Government Archives by R. Gosnold. Victoria. B. C. 1903."

1837. The fourth—and acceptable—candidate is cited thus in *The Encyclopedia of Canada:*

"McLeod, John (fl. 1821–1842), fur-trader, was a native of Scotland, and entered the service of the North West Company prior to 1821, when he was clerk on the Churchill river. From 1824 to 1830 he was in the Mackenzie river district; and later he was employed on the Columbia. He was promoted to the rank of chief trader[4] in 1834, and he retired from the fur-trade in 1842. Sir George Simpson . . . described him as 'a fine active fellow.' "

But there was still no actual proof that a John McLeod *made* the collection. This was obtained through correspondence with Dr. Willard E. Ireland, Provincial Librarian and Archivist of the Provincial Archives, Victoria, British Columbia. His first letter (September 23, 1948) was discouraging:

". . . While I am inclined to agree with Mrs. Maloney's suggestion that in all probability it was John McLeod, unfortunately none of our material here affords any clue to substantiate the idea, for our letters are considerably earlier and none in the period of 1837 . . . Similarly none of our Tolmie material throws any light on the identification . . ."

But a letter of a week later (September 30, 1948) brought good news:

"Further to my letter of September 23rd I am sure you will be pleased to know that I think we have established quite definitely that the source of the specimens mentioned as coming from 'friend of Mr. Tolmie' was John McLeod. When we went through some of our diaries of Tolmie for an earlier period we found that at the back there were many pages filled in with odd random notes of his, 5 pages of which, pages 317-321, listed 71 plants that he received from Mr. McLeod and the localities of their origin. For your information I enclose transcript of the pages in question. Unfortunately the list is not dated, but I feel quite sure that it was made in 1837, for just prior to this particular memo there are a series of dated ones in 1837, and shortly thereafter some for 1837 appear . . . I trust this information will be sufficient proof for your enquiry . . ."

Dr. Ireland very kindly sent me photostats of the Tolmie notes and I have checked their cited localities with those recorded as "(Tolmie)" collections in the "California Supplement" of *The botany of Captain Beechey's voyage.* The results of the comparison are as follows:

The Tolmie list, which is headed "Plants from Mr McLeod Localities &c—", cites seventy-seven numbers; of these numbers 27, 28A and 28B lack any data,[5] number 29 cites no locality but notes "Roots tuberose—edible"; we can reduce the list there-

---

4. Since the North West Company was amalgamated with the Hudson's Bay Company in 1821, McLeod was a chief trader with the proper organization!

5. Nos. 27, 28A, 28B and 29 may, like no. 26, have been from the Grande Ronde, but there are no ditto marks after no. 26.

fore (in so far as it might have been of service to Hooker and Walker-Arnott) to seventy-four.

The number of plants attributed to Tolmie's friend in the "Supplement" total, I believe, sixty-two, or twelve numbers less than the Tolmie list. Of the specimens sent Hooker a few may well have contained inadequate material or some may have been duplicates—the discrepancy in numbers was not unaccountably large.

Since the Tolmie list includes no plant names it could only be checked with the "Supplement" on the basis of cited localities. The list never mentions "Snake Country" (although it frequently cites Snake River and localities thereon), but the "Supplement" inserts "Snake Country" after practically all localities. The authors seem to have used it as a covering term and perhaps felt that they were accounting for some localities which they could not place with precision!

After finding that some forty localities of the Tolmie list were represented in the "Supplement" I brought the comparison to a close. Indeed, the similarity between certain citations seemed proof enough; for examples: "Bannock Defile between Snake and Bear Rivers" (p. 366, "Supplement") and "Bannock defile passing between Snake & Bear Rivers" (no. 7, Tolmie list); "Between Bruneau and Onyhee Rivers, Snake Country" (p. 333, "Supplement") and "Between Bruneau & Owyhee Rs" (nos. 43 and 44, Tolmie list); "Snake River, below the Salmon Falls, Snake Country" (p. 350, "Supplement") and "Snake R. below Salmon Falls" (no. 40, Tolmie list). I append the Tolmie record at the end of this chapter.

Having accepted John McLeod as collector, what route did he take to the rendezvous? The localities in the Tolmie list follow no route-sequence and are undated; but, after fixing the position of some of the more important, McLeod's travels become fairly clear. Whether he gathered the specimens going to or coming from the rendezvous is unimportant.

That the starting point was Fort Vancouver[6] and that the Columbia was ascended to Fort Walla Walla is probable. There is one record from "Columbia Falls" (no. 77). From Fort Walla Walla the direction was southeast, across northeastern Oregon: the Umatilla River was encountered in the county of the same name, "Utalla River" (no. 19), and the Blue Mountains (nos. 9, 15, 30, 31, 35, 42) crossed into the Grande Ronde valley (no. 26 and, perhaps, nos. 27, 28A, 28B, 29) in Union County. Whether McLeod proceeded from the Grande Ronde directly to the Snake River and ascended that stream to the mouth of Boisé River, or whether he went southeast across country I do not know but I believe the latter,[7] although the few localities mentioned are open

6. The "Supplement" lists three "(Tolmie)" collections from the "Walamet," or Willamette, River (pp. 395, 398, 401) which are not cited on the Tolmie list and these may well have been gathered by Tolmie himself.

7. The Oregon Trail went across country from the crossing of the Snake near the mouth of the Boisé River to the Grande Ronde. It was this route that Wyeth and Nuttall followed in 1834.

to either interpretation: "Burnt River" (no. 1) and "Between Burnt & Malheur Rs" (nos. 20 through 24).

The first Fort Boisé, erected in 1834 and called "Snake Fort" in the Tolmie list, was situated near the mouth of Boisé, or Reed's, River which enters the Snake from the southeast, in Canyon County, Idaho; nearly opposite this fort the Owyhee River enters the Snake from the southwest, in Malheur County, Oregon. There are numerous records from "Snake Fort" (nos. 3, 4, 14, 47, 52, 64, 65, 66, 69, 72, 74, 75).

From Fort Boisé, and ascending Snake River, McLeod crossed southern Idaho to Power or Bannock County, for there are records from "Between Bruneau & Owyhee[8] Rs" (nos. 43, 44), "Snake R. above Bruneau R" (no. 2), "Bruneau—sandy soil—ht 2 to 3 feet" (no. 38), "Snake R. below Salmon Falls" (nos. 40, 67), "Snake Riv: above Salmon Falls" (no. 45), "Snake R. below American Falls" (no. 71), "American Falls —Snake R." (no. 76). At the American Falls McLeod was in Power County.

At this point or somewhat farther up the Snake, McLeod must have left that river and turned southeast, across the southeastern corner of Idaho; he may have passed Fort Hall, built by Wyeth in 1834, but there is no record from there. There are records from "Bannock River" (no. 46), "Bannock defile passing between Snake & Bear Rivers" (no. 7), "Blackfoot River" (nos. 54, 62), "In the neighbourhood of the hot springs" (no. 63), which was probably the Soda Springs of Caribou County. Here he would have been on "Bear River . . ." (no. 11). Ascending Bear River he must have reached "Thomas's Fork" or "Thoma's Fork" (nos. 18 and 48, 61, respectively) which enters Bear River in extreme southeastern Bear Lake County, Idaho.

The records, "Between Henry's Fork & Smith's [Fork]" (nos. 57 through 60), indicate that McLeod crossed into Lincoln County, Wyoming, for Smith's Fork enters Bear River not far east of the Idaho-Wyoming boundary.

According to Russell, "The Thirteenth Rendezvous (1837)" was held on "Green River, 12 Miles south of Horse Creek" which would place it in central Sublette County, Wyoming. Most of the remaining localities of the Tolmie list which I have been able to identify satisfactorily, seem to have been in Wyoming—the specimens gathered as McLeod approached the meeting place of the trappers or while there: "Rock River [or Creek]" (nos. 37, 53), "Hams Forks" and "Hams Fork" (nos. 12 and 25, respectively), "Green River" (nos. 16, 32, 33, 39, 70), "Pine Creek near Green River"[9] (no. 5), Bonneville's Fort[1] (no. 68), and "Swamp at Green River" (nos. 50, 51).

8. Although the Owyhee River crosses southwestern Idaho, McLeod would not have gotten so far south in that state and the record must cover the country along the Snake between the mouth of Owyhee River, Malheur County, Oregon, and the mouth of Bruneau River, Owyhee County, Idaho.

9. Whether "Pine Creek," lacking qualifying phrase (nos. 13, 17, 34, 41, 49), is the same I do not know. Some useful maps examined list six Pine Creeks in the state of Oregon and four in Wyoming.

1. There is also a record: "Bonneville's Fort—Bear River" (no. 10). Where the Captain had his headquarters in 1837 I do not know, but presumably near the rendezvous.

Russell records that Sir William Drummond Stewart and the artist Alfred Jacob Miller were at the gathering; he does not mention the presence of John McLeod, but both W. J. Ghent and Bernard De Voto do so. Miller, as quoted by Russell, records a "Roman saturnalia" and I abandon Tolmie's "friend" in the midst of the revelry.

Between Fort Hall and Fort Boisé John McLeod followed a route which lay considerably south of the one which had been taken between the same points by Wyeth, Nuttall and Townsend; otherwise, from the rendezvous to Fort Walla Walla their paths differed little. Although David Douglas had collected more than once on the northern slopes of the Blue Mountains he had been unable to cross their heights and explore the Grande Ronde Valley. Nuttall seems to have been the first plant collector to enter the valley, in 1834; Meredith Gairdner is said to have been there in 1835; McLeod must have been the third, in 1837.

McLeod may well have learned how to prepare botanical specimens from John Kirk Townsend who, in his turn, must have learned the art during his months of association with Nuttall. Townsend's *Narrative of a journey across the Rocky Mountains, to the Columbia* . . . tells of a trip which he made with McLeod (I believe the McLeod of this chapter) who was just back from the rendezvous of, probably, 1835. Between September 1 and November 30 of that year (when Townsend left the Columbia for the last time) the *Narrative* includes a number of references to McLeod.

With the kind permission of Dr. Ireland I include a transcription of the Tolmie list.

Plants from Mr McLeod Localities &c—

No.   1. Burnt River
      2. Snake R. above Bruneau R
      3. Snake F. on Reeds River
      4.   do.   do.
      5. Pine Creek near Green River
      6. In the most arid plains—as 2
      7. Bannock defile passing between
          Snake & Bear Rivers
      8. Snake River
      9. Near Blue Mts
   10. Bonneville's Fort—Bear River
   11. Bear River, but common throught
   12. Hams Forks.
   13. Pine Creek
   14. Snake Fort
   15. Blue Mts—common throughout
   16. Green River
   17. Pine Creek

18. Thomas's Fork
19. Utalla River
20. Between Burnt & Malheur Rs
21.   "   "
22.   "   "
23.   "   "
24.   "   "
25. Hams Fork
26. Grand Rond
27.
28. A—
28. B—
29. Root tuberose—edible
30. Blue Mts
31.   "   "
32. Green River
33.   "   "
34. Pine Creek
35. Blue Mts clayey soil
36. Snake River
37. Rock River.
38. Bruneau—sandy soil—ht 2
     to 3 feet
39. Green River
40. Snake R. below Salmon Falls
41. Pine Creek
42. Blue Mts—Abundant
43. Between Bruneau & Owyhee Rs
44.   "   "
45. Snake Riv: above Salmon Falls
46. Bannock River
47. Snake Ft
48. Thoma's Forks
49. Pine Creek
50. Swamp at Green River
51.   "   "   "   "   sandy soil
52. Snake Ft
53. Rock River
54. Blackfoot River
55. Henry's Fork

56. Green River—Bitter Root com:
    throught Interior
57. Between Henry's Fork & Smith's
58. do.
59. do.
60. do.
61. Thoma's Fork
62. Blackfoot Rivr
63. In the neighbourhood of the hot springs
64. Snake Fort
65.   "   "
66.   "   "
67. Snake R. below Salmon Falls
68. Bonneville's Fort
69. Snake Ft
70. Green River
71. Snake R. below American Falls
72. Snake Ft
74.   "   "
75.   "   "
76. American Falls—Snake R.
77. Columbia Falls

Anonymous. 1936. John McLeod (fl. 1821–1842).[2] Encycl. Canada 4: 209.

Brosnan, Cornelius James. 1935. History of the state of Idaho. rev. ed. New York. 70, 93.

De Voto, Bernard. 1947. Across the wide Missouri. Illustrated with paintings by Alfred Jacob Miller, Charles Bodmer and George Catlin. Boston.

Ghent, William James. 1931. The early far west. A narrative outline 1540–1850. New York. Toronto. 262.

---

2. Valuable references concerned with the various men by name John McLeod who were employed by the North West and Hudson's Bay Companies, were contained in a memorandum from Dr. Willard E. Ireland dated January, 28, 1955. Although his conclusion in regard to the identity of the man discussed in this chapter is not at variance with mine it seems of value to supply these references. They are: (1) "... bibliographical information supplied in 'Appendix B' to E. E. Rich (ed.), Journal of Occurrences in the Athabasca Department by George Simpson, 1820 and 1821, and Report (London: Hudson's Bay Record Society, 1938), pp. 455-56 ..." (2) "... Biographical Appendix to W. Stewart Wallace (ed.), Documents Relating to the North West Company (Toronto: Champlain Society, 1934), p. 481 ..." (3) "... manuscript Life of John McLeod, Sr (1788-1849) as prepared by his son, Malcolm McLeod." [This is in the Archives of British Columbia.] (4) "Further details concerning his service ... in E. E. Rich (ed.), Minutes of Council Northern Department of Rupert Land, 1821-1832 (London: Hudson's Bay Record Society, 1940) ..." (5) "Details concerning the service of [the John McLeod of this chapter] ... in E. H. Oliver (ed.), The Canadian North-West, its early development and legislative records, (Ottawa, 1915), volume 2 ..."

HOOKER, WILLIAM JACKSON & WALKER-ARNOTT, GEORGE ARNOTT. 1830–1841. The botany of Captain Beechey's voyage; comprising an account of the plants collected by Messrs Lay and Collie, and other officers of the expedition, during the voyage to the Pacific and Bering's Strait, performed in His Majesty's ship Blossom, under the command of Captain F. W. Beechey, R.N. in the years 1825, 26, 27, and 28. London. ["California.—Supplement." 316-409. Issued from 1840–1841.]

MCLOUGHLIN, JOHN. 1941. The letters of John McLoughlin from Fort Vancouver to the Governor and Committee. First Series 1825–38. Edited by E. E. Rich . . . With an introduction by W. Kaye Lamb . . . Toronto. Publications of the Champlain Society 4: 208.

RUSSELL, CARL PARCHER. 1941. Wilderness rendezvous period of the American fur trade. *Oregon Hist. Quart.* 42: 37-39.

TOLMIE, WILLIAM FRASER. 1837. Plants from Mr McLeod Localities &c—[Photostat of this list in the Library of the Arnold Arboretum. Original in Provincial Archives, Victoria, British Columbia].

TOWNSEND, JOHN KIRK. 1839. Narrative of a journey across the Rocky Mountains, to the Columbia River, and a visit to the Sandwich Islands, Chili, &c., with a scientific appendix. Philadelphia.

CHAPTER XXXI

HINDS, SURGEON-NATURALIST WITH BELCHER IN THE SULPHUR, *VISITS THE COLUMBIA RIVER REGION AND, ON REACHING CALIFORNIA, MAKES A TRIP UP THE SACRAMENTO RIVER AND STOPS AT THE PORTS OF SAN PEDRO AND SAN DIEGO*

ON December 24, 1835, two ships of the British Navy—the *Sulphur* commanded by Captain Frederick William Beechey and the *Starling* commanded by Lieutenant H. Kellett—left Plymouth, England. On arrival at Valparaiso, Chile, Beechey found that ". . . his constitution was too much shattered to allow of his continuing the command, subject to such changes of climate as it would necessarily entail . . ." and he was "invalided" and returned to England. The Lords of the Admiralty finally appointed Captain Edward Belcher to take his place; he joined the ships at Panama and took command of the *Sulphur* in February, 1837. Kellett was still in charge of the *Starling*.

The expedition was sent out to verify existing surveys, coastal and other, and to obtain miscellaneous facts of interest to the British Admiralty. Another objective, desirable but of secondary importance, related to the natural history of the countries visited.

"Large collections of natural history cannot be expected, nor any connected account of the structure or geological arrangement of the great continent which you are to coast; nor indeed would minute inquiries on these subjects be at all consistent with the true objects of the survey. But at islands, and even along the coast, to an observant eye, some facts will unavoidably present themselves, which will be well worth recording, and the medical officers of both vessels will no doubt be anxious to contribute their share to the scientific character of the Survey."

Richard Brinsley Hinds, attached to the *Sulphur,* was surgeon-naturalist. Although his task included botany, a second individual was to concentrate upon that subject, and in reading the orders one realizes that there were class distinctions in botanical as well as in naval circles:

"You are to prepare a berth for the botanical collector for plants and seeds for his Majesty's garden at Kew, who is to be borne on the book of the Sulphur for victuals only, and who will mess with the warrant officers; you will furnish him with the means of landing on such parts of the coast of the shores you may visit, to make his collections, when it will not interfere with the Survey."

The name of this collector is only supplied on the last page of Hinds's and Bentham's volume upon the botany of the expedition to which I refer later. He was a "... Mr. Barclay ... sent out by the Royal Garden at Kew ..." In *The botany* Barclay is occasionally cited as the collector of a plant and Belcher quotes him in regard to the flora of the Bay of Magdalena, Lower California, where in November of 1839 the ships remained for eighteen days: "In the vegetable kingdom our botanical collectors were not idle. The following is from Mr. Barclay's report." "Collectors" is plural and *The botany* also credits specimens to a "Dr. Sinclair."

Captain Edward Belcher's *Narrative of a voyage round the world, performed in Her[1] Majesty's ship Sulphur, during the years 1836-1842* was published in 1843 in two volumes. This supplies the route taken by the vessels: after touching at various Central American and Mexican ports and at the Hawaiian Islands they approached their northernmost point at Port Etches in Prince William Sound (where the position of Mount St. Elias was determined) and then, moving southward, reached Sitka, or "New Archangel," on Baranof Island, on September 12, 1837, where they remained until the 27th. Belcher notes that he was received by the Governor at Sitka, Captain Kouprianoff, "... in the warmest manner, and tendered all facilities which the port or arsenal could afford ... he requested I would ... make my own arrangements ... for my observatory or any other pursuits ... his civilities were overpowering. The Sulphur is the first foreign vessel of war that has visited this colony."

Belcher's report on the Russian colony was on the whole a favorable one. He was considerably impressed by the ladies who "... acquitted themselves with all the ease, and I may say elegance, communicated by European instruction ... I believe that the society is indebted principally to the Governor's elegant and accomplished lady for much of this polish ... The lady of Baron Wrangel, I think, was the first Russian lady who ventured so far. The whole establishment appears to be rapidly on the advance, and at no distant period we may hear of a trip to Norfolk Sound (through America) as little more than a summer excursion."

1837

On October 3 the ships reached Nootka Sound. Belcher was disappointed. "At first I doubted my senses, that so small a space could have occupied so much type ..." After he had examined the Sound he reached the conclusion that it could not afford shelter to two vessels. The Indian chief was invited aboard and proved to be married to a descendant of "... the Macquilla or Macquinna of Vancouver ..." Belcher regretted that he had no presents to bestow on the daughter, "Government, or Captain Beechey, not having considered the ladies of sufficient importance, to provide the

1. The *Narrative* records that in December, 1837, on arrival at San Blas, "... a large packet of newspapers, affording us the official intelligence of the accession of our Maiden Queen, Victoria, was on our arrival despatched to us, with dates to September."

Victoria succeeded to the throne on June 20, 1837, and was crowned Queen on June 28, 1838.

presents necessary for their gratification. This is bad policy . . . if *they* are not gratified, we well know the result the world over."

At Nootka "No vestige remains of the settlement noticed by Vancouver . . . not a trace remains of an European." Belcher sailed away on October 9, intending to enter the Columbia River, but the weather ". . . proved boisterous, attended by a long westerly swell . . ." so he continued on to San Francisco and, on the morning of October 19, ". . . made Punta de los Reyes and the Farallones . . . and shortly after midnight, as the moon cleared the eastern hills, we dropped anchor in Yerba Buena Bay."

Belcher, who rarely found *anything* to his satisfaction *anywhere,* was up at daylight ". . . to take a peep at our old ground . . ."[2] and ". . . was surprised to find everything going to decay, and infinitely worse than we found them ten years before.

"Of the revolution, of which we heard much and expected more, not a trace could be observed; it was a sore subject . . . they were evidently aware of their inability to govern themselves: no one stepped forward to attempt it, and they quietly fell back under the Mexican yoke. Another fate attends this country; their hour is fast approaching; harassed on all sides by Indians, who are stripping them of their horses, without which their cattle are not to be preserved; pestered by a set of renegado deserters from whalers and merchant ships, who start by dozens, and will eventually form themselves into a bandit gang, and domineer over them; unable, from want of spirit, to protect themselves; they will soon dwindle into insignificance."

Armed with a letter of "special introduction" to the Padre Presidente of the mission at Santa Clara, Belcher, with Kellett and Hinds, set out to visit him; it was hoped to obtain supplies "instead of taking our chance at the beach."

". . . After much toil, and a night spent in the marshes by the fault of a bad pilot, we reached Santa Clara to breakfast, but were miserably disappointed, the padre being absent at San Josef. The mission is fast falling to decay, and scarcely common civility was shown to us."

On October 24 Belcher left with the *Starling,* a ". . . pinnace, two cutters, and two gigs, to explore the navigable limit of the Rio Sacramento; one of three streams, diverging about thirty miles up the north-western arm of Estrecho Karquines [Strait of Carquinez], where the Blossom's survey terminated."

They started at dawn and made thirty-six miles. On October 26, ". . . the pilot assuring us that she could not be carried further, we stored our boats with as much provision as they could stow, and moved on. We soon found our pilot mistaken, but it was now too late . . . several boats [were] twenty miles in advance . . .

"From former descriptions of the river, I was greatly disappointed at not meeting with either the San Joachim [San Joaquin] or Jesus Maria, equally large streams, said to trifurcate north and south with the Sacramento. These streams may possibly be found upon a closer examination, but no such idea is conveyed . . . on entering the

2. Belcher had served under Beechey in the *Blossom* in 1825–1827, and that ship had been in the Bay of San Francisco in November of 1826.

mouth of the Sacramento which becomes a stream about twenty miles above . . . where we left the Starling . . . our guide appeared quite as much in the dark as ourselves . . . This guide was one of those trained in former days to *hunt for Christians!*[3]

". . . Having entered the Sacramento, we soon found that it increased in width as we advanced, and at our noon station of the second day was about one-third of a mile wide. The marshy land now gave way to firm ground, preserving its level in a most remarkable manner, succeeded by banks well wooded with oak, planes, ash, willow, chestnut, walnut, poplar, and brushwood. Wild grapes in abundance overhung the lower trees . . . All our efforts were directed to reach the head of the stream without delay . . ."

On October 30, at about four in the afternoon, they ". . . found the deep boats stopped at a point where the river forked." After examining the main stream Kellett reported ". . . 'no water for our lightest boats.' The natives also assured us that this was the ford where the hunters cross.

"I landed at 'the Fork,' which was named Point Victoria . . ."

October 31: ". . . the more arduous part of our duty commenced, viz. the trigometrical survey from hence to the junction with the Blossom's Survey at the mouth of the San Pablo. By these observations Point Victoria was found to be in latitude 38° 46′ 47″ north, longitude 0° 47′ 31″ 5 east of the observatory on Yerba Buena; traversing in its meanderings about one hundred and fifty miles."

The present city of Sacramento is situated at the confluence of American and Sacramento rivers (which *may* be the "fork" referred to by Belcher on October 30); its latitude is now given as 38° 34′ North and its longitude as 121° 26′ West. It is said to be situated at the head of navigation for large steamboats but, throughout the year, can be approached by smaller craft. There have been various estimates of the distance traveled up the Sacramento River by Belcher and his party. Brewer merely states that they went "some distance up." Stillman states that von Kotzebue in 1824[4] (on his second visit to California) ". . . in the month of November . . . actually ascended the Sacramento to the latitude of thirty-eight degrees thirty-seven minutes, nearly as high

3. In a footnote Belcher explains: "Boats with soldiers were sent under the direction of the padres to capture Indians and bring them to the missions, where they were made Christians *nolens volens.*"

4. In 1824 Otto von Kotzebue, then commanding the frigate *Predpriatie,* or *Enterprise,* arrived again at the Bay of San Francisco, staying two months—according to Bancroft (using the Russian calendar presumably) from October 8 to December 6, and according to Essig from September 27 to November 25. Among other scientists in the party was Dr. Johann Friedrich Eschscholtz who, on this occasion, seems to have confined his attention to zoology. And apparently Eschscholtz traveled more than on his visit in 1816. Essig records that he and von Kotzebue went to the Santa Clara mission the day after arrival, staying three days (September 28 to 30); that they left for Fort Ross on October 3, traveling to the San Rafael mission in small boats and from there moving on on horseback to Sonoma and Bodega Bay and "northward along the coast." Eschscholtz remained at Fort Ross for a week and then returned to the Bay of San Francisco by boat, putting in at Point Reyes during a storm, and reaching his destination October 12. From November 12 to 23 he traveled by boat up the Sacramento River (it had been intended to explore both that stream and the San Joaquin) but they reached, according to Essig, only "approximately

as where the state capital now stands . . ." and that ". . . the farthest limit of Belcher's explorations fell short of those of Kotzebue, about one-fifth of a mile, allowing the observations of both navigators to have been correct . . ." Essig states that von Kotzebue and Eschscholtz ascended the Sacramento River ". . . approximately to the present site of Rio Vista." If the reference is to Riovista, Solano County, about sixty-five miles northeast of San Francisco, on the west side of the Sacramento River, then von Kotzebue had a considerable way to go before reaching what is now the site of Sacramento.

This was the only inland trip of any importance made by Belcher's party while in California. I quote, therefore, two descriptions of the country: first the one supplied by Belcher in the *Narrative:*

"Throughout the whole extent, from Elk station to the Sacramento mouth, the country is one immense flat, bounded in the distance N. W. by Sierras Diavolo, W. Sierras Bolbones, and E. N. E. to E. S. E. by the Sierras Nievadas, from whence no doubt this river springs, and rises in proportion to the rains and thaws. Our course lay between banks . . . produced by some giant alluvial deposit. Sand did occur at times, but not a rock or pebble varied the sameness of the banks. These were, for the most part, belted with willow, ash, oak, or plane, (platanus occidentalis,) which later, of immense size, overhung the stream . . .

"Within, and at the verge of the banks, oaks of immense size were plentiful. These . . . form a band on each side, about three hundred yards in depth, and within . . . were to be seen disposed in clumps . . . Several of these oaks were examined, and some of the smaller felled. The two most remarkable measured respectively twenty-seven feet and nineteen feet in circumference, at three feet above the ground. The latter rose perpendicularly at a (computed) height of sixty feet before expanding its branches, and was truly a noble sight.

to the present site of Rio Vista." The *Predpriatie* left San Francisco on November 25. Essig records that ". . . Many interesting animals were encountered on the various excursions . . . Eschscholtz took nearly a hundred species of Coleoptera, all new excepting one.

"In summarizing the zoological collections made during the entire voyage round the world, Eschscholtz states that of the 2400 species or kinds of animals taken, 1400 were insects. The insects were deposited in the collections of the University of Dorpat and a considerable number of beetles and a few others were described by Eschscholtz, but his early death, at the age of thirty-eight, cut short his work and left his fine collection to the mercies of his fellow entomologists."

BANCROFT, HUBERT HOWE. 1885. The works of . . . 19 (California II. 1801–1824): 522-525.

BRETSCHNEIDER, EMIL. 1898. History of European botanical discoveries in China. London. 321.

ESCHSCHOLTZ, JOHANN FRIEDRICH. 1836. Histoire naturelle d'un voyage autour du monde, fait avec le Capitaine de Kotzebue, 1823–1826. *Bull. Soc. Imp. Nat.* 1: 19-24.

ESSIG, EDWARD OLIVER. 1933. The Russian settlement at Ross. *Quart. Cal. Hist. Soc.* 12. 208, 209. Reprinted: *Cal. Hist. Soc.* Spec. Publ. 7: The Russians in California. (1933.)

KOTZEBUE, OTTO VON. 1830. A new voyage round the world, in the years 1823, 24, 25, and 26. By Otto von Kotzebue, Post Captain in the Russian Imperial Navy. 2 vols. London. [Volume 2 contains a chapter on California (71-150).]

STILLMAN, JACOB DAVIS BABCOCK. 1869. Footprints of early California discoverers. *Overland Monthly* 2: 260-262.

"All the trees and roots on the banks afford unequivocal proofs of the power of the flood-streams, the mud line on a tree we measured exhibiting a rise of ten feet above the present level, and that of recent date.

"At the period of our examination the river was probably at its lowest . . . at times almost still water; and yet up to our highest position the Sulphur might have been warped or towed by a steamer . . ."

Hinds, writing in the *Botany* of the same trip up the Sacramento, is the first person to mention an interesting tree, the California walnut:

"It was late in the autumn of 1837, when an Expedition up the Rio Sacramento penetrated from San Francisco some distance up the interior. The country exhibited a vast plain, rich in a deep soil, and subject to periodical submersion. Occasional clumps of fine oaks and planes imparted an appearance of park-land. They were already shedding their leaves; a small grape was very abundant on the banks; and we sometimes obtained a dessert from the fruit of a juglans.[5] We had scarcely returned, when a storm covered the maritime range of hills with snow; and this set the final seal on the year's vegetation. On quitting the coast for the interior, we exchanged the evergreen oaks for deciduous species. The latter grow to fine trees with wood of great specific gravity. But the natives have a very pernicious practice of lighting their fires at the bases; and as they naturally select the largest it was really a sorrowful sight to behold numbers of the finest trees thus prematurely and wantonly destroyed. And it is not a country where wood is superabundant; for no sooner is the Oregon crossed than the spruce forests disappear, and the prevailing trees are oaks, which towards the South become gradually less abundant."

On November 18 the party got back to the *Starling* which they had left on October 31 and, writes Belcher, ". . . had the satisfaction, after twenty-three days confinement in the boats, of again luxuriating in a wholesome bed. As the work of each day was entirely completed on paper before we retired to rest, (sometimes at four a.m.) the *severe* part of our labours was here ended."

He wrote further on November 24:

"Having completed our connexion with the Blossom's Survey up to Yerba Buena, we reached the Sulphur . . . having been absent altogether on this interesting but harassing service thirty-one days.

"As far as navigation is concerned, the Sacramento affords every facility for small

5. Ralph E. Smith wrote in 1909:

"According to the latest arrangement by Professor W. L. Jepson, the southern California walnut is to be regarded as *Juglans californica* Wats. . . . The northern nut averages fully twice as large, and the tree has typically a tall regular form approaching that of the eastern black walnut. The origin of the northern California walnut is much in doubt . . . there are only a few places where there is any indication of the tree having been indigenous . . ."

Sereno Watson's description of *Juglans californica* is found in his "Revision of the genus *Ceanothus*, and descriptions of new plants, with a synopsis of the western species of Silene." *Proc. Am. Acad. Arts Sci.* 10: 349, 1875.

craft as high up as the 'Fork;' but I cannot at present perceive any advantage to be derived from taking large vessels above the Starling's position, or even above the creek at the mouth of the Estrecho Karquines, which communicates with the mission of San José, and which, until settlements are made above, will be the extent of traffic, except for timber.

"Taking into consideration the whole port of San Francisco, the Sacramento, and minor streams, there is immense field for capital, if the government could protect its citizens or those inclined to reside. At this moment (December, 1837) they are reduced to almost their extreme gasp . . . they sadly want the interposition of some powerful friend to rescue them. To Great Britain their hopes are directed; why, I cannot learn, but I am inclined to think that it is rather from a pusillanimous fear, and want of energy to stand by each other and expel their common enemies, than from any friendly feeling to Great Britain.

"Besides this, they look with some apprehension upon a power daily increasing in importance—an organised independent band of deserters from American and English whalers . . . These men . . . will, whenever it suits their purpose, dictate their own terms and set all law at defiance. It is distressing to witness the downfall of this splendid port, all the forts in ruins, not even a signal-gun mounted! Such are the blessings of revolution!"

Before departing with the *Sulphur* from San Francisco I quote what Belcher has to tell of the lower reaches of the Sacramento River:

"Near the mouth of the river the soil is entirely peaty . . . The spring-tides overflow all the lower lands, which are well stored with long flag grass, and rushes of great size . . . The ground does not assume a substantial bearing until the flood is overcome by the fresh water; and there the soil is of the finest kind. Roses, arbutus, and other small shrubs flourish luxuriantly, and wild grain[6] produces and re-sows itself, affording perpetual pasturage for deer, &c. During the dry season the natives burn this down, and probably by such means destroy many oak plantations which otherwise would flourish.

"The oak of California does not bear a high character, although it is the same as that used generally on the eastern coast of America, about the same parallel. The ash is excellent, but does not attain any great size. Wild grapes generally prefer it . . . Our friend the plane, however, will not be eclipsed. The timber of this tree is solid, and does not swim; when green it seasons well, and I have found it made good gunwales and timbers for light boats. Laurel, varieties of oak, sumach, pine, &c., we noticed; also the bulbous root termed *ammoles* by the Spaniards, and generally used as a detergent in washing. It is roasted, and used by the natives as food. It has a sweetish taste.

"The grapes are abundant and well-tasted, but small in size and large in seed, therefore not very great luxuries. Some of the acorns were as sweet as chesnuts. The fruits of the hiccory and walnut we occasionally met with, and not having better, we thought

6. Probably *Avena fatua* L., the wild oat; introduced from Europe; common on Pacific coast.

them excellent; but the shells being very thick, and the fruit small, they were as little prized as the grapes . . ."

On November 30 the ships left San Francisco Bay for Monterey where they arrived on December 2. Belcher found "Monterey . . . as much increased as San Francisco had fallen into ruin. It was still, however, very miserable, and wanting in the military air of 1827 . . . There were guns it is true, (about seven,) but they were in a state infinitely more dangerous to those using them, than to those against whom they might have been used . . .

"Yet I find in the restrictions enjoined on poor Douglas that he was in honour bound 'to desist from entering or *taking plans of the fortifications!*' which consisted of a mud wall of three sides . . ."

On December 6 the *Sulphur* quitted Monterey for San Blas [Jalisco]. The *Starling* had been sent ahead for mail. On reaching there they ". . . found no letters; that dreadful damper after long-cherished expectation, and particularly on such a service as the present, where year after year fate may send them without a chance of reaching us."

1838

After surveys along the coast of Central America and elsewhere the ships got back to Panama some ten months later, or in October of 1838. "Here," states Belcher's preface, "may be said to have ended her first cruize . . ." Between October and March of 1839 other surveys were made, ". . . after which the ship moved northerly, repeating her cruize of 1837."

1839

From May 29 to June 16 the ships visited the Hawaiian Islands for a second time. From there the *Starling* was sent to the Columbia River to await the arrival of the *Sulphur.*

Belcher was again at Mount Edgecumbe on July 16 where he found the same governor, and so on. A certain amount of "reserve" was apparent because the Russians ". . . were anticipating hostilities . . ." On July 19 he left for the Columbia River; on July 28 the *Sulphur* ". . . reached the mouth of the Columbia, when Lieutenant Kellett, having descried us, weighed and stood out with the Starling to conduct us in . . . the weather admitted of our entering, otherwise the very imperfect sailing directions might have led us into danger. The shoals in the entrance . . . have most materially changed their features within the last two years . . ."

Both ships went aground, the *Sulphur* floating again at the next tide, the *Starling* losing her rudder and needing repairs. On July 31 ". . . we started . . . for Fort George . . . Off this fort, the well-known 'Fort Astoria,' of Washington Irving, we anchored for the night. It has dwindled considerably since the Hudson's Bay Company took charge, who removed their chief establishment to Fort Vancouver, and allowed it to run to utter ruin. Not a vestige remains . . .

". . . we examined the great fir mentioned by Douglas[7]. . ."

The Belcher *Narrative* continues:

August 1: ". . . we proceeded on our voyage through the intricacies of Tongue Point Channel, and after grounding occasionally, which I take to be according to practice, managed by sunset to find a soft berth for the night, 'on an unknown spot where no bank ought to have been,' according to our pilots. This delayed us one day . . . Our detention occurred close to the 'Pillar Rock,' . . . There is but little to interest one here . . ."

August 5: ". . . we passed round the southern side of Puget's Island, and . . . reached Oak Point, where we anchored for the night. At Puget's Island the scenery may be said to change, the foliage being mixed with ash, willow, alder, maple, &c. I noticed the cypress also amongst the pine, but its timber here is of no value. Indeed I have not heard of the cypress of Norfolk Sound below that latitude, although the same tree in leaf, bark, and other characters, occurs here . . ."

August 7: ". . . we passed Corpse Island and Coffin Mount of Vancouver . . . In the year 1836, the small-pox made great ravages, and it was followed a few years since by the ague . . . skulls and skeletons were strewed about in all directions . . ."

August 9: ". . . after being nearly devoured by mosquitoes, we reached Fort Vancouver . . . as the crow flies, eighty-two miles from Cape Disappointment . . . As to the appellation of Fort Vancouver, it is clearly a misnomer; no Fort Vancouver exists; it is merely the mercantile post of the Hudson's Bay Company . . .

"The attention of the chief[8] to myself and those immediately about me . . . I feel fully grateful for, but I cannot conceal my disappointment at the want of accommodations exhibited towards the crew of the vessels under my command, in a British possession.[9]

7. Possibly this is the same fir which is described by Douglas in "Some American pines" which was published as Appendix VIII of the Royal Horticultural Society's volume, *Journal kept by David Douglas during his travels in North America, 1823–1827* (1914). The tree was a Douglas fir.

"Behind Fort George, near the confluence of the Columbia River . . . there stands a *stump* of this species which measures in circumference 48 feet, 3 feet above the ground, without its bark. The tree was burned down to give place to a more useful vegetable, namely potatoes."

8. The Chief Factor, presumably.

9. At this point it should not be amiss to quote a few comments about Belcher and to refer to the attitude of the Hudson's Bay Company.

Longstaff and Lamb state that "Belcher was unfair to [Chief Factor James] Douglas . . . Nor is it likely that Belcher took much account of the many years that McLoughlin had devoted to building up the herds he saw about the fort . . ." Further: "One wishes that Captain Belcher had been a more attractive personality, for even the *Dictionary of national biography* goes so far as to say that 'Perhaps no officer of equal ability has ever succeeded in inspiring so much personal dislike . . .' It is true that one of the purposes of his

"We certainly were not distressed . . . or I would have made a formal demand . . . I enquired in direct terms what facilities her Majesty's ships of war might expect . . . I certainly was extremely surprised at the reply, that 'they were not in a condition to supply.' . . . As . . . I well knew this point could be readily settled where authority could be referred to, I let the matter rest . . . no offer having been made for our crews, I regretted that I had been led into the acceptance of private supplies; although . . . other officers of the establishment had told my officers that supplies would of course be sent down . . ."

According to Belcher, the relationship of British and Americans was none too good at this period, especially along the Willamette and other rivers; the former, instead of selecting their own countrymen ". . . to afford them spiritual assistance, [had] recourse . . . to Americans,—a course pregnant with evil consequences . . . No sooner had the American and his allies fairly 'squatted,' (which they deem taking possession of the country,) than they invited their brethren to join them, and called on the American government for laws and protection! . . . They are now loud in their claim of right to the soil, and a colony of American settlers was *en route* in the plains when we quitted.

"On the . . . Catlamet [or Cathlamet], they have two missionaries, one Protestant, the other Roman Catholic; but as this is a company's farm, and on the north side, I believe that there is no present fear of intrusion.

"The territory has at length, by dint of British capital and perseverance, attained

expedition was to check the longitudes recorded by Vancouver, but the obvious relish with which he remarks upon the discrepancies uncovered is not much to his credit . . ."

A letter from James Douglas "To the Governor, Deputy Governor and Committee Hon. Hudson's Bay Compy.", dated "Fort Vancouver 14*th Octr.* 1839," which is included in *The Champlain Society Publications,* pictures the Belcher visit from the Company's angle:

"Her Majesty's surveying ships the *Sulphur* and *Starling* visited this River and were employed in its survey from the 16th July to the 12th September. They have not made any discovery of importance, if we except a hitherto unexplored passage from Ft. George leading through the south sands and uniting with the regular channel about a mile west of Cape Disappointment . . . the old passage is preferable . . . The *Starling* met with several accidents in this River, and on different occasions lost two rudders, which we furnished means to replace. We aided their professional researches to the utmost of our ability, at the same time Captain Belcher and Officers were entertained with the kind of attention, pointed out to us in your instructions and by their own peculiar situation."

The literature of the period makes it apparent that the British Company, in addition to conducting its own tremendous organization, was conducting as well what amounted to hotels, travelers' aid societies, banks, hospitals, etc., etc., for the benefit of each and every individual or group arriving at Fort Vancouver or any other Company post. Beneficiaries usually expressed appreciation, but Belcher—obviously impressed with his own importance as an officer in the British Navy—was an exception to this rule. Reading the lengthy correspondence of the Chief Factors, one marvels at their patience in explaining to their superiors in the home office just how each arrival had been treated and so on, *ad infinitum.*

Robert Greenhow's opinion of Belcher's *Narrative* was by no means favorable:

"This large and extensive work, though very amusing to the general reader, abounds in misstatements and inconsistencies, and contains scarcely a single fact or observation of importance with regard to the different places visited. The results of the scientific investigations, especially the geographical positions of many important points, which were determined, doubtless, with the utmost accuracy during the voyage, are omitted."

such importance, that America doubtless is anxious to open the field to her sub-jects; and having her eyes open to what has been virtually lost in an over-reaching at-tempt by the Astor Company, to obtain a value by negociation, for that which in a few hours would have been *public* British property *by capture,* she attempts to disavow the legality of the transfer. This renders the matter still worse, as, had the capture taken place, they might, with some plea of propriety, have looked for its restoration at the peace. But if *bona fide* purchasers are, at this late date, to be marred in their specula-tion . . . then must we bid farewell to good faith in mercantile transactions; for in no other light than that of a plain transfer of private property can it honestly be con-templated.

"The possession of the trading post at Fort Vancouver (not the Fort Astoria, pur-chased from the Americans by the British Company) cannot in any way be ques-tioned in reference to the Columbia territory. That boundary line must first be defined before America ventures to claim any point of the head-quarters of the Hudson's Bay Company."

American settlers on the Willamette—the first had gone there in 1830—numbered, at the time of Belcher's visit, "fifty-four souls, or fifty-four farms," including

"24 Canadians, H. B. C.
20 American stragglers from California.
10 Clergymen, teachers, &c., American Methodist Mission

There are also four other missionary stations, (all American,) [viz.,]
One at the Dalles of this river.
— — Walla Wallah, 25 miles S. of Fort Nez Percez.
— -- Clear Water River.
— — Spokane."

In regard to the relationship existing between the Russians and the British, farther north, Belcher tells us that the latter in 1834 had attempted ". . . to establish a trading post on the river Stikine, which falls into Clarence Straits . . . but the Russians having notice of their intention, had erected a blockhouse, and placed one of their corvettes at the mouth of the river, to prevent their effecting their object."

He reviews the treaty of 1825[1] which reserved to British traders the right to ". . . *'freely navigate all rivers which crossed the line of demarcation.'* " There was too, an-other interesting provision: ". . . neither party under any circumstances should have recourse to force, without first transferring the dispute to their government; a formal appeal was made to Baron Wrangel, at that period Governor of Sitka, but without success."

Before leaving the Columbia, Belcher comments upon the region's trees:

1. The terms of this treaty are given in Belcher's Appendix 2: "Convention between His Majesty, and the Emperor of Russia, respecting the free navigation, commerce, and fisheries in the Pacific Ocean, and the limits on the N. W. coast of America. Signed at St. Petersburgh, February 28th, 1825 . . ."

"The timber of the Columbia, either for spars or plank, cannot be compared to that of the higher latitudes; for topmasts and topgallant masts it is probably as tough though heavier; oak and ash are better. Probably no part of Western America can produce timber of the dimensions grown in the regions of the Columbia and the northern confines of California. Amongst the *drift* trees on the banks of the Columbia, we measured one, one hundred and seventy-four feet in length by twenty feet circumference; many one hundred and fifty by thirteen to eighteen. These of course were washed from its banks and therefore not the largest, which grow invariably in the thickest parts of the wood . . ."

Having been reconditioned, the ships started down the Columbia in early September.

"Having completed the Starling's refit, we commenced our return, surveying the river downwards. We had reached Puget's Island, when she . . . drifted on a snag, (or stump of a tree under water,) and broke her rudder short away . . . On this occasion I merely despatched the requisite officers to Fort Vancouver with fresh demands, and moved downwards with the Starling to Fort George . . ."

When the *Starling* was again in order, the ships ". . . dropped down to Baker's Bay . . ." They "quitted" this on September 14 but ". . . did not clear the heads until the wind failed, compelling us to anchor . . . the strength of the breeze parted our cable . . . I was heartily glad . . . to leave the anchor, and get clear of this dangerous port . . . The Starling . . . met with a similar accident, leaving also her last anchor but one. Heartily sick of this nest of dangers, we took our final look at Cape Disappointment, and shaped our course for Bodega."

On the morning of the 20th they arrived ". . . in a thick fog, off the settlement of Ross, (of which Bodega is the port, distant about thirty miles southerly,) and having stood into fifteen fathoms, we suddenly observed the trees on the heights over the fog, and . . . found ourselves within gun-shot of the settlement."

After acquiring two pilots, the ships "bore away for Bodega." There the *Starling* was left to complete a survey and the *Sulphur* moved on to San Francisco. Entering the harbor on September 22, Belcher stopped at his "old anchorage." The governor of Fort Ross was at San Francisco and Belcher, finding that ". . . the same feeling of distrust, arising from the anticipated rupture between Russia and Great Britain, appeared to produce more than ordinary reserve . . . sailed again that evening . . ."

On September 23 he was back at Bodega ". . . after an absence of only forty-eight hours . . ." The bay was surveyed but Belcher did not attempt to visit Fort Ross. His description of the settlement was based upon information derived from a friend who resided there and from his own "observation by telescope." He then started back for San Francisco. On the 24th the fog ". . . cleared sufficiently to round Punta de los Reyes the same afternoon, and to enable us to get well towards the mouth of San Francisco . . ."

September 25: ". . . About ten we got sight of the land, and ran in . . . we passed on to our anchorage at Yerba Buena. Our stay here merely enabled us to take in a supply of bullocks and vegetables, and complete observations confirming those of former visits, when we moved on to Monterey."

On arrival at Monterey on October 5 Belcher ". . . had no hesitation in anchoring." On October 6 they had completed their observations and left for Santa Barbara.

October 9: "We reached our anchorage in the bay of Santa Barbara . . . The customary guide in approaching the coast is the 'kelp line,' which generally floats over five to seven fathoms . . . on its verge there is no danger . . . It is the fucus giganteus, and sufficiently strong to impede steerage, if it takes the rudder . . .

"We were fortunate in landing comfortably, and by four o'clock . . . we moved on towards San Pedro . . .

"Having been requested to afford surgical assistance to a young man of one of the first families, who had fractured a limb, and was unable to be moved from Buenaventura, the Starling was despatched for that purpose, as well as to obtain its position, the ship standing close along the coast towards the bay of San Pedro . . ."

October 11: ". . . on opening the bay of San Pedro, we noticed several lights . . . vessels at anchor . . . we anchored amongst the vessels . . . masters of two American ships immediately came on board to offer their services . . . Inside of the small island in the bay is a very snug creek, but only accessible to small craft . . .

"Having completed this bay, we moved to the bay of San Juan, despatching the Starling . . . to examine the island and anchorage of Santa Catalina[2] rejoining us at San Diego."

October 13: ". . . we dropped anchor in the bay of San Juan . . . This bay was examined and surveyed. The anchorage is foul under five fathoms, is unprotected, and the landing bad.

"The mission is situated in a fruitful-looking sheltered valley, said to abound in garden luxuries, country wines, and very pretty damsels, whence the favourite appellation Juanitas . . .

"At four we quitted San Juan . . ."

On October 17 they reached San Diego, ". . . having anchored within the kelp the night previous . . . The chief drawback is the want of fresh water, which even at the presidio, three miles from the port, is very indifferent . . . The mission is situated up a valley, about seven miles from the presidio. There they have . . . the finest water . . . Several varieties of cacti, particularly of the Turk's head variety, are abominably abundant . . .

"Since the missions have been taken from the padres, and placed under the ad-

2. Had the surgeon-naturalist of the expedition, Richard Brinsley Hinds, taken this opportunity to visit Catalina Island on the *Starling* he, rather than William Gambel, would have been the first botanist to do so.

ministradores, they have fallen entirely into decay and ruin . . . it is not improbable that the whole country will ere long either fall back into the hands of the Indians, or find other rulers . . . The garden . . . famed in former days for its excellence, has now fallen entirely into decay, and instead of *thousands* of cattle, and horses to take care of them, not twenty four-footed animals remain!"

On October 22 Belcher left San Diego[3] ". . . intending to land and fix the position of Cape Colnett," in Baja California.

After further surveys off Mexico, etc., the expedition started homeward across the Pacific in January of 1840, reaching Singapore in October of that year. Here Belcher received orders to proceed to China where his ship "took an active part in the operations against the Chinese, till nearly the close of the year 1841, when she sailed for England . . ." She went around the Cape of Good Hope and reached Spithead, England, on July 19, 1842.

Before closing the outline of Belcher's voyage, it may be of interest to note that, in June of 1840, while involved in a none too enjoyable survey of the "Feejee" Islands—with cannibals in evidence—Belcher spent eighteen interesting hours with Captain Charles Wilkes of the United States Exploring Expedition, who had reached the same islands from Antarctica.

The Appendix to Belcher's *Narrative* contains a long dissertation by the surgeon-naturalist Richard Brinsley Hinds, entitled "The regions of vegetation; being an analysis of the distribution of vegetable forms over the surface of the globe in connexion with climate and physical agents." This is the first floristic analysis[4] of the vegetation of our west coast made as a result of actual field-observation and, as such, a landmark in the botany of the region. Hinds's opening "Advertisement" explains that

"Her Majesty's ship Sulphur was the school in which I more particularly studied geographic botany. Preconceived views, and results drawn from the perusal of the writings of scientific travellers, were here practically tested. Her extensive voyage, and rapid transition from one portion of land to another, afforded rich and most favourable sources of comparison . . . Climate is the basis on which the earliest data must be founded, and with the liberal use of instruments, observations on temperature and humidity were in time collected. These, with observations on the physical condition of the surface, furnish us with many of the circumstances which govern the distribution of the flora of the world. What I have accomplished under these heads has been collected together, and form the subject of a lengthened paper, which, through the

3. Hinds, attached to the *Sulphur*, seems to have been the fourth botanist to visit San Diego; he had been preceded there by Menzies (1793), by Coulter (1832) and by Nuttall (1836). Deppe, a German botanist, had reached San Diego overland from Lower California but the year of his arrival, 1831 or 1832, is uncertain and, moreover, he is reported to have confined his collecting in California to birds and mammals.

4. A "floristic analysis" being merely a comparison of the plants of one region with those of another, or a study of the vegetation in relation to climate and general environment.

liberality of the proprietors of the Annals of Natural History, have been already published[5]. . .

"The result of these investigations was the developement of regions of vegetation, and which had their origin and stability in previously established views. At the same time, I do not insist that these are natural, but that taken in their entireness, they present in situations circumstances of remarkable individuality. In the meantime they will be found eminently useful in studying the features of vegetation, and more particularly in leading the subject to the naturalization of plants—the great end and aim of geographic botany."

Hinds's paper is divided into regions: "America, North," "America, South," "Australia," "Africa," and "Europe," and these in turn are subdivided. Our interest centers in "America, North," and its subdivisions "II. The North-West America region" (pp. 331-335) and "V. The California region" (pp. 345-348). I found much that he has to say of interest and shall quote from his analyses of both these subdivisions; first in regard to the northwest:

"Extent.—The rocky mountains and Pacific Ocean on the east and west, and 68° N. lat., and the Columbia river to the north and south, enclose this region.

"Physical characters.—The surface is irregular, consisting entirely of mountain and valley, without the least pretensions to plain . . . The soil is often rich, from the great accumulation and rapid decomposition of vegetable remains.

"Climate.—Being freely exposed to winds from the ocean, and westerly winds prevailing, the climate is considerably modified. Compared with Europe, it is far cooler for the latitude . . . rainy days are very frequent . . . At the Columbia river in 46° N. lat., being the southern limit . . . the quantity of rain 53.6 inches, and snow is rarely seen.

"Flora.—. . . soil is abundant, and the investing vegetation vigorous. The constant moisture favours premature decay, and thus the trees are early undermined, and falling . . . in the forest, cover the ground in vast numbers . . . Within the tropics I have never seen anything equal to the scene of devastation the northern part of this region presents; trunks of trees, of great length and clear of branches, are seen on all sides strewed in tiers, and covered with a dense agamic vegetation. It would often seem that they were unable to attain a good old age, as, always exposed to moisture from the repeated rains, they have yielded to its influence immediately that period of life arrived when the activity of vegetation diminishes. Here everything is moist, the soil is completely saturated, mosses and lichens are in their liveliest vigour, and much of the surface is swampy.

---

5. In 1842 Hinds published "The physical agents of temperature, humidity, light and soil, considered as developing climate, and in connexion with geographic botany" (*Ann. Mag. Nat. Hist.* 9: 169-189, 311-333, 469-475. *Suppl.* 521-527). And, in 1846, in the same journal, "Memoirs on geographic botany" (15: 11-30, 89-104). In 1842 he had published three papers on similar subjects (*London Jour. Bot.* 1: 128-135; 312-318; 669-676).

"Tracing the regions from Prince William's Sound in 6° north latitude[6] to the east, and then to the south, the whole will be found to be covered with one vast forest. It extends to the north as far as the boundary line,[7] and to the south . . . to the Columbia river, where a sudden change occurs, and which is a very decided line of demarkation between this and the California region . . .[8] Elsewhere the forest, though dense, consists of but few species; abies has three, which, with cupressus thyoides, constitute all the larger trees, whilst some smaller are contributed by cratægus, salix, cerasus, betula, and to the south diospyros.[9]

"The undergrowth of shrubs is so extremely luxuriant, that it appears a chief characteristic, and, regardless of the shade of the forest, flourishes in great vigour. These shrubs are chiefly the species of vaccinium, menziesia, rubus, and ribes, which, though numerous in species, have a multitude of individuals. Towards the south, lonicera involucrata, mahonia glumacea, symphoria racemosa, gautheria [sic] shallon are superadded, and particularly aspidium munitum, a handsome fern, very social, and covering portions of the surface to the exclusion of others. Another peculiarity is, that though some of the genera appear through several degrees of latitude, they are continued by new species; thus ribes, rubus, rosa, and lupinus, are seen everywhere in the region, yet each species had but a small range, and is immediately succeeded by another.

"Relations.—Two plants are common which are eminently distinguished for their large foliage, and as members of families of a warmer climate; panax horridum, a fine shrub with large showy leaves . . . and dracontium camtschaticum, with a very different habit . . . Mimulus guttatus has a wide habitat . . . The herbaceous plants are of families common to these latitudes, though both cruciferæ and umbelliferæ are scarce . . . The southern part mixes but feebly with the California region, and the features are preserved singularly intact even to the banks of the Columbia. Here quercus commences with many others, abies ceases suddenly, and pinus partly supplies its place, nor disappearing from the elevated lands till it arrives in the vicinity of Panama. A collection of plants from its northern part contained about one half common with the north of Europe, and a similar number with Siberia."

Turning now to Hinds's analysis of "The California region":

"Extent.—After crossing the Columbia river from the north, an entirely altered vegetation commences. The dense compact forests of abies cease suddenly, and are supplanted by an open country, spotted by occasional clump[s] of oaks, and the river lines fringed by platanus, fraxinus, juglans, and salix. The outline of the region may

6. Obviously a typographical error, for the sound is north of latitude 60°.

7. Presumably the southern Russian boundary line cutting across "New Caledonia" at 54° 40′ north latitude.

8. Here there follows a description of the vegetation of Prince William Sound which I omit.

9. The persimmon does not occur spontaneously north of southern Lower California.

be traced up the Columbia river to the Rocky Mountains, which it meets in about 50° N. latitude, and is continued along them to the south, till approaching the commencing waters of the Colorado, it runs along its course to the gulf of California. The remaining portion is circumscribed by the Pacific Ocean.

"Physical characters.—In the northern part the surface is regular . . . there are some well-watered fine alluvial plains, without a rock or stone. Occasionally ranges of low mountains traverse it . . . not of sufficient elevation to affect materially the vegetation, but support some groves of pinus lambertiana and abies religiosa; pinus rigida prefers the plains. The broad plains which separate them are often overflowed in the winter, which with their deep rich soil renders them very fertile. To the south, the scenery is wild and rugged, nearly altogether mountainous, the ranges running from north to south. Not a tree is to be seen, but there is a moderate sprinkling of a more lowly and interesting vegetation . . . There is no soil nor fertilizing streams, water being very scarce.

"Climate.—To the north the climate is even and temperate; the winters are mild and of short duration, and snow appears on the loftier hills; and the summers have an agreeable warmth, with the atmosphere clear and transparent . . . The rains are soon over, but during their continuance deluge the country . . .

"Flora.—The finest part of the region is to the north, where an open country prevails, varied by patches of trees of noble growth. Of the oaks, two species are deciduous, and two evergreen. The latter are confined to the neighbourhood of the sea coast between 38° and 34° N. latitude.[1] The other trees are not numerous, and are chiefly comprised under platanus, acer, pavia, juglans, cornus, laurus regia, and the aromatic tetranthera californica . . . The undergrowth consists of several species of rubus, ribes, lupinus, rhus, vaccinium, arbutus, and lonicera; and such is the variety of some of these, that a new species may be met with almost every hundred miles. Vitis, scarcely expected, grows abundantly on the margins of some of the rivers. Shrubby compositæ prevail throughout, but are in the greatest intensity towards the centre of the region; and in the more arid parts cacteæ and euphorbiaceæ are particularly numerous, with a few leguminosæ. Cacteæ are not seen further north than 34°; here also is the limit of ricinus communis, of course introduced, as is phoenix dactylifera, a few large trees of which may be seen about San Diego, but only yielding a sour fruit.

"As characteristic peculiarities of the region may be mentioned, its great aridity, general scarcity of trees, superior prevalence of cacteæ, compositæ and euphorbiaceæ, great number of plants with lactescent juices, and with fragrant foliage, the frequent developement of the flowers and leaves at different periods, and the general small range of its species. The negative features consist in the scarcity of ferns, mosses, and fungi, none of which exist in the southern part, except perhaps the latter during the rains. Lichens, with sickly aspects, occasionally cling to the trees or rocks.

1. The latitudes, respectively, of San Francisco and of Los Angeles.

"Relations.—... In establishing a comparison between the western and eastern parts of the American continent, a superiority must be assigned in the forest trees to the east, and in the herbaceous vegetation to the west."

Reading the above, one remembers that Hinds was attached to a coastal survey and got no farther into the interior of our Pacific northwest than the region of Fort Vancouver on the Columbia River and only into the interior of California on his short trip up the Sacramento River.

The botany of the voyage of H. M. S. Sulphur, under the command of Captain Sir Edward Belcher, R.N., C.B., F.R.G.S., etc. during the years 1836–42 was, according to the title page, "Published under the authority of the Lords Commissioners of the Admiralty," and "Edited and superintended by Richard Brinsley Hinds, Esq., Surgeon, R.N., attached to the expedition." The same page records "The botanical descriptions by George Bentham, Esq." It appeared in six numbers, issued from 1844–1845:[2] I. (pp. 1-16, plates 1-10); II. (pp. 17-48, plates 11-20); III. (pp. 49-72, plates 21-30); IV. (pp. 73-96, plates 31-40) in 1844; V. (pp. 97[3]-144, plates 41-50) in 1845; VI. (pp. 145-195, plates[4] 51-60) in 1846. The publisher was Smith, Elder and Company, London.

Three regions are considered: "I. North-west America" (pp. 1 and 2), "II. California" (pp. 2-57) and "Tropical America, from Mexico to Guayaquil" (pp. 58-181) and, of these, only the first two relate to the region of my story.

No collections are described from "North-west America," and the reason is explained, by Bentham presumably, in a paragraph following introductory remarks by the editor Hinds:

"The number of species collected at these several places was altogether about 200; many of them are as yet very scarce in herbaria, but none entirely new, the whole line of coast having been already pretty well explored by English, American, and Russian botanists. They have also been well described in several works, amongst which we may particularly refer to Hooker's Flora Boreali-Americana, Bongard's Végétation de l'Isle de Sitcha and to two general works now in the course of publication, Ledebour's Flora Rossica for the Russian possessions, and Torrey and Gray's Flora of North America, which includes the whole of the territory visited. An enumeration of the species collected by the Expedition would therefore be superfluous."

2. *See: Jour. Arn. Arb.* 11: 243, 244 (1930).

3. Bentham records (p. 182): "The Editor having left Europe shortly after publication of the Fourth Part of this Work, the last two parts, commencing at p. 97, have been completed by the Author of the Botanical Descriptions ..."

4. The handsome plates of *The Botany* are inscribed: "Drawn from Nature and on Stone by Miss Drake." Wilfrid Blunt comments that "Of Miss Drake 'of Turnham Green' (*fl.* 1818–47) almost nothing seems to be known beyond her work ..." He thought highly of her ability and devotes pp. 214-216 of *The art of botanical illustration* to comments upon her contributions to a number of botanical books and periodicals.

Although the conclusions reached by Hinds in his paragraph upon "North-west America" do not differ greatly from those supplied in "The regions of vegetation . . ." already quoted, they are differently expressed and I quote his comments in part:

"The portion of country visited, and which may be accepted under this popular denomination, extends from 60° 21′ to 46° 19′ N. lat. Port Etches, in King William's Sound, and the Columbia River are situated at the extremes, and Port Mulgrave, Sitka, and Nootka Sound, are intermediate. These places were severally visited during the autumn of 1837, with the exception of the Oregon or Columbia River, the examination of which was deferred till the months of August and September in 1839. The whole territory, though extensive, is remarkably uniform in its physical character and natural productions. The climate is far more moderate than on the eastern coast, not being liable to those great vicissitudes, nor ever known to display any great range of temperature. The number of rainy days in the year is very great . . . The whole country is bold and mountainous, intersected by deep and moist valleys, and is every where covered by a gloomy forest of spruce. These vast forests offer a scene which powerfully arrests attention. The trees are often of enormous dimensions; stretching upwards, with scarcely a branch, to where the eye almost fails to follow them, with enormous trunks, very deceptive till brought within the scope of our experience by means of the tape-line; beneath, a most luxuriant undergrowth everywhere abounds . . . But over these the influence of the moist climate is unceasing. It most probably hurries through a rapid existence the more lowly shrubs, and its effect on the trees is very marked. None are seen to attain any great age, that is, none have that appearance; but when the vigour of life is past, they rapidly yield to the constant influence of the moist atmosphere, soil, and investment of mosses and lichens, and soon fall to the ground, which in some places they occupy in great numbers . . . the variety in species is not great; and some will occur with multitudes of individuals covering a very large space. It is curious to observe how tenaciously some genera extend throughout this territory, though continually represented by a different species. This is particularly conspicuous with *Vaccinium, Rubus, Ribes, Rosa,* and *Lupinus.* The former has several deciduous species towards the northern portion, but towards the south they become neat evergreen shrubs, with a myrtle-like foliage. To a European, the general features of the vegetation are entirely such as he is familiar with, only modified by the character of the climate and country; with the exception, that there are two common plants, *Panax horridum* and *Dracontium camtschaticum,* which differ so entirely from the surrounding vegetation, as to exert a very considerable influence on the physiognomy."

Hinds's introduction to "II. California." begins thus:

"It usually happens, that the season most favourable for the nautical examination of a country, is that which is least so from acquaintance with its vegetation; and this proved particularly the case with California. Indeed, the residents would almost check our pursuits, by representing that the season was past when Botany should be pursued, whilst they dwelt with much animation on the rich vesture the country assumed after the period of the rain.[5] It was late in the autumn of 1837, when an Expedition up the Rio Sacramento penetrated from San Francisco some distance into the interior[6] . . .

"But Upper Californa had already been tolerably examined; and it was our good fortune to touch rapidly at several places on the coast of Lower, or New California, during October and November, 1839, and here we trod in no footsteps, as none had preceded us. I shall confine myself to a few brief abstracts of notes written at the time, as they may convey fresher and more correct impressions; merely premising, that the two Californias are countries differing in many essential particulars, and that San Diego is their political place of separation."

The remainder of Hinds's introduction to "II. California" is divided into a discussion of particular regions: *"New California,* Oct. 15th.", *"San Diego,* 32° 29′ N. lat.," *"San Quentin,* 30° 21′ N. lat.," *"San Bartolomè,* 27° 40′ N. lat., Oct. 29th.," and *"Bay of Magdalena.* 24° 38′ N. lat., Nov. 2nd." Obviously, from dates and latitudes mentioned, both the paragraph on *"New California,"* and the one on *"San Diego,"* concern regions which are now included in the present state of California although the "political place of separation" demanded their inclusion in the "Lower, or New California" of Hinds's day.

Hinds wrote of *"New California":*

". . . We have touched already at several places on the coast. Everywhere it has much the same character, being almost destitute of wood or even of shrubs; where there happens to be any of the former, consisting of evergreen oak. At this season the soil is dried and cracked, and the vegetation extremely arid. Yet even here after a day's wanderings we return with a dozen or more different species in flower . . . The prevalence of *Compositæ* is truly great, and one is surprised at the variety of aspect the flowers are capable of assuming. They are not all blue Asters, or yellow Coreopsis, as is seen in an English garden, but have many varied colours and tints. At different places on the coast the species vary, and their total is perhaps considerable. My attention has been directed to the distribution of *Cacteæ,* by meeting with two species for the first time at San Pedro, as we were descending the coast. Their limit here then may be stated as 34° N.; but in the Rocky Mountains, a species has been recorded at 49°. On the plains of the Missouri four species attain to 48°. In Europe we have a representative in 44°. And in Chiloe [an island off the west coast of Chile] they probably cease at about 42° S."

Hinds wrote of "San Diego":

". . . The vegetation generally is highly aromatic, not certainly always fragrant or

5. Still a popular comment a century later; not only in California, but in all parts of our southwest where the desert bursts into bloom after spring rains.

6. I have already quoted Hinds's report upon the trip.

agreeable, but abounding in strongly scented properties. It continues to consist of a low shrubby character, amongst which multitudes of quail, rabbits, and hares love to nestle. *Compositæ* greatly prevail, and are numerous even as species. *Cacteæ* are now common, and three species have been noticed; there are a few lactescent plants, and many of the shrubs have tough leathery leaves—intelligible indices of the prevailing climate. *Ricinus communis* is seen for the first time, and a few trees of *Phoenix dactylifera*. The latter bears no fruit fit for eating,[7] it being very sour. One of the trees was tall and fine grown, and stood a solitary monument amid a lowly growth. None of the stunted evergreen oak have been seen below Santa Barbara, and their Northern limit is the waters of San Francisco; thus ranging on the coast between 38° and 34° N. lat."

The remainder of Hinds's comments upon "California" relate to what is now understood as Lower California. Bentham adds a paragraph:

"The flora of this country is particularly interesting, as forming the connecting link between the north-west and the tropical vegetation. The species collected in Upper California are for the most part already published in the Botany of Captain Beechey's Voyage, or in Torrey and Gray's Flora. As species they are generally peculiar to the districts where they are found, but belong chiefly to the same genera or groups as the north-western plants. But the Lower Californian plants . . . are almost all either tropical or Mexican forms, but chiefly new as to species . . . I am not aware of any South Californian plants having been yet published." [8]

Bentham prefaces his treatment of the botanical collections with the comment that ". . . the species already described are merely mentioned, with a reference to some work where their character may be found. The new species will be described as fully as the state of the specimens admits of." At the end of his enumeration he has this to say:

"The principal collection . . . was that made by the Editor himself, Mr. Hinds, through whose liberality the original specimens have been deposited in the subscriber's herbarium. This extends over the whole of the stations mentioned in the work. A second collection was made likewise at the whole or the greater part of the stations, by Mr. Barclay, the collector sent out by the Royal Garden of Kew, and, through the

---

7. *Phoenix dactylifera* Linn. is the date palm, native of Arabia and North Africa, and the specimen seen by Hinds was of course introduced. As one who has enjoyed the delicious fresh dates of southern California, I comment on Hinds's adjective "sour"! Since the date palms growing in California in his day were all seedlings—no two alike—the fruit was doubtless as described. The date palms now cultivated in California are all from imported, selected varieties, and propagated from "suckers" or offsets. They were introduced through the instrumentality of the United States Department of Agriculture within the present century.

8. Nor, so far as I know, was anything more published upon the flora of Lower California until, in 1860, John A. Veatch brought out his paper entitled "About Cerros Island." The island is usually called Cedros at the present time.

kindness of Sir William Hooker, the subscriber has been enabled to avail himself of a set of these plants deposited in Sir William's herbarium."

After referring to the specimens from "Western Tropical America," a "considerable portion" of which were gathered "by Dr. Sinclair," Bentham ends thus: "The original specimens of the Orchidaceæ are in Dr. Lindley's herbarium, and those of the *Ferns* in Sir William Hooker's. The total number of species gathered in the voyage amounts to near two thousand, of which above four hundred were previously undescribed."

The Bentham and the Hooker herbaria are now a part of the general herbarium at Kew. The Lindley orchid herbarium is also at Kew.

BANCROFT, HUBERT HOWE. 1886. The works of . . . 21 (California IV. 1840–1845): 142-147.

———— 1884. The works of . . . 28 (Northwest coast II. 1800–1846): 611, 612.

———— 1886. The works of . . . 29 (Oregon I. 1834–1848): 232-234.

BELCHER, EDWARD. 1843. Narrative of a voyage round the world, performed in Her Majesty's ship Sulphur, during the years 1836–1842, including details of the naval operations in China, from Dec. 1840, to Nov. 1841. Published under the authority of the Lords Commissioners of the Admiralty. By Captain Sir Edward Belcher, R.N., C.B., F.R.A.S., &c. Commander of the expedition. 2 vols. London.

BREWER, WILLIAM HENRY. 1880. List of persons who have made botanical collections in California. *In:* Watson, Sereno. Geological survey of California. Botany of California 2: 555.

BRITTEN, JAMES & BOULGER, GEORGE SIMONDS. 1931. A biographical index of deceased British and Irish botanists. ed. 2. London. 149.

ESSIG, EDWARD OLIVER. 1933. The Russian settlement at Ross. *Quart. Cal. Hist. Soc.* 12: 191-209. *Reprinted: Cal. Hist. Soc.* Spec. Publ. 7: The Russians in California.

GREENHOW, ROBERT. 1845. The history of Oregon and California, and other territories on the north-west coast of North America: accompanied by a geographical view and map of those countries, and a number of documents as proofs and illustrations of the history. ed. 2. Boston, 377, *fn.*

HINDS, RICHARD BRINSLEY. 1843. The regions of vegetation; being an analysis of the distribution of vegetable forms over the surface of the globe in connexion with climate and physical agents. By Richard Brinsley Hinds, Esq. Surgon [*sic*] to the expedition. *In:* Belcher, E. 1843. Narrative of . . . 2: 323-460.

———— & BENTHAM, GEORGE. 1844. The botany of the voyage of H. M. S. Sulphur, under the command of Captain Sir Edward Belcher, R.N., C.B., F.R.G.S., etc., during the years 1836–42. Published under the authority of the Lords Commissioners of the Admiralty. Edited and superintended by Richard Brinsley Hinds, Esq., Surgeon, R.N. attached to the expedition. The botanical descriptions by George Bentham. London.

LAUGHTON, J. K. 1885. Edward Belcher. *Dict. Nat. Biog.* 4: 121, 122.

LONGSTAFF, F. V. & LAMB, WILLIAM KAYE. 1945. The Royal Navy on the northwest coast, 1813–1850. *Brit. Col. Hist. Quart.* 9: 11-16.

McLOUGHLIN, JOHN. 1943. The letters of John McLoughlin from Fort Vancouver to the Governor and Committee Second Series 1839–44. Edited by E. E. Rich . . . With an introduction by W. Kaye Lamb . . . Toronto. *Publications of the Champlain Society* 6: Appendix A. 227, 228.

SARGENT, CHARLES SPRAGUE. 1891. The Silva of North America. Boston. New York. 2: 44, *fn.* 1.

SMITH, RALPH ELIOT. 1909. Report of the plant pathologist and superintendent of southern California stations, July 1, 1906 to June 30, 1909. *Cal. Agric. Exp. Sta. Bull.* 203: 27-29.

STILLMAN, JACOB DAVIS BABCOCK. 1869. Footprints of early California discoverers. *Overland Monthly* 2: 263.

CHAPTER XXXII

GEYER ACCOMPANIES NICOLLET ON HIS SURVEYS OF THE BASINS OF THE UPPER MISSISSIPPI AND MISSOURI RIVERS

UNDER earlier decades I have told how Cass and Schoolcraft in 1820, and Schoolcraft in 1831 and 1832, searched for the sources of the Mississippi River. Also how Keating and Long in 1823 ascended St. Peter's, or Minnesota, River to its headwaters and, after descending the Red River of the North to the recently established boundary between British and American domains, raised the American flag on the 49th parallel at Pembina, North Dakota. Each of these exploratory expeditions had included a plant collector.

It still remained to chart the territory drained by the aforementioned and other great rivers and the United States government selected a Frenchman, Joseph Nicolas Nicollet, who is said to have emigrated to North America in 1832, to perform the task. Nicollet was well qualified by training and, after his arrival in this country, had spent five years before his appointment in 1838, in exploring these very regions. In his own words, after ". . . having explored the Alleghany range . . . and having ascended the Red river, Arkansas river, and to a great distance the Missouri river . . ." he had determined to ". . . undertake the full exploration of the Mississippi river from its mouth to its very sources." It was after ". . . the expiration of this long (and, as I found it, arduous) journey . . ." that he was asked by the War Department to command an expedition enabling him ". . . to complete, to the greatest advantage . . . the construction of a geographical and topographical map of the country explored." His instructions, dated April 7, 1838, were from J. J. Abert, Lieutenant Colonel in the Corps of Topographical Engineers.

The account of Nicollet's travels—"Report intended to illustrate a map of the hydrographical basin of the upper Mississippi River"—is not long but is rich in substance; it was published in 1843 in the Senate Documents (2nd Session, 26th Congress, no. 237). The accompanying map (said to be reduced, but nonetheless large) is extremely valuable to anyone interested in the region and period. In his "Memoir" of 1859 G. K. Warren refers to it as ". . . one of the greatest contributions ever made to American geography." When Nicollet died in September, 1843—aged, according to an obituary notice, ". . . it is supposed about forty eight . . ."—he had not completed the introduction nor revised the "Report" in essential ways. It is a descriptive account of his travels, not a journal, and Nicollet's itinerary cannot be followed with precision; nor do the routes traced on the map indicate the direction taken, and with exceptions, dates are omitted. Nicollet's assignment covered the years from 1836 through 1839[1] and portions of the map seem to be applicable to that period.

Nicollet's travels during 1838 and 1839 have a botanical interest for he engaged at his own expense ". . . the services of a practical botanist, Mr. Charles Geyer . . ." in order that he might make a collection of plants.

He was Karl Andreas Geyer, a German[2] botanist, who was born, according to Coville, in Dresden, Saxony, November 30, 1809, the son of a market gardener. He reached North America in 1835 and remained nine or ten years, accomplishing his most important work in the decade of 1840–1850. On arrival in this country he had worked independently from 1835 to 1838 but seems to have accomplished little. When W. J. Hooker undertook the publication of Geyer's later collections (of 1843 and 1844) he evidently lacked information as to the man's early years in this country; Geyer sets this information forth in a series of letters[3] and these obviously formed the basis for Hooker's editorial comments. I shall quote from the letters themselves, rather than from Hooker's paraphrased version, altering the paragraphing only. He wrote Hooker from Kew, August 13, 1845:

"I made a first botanical tour from New York to the Missouri territory in 1835, as far as the Big Nemahaw, lower Platte. It turned out abortive on account of fever and maltreatment by a party of Indians. I ventured alone among them with one man only.

"On my way down to St. Louis from that unlucky excursion I met Chev. J. N. Nicollet on board the steamer of the Am. Fur Company, and here I got acquainted with that gentleman now probably no more. He called on me to join him on his expedition to the source of the Mississippi in 1836 & 37 but I was so discouraged that I did not like to venture again among the Indians . . ."

There is a slightly fuller account of this trip in a letter which Geyer wrote George C. Thornburn on December 28, 1845:

"When I left New York on the 1st of March 1835, my course was directly to the far west, and on the 18th of April I was already on the western borders of the state of Missouri; I equipped myself and made a journey to the Pawnee-loups Indians on the Big Nemahaw and lower North Fork of the Platte river, got in some difficulty with the Indians, left them almost destitute of every thing, sick . . . and barely did I bring my life back to Missouri, it took me a long time to recover and I embarked from the

1. Warren wrote: "The years 1838 and 1839 were spent in explorations in Minnesota . . . Mr. Nicollet had nearly completed the map, and written a portion of his report when death put an end to his labors before he was enabled to finish it, or to revise what had previously been written. The report does not, therefore, do justice to the surveys, and it is impossible to specify the routes he pursued except for the years 1836, 1838, and 1839, and somewhat imperfectly for these, even though I have consulted his original notes in the Topographical Bureau . . ."

2. In his *Rocky Mountain naturalists* (p. 214, 1950. Denver) Joseph Ewan refers to him as an "Austrian botanical collector . . ."

3. For Geyer's letters see p. 770, *fn.* 1, where they are enumerated.

mouth of the Kansas River for St. Louis in the Septbr, in company with the celebrated astronomer and topographer Nicollet and Sir Charles Murray Viscount Dunmore from Scotland. In St. Louis I found to my displeasure that all my letters had failed me, and to occupy myself with advantage and to make my living also I applied myself to printing, to acquire at the same time the English language of which I was much in need."

Geyer, I regret to say, appears to have had a propensity for getting into "some difficulty" with almost everybody, not only Indians. His letter to Hooker quoted above now turns to the years spent with Nicollet:

". . . in 38 & 39 I was with him and we surveyed the Missouri as high up as the Little Missouri and almost the whole of that immense country (now Dacotah and Ioway Territory) between the Missouri and Mississippi. Had not the principal collection I made been lost I believe I would have gained a good deal of credit from these travels. Fremont[4] was also with Nicollet during that time."

Geyer's letter to Thornburn is again a trifle more explicit:

"I had enough for a long time on my first journey to the wilderness and remained stationary at St. Louis until March 1838, then I got a letter from . . . Nicollet from Washington city, inviting me as an officer of the western exploring expedition for the botanical department, under order of the Secretary of War. This I accepted and I filled, (I think) my place to the satisfaction of my patrons for several consecutive expeditions; to the sources of the Mississippi, throughout the whole territory between the Missouri & Mississippi up to the line of British America, and the Missouri River to the Yellow Stone: I held my place up to the 1st of July 1840 . . ."

I shall quote Warren's outline of Nicollet's travels in the years 1838 and 1839; the outline is concentrated into two paragraphs (p. 41 of the "Memoir"). Warren does not mention the presence of Geyer. First the journey of 1838:

"In 1838 Mr. Nicollet . . . started from Fort Snelling, accompanied by Lieutenant Frémont; they ascended the St. Peter's or Minnesota river to the mouth of the Waraju, and passed up the valley of this river to its source in the Côteau du Prairie. Continuing to the westward they examined the source of the Des Moines river, and the Indian red pipe stone quarry; thence turning north they examined Lake Benton; traveling west from this point they struck the Big Sioux river, crossed over and examined Lakes Ti-tanka-he, Preston, Poinsett, and Abert; and leaving the Big Sioux again at the mouth of Redwood creek, took a northeasterly course to Lake Tizaptonan; thence they proceeded down the Intpah to Lac qui Parle. They now ascended the St. Peter's to the Manka Re Osey river. Here they turned westward along this stream, passing from it to the sources of the Second and Third forks; the latter they followed to the Izuzah,

4. According to Abert's orders to Nicollet, "Mr. Fremont will be assigned to you as assistant." Nicollet turned over to Frémont the ". . . reconnoissance of the country traversed each day, or rather the survey of our route, by land or by water . . ." and pays tribute to ". . . the talents which he displayed for this branch of service, and the activity and accuracy which have always characterized whatever he has had occasion to perform under my directions . . ."

whence they crossed over to Big Stone lake. They also examined Lac Traverse, and returned to Fort Snelling by the St. Peter's river. Mr. Nicollet placed the source of the Big Sioux (which he did not visit) about twenty-five miles too far north; it is now known to head in Lake Kampeska."

Interpreting the above journey in terms of a modern map, it is evident that, starting from Fort Snelling, near the confluence of the Minnesota and the Mississippi, Nicollet ascended the Minnesota to the "Waraju river"—which appears to be the Big Cotton-wood River, entering the Minnesota from the west about opposite New Ulm—and followed that stream to the "Côteau du Prairie." [5] He was still in the state of Minnesota (its southwestern corner) when at "the source of the Des Moines river" and when at "the Indian red pipe stone quarry"—now the Pipestone National Monument, Pipestone County—and when at Lake Benton, Lincoln County.

When Nicollet reached the Big Sioux River he must already have moved into southeastern South Dakota where the "Lakes Ti-tanka-he, Preston, Poinsett, and Abert" are situated in Brookings, Kingsbury and Hamlin counties. When he left the Big Sioux "at the mouth of Redwood creek" [in southeastern Hamlin County?] and "took a northeasterly course to Lake Tizaptonan," he must have crossed Deuel County and reëntered the state of Minnesota in Lac qui Parle County.

The "Intpah" River down which Nicollet now proceeded is the Lac qui Parle River which empties into the southern end of the lake of the same name (which is merely an enlargement of the Minnesota River). The Minnesota was ascended to "the Manka Re Osey river"—from its location on Nicollet's map this appears to be the Yellow Bank River—and when he turned "westward along this stream" Nicollet again entered South Dakota, in Grant County.

The "Izuzah river" and its "Second and Third forks" appear from their positions on Nicollet's map to be the Whetstone River (with its branches); this enters the southern end of Big Stone Lake (drained by the Minnesota River) which forms the boundary between Grant County, South Dakota, and Bigstone County, Minnesota.

From Big Stone, or Bigstone, Lake, Nicollet proceeded north and "examined Lac Traverse," from which the southernmost tributary of the Red River of the North begins its journey northward. From Lake Traverse Nicollet returned to Fort Snelling by way of the "St. Peter's River."

Nicollet's journey of 1839 is described by Warren thus:

"Mr. Nicollet and Lieutenant Frémont again started from St. Louis on board the American Fur Company's steamer Antelope, April 4, 1839, bound for Fort Pierre, at which place they arrived June 12. The course of the river was sketched throughout most of the distance as the boat ascended. Their design being to explore Mini-

---

5. This elevated region, usually called the Coteau des Prairies, is included in the states of North and South Dakota and separates the basins of the Mississippi and the Missouri rivers.

Its northern portion, as indicated on Nicollet's map, lay between St. Peter's River on the east and the "Riv. à Jaques," or James River, on the west.

wakan lake, a party was organized at Fort Pierre. They took the field on the 2d of July, and proceeded in a northeast course, striking the Rivière à Jacques, or James river, at the old trading-houses called the 'Oakwood Settlements.' They explored the valley of James river as far north as Butte aux Os, or Bone Hill; thence they struck northeastwardly to the valley of the Shayenne Oju river. This valley they followed as far as the parallel of 47° 45′, when they crossed the stream and traveled northwest to Miniwakan or Devil's lake. Having examined all its shore except the northwest extremity, they returned to the Shayenne river, and crossed the high divide separating it from the Red river valley on the east. Traveling south, near the sources of the west branches of Red river, they recrossed the Shayenne near latitude 46° 30′, and continuing in nearly the same direction passed the sources of Wild Rice river, and examined the numerous lakes about the head of the Côteau du Prairie. Falling upon the sources of the Izuzah river, they returned, by that stream and the St. Peter's river, to the settlements."

Nicollet's explorations of 1839 were, for the most part, considerably west of and reached much farther north (or beyond the 48° parallel) than those of 1838; they are clearly indicated on his map and, from July 4 to August 19, the route is dated at intervals.

Fort Pierre[6] where Nicollet spent from June 12 to July 2 was situated on the west side of the Missouri River and to the north of the Bad, or Teton, River, which enters the larger stream from the west, in Stanley County, South Dakota. On his arrival at the post Nicollet reported that, from St. Louis, they ". . . were sixty-nine days in ascending a distance of 1,271 miles, which, on the Mississippi, with a steamboat of the same power, could have been accomplished in twelve days."

"Miniwakan lake," which it was planned to explore, was Devil's Lake, lying between Benson and southern Ramsey counties, North Dakota. Starting on July 2 they proceeded northeastward, reaching the James River ("Rivière à Jacques") "at the old trading-houses called the 'Oakwood Settlements' " in southern Brown County, South Dakota. According to Nicollet's map, the site was on the west side of James River, between the Muddy (entering from the east) and the "Mocasin" (entering from the west).

From there, proceeding up the James River, they crossed into North Dakota, and on July 18 were "as far north as Butte aux Os, or Bone Hill," which is shown on Nicollet's map on the eastern side of the James River, not far north of Bone Hill River. On modern maps a creek of this name enters the James in Lamoure County. From there Nicollet—according to Warren—moved "northwestwardly"[7] to the valley of the "Shayenne Oju river," which usually appears as the Shayenne, or Cheyenne, on

6. "Fort Pierre Chouteau, or simply Fort Pierre, is the upper limit of my navigation of the Missouri . . ." (Nicollet, "Report," p. 41.)

7. This is an error; they moved *northeastward*.

modern maps; the map of the "Report" indicates that he moved up the western side of this river until, at 47° 45′ north latitude, he crossed to the other side—the north side, for here the river runs almost from west to east; the crossing must have been in Griggs County. From there Nicollet proceeded "northwest to Miniwakan or Devil's Lake" and, from August 1 to 6, examined "all its shore except the northwest extremity." He then returned to the "Shayenne Oju" and crossed "the high divide separating it from the Red River valley on the east." He then proceeded almost due south, crossing the upper waters of streams entering the Red River from the west and on August 24 crossed the "Shayenne Oju" near latitude 46° 30′, in what is now Ransom County. Still moving south Nicollet passed "the sources of Wild Rice river," and "examined the numerous lakes about the head of the Côteau du Prairie."

While examining "the numerous lakes about the head of the Côteau du Prairie," he must have crossed from extreme southeastern North Dakota (Sargent or Richland county) into extreme northeastern South Dakota (Marshall, Day or/and Roberts counties). Having done this, the map of the "Report" indicates that he turned east—between latitude 45° 30′ and latitude 45°—and reached the Big Stone Lake; he then followed its western shore southeastward through Roberts County.

Near the southern end of this lake he encountered "the sources of the Izuzah river," or the Whetstone River, and followed it downward to its junction with the Minnesota, or "St. Peter's," River. He must have reëntered Minnesota in Big Stone County. He then descended the Minnesota "to the settlements" which, presumably, meant the region of Fort Snelling. At the Whetstone River his explorations of 1839 seem to have connected with his more southerly ones of 1838.

Nicollet's "Report," detailing the physical, geological and other aspects of the country examined, contains occasional references to the flora of particular regions but the important treatment of the subject is contained in Appendix B of which he wrote in his introduction:

"Appendix B is a catalogue of plants, for which I am indebted to that eminent botanist, Dr. James [sic] Torrey, to whom I am proud of an opportunity—brought about by the liberal and disinterested intercourse which characterizes American savans, but perhaps more especially the gentleman whom I have now the honor to name—of submitting my collection. The catalogue has been arranged in accordance with the system adopted in the publication of the American Flora, by Drs. Torrey and Grey [sic]."

"In order to obtain this collection, I engaged, at my own expense, the services of a practical botanist, Mr. Charles Geyer. It does not appear now as complete as it was at one time, owing to the loss of a case containing nearly one-half of my original collection. As it is, I have still reason to believe that it will be no invaluable contribution to the natural history of the American territory. I owe my thanks to Mr. Geyer for the fidelity with which he served me."

The title of the paper is "Catalogue of plants collected by Mr. Charles Geyer, under the direction of Mr. I.[8] N. Nicollet, during his exploration of the region between the Mississippi and Missouri rivers: by Professor John Torrey, M.D." Torrey's opening paragraph describes the region where the collections were made:

". . . It lies between the Mississippi and Missouri rivers, embracing two extensive tracts: —

"1. The Coteau des Prairies, and the Mississippi and the Missouri valleys; consisting of prairies throughout, interspersed with woods and lakes, and the soil of which is alternately sandy, gravelly, or clayey, in the character of an erratic deposit. Its elevation above the sea is from 1,000 to 2,000 feet, and it extends from the 39th to beyond the 48th degree of north latitude.

"2. The extensive basin of the *rivière Jacques* . . . and the prolongation of the Coteau des Prairies, north of the Shayen-oju that empties into the Red river of the North. The basin of the *rivière Jacques* is a vast prairies, situated between the Coteau des Prairies, and the Coteau du Missouri,[9] and is from sixty to eighty miles west of the former. It is sparingly wooded, and its level above the sea is from 1,200 to 1,400 feet, or nearly 600 feet lower than the two Coteaux between which it lies. The northeastern portion is a region of salinas, including the great salt lake called Mini-wakan, or Devil's lake, together with all the headwaters of the rivers that empty into the Red river of the North, on its west side. The soil is sandy in every part, and the timber is found only along the water-courses and the borders of the lakes.

"It is well to remark, that there are two Shayen rivers—one emptying into the Red river of the North, and properly named *Shayen-oju;* the other pouring its waters into the Missouri a little below the 45th degree, and is called by the French, without any adjunct, *Shayen river.* It is the Washtey, or Good river of the Sioux.[1] But, in reference to the *habitale* of the plants mentioned in the catalogue, they have been principally collected within the region of the former river."

After a paragraph concerned with certain geologic formations, the introduction to the paper ends:

"Finally, although this collection is not sufficiently extensive, owing to the loss of a large portion of it between the rapids of the Des Moines and St. Louis, to justify any general views of the distribution of the principal families of plants in the region explored by Mr. Nicollet, its catalogue will, nevertheless, I trust, be a valuable contribu-

8. The title-page of the "Report" uses the same initial, as does Warren in his "Memoir."

9. On Nicollet's map the "Plateau du Coteau du Missouri" is shown running in much the same direction as the "Plateau du Coteau des Prairies"—that is from northwest to southeast; the portion included on the map extends from latitude 47° 30′ to 44° 30′; to its west is the Missouri and to its east the James River. The "Plateau du Coteau des Prairies" extends from 46° north latitude to 42° on Nicollet's map.

1. The river "Sheyenne" (of the modern map) enters Red River of the North from the southwest, in Cass County, North Dakota. The "Cheyenne" enters the Missouri from the southwest, between Armstrong County (north) and Standley County (south), South Dakota; Nicollet's map shows its lower reaches only.

tion to the geographical distribution of American plants, as well as for the number of new species it adds to the Flora."

The catalogue does not cite the year of collection although month and day are cited for each species. A few species were noted as undetermined and a *Helianthus* and two *Cirsium* are noted as new species with a question. *Alisma Geyeri* from "Muddy margins of ponds near Devil's lake," is noted as new, also *Muhlenbergia ambigua* from "Stony banks of Okaman lake, Sioux country," and *Elymus,* "(or a new genus between *elymus* and *hordeum.*)" had been found in "Heavy ferruginous loam on the Missouri, Jacques, and Shayen-oju rivers."

Geyer's letters to Hooker refer with admiration to Nicollet. On August 17, 1845, he wrote:

"Nicollet was an extraordinary man. In his person he resembled Michelangelo to a remarkable degree.[2] He was capable of any great undertaking, he fondled his friends like a grandfather, had an uncommon knowledge of the human heart, ever kind, never losing sight of his dignity and dangerous when touched in the heart, like every Parisian gentleman.

"The literary world ought to call out for the publication of Nicollet's manuscripts . . . No doubt they are in the hands of Dr. J. J. Ducatel of Baltimore, or perhaps at the Jesuit College of St. Mary's, Baltimore. If they are at the latter place[3] they will sleep for ever if nobody calls for publication . . ."

Strange it is, but true, that about two years after it was published, Geyer had not yet seen a copy of Nicollet's "Report" (nor, presumably, the Torrey "Catalogue" enumerating the plants which he had collected). He wrote Hooker on August 13, 1845:

"If the report is published and in your possession of Nicollet's travels, you would highly oblige me by allowing me a few hours of perusal . . . One evening would be enough to gratify myself."

Hooker evidently complied promptly with Geyer's request, for five days Geyer wrote him again:

"I am very much obliged to you for the perusal of these documents, especially the map. This is perhaps the most correct map that ever has been laid down. I spent a whole night and travelled in my mind the whole region over again, partly with delight, partly with grief at the thought that the principal reason of that expedition is no more. Many a new plant which I collected in that vast region got lost with that collection in 38 and will be found again by another. The most remarkable was a species of Botry-

2. Geyer had every opportunity to know what Nicollet looked like but his familiarity with the appearance of Michelangelo (1475–1564) is less understandable!

3. Grace L. Nute wrote in 1945:

"A hundred years have passed, and Nicollet's great collection of scientific papers are still unpublished. They were lost for many years. Then, during World War I housecleaning days, they were discovered in a corner of the State Department of Washington, D. C. They are now in the Library of Congress."

chium, 2 inches high, with a delicate triternate frond, in the valley of St. Peter's River, several Trifolia, Psoraleae, Pentstemons, etc. . . . Of that map only three hundred have been printed for members of Congress and different departments. This I mention to you as to the value of the same. The others are on a small scale."

Anonymous. 1843. Miscellanies 8. Death of Mr. J. N. Nicollet. *Am. Jour. Sci. Arts* 45: 404.

Coville, Frederick Vernon. 1941. Added notes on Carl A. Geyer. *Oregon Hist. Quart.* 42: 323, 324.

Dellenbaugh, Frederick Samuel. 1914. Frémont and '49; the story of a remarkable career and its relation to the exploration and development of our western territory, especially of California. Maps. New York. London. [Chapter 1 tells something of Frémont's early work as a topographer and includes a short sketch of Nicollet.]

Drury, Clifford Merrill. 1940. Botanist in Oregon in 1843–44 for Kew Gardens, London. *Oregon Hist. Quart.* 41: 182-188.

Geyer, Karl Andreas. [For his manuscript letters *see* p. 770, *fn.* 1.]

Hooker, William Jackson. 1845. [Editorial comments prefacing] Notes on the vegetation and general character of the Missouri and Oregon Territories, made during a botanical journey from the state of Missouri, across the south-pass of the Rocky Mountains, to the Pacific, during the years 1843 and 1844; by Charles A. Geyer. *London Jour. Bot.* 4: 479-482, and *fn.* (p. 479).

Nicollet, Joseph Nicolas. 1843. Report intended to illustrate a map of the hydrographical basin of the upper Mississippi River, made by I. N. Nicollet, while in employ under the Bureau of the Corps of Topographical Engineers. *U. S. 26th Cong., 2nd Sess., Sen. Doc.* 5: pt. 2, No. 237, 142 pp. Map. Washington.

Nute, Grace Lee. 1945. "Botanizing" Minnesota in 1838-39. *Conservation Volunteer* 8: no. 44, 5-8.

Torrey, John. 1843. Catalogue of plants collected by Mr. Charles Geyer, under the direction of Mr. I. [*sic*] N. Nicollet, during his exploration of the region between the Mississippi and Missouri Rivers. *In:* Nicollet, J. N. 1843. Report intended to . . . Appendix B. 143-165.

Warren, Gouveneur Kemble. 1859. Memoir to accompany the map of the territory of the United States from the Mississippi River to the Pacific Ocean, giving a brief account of each of the exploring expeditions since A.D. 1800 . . . *U. S. War Dept. Rept. expl. surv. RR. Mississippi Pacific* 11: 39-41.

1840-1850

Introduction to Decade of

1840–1850

IN the early years of this decade American settlers, in ever-increasing numbers, were moving into lands west of the Mississippi River. Those participating in the great overland migrations had, at first, the Oregon region as their goal but soon, and before the discovery of its gold in 1848, were turning southward into California. Near the end of the decade, or in 1847, the Mormons established their colony between Great Salt and Utah lakes—an advantageous situation since, as Frémont pointed out, it was ". . . intermediate between the Mississippi valley and the Pacific Ocean, and on the line of communication to California and Oregon." And, to the south and southwest, Americans were pushing ever westward into Texas.

During the opening years of this decade those Americans who proceeded beyond the regions acquired by the Louisiana Purchase were entering territory which was owned, or by virtue of occupation had long been controlled, by foreign powers—and I omit any mention of Indians and their tribal lands. Understandably, the advent of an extraneous population caused concern to those with established interests and friction between these elements was on the increase. In the northwest the "Oregon question"— relating to the boundary between British and American possessions—was becoming acute; in the south and southwest and in California—where Spain, and after Spain Mexico, was in possession—the situation was leading gradually towards war.

In 1846 the "Oregon question" was adjusted and our national domain was enlarged by the acquisition of the present states of Washington, Oregon and Idaho, as well as by portions of northwestern Montana and western Wyoming. From the Lake of the Woods westward to the Pacific Ocean, the forty-ninth parallel of north latitude was henceforth to constitute the boundary between British (Canadian) and United States possessions.[1]

The annexation of Texas by the United States in 1845, was, to a large extent, the cause of the Mexican War of 1846–1848. By this annexation, and in terms of the modern map, the United States government gained possession of Texas, the Oklahoma "Panhandle," a southwestern portion of Kansas, New Mexico east of the Rio Grande, as well as portions of southeastern and central Colorado, and south central Wyoming. In 1848, after the Mexican War, its territorial possessions were further increased by the addition of New Mexico west of the Rio Grande, and Arizona—with the exception of southern portions of both states which were purchased from Mexico in 1853—as

1. It was only in 1872 that the extension of this boundary into the coastal waters of the Pacific Ocean was finally adjusted.

well as by Utah, Nevada, California, western Colorado and southwestern Wyoming.[2] By 1853 Mexico no longer had any territorial possessions west of the Mississippi River.

By the end of this decade therefore, or actually by 1853, the exterior boundaries of the trans-Mississippi west existed as of today and those Americans who advanced westward did so, not as "foreigners," but as citizens seeking homes within the territorial possessions of the United States.

Only belatedly did the United States government demonstrate an interest in the welfare of those of its citizens who, of their own initiative,[3] had migrated to the far west. For only in 1841, or nearly forty years after the memorable journey of Lewis and Clark, did the United States Exploring Expedition reach by sea what is now our Pacific coast, visit posts of the Hudson's Bay Company in the northwest, take a look at Mexican possessions in the San Francisco region and, on its eventual return to Washington, report its observations. The aims of the expedition were purportedly scientific but Wilkes's reports certainly indicate that other matters of perhaps greater national interest were taken into consideration as well. Also, beginning in 1842 and intermittently until 1847, the government employed the topographer Frémont to explore overland routes to Oregon and to California. His reports and maps were put to immediate service by the emigrants—the Mormons are said to have utilized both on their westward journey. When Frémont reached the Columbia River in 1843 he had fulfilled the assigned task of connecting his overland journey with that of the United States Exploring Expedition which had reached the same point by sea two years before.

The United States government had, however, demonstrated a greater interest in the territory lying west of southern portions of the Louisiana Purchase. Even before the Mexican War it had sent topographers to explore and to map routes into what is now northern New Mexico, and during and subsequent to the conflict it had extended these surveys eastward into northern Oklahoma and Texas, southward along the Rio Grande, and westward, following the Gila River, into southern California. After 1845 many military expeditions, both topographical and protective, were sent into Texas. Indeed, by the end of my story, or at the opening of the decade of 1850–1860, routes

2. *See:* Paullin, Charles O. 1932. *Atlas of the historical geography of the United States.* Wright, John K., *editor.* Carnegie Institution of Washington Publication No. 401. Plate 46C ("Territorial acquisitions 1783-1853."). *Also:* Ghent, William James. 1931. *The early far west. A narrative outline 1540-1850.* New York. Toronto. Map, opp. p. 82 ("The original United States, with territorial accessions. From *Geological Survey Bulletin* No. 817.") *Also:* Goodwin, Cardinal. 1922. *The trans-Mississippi west (1803-1853).* New York. London. Map, opp. p. 504 ("Land acquired by the United States").

3. In his *History of the American frontier 1763–1893* (p. 350) F. L. Paxson comments that "Oregon and Deseret were two spontaneous colonies beyond the Rocky Mountains that owed nothing to Congress for their foundation. They were joined in the autumn of 1849 by a third, California, which framed a constitution for itself at the old Spanish village of Monterey, and demanded immediate admission as a State."

into and across the trans-Mississippi west were already mapped and being used.[4] They were not highways in the modern sense and to follow them required courage and intelligence; many of those who lacked these qualities fell by the way or returned whence they had come—Stansbury, among many, paints a vivid picture of what such journeys meant in 1849. Nonetheless the topographers were making headway throughout the far west whether north, south, east or west.

The emissaries of the descriptive botanists—the plant collectors—were participating in many of these westward movements. Brackenridge, assigned to the United States Exploring Expedition, visited portions of the northwest (much of it already explored by Douglas), and between Oregon and California entered regions where no botanist had hitherto penetrated. Frémont is said to have collected plants on all his expeditions before 1850, and these took him across the Sierra Nevada at more than one latitude, around and into the arid lands of the Great Basin, and into many other regions where the plants were still unknown. Men whose names are famous in botanical annals were in the field in the decade of 1840–1850: Gambel, Wislizenus and Fendler, all of whom reached Santa Fe; Geyer, who crossed the continent and on his way collected in the Rocky Mountains and in the "Oregon territory"; Hartweg, who worked in California; and Lindheimer who, because of his botanical knowledge and because he concentrated upon a limited field in Texas, accomplished great things. There were many other collectors; of some little is known along botanical lines; but because their names appear in the records they doubtless contributed something. Certain it is that the botanists on the home front were receiving, from most sections of the country west of the Mississippi River, as much material as they could handle—probably more.

Although written at a later date, a suggestion—which has also the ring of a protest—by the overworked Asa Gray might well have been applied, I surmise, to what he was experiencing in the 1840's. After explaining the labor involved in writing a flora, Gray continues:

"Of course we rely, very much indeed, upon the continued coöperation of all the cultivators of botany in the country; and it is gratifying to know that their number is increasing ... All can help on the work, and all are doing so, by communication of specimens and of observations. Those within the range of the published Manuals and Floras get on—or should get on—with only occasional help from us. They should send us notes and specimens to any amount; but they should not ask us to stop and examine and name their plants, except in special cases, which we are always ready to take up. Those who collect in regions as yet destitute of such advantages may claim more aid, and we take great pains to render it: partly on our own account, that we

4. In his biography of John Torrey, Rodgers quotes a letter written by George Engelmann dated from St. Louis on May 13, 1845: " 'You can have no idea how near we here consider ourselves now to Oregon & California: we mentally travel with those thousands of emigrants, and begin to think the Rocky Mts not much further off than the Alleghanies.' "

may assort their contributions into their proper places, partly for the encouragement of such correspondents, who otherwise would not know what they have obtained, and who naturally like to know when they have made interesting discoveries.

"But the scattered and piecemeal study of plants is neither very satisfactory nor safe. And it involves great loss of time, besides interrupting that continuity and concentration of attention which the proper study of any group of plants demands . . ."

Although, over the years, the United States government had permitted plant collectors to accompany some of the expeditions which it had sent into the field, it had not approached the problem of scientific participation (botanical participation certainly) in what might be called a generous spirit. We have seen examples of this more than once. Men such as Gray and Torrey exerted pressure in Washington and, as a result, their field-workers were usually given military protection but scarcely more than that; even in 1849 Wright, accompanying an army contingent, was obliged to proceed on foot[5] from San Antonio to El Paso. Yet Wright exemplifies, in a sense, a change of attitude on the part of the government toward the pursuit of botany. For, having acquired a great many collections during the progress of the United States Exploring Expedition and having established a place to house them, the Washington authorities began to take an interest in acquiring more; and the Smithsonian Institution contributed one hundred and fifty dollars to Wright's trip, and Gray's "Plantæ Wrightianæ texano-neo-mexicanæ" of handsome format was issued as one of its *Contributions*. A set of Wright's plants seems to have been the *quid pro quo*.

In this decade the botanical papers which were being published were somewhat different in content from those which had first recorded collections from our region; the taxonomists now possessed sufficient material to draw conclusions relative to plant distribution and their treatises are no longer mere lists of genera and species and their description—as many earlier ones had been—but are replete with citations of comparable material, often from regions far removed from those which their papers purport to consider. The authors, who had always had a knowledge of plant-ranges as an objective, were now in a position to present facts thereon. The need for such a compilation as the *Index Kewensis* (not to appear until 1895,[6] but now one of the great labor-saving essentials of any botanical institution) was already apparent.

Two great floras treating of North American plants were published in this decade, one in its entirety, the other never completed. They brought into convenient compass the sporadically published plant data which had been issued over the years since my story began.

The first was Sir William Jackson Hooker's *Flora boreali-americana; or, the botany of the northern parts of British America: compiled principally from the plants col-*

---

5. As a matter of fact, the best way to collect plants *is* on foot!

6. The title-page of the first issue (volume 1) records an important sponsor: "Compiled at the expense of the late Charles Robert Darwin under the direction of Joseph D. Hooker by B. Daydon Jackson."

lected by Dr Richardson & Mr Drummond on the late northern expeditions . . . to which are added . . . those of Mr Douglas, from north-west America, and of other naturalists . . . Publication of the work began in 1833 and was completed in 1840. Of its importance Asa Gray wrote in 1866:

". . . Although denominated 'the botany of the Northern parts of British America,' it embraced the whole continent from Canada and Newfoundland, and on the Pacific from the borders of California, northward to the Arctic sea. Collections made in the British arctic voyages had early come into his [Hooker's] hands, as afterwards did all those made in the northern land expeditions by the late Sir John Richardson, Drummond, etc., and the great western collections of Douglas, Scouler, Tolmie, and others, while his devoted correspondents in the United States contributed everything they could furnish from this region. So that this work marks an epoch in North American botany, which now could be treated as a whole."

The second great flora of this decade appeared in 1843, when Torrey and Gray published the second, and what was to be the last, volume of *A flora of North America, containing abridged descriptions of all the known indigenous and naturalized plants growing north of Mexico, arranged according to the natural system.* Publication had begun in 1838, and more had appeared in 1840. A multitude of reasons—all explained by Gray at a later date—contributed to its noncompletion. Five hundred pages of volume two were devoted to the *Compositae;* although his letters complain of being "half dead with Aster" (Mrs. Gray refers to his being " 'in the valley of the shadow of the Asters' "), there can be no doubt that this great family of plants had been meat and drink to Asa Gray for many years.

I know no better picture of what had been accomplished by the plant collectors—and by the systematists who had the knowledge, the industry and the indomitable will to bring unity and clarity out of the chaos of material which their field-workers had amassed—than Gray's statement of what his, and Torrey's, *Flora of North America* signified, from a number of angles. For, although issued at the beginning of the present decade, the treatise summarized the accomplishment of the sixty years considered in my story and presaged the work of many years to follow. Gray's paper upon the *Flora,* largely retrospective, was read in 1882. The author begins:

". . . I am to speak of the attempts made in my own day, and still making, to provide botanists with a compendious systematic account of the phaenogamous[7] vegetation of the whole country which the American Association [for the Advancement of Science] calls its own . . .

"Only two Floras of North America have ever been published as completed works, that of Michaux and that of Pursh[8] . . .

7. The phanerogams are plants with manifest flowers.

8. Under the decades in which these were published I have mentioned what Gray had to say of André Michaux's *Flora boreali-americani* (1803) and of Frederick Pursh's *Flora americae septentrionalis* (1814).

"I must omit all mention of more restricted works, even such as Nuttall's 'Genera of North American Plants,' . . . also the 'Flora Boreali-Americana' of Sir William Hooker, which . . . was restricted to British America . . ."

Gray then comments upon what the task of writing such a book had involved:

". . . the undertaking has become more and more formidable with the enlargement of geographical boundaries and of the number of species discovered. As to the increase in the number of species to be treated, we have by no means yet reached the end.[9] The area, that of our continent down to the Mexican line, we trust is definitely fixed, at least for our day . . .

"The area which Pursh's Flora covered was, we may say, the United States east of the Mississippi, with Canada to Labrador, to which was added a couple of hundred species known to him outside these limits northeastward.

"Torrey and Gray's Flora took the initiative in annexing Texas, ten years before its political incorporation into the Union;[1] although the only plants we then possessed from it were certain of Drummond's collections. California was also annexed at the same time, on account of Douglas's collections, and those of Nuttall, who had just returned from his visit to the western coast, which he reached by a tedious journey across the continent over ground in good part new to the botanist. Douglas had already made remarkably full collections along a more northern line. The British arctic explorers, both by sea and land, had well developed the botany of the boreal regions, and Sir William Hooker was bringing out the results in his Flora of British America. Of course our knowledge of the whole interior and western region was small indeed, compared with the present; and the botany of a vast region from the western part of Texas to the California coast was absolutely unknown, and so remained until after the publication of the Flora was suspended.[2]

"As to the number of species which Torrey and Gray had to deal with, I can only say that a rapid count gives us for the first volume about 2200 *Polypetalæ;*[3] that there are one hundred and nine species in the small orders which in the second volume precede the *Compositæ*; and that there are of the *Compositæ* 1054. So one may fairly conclude that if the work had been pushed on to completion, say in the year 1850, the 3076 species of Pursh's Flora of the year 1814 might have been just about doubled.

9. Alaska had been purchased by the United States fifteen years before Gray read his paper and was presumably included in this estimate; I am told that its flora is now (1954) better understood than was the flora of California at the turn of the century.

1. Elsewhere Gray makes it clear that he was referring to plant distribution, or phytogeography: ". . . As they clearly belonged to our own phyto-geographical province, Texas and California were accordingly annexed botanically before they became so politically."

2. The second and last volume was issued, as stated, in 1843, or before many of the collectors of the decade of 1840–1850 had gone into the field: Frémont had not yet made any trip to the west coast, nor had Geyer; Fendler had not visited Santa Fe, nor had Emory crossed New Mexico and Arizona into California; Wright had still to make his trip across Texas to El Paso, and so on.

3. Gray included, in the large group of many-petaled plants, the family of composites.

Probably more rather than less; for if we reckon from the number of the *Compositæ,* and on the estimate that they constitute one-eighth of the phaenogamous plants of North America, instead of 6150, there would have been 8430 species known in the year specified."

This estimate is in a sense a summary of what the technical botanists, basing their facts upon the collections of their field-workers, had learned of the plants of the North American continent up to 1843. I believe that I am correct in assuming that a large proportion of the increase in new genera and species after 1790 should be credited to the trans-Mississippi west, the region upon which the great taxonomists had been focusing their attention for more than fifty years.

I have many times referred to the discovery by our collectors of "new genera" and "new species." To determine what plants constituted such novelties had not been the simple task that one might suppose. Gray, in the paper just quoted, has something to say of the difficulties involved:

". . .The incoming of additional specimens may at a glance settle doubt as to the validity of a species; but new specimens are as apt to raise questions as to settle them; more commonly they raise the question as to the limitation and right definition of the species concerned, not rarely, also, that of their validity. When one has only single specimens of related species, the case may seem clear and the definition easy. The acquisition of a few more, from a different region or grown under different conditions, almost always calls for some reconsideration, not rarely for reconstruction. People generally suppose that species, and even genera, are like coin from the mint, or bank notes from the printing press, each with its fixed marks and signature, which he that runs may read, or the practised eye infallibly determine. But in fact species are judgments—judgments of variable value, and often very fallible judgments, as we botanists well know. And genera are more obviously judgments, and more and more liable to be affected by new discoveries. Judgments formed to-day—perhaps with full confidence, perhaps with misgiving—may to-morrow, with the discovery of new materials or the detection of some before unobserved point of structure, have to be weighed and decided anew. You see how this all bears upon the question of time and labor in the preparation of the Flora of a great country . . . where new plants are almost daily coming to hand . . ."

With this decade my story ends. In the space of sixty years the vast territory west of the Mississippi River which, to Americans certainly, was unknown in 1790, had been added to the domain of the United States and its boundaries had been established along the lines which exist today. Settlement of the region, even of portions geographically remote,[4] had begun—another forty years and the Director of the Census of 1890 was to consider that, on the basis of population *per* square mile, the country was "settled."

The plant collectors who went into the field before 1850 had participated to some

4. *See:* Goodwin, Cardinal. 1922. *The trans-Mississippi west (1803-1853).* Map, opp. p. 456 ("Distribution of population west of the Mississippi—1850").

extent in the exploration, conquest and settlement of the area. The technical botanists, to the best of their abilities, had kept apace of the wealth of plant material which
their field workers had collected. One great flora, including the plants discovered in
that region, had been published, although not in its entirety.

The partnership of descriptive botanist and field-worker was to continue and in the
ten years from 1850 to 1860 was to accomplish great things—as is demonstrated in
the botanical Reports issued in the course of the United States and Mexican Boundary Surveys and of the Pacific Railroad Surveys. We have seen in the final years of
the decade of 1840 to 1850—in the explorations of Parry and of Stansbury—fore-
shadowings of these fruitful years immediately ahead.

CHAPTER XXXIII

VOSNESENSKY, WHEN VISITING FORT ROSS IN 1841,

MAKES A COLLECTION OF PLANTS WHICH IS RETURNED

TO CALIFORNIA FOR IDENTIFICATION IN THE 1930'S

THE Russian settlement at Fort Ross, situated north of Bodega Bay, Sonoma County, was established in 1812, when Spain was in possession of California. The Russian American Company had its headquarters at New Archangel— or Sitka—and the advantages of possessing an outpost to the south had been portrayed to the Russian authorities at St. Petersburg by von Resanoff and by von Langsdorff after their visit to California in 1806, and later Russian expeditions which reached California in 1808 and in 1811 had reached similar conclusions. Eventually, when the Russians had become convinced that the California settlement was a liability[1] rather than an asset, they sold their establishment to John A. Sutter and took their departure from the colony in December of 1841; in the words of Essig, they ". . . set sail from San Francisco on January 1, 1842, after twenty-nine years of occupation."

H. H. Bancroft's *History of California* had much to say about ". . . the Russian right to territorial possessions in California . . ." and reached the following conclusions:

"Russia never made any pretension to sovereignty over the Bodega region or any portion of the California territory . . . It is absurd . . . to defend a Russian title never claimed by Russia or recognized by any other power. Not even the Russian American Company ever advanced a claim for territorial possessions in California. Their aim was to establish a post for fur-trading and for trade. Their efforts were to conciliate the Californians, and to maintain friendly relations. They wished to be let alone . . . The strongest claim in equity—though of no legal force in Spanish or Mexican[2] law— which the company could have set up to the lands actually occupied at Bodega would have been one of individual ownership, based on purchase from the natives, and an uninterrupted possession for thirty years . . . The company expressly excepted the land in their bargain with Sutter, and Sutter did not suppose that he had purchased any land . . ."

1. Essig explains why this was so: "Because of the encroachments of American settlers, the continued antagonism of the Mexican rulers in California, the failing supply of sea-otter furs and other skins, and the failure of the agricultural activities, it was apparent that Russian expansion in this part of North America was futile."

2. After Spain had relinquished California in 1821, all dealings of the Russian American Company had been with the Mexican authorities.

In these pages I have told of three botanists who came to California with Russian expeditions; von Langsdorff in 1806 (six years before the establishment of the Russian colony), and von Chamisso and Eschscholtz in 1816. The latter came again in 1824 but on that later occasion worked on other lines than botany. Essig refers to his having taken ". . . nearly a hundred species of Coleoptera, all new excepting one."

W. H. Brewer, writing of men who collected plants in California, records that while the Russians were in occupancy at Fort Ross, ". . . many botanical specimens were sent to St. Petersburg. Precisely how early these collections began, or who were the collectors, other than Wrangel and Wosnessensky, I have no information. But various Californian species were first described from specimens sent from this colony, or from plants grown in the botanic gardens of Europe from seeds collected here."

Of "Wrangel" above—Baron (or Admiral) Ferdinand Petrovitch Wrangel or Wrangell—the same author reports that he had ". . . arrived at Bodega about 1829, and lived there[3] as governor of the Russian possessions in America. He spent a number of years here, and collected many plants and seeds, which were sent to the Botanic Garden of St. Petersburg."

David Douglas had been in communication with the Baron both before and during his last trip to the northwest coast; at that time he had hoped to return to England by way of Russia's Alaskan possessions and from there cross to Siberia, but Indian unrest in New Caledonia (as British Columbia was then called) had made the journey inadvisable and his plan was abandoned. In his letters to William Jackson Hooker, Douglas always referred to von Wrangel with great respect: on August 6, 1829, as ". . . a man of vast information, and joins heart and hand with all those who have scientific views . . ." and on November 23, 1831, as ". . . the Captain Parry of Russia, keenly alive to the interests of Science, and anxious to assist in every way, those who labour in this field." [4]

A very complete account of the Russian settlement at Ross is supplied in five documented articles by T. Blok, E. O. Essig, Adele Ogden and Clarence J. DuFour, first published in 1933 in the *Quarterly* of the California Historical Society and reprinted

---

3. Brewer may have been correct, that von Wrangel's official residence was at Bodega as against New Archangel. However, Adele Ogden refers to his having ". . . appeared at Fort Ross in person in 1833 and again in 1835." He was Russian America's sixth governor and held office from 1831 to 1836.

4. There is a reference to the man's helpfulness in other directions in Dana's *Two years before the mast* (ed. 1. New York. 1840, pp. 291, 292). Von Wrangel had left Alaska after resigning his post and had reached Monterey on his way to Mexico and elsewhere:

"The only other vessel in the port was a Russian government bark, from Asitka, mounting eight guns, (four of which we found to be Quakers,) and having on board the ex-governor, who was going in her to Mazatlan, and thence over land to Vera Cruz. He offered to take letters and deliver them to the American consul at Vera Cruz, whence they could be easily forwarded to the United States. We accordingly made up a packet of letters, almost every one writing, and dating them 'January 1st, 1836.' The governor was true to his promise, and they all reached Boston before the middle of March; the shortest communication ever made across the country."

in September of that year as a special publication of the same society under the title "The Russians in California." Since our interest lies in the botanical aspects of the colony, it is regrettable that Essig's chapter on the Russian settlement lays emphasis upon such scientific matters as "beetles," "insects" and so on—very naturally so, however, as he was Professor of Entomology at the University of California. He does give considerable information upon the agricultural, gardening, and fruit-growing accomplishments of the Russians at Fort Ross. He also reports upon the visitors who reached the settlement:

"During the last few years of the Russian occupation of Ross a number of noted entomologists and collectors visited the place and added much to the insect collections in the museums of St. Petersburg, Moscow and at other cities in Russia and Europe. Among these were Ferdinand P. Wrangell, Governor of Russian America; Dr. F. Fischer, a physician of the Russian American Company and a collector of insects in Alaska and at Ross; Dr. Edward L. Blaschke, also a physician of the company and an ardent collector of beetles in Sitka and California; George Tschernikh, an agriculturist and overseer of the Tschernikh (George's or Gorgy's) Ranch, five miles north of Bodega Bay, and a most industrious and successful collector of beetles in Alaska and in California; and I. G. Vosnesensky, naturalist and curator of the Zoological Museum of the Academy of Natural Sciences, St. Petersburg. The last named was the only trained entomologist, being sent out by the museum to collect insects in California. He collected extensively over the territory occupied by the Russians from Bodega Bay to Ross and also around San Francisco, at New Helvetia (Sacramento) and the area between Ross and the Upper San Francisco Bay region. With Tschernikh, on June 12, 1841, he was the first to climb Mount St. Helena, which he named for the Empress of Russia . . . the copper plate[5] placed on the top of this mountain by him was removed years afterwards, but another commemorating the event was erected there in his honor in 1912."

Brewer's "List" supplies but little information about Vosnesensky nor does Jepson's paper of 1899 upon the "Vegetation of the summit of Mt. St. Helena." I quote their comments in succeeding paragraphs:

"Dr. Wosnessensky was sent, by the Academy of Sciences of St. Petersburg and the Zoological Museum of that place, to the North Pacific, and spent ten years on the coast. He spent some time collecting in California, but when he first came, how often or how long here, or how extensive his botanical collections were, I do not know. He was on Mount St. Helena June 12, 1841, which is the only date I have of his visit."

5. Essig supplies its inscription which, translated, read: "Russians, 1841, June, E. L. Voznisenski iii, E. L. Chernich." It was evidently discovered on the mountain's top in 1853, but Essig points out that there exists ". . . much confusion and misunderstanding regarding this event." The original plate was destroyed in the San Francisco fire but fortunately a paper copy had been made and was preserved.

"The first explorer to reach the summit of Mt. St. Helena[6] was the Russian botanist Wosnessensky, and his companion Tschernich, who formally named the mountain in honor of the Russian empress on the date of the visit, June 12, 1841. I find no record of any plants collected by them on or near the mountain."

More light has been thrown in recent years upon the botanical accomplishments of I. G. Vosnesensky: in 1937 Mr. J. T. Howell, of the California Academy of Sciences, published upon a collection which, ". . . originally sent to St. Petersburg *via* Alaska and Siberia . . .", was returned to California, on the way crossing ". . . western Europe, the Atlantic Ocean, and North America." After the specimens had been determined ". . . all were returned to Leningrad except a few duplicates . . . now in the Herbarium of the California Academy of Sciences."

Mr. Howell records that, with the exception of eight specimens gathered "by Kuprianov, governor of Russian America at the time Fort Ross was abandoned," the collection was obtained by "E. Voznesenski, scientist-naturalist from the zoological museum of the Russian Academy in St. Petersburg." The collection

". . . contained 346 specimens representing 214 species and varieties . . . collected in 1840 and 1841, in general a fairly representative collection . . . from the flora of present-day Sonoma County: a number obviously from the coastal hills and mesas in

6. A later visitor to Mount St. Helena, more famous than Vosnesensky, was Robert Louis Stevenson who spent the summer of 1880 in a deserted shack upon its slopes. With him went his bride, Fanny Van de Grift, and his stepson Osbourne. Their strange honeymoon abode was chosen for reasons of health but Stevenson's biographer, Graham Balfour, states that both bride and stepson became ill with diptheria during their sojourn—this fact is not mentioned in *The Silverado Squatters*. According to Balfour's bibliography, this little volume was first published in book form in December of 1884—in Boston by the firm of Roberts Brothers and in London by Chatto & Windus.

This brief tale—as a picture of an abandoned mining region and of the strange examples of humanity with whom they came into contact—is very readable. Stevenson was obviously interested in the vegetation which he saw about him but could derive little information thereon by questioning—merely what he calls in one instance "false botany"—because his lot had ". . . fallen among painters who know the name of nothing, and Mexicans who know the name of nothing in English . . ."

It had once been a redwood region. In Stevenson's day one money-wise rancher who knew more about "putrefaction" than "petrifactions" was showing their petrified trunks "at the modest figure of half a dollar a head, or two-thirds of his capital when he first came there with an axe and a sciatica." I quote one passage, exemplifying Stevenson's readable comments upon the vegetation: ". . . The oak is no baby; even the madrona, upon these spurs of Mount Saint Helena, comes to a fine bulk and ranks with forest trees; but the pines look down upon the rest for underwood. As Mount Saint Helena among her foot-hills, so these dark giants out-top their fellow-vegetables. Alas! if they had left the redwoods, the pines, in turn, would have been dwarfed. But the redwoods, fallen from their high estate, are serving as family bedsteads, or yet more humbly as field fences, along all Napa Valley."

In some short notes prefacing his story Stevenson refers to Mount St. Helena as ". . . the Mont Blanc of one section of the Californian Coast Range, none of its near neighbors rising to one-half its altitude . . ." and notes that three counties, ". . . Napa County, Lake County, and Sonoma County, march across its cliffy shoulders . . ." The volume *California* ("American Guide Series"), p. 411, mentions that a Robert Louis Stevenson Monument now marks the site of the "bunk-house."

the immediate vicinity of Ross,[7] and others from the interior, some certainly from Mt. St. Helena. But whether the specimens originated on the coast or in the interior, most of the labels accompanying the specimens carry only the printed data, 'California boreal. Ross.—leg. Wossnesensky,' and on only a few appear more definite designations of locality in script . . ."

Mr. Howell points out that, had the collection been reported upon promptly, ". . . what might have proved a classical collection of Californian plants is noted here as something of merely botanico-historical interest." For many of the plants at the time of collection were then ". . . unknown to science . . . and . . . had they been sought out and named, the collection would now rank with those obtained by Nuttall, Douglas, and Hartweg, and by the earlier Russian collectors, Langsdorff, Chamisso, and Eschscholtz."

It remains, perhaps, for a botanist to examine the collections in Russian herbaria before any satisfactory picture of botanical accomplishment during the occupation of Fort Ross can be obtained. William Jackson Hooker, writing in 1836, called attention to some twenty Californian plants which had been described by F. E. L. von Fischer and C. A. Meyer:

"In the 'Index Seminium, quæ Hortus Botanicus Imperialis Petropolitanus pro mutua commutatione offert,' recently printed by Drs. Fischer and C. A. Meyer, those excellent Botanists have given the characters of several lately-discovered species of plants of New California, a country peculiarly interesting to us, in consequence of the researches of Mr. Douglas and the Naturalists of Captain Beechey's voyage, in that and the neighbouring western shores of North America. As Indices of this description are not in general circulation, nor preserved with the care that many of them merit, we are anxious to afford greater publicity to these descriptions, and to give them a place in our Journal."

No dates of collection and no names of collectors are cited. Most species were from the region of Bodega and of Fort Ross—"Hab. in Nova California, circa Coloniam Ruthenorum Ross.", "Hab. Circa coloniam Ruthenorum Ross, in sinu Bodega Nova California.", "Hab. in Novæ Californiæ portu Bodega." and so on.

7. Fort Ross, as distinguished from Ross in Marin County.

BANCROFT, HUBERT HOWE. 1885. 1886. The works of . . . 19 (California II. 1801–1824): 628-652; 21 (California IV. 1840–1845): 158-189.

———— 1884. The works of . . . 27 (Northwest coast I. 1543–1800): 526; 28 (Northwest coast II. 1800–1846): 330, 664.

———— 1886. The works of . . . 33 (Alaska 1730–1885): 476-489.

BREWER, WILLIAM HENRY. 1880. List of persons who have made botanical collections in

California. *In:* Watson, Sereno. Geological survey of California. Botany of California 2: 554, 555.

ESSIG, EDWARD OLIVER. 1933. The Russian settlement at Ross. *Quart. Cal. Hist. Soc.* 12: 191-209. *Reprinted: Cal. Hist. Soc.* Spec. Publ. 7: 3-21 (1933).

HOOKER, WILLIAM JACKSON. 1836. Botanical information. *Comp. Bot. Mag.* 2: 6-12.

———— 1836–1837. A brief memoir of the life of Mr. David Douglas, with extracts from his letters. *Comp. Bot. Mag.* 2: 151, 155.

HOWELL, JOHN THOMAS. 1937. A Russian collection of botanical plants. *Leafl. West. Bot.* 2: 17-20.

JEPSON, WILLIS LINN. 1899. Vegetation of the summit of Mt. St. Helena. *Erythea* 7: 109.

OGDEN, ADELE. 1933. Russian sea-otter and seal hunting on the California coast, 1803–1841. *Quart. Cal. Hist. Soc.* 12: 217-239. *Reprinted: Cal. Hist. Soc.* Spec. Publ. 7: 29-51 (1933).

CHAPTER XXXIV

BRACKENRIDGE, "HORTICULTURIST" OF THE UNITED

STATES EXPLORING EXPEDITION, MAKES TWO TRIPS

IN THE STATE OF WASHINGTON, AND TRAVELS OVER-

LAND FROM THE COLUMBIA RIVER TO THE BAY OF

SAN FRANCISCO

ALTHOUGH Lewis and Clark had reached the mouth of the Columbia River in 1805, nearly four decades elapsed before the United States government demonstrated sufficient interest in the Pacific northwest to send to that region two expeditions with an interrelated purpose: one was the United States Exploring Expedition which reached there by sea, the other was an overland expedition led by John Charles Frémont.[1]

Charles Wilkes,[2] then a Commander in the United States Navy and head of the United States Exploring Expedition, had set out in 1838 to examine and survey the South Seas and the Pacific Ocean. The ships arrived off the coast of Washington in May of 1841 and proceeded eventually to the Columbia River and to the Bay of San Francisco.

The literature concerned with the United States Exploring Expedition is voluminous, as a glance at the 188-page bibliography compiled by Daniel C. Haskell will indicate.

Authorization for what has more than once been termed "our first national exploring expedition," [3] was given in May, 1836, and governmental discussion of the

1. Frémont had accomplished his task when he arrived at the Dalles of the Columbia River in November of 1843. "The object of my instructions had been entirely fulfilled in having connected our reconnoissance with the surveys of Captain Wilkes . . ."

2. Wilkes may be remembered for two achievements unrelated to my story: in 1840 he sighted the Antarctic continent and for two months coasted its shores for some one thousand miles, and, during the Civil War period, it was Wilkes, then commanding the *San Jacinto,* who, on November 8, 1861, removed from the British steampacket *Trent,* the two Confederate Commissioners to France and to Great Britain, John Slidell and James Murray Mason.

3. As the opening paragraph of Wilkes's introduction to his *Narrative* explains:
"The Expedition . . . was the first, and is still the only one fitted out by national munificence for scientific objects, that has ever left our shores."
In other words, the first to travel *by sea.* Although little "national munificence" had been involved, the Lewis and Clark and the "Yellowstone" expeditions had both been instructed to concern themselves with "scientific objects."

project—all in print—had begun much earlier. The last scientific memoir was not issued until 1874 and publications concerned with the expedition and its members continue to appear. Haskell reported in 1942 that both a biography and an autobiography of Wilkes were in preparation.

Despite the voluminous literature, some of the scientific memoirs, although ready for the press, were never published: Asa Gray's second volume on botany (XVIII of the contemplated official set) still remains in manuscript in the Gray Herbarium, for scientific treatises soon become of mere historical interest.

The expensive editions which the government ordered printed were extremely small and were unintelligently designated for distribution; none, for example, were assigned scientific institutions (the suitable repository for scientific works) and because no copies were available for purchase, these institutions were unable to acquire them. Contributing to the reduction of the small number of copies issued, there were a series of fires in which many of the few existing copies were destroyed. It has been suggested by Frank S. Collins that some of the volumes ". . . might be unearthed in Washington by one who had time and energy for the work."

Of the twenty-four volumes which Collins enumerates as the "official set," four were never published; one, Gray's second volume on botany, I have mentioned. Writes Collins:

"To explain the non-publication of Vols. XVIII, XXI, XXII, and XXIV would take much space and would bring up interesting but not specially edifying stories of inefficiency, extravagance and plunder, which when compared with present conditions, indicate that the standards in such matters are higher now than in the days of the fathers. At the time the publication was authorized there was no National Museum, no Smithsonian Institution, practically no organ that could deal with scientific matters; Congress, probably with a vague feeling that literature had something to do with science, and that a library had something to do with literature, put the matter in charge of the Joint Committee of the Library of Congress, with which it remained from 1852 to 1874, the date of the last issue. The committee, its membership continually changing, left the whole matter practically in the charge of Wilkes, who dealt with authors and printers much as he was accustomed to do with sailors and marines. The qualities that made him an excellent commander of an exploring expedition, accomplishing a work of lasting credit to the navy and the nation were not suited to estimating the value of scientific memoirs . . .

"There is no evidence of anything dishonest on the part of Wilkes himself but the report of the Library Committee in 1876 shows that the greater part of the money appropriated, about $350,000.00 in all, had gone for 'superintendents,' the amounts for the work itself being in some years absurdly small. The disgust of Congress at these revelations led to an immediate stoppage of the work, without regard to the memoirs ready and waiting. The money to publish them had been appropriated over and over again but had been squandered."

Eventually some of the scientists were permitted to publish their own memoirs and a number of printers brought out editions. These miscellaneous issues (constituting most of the volumes readily available for examination) contributed to the bibliographical complexity which has been depicted by Collins, by Haskell, and by H. H. Bartlett. Persons interested in this aspect of the expedition's accomplishment should turn to what these men have written.

There exists also, still unpublished, a considerable amount of manuscript material, all meticulously listed as to content and repository in Haskell's bibliography.

In 1940 a symposium, commemorating the centenary of American polar exploration which had had its inception in the work of the United States Exploring Expedition, was held in Philadelphia, at the American Philosophical Society. Appropriately so, since that institution had advised the expedition upon all important phases of its contemplated scientific work. In 1940 the Society published in its *Proceedings* eighteen papers read at the meetings. I found the following four of special interest:

(1) "Connection of the American Philosophical Society with our first national exploring expedition," by Edwin G. Conklin.
(2) "Connection of the Academy of Natural Sciences of Philadelphia with our first national exploring expedition," by James A. G. Rhen.
(3) "The purpose, equipment and personnel of the Wilkes expedition," by Captain G. S. Bryan.
(4) "The reports of the Wilkes expedition, and the work of the specialists in science," by Harley Harris Bartlett.

The contribution of Charles Wilkes—*Narrative of the United States Exploring Expedition during the years 1838, 1839, 1840, 1841, 1842*—appeared promptly, in official issue, in 1844, in five volumes and an atlas. Volumes IV and V contain an account of the months spent along the coast and in the interior of the Pacific northwest as well as of the visit to California and the Bay of San Francisco.

A manuscript in the archives of the Maryland Historical Society is cited thus in Haskell's bibliography—his item 463:

"Brackenridge, William Dunlop. Journals kept while on the United States Exploring Expedition, 1838–1842. 6 v., consisting of two large folio volumes covering the periods Sept. 10, 1838–Sept. 9, 1839, and Nov. 29, 1839–Oct. 28, 1841, respectively, and four small note books containing rough notes for the preceding, covering April 14–August 29, 1840, April 28–June 2, 1841, June 4–July 14, 1841, and Sept. 1–Oct. 27, 1841, respectively."

Two transcripts of these manuscript journals, covering different periods, have been published:

Transcript (1), entitled "Our first official horticulturist The Brackenridge journal," was issued in five installments in *The Washington Historical Quarterly* for 1930 and 1931, and was edited by O. B. Sperlin. It begins on April 28, 1841, when the ships of the expedition were off Cape Disappointment and describes the voyage through the Straits of Juan de Fuca and portions of Puget Sound to the anchorage opposite Fort Nisqually, reached on May 11. From that date to May 17 it concerns the region about Fort Nisqually. It then tells of two inland trips made from Nisqually and of a third which started from Fort Vancouver. Brackenridge participated in all three trips.

The first trip (May 17–July 15) was into the interior of the present state of Washington, with a very short excursion into Idaho. The second trip (July 19–August 31) was from Fort Nisqually to the Pacific coast, to Grays Harbor and Willapa Bay and thence to Cape Disappointment where the Columbia River was crossed to Fort George (once Fort Astoria), Oregon; from there the Columbia was ascended to Fort Vancouver, Washington. The third trip (extending in its entirety from September 1 to October 28) started from Fort Vancouver and ended at the Bay of San Francisco and, as far as botany was concerned, was the most important of the three. The Sperlin transcript, covering Brackenridge's work in Washington and Oregon, ends on October 2 when the party crossed what is now the northern border of California. Plant identifications are supplied by Sperlin and appear in footnotes—for them the editor acknowledges his indebtedness to a number of authorities.

Transcript (2), entitled "Journal of William Dunlop Brackenridge," was published in the *California Historical Society Quarterly* in 1945 and was edited by Alice Bay Maloney. It begins on October 1, 1841, when the party reached California and ends October 28, when it arrived at the Bay of San Francisco. Associated with this transcript is an article by Alice Eastwood, entitled "An account and list of the plants in the Brackenridge journal." Here the plants mentioned in the journal are identified by numbers with the Eastwood determinations.

I turn now to the expedition and, in particular, to its botanical activities during its visit of approximately six months to regions in western North America which are included in the locale of my story.

Wilkes's introduction to his *Narrative* explains that, in March, 1838, the command of the expedition had devolved upon him. By April he was informed that four vessels had been appointed for the service: ". . . the sloops of war Vincennes and Peacock, the brig Porpoise, and store-ship Relief. The tenders Sea-Gull and Flying-Fish were subsequently added." [4] This was evidently a reduction from the number of ships originally planned and, with less accommodation available, there had had to be a ". . . reduction of those departments that were placed under the corps of civilians, including naturalists as well as artists. As many of these were taken as could be accommodated. The selection was made with much deliberation, and with great impartiality. Reference was had to the departments in which results were most to be expected, and

4. Only the *Vincennes* and the *Porpoise* made the entire trip; the *Sea-Gull* was lost off Cape Horn, the *Relief* was sent home as too slow, the *Peacock* met its end on the bar of the Columbia River—as had many ships before it—and the *Flying-Fish* was sold as "unseaworthy." The *Sea-Gull* and the *Flying-Fish* had only been purchased two weeks before the expedition's departure from the east coast.

most desired by the country. The only new one added was the Horticulturist and Assistant-Botanist, Mr. Brackenridge."

The "Instructions" issued to Wilkes enumerated the "corps of scientific gentlemen" under his direction. Those whose task was related to botanical work were: Charles Pickering, "Naturalist," attached to the *Vincennes* commanded by Wilkes; William Rich, "Botanist," attached to the *Relief,* and William Dunlop Brackenridge,[5] "Horticulturist," who, like Pickering, traveled in the *Vincennes.* There was some shifting about during the long voyage but such were the original assignments. Pickering and Brackenridge were both on the *Vincennes* when it reached the northwest coast of North America.

Not until volume four of the Wilkes *Narrative* do we read of the expedition's arrival along our Pacific coast. This was on April 28, 1841, when they ". . . made Cape Disappointment . . ." off the mouth of the Columbia River. Wilkes could not conceive how Vancouver, for one, had been unable to recognize the ". . . mouth of the mighty river . . ." He, himself, was unfavorably impressed with "the terrors of the bar" at the river's mouth and decided to put off its crossing and proceed instead to the Straits of Juan de Fuca and there begin his coastal surveys. By May 1 the ships were ". . . well into the straits . . . we hastened to reach Port Discovery, where we anchored . . . on the 2d of May; just forty-nine years after Vancouver . . . had visited the same harbour." Obviously summarizing botanical information derived from Brackenridge (although no credit is given), Wilkes wrote in his *Narrative:*

"We remained at Port Discovery until 6th May . . . Our botanists had a large and interesting field opened to them, and there are few places where the variety and beauty of the flora are so great as they are here. Dodecatheon, Viola, Trifolium, Leptosiphon, Scilla (the cammass of the natives), Collinsia, Claytonia, Stellaria, &c., vied with each other in beauty, and were in such profusion, as to excite both admiration and astonishment. According to Mr. Brackenridge, the soil in which the plants grow consists of a light-brown loam . . .

"The trees grow so closely that in some places the woods are almost impenetrable. The timber consists principally of pine, fir, and spruce. Of the latter there are two species, one of which resembles the hemlock-spruce of the United States: it has a very tall growth, and puts out but few, and those small lateral branches. Some maple-trees grow in the open grounds and on the banks but they are too small to be of any service to the settler."

On May 6 Wilkes was ready to sail for Fort Nisqually. This Hudson's Bay Company post had been established in 1833 and was situated about seventeen miles south of present Tacoma, Pierce County, Washington. The ships arrived at the anchorage on May 11, 1841. Wilkes commented: "Nothing can exceed the beauty of these waters, and their safety: not a shoal exists within the Straits of Juan de Fuca, Admiralty Inlet,

5. In the "List of officers and men . . .", as throughout the *Narrative,* the man's name appears, incorrectly, as "J. D. Brackenridge."

Puget Sound, or Hood's Canal, that can in any way interrupt their navigation by a seventy-four gun ship. I venture nothing in saying, there is no country in the world that possesses waters equal to these."

Although the Brackenridge journal covers the period from April 29 (arrival at Cape Disappointment) to May 11 (arrival at Nisqually) and has something to tell of the environs of Nisqually, I turn at once to the botanist's first journey into the interior. I quote three paragraphs in which Wilkes describes the personnel of the party, outlines the route in detail (even supplying a map), and specifies some of his orders:

"The land party intended to explore the interior was placed under the command of Lieutenant Johnson[6] of the Porpoise. With him were associated Dr. Pickering, Mr. T. W. Waldron . . . Mr. Brackenridge, Sergeant Stearns, and two men. Eighty days were allowed for the operations of this party, which it was intended should cross the Cascade range of mountains, towards the Columbia, proceed thence to Fort Colville, thence south to Lapwai the mission station on the Kooskooskee river, thence to Walla-walla, and returning by way of the Yakima river, repass the mountains to Nisqually.— (The orders are given in Appendix XII.)"[7]

"You will leave this place on the 17th, and proceed . . . across the mountain range north of Mount Rainier . . . Thence you will pursue a route to the northward, keeping to the west of the Columbia river until you reach Fort Okonagan, where you will doubtless obtain much information relative to the country, and proper guides . . . From thence I should desire you to push to the northward, if possible, and your time will permit, making a detour so as to stop at Fort Colville; thence across the Saptin or Lewis [Snake] river, and down to Fort Nez Percé [Fort Walla Walla], and thence, in a direction to Mount Rainier, to Nisqually, where you will join me or receive orders to govern your farther movements." Further:

"It may be desirable for Dr. Pickering and Mr. Brackenridge to make occasional short excursions from your direct route; you will in that case afford them all the facilities in your power to promote their researches. [Finally] . . . you are not to deviate from the route pointed out unless insurmountable difficulties should render it impossible to pursue the course specified, and in no case are you to go to the southward of the limits pointed out."

Wilkes has something to tell of the movements of the above party (following an account of an overland journey which he himself made to the Columbia River during their absence). His accounts of these trips contain an amazing amount of information about the character of the country and the activities of the Hudson's Bay Company. This was a period—just five years before the extinguishment of British claims south of the forty-ninth parallel—when such information would be extremely interesting

---

6. Lieutenant R. E. Johnson who, as we shall see, proved to be a thoroughly irresponsible person and an incompetent leader.

7. They were issued on May 13, 1841, at "Nisqually Harbour." A map was included with the orders.

to those in Washington, D. C. It would, therefore, require some credulity to read the *Narrative* in the light of an ingenuous travelogue or (despite the enumeration of botanical and other scientific accomplishment) in the light of a scientific investigation pure and simple.[8] Since the perspicacious Factors of the British company were already disturbed by American interest in the Oregon country it is much to their credit that the expedition—at every Company post—received great courtesy and much wise advice—indeed it would be difficult to see how the Johnson party could have made their excursion without its assistance and certainly not in the time allotted.

Comparing the Wilkes and the Brackenridge accounts of the inland journey, it is obvious—as it is on other occasions—that, as far as botanical matters are concerned, the commander's *Narrative* is based largely upon Brackenridge's information. Johnson's report[9] can have been of little value. I therefore omit the version of the trip supplied by Wilkes and turn at once to the first-hand account in the Brackenridge journal, quoting from Sperlin's carefully edited transcript; the editor's footnotes (plant identifications, comments upon the route, interpretation of old place-names and so on) are extremely helpful and I include many between brackets in the narrative. The journal bears reading in its entirety but I omit much that is unrelated to botany. I have italicized the scientific plant names mentioned by Brackenridge, however strangely spelled, and have sometimes altered the paragraphing.

On May 15 Brackenridge noted that he "Remained on board all day put[t]ing the Botanical collections in order and transferred the same *by order* over to Mr. Jont. Dyes[1] to look after." On May 19 they started on their way and, with the exception of June 25 and 26, when a short excursion was made into present Idaho, the route from

8. G. Brown Goode (1888) makes this interesting comment: "... It is customary to refer to the Wilkes expedition as having been sent out entirely in the interests of science. As a matter of fact it was organized primarily in the interests of the whale fishery of the United States."

9. Sperlin states in his introductory words: "The publication of the Brackenridge journal shows that only a minor quantity of the substance was Johnson, the major quantity being Brackenridge and Pickering."

1. In the "List of officers and men ..." given in the *Narrative,* John W. W. Dyes is mentioned as "Assistant Taxidermist," assigned to the *Vincennes.* Sperlin suggests that Brackenridge, in underlining the phrase *"by order,"* "... seems to hint that he suspected that all was not 'shipshape' after the collections had left his hands." Perhaps so, and he may not have considered Dyes a suitable custodian.

In Piper's "Flora of the state of Washington" appears this comment: "Unfortunately the original labels of the specimens seem in some way to have become intermixed, with the result that a good many plants confined to eastern Washington bear such labels as 'Port Discovery' and 'Nisqually,' while other species confined to western Washington are labeled 'Walla Walla,' or 'North Fork of the Columbia.' On some sheets eastern and western Washington species are mixed, and mounted over a single label. With the details of the party's itinerary known, it is possible, however, to tell with some accuracy where the specimens must have been gathered."

To associate Dyes with this mixing of labels is, of course, only a matter of inference.

It seems to me possible that in his phrase *"by order,"* Brackenridge may merely have been indicating his resentment to naval discipline. Another entry made two days later suggests this interpretation; the party left the ship on the date set by Wilkes. "Our tents were pitched for the night outside the Fort, & our luggage piled up in a heap; which obliged us to stand watch, when we could just as easy have placed the whole

day of departure to July 14 (when they were back at Nisqually) lay through the state of Washington.[2]

May 20: ". . . remooved about 5 miles farther along the plain in the direction of Mount Ra[i]nier . . . On the plains I observed several specimens of a very large Pine, the height of many I estimated at 130 ft. . . . in Co. with Dr. P. we wandered from the camp . . . and found . . . a very handsome yellow *Ranunculus,* described by Dr. Hooker, in thickets, *Trillium* sp: with large leaves & small flowers; *Lupinus polyphyllus* ? not in flower. Solitary specimens of a rich orange colour'd *Cruciferae,* annual plant, were observed here for the first time."

May 21: ". . . in the early part of the forenoon crossed a large river ['The Puyallup River.'] about 70 feet wide . . . we began a gradual ascent . . . Among dense masses of *Gaultheria Shallon,* Hazel, *Spiraea, Vaccinium, & Cornus.* Towards eavening we came upon the Smalocho river ['. . . The party did not reach the Smolocho (White) till two days later. The river . . . is the Puyallup . . .'] and encamped at the junction of the Upthascap ['The Carbon River, and its branch, South Prairie Creek.'] with the former. Though deserted of inmates, I here saw a very snug and perfectly water tight house built from plank split out of *Thuja* or Arbor Vitae, a tree which attains a great size on these mountains. The planks were as smooth as if cut by a saw & many of them three feet wide."

According to Sperlin, the party by May 23 was ". . . four days out from Nisqually on the Naches Trail across the Cascade Mountains to the eastern part of Old Oregon now Eastern Washington and Idaho . . ." The route seems to have led southeastward, across present Pierce County. Brackenridge wrote on May 23:

"Our route [a]cross the range lay somewhat to the north of Mt. R[ainier] where the finest timber exists that I ever beheld. For several days our route lay through dense forests of Spruce the stems so straight and clean that it was seldom you could find a branch closer than 150 feet to the ground. A prostrate trunk of a Spruce which we took with a tape line measured—length 265 ft. circumfe: (10 ft. from base) 35 ft. . . . the whole height of Said tree when standing would be 285 ft. In deep moist valleys I have seen the *Thuja,* or Arbor Vita at least ⅓ more in circumference, but not so high by

inside the Fort,—an offer which was made us by Mr. Anderson, but no—we must stand guard, while we could have reposed . . ."

Wilkes was an efficient commander and it was doubtless necessary to enforce discipline but some of his methods seem unduly harsh. Read the general order which he issued (included in Part V of the Appendix, volume V of the *Narrative*) for the punishment of sailors guilty of an "intemperate use of intoxicating liquors." Menzies had considered, as had the mission padres, that Vancouver might have accomplished more had he shown greater leniency to a deserter. And what was resented by a British civilian in the 1790's was doubtless even less acceptable to an American civilian of the 1840's!

2. Bancroft's *History of the northwest coast* includes a map captioned "Johnson's Excursion," which is extremely helpful.

100 ft. A *Populus*, or Cotton tree which we measured was upwards of 200 ft. high. Many of the Spruce stems which lay prostrate were so stout that when on horse back we could not see over them. On the decayed bark of such seedlings of the Spruce vegetated freely, forcing their roots through the bark, over the body of the trunk till the[y] reached the ground so that when said trunk became entirely decayed, the roots of the young trees became robust [and] formed a sort of arch way, under which we occasionally rode."

May 25: "The banks of the Streams and Rivers . . . afforded the greatest variety of trees and Shrubs, these consisted of *Populus, Rhamnus,* 50 ft. h:, *Cornus* 30 to 40 ft: h:, several species of *Salix, Alnus, Acer* 2 species, & occasionally a solitary Yew. The Shrubs or under gro[w]th in the forest were chiefly—Hazel, *Vaccinium, Gaultheria,* a prickly species of *Aralia* with large peltate leaves. *Euonymus,*[3] a little evergreen shrub, was also common, The herbaceous plants in such places were—*Goodyera, Neottia,* several species, *Claytonia, Corallorrhiza, Aquilegia*—the majority of such plants have not yet come into flower, so that we left many of such till our return."

It was on this day, according to Sperlin, that they crossed White River. On the 26th they set the woods on fire while cooking supper and ". . . Dr. P. came near loosing the tails of his Coat by a brand of fire that had accidentally fallen into his pocket." On the 27th they ". . . came upon an open space . . . which I estimated to be the summit of the range . . ." Sperlin identifies this as the "Naches Pass." It lies close to the meeting of the four counties of Pierce, King, Kittitas and Yakima.

". . . Here . . . the spruce trees (there being no other) were more dense in foilage . . . and stunted in habit . . . we began to descend . . . At three in the afternoon when we had got fairly clear of all snow we came upon the head of the Spipe ['The Naches River.'], where we incamp'd pretty well fatigued . . . I am not aware what our height was here, but . . . the vegetation was farther advanced on the east side of the Mts. than it was at what I considered the same height on the West Side. The *Pulmonarias* and several small annuals were in a more advanced state. On our descent we fell in with a good many fine Larch trees . . . During the trip over the range I saw one real *Pinus*— (Pine) which consisted of a few small trees in an open place on the west side of the range, the Alpine plants which we expected to meet with on the high ridges being covered with snow."

While the luggage was being brought up on May 28, Brackenridge and Pickering ". . . went collecting & found some pretty curious plants particularly a glaucus leaved species of *Pyrola,* and a little fern—perhaps *Cheilanthes*—on a rocky situation." On the 29th ". . . Mr. J himself came on, having left four of the Horses behind, and came near loosing himself by getting out of the path when the Indians had left him." At this point some of the Indians were dispensed with as no longer necessary; Brackenridge

3. Sperlin queries: ". . . Could this have been *Pachistima myrsinitis?* The only species of Euonymus known here is deciduous . . ."

evidently sent back his collections made to date. "All the specimens that were dry, of my collecting, were neatly stitched up in a canvass cover by Mr. J. to be taken back to the Ship accompanied by a note of instructions from myself to Mr. Dyes, relating to their preservation." On May 30 the two botanists and a sergeant were allotted a horse ". . . to be used alternately, but [writes Brackenridge] as the Doctor and myself were prooved pedestrians & had plants to pick up, we started to walk . . . keeping close to the river . . . our path lay on the north side." The character of the country was changing. Instead of flowing between flats and meadows, the river ran between rocky bluffs for some twenty miles.

"Towards the eavening we could dis[c]ern a particular difference in the character of the vegetation—every thing was in a more advanced stage & several interesting genera of Plants made their appearance, as *Paeonia Brownii* ? *Cypripedium Oreogonium,*[4] *Pentstemon, Ipomopsis elegans,* and several neat little *Compositae.*"

On May 31, keeping along the river, they ". . . fell in with *Purshia tridentata,* a very handsome flowering shrub . . ." and, ascending a steep hill, found ". . . a beautiful little species of *Polemonium,* about three inches high."

On June 1, three miles "farther down the River," they turned north, away from the stream, and doing so ". . . had a range of Mts. to cross." Sperlin notes that ". . . The party here, considerably above the confluence of the Tieton, reach the chief arterial trail from Celilo (near the Dalles) to Chelan . . ." and identifies the range of mountains as the Cleman Mountains.[5] After a climb of some six hundred feet the vegetation assumed a different appearance and on top of "a bare ridge" (which Sperlin identifies as "Umtanum Ridge") they met some ". . . Spipen Indians, collecting Cammass and other roots." They purchased cakes made from various roots and Brackenridge describes how they were prepared and ends with the comment that, while digging the roots, the Indian women worked so diligently that they ". . . seldom pay any attention to a passer by." He ends his report of the day with the information that he had had the misfortune of losing his ". . . Note Book, so that what I have said of our route from Nesqually to this place is mostly from memory and partly from those who were in the party, but I believe it to be correct." On the 2nd the "Eyakema," or Yakima, River was reached:[6]

". . . The character of the country between the Spipe and the Eyakema River is mountaneous very thinly wooded with about equal proportions of Spruce and Pine.

4. The "Cypripedium" and the "Oreogonium" should certainly have been separated by a comma. Throughout his narrative Brackenridge uses "Oreogonium" for *Eriogonum.* At this point in his transcript Sperlin considered that the word must have been used as the specific name of a *Cypripedium.* But at the end of his transcript he devotes two pages to the word and reaches the conclusion that Brackenridge merely misspelled *Eriogonum:* " 'Oreogonium' is Eriogonum, with the accent on the *og*!"

5. The Tieton River enters the Naches in Yakima County and the Cleman Mountains are in the eastern portion of the same county.

6. This was, according to Sperlin, "near the site of Ellenburg." Or in southern Kittitas County.

A great many of the latter I observed to be forked at top—a feature rather rare in this tribe. The soil is a poor obstinate yellow Loam producing very indifferent feed for Cattle—in the plant way it is rich. I mention a few of the finest—*Lupinaster*[7] sp: with flesh colour'd flos:, *Viola* flos: white and purple, *Parnassia,* a very dwf. sp:, *Dodecatheon* like *integrifolia, Trollius* sp: *Sisyrinchium* sp: the five last in marshy grounds. A sp: of *Balsamoriza* different from the one on Nesqually plains.—"

The Yakima had been crossed by inflating ". . . the Gum elastic Bolsters in order to form a raft . . ." This carried a greater weight than a canoe and worked successfully according to Brackenridge. On June 3 they kept along the sandy flat lands bordering the river and, after crossing ". . . a sandy prairie with a number of small shrubs on it . . . immediately a group of Mts. set in by degrees upon us . . ." Sperlin states that, from near the site of Ellenburg, they had turned north, ". . . up Wilson (Kittitas) Creek and its tributary the Naneum. The crossing was made by Colockum Pass, 5323 feet elevation." The "group of Mts." he identifies as the Wenatchee Range.

June 4: ". . . Our road for the first 5 miles was over swampy ground on the brow of a Mt. A few patches of spruce trees stud[d]ed the crest and the[y] began to thicken as we descended a small stream . . . at 12 o'clock we haulted to take some breakfast; which was ready about one and served us at the same time as dinner also. In all our opperations we wanted system, but with such a leader we had all we could expect, in fact fared and got farther along than I at first anticipated we could with our journey."

Sperlin notes that, in referring to "the brow of a Mt.," "Brackenridge here passes lightly over one of the hardest days of the trip, over the Wenatchee Mountains. Swamps in the summit region are characteristic of the range." On this day the "Baromiter tube" had been broken and Brackenridge comments that this ". . . put an end to its use for the Cruize."

"At three in the afternoon we struck the north branch of the Columbia, and by keeping along its banks, shortly arrived at the Piscouas[8] River, which takes its rise in the range of snowy Mts. which lay in a N. West direction from us . . . About ½ mile from its junction with the Columbia . . . we camped for the night . . .

"Plants: two species of *Calochortus* was found today. A shrubby *Phlox,* a *Marrubium,* & four or five *Compositae;* the first Birch observed since leaving Nesqually was found on the banks of a small run overrun by a *Clematis* of which we found two species, perhaps the smallest one *Atragene.*"

On June 5 the "Piscouas," or Wenatchee, River was crossed and they started to follow the "North"—but actually west—bank of the Columbia. The route was "more rich

---

7. Surely *Lupin* and *aster* should be separated by a comma.

8. According to Sperlin: "The Wenatchee River. Johnson, following instructions from Wilkes to use Indian names, records *Wainape,* his best attempt at *Wenatchee.*" The Wenatchee River enters the Columbia in southern Chelan County.

in plants" than the day previous: "A *Cupressus* tree,[9] and a *Cruciferous* plant on rocks, was among the number."

On June 6 they reached a river which Sperlin identifies as the Entiat River. After crossing it Johnson, in paying some Indians for their assistance, offended ". . . the Chief of the Tribe very much, in the manner in which he proffered him some Tobacco, which was not accepted of." For some reason, unexplained, they crossed to the east bank of the Columbia. On the 7th they traveled inland from the river and the guides cautioned them regarding the scarcity of water; Brackenridge and a companion stopped near a spring, the last they would meet, but Johnson ordered them to proceed ". . . without knowing where water was to be found, or where he was leading the party to. He was by far to[o] consequential & ignorent of such undertakings, to think of consulting any one. $60 for a guide and 50 more per month for a boy to wait upon him day and night, is rather a round sum to go for nothing . . . From this spring Mr. J rode a head of the party and . . . in place of taking a north direction rode straight on. There being no marked path we all stuck to our Indian guide, who was leading us in the direction of Okanagan . . ."

Johnson was missing at sundown and had to be found and eventually got to camp, where the party had neither fire nor water. On the 8th day they reached "Okanagan Fort" and a canoe was sent to ferry them across. The fort was in charge of a "Canadian Frenchman," by name La Pratt, who rendered all the assistance in his power.

". . . Provisions was the principal thing we wanted, and tis a fact as singular as tis true, that the party after starving for 10 days, should bring into this place not less than 25 lbs of Pork, 3 whole Cheeses, 3 cases of Sardinias, with some Butter—(the Sardinias & 2 of the Cheeses were afterwards made presents off to individuals). Had I then[1] had the least idea that such conduct would have been approved off by the commander, or that he had direct orders to act as he did, I would certainly have taken the shortest way for the U. States, viz: across the Rocky Mountains."

Brackenridge spent most of the 9th caring for the "Collection of Plants" which, ". . . by being carried in a bag on horseback, were moist and a good deal bruized . . ." In the evening, however, he visited the junction of the Okanagan and Columbia rivers, about two miles below the fort. On June 10 he made a trip of eight miles along the Columbia. That day Johnson did not return from one of his excursions, although his horse did. On the 11th he was found, asleep, about three miles from camp: ". . . this frolick[2] of his kept the party ⅔ of a day back."

9. Sperlin makes this comment: "This genus (cypress) is not reported as native for Washington. What tree was it that he found? The Alaska cedar, *Chamaecyparis nutkaensis*, is popularly called a Cupressus, of which it has the main characteristics. Piper says the Alaska cedar is found as far east as Idaho."

1. "This passage [writes Sperlin] was almost certainly written long after, probably after the court-martial trial of Captain Wilkes."

2. Bancroft's *Northwest Coast* (2: 678, *fn.* 10) explains that, when he left the fort on the 10th: "Indulging in somewhat too liberal potations at parting, or else overcome by his private bottle, Johnson became separated from his party, and lay the first night out upon the ground, alone, dead-drunk."

On June 12 the party reached the Grand Coulee, situated near the junction of four counties: Douglas, Grant, Lincoln (south) and Okanagan (north). It was ascending the Columbia River. Travel was rapid. On the 13th camp was about two miles below the confluence of the Spokane which enters the larger stream from the east and now forms the boundary between Stevens County (north) and Lincoln County (south). That day Brackenridge had found "a singular species of *Trillium* almost stemless." On the 14th Johnson went off to visit the missionaries Cushing Eells and Elkanah Walker but, finding them away, rejoined the expedition two days later. In the meantime the others kept up the Columbia and on the 15th ". . . reached Colville . . . where [notes Brackenridge] we were kindly welcomed by Dr. A. McDonald the Superior." On the journey he had kept near the river banks, through ". . . a loose sandy Pine forest . . ." and had discovered a ". . . *Campanula* & several fine Polemonaceous Annuals . . ." among other plants. He spent June 16 ". . . reducing the size of our specimens bag, also drying some of the recently collected ones & picking up a few fresh ones in the vicinity of Fort . . ." On the 17th, with Pickering, he walked to "Kettles Falls" less than a mile below the Fort. One of Johnson's last actions before departure, was to make ". . . a present to Mr. Maxwell—one of the Ships Bowie Knife Pistols, which Mr. P[ickering] carried to protect his person—the rest of us had our Fowling pieces." When the fort was left on the 19th, the course was "due south by comp[as]s" through a marshy meadow.

"Met with today *Potentilla fruticosa, Lilium* sp: in woods. Saw also a few Larch trees intermixed with Pines on the hills. The Spruce was very scarce, no hard wood . . ."

On June 20 the party ". . . reached the Mission Station, of Messrs Walker & Eel[l]s, who belong to the American Board of Missions . . ." Johnson and others were to visit ". . . the Pointed heart Lake [Lake Coeur d'Alene] . . . the remainder of our party were to make the best of their way to Fort Walla Walla, by way of Mr. Spauldings Station on the Kus Kutskii River, to be under Charge of Mr. T. Waldron. This Mission is distant from Colvile 60 miles, between which is the valley that stretches from the one to the other . . ."

June 21: "Attended family worship in the House of Mr Eel[l]s . . . after which made a Botanical excursion . . . a warm discussion ensued between Dr. P. & Mr. J. respecting the moovements of the party, and from Mr. J.'s reasoning the old adage to me was made clear—That a Sailor on shore was a fish out of Water.—

"Our party left Mission station . . . Mr. & Mrs. Walker accompanying us several miles . . . After riding 10 miles made the Spokane and before dark both Horses & luggage were landed on the opposite bank where we pitched our tents. The track of country from the Mission to the River Sp[okane] is poor & thinly wooded with Spruce, Larch, & Pine, neither of which are of great size. The banks or margin of river for a good distance back is sand & gravel with a few Alder and Willow bushes . . ."

June 22: "Travelled today about 30 miles, in an E S East direction . . . the country is hilly, with Lakes and open glades intervening—the soil is poor sand and stones, a few scattered Pines on the Hills, while arround the Lakes Cotton wood and Willow bushes were seen. In water *Nuphas lutea,* & *Menyanthes trifoliata,* were abundant."

On the 23rd they rode for twenty miles in the afternoon ". . . through a rich fertile valley . . . our horses walking up to the knees in Clover . . ." On the 24th they moved ". . . over a fine rolling prairie country producing as fine pasture as I ever beheld in my life, well watered, though destitute of wood . . . Plants observed—*Coronilla* sp: *Frazera, Habenaria, Calochortus, Baptisia, Trifolium* sp: a good plant for Cattle."

June 25: "Made the Kus-Kutskii river [the Kooskooskee, or Clearwater] about mid-day; had to ride two miles up along its banks before we came opposite Mr. Spauld-ings[3] Station, when a boat was immediately sent over for us. We found Mr. S. and fam-ily living in a snug and comfortable House into which we here heartily welcomed. Mr. S. took me out in the afternoon to shew me farm and Cattle. The former consisted of 20 acres of fine Wheat, a large field in which were Potatoes, Corn, Melons—Musk and Water—Pumpkins, Peas Beans &c &c the whole in fine order . . . Yews breed with him twice every year . . . He has also built a Saw and Grist Mill, both of which is at the use of the Natives when they apply . . . Mrs. S. has regularly about her a number of young feamales, which she is teaching to Card, Spin, Weaves Blankets & Knit stock-ings. Mr. S. has made a great many hand Looms . . . He has also given them Cattle, Sheep, Seeds of Wheat, Corn, Potatoes &c and made them Plows & other empliments of Agriculture.

"We had a large Meeting of Natives in his House, when we endeavoured to impress on their minds the utility of taking the advice and following the example of Mr. S. in cultivating the Soil & raising Cattle, to which the[y] all agreed . . ."

On June 26, after Brackenridge had reduced the bulk of plant specimens by parting with some of the collecting paper, the party started on again, accompanied part of the way by the Spaldings and two visiting missionaries.

". . . At 3 P.M. came upon the banks of the *Snake River* . . . The Snake River where we crossed was about one mile above its junction with the Kus-Kutskii, its breadth about 250 yards, destitute of bushes or trees on its banks. This river abounds in Salmon, in Kus-Kutskii the water is so clear that the Natives say that the[y] cant spear them. These two rivers form the south branch of the Columbia by some called Lewis River.[4] A few miles below the mouth of the Snake the Lewis is bounded on the side

---

3. Henry Harmon Spalding, whose mission post was in the Lapwai valley of Idaho. He had settled there in 1836 according to Sperlin. I have devoted a chapter to Spalding who did some plant collecting for Asa Gray.

4. Sperlin states: "Now generally called the Snake, with Clearwater merely a tributary."

by a range of high basaltic columns, under the shelter of which we camp'd for the night . . ."

The excursion of the 25th and 26th had led the party into what is now the state of Idaho. On the 27th they reëntered Washington, presumably in Asotin County.

June 27: ". . . our course at first west, in the afternoon S. W. by West . . . Travelled . . . 40 miles. Country hilly with deep valleys in which we generally found water and abundance of good pasture. No Pine trees, a few Willow and Alder bushes in moist places. Soil a stiff yellow Loam. Plants: saw a *Pentstemon* with serrate leaves; *Monarda* sp: *Lupinus* sp: with white flos: very common. Camp'd at dark on top of hill."

June 28: ". . . came on a spur of the Walla Walla river, lined with a few scattered Pine & Willow bushes. At mid-day . . . we rode ahead in order to reach the Fort[5] before dark, which after riding 20 miles over a waste sandy prairie we effected. Mr. McLean one of the Companys Clerks (since the death of Mr. Pombran[6]) was in charge of this fort, which is generally called F. Nez perce, welcomed us in a very friendly manner . . ."

On June 29, after "regulating our collections," Brackenridge visited the gardens and farms belonging to the fort. On the 30th, in company with Pickering, he started for "Lewis," or Snake, River, ten miles away; but the country was "so destitute of interest in a plant way" that the Columbia was visited instead. "Of Plants we found a *Salsola*,[7] *Opuntia* sp: *Dalea*[7] sp: *Oberonia* sp: a handsome plant. *Verbena* sp: with several *Compositae*."

July 2: "Dr. Whitman[8] came down to pay us a visit; found him a very intelligent man. I was very anxious to visit his place but could get no satisfaction from Mr. Johnson as to when he intended leaving this fort, so I occupied my time in put[t]ing my collections in order to be ready to start on the shortest notice."

On the 3rd Brackenridge "Kept lingering about all day expecting every minute to

5. According to Sperlin, "Sometimes called Fort Walla Walla, but more properly Nez Perce. It was built first in 1818 by Alexander Ross."

6. Pierre Chrysologue Pambrun, who had long been a trader of the Hudson's Bay Company.

7. Writes Sperlin: "Since the Russian thistle did not reach us until after statehood and since it is the only member of Salsola in the Northwest, what plant should this be? The *Dondia depressa* or *Suaedia depressa* was called Salsola depressa by Pursh (Piper, p. 240). This blite is found in Eastern Washington, and its old name seems to identify it with the Salsola here mentioned." And of *Dalea:* "This large genus of the bean family is in Gray's Manual but not in Western Flora. Brackenridge probably found a similar genus, Petalostemum, of which a Species, *P. arnatum,* was collected earlier by Douglas on the arid plains near the Blue Mountains."

8. This was the missionary Marcus Whitman who had established his post at Waiilatpu about twenty-five miles above the mouth of Walla Walla River.

set out on our journey." On the 4th, having borrowed horses from their hosts, the party, after being taken across the Columbia River in canoes, started on their way and, when they got opposite the mouth of the Snake River, turned westward. They were now on the final lap of their inland journey and bound for Nisqually. The crossing had been made in Benton County. Brackenridge notes that on the 5th they had a glimpse of Mount St. Helens—which Sperlin states was really Mount Adams—and reached and crossed the Yakima River; this they followed westward and northwestward and must soon have entered Yakima County. The route, which kept close to the stream, was ". . . through one of the most barren countrys[9] it has ever been my lot to witness. A few scattered wormwood bushes with an occasional tuft of grass is all that releaves the eye from the open almost unbounded waste. We encamp'd near a marsh where the Moschetos were so plentiful that we could get no rest. Rattle Snakes is here also very abundant."

July 7: "In the early part of this day came up to the junction of the Spipe [Naches] river with the Eyakema [Yakima] which we foorded, but previous to this another river flowing from the S. West had joined it . . . cross'd over some high hills . . . then Came into the Spipe valley . . . here we for the first time from Walla Walla came upon Pine trees, and a small river coming in from the S.S. West[1] here joined the Spipe . . . Along its banks was a range of Basaltic columns. For the last two days we had found little new in the plant way, if we except two *Oreogoniums* and several *Compositae,* a number of small Oaks, of the same species as observed at Nesqually . . . These appeared local, as we observed them nowhere else."

On July 8 they crossed to the south side of the Naches River, then back to the north side and reached ". . . the path where we struck off on our way for Okanagan . . ." This had been on June 1, at what Sperlin called the "north-and-south arterial trail."

July 9: ". . . The pack Horse carrying the Botanical specimens got into the river and before we could drive him back swam to the opposite side. One of our Indians . . . by a good deal of trouble succeeded in getting him over again to the party. The plants being in Gum elastic bags were not the least injured, which was fortunate.

"At one Oclock P.M. arrived at the place on the Spipe at the base of the range of Mts. where the first party encamped on the 17th of May[2] to await Mr. Johnsons arrival; here we haulted . . . and then set forward to reach if possible the summit of the ridge before dark, which we effected. The snow had entirely dissapeared with the ex-

9. States Sperlin: "Even a horticulturist, as he looks upon the sage brush plain, does not always envision the marvelous change that irrigation will make, as Yakima, the sixth richest agricultural country in the United States, could prove to Brackenridge could he revisit this fair valley."

1. According to Sperlin, the river entering from the "S. S. West" was the Ahtanum River, or more properly Creek; the one from the "S.S. West" was Cowiche Creek.

2. Brackenridge must have meant May 27, *not* 17, for they were at the eastern end of Naches Pass.

ception of a few lumps under Spruce trees. The plants were now in all their beauty, *Ranunculus, Cla[y]tonia, Caltha, Menziesia, Vaccinium.* and *Helonias tenax,*[3] all vied with each other."

July 10: "On the west side of the range we found the snow more plentiful than on the east, although vegetation on both sides appeared much in the same stage of advance. it rained very hard . . . we were all drenched to the Skin."

On the 11th the party did not get started until midday which gave Brackenridge an opportunity of ". . . picking up a few plants, there being no opportunity to do so on the march from the rapidity with which we proceeded, all having enough to do taking care of his person and Horse among the multitude of Logs, Rivers and precipices . . ." Rain continued the next two days—"every thing except what was in Gum elastic bags was wet through, so that at night we all fared very uncomfortable."

July 15: "On the afternoon of this day the party reached Fort Nesqually, being absent exactly sixty days from the Ship. And casting a look back on that part of the Oregon that we had traversed it appears to me that we certainly must have viewed it in a very different light from the majority of writers that have come out so boldly in its favour. As an agricultural country to me it appears almost posetive that to take the upper lands (or those above Walla Walla) on an average, that Ten acres out of a Hundred would not produce Rye enough to cover expence of Seed and Labour. I have chosen Rye as being a grain that succeeds better on poor soil than any other that I know. And that not more than Two acres out of one Hundred would produce Wheat that would pay the farmer for his trouble. The valley extending from Colville to Chimekane (Messrs Walker & Eel[l]s station), at Lapwai on the Kus-Kutskii, (Mr. Spauldings station) and at Waiiletpu on the Walla Walla are the only three good tracts of land that we saw, or could learn anything of. Much has been said of the Willamette as a wheat country but of that I can say nothing.—Nature seems to have designed the upper part of Oregon more as a pastoral, or country for the raising of Cattle. That part of it pass'd over by us between Okanagan, by the way of the Coule[e] to the River Spokane, & from that over to the Kus-Kutskii could not be surpassed as a sheep Country by any in the world, although I have not the least doubt but that the incursions of the Woolves[4] and Indians on the herds might proove an obsticle to the Sheep farmer at the Commencement."

Wilkes's account of the foregoing trip ends with the following comment, and italics are mine:

". . . they . . . reached Nisqually, all well; having performed a journey of about one

3. The reference is to the bear grass, *Xerophyllum tenax* (Pursh) Nuttall.

4. To this Sperlin adds: "Dr. Pickering says, 'In the course of a journey of eight hundred miles, the only large quadruped we saw was a solitary wolf.' (*Quarterly*, XX: 60 . . .)"

thousand miles . . . They traversed *a route which white men had never before taken,* thus enabling us to become acquainted with *a portion of the country about which all had before been conjecture.* They had also made a large addition to our collection of plants."

If officials of the Hudson's Bay Company read this statement, they must have smiled! From the time that Wilkes's ships had sailed into the Straits of Juan de Fuca until they reached Nisqually the party had been in regions surveyed by Vancouver, and where the botanist Menzies went ashore on many occasions. Tolmie had spent six months at Nisqually in 1833. As far as the trip itself was concerned, it seems unlikely that any portion of the route was untraversed by white men. Most of it was certainly familiar ground to the British Company's traders and trappers. On the whole it would seem to have been a simple (though at times an uncomfortable) journey, and it took only fifty-eight days of the eighty allotted for its completion.[5] Wilkes had supplied Johnson with a map and had even told him on which side of the Columbia to travel, certainly indicating that something was known of the country. The party had been well received at Hudson's Bay Company posts and at missionary stations, ferried across rivers, supplied with food, given directions about routes ahead (even started in the proper direction at times), equipped with fresh horses, and so on. The only person who ever got lost was the incompetent Johnson.

What portions of the route were unknown to plant collectors? Tolmie in 1833 had made a trip from Nisqually to Mount Rainier and had collected plants on its lower slopes. David Douglas had collected along the Columbia River from its mouth far into what is now British Columbia; he had been to Fort Colvile, as had Wyeth, and to Kettle Falls; he had entered Idaho in much the same region as the Wilkes party but had gone farther, or into the Craig Mountains. So far as I know the portions of the route which were botanically untouched lay between the Mount Rainier region and the Columbia River which was marked by the crossing, as Sperlin has stated, of two mountain ranges, the Cascades (by way of Naches Pass) and the Wenatchee Mountains (by way of Colockum Pass); and, after leaving Fort Walla Walla on the return trip to Nisqually, the route westward along the Yakima River.

The journey just described had hardly ended before Wilkes ordered another—this time from Nisqually to the Columbia River. Pickering is not named as a participant but Brackenridge made the trip and it is again from his journal—and its transcription by Sperlin—that we derive our information thereon. Wilkes, as usual, described the journey in his *Narrative* (volume V), but more briefly than the earlier one. The orders,[6] dated July 17, 1841, were issued at "Nisqually Roads" and addressed to Johnson, but in an appended note his name is "erased" and that of Henry Eld "inserted";

5. Although Johnson's party left the ship on May 17 it did not start until the 19th.

6. They appear as part XIV of the Appendix in volume IV of Wilkes's *Narrative*.

he was a "Past Midshipman." It seems that Johnson had displayed insubordination to Wilkes himself. The party, briefly, was to proceed from Nisqually ". . . to Gray's Harbour, by the portage of Shaptal, and through the lakes of the same name thence down the Chickeeles [Chehalis] river, which empties itself into the ocean, forming Gray's Harbour . . . Your departure will take place . . . on the 19th, proceeding up Puget Sound to the portage . . ."

After a survey of Grays Harbor ". . . you will proceed along the coast in your canoes (choosing a smooth time[7]) . . . to Shoalwater Bay, from whence you will despatch a letter for me, directed to Mr. Birnie at Fort George . . ." After the survey of Shoalwater Bay they were to survey the shores ". . . around to Cape Disappointment . . ." In so doing ". . . you will pass over a small portage, when you will enter a lake that has its outlet in Baker's Bay, where you will join the ships, or await my arrival there . . . *Wishing you a pleasant time* [italics mine] . . . I am, &c. . . ." The party had been equipped with two canoes and four "balsas," together with ". . . provisions for twenty days, which is considered by me [Wilkes] ample time to effect the . . . objectives . . ." Until August 28 (when the party crossed to Fort George) the route was in the state of Washington. I now resume Sperlin's transcript of Brackenridge's journal:

July 19: "Having spent three days at Nesqually part[l]y on board Ship and partly on the plains, collecting a few of the summer plants; during this time a party was organized (in which I was included) which was to pass up to the head of one of the branches of Puget Sound cross overland to find the source of the Satchell[8] river, one of the tributaries of the Chekilis [Chehalis], follow this last down to the Coast where it terminates in what is called Grays Harbour. then to proceed along the Coast by way of Chennook b[a]y and join the Ship at Fort George on the Columbia . . . The party started from Nesqually . . . every one took a paddle and did his best, but night had closed in . . . before we reached the head[9] of the Sound, where we pitched our Tents for the night, without knowing where a drop of water was to be found for Supper. I had advised the propriety of haulting a few miles back at a spring & resuming our journey early in the morning, but this was not listened to, and it was only when the[y] had run themselves into difficulty that I was consulted . . . Young men from a Ship are about the poorest hands to conduct an expedition of this sort that I have ever fallen in with . . ."

While two members of the party went overland to the "Lakes[1]" to procure Indians,

7. More than once Wilkes's orders leave nothing to the intelligence of his subordinates!

8. According to Sperlin "(Called Shaptal in Wilkes's order), now the Black River, from the color of its water . . . Brackenridge is following Wilkes's instruction in using the Indian names . . ."

9. Sperlin identifies "Chennook b[ay]" above with Willapa Harbor or with "Shoalwater Bay of the *Narrative*"; and "the head of the Sound" with "Eld Inlet, more often called Mud Bay."

1. The Shaptal Lakes mentioned above in Wilkes's orders.

horses and assistance in making the portage, Brackenridge went ". . . a collecting plants, and in the woods found an *Epimedium,* the first observed by us in the Oregon. Saw Trees of *Cornus Nuttalliensis* 40 feet high [,] at 2 ft. from the ground stem 11 inches in diam . . ." The needed assistance arrived on the 21st and the portage was made to the Shaptal Lakes, crossing a series of prairies:

". . . The portage overland to Lake is a distance of about 4 miles through a forest of high trees of Spruce and Maple, with *Cornus,* Hazel, & *Spiraea* as underwood. When within about 1 mile of Lake we had to cross the end of a small prairie . . . on leaving this . . . passed through a thicket of trees . . . and another small prairie . . . surrounded on all sides by low Willow & Alder bushes . . . in it I found *Nuphar lutes,* two species of *Potamogeton,* and a little pink flowered plant in habit of *Nymphaea,* but in its vicinity plants were very scarce . . . Came into Camp taken bad with a cramp in the Stomach . . ."

On the 22nd they started on the "River excursion," which Sperlin states was on Black River or "from Black Lake to the Chehalis." It was ". . . nearly chocked up with flow[er]s: of *Nup[h]ar, Potamogeton, Comarum,* & *Sparganium* . . . Towards eavening the Moschitos became so thick that we could scarcily find the channel among the weeds . . . there being no place to land we had to keep on till 9 P.M. . . . (Near to the Lake are a good many scattered Oaks . . .)" On the 23rd another portage was necessary because the river was "chocked by brush"; Brackenridge noted "a good many Ash trees." All this time the weather had been pleasant.

July 25: "About 10 O'Clock got into the Chekilis [Chehalis] River, a little below the mouth of the Satchell (the name of the river and Lake that we approached it by) . . . with a gravely bottom and smooth water except where gravel bars set across the Channel, the soil on both sides . . . is a rich deep alluvial Loam, overgrown with Poplar, Alder, Willow, Dogwood & Raspberry bushes . . ."

July 26: ". . . no fish to be had . . . reached the mouth of the Satchap [Satsop] River, where we encampd. & made preparations to ascend this as far up as the Lake where it takes its rise . . ."

While Eld and others started on the 27th for a lake (Nawatzel by name according to Sperlin) Brackenridge remained behind and ". . . penetrated back into the Spruce forest 3 or 4 Miles during which I did not find a single new plant." The next day he went four miles up the Satsop River, finding it much "chocked up" with driftwood. On July 29 he comments that ". . . the whole of this trip so far is the poorest for a Botanist that could be picked out of the whole Oregon Territory . . ." Eld having returned with his party from their trip, the descent of the river was begun on the 30th. They had to fight the tidal waters setting in from the sea. On the 31st wind and tide made "a pretty round ripple on the water," and the canoes shipped water; ". . . by the time we reached the upper end of the Harbour (Grays), we had to make for the North Shore . . ."—some Indians had come to their assistance and Eld had a dispute with them about a canoe. Finally they camped on the bank of a small creek which Sperlin states was "Charlies Creek." On August 1 Eld began his surveying, while Brackenridge occupied himself ". . . in searching for plants and preserving the same . . . Our . . . stock of provisions was now getting low and little was procurable from the Natives. Game . . . very scarce. Of Plants the most interesting found was *Nuttallia, Epilobium, Scrophularia, Stellaria,* the former in great beauty and abundance on the meadow land behind our Camp, but on the whole a more same and meagre vegetation could scarcely cloathe any Country."

After five days in the same camp they moved, on the 6th, five miles farther down the bay ". . . to a nice patch of Meadow close to a stream of water[2] . . ." On the 7th Brackenridge ". . . penetrated the forest behind our tents to the distance of 6 Miles in search of Game & Plants, of the former I got a couple of Pigeons, of the latter only a few ferns, a *Menziesia, & Tanacetum.*" The party was hungry and on the 8th Brackenridge went, unsuccessfully, in search of clams; instead he filled his ". . . Collecting Case with Shallon berrys—*(Gaultheria Shallon).*" On August 10 they moved farther down the shore of the bay—to ". . . the S. head (or Chehilis point.)" identified by Sperlin as Point Hanson. On the 11th Brackenridge refers to the fact that the party was in "a state of Starvation." However, on the 13th, a party sent out by Wilkes arrived, bringing news of the loss of the *Peacock;* they left at once ". . . to send on as quick as possible a stock of provisions as we were in a state of Starvation, subsisting on dead Fish cast on shore by the surf, a few Clams and now and then a Grouse . . . very scarce. Our men from bad fare and the moisture of the climate with their feet always wet by dragging the Canoe over the mud flats became feable and sick, so that little progress was made in the survey . . ."

Brackenridge ("nothing of interest occur[r]ing") now passes on to the 18th, on that day giving a gloomy description of the whole region:

"The Character of the land arround Grays Harbour and up the Chehilis as far as the entrance of the Satchap is of the poorest description, the spruce forest along the most of it setting down to the waters brink, with occasionally a small patch of Salt marsh or meadow, producing coarse grass and Cat Tails . . . The only patch of land that could be cultivated as a farm is that immediately within the South Head, in extent about 100 acres, but perhaps better calculated for Cow pasturage. The coast from this point to Cape Shoalwater[3] is a smooth sand beach, with behind a line of low sand hills about 20 feet high, inclining gradually to descend in towards the Pine trees; a belt of which skirt the Spruce forest which extends inland. The coast vegetation is same all along, consisting of the following *Plants*[4]: *Obronia,* sp: flos: orange, stem prostrate, leaves alternate, Sand Hills. Umbelliferous plant, Dwf: habit, an *Hieracium* &c &c *Lathrus,*

2. This was O'Leary Creek according to Sperlin.

3. Sperlin notes that this is the ". . . north cape at entrance to Willapa Harbor, Brackenridge's Chenook Bay."

4. I run this, and other lists of plants, into a single paragraph.

an *L. Maratima,* Common Sand Hills *Sisyrinchium* sp: flos: yellow, leaves glaucus, Salt marshes *Neottia* Sp: Salt Marsh, a *Herbinaria* with green flos: same locality *Ambrosia* sp: Sand Hills *Tanacetum* same locality Aster two sp: with several *Juncaea* and *Graminaea,* an *Armeria,* with a number of Saline plants. The *Gaultheria Shallon* which produces a very palatable[5] fruit was abundant among the Pine trees."

The party had reached Grays Harbor on August 1 and had been there twenty-four days, the weather "wet and foggy" during that time. On August 24 ". . . each man with his burden on left the miserable Grays Harbour station . . ." and moved to Willapa Harbor, which Brackenridge calls "Chenook bay." On the 26th they reached a "lake" which Sperlin states was "Whealdon's pond, sometimes dignified with the name Black Lake."

". . . With the exception of the belt of Spruce in which we campd, the whole up from the bay to the Lake is one Continued Marsh or Swamp: in which I found *Vacciniums, Ledums,* and a *Myrica* . . ."

On August 27 they reached Baker's Bay where the schooner *Flying Fish* was waiting and were taken across to Fort George (the earlier Fort Astoria); Eld, the leader of the party, who had disappeared the day before, was found aboard. They were now in Oregon. From Fort George they moved up the Columbia River on a barge, leaving on the 29th and arriving at Fort Vancouver on the 31st. The entire journey had taken forty-four days as against the twenty which Wilkes had considered ". . . ample time to effect the . . . objectives . . ."

From Nisqually, Pierce County, Washington, the party had crossed northern Thurston and southern Grays Harbor counties and had reached Grays Harbor on the Pacific; from there had moved southward to Willapa Bay and to Cape Disappointment, both in western Pacific County; from Cape Disappointment had crossed the Columbia River to Fort George in Clatsop County, Oregon, and at Fort Vancouver were in Clark County, Washington.

In October and November of 1825, David Douglas had made an overland trip from Cape Disappointment to Willapa Bay and to Grays Harbor; from there he had gone up the Chehalis River, crossed to the Cowlitz which he descended to the Columbia River and then had ascended that stream to Fort Vancouver. He had an abscessed knee at the time, had traveled with only occasional assistance and at an extremely poor time of year; as far as food was concerned he had none for the most part. He liked the country as little as did Brackenridge. Although the Eld contingent did not have the "pleasant time" wished them by Wilkes, I doubt that its difficulties equalled those of Douglas sixteen years earlier.

Brackenridge, thirty-one years of age and evidently of strong physique, was no sooner back from this journey before he started immediately upon a new one.

5. Sperlin commented, under date of August 8, that "The salal berry is eaten, but it cannot be called a favorite."

The story of this trip—from the Columbia River to the Bay of San Francisco—is told in volume V, chapter VI, of Wilkes's *Narrative;* Wilkes's orders[6] were to Lieutenant George F. Emmons[7] who had been attached to the *Peacock* (wrecked some six weeks earlier). The party's ". . . absence is limited to the 10th of September . . ." The "Scientific gentlemen" were to be permitted "occasional short Excursions" while on the way. The route ". . . to be pursued by the party, is up the Willamette Valley in a southerly direction, crossing the Umpqua river and mountains, thence south and west of the Shaste Mountains to latitude 42° N." Additional orders were issued on September 1 and included more directional instructions; after passing, "if possible," west of the "Shaste Mountains" the party was ". . . to strike the waters of the Sacramento, passing over to headwaters of various streams that empty into the ocean, viz., the Umpqua, Klamet, and their branches . . . After you start, which must not be later than the 5th or 6th, I give you twenty-five days to reach the forks of the Sacramento, where the boats of the Vincennes or squadron will be on the 30th of September . . . If you should fall upon the Sacramento, taking a more easterly route, you will, if you find it difficult to proceed with your horses, abandon them, and proceed in canoes down the river . . . The route you will probably follow, is that usually taken by the [Hudson's Bay] Company's party . . ."

Under Emmons' command, of "scientific gentlemen," were Brackenridge, Titian Ramsey Peale, a "naturalist," the mineralogist James D. Dana, the botanist William Rich—who is rarely mentioned by Brackenridge—and the draughtsman or "Artist" Alfred T. Agate;[8] Pickering does not seem to have gone on this journey. Traveling with the party were settlers bound for California.

A first-hand account of the trip is, again, supplied in the journal kept by Brackenridge; Sperlin's transcript of this record covers the period from September 1 through October 2—or from the Columbia to the present Oregon-California border. Here the transcript of Alice Bay Maloney takes up the record, from the California line to Yerba Buena (San Francisco) and covers the period from October 2 to 28, inclusive. My citations as before are, for a time, from Sperlin.

September 1: "Had received orders from the Commander to prepare by the following morning to join a party of the Peacocks officers who had been laying encampd. up the Willamette for several weeks, this party under the guidance of Lt. Emmons was to proceed overland to California & join the Squadron at San Francisco. We had already had a pretty hard Campaign, but as it was a new field we were anxious to get into it.

"Of F. Vancouver and its environs I can say but little, having only spent a few

6. They appear in part IV of the Appendix, volume V of the *Narrative* and for some reason were "(Confidential)."

7. George Foster Emmons who, according to Maloney, later became an Admiral in the U. S. Navy.

8. Barnhart refers to Agate as ". . . particularly engaged in making drawings of living plants."

hours with the principal, Dr. M. Loughlin, who in the most friendly manner showed me round his gardens, under the keeping of Mr. Bruce,[9] a Scotch Highlander by birth. The Apple Trees bore remarkable heavy crops of fruit and were invariably in a healthy Condition, there were from 4 to 500 of these in a bearing state, and with the exception of a few approved varieties imported from England the whole stock had been raised from Seeds at Vancouver, and to my taste the majority were better adapted for bakeing than for a dessert, but in a new Country certainly a great acquisition. Gooseberrys and strawberrys the[y] had of the finest sorts. Peaches and Nectarines the same. Grapes I was told had succeeded well but of late years their cultivation had been neglected. Melons, Musk & Water, do well. Of Vegetables the[y] Can raise any quantity, all of which produce good Seeds . . ."

On September 2 Emmons, who had been kept at Fort Vancouver with the "Tertian Fever" but had now recovered, started overland for the Willamette, while Brackenridge, Eld, and "Henery" (always so spelled in the journals) Walton, traveled thither by canoe. Writes Brackenridge, ". . . as this river has been denominated the paradise of the Oregon, I kept my eyes about in order to catch some of its beauties. After leaving the Columbia, the first 10 miles of the Willamette, the banks is covered with Willow, Alder, & Dogwood, behind which rises spruce trees, *(Abies Douglasii),* and as you proceed farther up towards the *Falls,* the same kind of brush wood line its banks, but behind this occasional patches of open prairie which support solitary Oaks . . . are observed . . ." On the 3rd they passed the Falls—situated at present Oregon City— and kept on to ". . . what is called Champooi [usually spelled Champoeg], or the first of the Settlement where we camped in a Barn yard . . ." This proved to be on the farm of Thomas McKay.[1] He provided the party with horses and a cart and they moved on to a point opposite "the Mission Station." This, according to Sperlin, was on the east side of the Willamette, south of Wheatland. Brackenridge had walked this distance and offers a number of opinions about the settlers: the "Canadian Frenchmen" lacked "both means and energy ever to become very formidable competitors in the farming line"; the Methodist Mission "by some claim or another have taken possession of 8 Square Miles of the best of the land," cultivating part and preventing settlement on

9. Wilkes also refers to the gardens and to the fact that they contained ". . . all kinds of vegetables and many kinds of fruits, with which the tables are abundantly supplied by the gardener, 'Billy Bruce.' After William Bruce's first term of service had expired, he was desirous of returning to England, and was accordingly sent." While in England, Wilkes continues, he was employed at ". . . Chiswick, the seat of the Duke of Devonshire . . . but no place was like Vancouver to him, and all his success here continues to be compared with Chiswick, which he endeavours to surpass . . ." There is further reference to this interesting character: "Even Billy Bruce the gardener made us his debtor, by sending us repeatedly some of the fine fruit and vegetables grown under his care. I have endeavoured to repay him, by sending him seeds; but the route is so long and circuitous, that it is questionable whether they ever arrive, and when they come to hand, if I shall not be classed by him with those who have sent 'trash' to Vancouver, for him to waste his time and experience on, in attempting to cultivate."

1. Sperlin states that he was the son of Alexander McKay, and a stepson of John McLoughlin.

the remainder; the Americans, for the most part, had settled on the Tualatin plains which was said to be good farming land but, "from the Careless manner that the[y] plow and dress their land," probably did not get as high a yield in wheat as they otherwise might. The party evidently remained here for two days and Brackenridge did some collecting.

"... The season for flowers being now past, the Seeds were in good condition, particularly Annuals: *Oenotheras, Gilias, Escholzia Californica, Clintonia, Mimulis,* and several other good plants I secured ... I did not visit any of the Mission Gardens, but I understood the[y] are well managed and produce good Vegetables."

After getting "the luggage devided out into Horse loads" on the 7th, they moved on to another farm where they were obliged to remain on the 8th while a search was made for two lost horses.

September 9: "The party again made a start, mooving slowly along over an open prairie country, swelling gently into round hills with a few scattered Oaks along their Summits. In moist low ground a Species of Ash reaching to the height of 30 or 40 feet, the trunks of which are to[o] small ever to be of much utility. Campd. on Ignat Creek[2] ..."

The mileage from the 9th through the 13th totalled sixty miles. Brackenridge does not refer to the Willamette River during this period but presumably they were following its course; on the 11th he mentions that on the banks of "the Lamale River"— which Sperlin states has "been modernized to Long Tom River"—grew "... Dogwoods, *Spiraea,* Willows, Alder, and Close by Clumps of a large *Pinus,* near to *P. ponderosa,* procured Seed of *Madia elegans,* but the rascally Indians by setting fire to the prairies had deprived us of many fine plants ..."

On the 15th they came "to the base of the Elk mountains," which Sperlin identifies as "the divide between the Willamette and the Umpqua."

"... Their ascent was gradual on both sides, breadth of the range where we crossed, 10 Miles. The whole is a broken Chain of round [k]nobs the highest of which we estimated at 1600 feet.—the summits are clothed with Spruce and Oak trees, with an undergro[w]th of *Cornus,* Hazel, *Arbutus, Rubus, Ceanothus,* also a species of *Castanea* with lanceolate leaves very rusty beneath, the cups of the nut very prickly, in moist places at the base a sp: of *Rudbeckia* four feet high ..."

On September 16 they reached Elk River, according to Sperlin a "... northern affluent of the Umpqua [which] enters about three miles below Fort Umpqua." The Fort was situated "opposite Elkton [northern Douglas County] on the Umpqua River." Here they encamped.

"... Mr. R[ich] and myself examined our collections, and in the afternoon walked out, where we found a species of Oak, new to us, its size and habit is near that of the one on the Willamette (and which continued plentiful all along our route), but the

2. According to Sperlin, "The name does not seem to have survived."

lobes of the leaves have a spine at their termination, the acorns are larger and sit deeper in the Cup. A *Lonicera* with yellow flos: was also got on the banks of the River . . ."

On the 17th Brackenridge was busy "collecting Seeds and plants." Others visited Fort Umpqua, returning with fruits and vegetables. On the 18th they proceeded, through a country which for a time was "rather mountaneous in its character,"

". . . the tops of the ridges *stud[d]ed* with Pine and Spruce trees, flanked beneath with Oaks (2 sp:) Hazel, *Ceanothus, Rhamnus,* & *Arbutus.* The flats between the rising ground is rich deep soil with Clumps of Ash and Dogwood, the grass had been burnt up by a fire which we saw rageing ahead of us and were compelled to urge our horses through it. Campd. on Billys River . . . *Plants, Asclepias,* sp: in moist places. *Compositae* small white fld. Annual. *Epilobium,* flos: large purp[l]e, a very handsonme annual."

September 19: ". . . the country swelling gently today into Mts: 6 & 800 feet high covered with Pine and Oak, with solitary large trees of *Arbutus procera.*[3] The natives appear to use the bark of this tree for some purpose or another, as the base of the trunks were in general partly stripd. of it. . . . At mid-day made the north branch of the Um[p]qua River . . . *Plants: Ceanothus* sp: flower beautiful Sky blue, a shrub 6 feet high—A Shrub 15 feet high, related to *Dalabardia.* Seeds like *Pelargonium. Oreogonium* 2 species on sand flats, both with orange blossoms—*Laurus*[4] *Ptolemii?*—of Hooker, a very handsome evergreen tree. *Necotiana* sp: flos: white & tubular—also a small annual new. The banks of the river are steep and thickly lined with *Cornus, Dalibardia?,* Ash, *Lonicera,* Alder, & Hazel bushes . . ."

Throughout this part of the journey the party seems to have been wary of the Indians, keeping them away from camp, tying up the horses at night, and so on.

September 20: ". . . in the afternoon crossed the South branch of the Um[p]qua River, which is much smaller than the north one . . . the party encamped on its banks . . . *Plants: Madia* sp: (1) leaves linear. flos: yellow, plant 2 ft. *Lupinus* sp: (1) flos: yellow, solitary on plains. *Monarda* sp: Same as seen at Fort Okanagan. *Compositae* flos: white, seen at the Willamette. *Vitis,* A kind of Fox Grape, fruit edible, in considerable abundance here. *Fraxinus,* sp: leaves wooly, the *Laurus* was found here in large masses."

September 21: ". . . had to ford the Um[p]qua three times during the day . . . Campd. on a small stream at the base of the Um[p]qua Mts. . . . *Plants: Prenanthes,* flos. yellow 4.—*Epipactus* sp: moist places. *Escholzia Californica* san[d]y plains. *Acer* sp: like

3. This is now *Arbutus menziesii* Pursh, the madroña. I am told that in California the preferred spelling is madroño.

4. Writes Sperlin: "Since neither of the two species of Laurus is native to the Northwest, the plant here is probably *Umbellularia californica,* California laurel."

A. Macrophylla, a large tree. Azalea sp: fol. glutinosa, 10 feet: Mts. *Compositae* sp: 4 flos: yellow sand banks."

September 22: "Began to ascend the Mountains . . . which at first was gradual and easy . . . we descended occasionally nearly as much as what we had risen . . . on gaining the summit we kept along the ridge for a short way and descended rapidly into a valley . . . soon after this came to the highest and most difficult pass of all, which took us at least three hours to accomplish, the woods [had] lately been on fire and before we got over wer[e] as black and as unc[h]ristian like as so many Negroes from the coast of Africa . . . arrived at small prairie where we campd. . . . The whole range is densely wooded with the following trees & Shrubs. *Pinus Lambertiana*—120 feet in height, *Pinus* sp: an *P. ponderosa Quercus* 2 sp: large trees—*Andromeda* sp: same found at Nesqually. *Arbutus* sp: frt. red, fol: glaucus, a bush 6 to 8 feet high. A Shrub like *Beurhamia,* 20 feet high. *Gaultheria Shallon Prunus* sp: *Cornus Nuttalliana. Mahonia* sp: perhaps new. Yews. Dogwoods. Hazel. *Spiraea & Castanea* formed the vegetation."

I pause to mention another botanist who had reached much the same region, and by a similar route, fifteen years earlier. David Douglas, in September and October of 1826, from the Columbia River had traveled up the Willamette, crossed to the Umpqua River and, in the vicinity of what is now Roseburg, Douglas County, had acquired the pine cones which were the objective of his journey. More than one of the plants mentioned by Brackenridge had been observed and collected by his predecessor-botanist. I mention but two. The ". . . species of Castanea with lanceolate leaves very rusty beneath" which Brackenridge had noted in the Elk Mountains on September 15 was the golden-leaved chestnut or chinquapin of the northwest, first discovered by Douglas and named by Hooker *Castanea*[5] *chrysophylla;* every collector sent out from Britain at this period was instructed to obtain propagating material of this tree. The second species noted by Brackenridge on September 22 was the sugar pine— *Pinus Lambertiana*—collected by Douglas at considerable risk to his life on October 26, 1826. Douglas named and described this tree in 1827.

Travel over this route appears to have presented some of the same dangers in 1841 that it did in 1826. The Indians of the region were still imperiling the lives of white men; the "Tertian fever," soon to attack the members of Brackenridge's party, was probably identical with the "intermittent fever" of Douglas' day. But we note a difference. In the 1820's all those who wished to move from Oregon to California, or *vice versa,* traveled by sea, usually by way of the Hawaiian Islands; in the early 1840's even settlers were making their way overland.

To move many years forward: much of the route from the Columbia River (followed as far south as Roseburg by Douglas, and all the way to the Bay of San Francisco by Brackenridge) is now closely approximated by the Shasta Route of the "Pa-

5. Subsequently placed in the genus *Castanopsis;* although a further change has been suggested in recent years, many botanists still adhere to this classification.

cific System" of the railroads. For, in Oregon, this follows the lower elevations of the river valleys to the west of the Cascade Range, crosses the Siskiyou Mountains near the Oregon-California border and, on entering California, continues down the Sacramento River valley between the Klamath Mountains and the Coast Range (to the west) and the Cascade Mountains and the Sierra Nevada (to the east).[6]

September 23: "Mr. Peales note book and several other things belonging to him having got lost yesterday, a party went back in search of them, which detained the main party in Camp all day: in the vicinity a good many seeds were collected: of Plants we found a bulb in habit of an *Anthericum* which the natives of California made use of in place of Soap. *Lilium* sp: with large Orange flowers. *Pyrola* sp: with glaucus leaves. Also a *Nuttallia* which I thought different from the one on the Willamett . . ."

September 24: ". . . had . . . some difficult passes to make with dense thickets of brush wood on all hands. The country was of a mountaneous nature, well wooded, with tracts of Prairies between . . . pitched our Tents . . . on Youngs creek. *Plants: Actea* sp: on the banks of creeks. *Arbutus* sp: leaves wooly."

September 25: ". . . At 3 P.M. cross'd Rogue River, so named from the Rascally Indians that live on its banks. The country around is hilly with numerous Oaks & *Pinus Lambertiana* growing on them. Several of the Cones of this species that we examined, measured 15¾ in length by 18½ in circumference,—that is when the scales were reflexed . . . *Plants: Pentstemon* sp: leaves lineare. A second sp: on prairies. *Coronilla* sp: flos: yellow (same at the Spokane. *Eryngium* sp: flos: blue. *Oreogonium* sp: in dry situations, very neat and dwf: several small compos:"

On the 26th they got to "what is called Turners encampment," where a man of that name and his party had had a battle with some Indians, with losses on both sides. On the 27th careful watch was kept, for the savages were known to be in the vicinity, ". . . the party . . . wanted nothing better than to get a sight on one of the rascals." On the 28th "Several of the Gentlemen who had already suffered from the Tertian fever, were again attacked today, but the leader of the party in place of holding on for them, pushed ahead, but five of the party who had a little more sympathy for their fellow beings remained behind to protect the sick individuals from the assaults of the Savages . . . the[y] did not reach the Camp till after dark . . . *Plants,* nothing new to the list—"

On September 29 the party was ". . . at the base of the Shaste mountains, which are in general considered the boundary line between the Oregon & California Territories . . ." Sperlin mentions that Wilkes's *Narrative* refers to them as ". . . Boundary Mountains, i.e., boundary between Oregon and Mexico (California). Now the Siskiyou Mountains." Brackenridge continues:

". . . When on the summit we got a sight of a high Snowy mountain in form some-

6. *See* Diller, J. S. and others. 1916. "Guidebook of the western United States. Part D. The Shasta Route and Coast Line." (*Dept. of the Interior, U. S. Geological Survey* Bull. 614.)

what like Mt. St. Helens, & soon descended into an extensive valley where we encamped on the bank of a small stream for the night. The height of the range where we
crossed I estimated at 1500 feet. In vegetation it is very poor and the only things new
in the plant way was, *Compositaea,* fol. lineare silky, flos: yellow, shrub 4 ft. *Cupressus*
sp: a tree 30 ft supporting a curious parasite: near allied to *Viscum. Baptisia* sp: in
seed. *Cassia* sp: no flos: in seeds. *Oreogonium* sp: flos: orange 18 inch high. No *Pinus
Lambertiana* for the last three days. Soil very poor . . ."

Sperlin says that the camp of the 29th had been one mile from the point where "the
small stream," Otter Creek in Wilkes's *Narrative,* joins the Klamath River. The party
appears to have left the range of the sugar pine near Rogue River, or on the 25th.

September 30: "The Physician having recommended the propriety of the party remaining in Camp all day, in order to recruit the Sick. On which Mr. E. condescended
so far as to permit those who were well, to leave the Camp on duty. While I went on
an excursion three of the party were s[e]ized with the fever . . ."

The Sperlin transcript of the Brackenridge journal continues through October 2.

Sperlin ends his paper with two lists of plants, one entitled "Fruits Indigenous to
the Oregon Territory That Are Eaten by the Natives," the other, "Roots of Indigenous
Plants Eaten by the Oregon Indians." To the first the editor appends the following
note: "This list of edible fruits, seeds, and roots is inserted by Brackenridge immediately following the Inland Expedition;[7] but as it lists a large number of these that
were not found until later, I have transferred it to the end of the Oregon exploration."

The Maloney transcript begins on October 1 and I shall now follow the Brackenridge journal therein.

As already stated the California plants mentioned by Brackenridge in his journal
are associated by number with the identifications supplied by Alice Eastwood, who
comments:

". . . He recognized well-known genera along the route, and the families to which
the plants belonged; but, with few exceptions, the species were unfamiliar to him,
since the flora of California is unique among world floras. His descriptions of those he
did not know, however, and his likening them to plants with which he was familiar,
were of much help in making the identifications;[8] and, besides, I have made many collecting trips in the area through which the expedition passed."

October 1: "Mooved from camp ground . . . crossed the Chaste River soon after . . .

7. The journey from Nisqually to Fort Covile and back.

8. Miss Eastwood states later in her paper: "Not having access to Brackenridge's collection at the National Herbarium, nor to Torrey's *Phanerogamia of Pacific North America* (printed in 1874 in Philadelphia as a part of Volume XVII of the Expedition's publications) the identifications are in certain instances doubtful . . . Although there may be a few that are incorrectly identified, all the species in the appended list are to be found along the route . . ."

passd over during the day a gravely sandy desert which continued 12 miles and bounded by conical low hills, came again on to the Shasty & camped by it . . . Plants,[9] *Compositae*—in habit of *Ephedra*. Salt Marsh. *Cruciferaea*—flor. in spikes of a cream colour, 15 to 18 inch. *Dalibardia*[1]—a smaller species than the one on the Umpqua River. *Fournfortia*—habit of, leaves glutinose, *Rhus* sp.—leaves trilobate. *Helianthus* sp.—flor. yellow, *Scilla* looking plant in meadows near river. *Lychnus* sp.—flor. Lilac fringed. Weather very warm, no water for 15 miles, miserable country, the Shaste Valley—"

October 2: "The country was a trifle better than yesterday, a deal of crusted salt was found in low places on the prairies, with patches of *Spiraea* & Dogwood & a better supply of grass . . . Plants. *Oreogonium* sp.—fol. spatulate, flor. pale rose, stem 4 inch —salt marsh. *Parnassia* sp.—an *P. palustris* of Europe—stem 6 inches, flor. white— marsh. *Gentiana* sp.—annual, flor. pale lilac, fol. glaucus, 8 inches—borders of stream, a second sp. was found in marsh but no flowers. *Bartonia* sp. flor. orange— local on prairie. *Cleome* sp.—flor. yellow very handsome, *Lupinus* sp. flor. blue—stemless, three or four different *Compositaea,* & a *Campanula*. Also a sp. of *Pinus* like *Pinus Sylvestris*[2]. . ."

October 3: "We had now to ascend and cross the California range of Mountains which according to our guides opinion was to take us at least seven days, but as neither our Horses & many of the party being in the best condition we did not reach the head of the valley of the Sacramento till the afternoon of the 10th, the route which we took was to the west of the Bute or snowy Mountain,[3] but tis my belief that had we kept to the eastward of it our route would have been shorter and easier . . . On the afternoon of the first day . . . to the west of the Bute we came upon the head waters of the Sacramento— a small stream about three yards broad, we were told by our guide that this was the principal branch & off and on we kept to this river till we reached the valley on the opposite side, of fine flowers and shrubs in the proper season there must be a great abundance & and we had still I think the good luck to find some plants that have not yet been known to Botanists, and in the following list theese will be found,—

9. As in the Sperlin transcript, I run Brackenridge's plant lists into a single paragraph and italicize the scientific names. Many are misspelled but the errors are corrected in Miss Eastwood's list.

1. According to Eastwood, ". . . One species on the Atlantic coast, *Dalibarda repens,* is a low plant related to the raspberry *(Rubus)*; but nothing similar, nor one described as having glutinous leaves, has ever been found in the Pacific coast area. However, a common, shrubby, wild currant, *Ribes cereum,* has glutinous leaves somewhat similar in shape to those of *Dalibarda,* and at the time Brackenridge saw it would have been without flowers or fruits. What he saw was probably this plant."

2. Identified by Eastwood as "*Pinus Murrayana* Balf., lodge-pole pine and tamrac."

3. Maloney cites George Gibbs to the effect that this is " 'Shaste,' or as it is usually called the 'Shaste Butte,' . . ." not part of a connected chain but rising " 'near the connection point of several; the headwaters of the Sacramento separating it from the great range bounding the western side of its valley, and from the peaks which form the source of the Trinity.' "

"Plants observed, and specimens collected on California Mountains—*Chimonan-thus* sp.—a shrub 10 ft.—leaves fragrant, hab. moist banks. *Sarracenia*[4] sp.—leaves 3 ft. long—flower stem exceeding the leaves in length, hab. wet places. *Diospiros*[5] sp.—? a shrub 4 ft. high—on dry banks. *Pavia* sp.—a small tree or bush—fruit the size of a Peach, hab. on dry banks. *Zauchnina* of Presl, a splendid plant with the fruit of an *Epilobium* & flower of a *Fuchsia*—hab. in dry sandy places—*Cercis* sp.—in fruit, a shrub 10 feet high—found afterwards in the valley of Sacramento *Paruassia* sp.—flor. white, hab. in marshes. *Mitella* sp.—large broad leaves like *Gunnera,* leaves 15 inches broad, hab. banks of river. *Cypripedium* sp.—moist places—out of flower. *Convolvulus* sp.—a neat little creeper with white flowers—hab. in dry rocks. [*Convolvulus*] sp:— leaves hastate—flor. a greenish white—plant prostrate woody, hab. rocks. *Quercus*[6] sp.—acorns long, cups shallow, leaves entire, evergreen—a tree 60 ft., bushy —[*Quercus*[7]] sp.—acorns slender very pointed, a scrub oak much branched, hab. on top of Mts. *Ceanothus* sp.—leaves lanceolate, green above & whiteish beneath, stem enclined. frt. in racemes. [*Ceanothus*] sp.—leaves dentate, stem trailing on the ground, habit that of *Dryas octopetala. Rhamnus* sp.—flor. small green, frt. a black berry the size of a Pea, shrub 8 ft.—handsome. *Polygala* sp.—flor. pale blue, whole habit that of *P. Senege,* hab. the dry rocky banks. *Pentstemon* sp.—leaves glaucus and pointed, hab. dry sandy places. *Viola* sp.—a very singular plant growing in company with an *Aurum*—no flor. on either. *Thuja*[8] sp.—a tree 40 ft.—different from the one on the Columbia. Two trees related to *Castanea* with prickly capsules. *Oreogonium* sp.— stem herbaceous—simple, flor. globose, leaves spatulate. [*Oreogonium*] sp.—stem shruby—flor. deep orange, hab. dry sandy places. *Cephalanthus* sp.—6 ft. high— banks of the river. *Dendromicon rigidum*. Hooker; related to *escholtzia*—a shrub on the tops of open ridges. *Audibertia incana*[9]—flor. bright scarlet—a very handsome plant. *Scrophalarinea*—a handsome lilac flowered plant on dry rocks. *Psoralea* 2 sp. —both now out of flower—*Cheilanthes*[1] sp. a neat little fern on the face of dry rocks. *Woodwardia*[2] sp.—found in moist shady places. *Mahonia* sp.—leaves more glaucus—

4. This was the *Darlingtonia californica* Torrey, the California pitcher plant. Barnhart records that the date of its collection "... was the fifth of October (1841, not 1842 as has been stated erroneously in all accounts of the discovery of this plant) ..."

5. Eastwood notes that "... there is no native persimmon on the Pacific coast ..."

6. According to Eastwood, "*Quercus chrysolepis* Liebm., golden oak."

7. According to Eastwood, "*Quercus Brewari* Engelm., Brewer's oak or mountain oak."

8. States Eastwood: "*Chamaecyparis Lawsoniana* Parl., Lawson cypress or Port Orford cedar." She also comments that its foliage "... somewhat resembles that of *Thuya,* Lawson's cypress can be seen today near Shasta Springs."

9. Eastwood notes: "*Salvia carnosa* Dougl. (*Audibertia incana* Benth.)"

1. "This is probably what he later named *Onychium densum* Brack., renamed by Hooker *Pellaea densa,* and later renamed by Maxon *Cheilanthes siliquosa.*" (Eastwood)

2. "*Woodwardia Chamissoi* Brack., perhaps named by him from these specimens ..." (Eastwood)

prickly & upright than the Oregon ones. *Aristolochia* sp.—a climbing shrub—but now out of flower. *Photinia arbutifolia* Lindl. a very handsome shrub. 15 ft. high. *Crucifera*—flor. purple, pod hook shaped, leaves ovate, amplexicaule, glaucus. *Crucifera* sp.—on rocks of an upright habit. *Composita*—4 or 5 very neat flowering annuals. plant related to *Ilex,* with red fruit and much branched. *Pinus*[3] sp.—cones arranged in bunches on old stem, leaves in threes, stem 40 ft. high. [*Pinus* sp.]—cones globular oval large, leaves long, 6 inches, tree much branched, this species as also that of *Lambertiana* furnish large edible seeds, which the Indians are fond of. *Mimulus* sp.—perhaps *M. Cardinalis,* but it struck me that our plant was much finer, the flowers being larger and of a deeper crimson."

On October 11 Brackenridge comments that, in crossing the mountains, it was too late in the season to see much of the "annual vegetation" but, from the "dried fragments" observed, he felt that the plants must be beautiful and varied. The party passed "an extensive Soda spring," identified by Maloney as "Lower Soda Springs on Soda Creek, which flows into the Sacramento from the east near Castle Crags." They are on the northern border of Shasta County. Brackenridge, who seems rarely to find much to praise, evidently thought well of Soda Springs ". . . the effervescence of whoos waters were as agreeable as any manufactured in our large citys—and to be in possession of such a fountain in the U. States would be having of a fortune.

"At the head of the valley we met with a great many Indians, who were very friendly and docile in there manner, theese people subsist principally on fish, Nuts and other seeds of plants, the[y] shewed us cakes of bread as black as coal, made of Acorns pounded into a meal between two stones, this bread was sweatened by the berrys of the *Arbutus* which have a very pleasant tartish flavour. Grapes of the Chicken sort were now in great profusion on the sacramento, and we found the Indians busy collecting the acorns from the trees & drying them in the sun for a winters store, theese acorns in a raw state are very agreeable to the palate, being altogether free of that bitterish tannin-principle peculiar to most of the Oaks.

"By a calculation made each day of the number of miles gone, I find that the breadth of the range where we crossed—was—108 miles—

"The first days journey down the valley was on the north side, the country descending very gently—soil, sand and gravel, supporting a few scraggy bushes. I observed a great many decayed stems of several kinds of bulbs, which I took for *Calochortus* or *Scilla,* this country appears to be rich in Bulbous plants . . . camped on the banks of the Sacramento, where we found a fine species of *Platanus* of very graceful groth."

On the 12th they kept close to the north bank of the river for some time and then crossed it. They had seen several grizzly bears— ". . . of Antelope, Dear, Elk & Bears I never [saw] such number in my life . . . no variety of plants—the prairies having been burned over—"

3. According to Eastwood, "*Pinus Sabiniana* Dougl., digger pine . . ."

October 13: "We found the valley on the South side of the River flater and the soil richer, the most of the good land was covered with stately Oaks of two different species.[4] I calculated 20 good trees to the acre, saw a number of the Indians collecting the Acorns ... Plants observed—*Gnaphalium* sp.—2 inches high, whole plant downy with short spines, prairie ground. *Fraxinus* sp.—leaves broad—much devided, appearance of *Acer,* 30 ft. *Umbrosia* sp.—banks of the Sacramento—*Rosa* sp.—flowers much clustered—handsome—*Dipsacus*[5] sp.—banks of river—*Platanus,* Willows & Ash line the banks of the river—"

Following the river on the 14th they got into mud and had to make an eight-mile detour; the only plants recorded are "*Valerianella* sp.—not in flower—*Baccharis* sp.— a shrub 10 feet high—banks of the River—" On the 15th there were no oaks on the prairie, only along the river. The valley had been flooded for most of the summer. There were "... a few annuals related to *madia*—of very stiff habit—nothing in the way of plants new to us was observed ..." On the 16th they came in sight of what their Canadians called "the Bute mountains," barren peaks with rocky rugged summits and with "a few scattered bushes & trees vegetating in the cliffs." On the 17th they got through the mountains after experiencing "a complete hurricane" during the night, emerged upon "an extensive prairie thickly covered with Oak trees," reached Feather River and, since no good fording place was found, kept down it for a time. Brackenridge noted that "... near this river I saw some of the finest Oak timber observed during the whole trip through from the Columbia ... of Plants new to us Heliotrope, on salt marsh. Ariodeous plant, in habit of *simplocarpus,* margin of Ponds. *Cassia* sp.— pods bristly—common on dry prairie."

On October 18[6] the party "... forded the Feather River about ¼ of a mile above

4. Identified by Eastwood as *Quercus lobata* Née, valley oak, and *Quercus Wislizenii* A. DC., interior live-oak.

5. Eastwood calls this "*Dipsacus fullonum* L., fuller's teasel. This is surprising, as it is an introduced species in California—perhaps by the Mission fathers for the purpose of carding wool (?)."

6. It was on October 18, 1841, according to Bancroft's *History of California,* that Wilkes, commanding the United States Exploring Expedition, met "outside the heads" of the Bay of San Francisco, a French traveler by the name of Eugène Duflot de Mofras who was bound for Fort Vancouver in the ship *Cowlitz.* This visitor to our west coast deserves mention.

Bancroft states that De Mofras was then "... attached to the [French] legation at Mexico, with a special mission to visit the north-western provinces of the [Mexican] republic, and the American, English, and Russian posts beyond, 'in order to ascertain, independently of a political point of view, what advantage might be offered to our commerce and to our navigation by mercantile expeditions, and the establishment of trading-posts in those regions still little known to France.' ..." Bancroft supplies the visitor's itinerary and I quote therefrom: "... In April 1841 he came up from Mazatlan ... on the *Ninfa,* touching first, perhaps at San Pedro, and arriving at Monterey in May. Before June 11th, he had visited Sonoma ... and probably went to Ross before returning to the capital. In July he was at Monterey ... September 1st he arrived at Sutter's fort; and during the same month was at San José and Santa Cruz. October 18th ... had embarked at San Francisco on the *Cowlitz* for Fort Vancouver, meeting Wilkes outside the heads; and on December 30th he came back on the same vessel to San Francisco ... immediately took passage on the *Bolivar* for Monterey ... on January 3d [1842] ... sailed ... for Mazatlan ... on the *Maryland,* which touched at Santa Bárbara, and remained for nine days, January 18th to 27th, at San Diego. During the travels of which I have presented this fragmentary record, Mofras visited probably every mission and other settlement in California. I suppose that the Santa Bárbara district was explored in April, as the *Ninfa* came up the coast; those of Monterey and San Francisco from May to October, the explorer making his headquarters at the capital and Yerba Buena; and that of San Diego in January 1842, while the *Maryland* was disposing of her cargo ..." Bancroft's *History of the northwest coast* has considerable to tell of the personality of De Mofras, and his *History of Oregon* mentions that he reached Fort Vancouver from the Hawaiian Islands, almost "simultaneously" with the return visit of Sir George Simpson (which had been on October 22, 1841) and that he sailed with Sir George "... from Oregon for San Francisco Bay, in the bark *Cowlitz* ..." about "... the end of November [1841] ..." So much for the man's itinerary.

Brewer's "List of persons who have made botanical collections in California" mentions De Mofras' visit of 1841 and states: "How much he collected in this State it is not easy to say; but in an Appendix to his 'Explorations du Territoire de Oregon des Californies,' etc. ... there is a catalogue of the principal plants of the Northwest Coast, which enumerates about two hundred and ninety species, without even the usual specific authorities, and with so many errors of one kind and another as to be of little scientific value."

De Mofras' account of his trip was published in Paris in 1844, in two volumes with an atlas: *Exploration du territoire de l'Orégon, des Californies et de la mer Vermeille* [the Gulf of California], *exécutée pendant les années 1840, 1841 et 1842;* the included map, in color and extremely elaborate, is reproduced (small scale and in black and white) in Bancroft's *California;* the Spanish missions which are described in detail in De Mofras' text, are all indicated thereon. The Frenchman's botanical contribution is found in the second volume of the *Exploration*—"Catalogue des principales plantes de la côte nord-ouest de l'Amérique, du territoire de l'Orégon et de la Californie"—and lists approximately the number of species named by Brewer: we find "Rubies" for *Ribes,* or perhaps *Rubus,* "Sycomorus (sycomore)," and so on; even in the 1840's most persons did "better than that"! Nor, throughout his volumes, does the author make any reference worth mentioning to the amazing trees which he must have seen on his travels.

I found that the *Exploration* provided one of the most interesting contemporaneous accounts of the missions and presidios which it has been my fortune to read. Stillman considered that "... De Mofras gave undoubtedly the most thorough history of California, both political and physical, that had ever been condensed into one work." Bancroft accords it "a high degree of praise," and states that "... Forbes' work [Forbes, Alexander. 1839. *A history of upper and lower California.* London] is the only one of the time that can be compared with this; but ... it is very much less extensive and complete ..." On the other hand Robert Greenhow, in the preface to the second edition (1845) of *The History of Oregon and California,* credits De Mofras with plagiarism: "... the greater portion of the work is extracted from the present History and the preceding Memoir on the same subject, and it contains scarcely anything which might not have been produced by one who had never quitted the barriers of Paris. The errors and misstatements of M. de Mofras are indeed innumerable, particularly in all that relates to the United States, towards which he appears to entertain feelings of aversion even stronger than towards Great Britain ..."

BANCROFT, HUBERT HOWE. 1886. The works of ... 21 (California IV. 1840–1845): 248–255. Map ["Mofras' Map of California." 254].

———— 1884. The works of ... 28 (Northwest coast II. 1880–1845): 250, 251.

———— 1886. The works of ... 29 (Oregon I. 1834–1845): 250, 251.

BREWER, WILLIAM HENRY. 1880. List of persons who have made botanical collections in California. *In:* Watson, Sereno. Geological survey of California. Botany of California 2: 255.

DUFLOT DE MOFRAS, EUGENE. 1844. Exploration du territoire de l'Orégon, des Californies et de la mer Vermeille, exécutée pendant les années 1840, 1841 et 1842, par M. Duflot de Mofras, attaché à la légation de France à Mexico; ouvrage publié par ordre du roi, sous les auspices de M. le maréchal Soult, Duc de Dalmatie, président du conseil, et de M. le ministre des affaires étrangères. Atlas. 2 vols. Paris.

GREENHOW, ROBERT. 1845. The history of Oregon and California, and the other territories on the northwest coast of North America: accompanied by a geographical view and map of those countries, and a number of documents as proofs and illustrations of the history. ed. 2. Boston, xi, xii.

STILLMAN, JACOB DAVIS BABCOCK. 1869. Footprints of early botanical discoverers. *Overland Monthly* 2: 263.

its junction with the Sacramento . . ." and camped on the last. "Plants, *Mimulus* sp.—flos. brownish—habit that of *Latea,* banks of river. *Scirpus* 2 species—and a number of other Grasses."

October 19: "We were now 15 miles from the first settlement, viz—that belonging to Captain Sutter, one of our party had gone on yesterday before us, and as we had got fairly over the Rio de Los Americanos—or the American River, Capt. S. with a number of his attendants met us and gave the party a very friendly invitation to make a stay at his place for a few days.—Capt. S. is a Swiss by birth & has served for a considerable number of years in the French Army . . . On the termination of the Russian grant in California, he purchased of them, all there moveable stock—which consisted principally in Cattle, Horses—Guns &c &c. He has succeeded in procuring from the Mexican Government a grant of Land on the Sacramento—of 30 Square leagues in extent—which he calls Nova Helvetica. His Hacendo is situated on a plain about half way between the Sacramento & the Americans rivers, a little way below there junctions . . . The Sacramento is navagable for vessels of 20 Tons up as far as opposite his place which is about 115 miles inland from San Francisco . . . At Boteka [Bodega] one of the Russian old Settlements, he told me that there is large Gardens well stocked with Grapes, Apples, Pears and other small fruits—which he is about to remove in spring over to the Sacramento where he intends to cultivate the Grape, in order to make wines from it—and I doubt not but he will succeed."

Since, according to DuFour, it was on December 13, 1841, that the Russians completed arrangements to sell their property to Sutter, some of Brackenridge's comments under date of October must have been inserted in his journal at a later period.

October 20: "Like his countrymen he [Sutter] certainly talks a little largely—but granting him a little license in this respect, with the stock he has begun on—and the intelligence and perseverance which he possesses, in a few years there will be nothing in California to compete with him in point of strength, wealth, & influence. Our party was encamped on the banks of the Sacramento—and as the Captain gave the use of his Launch to the party, a number of those who were sickly preferred going down to S. Francisco by water with Mr. E. [Emmons] at there head, while a party with the Horses should proceed by land, the last route suiting my pursuits better—I joined, Mr. Eld being appointed the leader."

This overland party got started on the 21st: ". . . the land appeared rich but rather thinly clad with trees." On the 22nd Brackenridge noted that "A number of evergreen species of Oak made there appearance . . . along the banks of a small river called the Moqueles . . . a species of *Pavia* (perhaps the same as that found on the upper Mts) was found in large quantities . . ." The next day Brackenridge indulges in a few (final) comments about Henry Eld:

October 23: ". . . we came to the River called Rio San Joaquin . . . our guide advised us to camp here for the night as there was no water within 30 miles of us if we went on, but Mr. E. wishing to bring into play the very little power with which he was invested rejected the advice of the Guide and ordered the party to proceed, when night closed arround we were not in sight of any habitation, but after riding several miles we came upon a miserable Rancho where not a blade of grass to be had for our horses . . . we brot up a few minutes before Ten p.m. on the margin of a morass where not a drop of water was procurable for Supper—after fasting all day, and obliged to lay down at last to rest among the wet flags . . . of Plants—*Zauschina* sp.—a very handsome scarlet perennial. *Astragalus* sp.—pods very much inflated—on high hills. *Monarda* sp.—*Compositae*—a neat shrub—on rocky situations."

October 24: "A bare and bleak range of Mountains set in this morning soon after leaving the morass . . . At 2 p.m. we reached the Mission of San Jose which is situated on an enclined plain at the base of a range of low mountains, and surrounded on one side by Gardens and the rest by fields . . . the whole in a delapidated state. We had a letter of introduction to Don Jose Antonio Estrado (one of the Mission) who received us coldly and in a sort of uncerimonious manner invited us into a large ruinous sort of Hall which like our host I have not the least doubt seen better times, at least judging from the present mean appearance of them both. Our own external appearance was *certainly* anything but preposessing, but he might *certainly* have detected beneath the Buckskin dress & from the conversation that insued—some faint traits of Gentlemen; —which Mr. Forbs, Agent for the Hudson B. Co. who happened to come in at the time was not long in discovering, and kindly invited us out to spend the night at his House which was on our way—The pack Horses being drove on ahead, Mr. R. and myself followed taking leave of this miserable fallen place and its unhospitable occupants. Messrs. P. & E. remained to see the Chapel And the Old Padre whoom we left asleep—The whole of our party reached Mr. Forbs's house at dusk . . . no plant of any consequence was seen—every particle of vegetation (but trees) being browsed down by the numerous herds of cattle everywhere to be found in California."

October 25: "Mr. F. rode out with us a considerable distance . . . giving us all the information in his power relating to our route through to Yerbo Bueno. The Mission of Santa Clara laying on our way we concluded to visit it . . . at the Mission . . . the Padre or Principal . . . received us in a courteous manner . . . we were afterward shewed by another individual into a small fruit Garden . . . principally stocked with Grape vines . . . a number of Fig, Peach, Olive, & Almond trees . . . Apples, Pears, and small fruits in abundance, the fruits produced here are of fine quality, the Prickly Pear is here cultivated to a great extent and produces during the whole year a profusion of fruit . . . on leaving the old Padre generously made us a present of some fruit—his conduct towards us strongly contrasting with that experienced at the neighbouring Mission of San Jose . . . we came up to our party . . . encamped on the margin of creek . . . influenced by the tides . . . Plants. *Verbena* sp.—in habit of *V. venosa*, flor. blue.

Salicornia sp.—abundant on salt marshes near St. Clara. *Stachys* sp.—in thickets near St. Clara."

On the 26th the party halted a day to rest the horses. Everyone was anxious to get to the ship; Brackenridge notes that he had ". . . only been on board of her once—for the last 6 months . . ."

". . . this day was spent in collecting Seeds & Botanical specimens, to the last we added—*Nicotiana* sp.—flos. white—leaves very sticky—plant annual. *Mimulus* sp.—stem shruby, leaves lanceolate. flor. pale orange, this I believe to be the same plant that Mr. Nuttall introduced into the U. States from this country. *Quercus* sp.—a tree 60 feet—stem smooth much branched at top, leaves prickly, acorns large with a sharp point. *Rhamnus* sp.—a shrub 10 feet high—common in thickets. *Leptosiphon* sp.—a slender annual with white flowers, dry places. *Solanum* sp.—stem shruby—flos. a fine sky blue—a handsome plant. *Laurus* sp.—perhaps *L. Ptolmi* of Hooker, this tree reaches the height of 40 feet—and covered with a profusion of fruit in size and appearance of a Damson plum. Two species of *Composita* related to *Grindelia,* from Salt Marshes."

It was expected that they would reach their destination on the 27th but many of the horses gave out. On October 28, after a halt at the "Mission of the Nuestra Fra—de los Dolores" which was in a dilapidated state, they finally reached Yerba Buena where the boats from the ship were awaiting their arrival.

"The country passd over today was mostly of a barren & sandy nature, particularly that part between the last Mission and the village of Yerba Buena, which is almost of a pure sand, producing a variety of scraggy shrubs, many of which were valuable acquisitions to our overland collection. Among them were the following—*Lotus* sp.—flos. yellow—on sand banks—10 inches high. Cruciferous plant like *cheranthus,* flor. cream colour—sandy places. *Lupinus* sp.—a shrub 6 feet high. flos. yellow—very common. A *Croton* looking shrub—prostrate on sand banks. *Ceanothus* sp.—shrub 15 ft. high—sand hills near Yerba Buena. *Echiveria* sp.—in rocky situations."

To look back, and then forward, over the route followed by Brackenridge: from southern Oregon southward to the region of what is now the city of Sacramento he had been the first botanist in the field. Four years before, or in 1837, Richard Brinsley Hinds had made a trip up the Sacramento River which had taken him from the Bay of San Francisco to a point probably not far from Sacramento, so that he had antedated Brackenridge in that region. In another five years, or in early 1846, Frémont was to ascend the Sacramento River valley to Klamath Lake; and in 1847 Theodor Hartweg was to reach Marysville on Feather River and to collect in the foothills of the Sierra Nevada.

Two reports of the United States Exploring Expedition concern plants collected on the three explorations in which Brackenridge participated:

(1) The most comprehensive was by John Torrey, but edited and altered (to some

not clearly defined extent) by Asa Gray. It had been ready for publication in 1861 or 1862, but by the time it was called for by the authorities Torrey had died (in March, 1874) and his friend and fellow-worker took up the task of putting it through the press. Gray noted in his preface to the volume:

"About twelve years have elapsed since the late Dr. Torrey prepared this Report upon the Phanerogamous plants collected by the naturalists of the Expedition upon the Pacific coast of this country. Circumstances beyond his control have delayed the printing of it until now. The plates only have been published, a small edition having been distributed among botanists, both in this country and in Europe . . . Foreseeing that he would probably not be able to revise this Report himself, he committed the manuscript to my hands and desired me to edit it . . . There has, of course, been no attempt to bring the Report down to the present day. It represents our knowledge in respect to the subject-matter as it was a dozen years ago . . . It is much to be regretted that this Report was not printed and published, as was expected, in the year 1861 . . ."

The Torrey Report[7] appeared as part II of volume XVII of the Expedition's reports which contained: "Botany. I. Lower cryptogamia. II. Phanerogamia" [8] of the Pacific coast of North America. A "Note" following the title-page states: ". . . the late Dr. Torrey's Enumeration of the Phanerogamia collected on the Pacific Coast of the United States is now issued (July 1874), pages 205-514, with seventeen plates." No expense seems to have been spared in publication.[9]

Despite the bulk of the quarto volume and the large number of plants described, but seventeen new species and but two new genera were included—and none of them what might be considered conspicuous, or "showy," plants. Because of the long delay in publication the botanists, eager to describe the more unusual plants before others did so, had taken matters into their own hands and had published descriptions elsewhere. Torrey, for example, had described the California pitcher plant—perhaps the most interesting of Brackenridge's discoveries—in the *Smithsonian Contributions to Knowledge* in 1853 and had included a fine plate.[1] He named the plant *Darlingtonia californica,* dedicating the genus to his ". . . highly esteemed friend Dr. William Darl-

7. The title-page of the copy which I examined bore the imprint "Philadelphia: Printed by C. Sherman. 1862, 1874."

8. A phanerogam is a plant with manifest flowers or with stamens and pistil distinctly developed.

9. Preceding the Torrey Report are two more title-pages: the first inscribed "Phanerogamia of Pacific North America"; the second, "Phanerogamia of the United States Exploring Expedition." The Report itself is headed, "Phanerogamous plants collected in Washington Territory, Oregon and California"!

1. When Brackenridge had collected the plant it had not been in flower and he mistook it for one of the eastern pitcher plants *(Sarracenia)*; but Torrey states that, at his urgency, a friend ". . . Dr. G. W. Hulse [Gilbert White Hulse], of New Orleans . . . found it in flower in May, 1851, in the same region, and perhaps in the very spot where it was discovered many years before by Mr. Brackenridge."

In the Brackenridge journal the "Plants observed, and specimens collected on California Mountains," or from October 3 to October 11, appear in one list, without specific dates. Barnhart, as already noted,

ington,[2] of West Chester in Pennsylvania, whose valuable botanical works have contributed so largely to the scientific reputation of our country."

Moreover, as the years passed, the descriptive botanists were finding fewer new genera and fewer new species in the collections received from the field-workers and many of the plants found by Brackenridge in 1841 had already been described from material gathered by such men as Née, von Chamisso, Nuttall, Douglas and others whose stories have been told in my pages. And when, thirty-three years after Brackenridge had been in the field, the Report was finally published,[3] what was known of the plants indigenous to the trans-Mississippi west had grown by leaps and bounds. This is shown in part by the Report's references to plant distribution the knowledge of which was based on actual specimens; it cites such plant ranges as "Interior of Oregon and Washington Territory; eastward to the Great Bend of the Missouri, and southward along the mountains to New Mexico," or "Upper Columbia, Washington Territory.— This species is pretty widely diffused over the country west of the Mississippi. It occurs as far south as Eagle Pass on the Rio Grande." Cited as proof of such distribution are specimens collected on such expeditions as Whipple's, and by men such as Newberry, Kellogg and others who worked in the field after my story ends in 1850.

Some of the plant-localities cited in the Report reflect the fact that historical changes had taken place between 1841 (the year of Brackenridge's journey) and 1874 (the year the Torrey Report was published): for British claims in the Oregon country had been relinquished in 1846 and, although what is now the state of Washington was still "Washington Territory," Oregon had become a state in 1859 and is referred to as such; as is true of California which, relinquished by Mexico in 1848, became a state in 1850. And, although Idaho—into the western portion of which Brackenridge made a short excursion—was not to become a state until 1890, it was in 1874 recognized as part of the national domain having been relinquished by the British (along with Oregon and Washington) in 1846.

(2) Printed earlier than the Torrey Report, the second publication concerned with

gives the date of collection as October 5, 1841. Eastwood states that Brackenridge found it "... growing in marshy areas along the main branch of the Sacramento River on the way to Mount Eddy."

D. C. Gilman relates: "The story is told ... that on his way from Mount Shasta to San Francisco, an alarm from the Indians caused the party of explorers to run. Brackenridge saw a strange-looking plant, grabbed a clump of it and carried it to camp. This was the *Darlingtonia Californica*."

2. To name plants for friends was a common practice at this period. Darlington doubtless was deserving of honor, but a name of more significance might well have been chosen. Eastwood states that "Several genera had already been named in Darlington's honor, so that *Darlingtonia* was replaced by *Chrysamphora*, from the Greek meaning golden pitcher, being so named by Dr. E. L. Greene. However, in general usage, *Darlingtonia* persists." *Darlingtonia* Torr., 1853, was proposed for conservation in 1950.

3. Gray's preface to the Torrey Report refers to "... the understood wishes of the Library Committee that the Report should be printed as nearly as possible in the state the author left it as well might be." As noted earlier in this chapter the Library Committee is said to have been poorly qualified to advise on scientific matters and any report to be of other than historical interest had to be up to date.

plants from the region of my story was written by Brackenridge himself. It described a few collections from western North America but by far the greater number were from such distant lands as the Hawaiian Islands, the Philippines, and so on. Issued as volume XVI of the Expedition's reports,[4] it bore the title *Botany. Cryptogamia. Filices, including Lycopodiaceae and Hydropterides*[5] ... *With a folio atlas and forty-six plates.* I quote from Bartlett's comments upon this publication:

"On the cruise William D. Brackenridge ... devoted himself to the ferns, and the report on this group was his only contribution to scientific botany. Unfortunately it is such an exceedingly rare volume that it has been inaccessible to many students of ferns in this country and abroad.[6] Twenty copies of the small official edition were destroyed by fire in 1856, and although there was an unofficial edition, it is even more excessively rare than the official one, for it was destroyed by fire ... after only ten of presumably a hundred or more copies had been sold ..."

Barnhart mentions that "... the folio atlas, with its 46 beautiful plates, seems to be even rarer, and was not to be found in any of the great libraries of New York City, until a copy was recently secured by purchase for the library of the New York Botanical Garden.[7] He comments upon Brackenridge's qualifications for writing his report:

"Brackenridge was a good field-botanist, with the advantage of four years of intimate association during the voyage with the scholarly Pickering; and he was by no means illiterate ... But he was not well versed in the technical forms of descriptive plant taxonomy, and his knowledge of Latin was very limited. The rules laid down for the monographs of the report series demanded that every description should be printed in both Latin and English, and Brackenridge appealed to ... John Torrey ... to help him out of his difficulty. Torrey revised much of the fern manuscript, supplying Latin translations as he went along ... subsequently it was placed in the hands of Asa Gray, who completed the preparation of the work for the press ..."

Before the great amount of material[8] gathered by the United States Exploring Ex-

4. The title-page of the volume examined bore the imprint "Philadelphia. Printed by C. Sherman. 1854."

5. A Cryptogam is a plant with no flowers, reproduced by means of spores. *Filices* are ferns, a *Lycopodium* is a nearly allied plant and the *Hydropterides* are water ferns.

6. The copy in the library of the Gray Herbarium bears a bookplate recording it as "The gift of Asa Gray."

7. The Gray Herbarium possesses two copies of this remarkable atlas. One, according to an attached letter, was sent as a gift to Asa Gray on July 5, 1882. Three days later Gray wrote on the back of this letter: "This is exceedingly rare. I had the letter press, fortunately. Nearly all the Atlas & letter press too, was destroyed by a fire. I despaired our ever having it—till it has now come as a gift ..."

8. In a brief, unsigned article describing the results of the Expedition which appeared in 1843 in Silliman's *American Journal of Science and Arts* (44: 395-408) it is noted that "Ten thousand species of plants, and upwards of fifty thousand specimens, constitute the herbarium of the expedition. The following catalogue gives the number of species collected at the several places visited." From Oregon 1218 are cited, from California 519. Brendel (1880) names the same figures.

pedition had begun to reach Washington it was realized that some housing place would be essential. Perhaps the most important result of the expedition's work—from the angle of the natural sciences including botany—was the establishment of a national repository for the collections: for, from the National Institute to which their care was first assigned, there developed, through a series of complicated changes, the National Museum and the Smithsonian Institution. H. H. Bartlett and G. Brown Goode have described the beginnings of these great establishments; Bartlett states:

"If it had not been in anticipation of the enormous collections in natural history that would have to be taken care of, there would have been no immediate need to plan and build up the National Institute. Without the Institute, no place would have been provided for the collections, which, if dissipated to institutions throughout the country for study, would never have been reassembled to form the substantial basis for a National Museum. Wilkes was eventually to be harshly criticized for wanting reports written in Washington,[9] then 'a city without books', but by adhering as inflexibly as he could to an arbitrary policy that the collections were to remain in Washington he performed a great service to the nation."

I have limited my comments upon the plant collections to those from my chosen locale, but there were vast numbers from many other parts of the globe. This wide coverage enhanced the importance of the collections and started the institution in which they found a home on its way to becoming one of the great representative herbaria of the world. The famous herbaria of Europe and of Great Britain had long aimed at such a compass; we have seen how many of the earliest collections from west of the Mississippi were destined for other lands. Before the United States Exploring Expedition went into the field the great American botanists such as Torrey and Gray had been primarily concerned with assembling plants from the North American continent. Gray appreciated fully that to properly determine and describe plants from parts of the world with which he was not so familiar[1]—as he was asked to do—would necessitate visiting the comprehensive herbaria of other lands; only, probably, because he was regarded as indispensable and only because he was, so it seems, a law unto himself, was he permitted to do so—for the Washington authorities, unversed in such matters and apparently sensitive in regard to anything belittling the nation's standing, scientific or other, were determined that all phases of the Expedition's work should be "American" pure and simple. There are humorous aspects to the many fights (as good a name as any) that marked the progress of the Expedition from start to finish.

Brackenridge made another contribution towards the botanical accomplishment of

9. See: *Letters of Asa Gray*, edited by his wife Jane Loring Gray, for the botanist's acidulous comments upon this stipulation—as well as upon other aspects of the Expedition.

1. The plants described by Asa Gray in part I of his *Botany* (volume XV of the original issue) were all from regions other than those which I consider here. Part II of the work (which was to have been issued as volume XVIII of the reports) was never published as already stated. Gray had been retained to go as botanist on the voyage but had withdrawn for reasons of his own.

the Expedition and this concerned its living collections. *The magazine of horticulture, botany, and all useful discoveries and improvements in rural affairs* mentions two visits paid by its editor, C. M. Hovey, to the "Experimental Garden of the National Institute" in Washington, D. C. One was on October 19, 1843:

". . . the only particularly new movement in gardening, is the establishment of the garden of the National Institute . . . Pickering and Breckenridge, the botanists attached to the Expedition . . . brought with them, in addition to great quantities of seeds, bulbs, &c., upwards of one hundred species of live plants . . . a small piece of ground was selected, in the rear of the new Patent Office, and . . . a house was completed about thirty feet long, and the plants removed thereto, where they were under the charge of Mr. Breckenridge . . ."

The plants grew and multiplied and the "house" had to be enlarged and in addition a ". . . square of ground, containing perhaps half an acre . . . was to be fenced off for an experimental garden." When Hovey paid a second visit two years later (October 17, 1845) he found that ". . . a great accession had been made to the Collection, through the untiring exertions of Mr. Breckenridge . . ." He reported that

". . . an additional wing had been added to the house so that it now extends upwards of one hundred and fifty feet, built in the most thorough and substantial manner . . . We are gratified to record the improvements which have been made in this department of the Institute . . . they cannot fail to be of much benefit to the spread of a taste for plants . . . the representatives, who assemble at the capitol a large portion of every year, will be occasionally induced to visit the gardens, and become better acquainted with the floral productions of the globe."

After quoting from a report which Brackenridge presented to the National Institute in November, 1842—it outlined his ambition to have the small living collection develop into one which should ". . . vie with the most celebrated establishments of the same kind in Europe . . ."—Bartlett wrote:

"We find in the passages just quoted the ideas which led to the development of two of our national establishments, the United States Department of Agriculture and the United States Botanic Garden . . . The present vast work of the Department of Agriculture's Plant Introduction Office is just as clearly to be interpreted as an outgrowth of what Brackenridge started at the little Patent Office greenhouse . . .

"It may be added that we find in Brackenridge's early and most meritorious efforts the germ of the later Congressional Free Seed Distribution, of malodorous memory . . ."

When Hovey made a third report upon the progress of the garden of the National Institute—on March 10, 1848—he thought well of what was to become later a questionable and certainly overdone practice, the distribution of seeds under the franking privilege:

"This garden, under the superintendence of . . . Mr. Breckenridge, is well repaying the expense attending its management: through the exertions of Mr. Burke, the Commissioner of Patents, thousands of papers of seeds, raised in the garden, are distributed, through the representatives of the several states ... and handsome, even rare and choice, kinds of seeds find their way where, through the ordinary course of trade, they would not be probably introduced for years ... this medium of sending seeds is of the greatest importance ...."

There exist only brief biographical sketches of Brackenridge; by Barnhart, by Meehan, by Peattie, and so on. Barnhart records that he was born at Ayr, Scotland, on June 10, 1810, tells of his early training as a gardener in Edinburgh and on the continent (in Poland and at Berlin) and of his arrival in America in 1837 where he was employed by the nurseryman Robert Buist[2] of Philadelphia. In 1838 he went on the Exploring Expedition. Barnhart notes that Brackenridge's modesty was "so excessive" "... that he could never be induced to furnish biographical information to those who sought to secure it directly from him as the most authoritative source, and the world is indebted to the late Thomas Meehan for the preservation of not a few of the facts of this interesting career. Meehan published an account of Brackenridge, with a portrait, in the number of his *Gardeners' Monthly* for December, 1884, and recorded his death February 3, 1893, in the number of *Meehan's Monthly* for March, 1893 ...."

Brackenridge in 1855 bought a farm near Baltimore, Maryland, where he spent the remaining years of his life. "For some years he was horticultural editor of the *American Farmer* ... His death occurred on the third of February, 1893."

When naming the genus *Brackenridgea*[3] Asa Gray wrote that "The name selected for this genus is intended to commemorate the important scientific services of William D. Brackenridge, the Assistant Botanist of the Expedition, through whose indefatigable zeal and industry this botanical collection was principally made, of which he had himself elaborated the Ferns and allied orders." Bartlett mentions that "There were also species named 'Brackenridgei' in seven genera, viz., *Astragalus, Clidemia, Cupania, Draba, Eugenia, Hibiscus,* and *Pittosporum,* these in vol. IV; subsequently there were species in several other genera."

2. Buist had learned his trade, according to Harshberger, under James MacNab, Curator of the Edinburgh Botanic Garden, and elsewhere. He started a florist business— "... probably the first florist's establishment [in Philadelphia presumably]."

3. In volume XV (1854) of the reports of the Exploring Expedition (*Botany. Phanerogamia* ... pp. 361, 362).

BANCROFT, HUBERT HOWE. 1886. The works of ... 21 (California IV. 1840–1845): 240-248. Map ["Wilkes' Map. 1841." 244].

———— 1884. The works of ... 28 (Northwest coast II. 1800–1846): 668-684. Map ["Johnston's Excursion." 678].

———— 1886. The works of ... 29 (Oregon I. 1834–1848): 246-249.

BARNHART, JOHN HENDLEY. 1919. Brackenridge and his book on ferns. *Jour. N. Y. Bot. Gard.* 20: 117-124.

——— 1921. Gilbert White Hulse. *Jour. N. Y. Bot. Gard.* 22: 132, *fn.* 1.

BARTLETT, HARLEY HARRIS. 1940. The reports of the Wilkes expedition, and the work of the specialists in science. *Proc. Am. Philos. Soc.* 82: 601-705.

BELCHER, EDWARD. 1843. Narrative of a voyage round the world, performed in Her Majesty's ship Sulphur, during the years 1836–1842, including details of the naval operations in China, from Dec. 1840, to Nov. 1841. Published under the authority of the Lords Commissioners of the Admiralty. By Captain Sir Edward Belcher, R.N.C.B., F.R.A.S., &c. Commander of the expedition. 2 vols. London. 2: 46-48.

BRACKENRIDGE, WILLIAM DUNLOP. 1855. Botany. Cryptogamia. Filices, including Lycopodiaceae and Hydropterides. By William D. Brackenridge. With a folio atlas of forty-six plates. *In:* United States Exploring Expedition. During the years 1838, 1839, 1840, 1841, 1842. 16: Philadelphia. C. Sherman, printer. i-viii, 1-357. [*See:* No. 66 of Haskell, D.C. 1940–1942. The United States Exploring Expedition . . . and its publications . . .]

——— 1930. 1931. Documents—Our first official horticulturist—The Brackenridge journal. Edited by O. B. Sperlin. *Wash. Hist. Quart.* 21: 218-229; 298-305 (1930): 22: 42-58, 129-145, 216-227 (1931).

——— 1945. Journal of William Dunlop Brackenridge October 1-28, 1841. [Edited by Alice Bay Maloney.] *Cal. Hist. Soc. Quart.* 24: 326-336.

BRENDEL, FREDERICK. 1880. Historical sketch of the science of botany in North America from 1840 to 1858. *Am. Naturalist* 14: 27.

BREWER, WILLIAM HENRY. 1880. List of persons who have made botanical collections in California. *In:* Watson, Sereno. Geological survey of California. Botany of California 2: 555.

BRYAN, GEORGE SMITH. 1940. The purpose, equipment and personnel of the Wilkes Expedition. *Proc. Am. Philos. Soc.* 82: 551-560.

COLLINS, FRANK SHIPLEY. 1912. The botanical and other papers of the Wilkes Exploring Expedition. *Rhodora* 14: 57-68.

CONKLIN, EDWIN GRANT. 1940. Connection of the American Philosophical Society with our first national exploring expedition. *Proc. Am. Philos. Soc.* 82: 519-541.

DILLER, J. S. & others. 1916. Guidebook of the western United States. Part D. The Shasta Route and the Coast Line. *U. S. Dept. Int. Geolog. Surv. Bull. 614.* Washington.

DUFOUR, CLARENCE JOHN. 1933. The Russian withdrawal from California. *Quart. Cal. Hist. Soc.* 12: 240-249. *Reprinted: Cal. Hist. Soc.* Spec. Publ. 7: 52-61 (1933).

EASTWOOD, ALICE. 1945. An account and list of the plants in the Brackenridge journal. *Cal. Hist. Soc. Quart.* 24: 337-342.

GILMAN, DANIEL COIT. 1899. The life of James Dwight Dana . . . Scientific explorer, mineralogist, geologist, professor at Yale University. New York. London. 62, 63.

GOODE, GEORGE BROWN. 1889. The beginnings of American science. The third century. *Proc. Biol. Soc. Washington* 4: 45, 46.

——— 1892. The genesis of the National Museum. *Smithsonian Inst. Ann. Rep. 1891:* 272-380.

GRAY, ASA. 1854. Botany. Phanerogamia. By Asa Gray, M.D. With a folio atlas of one hundred plates. vol. 1. *In:* United States Exploring Expedition, during the years 1838, 1839, 1840, 1841, 1842. Under the command of Charles Wilkes, U.S.N. 15. New York. George P. Putnam & Co. 361 [*Brackenridgea* described]. [*See* Haskell, D.G. 1940–1942. The United States Exploring Expedition . . . and its publications . . . No. 61.]

——— 1893. The letters of Asa Gray. Edited by Jane Loring Gray. 2 vols. Boston. New York.

HARSHBERGER, JOHN WILLIAM. 1899. The botanists of Philadelphia and their work. Philadelphia. 193-195 [Robert Buist].

HASKELL, DANIEL C. 1940–1942. The United States Exploring Expedition, 1838–1842, and its publications 1844–1874. A bibliography by Daniel C. Haskell, Bibliographer of the Library. With an introductory note by Harry Miller Lydenberg. *Bull. N. Y. Public Library.* J 1940 and J J1 & O 1941. Reprinted 1942. New York.

HOVEY, CHARLES MASON. 1844. Notes and recollections of a tour through Hartford, New Haven, New York, Philadelphia, Baltimore, Washington and some other places, in October, 1843. *Mag. Hort. Bot.* (Hovey) 10: 81-83.

——— 1846. Notes of a visit to several gardens in the vicinity of Washington, Baltimore, Philadelphia, and New York, in October, 1845. *Ibid.* 12: 241-243.

——— 1848. Notes on gardens and nurseries in the vicinity of New York, Philadelphia, Baltimore, and Washington. *Ibid.* 14: 241, 242.

KELLY, HOWARD ATWOOD. 1929. Some American medical botanists commemorated in our botanical literature. New York. London. 151-153 [Charles Pickering].

MALONEY, ALICE BAY. 1945. A botanist on the road to Yerba Buena. *Cal. Hist. Soc. Quart.* 24: 321-325. [Introductory notice to Brackenridge, W. D. 1845. Journal of . . .]

MEEHAN, THOMAS. 1884. Editorial notes. *Gard. Monthly* (Meehan) 26: 375, 376.

——— 1894. Biography and literature. *Meehan's Monthly* 3: 47.

PEATTIE, DONALD CULROSS. 1929. William D. Brackenridge. *Dict. Am. Biog.* 2: 545, 546.

PIPER, CHARLES VANCOUVER. 1906. Flora of the state of Washington. *Contr. U. S. Nat. Herb.* 11: 15.

REHN, JAMES ABRAM GARFIELD. 1940. Connection of the Academy of Natural Sciences of Philadelphia with our first national exploring expedition. *Proc. Am. Philos. Soc.* 82: 543-549.

TORREY, JOHN. 1853. On the Darlingtonia Californica. A new pitcher-plant, from northern California. *Smithsonian Contr. Knowledge* 6: 1-8, pl. 12.

——— 1874. Botany . . . II. Phanerogamia of the Pacific coast of North America. *In:* United States Exploring Expedition, during the years 1838–1842, under the command of Charles Wilkes, U.S.N. 17. Philadelphia: Printed by C. Sherman. 1862, 1874. 17 plates, 205-514. [*See* Haskell, D. G. 1940–1942. The United States Exploring Expedition . . . and its publications . . . No. 70.]

WARREN, GOUVENEUR KEMBLE. 1859. Memoir to accompany the map of the territory of the United States from the Mississippi River to the Pacific Ocean, giving a brief account of each of the exploring expeditions since A.D. 1800 . . . *U. S. War Dept. Rept. expl. surv. RR Mississippi Pacific* 11: 38, 39.

WILKES, CHARLES. 1845. Narrative of the United States Exploring Expedition. During the years 1838, 1839, 1840, 1841, 1842. By Charles Wilkes, U.S.N. Commander of the expedition, member of the American Philosophical Society, etc. In five volumes, and an atlas. Philadelphia. Lea & Blanchard. 1. 4. 5. [*See* Haskell, D. G. 1940–1942. The United States Exploring Expedition . . . and its publications . . . No. 2.]

CHAPTER XXXV

GAMBEL PASSES THROUGH SANTA FE AND, WITH THE

WORKMAN PARTY, REACHES SOUTHERN CALIFORNIA

OVERLAND FROM THE EAST, FROM THERE VISITING

THE ISLAND OF SANTA CATALINA IN SAN PEDRO BAY

WILLIAM GAMBEL, a doctor of medicine and an ornithologist, is credited with being the "first botanist" to collect plants about Santa Fe, New Mexico. He was also the first to collect plants on the island of Santa Catalina off the coast of California and lying southwest of San Pedro Bay. He was also the first of the botanical persuasion to enter California overland from the east.

Statements have differed as to the year he was in Santa Fe.[1] And, as far as I know, no one has reconstructed his route in any detail and published the findings. It seemed of interest to do so.

The trail which I pursued through the literature was somewhat circuitous and *began* at the *end* of Gambel's journey with his arrival in California with the "Workman party." And—after I learned that the "Workman party" had taken the route followed by "Wolfskill" in 1831—Gambel's journey from Santa Fe to Los Angeles was fairly clear, certainly for the greater part of the way.

My first clue was found in the "Pioneer register and index 1542–1848" included in H. H. Bancroft's *History of California* (3: 751, 752). The italics are mine:

"Gamble (Wm), *1841,* a young naturalist sent out from Phila. by Nuttall to collect specimens; *came from N. Mex. in the Workman party.* IV. 278-9. Being financially crippled, he was employed by Com. Jones in '42 as clerk on the *Cyane,*[2] and perhaps went away on that vessel; in '44 at Callao;[3] said by Given to have ret. to Cal. about '49."

Bancroft's *History of California* (4: 276-279) has considerable to tell about the Workman party which Gambel accompanied from New Mexico. Again the italics are mine.

---

1. Brewer (1880) mentions the year of his trip as 1842. Standley (1910) refers to his passing through Santa Fe in 1841 or 1842. And Pennell (1936) places the trip in "the autumn of 1844."

2. The *Cyane,* of the United States Navy, was serving in Pacific waters when, in 1842, Commander T. A. Catesby Jones became prematurely involved in the seizure of Monterey from the Mexican authorities. Bancroft, on occasion, mentions her presence off Mexico, off California, etc., and his *History of California* (4: 322) refers to her departure from the harbor of San Pedro on January 21, 1843. Brewer states that Gambel returned to the east in 1843 and this suggests that, as Bancroft thought possible, he may have left California in the *Cyane;* but he did not leave for the east coast in 1843, nor did the *Cyane*—not apparently until 1845.

3. Situated north of Lima, Peru.

"Another party of emigrants, twenty-five in number, came this year [1841], arriving at Los Angeles nearly at the same time that the Bartleson party reached San José. This company was organized in New Mexico, where most of the members had for some time resided; but a few men, including Given and Toomes, had come to Santa Fé from Missouri with the intention of going to California . . . There were political reasons which influenced the departure of [William] Workman and [John] Rowland, the organizers and leaders of the company, and probably of some others . . . They started from Abiquiu [situated about sixty miles northwest of Santa Fe, and in Rio Arriba County, New Mexico] in September, crossed the Colorado, and *followed the same route as that taken by Wolfskill in 1831*, which had often been chosen by the New Mexican traders. They drove a flock of sheep for food; met with no adventures and few hardships; and arrived at San Gabriel [not far north of Los Angeles, Los Angeles County, California] early in November . . ."

Bancroft notes further that Workman and another of the party brought their families intending to settle permanently, that some were seeking adventure, and that ". . . three, Gamble, Lyman and Mead, were men of scientific proclivities, and spent but a short time in California . . . Rowland on his arrival furnished the authorities a list of his companions, with a statement of their intention to obey all legal requirements. I append in a note a complete list of the company."

This list, supplied by Bancroft (p. 278, *fn*. 38), includes "Wm Gamble," the name marked by an asterisk to indicate that he did not remain in California permanently.

Having learned that Gambel had left Santa Fe in September, 1841, and that he had reached Los Angeles in November, it remained to trace the route which Wolfskill followed between these points in 1831. In the travel-annals of our southwest his journey seems to have been an important one; although the greater part of the route is clear, certain portions are still obscure and promise to remain so. I shall first quote Bancroft's account of the journey as supplied in his *History of California* (3: 386):

"In the autumn of 1830, William Wolfskill fitted out a company in New Mexico to trap in the great valleys of California. He was a Kentuckian by birth, thirty-two years of age, with some eight years' experience of trapping and trading in the broad territories surrounding Santa Fé from the north to the south-west. He had been a partner of Ewing Young, then absent in California . . . There is extant neither a list of the company nor diary of the trip;[4] but the expedition took a route considerably north of that usually followed, left Taos in September, crossed the Colorado into the great basin, and pressed on north-westwardly across the Grande, Green, and Sevier rivers, then southward to the Rio Virgen, trapping as they went. It seems to have been the intention to cross the mountains between the latitudes 36° and 37°; but cold weather, with symptoms of disorganization in the company, compelled the leader to turn south-

4. Some chronicles were published in 1923 which include facts about the journey as told by participants. I refer to this publication below.

ward to Mojave. Thence he crossed the desert westward, and arrived at Los Angeles early in February 1831 . . ."

William J. Ghent in 1931 provided a slightly more explicit outline of Wolfskill's route, basing this upon "The Chronicles of George C. Yount California pioneer of 1826," edited by Charles L. Camp and published in the *Quarterly* of the California Historical Society in 1923.[5] I quote from Ghent.

". . . It is to William Wolfskill that the credit is generally given of having broken the Old Spanish Trail through to California. It was a route that had often been used by Spaniards, Mexicans, and Indians between Sante Fé or Taos and Utah Lake, but as far as known had never been travelled by white men farther westward. His party, of some twenty-two men, and including George Yount, left Taos about the end of September 1830. They struck the Grand (now the upper Colorado) below the mouth of the Dolores [Grand County, eastern Utah], swung southwestward to near the junction [in northwestern San Juan County, Utah] of the Grand and Green, and after crossing both streams, appear to have moved northeastward to Sevier Valley.[6] Further south they became lost in the high mountains of the lower Wasatch chain where they encountered deep snows and extreme cold. Later they made their way to the Virgin River, which they descended to the Mohave villages where they struck across the desert to Cajon Pass. At San Gabriel Mission, which they reached in February 1831, they were hospitably entertained, and after resting for a week or two they went on to Los Angeles where the party broke up. It is only by a liberal stretch of the imagination that this expedition can be said to have followed the eastern half of the Spanish trail, or to have broken what later became its western extension.[7] For the greater part of the journey the two routes were widely apart."

Camp states that the Clark narrative begins only after the crossing of the Green

5. The introductory remarks explain that the "detailed but disconnected and unarranged reminiscences" were taken down by the Reverend Orange Clark as they were told him by Yount. Much of "The Chronicles" has nothing to do with Wolfskill's journey but some portions (of pp. 36-41) supply a very few details on the route over and above what is related by Ghent. Other participants are quoted besides Yount.

6. The Sevier River is about two hundred miles long and the portion of its course which seems to be designated on the maps as "Sevier Valley" lies in southern Sanpete County, or in about the center of the state of Utah.

7. In his *History of Nevada, Colorado and Wyoming* (p. 39) Bancroft refers to the fact that Wolfskill, from Taos had ". . . followed one old Spanish Trail towards Salt Lake, and another away from that region towards Los Angeles . . ." There was an old Spanish Trail which lead from Taos northwestward to Sevier River and Wolfskill certainly took it.

It has been stated by T. S. Palmer (interested in Gambel's ornithological work) that ". . . Gambel . . . crossed the continent in 1841 via the Santa Fe Trail and *then from Utah to California via the Mormon Trail* with the Workman Party [italics mine]." Although the Mormons eventually had an outpost in southern California whence overland travel was easier and shorter to Salt Lake City, the fact that their advance guard only got to Great Salt Lake in 1847 does not suggest that they had already established a trail between that lake and the west coast by 1841. Unfortunately Palmer does not cite his authority for the statement.

River. From "The Chronicles" we glean a few meagre records of the route not mentioned by Ghent:

(1) That, before reaching Sevier Valley (which they called "Pleasant Valley"), Wolfskill's party had, two days of short marches before, been in "St-Joseph's Valley" (perhaps Castle Valley, in Carbon County, Utah).

(2) That it was near "the junction of the forks of Sevier River" (or in Piute County, Utah) that they had become confused and ". . . went due south or southeast into one of the high plateaus, the southern continuation of the Wasatch Range . . ."

(3) That from the mouth of the Virgin River they had proceeded *down the Colorado* and ". . . struck across the desert to the Mohave River . . ." and *followed that stream,* which brought them to the vicinity of Cajon Pass.

Further information on the trip was brought to my attention by Dr. F. W. Pennell of the Academy of Natural Sciences of Philadelphia. On April 20, 1949, he sent me a memorandum prepared by Dr. Francis Harper of the same Academy:

"Dr. Witmer Stone left a partly finished MS of perhaps 1000 pages on the early annals of North American ornithology. This MS belongs to Mrs. Arthur C. Emlen, formerly of Awbury, Germantown, Pa., but now living somewhere in California. She has entrusted the MS to me, but I feel sure she would have no objection to Mrs. McKelvey's use of that portion of it pertaining to William Gambel.

"This portion consists of about 60 pages. Some of the material, of a biographical nature, would doubtless be of very considerable value to Mrs. McKelvey, but part of it consists merely of lists of Gambel's birds. There are several letters from T. George Middleton . . . a nephew of Gambel's who was born in 1872 and still survived in 1938. Also a couple of letters from a historian, Harry N. M. Winton . . . (including a map of Gambel's probable route to California). Among the various papers are some valuable references to published sources of information.

"This Gambel material of Dr. Stone's, like the rest of his MS, is in very scrappy condition. Some of the sheets would be scarcely identifiable if they got out of sequence . . ."

Dr. Harper sent me ". . . a copy of Mr. Winton's sketch map of Gambel's probable route . . ." with the qualification that he could not ". . . vouch for the accuracy of the original." Also a memorandum of June 3, 1949, listing "Some of the more important information" included in Dr. Stone's files. This related to published sources which I had already examined, and to three letters, two written in 1938. These I have not seen.

The Winton "sketch-map" shows the "Probable route taken by William Gambel to California, 1841." Starting from Independence, Missouri, the route crossed Kansas southwestwardly to Dodge City, and from there continued southwest—across southwestern Kansas, across the extreme western portion of the "Panhandle" of Texas into northeastern New Mexico, and on to Santa Fe. From there a side-trip to Taos is indicated. From Santa Fe westward to Los Angeles it follows closely the Wolfskill route outlined above—as closely indeed as the known facts permit.

Upon what records Winton based the eastern portion of Gambel's route—between Independence and Santa Fe—I do not know. The route westward from Independence for about four hundred miles was usually the same for all; but about twenty miles above Dodge City travelers bound for Santa Fe came to the first of two routes leading to that city. After crossing the Arkansas this first route followed the Cimarron River— usually waterless—for several days, crossed to the Canadian (in the extreme western end of the "Panhandle" of Texas) and, entering what is now New Mexico, turned southwestward and then northwestward to Santa Fe. Travelers who wished to avoid this "thirsty" trail continued along the Arkansas to about four miles above Bent's Fort (or to near present La Junta), where they made a crossing, traversed Raton Pass in the mountains of the same name and then moved on to Santa Fe. The Winton map indicates that Gambel followed the first of these trails, the so-called Cimarron route. Wistar, who traveled with Gambel at a later date, states that on an earlier journey Gambel had been ". . . to the Raton Mountains and Santa Fe . . ." and he may have derived this information from his companion. Witmer Stone refers to Gambel's ". . . exploring the Raton Mountains of northern New Mexico and passing thence from Santa Fé to the Colorado River . . ." Stone may have derived *his* information from some of Gambel's bird-records although those which I have seen supply nothing certain as to his route.[8] However, despite the Winton map, the evidence seems to point to the Raton Pass route.

The plants collected by Gambel on his trip to California were described by Thomas Nuttall who took the opportunity to include a considerable number which he himself had gathered[9] on his trip to the west coast in 1834. His paper appeared twice, in slightly different forms:[1] first in the *Journal* of the Academy of Natural Sciences of Philadelphia (new ser. 1: 149-189, 1848), under the title "Descriptions of plants collected by William Gambel, M.D., in the Rocky Mountains and Upper California"; and next (the preliminary diagnosis) in the *Proceedings* for 1848 and 1849 of the same Academy (4: 7-26, 1850); the title varies to the extent that the plants were "collected

8. In a paper on our western birds, published in the *Journal* of the Philadelphia Academy, Gambel mentions the "Common Turtle Dove" as breeding ". . . on the prairies near the rivers Arkansaw and Cimaron, in June . . ." This offers no enlightenment since "the plains near" these rivers would have been encountered on either journey.

The reference to the month of June supplies some indication, however, of the length of time that Gambel spent at Santa Fe. If he was some distance from that town in June and left it in September, his stay—at most—was not longer than two or three months.

9. These Nuttall specimens were from widely scattered regions such as ". . . the dividing ridge of the Rocky Mountains," "Big Sandy Creek of the Colorado of the West," ". . . hills near Scott's Bluffs of the Platte," "Near the first range of the Rocky Mountains of the Platte," ". . . on the Oregon near the outlet of the Wahlamet [Willamette]," "Near San Diego, Upper California," "About Ham's Fork of the Colorado of the West," and so on: for the most part they are distinguished from those of Gambel by the notation "(Nuttall)."

1. According to a record dated February 29, 1848, "The Committee on Mr. Nuttall's paper, read 1st and 11th insts., reported in favour of publication in the Journal and Proceedings."

by Mr. William Gambel" for he had only received his medical degree in March, 1848, or after the paper had been read. The dedications to the two papers also vary slightly.[2]

In his paper Nuttall described the new genus *Gambelia* (with type species *G. speciosa*) but, although still known as "Gambel's snap-dragon," the plant was later reduced to synonymy under the genus *Antirrhinum,* so that the honor conferred on Gambel is now only a matter of the record. Meehan tells the story and includes a colored plate:

"It is only by rare good chance that we have been able to illustrate this plant, which is one of the rarest as well as of the most beautiful of the wild flowers of the United States. It was first collected by Mr. William Gambel, in 1842, on the island of Catalina, off the coast of California, and does not appear to have been found again till 1875, when it was gathered on another island off the same coast—Guadalupe—by Dr. Edward Palmer;[3] and it is from seeds collected by him that the plants were raised in the gardens of the Arnold Arboretum of the Bussey Institute, near Boston, from which our artist made the drawing for this plate.

"The plants collected by Mr. Gambel were examined by Mr. Nuttall, and described by him . . . under the title of 'Plantæ Gambelianæ,' [4] and this particular one believed to constitute a new genus, and which he called *Gambelia;* but which has since been decided by Dr. Gray not to be distinct from Antirrhinum, the well-known 'Snap-dragon.' Mr. Nuttall himself seems to have perceived the close relationship to *Antirrhinum*."

After quoting Nuttall's description Meehan adds:

"It is to be regretted that the attempt to honor Mr. Gambel in this genus could not be sustained, for he seems to have been one of those very meritorious persons, who, triumphing over early obstacles, succeed by their perseverance in serving their fellows, and often, as in this instance,[5] at the expense of their lives."

Nuttall conferred upon Gambel what was perhaps a lesser honor but to date a

2. In the *Journal* the dedication reads: "In honor of Dr. William Gambel, a naturalist, who has explored Upper California, and particularly elucidated the ornithology of that country," and in the *Proceedings:* "In honor of Mr. William Gambel, a naturalist, who has explored Upper California, and made an interesting collection of the plants of that country."

The small variations suggest that Gambel's title and qualifications to fame had undergone a slight revision, recorded with the accuracy that the scientific approach demands!

3. Meehan comments on the flora of the Catalina group:

"Our very pretty wild flower seems confined to these small islands along the California coast . . . Guadalupe Island, where Mr. Palmer found this plant, is only twenty-six miles north and south, and only ten miles across on the widest line . . . of the flowering plants that Dr. Palmer collected there, one-fifth have so far not been found in any other part of the world! . . ."

4. Although both Meehan and Brewer cite this as the title of Nuttall's paper I have found no authority for this. Both men published in the same year (1880) and one may have quoted from the other.

5. Gambel's death did *not* occur when collecting the snapdragon, as might perhaps be inferred from the wording.

permanent one, when he named for him a shrubby, deciduous oak, *Quercus Gambellii*, ". . . with the aspect of our northern oaks, but very distinct . . . On the banks of the Rio del Norte, but not abundant." Gambel had been near the Rio Grande when at Santa Fe and when he left that city for the west.

The localities of collection cited by Nuttall for Gambel's plants give some indication of where the collector went, and the dates give, perhaps, a suggestion of when he visited a particular region.

Eight plants are cited from Santa Fe[6] (two flowering in August); a few are from "Rio del Norte, New Mexico," (flowering in September). Gambel left Santa Fe on the first of September, as we shall see later.

From California, collected in late 1841 or in 1842 and subsequent years, are the following: eight from Santa Catalina (five flowering in February); three from "San Pedro, Upper California" [7] (one flowering February); a number from "Pueblo de los Angeles" (flowering April); others from Santa Barbara (one flowering April, one May); some from "Monterey, Upper California" (one flowering September and October); still others from "St. Simeon, Upper California," San Luis Obispo County. According to these plant records, Monterey seems to have been the most northernly coastal point reached by Gambel, and, except for a few records from San Diego, which were possibly Nuttall's own, the most southernly point was San Pedro, Los Angeles County, the usual point of departure for Santa Catalina. Collections from inland California are few and the localities cited are vague; they were probably made when Gambel entered California in 1841. Such are: "Mountains of Upper California," "Rocky Mountains of Upper California," "On the Sierra of Upper California," and so on. The *Proceedings* of the Philadelphia Academy (2: 276, 279, 1846) record two gifts from Gambel but whether of his own collecting is not stated: one, "Cone and seeds of Pinus Lambertianus Douglass; from the mountains of Santa Lucea, Upper California . . ." received September 23, 1845; the other "Leafy branch of Pinus Lambertianus Douglass . . ." from the same locality, received October 14, 1845.

From the above it is evident that Gambel did no great amount of plant collecting, and least of all when on the march; or, if he did, Nuttall recorded but a few specimens, what he considered noteworthy. His claim to fame rests largely, it would seem, upon the fortuitous fact that he collected in two hitherto unexplored regions (Santa Fe and Santa Catalina) with an untouched flora. I have noted also that he was the first

6. Gray's "Plantæ Fendlerianæ" cites three Gambel specimens, but only one mentions Santa Fe specifically. *Erigeron cinereum*, a new species, is described from a Fendler collection, but there is a notation that "From Mr. Lowell's herbarium I find that Dr. Gambell gathered the same species in the vicinity of Santa Fé."

7. States Robert Greenhow: ". . . Upon the overthrow of the Spanish power in 1822, California was divided politically into two *territories* of which the peninsula formed one, called *Lower California;* the other, or *Upper California*, embracing the whole of the continental portion . . ."

"botanist" to enter California overland from the east, crossing the Sierra Nevada, in their Sierra Madre extension, through Cajon Pass.[8]

Gambel's work in ornithology should be mentioned. His important publication on our western birds was entitled "Remarks on birds observed in Upper California, with descriptions of new species" which appeared in two parts (December, 1847 and August, 1849) in the *Journal* of the Philadelphia Academy; there is no mention of a continuation. Preliminary diagnoses were published in the *Proceedings* in 1843, 1845 and 1848 after portions had been read.

T. S. Palmer tells that Gambel ". . . spent several years in California collecting at various points along the coast chiefly in the vicinity of the Missions, as far north as San Francisco Bay." Of bird habitats mentioned in the *Journal* a hawk *(Elanus laucurus)* is cited. "At the Mission of St. John, between Monterey and the Bay of San Francisco, I procured three specimens in one day." And the "Roseate Spoonbill" is said to have extended in flocks ". . . up the coast even as far as San Francisco."

A few places along the route to California are mentioned: the "Common Turtle Dove" was found breeding ". . . on the prairies near the rivers Arkansaw and Cimaron, in June . . ."; "Harris's Woodpecker" was ". . . killed on Green River, in September . . ."; the "Ruffed Grouse" was met with in flocks ". . . along the solitary and desolate Rio Severo[9] (Nicolet's river of Fremont) . . ." Most localities are, however, comprehensive ones, as, ". . . from New Mexico to California . . .", ". . . on the eastern side of the California range of mountains . . ." and so on.

Gambel's observations on birds are delightfully specific and indicate where the man's real affection lay. I quote two: first his description of the "Ground Tit," which, on plate VIII, appears to have been a vivacious little creature; and then the one of the "American Barn Owl." They appear in his "Remarks on the birds observed in Upper California."

"For several months before discovering the bird, I chased among the fields of dead mustard stalks, the weedy margins of streams, low thickets and bushy places, a continued, loud, crepitant, grating scold, which I took for that of some species of wren, but at last found to proceed from this Wren-Tit, if it might be so called. It is always difficult to be seen, and keeps in such places as I have described, close to the ground; eluding pursuit by diving into the thickest bunches of weeds and tall grass, or tangled bushes, uttering its grating wren-like note whenever an approach is made towards it.

8. Brackenridge's overland trip into northern California, also made in 1841, had taken him from the Columbia River southward through Oregon. Deppe had entered California overland from Mexico in 1831 or 1832, but is said to have made no plant collections while there. All other plant collectors, so far as I know, had reached and left California by sea.

Gambel might well have been antedated in the Catalina region by R. B. Hinds who records that the *Starling* had been dispatched from San Pedro Bay ". . . to examine the island and anchorage of Santa Catalina; rejoining us at San Diego." The date was October 9, 1839. But evidently Hinds had not made the trip.

9. "Rio Severo" is Spanish for Sevier River.

But if quietly watched, it may be seen, when searching for insects, to mount the twigs and dried stalks of grass sideways, jerking its long tail, and keeping it erect like a wren ... At the same time uttering a very slow, monotonous, singing, chickadee note, like *pee pee pee pee peep;* at other times its notes are varied, and a slow, whistling, continued *pwit, pwit; pwit, pwit, pwit,* may be heard. Again, in pleasant weather, towards spring, I have heard them answering one another, sitting upon a low twig, and singing in a less solemn strain, not unlike a sparrow, a lively *pit, pit, pit, tr r r r r r r r;* but, if disturbed, at once resuming their grating scold." (p. 35)

"This delicate feathered and familiar owl, is in California sufficiently abundant ... Its favorite resort is in the neighborhood of the towns and Missions ... Its nesting place is under the tiled roofs of the houses of the towns, numbers under one roof, and shows but little fear when approached. I have scarcely ever visited a Mission without disturbing some of these birds, which were roosting about the altar, chandelier, etc., of the chapel, and hearing the bendition of the Padre for drinking all the oil out of the lamps. Every where in California, when speaking of it, we are sure to be told of its propensity for drinking the sacred oil ..." (pp. 29, 30)

Gambel was familiar with the literature on birds; there are references to the opinions of the "Prince of Wied," to a bird noticed by "La Perouse" at Monterey and figured in the "atlas to his voyages," many to Audubon, to Nuttall, and so on.

T. S. Palmer thought highly of Gambel's contribution:

"He was the first ornithologist to spend any length of time in the state and his papers are the most important of the early publications on the birds of California. He described several characteristic birds including the Elegant Tern, Nuttall's Woodpecker, California Thrasher,[1] Plain-crested Titmouse, Wren-tit [the 'Ground Tit' above], Cassin's Auklet, Blanding's Finch or Green-tailed Towhee and Mountain Chickadee. Three of his names have proved to be synonyms ... In addition to the Mountain Chickadee, four other birds have been named in his honor: a Goose ... a Quail[2] ... a White-crowned Sparrow ... and a Shrike ..."

Zoologists were as unsuccessful as botanists in their attempts to name a genus after Gambel; Palmer notes that the "sub-generic name" *Gambelia* was proposed for a group of lizards, "but apparently has never come into use."

According to Witmer Stone, Gambel's bird-specimens met with an enthusiastic reception on arrival in Philadelphia; John Cassin who, we are told, later showed some animosity towards Gambel, wrote an associate:

" 'Eureka! Gambel is here with his California birds and others—not very many,

1. Alden and Ifft refer to "the California thrasher" as having been beautifully figured, and "easily identified" in the Atlas of the Lapérouse expedition.

2. No one traveling extensively in the Lower Sonoran deserts of our southwest can fail to encounter and to be charmed by the flocks of "Gambel quail" with their little plumed heads. F. M. Bailey's *Birds of New Mexico* supplies their scientific name which would have been after Gertrude Stein's own heart: *Lophortyx gambeli gambeli* Gambel!

but some of the most magnificent specimens I ever saw . . . Unfortunately he has made it an object merely to make one good series . . . Of many birds he has but one specimen,[3] though of several species he has duplicates . . ."

Gambel made a second transcontinental trip in 1849 which ended with his death on arrival in California. Although it seems highly doubtful that any plants were saved —even if collected—I shall, nonetheless, tell what is known of his journey.

The *Autobiography of Isaac Jones Wistar* was published in 1914 and describes the first part of the trip. Wistar and Gambel traveled together from Philadelphia to St. Louis; although an unusually fine map shows General Wistar's trip across the continent, the two men parted company on June 2, 1849, after they were five weeks out from that city. By that date, with the breaking-in of mules, with cholera, pneumonia and smallpox in the party, they had only, following the Platte River, gotten slightly beyond Grand Island, Hall County, Nebraska. Wistar mentions his companion more than once up to the time they separated and tells what he was later to learn, not very much, of Gambel's death.

Wistar was twenty-two years old and Gambel his senior by five or six years. According to Meehan, Gambel had graduated from the medical school of the University of Pennsylvania in March, 1848[4] and during his weeks with the Wistar party evidently began an extremely active practice.

The news of the discovery of gold in California had reached Philadelphia and the fascination of "prairies and mountains" and of "the golden prospects beyond" had been enough to start Wistar on his way, with one hundred dollars and "a fine half ounce calibre rifle." I quote from the *Autobiography:*

". . . in company with Dr. William Gambel, assistant curator of the Philadelphia Academy of Natural Sciences, a young naturalist and author already of some distinction, I started at 11 p.m., April 5, 1849 . . . Gambel and I traveled by rail via Baltimore to Cumberland . . . thence by stage-coach over the national turnpike . . . to Wheeling . . . We took a steamer to Cincinnati, where the uncertainty of the steamboats obliged us to remain a day or two, and where I purchased a wagon actually in street use, which struck my fancy as a light, strong shortcoupled vehicle adapted to the purpose . . . I made no mistake, for that wagon proved to be one of the only two of our entire outfit which . . . actually reached the Pacific coast . . .

"At St Louis . . . I . . . took passage . . . by steamer up the Missouri, Gambel re-

3. Was Gambel an early conservationist? What he did collect was "good," even "magnificent," although insufficient for the ornithological Oliver Twists of his day. This desire for "more" is, however, understandable, for classifications—whether of birds or plants—may be modified on the basis of full material.

4. The *Proceedings* of the Philadelphia Academy, under date of May 27, 1849, records:

"The resignation of Dr. William Gambel as recording secretary, and as a member of the Publication Committee, was read, Dr. Gambel having left Philadelphia for California, where he proposes remaining for one or two years. On motion the same was accepted . . ."

maining a day or so longer for some purposes of his own or because he preferred more leisurely traveling . . ."

On board the boat cholera broke out, twenty persons died, and Wistar had a short but severe attack.

". . . when five or six days later we arrived at Independence landing . . . I was already sufficiently recovered to go about my affairs, after a fashion . . .

"We . . . made a start on April 25th . . . sixteen in all, with thirty-five mules . . . Dr. Gambel had joined himself to five Virginians who with their one wagon and eight mules traveled with us . . . it was not till May 2d that we at last crossed the State Line into the Indian country . . ."

As the map shows, they were, on that date, on the south side of the Kansas (or "Kaw") River, in eastern Kansas. A Negro slave, one of the party, soon came down with smallpox and was left behind.[5] Members of the party are usually referred to by initials. When another man came down with smallpox he was left ". . . with his friend G. to care for him, at Lipscombe's—the 'last house.' " When they caught up with the advance party on May 12 they were on ". . . a small tributary of the Kaw, or Kansas, and within a few miles of that river, which must be crossed tomorrow." According to the map this must have been Wakarusa Creek and they had not gotten far. By May 14 "H." was ill.

"Spent this day in camp owing to the serious illness of H. which Gambel now pronounces to be pneumonia, and no great wonder, considering the weather, labor and exposure we have had. It has been severe on all of us . . . with more experience and less haste, we might have avoided a considerable part of it . . ."

Gambel was certainly not accomplishing any ornithological or botanical work:

"A fair march was made till shortly after noon, when J., another of our invalids, suddenly became so much worse that Dr. G. advised a halt, to let him die in peace . . . J. and H. the Doctor says, are both dying, and G. also is very ill, thus increasing our labor, while seriously reducing our effective force. However, the Virginians came nobly to the breach, in regard to the extra guard duty."

On May 19, with the three sick men lashed "down solid" into a wagon, they crossed what was presumably the Kansas and camped on Big Vermilion. They followed the north side of the Little Blue River—Wistar remarks that "The Little Blue, notwithstanding its name, seems to be much larger than the Big Blue"—until May 30 and crossed from its headwaters to the Platte, reached near Grand Island. For a number of days Wistar had had ". . . a painful and troublesome return of the Missouri River cholera . . ." On May 30 they ". . . plodded up the Platte bottom . . . The rain continues . . . The road is soft and slippery, and sixteen miles were enough for us. That distance, however, cleared us of Grand Island, and gave us the whole width of the

5. For those who may be interested, the sick man recovered and was ". . . shipped by express, with a label sewed to his breast, from Independence to Georgia, and arrived safely!"

river [the Platte] to look at, but as it is the ugliest one I ever saw, the view is not an un-mixed joy . . ."

Under date of June 2 Wistar tells of parting from Gambel and then tells of what he learned of his death a few months later:

"Gambel being desirous of traveling more leisurely and comfortably, left us today and joined the large ox train led by Captain Boone of Kentucky, who is anxious to have him and will dispense with any aid from him in driving or working, in return for his medical services. We gave him a mule with his proportion of the tools and provisions. He is an amiable, excellent fellow and very pleasant in conversation, having formerly made a similar journey to the Ratone Mountains and Santa Fé for the Philadelphia Academy of Natural Sciences, of which he is a prominent member. But he is averse to camp duty and hard work, and fond of taking things easy, and there is no doubt that Boone's large train with plenty of men and animals, and leisurely rate of traveling will suit him better than our headlong methods, especially as he has formed a warm friendship with Boone. (I never saw Gambel after that separation, and may as well state here what I did not learn till long afterward, and then only by hearsay. Boone's train after losing many teams and wagons in the Humboldt River desert, arrived late in the season at the Sierra, where they encountered more obstacles and losses, reaching California after the beginning of the rains. Gambel personally made his way as far as Rose's Bar[6] on Feather River, where he died almost immediately from typhoid fever resulting from the extreme privations suffered during the latter part of his journey. Either Boone himself or some of his party, among whom Gambel was a great favorite, were with him at the time of his death.)"

In a letter to C. S. Sargent which is included in *The Silva* Wistar repeats much the same story, but with a few additional details which I quote:

" 'In the year 1850, I met two men of Boone's train at Foster's Bar, who gave me the first information I had received of the fate of the majority of the overland party. Being well furnished and provisioned and mostly older men than we, they traveled leisurely and reached the Sierras only in October. After the loss of most of their cattle and consequent abandonment of many wagons in the Humboldt desert, they were caught by snow in the mountains; and instead of abandoning the remainder and pushing through they camped to await better weather, which did not come. The snow constantly accumulated, all the cattle died, provisions were consumed, and when too late they made snowshoes and tried to save themselves. But few got across the range, in-

6. Lest anyone should infer—as I did for a brief second—that Gambel in his last hours was given shelter in a "bar" (modern sense) and that a lady by the name of Rose was his ministering angel, I quote from Bancroft:

"The Yuba revealed gold as far down as Marysville [Yuba County] . . . Above lay the bars . . . known as Sand, Long . . . Spect . . . [and] the richer and enduring Rose Bar . . ."

The three main branches of Yuba River unite about nine miles northwest of Nevada City and empty into Feather River at Marysville.

cluding Gambel, and those saved little but what they stood in. With numbers rapidly diminishing, the remnant pushed on down to Rose's Bar, where several, including Gambel, died almost immediately of typhoid fever. Gambel was buried on the Bar, which, however, I have understood, has since been entirely removed by hydraulic mining. His death occurred in the latter part of November,[7] 1849, and I have never since seen any of the survivors of his party or heard further particulars . . .' "

A possible clue to the route taken across the Sierras by Gambel and his companions appears in an extract from a Philadelphia newspaper, the *North American,* which is quoted in Witmer Stone's account of Gambel:

". . . he . . . joined a company commanded by Capt. Boone,[8] of Kentucky, which followed the trail opened by Hudspith's Company, crossing the Sierra near the head of Sacramento Valley . . ."

"Hudspith" might well be the James M. Hudspeth whose name is associated with Lansford Hastings, of "Hastings Cut-Off" fame. If so, Gambel and his party may have tried the same disastrous route which Hastings is said to have recommended to the Donner party. Certainly what befell the two expeditions sounds remarkably alike.

It is unfortunate that a journal kept by Gambel should have been lost. Stone, enumerating in 1911 the meagre Gambel records at the Philadelphia Academy, refers to ". . . a letter from his young widow thanking the Academy for their resolutions upon his death, and a record showing that she loaned the Curators Dr. Gambel's journal of his last trip—a manuscript which I have failed to trace, but which if still extant must be fascinating reading."

In the Historical Society of Pennsylvania is a scrapbook relating to William Gambel. Dr. Francis Harper was good enough to bring this to my attention in February of 1951, or after this chapter was written, and the Society kindly provided me with a microfilm. There are six letters written by Gambel to his mother during the period of his first trip to the west coast, and three letters written to Gambel by Nuttall who was then living in England on his inherited estate. Although most of the content of these letters is unrelated to botany, I found them interesting, largely because of what they disclose of Gambel's personality—of which the literature reveals but a pittance—and of what

7. On December 13, 1849, according to Meehan.

8. The disappointments of research are many. I cite one example. In the "California emigrant letters" compiled by Walker D. Wyman and published in 1945 in the *Quarterly* of the California Historical Society, a few quoted from the Missouri *Statesman* are signed "Old Boone." One letter so signed describes "Scenes around St. Joseph and Independence" under date of May 3, 1849; and another, describing "Life Enroute," was dated June 3, 1849, the day after Gambel joined "Captain Boone." I hoped that "Old Boone" might prove to be Gambel's companion. But "Old Boone" also wrote from the gold-fields of California on August 15, 1849, and—according to Wistar—Gambel, and his particular Boone also, reached the Sierras only in October. Belatedly discovered, but conclusive and disheartening, "Old Boone" was recorded in the index of the *Quarterly* as a pseudonym! I have been unable to discover with which particular Boone Gambel traveled.

they disclose of Nuttall in the capacity of friend. Considerable has been published about his peculiarities! They also indicate how difficult it was for him, remote from botanical contacts, to carry on his work while in England. I believe that some may find the letters interesting, in part at least.

Curiously, although the son signs his family name "Gambel," two of the letters to his mother are addressed to "Mrs. Elizabeth Gamble," and apparently in the writer's own hand. I quote the letter-headings and hereafter cite them as (1), (2), and so on:

(1) "Independence Missouri May 4th 1841."

(2) "Santa Fé, New Mexico, July 25th 1841."

(3) "Pueblo de los Angeles Upper California January 14th 1842."

(4) "U. S. Frigate 'United States' Callao-Bay January 18th 1844."

(5) "U. S. Frigate Savannah Callao Bay July 15th 1844."

(6) "U. S. Frigate Savannah At Sea January 13th 1845."

Letters (1), (2) and (3) are the only ones which include any information about Gambel's trip of 1841 and this information is of no great importance, in the sense that the writer mentions what most other travelers of the period do—prairies, Indians, lack of water and food, buffalo, etc. They do, however, outline his route from the east to St. Louis (down the Ohio and up the Mississippi) and supply a few dates. Gambel was in Santa Fe on July 25 and left September 1, so he was there, certainly, for a little more than a month. Punctuation and capitalization are virtually nonexistent but the writing for the most part is clear and neat. Letter (1) follows, with unimportant omissions:

"I shall start from here to morrow morning [May 5, 1841] for Santa Fé and this is the last chance I shall have for writing till I get back which will not be for a year or more as there is no party coming back till that time you must excuse me for not writing before, but you know I dont like to write and when I do it is as much as you can do to read it I suppose. tell my sisters not to be frightened about me for I will get back some of these days. and perhaps before I do I will find Eliza has got married and Maggy thinking about it. I hope you will get along well while I am gone and not be anxious about me for you know I am a careful fellow and sooner run than fight. I have not room in this letter to tell you all about the places I have been in but can mention them in 5 days after I left I got to Pittsburg it snowed and hailed the whole way. I then took steamboat down the Ohio river stoped at Cincinnati Louisville Ky. and other places on the river . . . I then came up the Mississippi river to St Louis staying there a little while I came up the Missouri river to this place which is the last settlement before you get in the prairies which stretch for hundreds of miles without seeing a plant 6 inches high except grass so that it appears like the sea. through these I have to travel now for the next 3 months before I get into Mexico. I am in a hurry now as we are about starting. there will be in the party about 60 men all mounted on mules, horses not being able to do without water and corn like a mule . . ."

Letter (2):

"This is the first opportunity I have had of writing to you, by the Company which returns this fall. it has been so long since you have heard from me that you must nearly have forgotten me, but tis different with me when standing Guard or in among Indians. I would often think of home [and] the dangers I was in. & sometimes feel sorry to have left it. We met about half way the whole body of the Arrapaho Indians consisting of about 500 warriors but we by giving them a great many presents got clear of them without fighting, and when approaching the Rocky Mountains we met with the [Enturo ?] Indians about 400 warriors which attackted us and us being only 90 men in all had a great deal of trouble to get clear of them, they fired on us nearly all the forenoon from the hills that were round us but at too great a distance to do any harm although the ball would fall within a few feet of us sometimes. We also suffered much for want of water sometimes having to do without it for two days however I have got through with safety not even with a cold from laying on the ground with nothing but the sky above the earth below and perhaps Indians around us for nearly 3 months. but now I am only half way to where I am going (but you must not mind it I shall get home safe and in due time if God is willing) it is to California, I am going straight across the country to there and from there to Sandwich Islands and from there a five months voyage round Cape Horn to the U. States. When I get to California I will write to you but will take the letter 5 or six months to go to you, besides the time I am going from here to there which will be 3 perhaps, so you will hear from me again in about 8 or 9 months and se me back in perhaps 10 month. It is a very long while but I cannot help it, as I cannot go by railroads but jog jog on a mule from morning to night. I shall not start from here for a month yet . . . I am now in a different country now having to speak the Spanish language altogether. I have studied it a [great ?] deal and now can speak pretty well . . . manners and customs are also very different but any thing suits me. I would as soon live here nearly as in Philadelphia. On the road here the Buffalo were very abundant on the Arkansaw river I saw at least ten miles square black with them as thick as sheep. it is the finest meat in the world and we lived on it all the way out . . . so until you hear from me again . . ."

Letter (3) was written almost six months later:

". . . you have perhaps given me up for lost, but thank God I am yet safe, though far, far from home . . . though it was a long and dangerous journey, I have got through without sickness or accident. We left Santa Fé on the 1st of September and arrived here the last of November being three months travelling over Rocky mountains, barren deserts, worse than those of Arabia, sometimes having to do without water 2-3 days at a time, and towards the last almost starving for want of provisions, suffering also innumerable other difficulties which I have not time to mention, but am glad I have got through them safe, and am in California, on the banks of the broad Pacific Ocean, from where I shall go home in a vessel perhaps round Cape Horn, if not I will go to Panama and from there across to the west India Islands and from there home. but it is some time yet to that, for I most probably shall stay here until August or September next before I leave.[9]

"I have not time at present to write you a long letter as a vessel has arrived in the port of St Pedro about 30 miles from here and sails in the morning so that I have but just time to write a few lines to you, Mr. Nuttall, and my friend James Carlin, to let you all know that I am yet in the land of the living . . . I have enjoyed the best of health, not having been sick a day since I left though I have knocked about in all kinds of weather, and lived almost as rough as an Indian.

"California is a fine rich country and to which many western people are now commencing to emigrate, and in a few years I expect it will be under the Government of either the British or Americans. They raise an immense number of cattle and horses here which you can buy for about nothing a good fat ox only costing 3 dollars and you can take the hide to a store and sell it for 2 dollars so that the meat only costs a dollar. the best horses also can be bought for 5-10 dollars and mares from 1 dollar to 2 dollars If our Yankee farmers were here they would soon make their fortunes instead of cultivating the poor rocky soils of Connecticut and Vermont . . ."

The letter ends with an admonitory postscript to his sisters:

"Give my love to my sisters, and tell them not to give theirs too quickly to any young fellows they may come across. Yours W. G."

Letters (4), (5) and (6) show that Gambel served, in some capacity, on the frigates *United States* and *Savannah* and moved about with these vessels—to Peru, to the Hawaiian Islands, to Mexico, to California probably (for he mentions the possibility of getting to San Francisco and Monterey), and elsewhere. Bancroft, as noted already, mentions that, being "financially crippled," Gambel had taken service on the *Cyane*. Letter (5) refers to the fact that that ship had departed for the United States in June of 1845. The same letter indicates that he liked the sea and was thinking of entering the navy as a means of supporting his family.

"I expect it somewhat surprises you to see how I talk about such long voyages as if they were nothing, but I am now an 'old Sea Dog' as the sailors say, and feel as safe and as much at home on board ship, as if I was on shore. I have now sailed further than twice round the world! . . .

". . . perhaps it is better for my future prospects in life, that I have been so long in my present situation in the Navy, for I have made friends of influential men in our country,—and there is hardly a doubt that on my return I may if I wish, be able to get a Pursers appointment in the Navy, which will make me an independant man, and able to support both myself and you is [in] easy and decent circumstances. The pay of a Purser in the Navy varies according to the duties he is ordered upon, it is from $1,500. to $3,500 a year, and he receives $1,000 when he is doing nothing. I do not

9. Gambel's plans for returning east seem to have been as variable as the wind. Whether they were altered by necessity or by *wanderlust* is hard to tell.

wish you to mention anything about my prospects to even your most intimate friends. I have told no one but you, and I want it to go no further . . ."

Gambel's admonitions to his mother are curious. One concerned with the addressing of letters ends thus: "Get some one to direct them outside in a mercantile hand and they will be more certain to come." But advice to his mother is as nothing compared to what he bestowed upon his sisters. Perhaps he had been asked to send it, or perhaps he felt he was not living up to the responsibilities of a brother. It would be interesting to know how sisters of the 1840's relished it! I quote from letter (6):

"I often think of my sisters, and what they may be doing; and I hope that you are able to live without difficulty, though no doubt you have seen hard times since I left home. May God preserve you until I get back. But I hope that in every difficulty, you have taught my sisters to look to their heavenly Father, as the protector of the father-less children, and widows and those who are desolate, and oppressed. Oh! how happy should I be, to hear that they have indeed become his children. I wish my dear Mar-garet & Elizabeth that for my sake your only brother who is far away, thousands of miles on the broad Pacific, that you would read one chapter in your Bible at least, every day, until I get back; and attend the Episcopalian church regularly on Sundays. As I told you before, beware of the company you keep, and do not think because you are poor, that you are obliged to mingle with all those who are [word illegible] it is better to keep aloof and be called stiff and proud, and whatever else they choose, than to fall into the bad habits and ways of those who may want to associate with you be they rich or poor. Respectability and virtue depend neither upon riches or poverty; and you may be both, in any situation of life, by 'taking heed unto your ways.' Solomon says, 'He that walketh with wise men shall be wise: but a companion of fools shall be destroyed.'

"I hope also, that you read and study all that is in your power; I do not mean novels, tales, travels, and the such like trash that is publishing every day in Phila; but good, sound, sensible books, that will teach you common sense, and how to live, and make the most of life; for, as the wise man said again 'Every *wise woman* buildeth her house: but the foolish plucketh it down with her hands,' that is, that by their ignorance and carelessness in the management of affaires, they soon go to destruction . . ."

Precisely how and when Gambel got back to the east coast I do not know, but a letter written by Nuttall from England on September 8, 1845, mentions that he is glad to hear of his safe return. Gambel's letter (6), written in January, 1845, mentions that he was expecting to reach Valparaiso ". . . by the 1st of March, and we shall then start immediately . . ."

The three Nuttall letters in the Historical Society of Pennsylvania collection are full of affection and of eagerness to help Gambel in some concrete way; the fact that, with much material calling for publication, he described Gambel's plants as soon as he got back to the Philadelphia Academy on his last visit, may have been motivated by such a desire. Gambel's letters indicate that he wrote Nuttall rather frequently, but it seems unlikely that his side of the correspondence is still in existence. Nuttall's letters are all from his estate in England. The first is dated July 2, 1845, from "Sutton, nr Prescott, Lancashire England":

"Your last letter, 'Bay of Mazatlan Feby 27th 1843,' and the place of its date gave me no small astonishment as well as inward delight, because my eyes again dwelt on your hand-writing and ran over your interesting adventures. The time appears very long since we parted and both of us have gone thro' unexpected vicissitudes and changes of circumstances. By the death of my aunt Nuttall I have been called here to dwell in England on a pretty good estate, and where I wish to see you as soon as you possible can . . . I a good while since got a letter from you from 'Angelos' stating your desire and as I thought intention of returning in my good old ship the Alert; I had previously written to Bryant and Sturgis thro' our friend Brown desiring them on my account to give you a passage home and advance you what money you might immediately want, but alas all my efforts have yet been in vain to relieve my favourite disciple from difficulty and embarrasment. I have also written Mr. Wright Boott of Boston to advance whatever you might want on landing before arriving in Philad[elphia] at Carpenters when I directed him also to advance whatever you might want. As to writing to you, I was thrown into despair of any thing ever reaching you wandering about as you were. I have even now but a faint hope of ever reaching you by letter considering the long lapse of *weary time* wh must arrive before there is any chance of this epistle meeting you. As it is, to use a canting tho' scriptural phrase, my wishes to you on paper perhaps may prove like 'bread cast upon the waters wh may be found after many days.' Write you say, if only a line ! my dear friend, do you suppose for a moment, that I can ever forget you. No, while this pulse beats wh now allows me to communicate with you you are seldom out of my anxious thought, and if good *wishes* would have been of any avail, you would this moment have been as rich as Croesus.—I have a good garden, greenhouses, &c so that what plants you may bring living can be well cared for. Your collections of dried plants I can get sold for you here as well as possible, as also skins of birds, &c and I wish the snails to be *wholly* your own. Come to me and write out your journal, describe your novelties, and suit yourself. I only got one of your papers on birds, no synopsis of birds, but the descriptions containing also the account of the *Couri camena,* wh[ich] for you I called *Eudromus* (good traveller) but strange to say, the Academy or their committee remark in a note appended to the publication of your species that they have *omitted* the description of the Eudromus untill they should see the specimen, the thing of all others I most anxiously wished to have published for your sake.[1] The name, however,

---

1. Among the letters sent me by the Historical Society of Pennsylvania is one written by Gambel to some member, not named, of the publications committee of the Philadelphia Academy in regard to the bird just mentioned. If it was sent (for the one seen may well have been a copy retained by Gambel), it may still be in the unknown recipient's files. It is dated "U. S. Frigate Savannah Callao Bay Peru July 6th 1844."

"Dear Sir,

"An officer of the Frigate 'United States' which sails for the United States to day, offering to take any

thus incidentally given must stand good, and I shall have an eye to any thing wh may arrive in England so as to identify your curious genus.

"You remember that I sold my specimens of plants at 6$ a 100. While Cumings and others for tropical Asiatic plants (those of the Phillipines) got 50 shillings sterling or double what I obtained for mine. As to Jones you astonish me to find that he could have been so unfriendly towards you. For my part I never trusted his friendship tho' I may say, however, that Townsend[2] abused it, wh might perhaps account for his coldness towards you. How important it would have been to you if he would have assisted you in your extremity.

"I am still slowly progressing with my descriptions of the plants collected in my tour, many of wh yet remain to be described, and in many I have been anticipated. I have just returned from a visit to London to Sir Wm J. Hooker who for 3 or 4 days at his house at the Queen's Botanic Gardens at Kew, assisted me in comparing my specimens with his *stupendous herbarium!* His uniform kindness, frankness, and honor I shall never forget.

"Your mother and sister as far as I know last from James Carlin are well.

"I remain, your affectionate friend

"Thos Nuttall

"N.B. I have never had a word of any of your collection arriving with Dr. Engelman or in St. Louis. Collect all the Carices you can for Dr. Boott of London."

The second Nuttall letter was written from "Nutgrove nr Rainhill. Sept. 8th 1845":

"Your letter of July 31st came duly to hand, and gave me much pleasure to hear of your safe return after so long an absence indeed laterly I began to doubt whether you ever would return. I am sorry you met with so cool a reception from Mr. Browne for

thing I might wish to send home, for me among other things, I happened to have some few of the land Shells which I have collected handy, and although I donot know whether they are worth it or not, yet I concluded to send them to you knowing how fond you are of *land* Shells.

"Accidentally finding a volume of the transactions of the Academy in the Squadron, I discovered my paper describing some new birds in it, and saw that you were one of the committee to whom it was referred for publication,—I cannot but think it strange that those gentlemen should credit the descriptions of my other new birds and not that of the *Eudromus variegatus,* the existence of which bird was well known to several ornithologists, even before I left Phila, but I was the first one to procure a specimen and describe it, and I would have considered myself in some measure repaid for my weary wanderings by having been the first to make so valuable and interesting an addition to our Fauna as the *Correcamino.* The fine specimen which I have I think too much of to risk sending home until I go myself.

"Should you like to write to me, or if you have any commands you would like executed in this ocean and express them in a letter I shall be most happy to attend to them ... I expect to be in the U. S. in about 8. or 9. months ... I am Sir

"Very Resptfly   Your Obdt Servt

    "Wm Gambel."

This bird, in which both Gambel and Nuttall displayed such interest, is, I believe, the road-runner, *Geococcyx Californianus.* Both men were apparently unaware that C. E. Botta had collected the bird and that it had been described, before 1835, by Lesson.

2. Whether "Jones" was Commander T. A. Catesby Jones of the *Cyane,* and whether "Townsend" was John Kirk Townsend, the ornithologist, I have no way of knowing.

wh I can not account a letter you have by this time found for you long left with Mr. Carpenters, and I hope ere long to see you in England where I know your collections will be well appreciated indeed in my every letter from Sir Hooker now director of the Royal Garden at Kew you are almost always enquired of. Geyer has arrived already and has paid a visit to Sir William who has most kindly assisted him in arranging his collections for sale (now offered at 2 £ sterling for 100) in wh way he will make something handsome. In regard to Botany or Natl History I have done very little since coming to England. All I have done is to arrange my herbarium and place it in a convenient cabinet. The fact is, there are in this part of England scarcely any who cultivate any branch of Natural Science. we have indeed not far off a kind of society so named but little or nothing is done in this subject but in the vicinity of the metropolis. The royal botanic garden at Kew is however, rising to be the greatest establishment of the kind in existence but unfortunately I am situated a great way from it. When you come we can visit it together, and you can make some arrangement to dispose of or publish your things as you may think proper. . . .

"Mr. Carpenter will render you on my credit what assistance you may require . . . Dobson has wofully [deceived ?] me. The continuation of Michaux has as good as stood still since I left. I dont know what he will say to you about it but I wish my manuscript were published in some way. . . ."

The last letter is dated "Sutton, nr Rainhill March 12th 1846.":

"I hope you have not been waiting for me to write, by your long silence any thing you choose to communicate I shall always be glad to hear and answer. From Mr. Carpenter I now learn that you are about to study medicine[3] wh I was pleased to hear and hope you may meet success It was time you had some profession to depend upon and I trust you have chosen that wh will suit you. I shall be much pleased to hear from you, and more so to see you whenever you can come over. I had thought of visiting America last autumn but, if I live, I shall I think certainly do so this fall.[4] I have desired Mr. Carpenter to offer my cabinet of minerals for sale, but I hardly know how it can be effected as they will all more or less want labelling as to character and locality.

"The balance of minerals wh I expected in exchange from Mr. Clay's brother at Vienna have never yet been forwarded. Ask Mr. Clay if his brother is still in Austria?

"My continuation of Michaux will I suppose now remain unfinished as Dobson has it seems given it up to some one from whom I hear nothing whatever. Torrey and Gray appear to have given up the Flora for the present but on what account, more than the want of a remunerative sale of the work, is more than I can conjecture. For my own part since living in England I have, in fact, done absolutely nothing more than arrange my herbarium. Many of my plants yet remain unpublished. I have next to no botanical acquaintance except Sir Wm Hooker. There is a good botanic Garden in Li[ver]pool,

---

3. This would account for Gambel's interim years between his two western journeys.

4. According to Pennell, "In the last months of 1847 and the first of 1848 Nuttall again visited America . . ."

but it is much neglected by the public and its support is precarious. All my amusement is seeing to the cultivation of a small garden. I see very little of society, as usual, for I value retirement more . . ."

ALDEN, ROLAND HERRICK & IFFT, JOHN DEMPSTER. 1943. Early naturalists in our far west. *Occas. Papers Cal. Acad. Sci.* 20: 10, 11.

ALLEN, JOSEPH ASAPH. 1911. Recent literature. *Auk,* new ser. 28: 288, 289.

BANCROFT, HUBERT HOWE. 1889. The works of . . . 17 (Arizona and New Mexico 1530–1888): 338.

—— 1885. 1886. 1888. *Ibid.* 20 (California III. 1825–1840): 386, 751, 752 (1885); *Ibid.* 21 (California IV. 1840–1845): 276-279, 322 (1886); *Ibid.* 23 (California VI. 1848–1859): 72, 359, *fn.* 7 (1888).

—— 1890. *Ibid.* 25 (Nevada, Colorado and Wyoming 1540–1886): 38, 39.

BREWER, WILLIAM HENRY. 1880. List of persons who have made botanical collections in California. *In:* Watson, Sereno. Geological survey of California. Botany of California 2: 556.

CLELAND, ROBERT GLASS. 1950. This reckless breed of men. The trappers and furtraders of the southwest. New York. [Map 3, "From Santa Fe to California . . .", shows the route taken by Wolfskill.]

GAMBEL, WILLIAM. 1843. 1845. 1846. 1847. Descriptions of some new and rare birds of the Rocky Mountains and California. *Proc. Acad. Nat. Sci. Phila.* 1: 258-262 (1843); 2: 263-272 (1845); 3: 44-48, 110-115 (1846); 154-158, 200-205 (1847). [The titles vary slightly in the different installments.]

—— 1847. 1849. Remarks on the birds observed in upper California, with descriptions of new species. *Jour. Acad. Nat. Sci. Phila.* ser. 2, 1: 25-56 (1847); 215-229 (1849).

GHENT, WILLIAM JAMES. 1931. The early far west. A narrative outline 1540–1850. New York. Toronto.

GRAY, ASA. 1849. Plantae Fendlerianae Novi-Mexicanae: an account of a collection of plants made chiefly in the vicinity of Santa Fé, New Mexico, by Augustus Fendler; with descriptions of new species, critical remarks, and characters of other undescribed or little known plants from surrounding regions. *Mem. Am. Acad. Arts Sci.* new ser. 4: 1-116.

GREENHOW, ROBERT. 1845. The history of Oregon and California, and other territories on the north-west coast of North America: accompanied by a geographical view and map of those countries, and a number of documents as proofs and illustrations of the history. Boston. ed. 2. 365.

HARSHBERGER, JOHN WILLIAM. 1899. The botanists of Philadelphia and their work. Philadelphia. 157, 231-233.

MEEHAN, THOMAS. 1880. The native flowers and ferns of the United States in their botanical, horticultural, and popular aspects. Philadelphia. ser. 2, 2: 61-64.

NUTTALL, THOMAS. 1848. Descriptions of plants collected by William Gambel, M.D., in the Rocky Mountains and upper California. *Jour. Acad. Nat. Sci. Phila.* new ser. 1: 149-189. 3 plates.

———— 1850. Description of plants collected by Mr. William Gambel in the Rocky Mountains and upper California. *Proc. Acad. Nat. Sci. Phila.* 4: 7-26. 3 plates. [Read in February, 1848. Preliminary diagnosis.]

PALMER, THEODORE SHERMAN. 1828. Notes on persons whose names appear in the nomenclature of California birds. A contribution to the history of west coast ornithology. *Condor* 30: 278.

PENNELL, FRANCIS WHITTIER. 1936. Travels and scientific collections of Thomas Nuttall. *Bartonia* 18: 43, *fn.* 101.

SARGENT, CHARLES SPRAGUE. 1895. The Silva of North America. Boston. New York. 8: 35, *fn.* 2.

STANDLEY, PAUL CARPENTER. 1910. The type localities of plants first described from New Mexico. *Contr. U. S. Nat. Herb.* 13: 167.

STANSBURY, HOWARD. Exploration and survey of the valley of the Great Salt Lake of Utah, including a reconnoissance of a new route through the Rocky Mountains. *U. S. 32nd Cong., Spec. Sess., Sen. Exec. Doc.* 2: No. 3, 1-267.

STONE, WITMER. 1911. William Gambel, M.D. *Cassinia:* No. 14: 1-8.

WISTAR, ISAAC JONES. 1914. Autobiography of Isaac Jones Wistar 1827–1905. 2 vols. Philadelphia. [Printed by The Wistar Institute of Anatomy and Biology.]

WYMAN, WALKER D., *compiler.* 1945. California emigrant letters. *Quart. Cal. Hist. Soc.* 24: 37-39, 43-45, 126, 235, 236, 363.

YOUNT, GEORGE C. 1923. The chronicles of George C. Yount California pioneer of 1826. Edited by Charles L. Camp. *Quart. Cal. Hist. Soc.* 2: 36-41.

CHAPTER XXXVI

FRÉMONT CROSSES THE ROCKY MOUNTAINS THROUGH

SOUTH PASS AND GETS AS FAR WEST AS THE WIND

RIVER MOUNTAINS OF WYOMING

JOHN Charles Frémont began work as a topographical engineer and his first field assignments were in our eastern states. Next, in 1838 and 1839, he served under Joseph Nicholas Nicollet, who was then completing surveys for a map of the sources of the Mississippi River and adjacent regions. In 1841 he surveyed the lower portions of the Des Moines River from its confluence with the Missouri to Raccoon Forks,[1] in Polk County, Iowa. Soon after his return from this task Frémont married Jessie Benton—daughter of Thomas Hart Benton, influential United States senator from Missouri—a lady who sometimes took matters into her own hands and not always, I must believe, to her husband's eventual benefit.

Frémont's ability as a map maker, his love of exploration and his marriage to the daughter of a strong advocate of westward expansion all played a part in securing his appointment to head three government expeditions associated with the decade of 1840–1850—the first in 1842, the second in 1843–1844 and the third in 1845–1846. On the first two of these assignments Frémont adhered largely to topography, but on the third he played a part in the seizure of California. Whether he was justified in assuming the rôle of conqueror is for historians to decide. He was charged with mutiny and convicted and, although President Polk set aside the sentence, Frémont resigned from the United States Army. On his own initiative Frémont undertook still a fourth expedition in this decade. In 1848–1849, after a foolhardy and disastrous attempt to make a shortcut, in midwinter, through the mountains of Colorado, he eventually, by a southern route, reached the junction of the Gila and Colorado rivers and from there moved northward through California. He had acquired a piece of property through which—by mere good luck—ran one of the richest of the gold lodes of Mariposa County,[2] and Frémont's topographical (and botanical) work appear to have ended with his entry into California in 1849.

Frémont's later life included, among other activities, two years as United States Senator from California (1850–1851), a campaign as Presidential candidate of the Republican party (1856), and the Governorship of the Arizona Territory (1879–

1. I have referred to this summer's assignment when writing of Karl Andreas Geyer, the German botanist and plant collector.

2. Because this land had been acquired from the Mexican authorities, Frémont's title to the property was under dispute for some years, but his claim was finally allowed by the Supreme Court of the United States.

1882). Although for a time extremely rich, it was not long before he lost his property with the collapse of a railroad enterprise.

Frémont once loomed large in public estimation, but his stature with the passage of time seems to have diminished. Certainly many of his activities along lines other than topographical are even now subjects for controversy. During his lifetime he received a vast amount of publicity and, although the eulogies were not entirely of his own composing, there can be no doubt that he made some unjustifiable statements about his accomplishments. H. H. Bancroft[3] comments more than once upon the man's propensity for self-eulogy. I quote his appraisal of Frémont's accomplishments on his expedition of 1842 and of the explorer's claims on arrival at Great Salt Lake during the course of his journey of 1843–1844:

"He was to connect his explorations with those of Wilkes on the Pacific coast, but did nothing further this year [1842] than to make a summer jaunt to the South pass, which, being a military officer and not a private citizen, trader, trapper, missionary, immigrant, or what not, he 'discovered,' naming its altitude, and ascending the highest peak in the Wind River range, 13,570 feet, planted thereon the American flag. This mountain he named Frémont's peak; and considering that the government paid all the costs, and that he had an experienced mountain man, Kit Carson, for a guide, it must be admitted that the eternal mountains might be put to nobler use than to perpetuate such achievements. He did, however, in his subsequent expeditions actually explore some new territory."

"In 1843. . . . Frémont followed the emigrant trail through the south pass, and on the 6th of September stood upon an elevated peninsula on the east side of Great Salt Lake, a little north of Weber River, beside which stream his party had encamped the previous night. Frémont likens himself to Balboa discovering the Pacific; but no one else would think of doing so. He was in no sense a discoverer; and though he says he was the first to embark on that inland sea, he is again in error, trappers in skin boats having performed that feat while the pathfinder was still studying his arithmetic, as I have before mentioned."

Our interest in Frémont's travels arises from the fact that on his first three journeys he collected plants which were described by John Torrey. There is reason to suppose that he made some collections on his fourth expedition as well but, as far as I know, no report was published upon them. I shall tell of Frémont's expeditions in four separate chapters.

The man's interest in plants may have been stimulated by association with the bot-

---

3. Hubert Howe Bancroft's historical collections and writings are proving their importance with the passage of time. De Voto wrote: "I cannot imagine anyone's writing about the history of the West without constantly referring to Bancroft. His prejudices are open, well known, and easily adjustable. A generation ago it was easy for historians to reject much of what he wrote; in the light of all the research since done, it is not so easy now . . . I have found that you had better not decide that Bancroft was wrong until you have rigorously tested what you think you know . . ."

anist Geyer in whose company he had passed the summer of 1841, as I have stated. Geyer, somewhat prone to belittle, was not impressed by his associate's botanical abilities, writing William Jackson Hooker on August 13, 1845:

"In 1841 I made a tour with Fremont in the Desmoines River, lower Iowa territory in which . . . survey was the principal occupation . . . I did not [in 1842] accompany Fremont. The collections he had were made at and on the Wind River Mountains at random, neither he, himself, nor any one of his party understood anything about Botany; he found a new field where no botanist had collected before on account of the great number of men requisite for protection against the Indians. This tour Fremont made in 42 of which Dr. Torrey published the report . . ."

Frémont's account of the expedition of 1842 was printed at least twice by the United States government: first, in 1843, under the title "A report on an exploration of the country lying between the Missouri River and the Rocky Mountains, on the line of the Kansas and Great Platte Rivers" (Senate Document No. 243, 27th Congress, 3rd Session); again in 1845, when it was reprinted in conjunction with the account of Frémont's journey of 1843–1844 (Senate Document No. 174, 28th Congress, 2nd Session). Here the main title-page is inscribed "Report of the exploring expedition to the Rocky Mountains in the year 1842, and to Oregon and North California in the years 1843–'44," while the title-page of the reprint of the expedition of 1842 is similar to that of the printing of 1843, with minor exceptions.[4] Appended to both printings is John Torrey's "Preface" to, and "Catalogue of plants collected by Lieut. Fremont . . ." and to these I refer again at the end of this chapter. The map included in the first printing of the "Report" (1843) was, quoting G. K. Warren, ". . . on a scale of 1 to 1,000,-000, (nearly sixteen miles to an inch,) embracing the country from the forks of the Platte to the South Pass, between the forty-third and forty-fifth parallels." Of the one included in the second printing (1845) Warren wrote:

". . . the accompanying map exhibits the routes followed during [1842] . . . as well as during the years 1843 and 1844. The longitudes given on this map and in this report (pp. 100 and 101) differ materially from those of the first report and map, the reason for the change being explained on page 321. The new map is on a scale of thirty-two miles to an inch, and is 'strictly confined to what was seen and to what was necessary to show the face and character of the country.' It was drawn by Charles

---

4. Editors, printers, or whoever, had difficulty with the name "Frémont," sometimes omitting, sometimes inserting, the accent. Since Frémont's *Memoirs of my life* (1887) adopts the accent it is clear that the French form is the one in which the author wished his name to appear. Comparing the two printings of the "Report" we find that the reprint (1845) attempts to insert the accent omitted throughout the first printing (1843). The reprint of Torrey's "Catalogue of plants" (1845) also attempts to incorporate the accent but here again the typesetters could not quite cope with the problem.

Dellenbaugh uses "Fremont," as does Nevins, who states: "Family tradition holds that the name was always Frémont . . ." But usage is not consistent, for some authors still include, others ignore, the accent. It is always omitted on the maps and in geographical names. I include the accent except in the case of place names.

Preuss whose skill in sketching topography in the field and in representing it on the map has never been surpassed in this country . . . After the investigations necessary in compiling the map which accompanies this memoir [Warren's own], I may be permitted to add my testimony to the truth of Captain Frémont's assertion . . . 'that the correctness of the longitudes and latitudes may well be relied upon.' They contain only such errors of longitude as are inherent to results obtained from observations made with the instruments employed. A mercurial barometer was carried across the continent on the road to Oregon as far as the Blue Mountains, where it was broken. The subsequent elevations on the route were determined by the temperature of boiling water."

So much for the government's printings of the "Report." In 1887, or more than forty years after these had been issued, Frémont published the first—and what was to be the last—volume of his *Memoirs of my life,* an extremely heavy, bulky and disagreeable book to handle. The title-page announces ". . . five journeys of western exploration, during the years 1842, 1843–4, 1845–6–7, 1848–9, 1853–4." Unfortunately the *Memoirs,* as published, only included the first three expeditions.[5] The greater part of the subject matter dealing with the journey of 1842 is taken verbatim from the "Report" issued by the Senate although there are a few interpolations. The included map is extremely helpful. When Frémont *really* began to "get about" the country, as he did on his later journeys, one needs many good maps!

I shall follow Frémont's expedition of 1842 in the Senate reprint of 1845 and shall attempt to translate the route into the parlance of the modern map. Warren's outline of the trip is sometimes of help, as are Joseph Ewan's comments upon the explorer's routes through Colorado and the plants encountered there. The "Preface" to Torrey's "Catalogue of plants" contains an outline of Frémont's journey accompanied by references to plants found along the way. These references I mention in footnotes or otherwise at the appropriate places in my narrative.

Frémont's orders (issued by Colonel J. J. Abert, Chief of the Corps of Topographical Engineers) were ". . . to explore and report upon the country between the frontiers of Missouri and the South Pass in the Rocky Mountains, and on the line of the Kansas and Great Platte rivers . . ." On May 2, 1842, he left Washington for St. Louis, Missouri. Twenty days later he states that he had reached his destination ". . . where the necessary preparations were completed, and the expedition commenced . . ." From there he proceeded by steamboat to ". . . Chouteau's landing, about four hundred miles by water from St. Louis, and near the mouth of the Kansas river,[6] whence we proceeded twelve miles to Mr. Cyprian Chouteau's trading house, where we completed our final arrangements . . ."

5. The publishers, Bedford, Clarke & Company, are said to have lost money on the expensive volume, and evidently thought it the part of wisdom to issue no more. Moreover, by the time the *Memoirs* appeared, much of the content was an old story to the public.

6. The Kansas River enters the Missouri in Wyandotte County, Kansas, and Chouteau's Post, sometimes called Kansas Post, was about twelve miles up that stream, on the south bank.

Frémont's party consisted of about twenty-five persons, traveling on horseback or in wagons. Charles Preuss was assistant topographer. Christopher ("Kit") Carson served as guide, as he was to do on Frémont's next two journeys. He had had sixteen years of experience in the west and had traveled with such a wise "mountain man" as Thomas Fitzpatrick; one may be sure that a large part of Frémont's success was attributable to his presence, for when he did not accompany Frémont's expedition of 1848–1849 matters went disastrously.

For eleven days (June 10 to 21) the party traveled through Kansas. June 10 ". . . happened to be a Friday—a circumstance [states Frémont] . . . our men did not fail to remember[7] . . . " On the 14th they reached ". . . the ford of the Kansas . . . where the river was two hundred and thirty yards wide . . . swollen by the late rains . . . an angry current, yellow and turbid as the Missouri . . . By our route, the ford was one hundred miles from the mouth of the Kansas . . ."

We hear of a famous transcontinental caravan on the 17th: "A party of emigrants to the Columbia river, under the charge of Dr. White,[8] an agent of the Government in Oregon Territory, were about three weeks in advance of us . . . There were sixty-four men, and sixteen or seventeen families . . . a considerable number of cattle . . . [they] were transporting their household furniture in large heavy wagons . . ."

On June 18 they reached the ford near the mouth of the Vermilion, which enters the Kansas River from the north in Pottawatomie County, and camped on its western side. The next objective was the Platte River to the northwest; the party went across country. Frémont relates on the 19th ". . . many beautiful plants in flower, among which the *amorpha canescens* was a characteristic, enlivened the green of the prairie . . . I remarked, occasionally, thickets of *salix longifolia,* the most common willow . . ." On June 20 they crossed "the Big Vermilion," and after marching twenty-four miles camped on "the Big Blue."

On the 21st: ". . . the *amorpha* has been very abundant . . . Every where the rose is met with . . . scattered over the prairies in small bouquets . . . The *artemisia,* absinthe, or prairie sage . . . increasing in size, and glitters like silver . . . The *artemisia* has its small fly accompanying it through every change of elevation and latitude . . . wherever I have seen the *asclepias tuberosa,* I have always remarked . . . on the flower a large butterfly . . ."

From June 22 to July 8 the route was across Nebraska. On the first day at ". . . Wyeth's creek[9] . . . a pack of cards, lying loose on the grass, marked an encampment of our Oregon emigrants . . ." and camp was near the Little Blue. From the 23rd to the 27th the dates are confused in the "Report," but during that period they came to ". . .

7. This and other forebodings never came to anything.

8. In 1842 Dr. Elijah White had been appointed Indian Agent for the United States government, the settlers in Oregon having asked for protection. He was on his way to his post, leading the second emigrant train into that territory.

9. The volume *Nebraska* ("American Guide Series") identifies this as Rock Creek, near Fairbury in Jefferson County.

Sandy creek . . ." sometimes called "the Ottoe fork," ascended the valley of the Little Blue and reached a fork ". . . where the road leaves that river, and crosses over to the Platte . . . The road[1] led across a high and level prairie ridge . . ." Twenty-one miles beyond they came to ". . . the coast of the Nebraska, or Platte river[2] . . . From the foot of the coast, a distance of two miles across the level bottom brought us to our encampment on the shore of the river, about twenty miles below the head of Grand Island. From the mouth of the Kansas . . . we had travelled three hundred and twenty-eight miles." [3]

Frémont had mentioned various plants observed from the Little Blue to this point: "The country has become very sandy . . . the plants less varied and abundant, with the exception of *amorpha,* which rivals the grass in quantity . . . *cacti,* for the first time, made their appearance." In the valley of the Little Blue ". . . an abundance of prêle *(equisetum)* afforded fine forage to our tired animals." At an old encampment of some Pawnees was found, mingled with ". . . the usual plants, a thistle *(carduus leucógraphus)* . . . along the river bottom *tradescantia (virginica)* and milk plant *(asclepias syriaca)* . . ." Near the head of Grand Island (June 27) "The soil . . . was light but rich . . . in some places rather sandy . . . the timber, consisting principally of poplar, *(populus monilifera,)* elm, and hackberry *(celtis crassifolia,)* is confined almost entirely to the islands."

Frémont's next objective was the forks of the Platte, where the party arrived July 2. He had met (June 28) some fur traders on their way from Fort Laramie to St. Louis; they had started downstream sixty days before, their barges laden with furs, but had been obliged to abandon their entire cargo because of low water and now possessed only what they could carry on their backs: "We laughed at their forlorn . . . appearance, and, in our turn, a month or two afterwards, furnished the same cause for merriment to others." For the first time (June 30) buffalo were observed ". . . swarming in immense numbers over the plains, where they scarcely left a blade of grass standing . . ." Camp was at "Brady's island," Lincoln County, on July 1, and on July 2 the Platte was crossed and the party ". . . encamped at the point of land immediately at

1. Indicated on the map as "Pawnee Trail."

2. According to Warren, they reached the Platte ". . . near the present location of Fort Kearney . . ." *Nebraska* ("American Guide Series"), p. 58, includes this note: ". . . old Fort Kearney (now Nebraska City) on the Missouri 50 miles south of Omaha, established 1847 and abandoned 1848; new Fort Kearney on the south side of the Platte opposite the upper end of Grand Island, established as Fort Childs 1848 and abandoned 1871 . . ." Warren, writing in 1859, referred therefore to the "new" Fort Kearney, situated in present Buffalo County. W. J. Ghent mentions that "To protect the Oregon Trail the Government, in 1848, built Fort Kearney, on the Platte . . ." The spelling of the name differs: the volume *Nebraska,* quoted above, states (p. 334) that the town of Kearney ". . . was named for Fort Kearney . . . the name honored Gen. Stephen Watts Kearny. (The misspelled name of the town and fort became statutory.)"

3. Torrey's "Catalogue" records: "Crossing over the waters of the Kansas . . . Frémont arrived at the Great Platte, two hundred and ten miles from its junction with the Missouri . . . The valley . . . is rich, well timbered, and covered with luxuriant grasses. The purple *liatris scariosa,* and several *asters,* were here conspicuous features of the vegetation."

the junction of the North and South forks[4] . . . in the vicinity . . . there was a bluish grass, which the cattle refuse to eat, called by the voyageurs 'Herbe salée,' (salt grass.)"

At this point the party was at the present city of North Platte, in Lincoln County. A cache of pork was made and they took pains that the Indians should see the content, for they would leave that particular form of meat undisturbed. They traveled up the South Fork on July 3 and on the 4th came to a road which led to the North Fork. Here Frémont divided his party. Charles Preuss joined the section making a detour to the north;[5] Frémont chose to ascend the South Fork and its tributaries to St. Vrain's Fort, two hundred miles away and nearly east of Long's Peak. ". . . Crossing the country northwestwardly from St. Vrain's fort, to the American company's fort at the mouth of the Laramie, would give me some acquaintance with the affluents which head in the mountains between the two . . ." Moreover, "In a military point of view, I was desirous to form some opinion relative to the establishment of posts on a line connecting the settlements with the South pass of the Rocky mountains, by way of the Arkansas and the South and Laramie forks of the Platte."

They moved south and southwest (July 5 and 6) following the South Fork.[6] When, on the 6th they reached the confluence of Lodgepole Creek, they had crossed from Nebraska into Colorado. The two streams unite near Julesburg, Sedgwick County. The route was to continue through Colorado until July 14.

On July 8 they came to the mouth of ". . . one of the most considerable affluents of the South fork, *la Fourche aux Castors,* (Beaver fork,) heading off in the ridge to the southeast" and were now in Morgan County.

On the 9th ". . . we caught the first faint glimpse of the Rocky mountains, about sixty miles distant . . . we were just able to discern the snowy summit of 'Long's peak,' ('*les deux oreilles*' of the Canadians,) . . . I was pleased to find that among the traders and voyageurs the name of 'Long's peak' had been adopted . . ."

They reached and crossed Bijou Creek, still in Morgan County, that day, and on July 10 arrived at St. Vrain's Fort in southwestern Weld County. It was situated ". . . on the South fork of the Platte, immediately under the mountains, about seventeen miles east of Long's peak. It is on the right bank, on the verge of the upland prairie, about forty feet above the river . . . Mr. St. Vrain . . . received us with much kindness

4. Torrey records: "I was pleased to recognise, among the specimens collected near the forks, the fine large-flowered asclepias, that I described many years ago in my account of James's Rocky Mountain Plants, under the name *A. speciosa,* and which Mr. Geyer also found in Nicollet's expedition. It seems to be the plant subsequently described and figured by Sir W. Hooker, under the name *A. Douglasii.*"

5. This party rejoined Frémont at Fort Laramie on July 13. An extract from Preuss's journal (July 6-13) is included in Frémont's "Report." They had met a party led by James Bridger, who reported that the Sioux, Cheyenne and Gros Ventres Indians were out with war parties and were in the vicinity of Red Buttes, Natrona County, Wyoming, a region Frémont was planning to cross. Although warned to proceed no farther, so he says, he did not follow advice. Nothing disastrous happened!

6. Torrey mentions that "On the Lower Platte, and all the way to Sweet Water, the showy *cleome integrifolia* occurred in abundance."

and hospitality . . . Pike's peak is said to be visible from this place, about one hundred miles to the southward . . ."

St. Vrain's post[7] was supplied with necessities from Taos, New Mexico. Frémont now turned north, bound for Fort Laramie, one hundred and twenty-five miles away. He noted that for a short distance the road lay down the valley of the South Platte which ". . . resembled a garden in the splendor of fields of varied flowers which filled the air with fragrance. The only timber I noticed consisted of poplar, birch, cotton-wood, and willow." [8] Soon they passed Thompson's Creek, entering the South Fork in Weld County, and ten miles beyond reached *"Cache à la Poudre"* River. After traveling about twenty-eight miles (July 12) they camped on a creek which Frémont named Crow Creek, with the statement: ". . . I had great difficulty in ascertaining what were the names of the streams we crossed between the North and South forks of the Platte[9] . . ."

On July 13 they came to what was supposed to be a branch of Lodgepole Creek and at their camp Frémont amused himself "hunting for plants among the grass." [1]

From July 14 (or possibly a day or so earlier) until September 15 the party was in Wyoming, entering that state in Laramie County. Camp on July 14 was near what was said to be "Goshen's hole." [2] On the 15th they reached the North Fork of the Platte about thirteen miles below Fort Laramie and soon arrived at ". . . the post of the American Fur Company, called Fort John, or Laramie . . ." It was situated where Laramie River enters the Platte,[3] in western Goshen County. The post was in charge of "Mr. Boudeau." The section of Frémont's party that had taken a more northern route was already there.

They stayed at Fort Laramie for five days and then set out for South Pass which lay

7. Ewan tells us that "St. Vrain's Fort, [was] located between the confluences of Thompson's Creek and St. Vrain Creek with the South Platte, south of Greeley . . . This fort was owned by Marcellus St. Vrain and Col. William Bent and served as a base of operations for independent trappers who worked the region of the Front Range."

8. Ewan identifies the poplar as "Likely the Populus angustifolia, a poplar type, as distinguished from the cottonwood type." The birch as "Certainly 'alder,' that is, genus Alnus, probably A. tenuifolia." The cottonwood was "Certainly Populus Sargentii, a characteristic cottonwood." The willow Ewan does not—very wisely—attempt to identify!

9. Even Ewan does not interpret the route from Cache de la Poudre to the Platte, reached on July 15, merely noting that they were "Threading their way across the complex of hills along our Wyoming border . . ."

1. According to Ewan, "No actual specimens taken on July 13th are listed in Torrey's 'Catalogue of Plants.' "

2. The volume *Wyoming* ("American Guide Series"), p. 14, notes that "Edged in some places by chalk bluffs, Goshen Hole, a depression in the great plateau of east-central Wyoming, looks like a scooped-out bed of an ancient lake . . ." It is not far from Hawk Springs, southeastern Goshen County.

3. According to Torrey, it was at Laramie's Fork that ". . . that singular leguminous plant, the *kentrophyta montana* of Nuttall was first seen, and then occurred at intervals to the Sweet Water River."

almost due west, in southwestern Fremont County.[4] They reached the pass on August 8. Frémont noted on July 21 that "... the road led over an interesting plateau between the North Fork of the Platte on the right, and the Laramie river on the left." [5] He recorded on the 22nd:

"One of the prominent characteristics ... of the country is the extraordinary abundance of the *artemisias*. They grow every where ... in tough, twisted, wiry clumps ... the whole air is strongly impregnated ... with the odor of camphor and spirits of turpentine which belongs to this plant. This climate has been found favorable to the restoration of health, particularly in cases of consumption ..."

On July 22 they camped on "... the Fer-à-Cheval, or Horse-shoe creek ..." and on the 23rd on the Platte. Frémont had "... occasionally remarked among the hills the *psoralea esculenta,* the bread root of the Indians." They camped on the right bank of the Platte on the 24th and kept to that side on the 25th. On the 26th they crossed "... *La Fourche Boisée* ... well timbered ..." Frémont had "remarked" several asters. Five miles beyond grew "... thickets of *hippophaæ,* the *grains de bœuf* of the country. They were of two kinds—one bearing a red berry, (the *shepherdia argentia* of Nuttall); the other a yellow berry, of which the Tartars are said to make a kind of rob ..."

They camped at "... the mouth of Deer creek ... It is the largest tributary of the Platte, between the mouth of the Sweet Water and the Laramie." This stream enters the Platte from the south in western Converse County. On July 27 "The main chain of the Black [Laramie] Hills was ... only about seven miles to the south, on the right bank of the river ..." On the 28th they came to "... the place where the regular road crosses the Platte ..." and led to Independence Rock, on the Sweetwater. Frémont followed the Platte instead, reaching the Red Buttes of Natrona County, and on the 29th "Goat Island," near the mouth of the Sweetwater. On the 31st they left the Platte and crossed to the Sweetwater and on August 1 came to Independence Rock in southwestern Natrona County. On August 2, five miles above the Rock they came "... to a place called the Devil's Gate, where the Sweet Water cuts through the point of a granite ridge. The length of the passage is about three hundred yards, and the width thirty-five yards." On the 3rd they "... caught the first view of the Wind river mountains ..." which were seventy miles away. On August 4, 5, 6, and 7 Frémont kept along the Sweetwater[6] and on the 8th reached and crossed South Pass. On the ap-

4. Frémont's name is perpetuated, not only in the name of this county, but in various geographical features of Sublette County to the west of South Pass.

5. According to Torrey, "The route along the North fork of the Platte afforded some of the best plants of the collection. The *senecio rap[i]folia,* Nutt., occurred in many places, quite to the Sweet Water; *lippia (zapania) cuneifolia* (Torr. in James's plants, only known before from Dr. James's collections) *cercocarpus parvifolius,* Nutt.; *shepherdia argentea,* Nutt., and *geranium Fremontii,* a new species, (near the Red Buttes,) were found in this part of the journey."

6. Torrey's "Preface" to the "Catalogue of plants" states: "Along the Sweet Water, many interesting plants

proaches thereto Frémont had noted that "... A variety of *asters* may now be numbered among the characteristic plants, and the artemisia continues in full glory; but *cacti* have become rare, and mosses begin to dispute the hills with them." His description of the pass does not picture it as a very formidable affair:

"... About six miles ... brought us to the summit. The ascent had been so gradual ... we were obliged to watch very closely to find the place at which we had reached the culminating point ... I should compare the elevation which we surmounted immediately at the Pass, to the ascent of the Capitol hill from the avenue, at Washington ... the travller, without being reminded of any change by toilsome ascents, suddenly finds himself on the waters which flow to the Pacific ocean. By the route we had travelled, the distance from Fort Laramie is three hundred and twenty miles, or nine hundred and fifty from the mouth of the Kansas." [7]

Having crossed South Pass Frémont now turned northwest, into the Wind River Mountains of western Wyoming; the range culminates in a peak which now bears Frémont's name. The explorer was entering an untouched and interesting botanical field. [8]

were collected, as may be seen by an examination of the catalogue; I would, however, mention the curious *œnothera Nuttallii*, Torr. and Gr.; *eurotia L[a]nota*, Mocq.; (diotis lanata *Pursh*,) which seems to be distinct from *E. ceratoides; thermopsis montana*, Nutt.; *gilia pulchella*, Dougl.; *senecio spartioides*, Torr. and Gr.; a new species, and four or five species of wild currants, (*ribes irriguum*, Dougl., &c.) Near the mouth of the Sweet Water [Frémont had been there on July 31, and was to be there again on August 25] was found the *plantago eriophora*, Torr., a species first described in my Dr. James's Rocky Mountain Plants. On the upper part [of the Sweetwater], and near the dividing ridge [South Pass], were collected several species of *castilleja; pentstemon micrantha*, Nutt.; several *gentians;* the pretty little *androsace occidentalis*, Nutt.; *solidago incana*, Torr. and Gr.; and two species of *eriogonum*, one of which was new."

7. According to Rodgers: "Frémont proceeded onward into the mountains to South Pass, the famous locality through which the great Oregon emigration crossed the Rocky Mountains. Frémont was the first to explore this region for scientific ends."

It seems to be an accepted fact that N. J. Wyeth, about ten years earlier or in mid-June of 1834, had led his party through South Pass; Thomas Nuttall and the ornithologist John Kirk Townsend were members of Wyeth's party. Ewan, when writing of Nuttall's travels along the Sweetwater at this period, mentions that "... several 'novelties' had been described from his collections mostly by Nuttall himself or at least named in mss. by Nuttall and published by Torrey and Gray in their *Flora of North America* ..."

8. K. A. Geyer's "Notes" express regret that he had been unable to collect in these and other mountains on his transcontinental journey of 1843: "... It was not in my power to visit the Wind River and Rocky Mountain Chain; for a description of which, I refer to the account given by Captain Frémont, in his report of 1842, with the notes of Dr. Torrey." And again; in a footnote: "It may safely be asserted that much more is known about the vegetation of Oregon and California than that of this mountain region [the reference here is to '... the Black Hills, or Black Mountains ...' of northeastern Wyoming, and so on]. Most botanists have not the means to equip expeditions of their own, and are obliged to attach themselves to caravans of traders, missionaries, or emigrants, who move onward without loss of time, passing rapidly most striking portions, to stop, perhaps, for a day or so only, at the least interesting. The great number of new plants which Captain Frémont found on the Wind-river Mountains prove sufficiently that he must have been the first explorer. Not less a rich collection of new plants could be made by an exploration of the 'Black Hills,' difficult and perilous though the undertaking would be ..."

On leaving South Pass the party reached after eight miles ". . . the Little Sandy, one of the tributaries of the Colorado, or Green river of the Gulf of California." On August 9 they came to the ". . . Big Sandy, another tributary of Green river . . ." and camped that night ". . . on the first New Fork. Within the space of a few miles, the Wind mountains supply a number of tributaries to Green river, which are all called the New Forks. Near . . . were two remarkable isolated hills . . . called the Two Buttes . . ."— presumably the Twin Buttes of eastern Sublette County. On August 10 Frémont wrote:

"We are now approaching the loftiest part of the Wind river chain . . . and I left the valley . . . to penetrate the mountains . . . we came . . . in view of a most beautiful lake, set like a gem in the mountains . . . I determined to make the main camp here . . . and explore the mountains with a small party . . ."

Frémont named the lake ". . . Mountain lake . . . This was the most western point at which I obtained astronomical observations . . . 110° 08′ 03″ west longitude from Greenwich, and latitude 43° 49′ 49″." This was, I believe, the lake in northeastern Sublette County which now bears the explorer's name. From there, from August 12 to 16, Frémont made a trip into the mountains and climbed a peak ". . . which from long consultation as we approached . . . we had decided to be the highest of the range." At this crucial moment the barometer, which had been transported safely for one thousand miles, was broken but Frémont was ingenious enough to produce a substitute from a powder horn and buffalo glue. Early on the morning of the 12th they left camp, ". . . fifteen in number, well armed, of course, and mounted on our best mules. A pack animal carried our provisions . . ." The air as they climbed made them feel as if they ". . . had been drinking some exhilarating gas. The depths of this unexplored forest were a place to delight the heart of a botanist. There was a rich undergrowth of plants, and numerous gay-colored flowers in brilliant bloom." Having reached a high point among the pines on August 13 the mules were abandoned and the party continued on foot, reaching ". . . the upper limit of the piney region . . . above this point, no tree was seen . . . The flora of the region . . . was extremely rich, and, among the characteristic plants, the scarlet flowers of the *dodecatheon dentatum* every where met the eye in great abundance. A small ravine . . . was filled with a profusion of alpine plants in brilliant bloom."

With five men, including Preuss and Carson, Frémont reached the top of the peak on the 15th and makes clear that, on arrival, *he* was the first to do so.

"I sprang upon the summit, and another step would have precipitated me into an immense snow field five hundred feet below . . . As soon as I had gratified the first feelings of curiosity, I descended, and each man ascended in his turn . . . We mounted the barometer in the snow of the summit, and fixing a ramrod in a crevice, unfurled the national flag to wave in the breeze where never flag waved before . . . Here, on the summit . . . while we were sitting on the rock, a solitary bee *(bromus, the humble bee)* came . . . and lit on the knee of one of the men . . . seizing him immediately [we] put him in at least a fit place—in the leaves of a large book, among the flowers we had collected on our way . . . It is certainly the highest known flight of that insect."

Frémont gives the altitude of the peak as ". . . 13,570 feet above the Gulf of Mexico . . ." and states that they ". . . had climbed the loftiest peak[9] of the Rocky mountains . . ."

By late afternoon of August 16 the party got back to the main camp on "Mountain lake," and we are told that "All heard with great delight the order to turn our faces homeward . . ."

Torrey's "Preface" to the "Catalogue of plants" records of the collections made between August 10 and 17:

"On the borders of a lake, embosomed in one of the defiles, were collected *sedum rhodiola* DC., (which had been found before, south of Kotzebue's sound, only by Dr. James;) *senecio hydrophilus*, Nutt.; *Vaccinium uliginosum; betula glandulosa*, and *B. occidentalis*, Hook.; *eleagnus argentea*, and *shepherdia Canadensis*. Lieutenant Frémont . . . ascended one of the loftiest peaks on the 15th of August . . . The vegetation of the mountains is truly alpine, embracing a considerable number of species common to both hemispheres, as well as some that are peculiar to North America. Of the former, Lieutenant Frémont collected *pleum alpinum; oxyria reniformis; Veronica alpina;* several species of *salix; carex atrata; C. panicea;* and, immediately below the line of perpetual congelation, *silene acaulis,* and *polemonium cœruleum. β* Hook. Among the alpine plants peculiar to the western hemisphere, there are found *oreophila myrtifolia*, Nutt.; *aquilegia cœrulea*, Torr.; *pedicularis surrecta*, Benth.; *pulmonaria ciliata*, James; *silene Drummondii*, Hook.; *menziesia empetriformis, potentilla gracilis*, Dougl.; several species of *pinus; frasera speciosa*, Hook.; *dodecatheon dentatum*, Hook.; *phlox muscoides*, Nutt.; *senecio Fremontii*, n. sp., Torr. and Gr.; four or five *asters*, and *vaccinium myrtilloides*, Mx.; the last seven or eight near the snow line. Lower down the mountain were found *arnica angustifolia*, Vahl.; *senecio triangularis*, Hook.; *S. subnudus*, DC.; *macrorhynchus troximoides*, Torr. and Gr.; *helianthella uniflora*, Torr. and Gr.; and *linosyris viscidiflora*, Hook."

The return journey, in brief, was through South Pass, down the Sweetwater to the Platte, and down the Platte to Fort Laramie; thence, down the Platte to the Missouri, and down the Missouri to St. Louis. As they started on August 17 the barometer was again broken, this time "past remedy." Frémont commented that ". . . it had done its part well, and my objects were mainly fulfilled." They were back at the "Two Buttes" on the 17th, camped on "Little Sandy river"[1] on the 18th, went through South Pass and camped on the Sweetwater on the 19th. "Our coffee . . . was expended . . . we

9. Ewan suggests the possibility that the explorer may have climbed a peak other than the one now bearing his name when he comments: "Fremont Peak (or possibly Gannett Peak) was reached August 15, 1842, and constituted the literal high spot of the First Expedition."

1. Notes Torrey, "Along Little Sandy . . . were collected a new species of *phaca* (*P. digitata,*) and *parnassia fimbriata*."

made a kind of tea from the roots of the wild cherry tree." On August 22 they reached Independence Rock and Frémont engraved thereon ". . . a symbol of the Christian faith . . . a large cross . . . covered with a black preparation of India rubber, well calculated to resist . . . wind and rain." [2] Arriving at the confluence of the Sweetwater with the Platte on the 23rd, they found the river flowing ". . . in a broad, tranquil, and apparently deep stream . . . from its appearance . . . considerably swollen . . . the afternoon was spent in getting our boat ready for navigation the next day." This boat, made of rubber and inflatable, did well through various rapids and cataracts but was finally upset, throwing men and equipment into the water. Frémont refers to the spot as "this Thermopylæ of the west," and notes:

"Favored beyond our expectations, all of our registers had been recovered, with the exception of one of my journals, which contained the notes and incidents of travel, and topographical descriptions, a number of scattered astronomical observations . . . and our barometrical register west of Laramie . . ."

The party continued downstream, some on one, some on the other side of the river, but were united again at Goat Island. Frémont recorded that he slept soundly on the 24th, ". . . after one of the most fatiguing days I have ever experienced." On August 28 they camped at the ford of Platte River (where they had been last on July 28) and on the 31st ". . . reached Laramie Fort . . . after an absence of forty-two days . . . and felt the joy of a home reception in getting back to this remote station, which seemed so far off as we went out." Frémont was fond of traveling on schedule and he mentions that "The fortieth day had been set for our return."

Fort Laramie was left on September 3 and they continued down the Platte, reaching the forks on the 12th. A bullboat was constructed at this point, but when they attempted to use it on the 15th the water was too low and the party continued on foot. Frémont had sent a message ahead to ask "Mr. P[eter A.] Sarpy" at the American Fur Company station (Bellevue Post in Sarpy County, Nebraska) to build a boat for the descent of the Missouri; and when the party arrived on October 1 they found it on ". . . the stocks . . . a few days sufficed to complete her . . ." After selling "All our equipage—horses, carts, and the *materiel* of the camp . . . at public auction at Bellevue . . ." the descent of the Missouri was a rapid one and the party reached St. Louis on October 17 where ". . . the sale of our remaining effects was made . . ." Frémont left St. Louis on October 18 by steamboat and on October 28 reported to Colonel J. J. Abert in the city of Washington. He signed his "Report" as a Second Lieutenant.

At appropriate places in Frémont's narratives I have referred to comments made by John Torrey in his "Preface" to his "Catalogue of plants collected by Lieut. Fremont, in his expedition to the Rocky Mountains"; this was published in both printings

2. Should Frémont be remembered as the first to introduce into the trans-Mississippi west the horrible painted inscription which now defaces so many features of the American landscape? *Wyoming* ("American Guide Series"), p. 387, states that "Frémont's cross has disappeared, nevertheless." Someone, I know not who, once made the comment that only by the introduction of ships had humanity added to the beauty of nature!

(1843 and 1845³) of Frémont's "Report" and was dated "New York, *March,* 1843."
It ends thus:

"As the plants of Lieutenant Fremont were under examination while the last part
of the Flora of North America was in the press, nearly all the new matter relating to the
Compositæ was inserted in that work. Descriptions of a few of the new species were
necessarily omitted, owing to the report of the expedition having been called for by
Congress before I could finish the necessary analyses and comparisons. These, however, will be inserted in the successive numbers of the work to which I have just
alluded."

Torrey's "Preface" is evidently based, to some extent, upon a long letter written him
by Frémont on March 11 [? 1843] which is quoted in Rodgers' biography of that
botanist (pp. 155-157). The poor quality and small amount of material are explained
by the difficulties of the journey. However, Frémont's "Report" reveals that none of
the anticipated dangers ever materialized and that, by comparison with many other explorations, Frémont suffered little hardship; although he received numerous warnings
about Indians, he was never more than "sometimes annoyed" by them. His worst
experience—the shipwreck on the Platte River—resulted from his own ill-advised attempt to navigate that stream in his rubber boat.

Torrey's "Preface" refers to Frémont's collection of 1842 as ". . . a very interesting
contribution to North American botany . . ." although made under unfavorable circumstances, for after the wagons had been abandoned at the Red Buttes, transportation was on horseback and so rapid ". . . that but little opportunity was afforded for
collecting and drying botanical specimens."

For an inexperienced collector, Frémont probably did a creditable piece of work.
It was becoming ever more difficult to find plants that had not been collected before
and many of those which he brought back had already been described by Pursh, Nut-
tall, Hooker and others; except for his few days in the Wind River Mountains of
Wyoming and for insignificant short trips such as the one from St. Vrain's Fort to
Fort Laramie, Frémont had traveled only in regions already explored by plant col-
lectors. Nonetheless, on the basis of Frémont's collections, Torrey distinguished some
sixteen new species⁴ and the new genus *Fremontia.* Among the new species were:
Phacelia leucophylla, from "Goat island, upper North fork of the Platte [Wyoming].
July 30"; *Ipomoea leptophylla,*⁵ from the "Forks of the Platte to Laramie river. July

---

3. This "Preface" was reprinted in Frémont's *Memoirs of my life* (1887) under the heading "Sketch of
the vegetable and geological character of the region covered by the first report" although Torrey's "Catalogue" itself was not reprinted.

4. Brendel states that the collection of plants ". . . consisting of 352 species, contained fifteen new ones
. . ."

5. Ewan wrote: ". . . Fremont records such a succession of encounters with the Indians, all to no harm,
that he seems not to have collected many plants on this his first Colorado trip. The only [? new] plant
that I find Fremont definitely collected in Colorado on the First Expedition was the Bush Morning glory
(Ipomoea leptophylla). The type specimen of this Ipomoea was taken, according to Torrey's description,

4–September 3"; *Gentiana Fremontii,* from the "Wind river mountains [undated but from Wyoming]"; *Abronia (tripterocalyx) micranthum,* from "Near the mouth of Sweet Water river [Wyoming]. August 1." Of the new genus *Fremontia* Torrey wrote in his "Preface":

"In saline soils, on the upper Platte, near the mouth of Sweet Water, were collected several interesting Chenopodiaceæ, one of which was first discovered by Dr. James, in Long's expedition; and although it was considered a new genus I did not describe it, owing to the want of the ripe fruit. It is the plant doubtfully referred by Hooker, in his Flora Boreali Americana, to Batis[.] He had seen the male flowers only. As it is certainly a new genus, I have dedicated it to the excellent commander of the expedition, as a well-merited compliment for the services he has rendered North American botany."

The type of the new genus was *Fremontia*[6] *vermicularis: "(Batis* ? *vermicularis,* Hook.) . . .* Upper north fork of the Platte, near the mouth of the Sweet Water [Wyoming]. July 30."

In the Gray Herbarium is a letter written by Torrey to Asa Gray under date of November 18, 1842, which explains how Frémont's collections had become available to him:

"A few days ago I recd a letter from Jaeger—formerly of Princeton, giving me an account of some plants collected towards the Rocky Mountains by a Lt. Fremont in the U. S. service. He advised the gentleman to send the whole to me—& this morning a letter arrived from the gentleman himself—informing me that the box was dispatched from Washington on the 16th . . . The specimens were collected, he says 'the present year in the course of a geographical exploration to the Rocky Mountains. The region over which the collection was made, extends from the 39th to the 43rd degree of N. Latitude—& from the 95th to the 112 deg. W. Longitude. The labels which are affixed to the specimens, will enable us to assign these to their exact localities on a topographical map of the country which I am now engaged in constructing . . .' He writes something like a foreigner—but he signs himself J. C. Frémont, Lt. Topog. Engineers—He expects next year, to continue the explorations to the Pacific & offers me what he collects—So here is a chance for you to get seed &c—How would it do to send a collector with him, Leavenworth wishes to go somewhere—& this place might suit *him*—but not *us* in all respects—When I get the box, I will send you the Compositae. & such duplicates of the others (if there be any) as you may desire for your own herbm."

Of the publication of Torrey's "Catalogue of plants" we are told by Rodgers that

at an unrecorded station above the 'Forks of the Platte'. Since this man-in-the-ground grows, not in river bottoms, but on the 'high hills along the Platte', to borrow Edwin James's phrase for other prairie plants, Fremont must have left the Platte bottoms for some distance at least once."

6. Torrey was to discover at a later date that his new genus had already been described in 1841 (from a plant in Maximilian's collection) by Nees von Esenbeck under the name *Sarcobatus,* and that the name *Fremontia* (1843) was invalid. *See* pp. 555, 566.

"Torrey worked with Carey[7] on the Frémont plants but as Carey had suffered financial losses and was despondent about them, Torrey, after sympathizing for a while and trying to help him, lost patience with him, and by March 9, 1843, finished the account himself of the plants of Frémont's first expedition, and he sent off a copy to Gray. In many respects he was disappointed with the printing; in many respects he was disappointed in the material."

The same author also quotes a letter from Torrey to Gray (no date is supplied) which tells more about the make-up of the "Catalogue" which was evidently not to Torrey's satisfaction:

" '. . . My catalogue of his plants will, I fear be shockingly printed. I have only received one proof sheet, & that was as bad as it could be. The whole style of the thing was changed from my Mss. I wished it set up like my Rocky Mo paper but they made it purely Etonian, & employed a very fine type. The extra copies that I requested have not been sent to me & if they are as bad [as] I fear they will be I shall destroy the whole.' "

7. John Carey.

BANCROFT, HUBERT HOWE. 1890. The works of . . . 20 (Nevada, Colorado, and Wyoming 1540–1888): 356, 357, *fn.* 19, 688.

———— 1886. *Ibid.* 21 (California IV, 1840–1845): 3-436. Map ["Frémont's Map, 1844." p. 442].

———— 1889. *Ibid.* 26 (Utah 1540–1886): 32.

BRENDEL, FREDERICK. 1880. Historical sketch of the science of botany in North America from 1840 to 1858. *Am. Naturalist* 14: 28.

DELLENBAUGH, FREDERICK SAMUEL. 1914. Frémont and '49; the story of a remarkable career and its relation to the exploration and development of our western territory, especially of California. Maps. New York. London. 46-105. [Map, opp. p. 68, shows routes of all Frémont's expeditions.]

DE VOTO, BERNARD. 1943. The year of decision 1846. Boston. 525.

EWAN, JOSEPH. 1943. Botanical explorers of Colorado III. John Charles Fremont. *Trail & Timberline* No. 291: 31-34, 37.

———— 1950. Rocky Mountain naturalists. Denver. 21-27 [this is largely a reprint of his earlier account of Frémont's first expedition of 1842], 211.

FRÉMONT, JOHN CHARLES. 1843. A report on an exploration of the country lying between the Missouri River and the Rocky Mountains, on the line of the Kansas and Great Platte Rivers. By Lieut. J. C. Fremont, of the Corps of Topographical Engineers. *U. S. 27th Cong., 3rd Sess., Sen. Doc.* 4: No. 243, 1-76. [The included map shows the route followed by Frémont in 1842.]

———— 1845. Report of the exploring expedition to the Rocky Mountains in the year 1842, and to Oregon and north California in the years 1843–'44. By Brevet Captain J. C. Frémont, of the Topographical Engineers, under the orders of Col. J. J. Abert, Chief of the Topographical Bureau. *U. S. 28th Cong., 2nd Sess., Sen. Doc.* 11: No. 174, 1-290.

[The included map shows the routes followed by Frémont in 1842 and in 1843–1844. The portion of the report (pp. 7-79) concerned with Frémont's expedition of 1842 bears the same heading as the printing of 1843. Frémont's "Notice to the reader" (p. 3-6) refers to his expedition of 1843–1844.]

——— 1887. Memoirs of my life, by John Charles Frémont. Including in the narrative five journeys of western exploration, during the years 1842, 1843–4, 1845–6–7, 1848–9, 1853–4 . . . A retrospect of fifty years, covering the most eventful period of modern American history . . . with maps and colored plates. Chicago. New York. 73-168. [The map does not show Frémont's route of 1842. The volume extends only through the third expedition and no second volume was ever published.]

GEYER, KARL ANDREAS. 1845. Notes on the vegetation and general character of the Missouri and Oregon territories, made during a botanical journey from the state of Missouri, across the South-pass of the Rocky Mountains, to the Pacific, during the years 1843 and 1844, by Charles A. Geyer. *London Jour. Bot.* 5:35 *fn.*, 40.

GHENT, WILLIAM JAMES. 1931. The early far west. A narrative outline 1540–1850. New York. Toronto. 321-323.

GOODWIN, CARDINAL. 1922. The trans-Mississippi west (1803–1853). A history of its acquisition and settlement. New York. London. 233-235.

NEVINS, ALLAN. 1931. John Charles Frémont. *Dict. Am. Biog.* 7: 19-23.

——— 1939. Frémont. Pathmarker of the west. New York. 89-126.

RODGERS, ANDREW DENNY 3RD. 1942. John Torrey. A story of North American Botany. Princeton. 154, 155, 158.

TORREY, JOHN. 1843. Preface [to Torrey, John. 1843. Catalogue of plants . . .]. *In:* Frémont, J. C. 1843. A report on . . . 79-81. *Reprinted:* Frémont, J. C. 1845. Report of . . . 83-85. *Reprinted:* Frémont, J. C. 1887. Memoirs of my life . . . [under the heading: "Sketch of the vegetable and geological character of the region covered by the first report."]

——— 1843. Catalogue of plants collected by Lieut. Fremont, in his expedition to the Rocky Mountains. *In:* Frémont, J. C. 1843. A report on . . . 83-94. *Reprinted:* Frémont, J. C. 1845. Report of . . . 87-98.

WARREN, GOUVENEUR KEMBLE. 1859. Memoir to accompany the map of the territory of the United States from the Mississippi River to the Pacific Ocean, giving a brief account of each of the exploring expeditions since A.D. 1800 . . . *U. S. War Dept. Rept. explr. surv. RR Mississippi Pacific* 11: 41, 42 .

CHAPTER XXXVII

GEYER CROSSES THE "MISSOURI AND OREGON TERRI-

TORIES" AND COLLECTS A BEAUTIFUL HERBARIUM

FOR THE ELDER HOOKER

I N an earlier chapter I told of the collecting done by the German botanist Karl Andreas Geyer through the year 1839. I now turn to his work between 1840 and 1845, touching briefly upon the years 1840, 1841 and 1842. His important journey in this decade took place in 1843 and 1844.

1840

A letter[1] which Geyer wrote William Jackson Hooker on August 13, 1845, contains a short reference to his activities during 1840: "In 1840 I collected about St. Louis, waiting orders till July, when it was too late to undertake any tour." An earlier letter to George C. Thorburn, dated December 28, 1843, is more explicit:

". . . I held my place [awaiting orders from Nicollet] up to the 1st of July 1840 when I attented your brothers establishment at St Louis during his journey and voyage to Scotland. I will not dwell upon the unfortunate catastrophe which brought ruin on his fine establishment, nor will I name the causes . . . Later I had reason to disapprove of the course your brother took, and had no more intercourse with him . . .

1. In the Royal Botanic Gardens, Kew, are eighteen letters written by Geyer to Hooker. Typewritten copies of these letters were in the State College of Washington and through the courtesy of that institution and of Dr. F. Marion Ownbey I was permitted to have copies made. A few were published in an article by Clifford M. Drury in 1940.

In the New York Botanical Garden are two Geyer letters: one to "Messrs Torrey & Gray Professors Drs etc etc," and another, the more important, to George C. Thorburn, dated December 28, 1843. Dr. H. W. Rickett kindly sent me photostats of both. The Thorburn letter was published in part by Grace L. Nute in 1946.

In the Missouri Botanical Garden are thirteen letters in German script written by Geyer to George Engelmann. Miss Nell Horner, Librarian, was good enough to forward photostats and they were read on my behalf by Professor Alfred Rehder. The one quoted in this chapter was written from Fort Colvile on April 6, 1844; portions of it were published by Dr. Nute in 1946 and I shall quote from her translation.

Since Geyer's letters are often mentioned and quoted in this chapter, I comment upon them briefly. I found them wearisome, cumulatively so, and suspect that the recipients had a similar reaction, for Geyer more than once complains that he has received no reply. They give the impression that Geyer was a self-satisfied individual and inclined to be critical of those in his own station in life. When, on the other hand, his assurance was taken to task by such a personality as Hooker—at this period an all-powerful figure in a British botanical world of well-defined hierarchies—he reveals himself in the light of a sycophant, signing himself thereafter as "Your most obedient humble servant," "Humbly," "Yours humbly,"—ever

"... By remembering me to your old father I would beg you to give excuse for me for any improper conduct which then, at the time want of good sense and vanity made me guilty of ..."

1841

Writing of John Charles Frémont, John Bigelow relates that "During the summer of 1841 ... he received a mysterious but inexorable order[2] to make an examination of the river Des Moines, upon the banks of which the Sacs and Fox Indians still had their homes. Iowa being at that time a frontier country. He set out to the discharge of this duty ... finished it, and returned to Washington ..."

Geyer wrote Hooker on August 13, 1845: "In 1841 I made a tour with Fremont in the Desmoines River, lower Iowa territory in which however, survey was the principal occupation. In my botanical collection, which suffered much by the filling of a canoe, I had several new plants which are in the herbarium of Dr. Engelmann[3] ..."

Frémont, in the *Memoirs of my life,* explains that Nicollet, whose health was failing rapidly, was desirous that his surveys of "some of the larger rivers" should be completed:

humble in some form. However, Geyer's means of livelihood depended upon patronage. He did not lack good intentions, but his approach was Germanic and not too tactful. One letter to Hooker dated April 15, 1846, should serve as an example. After a long and tiresome account of a heating apparatus Geyer turns to the raising of plants:

"I have also to add that I have raised a few seedlings of Lewisia rediviva ... Singular enough it is that I have grown from the few stray seeds which I accidentally gathered from specimens, or out of the rummage parcels, almost every grain has come ... Mr. Pierce who got all my seeds complained that they did not grow. This I think rests a little on the faulty way seeds and cuttings are treated in England under a far [more] favourable sky for such purposes. So much I find to be true that the German gardeners have deeper enquired into the life and contingencies to the life of vegetable than the English, the astounding easy manner of propagation throughout Germany, stands as a satisfactory proof."

Granted the ridiculous assumption that the British did not understand plant-culture, it would have been wiser not to explain this to Hooker, Hooker in particular!

I conclude, after reading Geyer's letters, that Frémont's decision not to take Geyer on his 1842 journey (according to Geyer "I did not accompany Fremont," according to Gray writing Engelmann, "... Frémont will not take Geyer, but I believe he wants some one") and Sir William Drummond Stewart's failure to employ him as a gardener—a matter which seems to have hung fire for a long time—may have denoted wisdom on the part of these men. Both had traveled with Geyer and there is no better may of becoming acquainted. As our story progresses, it becomes evident that most people wearied of the fellow.

Frémont's *Memoirs of my life* refers pleasantly enough to Geyer, but he was writing fifty years later: "No botanist had been allowed to Mr. Nicollet by the Government, but he had for himself employed Mr. Charles Geyer, a botanist recently from Germany, of unusual practical knowledge in his profession and of companionable disposition."

2. Perhaps instigated by his future father-in-law Thomas Hart Benton, who was averse to Frémont's marriage to his daughter Jessie.

3. Exchange of civilities was customary at this period and Hooker adds: "... and they cannot be in better hands, for that gentleman has himself successfully explored the botanical riches of a great part of the state of Arkansas, and is familiar with the flora of St. Louis." Hooker mentions that a few of the plants were new.

"The Des Moines was one of these; and at his request I was sent, in July, to make such a reconnoissance of its lower course as would nearly complete it. Whether or not this detachment of myself from Washington originated with Mr. Nicollet I do not know, but I was loath to go.[4]

"I had again with me on this survey one of my companions of the former expedition in Mr. Charles Geyer, who accompanied me as a botanist. I established the course of the river upward from its mouth about two hundred miles, which brought the survey to the Racoon Forks; and Mr. Geyer did all that the season and time allowed for botany. It was here that Geyer found the snake under his flowers. There were many snakes along the river, and botany became a hazardous pursuit . . . our examination was confined to the immediate valley of the river, but we frequently ranged into the woods . . ."

The Des Moines River, which rises in southwestern Minnesota, empties into the Mississippi River between Lee County, Iowa, and Clark County, Missouri. Raccoon Forks is in Polk County, Iowa.

1842

Geyer spent the year 1842 gathering plants in Missouri and Illinois for sale to the "botanical public." Early in 1843 Asa Gray advertised sets of these plants, obtainable on application to himself or to George Engelmann:

". . . a selection from his collections of the last year in Illinois and Missouri; consisting of twenty sets of one hundred and fifty species of plants, which are offered at six dollars per set. A list of this collection, with critical remarks, and descriptions of some new species it contains, received from Dr. Engelmann too late for present insertion, will find a place in the ensuing number of this journal."[5]

Engelmann enumerated and advertised the same collections in his "Catalogue of a collection of plants made in Illinois and Missouri, by Mr. Charles Geyer . . .", published in 1843. According to his introductory words:

"Mr. Geyer, who is an excellent collector, is now absent on an expedition to the Rocky Mountains and Oregon . . . Being unwilling to adopt the common plan of selling his collections to subscribers before they are actually made,[6] he prefers to seek

---

4. Another reference to the romance. The trip took less than four months, for Frémont was married to Jessie Benton on October 19, 1841.

5. *American Journal of Science and Arts.*

6. It seems very possible that selling collections before they were made—not unlike, for the purchaser, buying a pig in a poke—had not proved too successful; but there are different ways of expressing the same fact!

Writing Hooker from Dresden on December 14, 1845, Geyer had not yet disposed of his 1842 sets: ". . . I would feel greatly obliged to you should you be willing to bring once more to notice the Illinois flora sets, for sale at New York, they are preserved in the best possible manner, at least they were so when they came out of my hands." It was probably in answer to this plea that Hooker mentioned the sets, casually, in 1845.

some needful aid in the prosecution of his arduous undertaking, by offering to the botanical public sets of the following plants, collected in 1842 near St. Louis, Missouri,
and around Beardstown on the Illinois River. This collection . . . consists of the following species . . ."

No localities, other than general ones, are given in the Engelmann "Catalogue," nor
any precise dates of collection. I do not know, therefore, and it is perhaps unimportant,
how many plants come from east, how many from west, of the Mississippi River.
Geyer wrote Hooker (August 13, 1845) of having been in ". . . the upper Illinois
country, especially about Songamon River . . ." The Sangamon, formed by the unison
of a north and south fork, flows into the Illinois River about ten miles above Beardstown, Cass County, Illinois.

It is interesting that Engelmann considered it advisable to report in the "Catalogue"
upon the great numbers of introduced plants that had become naturalized about the
Mississippi River settlements:

"We now ought to be careful observers of such plants as are apparently common to
both continents; in after years it will be much more difficult to decide which are natives
and which introduced. Many European plants, now common weeds of the Alleghany
Mountains, have not yet found their way to the Mississippi valley, but undoubtedly
will arrive in a short time. Other plants are here already as common as they are in
Europe, from whence they were derived, or in middle Asia, perhaps their original
home. It behooves us therefore to note the progress of these intruders, and distinguish
them from the true natives.

"We are able to distinguish several different classes of such plants: —

"1. *Nearly allied geographical species,* where one takes the place of the other in
the other continent . . .

"2. *Geographical varieties,* where no specific distinction can be discovered between
the natives of both continents, but where the American and European variety can always be distinguished by some points of minor importance . . .

"3. *Identical plants,* true natives of both continents, especially arctic or at least
northern plants; also marine species and cryptogamic plants . . .

"4. *Naturalized plants,* spreading with the progress of civilization . . ."

1843

Geyer's collections up to now had been described by Drs. John Torrey and George
Engelmann. Geyer had hardly started west in 1843 when Asa Gray wrote an enthusiastic prospectus of the plants he was expected to collect and hoped subsequently to
sell to subscribers. Engelmann also mentioned them when writing of Geyer's Illinois
plants. Obviously, these two botanists expected to be the distributors of the sets. I
quote from Gray, whose stipulations in regard to payment are cautious to say the
least:

"We take much pleasure in announcing that three enterprising botanists are now engaged in exploring the most interesting portions of the *far West* and that their collections of dried plants will be offered to subscribers, in sets, as they come to hand. Two of these collectors, Mr. *Charles A. Geyer* . . . and Mr. *Lüders,* who are for the present attached to Sir. Wm. Stewart's party, have by this time reached the Rocky Mountains. The particular field of Mr. Geyer's operations, and the extent of his journey, were undecided at the time of his departure from St. Louis . . . [Lindheimer, who was contemplating a trip to Texas, is also mentioned.] [They] will doubtless avoid the common and better known plants of this region; and thus their collections may be expected to prove unusually choice and valuable . . .

"These several collections will be assorted and distributed, and for the most part ticketed, by Dr. Engelmann of St. Louis; assisted, as far as need be, by the authors of the *Flora of North America,* who promise to determine the plants, so far at least as they belong to families published in that work; and, for the information of subscribers, particular notices of the *centuria* offered for sale, will probably appear in this Journal, as they come to hand. The number of sets being limited, earlier subscribers will receive a preference. The three explorers are entirely independent of each other; and their collections are to be separately subscribed for.

"The price of the Rocky Mountain collections of Geyer or of Lüders, is fixed at ten dollars (or two guineas) per hundred; that of Dr. Lindheimer's Texan collections at eight dollars (or £ 1, 13s. 6d. sterling) per hundred—payable on delivery of the sets at St. Louis, Missouri, by Dr. George Engelmann . . . and Prof. A. Gray, of Harvard University, Cambridge, Massachusetts, to either of whom subscribers may address themselves *(post paid)* by mail. The additional expense of transportation, doubtless trifling in amount, will be charged upon the sets deliverable in London.

"The writer of this notice cheerfully states that the dried specimens made by these botanists which have fallen under his observation, are well selected, very complete, and finely prepared; and he cordially joins Dr. Engelmann in recommending the enterprise to the patronage of botanists . . ."

This "notice" I have quoted for several reasons: first, because it indicates the increased amount of plant collecting[7] in the 1840's; second, because it shows that botanists were already distinguishing between "common and better known plants"; and finally, because of the curious fact that, despite Gray's "notice," Geyer's collections went to William Jackson Hooker, were distributed to subscribers through his instrumentality, and were described by the British botanist. The change in recipiency, etc., is to some extent explained in the Geyer letters—Hooker's replies can only be surmised.

In a letter to Sir William, dated May 16, 1845, Geyer explains who he is, tells of his work in the field, and mentions that he has shipped him his collections: "The writer . . .

7. Geyer wrote Hooker on August 13, 1845: "Local collectors are many in the North U. States but without collecting voluminous or in any way satisfactory."

is the botanist who set out at your desire, as expressed in a letter to my friend Dr. Geo. Engelmann at St. Louis Missouri, in February 1845 [correctly, 1843], to explore the upper Missouri country . . ." The next letter, of July 14, 1845, contained criticism of Hooker and ends: "Dispose of the collections as you like, they were made at your desire and I am summarily tired of them." The third letter, undated, is an explanation of why the previous letter (". . . that unfortunate letter of mine . . .") was written:

". . . When your letter reached Dr. Engelmann, the latter mentioned to me your offer as something so desirable of which I ought to avail myself, which I did.

"Both yourself and Dr. Engelmann were interested as to the result of my expedition. Before I set out I had to sign a paper for Dr. Engelmann by which I gave him the desired right for the disposal of my collection, and as I returned this way and not to St. Louis I naturally thought that that right devolved upon you.

"In this light I looked upon the collection as not strictly my own, at least the disposal of it; but, aside from every above consideration, had I sent specimens elsewhere, —after having already received a patronage and divers kindnesses,—I am sure dear Sir that you would never have forgiven such as act, little as the value of the collection may be.

"As however, my trespass does not implicate my honesty, not more than you meant to place the same in doubt, so I hope that . . . it may be forgotten by you, if you think it compatible with your rank . . ."

The reason for the abrogation of his *written contract*—Geyer's return to London instead of to St. Louis—is curious at least! Hooker undoubtedly felt that Geyer had placed him in a difficult position. How he adjusted the affair with Engelmann is not disclosed but it is interesting that, although Gray's name appears as one of the subscribers to the collector's "sets," the name of the St. Louis botanist is conspicuous by its absence. Geyer seems never to have forgotten his mistake and his humility is in evidence from this time onward.

Hooker in his turn now advertises the Geyer "sets." This was in 1845, in an introduction to the Geyer "Notes":

"(It is with no small satisfaction we are able to announce to our scientific friends that Mr. Charles Geyer . . . has recently arrived in England with a very valuable and beautifully preserved collection of Plants, gathered in the Upper Missouri, on the Rocky Mountains, and in the Oregon Territory, during the years 1843 and 1844. Mr. Geyer is . . . honourably mentioned by Drs. Torrey and Gray in the 2nd volume of their admirable 'Flora of N. America'[8]. . .

"The collection is peculiarly interesting, as illustrating the Botany of a region lying considerably south of those countries so successfully explored by the intrepid travellers, Douglas and Drummond, extending as it does from the 29° to the 48° of lat.; and, in conjunction with the discoveries of the talented and indefatigable Nuttall, and

8. Hooker fails to mention Gray's prospectus!

of Lieut. Fremont . . . must render our knowledge of the vegetation of these extensive wilds very considerable.

"Mr. Geyer has now divided his ample collection into 20 sets; the fullest of which amounts to 600 species; the lowest to 2 or 300;[9] but the species wanting in these lower sets are not generally the scarcest kinds, for of such Mr. Geyer was careful to collect abundantly:[1] and the sets are now offered at the rate of £2 the 100 species, all expenses included. Orders may be sent to Mr. C. A. Geyer, at the Royal Botanic Gardens, Kew, or . . .

"It will be our agreeable task to publish a Catalogue of this collection, with remarks and descriptions of the new species; this Catalogue to be prefaced by some account of the journey detailed by Mr. Geyer himself.).—Ed."

Apart from Geyer's letters and a few miscellaneous records, my information about Geyer's journey is derived from his "Notes on the vegetation and general character of the Missouri and Oregon Territories, made during a botanical journey from the state of Missouri, across the South-pass of the Rocky Mountains, to the Pacific, during the years 1843 and 1844," which was published in Hooker's *London Journal of Botany* in 1845 and 1846.

This paper abounds in references to plants and their distribution over large geographical and geological areas. One or often several collecting numbers follow references to particular species, indicating that the same plant has been gathered in more than one region. It would be only by arranging the specimens themselves in numerical sequence—and Geyer claims to have collected nine or ten, in one instance twenty, thousand—that they might serve (always provided that the specimens cite precise localities) as a means of tracing his route. This would be an impossible task at this date. By contrast with David Douglas' day-by-day journal, which is only complicated by variations in its different transcripts, Geyer's form of presentation is far less useful for it gives no indication of the man's whereabouts at a particular time. Occasionally the "Notes" name a river, a plateau, or a mountain range, but just when Geyer was there we are not told. If he kept a journal it goes unmentioned.

Hooker's preface to Geyer's "Notes" contains a brief footnote account of the collector's work before 1843; he had evidently asked for this information and Geyer supplied it in his letters—the content of the Hooker preface and the Geyer letters are very similar. Some "Preliminary Remarks" by Geyer are largely devoted to expressions of appreciation for kindnesses received and suggest that they were inserted (certainly phrased) by Hooker.

Geyer's account of the "Missouri Territory" included in his "Notes" is preceded by a short paragraph outlining his route as far as the sources of the Missouri River:

9. Some of the last sets to sell were even smaller, ". . . of 140 species each . . ." according to a Geyer letter dated March 18, 1846.

1. Comments such as this are doubtless responsible for threadbare jokes about the part played by botanists in exterminating rare or localized species.

"Passing up the Platte River to Fort Laramie, thence through the most northerly narrow range of the 'Black Hills' across the Saline desert to the 'Red Butter [Buttes],' and 'Rock Independence' at Sweet-water, or Eau Sucrée River—Thence to 'Wind River Mountains,' and across the 'Upper Colorado,' near the mouth of 'Grand Sub-leuse,' to 'Fort Hall,' and 'Boiling Springs' of 'Lewis River'—and finally to the sources of Missouri, across Madison's fork at Beaver-head, on the central chain of the Rocky Mountains: connected with previous observations up the Missouri as far as the Little Missouri, in 1839."

In other words, Geyer had moved westward across Nebraska and on to Fort Laramie in western Goshen County, southeastern Wyoming; thence westward across Wyoming to the Red Buttes (near Casper) and to Independence Rock on the Sweetwater, both in Natrona County; thence—certainly through South Pass, Fremont County—to the Wind River Mountains which, with other ranges, form part of the Continental Divide in western Wyoming. After reaching Green River (Geyer's "Upper Colorado" or "Upper Colorado of the West") at the mouth of the Big Sandy (Geyer's "Grand Sableuse") in northwestern Sweetwater County, Wyoming, Geyer parted—I do not know the date[2]—from the party with which he had been traveling—it had been organized by Sir William Drummond Stewart[3]—and joined a group of Jesuits bound for

2. Possibly on or about August 2, 1843, for on that day Sir William wrote for Geyer a letter of introduction to Chief Factor John McLoughlin at Fort Vancouver.

3. Sir William Drummond Stewart of Murthly Castle, Scotland, was traveling in the Rocky Mountains for his own delectation. De Voto's *Across the wide Missouri* refers to his trip of 1843 thus:

"... Before he left Scotland he sold Longiealmond, the Drummond estate which was not entailed, for something over a million dollars and so could shoot the moon. As Western sportsman he had been first in a good many ways and this was the West's first dude expedition, with staggering luxuries, a handful of rich American bloods as paying guests, full newspaper publicity—the grand style ornate ..."

According to C. M. Drury: "A member of Stewart's party, Richard Rowland, says that this was a party of pleasure, the first pleasure party that ever went to the Rocky Mountains; ms. in Oregon Historical Society."

Audubon relates that he had been asked to join the expedition at St. Louis but had declined, stating (according to his biographer S. C. Arthur) that "... 'Sir William Stewart and his gang cannot however go off until the 10th or 11th of May.' ..." and Audubon evidently got off sooner.

In addition to Geyer there were other plant collectors in the party: Alexander Gordon, "Luders" [Friedrich G. F. Lüders] and "Burke" [Joseph Burke]. Presumably these men had all availed themselves of an opportunity to travel in the safety of a large group.

Geyer's "Preliminary Remarks" to the "Notes" acknowledge his indebtedness to Stewart, among other things for a letter of introduction to Chief Factor John McLoughlin at Fort Vancouver. He sent this letter ahead from Fort Colvile but at once began to wonder whether he had done the wise thing—whether it would be injurious or helpful. G. L. Nute translates and quotes from a letter which Geyer wrote Engelmann on April 6, 1844: " 'As soon as I became acquainted with some of the Hudson's Bay Company factors ... I learned that Sir Wm. Stewart did not stand in the highest estimation here, on account of crude violations of hospitality at Vancouver.' "

One reads a good deal in Geyer's letters about the possibility that he should take the position of gardener at Sir William's estate in Scotland. He even went to see Murthly Castle, Pertshire, on his return—Sir William was not there—and wrote Hooker that he had found it "beautiful beyond expectation." It may merely have been a light-hearted suggestion on the part of the owner and, as late as July 5, 1846, Geyer had

their mission station among the Flathead Indians who inhabited lands in western Montana, west of the Continental Divide. To get to this mission Geyer, and his companions, must have crossed southeastern Idaho (for he was at Fort Hall in southern Bingham County) and then turned northward into southwestern Montana. His phrase, "connected with previous observations up the Missouri as far as the Little Missouri, in 1839," meaning apparently that, except between the headwaters of the Missouri (in southwestern Montana) and the confluence of the Little Missouri with the Missouri (in Stanley County, South Dakota), Geyer had traveled the length of the Missouri River.

Having outlined his route through the "Missouri Territory," Geyer then divides the region into major tracts, some with numerous subdivisions, and enumerates their distinguishing characteristics, in particular the plants indigenous to each. His long dissertation had best be read in its entirety by those interested in the region or in specific portions thereof.

Geyer explains in a letter to Hooker (May 16, 1845) that he had not traveled by the route which he had hoped to take:

"I failed to get a passage up the Missouri to the Yellow-stone River, which I thought was the country which you wished to see further explored. I was in those regions in the early part of summer 1839 and was well aware that that country is not half explored yet, chiefly owing to the dangerous tribes of Indians. Nevertheless I resolved to lead even a perilous existence and to go there, but could not prevail on the Amer. Fur Company to grant me a passage in their steamboat, by claims of friendship and obligation even it was indirectly rejected. The reason is plain to those who know their illegal dealing with intoxicating liquors, to which they of course, want as few witnesses as possible. The venerable Audibon also had hard work to get a passage[4] for a short distance to examine the quadrupeds of that region."

The account of the "Oregon Territory" in Geyer's "Notes" is not preceded by a brief outline of the route, as was his account of the "Missouri Territory." Instead it is prefaced by general opinions and conclusions about the region.[5] His itinerary and dates can only be pieced together from his letters (to Thorburn and to Hooker which are not always in complete agreement) and from comments (often included as footnotes) in

heard nothing further about the position. He evidently wanted it and doubtless needed employment, but did not want to appear too eager, or too subservient. He wrote Hooker:

"Till the middle of July I waited for Sir Wm. Stewart ... I have seen nothing of Sir Stewart as yet and think myself now quite absolved from my obligation and promise, to go and live at his place ... I would be quite willing to go thither but not to wait in uncertainty as I have done till now.—I express this to you, should you perhaps come together with Sir William Stewart ..."

4. Audubon was sixty-three years of age and, I feel sure, did not consider himself venerable. Geyer was but thirty-four, however! The ornithologist, in 1843, had had no difficulty as far as I know in obtaining a passage up the Missouri, nor had Maximilian had any in 1833. Why Geyer was refused passage is not explained. It might well be that the American Fur Company officials had some reason to dislike him.

5. I quote from these later in this chapter.

his "Notes." To reconstruct the journey with precision is therefore baffling, and the fact that Geyer rarely refers to an Indian tribe, a mission station or a geographical feature by the same name twice is not helpful. Actually, although certain points such as the missions can be placed, his excursions between them, and his trips in their vicinity, are so casually mentioned that, except in a general way, he perhaps did not quite know where he was on occasion! [6]

However, two brief outlines of Geyer's route have been published and, since so little is known about it, I shall quote from both although they are very similar. The journey consumed more than one year—we can only surmise when he parted from Stewart's party[7]—for he left the country of the Flatheads in November, 1843, and sailed for England from the Columbia River in November, 1844.

The first outline is supplied by Charles V. Piper (1906). He states that Geyer

". . . in November, 1843, crossed a high spur of the 'Green' (Bitterroot) Mountains from the Flathead to the Spokane or Coeur d'Alene River, and passed the winter at Chamokane Mission . . . on Chamokane Creek . . . During the season of 1844 . . . made excursions northward to Old Fort Colville on the Columbia, southeastward up the Spokane River and into the mountains about Lake Coeur d'Alene, and southward to the Palouse River and to Lapwai Mission, near the mouth of the Clearwater. From here . . . explored the Craig Mountains of Idaho . . . overland to Fort Walla Walla he descended the Columbia . . . reached Fort Vancouver November 13, 1844, whence he sailed to England . . ."

The second outline is supplied by Frederick V. Coville (1941):

". . . In November 1843 . . . crossed a 'high spur' of the Bitterroot Mountains (which he called the Green Mountains) to the Coeur d'Alene River, Idaho . . . on Christmas Day, 1843, reached the Chamokane . . . Mission, on Chamokane Creek, State of Washington . . . where two American Board missionaries, Eells and Walker, were located . . . passed the winter in this mission. During the season of 1844 . . . made several excursions: northward to old Fort Colville . . . eastward up the Spokane River, across Coeur d'Alene Lake, Idaho, and into the surrounding mountains; southward to the Palouse; and southeastward to the junction of the Clearwater and Snake, from which point he explored the valley and adjacent plateaus of the Clearwater, apparently well into the mountains, and the valley of the Snake as far as the mouth of Salmon River. It was during this season, 1844, that Geyer became acquainted with H. H. Spalding, who had charge of the mission at Lapwai . . . closed the year's work at old Fort Walla Walla and descending the Columbia to Fort Vancouver, on November 13, 1844, sailed for England . . ."

6. Hooker may have become a trifle confused himself and have advised against including the outline of the journey!

7. It may have been in early August for John McLoughlin mentions in a letter (see p. 784, *fn.* 8) that Geyer's letter of introduction to him had been dated August 2, 1843, and it seems probable that Sir William wrote it when Geyer was about to take his departure.

Geyer, we know, was at the headwaters of the Missouri River when he joined the Jesuits who were bound for their mission among the Flatheads. Presumably the reference contained in his letter to Thorburn (". . . passing Lewis River and the sources and upper forks of the Madison, from there to Clark River, through the country of the terrible Blackfeet Indians . . .") refers to the early part of this journey. According to Geyer's "Preliminary Remarks," he ". . . enjoyed their hospitality, and finally accompanied a mission party to the Coeur d'Aleine Indians, an entirely new field for my researches on the upper waters of the Spokan and Kalispell Rivers." Writing to Thorburn, he is a trifle more specific, for he refers to leaving the territory of the "Salesh" (or Flathead) tribe in November, 1843, and moving westward to the territory of the Coeur d'Alene Indians (whom he calls as well the "Skitsoe" or the "Skitsowich" Indians). According to Chittenden's map, the Flatheads occupied lands west of the Continental Divide and the country occupied by the Coeur d'Alene Indians was still farther westward, or beyond the Bitterroot Mountains, in northern Idaho.

It was when he moved from the Flathead to the Coeur d'Alene country that Geyer crossed ". . . a high spur of the Green Mountains . . ."—doubtless a northern extension of the Bitterroots. In writing Thorburn, he refers to the trip as ". . . one of the most terrible journeys I ever made, especially in the midst of winter, crossing 76 times streams, (tributaries of Clark River). Some we had to swim . . ."

The trip is also mentioned in Geyer's "Notes":

"Owing to the difficulties of this crossing, which cost us most of our horses, (having for nearly five days nothing to feed on) I could not pay proper attention to the vegetation. But this much I do know, that I saw *Taxodium sempervirens*[8] growing with the *Thuja gigantea,* on the borders of these woods, and that my hungry animal fed upon its branches; they were all low slender trees, of somewhat naked aspect; but I never met with them again on any of my future excursions in similar regions."

How long Geyer remained among the Coeur d'Alene Indians is not disclosed. He merely refers to ". . . my stay with the Skitsoes, in November 1843 . . ." Since he was with Jesuit missionaries he probably stayed at the first Coeur d'Alene Mission, established in 1842.[9] To Thorburn he mentions that ". . . the hospitality which the Jesuits showed me was scanty and beggarly . . ." It seems probable that the missionaries themselves were living in no regal manner, having just established their mission the year before Geyer was their guest.

Geyer's next move was into the state of Washington. He planned ". . . in the beginning of December 1843, to go to Fort Colville on the Columbia River, a distance of

8. This is an interesting misidentification. *Taxodium sempervirens* (the name bestowed upon the coast redwood by Lambert in 1824) certainly does not grow where Geyer claims to have found it. Geyer was no botanical amateur so that what he mistook for the redwood is hard to imagine; I hazard no guess. The *Thuja gigantea* which he also mentions was doubtless *Thuja plicata* D. Don.

9. According to Brosnan's extremely helpful map of the "Early fur-trading posts and missions" of Idaho and adjacent regions, this mission was situated between Coeur d'Alene and St. Joe rivers, near the southern end of Lake Coeur d'Alene, in present Kootenai County, Idaho.

about 180 or 200 miles on the winter road . . . I . . . came to the resolution to go alone, though utterly ignorant of the route and the country generally . . ."

There is a long account of this hazardous journey in a footnote to Geyer's "Notes." The trip took approximately two weeks, and since he arrived at the Chamokane Mission on Christmas Day, he must have left the Coeur d'Alene Mission about December 10, 1843. He traveled in deep snow, met an occasional Indian without deriving much information as to where he was, but finally he ". . . recognized Spokan River in a rather broad valley. Both rivers [Chamokane Creek and Spokane River presumably] joined a short distance below, and enclose a point of land of classic reputation in Oregon; namely, the place where the trading-company led by Wilson P. Hunt, of St. Louis, built their first trading post . . . Nothing remains but a little elevation of the place where the chimney stood." Geyer camped ". . . under a gigantic . . . *Abies balsamea* near the river [Spokane]. I made a shelter of a blanket, and stretched on my bear skin, I mused over the changes of the day, and over past times, for it was Christmas Eve." The tree bore testimony to an earlier visitor: "In this tree were sundry marks hewn and cut. Amongst others, I found the initials (D. Dgls.), which I take to be those of the late Mr. David Douglas, who made a summer excursion to this place."

Geyer's reflections were interrupted by an Indian boy who explained that ". . . to go to the Mission establishment I had to cross a high snowy mountain . . ." With the boy as guide Geyer reached the ". . . valley Tshimakain. Soon I shook hands with Messrs. Eells and Walker, and accepted the permission joyfully to make myself at home in their residence . . . I shall remember their kindness throughout the whole of my life . . . I could enjoy the luxury again of sleeping under a roof, of which I had not had an opportunity for eight long months in succession."

The Chamokane Mission was on Chamokane Creek, some ten miles above its confluence with Spokane River, in present Stevens County, Washington. Geyer spells both river and mission "Tschimakain"[1] and sometimes refers to the last as "Spokan Mission." The missionaries were Cushing Eells and Elkanah Walker and side lights on Geyer's prolonged visit have been supplied by Drury,[2] who quotes a letter written by Mrs. Eells on January 4, 1844:

" 'This winter finds me fully occupied as ever & what I write must be done in a hurry amidst all the noise and confusion of a kitchen filled with children & Indians besides a German agent who is spending a few weeks with us. He is a professional Botanist. So far as we are able to judge he understands his profession . . . He wishes to spend

1. Both spellings appear on modern maps.

2. The Reverend Clifford Drury wrote a biography of the missionary H. H. Spalding, and another of Elkanah Walker and his wife Mary. Both mention Geyer but contain no more information about him than is found in a short article by the same author. My quotations are from Drury's short paper.

The article includes a reproduction of a drawing of the Tshimakain Mission made by Geyer in December, 1843, and despite the fact that Drury states that "The proportions in the picture are not good," I should say that it displays considerably better draughtsmanship than most drawings found in narratives of the period.

the winter here & at Colville—in the spring to go to Vancouver by land for the purpose of botanizing through the lower country . . .' "

The oft-quoted Geyer letter to Thorburn gives a picture of "home life" with the Eells and Walker families:

". . . arrived at Christmas-day at the missionary station of the American board of Missions at the Spokan Indians, at the house of Mrss Eells and Walker Revds., where I was welcomed like a Christian, and indeed, I was in want of hospitality, 7 months had I not slept under a roof, constantly exposed to all the changes of weather like a savage; the hospitality which the Jesuits showed to me was scanty and beggarly, readily did I brave 2 feet snow in the mountains and the rains of the plains for 10 days and nights, with scarcely anything to eat, to arrive at a hospitable house of Protestant friends. Rev. Mr. Eells and Walker are from Massachusetts and since 5 years established at the Spokan River, on the road from Wallah-Wallah to the Nez-perçez Mission. Their labors among the Spokan Indians are prodigious and they beginn to make progress in civilization and christian devotion. We have divine worship every morning and every evening; Rev. Mr Eells and his lady sing excellent and Mr Eells guides the singers by playing on the Violoncello; Mrs. Walker also, a highly educated lady sings, and Rev. Mr Walker, also is like his Rev. Brother a man of high education & social & litterary acquirements. Never could I appreciate company more highly than now, never did I feel so convinced that ruling providence guards our steps so parental as at present."

1844

Geyer was with the Walkers on March 1, 1844, for Drury quotes an entry which he wrote that day in ". . . Mrs. Walker's memory book or album . . . the original . . . in the Walker collection at the State College of Washington."

He was there in July! Drury, in his short paper on Spalding, quotes from Mrs. Walker's diary:

"On July 7, 1844, she wrote: 'We are tired on account of Mr. Geyer but do not know what to do with him; we are determined to be rid of him.' The next day the Walkers told Geyer that they did not have the means to entertain him longer. They met him again on the trail between their home and Fort Colville on Monday, September 16, 1844. The entry in Mrs. Walker's diary for that day states: 'We met Mr. Geyer of whom we took final leave.' "

Despite Mrs. Walker's understandable impression that the visit had been unduly prolonged, Geyer must have been away on trips for some of the approximate six months between December 25, 1843, and July 7, 1844:

He was at Fort Colvile on April 6, 1844, for he wrote Engelmann from there and the letter gives a suggestion of the length of his stay—"During the month I have been here . . ." We know that from Fort Colvile he visited Kettle Falls, Stevens County, no great distance northwest, for the "Notes" refer to his collection (no. 460) of "A very rare plant . . . the *Hedyotis* . . . which I picked on the rocks at Kettle-falls, near Fort Colville, and of which I found only one specimen." And Drury tells us that "Geyer spent several weeks with the Reverend H. H. Spalding at Lapwai in June, 1844, and interested Spalding in botany."[3] So that this must have been another occasion when Mrs. Walker had a well-merited respite.

The Lapwai mission was on Lapwai River, in the triangle formed by that stream's confluence with Clearwater River, Nez Percez County, northwestern Idaho, so that Geyer was entering that state for the second time. This would explain his references (in his letter to Hooker of May 16, 1845) to having ". . . also visited the high plains of the Nez-Percez Indians [he also calls them the 'Saptonas'], Salmon and Koos Kooskii [Clearwater] Rivers . . . with few exceptions new fields, though close to regions explored by Messrs. Douglas & Nuttall."[4]

In his "Preliminary Remarks" Geyer refers to visiting with Spalding: ". . . another new field, the Highlands of the Nez-Perçez Indians, where he accompanied me on my excursions and also afforded facilities to investigate the flowery Koos Kooskee [Clearwater] valley over again, where previous botanists had but cursorily passed . . . It must be gratifying to the lovers of natural history that such assistance is rendered to scientific travellers . . . it . . . shows that the necessity for extending our knowledge of the productions of nature is felt and cheerfully aided, even in the recesses of that vast western wilderness[5] . . ."

Geyer thought highly of the missionary work being done by Spalding; indeed, only here, of all the stations visited, did he find the work ". . . propitious, beyond my expectation . . ." In a long footnote he praises[6] the man's accomplishments: the Nez-Percés Indians were ". . . under the surveillance of an American Missionary, belonging to the American Board of Foreign Missions, the Rev. Mr. Spalding, who resides at Lapwaï, on the Koos-Kooskee . . . Mr. S. is by far the most successful Indian missionary deputed by the American Board of Foreign Missions . . . He boldly left off the absurd custom of teaching the Indian to pray, before endeavouring to fill his hungry stomach . . . The American Board of Foreign Missions has committed an error in not aiding Mr. Spalding, or giving and entrusting to his hands the surveillance of all the Missions of that Board in Oregon. They leave him to struggle alone, and conse-

3. Spalding's botanical endeavors are told in a later chapter.

4. Douglas had been in the Craig Mountains of Idaho in July, 1826; and Nuttall had moved from the Grand Ronde across the Blue Mountains into Washington in 1834.

5. The last sentence has the ring of a Hooker interpolation, definitely!

6. J. Orin Oliphant states ". . . that a close relationship existed between Geyer and Spalding, and that each became interested in the work of the other. Although by the time of Geyer's arrival at Clear Water, Spalding had received word that the order of the American Board recalling him from the Oregon Mission had been rescinded . . . nevertheless he was still suffering from a sense of personal injury . . . it is not unlikely that he imparted to the botanist his version of the controversy . . . At any rate, Geyer begged the indulgence of 'the scientific reader' for a lengthy digression made for the purpose of doing 'justice' to Mr. Spalding . . ."

quently the credit and praise belong solely to him . . . I have longed to do justice to Mr. S. . . . Those who have travelled in North America, and visited Indian Missions, will be . . . aware of their fruitless efforts to civilize the Indians, and of the immense sums squandered liberally by the American citizens for that laudable object . . ."

On what date Geyer reached Fort Walla Walla I do not know, nor how long he remained.[7] His "Notes" mention that he arrived safely at Fort Vancouver with the brigade of the Hudson's Bay Company and refers to his ". . . rapid passage from Fort Walla-Walla to Fort Vancouver . . . Many interesting trees and shrubs I noticed on my rapid passage, but constant rain forbade . . . collecting specimens . . . the reader must remain satisfied with the information which Mr. Douglas has given." He was well received at Fort Vancouver[8] and given every assistance in getting off for England. The "Notes" record that "The vessel of the Hon. Company, the barque Columbia, Captain A. Duncan, left Fort Vancouver on the 13th of November, 1844, and arrived safely

7. He was there when Joseph Burke arrived on October 27, 1844.

8. It seems to me appropriate to mention at this point some of the problems confronting the Chief Factors of the Hudson's Bay Company!

The *Champlain Society Publications* include a letter from John McLoughlin to the Company's officials, dated "Fort Vancouver, Columbia River, 20*th November* 1844." It contains a reference to Geyer's departure:

"Mr. Gyer goes to England as a passenger in the *Columbia*. He was recommended to me by Sir W. D. Stewart, who wrote me that he would pay his Passage at fifty Pounds, but as I presume this rate was intended only for officers in the Service, I wrote Sir W. D. Stewart that he would have to pay one hundred Pounds for Mr. Gyer's Passage."

McLoughlin also wrote Archibald Barclay, Secretary of the Company on November 25, 1844:

"There is a gentleman of the name of Gyer who goes to England on the *Columbia* he came to the Country Summer 1843 has been till a few days in the Rocky Ms. and was recommended to me by Sir W. D. Stewart of Murthly Castle north Britain who requested me to furnish necessary supplies and send him to London in the Company's ship and that he would pay the expense & accordingly goes home But Sir W. D. Stewart has forgot to write me on whom to draw for the money will you please charge it to my a/c and when Sir W. D. Stewart pays it in put it to my credit."

Settlement of this debt was still under discussion with Company officials on July 1, 1846—or some eighteen months later—for on that date McLoughlin wrote:

". . . I am censured for giving Debts to the Immigrants . . . But as soon as circumstances permitted I stopped the Credit System—But after overcoming the Difficulties with which I had to contend, saving the Companys and Puget Sound Business and Bringing them to order and making our conduct and the manner the Business was managed the subject of praise by every Respectable person who came to the Country and now to be treated as I am Is what I could not expect . . . I see by my account that you have ordered Mr. Geyer passage amounting to one hundred pounds to be charged to me and also twenty two pounds nineteen shillings and nine pence advanced him from the Vancouver Stores Mr. Geyer was recommended to me by Sir W. D. Stewart of Murthly Castle Perthshire who wrote me from the Rocky Mountains 'From Great Sandy Creek under the Windy Mountains 2d Aug. 43 Mr Geyer a German of considerable Eminence as a Botanist will Visit Vancouver I Believe next Autumn and I am sure will Recommend himself to you by his knowledge of Gardening and Specimens of plants he can give you of which I have spoken to him and of your taste for Gardening I think of hireing him to superintend my Garden in Scotland and if he has any Difficulty in settling for a passage home to England he may have what assistance you may consider Reasonable on my account.' Mr. Douglass writes me that Sir W. D. Stewart informed you he had sent the money. If he did it never came to me I wrote him on the subject and

in London, viâ Sandwich Islands and Cape Horn, on the 25th of May, 1845." Geyer's letter to Hooker (May 16, 1845) includes a few more dates: "I . . . arrived at Fort Vancouver the 29th of October, embarked the 12th of November . . . we arrived in the port of Honolulu on the last of December we left again on the 12th of January [1845] without touching at any other port."

Before turning to Geyer's collections it may be of interest to tell what has been learned of a man whom Geyer refers to as ". . . Dr. Mersh, of Luxemburgh in the suite of Sir Wm. Stewart . . ." Geyer mentions him twice in volume five of *The London Journal of Botany*. First (p. 25 *fn.*) in connection with a cactus:

"This *Mammillaria* is one of the largest in size among the *Conothelae,* and like another species of the same division from the sandy plains at the uppermost forks of the Platte, always solitary and with yellowish-white spines. Both species were lost, I regret to say, but I hope that future travellers will again find them . . . The one from the Platte-plains . . . was first discovered by Dr. Mersh, of Luxemburgh, in the suite of Sir Wm. Stewart; as I noted it down in my journal, as *M. Mershii . . .*"

The second reference (p. 36, *fn.* 2) mentions that, when encamped nearby, ". . . Sir Wm. Stewart desired Dr. Mersh and myself to take a special survey of Scott's Bluffs, where we had an opportunity of witnessing a curious meeting between the grizzly bear and antelope . . ." It seems doubtful if any "survey" was made for, according to Geyer's long story, they derived the impression that they were being pursued and that "Mersh's" life was preserved by Geyer's wisdom:

"It was with difficulty that I prevented Dr. Mersh from firing, he being unacquainted with the revengefulness and ferocity of that animal; and an attack on our part would have been almost certain death, for Dr. Mersh had but one gun, and myself a brace of pistols . . ."

if he hands the amount and Interest will you please order it to be Received and place to my Credit. Sir William D. Stewart was Introduced to us by a circular from the Right Honble. E. Ellice addressed to Several Gentlemen in Canada Sir G. Simpson and the Chief factors and Chief traders of Hudson Bay Company, and we never had any Difficulty before about his Draft—But they were drawn on his Agent by himself he has made several trips from Scotland to the Rocky Mountains . . ."

In a footnote to the above the editor of the *Champlain Society Publications*, Mr. E. E. Rich, offers further data on the matter:

". . . A. 6/27, fo. 8d., Governor and Committee to McLoughlin, Ogden and Douglas October 8, 1845, 'The Bills advised in the 80th par[agraph] of C. Factor McLoughlin's Despatch of Novr. 20th have been honored with the exception of one for £22.19.9 drawn by Mr. Geyer on Sir W. D. Stewart, who, on application being made to him for payment, stated that he had remitted the money to Mr. McLoughlin. It has therefore been charged to that gentleman's account.' . . ."

When one notes that McLoughlin's despatch had devoted seventy-nine paragraphs to Company matters before reaching Geyer—"G" perhaps!—one can readily understand that a Chief Factor's correspondence must have been something of a problem. One can only hope that "Sir W. D. Stewart's" remittance reached McLoughlin ultimately and that he was not out of pocket for twice "£22.19.9."

The Chief Factors were in a vulnerable position—open to criticism if they did too little or if they did too much for the many visitors who were "wished upon" them from London or elsewhere.

It developed, however, that the two bears ". . . had no ill intentions, but seemed impatient to get out of our way . . ."

Even the well-informed have been unable to supply information regarding "Dr. Mersh" although Dr. Rickett supplied the information that ". . . Torrey mentions a 'Dr. Mersche of Baltimore, Md.' as having been on that same expedition of Stewart's. This is probably the same man. About the only place to look would be some ancient medical directories, if such there be; I can't find any here . . ." This was in a letter to Mrs. Lazella Schwarten, Librarian of the Arnold Arboretum, who then wrote to our Minister to Luxembourg, Mrs. Perle Mesta, on February 17, 1953, and with what seems to have been remarkable efficiency, acquired the desired information from the American Consul at Luxembourg, Mr. J. Tuck Sherman. His letter was dated February 26, 1953! It enclosed a copy of a letter from the Director of the Bibliothèque Nationale of Luxembourg, Mr. Alphonse Sprunck ("En reponse a votre demande telephonique de ce matin [February 25]"), which transmitted a biographical notice upon "Charles-Mersch-Faber." All accents are lacking in the copy.

". . . D'apres la Bibliographie Luxembourgeoise de Martin Blum, Mersch est ne a Luxembourg le 30 avril 1810 et decede a Bissen dans le Grand Duche, le 25 aout 1888.

"D'apres ce bibliographie il publia dans les 'Publications de la Societe des Sciences Naturelles et Mathematiques du Grand Duche de Luxembourg'—Sur la representation algebrique des mosaiques geometriques. Tome 19, annee 1883, page 153-208, avec 7 planches.—

"Dans le 'Rapport de la Commission speciale chargee d'emettre son avis etc.' Rapport de la Commission speciale . . . Guillaume-Luxembourg p. 3-26 (en collaboration avec Majerus Francois-Emile) . . ."

The second enclosure is in German and headed: "Porträt-Gallerie hervorragender Persönlichkeiten aus der Geschichte des Luxemburger Landes von ihren Anfängen bis zur Neuzeit. Mit biographischen Notizen. Von K. Arendt, E.—Staatsarchitekt, wirkl. Mitglied der histor. Abtg. des grossh. Institutes. IVter Band, Seite, 35." This is followed by the heading *"Eisenbahn-Kommissar K. Fr. Mersch-Faber. (1810–1888)."* I offer a translation of what follows:

"In the roster of Luxemburg scholars who undertook to turn varied information into practial use, Karl Friedrich Mersch, born in 1810 in Luxemburg, occupied a prominent place. He received his first appointment 1834–1837, as Professor of Chemistry in the local Atheneum. Then he accepted the invitation of a friendly rich American, by the name Wart, of New York, to try his luck in the New World. (Mersch had been a private tutor to this young man during his student years at the Polytechnic.) In the Rocky Mountains of North America there were both blast furnaces and steel mills, whose stock after but a few years paid up to forty percent dividends. Still greater luck awaited the zealous engineer in San Francisco. This brisk booming port city lacked natural drinking water. Here Mersch studied the question of providing fresh water and he succeeded in completely solving this with lucrative results.[9] At the same time he became interested in a profitable import business of Parisian manufactures. He returned to his home at the beginning of his sixtieth year as a rich man. In acknowledgment of his technical and business knowledge, the Luxemburg Government named him Railroad Commissioner, Councilor, and Curator of the Atheneum (1865–1870). From 1882 until March 1888 Mersch lived in Bissen in retirement. He was an Officer of the 'Eichenlaubkrone' (Oak-leaf-Cross) and 'Comtur des Preussischen Kronenordens' (Order of the Prussian Crown). Moral seriousness, strong devotion to duty and charity were the principal characteristics of this gentleman."

Hooker described Geyer's plants in his "Catalogue of Mr. Geyer's collection of plants gathered in the upper Missouri, the Oregon territory, and the intervening portion of the Rocky Mountains," which was published in *The London Journal of Botany* in 1847 and in its successor-periodical, Hooker's *Journal of Botany,* in 1851, 1853, 1855 and 1856. Hooker refers to the collection as a "beautiful Herbarium."

The author cites only general regions of collection and months, not precise dates. There are some interesting remarks upon certain species and many references to earlier collections made by such men as Wyeth, Tolmie, Nuttall, Douglas, etc.; references too to collections of Alexander Gordon. Geyer's much-quoted letter to Hooker (May 16, 1845) mentions the number of his specimens:

"The collections made amount to between 9-10,000 specimens and fall far short of my expectations,[1] chiefly owing to my limited means which compelled me to spend valuable time for other necessary labours.[2] The collections were at Fort Vancouver in good order and being placed at the best possible place on leaving the ship, I have no doubt that they will be in good order. Those made across the plains are not so good as could be desired owing to the many perilous crossings of Rivers. They amount to about 3,000 specimens and include, I have no doubt, something new . . .

"P.S. I marked the boxes containing the collection with your name, thinking that it would be a passport with regard to the customs house inconvenience . . .

"I possess also a collection of seeds (carefully kept) of about 340 or 50 species which

9. That "blasts furnaces" and "steel mills" existed in the Rocky Mountains in 1845 seems impossible. Since Mersh is said to have returned to his native land at the age of sixty (or in 1870) his investments in such lucrative enterprises must have been much later than his meeting with Geyer. And as far as his "fresh water" enterprises about San Francisco are concerned I do not find his name mentioned by H. H. Bancroft although the historian has considerable to tell about the so-called "water-grabbers" of California. Let us hope that Dr. Mersh did not belong in that category!

1. He had written Thorburn on December 28, 1843: "My collection of Botanical Specimens, which will probably amount to 20,000 Specimens is destined for Sir Wm Jackson Hooker in Royal Kew Gardens . . ." The over-estimate is interesting; it is also interesting that, *almost a year before Geyer sailed from the Columbia, he was planning to send his collections to Hooker,* despite his written contract with Engelmann.

2. Just what these "labours" could have been is hard to imagine; possibly he felt that he should have had an assistant of some sort when in the field.

are all destined for Sir Stewart[3] and of each species may be planted some as you may desire at your gardens."

Hooker's office managed the sale of Geyer's sets and the collector expresses his appreciation for this assistance. A number of the recipients or purchasers are mentioned in his letters to Hooker and some individuals are named more than once: Bentham, Brown, Baron Delessert, Fielding, Webb, and so on; Geyer himself took sets for ". . . Messrs. Klotsch, Lehmann, Reichenbach and Endlicher and others[4] . . ." One set he gave to the King of Saxony, and this despite the fact that he feared ". . . they would be laid by . . ." The King, who had bestowed on Geyer the honor of a private audience, was "an amateur botanist," and Geyer wrote Hooker that ". . . he wants to see me, on account of the names . . ." ! On August 6, 1845, he gratefully acknowledges to Hooker ". . . most liberal pay for the collection . . ." and adds: "I was surprised at the offer of £20 for the collection as I always understood that the complete set belongs as a matter of course free of charge to the publisher [Hooker in this instance]." Despite Geyer's sense of having been liberally treated, plant collecting could hardly have been a remunerative profession. More than once Geyer was in need of funds. It would be interesting to know his net return for two years of hard field-work. A receipt (among the letters at Kew) records one payment (*minus* an advance):

"Received from Sir William Hooker on account of Mr. C. A. Geyer for the sale of his plants forwarded to Baron Delessert, Dr. Asa Gray, Dr. Harvey, Mr. Bentham and Mr. Fielding (deducting £20 paid in advance) the sum of £21 (twenty-one pounds) 6/-. Kew, February 18, 1845."

Geyer's letter of August 17, 1845, to Hooker, includes this curious comment: "I have sent off parcels to Mr. Bentham, Mr. Fielding, M. Delessert. The latter I put in a neat box, Mr. F's in oil cloth and Mr. B's only in strong papers." For the Baron in France a box perhaps, but just why oil cloth for "Mr. F" and strong papers only for "Mr. B" ?

Geyer would have been pleased by Coville's commendatory words, written in 1941: "Geyer's botanical work in the western United States was of such importance that no less than thirteen American plants have been named in his honor, as follows: *Alisma geyeri   Allium geyeri   Antennaria geyeri   Aster geyeri   Astragalus geyeri   Carex geyeri   Coggswellia geyeri   Delphinium geyeri   Euphorbia geyeri   Melica geyeri   Mimulus geyeri   Physaria geyeri   Salix geyeriana.*"

3. This is the only reference to collections made on behalf of Sir William Stewart; since Stewart was a horticulturist, with a garden, he was probably only interested in propagating material. Geyer may have been making a present to one whom he then believed was to be his future employer.

4. These recipients, in the order named, and the present repositories of their herbaria are as follows: George Bentham (at Kew); Robert Brown (British Museum); Baron Delessert, then of Paris (at Geneva); Henry Barron Fielding (at Oxford); Philip Barker Webb (at Florence, Italy); Johann Friedrich Klotzsch (at Berlin but destroyed in World War II); Johann Georg Christian Lehmann (once at Berlin); Heinrich Gottlieb Ludwig Reichenbach and his son Heinrich Gustav Reichenbach (once at Berlin); Stephen Ladislaus Endlicher (at Vienna).

Geyer was not optimistic about the prospect of opening up the Oregon country either by railroads from the east or by sea through the mouth of the Columbia River. He wondered why ". . . so many American citizens of the United States leave their desirable residences in the back parts of their own country . . . in order to search for a new home in Oregon territory! Not less wonderful is it that the final possession of that country causes, and has already caused, so many violent demonstrations in the councils of several nations. Certainly Oregon will make a stronghold, for it is already so strong, that it is equally difficult and perilous to get out, as to get in!"

I quote from the end of his long dissertation which makes interesting reading in the 1950's—predictions as it were of things to come:

". . . we must take another and wider view of Oregon, by adding to that territory the whole north-west coast, and even more; from Mount Elias down to the gulf of California; or from the limits of vegetation down to the centre of the region of palms . . . we have . . . all the elements necessary for the formation of a mighty future empire . . . the final possession of the entire coast! . . . Wherever he [the American] has selected a home, he tries to establish his political principles . . . with them he is contented, without them, unhappy. Trusting his rights to no other hands but his own, he despises over-careful governments, and hired military power; but is ever ready for defence, if compelled by necessity, and to handle the rifle instead of the plough. Such are the main features of an American countryman . . . and those individuals in foreign countries are much mistaken, who believe that American politics, affecting the material welfare of that nation, will ever be subject to a single impulse, as in Europe. Such . . . are those thousands of men who have already emigrated to Oregon and California . . .

"Ere long, the hardy scattered emigrants both in Oregon and California will consolidate a government and appear on the theatre of nations, independent of all others. They will, by their enterprize and unceasing civil conquests, overcome successfully the heroic indolence of their Mexican neighbours, regenerate their political and social institutions, and form, in connection with the mother country, on that coast, a great western empire; an outpost of civilization, which, in time, will be the doom for the reckless despotism in the Old World, opposite the great Pacific. Of that future great empire, the present limited Oregon territory is only the nucleus."

The best biographical account of Geyer is contained in Grace Lee Nute's " 'Botanizing' Minnesota in 1838-39," [5] and I quote therefrom:

"In 1826 he entered the garden at Zabelitz as an apprentice. In 1830 he was back in Dresden as an assistant in the botanical garden and in other horticultural establishments. The members of a famed family of horticulturists, the Reichenbachs,[6] were his

5. The author expresses her indebtedness to ". . . Mr. Harold Russell, reference librarian of the University of Minnesota Library, and his staff for many references . . ."

6. Geyer's letters to Hooker contain many references to the Reichenbachs, father and son. Evidently he

friends . . . He left Dresden in February, 1834, for North America . . . Geyer was back in Dresden in the fall of 1845, when one of his friends described him as looking 'at least 20 years older.' " Not being able to secure a botanical position to his liking in Germany, he bought a piece of land in Meissen[7] and began a nursery. In his leisure hours he gave lessons in botany and English. He married Emma Schulze on August 24, 1846. During the last three years of his life he edited *Die Chronik des Gartenwesens,* a horticultural journal of merit, for which he himself wrote the leading articles. He died on November 21, 1853.''

was in close touch with both men on his return from North America. Dr. H. G. L. Reichenbach was ". . . very busy writing a great zoological work, besides his 'Icones Florae Germanicae' of which there are now again printed Vol. I–VII.–VIII. 1846."

7. Fifteen miles northwest of Berlin.

ARTHUR, STANLEY CLISBY. 1937. Audubon. An intimate life of the American woodsman. New Orleans.

BIGELOW, JOHN. 1856. Memoir of the life and public services of John Charles Fremont. New York. 34.

BROSNAN, CORNELIUS J. 1935. History of the state of Idaho. rev. ed. New York.

CANDOLLE, ALPHONSE DE. 1880. La phytographie ou l'art de décrire les végétaux considérés sous différents points de vue. Paris. 415.

CHITTENDEN, HIRAM MARTIN. 1902. The American fur trade of the far west. A history of the pioneer trading posts and early fur companies of the Missouri valley and the Rocky Mountains and of the overland commerce with Santa Fe. Map. 3 vols. New York.

COVILLE, FREDERICK VERNON. 1941. Added botanical notes on Carl A. Geyer. *Oregon Hist. Quart.* 42: 323, 324.

DE VOTO, BERNARD. 1947. Across the wide Missouri. Illustrated with paintings by Alfred Jacob Miller, Charles Bodmer and George Catlin. Boston.

DRURY, CLIFFORD MERRILL. 1936. Pioneer of Old Oregon: Henry Harmon Spalding. Caldwell, Idaho. 312, 313.

——— 1940. Botanist in Oregon in 1843–44 for Kew Gardens, London. *Oregon Hist. Quart.* 41: 182-188.

——— 1940. Elkanah and Mary Walker. Pioneers among the Spokanes. Caldwell, Idaho. 188-190, 254. Pl. ["Drawing of the mission at Tshimakain made by K. A. Geyer in 1844 . . . Original at Washington State College . . ."]

ENGELMANN, GEORGE. 1843. Catalogue of a collection of plants made in Illinois and Missouri, by Mr. Charles Geyer; with critical remarks, &c. by George Engelmann, M.D., of St. Louis. *Am. Jour. Sci. Arts* 46: 94-104. *Reprinted:* Trelease, William & Gray, Asa. 1887. The botanical works of the late George Engelmann. Cambridge, Mass. 506-510.

FRÉMONT, JOHN CHARLES. 1887. Memoirs of my life . . . Including in the narrative five journeys of western exploration, during the years 1842, 1843–4, 1845–6–7, 1848–9, 1853–4 . . . A retrospect of fifty years . . . Map. Chicago. New York. 34.

GEYER, KARL ANDREAS. 1845. 1846. Notes on the vegetation and general character of the Missouri and Oregon territories, made during a botanical journey from the state of Missouri, across the South-pass of the Rocky Mountains, to the Pacific, during the years 1843 and 1844, by Charles A. Geyer. *London Jour. Bot.* 4: 479-492, 653-662 (1845); 5: 22-41, 198-208, 285-310, 509-524 (1846). [*See* Hooker, W. J. 1845, for an account of Geyer's work before 1843, a short account of his life, and comments on the genus *Fremontia.*]

—— For manuscript letters *see* p. 770, *fn.* 1.

GRAY, ASA. 1843. Notice of botanical collections. *Am. Jour. Sci. Arts* 45: 225-227. Reprinted: *London Jour. Bot.* 3: 139-141 (1844).

—— 1893. Letters of Asa Gray. Edited by Jane Loring Gray. 2 vols. Boston. New York. 1: 298.

HOOKER, WILLIAM JACKSON. 1845. [Editorial comments on Geyer, K. A. 1845. 1846. Notes on the vegetation . . .] *London Jour. Bot.* 4: 79-82, and *fns.*

—— 1847. Catalogue of Mr. Geyer's collection of plants gathered in the upper Missouri, the Oregon territory, and the intervening portion of the Rocky Mountains. *London Jour. Bot.* 6: 65-69, pl. 5, 6; 206-256.

—— 1851. 1853, 1855. 1856. *Ibid.* contd., *Jour. Bot.* 3: 287-300 (1851); 5: 257-265 (1853); 7: 371-378 (1855); 8: 16-19 (1856).

McLOUGHLIN, JOHN. 1944. The letters of John McLoughlin from Fort Vancouver to the Governor and Committee Third Series 1844–46. Edited by E. E. Rich . . . With introduction by W. Kaye Lamb. Toronto. Publications of the Champlain Society 7: 53, *fn.* 1, 66, 67, 160, 161, *fn.* 1.

NUTE, GRACE LEE. 1945. "Botanizing" Minnesota in 1838–39. *Conservation Volunteer* 8: no. 44, 5-8.

—— 1946. A botanist at Fort Covile. *Beaver, Outfit* 277: 28-31.

OLIPHANT, J. ORIN. 1934. The botanical labors of the Reverend Henry H. Spalding. *Washington Hist. Quart.* 25: 93-102.

PIPER, CHARLES VANCOUVER. 1906. Flora of the state of Washington. *Contr. U. S. Nat. Herb.* 11: 16.

SPAULDING, PERLEY. 1909. A biographical history of botany at St. Louis, Missouri. *Pop. Sci. Monthly* 74: 124, 125.

CHAPTER XXXVIII

DESPITE THIRTY-EIGHT MONTHS OF ARDUOUS WORK, BURKE IS VIRTUALLY A FORGOTTEN COLLECTOR

ACCORDING to Britten and Boulger's *Biographical index of deceased British and Irish botanists,* Joseph Burke collected for Lord Derby in South Africa from 1839 to 1842 (in company with Zeyher[1] from 1840 to 1841) and in North America from 1844 to 1846. From the very little that has been published about Burke one derives the impression that he did creditable work in South Africa. In a biographical note the editor of the *Champlain Society Publications* states (on information derived from the Director of Kew) that Burke's accomplishment there ". . . was so satisfactory he was commissioned to undertake a similar errand in North America and California for the Earl of Derby and the Royal Botanic Gardens." And William Jackson Hooker, in 1843, mentions that ". . . it has been my good fortune, under the liberal patronage of the Governor and Directors of the Hudson's Bay Company, to unite with the Earl of Derby in sending out an able collector (Mr. Burke), to North-western America and California . . ."

In view of these favorable comments it seemed strange that nothing further was published about the man for his mission to North America certainly lasted more than three years. He left Gravesend in June of 1843 and his last letter from America was written from Fort Walla Walla in October of 1846; this tendered his resignation to Hooker. I do not know the date when he sailed for England but the first letter in the Hooker correspondence indicating that he was back in London is dated "Dover St Novr 6th 1847."

Through the kindness of Sir Edward Salisbury I obtained a microfilm of Burke's letters preserved at Kew. They included sixteen letters to Hooker, a portion of a letter to Lord Derby (dealing mainly with specimens of birds and animals obtained in the summer of 1844) and a long memorandum, probably fragmentary, concerned with plants and seeds collected in the summer of 1845. From these letters I derived the impression that Burke was honest, hard-working, and clearly desirous of fulfilling his obligations to his employers. But he evidently "got off on the wrong foot" as far as collecting-seasons, transportation by Hudson's Bay Company brigades, and much else was concerned. Since Burke's letters often refer to his attempts to follow "instructions," this suggests poor planning on the part of those who sent him into the field. Moreover, there were miscellaneous misfortunes for which he could hardly be held responsible: a cold, wet, snowy summer in 1844, followed by a river accident which

---

1. Lasègue mentions that Jean-Michel Zeyher, director of the garden at Schwetzinger, left a rich herbarium which was bought by the Grand Duke of Baden and deposited at Carlsruhe. Zeyher died in 1845.

did his specimens no good. Most serious of all, his letters and collections seem to have been greatly delayed in reaching London. Perhaps it is not surprising that Hooker— lacking evidence of any accomplishment by August of 1845—acquired the impression that Burke was not fulfilling his obligations and wrote him—or so it would seem—to that effect.[2] Burke, I may add, received the letter ". . . one year & 47 days after the date . . ." so that mail between England and America moved slowly whatever its direc- tion. Burke is not the only collector whose problems were underestimated (if they were understood) by sponsors thousands of miles away. That Burke's collections contained much, if anything "new" is questionable for they were never described as an entity. His journeys were usually hurried ones for he traveled with the Hudson's Bay Com- pany traders for the most part and at a period and in regions of Indian unrest. Fort Hall was his headquarters for most of the time between November, 1844, and August, 1846, and although other collectors—Nuttall, McLeod and Geyer—passed through some of the same country, Burke did get into botanically untouched regions at times and must have contributed to the knowledge of plant distribution if no more. Per- force, Burke spent considerable time in Canada before he was able to cross the Rockies and start down the Columbia River. I shall abridge considerably the content of his early letters and follow his journey in more detail after he reached Fort Walla Walla and the regions of my story.

After his return from South Africa Burke had been employed on Lord Derby's estate, Knowsley,[3] as under-gardener to "Mr. Jennings." He writes that he had sent there "a great many good plants from Algoa Bay, but scarcely any of them are to be found—It is very disheartening . . ." Lord Derby was close to seventy, his memory was

2. Whether Hooker's letter to Burke still exists I do not know. That it was forcefully worded is certain. Burke replied that ". . . Had I any reason to consider myself the neglectful person that I am accused as, in that letter, I would never show my face where I am known again . . ."

3. This was Edward Smith Stanley, thirteenth Earl of Derby (1775-1851). The *Dictionary of National Biography* states that he was president of the Linnean Society from 1828 to 1833 and president of the Zoological Society. He had a "private menagerie" at his estate, Knowsley, in Lancashire, and left his "museum" to the city of Liverpool.

The second chapter of Angus Davidson's *Edward Lear, landscape painter and nonsense poet (1812- 1888)* casts delightful side lights on Lord Derby and on life at Knowsley during the late 1830's and early 1840's. Edward Lear, born in 1820, had begun when a lad of fifteen to make detailed drawings of birds, butterflies, and so on and his work was so good that he was employed at the age of eighteen by the Zoologi- cal Society to make drawings of the parrots "at the zoo in Regent's Park." "The fruits of his labour ap- peared in a noble folio volume entitled *Illustrations of the Family Psittacidae*, with forty-two lithographic plates—the first complete volume of coloured drawings of birds on so large a scale to be published in England." The fine quality of his work met with the approval of the "eminent zoologists" at the British Museum and for two or three years Lear was employed by Dr. John Edward Gray ". . . to make the plates for his volume ... *Tortoises, Terrapins and Turtles* (. . . not published till 1872) . . ." as well as drawings ". . . for Bell's *British Mammalia,* and for some of the volumes ... of the *Naturalist's Library* . . . [and the ornithologist John] Gould had employed him to make drawings for his *Indian Pheasants, Toucans* and *Birds of Europe* . . ." This brings us down to the period of Lear's work at Knowsley:

". . . A very fine collection of animals and birds had been brought together by the 13th Earl of Derby, a keen naturalist, who had formed a unique private menagerie at his country house, Knowsley Hall . . .

"not so good as it used to be," and he was subject to "a sort of fainting." Nonetheless, the old gentleman kept ". . . very busy looking over his different works and writing lists of things . . ." he wished Burke to collect for him. The trip to North America had evidently been planned. Burke sailed from Gravesend on June 4, 1843, in the *Prince Rupert*. They stopped at the Orkney Islands off the coast of Scotland and eventually reached York Factory on Hudson Bay.[4]

1844

Burke wrote from Edmonton House (post of the Hudson's Bay Company), Alberta, on March 1, 1844. He was to leave there in a few days for Jasper House in the same province and expected that he would be obliged to wait there until June before making a crossing of the Canadian Rockies.

". . . If I can find sufficient game to supply myself & those who accompany me with food, I intend spending some time in collecting on the hight of land and the west side of the Mountains—The other side I believe has been much examined by Drummond— He spent a whole season at Jaspers house & made several excursions amongst the Mountains—I am quite tired of the long winter—I am very glad to be able to continue my journey . . . I am told I shall not be able to do much in the way of collecting until

Lord Derby was now contemplating the publication of a book to illustrate his collection . . . Naturally enough, he applied to Dr Gray of the British Museum, and Dr Gray mentioned the name of Edward Lear . . . Lord Derby went off to the Zoo and watched Lear at work there on his parrot drawings . . . and engaged him on the spot to come and draw for him at Knowsley . . .

". . . When he arrived at Knowsley as a humble and unknown artist he was a figure of very little importance in the vast and complex organization of the great house, which was a sort of small town in itself: he came as an employee rather than as a guest, and took his meals in the steward's room. But, though he saw little of Lord Derby and was occupied all day with his drawings in the menagerie, he soon made the acquaintance of Lord Derby's grandsons. The latter were in the habit of dining every day with their grandfather, but soon Lord Derby remarked that they now seemed anxious, instead of sitting with him during the evening, to make their escape as soon as possible after dinner was over . . ."

The Earl was told, "with the candour of youth," that they derived more pleasure in the company of "that young fellow in the steward's room." Davidson notes that ". . . Old Lord Derby was a wise man, and was not offended: instead he invited Lear to dine with him upstairs, and thenceforth the young draughtsman became an intimate friend of the family . . ." A pleasant story I found it! We are told that "The results of Lear's four years of work at Knowsley were published in 1846 in a magnificent volume (privately printed, and now a rare and valuable book) entitled *Gleanings from the Menagerie and Aviary at Knowsley Hall* and containing a large number of lithographs . . . The volume was edited by Dr Gray . . . But those four years also saw the beginnings of a very different book, a book that was to bring Lear a fame and an immortality that all his drawings of animals and birds could never have given him—*The Book of Nonsense*."

4. George Bryce states: "The full name of the Company given in the Charter [of May 2, 1670] is, 'The Governor and Company of Adventurers of England, trading into Hudson Bay.' They have usually been called 'The Hudson's Bay Company,' the form of the possessive case being kept in the name though it is usual to speak of the bay itself as Hudson Bay." *See: The remarkable history of the Hudson's Bay Company.* ed. 3. Toronto. 14 (1910).

the latter end of May—I now find it would have been better to have gone by way of Cape Horn.[5] But it is impossible to foresee these things—I have seen little else but winter since I landed at York so that I hope when the season comes I shall be able to make up for lost time—I am not at all satisfied but I could do no more—it is impossible to work against such a climate as this . . ."

Burke had been able to send a few seeds, evidently collected between York Factory and Edmonton House: ". . . It is such a trifle am almost ashamed to send."

The next letter is dated "Jaspers House May 1st 1844." He had been obliged to abandon the plan outlined in his earlier letter:

". . . The information I received at Edmonton was from a gentleman, I now find had never been here & was not aware of the difficulties of the seasons & other causes which prevent my proceeding as I intended. Mr. Colin Fraser who has charge of Jaspers House has been here eight years & is perfectly acquainted with boath sides of the Mountains—He says there will be at least eight feet of snow on the hight of land at Midsummer—& when I do cross I must wait at boat encampment until the canoe arrives in the fall from Fort Colville to meet the party going to the Columbia . . . There is nothing at Jaspers House at present . . . I intend moving about all the summer & return to Jaspers house in time to be ready to cross to the west side by the 1st of Septr— I shall then have six weeks to collect on the west side, before the Columbia party arrives . . ."

This is followed by a description of the trip between Edmonton and Jasper Houses (March 5–20). After an attempt to proceed farther, all the time in snow, he had returned to the last-named post.

"I hope the plan I intend to adopt during the summer will meet with the approbation of yourself & the Earl of Derby—I shall arrive at Fort Van Couver about the same time as is mentioned in my instructions—the only differance will be that I shall spend the whole season amongst the Mountains instead of the Columbia River—I hope—when I arrive at Fort Van Couver, it will be the season the trappers visit the Sacramento— Although winter it being much farther south than the Columbia—there will be a good chance of collecting seeds—Afterwards I should like to visit the Snake country—Mr Drummond spent the summer at the same place we intend forming our encampment . . . My guide was well acquainted with Mr Drummond—he says he never made an excursion so long as to prevent his returning to the encampment at night—I intend moving about in different directions, & return to the encampment with what I collect about once a week By that means, I hope to find much that was not seen by Mr Drummond."

5. When referring to the disastrous journey of Robert Wallace and Peter Banks (p. 797, *fn.* 8) I mention that both the Hudson's Bay Company and John Lindley had advocated that they travel to the Columbia River by way of Cape Horn, but this advice was not followed, their sponsors, Hooker and others, determining otherwise. Their trip had been made in 1838 and 1839.

The same letter is continued on May 14, after the Company's brigade had arrived. Peter Skene Ogden traveled with the party and was on his way to England. The collector mentioned to Burke by Ogden was Karl Andreas Geyer:

". . . He tells me, the whole of the Country on the south side of the Columbia, is taken up by the American emigrants—No more trapping parties will go to California— He tells me, there is a German collector at Fort Coville who sends part of his collections to you—I hope to see him in the Fall at Fort Van Couver that I may know what he has done—He spent the whole summer in the Snake Flathead countries & intends to do the same this summer . . . I have been nearly twelve months from England & have not done any thing—I hope when the summer sets in to be able to make up for lost time . . ."

The next letter was also from Jasper House, September 11, 1844. Burke's summer work had been unremunerative.

"The past summer has been the most unfavourable, the oldest persons about this place ever remember witnessing . . . Mr Frazer . . . has been upwards of seventeen years in N. W. America—& eight years of that time at Jaspers House . . . has seen unfavourable seasons before—but not at all compared to the past summer—Whe have had frost, snow, or coald[6] rain, nearly the whole time—I expect in a few days to be able to cross the Athabasca, on my way to the Columbia—I fear I shall find few ripened seeds, on account of the unfavourable season . . ."

There is no reason to doubt Burke's story of misfortunes. But such fungi as he was able to collect rotted, the pines (one of which he considered especially interesting) failed to produce fruit ("Weather this has been a season of rest—or weather the unfavourable season des[troyed[7]] the Pine blossoms, I cannot tell . . ."); the weather, mentioned in detail and almost daily, was uniformly unpleasant in some way or other! Hooker—and he is not alone in this—may have grown weary of calamities, but Burke himself could not have enjoyed them.

Burke's next letter is dated "Fort Hall, Snake country Decr 28th 1844." By the time it was written the collector had made his trip to Boat Encampment and had waited there for the westward-bound brigade until his food gave out. Fortunately, soon after he had started back to Jasper House, the party arrived and on October 10, 1844, he began the journey to Fort Vancouver. At Boat Encampment, four days later, they found ". . . Mr. McDonald & his family . . ." This was Archibald McDonald.

"I had written to Mr McDonald in the spring, that I should like to go to the Snake Country—He advised me to go there as quick as possible—& not go to Van Couver, as I had intended—He gave me an introductory letter to his nephew—who he ex-

6. Burke wrote a good and, for the most part, legible hand. His spelling is often peculiar—as far as some words are concerned, uniformly so.

7. Burke's letters may have been bound into a book for often the beginning or end of a line is cut off in the microfilm. Where the lost portion is obvious I have supplied it.

pected I should find at Fort Hall—but I arrived at Walla Walla time enough to go with him to the Snake Country . . ."

The trip down the Columbia with the brigade began October 15 and that evening they encamped ". . . below the Dalles des morts, Where Wallice[8] and Banks, the collectors

8. Robert Wallace and Peter Banks were two young gardeners sent out from England to collect plants. They had been employed at Chatsworth, the estate of William George Spencer Cavendish, sixth Duke of Devonshire. The Duke was a horticulturist, much interested in embellishing his grounds and greenhouses with rare and interesting plants; in this he was abetted by Joseph Paxton who had risen from the position of gardener at Chatsworth to that of friend and adviser to the Duke. They had already sent one collector, John Gibson, to India and the experiment had been rewarding but extremely expensive, for the Duke was in the habit of doing things in the grand manner.

The Duke and Paxton, together with other horticulturists, nurserymen and the scientific botanists (William Jackson Hooker in particular) agreed to share in the costs of sending out Wallace and Banks, both unmarried when they left Britain. Violet Markham, in her entertaining volume *Paxton and the bachelor Duke*, tells that the young men were ". . . bidden to beware of bears and women, both hinderances to the quiet life of a plant collector." The envoys may have been careful about the bears but Wallace paid no heed to the last peril!

Against the advice of the Hudson's Bay Company and of John Lindley, who advocated the slower trip around the Horn, they were sent to the mouth of the St. Lawrence and from there, following a route by the Great Lakes and the interior of Canada, eventually crossed the Canadian Rockies through Athabasca Pass and arrived at Boat Encampment in October of 1838. It was while descending the Columbia River, on October 22, that the overcrowded boat in which they were traveling struck a rock and twelve, or some say thirteen, persons aboard were drowned, among them "Mr. and Mrs. Wallace," and "Mr. Banks." Accounts of the accident are found in H. H. Bancroft, in the *Champlain Society Publications,* and elsewhere. According to J. A. Stevenson's "Disaster in the Dalles," the wreck took place in ". . . the Little Dalles . . . just above the present city of Revelstoke [British Columbia] . . ." We are told by Markham that news of what had happened reached Chatsworth only in May of 1839.

Disasters to plant collectors are almost commonplace. But since romance in the field is unique—certainly in the period of my story—I quote a letter written by Chief Factor John Rowland, stationed at Fort Edmonton, to James Hargrave. It is found in Stevenson.

" '. . . I have a piece of news to tell you which will surprise you, you must know that Miss Maria Simpson was married here last fall to a Mr. Robert Wallace during their stay at Norway House last summer they did not only get acquainted but madly in love who promised their hand & heart to each other before leaving that place & all what I could say to Miss I could not persuade her to wait and get a Fathers consent no she would be Mrs. Wallace *coute que coute*. That business has caused me much uneasiness and yet I do not feel comfortable, God knows I acted for the best still perhaps our friend the Governor will blame me, however he never was more fond of his amiable wife himself, than those two seemed to be between you and me. N. H. is not a fit place to have young ladies under the same roof with young Batchelors as they were fixed here last summer. When we meet if you wish to hear more about this it will be time. All I have to tell you is that I was very sorry to see her attach to a man she has seen for a short time only. All the Gentlemen who were here this fall were of one opinion it was better they should be united & so they were now god bless them Both if I am to be found fault with gods will be done.' "

Stevenson explains that the identity of the bride is difficult to determine, but concludes that she ". . . was the daughter, *sub rosa*, of George Simpson and 'Betsey'—who later became Mrs. [Robert S.] Miles— and that she was consequently referred to by some as Maria Simpson and by others as Maria Miles." The editor of the *Champlain Society Publications* supplies the same history. The bride was sixteen years of age.

Stevenson also quotes from a letter written by Wallace to Paxton in July of 1844 which indicates that little collecting had been done even up to the time the young men had arrived at Norway House on June 28:

" 'We have done but little in the way of botany as yet . . . Our mode of travelling will not admit of such,

were lost, with several others a few years ago . . ." They reached Fort Colvile on the 18th, crossed the portage at Kettle Falls on the 19th and rested the next day, a Sunday. On the 22nd they camped above Spokane River, on the 24th were at Okanagan, on the 26th at Priests Rapids and at Fort Walla Walla on October 27.

Burke's letter refers to finding Geyer at the post:

". . . There I saw Mr Geyer the collector—I had the pleasure of his company but a short time, as he left the same day by the boats that I came on.

"I was also engaged as Mr. McDonald[9] was quite prepared, & wished to commence his journey to the Snake country that evening—He however put it off until the next day as I could not possibly be ready . . . all my things were soaked with water taken in when running the rapids—Mr Geyer will describe[1] the state my things were in better than I can write it—I packed two reams of [? the dozen] of my papers & a few other things that I could not dispense with & the next day Octr 28th commenced my journey

as we are on the water from daybreak until sunset except at the portages, or when ashore for breakfast or dinner.' "

To this Stevenson adds that ". . . they stayed some time at Norway House, and Wallace would have had more time to study the botany of the region, if his interests had not been otherwise engaged . . ."

BANCROFT, HUBERT HOWE. 1884. The works of . . . 28 (Northwest coast II. 1880–1846): 338.

———— 1886. The works of . . . 29 (Oregon I. 1834–1848): 316, fn. 2.

HOWAY, F. W., LEWIS, WILLIAM S. & MEYERS, JACOB A. 1917. Angus McDonald; a few items of the west. Washington Hist. Quart. 8: 215, 216.

KANE, PAUL. 1859. Wanderings of an artist among the Indians of North America; from Canada to Vancouver's Island and Oregon through the Hudson's Bay Company's territory and back again. London. 332.

McLOUGHLIN, JOHN. 1941. The letters of John McLoughlin from Fort Vancouver to the Governor and Committee First Series 1825–28. Edited by E. E. Rich . . . With an introduction by W. Kaye Lamb. Toronto. Publications of the Champlain Society 4: Appendix A, 293, 294.

———— 1943. Ibid. Second Series 1839–44. Publications of the Champlain Society 6: 1-3, fn. 2.

MARKHAM, VIOLET. 1935. Paxton and the bachelor Duke. London.

STEVENSON, J. A. 1942. Disaster in the Dalles. Beaver, Outfit 273: 19-21.

9. Burke's journey to Fort Hall was with Angus McDonald, nephew of Archibald.

1. Geyer wrote Hooker of seeing Burke; the letter is dated from "Barque Columbia of the Honbl, Hudson Bay Company, before the entrance of the British Channel May 16th 1845:"

"I met Mr. Burke at Fort Walla-walla, he was much disappointed at his summer expedition at Jasper's House, the summer being cold, stormy and snowy. There was no time to write. He requested me to state to you his part ill luck, and also to his noble employer the Earl of Derby. He had just escaped wrecking, his boat struck at the rocks in the rapids where Messrs. Banks & Wallis lost their lives. His papers and books got wet and during the two hours only which we spent together he was busily employed in getting horses and saddles ready to start next day for Fort Hall on the upper Louis [Lewis or Snake] River, his intended winter quarters, to descend from thence next spring, down towards the Sacramento River, into California . . ."

Geyer also mentions Burke in a letter which he wrote Hooker from "Dresden, Decr,. 14th 45:"

"Mr. Burke is now in new fields for botanists. You may look for numbers of new species, for he, doubtless will have opportunity to visit some of the Black-mountain ranges, southwardly, rich in Eriogona etc. In that region exists a Melocactus (at all descriptions from hunters) which has large, long, flat, spirally twisted spines in the shape of a ram's horns, growing on the tops of calcareous cliffs; also a spiny, shrubby Oenothera in the desert. I spoke of both to Mr. Burke at Walle-walle."

to the Snake country—Mr McKinley . . . in charge of Walla Walla promised to dry my paper & other things I left with him . . . The Brigade was of about thirty loaded horses and about twenty loaded [? mules] . . ."

In making the trip to Fort Hall, situated in southeastern Idaho, Burke and the Hudson's Bay Company brigade were, starting from Walla Walla, to move from southeastern Washington into northeastern Oregon (where the Blue Mountains were crossed and the Grand Ronde valley traversed for its entire length); from there the route was along the so-called Oregon Trail to Fort Boisé and thence, across southern Idaho, to Fort Hall. This was much the route taken by Wyeth and Nuttall in 1834 (although in reverse) and by McLeod in 1837.

Burke's letters are long and, since the content is rarely differentiated by paragraphing or other means, they make difficult reading. I have, therefore, at times, broken them down according to dates and have sometimes substituted other punctuation for the dash which he uses with great frequency.

Leaving Fort Walla Walla on October 28 the party reached "the foot of the Blue Mountains" on October 30 and camped on the "Umtallow," or Umatilla, River. They were then in Oregon. The migrations of the 1840's were in progress and on the 30th Burke mentions that they ". . . met several American emigrants, with their families & waggons, going to the Wallamette [Willamette] settlements."

By November 1 they reached the top of the pass over the Blue Mountains where "it was rather coald," and camped ". . . on the Grand Rond R The Grand rond is the prettiest plain I have seen in the N.W. It is of a circular form, about 40 miles in breadth, & about [?] in length—A small river runs through it with willows on its banks, & a few poplars, the only wood on the plain—There are hundreds of Pines on the mountains that surround it on all sides." That day more caravans were encountered.

"During the day, we met several other American waggons. There were many carcases of oxen belonging to the emigrants which had died on the road—The emigrants themselves appeared very ill off—they were ragged, dirty, & looked half starved—Many of them were short of provisions—They told me they had been six, & some seven months on the journey . . ."

Camp of November 2 was at the southern end of Grande Ronde valley and that of the 3rd at ". . . the lone tree encampment The stump of a large Pine is all that now remains—The American emigrants last year, cut it down. It was a great pity—it being the only large tree, on the whole journey from the blue mountains to Fort Hall." On November 4 Burke refers to being "nearly smothered with dust." On the 5th they were following up ". . . a small stream called Burnt river, running amongst the high hills," and on the 6th reached, but left immediately, "the bank of the Snake river . . ." On the 7th they were near ". . . Fort Boisee a small Fort on the east side of the Snake river," and the next day crossed to the fort: ". . . the packs were taken across the river in a canoe."

On November 9 and 10 the route was along the Boisé River; on the 11th and 12th the way is only described as very rough, "the country being [covered] with stones." On November 13 they ". . . passed a cluster of hot springs[2]—the water was . . . too hot to keep the fingers an instant in the bubble . . ." and again reached Snake River which they followed, or kept close to, the rest of the way to Fort Hall where they arrived on the 21st. Burke had commented upon the salmon in the Snake: "The shore was strewed with dead salmon. I had been told, none of the Salmon that ascend the river [live] to return—Their fins & tails are nearly worn away, ascending the rapids & falls." He had also mentioned (on the 16th) an accident which had involved his collections. They were leaving the Snake River at one point:

". . . On ascending the bank which was very steep & rocky & about a hundred yards high—When nearly at the top a horse [bit] another that was before him. he kicked until he threwed his load which although on the ground was still attached to the saddle—he then took fright & ran down towards [the] river, dashing his load from rock to rock—when at the bottom [he] fell & layed there until he was taken—It chanced to [be] my luggage—It was quite distressing to examine it—My insects & bottles filled with spirits of wine were all broken to pieces—Other tin boxes that I had were beaten out of all shape—In short I have nothing left that it was possible to destroy by dashing against rocks . . ."

Fort Hall was ". . . on the south bank of the snake river—in Lat 43 N & Long 112-30 W. The altitude being very great, it is rather coald at present—The climate in Winter, is about the same as London, but . . . more snow—The Summer is said to be very hot & dry . . ." The 15-page letter describing the foregoing journey contained a brief description of the character of the country traversed and one wonders what Hooker thought of it as a collecting ground:

". . . After leaving the Blue mountains the whole journey is through a [? wilderness] with mountains on each side, topped with snow—The appearance of the country is what is called in Southern speech a half desert—a loose sandy soil & in places stony—covered with low stunted bushes & abundance of grass, growing in tufts—It is quite destitute of trees . . ."

1845

Burke's next letter was also from Fort Hall, and dated "June 4th 1845." The winter had been "very severe" but by April 7 Burke had started off with a "trading party going to the [Yo]uta country." This excursion was to last until May 27. From Fort Hall the route led southeast (across southeastern Idaho) into northeastern Utah, and Burke, who for a time traveled alone except for one companion, seems to have gotten

---

2. Possibly Twin Springs in Elmore County. *The Idaho Encyclopedia. Compiled by the Federal Writers' Project of the Works Progress Administration* (51-56, 1938) lists many springs in the Boisé River region and states that "There are hundreds, possibly thousands of hot and cold springs in the State with more than ordinary mineral content. . . ."

as far south as the Provo³ River, for that stream and his "Provosts river" were one and the same, and to have, by way of that river, reached Utah Lake (which he calls "Little Youta lake"). The Utah Indians, from what Burke has to tell—and on the evidence—were a tribe to reckon with at this time and the amount of plant collecting that he was able to do was probably small. Certainly he left Utah Lake with some haste and started north again. I quote some of his story of the journey.

Leaving Fort Hall on April 7 the party ". . . crossed the plain in which Fort Hall is situated in a southerly direction & encamped in the evening on the bank of Port Neuf river." They kept up that stream on the 8th and 9th and crossed it on the 10th. On leaving the fort, ". . . the grass had not begun to shoot—after entering the Mountains vegetation was much more forward in the warm vallies . . ." On April 11 they camped ". . . by a long narrow [piece] of water called rush lake . . ." and on the 12th, after passing "the Bear river . . . encamped on its bank [at] the commencement of Cache valley . . ."

They must have entered Utah and the fertile valley in Cache County through which flows the Bear River. Following this valley, they reached its southern end on the 20th and, on the 22nd, left it by ascending a mountain on the top of which they encamped ". . . amongst the snow banks . . ." Burke comments on the 23rd that ". . . We had scarcely anything to eat for three days. The people killed a dog & ate it. I was not quite hungry enough to join them . . ."

On April 24 they moved the camp "to Ogdens hole," known also as Ogdens Valley.⁴ On the 26th they left this valley and on the 27th camped on ". . . the bank of the Weaver river⁵. . ." They were still in Weber County, to the east of Great Salt Lake; the Weber River enters the lake about opposite Fremont Island.

I have omitted Burke's daily reports upon the weather. Suffice it to state that it had been cold, wet and snowy for most of the time. Hunting was poor and the party was ". . . in a starving state & nothing to be got to eat—In some places we found a sp of Alium [*Allium*] which made us ill if we eat much of it . . ." On April 29 they were able to acquire a fat mule from some Indians and had a good meal.

April 30: ". . . camp was removed a short distance up the Weaver—Myself & [? Michaux] the Canadian that joined our party in Cache valley passed through an opening in the mountains, following the course of the Weaver—to see the Great Salt lake—All the waters from the Bear river to Provosts river begin in the great Salt lake—& no

3. Named for Étienne Provot, a French Canadian, who explored the region about 1825.

4. This and Ogden River were named for Peter Skene Ogden of the Hudson's Bay Company; the city of Ogden, Ogden Peak—all in Weber County—suggest that his visit to this region had been deemed worthy of commemoration.

5. According to the volume *Utah* ("American Guide Series"), pp. 357, 358: "The Weber (pronounced WEE-ber) is presumed to have been named for Captain John G. Weber, who claimed to have joined General Ashley as a trapper in 1823. The Weber was also called Weaber and Weaver, and may have been named for Pauline Weaver, Arizona frontiersman."

river is known to run from it. The water of the lake is as salt as brine—The shore is covered with salt."

May 5: ". . . We moved our camp to Provosts river—Several years ago upon this river —a party of twenty five men under one Provosts [were] surprised by the Snake Indians —Twenty of them were treacherously murdered, the Indians at the time pretending to be friendly with them—Five of the party escaped."

May 7: ". . . Early in the morning myself the Interpreter & another—left the encampment to visit the Little Youta lake . . . Following up the course of Provosts river through the Mountains—after a rough days journey we arrived at the lake . . . At the little lake we found an Indian, called the little chief & his party—They were not inclined to be very friendly with us . . ."

The "little chief" had been fighting other "Youtas" and had robbed a party belonging to the American Fur Company. Attempts to trade and to obtain trout, of which the Indians had an abundance, were fruitless. "The little chiefs brother, who has always been friendly with the whites . . . told us unknown to the others, to saddle our horses & retreat as fast as possible . . ." This advice was acted upon promptly and the next day Burke and his two companions rejoined the main party which Burke tells us (on May 9) had now removed some distance up the river to a small plain called ". . . Camass prarie—A small party of Snakes had encamped with them—In the afternoon two Youtas arrived . . . One of them (as evil a looking fellow as I have ever seen), three years ago murdered a Canadian trapper; his wife & two children in a most cruel manner[6] . . . The murderer has hid himself from the whites until lately—Mr. Smith, an American fur trader offered a reward of horses to the Youtas if they would deliver him up—he . . . left his tribe & has only lately returned He offered me a piece of meat— although I was hungry, I turned from the wretch in disgust—I believe if I was starving I could not have taken food from him. I felt a greater inclination to shoot him than any animal I have ever seen—but upon second thoughts, I thought it better to leave him to the friends of the murdered parties, who will certainly meet with him some day or another. Our Interpreter wished to shoot him—he asked me if he should do so—I would not sanction it—I told him if I had thought it best to kill him at that time, I would have done it myself . . ."

Instead the party started back towards Fort Hall the following day (May 10) and followed the Weber River until the 13th. Hearing (May 15) that ". . . a party of American fur traders . . . were encamped on the Bear river . . .", Burke with the interpreter ". . . went to see them—We there found Mr Smith[7] who sends his furs to Fort Hall . . ." Burke breakfasted with him on the 16th and was told that a party

6. I omit the horrible details.

7. This was certainly Jedediah Strong Smith, famous trapper and explorer.

of American trappers who had left Fort Hall for the Missouri had had their horses stolen and been obliged to walk back to the fort. Burke found that his party had moved to "Sand creek" where he rejoined them. On the 17th they moved to Bear River (when crossing it "most of our things were soaked"), proceeded down that stream on the 18th, camped on "Smith's Fork" on the 19th, and on "Thomas's fork" on the 21st. The references to Smith's and to Thomas' Forks indicate that, on leaving Utah, the party crossed into Lincoln County, in southwestern Wyoming, for a short time. But by May 23 or 24 they were back in extreme southeastern Idaho, for on the latter date they camped ". . . about half distance between the Big timber & the White clay . . ." On May 25 they camped by ". . . a number of Soda springs, near the White clay—Near the Soda springs is a [?] on the bank of the Bear river—It throws its water through a hole in a rock, making a great noise—The Mountain men call the stream bout spring . . ." Soda Spring was, doubtless, the famous Beer or Soda Springs of southwestern Caribou County, Idaho. And "bout" was a corruption of the French word for boiling. I have found no reference in books or maps (though such probably exist) to either the "Big timber" or the "White clay." Burke had mentioned that the "Big timber" was so named ". . . on account of the Poplars being more plentiful than in other parts of the [? Bear] river . . ." The trees may have been aspens and, on one map, an "Aspen Range" is located directly west of Soda Springs.

May 27: ". . . We arrived at Fort Hall in the afternoon . . . Our whole journey has been amongst mountains—The season was rather too early for collecting plants, but I got several, perhaps a hundred sps & think this country will prove very rich in Perennials & perhaps annuals. but trees, & shrubs are very scarce—The Indians by continually burning the country, have destroyed nearly all the wood—The only Pine I have seen is Douglasii & that is very high on the mountains where it has been saved from the fire by the snow. There is also a dwarf stunted sp of oak which has not begun to put forth its leaves . . ."

Burke adds to his letter (under date of June 18) that a brigade had arrived at Fort Hall from Fort Vancouver, bringing him letters from Hooker and Lord Derby. He states that he had obtained a few seeds for the Earl; also, that, since his return to Fort Hall, he had been doing some collecting, but was about to depart on another trip.

"The country is just getting beautyful—I leave here directly with a party going to the head of the Missouri to hunt buffalo—The whole country is quite in a ferment—We hear of nothing but fights & murders—I expect there will be sad work when the emigrants cross this summer . . . The brigade will have left Fort Hall before our return from the Missouri—which will prevent my sending my collections by the ship this year —I expect to send a large collection of seeds across the mountains in the spring . . . (I am stung out of all patience by the Moschett[os])—I shall forward by the ship the few seeds I got from the trappers that came from the Colorado."

Burke now set out with the traders and hunters bound for "the head of the Missouri." The journey lasted from June 25 to August 30, 1845, and is described in a 19-page letter to Hooker, dated from Fort Hall, September 3, 1845. It must have taken an immense amount of time to write such an epistle. The weather, the party's whereabouts, and some of its activities, are recorded daily. I shall not follow the journey in any detail.[8] Until July 5 or 6 the party was in Idaho; from then until into the end of the third week of August in Montana; and from August 21 (certainly) to September 3 back in Idaho. The party ". . . mustered twenty seven, all well armed & plenty of ammunition." They visited territory frequented by Nez Percés (traveled with lodges of this tribe from July 4 to August 3), Flatheads, Blackfeet and Crows; even some Pend Oreilles put in an appearance. The Indians were hunting and moving about, keeping watch of the white men as well as of enemy tribes of their own race.

On June 25 Burke overtook the brigade ". . . at the Blackfoot Bute, about seven miles from the Fort . . ."; on the 26th ". . . proceeded up the Snake river . . .", crossed it on the 27th, and followed it until June 30 when it was left, the party then proceeding ". . . in a northerly direction . . ." On July 1 they ". . . encamped by a small river called Camass creek . . .", and on the 3rd "close to a range of Mountains, on a small stream called Medicine river—it takes its name from a very large Medicine lodge being made there several years ago by the Black feet . . . The country between Fort Hall & this place, is a half desert, very sandy, & in places . . . covered with low shrubs & grass in tufts." On July 4 they ". . . entered the Mountains, following up the Medicine river . . ."

I believe the brigade must have left Snake River at the point in Jefferson County where that stream makes a bend to the south, and kept north or northwest until, in Clark County, it encountered Medicine Lodge Creek and followed it into that portion of the Rocky Mountains situated along the boundary between Idaho and Montana. It was in these mountains, on July 5, that Burke found a plant that filled him with enthusiasm:

". . . we encamped in a beautiful spot amongst the hills, sloaping from the Mountains was quite a flower garden—all the sp[ecime]ns marked with that date, were collected within a few hundred yards of the encampment—In that place I found a most beautyful Columbine, I did not see it before, & have not seen it since—It was growing at the foot of a hill, in a rich loamy soil—The flower stems are about ten inches in hight —The flowers very large & beautyfully white, with varieties shaded a clear light blue— In my opinion It is not only the Queen of Columbines, but the most beautyful of all herbaceous plants—I never felt so much pleasure in finding a plant before—I sent a few seeds of a red sp from Jaspers house—I also found the same near this one— Amongst the spns you will find a pretty pink sp which I think a variety between the red & white—I only saw one plant of it."

8. Photostats of Burke's letter are filed in the library of the Arnold Arboretum. I believe that persons familiar with the terrain of Idaho and Montana should be able to follow the route from start to finish.

On July 6 they camped on a stream ". . . forming one of the heads of the Missouri . . ." and were now in Beaverhead County, Montana, for on the 7th they reached ". . . a small valley called the horse prarie . . ." Horse Prairie Creek unites with the upper waters of Beaverhead River near Armstead, Beaverhead County.

Finding Burke's outline of the journey into Montana extremely difficult to follow, I appealed for assistance to Dr. Paul Chrisler Phillips of Montana State University, Missoula. He very kindly read Burke's record and supplied me with his interpretation thereof. Of the route taken from Idaho into Montana he wrote (March 18, 1953) that it was ". . . by way of Monida Pass or Bannock Pass, and then to Rock Creek and to Horse Creek, near the present Armstead, Montana. The term Horse Prairie is used today as it has been since very early times . . ."

I shall first quote Burke's comments upon the route, italicizing his dates:

On *July 9* they ". . . entered a large valley called the Grand True . . . The *10th* . . . encamped by a small tributary of the Missouri in the Grand True. The *11th* . . . arrived at a bend in the Grand True, where it suddenly turned to the eastward . . . The *12th* . . . encamped . . . at the eastern end of the Grand True. The *13th* . . . came up with the Nezperces camp . . . about fifty lodges & several thousand horses . . . encamped on the Missouri fork. On the *14th* we crossed a high mountain with a great deal of Pinus Douglasii growing on it. After descending the east side . . . encamped on the bank of the river we had left in the morning . . . The next day . . . encamped again on . . . the same river. The *16th* encamped near a rock called the Beavers head. The *17th* . . . crossed the Missouri fork, & encamped . . . The *18th* . . . continued following the same river. The *19th* . . . we did not travel . . . many of the horses being much fatigued. The *20th* . . . proceeded, still following the river . . . The *21st* . . . left the Missouri fork & proceeded in a southeasterly direction. We travelled but a short distance that day . . . The *22nd* but a short distance There being a great many sick persons in the Nezperces camp chiefly children[9] . . . The *23rd* . . . we removed our camp but a short distance . . . the *24th* we encamped at Tobacco root river, a branch of the Missouri . . . The *25th* . . . we did not travel on account of the sick. On the *26th* we encamped at the Three forks, also tributaries of the Missouri . . . The *27th* . . . we did not travel . . . On the *31st* we encamped by the head waters of the southwest branch of the Muscle shell river . . .

9. Burke thought very little of medical procedure among these Indians: "The Indian doctors were all busily employed, & a great noise they made—It was enough to give a person in health the headache, to be in the same lodge with them—The Medicine man as the Doctor is called has several helpers, who beat with sticks on a piece of wood laid across the lodge—They all sing at the top of their voices—They say it is to frighten away the sickness—I hear Medicine men are often shot—When the patient gets worse under one of them, another is sent for—If he sees the patient cannot live, he sais he can do nothing for him the discarded Doctor has thrown bad Medicine on the sick person, which is too powerful for his good medicine In many cases the first doctor is shot by the friends of the sick person (that is, if he should die) & often, his friends will shoot the second doctor—Although it is a dangerous calling, it is so profitable that there is no great want of doctors . . ."

"On *August 1* . . . we did not proceed . . . The *2nd* . . . We proceeded down the Muscle shell . . . On the *3rd* . . . we parted company with the Nezperces they proceeding towards the Missouri, & us towards the Flathead camp. We found them on another branch of the Muscle shell, just outside the Mountains—It being the first time we were clear of them, since . . . the 4th of July . . . On the *5th* we removed, proceeding in a North easterly direction, in company of the Flat heads, following the Buffalo. On the *6th* we removed to the south . . . On the *7th* we proceeded towards the south west . . . On the *9th* . . . a small party of Flat heads left the encampment to go to the American Fur companys Crow Indian Fort, on the Yellowstone—which was about two days journey . . . On the *11th* we left the Flat heads, on our return to Fort Hall. We encamped on the Otter river, A branch of the Yellowstone . . . The . . . *12th* . . . we encamped on the Bank of the Yellow stone, after crossing that river . . . The *13th* . . . we did not travel. The *14th* . . . we encamped on the Yellow stone a short distance below the 25 yards river on the opposite side . . . The *15th* . . . we crossed the Yellow stone & encamped. The *16th* . . . we still proceeded up the Yellow stone. The *17th* . . . we passed over a narrow plain about twenty miles in length, covered with Opuntia. We encamped at a bend of the Yellow stone, when we left that river. To follow it any farther would have been too much to the South for our course. The *18th* . . . we crossed a high mountain, the descent [on] the opposite side, towards the west, was very difficult. We encamped in a pretty valley at the head of the three forks . . . The next day [the *19th*] . . . we travelled a good deal through Pine forests. We encamped in a pretty little prarie surrounded on all sides by steep Pine clad Mountains. The *20th* . . . We travelled through thick Pine forests & encamped on the bank of Tobacco root river, in a large Prarie called the Burnt hole. In a forest at the foot of a mountain at the edge of the Burnt hole are several boiling & Sulphur springs. One of them I am told, throws the water to the hight of one hundred feet. It was too far out of my way to visit them. The Tobacco root river in the place, is very full of fine trout. The *21st* . . . we encamped on the waters of Snake river, near Henry's lake . . ."

This ends the portion of Burke's letter which describes the trip into Montana, for the party was back in the northernmost corner of Fremont County, Idaho, on arrival at Henry Lake. Burke's letter had had a good deal to tell of the Indians with which they came in contact, but I have confined my quotations to geography. Dr. Phillips refers to his own notes upon the route as "rough" and I have rephrased some of them, as the writer evidently intended should be done.

Burke's "Grand true" (July 9 to 12) was the Grand Trou, so named by the French trappers, and called by the English "Big Hole"; "evidently the party kept north from near present Dillon to Reichle on Big Hole [July 11]," and were at the "east end of Big Hole near Rochester Creek," on the 12th. "The 'Missouri fork' [of July 13] should equal Jefferson River but probably the party crossed south of the Beaverhead, so named by Lewis and Clark." Beaverhead Rock, of the 16th, "is now known as Point of Rocks—a short distance above the forks of the Jefferson." On the 21st they were "on the Jefferson near present Loomont" and moved "southeast to the Madison River —Burke's 'Tobacco root river'—thirty-five miles from the Jefferson, on the 21st." On the 26th the route was "down the Madison to Three Forks."[1] The route of July 27 to 31 was "up Sixteen Mile Creek to the South Fork of the Mussellshell River."

Burke's phrase of August 3—"just outside the Mountains"—is interpreted by Phillips as "south of Little Belt Mountains or east of Castle Mountains." Under date of August 9 Phillips notes that "Crow Indian Fort on the Yellowstone was either Fort Cass built in 1832 at mouth of Bighorn River or Fort Van Buren[2] built in 1835 at mouth of Rosebud River."[3] Otter River, of August 10, "flows into the Yellowstone near Big Timber." Burke's "25 yards river" of the 14th was "Shields River." The route of the 17th and 18th "crossed the plain at Livingston, thence over Bozeman Pass, and down the Gallatin to Three Forks." Of the remainder of the journey, from Three Forks to Henry's Lake, Idaho, Phillips notes: "From Three Forks the route was up the Madison [Burke's 'Beaverhead'] to Burnt Hole . . . Boiling springs may refer to Old Faithful which throws water 100 feet. There are 'boiling fountains' west of Burnt Hole. The party crossed over a rugged pass to Henry's Lake and down Henry's Fork a short distance then turned west towards Camas Creek."

Translating the above into terms of present-day Montana counties, Burke's party appears to have crossed Beaverhead County from southwest to northeast, crossed northern Madison County into northeastern Gallatin County (Three Forks), and from there moved into southern Meagher County and Wheatland County (where they were "just outside the Mountains"). Whether Burke proceeded from there southeast to Fort Cass (Treasure County) and to Fort Van Buren (Rosebud County) I do not know, as stated in my last footnote. But on Otter River (August 10) the party was in Sweetgrass County. To reach Three Forks it then moved westward across Park and Gallatin counties. From Three Forks the party followed a more easterly route than they had taken on entering Montana, proceeding southward up the Madison River through Madison County (keeping to the north and northwest of what is now the northwestern corner of Yellowstone National Park) and then crossing into Fremont County, Idaho. From the Three Forks to Henry's Lake the route had led almost due south. For the location of "Burnt hole" (August 20) Dr. Phillips referred me to the

1. Burke refers to "Three forks, also tributaries of the Missouri." Phillips corrects the phrase "tributaries of the Missouri" to "which form the Missouri."

2. Chittenden refers to Fort Cass as the first American Fur Company post in the Crow country and states that it was abandoned in 1835. Also that Fort Van Buren was the second, built in 1835 and abandoned in 1843. He places it at the mouth of Tongue, rather than Rosebud, River—which is not in accord with Dr. Phillips' statement as to the site.

3. Whether Burke accompanied the "small party of Flat heads" which left on the 9th "to go to the "American Fur companys Crow Indian Fort, on the Yellow stone," or remained with the main party of that tribe, is not clear in his letter.

map contained in William F. Raynolds' "Report" upon the exploration of the Yellowstone River.[4] Burnt Hole is indicated between two of the uppermost branches of Madison River, and lies nearly due east[5] of Henry's, or Henry, Lake.

Back in Idaho on August 21 Burke was filled with a desire ". . . to leave the Brigade & return to the place I collected the beautyful Columbine on the 5th of July to collect the seeds.

"I was obliged to leave my man in charge of the collections, & none of the Companys people would volunteer to accompany me—Even those that a short time before said they would go, & were well acquainted with the mountains. When it came to the point [they] said they did not know the way . . . on my return they could tell me the point I had gone wrong . . . the Banak that joined us on leaving the flat head camp . . . did not hesitate a moment—he was very willing to go. He received his instructions through such a very awkward interpreter, that it was almost useless. Myself & the Banak left the Brigade early on . . . the 25th. We had three good horses. We . . . encamped . . . in sight of the Medicine river . . . The 26th . . . We were wandering amongst the Mountains . . . without striking on the track I wanted—I then found my guide had not been rightly informed as to the place I wanted to go—He made signs to that effect, & also for me to take the lead—Neither of us could speak one word that the other could understand—I went in the direction I knew the Medicine river to be & struck it at the place we encamped on the 4th of July I then knew where I was—About an hour before sun set we arrived at the place I desired   It had a very different appearance to what it had two months before—The whole country was dry & barren—Many plants that I wished to collect seeds from, had disappeared altogether—I found I was too late for the Columbine I so much desired—After a short search I found it amongst the dried grass—The seed vessels were about half emptied by the wind—I collected seeds until it became dark, & commenced again the next morning as soon as it was sufficiently light—I collected all I could find, which is rather a large packet—I collected all the other sps that had not shed their seeds . . . we . . . commenced our return journey very much satisfied with my success—I had been thinking some time, I would

4. Raynolds, William F. 1868. "Report on the exploration of the Yellowstone River, by Bvt. Brig. Gen. W. F. Raynolds. Communicated to the Secretary of War, in compliance with a resolution of the Senate, February 13, 1866." *U. S. 40th Cong., 1st [? 2nd] Sess., Senate Exec. Doc.* No. 77, 174 pp. Map ["U. S. War Department. Map of the Yellowstone and Missouri Rivers and their tributaries, explored by Capt. W. F. Raynolds Topl Engrs and 1st Lieut. H. E. Maynadier 10th Infy Assistant. 1859–60. To accompany a report to the Bureau of Engineers. 1867."].

5. Whether it lies within the confines of Yellowstone National Park, Wyoming, or in the southernmost portion of Gallatin County, Montana, I have been unable to determine. Although there is a reference to Burnt Hole in another interesting book mentioned by Dr. Phillips (Ferris, Warren Angus. 1940. *Life in the Rocky Mountains. A diary of wanderings on the sources of the rivers Missouri, Columbia, and Colorado from February, 1830, to November, 1835. By W. A. Ferris then in the employ of the American Fur Company. And supplementary writings by Ferris. With a detailed map of the fur company, drawn by*

be too late for the seeds. It takes off a great deal of the pleasure of collecting knowing as I do that Mr. Geyer passed through this country in the seed season—Mr. Gordon[6] I suspect, came on Mr. Geyer's [? trail] with the Emigrants, or the American Fur company belonging to Fort Larime—I hope neither of them have found my tall shrubby OEnothera . . ."

On August 29 Burke rejoined the brigade and on the 30th the expedition was back at Fort Hall. "Shortly before we arrived, between 450 & 500 waggons had passed, for the Walla amette settlement, & several for California."

In the memorandum included with Burke's letters—headed "References to particular sp[ecime]ns & seeds, Collected during the summer of 1845"—the Columbine is mentioned and three *Oenothera,* two described as shrubby. The memorandum is evidently incomplete, but includes notes on fifteen plants and should have been a useful supplement to Burke's specimens.

Burke's next letter—like the two previous ones—was from Fort Hall, and dated December 15, 1845; he was off again to Cache Valley, Utah, where he had been in April and in May.

"After staying at Fort Hall until I had secured the seeds I had collected, & examined my specimens I prepared to depart on a journey to the South end of Cache Valley—My main object for going there, was to collect a beautyful Liliaceous plant I had found there in the spring Spn No 6 & also an Oak which grows there in abundance on the Mountains—I was aware it would be dangerous, on account of the Youtas. But little is to be got in this country at present, without some risk—Mr. Grant & others advise me not to go, if it could be avoided—To avoid going was impossible, as I did not know where else to get the plants I wished. My man did not hesitate a moment about going— He is exceedingly brave, & that is about the only good quality I have seen in him. He is a sad rogue & never missed a chance of cheating me when it was in his power . . . with all his faults, if not the best, I believe he is as good as any freeman in this country—I am now pretty well acqainted with most of them, & never met their equals—I thought the Hottentots were the most degraded set on earth—but I now find they are superior to these people—It is said, there is honour amongst [? thieves] but such is not the case amongst these mountain free men—for they cheat each other every opportunity—Not

Ferris in 1836. Edited, with a life of Ferris, and a history of explorations and fur trade, by Paul C. Phillips. Denver.), this merely refers to the Raynolds map.

The importance of the mountain valleys which were called Holes in the days of the fur trade is discussed at some length by Chittenden (2: 743-751) and the locations of many are described. Burnt Hole is not mentioned although he does say that ". . . Near the sources of the Madison there were also numerous parks, which are still without distinctive names; but which were familiar resorts to the hunters and trappers. All of these valleys were connected by easy passes over the Continental Divide with the equally important basin of Henry Lake, one of the ultimate sources of the Columbia."

6. Alexander Gordon who, like Geyer, had traveled to the Rocky Mountains with Sir William Drummond Stewart in 1843. Any references to his presence west of the Rockies are extremely rare.

being acquainted with their real character—I thought I should by their means, make a compleat sweep of the seeds of the Snake country—But I was greatly mistaken—I got but two very small packets . . ."

Burke left Fort Hall on September 6 and reached the Portneuf where he relates that he saw on the 7th ". . . quantities of Ipomipsis [*Ipomopsis*], the same as the specimens[7] but the seed season was passed—Mr McDonald told me [that he] saw this summer, towards the head of Green River—a white variety of this plant—I hoped to have got many seeds from him—I am shure [he] would be willing to oblige me in that, or any other way in his power—But whither he does not know the seeds, or whither he passes by without noticing them, I cannot tell—at any rate, I have not been so fortunate as to get any from him—it is the same with all I have met with—They either know nothing about it, or forget it altogether which is not surprising, as it is a branch that the traders think very little of—I will nonetheless lose no opportunity of adding to the collections for the want of asking, or by little expenses—At the same time, I fear I shall get but little, but by my own bodily exertions . . ."

The route to Cache Valley was the one taken in April and May of 1845; the first night (September 6), as noted, was spent on the Portneuf; the south end of Rush Lake was reached (the 10th), Bear River (the 11th) and Cache Valley descended (the 12th through the 14th); on the 15th Burke ". . . ascended the Mountain which divides cache valley from Ogdens hole, & arrived at the top about midday.

"Not a vestage was to be seen, of the Liliaceous plant I was seeking—I opened a trench with my pickaxe about fifteen inches deep—I continued working until night fall—by that time I had found several bulbs—I broke many of the finest—The bulbs being very brittle they broke when removed [from the] earth—My man having his children [with] him, felt very uneasy concerning the Youtas, as it was expected they were in Ogdens hole—I had secured the Liliaceous plant, but not the Oak—I had ex-amined the trees on ascending the Mountain, but could not find a single acorn—We met a party of Snakes in cache valley who told us, it was an entire failure amongst acorns, which has been the case with almost all [? trees] this season—probably caused by the unfavourable spring—I thought I might get a few where [were] it only speci-mens, on the south side—Accordingly I departed . . . at day break—I returned with-out succeeding—I collected a few branches, which are amongst the specimens."

Burke and his companion left promptly—the fires of the Utahs could be seen at the spot where the previous night had been spent. They returned through Cache Valley (September 17), reached Trout Lake (the 18th) and its north end (the 21st), Bear River (the 22nd); on reaching Thomas' Fork, Wyoming, on the 24th Burke spent a day collecting seeds in the mountains. He was back at Fort Hall on the 27th. Here he learned that some Snake Indians had brought word that members of the Utah tribe

7. In Burke's memorandum ("References to particular sps & seeds . . .") this is his number 9: "Ipomipsis elegans ? . . ." It had been seen east of Fort Hall, both northward and southward, but on his return from the trip to Montana ("to the Yellow stone") the seed season was over. "By examining a great many seed vessels, I found four seeds."

had murdered a party of some thirteen persons traveling from California, had bound the leader and thrown him in the river. This day he had met ". . . Dr. White[8] & three other Americans from the Walla amette & going to the States.

"The Doctor had been three years in Oregon as Indian agent for the American government— I halted to dine with him—Amongst other things, I enquired concerning the chestnut[9] you are so anxious to procure   I made all the enquiry in my power before but he was the first that could give me any information—He sais it is to be found between the Columbia river & California by the way of Umpqua—About Umpqua it is very abundant on the mountains—He told me attempts have been made to take it to the States, but without success—The nuts would not keep—That is no more than I expected although I intended & still shall try the same the first opportunity—But the plan by which I expect to succeed, is to plant the nuts as soon as possible after they are collected, & bring home living plants in cases—I have since heard that the poor Doctor & his little party were cut off by the Siux between Green river & Fort Larime—It could scarcely be expected that such a small party could get through—The Snakes & Siux have been at war all this summer . . ."

A few lines later comes a break in the letter. The missing portion must have included the beginning of an interesting discussion concerned with Burke's duty to his two patrons, Hooker and Lord Derby. The portion that remains speaks well for Burke's generous and disinterested approach to his obligations:

". . . Perhaps a collector was never so situated before—I would wish strictly to do my duty by both parties, but everything conspires against my doing it in the manner it was planned—Any arrangement that can be made so far as to give satisfaction, will be pleasing to me—As you have the arrangement for every thing connected with this undertaking—I beg Sir William, as a great favor—that in the arrangement that may be made—That any thing which may concern my own private interest, be considered as nothing—for under present circumstances, it is so with me."

1846

Burke remained at Fort Hall for the early part of the winter of 1845–1846. In a letter dated from Fort Walla Walla, May 3, 1846, he tells that on January 11, 1846, he left Fort Hall with the trappers, carrying his collections with him. ". . . We crossed

8. Dr. Elijah White. Frémont, traveling west in 1842, records that he traveled behind White, who was then leading the second emigrant train to Oregon.

9. From the time that Hooker had first learned of *Castanopsis chrysophylla* (Hook.) A. DC. from its discoverer David Douglas, not a collector from Britain went into the far west without instructions to acquire propagating material; I have commented more than once upon Hooker's persistence in this matter. Even the officials of the Hudson's Bay Company had been called upon to exert themselves on his behalf! Douglas had first observed it on October 9, 1826, but had been unsuccessful in procuring satisfactory propagating material, as were other collectors who attempted the task. Burke was to acquire fruit on a later journey as we shall see.

the Blue Mountains without any difficulty, which is a very rare occurance as the snow is very deep and hard at that season." They reached Fort Walla Walla on February 2.

On February 19 Burke left for Fort Vancouver, arriving there nine days later, February 28. After stating that, while there, he went about as much as possible—"to see the field that Douglas had been so successful in"—it was early in the season and but few ". . . of Douglas's plants were in flower, the most abundant & the most beautyful was Ribes sang [uineum]." He has something to say of the boundary dispute, then at its height:

"The country is fast settling by the Americans—Oregon city considering the short time it has emerged from a forest, has quite a respectable appearance & a great number of British subjects from Red River & retired servants of the H B company are settling amongst them—The boundary question is so [long] in settling, & the Americans taking possession of the best lands on boath sides of the Columbia British subjects have no alternative but to join with them—The speech of Sir Robert Peel has given them a little courage, as it is expected by them that the Columbia will be the boundary—not so with the Americans   They expect to get to lat 49 which includes all the country that is of any use as it [lies] in the best lands, the Columbia river & the [most] of Frazer's river, which will cut off all the water communication to the north of 49 & if the Americans at any time think proper they can with very little difficulty cut off the Salmon fisheries which are the chief support of the interior."

Burke left Fort Vancouver on March 25 with the company express and reached Walla Walla April 3. Their boat was nearly lost in the rapids. On April 7 he went collecting in the Blue Mountains but returned May 2 so that he might accompany Angus McDonald to the Snake country! He comments upon his work:

". . . I have added several plants to the collections, from the Blue Mountains, but not so much as I expected, as the general face of the vegetation is the same all through the country that I have travelled on this side of the Rocky Mountains, where the soil is the same. I hoped to have received letters from England by the ship, but she had not arrived up to the time that Mr. McDonald left Van Couver."

Burke's next—and last—letter from America is dated from Walla Walla, October 17, 1846. In it he tells that he had left Fort Walla Walla on May 11 and had reached Fort Hall on June 12:

"I collected many plants by the way & some seeds—I intended collecting during the remainder of the summer on the head waters of the Snake & Green rivers. The Blackfeet having destroyed several lodges of Snakes in that quarter no one would engage to go with me. In short no free man would engage atall . . ."

Burke evidently got into the mountains lying southeast of Fort Hall and from July 8 to 16 seems to have made a visit to Bear River. But he had just started on a trip with McDonald on July 22 when they were called back:

"We had scarcely left the Fort when the first of the Emigration arrived—an express was immediately sent after us with orders to return—I could get no person to accompany me—all were anxious to trade with the emigration—Mr. Grant had scarcely time to take his meals—I could not leave the Fort—My situation was most an[noying]."

Burke next participated in a journey of some historical interest, for he accompanied Jesse Applegate and a party of young men who prepared (cut out and marked) a road from Fort Hall to the Willamette Valley of Oregon. This party preceded a wagon train of some ninety to one hundred wagons—the first emigrant train over a route which came to be known as the "Applegate Cut-off." [1] Burke describes the trip in his letter to Hooker of October 17, 1846, but unfortunately in no great detail. He had disturbing matters on his mind, for his epistle ends with his resignation from Hooker's service.

I shall outline the route as I have been able to reconstruct it from a number of sources[2] and then quote Burke's account of the trip:

From Fort Hall, Idaho, the route went southwest to Raft River and thence to Goose Creek; continuing up that stream, it crossed from Cassia County, Idaho, into the extreme northeastern corner of Elko County, Nevada, continued southwestward down the valley of Rock Springs River and thence to the upper waters of the Humboldt River (which Burke calls "Ogdens river"). The Humboldt was followed to the point in southeastern Humboldt County (probably in the vicinity of present Winnemucca) where the river bends southwestward; here the route left that stream and, trending west and northwest, crossed the Black Rock Desert (Camp McGarry was later built on the cut-off) and reached the border of Nevada in northwestern Washoe County, crossing into eastern Modoc County, California, between Upper and Middle lakes. From there it turned northward, crossed Fandango (or Lassen) Pass at Fandango Mountain, turned southward through Goose Lake Valley, went around the southern end of Goose Lake—in so doing crossing Pit River, northern tributary of the Sacramento River. It then turned westward, crossed the lava bed region to the south of Clear and Tule lakes, and entered Klamath County, Oregon, south of Upper Klamath Lake. Once in Oregon, the route appears to have gone west and north—since it reached the Rogue and Umpqua rivers—and came eventually to the upper waters of the Willamette. It followed that stream to the Columbia.

I break down Burke's description of the trip into several paragraphs:

"Late on the evening of the 8th of August Mr. Applegate from the Walla Amett settlement arrived. He had discovered a south route from the Walla Amett valley to Ogdens river and then east to Fort Hall. He gave such a fine description of the country between the California line & the Walla Amette valley that I felt most anxious to ac-

1. Bancroft's *History of Oregon* (1: 555-565) cites the date of departure from Fort Hall as August 9 [1846]. Burke states that Applegate arrived at Fort Hall on August 8 but put off his departure until the 11th so that Burke might prepare to accompany him.

2. Bancroft's *History of Nevada, Colorado, and Wyoming* (pp. 62-64) as well as his *History of Oregon* (already cited) refer to the cutoff. Bernard De Voto's *Year of decision 1846* mentions it more than once and stresses its importance to the emigrants bound for either California or Oregon. He notes that ". . . use and further exploration changed details of it."

company him & his party on their return—The next day being Sunday he allowed me the time to prepare for the journey—I packed a small quantity of paper, a quantity of seed bags & forty days provisions—

"On the 11th we left the Fort—after following the Snake river about 45 miles we turned up the raft river, then striking over to the head of Goose creek, then crossing to the hot spring valey & from there to Ogden's river. It was then the 17th of August—We were . . . about 200 miles from Fort Hall—We followed Ogdens river until the 26th of August—The river & California trail which we had been following turns with a sharp bend to the S a little inclining to E [? West]—& about 400 miles from Fort Hall—After leaving the river we passed over about 60 miles of most miserable volcanic region, with many boiling springs   With Mr. Applegates party, & some young men from the waggons which came to clear the road when it was absolutely required we numbered 24—On the 8th of Septr we entered the Sacramento valey—after crossing the valey—about 20 miles, we crossed the Sacramento [the Pit] river a short distance below the lake which is divided from the Clamet lake by a low narrow ridge—that evening we encamped by the Clamet lake—On the 14th we entered the Rogue river or Shasty valey—There I saw abundance of Eschscholtzia californica, the ground had been burnt, & it was flowering the second time—The next day we [?] the Rogue river, here for the first time I saw the vine in its wild state. It appears to be a very shy fruiter. In this valey the soil is generally good, & continues so all the way to the Walla Amett. The country is well timbered with Pine & Oak—On the 19th we entered the Umpqua Mountains—On the 20th we encamped on the Callipua [Calypooa] Mountains—From these Mountains back to the Mountains bordering the Clamet lake, the chestnut you were so anxious to get is very abundant—It is a very shy fruiter, & not ripe when [we] passed—I collected the fruit leaving the [? nut] in the husk, with a small piece of the [stem] hoping they may ripen sufficiently to grow   It is generally a dwarf shrub, but in very sheltered places, it grows to a beautyful tree about 20 feet high[3]

"On the 26th we arrived at Mr. Applegates farm in the Yam hills [Yamhill] about 50 miles from Oregon city—Our horses were completely worn down & ourselves much fatigued—By great attention to watching boath day & night, we passed through the different tribes of Indians without any mishap what has seldom occured to a party before

"After allowing the horses to rest three days, I proceeded towards Oregon city—In the evening I arrived at Mr. J. McLoughlins farm—To reach his place I had to cross the Yam hill river. although the distance I had travelled was but 25 miles, the horses were so weak they were nearly drowned in swimming it   The next morning Mr.

---

3. Burke had been successful, therefore, in acquiring the golden-leaved chestnut (*Castanopsis chrysophylla* A. DC.) in which, as I have stated, Hooker had long been interested; not in the condition that Burke could have wished, but that was scarcely his fault, misfortune rather. It would be interesting to know whether his propagating plan was a success.

McLoughlin lent me fresh horses to take me about 5 miles to the companys station at Champouic [Champooic] where a boat was to start from that morning to Oregon city —I arrived one hour too late for the boat   The next morning Mr. McDonald who is in charge at Champouic, engaged a canoe which took me to Oregon city about an hour after my arrival Mr. A. McDonald of the Snake country arrived on a visit to Mr. McKinley, who is at present in charge of the Oregon city establishment—I arrived at Van Couver on the evening of the 5th [of October]. On the 7th I left with Mr. A. McDonald for Walla Walla—On the 10th we arrived at the dalles where horses were waiting to take us to Walla Walla & arrived yesterday [the 16th]."

Burke's letter states that on arrival at Fort Walla Walla he found two letters from Hooker: "One of June 6th & the other of August 29th 1845—with the report concerning the Royal Botanic Gardens. The latest of these letters being one year & 47 days after the date—Had I any reason to consider myself the neglectful person that I am accused as, in that letter, I would never show my face where I was known again— But as I am innocent, I fear nothing—" He then reviews his years of work:

"I have done my utmost to give satisfaction—I left England in June 1843 & arrived at York Factory in August. I then proceeded up the Saskatchawan in the fall boats, which is the most severe season to travel in this country—much worse than the dead of winter & March I went to Jaspers house, over the Ice & snow—The Ice had but partly disappeared on the lakes when the express passed in May no one could have collected much during [the] long dreary winter which was then just passing away & which nearly compleated my first year—I spent the summer amongst the Mountains doing all I could for my employers—In the fall I returned to Jaspers house, packed my collections & wrote my letters—I was detained a few days from commencing my journey across the Mountains by the heavy snows. I felt anxious to get such seeds I could as winter was just approaching—I left two boxes at Jaspers house. One for the Earl of Derby & one for yourself, & also my letters which I gave in charge of Mr. Frazer. The whole to be forwarded by the first opportunity—Mr. Frazer tried to dissuade me from entering the Mountains as the waters were unusually high & would be attended with a great deal of danger   My answer to him, was, did I collect for myself I would not by any means refuse his advice but as I was collecting for others, danger would be but a poor excuse—I narrowly escaped the mountain torrents but with the destruction or injury of such things as could be destroyed or injured by water. I saw but few seeds on that rough journey, & to have sent them by Cape Horn would have been acting quite contrary to your orders. On seeing Mr. McDonald at Boat encampment he advised me to go to the Snake country. he told me his Nephew would have left Walla Walla but I could get Indian guides. On my arrival Mr. Angus McDonald was about leaving for the Snake country, but he told me he would wait a short time for me but not the whole of the day   I had but a few hours to prepare for a journey on horse back of six hundred miles in the fall of the year. I had to take such things as I wanted the ensuing summer, but at the same time received orders to take as little as possible. The only opportunity I had of speaking to Mr. Geyer was his [standing] by me, when preparing for the Snake Country—I told him I had not one moment to spare—to mention at home how I was then situated. I also told him as much as I could remember concerning the summer before—As I had written full letters from Jaspers house & [several] times before. I thought my journey across the mountains & down the Columbia river at that late season of very little consequence—During the spring & summer of 1845 I collected in the Youta, Blackfoot, & other countries. This past winter soon after new year I brought my collections to Van Couver—The seeds were sent across the mountains by the express & the specimens will go by the ship this fall from Van Couvers Island—As far as I know, nothing has been detained in the country when it was possible to send them home—I think Sir William, it is a very hard case if a collector is sent from the Royal Botanic Gardens to a country where he cannot send his collections by any means by the time mentioned in your letter that his funds are to be stopped—If it only means my salary, I think very little of that but should it mean that my supplies are to be stopped in this country, It would dishonor me for life—I must now go to the Snake country as I left my summers collections there—I have a fair collection of seeds from the present journy—I shall bring my collections down in the winter—leave the specimens to go round the horn, & return to England by the express across the rocky Mountains—I trust Sir William you will forgive my retiring from the service without waiting an answer, as it would be two years or upwards before I could receive one—

"I am Sir William with the greatest respect your most obedient & very humble servant Joseph Burke"

Hooker noted on this letter of October 17, 1846: "Only received in *January* 1848! By Post W.J.H."

The Hooker correspondence includes two short letters written by Burke after he had returned to England. They are of no great importance but do indicate that the collector completed his end of the agreement with his sponsor. Apparently the two men did not meet and Burke gives no indication that he wanted to see Hooker again. The first letter is dated "Dover St   Novr 6th 1847":

"I have been every day to the Docks since the Prince Rupert arrived & could not get my things cleared until last evening—I forward to Kew a box containing four sp of Orchidaceous plants. specimens of the same, you will find in Dr. Smillies parcel, which I allso forward—I would have brought them myself but the great deal of walking I have had lately has caused my thigh which was injured to swell, & I require to be as quiet as possible for a short time—"

The last letter cites no year but is dated from Piccadilly, November 20th:

"I have just received a letter from Mr. Jennings[4] requesting me to go to Knowsley as soon as possible—Would you be pleased to forward to this place the papers I delivered to you on arrival as Lord Derby & Mr. Jennings wishes to see them—I was at

4. The head gardener at the Earl of Derby's estate.

Kew on Thursday, but did not see you. I could not wait as I had to meet Dr [? Kel-lack] at a certain hour—I must go to him to day at Stokes Newington, as I want further advice before I leave London—"

It would be interesting to know what Burke did in later years, but the curtain falls at this point on what seems (to me) to have been a sorry story, for the unfortunate Burke certainly.

Inquiry regarding the Burke collections at the Royal Botanic Gardens, Kew, brought the following prompt reply from the Director, Sir Edward Salisbury, dated "4th December, 1952":

". . . no list of specimens collected by Burke or Gordon[5] have been found in the Herbarium. [Follows a reference to the Burke-Hooker and Gordon-Hooker letters.]

. . .

"Burke was a disappointment as a collector of *seeds;* only one great woody plant was raised, a Juniper, which by 1884 was 5 feet high and appeared to be an unde-scribed species. Burke must have collected rare species for the Herbarium, however, since he visited the Suche[6] River region where Douglas had been so successful . . ."

5. The inquiry included a request for similar information regarding the collections of Alexander Gordon.

6. Probably this is a reference to the Deschutes River of Oregon which empties into the Columbia from the south (to the east of the Dalles) and forms in that region the boundary between the present counties of Wasco (west) and Sherman (east). Of course Douglas visited the Dalles of the Columbia more than once. The Deschutes River was called by the French the "Rivière aux Chutes."

BRITTEN, JAMES & BOULGER, GEORGE SIMONDS. 1931. A biographical index of deceased British and Irish botanists. ed. 2. London.

BURKE, JOSEPH. [His letters to William Jackson Hooker are at Kew. A microfilm and photostats of these letters are in the Arnold Arboretum.]

CHITTENDEN, HIRAM MARTIN. 1902. The American fur trade of the far west. A history of the pioneer trading posts and early fur companies of the Missouri valley and the Rocky Mountains and of the overland commerce with Santa Fe. Map. 3 vols. New York. 2: 743-751.

DAVIDSON, ANGUS. 1939. Edward Lear, landscape painter and nonsense poet (1812–1888). New York.

DE VOTO, BERNARD. 1943. The year of decision 1846. Boston.

HOOKER, WILLIAM JACKSON. 1843. Figure and brief description of Castanea chrysophylla; by W. J. H. *London Jour. Bot.* 2: 495-497. Pl. 16.

LASÈGUE, ANTOINE. 1845. Musée botanique de M. Benjamin Delessert. Notices sur les co-lections de plantes et la bibliothèque qui le composent; contenant en outre des docu-ments sur les principaux herbiers d'Europe et l'exposé des voyages entrepris dans l'in-térêt de la botanique. Paris. 207, 268.

McLOUGHLIN, JOHN. 1944. The letters of John McLoughlin from Fort Vancouver to the Governor and Committee. Third Series, 1844–46. Edited by E. E. Rich . . . with an in-troduction by W. Kaye Lamb . . . Toronto. *Publications of the Champlain Society* 7: 59, 60.

CHAPTER XXXIX

GORDON MIGHT HAVE GATHERED ADDITIONAL "EX-QUISITELY DRIED SPECIMENS" HAD HE RECEIVED FINANCIAL ASSISTANCE WHEN THAT WAS SORELY NEEDED

IN 1845 William Jackson Hooker wrote in the London *Journal of Botany* upon the "Proposed botanical journey of Mr. Alexander Gordon, to the mountains of Texas, &c." The "&c." was, I believe, the Santa Fe country to which Hooker and other botanists had long been desirous of sending a competent collector. On the maps Texas doubtless looked like the best approach to Sante Fe but actually the only feasible way to get there at this period was from St. Louis.

As far as I know, Gordon never got to Texas although he might have done so had he been given assistance when he was in financial distress. He eventually got to the vicinity of Santa Fe, but not until 1848; that he did so is evidenced by certain specimens cited in Gray's "Plantæ Fendlerianæ Novi-Mexicanæ."

Information about Gordon is very meagre. Britten and Boulger refer to him as a gardener who collected in the Rocky Mountains and South Carolina for the nurseryman and "Covent Garden seedsman," George Charlwood. This employment must have been both before and during the period of Gordon's journey to the Rocky Mountains.

It was after reporting the departure of Karl Andreas Geyer for the Rocky Mountains in 1843—he was then traveling with Sir William Drummond Stewart's party—that Hooker mentions that ". . . an equally indefatigable Scottish Botanist was of the party, Mr. Alexander Gordon, who had long been resident in the United States, and had thence transmitted many rare seeds and roots to Europe. On his return from that journey, he lost by shipwreck a large part of his collections soon after his embarkation at New Orleans for England. Among what remained, seeds of several rare plants have been reared, and a considerable collection of exquisitely dried specimens came into the possession of Mr. H. Shepherd, Curator of the Liverpool Botanic Garden, and Mr. Lawson of Edinburgh. Through their kindness, my Herbarium has been enriched with many of these plants, and I shall have occasion to notice several, when treating of those of Mr. Geyer in the present Journal . . ."

In a letter to Hooker dated August 13, 1845, Geyer mentions Gordon's presence on this trip, a trifle superciliously, as was his wont when referring to those in his own station in life:

". . . Gordon in his time collected only in Sir William H's[1] route. He had a long Journal, whether anything has been published of it or not? I found the *Oenothera guttata* of my collection before him and since that time he never communicated to me what plants he found, whether you received or Dr. Torrey, I should like to know. He evidently collected specimens for the first time,[2] what I could see, but he had good knowledge of the Flora of Lower Louisiana, Alabama, etc."

Just where Gordon went in 1843, or following the trip with Stewart, I do not know; one record suggests, no more than that, that he continued west.[3] And he must, from some point, have proceeded to England. For, having commented on Gordon's earlier work in this country, Hooker so states in writing of his "Proposed botanical journey" of 1845:

". . . Mr. Gordon embarked again for the United States in the autumn of last year [1844] . . . and his first letter to me conveyed the information that misfortune still attended his wanderings, so that he was detained at Mobile in Alabama much longer than he could have wished. The circumstances are . . . detailed in his letter from that place, dated December 23, 1844."

This letter recounts that, from New York, Gordon had traveled by way of the Ohio River to the Mississippi and was descending that "noble stream" to New Orleans in the ". . . stream-boat 'Belle,' a splendid new vessel . . ." when she was run into: ". . . I lost everything except my shirt and trousers,[4] and four dollars that were in my pocket." The unfortunate man seems to have met with misfortunes at every turn. Now destitute, Gordon began collecting plants which he believed would find ". . . a ready market at New York . . . I may naturally look for payment by February [1845], and so be in a position to pursue my route early in spring . . . I . . . hope that little time is lost by the delay, for whether I go . . . to the Texian Mountains; or what now seems more probable, to Santa Fè . . . I shall arrive soon enough for the spring Flora. If I decide on the latter course, I shall join the regular Traders at the City of St. Louis, and avail myself of their protection to Santa Fè; they go annually . . ."

Hooker then quotes from another letter which Gordon wrote him from Mobile on April 17, 1845. The writer expressed appreciation for the interest which Hooker had

1. In typing the Geyer letters, the copyist may have made a mistake. "H" must surely have been "S"—for Stewart. For Geyer letters see p. 770, *fn.* 1.

2. On entering what was for him a new field, Gordon may not have distinguished between common plants and rarities, already an important distinction at this period; to Geyer, who had worked eastward of the Rockies since 1835, he may not have appeared to be discriminating. Hooker had noted his "exquisitely dried specimens" so that he could have been no novice. He evidently had sufficient acumen to collect the *Oenothera,* and was astute enough to keep what he subsequently found to himself!

3. Joseph Burke wrote W. J. Hooker (September 3, 1845) from Fort Hall: ". . . it takes off a great deal of the pleasure of collecting knowing as I do that Mr. Geyer passed through this country in the seed season—Mr. Gordon I suspect, came on Mr. Geyer's [trail] with the Emigrants, or the American Fur company belonging to Fort Laramie—I hope neither of them found my tall shrubby Œnothera . . ."

4. In the original, "pantaloons," but Hooker made what he considered suitable emendations!

taken in his affairs. The "interest" seems not to have been accompanied by financial assistance and Gordon, with reluctance, had been obliged to put off his trip: ". . . poverty is a powerful check-rein, and at present there is no alternative . . ." Instead, he worked as superintendent on an estate, collecting plants in his spare time. It is obvious that he was not of a mercenary nature:

". . . I shall be enabled to transmit you so large a collection for the sum you specify, as will give you entire satisfaction. Indeed, I should be sorry to restrict my exertions to mere payment; I shall feel pleasure in sending all I can . . ."

Hooker makes this laudatory comment:

"It is impossible not to admire the ardour with which Mr. Gordon thus carries on his botanical investigations . . . he has hired himself out . . . earning, with the sweat of his brow, the scanty means for prosecuting his favourite pursuit . . . we trust that when his Alabama plants arrive . . . purchasers will be found . . . thus enabling him to collect the more extensively . . . in the mountains of Texas and of North Mexico."

I have found no published record of Gordon's activities between 1845, when he was in Mobile, and 1848 when some specimens cited in Gray's paper on Fendler's plants bear testimony to his presence in New Mexico. The localities named are as follows:

". . . flowers of this species, sent me by Dr. Engelmann from a specimen gathered on the Upper Canadian in April, 1848, by Mr. Gordon . . ." (p. 40)

". . . Raton Mountains, April, 1848, *Mr. A. Gordon* . . ." (p. 100, *fn.*)

". . . It is also in Mr. Gordon's collection, from towards the sources of the Canadian." (p. 101, *fn.*)

". . . Mr. Gordon also gathered, on the sources of the Canadian . . ." (p. 101, *fn.*)

". . . a flowering specimen . . . gathered by Mr. Gordon, near the sources of the Canadian . . ." (p. 116, *fn.*)

In his *Rocky Mountain Naturalists* Joseph Ewan refers to collections made by Gordon in the "Valley of Platte River on the east side of Rocky Mts.," which would be in Ewan's "present Nebraska." And he adds that he ". . . botanized in present Nebraska and eastern Colo[rado] at least as far south as the 'Raton Mountains,' which appear on the tickets as 'Ratoon Mts.,' evidently following the old route passed over by W. H. Emory . . . and others. His Raton coll[ection]s made in April, 1848, are preserved at K[ew] and at G[ray] H[erbarium]. Occasional coll[ection]s at G[ray] H[erbarium] carry collection numbers (e.g., *Astragalus missouriensis,* no. 21), which may have been added by W. J. Hooker when sent to Asa Gray from Kew. Some coll[ection]s are labelled, 'Valley of Platte River on the east side of Rocky Mts.' . . ."

It seems to me probable that the specimens, undated, from the Platte River and Nebraska were collected in *1843* when Gordon traveled with the Stewart party over the Oregon Trail—across Nebraska and Wyoming, through South Pass and so on. And that the specimens dated *1848* were made on another journey which, like Emory's in 1846, was over the more southern Santa Fe Trail from Fort Leavenworth, Kansas, to Bent's Fort, Colorado, and thence by way of Raton Pass into New Mexico where both men soon encountered the upper waters of the Canadian River. But Gordon's movements—both immediately before and after "April, 1848"—go unrecorded.

His collections are mentioned in a number of publications not specifically concerned with his work. I have cited Gray's references to certain ones from New Mexico. Hooker, as planned, refers to some in his "Catalogue" of Geyer's plants; and Nuttall mentions one at least (from the Rocky Mountains) in his paper on Gambel's collections.

Ewan ends his paragraph on Gordon with this query: "Can this be the same Alexander Gordon, New Orleans representative of a Liverpool cotton milling factor and son of Major William Gordon of Natchez, who became J. J. Audubon's brother-in-law?"

I feel sure, relying upon dates alone, that our impecunious Gordon was a different person. In S. C. Arthur's biography of Audubon we read that the ornithologist, on arrival at New Orleans in 1821, had inquired the way to the ". . . office of Gordon, Grant & Co., he was in search of Alexander Gordon, a cotton commission factor, who three years later [which must have been in 1824] became his brother-in-law by marrying Ann, the youngest of the four attractive Bakewell girls." If we accept the fact that our Alexander Gordon was born in 1813, as Ewan states, he could scarcely have been of marriageable age in 1824! The name "Alexander Gordon" has a ring of distinction, but I am inclined to believe that it is not an unusual one in the British Isles.

Through the kindness of Sir Edward Salisbury, microfilms of Gordon's letters to Hooker were sent to the Arnold Arboretum. They begin when the collector was in London, after he had returned from his trip to the Rockies with Sir William Drummond Stewart's party. Portions of these letters, written in 1844, are of some interest to the over-all picture of botanical collecting. For Gordon had the courage, or temerity, to protest what he considered was to be an inadequate recompense for his future services. The first letter is dated August 17, 1844:

"I hope Sir William will pardon me making one remark on the low price offered per package at the Royal Botanic Gardens. I must confess it appears to me a degree of penury unworthy of a great Nation towards a class of men who devote their lives to enrich our Gardens with the Gems of other climes and under such a variety of complicated obstructions. But it is only a Collector who can appreciate the difficulties, the dangers, the vicissitudes, and the privations which such men have to encounter. If my memory serves me the Collector who accompanied us to the Rocky Mountains (Mr Geyer) at the recommendation of Dr Engleman informed me that the Kew price was one *Shilling* pr packet.—Apropos to this subject. I have taken the liberty of perusing your kind letter of introduction of me to Dr. E . There you seem to think the Collector he had spoken of to you had failed in making a collection. It must have escaped your memory that I informed you, when I had the honour of your visit in Munster Street that Dr Englemans Collector accompanied us to the Mountains—&

then left us with some Jesuit Priests for the Nation of the Flat Head Indians—was to remain there some time & to proceed from thence to Fort Vancouver   [He] did not purpose returning untill 1845–6. Should Mr Geyer ever reach you, I am certain from his unwearied zeal, his general Knowledge of plants, & his admirable tact in displaying their respective parts that a rich treat is in store for you. Should he ever come to Kew you would particularly oblige me by giving him my kind regards. I am perfectly sensible of the great advantages which I should derive in the disposal of my plants &c by your occasionally noticing them in your Botanical work—& I can only say I trust you will ever find me deserving & grateful . . .”

By August 19th, when Gordon wrote again, he had evidently received a prompt reply:

“Sir William I am this moment honoured with yours of date yesterday & I fear some of my expressions in my last were couched in too strong terms but I can assure you nothing was further from my intention than to give the least offense. If I did so I deeply regret it. However it has been attended with one good effect—viz eliciting a thorough understanding respecting the wide latitude you intend for such collections of seeds as I may send to the Royal Gardens, & which ought & shall have its due effect with me when contrasted with a collection of Select & Showey plants. I therefore beg leave distinctly to state that a General collection including grapes, Euphorbias, &c &c I would put up at 3 £ per 100 in double papers (which I consider highly advisable)—Select collections from 5 to 10 £ per 100 according to their merits & Scarcity, Roots of plants carefully packed from 5 £ to 10 £ per 100—unless a number of . . .”

At this point the microfilm breaks off—a page, or perhaps merely a line, is lacking, and one is left in the dark about the compensation! The Gordon ultimatum ends with the statement that he would supply “Dried specimens at 2 £ per 100 species, carefully dried, put up & numbered.” Finally, “I am well aware that this scale of prices will scarcely more than remunerate & I think any one desirous of adding to their collection would not object to the prices wanted.”

Whether Gordon won a victory, suffered a defeat, or stood just where he had been before making his protest, I do not know. Perhaps he was glad to have spoken his mind, but, very possibly, he regretted having done so. The very few collectors of my period who protested to their sponsors what they considered unfair treatment seem, thereafter, to have been somewhat subdued in spirit!

Ewan records that three species were named in his honor: *Eriogonum gordoni* Bentham, *Philadelphus gordoni* Lindley, and *Pentstemon gordoni* Hooker.

In response to an inquiry regarding the Gordon collections at Kew, Sir Edward Salisbury wrote (December 4, 1952) that the Herbarium possessed no list of his specimens, but mentioned the “deeply interesting letters to W. J. Hooker” from which I have quoted.

Two paragraphs written by U. P. Hedrick come as something of a surprise! For Alexander Gordon is mentioned as both a botanical collector and as a “writer of some note, who lived at Baton Rouge, Louisiana." Hedrick quotes from his description of a garden north of New Orleans, which was written in 1849 and which displays some familiarity with landscape architectural styles, both French and English. It would be pleasant to believe that the problems of the Gordon of this chapter found their solution in a literary career!

ARTHUR, STANLEY CLISBY. 1937. Audubon. An intimate life ot the American woodsman. New Orleans.

BRITTEN, JAMES & BOULGER, GEORGE SIMONDS. 1931. A biographical index of deceased British and Irish botanists. ed. 2. London. 62, 125.

EWAN, JOSEPH. 1950. Rocky Mountain naturalists. Denver. 217.

GORDON, ALEXANDER. [His letters to William Jackson Hooker are at Kew. A microfilm and photostats of these letters are in the Arnold Arboretum.]

GRAY, ASA. 1849. Plantae Fendlerianae Novi-Mexicanae: an account of a collection of plants made chiefly in the vicinity of Santa Fé, New Mexico, by Augustus Fendler; with descriptions of the new species, critical remarks, and characters of other undescribed or little known plants from surrounding regions. By Asa Gray, M.D. *Mem. Am. Acad. Arts Sci.* new ser. 4: 1-116.

HEDRICK, ULYSSES PRENTISS. 1950. A history of horticulture in America to 1860. New York. 355, 356.

HOOKER, WILLIAM JACKSON. 1845. Proposed botanical journey of Mr. Alexander Gordon, to the mountains of Texas, &c. *London Jour. Bot.* 4: 492-496.

—— 1847. Catalogue of Mr. Geyer's collection of plants gathered in the upper Missouri, the Oregon territory, and the intervening portion of the Rocky Mountains. *London Jour. Bot.* 6: 209, 212, 215, 222, 224, 240.

NUTTALL, THOMAS. 1848. Descriptions of plants collected by William Gambel, M.D., in the Rocky Mountains and upper California. *Jour. Acad. Nat. Sci. Phila.* new ser. 1: 149-189.

—— 1850. Description of plants collected by Mr. William Gambel in the Rocky Mountains and upper California. *Proc. Acad. Nat. Sci. Phila.* 4: 7-26.

CHAPTER XL

INSPIRED BY A VISIT FROM THE BOTANIST GEYER,

THE MISSIONARY SPALDING COLLECTS PLANTS FOR

ASA GRAY

O N Christmas Day of 1843 the German botanist Karl Andreas Geyer arrived at the Chamokane Mission conducted by the Reverend Elkanah Walker and the Reverend Cushing Eells. It was situated on Chamokane Creek,[1] in present Stevens County, Washington. In June of 1844 he moved on to the Lapwai Mission in charge of the Reverend Henry Harmon Spalding and his wife and situated in the triangle formed by the junction of Lapwai and Clearwater rivers, in Nez Percez County, northwestern Idaho.

Geyer's "Notes on the vegetation and general character of the Missouri and Oregon Territories, made during a botanical journey from the state of Missouri, across the South-pass to the Rocky Mountains, to the Pacific, during the years 1843 and 1844" was published in *The London Journal of Botany* in 1845 and 1846; the author's "Preliminary Remarks" refer to having visited with Spalding ". . . another new field, the Highlands of the Nez-Perçez Indians, where he accompanied me on my excursions and also afforded facilities to investigate the flowery Koos Kooskee [or Clearwater] valley over again, where previous botanists had but cursorily passed . . ."

Geyer, more prone I believe to criticism than to praise, was most commendatory of the missionary work being done by the Spaldings; and he must have been to some extent responsible for arousing their interest in plant collecting. Clifford Merrill Drury's biography of Spalding relates:

"Spalding's interests were multitudinous. He was able to turn his hands to many tasks and there be efficient. He was preacher, teacher, doctor, farmer, horticulturist, mechanic, printer, lumberman, weaver, miller, carpenter, musician, translator, and author by turn and as occasion demanded. He kept weather reports with such faithfulness and accuracy that the members of the Wilkes expedition saw fit to include them in their report.[2] Among his interests was botany . . .

1. On modern maps the name appears as "Chamokane" or sometimes "Tshimakain."

The volume *Washington* ("American Guide Series") mentions that the site of the "Tshimakain Mission (pronounced Shim-ik-in)" is marked by a monument to the two missionaries Walker and Eells, who had come there in 1838: "The mission was situated on Chamokane Creek. ('Chamokane' is another form of the Indian word meaning 'plain of springs') . . ." (p. 442)

2. A contingent of the United States Exploring Expedition, in charge of Lieutenant R. E. Johnson and including the botanist Charles Pickering and the "Horticulturist" William Dunlop Brackenridge, had been sent from Nisqually on an inland excursion. On June 25, 1841, they had arrived at "Mr. Spauldings Station," remaining in that vicinity until the next day only.

"In June, 1844, a German botanist by the name of Karl Andreas Geyer spent a few weeks with Spalding, gathering specimens for a collection which was later sent to London. He interested Spalding in botany with the result that Spalding made a collection of several hundred varieties and sent them to Greene . . ."

This was the Reverend David Greene, Secretary of the American Board of Commissioners for Foreign Missions—commonly abbreviated to A. B. C. F. M. Drury states that he was "secretary in charge of Indian missions." Although Drury gives a brief account of Spalding's plant collections, a fuller one is found in a short article by F. Orin Oliphant which was published in *The Washington Historical Quarterly* in 1934: "The botanical labors of the Reverend Henry H. Spalding." The letters upon which Oliphant based his information insofar as Spalding's connection with botany is concerned are now deposited by the American Board in the Houghton Library of Harvard University and through the courtesy of Miss Carolyn Jakeman I was permitted to examine them and make transcriptions. My thanks are due the American Board for permission to publish them.

Mail moved slowly from east to west and vice versa. Spalding first mentions his interest in plants in a letter of June, 1846; but not until March of 1849 did instructions on the preparation of specimens, and the paper in which to place them, start on their way to the missionary. What is known of the man's collecting activities is contained in an exchange of letters (those examined by Oliphant) and I transcribe them chronologically.

The first was addressed by Spalding "To the Rev. David Greene Sect. of A. B. C. F. M." and is dated from "Clear Water, June 6, 1846.":

"My dear Sir: I know not what will be your feelings on learning the fact that I have taken tim[e] to make a collection of Flowers & Plants, but judge they will be no other than favorable. In fact I bestowed but little time directly upon the effort, but made the collections almost entirely when traversing the Plains, the vallies & hills looking after my cows, horses &c or as my duties called me to different stations of the Mission. Mrs Spalding did most of the drying in papers.[3] We found it a most pleasing relief to our monotonous labors in our lonely situation. And when examined by a Botanist there may be found some new plants, in fact, although I am not acquainted with the science & therefore can not judge, yet I confidently believe very many of this collection will be found to be new for several reasons; first no Botanist has ever spent a whole season in this vicinity or even in this country & therefore could not collect the flow[er]s which were not in existence at the time of his travels . . . But I found a new set of flowers showing themselves almost every week & many of very short duration, consequently a Botanist who might make collections in June only[4] would fail of the March & April flowers & could not obtain those which do not show themselves till Aug[ust].

3. Piper states: "According to the testimony of his son, the late H. H. Spalding, the specimens were largely gathered by his mother."

4. Geyer had been there in June, as noted elsewhere.

Again after visiting the stations of Cimakain, Waiilatpu, as also Walla Walla & several parts of the Nez Perces country & traversing almost every mile square in the vicinity of this station, I found this to be far the richest field for flowers also that very many of them are confined to comparatively small beds consequently a Botanist passing once or twice through a district would fail could not fall in with all the flowers or have the advantage of one stationary & going out almost every day in different directions. A German Botanist by the name of Geyer was here a fewe weeks in June in the year /44 but flowers had disappeared & besides he merely traversed this valley & did not visit the numerous vallies & plains around all of which afford new flowers. Mr. Geyer sent his collection to London. The Renowned Douglass did not . . . come as far up as this.[5] Mr. Breckenridge[6] of the American Exploring Squadron passed this place in /40 [actually in 1841] but too late in the season & too rapidly to make many collections.

"I am no botanist & therefore have made no attempt to describe the flowers except the date of collecting them & their locality. Care should be taken not to disarrange the short notes placed upon each parcel collected at the same date & place. My numerous labors and cares prevented me from besto[w]ing as much labor upon [them] as they justly required, in fact I could not command time to put them up last fall & now I have been compelled to put them up in to[o] much haste to arrange them properly. I send them to you with the expectation that you will dispose of them as may be thought best. Should they arrive uninjured they will be worth $5.00 or $6.00 a hundred i.e. the Botanical Gardens in London offer that price for flowers from this country. I know not how many kinds there are in this collection or how many specimens of each kind of some more & of some less according as they were easily obtained or easily dried. When sold e[a]ch specimen cou[n]ts one & there may be some instances of 50 specimens of one flower which of course counts 50.[7] Should you think best to sell the whole or a part, the profit of course belongs to the Board. I have taken the liberty to send a full set to the National Institute, City of Washington.[8] Should you think best to present a set of specimens to some institution or individual of course you are at liberty as I am a servant of the Board, but the Box should be overlooked & the Plants arranged by a Botanist. This letter although addressed to yourself should accompany the flowers.

"Should any Institution or Botanist desire me to make an other collection they will please give me some directions how to put up the plants, & send in a shall [? small] Box a specimen of the flowers required where I have not given the Native name as many of this collection of course will be found to be common.

5. David Douglas had been at the confluence of Snake and Clearwater rivers from July 24-30, 1826, and had then made a trip into the Craig Mountains. If he did not get "as far up" as the Lapwai Mission, he was certainly no great distance from it.

6. William Dunlop Brackenridge.

7. In all these somewhat detailed financial directions, I sense the advice of Geyer.

8. Spalding, in 1846, seems to have been the first of my plant collectors to send, of his own accord, a collection of plants to augment the government collections. He must have heard of the "National Institute" (later the Smithsonian) from Brackenridge or from Pickering.

"I made a good collection of seeds but deem it useless to send them by ship as all the seeds from home by sea failed to germinate & I conclude it will be so with seeds from this country by sea

"Last year there was a very luxuriant growth of plants in the early part of season & I gathered many flowers which have not shown themselves this season.

"With best wishes I am &c   H. H. Spalding" [9]

Twenty-two months later, or in the late winter of 1848, we learn that Spalding's letter and collection were in Boston. The news appears in a letter dated "(March 18, 1848.) Saturday, Cambridge," which Asa Gray wrote to the "Rev. Dr. Greene Missionary House Pemberton Square" in Boston: [1]

"I have to day your favor of the 16th—and, as I was coming to Boston hoped to see the plants collected by Mr. Spaulding, but was too late.

"I shall be in Boston again on Monday evening, but shall not be able to be in during business hours until the end of next week or near it.

"If you choose to send the box up to Mr. C. G. Loring's 8, Ashburton Place, addressed to me, I will look over the plants and report on them: or if there is no haste I will call as soon as I am in town at the proper hours.

"If they are pretty good specimens I shall be able to find a purchaser for them."

Dr. Greene's reply was prompt, dated "Missy House Boston 15th April. 1848." and addressed to Gray at Harvard College. [2] It expressed his regret to have missed Gray's visit of the day before and his pleasure ". . . that the plants and flowers sent on by Mr. Spalding perhaps have some interest and value to you . . ." and he forwarded ". . . copy of Mr. S.'s letter relating to the box and its contents, the statements in which may be of some use to you." So far so good! About *eleven months later*, or on March 1, 1849, Gray wrote to the Reverend Dr. Rufus Anderson, Secretary of the American Board, at Missionary House: [3]

"A few days ago I had the pleasure of handing you $35. received for two sets of specimens which I made from the collection of Mr. Spalding, at Clear Water, Oregon. I hope still to realise as much more from them, which I shall duly pay over to the Board. I could have turned these specimens to much better account, and they would have been intrinsically much more valuable had they been better prepared, and packed with more paper. As I trust Mr. Spalding will have opportunity, and be disposed, to make further collections (which I shall be happy to arrange, name, and dispose of for his benefit [or] of that of the Board) allow me to express to him a lively interest which I feel in the matter, and to say that attention to the following points will enable him to make excellent and durable botanical specimens.

"1. The thick roots and bulbs, &c, which abound in that region, should be thinned,

9. The letter is filed in volume 248, no. 133 of A. B. C. F. M., *Letters and Papers.*

1. Filed in volume 270, no. 293, of A. B. C. F. M., *Miscellaneous Domestic Letters 1846-1860.*

2. Filed in A. B. C. F. M., *Domestic Letter Book,* no. 29, p. 359.

3. Filed in volume 270, no. 294, of A. B. C. F. M., *Letters and Papers (Miscellaneous Domestic Letters, 1846–1860).*

or one half cut away with a knife, when the specimen is pressed.—So of thick and hard fruits, which, like the roots, it is very desirable to have.

"2. The specimens should be dried as quickly as possible between *numerous* thicknesses of soft, bibulous paper, and under *strong pressure,* as strong as can be applied without crushing.[4] Then they dry more rapidly, keep their color pretty well, and are not fragile; as they otherwise are. While drying the specimens should be changed into dry papers every day. When dry put away in fresh dry paper,—not crowded layer over layer—under gentle pressure.

"I hope you will send him a supply of cheap paper, by first vessel. If you wish I will select and purchase a proper article for the purpose.

"Mr. Spalding should also know, that I am able to make a small annual appropriation from the funds of the Botanic Garden for seeds, bulbs, & roots. There will now be overland communication by which packages of seeds can readily be sent home, in autumn. They should be gathered only when fully ripe, in a dry day, dried in the sun for a few hours, done up in paper, and kept in a very dry place until sent.

"When the means of communication (over the Isthmus) will allow, I am anxious that Mr. S. should gather, in autumn (or any time after the foliage decays) fleshy roots, bulbs, & tubers,—especially of the numerous plants used for food by the natives— pack closely and compactly in *dry moss,* in a close box, and send to me for cultivation. I shall be able to make a proper remuneration.

"Will you send Mr. Spalding the accompanying copy of a memoir, just printed,[5] which contains a few notices of interesting plants from his collection—as far as yet studied—and oblige . . ."

On March 13, 1849, the Reverend Selah B. Treat, who, according to Oliphant, ". . . had succeeded the Reverend David Greene as secretary to the American Board . . .", wrote[6] to Spalding and sent him a copy of Gray's long letter to Dr. Rufus Anderson noting that "In accordance with his suggestion, we have purchased four reams of paper, which will be sent to you *via* the Sandwich Island. The Memoir, mentioned in the last paragraph,[7] will also be sent by the first favorable opportunity . . ."

The letters concerned with Spalding's collecting of plants end at this point—certainly there are no more in the collection of the American Board. Although nothing startling took place in the life of Boston and of Cambridge at this period, much of a disturbing character was transpiring in the far northwest. Oliphant writes thus:

4. Just what Gray meant by "without crushing" I do not know, for, unless flattened I have found that plants wrinkle in drying.

5. This was, undoubtedly Gray's "Plantæ Fendlerianæ Novi-Mexicanæ," published in the *Memoirs of the American Academy* (new ser., 4: 1-116) in 1849. In this paper Gray mentions many specimens collected by other than Fendler and from regions far removed from New Mexico. Many of Geyer's specimens were included and some fifteen of Spalding's—of the last none are referred to as new species, however.

6. Filed as no. 12, p. 71, of A. B. C. F. M., *Indian Letter Book.*

7. Only the first paragraph of the Reverend Treat's letter refers to Spalding's collections.

". . . In the meantime evil days had come upon the Oregon Mission; the massacre at Waiilatpu[8] occurred on November 29, 1847. Months before Professor Gray had written the letter quoted above, all the stations of the mission had been abandoned, and Spalding, Eells, and Walker, together with their families, had sought refuge in Lower Oregon. It is probable that the massacre marked the end of Mr. Spalding's botanical labors in the Oregon country."

When, and if ever, Spalding received Gray's instructions, I do not know. He would have found them valuable and have been gratified to know—from Gray's paper— that his efforts on behalf of botany were already in print.

Piper records that, in the vicinity of the mouth of the Clearwater, ". . . Spalding collected a good many plants which are in the Gray Herbarium. Most of them are labeled 'Clearwater, Oregon,' but inasmuch as a number of them have not since been found near Lapwai it is not improbable that they were collected elsewhere. Spalding traveled quite extensively in the course of his labors, and doubtless gathered some of his specimens at other places than Clearwater, as, indeed, some few of the labels indicate. His notes on the Indian food plants are most interesting and often quite detailed . . ."

Miss Ruth Sanderson, Librarian of the Gray Herbarium, very kindly attempted to find among Asa Gray's papers some record of the plants which he received from Spalding, but was successful only in discovering an entry in a notebook (labeled "Dr. Gray's account books 1838–1850" and containing, in addition to monetary matters, many notes of a miscellaneous character). It was dated June 1, 1850, and was evidently a reminder: "Set of Spalding Oregon—with tickets—(to name at Ge[neva]."

8. When the missionary Marcus Whitman, his wife Narcissa and eleven others were killed by a band of Cayuse Indians. Whitman College, Oregon, was named in his honor.

Drury, Clifford Merrill. 1936. Pioneer of Old Oregon: Henry Harmon Spalding. Caldwell, Idaho. 312-314.

Oliphant, J. Orin. 1934. The botanical labors of the Reverend Henry H. Spalding. *Washington Hist. Quart.* 25: 93-102.

Piper, Charles Vancouver. 1906. Flora of the state of Washington. *Contr. U. S. Nat. Herb.* 11: 16.

CHAPTER XLI

AUDUBON TRAVELS UP THE MISSOURI RIVER AS FAR

AS THE MOUTH OF THE YELLOWSTONE

ALTHOUGH Audubon's fame is not based upon his contributions to botanical science I believe, nonetheless, that he deserves inclusion in this story, and for a number of reasons: he did bestow a name upon one beautiful small tree of our Pacific coast; the plants illustrated in his plates of birds and mammals are so accurately depicted that they merit comment; and he was in contact with many of the men whose names appear more than once in my narrative. By 1843 when Audubon traveled up the Missouri River—less than forty years after Lewis and Clark made their slow way up that stream—the trip was "in a good Steamer" and those making the journey were in touch with white men, and the doings of white men, during the entire period of their excursion.

Audubon had but slight training as an artist although he studied for a short time under Jacques Louis David, painter to the French court at the time of Napoleon. Of David's influence upon Audubon's style, one of his biographers, Francis Robert Herrick, has this to say:

". . . Buchanan remarked that the mannerism of David could 'still be traced in certain pedantries discernible in Audubon's style of drawing,' which is a fancy without any basis in fact. If it could be shown that drawing from the casts of antique statues could develop mannerisms in the careful delineation of birds and mammals, it would still appear that Audubon's style was really formed at a later period.

"This brief episode, which at most could have lasted but a few months, represented all the formal instruction which Audubon ever received[1] in drawing, although he enjoyed some private tuition at a much later day. As to the sciences now embraced in biology, that is, zoölogy and botany, which would have been most useful to him, the score was blank; even books on any of these subjects were rare in America at the beginning of the nineteenth century."

By far the greater number of the trees, shrubs, vines and herbaceous plants which are shown in the Audubon plates are recognizable to persons familiar with the plants of our eastern, western, northern and southern states, and even to the humblest representatives. Writes Herrick:

"Audubon began in the usual way, by representing his birds in profile, and often on

---

1. Given that Audubon's training was slight, he must have been naturally gifted as an artist for, apart from his skill in depicting the fauna, his portraits of his two young sons—Victor Gifford and John Woodhouse Audubon—which were painted in 1823 and are reproduced in Maria Rebecca Audubon's *Audubon and his journals,* have a professional and delightful quality. Audubon's younger son in particular seems to have inherited some of his father's skill as a painter as well as his interest in natural history.

a single perch, but gradually introduced accessories which eventually became such an important part of his plan that, after 1822, his plates took on more the character of balanced pictures, literally teeming with the characteristic fruits and flowers of America, as well as with insects and animals of every sort, suggestive of the food and surroundings of his subjects, not to speak of American landscapes drawn from many parts of the country."

After describing the controversy aroused by Audubon's plate 21, which included a much criticized rattlesnake—subsequently proven to be as depicted—Herrick had this to say:

"As I have not hesitated to speak of Audubon's real or supposed mistakes, I will give another and more striking instance of his tardy vindication. In his plate of the American Swan (No. ccccxi), which was published in 1838, there is represented a yellow water lily, under the name of *Nymphaea lutea* [=*flava* on the plate examined]. Since this lily was then quite unknown to botanists, it was ignored and treated as a fable, or as an extravagant vagary of the naturalist's imagination, until the summer of 1876, when it was rediscovered in Florida by Mrs. Mary Treat.[2] Audubon's lost lily was identified and acknowledged by Professor Asa Gray, the botanist, who, with poetic justice, proposed to rename it after the discredited enthusiast, in view of the fact that it had been originally discovered and faithfully depicted by him a generation before."

This water lily, of the family *Nymphaeaceae,* has never borne the name suggested by Gray although more than one has been bestowed upon it: *Castalia mexicana* Coulter, *Nymphaea flava* Leitner, and *Nymphaea mexicana* Zuccarini (by which it is now generally known). Stanley Clisby Arthur, recent Audubon biographer, quotes letters written by the ornithologist which indicate that he had learned of collections of birds and flowers made in the Florida everglades by a young German botanist and one was, presumably, the water lily which he wished to name for its discoverer: " '. . . I have represented a New Nymphea, which [if] unpublished by him, I should like, in my letter press to name after Docr Leitner's name; "Nymphea Leitneria!" . . .' " In spite of this wish Audubon's plate 411[3] bears this legend in the elephant folio edition: "Com-

2. For a time Mrs. Treat seems to have collected plants for C. S. Sargent and, during the years when he was editor of *Garden & Forest* (1888–1897) was a frequent contributor to that periodical, signing her articles from Vineland, New Jersey.

3. Desirous of seeing this particular plate, I was assisted in doing so by Mr. David M. K. McKibbin of the Boston Athenæum staff. "Assisted" is accurate for, as was to be expected, it was contained in the bottom volume of the well-named *elephant* folios! How close seems more than a century ago when one reads of these very Athenæum plates in a letter written Audubon by Edward Everett on May 19, 1831; it is quoted by Herrick: " 'The portions of your work, which arrived at Washington before I left it, were publicly exhibited in the library, and attracted great attention and unqualified admiration. The same is true of the copy received by the Boston Athenæum. The plates were especially exhibited in the great hall of the Athenæum, to the entire satisfaction and delight of those who saw them.' "

Such matters were carefully carried out in those days! The Boston Athenæum possesses a letter written by one of its Proprietors, Mr. Thomas Handasyd Perkins, to the Librarian, Dr. Bass: "Mr Elliot agrees

mon American Swan, Cygnus Americanus, Sharpless. Nymphea flava—Leitner." The author of the name was Edward F. Leitner, for whom the genus *Leitneria,* or cork-wood,[4] was named. According to Arthur, he was an able physician and botanist who met with a tragic death, having been killed and scalped by Indians in the Seminole War, on January 15, 1838.

Audubon did, however, bestow the name *Cornus Nuttallii* upon the flowering dog-wood of the west coast of North America. The plant was illustrated in the elephant folio edition of *The birds of America,* plate 367 which bore the legend "Band-tailed Pigeon. 1. Male. 2. Female. Columba Fasciata, Say. Plant Nuttall Cornel. Cornus Nuttallii. Aud." Audubon's *Ornithological biography* relates:

"In my plate are represented two adult birds, placed on a branch of a superb species of Dogwood, discovered by my learned friend, Thomas Nuttall, Esq., when on his march toward the shores of the Pacific Ocean, and which I have graced with his name! The beautiful drawing of this branch was executed by Miss Martin, the amiable and accomplished sister [sister-in-law] of my friend Dr. Bachman [the Rev. John Bach-man]. Seeds of this new species of *Cornus* were sent by me to Lord Ravensworth and have germinated, so that this beautiful production of the rich valley of the Columbia River may now be seen in the vicinity of London, and in the grounds of the nobleman just mentioned, near Newcastle-upon-Tyne."

Audubon then quotes "an interesting notice" with which he had been "favoured" by Nuttall and which mentions the feeding-habits of the band-tailed pigeon. It refers to ". . . swarming flocks feeding on the berries of the elder tree, those of the Great Cornel *(Cornus Nuttallii),* or, before the ripening of berries, on the seed-germs of the

with the other Gentlemen on the subject of exhibiting the Engravings in the Picture Room You will there-fore please to have the Rails put up on the East & South side—I think it will be well to have some Cards struck off, and sent to each Proprietor—as follows—Admit ——— Proprietor in the Boston Athenæum Corporation to the Picture Gallery, with his Friends, to see the splendid engravings of Audubon, belong-ing to the Institution. Signed Seth Bass   Open from 10 untill 1 Oclock   Show this notice to some of the Gentlemen and if they approve, get as many Cards (and a few to spare) as there are proprietors—Make what alterations you may think best after consulting Mr Jackes or other Gentlemen—Yrs Obed. Sir—T H Perkins   I would not send the tickets untill the Engravings are tacked to the Walls—You will want 4 tacks for each plate—those which are tinned will be the best, and they may be three quarters or half an inch long and drove about half the length in the wall—when the Room is ready I will find the Engravings and we can determine the heighth from the floor which will suit best—THP"

I suggest—in connection with size of books and plates—that the reader look at figure 51 (p. 215), in *The art of botanical illustration* by Wilfrid Blunt (1950), which depicts "The Librarian's Nightmare. Vignette by G. Cruikshank from J. Bateman, *The Orchidaceae of Mexico and Guatemala,* 1837-41," a work which Mr. Blunt characterizes as ". . . probably the finest, and certainly the largest, botanical book ever produced with lithographic plates . . ."

4. The name "corkwood" suggests one character of the tree which Sargent refers to thus in *The Silva of North America* (7: 110, *fn.* 2): "The wood of no other North American tree that has been examined is as light as that of Leitneria; the wood which approaches it nearest in lightness is that of the Florida *Ficus aurea* . . ." Despite its southern habitat, this tree was once growing happily in the Arnold Arboretum. The generic name is easy to remember if one is not superior to puns!

young pods of the Balsam poplar . . . They remain on the lower part of the Columbia nearly the whole year, late in the season (October and November) feeding mostly on the berries of the Tree Cornel . . ."

The name *Cornus Nuttallii* was taken up by Torrey and Gray in the *Flora of North America* (1: 649, 1840) and by Nuttall in *The North American Sylva* (3: 51, t. 97, 1852) and for more than a century has been acceptable to the botanists.[5]

Sargent mentions in *The Silva* (5: 69, 1895) that "Cornus Nuttallii was discovered on the banks of the Columbia River by David Douglas in 1825 or 1826; it was first mistaken for the Flowering Dogwood of the east, and was not distinguished from that species until several years later by Thomas Nuttall in his transcontinental journey."

Audubon's journals more than once record that he collected plants. I surmise that he did so not for botanical purposes as such, but in order that his creatures should appear in appropriate surroundings accurately depicted. And, such is the lifelike quality of the greater number, that it may be inferred that they were drawn from freshly gathered material rather than from the "dried specimens" with which the majority of our plant collectors were concerned. They have—if plants may have—the animated quality for which the man's birds are famous. We read, more than once, of Audubon's drawing plants when in the field.

In 1937 The Macmillan Company reproduced five hundred plates taken from Audubon's folio and first octavo editions of *The birds of America,* accompanied by "an introduction and descriptive text by William Vogt." For those to whom the orginal editions are not available this one-volume book serves a useful purpose and, moreover, is valuable to all persons for easy reference. Of the five hundred plates, nos. 1-435 are reproduced from the elephant folio edition, issued from 1827–1838 and, as Vogt states, ". . . generally found bound in four volumes . . ." Plates nos. 436-500 are reproduced from the octavo edition first published from 1840 to 1844. Vogt includes a "Transcript of the legends on the original plates . . ."

Examining these legends (for the 435 plates of the folio edition only) I find that volume one, plates 1-100, includes seventy identifications of plants (citing scientific name, "vulgo" or common name, or both); volume two, plates 101-200, includes seventy-five such identifications; volume three, plates 201-300, includes but one; and volume four, plates 301-435, but sixteen—making a total of one hundred and sixty-two identifications of plant species illustrated in four hundred and thirty-five plates. The obvious reason why the numbers of such identifications drop off noticeably in volumes three and four is that Audubon was illustrating water fowl and the backgrounds of his plates show mainly sea, rocks and distant landscapes. When, in volume four, he returns to land-birds, plants and their identifications begin again. I do not know who helped Audubon to name his plants; the spellings are often peculiar and,

5. In 1935 and again in 1948 changes in the generic name were suggested, but to date, and as far as I know, they have not been taken up with enthusiasm. In any event the specific name remains.

as is to be expected, many of the names are no longer used by the botanists. It might be an interesting task—and I am not aware that it has ever been attempted—to identify the plants in the complete series of plates, supplying the now-accepted names.

In 1837 Audubon spent about two months in Texas. This was the year, according to Samuel Wood Geiser, when ". . . the United States of America recognized the independence which the Texans had won the year before on the field of San Jacinto." In his journal of this trip (first edited by Buchanan in 1868) Audubon makes this entry on April 29: "John took a view of the rough village of Galveston, with the Lucida.[6] We found much company on board on our return to the vessel, among whom was a contractor[7] for beef for the army; he was from Connecticut, and has a family residing near the famous battle-ground of San Jacinto. He promised me some skulls of Mexicans, and some plants, for he is bumped with botanical bumps somewhere . . ."

Audubon's Texas journal has only this to tell of plant collecting; the author was on Galveston Island on May 2: ". . . We also found a few beautiful flowers, and among them one which Harris and I at once nicknamed the Texas daisy; and we gathered a number of their seeds, hoping to make them flourish elsewhere."

In 1843 Audubon traveled up the Missouri River by American Fur Company steamboat from St. Louis to Fort Union at the mouth of the Yellowstone River. The same trip had been made by Maximilian, Prince of Wied-Neuwied, in 1833. It took Audubon ". . . forty-eight days and seven hours from St. Louis." The party had started on ". . . April 25th, at noon; reaching Fort Union[8] June 12th at seven in the evening." Maximilian's trip had taken one month longer for he left St. Louis on April 10 and arrived at the post on June 24. In "The Missouri River Journals"[9] Audubon refers to the Prince's earlier visit:

"Our room was small, dark, and dirty, and crammed with our effects, Mr. Culbertson saw this, and told me that to-morrow he would remove us to a larger, quieter, and better one. I am glad to hear this, as it would have been difficult to draw, write, and work in; and yet this is the very room where the Prince of Neuwied resided for two months, with his secretary and bird-preserver . . . Harris and I made our beds up; Squires fixed some Buffalo robes, of which nine had been given us, on a long bedstead, never knowing it had been the couch of a foreign prince . . ."

Edwin T. Denig's "Description of Fort Union," included in the same book, tells that "In the upper story are at present located Mr. Audubon and his suite. Here from the pencils of Mr. Audubon and Mr. Sprague emanate the splendid paintings and

6. The camera lucida is mentioned more than once in the Audubon journals.

7. Geiser identifies him as ". . . John W. Moore, of Harrisburg, who was appointed to the position [of alcalde] Nov. 15, 1853. Frequent references to him are found in the Lamar and Austin papers . . ."

8. From his headquarters at this post, Audubon and his party worked in the region until August 16 when, by mackinaw boat, the return trip to St. Louis began; they were back at St. Louis on October 19.

9. Included in Maria R. Audubon's *Audubon and his journals . . .*

drawings of animals and plants, which are the admiration of all; the Indians regard them as marvellous, and almost to be worshipped."

"Mr. Sprague" was the gifted Isaac Sprague who was accompanying Audubon "as artist." His name is familiar to botanists; for Torrey and Gray employed him to illustrate many of their botanical papers. His name occurs frequently in my narrative. There are numerous references to Sprague in Asa Gray's letters edited by his wife and Mrs. Gray included the following short account of the man ". . . so long associated with Dr. Gray as illustrator of his works":

"Isaac Sprague was born at Hingham in 1811. He early showed a faculty for observation, and a gift for painting birds and flowers from nature. His talent was discovered, and he was invited by Audubon in 1843 to join his expedition to Missouri, and to assist in making drawings and sketches. President, then Professor, Felton . . . knowing Dr. Gray was looking for someone for his scientific drawings, recommended Mr. Sprague, and he began with the illustrations for the Lowell lectures and the new edition of the 'Botanical Text-Book.' Dr. Gray was delighted with his gift for beauty, his accuracy, his quick appreciation of structure and his skill in making dissections. Mr. Sprague was from that time his chief and almost only, illustrator for his books, both educational and purely scientific.[1] Dr. Gray is said to have stated that Mr. Sprague had but one rival,—Riocreux,[2] and that he considered that draughtsman's classical drawings inferior to Mr. Sprague's."

On the Missouri River journey Sprague is most often mentioned as shooting birds. Audubon named one for him, ". . . a Lark, small and beautiful . . .", first mentioned on June 19.[3] "Sprague's Missouri Lark, Male" (plate 486) is shown against a pink-flower-

1. Interesting, as carrying on this Gray-Sprague association of fifty years earlier, was a small bequest made to Harvard University in 1938: "Residuary bequest of Helen O. Sprague 'to be known as the Isaac Sprague Memorial Fund, in memory of my father-in-law Isaac Sprague, the income only to be used for the Gray Herbarium.' " (*See: Endowment Funds of Harvard University June 30, 1948*. Cambridge. 1948. p. 316.)

2. In his preface to *The Silva* Sargent mentions that the plates in that work were engraved ". . . under the general direction of Monsieur A[lfred] Riocreux . . ." whom he characterizes as ". . . the most distinguished European botanical artist." *See also* Wilfred Blunt, *The art of botanical illustration*. London. 1950.

3. According to Elliott Coues: "This is the first intimation we have of the discovery of the Missouri Titlark, which Audubon dedicated to Mr. Sprague under the name of *Alauda spragueii*, B. of Am. vii., p. 334, pl. 486. It is now well known as *Anthus (Neocorys) spragueii* . . ."

Sprague's own diary of this trip (mentioned later in this chapter) indicates that he was a naturalist as well as a botanical artist; and it has this to tell of his discovery of the bird and of its habits as he observed them in the field: "June 19 . . . Messrs Harris and Bell went out shooting and brought in . . . a small species of titlark which is probably new . . . June 23 . . . While I watched one of the new titlarks for nearly an hour—it sailed around over my head high in the air—singing its simple notes at intervals of about 10 seconds the song itself occupying about 5 seconds while singing they remain nearly still moving their wings in a rapid manner like a little hawk—and in the intervals between they sweep around in an undulating manner closing the wings to the body, like a goldfinch. 3 of these titlarks killed to day . . . June 24. Found the nest of the titlark—and shot the female as she rose from it it was built on the ground, in a small cavity so that the top of the nest was even with the surface and slightly shaded by a small tuft of grass—the eggs five in number are pale brown thickly spotted with darker."

ed cactus, certainly of *Mammillaria* affiliation. We read more than once that Audubon and Sprague made drawings of cacti on this journey.

What one would like to know, of course, is whether any of the plants in the Audubon plates were actually drawn by Sprague. I have found no reference which attests to collaboration although—to me—it seems highly probable that such existed.[4]

In the Boston Athenæum is a small manuscript diary kept by Isaac Sprague while accompanying Audubon on the trip up the Missouri River which took them from St. Louis to the mouth of the Yellowstone River and back. This neatly penned little document was presented to the library by the artist's grandson, Isaac Sprague, Junior, a member of the staff. Unfortunately it offers no *proof* that Sprague drew the plants in the Audubon plates made at this period; indeed, the writer rarely mentions Audubon —and always as "Mr. A."—and refers only in a general way to their mutual activities as collectors and artists. The diary contains more references to plants than Audubon's own journal of the same trip and it might be possible—although no useful purpose would be served—to identify, by means of dates and Sprague's very slight descriptions, some of the plants which he mentions with those in the plates. It is an interesting little diary and it mentions many of the places in the Missouri River valley described in earlier years by men whose travels are narrated in my story. Sprague refers to the trip to Fort Union as ". . . the quickest trip that has yet been made . . ." Perhaps so, but steamboat captains have a propensity for telling passengers something unusual to remember! Mr. Sprague, the grandson, tells me that plants grown from seeds collected on this journey still live in the gardens of Wellesley Hills, Massachusetts, where his grandfather later lived and where he died.

The last chapter of Arthur's life of Audubon refers to the journey of 1843 and mentions Sir William Drummond Stewart with whose party several of the botanists men-

---

4. This very matter—credit rendered for collaboration—came up in connection with a young man, Joseph Robert Mason, who traveled with Audubon in the 1820's and claimed that he had drawn many of the plants, even portions of birds, depicted in the Audubon plates of that period. Mason's claims were aired by John Neal, editor of a Boston newspaper, *The New England Galaxy*, in 1835 (January 3, February 7, and April 18).

Audubon's biographer, S. C. Arthur, tells the story in considerable detail and notes that ". . . an examination of these original drawings discloses that Joseph Mason did draw many of the floral designs . . . but when the plates were engraved and issued there was no mention of the talented boy's part in making many of these bird plates real pictures . . ."

Richardson Wright revives the matter in his *Gardener's tribute* (1949) and notes that Mason's story, as reported by Neal, ". . . seemed a pretty tall claim." Wright comments: ". . . The biographer quoted here [Herrick] states that among the originals are two plates on which the pencil credit line 'Plant by J. Mason' is discernible—'simple studies of the blue yellow-backed warbler with red iris and the pine-creeping warbler with pine spray.' A later and better documented biographer, Stanley Orsby [= Clisby] Arthur, feels that Audubon failed to keep his promises to young Mason. He is believed to have been responsible for the botanical part of the following plates in the *Birds of America*: 2, 3, 8, 12, 14, 17, 21, 24, 32, 38, 40, 42, 47, 51, 63, 66, 122 and 126."

See also Peattie, D. C. 1940. *Audubon's America* . . . for this collaboration. The author states: ". . . In my opinion no botanical illustration ever done in America has come up to Mason's." (p. 135) I should consider Isaac Sprague no mean competitor!

tioned in my narrative traveled to the Rocky Mountains in that year. I quote from Arthur:

"It was the first day of April before Lord Stewart and his party reached St. Louis. The Britisher and his friends had started from New Orleans on the *J. M. White,* then the finest and fastest boat on the Mississippi river, but when only a dozen miles below St. Louis the *White* struck a ledge of rocks, called the Grand Chain, and in a few moments was sunk in deep water. Although the boat was a total loss all the passengers and most of the baggage were saved.

"Sir William thereupon changed his mind about penetrating the wilds via a steamboat, and decided to organize a land expedition. He and Audubon held conference, 'Sir Wm is so desirous that we should accompany him & party, that he offered me 5 Mules and a Waggon for ourselves; but they will not Change my plans!' Audubon informed his wife . . . A week later he reported, 'Sir William Stewart and his gang cannot however go off until about the 10th or 11th of May.—No one here can understand that Man, and I must say that in my opinion he is a very curious Character. I am told he would give a great deal that we Should Join him. If So why does he not offer some $10,000; who Knows but that in such a Case I might Venture to leap on a Mule's back and trot some 7 or 8 thousand Miles.' . . ."

Arthur tells us that Audubon stuck to his plans to travel by boat and includes an amusing comment taken from a letter to his wife Lucy:

" 'Sir Wm Stewart goes off in about a week to Independence with his 70 followers of all Sorts.—They start from thence when the grass is sufficient to feed their beasts, and take with them 16 days provisions only; then comes the tug of War.—Buffalo meat, when it can be got, and dried when not fresh.—Water very scarce at times &c. How many of the *Young Gents* will return before they have a sight of the Mountain is more than I can say?—He has done all he could to persuade me to Join the party, but it was no go. he promises *now* to let me have all the Animals I do not procure, since I have procured a German Youth, to preserve Bird & Beasts for *him!* This is the way with this World!"

Who this collector could have been is a question. It might have been Frederick G. J. Lüders who was with Stewart's party, but whether a "Youth" I do not know. Karl Andreas Geyer traveled with Stewart at this time, but he was then thirty-four years of age. Geyer refers to Audubon in a letter to William Jackson Hooker, dated August 13, 1845: "The celebrated Audibon went up the Missouri in 43 and up the Yellow Stone River also to collect the quateripeds of that region. He also intended to collect plants and as he, in my judgment has passed through the region of Bostonia,[5] I think he must have found very fine and new plants."

5. It is not clear to me what Geyer meant by "the region of Bostonia." The only town of that name which I have been able to find in any gazetteer is in San Diego County, California, not far northeast of the town of El Cajon, but far inland from the city of San Diego. Possibly it was a reference to Boston, expressed in Geyer's often-curious terminology.

Fond of reporting bits of gossip, Geyer wrote Hooker on May 16, 1845, that he had been unable to obtain passage up the Missouri River in the American Fur Company steamboat because the company wanted no witnesses to "their illegal dealing with intoxicating liquors," and added that "The venerable Audibon also had hard work to get a passage for a short distance to examine the quadrupeds of that region." That Audubon had any great difficulty in obtaining passage seems doubtful for, in a letter written February 10, 1843, to his friend Spencer F. Baird (it is quoted in Herrick) he comments:

" '. . . I wish you would assure your good mother that to go to Yellow Stone River, in a good Steamer, as passengers by the courteous offers of the President of the American Fur Company who himself will go along with us, that the difficulties that existed some 30 years ago in such undertakings are now rendered as Smooth and easy as it is to go to Carlisle [Pennsylvania] and return to N. Y. as many times as would make up the Sum in Miles of 3000 . . .' "

The government was trying to prevent the sale of alcohol to the Indians. On May 10 the steamboat in which Audubon was a passenger had its cargo examined and while this was being done the passengers, to their great joy, were allowed ashore. Wrote Elliott Coues: "The two precious hours of Audubon's visit were utilized by the clever captain in so arranging the cargo that no liquor should be found aboard . . ." Sprague's diary too casts a side light on alcoholic matters: on October 9 he tells of a walk through the forests along the Missouri and expresses his delight in the "profound stillness," broken only by the "chatter of a squirrel" or the "scream of the parrots." "But in the midst of this—and quite out of place I thought we found that curse of the red and white man—a distillery! . . ."

In writing of the botanist Beyrich, I quoted the account of his death which was written by the artist George Catlin, who was present at the time. Catlin is often mentioned by Audubon in his journal of the Missouri River journey and he evidently did not think highly of the painter's accuracy, nor agree with his glowing accounts of the Indians. Herrick comments that "Audubon's opinion of the Indian was modified considerably after having seen him in the western wilderness, and his confidence in George Catlin's descriptions was completely shattered . . ." Audubon made comments such as the following—they appear in his Missouri River journals published by Maria A. Audubon. On May 17:

"Ah! Mr. Catlin, I am now sorry to see and to read your accounts of the Indians *you* saw—how different they must have been from any that I have seen!" [June 6:] "The Mandan mud huts are very far from looking poetical, although Mr. Catlin has tried to render them so by placing them in regular rows, and all of the same size and form, which is by no means the case. But different travellers have different eyes!" [June 10:] "We saw many very curious cliffs, but not one answering the drawings engraved for Catlin's work." [June 11:] "We have seen much remarkably handsome scenery, but nothing at all comparing with Catlin's descriptions; his book must, after all, be altogether a humbug. Poor devil! I pity him from the bottom of my soul; had he studied, and kept up to the old French proverb that says, 'Bon renommé vaut mieux que ceinture doré,' he might have become an 'honest man'—the quintessence of God's works." [June 21:] "Buffaloes never scrape the snow with their feet, but with their noses, notwithstanding all that has been said to the contrary by Mr. Catlin." [July 22:] "When and where Mr. Catlin saw these Indians as he represented them, dressed in magnificent attire, with all sorts of extravagant accoutrements, is more than I can divine, or Mr. Culbertson tell me."

When Audubon was in the eastern United States and when he traveled in Europe, he visited many centers of science and of horticulture, and got in touch with persons from whom he might derive help, not only in selling subscriptions to his books, but in acquiring knowledge.

Thomas Nuttall had reached Boston on September 20, 1836, from his trip across the continent, to the Hawaiian Islands and to California—regions which Audubon longed to visit. In the *Life of Audubon* edited by his widow—the remarkable and devoted Lucy—we read that the two men met at once after Nuttall's return:

"September 21 . . . heard that our learned friend Thomas Nuttall had just returned from California. I sent Mr. Brewer[6] after him, and waited with impatience for a sight of the great traveller, whom we admired so much when we were in this fine city. In he came, Lucy, the very same Thomas Nuttall, and in a few minutes we discussed a considerable portion of his travels, adventures, and happy return to this land of happiness. He promised to bring me duplicates of all the species he had brought for the Academy at Philadelphia, and to breakfast with us to-morrow, and we parted as we have before, friends, bent on the promotion of the science we study.

"September 22. This has been a day of days with me; Nuttall breakfasted with us, and related much of his journey to the Pacific, and presented me with five new species of birds obtained by himself, and which are named after him . . ."

Audubon was unable, however, to acquire immediate possession of—or even access to—all the specimens of birds that Nuttall and John Kirk Townsend had collected on their journey and he resented this. The *Ornithological biography* records his grievances:

". . . then returned to Boston, and as fortune would have it, heard of the arrival of Thomas Nuttall, Esq., the well-known zoologist, botanist, and mineralogist, who had performed a journey over the Rocky Mountains to the Pacific Ocean, accompanied by our mutual friend John Kirk Townsend, Esq., M.D. Mr Nuttall generously gave me of his ornithological treasures all that was new, and inscribed in my journal the observations which he had made respecting the habits and distribution of all the new and rare species which were unknown to me . . .

6. The ornithologist Thomas Mayo Brewer.

"Dr. Townsend's collection was at Philadelphia; my desire to examine his specimens was extreme; and I therefore . . . hurried off to New York . . . [and thence to Philadelphia] . . .

"Having obtained access to the collection sent by Dr. Townsend, I turned over and over the new and rare species; but he was absent at Fort Vancouver, on the shores of the Columbia River;[7] Thomas Nuttall had not yet come from Boston, and loud murmurs were uttered by the *soi-disant* friends of science, who objected to my seeing, much less portraying and describing, those valuable relics of birds, many of which had not yet been introduced into our Fauna . . . It was agreed that I might *purchase duplicates, provided* the specific names agreed upon by Mr. Nuttall and myself were published in Dr. Townsend's name. This . . . was perfectly congenial to my feelings, as I have seldom cared much about priority in the naming of species. I therefore paid for the skins which I received, and have now published such as proved to be new, according to my promise. But, let me assure you, Reader, that seldom, if ever in my life, have I felt more disgusted with the conduct of any opponents of mine, than I was with the unfriendly boasters of their zeal for the advancement of ornithological science, who at that time existed in the fair city of Philadelphia."

Herrick quotes some of the above passage with the following comment:

"For many years Audubon had expressed great contempt for all seekers after priority in the naming of species of animals, but now he began to find the pressure from without too strong to be resisted. Rivalry in this field had become keen on both sides of the Atlantic, and in a commendable desire to render his work as complete as possible, he was inevitably drawn into a struggle in which the higher aspirations of scientific men are all too apt to be obscured by petty vanities, suspicions and disputes."

Although "priority in the naming of species"—whether of animals or of plants—may seem a trivial or even "small" approach to those broad-minded and disinterested individuals who believe that it matters not *who* published *first* so long as it *was* published, priority was an important matter to the collectors of the period of which I write, and understandably so. The sponsor or financial "backer"—whether an individual or an institution—sending a collector into the field expected, and perhaps had a right to expect, some "return for his money," and this return was greatest when it consisted of some new or rare discoveries; for the sponsor, the collector and the expedition gained thereby a certain renown—but never, I must believe, any monetary recompense worth mentioning. When a man, as some did, adopted plant collecting as a profession and sold what he discovered to those assembling herbaria, it was essential that he should have something valuable (represented by something "new") among his wares, provided he hoped to sell his future collections. The poor—and I mean financially poor—botanist who expected to write up his findings in a saleable flora or narra-

7. Townsend did not get back to the east until November 17, 1837, for he had remained at Fort Vancouver in order to fill the position of Dr. Meredith Gairdner when that medical man had had to retire on account of illness.

tive, was irreparably harmed when his "new" material was published with greater promptitude by someone else. John Bradbury was one outstanding example of such a disappointed and thwarted man. When Townsend, far away, was not in a position to express his wishes regarding his collections, the institution which had sent him west was protecting his interests as well as their own when they refused Audubon's request for "those valuable relics of birds." The apprehension expressed by such a man as David Douglas—lest Drummond gain access to, even a glimpse of, the plants which he had struggled to acquire—makes the reader smile; but his fear connoted as well a praiseworthy ambition and an inward prayer that years of effort might not be rendered futile by an untrustworthy rival. Fortunately, Drummond was an honorable gentleman, as Douglas soon discovered. Audubon's "commendable desire to render his work as complete as possible" was commendable only insofar as the desire did not infringe upon the rights of others. I believe that Townsend eventually turned over his valuable collections to Audubon—who may have accomplished the task of publication better and with greater promptitude than Townsend could have done. But this is matter apart from the simple ethics of "priority"!

From Philadelphia Audubon went to Washington, fortified with letters of introduction to influential persons which had been given him by Washington Irving. There he was in touch with Colonel J. J. Abert and through him with Secretary of the Navy Dickinson. The United States Exploring Expedition had been authorized in May of 1836 and Lucy Audubon quotes her husband to the effect that Dickinson ". . . paid me some compliments, and told me the moment the expedition had been mentioned he had thought of me, of Nuttall, and Pickering—a glorious trio! I wish to God that I were young once more; how delighted I would be to go in such a company, learned men and dear friends . . ."

Irving's letters of introduction refer to Audubon's publications as ". . . a national work . . . highly creditable to the nation . . . particularly deserving of national patronage . . ." Did Audubon's birds and "quadrupeds" merely have a wider appeal than the dried plants of my collectors? Or, was the lack of understanding between scientist and layman becoming less pronounced? Whatever the reason, one wonders whether Audubon did not accomplish more towards arousing interest in the natural sciences— to make them matters of *national* interest to the American people—than the sum total of all the men of whom I have written in these pages. David Douglas might, perhaps, represent the customary exception.

I end with a comment made by Audubon himself which appears in his "European Journals" under the date of January 22, 1837. He had visited Sir Walter Scott whom he greatly admired, and his approach to the interview was one which, I suspect, might go far towards ensuring anyone's success: "There was much conversation. I talked little, but believe me, I listened and observed, careful if ignorant."

ARTHUR, STANLEY CLISBY. 1937. Audubon. An intimate life of the American woodsman. New Orleans.

AUDUBON, JOHN JAMES. 1827–1838. The birds of America, from original drawings by John James Audubon, Fellow of the Royal Societies of London & Edinburgh and of the Linnaean & Zoölogical Societies of London, Member of the Natural History Society of Paris, of the Lyceum of New York, &c. &c. &c. 4 vols. London. 435 plates. Double elephant folio. Published by the author.

—— 1831–1839. Ornithological biography, or an account of the habits of the birds of the United States of America; accompanied by descriptions of the objects represented in the work entitled The Birds of America and interspersed with delineations of American scenery and manners . . . 5 vols. Edinburgh.

—— 1840–1844. The birds of America, from drawings made in the United States and their territories . . . 7 vols. of text and plates. New York. Published by J. J. Audubon. Philadelphia. Published by J. B. Chevalier. 500 plates. Royal octavo.

—— 1868. The life and adventures of John James Audubon, the naturalist. Edited, from materials supplied by his widow, by Robert Buchanan. London.

—— 1883. The life of John James Audubon, the naturalist, edited by his widow [Lucy Bakewell Audubon]. With an introduction by Ja[me]s Grant Wilson. New York.

—— 1897. Audubon and his journals by Maria A. Audubon. With zoölogical and other notes by Elliott Coues. 2 vols. New York.

—— 1937. The birds of America. John James Audubon. With an introduction and descriptive text by William Vogt. New York. 500 plates. [Of these plates 435 are reproduced from Audubon's The birds of America (1827–1838) and 65 from The birds of America (1840–1844).]

—— See also under Peattie, D. C. 1940. Audubon's America.

GRAY, ASA. 1893. The letters of Asa Gray. Edited by Jane Loring Gray. 2 vols. Boston. New York.

GEISER, SAMUEL WOOD. 1930. Naturalists of the frontier VIII. Audubon in Texas. *Southwest Review* 16: 108-130.

—— 1937. Naturalists of the frontier. 317, 318, 330.

—— 1948. *Ibid.* ed. 2. Dallas. 79-94.

HERRICK, FRANCIS HOBART. 1917. Audubon the naturalist. A history of his life and time. 2 vols. New York. London.

PEATTIE, DONALD CULROSS.[8] 1940. Audubon's America. The narratives and experiences of John James Audubon. Edited by Donald Culross Peattie. Illustrated with facsimiles of Audubon's prints and paintings. Boston.

—— & DOBSON, ELEANOR ROBINETTE. 1928. John James Audubon. *Dict. Am. Biog.* 1: 423-427.

WRIGHT, RICHARDSON. 1949. Gardener's tribute. Philadelphia. New York. 193-201.

8. D. C. Peattie has written much about Audubon. His *Singing in the wilderness. A salute to John James Audubon* (New York. 1935) is a delightfully told story; and Chapter X ("Wilderness birdsmen: Wilson and Audubon") of his *Green laurels. The lives and achievements of the great naturalists* (New York. 1936) offers interesting pictures of the two ornithologists.

CHAPTER XLII

ON A JOURNEY FROM THE MISSOURI FRONTIER TO THE COLUMBIA RIVER AND BACK, FRÉMONT ENCIRCLES THE GREAT BASIN—IN SO DOING CROSSING THE SIERRA NEVADA AT A NUMBER OF LATITUDES AND MAPPING ROUTES THROUGH THE DESERTS OF NEVADA, CALIFORNIA AND UTAH AND THROUGH THE THREE PARKS OF NORTHERN COLORADO

FROM the point of view of geography, John Charles Frémont's second expedition of 1843–1844 was important, for during its progress the explorer investigated and mapped two routes to the Pacific coast—one into Oregon and one into southern California—as well as another from the Columbia River all the way to southern California, in the course of these journeys making several crossings of the Sierra Nevada.

When botanists learned of the proposed journey they recognized an opportunity to acquire plants. Asa Gray wrote John Torrey on December 5, 1842:

". . . I wish we had a collector to go with Frémont. It is a great chance. If none are to be had, Lieutenant F. must be indoctrinated, and taught to collect[1] both dried specimens and seeds. Tell him he shall be immortalized by having the 999th Senecio called S. Fremontii; that's *poz.,* for he has at least two new ones . . ."

The choice of a collector was still on Gray's mind on February 13, 1843, for he wrote Engelmann on the subject, suggesting Lindheimer—who went instead to Texas—and explaining that "Frémont will not take Geyer; but I believe he wants someone." No trained collector was a member of the party and Frémont himself acted in that capacity.

The account of Frémont's second expedition was published in 1845 (Senate Document No. 174, 28th Congress, 2nd Session), following a reprint of the report on the first expedition of 1842. The main title-page was inscribed "Report of the exploring expedition to the Rocky Mountains in the year 1842, and to Oregon and north California, in the years 1843–'44." The title-page of the expedition under discussion read, "A

---

1. Rodgers quotes a letter which Torrey wrote Gray on March 26, 1843, which states " 'I have already given him directions for collection & preserving specimens & he promises to pay attention to what we, of course, consider the main object of the expedition.' "

report of the exploring expedition to Oregon and North California,[2] in the years 1843–'44." It was dated "Washington City, *March* 1, 1845," was consigned to Colonel J. J. Abert, and signed by "Brevet Captain[3] J. C. Frémont, of the Topographical Engineers."

The opening lines of the report explain that Frémont's instructions had been ". . . to connect the reconnoissance of 1842 . . . with the surveys of Commander Wilkes[4] on the coast of the Pacific ocean so as to give a connected survey of the interior of our continent . . ."

Charles Preuss was again Frémont's topographical assistant and there can be no doubt that the presence of two experienced guides had much to do with the expedition's success. They were Thomas Fitzpatrick ". . . whom many years of hardship and exposure in the western territories had rendered familiar with a portion of the country it was designed to explore . . ."; and Christopher, or Kit, Carson who became a member of the party when it reached Pueblo, Colorado, in July. There were about forty men, and twelve carts, with camp equipment and provisions, and ". . . a light covered wagon, mounted on good springs . . ." carrying the surveying instruments.

The party ". . . was armed generally with Halls' carbines, which, with a brass 12-lb. howitzer, had been furnished . . . from the United States arsenal at St. Louis, agreeably to the orders of Colonel S. W. Kearney, commanding the 3d military division. Three men were especially detailed for the management of this piece, under the charge of Louis Zindel, a native of Germany . . ."

This howitzer served no purpose, was never used, and finally was abandoned many, many miles from St. Louis; actually it did more harm to Frémont than to anyone else![5]

Frémont explains that it was his intention to reach the Rocky Mountains by a dif-

2. In distinction to Lower California, presumably, for Frémont left California by way of Cajon Pass.

3. Warren states that "Frémont did not receive his promotion to the rank of brevet captain till the termination of his second expedition."

Of this promotion Frémont wrote in his *Memoirs of my life* (1887): ". . . I was appointed by President Tyler captain by brevet, 'to rank as such from the 31st day of July, 1844: for gallant and highly meritorious services in two expeditions commanded by himself; the first to the Rocky Mountains, which terminated October 17, 1842; and the second beyond these mountains, which terminated July 31, 1844.' This brevet has the greater value for me because it is the only recognition for 'services rendered' that I have received from my own Government."

4. Charles Wilkes and the United States Exploring Expedition (1838–1842) had reached the northwest coast of North America in May, 1841, by sea.

5. Frémont had moved to the small town of Kansas, where he was to complete the final arrangements for the journey, leaving his wife Jessie in St. Louis. When a letter from Colonel Abert reached St. Louis, Mrs. Frémont opened it: her husband was ordered to return at once to Washington and explain the presence of a twelve-pound howitzer on a "peaceful, scientific survey." Instead of forwarding the letter to her husband, Jessie sent him a note urging him to be on his way immediately; he took the hint and left at once. Since relations with both the British and the Mexicans were not too cordial at this period, the government was showing some wisdom, although at the present time one piece of artillery seems rather a joke.

Although Mrs. Frémont later confessed her part in the matter to both Kearny and Abert, the action

ferent route from the one followed in 1842 which had been ". . . up the valley of the Great Platte river to the South Pass, in north latitude 42° . . ." Instead, he planned to ascend the valley of the Kansas River ". . . to the head of the Arkansas, and to some pass in the mountains, if any could be found, at the sources of that river. By making this deviation from the former route, the problem of a new road to Oregon and California, in a climate more genial, might be solved . . .", and so on.

I shall follow the journey in Frémont's "Report" of 1845 and attempt to translate it into terms of a modern map.[6] Since the explorer was in eleven different states (in six of them twice, in one three times) and always in different regions, I preface the detailed account of his journey with a brief outline of his whereabouts at given times, which I believe may be useful for ready reference; the directions taken are, of course, general ones:

Frémont moved westward across Kansas (May 17–June 25, *1843*) and Nebraska (June 26–June 30); southward through Colorado (July 1–July 29, during that time making a trip from St. Vrain's Fort south to Pueblo on the Arkansas River and back from July 4–23); then north and northwest in Wyoming (July 31–August 22); west and south in Idaho (August 22–? 25); south and then north in Utah (August ? 25–September 15, during that time spending from September 6–12 at Great Salt Lake); north and then west in Idaho (September 15–October 9); north and northwest in Oregon (October 9–23); north and northwest in Washington (October 23–29). He was, for undeterminable periods, in Washington and Oregon (October 29–November 25, during that period traveling from Fort Walla Walla to Fort Vancouver and back to the Dallas of the Columbia River); south in Oregon (November 26–December 26); south in Nevada (December 26–January 26, *1844*); east, south and finally northeast in California (January 26–May 2); northeast in Nevada (May 2–9); northeast in Arizona (May 10, following the Virgin River); north and then northeast in Utah (May 11–June 7); northeast in Colorado (June 7–? 12); northeast and south in Wyoming (June ? 12–15); south and finally east in Colorado (June 15–July 8); and east across Kansas (July 8 or 9–July 31).

1843

From May 17, when Frémont arrived at the "little town of Kansas, on the Missouri frontier, near the junction of the Kansas river with the Missouri river," until June 25, when the party entered Nebraska, the route was across the state of Kansas.

was highhanded and made an unfavorable impression on these two officers. Frémont came into conflict with Kearny on his third expedition of 1845–1846, and his earlier act of insubordination—although a matter of the past and perhaps not committed knowingly—was undoubtedly remembered.

6. The map included in Frémont's *Memoirs of my life* ("Map showing the country explored by John Charles Frémont, from 1841 to 1854 inclusive. Drawn and engraved expressly for Frémont's Memoirs.") is helpful and includes, roughly speaking, an area lying between 102° and 124° of longitude and 32° and 46° of latitude. It is smaller than the map included in the "Report" and the reduction is helpful rather than otherwise.

Two weeks were consumed in making final arrangements, but the expedition started on May 29 and camped that night ". . . four miles beyond the frontier, on the verge of the great prairies." On May 31 it reached Elm Grove, in company with several emigrant wagons which were proceeding ". . . to Upper California, under the direction of Mr. J. B. Childs, of Missouri." In addition to much else Childs was transporting ". . . an entire set of machinery for a mill . . ." which was to be set up ". . . on the waters of the Sacramento river emptying into the bay of San Francisco." Frémont notes that the route until June 3 was nearly the one which he had taken in 1842 with the difference that a great migration[7] was now taking place: "Trains of wagons were almost constantly in sight; giving to the road a populous and animated appearance."

I shall not follow Frémont's route across Kansas in detail. He did not cross the Kansas River at the usual fording place (as he had done on June 18, 1842), but kept along its southern side. On June 4 he reached Otter Creek, on the 8th came to ". . . the mouth of Smoky-hill fork, which is the principal southern branch of the Kansas; forming here, by its junction with the Republican, or northern branch, the main Kansas river." These streams unite near present Junction City, Geary County. After building a raft, they crossed the river and on the 11th resumed the journey, up the Republican River. By the 14th they were ". . . 265 miles by our travelling road from the mouth of the Kansas . . . at an elevation of 1,520 feet . . . where we . . . encamped is called by the Indians the *Big Timber*."

Frémont found travel slow and divided his party, going ahead with Preuss and fifteen men, the howitzer and the instruments, and leaving Fitzpatrick to follow with the heavier equipment. Frémont bore west, according to Warren keeping on the divide between the Republican River and Solomon's Fork. On the 16th the party soon ". . . entered upon an extensive and high level prairie . . ."

". . . Among a variety of grasses which to-day made their first appearance, I noticed bunch grass *(festuca,)* and buffalo grass, *(sesleria dactyloides.)* Amorpha canescens *(lead plant)* continued the characteristic plant of the country, and a narrow-leaved

---

7. Robert Greenhow, writing of the spring migrations of 1843, comments that their success ". . . encouraged a still greater number to follow in 1844, before the end of which year the number of American citizens in Oregon exceeded three thousand." The same author supplies an interesting British comment upon the importance of these migrations:

"It may be remarked, that, on the 1st of July, 1843, while this crowd of men, women, and children, with their wagons, horses, and cattle, were quietly pursuing their way across the continent, to the regions of the lower Columbia, an article appeared in the Edinburgh Review . . . in which it was affirmed, *ex cathedra*, that—'*However the political questions between England and America, as to the ownership of Oregon, may be decided, Oregon will never be colonized overland from the United States.*' The Reviewer asserts that —'*The world must assume a new face, before the American wagons make plain the road to the Columbia as they have done to the Ohio;*' and he determines that—'*Whoever, therefore, is to be the future owner of Oregon, its people will come from Europe.*' This is not the first occasion, in which European predictions, implying doubts as to the energy of American citizens, and their capacity to execute what they have undertaken, have been contradicted by facts, as soon as uttered . . ."

lathyrus occurred . . . in beautiful patches. *Sida coccinea* occurred frequently, with a *psoralia* near *psoralia floribunda,* and a number of plants not hitherto met, just verging into bloom. The water on which we had encamped belonged to Solomon's fork of the Smoky-hill river, along whose tributaries we continued to travel for several days . . . the numerous streams . . . were well timbered with ash, elm, cottonwood, and a very large oak . . . occasionally, five and six feet in diameter, with a spreading summit. *Sida coccinea* is very frequent in vermilion-colored patches on the high and low prairies . . . it has a very pleasant perfume. The wild sensitive plant *(schrankia angustata)* occurs frequently . . . the leaflets close instantly to a very light touch. *Amorpha,* with the same *psoralea,* and a dwarf species of *lupinus,* are the characteristic plants."

On the 19th "the Pawnee road to the Arkansas" was crossed; the elevation was increasing. On the 21st travel was westward and "affluents of the Republican" were being crossed; plants were few: ". . . with the short sward of the buffalo grass . . . were mingled patches of a beautiful red grass, *(aristida pallens,)* . . .". They camped at a fork which was ". . . well wooded with ash-leaved maple, *(negundo fraxinifolium,)* elm, cottonwood, and a few white oaks." On June 23 camp was in "the valley of a principal fork of the Republican" which they named "Prairie Dog river." [8] The prevailing timber was ". . . a blue-foliaged ash, *(fraxinus,* near *F. Americana,)* and ash-leaved maple . . . cottonwood, and long-leaved willow." On the 25th they camped within ". . . a few miles of the main Republican . . . where the air was fragrant with the perfume of *artemisia filifolia,* which we saw here for the first time, and which was now in bloom." It must have been on the 25th or 26th that the party crossed into southern Nebraska. They were to travel in that state until June 30.

On the 26th Frémont notes that suddenly ". . . the nature of the country had entirely changed. Bare sand hills every where surrounded us . . . and the plants peculiar to a sandy soil made their appearance . . . we entered the valley of a large stream, afterwards known to be the Republican Fork of the Kansas . . . no timber of any kind was to be seen . . . the country assumed a desert character . . . We travelled now for several days through a broken and dry sandy region, about 4,000 feet above the sea, where there were no running streams . . .

"The soil of bare and hot sands supported a varied and exuberant growth of plants, which were much farther advanced than we had previously found them, and whose showy bloom somewhat relieved the appearance of general sterility . . . on . . . the 30th of June we found ourselves overlooking a broad and misty valley, where . . . the South fork of the Platte was rolling magnificently along . . ."

It must have been on July 1 that the party entered Colorado. Ewan states that this was ". . . in crossing from South Park Republican River, to the south of the present

8. Prairie Dog Creek empties into Republican River in Harlan County, southern Nebraska, and Frémont was not far from the Kansas-Nebraska border.

towns of Wray and Yuma [both in Yuma County], and on to the South Platte east of Fort Morgan [Morgan County]." The route was in Colorado until July 29.

On July 1 the party moved up the South Platte and camped at the mouth of Bijou Creek (Morgan County); on the 4th it reached Fort St. Vrain (Weld County) and on the 6th camped sixteen miles above; on the 7th it reached Cherry Creek (emptying into the South Fork of the Platte near Denver, Denver County); still traveling up the South Platte and twenty-one miles above St. Vrain's Fort, it came, on the 8th, to a point ". . . where the stream . . . divided into three forks; two . . . issuing directly from the mountains on the west . . .

"On the easternmost branch, up which we took our way, we first came among the pines growing on the top of a very high bank, and where we halted on it to noon; quaking asp *(populus tremuloides)* was mixed with the cottonwood, and there was excellent grass and rushes for the animals. During the morning there occurred many beautiful flowers, which we had not hitherto met. Among them, the common blue flowering flax made its first appearance; and a tall and handsome species of *gilia,*[9] with slender scarlet flowers, which appeared yesterday for the first time, was very frequent."

By July 9 the food supply was low and Frémont turned eastward ". . . along the dividing grounds between the South fork of the Platte and the Arkansas . . ." in the hope of finding buffalo, crossing, on the 10th, "a head water of the Kioway river," and camping on "Bijou's fork." That day Frémont had observed ". . . the characteristic plant . . . *esparcette,*[1] *(onobrychis sativa,)* a species of clover which is much used in certain parts of Germany for pasturage for stock . . ." On July 11, no buffalo having been found, the party turned south, ". . . up the valley of Bijou."

". . . *Esparcette* occurred universally, and among the plants on the river I noticed, for the first time during this journey, a few small bushes of the *absinthe* of the voyageurs, which is commonly used for fire wood *(artemisia tridentata.)* Yesterday and today the road had been ornamented with the showy bloom of a beautiful *lupinus,* a characteristic in many parts of the mountain region, on which were generally great numbers of an insect with very bright colors . . ."

Following the stream to its headwaters (about "7,500 feet above the sea"), they reached ". . . a piney elevation . . . from which the waters flow, in almost every direction, to the Arkansas, Platte, and Kansas rivers . . ." They were on the Arkansas divide, in El Paso County. They then turned southwest, reached the road to the settlements on the Arkansas River, and camped on the 12th on ". . . the *Fontaine-qui-bouit* (or Boiling Spring) river . . ." Frémont noted that day ". . . a tall species of *gilia,* with a slender white flower . . . and . . . another variety of *esparcette,* (wild clover) having

9. According to Ewan, ". . . surely Gilia aggregata . . ."

1. Ewan notes that "This may have been Onobrychis, that is Sainfoin or Sand-foin but this is open to some question. The racemes of Sainfoin resemble those of our Astragalus species so that this large and varied group of natives may have been the plants in question."

the flower white . . ." Descending Fountain Creek on the 13th, the animals luxuriated on ". . . rushes *(equisetum hyemale,)* . . ."

". . . A variety of cactus made its appearance, and among several strange plants were numerous and beautiful clusters of a plant resembling *mirabilis jalapa,* with a handsome convolvulus . . . not hitherto seen, *(calystegia.)* . . ."

July 14 they camped at the mouth of Fountain Creek, which empties into the Arkansas River near Pueblo (on the boundary between El Paso and Pueblo counties).

Writes Frémont: ". . . A short distance above our encampment, on the left bank of the Arkansas, is a *pueblo,* (as the Mexicans call their civilized Indian villages,) where a number of mountaineers, who had married Spanish women in the valley of Taos, had collected together, and occupied themselves in farming, carrying on at the same time a desultory Indian trade. They were principally Americans, and treated us with all the rude hospitality their situation admitted; but as all commercial intercourse with New Mexico was now interrupted, in consequence of Mexican decrees to that effect, there was nothing to be had in the way of provisions . . . By this position of affairs, our expectation of obtaining supplies from Taos was cut off . . ."

At this point Frémont had the good fortune to meet ". . . our good buffalo hunter of 1842, Christopher Carson . . ." and was able to secure his services again. With him, in addition to Fitzpatrick, Frémont had two of the best guides of the far west. Carson was dispatched immediately to acquire mules at Bent's Fort, ". . . on the Arkansas river about 75 miles below *Fontaine-qui-bouit* . . . ," and rejoin Frémont at St. Vrain's Fort.

On July 16 Frémont turned northward again, up Fountain Creek, with the intention of visiting ". . . the celebrated springs from which the river takes its name, and which are on its upper waters, at the foot of Pike's peak.

". . . *Ipomea leptophylla,* in bloom, was a characteristic plant along the river, generally in large bunches, with two to five flowers on each. Beautiful clusters of the plant resembling *mirabilis jalapa*[2] were numerous, and *glycyrrhiza lepidota* was a characteristic of the bottoms. Currants nearly ripe were abundant, and among the shrubs which covered the bottom was a very luxuriant growth of chenopodiaceous[3] shrubs, four to six feet high."

The "celebrated springs"—Manitou Spring—was reached on the 17th and the party remained there the next day.

"The trees in the neighborhood were birch, willow, pine, and an oak resembling *quercus alba.* In the shrubbery along the river are currant bushes, *(ribes,)* of which the fruit has a singularly piney flavor; and on the mountain side, in a red gravelly soil, is a singular coniferous tree, (perhaps an *abies,)* having the leaves singularly long, broad and scattered, with bushes of *spiræa ariæfolia* . . ."

On the 19th they descended the river ". . . in order to reach the mouth of the eastern

2. ". . . being Mirabilis multiflora." (Ewan)

3. ". . . of course, Atriplex canescens." (Ewan)

fork . . ." which Frémont planned to ascend; they traveled ". . . up the eastern fork of the *Fontaine-qui-bouit river* [? Monument Creek] . . ." crossing deep gullies and dragging along, or so it seems, the famous howitzer. A shaft of the gun carriage got broken.

July 20: ". . . We continued our march up the stream, along a green sloping bottom, between pine hills on the one hand, and the main Black hills on the other, towards the ridge which separates the waters of the Platte from those of the Arkansas. As we approached the dividing ridge, the whole valley was radiant with flowers; blue, yellow, pink, white, scarlet, and purple, vied with each other in splendor. Esparcette was one of the highly characteristic plants, and a bright-looking flower *(gaillardia aristata)* was very frequent; but the most abundant plant . . . was *geranium maculatum* . . . the characteristic plant on this portion of the dividing grounds. Crossing to the waters of the Platte, fields of blue flax added to the magnificence of this mountain garden; this was occasionally four feet in height, which was a luxuriance of growth that I rarely saw this almost universal plant attain throughout the journey . . ."

They camped at the three forks of the South Platte where they had been on July 8. Frémont comments upon the "*Achillea millefolium* (milfoil)" seen on the river bottoms:

"This was one of the most common plants during the whole of our journey, occurring in every variety of situation. I noticed it on the lowlands of the rivers, near the coast of the Pacific, and near to the snow among the mountains of the *Sierra Nevada*." On July 23 the party was again at St. Vrain's Fort, having (on their excursion) surveyed ". . . to its head one of the two principal branches of the upper Arkansas . . . and entirely completed our survey of the South fork of the Platte, to the extreme sources of that portion of the river which . . . heads in the broken hills of the Arkansas dividing ridge, at the foot of the mountains . . ." Fitzpatrick, with whom Frémont had parted on June 15, was already there and Carson also, with ten good mules and the necessary packsaddles, acquired at Bent's Fort. Engaged as a hunter was Alexander Godey, who in ". . . courage and professional skill . . . was a formidable rival to Carson . . ."

From the time of his arrival in Colorado (about three weeks earlier), Frémont had made no noteworthy departure from the route which had been followed by Stephen H. Long and Edwin James in 1820. Ewan notes that ". . . Frémont seems to have been unacquainted with the James narrative of this region; at least he makes no reference to it."

It was now a question of acquiring information about ". . . the passes in this portion of the Rocky mountain range, which had always been represented as impracticable for carriages . . ." It was necessary to discover ". . . some convenient point of passage for the road of emigration, which would enable it to reach, on a more direct line, the usual ford of the Great Colorado . . . It is singular that, immediately at the foot of the mountains, I could find no one sufficiently acquainted with them to guide us to the plains at their western base; but the race of trappers, who formerly lived in their recesses, has almost entirely disappeared—dwindled to a few scattered individuals[4] . . ."

Frémont divided his party again, placing Fitzpatrick at the head of one contingent[5] and keeping Carson and Preuss with his own. There are references to the dangerous Indians of the region who, however, and as in 1842, caused no trouble.

From Fort St. Vrain on July 26 Frémont and his party started northwestward, reaching the "Câche-á-la Poudre" River on the 28th and following that stream upward until the 30th. Frémont had commented on the 26th that ". . . the green river bottom was covered with a wilderness of flowers, their tall spikes sometimes rising above our heads as we rode among them. A profusion of blossoms on a white flowering vine, *(clematis lasiantha,*[6]*)* . . . was abundant . . . The stream was wooded with cottonwood, box elder, and cherry, with currant and serviceberry bushes." On July 31 they camped at night on the Laramie River; along the way the slopes and ravines ". . . were absolutely covered with fields of flowers of the most exquisitely beautiful colors . . . a new *delphinium,* of a green and lustrous metallic blue color, mingled with compact fields of several bright-colored varieties of *astragalus* . . ." It was on this day that they crossed from Colorado into Wyoming (Albany County)—and until August 22 traveled in that state.

They kept to the east of the Medicine Bow Mountains[7] and camped, on August 2, on ". . . the principal fork of Medicine Bow river, near to an isolated mountain called the Medicine Butte [Carbon County] . . ." In the course of the day Frémont ". . . became first acquainted with the *yampah, (anethum graveolens,)* . . . among the Shoshonee or Snake Indians . . . this is considered the best among the roots used for food . . ." On the 3rd they ". . . entered the pass of the Medicine *Butte,*[8] through which led a broad trail . . . recently traveled by a very large party . . ." On August 4 they reached

4. Chittenden has stated that "The true period of the trans-Mississippi fur trade . . . embraces the thirty-seven years from 1807 to 1843." Frémont was writing in the last year of that period, but it is hard to believe that he lacked advice—with Fitzpatrick and Carson in his party.

Again, referring to the traders and trappers, he notes that *"They* were the 'pathfinders' of the West and not those later official explorers [inclusive of Frémont presumably] whom posterity so recognizes. No feature of western geography was ever *discovered* by government explorers after 1840. Every thing was already known and had been for fully a decade."

5. Fitzpatrick was to cross the plains to the mouth of Laramie River (entering the North Fork of the Platte in western Goshen County, Wyoming) and from there proceed by "the usual emigrant road" to Fort Hall, ". . . a post belonging to the Hudson Bay Company, and situated on Snake river, as it is commonly called in the Oregon Territory, although better known to us as Lewis's fork of the Columbia. The latter name is there restricted to one of the upper forks of the river." Built, as we know, by N. J. Wyeth in 1834, Fort Hall was situated not far from the mouth of the Portneuf, in Bingham County, southeastern Idaho; its life as an American fort had been brief, for Wyeth had been obliged, for a number of good reasons, to sell it to the British in 1836. Frémont was to rejoin Fitzpatrick at that point.

6. ". . . here being Clematis ligusticifolia . . ." (Ewan)

7. Ewan states that they ". . . passed near present Centennial [Albany County] . . ."

8. Warren refers to this as "the Medicine Bow Butte Pass."

the "North Platte" at a good ford and remained two days—a shaft of the howitzer's carriage needed repairs. Keeping north they camped on the 8th ". . . about twenty miles above Devil's gate . . ." This was near Independence Rock, Natrona County, and they were on the Sweetwater. Here, states Frémont, ". . . passes the road to Oregon; and the broad smooth highway, where the numerous heavy wagons of the emigrants had entirely beaten and crushed the artemisia, was a happy exchange to our poor animals for the sharp rocks and tough shrubs among which we had been toiling so long; and we moved up the valley rapidly and pleasantly. With very little deviation from our route of the preceding year, we continued up the valley; and on the evening of the 12th encamped on the Sweet Water, at a point where the road turns off to cross to the plains of Green river . . ."

On the 13th—they were now traveling west—they crossed ". . . the dividing ridge which separates the Atlantic from the Pacific waters . . . by a road some miles further south than the one . . . followed on our return in 1842 . . . very near the table mountain, at the southern extremity of the South Pass . . . and already traversed by several different roads . . ." Frémont took a barometrical observation, correcting the one made in 1842.

". . . From this pass to the mouth of the Oregon is about 1,400 miles by the common travelling route; so that . . . it may be assumed to be about half way between the Mississippi and the Pacific ocean, on the common travelling route. Following a . . . slight and easy descent . . . we made our usual halt four miles from the pass . . . Entering here the valley of Green river—the great Colorado of the West—and inclining very much to the southward along the streams which form the Sandy river, the road led for several days over dry and level uninteresting plains . . . on the evening of the 15th we encamped in the Mexican territory, on the left bank of Green river, 69 miles from the South Pass . . . This is the emigrant road to Oregon, which bears much to the southward, to avoid the mountains about the western heads of Green river—the *Rio Verde* of the Spaniards."

According to Warren's "Memoir," Frémont had followed ". . . the course of the Big Sandy to its mouth . . ." It enters Green River in northwestern Sweetwater County. On the 16th he seems to have turned down Green River (or southeastward) until the road left the stream, and then moved westwardly, encountering "Black's fork of the Green river," and, on the 17th, "Ham's fork" (of the Black Fork). Plants had been "very few in variety" since leaving the pass, "the country being covered principally with artemisia." On the 19th Carson was sent ahead to Fort Hall to arrange for provisions. On August 21 they reached ". . . the fertile and picturesque valley of Bear river, the principal tributary to the Great Salt lake . . ." and were, presumably, in southwestern Lincoln County, very near the Wyoming-Idaho border. On the 22nd, descending Bear River, they came to "Smith's fork," which empties into Bear River from the north, in Lincoln County, and by afternoon reached "Thomas's fork," which also empties into Bear River from the north but in Idaho, in Bear Lake County.

Frémont was to remain in Idaho for about three days, following Bear River to its northernmost point at Soda Springs, Caribou County (where the river turns abruptly southward) and thence, with only slight deviations, following it southward through Bannock and Franklin counties, Idaho, into Cache County, Utah, and thence, through Boxelder County to Great Salt Lake. Just what day the Idaho-Utah border was crossed I do not know, perhaps August 25, but they were at the lake on September 6 and remained in Utah until September 15.

On August 25 the party reached "the famous Beer springs"—Soda Springs—and halted nearby. On the 26th, avoiding the road that led to Fort Hall, they moved southward, down Bear River.[9] On the 27th the country had an autumnal appearance, "in the crisped and yellow plants, and dried-up grasses." The "... *epinettes des prairies* (grindelia squarrosa) ... is among the very few plants remaining in bloom ..." On August 31 they were on "*Roseaux, or Reed*[1] river." On September 1 the country "... plainly indicated that we were approaching the lake, though, as the ground ... afforded no elevated point, nothing of it as yet could be seen ...

"... the grass was every where dead; and among the shrubs ... (artemisia being the more abundant,) frequently occurred handsome clusters of several species of *dieteria* in bloom. *Purshia tridentata* was among the frequent shrubs ... On the river are only willow thickets, *(Salix longifolia,)* and in the bottoms the abundant plants are canes, solidago, and helianthi, and along the banks of the Roseaux are fields of *malva rotundifolia* ..."

A large part of September 2-4 was spent in trying out one of the "useful things which formed a portion of our equipage ... an India-rubber boat, 18 feet long ..." with inflatable sides. It served well as a means of transporting the luggage across the river, but it "moved so heavily" that when Frémont made a trip in it he could not get back to camp by nightfall and had to cache it to be retrieved the next day.

September 5: "Before us was evidently the bed of the lake, being a great salt marsh, perfectly level and bare, whitened in places by saline efflorescences, with here and there a pool of water, and having the appearance of a very level sea shore at low tide. Immediately along the river was a very narrow strip of vegetation, consisting of willows, helianthi, roses, flowering vines, and grass; bordered on the verge of the great marsh by a fringe of singular plants, which appear to be a shrubby salicornia, or a genus allied to it ..."

They stopped at "a beautiful little stream of pure and remarkably clear water"

9. The narrative is broken during this period by dissertations upon the waters of Soda Springs, upon the roots eaten by the Indian tribes of this part of Idaho (several encampments were visited), and upon the scarcity of the buffalo west of the Rockies (where they had once been plentiful)—the usual story of ruthless and useless extermination.

1. Not the Boisé River, Idaho, which is often referred to in the narratives as "Reed" or "Reid" River. This Utah stream, since *roseau* is French for reed or rush, doubtless was named for the plant.

which they named *"Clear creek,"* and camped that night on a "large and comparatively well-timbered stream, called Weber's fork." From there, on September 6, after
crossing a "slough-like creek . . . wooded with thickets of thorn *(cratægus)* . . . loaded
with berries," they ascended a butte from the summit of which they beheld ". . . the
object of our anxious search—the waters of the Inland Sea, stretching in still and
solitary grandeur far beyond the limit of our vision. It was one of the great points of the
exploration; and as we looked eagerly over the lake in the first emotions of excited
pleasure, I am doubtful if the followers of Balboa felt more enthusiasm when, from the
heights of the Andes, they saw for the first time the great Western ocean. It was certainly a magnificent object, and a noble *terminus* to this part of our expedition . . . So
far as we could see, along the shores there was not a solitary tree . . . Weber's fork . . .
appeared to be the nearest point to the lake where a suitable camp could be found . . .
with good grass and an abundance of rushes, *(equisetum hyemale.)* . . ."

Frémont and his party remained at Great Salt Lake until September 12, making
a survey of the northern portion of the lake on the 7th and testing out their rubber
boat along the shore on the 8th—considered to be a perilous proceeding since the
craft was "pasted" rather than sewn together. On the 9th it was thought safe to embark for one of the islands: and, having landed thereon, Frémont made the comment:
". . . We felt pleasure . . . in remembering that we were the first who, in the traditional
annals of the country, had visited the islands, and broken . . . the long solitude of the
place . . ." It was this remark to which Bancroft took exception, pointing out that
"trappers in skin boats" had done so "while the pathfinder was still studying his
arithmetic." We are told of the plants which grew on the island (named *"Disappointment island"* because of its lack of fertility):

". . . the *Fremontia vermicularis* . . . was in great abundance . . . This is eminently
a saline shrub; its leaves have a very salt taste . . . It is widely diffused over all this
country. A chenopodiaceous shrub, which is[2] a new species of Obione, (O. rigida,
Torr. & Frem.,) was equally characteristic of the lower parts of the island. These two
. . . belong to a class of plants which form a prominent feature in the vegetation of
this country . . . also, a prickly pear of very large size was frequent. On the shore . . .
was a woolly species of *phaca;* and a new species of umbelliferous plant *(leptotæmia)*
was scattered about in very considerable abundance. These constituted all the vegetation that now appeared upon the island."

They had made themselves, out of driftwood, some rude shelters along the shore of
the lake (calling it *"Fisherman's camp"*). It "made quite a picture." Here ". . . *Lyno-
siris graveolens,* and another new species of Obione, (O. confertifolia—*Torr. & Frem.,*)
were growing . . . with interspersed spots of an unwholesome salt grass . . . with a few
other plants." On what would seem to have been September 10, those who had made

2. Although all Frémont's comments suggest a great familiarity with plants, one may be sure that all
the names were inserted after Torrey had made the determinations.

the excursion to the lake returned to the base camp on Weber's River and ". . . were received with a discharge of the howitzer[3] by the people, who . . . had begun to feel some uneasiness." On the 11th the salt content of the lake water was analyzed and on the 12th the party started back, on their way to Fort Hall, "returning [north] by nearly the same route which we had travelled in coming to the lake." Along the way, and while still in Utah, Frémont mentions certain plants observed: on September 12, on Clear Creek, there grew a variety of trees, among them ". . . birch *(betula,)* the narrow-leaved poplar *(populus angustifolia,)* several kinds of willow *(salix,)* hawthorn *(cratægus,)* alder *(alnus viridis,)* and *cerasus,* with an oak allied to *quercus alba,* but very distinct from that of any other species in the United States." On the 13th ". . . the irrigated bottom of fertile soil was covered with innumerable flowers . . . purple fields of *eupatorium purpureum,* with *helianthi,* a handsome solidago *(S. canadensis,)* and a variety of other plants in bloom." On September 15 they were again on Roseaux and Bear rivers and there, from a Snake Indian boy, Frémont learned the identity of "the *kooyah* plant, which proved to be *valeriana edulis.* The root, which constitutes the *kooyah,* is large, of a very bright yellow color . . ."

In the afternoon they ". . . entered a long ravine leading to a pass in the dividing ridge between the waters of Bear river and the Snake river, or Lewis's fork of the Columbia . . . The approach to the pass was very steep; and the summit about 6,300 feet above the sea—probably only an uncertain approximation . . . We descended, by a steep slope, into a broad open valley . . . coming down . . . upon one of the headwaters of the Pannack [Bannock] river . . ."

They now re-entered Idaho, where they were to be until October 9. On September 17 they reached the valley of the main Bannock River, probably in Power County, and on the 18th ". . . emerged on the plains of the Columbia,[4] in sight of the famous *'Three Buttes,'* a well-known landmark . . . distant about 45 miles . . . we were agreeably surprised, on reaching the Portneuf river, to see a beautiful green valley with scattered timber spread out beneath us, on which, about four miles distant, were glistening the white walls of the fort." Frémont described the location of Fort Hall thus:

". . . It is in the low, rich bottom of a valley, apparently 20 miles long, formed by the confluence of Portneuf river with Lewis's fork of the Columbia [the Snake River], which it enters about nine miles below the fort . . . Allowing 50 miles for the road from the *Beer springs* of Bear river to Fort Hall, its distance along the *travelled* road from the town of Westport, on the frontier of Missouri, by way of Fort Laramie and the great South Pass, is 1,323 miles. Beyond this place, on the line of road along the *barren* valley of the Upper Columbia, there does not occur, for a distance of nearly

3. This was, as far as I know, one of the two times the howitzer was fired—and without international repercussions!

4. The Snake River plain.

three hundred miles to the westward, a fertile spot of ground sufficiently large to produce the necessary quantity of grain, or pasturage enough to allow even a temporary repose to the emigrants[5] . . ."

Fitzpatrick, who had parted from Frémont at Fort St. Vrain, Colorado, on July 26, was at Fort Hall as planned. Since winter was approaching and supplies were a problem, it was decided to send home a considerable number of the men. Many, moreover, had decided that they were not adapted to privations, past or future. On September 22 both the eastward- and westward-bound parties started. After fording the Portneuf, Frémont and his men moved on three miles, camping at the mouth of "the Pannack river, on Lewis's fork . . ." Snow was falling on the 23rd—". . . an inauspicious commencement of the autumn, of which this was the first day." By noon of the 24th they were "half a mile above the *American falls* of Snake river," in Power County. Noted, growing among the willows, were "bushes of Lewis and Clarke's currant, *(ribes aureum.)*" They kept down the Snake River and on the 26th caught up with the party of "Mr. Jos. Chiles"—last met on the Missouri frontier on May 31 (and then referred to as "Mr. J. B. Childs"). The party, too large to travel comfortably, had divided into two sections.[6] They reached this day "a stream called Raft river, (Rivière aux Cajeux)," emptying into the Snake from the south, in Cassia County; on the 27th "Swamp creek" (perhaps Marsh Creek, not far west); on the 28th "Goose creek," also entering from the south in Cassia County; on the 29th "Rock creek," also from the south, in Twin Falls County. By October 1 they were a mile below "the Fishing falls, a series of cataracts . . . probably so named because they form a barrier to the ascent of the salmon . . ." They were still in Twin Falls County; Frémont devotes considerable space to describing the numerous falls. On October 7 they ". . . came suddenly in sight of the broad green line of the valley of the *Rivière Boisée,* (wooded river,) . . . At the time of the first occupation of this region by parties engaged in the fur trade, a small party of men under the command of —— Reid, constituting all the garrison of a little fort on this river, were surprised and massacred by the Indians; and to this event the stream owes its occasional name of Reid's river." [7] On the 8th they camped on the right bank of the Boisé, a mile above the mouth, and on the 9th reached Fort Boisé, ". . . a simple dwelling-house on the right bank of Snake river, about a mile below the mouth of Rivière Boissée . . . we were received . . . by Mr. Payette, an officer of the Hudson Bay Company, in charge of the fort . . ." They were now in Canyon County,

5. It was at this place that the farseeing Wyeth in 1834 had erected his Fort Hall. Unfortunately, he was a decade ahead of events and had been obliged because of maintenance problems to sell it in 1836 to the Hudson's Bay Company. Frémont in 1843 suggests that the spot would form "the *nucleus* of a settlement" on the road to Oregon.

6. Chiles took a northern route to the Sacramento valley, across the Sierra Nevada, and Joseph Walker led another group into the San Joaquin valley. States Frémont: "in the course of our narrative, we shall be able to give you some information of the fortune . . . of these adventurous travellers."

7. Chittenden refers to him as John *Reed* and states that the massacre occurred in the winter of 1813–1814.

but immediately crossed the Snake River into northeastern Malheur County, Oregon. They were to travel through that state until October 23.

On October 11 the party resumed their journey and reached that day ". . . the *Rivière aux Malheurs,* (the unfortunate or unlucky river,) . . ." From the west this enters the Snake River in Malheur County. On the 12th they came to ". . . the *Rivière aux Bouleaux,* (Birch river,) . . .", descended it for seven miles, crossed it and arrived at Snake River. On the 13th Frémont records:

"We were now about to leave the valley of the great southern branch of the Columbia river, to which the absence of timber, and the scarcity of water, give the appearance of a desert, to enter a mountainous region where the soil is good, and in which the face of the country is covered with nutritious grasses and dense forest—land embracing many varieties of trees peculiar to the country, and on which the timber exhibits a luxuriance of growth unknown to the eastern part of the continent and to Europe . . ."

That day they descended into the valley of *"Burnt river"* [8] and followed it the next; on the 15th the road improved after the ". . . dividing grounds between the *Broulé* (Burnt) and Powder rivers . . ." had been crossed. They camped on the last:

". . . we had looked in vain for a well-known landmark on Powder river, which had been described to me by Mr. Payette as *l'arbre seul,* (the lone tree;) . . . on arriving at the river, we found a fine tall pine[9] stretched on the ground, which had been felled by some inconsiderate emigrant axe. It had been a beacon on the road for many years past . . ."

On October 16 they were still on Powder River; and on the 17th traveled along its affluents and entered ". . . the *Grand Rond*—a beautiful level basin, or mountain valley, covered with good grass, on a rich soil, abundantly watered, and surrounded by high and well-timbered mountains; and its name descriptive of its form—the great circle . . ." The party encamped on the "Grand Rond river." In descending into the valley Frémont had noted ". . . mingled with the green of a variety of pines . . . the yellow of the European larch *(pinus larix,)* which loses its leaves in the fall . . . here a magnificent tree, attaining sometimes the height of 200 feet, which I believe is elsewhere unknown." [1] On the 18th Frémont continued north through the valley and on the 19th halted at the foot of the "Blue mountains, on a branch of the Grand Rond river . . ." and, after stopping "a few minutes," started over the pass, where night over-

8. Burnt River enters the Snake from the west, in southeastern Baker County.

9. Joseph Burke comments that it was the "only large tree, on the whole journey from the blue mountains to Fort Hall." But a stump remained when, on November 3, 1844, he passed "the lone tree encampment."

1. The reference is to our western larch, *Larix occidentalis* Nuttall. The other larch of our northwestern states, *Larix Lyallii* Parlatore, does not attain such a height nor, probably, is it found so far south.

Because of the rarity of this tree, eluding collectors, Professor Sargent enjoyed referring to it as "the lary Laxilli," a name which is easy—but unimportant—to remember. J. G. Jack found it at Piegan Pass, Glacier National Park in 1922, a region from which it was then unrecorded. *See:* Standley, P. C. 1921. "Flora of Glacier National Park Montana." *Contr. U. S. Nat. Herb.* 22.

took them. The ascent continued on the 20th, but on the 21st they reached ". . . one of the head branches of the *Umatilah* [Umatilla] river." The conifers had been remarked upon from the time they had begun to climb:

". . . Some of the white spruces which I measured to-day were twelve feet in circumference, and one of the larches ten; but eight feet was the average circumference of those measured along the road. I held in my hand a tape line as I walked along, in order to form some correct idea of the size of the timber. Their height appeared to be from 100 to 180, and perhaps 200 feet, and the trunks of the larches were sometimes 100 feet without a limb; but the white spruces were generally covered with branches nearly to the root. All these trees have their branches, particularly the lower ones, declining."

Descending the "western slope" of the Blue Mountains on the 22nd, they could see below them "the great *Nez Percé* (pierced nose) prairie . . . a considerable stream was seen pursuing its way . . . towards what appeared to be the Columbia river. This I knew to be the Walahwalah [Walla Walla] river . . ." On the 23rd they reached this stream; on the 24th crossed "a principal fork" of the river and after six miles came to ". . . the missionary establishment of Dr. Whitman,[2] which consisted, at this time, of one *adobe* house—i.e. built of burnt [sun-dried] bricks, as in Mexico. I found Dr. Whitman absent on a visit to the *Dalles* of the Columbia . . ." They had entered Washington on the 23rd:

On October 26 the party ". . . arrived at the Nez Percé fort, one of the trading establishments of the Hudson Bay Company, a few hundred yards above the junction of the Walahwalah with the Columbia river. Here we had the first view of this river, . . . about 1,200 yards wide . . . presenting the appearance of a fine navigable stream . . . The river is . . . a noble object, and has here attained its full magnitude. About nine miles above . . . is the junction of the two great forks which constitute the main stream —that on which we had been travelling from Fort Hall, and known by the names of Lewis's fork, Shoshonee, and Snake river; and the North fork, which has retained the name of Columbia, as being the main stream.

"We did not go up to the junction, being pressed for time . . .

"From the South Pass to this place is about 1,000 miles; and as it is about the same distance from that pass to the Missouri river at the mouth of the Kansas, it may be assumed that 2,000 miles is the *necessary* land travel in crossing from the United States to the Pacific ocean on this line . . ."

The journey down the Columbia seems to have been begun on the 28th; on the 29th they crossed the Umatilla River near its mouth (evidently following the south bank of the Columbia); on November 2 crossed John Day's River (emptying into the Columbia

2. Ghent states that Marcus Whitman's mission (as of the year 1837) was at ". . . a point called Waiilatpu, 'the place of rye grass,' twenty-five miles from the [Walla Walla] river's mouth . . . among the Cayuse Indians . . ." It was in southern Walla Walla County, not far west of the present city of the same name.

from the south between Gilliam and Sherman counties); and on the 3rd reached "...
the ford of the Fall river, *(Rivière aux Chutes,*[3]*) ...* We delayed here only a short time
to put the gun [presumably the howitzer!] in order ... The roar of the *Falls of the
Columbia* is heard from the heights, where we halted a few moments to enjoy a fine
view of the river below." They had traveled a little inland but on the 4th descended
to "... the *Dalles of the Columbia.* The whole volume of the river at this place passed
between the walls of a chasm ... We passed rapidly three or four miles down the level
valley and encamped at the mission[4] ... Our land journey found here its western
termination." For it was from here that Frémont decided to begin the homeward
journey.

He left Carson in charge of the men and equipment, ordered Fitzpatrick to join the
party there (he was then at Walla Walla), and with Preuss and five others, "a motley
group," started for Fort Vancouver on November 4 by canoe, which he found an enjoyable change.

"Being now upon the ground explored by the South Sea expedition under Captain
Wilkes, and having accomplished the object of uniting my survey with his, and thus
presenting a connected exploration from the Mississippi to the Pacific, and the winter
being at hand, I deemed it necessary to economize time by voyaging in the night, as is
customary here, to avoid the high winds ... which decline with the day."

They were assisted in portaging the Cascades on the 5th and on the 6th reached
Fort Vancouver where Frémont at once called on John McLoughlin and was told to
make himself at home. The place was full of American emigrants bound for the Willamette Valley. In two days—not taking time to visit the Pacific Ocean—Frémont was
ready to start home. He states that McLoughlin bestowed upon him "... a warm and
gratifying sympathy in the suffering which his great experience led him to anticipate
for us in our homeward journey ..." What the Chief Factor probably bestowed was
advice—not to try to cross the mountains so late in the season.

On November 10, availing himself of one of the Company's boats that was bound
up the Columbia, Frémont started back for the Dalles following the south or Oregon
side of the river; he and his party were to travel through Oregon until December 26.
When the Cascades were reached it was necessary to make a portage which seems to
have begun on the 13th and ended the 15th. While this was in progress, Frémont met
a botanist:

"A gentleman named Lüders, a botanist from the city of Hamburg, arrived at the
bay I have called by his name while we were occupied in bringing up the boats. I was
delighted to meet at such a place a man of kindred pursuits; but we had only the pleas-

3. Deschutes River comes into the Columbia from the south, between Sherman and Wasco counties, and
at no great distance east of the Dalles.

4. H. K. W. Perkins, with Daniel Lee, had started a station (on behalf of the American Board of Foreign
Missions) at the Dalles in 1838. H. H. Bancroft's volumes on Oregon refer to Perkins more than once.

ure of a brief conversation, as his canoe, under the guidance of two Indians, was about to run the rapids; and I could not enjoy the satisfaction of regaling him with a breakfast, which, after his recent journey, would have been an extraordinary luxury. All of his few instruments and baggage were in the canoe, and he hurried around by land to meet it in the Grave-yard bay; but he was scarcely out of sight, when, by the carelessness of the Indians, the boat was drawn into the midst of the rapids, and glanced down the river, bottom up, with the loss of every thing it contained. In the natural concern I felt for his misfortune, I gave to the little cove the name of Lüders bay." [5]

The entire expedition assembled at the Dalles, and was ready to start on November 25. Frémont records that he contemplated, in making the homeward journey, ". . . a great circuit to the south and southeast, and the exploration of the Great Basin be-

5. Bancroft moralizes upon this incident: ". . . The toils and dangers of this class of men occupy but little space in history, yet are none the less worthy of mention that they are not performed for gain or political preferment. It is a brave deed to dare the perils of the wilderness for these, in company of hundreds, but much nobler it is for the solitary student of science to risk life for the benefit of mankind."

Very little has been published about Friedrich G. J. Lüders. Asa Gray mentions him in a "Notice of botanical collections" which appeared in 1843 in Silliman's *American Journal of Science and Arts* and which was reprinted, with slight variations, in Hooker's *London Journal of Botany* for 1844. The prospectus was also briefly noted in 1845 in the *Beiträge zur botanischen Zeitung*. Gray was then advertising sets of plants which "three enterprising botanists . . . now engaged in exploring the most interesting portions of the Far West," were expected to collect. Two were traveling, in 1843, with the party of Sir William Drummond Stewart and bound for the Rocky Mountains—Geyer was one, Lüders another. ". . . Mr. Lüders expects to spend next winter, and perhaps the ensuing summer at a station of some Roman Catholic missionaries on the upper waters of Lewis and Clarke's or Great Snake River. These botanists, being well acquainted with the vegetation of the general Valley of the Mississippi and of the lower Missouri [evidently Lüders must have spent some time in the west before starting for the Rockies], will doubtless avoid the common and better known plants of this region . . . their collections may be expected to prove unusually choice and valuable . . . The price of the Rocky Mountain collections . . . is fixed at ten dollars (or two guineas) per hundred . . . The writer . . . cheerfully states that the dried specimens made by these botanists which have fallen under his observation, are well selected, and finely prepared . . . he cordially joins Dr. Engelmann in recommending the enterprise . . ."

Lasègue, writing erroneously of the year 1845, mentions that Geyer and Lüders had reached the Rockies and refers to Lüders' plans for visiting the missionaries, cited by Gray above. Delessert's herbarium in Paris contained none of his specimens, however.

One of Geyer's letters to Hooker (dated August 13, 1845, from Kew) must refer to Lüders although the typewritten copy of the letters uses the name "Lindon"—the copyist may have misread the name, or Geyer may have made a slip: "A Mr. Lindon from Hamburg, Scolar of Prof. Leimann [? Johann Georg Christian Lehmann] set out with me from the States but lost all he had in the Columbia River, barely escaped with his life [according to Frémont he had not been in the wrecked canoe]; went in the Hudson Bay Ship to Honolulu, thence to Valparaiso and joined a French expedition into Patagonia."

BANCROFT, HUBERT HOWE. 1886. The works of . . . 29 (Oregon I. 1834–1848): 420, 421.

EWAN, JOSEPH. 1950. Rocky Mountain naturalists. Denver. 254.

GEYER, KARL ANDREAS. For location of Geyer's letters see p. 770, *fn.* 1.

GRAY, ASA. 1843. Notice of botanical collections. *Am. Jour. Sci. Arts* 45: 225-227.

HOOKER, WILLIAM JACKSON. 1844. Botanical collections of north-west America. *London Jour. Bot.* 3: 139-141. [Briefly noted also in *Bot. Zeit.* 3: 42, 1845.]

LASÈGUE, ANTOINE. 1845. Musée botanique de M. Benjamin Delessert. Notices sur les collections de plantes et la bibliothèque qui le composent; contenant en outre des documents sur les principaux herbiers d'Europe et l'exposé des voyages entrepris dans l'intérêt de la botanique. Paris. 467, 505.

tween the Rocky mountains and the *Sierra Nevada* . . ." He wished to fix three geographical points recorded by report and on maps—Klamath Lake, Mary's Lake and the Buenaventura River. Only the first materialized.

A great part of his "projected line of return" was ". . . absolutely new to geographical, botanical, and geological science—and the subject of reports in relation to lakes, rivers, deserts, and savages hardly above the condition of mere animals, which inflamed desire to know what this *terra incognita* really contained. It was a serious enterprise, at the commencement of winter, to undertake the traverse of such a region, and with a party consisting only of twenty-five persons . . . of many nations . . . courage and confidence animated the whole party . . . the narrative will show at what point, and for what reasons, we were prevented from the complete execution of this plan . . . and how we were forced by desert plains and mountain ranges, and deep snows, far to the south and near to the Pacific ocean, and along the western base of the Sierra Nevada . . . the narrative . . . will first lead us south along the valley of Fall [Deschutes] river, and the eastern base of the Cascade range, to the Tlamath [Klamath] lake . . ."

The wagon which had transported the surveying instruments was abandoned. "The howitzer was the only wheeled carriage now remaining." On the 26th they turned away from the Columbia, and, after crossing a divide between that stream and the Deschutes River—which Frémont calls "Fall river"—camped on one of its main tributaries. For the remainder of November and for the first ten days of January Frémont traveled nearly south up the Deschutes River and its branches until, according to Warren, ". . . within a short distance of its head, where it turned into the mountains on the western side. Here he left it, and crossing a low and heavily timbered divide, came into the basin of Klamath lake." This was on December 10. At Upper Klamath Lake they were in southern Oregon, Klamath County. The lake was left on the 13th, and the party turned east and on the 16th (now traveling through deep snow) came upon another lake "bordered with green grass" upon which they bestowed the name "Summer Lake"; this is in central Lake County. Under the later date of January 29, 1844, Frémont states that it was on December 17 that they entered the Great Basin. On December 20, and still farther southeast, they reached "a handsome sheet of water" upon which Frémont bestowed the name ". . . of Lake Abert, in honor of the chief of the corps to which I belonged." This was also in Lake County. Other lakes, unnamed, are mentioned daily; the course was southeast. On December 26 Frémont states that "By partial observation to-night, our camp was found to be directly on the 42d parallel." This forms both the Oregon-California and the Oregon-Nevada boundary; the party was some ten to fifteen miles east of the first and, according to Warren, they ". . . entered the basin of the Mud Lakes . . ." situated in northern Washoe County, Nevada. They were to remain in Nevada until January 26.

For the remainder of December the party was traveling through snow and there was little food for the animals. "Artemisia was the principal plant, mingled with Fremontia and the chenopodiaceous shrubs. The artemisia was [on December 28] here extremely large, being sometimes a foot in diameter and eight feet high." It seems to have been the only firewood available.

December 31: ". . . our new year's eve was rather a gloomy one. The result of our journey began to be very uncertain; the country was singularly unfavorable to travel; the grasses being frequently of a very unwholesome character, and the hoofs of our animals were so worn and cut by the rocks, that many of them were lame, and could scarcely be got along."

1844

On January 1 the road was down a valley with ". . . a dry-looking black ridge on the left and a more snowy and high one on the right . . ."—presumably the Black Rock and Granite Ranges respectively.

January 3: ". . . Our situation had now become a serious one. We had reached and run over the position where, according to the best maps in my possession, we should have found Mary's lake, or river . . . the appearance of the country was so forbidding, that I was afraid to enter it, and determined to bear away to the southward, keeping close along the mountains, in the full expectation of reaching the Buenaventura river. This morning I put every man . . . on foot—myself, of course, among the rest—and in this manner lightened . . . the loads of the animals . . . Latitude by observation, 40° 48′ 15″."

On January 10 Frémont and Carson, making a reconnaissance of the route ahead, came to ". . . a hollow . . . several miles long, forming a good pass . . ." On reaching its summit they could see ahead ". . . a sheet of green water, some twenty miles broad . . . It was set like a gem in the mountains . . . we concluded it some unknown body of water; which it afterwards proved to be." The pass was to become known as Frémont's Pass and the lake, along the shores of which the party moved on the 14th, Frémont named Pyramid Lake because of a remarkable rock rising from its waters and resembling "the great pyramid of Cheops." On the 15th they came to the mouth of a fresh-water stream which, since they enjoyed some salmon from its waters, was named "Salmon Trout river." It was the Truckee River and they were in southeastern Washoe County. Meeting some Digger Indians they tried to obtain information about the country ahead:
". . . They made on the ground a drawing of the river, which they represented as issuing from another lake in the mountains three or four days distant, in a direction a little west of south; beyond which they drew a mountain; and further still, two rivers; on one of which they told us that people like ourselves travelled . . . I tried unsuccessfully to prevail on some of them to guide us for a few days on the road, but they only looked at each other and laughed."

January 17: "This morning we left the river, which here issues from the mountains on the west. With every stream I now expected to see the great Buenaventura; and Carson hurried eagerly to search, on every one we reached, for beaver cuttings, which he always maintained we should find only on waters that ran to the Pacific; and the absence of such signs was to him a sure indication that the water had no outlet from the great basin . . . We followed the Indian trail . . . which brought us, after 20 miles journey, to another large stream, timbered with cottonwood, and flowing also out of the mountains, but running more directly to the eastward."

This was later to be named Carson River. They followed it eastward for a time, but the animals were in such poor condition that Frémont decided that he had best attempt to reach California. "I . . . determined . . . to cross the Sierra Nevada into the valley of the Sacramento, wherever a practicable pass could be found. My decision . . . diffused new life throughout the camp." Retracing his steps along Carson River, he camped "close to the mountains." But, from a ridge, the country looked forbidding and he determined to make the crossing farther south; after twenty-four miles he reached ". . . another large stream, running off to the northward and eastward . . ." This was the Walker River, reached on January 21, and, according to Fletcher,[6] at a point ". . . a few miles north of the present site of Yerington . . ." in Lyon County, Nevada. From there, on the 22nd, he moved up the river, came to the point where the stream forked, and took the east branch which was followed for some days.

After entering Nevada the "Report" has little to say about plants although the omnipresent sagebrush, the *Fremontia* and *"ephedra occidentalis"* are mentioned. On the 24th Indians brought them a few pounds of seed from a pine tree ". . . which Dr. Torrey has described as a new species, under the name of *pinus monophyllus*;[7] in popular language, it might be called the nut pine." They saw it growing on the mountains on the 25th. There were ". . . heaps of cones lying on the ground, where the Indians had gathered the seeds."

On January 26 they must have crossed into Mono County, California. According to Fletcher, the party ". . . went into camp in the valley north of the present site of Bridgeport where it remained two days in order to recuperate the stock, and to allow Fremont to reconnoitre." The route was to be through California until May 2.

6. F. N. Fletcher's *Early Nevada* . . . supplies, as far as I know, the only detailed interpretation of Frémont's route from the time he reached Walker River, in Nevada, until he reached Sacramento, California. Although the explorer gives a fairly good description of the country, he, himself, did not know where he was for much of the time. Fletcher seems to have made a careful study of the route and I quote from his statements more than once.

7. This is perhaps a reference to the nut-pine rather than to the single-leaf pine, both of which are now considered (by some) to be varieties of *Pinus cembroides*—var. *edulis* and var. *monophylla*, respectively. Frémont did, however, collect the single-leaf pine on this journey, in 1844, at Cajon Pass, San Bernardino County, California, and Torrey described *Pinus monophylla* from that collection in 1845.

Making a reconnaissance on January 27, Frémont came to a pass between the east and west forks of Walker River, and from a height, could see a stream—the west fork —leading northwest and he determined to follow it. Travel was difficult and on the 28th the famous howitzer could not be dragged in to camp; on the 29th it was abandoned on Frémont's orders:

". . . in anticipation of the snow banks and snow fields still ahead, forseeing the inevitable detention to which it would subject us, I reluctantly determined to leave it there for the time.[8] It was of the kind invented by the French for the mountain part of their war in Algiers; and the distance it had come with us proved how well it was adapted to its purpose. We left it, to the great sorrow of the whole party, who were grieved to part with a companion which had made the whole distance from St. Louis, and commanded respect for us on some critical occasions, and which might be needed for the same purpose again."

The party now followed the west fork of Walker River and on the 29th learned from some Indian visitors that ". . . the waters on which we were . . . belong to the Great Basin, in the edge of which we had been since the 17th of December . . . we had still the great ridge on the left to cross before we could reach the Pacific waters." These Indians refused to serve as guides, indicating that snow would make the mountain crossing impossible. Fletcher states that [on the 30th] the party camped in Antelope Valley—in Alpine County presumably—and on the 31st ". . . went over a low pass into the southern end of Carson Valley, and camped . . . on the east fork of the Carson." At this point—after consulting some Indians who explained that before the snow fell ". . . it was six sleeps to the place where the whites lived, but that now it was impossible to cross the mountain . . ."—Frémont explained to his men that it was necessary to make the attempt, depicting that " . . . only about 70 miles distant, was the great farming establishment of Captain Sutter . . ." They ". . . received this decision with . . . cheerful obedience . . ." According to Fletcher, "The course was now up the Carson, and thence to Markleville Creek [in Alpine County] to the hot springs (Grover's) where camp was made on February 3. The next twelve days were occupied in a desperate effort to move the expedition through the deep snow up the mountain sides and across the ridge to what is now known as Hope Valley [in Alpine County] . . ." On that day Frémont commented for the first time upon ". . . a lofty cedar,[9] which here made its first appearance; the usual height was 120 to 130 feet, and one that was measured near by was 6 feet in diameter."

They were on their way up Carson Pass, in Alpine County, and, although they were not to reach the summit until February 20, they were cheered on the 6th by the sight, from an elevation, of what Carson recognized as "the mountains bordering the coast"

8. The volume *Nevada* ("American Guide Series"), p. 33, records that it was found after many years and removed to Virginia City to be used in "celebrations."

9. This was the incense cedar, *Libocedrus decurrens* Torrey. Frémont mentions it again on February 10 and other occasions.

and, nearer at hand, the Sacramento Valley. On the 10th Frémont comments upon the timber:

". . . The forest here has a noble appearance: the tall cedar is abundant; its greatest height being 130 feet, and circumference 20, three or four feet above the ground; and here I see for the first time the white pine,[1] of which there are some magnificent trees. Hemlock spruce is among the timber, occasionally as large as 8 feet in diameter four feet above the ground; but in ascending, it tapers rapidly to less than one foot at the height of 80 feet. I have not seen any higher than 130 feet, and the upper part is frequently broken off by the wind. The white spruce is frequent; and the red pine, (*pinus colorado* of the Mexicans,) which constitutes the beautiful forest along the flanks of the Sierra Nevada to the northward, is here the principal tree, not attaining a greater height than 140 feet, though with sometimes a diameter of 10. Most of these trees seem to differ slightly from those of the same kind on the other side of the continent."

On February 14 Frémont and Preuss climbed ". . . the highest peak to the right; from which we had a beautiful view of a mountain lake at our feet, about fifteen miles in length, and so entirely surrounded by mountains that we could not discover an outlet . . ." This was Lake Tahoe, through which runs the present Nevada-California boundary. The peak from which it was seen was, according to Fletcher, "probably" Stevens Peak, in northern Alpine County. "On *February* 20, 1844," writes Frémont, "we encamped with the animals and all the *materiel* of the camp, on the summit of the Pass in the dividing ridge, 1,000 miles by our travelled road from the Dalles of the Columbia . . . The temperature of boiling water gave the elevation of the encampment 9,338 feet above the sea." Fletcher states that "At Carson Pass the expedition again leaves the Great Basin, crossing the Sierra Nevada summit to the Pacific slope . . ."

The descent of Carson Pass, begun February 21, was as arduous and dangerous as the ascent had been and, although Frémont reached Sutter's Fort on the 6th, it was not until March 8 that the entire party finally arrived at their destination. Fletcher states that the course taken had ". . . led along the open ridges on the north side of the south fork of the American River to its junction with the main stream [the Sacramento], which Fremont erroneously mapped as the outlet of Lake Tahoe . . ." During this period the "Report" includes a number of references to the vegetation. Considering travel-conditions this would seem rather remarkable.

February 24: ". . . Green grass began to make its appearance . . . The character of the forest continued the same . . . the pine with sharp leaves and very large cones was abundant . . . We measured one that had 10 feet diameter, though the height was not more than 130 feet . . . oak trees appeared on the ridge . . . on these I remarked unusually great quantities of mistletoe . . . some beautiful evergreen trees, resembling live

---

1. *Pinus Lambertiana* Douglas. The identification of the tree at this point is made in Frémont's *Memoirs,* doubtless on the authority of Torrey.

oak . . . shaded the little stream. They were forty to fifty feet high, and two in diameter, with a uniform tufted top . . ."

February 25: "The forest was imposing . . . in the magnificence of its trees: some of the pines, bearing large cones, were 10 feet in diameter; cedars . . . abounded, and we measured one 28½ feet in circumference four feet from the ground . . . here . . . in its proper soil and climate. We found it on both sides of the Sierra, but most abundant on the west."

February 27: ". . . A new and singular shrub, which made its appearance since crossing the mountain, was very frequent to-day. It branched out near the ground, forming a clump eight to ten feet high, with pale-green leaves of an oval form, and the body and branches had a naked appearance, as if stripped of the bark, which is very smooth and thin, of a chocolate color, contrasting well with the pale green of the leaves[2] . . ."

March 1: ". . . We are rapidly descending into the spring . . . every thing is getting green; butterflies are swarming; numerous bugs are creeping out . . . and the forest flowers are coming into bloom. Among those which appeared most numerously to-day was *dodecatheon dentatum*."

March 3: ". . . At every step the country improved in beauty; the pines were rapidly disappearing, and oaks became the principal trees of the forest . . . the prevailing tree was the evergreen oak, (. . . the *live oak*;) and with these, occurred frequently a new species of oak bearing a long slender acorn, from an inch to an inch and a half in length . . . the principal vegetable food of the inhabitants of this region[3] . . . the low groves of live oak give the appearance of orchards[4] in an old cultivated country . . . *Dodecatheon dentatum* continued the characteristic plant in flower; the naked-looking shrub already mentioned [the manzanita] . . . beginning to put forth a small white blossom . . ."

March 5: ". . . we discovered three squaws in a little bottom . . . They had large conical baskets, which they were engaged in filling with a small leafy plant *(erodium*

2. In the *Memoirs*, Frémont makes clear that he was writing of one of the manzanitas. *Whatever the species*, I pay tribute to the beauty of this shrub which in its uniformly neat habit, polished mahogany-colored bark, tidy pale-green foliage and clusters of blueberry-like flowers, is handsome wherever met— from Texas westward to the Pacific. In certain florist shops of Boston it is now occasionally offered for sale. I can only hope that it may not become too popular for I must believe that the great branches displayed have taken a long time to grow! In many regions of the southwest it seems to constitute a great part of the chaparral.

3. This was undoubtedly the interior, or Sierra, live oak, *Quercus Wislizenii* described by A. de Candolle in 1864 from, states Eastwood, ". . . specimens collected by Dr. F. A. Wislizenus on the American River in 1851 . . ."

4. Traveling in our southwest, I have often noticed the orchard-like appearance of another plant, the mesquite. On arid hillsides, where soil and moisture can support but a given amount of vegetation, these small trees—not unlike well-pruned apple trees in form—stand out as individual specimens, regularly spaced, as it were in rows.

cicutarium[5]*)* just now beginning to bloom, and covering the ground like a sward of grass."

March 6: ". . . the valley [of the American River, and about ten miles above its union with the Sacramento] being gay with flowers, and some of the banks being absolutely golden with the Californian poppy, *(eschscholtzia crocea.)* Here the grass was smooth and green, and the groves very open; the large oaks throwing a broad shade among sunny spots . . ."

At this point, it was learned from an Indian who spoke Spanish, that Frémont's party had arrived ". . . upon the *Rio de los Americanos,* (the river of the Americans,) . . . the name of *American,* in these distant parts, is applied to the citizens of the United States . . . he went on to say that Capt. Sutter was a very rich man . . . his house . . . was just over the hill . . . in a few miles [we] were met . . . by Capt. Sutter himself. He gave us a most frank and cordial reception . . . and under his hospitable roof we had a night of rest, enjoyment, and refreshment . . ." I have said little of the difficulties endured by Frémont's party for many weeks, but, on going back to meet the contingent that was struggling to complete the journey, their leader notes that ". . . a more forlorn and pitiable sight than they presented cannot well be imagined. They were all on foot—each man, weak and emaciated, leading a mule as weak and emaciated as themselves. They had experienced great difficulty, in descending the mountains . . . many horses fell over the precipices, and were killed; and with some were lost the *packs* they carried. Among these was a mule with the plants which we had collected since leaving Fort Hall, along a line of 2,000 miles travel. Out of 67 horses and mules with which we commenced crossing the Sierra, only 33 reached the valley of the Sacramento, and they only in a condition to be led along . . ."

Frémont reports that on March 8 they ". . . encamped at the junction of . . . the Sacramento and Americanos . . . in the beautiful valley of the Sacramento . . ." The fort of John Augustus Sutter (or Johann August Suter) was situated within what are now the city limits of the city of Sacramento. He had obtained his first grant of land from the Mexican government and had purchased much of his farm equipment from the Russians when they took their departure from their Fort Ross colony in 1841. Joseph B. Chiles,[6] last encountered by Frémont at Fort Hall, had already arrived and had acquired a grant of land on the Sacramento from the Mexican authorities.

---

5. *Erodium cicutarium,* the red-stem filaree, is an introduced plant, native of the Mediterranean, and extensively naturalized. Frémont mentions it frequently. It is said to have arrived by way of the Mexican and Spanish missions.

6. Bancroft's *History of Nevada, Colorado, and Wyoming* mentions the man thus: "Joseph B. Chiles, of the Bartleson company of 1841, having returned to the States, organized a company which in 1843 followed the usual route to Fort Hall, where they divided, some of the men proceeding by a new route by way of Fort Boisé and the Malheur and Pit rivers to the Sacramento Valley, leaving the wagons and families in charge of Joe Walker, acting as guide, to be taken to California by a southern route, through Walker pass and by Owen Peak, the one by which he had returned from California to Great Salt Lake

Much had to be done before Frémont's party was ready to proceed but after two weeks, or on March 22, a preliminary start was made. On the 24th, while camped on the *"Rio de los Cosumnes,"* Frémont had outlined his proposed journey:

"Our direct course home was east; but the Sierra would force us south, above five hundred miles of travelling, to a pass at the head of the San Joaquin river. This pass, reported to be good, was discovered by Mr. Joseph Walker . . . whose name it might therefore appropriately bear. To reach it, our course lay along the valley of the San Joaquin—the river on our right, and the lofty wall of the impassable Sierra on the left. From that pass we were to move southeastwardly, having the Sierra then on the right, and reach the 'Spanish trail,' . . . which constituted the route of the caravans from *Puebla de los Angeles,* near the coast of the Pacific, to *Santa Fé* of New Mexico. From that pass to this trail was about 150 miles. Following that trail through a desert . . . until it turned to the right to cross the Colorado, our course would be northeast until we regained the latitude we had lost in arriving at the Eutah [Utah] lake, and thence to the Rocky mountains at the head of the Arkansas. This course of travelling . . . would occupy a computed distance of two thousand miles before we reached the head of the Arkansas; not a settlement to be seen upon it; and . . . but little trod by *American* feet. Though long . . . this route presented some points of attraction, in tracing the Sierra Nevada—turning the Great Basin, perhaps crossing its rim on the south—completely solving the problem of any river, except the Colorado, from the Rocky mountains on that part of our continent—and seeing the southern extremity of the Great Salt lake, of which the northern part had been examined the year before."

Frémont's route through California (after leaving Fort Sutter until it crossed into Nevada) has been described by the botanist S. B. Parish, in a short article entitled "Fremont in southern California." I shall quote therefrom on occasion and refer to some of Parish's plant identifications. W. H. Brewer's short outline of the route[7] is too brief to supply much detail.

From camp of March 24 the party moved on to the ford of the *"Rio de los Muke-lemnes,"* a rich, fertile country, and where a "showy *lupinus* of extraordinary beauty"

in 1834. This they accomplished, following down the Humboldt to the sink, then to Walker Lake, and over the Sierra; theirs being the first wagons to cross the state, as Bartleson's had been the first to enter Nevada."

On his *westward* journey, in 1833, Joseph Reddeford Walker had entered California by a route overlooking the Yosemite Valley and in a journal kept by one of the party, Zenas Leonard, appears the first known mention of the big tree *Sequoia gigantea.* (See p. 921, *fn.* 9) The various "Walker" rivers and passes mentioned in Frémont's "Report" were named in honor of this man's achievements. Walker joined Frémont's party in Utah, on May 12.

7. According to Brewer, "Captain John C. Fremont traversed the continent in 1843 by way of Great Salt Lake and Humboldt River [incorrect], crossed the Sierra Nevada in midwinter, just south of Lake Tahoe, and entered California in February, 1844, descending the South Fork of the American River. Late in March he set out on his return, passed southward along the eastern edge of the great valley, recrossed the Sierra Nevada at Tehachipi Valley and Fremont's Pass [? perhaps another name for Tehachapi Pass], thence to the Mohave and Virgin Rivers, and eastward, taking with him the earliest collection that had been made in any portion of the Sierra Nevada Range. In the report of this expedition (usually known as his second expedition), Dr. Torrey described about thirty species and four new genera . . ."

and four to five feet in height was in bloom. On the 25th camp was on the *"Arroyo de las Calaveras,* (Skull creek,) a tributary of the San Joaquin," where grew great quantities "of *ammole,* (soap plant,)" and a ". . . vine with a small white flower, (*melothria?*) called here *la yerba buena,* and which from its abundance, gives name to an island and town in the bay [of San Francisco] . . ." They were to keep up the San Joaquin valley until April 7. On March 27, traveling along the "Stanislaus" River, there is another reference to the beauty of the blue-flowered lupine, which was growing ". . . in thickets, some of them being 12 feet in height . . ." and, three or four plants together, forming clumps about ninety feet in circumference. The California poppy was plentiful, also the "*erodium cicutarium* in bloom, eight or ten inches high." By April 1, when the party stopped at the "*Rio de la Merced,* (river of our Lady of Mercy,)" the character of the country had changed from one of extreme fertility to one with a "more sandy," light soil. On the 2nd, after making a boat, they crossed to the east side of the San Joaquin; on the 3rd the foliage of the oak was "getting darker;" on the 4th bands of elk were observed and the prairies were "alive with immense droves of wild horses."

April 5: "Over the bordering plain were interspersed spots of prairie among fields of *tulé* (bulrushes,) which in this country are called *tulares* . . . On the opposite side, a line of timber was visible, which . . . points out the course of the slough, which, at times of high water, connects with the San Joaquin river—a large body of water in the upper part of the valley, called the Tulé lakes."

The next day they halted under some sycamore trees at a point where the San Joaquin came down ". . . from the Sierra with a westerly course . . . checking our way . . ." They found a good ford and crossed to the opposite, or west, side of the river. They were in Fresno County, where the San Joaquin, flowing from the northeast, turns northwestwardly. The Tulare Basin lay southeast.

April 7: "We made . . . a *traverse* from the San Joaquin to the waters of the Tulé lakes . . . over a very level prairie country. We saw wolves frequently during the day, prowling about after the young antelope, which cannot run very fast . . . we discovered timber . . . groves of oak on a dry *arroyo* . . . abundant water in small ponds . . . bordered with bog rushes *(juncus effusus,)* and a tall rush *(scirpus lacustris)* 12 feet high, and surrounded . . . with willow trees in bloom; among them one which resembled *salix myricoides.* The oak of the grove was the same already mentioned, with small leaves, in form like those of the white oak, and forming, with the evergreen oak, the characteristic trees of the valley."

On the 8th they reached ". . . a large stream, called the River of the Lake . . . the principal tributary to the Tulé lakes . . ." and some Indians, finding that they were not "Spanish soldiers," pointed out a good ford. On the 9th they were considerably nearer "the eastern Sierra" but these were still covered with snow. On the 10th camp was on a pretty stream principally timbered ". . . with large cottonwoods, (*populus,* differing from any in Michaux's Sylva.) . . ." The next day the face of the country was covered, not with grass, but with "... *erodium cicutarium,* here only two or three inches high ...", and, "... in bloom, for the first time since leaving the Arkansas waters," Frémont saw "the *mirabilis Jalapa.*"

April 12: "... the country was altogether sandy, and vegetation meagre. *Ephedra occidentalis* ... first seen in the neighborhood of the Pyramid lake, made its appearance ... very abundant, and in large bushes ... In the greater part, the vegetation along our road consisted now of rare and unusual plants, among which many were entirely new. Along the bottoms were thickets consisting of several varieties of shrubs, which made here their first appearance; and among these was *Garrya elliptica,* (Lindley,) a small tree belonging to a very peculiar natural order,[8] and, in its general appearance, (growing in thickets,) resembling willow. It now became common along streams, frequently supplying the place of *salix longifolia.*"

On April 13 the mountains were "very near" and many tracks of Indians and horses "imprinted in the sand" indicated that the creek on which the party traveled (and on the map of the "Report" called "Pass creek") issued therefrom. Parish states that "... It has been thought that Fremont left the San Joaquin valley through Tejon Pass, but an examination of his map, in connection with his itinerary, makes it evident that he crossed through Tehachipi Pass.[9] The summit was reached on the 13th [= 14th] ..."

April 14: "... we continued up the right-hand branch, which was enriched by a profusion of flowers and handsomely wooded with sycamore, oaks, cottonwood, and willow, with other trees and some shrubby plants. In its long strings of balls, this sycamore differs from that of the United States, and is the *platanus occidentalis* of Hooker —a new species, recently described among the plants collected in the voyage of the Sulphur ... Gooseberries, nearly ripe, were very abundant on the mountains; and as we passed the dividing grounds, which were not very easy to ascertain, the air was filled with perfume, as if we were entering a highly cultivated garden ... our pathway and the mountain sides were covered with fields of yellow flowers, which here was the prevailing color ... As we reached the summit of this beautiful pass ... we saw at

8. In such comments, doubtless of later insertion, one recognizes Torrey's knowledge.

9. Warren's "Memoir" states that Frémont "... crossed the Sierra Nevada mountains near their southern end by a 'beautiful pass' now called Tah-ee-cha-pah." Ewan, however, mentions the crossing of "... the southern Sierras at, I believe, Walker Pass in Kern County ..." Both Walker and Tehachapi Passes are in Kern County, the first some forty to fifty miles as the crow flies northeast of the last; both lead into the western margins of the Mojave Desert. Frémont, on March 24, had declared his intention of taking the pass "... discovered by Mr. Walker ..." and evidently thought that he had done so, for on May 12 he refers to "... our descent from Walker's Pass in the Sierra Nevada ..." On May 14 he states that the latitude and longitude of the pass "... may be considered as that of our last encampment, only a few miles distant ..." which had been at "... latitude 35° 17′ 12″ and longitude 118° 35′ 03″ ..." This would seem to point to Tehachapi Pass; however, as indicated, authoritative opinion differs; and, except for historical interest, the matter is unimportant. At the present day one finds the name spelled both *Tehachipi* and *Tehachapi.*

once that here was the place to take leave of all such pleasant scenes . . . The distant mountains were now bald rocks again . . ."

At this point Frémont reflects on his accomplishment to date. He had deviated from his original intention of going east from the Sacramento region and did not regret the change:

"It made me well acquainted with the great range of the Sierra Nevada of the Alta California, and showed that this broad and elevated snowy ridge was a continuation of the Cascade Range of Oregon, between which and the ocean there is still another and a lower range, parallel to the former and to the coast, and which may be called the Coast Range.[1] It also made me well acquainted with the basin of the San Francisco bay, and with the two pretty rivers and their valleys, (the Sacramento and San Joaquin,) which are tributary to that bay; and cleared up some points in geography on which error had long prevailed . . . No river from the interior does, or can, cross the Sierra Nevada . . . the Buenaventura . . . is . . . a small stream of no consequence . . . There is no opening from the bay of San Francisco into the interior of the continent. The two rivers which flow into it are comparatively short, and not perpendicular to the coast, but lateral to it, and having their heads towards Oregon and southern California. They open lines of communication north and south, and not eastwardly . . . this want of interior communication from San Francisco bay . . . gives great additional value to the Columbia, which stands alone as the only great river on the Pacific slope of our continent which leads from the ocean to the Rocky mountains, and opens a line of communication from the sea to the valley of the Mississippi."

Descending from the summit of the pass, still on April 14, they reached ". . . a country of fine grass, where the *erodium cicutarium* finally disappeared, giving place to an excellent quality of bunch grass . . . descending into a hollow where a spring gushed out . . ." they ". . . were struck by the sudden appearance of *yucca* trees,[2] which gave a strange and southern character to the country, and suited well with the dry and desert region we were approaching. Associated with the idea of barren sands, their stiff and ungraceful form makes them to the traveller the most repulsive tree in the vegetable kingdom . . ."

On the 15th and 16th the party moved south with an "illimitable" desert to the east —one from which even the boldest traveler "turned away in despair." The yucca trees still gave "a strange and singular character" to the country.

". . . Several new plants appeared, among which was a zygophyllaceous shrub (*zygophyllum Californicum,* Torr. & Frem.) sometimes 10 feet in height; in form, and in the pliancy of its branches, it is rather a graceful plant. Its leaves are small, covered with a resinous substance; and, particularly when bruised and crushed, exhale a singular but very agreeable and refreshing odor.[3] This shrub and the *yucca,* with many

1. Was Frémont the first to suggest this name?   2. *Yucca brevifolia* Engelm., the Joshua tree.

3. This was certainly the creosote bush, the *Larrea* to some, to others the *Covillea.*

varieties of cactus, make the characteristic features in the vegetation for a long distance eastward . . . in the afternoon emerged from the *yucca* forest . . . and came among . . . fields of flowers . . . which consisted principally of the rich orange-colored Californian poppy, mingled with other flowers of brighter tints . . . nightshade, and . . . buckwheat . . . attracted our attention . . ."

Parish states that Frémont was moving southward ". . . to the Sierra de la Liebra, probably somewhere near the site of the present village of Lancaster . . ." This is in northern Los Angeles County and the region in which, according to the records, the Joshua tree *once* grew to unusual proportions. Parish mentions that it was ". . . during this day's march[4] that *Dalea arborescens* was collected, if we may trust the date of the label." He identifies the nightshade and the buckwheat as ". . . probably, *Solanum Xanti* and *Eriogonum fasciculatum*."

On April 17 the party reached ". . . a small salt lake [Elizabeth Lake, according to Parish,] in a *vallon* lying nearly east and west . . ." Frémont comments that it was a ". . . most beautiful spot of flower fields . . . the hills were purple and orange, with un-broken beds, into which each color was separately gathered. A pale straw color, with a bright yellow the rich red orange of the poppy mingled with fields of purple[5] . . ." Leaving this region of "perfumed air," they passed through a defile overgrown ". . . with the ominous *artemisia tridentata* . . ." and came to another forest of yucca. On the 18th they continued, according to Parish, ". . . along the desert base of the San Gabriel mountains . . ." and camped ". . . at what must have been the wash of Big Rock creek . . ." or, it would seem, near present Littlerock. He identifies the stream which was being followed on the 19th as ". . . none other than Sheep creek . . ." Fré-mont records that there was nothing there "but rock and sand" and a few trees, among them ". . . the nut pine, *(pinus monophyllus.)*[6]" On this day they were ascending Cajon Pass, in southwestern San Bernardino County, and on the next, April 21, reached its summit.

April 20: ". . . a general shout announced that we had struck the great object of our search—the Spanish Trail—which here was running directly north . . . Since the mid-dle of December we had continually been forced south by mountains and deserts, and now would have to make six degrees of *northing,* to regain the latitude on which we wished to cross the Rocky mountains . . . A *road* to travel on, and the *right* course to go, were joyful consolations to us . . . our animals enjoyed the beaten track like our-selves . . . in 15 miles we reached a considerable river . . ."

On the 21st the party rested for a day and on the 22nd started along the trail, and

4. Either on April 15 or 16.

5. Parish offers a guess: the "pale straw-color" *Platystemon,* the "bright yellow" *Baeria,* and the "purple" *Orthocarpus purpurascens.*

6. It may have been on this occasion that Frémont collected the specimen upon which Torrey based his description of *Pinus monophylla.* See under January 24, *ante.*

the river (which they followed until the 25th). They used the camping places of the Santa Fe caravans which had not yet made their yearly passage. "A drove of several thousand horses and mules would entirely have swept away the scanty grass at the watering places . . ." The river, ". . . instead of growing constantly larger, gradually dwindled away, as it was absorbed by the sand." On April 23 the "Indian who spoke Spanish"—he had joined the party on April 13 as guide—informed them that ". . . this river finally disappeared. The two different portions in which water is found had received from the priests two different names; and subsequently I heard it called by the Spaniards the Rio de las Animas, but on the map we have called it the Mohahve river." Parish notes that Frémont may be ". . . considered as the first bestower of the present name of the river [the Mojave]. They must have encountered it near its exit from the mountains, probably at the present Las Flores ranch." [7] He adds that they "travelled down the Mojave river, apparently as far as the subsequent site of Camp Cady." [8] On the 23rd Frémont mentions that, although the country was a desert, it afforded much "to excite the curiosity of a botanist." On that occasion "limited time" and so on prevented "all extended descriptions." On the 24th, however, he mentions two interesting plants. They were crossing a plain, which a Mexican explained extended for some forty or fifty miles, and the Mojave River contained no water:

". . . Here a singular and new species of acacia, with spiral pods or seed vessels, made its first appearance; becoming henceforward, for a considerable distance, a characteristic tree. It was here comparatively large, being about 20 feet in height, with a full and spreading top, the lower branches declining towards the ground. It afterwards occurred of smaller size, frequently in groves, and is very fragrant. It has been called by Dr. Torrey *spirolobium odoratum*. The zygophyllaceous shrub [the creosote bush] had been constantly characteristic of the plains along the river; and here, among many new plants, a new and remarkable species of eriogonum *(eriogonum inflatum,* Torr. & Frem.) made its first appearance."

The small tree with the "spiral pods" is now *Prosopis odorata,* with a number of common names—screwbean mesquite or screwpod mesquite, tornillo and so on; it is not so common as *Prosopis juliflora,* the tree called simply mesquite. Referring to the period spent on the Mojave River, Parish states that there ". . . Frémont collected the type specimens of *Nicolettia occidentalis, Franseria dumosa, Anisocoma acaulis, Hymenoclea salsola, Coelogyne ramosissima, Lepidium Fremontii,* and the imperfect fragment which is the partial type of *Chaenactis Fremontii.* They are all still abundant in the region there traversed. The type of *Ampiachyris Fremontii* is also reported from

7. I do not know the site of this ranch but it may well have been near Victorville, in eastern San Bernardino County.

8. The map of Coville's "Botany of the Death Valley expedition" places Camp Cady on the south side of the Mojave River, at the northern end of a cluster of mountains which appear on more recent maps as the Cady Mountains, San Bernardino County, and in the heart of a triangle formed by the towns of Barstow (west), Ludlow (southeast) and Baker (northeast).

the Mojave river, but recent collections suggest that it may have been collected somewhat further east . . ."

On April 25 Frémont left the Mojave River and regained the trail which had been left for a time and moved northeastward, reaching *"Aqua de Tomaso."* This appears on the Coville map, north of the "sink of Mohave," which is in central San Bernardino County. During the day the country had assumed ". . . the character of an elevated and mountainous desert . . . But, throughout this nakedness of sand and gravel, were many beautiful plants and flowering shrubs, which occurred in many new species, and with greater variety than . . . in the most luxuriant prairie countries; this was a peculiarity of this desert. Even where no grass would take root, the naked sand would bloom with some rich and rare flower . . . Scattered over the plain, and tolerably abundant was a handsome leguminose shrub, three or four feet high, with fine bright-purple flowers. It is a new *psoralea,* and occurred frequently henceforward along our road . . ."

Parish identifies this as ". . . of course, a Dalea, and probably *D. Fremontii,* a species collected in about the same region by the Death Valley expedition."

On the 26th Carson and Godey, who had an encounter with some Indians, returned with two scalps: Frémont comments that "The time, place, object, and numbers, considered . . . this expedition . . . may be considered among the boldest and most disinterested which the annals of western adventure . . . can present . . ." On the 27th the party reached ". . . a large creek of salt and bitter water . . . called by the Spaniards *Amargosa*—the bitter water of the desert." For the past two days the creosote bush and the screwbean mesquite had been common—at the Amargosa Frémont noted ". . . a bed of plants, consisting of a remarkable new genus of *cruciferæ* . . .", which, according to Parish, was ". . . described by Torrey in the appendix to Fremont's Report as *Oxystylis lutea,* and assigned to the Capparidaceae. It is an exceedingly localized plant, and was not again collected until 1891, when it was rediscovered by Mr. Coville, at about the type station." On the 30th Frémont comments upon a plant ". . . with showy yellow flowers *(Stanleya integrifolia)* . . ." which had been abundant for two days.

Frémont and his party had reached a point not far south of the boundary between San Bernardino County (south) and Inyo County (north) and were close also to the California-Nevada line, which was probably crossed on May 2. Parish ceases to follow the journey at that point. Traveling northeastward, the expedition must have entered Nevada in southern Clark County, likely in the region east of the Spring Mountains, of which the Charleston Mountains appear to form a northern extension. The Las Vegas valley lies to the east of both these ranges. From May 2 to May 9 the route crossed Nevada, and in a northeasterly direction.

May 3: ". . . we encamped in the midst of another very large basin, at a camping ground called *las Vegas*—a term which the Spaniards use to signify fertile or marshy plains, in contradistinction to *llanos,* which they apply to dry and sterile plains."

On May 3 Frémont comments that the screwbean mesquite was common and now in bloom with fragrant blossoms, and that they occasionally ate the "bisnada"—the visnaga or barrel cactus—and moistened their mouths "with the acid of the sour dock, *(rumex venosus.)*" After some sixteen hours of travel they reached a "bold running stream" which proved to be ". . . the *Rio de los Angeles* (river of the Angels)—a branch of the *Rio Virgen* (river of the Virgin)." The branch was, likely, the Muddy River. On the 3rd, remaining in camp, they were much annoyed by visits from Indians, speaking "a dialect of the Utah," and generally known by "the name of *Diggers.*"

May 6: ". . . we left the *Rio de los Angeles,* and continued our way through the same desolate and revolting country, where lizards were the only animal, and the tracks of lizard eaters [the Digger Indians] the principal sign of human beings . . . After twenty miles' march . . . we reached the most dreary river I have ever seen . . . The banks were wooded with willow, acacia, and a frequent plant of the country . . . *(Garrya elliptica,)* growing in thickets . . . and bearing a small pink flower . . . Crossing it, we encamped on the left [north] bank . . . The stream was running southwest . . . It proved to be the *Rio Virgen*—a tributary to the Colorado."

The Virgin River[9] was encountered presumably in extreme northeastern Clark County, Nevada, and, since "for several days" the expedition continued "up the river," it must have crossed the extreme northwestern corner of Mojave County, Arizona—a transit of, perhaps, one day's duration—and then moved into the southwestern corner of Washington County, Utah, the region of the Santa Clara Valley. The crossing of Arizona was made, I believe, on May 10 for it was on that date that the "deep cañon" through which the Virgin River flows (between the Beaverdam Mountains, north and the Virgin Mountains, south) was encountered. On May 10 Frémont wrote:

"Our camp was in a basin below a deep cañon—a gap of two thousand feet deep in the mountain—through which the *Rio Virgen* passes, and where no man or beast could follow it. The Spanish trail, which we had lost in the sands of the basin, was on the opposite side of the river. We crossed over to it, and followed it northwardly towards a gap which was visible in the mountain. We approached it by a defile . . . here the country changed its character. From the time we entered the deserts, the mountains had been bald and rocky; here they began to be wooded with cedar [juniper] and pine, and clusters of trees gave shelter to birds—a new and welcome sight . . . Descending a long hollow, towards the narrow valley of a stream, we saw before us a snowy mountain . . . Good bunch grass began to appear on the hill sides, and here we found a singular variety of interesting shrubs. The changed appearance of the country infused among our people a more lively spirit, which was heightened by finding a halting place . . . on the clear waters of the *Santa Clara* fork of the *Rio Virgen.*"

9. From the northeast the Virgin River enters the northern arm of what, since the erection of Boulder Dam, is now Lake Mead.

The party was now not far southwest of St. George—where the Santa Clara enters the Virgin from the northwest—in Washington County, Utah. From May 11 to June 7 the route was in that state.

On May 11 they moved along the Santa Clara River, wooded with "sweet cotton-wood trees" of a "different species from any in Michaux's Sylva," and when the stream forked on the 12th they took the right-hand fork and camped on a ridge believed to be ". . . the dividing chain between the waters of the *Rio Virgen,* which goes south to the Colorado, and those of Sevier river, flowing northwardly, and belonging to the Great Basin. We considered ourselves as crossing the rim of the basin . . . It was, in fact, that *las Vegas de Santa Clara,* which had been so long presented to us as the terminating point of the desert, and where the annual caravan from California to New Mexico halted and recruited for some weeks . . . Counting from the time we reached the desert, and began to skirt, at our descent from Walker's Pass in the Sierra Nevada, we had travelled 550 miles, occupying twenty-seven days, in that inhospitable region . . .

"After we left the *Vegas,* we had the gratification to be joined by the famous hunter and trapper, Mr. Joseph Walker . . . who now became our guide. He had left California with the great caravan; and perceiving, from the signs along the trail, that there was a party of whites ahead, which he judged to be mine, he detached himself from the caravan, with eight men, (Americans,) . . . and succeeded in overtaking us."

Keeping evidently to the west of the mountainous regions which traverse central Utah from north to south, the party, on May 13, started northeastward, descending into ". . . a broad valley the water of which is tributary to Sevier Lake." On the 14th they ". . . came in sight of the Wah-satch range of mountains on the right, white with snow, and here forming the southeast part of the Great Basin . . . We travelled for several days . . . within the rim of the Great Basin, crossing little streams which bore to the left for Sevier lake; and plainly seeing . . . that we were entirely clear of the desert, and approaching the regions which appertained to the system of the Rocky mountains."

May 17: "The Spanish trail had borne off to the southeast, crossing the Wah-satch range. Our course led to the northeast, along the foot of that range, and leaving it on the right . . . Mr. Joseph Walker . . . who has more knowledge of these parts than any man I know, informed me that all the country to the left [west] was unknown to him, and that even the *Digger* tribes . . . could tell him nothing about it."

On May 23 they reached Sevier River, ". . . the main tributary of the lake of the same name—which, deflecting from its northern course, here breaks from the mountains to enter the lake . . . we made little boats (or, rather, rafts) out of bulrushes, and ferried across . . ."—presumably to the eastern side of the river. They were near the present town of Parley, Juab County, where the Sevier River makes an abrupt turn from its northwestern course and flows southwest towards the northern end of Sevier Lake. Frémont comments upon the name of the lake: "It was probably named after some American trapper or hunter, and was the first American name[1] we had met with since leaving the Columbia river.

". . . From the *Dalles* to the point where we turned across the Sierra Nevada, near 1,000 miles, we heard Indians names, and the greater part of the distance none; from Nueva Helvetia (Sacramento) to *las Vegas de Santa Clara,* about 1,000 more, all were Spanish; from the Mississippi to the Pacific, French and American or English were intermixed; and this prevalence of names indicates the national character of the first explorers."

Evidently no visit was made to Sevier Lake for on the 24th they were making their way ". . . towards a high snowy peak, at the foot of which lay the Utah Lake." They were now in Utah County. On the 25th camp was ". . . on the Spanish fork . . . one of the principal tributaries to the lake . . ." They visited Utah Lake on the 26th where Frémont noted that there were two "principal" plants in bloom—". . . the kooyah plant, growing in fields of extraordinary luxuriance, and the *convollaria stellata,* which, from the experience of Mr. Walker, is the most remedial plant known among these Indians." The first was ". . . an abundant supply of food, and the other the most useful among the applications which they use for wounds."

Frémont observes that "We had now accomplished an object we had in view when leaving the Dalles of the Columbia in November last: we had reached the Utah lake; but by a route very different from what we had intended . . . Its greatest breadth is about 15 miles, stretching far to the north, narrowing as it goes, and connecting with the Great Salt Lake. This is the report, and which I believe to be correct; but it is fresh water, while the other is not only salt, but a saturated solution of salt; and here is a problem which requires to be solved."

What Frémont learned was correct in the sense that Utah Lake empties into Great Salt Lake by the Jordan River. At this point the "Report" gives a résumé of the geography of the great territory covered between September, 1845 (when the explorer had been last at Great Salt Lake) and May, 1844 (when he was again in that region). The Great Basin had been circled but its "contents are yet to be examined."

Frémont was now to travel east to Brown's Hole, or Park, in the eastern portion of Daggett County, Utah, in so doing crossing a valley known as Uinta Basin. In E. H. Graham's "Botanical studies in the Uinta Basin of Utah and Colorado" (1937) it is stated that "The first botanical collections which were made in the Uinta Basin were those of two of the Fremont expeditions." That is, on the one under discussion and on the expedition of 1845–1847. Graham, writing of botanical expeditions sent out by the Carnegie Museum in 1931, 1933 and in 1935, makes no attempt to follow Frémont's route of 1844 in detail, merely quoting from Warren's "Memoir" which, as far as the crossing of the Uinta Basin is concerned, merely states that "Turning now [that is from the Utah Lake region] to the east, up the Spanish Fork, he crossed the Wasatch

1. This statement Frémont alters in his "Geographical memoir" (1848): "The river and lake were called by the Spaniards, *Severo,* corrupted by the hunters into *Sevier.*"

mountains at the source of White river. Continuing on toward the east, he passed north of the sources of Uinta river, between it and the Uinta mountains, arrived at Green river and crossed it at 'Brown's Hole.' " The two rivers mentioned by Warren—the White and the Uinta—are identified by Graham as, respectively, Price and Strawberry rivers. I might add that Frémont's botanical specimens were not only the *first* collected in the region but for a considerable number of years the *last;* for Graham, after reviewing the many explorations into, and in the environs of, the Basin, states that "This brings us to the turn of the century [1900], with only Fremont and [Marcus E.] Jones[2] having made botanical collections in the Basin proper and crossing it, and [Sereno] Watson having collected [in 1869] on the northwest rim."

With the data provided in the "Report," on its map, and on a more recent map of Utah (U. S. Geological Survey, Dept. of the Interior, 1921, 1922) it is possible to trace Frémont's route from Utah Lake to Brown's Hole with somewhat more precision, from the time certainly that he reached the upper waters of Strawberry (his "Uintah") River. *Without* a map the route until June 8 (as interpreted below) would mean little, if anything.

Leaving Utah Lake on May 27, he moved southeast, up the Spanish Fork[3] (crossing from Utah County into Wasatch County); he states that the stream was ". . . dispersed in numerous branches among very rugged mountains, which afford few passes . . ." On the 28th, after crossing ". . . by an open and easy pass, the dividing ridge which separates the waters of the Great Basin from those of the Colorado . . .", Frémont reached the headwaters of one of its ". . . larger tributaries . . . White [what is now called Price] river . . ." Continuing up one of its branches for a few miles, the party ". . . crossed a dividing ridge between its waters and those of the Uintah [Strawberry River] . . ." and, by a narrow ravine, descended from the summit [? Soldier Summit] and after the descent camped ". . . by the boiling point . . . 6,900 feet above the sea." On May 29 following the "Uintah," they reached a point where ". . . three forks came together . . ." and continued up ". . . the middle branch . . . named the Red river . . ." and camped ". . . on another tributary of the *Uintah,* called the *Duchesne* fork." In eastern Duchesne County, Red River (from the north) and Avintaguin River (from the south) unite with Strawberry River (which would appear to be Frémont's "middle branch," rather than Red River). Duchesne River (flowing from the north) and Strawberry River (one of Duchesne River's western affluents) unite no great distance east of Frémont's "three forks."

On June 1, the party left "Duchesne fork" and came to ". . . another considerable branch, a river of great velocity, to which the trappers have improperly given the name of Lake Fork." Lake Fork enters the Duchesne from the northwest, not far west of

2. Graham mentions that ". . . all of Jones's specimens which we have seen from the Uinta Basin were collected in the year 1908."

3. The volume *Utah* ("American Guide Series"), p. 52, states that from Utah Lake, Frémont ". . . went east by way of Spanish Fork Canyon . . ."

Myton, eastern Duchesne County. The 2nd was spent in crossing this stream. Although the daily progress of the expedition is indicated by its arrival at named rivers the map of the report indicates that these were reached, not at their confluence with Strawberry, Duchesne and, later, Green rivers, but somewhat upstream.

On June 3 the expedition arrived at ". . . the Uintah fort, a trading post belonging to Mr. A. Roubideau, on the principal fork of the Uintah river . . . a short distance above the two branches which make the river . . ." and remained until June 5. The site was not far north of present Randlett, western Uinta County, and on what is still called the Uinta River.[4] Leaving this post on the 5th, they crossed ". . . a broken country, which afforded, however, a rich addition to our botanical collection . . ." The next stream encountered was "called Ashley's fork." They were now traveling to the west of Green River, through northeastern Uinta County. On the 6th they forded Ashley Creek[5] and encamped ". . . high up on the mountain side, where we found excellent and abundant grass, which we had not hitherto seen. A new species of *elymus,* which had a purgative and weakening effect upon the animals, had occurred frequently since leaving the fort . . . we had a view of the Colorado below, shut up amongst rugged mountains, and which is the recipient of all the streams we had been crossing since we passed the rim of the Great Basin at the head of the Spanish fork."

Frémont's "Colorado" was, of course, Green River. On June 7 the party ". . . descended to 'Brown's hole.'[6] This is a place well known to trappers in the country . . ." It was on this same day (the 7th it seems[7]) that the party camped on "Vermillion creek," which provided brackish water and "indifferent grass." They had entered northwestern Colorado, Moffat County, and were to remain in that state until June 13.

On the 8th they made "a very strong *corál*" on the banks of ". . . the Elk Head river, the principal fork of the Yampah river, commonly called by the trappers the Bear river." They seem to have been traveling almost due east, for this stream enters Yampa River from the northeast, on the boundary between Moffat and Routt counties. On the 11th, following this stream on its northeasterly course, Frémont noted that it was ". . . handsomely and continuously wooded with groves of the narrow-leaved cotton-

4. The Frémont party was not far northwest of Green River, for the explorer's "two branches" (Duchesne River and Uinta River), uniting near Randlett, and flowing southeast, soon enter Green River at what is now Ouray.

5. Also Brush Creek, not mentioned by name in the "Report" but shown on its map. Both Ashley and Brush creeks enter the Green River from the northwest, the first at a point south of Jensen, the last at a point north of the same. Jensen is a point of approach (from Utah) to the Dinosaur National Monument, established in 1915, and located (as is Brown's Hole) in both northeastern Utah and northwestern Colorado. It is now said to comprise three hundred and twenty-seven square miles. Whether Brown's Hole and Dinosaur National Monument overlap in acreage I do not know.

6. The volume *Colorado* ("American Guide Series"), p. 290, refers to Brown's Park, or Hole, as ". . . an almost level valley 30 miles long and 5 miles wide, lying in both Colorado and Utah . . . the park became a rendezvous for the Mountain Men about 1830, when Baptiste Brown, a French-Canadian fur trader settled there . . ."

7. For several days the dates in the "Report" are not clear.

wood *(populus angustifolia);* with these were thickets of willow and *grain de bœuf.* The characteristic plant along the river is *F[remontia] vermicularis* . . . mingled with . . . saline shrubs and artemisia . . ." At night camp was ". . . below a branch of the river, called St. Vrain's fork." Ewan states that Frémont was ". . . north of Hayden [Routt County], reaching what he calls 'St. Vrain's Fork,' which must represent the Little Snake Basin." On the 13th they left that stream and crossed a "dividing ridge" which they had seen ahead for several days and which Ewan identifies as ". . . the Sierra Madre of southern Wyoming, down a canyon of that range called by him Pullam's Fork, and arrived at the present site of Encampment [southern Carbon County, Wyoming]." The party was to travel in Wyoming for two days or until June 15. Frémont wrote of a change in plan:

June 13: ". . . Leaving St. Vrain's fork, we took our way directly towards the summit of the dividing ridge . . . an elevation of 8,000 feet. With joy and exultation we saw ourselves once more on the top of the Rocky mountains, and beheld a little stream taking its course towards the rising sun. It was an affluent of the Platte, called *Pullam's* fork, and we descended . . . upon it . . . in the afternoon, we saw spread out before us the valley of the Platte, with the pass of the Medicine Butte beyond, and some of the Sweet Water Mountains. . . . We were now about two degrees south of the South Pass, and our course home would have been eastwardly; but that would have taken us over ground already examined . . . Southwardly there were objects worthy to be explored, to wit : the approximation of the head waters of three different rivers—the Platte, the Arkansas, and the Grand River fork of the Rio Colorado of the gulf of California; the passes at the heads of these rivers; and the three remarkable mountain coves, called Parks, in which they took their rise . . . We therefore changed our course, and turned up the valley of the Platte instead of going down it . . ."

On the 14th, therefore, they turned southward, up the North Platte River and on the 15th of June re-entered Colorado, finding themselves ". . . in the New Park—a beautiful circular valley of thirty miles diameter, walled in all round with snowy mountains, rich with water and with grass, fringed with pine on the mountain sides below the snow line, and a paradise to all grazing animals . . ." Ewan states that Frémont entered North Park ". . . which he called 'New Park' in contradistinction to 'Old Park,' that is, Middle Park." The party was to travel in Colorado until July 8.

They kept south through the park on the 16th and on the 17th ". . . crossed the summit of the Rocky mountains, through a pass which was one of the most beautiful we had ever seen . . . over an elevation of about 9,000 feet above the level of the sea . . . Descending . . . we found ourselves again on the western waters . . ." They had, according to Ewan, crossed Muddy Pass, the summit of which is given on modern maps as 8,772 feet.

They were now ". . . on the edge of another mountain valley called the Old Park in which is formed Grand river, one of the principal branches of the Colorado of California . . . The appearance of the country in the Old Park is . . . of a different character from the New . . . more or less broken into hills and surrounded by high mountains, timbered on the lower parts with quaking asp and pines." Middle Park is in Grand County. Frémont moved south on the 18th, into "the bottoms of Grand river," the Colorado River, and on the 19th stopped above a canyon where a ". . . southern fork of Grand river . . . makes its junction . . ." and traveled up it that day and the 20th. This was Blue River which enters the Colorado from the south, near Kremmling, southern Grand County. Reaching a point on June 21 when the river forked ". . . into three apparently equal streams . . ." [near Dillon, central Summit County], Frémont took the middle branch:

". . . The mountains exhibit their usual varied growth of flowers . . . I noticed, among others, *thermopsis montana,* whose bright yellow color makes it a showy plant. This has been a characteristic in many parts of the country since leaving the Uintah waters. With fields of iris were *aquilegia cœrulea,* violets, esparcette, and strawberries."

On June 22 the party was still traveling in a southeasterly direction. Frémont states that they crossed ". . . the summit of a dividing ridge, which would . . . have an estimated height of 11,200 feet . . . descended from the summit of the Pass into the creek below . . ." They turned up this creek and camped near its head. Because Frémont was not ". . . able to settle . . . satisfactorily . . ." where he was, and since even Ewan found the day's route "difficult to follow," I shall accept his ultimate conclusion: that Frémont did not actually cross Fremont Pass but, from the valley of Blue River, ". . . crossed Hoosier Pass [on the border between Summit County (north) and Park County (south)] to the Bayou Salade or South Park . . ." in Park County.

On the 23rd Pike's Peak was clearly to be seen. It was "a familiar object, and . . . had for us the face of an old friend." The party had camped at its base on July 16 and 17, 1843, when at Manitou Springs, El Paso County. On the 24th they ". . . crossed a gentle ridge, and, issuing from South Park, found ourselves involved among the broken spurs of mountains which border the great prairie plains." On the 28th they camped on a tributary of the Arkansas, near the main stream, and on June 29 were at Pueblo, ". . . near the mouth of the Fontaine-qui-bouit river . . ."—where they had been last on July 14, 1843. Of the route by which Frémont left South Park, Ewan states that it was *not* by Ute Pass but accepts (on Dellenbaugh's authority) that " '. . . he appears to have cut across to Oil Creek, then to the head of Ute Creek, and finally to Beaver Creek, which he descended to the valley, a very difficult and laborious road.' " "In any case," states Ewan, "Fremont reached Pueblo on July [= June] 29th, 1844 . . ."

On June 30 Frémont's "cavalcade" moved quickly down the Arkansas, and on July 1 reached Bent's Fort. Here lived the "mountaineers" who had taken part in the exploration and four, including Carson and Walker, left the party. On July 5 the journey was resumed down the Arkansas and on the 7th a large stream was crossed which ". . . we are inclined to consider . . . a branch of the Smoky Hill river, possibly it may be the Pawnee fork of the Arkansas." According to Warren, this was ". . . probably Sandy creek of the Arkansas." On the 8th or 9th the party must have crossed from Colorado into Wallace County, Kansas, for on the first of these dates camp was ". . . on the banks of a sandy stream bed . . . which afterwards proved to be the Smoky Hill fork of the Kansas river." The remainder of the journey, or until July 31, was through Kansas.

After continuing down Smoky Hill River "for two hundred and ninety miles," they left it, at the point where the river suddenly bore off to the northwest to join the Republican fork of the Kansas, and continued east; after twenty miles they reached the wagon road from Santa Fe to Independence, and ". . . on the last day of July encamped again at the little town of Kansas, on the banks of the Missouri river." Frémont ends the "Report" with the comment that,

"During our . . . absence of fourteen months . . . no one case of sickness had ever occurred among us. Here ended our land journey; and the day following our arrival, we found ourselves on board a steamboat rapidly gliding down the broad Missouri . . . On the 6th of August we arrived at St. Louis, where the party was finally disbanded, a great number of the men having their homes in the neighborhood."

The plant collections made on this long journey had met with two disasters. One, involving specimens collected between Fort Hall, Idaho, and Carson Pass, California, had occurred between February 21 and March 6, 1844, when a pack animal in descending the pass fell over a precipice. A second occurred on July 13, 1844, when the party was camped on the Kansas River. Frémont mentions that when they halted at night the stream was "less than a hundred yards broad" but that in the morning, after thunderstorms and heavy, continuous rain, ". . . the water suddenly burst over the banks . . . becoming a large river, five or six hundred yards in breadth . . . the baggage was instantly covered, and all our perishable collections almost entirely ruined, and the hard labor of many months destroyed in a moment." In a letter to John Torrey, dated from Washington on September 15, 1844 (it is quoted by Rodgers), Frémont described this flash flood[8] and its consequences in considerable detail and I quote a part:

" '. . . Everything we had was thoroughly soaked. We were obliged to move camp to the Bluffs in a heavy rain which continued for several days and one fine collection

8. Such avalanches of water come, as Frémont notes, "in a moment" and are soon over. The southwest abounds in precautionary reminders of the danger—such as markers warning the motorist not to cross dips in the highway after water reaches a certain height and in the visible evidence along dry streams of brush and débris hanging high in trees along the banks. But newcomers must learn that, although it may be clear and dry on the desert, rain in the distant mountains may with little warning alter the picture on the lowlands.

was entirely ruined . . . I brought them along and such as they are I send them to you. They are broken up & mouldy and decayed, and to-day I tried to change some of them, but found it better to let them alone. Perhaps your familiarity with plants may enable you to make something out of them. You will find them labelled with numbers which correspond to the numbers of notes in my books, which I will copy & send to you in case you can do anything with them . . . From the wreck of our Fossil collection I saved some in which the vegetable impressions seem to me to be very plain & beautiful . . . From the moment the plants were lost, I had formed a determination which has been strengthened by your letter—to return immediately to the interesting regions I have described to you, with the main and leading object of making anew such a collection as will enable us to give a perfect description of the vegetable character of the whole region. . . . we shall run an entirely new line in going out . . . I . . . shall certainly be again on the frontier early in October of next year (1845). In order to have efficient assistance in preparing & changing the plants &c. I take with me a young German gardener who has had the botanical education which they usually receive. We shall also have colored figures of the plants.[9] I trust that you will enter warmly into my enterprise & give me in the course of the winter whatever suggestions may offer themselves to you, tending to ensure our success . . .' "

Frémont's "Report" is prefaced by a "Notice to the reader" which, although it precedes the 1845 printing of both his first and second expeditions, refers only to the one under discussion (1843–1844):

"In the departments of geological and botanical science, I have not ventured to advance any opinions on my own imperfect knowledge of those branches, but have submitted all my specimens to the enlightened judgment of Dr. Torrey, of New Jersey,[1] and

9. On the expedition of 1845–1847, Frémont referred to the fact that ". . . Mr. Edward M. Kern . . . went with me as topographer. He was besides an accomplished artist; his skill in . . . drawing and coloring birds and plants made him a valuable accession . . ." What "young German gardener" Frémont was considering I have not discovered.

1. The association of Dr. Torrey with "New Jersey" doubtless stems from the fact that he owned a place called "Torrey Cliff" which was just north of the New Jersey-New York line, not far from Palisades, Rockland County, New York.

Dr. Cornelius Rea Agnew, an eminent eye specialist of New York, whose close friendship for Torrey began when he was his chemistry pupil at Columbia College, bought property adjacent to "Torrey Cliff" and later added the Torrey property to his own. It was later sold to the C. Philip Coleman family who, in turn, sold it to the Thomas W. Lamonts. Mrs. Lamont gave the estate to Columbia University and on May 28, 1951, the Lamont Geological Observatory was formally dedicated. In the course of these changes both the Torrey and the Agnew houses were torn down. Professor Maurice Ewing, Director of the Observatory, informs me that the main building was once the Lamont residence and that the Director's house is being built on the site of the Torrey house—to his great satisfaction evidently.

Dr. Agnew's daughter, the late Miss S. O. Agnew, has often told me how she attended a small school at "Torrey Cliff" conducted by two of Dr. Torrey's daughters: Miss Elizabeth, or could it have been Eliza, who was intellectual, somewhat terrifying and, since she suffered from neuralgia, with her head enveloped in a shawl; and Miss Margaret, possessed of more "human" qualities. "Sally" Agnew, during class, was alway seated on the protruding legs of a large table.

Dr. Hall, of New York, who have kindly classified and arranged all that I have been able to submit to them. The botanical observations of Dr. Torrey will be furnished in full hereafter, there not being time to complete them now. The remarks of Dr. Hall, on the geological specimens furnished to him, will be found in an appendix[2] to the report . . . Unhappily, much of what we had collected was lost in accidents of serious import to ourselves, as well as to our animals and collections . . . Still, what is saved will be some respectable contribution to botanical science, thanks to the skill and care of Dr. Torrey; and both in geology and botany the maps will be of great value, the profile view showing the elevations at which the specimens were found, and the geographical map showing the localities from which they came."

Section C of the Appendix to the Frémont "Report" relates to the plant collections made in 1843–1844. It opens with a "Note concerning the plants collected in the second expedition of Captain Fremont," written by John Torrey:

"When Captain Frémont set out on his second expedition, he was well provided with paper and other means for making extensive botanical collections; and it was understood that, on his return, we should, conjointly, prepare a full account of his plants, to be appended to his report. About 1,400 species were collected, many of them in regions not before explored by any botanist. In consequence, however, of the great length of the journey, and . . . numerous accidents . . . more than half of his specimens were ruined before he reached the borders of civilization. Even the portion saved was greatly damaged; so that, in many instances, it has been extremely difficult to determine the plants. As there was not sufficient time before the publication of Captain Frémont's report for the proper study of the remains of his collection, it has been deemed advisable to reserve the greater part of them to incorporate with the plants which we expect he will bring with him . . . from his third expedition, upon which he has just set out . . . much of the same country will be passed over again, and some new regions explored . . ."

Following Torrey's "Note" are four fine plates and ten pages of text, entitled "Descriptions of some new genera and species of plants, collected in Captain J. C. Frémont's exploring expedition to Oregon and North California, in the years 1843–'44; By John Torrey and J. C. Frémont." The new genera *Arctomecon* (illustrated in plate II), *Oxystylis,* and *Thamnosma* are still acceptable to the botanists. Certain plants now familiar to travelers in our far west were, in 1845, of sufficient interest to discuss at some length, although not all are asserted to be new species: as examples, a "shrubby species" of *Krameria,* possibly the ". . . *K. parvifolia* of Bentham, described in the voyage of the Sulphur. His plant was only in fruit . . . our specimens are only in flower . . ."; *Prosopis odorata* Torr. & Frém. (illustrated in plate I), the screwbean mesquite,

2. The two sections A and B of the Appendix, contributed by "James Hall, paleontologist to the State of New York," relate respectively to "Geological formations" and to "Organic remains" or fossils (some of plants) and concern material collected on the expedition of 1843–1844, from ". . . Oregon and North California . . ."

". . . we at one time regarded it as a distinct genus, to which we gave the name Spirolobium . . ."; *Cowania plicata* (now *C. Stansburiana* Torr.), ". . . without a ticket . . . It may prove to be a distinct species from the Mexican plant . . ."; and *Pinus monophyllus* (illustrated in plate IV), ". . . extensively diffused over the mountains of Northern California . . . it is alluded to repeatedly, in . . . the narrative, as the *nut pine*." Torrey notes that "The kernel is of a very pleasant flavor, resembling that of *Pinus Pembra* [*Cembra*]." I might add that the artist who drew the plates is not mentioned.

Another paper based upon the plants collected on this journey was Asa Gray's "Characters of some new genera and species of plants of the natural order Compositæ, from the Rocky Mountains and Upper California," published in the *Boston Journal of Natural History* in 1845. I omit Gray's outline of the journey, which may be described as "sketchy" and not always accurate, as well as his references to the losses of the collections, all of which is supplied in prefatory remarks to his enumeration of plants:

"The plants here described are selected from a collection made by Lieutenant Fremont . . . during his recent exploring tour . . . His botanical collections were doomed to . . . mishaps . . . so that his friend and our distinguished associate, Dr. Torrey, received only the *débris* of a collection of dried plants, which, considering the circumstances of the undertaking and the fact that researches in natural history were merely incidental to the main design of Mr. Fremont's tour, was originally of wonderful extent as well as richness. The *Compositæ* of this collection have been kindly submitted to me for examination.[3] But, as some months must elapse before a description of the new species will appear in the forthcoming part of the Flora of North America, by Dr. Torrey and myself, I have deemed it proper to present, in the present form, a few novelties in this family, which, in the course of a hasty examination, have arrested my attention.

"Among the plants of the Asteroid tribe, I notice a new species of a very marked

3. Torrey's "Descriptions of some new genera and species" had prefaced the family *Compositae* with the statement: "The plants of this family were placed in the hands of Dr. Gray . . . and he has described some of them (including four new genera) in the Boston Journal of Natural History for January, 1845. He has since ascertained another new genus among the specimens; and we fully concur with him in the propriety of dedicating it to the late distinguished I [= J]. N. Nicollet, Esq., who spent several years in exploring the country watered by the Mississippi and Missouri rivers . . . This gentleman exerted himself to make known the botany of the country which he explored, and brought home . . . an interesting collection of plants, made under his direction, by Mr. Charles Geyer, of which an account is given in the report of Mr. N[icollet]. The following is the description of this genus by Dr. Gray." Then follows the description of *Nicolletia*—"A humble, branching (and apparently annual) herb"—and of the species *N. occidentalis* Gray which Frémont discovered on the banks of "the Mohahve river," and which (should the reader be interested, and since Gray seems to have reached a considerable number of conclusions from "imperfect materials"!) is a species of the "interesting genus" *Nicolletia*, belonging ". . . to the tribe Senecionideæ, and the sub-tribe Tagitineæ . . . [with] the habit of Dissodia, and . . . both the chaffy pappus of the division *Tageteæ* and the *pappus pilosus of Porophyllum.—Gray*." Torrey states in a footnote that ". . . the notice of this genus by Dr. Gray was drawn up in Latin, but we have given it in English . . ." No one, to date, seems to have disputed Gray's findings!

and characteristic genus of the Rocky Mountain region, viz. Townsendia;[4] a genus founded by Hooker upon a single species, but to which Nuttall has since added four more . . ."

I turn, with a certain amount of relaxation, to Frémont's *Memoirs of my life,* where some forty years later, the explorer—whose botanical comments in the reports had certainly been subjected to the approval of Torrey—does a little "botanical" reminiscing on what may be called "his own." He had planned, so he states, to exclude from his *Memiors* all scientific data, but changed his mind, believing that travelers of scientific bent ". . . would, in travelling over the pages of the book, find a guide to show them the way to the objects they had in mind. A man reading to find something of interest in his particular science would find, and perhaps have lively pleasure in finding, that in the central ridges of the Sierra Nevada is the same rock of which his house is built on the shore of the Atlantic. And, sitting in some English home . . . another would be delighted to find on the plains of the San Joaquin the little golden violet—his Shakespeare's 'Love-in-idleness'— or, in the foothills of the Sierra, in the shade of the evergreen oaks, little fields of the true English crimson-tipped daisy . . . straightway some associations would cluster round the page. It is true that in the lapse of time the face of these regions has changed, but the change is only in degree."

Frémont pays tribute in his *Memoirs* to the disinterested help of those who had studied his collections—to ". . . such men as Torrey and Hall, whose only reward was in the delight they found in extending the confines of knowledge, and in their feeling of satisfaction at the reciprocated pleasure this contribution would give to their *confrères* in other parts of the world. For the men of science are the true cosmopolitans."

The *Memoirs* reprinted the Torrey and Frémont "Descriptions of some new genera and species" (section C of the Appendix of the "Report") and reproduced the four plates, but extremely poorly.

On his long journey Frémont had crossed many regions hitherto untrodden by botanist or plant collector. I mention a few which, from this aspect, were the more important.

He had, traveling west, visited the Great Salt Lake and one of its islands; had traveled through the deserts of western Nevada from the Oregon border as far south as Walker River (passing Pyramid Lake on the way); had reached the Sacramento River valley of California by crossing the Sierra Nevada through Carson Pass and descending the south fork of the American River; had ascended the San Joaquin River, examined the region of the Tulé Lakes, and had entered the western margins of the Mojave Desert by a second crossing of the Sierra Nevada through Tehachapi Pass; had traveled southward through the Mojave Desert and crossed—the third time—the Sierra Nevada, in their Sierra Madre extension, by Cajon Pass and then, turning west and northwest had entered southwestern Nevada and, by way of the Las Vegas valley, had reached the Virgin River; he had followed that stream through the extreme north-

4. Gray's new species was *Townsendia Fremontii,* an unpublished name selected by Torrey and himself.

western corner of Arizona; had entered southwestern Utah in the Santa Clara valley and had traveled north to Utah Lake and then northeast—crossing the Uinta Basin— and had reached the northeastern corner of that state in the region of Brown's Hole;[5] he had then entered the extreme northwestern corner of Colorado, diverged for a short time into southern Wyoming, and had then reëntered Colorado and had traveled south through its three Parks. This route had included, in its overall picture, an encirclement of the Great Basin. The entire journey had consumed but fourteen months.

Reading the "Report," one is conscious of the pressure under which Frémont traveled. Emigration from the United States was well under way and a knowledge of more direct routes to the far west was desirable. But was his haste essential or was the man actuated by the belief that speed, of itself, was praiseworthy? I believe the last and that it was inherent in his makeup. On his fourth journey west, in 1848 and 1849— when the only imaginable reason for haste was the desire to join his wife in California —this same characteristic, supplemented by overconfidence in his own ability to accomplish the impossible, resulted in the death of a third of his companions.

Had Frémont traveled more slowly and, in certain regions at least, at a more seasonable time of year, his accomplishments in natural history—in botany certainly— would have been greater. But, as Gray commented, plant collecting was a side issue to his main task, purely a voluntary contribution, and the eastern botanists who stimulated such endeavors were on the whole fortunate to get as much material as they did!

5. The botanist Wislizenus had visited Brown's Hole in 1839, on what was purely a pleasure excursion.

BANCROFT, HUBERT HOWE. 1886. The works of . . . 21 (California IV. 1840–1845): 434, 438, 444. Map ("Frémont's Map, 1844." p. 442).

——— 1890. *Ibid.* 25 (Nevada, Colorado, and Wyoming 1540–1888): 55-59, 357, *fn.* 19. Map ("Frémont's Route, 1843–4." p. 56).

——— 1889. *Ibid.* 26 (Utah 1540–1886): 32, 33.

——— 1884. *Ibid.* 28 (Northwest coast II. 1800–1846): 694-696.

——— 1886. *Ibid.* 29 (Oregon I. 1834–1848): 419-421.

BREWER, WILLIAM HENRY. 1880. List of persons who have made botanical collections in California. *In:* Watson, Sereno. Geological survey of California. Botany of California 2: 556.

CHITTENDEN, HIRAM MARTIN. 1902. The American fur trade of the far west. A history of the pioneer trading posts and early fur companies of the Missouri valley and the Rocky Mountains and of the overland commerce with Santa Fe. Map. 3 vols. New York.

COVILLE, FREDERICK VERNON. 1893. Botany of the Death Valley Expedition. *Contr. U. S. Nat. Herb.* 4. Map ("Map of parts of California, Nevada, Arizona and Utah . . .").

DELLENBAUGH, FREDERICK SAMUEL. 1914. Frémont and '49; the story of a remarkable career and its relation to the exploration and development of our western territory, especially of California. New York. London. 106-281. Maps. [Map opp. p. 68 shows routes of all Frémont's expeditions.]

EASTWOOD, ALICE. 1939. Early botanical explorers on the Pacific coast and the trees they found there. *Quart. Cal. Hist. Soc.* 18: 344.

EWAN, JOSEPH. 1943. Botanical explorers of Colorado III. John Charles Fremont. *Trail & Timberline* 291: 31-37.

―――― 1950. Rocky Mountain naturalists. 27-33, 211. Denver.

FLETCHER, F. N. 1929. Early Nevada. The period of exploration 1776–1848. Reno. Map. 112-140. [Useful for its interpretation of Frémont's routes through Nevada.]

FRÉMONT, JOHN CHARLES. 1845. Report of the exploring expedition to the Rocky Mountains in the year 1842, and to Oregon and north California, in the years 1843–'44. By Brevet Captain J. C. Fremont, of the Topographical Engineers, under the orders of Col. J. J. Abert, Chief of the Topographical Bureau. *U. S. 28th Cong., 2nd Sess., Sen. Doc.* 11, No. 174. 1-290. [The portion of this report concerned with Frémont's second expedition bears the sub-title: A report of the exploring expedition to Oregon and north California, in the years 1843–'44. 103-390. Frémont's "Notice to the reader" (3-6) refers to Frémont's expeditions of 1842 and 1843–1844.]

―――― 1848. Geographical memoir upon upper California, in illustration of his map of Oregon and California, by John C. Frémont: addressed to the Senate of the United States. Washington. *U. S. 30th Cong., 1st Sess., Sen. Miscel. Doc.* No. 148, 1-67. Map. [Contains many references to the expedition of 1843–1844.]

―――― 1887. Memoirs of my life, by John Charles Frémont. Including in the narrative five journeys of western exploration, during the years 1842, 1843–4, 1845–6–7, 1848–9, 1853–5 . . . A retrospect of fifty years, covering the most eventful periods of modern American history . . . with maps and colored plates. Chicago. New York. 169-410. [Map shows route taken by Frémont in 1843–1844.]

GHENT, WILLIAM JAMES. 1931. The early far west. A narrative outline 1540–1850. New York. Toronto. 262, 329-334.

GOODWIN, CARDINAL. 1922. The trans-Mississippi west (1803–1853). A history of its acquisition and settlement. New York. London. 235-241.

GRAHAM, EDWARD HARRISON. 1937. Botanical studies in the Uinta Basin of Utah and Colorado. *Annals Carnegie Mus. Pittsburg* 26: 12, 18, 20. Map ["Map of Uinta Basin and environs"].

GRAY, ASA. 1845. Characters of some new genera and species of plants of the natural order Compositae, from the Rocky Mountains and upper California. *Boston Jour. Nat. Hist.* 5: 104-111.

―――― 1893. Letters of Asa Gray. Edited by Jane Loring Gray. 2 vols. Boston. New York.

GREENHOW, ROBERT. 1845. The history of Oregon and California, and the other territories on the north-west coast of North America: accompanied by a geographical view and map of those countries, and a number of documents as proofs and illustrations of the history. ed. 2. Boston. 391, 392, & *fn.*

NEVINS, ALLAN. 1939. Frémont. Pathmarker of the west. New York. London. 127-289.

PARISH, SAMUEL BONSALL. 1908. Fremont in southern California. *Muhlenbergia* 4: 57-62.

RODGERS, ANDREW DENNY 3RD. 1942. John Torrey. A story of North American botany. Princeton. 163, 164.

SUDWORTH, GEORGE BISHOP. 1917. The pine trees of the Rocky Mountain region. *U. S. Dept. Agric. Bull.* 480: 21.

TORREY, JOHN. 1845. Note concerning the plants collected in the second expedition of Captain Fremont. *In:* Frémont, J. C. 1845. Report of . . . *U. S. 28th Cong., 2nd Sess., Sen. Doc.* 11: No. 174. Appendix C. 311.

———— & FRÉMONT, JOHN CHARLES. 1845. Descriptions of some new genera and species of plants, collected in Captain J. C. Frémont's exploring expedition to Oregon and north California, in the years 1843–'44. By John Torrey and J. C. Frémont. *In:* Frémont, J. C. 1845. Report of . . . *U. S. 28th Cong. 2nd Sess., Sen. Doc.* 11: No. 174. Appendix C. 311-319. 4 plates.

WARREN, GOUVENEUR KEMBLE. 1859. Memoir to accompany the map of the territory of the United States from the Mississippi River to the Pacific Ocean, giving a brief account of each of the exploring expeditions since A.D. 1800 . . . *U. S. War Dept. Rept. expl. surv. RR Mississippi Pacific* 11: 45-48.

CHAPTER XLIII

LINDHEIMER, RESIDENT IN TEXAS, SPENDS ABOUT NINE YEARS COLLECTING PLANTS FOR ASA GRAY AND GEORGE ENGELMANN

THE name of Ferdinand Jakob Lindheimer connotes, to botanists, the state of Texas. He arrived there in 1836 and, except for brief periods elsewhere, remained until his death in 1879. Approximately nine years (1843–1851) of the forty-three which he spent in the state were devoted to collecting plants as a profession. Although the greater part of his life in Texas was dedicated to promoting the establishment of the German colonists who, at this period, were turning thither for a life of greater freedom, I shall, in this chapter, confine myself mainly to the botanical aspects of his accomplishments. These are recorded in six publications which I cite in chronological order. Although there is some repetition, each has something different to offer.

(1) The first is Ferdinand Roemer's *Texas; mit besonderer Rücksicht auf deutsche Auswanderung und die physische Verhältnisse des Landes nach eigener Beobachtung geschildert,* published at Bonn in 1845. The author, best known as a geologist, became a friend of Lindheimer when on a visit to Texas. The volume contains a contribution by Adolf Scheele entitled "Verzeichniss der von Ferdinand Roemer aus Texas mitgebrachten Pflanzen." Roemer's book was translated into English by Oswald Mueller and published at San Antonio in 1935 as: *Texas with particular reference to German immigration and the physical appearance of the country described through personal observation by Dr. Ferdinand Roemer.*

(2) Part I of "Plantæ Lindheimerianæ; an enumeration of the plants collected in Texas, and distributed to subscribers, by F. Lindheimer, with remarks, and descriptions of new species, &c."[1] This is always cited as the work of George Engelmann and Asa Gray and was published in volume five of the *Journal* of the Boston Society of Natural History in 1845 (pp. 210-264).

(3) In addition to the chapter contributed to Roemer's *Texas,* cited under (1) above, Adolf Scheele[2] published a paper entitled "Beiträge zur Flora von Texas," which came out in seven installments in *Linnaea; ein Journal für die Botanik* from 1848 to 1852 (21: 453-472, 576-602, 747-768, 1848; 22: 145-168, 339-352, 1849;

---

1. The title of the reprint, issued in Boston in 1845, reads: "Plantæ Lindheimerianæ; an enumeration of F. Lindheimer's collection of Texan plants, with remarks, and descriptions of new species, etc." The pages are numbered 1-56. As Blankinship points out, ". . . it is necessary to add 208 . . . to make the page correspond with the same in the Boston Journal of Natural History (5: 210-264) . . ." (p. 207).

2. Georg Heinrich Adolf Scheele (Rehder, *Manual,* 918, 1940).

23: 139-146, 1850; 25: 254-265, 1852). This contained descriptions, not only of plants which Roemer had collected in Texas, but of many which Lindheimer had collected and turned over to Roemer before his departure for Germany—with no idea, certainly, that they would be transferred for publication to anyone, for he was working for Engelmann and Gray.

(4) "Plantæ Lindheimerianæ, Part II. An account of a collection of plants made by F. Lindheimer in the western part of Texas, in the years 1845–6, and 1847–8, with critical remarks, descriptions of new species, &c. By Asa Gray, M.D." [3] This was published in volume six of the *Journal* of the Boston Society of Natural History in January, 1850 (141-235). It ends with Gray's enumeration of the family of Composites for, as in others of his papers, Gray relaxed—to the point of stopping—when he had completed the family in which his interest lay; however, his intentions were good, for this, like his other incompleted papers, was "To be continued." Following Gray's enumeration is a contribution by Engelmann (234-240) which contained ". . . a brief account of the region in which the present collection was made . . ." It had been intended as a preface to Gray's paper, but had been received ". . . too late to take its proper place . . ."

(5) "Plantae Lindheimerianae. Part III." This was the work of Joseph William Blankinship and was published in the eighteenth annual *Report* of the Missouri Botanical Garden in 1907 (123-223). It opens with a discussion of the fascicles of Lindheimer plants which were distributed to subscribers and explains the intricacies of their numbering; then, under the heading "Lindheimer, the Botanist-Editor," Blankinship supplies an account of the collector's life, enumerates the plants described in his own paper, and ends with a "Bibliography of Texas botany" (which lists "Some publications with reference to the botany of this region") and an "Index to Plantae Lindheimerianae Parts I-III." Blankinship states that he had had access to an ". . . unpublished MS. containing Gray's notes on Scheele's 'Flor[4] von Texas,' " undated; to unpublished letters written by Gray to Engelmann; to letters written by members of Lindheimer's family and others, etc.

(6) Lastly, Samuel Wood Geiser's paper, "Ferdinand Jakob Lindheimer," first published with documentation in the *Southwest Review* (15: 245-266, 1930) and subsequently included, with slight revisions and without full documentation, in a chapter in the two editions (1937 and 1948) of his *Naturalists of the frontier*. Geiser's paper on Ferdinand Roemer, similarly issued, contains some references to Lindheimer.

I shall tell first how Lindheimer's association with Engelmann and Gray began. It is recorded in Gray's letters to Engelmann and in the advertisements to the "sets," or "fascicles," of the collector's plants which were offered for sale to interested individ-

---

3. In his "Autobiography" (chapter 1 of the *Letters of Asa Gray*) Gray comments that his medical degree ". . . gives me my place high enough on the Harvard University list to entitle me to a free dinner at Commencement."

4. Correctly *Flora* von Texas.

uals or institutions. I shall then tell of Lindheimer's relationship to Roemer and its unfortunate outcome. Next, tell where Lindheimer did his collecting during the years when he worked for Gray and Engelmann, quoting from Engelmann's and from Geiser's comments upon his whereabouts. And finally discuss the collections themselves.

According to Blankinship, Lindheimer from 1840 to 1843 had engaged in truck farming near Houston, but this had been an unprofitable venture and at the suggestion of "his friend, Dr. George Engelmann of St. Louis," he decided to turn to ". . . collecting the largely unknown flora of Texas and depend upon the sale of his specimens for a living." He had always enjoyed the study of botany and had "devoted much time" to it ". . . while in the university with Engelmann and other botanists.

". . . Moreover, the region in which he was situated was largely unknown botanically, only a few collectors having visited it[5] and the results of their work not having been published. The scattering collections already sent to Engelmann showed clearly the need of a scientific investigation of the plants of this borderland between the American and Mexican floras, and he urged Dr. Gray . . . to join with him and Lindheimer in the exploitation of this unique flora. Accordingly advertisements were inserted in several botanical journals, and in the spring of 1843 Lindheimer began collecting plants in quantity for distribution . . ."

Gray, in Cambridge, wrote Engelmann (July 26, 1842) that he had hopes of working upon *Compositae* during the next month and of spending a part of the time collecting roots and seeds ". . . for our Botanic Garden here; which I wish to renovate, to make creditable to the country and subservient to the advancement of our favorite science. I wish to see growing here all the hardy and half-hardy plants of the United States (as well as many exotics, etc.), and shall exert myself strenuously for their introduction. The Garden contains seven acres; the trees and shrubs are well grown up; we are free from debt, and have a small fund. The people and the corporation are anxious that we should do something, and I trust will second our efforts.

"Allow me therefore to say that yourself and your friend Lindheimer in Texas would render me, and also the cause of botany in this country, the greatest aid . . . if you will send me roots or seeds of any Western plants, especially the rarer, and those not yet figured or cultivated abroad. But nothing peculiar to the West and South will come amiss . . . I shall not be idle myself. I will defray all expenses of collection and transportation . . ."

About six months later he wrote Engelmann again (February 13, 1843):

"I note with interest what you propose in regard to Lindheimer's collections for sale in Centuriæ, fall into your plans, and will advertise in 'Silliman's journal' (and in 'London Journal of Botany') when all is arranged. Pray let him get roots and seeds for me.

---

5. The only important collections antedating those of Lindheimer had been made by Berlandier and by Drummond. Wright, on arrival in Texas in 1837, had done some collecting when his work as a surveyor and teacher permitted; but not until 1849 did he make the journey from San Antonio to El Paso with which his employment by Asa Gray began.

I will do all I can for him. But if the Oregon bill passes, a party under Lieutenant Frémont, or some one else, will go to the Rocky Mountains to Oregon . . . Now why not
send Lindheimer in some of these? . . . The interesting region (the most so in the
world) is the high Rocky Mountains about the sources of the Platte, and thence south.
I will warrant ten dollars per hundred for every decent specimen. If he collects in
Texas, eight dollars per hundred is enough . . . Let me know at once. The opportunity should not be lost. Do send Lindheimer to the Rocky Mountains if possible."

As agreed, Gray advertised the sets in Silliman's *American Journal of Science and
Arts* in July of 1843.[6] After reporting that Geyer and Lüders were also engaged ". . .
exploring the most interesting portions of the *far West* . . ." he mentions that a ". . .
third collector, Dr. Lindheimer, a very assiduous botanist, intends to devote a few
years to the exploration of Texas; and he pledges himself to exclude from his sets all
the common plants[7] of the southwestern United States.

"These . . . collections will be assorted and distributed, and for the most part
ticketed, by Dr. Engelmann of St. Louis; assisted, as far as need be, by the authors of
the *Flora of North America,* who promise to determine the plants, so far at least as
they belong to families published in that work . . . The number of sets being limited,
earlier subscribers will receive a preference . . ."

Presumably the extra recompense of two dollars for Rocky Mountains collections
bore some relation to Gray's concept of the difficulties involved. The advertisements
have a somewhat "canny" ring: the sets were to be paid for when they reached St.
Louis—not when they reached the subscriber—and any "additional expense of . . .
transportation, doubtless trifling in amount," was to be charged for delivery to London, or elsewhere!

The collections made in 1843 and in 1844 were smaller than had been expected, but
were fine material and by April 8, 1846, Gray was anxious for more and wrote Engelmann:

"What is Lindheimer about? Why is not his last year's collection yet with you? We
have just got things going, and we can sell fifty sets right off of his further collections,
and he can go on and realize a handsome sum of money, if he will only work now!
And he will connect his name forever with Texan Flora! . . ."

By July 2, 1846, Gray's hopes of sending a collector to Santa Fe were approaching
fulfillment, for it had been arranged that Fendler was to go there; his ability, however, was still untested[8] and possibly Gray thought it might be well to have two irons
in the Santa Fe fire. His obliviousness to Indians and hostile Mexicans was complete:

"It is said that a corps of troops is to be sent up through Texas towards New Spain.

6. A very similar notice was published by Hooker in 1844 in the *London Journal of Botany.*

7. Gray is so careful to specify this fact in his various notices that one wonders whether there had not
been complaints from subscribers about some of the material already received.

8. On July 15, 1846, Gray wrote Engelmann: ". . . Fendler has money enough to begin with. As soon as
he is in the field, and shown by his first collecting that he is deserving, I can get much more money ad-

Lindheimer ought to go along, and so get high up into the country, where so much is new, and the plants have really 'no Latin names.' "

In 1846 Lindheimer's collections were advertised again by W. J. Hooker in the *London Journal of Botany:*

". . . The first series have been distributed, and we can confidently say that finer and better prepared, or better selected specimens, have seldom come under the notice of Botanists, and Mr. Lindheimer has, as he pledged himself to do, excluded from them the common plants of the South-western States. The species are all labelled and numbered, and a list of names has been already published by Dr. Asa Gray, and copies of this Catalogue have reached this country.

"One object of the present notice, is to give the opportunity for saying that, together with the sets ordered by ourselves and friends, there have come two which are undisposed of, and which can be had by applying to the Editor of this Journal, at the price mentioned above (Vol. III. p. 140) with the addition of the share of freight. One of the two contains 186 species, the other 181 species."

One more reference to Lindheimer's collecting plans appears in Gray's letters to Engelmann of January 5, 1847. Engelmann had evidently decided where it was feasible, all things considered, for Lindheimer to collect and, from the vantage point of St. Louis, was certainly in a better locality than Gray, in Cambridge, to make the decision. Texas offered a good field and Lindheimer, concentrating on its flora, was to make a name for himself by so doing, whatever the height of the "Mountains"! With reservations, Gray was satisfied with the choice of locale:

"I am glad so fine a collection is on the way from Lindheimer, and greatly approve his going to the mountains on the Guadaloupe. How high are the Mountains? If good, real mountains, and he can get to them, and into secluded valleys, he will do great things . . ."

Nothing derogatory of Lindheimer's abilities as a collector seems ever to have found its way into print. Only Geyer—who once expressed the belief, subsequently modified, that he himself had amassed twenty thousand specimens on his transcontinental trip —found a belittling adjective: "A Mr. Lindheimer resident in Texas, is travelling and collecting there for some years but his collections are not very voluminous."

Roemer's *Texas* (chapter eight) gives a pleasant, firsthand picture of Lindheimer, describing him in the field, at his home on the Comal, and so on. I quote from Mueller's translation:

"During the first days of my stay in New Braunfels, I made an acquaintance which proved to be pleasant and valuable to me during the entire time I was there . . .

"At the end of the town, some distance from the last house, half hidden beneath a group of elm and oak trees, stood a hut or little house close to the banks of the Comal

vanced for him . . . If he only makes as good and handsome specimens as Lindheimer, all will be well . . . instruct him to get into the high mountains, or as high as he can find . . . The mountains to the north of Santa Fé often rise to the snow-line, and are perfectly full of new things . . ."

... When I neared this simple, rustic home, I spied a man ... busily engaged in splitting wood. Apparently he was used to this kind of work. As far as the thick black beard, covering his face, permitted me to judge, the man was in his early forties. He wore a blue jacket, open at the front, yellow trousers and the coarse shoes customarily worn by farmers in the vicinity. Near him were two beautiful brown-spotted bird dogs, and a dark-colored pony was tied to a nearby tree. The description fitted the man I was looking for. His answer to my question, given in a soft, almost timorous voice, which did not seem in harmony with the rough exterior of the man, confirmed my conjecture. It was the botanist, Mr. Ferdinand Lindheimer, of Frankfort-on-the-Main ..." (p. 107)

"... He bought a two-wheeled covered cart and a horse, loaded it with paper necessary to pack his plants, and a supply of the most necessary articles of food, such as flour, coffee and salt. Thereupon, he sallied forth into the wilderness armed with a gun and no other companions but his two hunting dogs. Here he busied himself with gathering and preserving plants. At times he was solely dependent upon the hunt for food sometimes not seeing a human being for months ..." (p. 109)

"... He ... began to work the rich and largely unknown flora of the surrounding country systematically with a leisure and convenience hitherto not enjoyed in Texas. He was soon convinced, however, that he could not collect properly and at the same time meet the demands which housekeeping imposed upon him. When he came home at night, tired from collecting plants, he was obliged to prepare his own meals first; when he tore his clothes in the thick underbrush of the forest ... he had to take needle and thread to repair the damage done; when he needed a clean shirt, he had to go to the river and wash it himself.

"He took the proper steps to remedy this situation thoroughly. He sought and found a companion among the daughters of the newly arrived emigrants ... The wife[9] not only takes care of the household duties, but also thoroughly understands the different processes of drying plants. She should rightly share in the praise which the botanists, who receive Lindheimer's plants from Texas, heap upon him on account of their keeping quality[1] and their careful preparation. Accompanied by Lindheimer, on foot and on horseback, I learned to know the near and distant surroundings of my new home through daily excursions. One of the first of this kind was the springs of the Comal ..." (p. 108)

Despite Roemer's reference to daily excursions with Lindheimer, only in the chapter quoted have I found specific references to their companionship: they were together at Mission Hill on February 9, 1846, and on another occasion "... followed the

9. According to Blankinship, Lindheimer married "Eleonore Reinarz of Aachen ... in 1846 ..."

1. Blankinship, when referring in 1907 to Lindheimer specimens in the Engelmann herbarium, comments that they "... are in a fairly good state of preservation, considering the lapse of more than half a century since their collection, the ravages of the usual herbarium pests and the accidents of transportation and storage during this time."

Guadaloupe several miles in its downward course." Although, when on occasion a number of plants are mentioned, one suspects the presence of the botanist, Roemer's references to their joint excursions end with those cited.

The results of these excursions, which were certainly detrimental to Lindheimer's efforts and to those of his sponsors, are described by Blankinship. Scheele's breach of botanical ethics was scarcely to be expected in a member of the clergy!

"Lindheimer and Roemer made many botanical excursions together during 1846 and the value of the latter's collections is largely due to Lindheimer's aid in the work. At the end of the season they appear to have exchanged a set of the collections made by each during the year, and Roemer, on his return to Germany, placed Lindheimer's with his own botanical specimens in the hands of Adolph Scheele, Pastor at Heersum near Hildesheim, who prepared a list of the species for Roemer's 'Texas,' and published the descriptions in Linnaea from 1848 to 1852 in his 'Beiträge zur Flor von Texas.' Not only did he publish the 'new species' of Roemer's collecting, but also those found among Lindheimer's duplicates, though he knew that Engelmann and Gray had already undertaken to describe these collections in their Plantae Lindheimerianae, and so industriously did he continue his work that he soon completely outdistanced his American competitors and left little for them to describe. This may have had something to do with the discontinuance of the Plantae Lindheimerianae, but not the slightest blame can be attached to Lindheimer, for he doubtless had no idea that any publication on his own collections was intended at the time the exchange was made . . ." (pp. 136, 137)

According to Blankinship, "Of the species from Texas described as new by Scheele, 73 were collected by Lindheimer and 66 by Roemer."

In his biographical sketch of Lindheimer already mentioned, Blankinship refers to the fact—interesting to read in 1949 when the bicentenary of the poet's birth was being celebrated—that Goethe was one of his ancestors. We are told that Lindheimer was born ". . . at Frankfort-on-the-Main, May 21, 1801, and died at New Braunfels, Texas, Dec. 2, 1879." His family was evidently prosperous and consequently he had been given ". . . the best education obtainable, attending a preparatory school in Berlin and finishing his education at Wiesbaden and Bonn, taking his degree at the latter university in 1827, after which he accepted a position in the Bunsen Institute . . . in his native city and taught there till 1833, when it was closed by the government and both he and George Bunsen were compelled to emigrate . . . young Lindheimer soon after . . . taking his patrimony, sailed for America early in the spring of 1834. He landed at New York . . ."

From there he eventually reached St. Louis and moved on to the German settlement at Belleville, Illinois, where the Engelmann family was living. Next he went to New Orleans, hoping to get into Texas, which Blankinship describes as at this time ". . . a terra incognita, the borderland between two hostile civilizations and ravaged alternately by bandits and Indians, and not even a map of the country could be found." Such being the case, he went instead to Mexico where he remained "some sixteen months" and then sailed for New Orleans. The boat was wrecked off Mobile, Alabama, and there Lindheimer ". . . enlisted at once in a company of volunteers forming to aid the Texas revolutionists . . . This company . . . on its arrival in Texas was stationed on Galveston Island . . ." The volunteers, ordered to join General Houston, did not get there until the day after the battle of San Jacinto (April 22, 1836). With the disbanding of the army ". . . Lindheimer seems to have come north to St. Louis and spent the summer of 1839 and probably the following winter here,[2] but the climate was too severe for his lungs and he again took up his residence in the new republic of Texas. He located near Houston and engaged in truck-farming (1840–1843) . . ."

This brings us to the period in which Lindheimer's work for Asa Gray began. Woven into the overall story of Lindheimer's life as told by Engelmann, by Blankinship and by Geiser, are references to the regions where he did his collecting from 1844 through 1851. These records vary slightly, or perhaps are merely expressed in differing geographical landmarks, and I shall quote from the two which seem to contain the most "meat," those of Engelmann and of Geiser.

Engelmann's records are found in Part II of "Plantæ Lindheimerianæ" where, as noted under (4) above, he supplies (pp. 234, 235) a "brief account" of the regions visited by Lindheimer. I italicize the years:

"In November, *1844,* Mr. Lindheimer left the neighborhood of the Brazos River, where he had made his collections in 1843 and 1844, and reached in January, *1845,* the shores of Matagorda Bay. In this and the following month he collected on the lower Guadaloupe. From thence he went up this river about one hundred miles. Here, where the Comale Creek empties into the Guadaloupe, the Association of German emigrants, with whom he had for the present joined his fortunes, selected a place for settlement and laid the foundations of New Braunfels, now [1850] a flourishing town, and the county seat of Comale County. The year *1845* was spent in exploring the country and making excursions in the mountainous region to the west and northwest, at that time very insecure, being the haunts of wild Indian tribes.[3] In . . . *1846,* Mr. Lindheimer made large collections in the interesting country about New Braunfels . . . The explorations of the year *1847* were extended northwest to the country watered by the Pierdenales River, where another German settlement, Friedrichsburg (or Fredericksburg), had been founded. Collections were made partly here and partly near New Braunfels. Late in the fall an excursion in a northern direction into the granitic re-

2. Blankinship bases this qualified statement as to Lindheimer's whereabouts on the fact that "A number of specimens in the Engelmann herbarium are labeled 'St. Louis. 1839. Lindheimer,' while similarly we find he was at San Felipe, Texas, in March and New Orleans in April of that year on his way up."

3. Blankinship mentions that Lindheimer ". . . was perfectly fearless of danger . . . and his immunity from the Indians is largely due to this fact, though he appears to have been held by them in extreme reverence as a 'medicine man,' who wandered aimlessly about securing herbs for his decoctions and incantations, and many are the stories told of his adventures with them during these troublous times."

gion of the Liano [Llano] river furnished some interesting plants not observed before. The year *1848* was spent principally on the Liano, where several new German settlements had been formed. But the country appeared to be less rich in botanical treasures than had been expected; the burning sun of the summer months had almost destroyed the vegetation on the granitic soil . . . Indian tribes . . . became troublesome, and the frontier settlements had to be abandoned. The spring of *1849* found Mr. Lindheimer farther south, at Comanche Spring, one of the headwaters of San Antonio River. He has now (in the spring of *1850*) returned to New Braunfels, where he intends again to go over the as yet insufficiently explored country, the most diversified and richest in botanical treasures as yet seen by him in Texas."

Following the above, Engelmann gives ". . . a short geographical and topographical sketch of the country explored by Mr. Lindheimer . . ." (pp. 235-238) and then adds ". . . a few details of localities and distances, which may not be found on the common maps." (pp. 238, 239) These paragraphs, as far as I know, contain the first published overall picture of the vegetation of any portion of the state of Texas[4] and for that reason, if for no other, are of interest. I omit a few references to climate, to geological formations and so on.

"Matagorda Bay, with its numerous branches, receives to the northeast the Colorado, one of the largest rivers of Texas. Southwest of the Colorado the smaller Guadaloupe River empties into the same bay after receiving not far from its mouth its southern branch, the San Antonio River. The headwaters of these rivers, together with the southern branches of the Colorado, drain the country investigated by Mr. Lindheimer since 1845.

"The coast of the bay itself forms a level saline plain, sandy . . . Cakile, Œnothera Drummondi, and Teucrium Cubense are characteristic plants: a little farther off are found Berberis trifoliata, Acacia Farnesiana, a shrubby Erythrina, groves of Sophora speciosa, Condalia, some large Yuccas, and large Opuntias with humbler Cactaceæ beneath them.

"Some miles higher up the rivers, on the clayey soil, solitary Elms and Palm trees are seen; the prairies have a stiff black soil thickly matted with grass. The prevalent tree now becomes the Live Oak along the rivers, as well as in small groves on the prairies, higher up on the rivers the Water Oak and the Spanish Oak (Q. falcata) are found mixed with the Live Oak. Swampy places are often densely covered with Marsilea macropoda, like fields of clover.

"Ten to twenty miles from the coast the country rises into the 'rolling prairies.' Along the rivers Quercus macrocarpa, Taxodium distichum, and Carya olivæformis constitute large forests of vigorous growth. The groves of the prairies are principally formed by Sophora speciosa, Condalia obovata, and Diospyros Texana. The prairies

4. See also Engelmann's paper "On the character of the vegetation of southwestern Texas" *Proc. Am. Assoc. Advancement Sci.* 5: 223-228 (1851).

themselves are richly studded by flowers, among which the blue and fragrant Lupinus Texensis and different species of red and yellow Castilejas are most conspicuous.

"About one hundred miles from the coast the country becomes hilly . . . the streams are more rapid and clear . . . Elm and Cypress are the principal trees along the rivers; Sycamores, Linden, and Hackberry are sparsely mixed with them. Many curious shrubs, among them the Ungnadia, are found in these river-forests. Here, also, the Muskit trees (Algarobia) make their first appearance, indicating the region of the Arborescent Mimoseæ; they form open woods, where the level ground, often over-flowed in the rainy season, brings forth abundance of the thin and wiry but nutritious 'Muskit grasses' (Aristida, Atheropogon, and others). Many other interesting plants are found in these 'Muskit-flats.'

"In this region, and at the base of the first plateau, are located the towns of San Antonio, New Braunfels, and Austin . . .

"A short distance north of this region, steep and sterile declivities . . . rise to the first plateau just mentioned. The high plains which are now reached are mostly sterile and stony . . . Many interesting plants mentioned in this catalogue, are peculiar to these plains: the smaller Cactaceæ, Echinocactus setispinus, Cereus cæspitosus, several Mammillariæ, and prostrate Opuntiæ grow here; different species of Yucca are common; the curious and stately Dasylirion[5] is here first met with. The trees of this region are Elms and Cedar among the rocks, and Cedar again, finely developed, along the banks of the streams, where Cercis occidentalis, the shrubby Red Bud, forms thickets. Juglans fructicosa and Morus parvifolia are here found; the Live Oak dwindles down to a shrub; and low bushes of Vitis rupestris, the mountain grape, cover large tracts of these plains.

"Twenty to thirty miles farther northwest the country rises again and becomes more hilly . . . producing many peculiar plants. The valleys between them are often wide, with a thin soil, covered with grass and often with sparse Post Oaks; or they are narrower, without any timber, but more fertile . . . The springs are here numerous . . . the streams clear and rapid. The beds of the larger watercourses are often entirely dry in summer, leaving a wide, stony or pebbly bed or naked rocks, abounding with interesting plants. The banks of the deeper streams are thickly covered with stately Cypress trees.

"A few miles north of the Pierdenales the first outlier of the granitic formation is seen, which is found extensively developed on the Liano. The vegetation here begins to show analogies to that of New Mexico . . .

"Green Lake and Caritas River are in the low lands near Matagorda Bay. Victoria is a town a little higher up on the Guadaloupe. New Braunfels on the Comale

5. In the windows of eastern florists the foliage of the *Dasylirion*, under the name of "spoon plant," is now much displayed, turned downside up in vases! The glossy, yellowish base of the leaf is presumably the "spoon," and the slender sage-green blade (cut off at the point where it ceases to be stiff) is what might be called the handle!

Creek and Guadaloupe River, is about one hundred miles to the northwest of the Bay, twenty-five miles northeast of San Antonio, and forty-five miles southwest of Austin, the present capital of Texas. The road from New Braunfels to San Antonio crosses the Cibolo, one of the confluents of San Antonio River ... The Salado, one of the heads of which is the often-mentioned Comanche Spring, is another branch of San Antonio river, and such, farther south, are the Leona and the Medina.

"In going west from New Braufels we reach, thirty-five miles from that town, the upper waters of the Guadaloupe, the so-called Guadaloupe crossings on the Pinto-trail. Several small streams in this neighborhood, Spring Creek, Wasp Creek, Three Creeks, and Sabinas (or Cypress Creek) are often mentioned as localities of different plants.

"North of this the road crosses several high ridges, (where, among other plants, Guajacum angustifolium, and in deep, clear ponds Chara translucens, were discovered), and reaches, sixty miles from the Guadaloupe, the Pierdenales, one of the branches of Colorado River. The town of Friedrichsburg is built near the Pierdenales in a rather barren, sandy region, thinly scattered with Post Oaks.

"About thirty-five miles north of this granitic region of the Llano or Liano is reached. The San Saba runs thirty miles farther north.

"The Flora of the country east of the Brazos River bears considerable resemblance to that of the southern United States. But south of the Brazos, and still more south of the Colorado, the character of the vegetation changes; it assumes the peculiarity of the flora of the Rio Grande valley, which I have tried to characterize in Wislizenus' Report. The flora of the Rio Grande connects the North American with the Mexican flora, and has also many peculiar plants of its own, some of which have for the first time been distributed in Lindheimer's collections: such are the interesting Rutosoma, the only American Rutacea known; Galphimia linifolia, the most northern Malpighiacea; several shrubby Mimoseæ; an evergreen Rhus; Sophora speciosa; the Eysenhardtia; a number of Nyctaginaceæ; the Dasylirion, and many others enumerated in this catalogue. The ligneous plants become shrubby and often thorny, and here the chaparals, so famous in northern Mexico, make their first appearance.

"Towards the northwest the granitic soil produces a number of plants, which indicate a connection with the flora of New Mexico, and again with that of our western plains."

Engelmann ends with a paragraph on "deserted ant-hills" (which I omit) and another of greater interest to botanists:

"In the neighborhood of New Braunfels the effects of cultivation on the distribution of plants are already apparent. Helianthus lenticularis, Verbesina Virginica, Croton ellipticum, Nycterium lobatum, different Cenopodiaceæ and Amaranthaceæ are becoming very common in cultivated places; but others, Digitaria sanguinalis, for example, so common in eastern Texas, have not yet made their appearance. In Cedar woods, Leria nutans, in damp bottom woods Dicliptera brachiata, on dry prairies the small blue Evolvulus, are getting much more abundant; while Pinaropappus roseus, Fedia stenocarpa and others are much rarer than they used to be in the first years of the settlement of the country."

I might add that Engelmann's remarks (pp. 234-240) quoted above are omitted by the editors of his botanical papers with the comment (p. 510, *fn.*) that they are "of little botanical interest."!

Geiser's chapter on Lindheimer contains many interesting facts about the collector and about the period in which he worked and I shall refer to some of these again. Here I shall merely extract what he has to tell of Lindheimer's whereabouts in the years he collected for Gray. I quote from the second edition of *Naturalists of the frontier* (pp. 136-140), occasionally altering the paragraphing and italicizing the years:

"Little is known of Lindheimer's life during the three years following the battle of San Jacinto . . . After the war for freedom, Lindheimer seems to have been completely submerged in the flux of incoming settlers . . . there is no scrap of evidence concerning his activities.

"During this period Lindheimer no doubt was collecting plants and corresponding on things botanical with his intimate friend Engelmann, who was at St. Louis. But the first definite information . . . concerning him after the war is found in specimens collected for Engelmann at San Felipe in March of 1839, as Lindheimer was on his way to New Orleans and St. Louis. From this time on, the record is ample."

The years of truck farming near Houston followed and Geiser states: "Fortified by this arrangement Lindheimer . . . left for Texas early in March of 1843. He was never to venture north again . . .

". . . he arrived in Galveston by the end of March [*1843*]; the herbarium of the Missouri Botanical Garden has specimens . . . labelled by Lindheimer at Galveston during that month . . .

"Until the first of June, *1843,* Lindheimer was occupied in making extensive collections around Houston. Early in that month he left Houston for the Brazos bottom in present Waller and Austin counties, and collected in the bottoms during the major part of June and July. In early August Lindheimer crossed the Brazos, and for a few days collected west of the river, probably near the present towns of Sealy and Bellville. After returning to Houston in the middle of August, at the end of the month he made a collecting trip to Chocolate Bayou, fifty miles to the south, and collected there in late September and early October. He returned to Houston by way of Galveston: his return was slow, for we find plant records reading 'Galveston Island, Oct., Nov., 1843' on his herbarium labels.

"In *1844* Lindheimer spent the whole season west of San Felipe, between the Brazos and the Colorado. He left San Felipe in February, and probably did not go beyond the confines of present Austin and Colorado counties. During this trip Lindheimer lived for a short time . . . at Cat Spring in Austin County . . . Lindheimer's collections also show that he spent three months of this season at Industry, in Austin County . . . It is probable that Lindheimer and Ervendberg[6] went together to Port Lavaca in December of 1844 to meet the first group of immigrants . . . they . . . accompanied the party on the slow journey up the Guadalupe to New Braunfels. Lindheimer's collections on this trip show specimens from Matagorda Bay and the Guadalupe bottom in Victoria County . . ."

Geiser does not actually mention the year *1845* but, since his last date was December of 1844, his reference to the fact that Lindheimer ". . . also explored the wild mountainous region to the northeast [of New Braunfels], a country still occupied by Indians," doubtless refers to 1845. Moreover, it conforms to Engelmann's statement.

6. Louis Cachand Ervendberg, a German Protestant pastor. Geiser devotes a chapter to the man. His years of useful work in Texas had a sad termination, and he went from there to Mexico. On the evidence found in six letters which he wrote to Asa Gray, Geiser comments on ". . . a tendency, occasionally exhibited by Gray, to leave his collectors in the lurch . . . As is apparent in the Ervendberg letters, to the very end Gray was remiss in rewarding Ervendberg's labors, although there is no indication that he was dissatisfied with the collector's work." Ervendberg collected for Gray from 1856 to 1860, or after the period of my story, but as a picture of the complexities in the collector-sponsor relationship has interest.

By March 4, 1857, he had evidently been told ". . . to go ahead and collect plants . . ." and Gray had promised ". . . to sell ten or twelve sets of any specimens he collected. Gray also advised Ervendberg to collect mosses for Dr. Sullivant . . . to be shipped in care of Dr. Gray. He promised to send plenty of drying paper . . ." In January of 1858 the collector wrote that he had already obtained ". . . a good deal but for want of paper and money I am bound to stop very soon." He begged for an advance of one hundred and fifty dollars on these plants—it would be difficult to forward them without buying a horse to carry them and without purchasing boards for a box in which to ship them: ". . . all that makes much expenses and I would hardly be able to send on plants without that favour . . ." The pleas for help are pathetic and are reiterated in October, 1858; by that time Ervendberg had been able to ship 226 specimens out of a larger number. Geiser states that Gray received these plants, but nevertheless Ervendberg received no acknowledgment nor assistance, despite the agreement. On April 12, 1860, he wrote his sixth letter ". . . *without having received any answer* . . ." and explaining again his very difficult problems; Gray endorsed the letter "Ansd June 26th," and ". . . evidently set a price on the plants sent in the first box, and terminated any arrangement existing between him and Ervendberg. On October 24th, 1860, Ervendberg made a slight draft on Professor Gray for ninety dollars—somewhat meagre pay for several years work. The draft was duly paid." Geiser then mentions that a twenty-five-page paper published by Gray in the *Proceedings* of the American Academy of Science and Arts (volume 5, 1862) ". . . contained nine new species and three new varieties of higher plants . . ." and was prefaced by a statement which I quote in part:

" 'This collection, being made by a person of limited botanical knowledge, contains a number of plants which are common weeds in most warm regions, but also a fair number of newer little-known species,— enough to show that this district of country . . . would well reward a proper botanical exploration, which is the object of this notice to encourage him to undertake . . . Supplied with proper appliances and facilities, Mr. Ervendberg would make a good, as he is a zealous collector . . .' "

Geiser comments: "Surely no enemy could have more effectively damned Ervendberg with faint praise than Gray did in this notice. Thus did the great botanist assist in the advancement of science in the ancient Mexican province of Huasteca." Accepting the fact that "Ervendberg cannot in any sense be said to have had adequate fundamental training for scientific exploration," Geiser concludes that ". . . if he had received from Professor Gray even the minimum of advice and counsel that simple humanity would seem to have dictated, he would have developed into a good and useful collector . . . the fiasco of Ervendberg's career in Mexico rests not on his shoulders, but squarely upon those of Gray . . ."

[In *1846*] ". . . Lindheimer kept on with his collecting in the New Braunfels region, sometimes in company with Roemer . . ."

". . . the following year [*1847*] he traveled up the Guadalupe to the new town of Fredericksburg, near the Pedernales River. After some time spent in this locality he joined the Darmstadt group who were on their way to establish their colony, Bettina, between the Llano and San Saba rivers. It appears that he remained at this colony through the winter of 1847–48 . . ."

[He returned to New Braunfels in February.] ". . . From February to June, *1848,* he collected at New Braunfels; and his plant labels indicate that in July and August he returned to the Pedernales and the Llano. No later records for this year are to be found among Lindheimer's collected plants . . .

"From Lindheimer's plant-labels it appears that he spent the whole of the collecting season of *1849* in the neighborhood of Comanche Spring (later known as 'Meusebach's Farm'), a camping place about twenty miles north of San Antonio on the Fredericksburg road. He collected there from February to November, and then, after a short trip up the Cibolo, returned to New Braunfels.

". . . His last two years of collecting for Engelmann and Gray, *1850* and *1851,* were spent at New Braunfels, and his collecting arrangement with them terminated that year . . ."

Geiser mentions that, with Lindheimer's ". . . assumption of the editorship of the newly-founded *Neu-Braunsfelser Zeitung* in *1852,* his active career as a botanical collector was brought to a close; but botany remained his avocation to the end of his life."

The three papers published in America in which Lindheimer's collections were enumerated, named and described were Parts I, II and III of "Plantæ Lindheimerianæ." Parts I and II, cited under (2) and (4) above, represented the joint work of Gray and Engelmann and Part III, cited under (5) above, the work of Blankinship. Any study of the included plants is complicated by two facts: first, that Gray ignored Lindheimer's field numbers (and indeed all data on his labels, to the extent certainly of distinguishing such data from his own) and bestowed his own series of numbers on the plants described—following, according to Blankinship, the botanical sequence of Bentham and Hooker; second, that the plants described were distributed in fascicles to subscribers, not only in a straight numerical sequence, but (beginning with Part II) with additional numbers "intercalated" or (if unnumbered) marked by a symbol as Gray explains,[7] and, until Blankinship supplied the numbers contained in the

7. ". . . Those inclosed in ( ) belong to the collection of 1847–8; for greater convenience in describing them, they are here intercalated. The few numbers in brackets below 319 belong to species which occurred in the former distribution. Those marked with a [symbol] in place of a number have not been distributed at all . . ."

Actually, in studying the enumeration, I found no numbers "in brackets," as distinguished from Gray's parentheses, and the two terms may have been interchangeable.

four fascicles distributed by Gray, it was impossible to tell just what numbers *were* included in fascicles three and four and described in Part II of the "Plantæ." [8] To construct, from the numbers cited by Gray (in Parts I and II) and by Blankinship (in Part III), a numerical sequence from 1 through 1283 (with which Blankinship's numbers end) has much the complexity, and some of the fascination, of the double acrostic, and even Gray got mixed occasionally.[9] The student of Lindheimer's specimens must associate the numbers cited by Gray with those supplied by the collector[1] himself, although in the numerical sequence continued by Blankinship, where Lindheimer's own numbers are also cited, the task of the student is presumably somewhat simplified. I cannot see that anything is to be gained at this late date by discussing the numerical complexities of the Gray fascicles at greater length and shall turn to the plants themselves.

Part I of "Plantæ Lindheimerianæ" included Gray's numbers I through 318. He notes that the first 214 numbers were collected in 1843, ". . . on Galveston Island, around Houston, on the Brazos, &c.," and numbers 215 through 318 were collected in 1844, ". . . between the Brazos near San Felipe, and the Colorado River,[2] in the neighborhood of Cat Spring of Mill Creek, the settlement of Industry, and thence westward towards the Colorado and along its bottom lands[3] . . ." The collections of 1843 and of 1844 had not been as large as anticipated for in 1843 the specimens had met with "various misfortunes" and some of the specimens of 1844 had been ". . . lost in the course of transmission to St. Louis." However, in Part I, Gray described about thirty new species—four were named for Lindheimer—and the new genus *Brazoria* (proposed, it is stated, by Engelmann). Engelmann's contribution upon the *Cactaceae* is included in extremely small type in a footnote—his paper

8. According to Blankinship, the results of Lindheimer's work were ". . . four fascicles of plants bearing the numbers 1 to 754, and the publication of the first two parts of Plantae Lindheimerianae describing a part of them. Fascicle I contained 214 species collected in 1843; Fascicle II represented the 1844 collection with nos. 215-318; Fascicle III consisted of nos. 319-574 of 1845-6; and Fascicle IV, comprising nos. 575-754, was collected in 1847-8. The specimens of the collection of 1849-51, here treated ["Plantae Lindheimerianae Part III"], were probably intended to form Fascicle V. It appears that the first two fascicles were issued in about 20 sets, only some 9 of which were at all full, while Fascicle IV contained about 40 sets. The collection of 1849–1851 contains about 650 numbers and there will be some 50 sets for distribution, of which about 35 are fairly full. This last collection of Mr. Lindheimer is therefore about as large as all the others together and duplicates a considerable number of their species . . ."

9. Gray's numbers 420, 421 and 422 (of his straight series) appear between parentheses when—if I understand his system—they should not be so inclosed!

1. Blankinship wrote: "Lindheimer gave a number to each collection in the field, usually with more or less data in German as to habitat, locality, date, etc., his numbers following in order of collection."

2. The Colorado River of Texas, of course.

3. Geiser states: "Lindheimer's collections . . . do not give any evidence of his having collected in the Colorado bottoms, as Engelmann and Gray state in one of their papers. Perhaps the data on which the statement was based were contained in the collections of Lindheimer, mentioned by Engelmann, which were lost in transmission to St. Louis during this year."

probably arrived too late for inclusion in Gray's text. He described seven (unnumbered) species, of which five were recorded as new. One was named for Lindheimer Part I cited few specimens made by other collectors although there are occasional references to those of Drummond, of Wright and others.

Part II of "Plantæ Lindheimerianæ" included, according to the title, collections made by Lindheimer ". . . *in the Western part of Texas*,[4] *in the Years* 1845–6, *and* 1847–8 . . ." The specimens apparently met with no mishaps during this period. Gray's enumeration began where it left off in Part I and included (in a straight series) numbers 319 through 448, with, as already noted, many other "intercalated" numbers and many other collections, unnumbered but marked with Gray's symbol. About twenty new species were described, of which seven were named for Lindheimer. The new genus *Lindheimera* is still accepted by the botanists, although two others *(Daucosma* and *Agassizia)* have, I believe, fared less well. Engelmann's contribution upon the *Cactaceae* (here inserted in the text rather than in a footnote) included seven new species, one of which was named for Lindheimer.

Part II cited many specimens collected by such men as Drummond, Wright (there is more than one reference to plants growing in the Botanic Garden at Cambridge which were raised from seed which he supplied), Berlandier, Gregg, Wislizenus, Fendler, Coulter, Gordon, Hartweg and so on.

By the time Part II was written many of Scheele's descriptions of Texan plants had already appeared and it is evident that Gray was not impressed by the determinations. Many of Scheele's names are reduced to synonymy without comment, but Gray sometimes says more. Of "Galactia Texana," upon which Scheele had bestowed the name "Lablab Texanus": ". . . Mr. Scheele, with his usual wisdom, provisionally refers the plant (without fruit) to Lablab!" Of Scheele's *Vitis rupestris:* "Like his other species . . . is by no means well characterized . . ." And, in a footnote to *Filaginopsis multicaulis* Torr. & Gray: "It is hard to say upon what plants (from a Texan collection, made by Roemer,) Mr. Scheele has founded two new species of Filago . . . We know of no indigenous North American *Filago* this side of California, nor of any naturalized species except *F. Germanica.* It may be seen, moreover, that no great reliance can be placed on this writer's determinations." Nonetheless, Gray seems to have accepted six at least of Scheele's species, some, from his comments, rather halfheartedly.

In Part III of the "Plantæ Lindheimerianæ" Blankinship, as already indicated, clears up some of the numerical complexities of the fascicles distributed by Gray. He divides his treatment into three series of numbers. The first two series (449-573 and 652-754 inclusive) were distributed, we are told, ". . . without names, localities and dates . . ." and remain in herbaria without such data. Since these series (contained in fascicles three and four) ". . . contain many type collections, particularly those of

4. Not in the "Western part" of present Texas, but more nearly in the southeastern and central portions of the state.

Scheele . . .", Blankinship endeavored ". . . to supply this information as far as the specimens can be found in the Engelmann herbarium at the Missouri Botanical Garden, and thus round out the work of the previous publication." In the numerical series 449-573 Blankinship cites at least twenty-six type collections of species described by Scheele (from collections made by Lindheimer and by Roemer) and, of these, treated some sixteen as synonyms. In the series of 652-754, which is headed "Fascicle IV. 1847–1848," no Scheele types are cited—Roemer had left Texas before Lindheimer made these collections. However, this series contains type collections described by botanists such as Hooker, Alexander Braun, Engelmann, Ferdinand von Mueller and others.

Blankinship's third series of numbers, 652-1283 inclusive, is headed "Species collected in Comal County and region adjacent in 1849–1851." It is explained that, after Engelmann's private herbarium had been given to the Missouri Botanical Garden, it was found to contain ". . . an undistributed collection made subsequent to the specimens described . . ." (in Parts I and II) and ". . . represented the work of Mr. Lindheimer during the years 1849, 1850 and 1851 . . . these collections have been carefully studied . . . and this paper prepared to complete the work of the first two parts of Plantae Lindheimerianae and render the data there contained more accessible to those concerned with the flora of Texas and regions adjacent. . . . This final collection . . . proves to be of considerable importance, not only from its historical interest, but also from the fact that it contains a large number of the the type collections, since described in various publications and many more from the type locality, made by the original discoverer of the species, while the great majority of the species are relatively rare in many of our herbaria, the older distributions having gone largely to Europe . . ."

Like Blankinship's other series, this one (nos. 652-1283), cites type collections of a few of Engelmann's new species and some of other botanists. As in all Blankinship's lists there are references to co-types and type localities and he defines these terms.[5]

Blankinship enumerates twenty-nine persons and institutions appearing as subscribers ". . . for the whole or part of the first four fascicles of the 'Flora Texana Exsiccatae,' [6] as shown by Gray's unpublished letters to Engelmann." Some of these subscribers need no comment but others, after a century, are not easy to identify. Following the names cited by Blankinship I record what I have learned of these individuals and of the disposition of their herbaria. I have *not* attempted to trace the Lindheimer

5. ". . . I have used the term 'type collection' to signify the collection from which the original description of the species was made; 'co-type' or 'co-type collection' to indicate other collections quoted in this description after that first mentioned; and the term 'type locality,' to indicate other specimens collected later at the locality from which the type collection came."

6. "Exsiccatae," as the derivation of the word suggests, are *dried* plants, and Blankinship merely refers, under a different name, to the sets sold in "fascicles" to subscribers.

collections themselves to their present abodes but anyone interested in so doing would know at least where such a search should begin.[7]

Many of the individuals and institutions mentioned in the list subscribed to the sets of other collectors of the period, so that the Blankenship records suggest the repositories of other North American collections. For instance, Gray mentions that B. D. Greene ". . . subscribed to all the large purchasable North American collections, beginning with those of Drummond in the southern United States and in the Mexican province of Texas." That the dissemination of these plants to great botanical centers of the world served a useful purpose is indicated by Gray's comment that, ". . . being distributed under numbers, among the principal herbaria of the world, and named and referred to in monographs or other botanical works, they were of prime importance as standards of comparison."

"Alexander, Dr.; England." Since he traveled in Canada and may be presumed to have had an interest in North American plants, this was probably Richard Chandler Alexander (1809–1908), who took the name of *Prior* in 1859. His herbarium is at Kew and his "plants"—those of his own collecting doubtless—are in the British Museum.

"Bentham, George; England." (1800–1884). His herbarium is at Kew.

"Boissier Herbarium; Geneva." Pierre-Edmond Boissier (1810–1875). His herbarium is now at the Jardin Botanique at Geneva, Switzerland.

"Braun, Alexander; Berlin." (1805–1877). His herbarium was in the Berlin Botanic Garden and Botanical Museum, but was destroyed in World War II.

"British Museum; London."

"Buckley, S. B.; Texas." Samuel Botsford Buckley (1809–1884). Ewan states (*in litt.,* January 19, 1953) that "Buckley's collections are at Missouri Botanical Garden, Phila. Acad., and Torrey Herb. at N. Y. Bot. Gard. . . . possibly there may be some good Buckley material, even Lindheimer material, at State Teachers College West Chester, Pa. where the Darlington Herb. is located. Certainly there is a good set at Phila. but I do not know where Buckley's prime set came to rest; I do not recall particularly noticing many Buckley sheets at Washington . . ."

"Carey, S. T.; New York." Samuel Thomas Carey. Gray (writing an obituary of the brother John Carey) refers to him as "also addicted to botany," adding that he "re-

7. For British and Irish subscribers I have turned to Britten and Boulger's *Biographical index . . .* (ed. 2, 1931) and, for European subscribers, to De Candolle's *La phytographie . . .* (1880) and to Lasègue's *Musée botanique de M. Benjamin Delessert* (1845). Asa Gray's *Scientific papers,* compiled by C. S. Sargent (1889), refers to some of the Americans mentioned. Queries to individuals have been answered as helpfully as possible in every instance. I am particularly indebted to Joseph Ewan, who may be said to have all such matters "under control!"

Someone—young, and not averse to work and travel—might perform a useful task by recording the location of important botanical collections in the great herbaria of the world!

mained in the city of New York, in active business, and so was only an amateur bot-anist." Ewan (*in litt.,* January 19, 1953) states that ". . . John Carey's U. S. colls. went to Kew; I cannot find any mention of the fate of S. T. C.'s herbarium, however, which may or may not have gone with John's to England . . ."

"Cleaveland, Prof. P.; Brunswick, Me."   Parker Cleaveland (1780–1858). Dr. Alton H. Gustafson, Professor of Biology, Bowdoin College, wrote me (January 9, 1953) that ". . . Cleaveland's herbarium is here. It has not been touched for many years and is covered with a thick layer of dust as I found it when I came to Bowdoin in 1946 . . . I hope some day to put it in more presentable form . . ." He informs me further that "We have just completed a new chemistry laboratory now known as the Parker Cleaveland Laboratory of Chemistry . . ."

"Durand, Elias."   Elias Magloire Durand (1794–1873), native of the United States. According to De Candolle, he gave his herbarium of American plants to the Museum of Natural History in Paris. He stipulated that it must be kept intact, as has been done.

"Engelmann, George; St. Louis, Mo."   His herbarium is in the Missouri Botanical Garden at St. Louis.

"Fielding, H. B.; England."   Henry Barron Fielding (1805–1851). His herbarium is at the University, Oxford.

"Gray, Asa; Cambridge, Mass."   His herbarium is in the Gray Herbarium, Harvard University.

"Greene, B. D.; Boston."   Benjamin Daniel Greene (1793–1863). In his obituary no-tice of Greene, Gray states that his ". . . botanical library and collections have been, by gift and bequest, consigned to the Boston Society of Natural History, of which he was one of the founders and the first president, and by which they will be preserved for the benefit of New England botanists . . ." This famous old society is now the Museum of Science. Dr. John Patterson, Acting Director, Planetarium, wrote me (January 6, 1953) that when the move was made from Berkeley Street, Boston, to the Museum's new site, ". . . the herbarium was placed in the Metropolitan Storage Warehouse in Cambridge where it will remain until we have sufficient room . . . to place it along with our other scientific collections. I believe that some specimens from the Greene and Lowell herbaria are still with our collection but it is nearly impossible to check the material in its present position . . ." This is quite understandable to anyone who has utilized storage warehouses! See footnote under Oakes, William, below.

"Harvey, Prof.; Dublin."   William Henry Harvey (1811–1866). His herbarium is at Trinity College, Dublin, and his "plants" (those of his own collecting) are in the Brit-ish Museum.

"Jardin des Plantes; Paris."

"Kew Gardens; England."

"Lamson, Prof.; Kingston, Canada." From Queen's University, Kingston, Ontario, Dean R. O. Earl wrote me (January 12, 1953) that my query regarding "Prof. Lamson" ". . . contained a puzzle. There never was a Professor Lamson at this University as far as I can discover and I know nothing of any herbarium of his. There is no other institution of learning here in which Botany is or has been taught . . ." Ewan solved this mystery—on the basis of a Blankinship misspelling—when he wrote me (March 2, 1953): " 'Prof. Lamson' was surely Prof. George *Lawson,* 1827–1895, Prof. of Natural History, Kingston, Canada, at least in 1858 (cf. Britten & Boulger, ed. 2. 182)." Britten and Boulger do not mention the disposition of George Lawson's herbarium. A letter to Dr. A. E. Porsild, Chief Botanist, National Museum of Canada, Ottawa, brought me the following reply, dated April 29, 1953: "The Lawson Herbarium was presented to the National Museum of Canada in 1950 by Mount Allison University of Sackville, N. B., after having been in storage since the death of Professor Lawson. Except for ferns and some other groups of Canadian plants in which Lawson was particularly interested, much of the Herbarium has remained unorganized, and much of the exotic material was still in the original wrappers in which Dr. Lawson received it. Since 1950 we have been slowly sorting and mounting the material. Of Lindheimer's Fl. Texana exsiccata there appears to be all, or parts, of fasc. III and IV, all still unnamed. Besides these, the National Herbarium of Canada possesses a set of the supplement issued by Missouri Botanical Gardens . . ."

"Leman." No individual whose name is spelled "Leman" has come to my attention. The man may well have been Charles Morgan Lemann (1806–1852) of Britain, who gave his herbarium of thirty thousand specimens to Cambridge University, and whose "plants" are at Kew. However, if Gray (in his letters to Engelmann) was spelling phonetically, there are other possibilities such as Johann Georg Christian Lehmann (1792–1860), professor at Hamburg, Germany, whose herbarium, according to De Candolle, was sold piecemeal, some bought by individuals and what remained by the Museum at Stockholm, Sweden.

"Lowell, John A.; Cambridge, Mass." John Amory Lowell (1798–1882). According to his obituary notice, written by Gray, ". . . he presented all his botanical books which were needed to the herbarium of Harvard University; and the remainder, with his herbarium, to the Boston Society of Natural History . . ." *See* Greene, B. D., above; and Oakes, William, below.

"Oakes, William; Ipswich, Mass." (1790–1844.) A New England botanist for whom the Liliaceous genus *Oakesia,* now *Uvularia,* was named by Watson. Ewan wrote me (January 19, 1953): "Wm. Oakes' herbarium was a part of the Boston Soc. Nat. Hist. collections. These were given to Gray Herb. and Gray Herb. checked their own herb. for the representation of the Boston Soc. sheet and if present already at Gray Herb., as would be the case with Lindheimer material, the 'duplicate' was then sent out in inter-institutional exchanges . . . Dr. Robert Foster might recall where the Lindheimer material went but that is only a bare possibility."[8]

"Olney, S. T.; Providence, R. I." Stephen Thayer Olney (1812–1878), for whom the genus *Olneya* was named by Gray. Dr. J. Walter Wilson, Chairman of the Department of Biology, Brown University, wrote me (January 3, 1953) that Olney ". . . was an amateur botanist and an industrialist . . . in Providence. He endowed a Professorship of Natural History which still bears his name . . . He established an herbarium which is an important part of the collections of the University . . . and still bears his name. All of the material he collected is as far as I know in that herbarium . . ."

"Saunders, William; England." William Wilson Saunders (1809–1879). His herbarium is at the University of Oxford.

"Shuttleworth, R. J.; England." Robert James Shuttleworth (1810–1874). His herbarium is at the British Museum.

"Smithsonian Institution, Washington." The collections are now in the United States National Herbarium.

"Stevens." This might perhaps refer to Christian von Steven (1781–1863), a Russian, who published a number of botanical papers from 1817–1862. De Candolle states that he gave his herbarium of twenty-three thousand species ("esp.") to the University of Helsingfors and duplicates ("échantillons") to the herbaria of De Candolle, of Zuccarini (in the Royal Herbarium at Munich and in the Imperial Garden at St. Petersburg) and of the University of Leipzig. *Or* it might refer to Henry Oxley Stephens (1816–1881), a surgeon of Bristol. Britten and Boulger state that his herbarium is at the Bristol Naturalists Society.

"Sullivant, Wm. L.; Columbus, Ohio." William Starling Sullivant (1803–1873). Gray, who wrote his obituary notice, states: "In accordance with his wishes, his bryological books and his exceedingly rich and important collections and preparations of Mosses are to be consigned to the Gray Herbarium of Harvard University, with a view to their preservation and long-continued usefulness. The remainder of his botanical library, his choice microscopes, and other collections are bequeathed to the State Scientific and Agricultural College[9] just established at Columbus, and to the

---

8. Dr. E. D. Merrill tells me that according to a signed agreement (that if at any time the large general herbarium of the Boston Society of Natural History should be broken up) the Gray Herbarium should have first choice of any extra New England collections in exchange for extensive sets of modern collections from New England already provided by Professor Fernald. In 1941 the decision was reached to retain only New England material in the Boston Society (now the Museum of Science). Thus all the extra New England material was sent over to Cambridge for sorting and Professor Fernald selected what in his opinion would add to the value of the reference collections at the Gray Herbarium.

9. Now Ohio State University.

Starling Medical College, founded by his uncle, of which he was himself the senior trustee."

"Thurber, George." (1821–1890.) In his *Rocky Mountain naturalists* (1950) Ewan refers to Thurber's service as ". . . botanist and quartermaster and commissary on the Mexican-U. S. Boundary, 1850–1853 . . ." and mentions that the Herbarium of the California Academy of Sciences ". . . housed a valuable set of Thurber's plants which were destroyed by the San Francisco fire of 1906." In a letter (January 19, 1953) he adds that at the Academy ". . . must surely have been his set of Lindheimer plants."

"Torrey, John; New York." The Torrey Herbarium, owned by Columbia University, is now deposited at the New York Botanical Garden.

"Webb, Barker." Philip Barker Webb (1793–1854), of Britain. His very large herbarium is at the Instituto Botanico, Via Lamarmora, Florence, Italy.

De Candolle cites a number of European herbaria which, in 1880, contained numbered collections made by Lindheimer in Texas—whether subscribed for or otherwise obtained is not stated: the herbarium of "de Franqueville";[1] the "Cosson" herbarium, Paris; the herbarium of the Johanneum at Graz (or Gratz) in what was then Austria-Hungary; and in the herbaria of the Imperial Garden at St. Petersburg, of the "Musés palatin" at Vienna, and of the University of Leipzig.

We are told by Blankinship that, after the termination of his work for Gray and Engelmann, Lindheimer for twenty years was editor and publisher of the *Neu-Braunfelser Zeitung* and that, in addition, he assumed ". . . many public duties. He conducted a private free school for advanced pupils. He served as Superintendent of Public Instruction in his county for several terms and was first Justice of the Peace of New Braunfels, till increasing age forced him to rest from his labors." Of Lindheimer's contributions in the field of botany Blankinship has this to say:

"His botanical work can be best appreciated by remembering the difficulties and dangers, the poverty and hardships under which his collections were made. He discovered and made known to the scientific world an enormous number of new species of plants from central Texas and many of these will ever bear his name. The beautiful *Lindheimera texana* is already not infrequent in ornamental cultivation . . . many plants grown from seeds of his collection are found in the Missouri Botanical Garden at St. Louis, in the Botanical Garden at Cambridge, Mass., and elsewhere. His private herbarium at his death came into the hands of Prof. Emil Dapprich of Milwaukee, Wisconsin, and was on exhibition at the World's Fair at Paris. On Dapprich's death in 1903 it came into the possession of the German-English Academy of Milwaukee, where I understand it still remains.

1. A French Count with a large herbarium in Paris.

"Mr. Lindheimer was a careful observer and a patient collector, and the notes accompanying his collections add greatly to their value . . . A number of his new species he himself described and named, but many of the names he suggested were found preoccupied and others given.

". . . A number of his principal scientific, philosophical and historical essays . . . have been republished in Germany under the title: 'Aufsätze und Abhandlungen von Ferdinand Lindheimer in Texas,' but the greater part are unknown and inaccessible to the general reader . . . his simple, direct, philosophical style is always interesting and his meaning clear, quite different from the usual complicated, involved German sentence . . ."

Geiser, whose paper on Lindheimer seems to me to offer the best overall picture of the man in relation to the Texas of his period, has this to say of his botanical accomplishments:

"In the field of botany, Lindheimer is honored by having a round score of species of plants named in his honor by scientific specialists. His name, along with that of Charles Wright . . . is indissolubly connected with the botany of Texas, to which in his collections for Engelmann and Gray he made contributions of outstanding value . . . Lindheimer's name lives forever in the very nomenclature of the science he loved."

BLANKINSHIP, JOSEPH WILLIAM. 1907. Plantae Lindheimerianae, Part III. *Missouri Bot. Gard. Ann. Rep.* 18: 123-223.

BRITTEN, JAMES & BOULGER, GEORGE SIMONDS. 1931. A biographical index of deceased British and Irish botanists. ed. 2. London. 30, 107, 141, 182, 249, 269, 275, 320.

CANDOLLE, ALPHONSE DE. 1880. La phytographie; ou l'art de décrire les végétaux considérés sous différents points de vue. Paris. 397, 399, 409, 418, 428, 437, 452.

ENGELMANN, GEORGE. 1850. Cactaceae. *In:* Gray, Asa. 1850. Plantae Lindheimerianae Part II. 195-209. *Reprinted:* Trelease, William & Gray, Asa, *editors.* 1887. The botanical works of the late George Engelmann . . . Cambridge, Mass. 117-122.

—— 1850. [". . . brief account of the region in which the present collection of plants was made . . ."] *In: Ibid.* 234-240.

—— 1851. On the character of the vegetation of southwestern Texas. *Proc. Am. Assoc. Advancement Sci.* 5: 223-228. [Stated by author to have been based on ". . . the extensive and beautiful collections of . . . Lindheimer, together with his full notes . . ."] *Reprinted:* Trelease, William & Gray, Asa, *editors.* 1887. The botanical works of the late George Engelmann . . . Cambridge, Mass. 529-532.

—— & GRAY, ASA. 1845. Plantae Lindheimerianae [Part I.]; an enumeration of the plants collected in Texas, and distributed to subscribers, by F. Lindheimer, with remarks and descriptions of new species, &c. *Boston Jour. Nat. Hist.* 5: 210-264. *Reprinted* [as to Engelmann's descriptions of new species]: Trelease, William & Gray, Asa, *editors.* 1887. The botanical works of the late George Engelmann . . . Cambridge, Mass. 510, 511.

EWAN, JOSEPH. 1950. Rocky Mountain naturalists. Denver. 321.

GEISER, SAMUEL WOOD. 1930. Ferdinand Jacob Lindheimer. *Southwest Review* 15: 245-266.

—— 1933. Ferdinand Jacob Lindheimer. *Dict. Am. Biog.* 11: 273, 274.

—— 1937. Ferdinand Jakob Lindheimer. *In:* Naturalists of the frontier. Dallas. 159-180.

—— 1948. *Ibid.* ed. 2. Dallas. 132-147.

GRAY, ASA. 1843. Notice of botanical collections. *Am. Jour. Sci. Arts.* 45: 225-227.

—— 1850. Plantae Lindheimerianae, Part II. An account of a collection of plants made by F. Lindheimer in the western part of Texas, in the years 1845–6, and 1847–8, with critical remarks, descriptions of new species, &c. By Asa Gray, M.D. *Boston Jour. Nat. Hist.* 6: 141-195; 209-233. *Reprinted* [as to Engelmann's descriptions of new species]: Trelease, William & Gray, Asa, *editors.* 1887. The botanical works of the late George Engelmann . . . Cambridge, Mass. 511-513. [*See also* under Engelmann, G. 1850. Cactaceae; and ". . . brief account of . . ."]

—— 1863. Benjamin D. Greene. *Am. Jour. Sci. Arts,* ser. 2, 35: 449. *Reprinted:* Sargent, Charles Sprague, *compiler.* 1889. Scientific papers of Asa Gray. 2 vols. Boston. New York. 1: 310, 311.

—— 1880. John Carey. *Am. Jour. Sci. Arts,* ser. 3, 19: 421. *Reprinted: Ibid.* 1: 417.

—— 1882. John Amory Lowell. *Proc. Am. Acad. Arts Sci.* 17: 408-411. *Reprinted: Ibid.* 1: 421-424.

—— 1893. Letters of Asa Gray. Edited by Jane Loring Gray. 2 vols. Boston. New York. 1: 14, 297, 298, 340, 342, 345.

HOOKER, WILLIAM JACKSON. 1844. 1846. Lindheimer's plants of Texas. *London Jour. Bot.* 3: 140, 1884; 5: 12, 1846.

ROEMER, FERDINAND. 1935. Texas with particular reference to German immigration and the physical appearance of the country described through personal observation by Dr. Ferdinand Roemer. Translated into English by Oswald Mueller. San Antonio. [Translation of the German edition of 1849. Bonn.]

SCHEELE, ADOLF. 1848. 1849. 1850. 1852. Beiträge zur Flora von Texas. *Linnaea* 21: 453-472, 576-602, 747-768 (1848); 22: 145-168, 339-352 (1849); 23: 139-146 (1850); 25: 254-265 (1852).

WINKLER, CHARLES HERMAN. 1915. The botany of Texas. An account of botanical investigations in Texas and adjoining territory. *Bull. Univ. Texas* 18: 1-27.

CHAPTER XLIV

FRÉMONT CROSSES THE GREAT BASIN BY WAY OF
GREAT SALT LAKE AND THE HUMBOLDT RIVER, EN-
TERS CALIFORNIA BY A PASS NORTH OF LAKE TAHOE,
AND EXAMINES THE COASTAL REGIONS ABOUT SAN
FRANCISCO AND MONTEREY AS WELL AS THE VAL-
LEYS OF THE SAN JOAQUIN, SACRAMENTO AND PITT
RIVERS

MUCH has been published upon the historical and political aspects of John Charles Frémont's third expedition of 1845–1847, in the course of which he became involved in the conquest of California. While still serving in the United States Army he is said to have acted under naval orders and claimed that he was correct in so doing. Whether his actions were justifiable I leave to the historians; but certain it is that, after June 7, 1846, Frémont's topographical work stopped abruptly and his interest in scientific objectives such as botany also ceased, temporarily at least.

Frémont wrote no itinerary of the journey—such as were contained in the reports upon his first and second expeditions—but assembled some of the results thereof in a paper entitled "Geographical memoir upon Upper California, in illustration of his map of Oregon and California, by John Charles Frémont: addressed to the Senate of the United States," which was issued as a Senate Miscellaneous Document in 1848 (30th Congress, 1st Session, No. 148) and which was accompanied by a map, captioned "Map of Oregon and Upper California from the surveys of John Charles Frémont and other authorities" and "drawn by Charles Preuss . . . Washington City 1848 . . . Scale 1: 3,000,000." Warren states that this embraced ". . . all the country between the thirty-second and fiftieth parallel of north latitude . . . and was at the time of its publication (1848) the most accurate map of that region extant." In presenting it to the Senate, Frémont wrote:

". . . it may be assumed to be the best that has yet appeared, but is still imperfect and incomplete . . . This geographical memoir . . . is only a preliminary sketch in anticipation of a fuller publication[1] which the observations of the last expedition would

---

1. Warren wrote: ". . . There are probably many reasons why a complete account of this third expedition as well as Colonel Frémont's subsequent ones, have never been published; but this desideratum will soon

justify . . . The results of the previous two expeditions were published by order of the Senate, and disposed of according to its pleasure. No copyright was taken; and whatever information the journals of the two expeditions contained, passed at once into general circulation. I would prefer a similar publication of the results of the last expedition; but being no longer in the public service, an arrangement . . . would be necessary . . ."

Although the "Geographical memoir" is largely a discussion of the regional aspects of the country under consideration and includes few dates, a brief and accurate record of Frémont's whereabouts, between August 16, 1845 and June 1846, is supplied in "A table of latitudes and longitudes deduced from the aforegoing astronomical observations,[2] calculated by Professor Hubbard."

The best running account of the present journey is found in Frémont's *Memoirs of my life,* already mentioned, which includes a helpful map, and I shall follow the route as supplied therein, supplementing from the explorer's "Geographical memoir." The outline of Frémont's route included in the Warren "Memoir," and the extremely interesting and careful analysis of the journey in Fletcher's *Early Nevada,* are referred to more than once when these clarify matters upon which Frémont himself is somewhat vague.

Frémont wrote in the *Memoirs of my life* that, "Concurrently with the Report upon the second expedition the plans and scope of a third one had been matured."

". . . It was decided that it should be directed to that section of the Rocky Mountains which gives rise to the Arkansas River, the Rio Grande del Norte of the Gulf of Mexico, and the Rio Colorado of the Gulf of California; to complete the examination of the Great Salt Lake and its interesting region; and to extend the survey west and southwest to the examination of the great ranges of the Cascade Mountains and the Sierra Nevada, so as to ascertain the lines of communication through the mountains to the ocean in that latitude. And in arranging this expedition, the eventualities of war were taken into consideration.

"The geographical examinations proposed to be made were in greater part in Mexican territory. This was the situation: Texas was gone[3] and California was breaking off

be supplied." And he adds in a footnote: "In press [1859], Colonel J. C. Frémont's Explorations, prepared by the author, and embracing all his expeditions.—Childs & Peterson, publishers, No. 602, Arch street, Philadelphia."

Since this anticipated publication does not appear in the catalogue of the Library of Congress, nor in a number of other bibliographies, it seems unlikely that it was ever issued. Possibly it eventually took form in Frémont's *Memoirs of my life,* of which one volume—all that was ever published—came out in 1887. The publishers were Belford, Clarke & Company of New York and Chicago.

2. Part II of the Appendix: "A table of astronomical observations made by J. C. Frémont at the four principal stations determined in this third expedition, namely: 1. The mouth of Fontaine Qui Bouit, on the upper Arkansas. 2. Southeastern shore of the Great Salt lake. 3. Lassen's farm, Deer creek, in the valley of the Sacramento. 4. The Three Buttes, valley of the Sacramento."

3. It had been admitted to the Union as a state on March 1, 1845, or on the very date that Frémont had presented to the Senate his "Report" upon his second expedition of 1843–1844.

by reason of distance; the now increasing American emigration was sure to seek its better climate. Oregon was still in dispute; nothing was settled except the fact of a disputed boundary; and the chance of a rupture with Great Britain lent also its contingencies.

"Mexico, at war with the United States, would inevitably favor English protection for California . . ."

On arrival at St. Louis Frémont camped on the adjacent frontier where his party was quickly organized. "The animals . . . left on pasture were in fine condition . . . thoroughly rested . . ." The remarkable Charles Preuss did not accompany Frémont on this expedition. Edward M. Kern, took his place as topographer and possessed the additional asset of being ". . . an accomplished artist; his skill in sketching from nature and in accurately drawing and coloring birds and plants made him . . . valuable . . ." With the comments that it was getting late in the year and that ". . . the principal objects of the expedition lay in and beyond the Rocky Mountains . . ." and that ". . . no time could be given to examination of the prairie region . . .", Frémont opens his narrative of the journey on August 2, at Bent's Fort—"our real point of departure"— which was situated on the south bank of the Arkansas River about half way between the towns of Las Animas (Bent County) and La Junta (Otero County), Colorado.

1845

Frémont remained at Bent's Fort[4] until August 16. On arrival he sent ". . . an express to Carson at a rancho . . . which with his friend Richard Owens, he had established on the Cimarron, a tributary of the Arkansas . . ." Carson responded by selling ". . . everything at a sacrifice, farm and cattle; and . . . brought his friend Owens . . . This was like Carson, prompt, self-sacrificing and true." In addition Carson secured a "Creole Frenchman of St. Louis," Alexis[5] Godey. Frémont comments that these three men ". . . under Napoleon, might have become Marshals . . ." On leaving Fort Bent Frémont had ". . . a well-appointed compact party of sixty . . ."

After leaving Bent's Fort, until October 13 when the party camped ". . . at a creek on the shore of the Great Salt Lake . . .", few dates are supplied and the route for the most part must be followed on the map. The meagre details indicate that, after the fort was left on August 16, the party had reached the mouth of *"Fontaine qui Bouit River"* (now Fountain Creek which enters the Arkansas River from the north at Pueblo) and had camped "at the mouth of the Great Canyon," or at the eastern end of the Royal

4. It was while there that Frémont dispatched two of his Lieutenants, J. W. Abert and W. G. Peck, with Thomas Fitzpatrick as guide, on their exploration of the Purgatory River, etc., which is described in my chapter XLV.

5. In late July of 1843 an Alexander Godey had joined Frémont's party at St. Vrain's Fort and had accompanied the expedition westward, distinguishing himself in April of 1844 by returning with two Indian scalps acquired in the Death Valley environs of California. "Alexander" and "Alexis" Godey appear to have been one and the same.

Gorge by Canon City, Fremont County, on August 26. To avoid the gorge, they had then kept along the southern slopes of the mountains at the southern end of South Park, and leaving the river, had crossed over ". . . a bench of the mountain which the trappers believed to be the place where Pike was taken prisoner by the Mexicans.[6] But this side of the river was within our territory . . ." On September 2 they were ". . . on the head-waters of the Arkansas in Mexican territory . . .", it would seem near Buena Vista in Chaffee County. They kept north and northwest, to the west of South Park. Frémont notes that "The Utah Pass was several days' journey to the southeast, and this part of the mountain[7] was out of the way of ordinary travel." The party passed Twin Lakes in southern Lake County and evidently went through Tennessee Pass, on the boundary of Lake and Eagle counties, reached Eagle River, in Eagle County, and on September 4 ". . . Piny River, an affluent of Grand River,[8] of the Colorado of the Gulf of California . . ." Piney Creek enters the Colorado River from the south, in northern Eagle County. No dates are mentioned in the *Memoirs* between September 4 and October 2. From Piney Creek they moved northwest, crossing southern Routt County and reaching the headwaters of White River in either northern Garfield County or in eastern Rio Blanco County. White River was followed westward across Rio Blanco County to and across the Colorado-Utah border and onward to its confluence with Green River in southern Uinta County. The route through Colorado had been farther west than the one taken on Frémont's journey southward in 1844, and Frémont entered Utah at a point considerably south of the one by which he had left that state in the same year. He was now to keep north of the route taken on the earlier expedition. The party was to remain in Utah until October 26.

The map of the *Memoirs* shows that, having reached Green River, they crossed that stream and ascended Duchesne River (which enters the Green from the northwest about opposite the mouth of White River) for a time, crossed a northern extension of the Wasatch Mountains, and on October 2 came to the upper waters of Provo, which Frémont called the "Timpanogos," [9] River. The party was then north or northwest of Utah Lake, but the map indicates no visit to that lake, so that Frémont's reference to reaching, October 10, "the shore of the lake," must mean Great Salt Lake. They came upon it, on this visit, near its southern end—the map of the "Geographical memoir" (1848) indicates at the "Mormon settlement," [1] and the map of the *Memoirs* (1887)

6. Pike's capture had been on the Rio Conejos, Conejos County, not far north of the present Colorado–New Mexico line; it would *seem* to have been south of Frémont's position.

7. Like others of the period, Frémont refers to such ranges as the Rocky Mountains and the Sierra Nevada as "the mountain."

8. At that time the upper waters of the Colorado were called Grand River.

9. The volume *Utah* ("American Guide Series"), p. 516, states: "The Provo River was called Timpanogos, 'water running over rocks.' "

1. Since the Mormon colony reached its destination only in July of 1847, Frémont was keeping abreast of historical events.

at "Great Salt Lake City." They remained in the environs of the lake until October 25 or 26. The "Geographical memoir," in the section "Lakes in the Great Basin," discusses the Great Salt Lake in considerable detail. On this occasion the party was at its southern and southwestern end. The *Memoirs* mentions that, on October 15, ". . . it began to rain in occasional showers . . . Flowers were in bloom during all the month." When a large island[2] in the southern part of the lake was visited, "About the 18th," ". . . *helianthus,* several species of *aster, erodium cicutarium,* and several other plants were in fresh and full bloom; the grass of the second growth was coming up finely . . ."

Frémont wished, from the lake, to take a route westward, ". . . over a flat plain covered with sage-brush." Since neither Carson nor Walker, so Frémont states, knew anything of the country in that direction, he sent out a party of three on October 25:

". . . apparently fifty or sixty miles away, was a peak-shaped mountain. This looked to me to be fertile, and it seemed safe to make an attempt to reach it . . . I arranged that Carson . . . should set out . . . and make for the mountain. I was to follow . . . the next day and make one camp out in the desert. They to make a signal of smoke in case water should be found."

Having learned on the 26th that his surmise as to the essentials of water and grass was correct, Frémont gave to ". . . the friendly mountain the name of Pilot Peak . . . Some time afterward, when our crossing of the desert became known, an emigrant caravan was taken over this route, which then became known as *The Hastings Cutoff.*"[3] The Pilot Range lies along the boundary between Utah and Nevada, and Pilot Peak, in eastern Elko County, Nevada, still bears the name bestowed on this occasion. The expedition was to be in Nevada from November 1 through December 6.

On November 2 they camped at a spring to which Frémont ". . . gave the name of Whitton, one of my men who discovered it." This, according to Fletcher, is now ". . . known as Flowery Springs, near Shafter . . ." and the party had moved southwest. Here Frémont decided that because of the probability of snow in the Sierras, it would be imprudent ". . . to linger long in the examination of the Great Basin." He therefore divided his party. According to the map of the "Geographical memoir," the division took place at Whitton Spring. Kern was placed in charge of the main contingent, with ". . . instructions to follow down and survey the Humboldt River and its valley to their termination in what was called 'the Sink.' . . . Thence to continue along the eastern foot of the Sierra to a lake to which I have given the name of Walker,[4] who was to be his guide on this survey . . . The place of meeting for the two parties was to be the

2. The map indicates this to have been Antelope Island.

3. It was by taking this route, publicized and misrepresented by Warren Lansford Hastings, that the unfortunate Donner party started on its way to disaster.

4. Joseph Reddeford Walker, who had been with Frémont's second expedition in 1844 from the time of its arrival in Utah until he left the party at Bent's Fort.

lake." Walker Lake is in Mineral County, Nevada, and no great distance from the Nevada-California line. Frémont (in disagreement with the map of the "Geographical memoir") states in his *Memoirs* that the division of the party took place at ". . . a small stream which I have called Crane's Branch . . . Crane's Branch led into a larger stream that was one of two forks forming a river to which I gave the name of Humboldt. Both the river and mountain to which I gave his name[5] are conspicuous objects . . . Years after . . . I was glad to find that river and mountain held his name, not only on the maps, but in usage by the people."

On November 3 Frémont's contingent, consisting of ten, started ". . . westward across the Basin, the look of the country inducing me to turn somewhat to the south." The map of the "Geographical memoir," shows that the party kept south and southwest, crossed the Humboldt Mountains and, moving at times south, at others west, arrived at the southern end of Walker Lake. On November 27 the Kern party arrived from the north, having followed the Humboldt River for a great part of the way.

Here the party again separated. Kern was to proceed south and ". . . pass around the Point of the California Mountain into the head of the San Joaquin valley." Walker was again Kern's guide and the meeting place of the two parties was to be at ". . . a little lake in the valley of a river called the Lake Fork of the Tularé Lake." The "Lake Fork of the Tularé Lake" was King's River. Frémont with fifteen men was to cross the Sierra Nevada to Sutter's Fort.

Leaving Walker Lake, Frémont kept north along Walker River and, when the course of the stream changed abruptly, he left it and kept north to Carson River, crossed it, and on December 1 came to ". . . the river which flows into Pyramid Lake, and which on my last journey I had named Salmon-Trout River." This was Truckee River and Frémont ". . . struck it above the lower cañon . . ." Fletcher states that this was "near the present site of Wadsworth." Frémont records that, on December 4, he ". . . camped at its head on the east side of the pass in the Sierra Nevada . . . the only snow showing was on the peaks of the mountains." On the 5th the party was ". . . on the crest of the divide . . ." Following the Truckee westward, Frémont must have crossed into California near Verdi, Washoe County, and entered Sierra County, California where I follow his journey until May 14, 1846.

5. In the "Geographical memoir" Frémont comments that he had bestowed the name Humboldt ". . . as a small mark of respect to the *'Nestor of scientific travellers,'* who has done so much to illustrate North American geography, without leaving his name upon any one of its remarkable features. It is a river long known to hunters, and sometimes sketched on maps under the name of Mary's, or Ogden's [river] . . . This river possesses qualities which, in the progress of events may give it both value and fame. It lies on the line of travel to California and Oregon, and is the best route now known through the Great Basin, and the one travelled by emigrants . . . Its head is towards the Great Salt lake, and consequently towards the Mormon settlement, which must become a point in the line of emigration to California and the lower Columbia . . ." The earlier name of Ogden's River—recording Peter Skene Ogden's sojourn on the river in 1833—might more appropriately have been retained.

Frémont appears to have crossed the Donner Pass[6] for, on the 5th, he reached (after crossing "the crest") the emigrant road ". . . here following down a fork of Bear River, which leads from the pass into the Sacramento valley . . ." or, in other words, westward to Emigrant Gap on Bear River. Believing this "a rugged way," Frémont ". . . turned to the south . . . We had made good our passage of the mountain and entered now among the grand vegetation of the California valley." In turning south he must have come upon the upper waters of the Middle Fork of American River and descended it to the valley.

This was Frémont's second crossing of the Sierra Nevada in much the same region— Warren states that it was ". . . about fifty-five miles north of his pass of January 17, 1844 . . ."—which had been Carson Pass. In the present instance he had kept north of Lake Tahoe which, on the map of the *Memoirs*, bears the name "Lake Bonpland." [7]

On December 9 the party reached ". . . what was then still known as *Rio de los Americanos*—the American Fork, near Sutter's Fort." On December 14—leaving ". . . the upper settlements of *New Helvetia*, as the Sutter settlement was called . . ." —they started for the rendezvous with Kern and Walker on King's River, specifically at ". . . the little lake in the valley of a river called the Lake Fork of the Tularé Lake." Frémont had ascended the San Joaquin valley on his last journey. He now mentions various rivers entering the San Joaquin from the east, and the map of the "Geographical memoir" shows that they were encountered some distance above their unison with the main stream. On the 14th the "Cosumné River" was reached and camp was about eight miles above its meeting with the "Mokelumné River." On the 16th, after crossing Calaveras[8] River, they reached the valley of the Stanislaus, and on the 17th ". . . approached the Tuolumné River, one of the finest tributaries of the San Joaquin . . ." The 18th found them on the ". . . Auxumné River—called by the Mexicans *Merced*— another large affluent of the San Joaquin . . ." On the 19th they were ". . . at the Mariposas River . . ."

6. Whether "Truckee Pass" and Donner Pass are one and the same I do not know—the first-named I have been unable to locate on a considerable number of maps. In line with the possibility that they are identical is a comment by Fletcher upon the origin of the name Truckee, which he states was bestowed by a member of the Stevens-Murphy party in 1844: ". . . the first party known to have followed the Truckee River to the pass of the same name and to have crossed the Sierra there. It is said to have named the river and the pass for an Indian Chief whose name, oddly enough, bore no resemblance to Truckee. The party was delayed near Donner Lake [lying below and east of Donner Pass] and the Indian had befriended it. His odd manners reminded one of the party of an acquaintance named Truckee so he gave him that name. Subsequently the name was attached to the river and the pass."

7. The accomplishments of the Humboldt-Bonpland affiliation seem to have been in Frémont's consciousness at this time for the name was a tribute to Aimé Jacques Alexandre Bonpland.

8. A name which always arouses memories of "The celebrated jumping frog of Calaveras County." I was delighted to read in the volume *California* ("American Guide Series"), p. 493, that in May of every year ". . . Angels Camp holds a Jumping Frog Jubilee. In the main event, frogs, after rigid inspection to prevent loading with buckshot, . . . compete for a first prize of $500. Models of frogs appear in the windows of stores and hotels. One restaurant advertises a jumping frog pie—made of prunes and raisins; and local businessmen have organized a Frog Boosting Club."

I have not quoted what Frémont, writing in 1887 in the *Memoirs,* has to say about the plants observed on the present journey. His comments, based on Torrey's determinations, differ little from those in the "Report" of 1845. One hoped, as he reached regions now famous for stands of the big trees (the Calaveras, Stanislaus, Tuolumne, Merced, Mariposa Groves, etc.), that he would refer to them, if only as hearsay. John Bidwell, owner of the Rancho Chico, stated that he mentioned them to Frémont in 1845, and Frémont's guide, Joseph Reddeford Walker, had been leader of the party now credited with having been, in 1833,[9] the first to see these giant wonders —when they crossed the Sierra Nevada into California in the region of Yosemite Valley. Although (both in the "Geographical memoir" and in the later *Memoirs)* he refers to forests of ". . . a cypress *(taxodium)* of extraordinary dimensions, already mentioned among the trees of the Sierra Nevada . . ." he was then writing of stands of redwood near Santa Cruz and I have found no references during his crossings of the Sierras which would suggest that he had come upon the big trees unless his comments

9. Zenas Leonard, whom Chittenden refers to as "the historian" of the Walker expedition of 1833 (the first party to follow the Humboldt River route into California and to cross the Sierra Nevada from east to west), is now accepted as the first to have written of the big trees. According to W. F. Wagner, they were ". . . the giant redwoods of Mariposa . . ." and the party saw them from heights above the Yosemite Valley.

F. P. Farquhar's *Yosemite, the Big Trees, and the High Sierras* (1948) cites two editions of Leonard's narrative and a "verbatim reprint" of the first edition, published respectively in 1839, 1904 and 1905. Under the title *Narrative of the adventures of Zenas Leonard,* and under the editorship of Milo M. Quaife, it was reprinted in the "Lakeside Classics" in 1934, an edition which Farquhar refers to as "the handiest for reading" and one to which I assign many other delightful attributes. I quote Leonard's reference therefrom—"the mountain" was the Sierra Nevada:

"In descending the mountain this far we have found but little snow, and began to emerge into a country which had some signs of vegetation—having passed thro' several groves of green oak bushes, &c. The principal timber which we came across, is Red-Wood, White Cedar and the Balsom tree. We continued down the side of the mountain at our leisure, finding the timber much larger and better, game more abundant and the soil more fertile . . . In the evening of the 30th [October] we arrived at the foot or base of this mountain—having spent almost a month in crossing over. Along the base of this mountain it is quite romantic—the soil is very productive—the timber is immensely large and plenty . . . From the mountain out to the plain, a distance varying from 10 to 20 miles, the timber stands as thick as it could grow, and the land is well watered by a number of small streams rising here and there along the mountain. In the last two days travelling we have found some trees of the Red-wood species, incredibly large—some of which would measure from 16 to 18 fathoms round the trunk at the height of a man's head from the ground."

For years the discovery was credited to John Bidwell—by C. C. Parry (1883) and by C. S. Sargent (1896). He had reached California with the Bidwell-Bartleson party in November of 1841, in the vicinity of the Calaveras Grove. His *Echoes of the past about California* (published in 1928 in the "Lakeside Classics" and also edited by Quaife) tells of having come upon them at night and, in retrospect, supposed them to have been ". . . trees in the Calaveras Grove of *sequoia gigantea* or mammoth trees, as I have since been there, and to my own satisfaction identified the lay of the land and the tree. Hence I concluded that I must have been the first white man who ever saw *sequoia gigantea,* of which I told Frémont when he came to California in 1845 . . ." Jepson (1923) states that a "hunter, one A. T. Dowd," when chasing a grizzly in the same region, "discovered" the big tree in 1852. But all these claimants to fame now give place to Zenas Leonard.

To turn from reports of the tree to the collection of actual specimens. The first to have any in his possession (". . . before June, 1852 . . ." according to Jepson) was the botanist Albert Kellogg. Kellogg evidently discussed them with William Lobb, then collecting plants in California for the Veitch firm of

on February 10, 1844 (of his second journey) about "the red pine, (*pinus colorado* of the Mexicans,)" should be so interpreted. He was then descending the Sierras to Sutter's Fort and the big trees are recorded from their western slopes in Placer County; the redwoods, or so I believe, are confined to the coast.

It is perhaps of interest that, when at the "Mariposas River," Frémont was no great distance from the region where he subsequently (in 1847) acquired a large piece of property—seventy square miles (according to Nevins 43,000 acres)—for which he paid three thousand dollars. It had been purchased for him by the American Consul Thomas O. Larkin during his absence in the east and, although not in the San Francisco region where he had intended the purchase to be made, it yielded a large amount

Exeter, England. Lobb, it seems, wasted no time in procuring his own material and carried it back to England where John Lindley in 1853 published a description of the tree under the name *Wellingtonia gigantea,* thus inaugurating its somewhat complicated nomenclature and generic classification, which, as late as 1939, involved its separation from *Sequoia* on the basis of morphological differences. I am told that, with "the critical approval of the best botanists" of the present day, the big tree is now *Sequoiadendron giganteum* (Lindley) Buchholz. Farquhar, after stating that the name *Sequoia gigantea* "is now in official use," comments that it ". . . has come to prevail among American botanists—Abrams, Eastwood, Jepson, McMinn, among others. A little flurry over the name *Sequoiadendron* does not seem to have disturbed them . . ."!

BIDWELL, JOHN. 1928. Echoes of the past about California. By General John Bidwell ... Edited by Milo Milton Quaife, secretary and editor of the Burton Historical Collection. Map. ["Map of the mining district of California by Capt. W. A. Jackson ... 1951."] The Lakeside Classics, Chicago. [Quaife states that it was first published serially in *The Century Magazine* in 1890.]

BUCHHOLZ, JOHN THEODORE. 1939. The generic separation of sequoias. *Am. Jour. Bot.* 26: 535-538.

CHITTENDEN, HIRAM MARTIN. 1907. The American fur trade of the far west. A history of the pioneer trading posts and early fur companies of the Missouri valley and the Rocky Mountains and the overland commerce with Santa Fe. Map. 3 vols. New York. 1: 409 *et seq.*

FARQUHAR, FRANCIS P. 1948. Yosemite, the big trees, and the High Sierra. A selective bibliography. Berkeley. Los Angeles.

GHENT, WILLIAM JAMES. 1931. The early far west. A narrative outline 1540–1850. New York. Toronto. 317, 318.

GRAY, ASA. 1872. Sequoia and its history. *Proc. Am. Assoc. Adv. Sci.* 21: 1-31. *Reprinted:* Sargent, Charles Sprague, *compiler.* Scientific papers of Asa Gray. 2 vols. Boston. New York. 2: 142-164. [Address delivered at Dubuque, Iowa, August, 1872.]

JEPSON, WILLIS LINN. 1923. The trees of California. ed. 2. Berkeley. 25.

LEONARD, ZENAS. 1904. Leonard's narrative. Adventures of Zenas Leonard. Fur trader and trapper 1831–1836. Reprinted from the rare original of 1839. Edited by W. F. Wagner, M.D. With maps and illustrations. Cleveland.

——— 1945. Narrative of the adventures of Zenas Leonard. Written by himself. Edited by Milo Milton Quaife. Secretary and editor of the Burton Historical Collection. Map ["Map of Leonard Country ... Outward route of Leonard from Fort Osage to Monterey."]. The Lakeside Classics. Chicago. 135, 136.

PARRY, CHARLES CHRISTOPHER. 1883. Early botanical explorers of the Pacific coast. *Overland Monthly,* ser. 2, 2: 415, 416.

——— 1888. Rancho Chico. *Overland Monthly,* ser. 2, 11: 561-576.

SARGENT, CHARLES SPRAGUE. 1896. The Silva of North America. Boston. New York. 10: 147, *fn.* 3.

SHINN, CHARLES HOWARD. 1889. The great sequoia. *Gard. & Forest* 2: 614, 615.

SUDWORTH, GEORGE BISHOP. 1908. Forest trees of the Pacific slope. *U. S. Dept. Agric. Forest Service* 139-145.

of placer gold not only to the owner but to innumerable prospectors as well. The property lay southwest of Yosemite National Park,[1] and through it now runs the state highway from Merced to the Park. Nevins, after describing Frémont's numerous vicissitudes as a landed proprietor, which included legal battles with both state and federal governments as to ownership rights, comments that the Mariposa grant "... did more to govern the central part of his career, and in the large view to warp it, than any other element for it led him from the scientific pursuits for which he had been trained into the alien world of business."

From the Mariposa River Frémont and his party moved on to the "... upper San Joaquin River ..." on December 20 and 21, and on the 23rd reached "... the *Tulére Lake* River. This is Lake Fork ... called by the Mexicans the *Rio de los Reyes* ... This is the principal affluent of the Tulére Lake ... In time of high water it discharges into the San Joaquin River ..." They had reached King's River and, expecting to make rendezvous with the Kern-Walker party, remained there for several days, on December 31 mounting to "... a safe camp between 9000 and 10,000 feet above the sea."

1846

On January 1 Frémont's party began to retrace its steps and by the 7th was again on King's River. Having failed to meet the other contingent, Frémont left most of his men to continue the search and himself departed for Sutter's Fort, arriving there on January 15 and remaining until the 19th. Having obtained passports for himself and eight men, he left for Monterey, spending on the way several days at Yerba Buena (San Francisco) and visiting a quicksilver mine at "New Almaden." It was while at Yerba Buena that he bestowed an indelible name upon the strait uniting the Bay of San Francisco with the Pacific Ocean. The name appears on the map of the "Geographical memoir" and its significance is explained in a footnote (p. 32). In the *Memoirs* (p. 512 and footnote) Frémont repeats:

"To this Gate I gave the name of *Chrysopylæ* or Golden Gate; for the same reasons that the harbor of Byzantium (Constantinople afterwards), was called *Chrysocera,* or *Golden Horn.*"

"The form of the harbor and its advantages for commerce ... suggested the name to the Greek founders of Byzantium. The form of the entrance into the Bay of San Francisco and its advantages for commerce, Asiatic inclusive, suggested to me the name which I gave to this entrance and which I put upon the map that accompanied a geographical Memoir addressed to the Senate of the United States in June, 1848."

1. Nevins' biography of Frémont includes a map (made from "Official Surveys, 1936") showing the boundaries of the property "... as finally laid out." Within its confines are shown the towns of Guadaloupe (south of Mount Bullion), Bridgeport, Mormon Bar, Mariposa and Bear Valley. The Merced River cut across its northern boundary, at a point south of Bagby, and its southeastern portion was crossed by Mariposa and Agua Fria creeks.

Frémont left San Francisco on January 24 and set out for Monterey. He observes on the 25th that this was his first "... ride down the valley of San José." He traveled with a man named Leidesdorff[2] who was "... a lover of nature and his garden at San Francisco was, at that time, considered a triumph." Crossing the Salinas plains on the 27th, Frémont reached Monterey where he saw the Consul Thomas O. Larkin and was taken to call on "the commanding general, Don José Castro," and others. His presence evidently called for an explanation:

"I informed the general and other officers that I was engaged in surveying the nearest route from the United States to the Pacific Ocean ... that the object of the survey was geographical ... and that it was made in the interests of science and of commerce ... during the two days I stayed I was treated with every courtesy by the general ..."

Having acquired supplies, Frémont took his departure northward and "by the middle of February" was reunited with Kern, Walker, and the rest "... in the valley of San José, about thirteen miles south of the village of that name on the main road leading to Monterey, which was about sixty miles distant." They had taken a different route from the one contemplated.[3]

The work of the expedition was resumed by the third week in February and, writes Frémont, "... on the 22d March [February] we encamped on the Wild-Cat Ridge on the road to Santa Cruz, and ... on the 23d near the summit." This ridge, he explains, protected the valley of San José from the northwest winds of the coast and was also called the *"Cuesta de los Gatos."* It was now spring in California and, although all the reasons for his sojourn may not have been disclosed, he reports that "The varied character of the woods and shrubbery on this mountain, which lay between my camp and the Santa Cruz shore, was very interesting to me, and I wished to spend some days there, as now the spring season was renewing vegetation, and the accounts of the great trees in the forest on the west slope of the mountain had aroused my curiosity. Always, too, I had before my mind the home I wished to make in this country, and first one place and then another charmed me ... and so far I had not stood by the open waves of the Pacific ..."

I quote what Frémont says of the coast redwood from the "Geographical memoir;" it is repeated, nearly verbatim, in his later *Memoirs.* There are two passages about this tree:

"... The place of our encampment was 2,000 feet above the sea ... The mountains

2. This would seem to have been William Alexander Leidesdorff to whom Bancroft devotes a long paragraph in his "Pioneer Register and Index" (*California* IV. 1846–1848, 711). Nothing is said of his horticultural interests, but he was evidently a prominent citizen of San Francisco.

3. Instead of going to "Tuláre Lake Fork," they had taken a pass considerably south of the lake and, until driven out by snow, had spent from December 27, 1845, to January 17, 1846, in a "... cove, near the summit of the Sierra, at the head of the river ..." To this river Frémont gave "... the name of my topographer, Kern ..." and, to one of the lakes along the party's route, "Owen's name."

were wooded with many varieties of trees, and in some parts with heavy forests. These forests are characterized by a cypress *(taxodium)* of extraordinary dimensions, already mentioned among the trees of the Sierra Nevada, which is distinguished among the forest trees of America by its superior size and height. Among many which we measured in this part of the mountain, nine and ten feet diameter was frequent—eleven sometimes; but going above eleven only in a single tree, which reached fourteen feet in diameter. Above two hundred feet was a frequent height. In this locality the bark was very deeply furrowed, and unusually thick, being fully sixteen inches in some of the trees. The tree was now in bloom, flowering near the summit, and the flowers consequently difficult to procure. This is the staple timber tree of the country, being cut into boards and shingles, and is the principal timber sawed at the mills. It is soft, and easily worked, wearing away too quickly to be used for floors. It seems to have all the durability which anciently gave the cypress so much celebrity. Posts which have been exposed to the weather for three quarters of a century (since the foundation of the missions) show no marks of decay in the wood, and are now converted into beams and posts for private buildings. In California this tree is called the *palo colorado*. It is the king of trees." (p. 36)

". . . A forest of *palo colorado* at the foot of the mountains in this vicinity [they had '. . . descended to the coast near the northwestern point of Monterey bay . . .'], is noted for the great size and height of the trees, I measured one which was 275 feet in height and fifteen feet in diameter, three feet above the base. Though this was distinguished by the greatest girth, other surrounding trees were but little inferior in size and still taller. Their colossal height and massive bulk give an air of grandeur to the forest.

"These trees grow tallest in the bottom lands, and prefer moist soils and north hill sides. In situations where they are protected from the prevailing northwest winds, they shoot up to a great height; but wherever their heads are exposed, these winds appear to chill them and stop their growth. They then assume a spreading shape, with larger branches, and an apparently broken summit." (p. 37)

Why Frémont—like other visitors to the Monterey coast—has nothing to say about the Monterey cypress (*Cupressus macrocarpa* Hartweg) is to me a mystery. He does, however, comment at this time upon the madroña (*Arbutus menziesii* Pursh):

"Another remarkable tree of these woods is called in the language of the country *madrono*. It is a beautiful evergreen, with large, thick, and glossy digitate leaves, the trunk and branches reddish colored, and having a smooth and singularly naked appearance, as if the bark has been stripped off. In its green state the wood is brittle, very heavy, hard, and close grained; it is said to assume a red color when dry, sometimes variegated, and susceptible of a high polish. This tree was found by us only in the mountains. Some measured nearly four feet in diameter, and were about sixty feet high." (p. 37)

On March 1 Frémont resumed his journey "along the coast" and on the 3rd camped at "the Hartnell rancho[4] . . . well out on the plain . . . We were now passing Monterey, which was about twenty-five miles distant . . . I was on my way to a pass, opening into the San Joaquin valley, at the head of a western branch of the Salinas River." It was on this day that letters came from the Mexican general, Castro, ordering Frémont ". . . forthwith out of the department, and threatening force in the event that I should not instantly comply with the order." By return messenger, Frémont states, ". . . I peremptorily refused compliance to an order insulting to my government and myself." On the 4th camp was moved a few miles, ". . . to the foot of the ridge which separates the Salinas from the San Joaquin, at the house of Don Joaquin Gomez. A stream here issues from the mountain which is called the Gavilan Peak. The road from Monterey passes this place, entering the neighboring San Juan valley by way of a short pass called the Gomez Pass."

Having fortified this camp, Frémont raised the American flag on "a tall sapling"; but when, after three days, this fell to the ground, it was accepted as an omen that it was time to move and on March 10 they left Gavilan Peak, entered the San Joaquin valley on the 11th, and the Sacramento valley on the 21st, camping the next day near Sutter's Fort.

Knowing nothing of the Sacramento River valley above this point, Frémont set out on March 24 to examine it and, having left the American River ten miles above its mouth, traveled "a little east of north" towards the Bear River settlements. On the 25th they reached Bear River, ". . . an affluent of *Feather* River, the largest tributary of the Sacramento . . ." On the 26th they encountered Feather River ". . . twenty miles from its junction with the Sacramento near the mouth of the *Yuba* . . ." Traveling northward "up the right bank of the river," they reached Butte Creek on the 28th, Pine Creek on the 29th and, on the 30th, Deer Creek, ". . . another of these beautiful tributaries of the Sacramento."

". . . It has the usual broad and fertile bottom-lands common to these streams, wooded with groves of oak and a large sycamore *(platanus occidentalis,)* distinguished by bearing its balls in strings of three to five, and peculiar to California. Mr. Lassen[5] . . . has established a rancho here . . . Salmon was abundant in the Sacramento. Those which we obtained were generally between three and four feet in length, and appeared to be of two distinct kinds. It is said that as many as four different kinds ascend the river at different periods . . ."

Here they remained until April 5 when they again started up the Sacramento and camped on a creek with the "snowy peak of Shatl [Shasta] . . . directly north." They moved eastward towards it on the 7th, with, to the east, *"the Sisters"*; and ". . . nearly

4. Doubtless the abode of William Edward Petty Hartnell with whom Douglas had resided when in California in 1833–1834.

5. Peter Lassen, a Danish blacksmith. See Bancroft's "Pioneer Register and Index" (*California* IV. 1846–1848, 708), which states that "His memory is preserved in the name of Lassen peak and county." He was killed by an Indian, or by a white man disguised as such, in 1859.

opposite, the Coast Range shows a prominent peak, to which I gave the name Mount Linn,[6] as an enduring monument to recall the prolonged services rendered by him in securing to the country our Oregon coast." Leaving the Sacramento this day, "at a stream called Red Bank," they entered "a high upland" and camped on "a large stream called Cottonwood Creek." On April 8, camping on the Sacramento, Frémont states that they were now ". . . near the head of the lower valley . . . The valley of the Sacramento is divided into upper and lower—the lower two hundred miles long, the upper known to the trappers as Pitt river, about one hundred and fifty. The division is strongly and geographically marked. The Shastl peak stands at the head of the lower valley . . . the whole valley of the Sacramento [is] three hundred and fifty miles long."

On the 9th they ". . . descended into the broad bottoms of . . . Cow Creek . . ." and retracing their steps on the 10th and 11th, again reached ". . . Lassen's, on Deer River . . ." Here, states Frémont, ". . . I set up the transit . . . This was the third[7] of my main stations . . ." Warren states that on the above journey Frémont had ascended the Sacramento "as far as Fort Reading," [8] and had then crossed to the valley of "Pit" River.

It was during a two weeks' sojourn at Lassen's ranch that Frémont and Carson took part in a fight with some Indians (threatening American settlers). The episode was not to their credit[9] and is not mentioned in Frémont's *Memoirs.*

On April 24 Frémont again left Lassen's Rancho for the north, with the purpose of aligning his present journey with the one made in late 1843 through the Klamath Lake region. After traveling up the Sacramento, they came to ". . . the head of the lower valley . . ." on the 25th—or to the region of Warren's "Fort Reading"—and by the 27th were ascending Pitt River which led into ". . . a region very different from the valley of California . . . one resembling that of the Great Basin . . . but more fertile and having much forest land, and well watered." On the 29th camp was ". . . on the upper Sacramento [or Pitt], above Fall River, which is tributary to it." On April 30 they reached ". . . the upper end of a valley . . ." to which Frémont gave the name *"Round Valley."* This is shown on the map of the *Memoirs* to the southeast of Mount Shasta. On May 1 they came ". . . on the southeastern end of a lake, which afterwards I

6. In honor of Lewis Fields Linn, United States Senator from Missouri, whose "Oregon bill" advocated that Oregon Territory should be "saved" from the English. This peak is now known as Castle Crags.

7. Of the four stations mentioned in part II of the Appendix of the "Geographical memoir."

8. The volume *California* ("American Guide Series"), p. 436, notes that although present Redding ". . . stands within Pierson B. Reading's Rancho Buena Ventura, northernmost Mexican land grant, it owes its name not to Reading but to B. B. Redding, Central Pacific R. R. land agent." Warren seems to have had Reading's rancho in mind.

9. The story is told briefly in *Kit Carson's autobiography* ("The Lakeside Classics," edited by Milo Milton Quaife). Carson comments that "We attacked them, and although I do not know how many we killed it was a perfect butchery . . ." and adds that the enemy was given a lesson not to be forgotten. Quaife cites one comment to the effect that the affair was "too unnecessarily revolting to prompt repetition here."

named Lake Rhett in friendly remembrance of Mr. Barnwell Rhett,[1] of South Carolina, who is connected with one of the events of my life which brought with it an abid-
ing satisfaction.[2] This camp was some twenty-five or thirty miles from the lava beds
. . ." or the place where ". . . Major-General [E.R.S.] Canby was killed by the Modocs,
twenty-seven years later[3] . . ." Since a member of the party got lost at this point they
remained in camp for several days, or until his return. On May 6 they reached ". . .
the Tlamath Lake at its outlet . . ." On the journey of 1843 Frémont had entered Nevada from Oregon at a point considerably east of his present position. The "Table of
latitudes and longitudes" of the "Geographical memoir" records their position on May
14[4] as at "We-to-wah creek, (southeastern end of Tlamath lake.)"

Here Frémont's scientific activities ceased and his military activities began. For a
naval officer, Lieutenant Archibald Gillespie (who had followed the explorer's trail
northward from Sutter's Fort) brought letters which, Frémont states, informed him
"officially" that his country was at war with Mexico and that ". . . the President's plan
of war included the taking possession of California . . ." Frémont claims that he had
his "warrant" and determined to start south. "This decision was the first step in the
conquest of California."

In 1853, in the *Smithsonian Contributions to Knowledge* (6: 1-24) John Torrey
published a paper entitled "Plantæ Frémontianæ; or, descriptions of plants collected
by Col. J. C. Frémont in California." It is explained in a footnote that "An Abstract of
this memoir was read before the American Association for the Advancement of Science, at its meeting held in New Haven, August, 1850, and published in the volume of
its Proceedings." Turning to the *Proceedings,* we find that the paper "On some New
Plants discovered by Col. Fremont, in California" was read on August 22, 1850, and
published in 1851, in the report of the "Fourth Meeting" of the Association. The abstract contains neither the botanical descriptions nor the plates of "Plantæ Frémontianæ," but it makes clear that the included plants were gathered on the expedition of
1845–1847. In other words it mentions but one expedition, while the "Plantæ Frémontianæ" refers to four, and without stating on which the plants described were collected. According to the abstract, "Of the collections made in Col. Frémont's third

1. Bartholomew's "Orographical map of the United States and part of Canada" (undated) shows
"Rhett Lake" lying east of "Little Klamath Lake" and just south of the Oregon-California border. In his
Oregon geographic names (ed. 3, p. 610, 1952) Lewis A. McArthur gives its location as ". . . Klamath
County, Oregon, and Modoc and Siskiyou counties, California . . ." He states that the name Tule Lake
is now in general use, in preference to Rhett Lake.

2. This suggests Frémont's marriage but Nevins' biography of the explorer does not cite Rhett in its index.

3. Canby, friendly towards the Modocs whom he was under orders to subdue, was shot down on April
11, 1873, while attempting to arrange a truce. In 1925 the Lava Beds National Monument was established—in northeastern Siskiyou County, on the boundary of Modoc County.

4. The last position in the "Table . . ." is dated June 7, at the "Buttes of the Sacramento."

expedition, in the years 1845–6–7, no public notice has hitherto been given, except that two or three of the new plants were briefly characterized by Dr. Gray, in order to secure the priority of their discovery. In the memoir, of which this is an abstract, I have given descriptions of ten genera of Californian plants, all of them discovered in the passes and on the sides of the Sierra Nevada, by Col. Fremont."

Although the abstract lacks the fine plates—which it states, erroneously, were made from the drawings ". . . of Mr. Charles Sprague,[5] of Cambridge, who ranks among the most eminent botanical draughtsmen of our day . . ."—as well as the technical descriptions prized by the botanists, it has a more readable quality and, besides, contains matter related to the nomenclature of the plants described which is not contained in what was, doubtless, the more important publication. The ten genera enumerated were:

(1) ". . . a remarkable genus of the natural order Portulacaceæ . . . As an expression of the estimate placed on the valuable services rendered to botany by Mr. Sprague, this new genus is called *Spraguëa*."

(2) "The next genus is a still more remarkable one. It is a tree, nearly allied to the celebrated *Cheirostemon* of Humboldt, or Hand Tree of Mexico; but is nevertheless wholly distinct from it . . . Several years ago, I named a genus in honor of the distinguished traveler [Frémont] just mentioned; but it was shown afterwards that I was anticipated a few months by Nees, who published the same genus, under the name *Sarcobatus*, in the appendix of Prince Maximilian's travels, a rare and costly work, which was unknown in this country until several years after my description was published. As the law of nomenclature in natural history is as just as it is inexorable, the name of Nees must be adopted instead of mine, and I have called the new Bombacean genus *Fremontia*,[6] with the specific name of *Californica*."

(3) After mentioning the genus Darlingtonia[7]—Torrey states that because of nomenclatorial rules it had been difficult to name a plant in William Darlington's honor—he enumerates "Of the great natural order Rosaceæ . . . three undescribed genera in Col. Fremont's collections, all of them shrubby."

5. The name, Isaac Sprague, is cited correctly in "Plantæ Frémontianæ."

6. Torrey states in "Plantæ Frémontianæ": "In my memoir on Batis, published in the present volume, I have given the reason for relinquishing the former genus Frémontia, and my intention of bestowing the name on a new plant from California, first detected by the distinguished traveller himself, whose valuable services to North American Botany it is thus intended to commemorate."

Nees von Esenbeck's *Sarcobatus* (1841) invalidated Torrey's *Fremontia* (1943). But Torrey—firm in his determination to have "the punishment fit the crime" as it were—led the way into nomenclatorial mazes when, in accordance with the general practice of the time, he described a totally different genus of the *Sterculiaceae* under the name *Fremontia* in 1853. This was renamed *Fremontodendron* by Coville in 1893 but this name is now relegated to synonymy, along with *Fremontia* Torrey (1843). According to the long list of officially conserved generic names, Torrey's second use of *Fremontia* stands! See *International Code of Botanical Nomenclature*, 119, 1952.

7. First collected by Brackenridge in 1841 and described by Torrey in 1853.

(4) *Coelogyne,* the type *C. ramosissima.*

(5) *Emplectocladus,* "nearly related to the genus Prunus," the type *E. fascicula-tus.*

(6) ". . . belongs to the true Rosaceæ, and is allied to Purshia. It was first dis-covered[8] by Col. Fremont; but Mr. Hartweg, an English botanist, detected it after-wards in California. Mr. Bentham, who has described all the collections of Hartweg, offered to adopt any name I wished to give this plant; but I waived the privilege, and he described it under the name *Chamæbatia foliolosa* . . ."

(7) "Another fine shrub discovered by Col. Fremont,[9] belongs to the small order of Philadelphaceæ . . . I have named this genus *Carpenteria* . . ."

Torrey now notes that "The two genera of Compositæ, which I have had drawn and engraved, have already been briefly noticed by my friend Dr. Gray, in Plantæ Fend-lerianæ, that the merit of discovering them might be secured for Col. Fremont." These were:

(8) ". . . one, the *Hymenoclea,* is allied to *Franseria.* A second species of this genus was afterwards found by Maj. Emory on the river Gila, and is noticed by me in his Report as *H. monogyna.*"

(9) "The other Composita is Amphipappus, and is characterized by Dr. Gray in the *Boston Journal of Natural History.*[1] The accompanying engraving gives a perfect representation of the plant."

(10) "The tenth and last genus which is described in my paper belongs to the sub-order Monotrapeæ of Ercaceæ . . . I have called it *Sarcodes sanguinea.*"

8. In a footnote, in "Plantæ Frémontianæ," Torrey states: "The plant on which this genus was founded was first discovered by Colonel Frémont, in his second expedition . . . early in the year 1844, as well as in his third expedition . . ."

9. One notes that the accent (in the name Frémont) is omitted throughout in the abstract, but inserted in the "Plantæ Frémontianæ."

1. *Boston Jour. Nat. Hist.* 5: 104-111, 1844. Plate.

BANCROFT, HUBERT HOWE. 1886. The works of . . . 22 (California V. 1846–1848): 2-6, 101-105; 24 (California VII. 1860–1890): 440, *fn.* 47.

BREWER, WILLIAM HENRY. 1880. List of persons who have made botanical collections in California. *In:* Watson, Sereno. Geological survey of California. Botany of California 2: 556.

CARSON, CHRISTOPHER. 1933. Kit Carson's autobiography. Edited by Milo Milton Quaife. The Lakeside Classics. Chicago. 95, *fn.* 82.

DELLENBAUGH, FREDERICK SAMUEL. 1914. Frémont and '49; the story of a remarkable ca-reer and its relation to the exploration and development of our western territory, es-pecially in California. New York. London. 282-380. Maps. [Map opp. p. 68 shows routes of all Frémont's expeditions.]

DE VOTO, BERNARD. 1943. The year of decision 1846. Boston.

FLETCHER, F. M. 1929. Early Nevada. The period of exploration 1776–1848. Reno. Map. 141-167.

FRÉMONT, JOHN CHARLES. 1848. Geographical memoir upon upper California, in illustration of his map of Oregon and California, by John Charles Frémont: addressed to the Senate of the United States. Washington. *U. S. 30th Cong., 1st Sess., Sen. Miscel. Doc.* No. 148. 1-67. Map. *Reprinted: U. S. 30th Cong., 2nd Sess., House Miscel. Doc.* No. 5. 1849.

———— 1887. Memoirs of my life, by John Charles Frémont. Including in the narrative five journeys of western exploration, during the years 1842, 1843–4, 1845–6–7, 1848–9, 1853–4 . . . A retrospect of fifty years, covering the most eventful periods of modern American history . . . with maps and colored plates. Chicago. New York. 411-602. [Map shows route taken by Frémont in 1845–1847.]

GHENT, WILLIAM JAMES. 1931. The early far west. A narrative outline 1540–1850. New York. Toronto. 343-348; 356-359; 373-375.

GOODWIN, CARDINAL. 1922. The trans-Mississippi west (1803–1853). A history of its acquisition and settlement. New York. London. 241-243.

GRAHAM, EDWARD HARRISON. 1937. Botanical studies in the Uinta Basin of Utah and Colorado. *Annals Carnegie Mus. Pittsburg* 26: 12, 13. Map ["Map of Uinta Basin and environs."].

NEVINS, ALLAN. 1939. Frémont. Pathmarker of the west. New York. London. 206-342, 383-384, 393. Maps ["Frémont's five exploring trips, 1842–54," p. 211; "Frémont's Mariposa Estate as finally laid out," p. 379].

TORREY, JOHN. 1851. On some new plants discovered by Col. Fremont, in California. *Proc. Am. Assoc. Adv. Sci.* 4: xxix, xxx; 190-193. ["Abstract" of Torrey, J. 1853. Plantae Frémontianae . . . but contains additional matter.]

———— 1853. Plantae Frémontianae; or, descriptions of plants collected by Col. J. C. Frémont in California. *Smithsonian Contr. Knowledge* 6: Art. 2, 1-24. 10 plates.

WARREN, GOUVENEUR KEMBLE. 1859. Memoir to accompany the map of the territory of the United States from the Mississippi River to the Pacific Ocean, giving a brief account of each of the exploring expeditions since A.D. 1800 . . . *U. S. War Dept. Rept. expl. surv. RR Mississippi Pacific* 11: 48-50.

CHAPTER XLV

THE YOUNGER ABERT EXPLORES THE PURGATORY

RIVER, THE NORTH FORK OF RED RIVER OF THE SOUTH,

THE UPPER WASHITA RIVER AND MUCH OF THE CANA-

DIAN RIVER FROM ITS HEADWATERS TO ITS CONFLU-

ENCE WITH THE ARKANSAS

IN his *Memoirs of my life* John Charles Frémont mentions under date of August 2, 1845, that he had detailed two of his Lieutenants—James William Abert and William Guy Peck—". . . to survey the Canadian from its source to its junction with the Arkansas . . . the Purgatory River, and the heads of the Washita . . ." Abert, son of John James Abert, Chief of the Corps of Topographical Engineers, commanded the expedition which included some thirty persons, and the famed Thomas Fitzpatrick served as guide. The account of the journey, entitled "Journal of Lieutenant J. W. Abert, from Bent's Fort to St. Louis, in 1845," was published in 1846 (Senate Document No. 438, 29th Congress, 1st Session) and included a beautiful map.

Abert was obviously interested in all branches of natural history, but appears to have been better informed about the fauna than the flora. I shall outline his route and quote some of his comments upon the plants which he observed. The plant names are often peculiar but whether the author or the typesetters should be held responsible is hard to say,—such papers seem to have been demanded immediately by the authorities.

Abert's "Journal" is dated from "Bent's Fort, on the Arkansas River, *Saturday, August* 9, 1845," and until August 24 travel was in Colorado. It begins on the 9th with the comment that "In compliance with orders from Captain Frémont, I this day moved across the river, to take command of the party detached for the survey of Purgatory creek, the waters of the Canadian and False Washita . . ." Three days later Fitzpatrick joined the party and Abert was "at once delighted with the carefulness of our guide," and comments on the 15th that "The preservation of our party was due to his vigilance and discretion."

On August 16 they began the descent of the Arkansas, keeping along "the right bank of the river" (the southern bank), and ". . . arrived at the junction of the Purgatory (or Las Animas) with the Arkansas . . ." This was in Bent County, not far east of the town of Las Animas. Abert had observed ". . . a profusion of prairie sage, 'artemisia tridentata,' . . . Cacti were numerous, and a species of cucurbitaceæ, 'cucurbita aurantia,' bearing a small spherical gourd, orange-colored." On the 17th they moved ". . . in a southwesterly direction up the Purgatory . . ." and, from the table-land, caught sight of the Rocky Mountains. ". . . To the southwest, the 'Wah-to-yah,' or Spanish peaks . . . to the northwest the snowy summit of Pike's peak was faintly discernible." The peaks were the East and West Spanish Peaks, lying northwest of present Walsenburg, Huerfano County.

They now left the Purgatory "some miles" to the left, traveled "nearly due west . . . in a direction nearly parallel with the Arkansas . . ." and came to "the Timpa," or Timpas Creek, ". . . which falls into the Arkansas 12 or 14 miles above Bent's fort." By crossing to that stream and following it for a time, the party was taking a more direct and easily traveled route than the one along the Purgatory. When making the crossing to the Timpas on August 19 Abert observed that "The sandy hills were . . . covered with artemisia, 'yucca angustifolia,' and a species of cactus—'cactus Peruviana.' The latter plant has a hard woody stem and numerous branches covered with long and sharp spines. As the plant rises to the height of four or five feet, the Mexican Spaniards frequently set them in rows to form hedges . . ."

August 20: "The bluffs on the left hand were . . . crowned with cedars . . . The cacti were very numerous; among which we saw the 'cactus openetia,' a kind resembling a small cantelope half hidden in the ground 'C. melocactus,' and the 'C. Peruviana,' before mentioned, as used for forming hedges. The stems, when old, are extremely hard, and resemble the wood-work carved by the South Sea islanders. The foliage of the cedar trees around our camp presents the appearance of an oblate speroidal mass, not otherwise differing from the common 'juniperus Virginianus.' . . . A plant called Adam's needle, 'yucca angustifolia,' was very abundant. The remarkable beauty of its conspicuous spike of campanulate flowers has procured it the name of the 'prairie light-house;' the Spaniards call it the 'palmilla.' "

On August 21 they reached ". . . the summit of the dividing ridge between the waters of the Arkansas and Purgatory . . ." and entered "a beautiful level plain," camping at ". . . a place called 'Hole-in-the-prairie,' a low marshy spot . . ." Here they saw, and collected seed of, ". . . the 'mirabilis jalapa,' . . ." On the 22nd they were back on the Purgatory, and turned off it to follow Raton Creek, entering from the south.

August 23: "We continued to follow the valley . . . the approaching rocks forced the trail into a defile formed by an affluent of the Purgatory . . . Our road now became exceedingly rough, leading along a tortuous valley, sometimes passing on one side of the Raton fork, sometimes on the other . . . We . . . saw the species of pine, 'pinus monophyllus,' which contains in its cone a number of eatable nuts nearly as large as the kernels of the ground nut, I brought in some of the cone, which were remarkably resinous. In the appendix to Captain Frémont's report of 1843 and 1844, there is a botanical description of this tree. We found them generally from 30 to 40 feet in height . . . Gregg says that considerable quantities [of nuts] are exported annually to the southern cities, and that they are sometimes used for the manufacture of oil, which may be used as a substitute for lamp oil. They form the chief article of food of the Pueblos and New Mexicans. Among the plants we noticed the blue larkspur, ('delphinium,') which resembles the cultivated kind, except the interior of the corolla, which presents a pubescent appearance; the wild geranium, ('geranium maculatum;') the beautiful flowering flax, ('linum perennium;') also, a delicate hair bell, ('campanula linifolia,') which I have found as far north as the Manitou islands of lake Michigan; also, a beautiful scarlet flower, 'convolvulaceæ.' "

On August 24 they ". . . ascended a hill 3 miles, to the summit of the mountain spur known by the name of the 'Raton.' " The boundary between Colorado and New Mexico now crosses at, or near, the summit of Raton Pass and Abert and his party must have entered New Mexico, remaining in that state until September 3. The "Journal" records on the 24th:

"Where the ravine widened we found quantities of cherries, 'Cerasus Virginianus,' but the bushes dwarfed by the rocky soil . . . we emerged from the gorge and entered a small prairie . . . We pitched our tents in a bend of the stream which forms the principal source of the Canadian, or Goo-al-pa, as it is called by the Kioways and the Camanches . . . We have now finally crossed the Reton ('Retoño') of the Spaniards, which is the only difficult part of the regular route to Santa Fe by the way of Bent's Fort . . ."

From the 24th until the 28th they followed the Canadian River, crossing, on the 26th, to its eastern side; this appears to have been at a point about due east of Eagle's Nest, in Colfax County. Abert's map shows that the Canadian was followed to a little south of the "Moro," or Mora, River, which enters the larger stream in northeastern San Miguel County. On August 28 and 29 they examined a gorge formed by the Canadian—". . . an excursion to the river and great cañon through which it flows, and from which is derived the name Canada, or Canadian river . . ."—and then made a detour (August 31–September 3) to the north, returning to the Canadian at a point just east of the mouth of Ute ("Utah") Creek, in northwestern Quay County and no great distance west of the present New Mexico-Texas border. On this detour Abert mentions (August 31) that "The stream we first struck was the 'Arroyo de los Yutas' or (as the Camanches call it) 'Salt creek.' . . . The sandy plain . . . was strewed with numerous flowers . . . Blue, red, and yellow were common; the yellow . . . predominated—cacti, and the prairie sensitive plant, 'Schraukia Angustata.' We saw many new plants which had not been found on the northern prairies, and again had to regret our misfortune in not having been able to save specimens. In the vicinity of our camp the sunflower, 'Helintheæ,' grew abundantly, and a species of belladonna, 'Solaneæ,' which had passed the flowering season, and was thickly covered with its round yellow seed-balls. There were also several species of convolvulus, amongst which was the 'Ipromæa leptophylla.' "

By September 1 the detour was becoming more and more difficult until they entered a valley which, though sandy, appeared fertile, ". . . being covered with high grass and multitudes of yellow flowering plants . . . Among the varieties . . . we observed the Adam's needle, 'Yucca angustifolia,' the sensitive plant, 'Schraukia angustata,' species of convolvulaceæ, leguminoseæ, and solaneæ."

On September 2 they reached a ". . . shady grove of tall buttonwood, 'Platanus occidentalis,' mingled with cotton-wood . . ." and discovered a few grapes. Next they "found themselves" back on the Canadian, ". . . just at the point where it is joined by the 'Arroyo de los Yutas.' " They had completed the detour.

"The bottom in which we encamped is every where covered with various species of cactus, the sharp spines of which penetrated our moccasins, making it painful to walk about. There is a plant still more annoying, commonly called 'Sand-bur.' This is a diminutive plant, lying close to the sandy surface, loaded with a profusion of little burs, which attach themselves to our clothes and blankets by their sharp prickles, and adhere with great tenacity. Amongst the sylva, the hackberry, 'Celtis crassifolia,' is quite common, and we observed, for the first time, an extensive grove of the pride of India, 'Melia azederach.'[1] . . ." After a short description he continues:

"The trees were every where loaded with heavy masses of grape vines, 'Nitis [Vitis] aestivalis,' which afforded the whole camp a great abundance of fruit . . . we thought they equalled in flavor any of the cultivated varieties. It was now the fruit season of the broad-leafed[2] cactus, 'C. opuntia,' and they were every where in great abundance . . . some being 3½ inches in length. In their flavor the raspberry and water melon seem mingled . . . We frequently paid dear for handling them, the little spines being barbed like a fish hook. We saw . . . great quantities of the 'musquit,' or muskeet, covered with its long sabre-formed legumes . . . We noticed a variety of cactus here, 'Cactus elatior'; the joints were ovate and oblong, covered with exceedingly long

1. Abert tasted the fruit, "so pungent that it was a long time before I could get rid of the unpleasant impression it produced." Since *Melia Azedarach* is an Asiatic species one would scarcely expect to find it naturalized in the wilds along the Canadian River. Liberty Hyde Bailey's article upon the genus mentions that, in this country, "The first tree that came to notice is said to have been found near the battlefield of San Jacinto, Texas, but with no record of its intro[duction] there . . ." It would be interesting to know whether Abert's record is earlier than the one mentioned by Bailey. The battle of San Jacinto took place on April 21, 1836, but Bailey does not tell when the tree was found, or by whom. The town of San Jacinto is in Harris County, Texas, on Galveston Bay, in a region where the tree's introduction would seem less remarkable than along a mountain gorge of the Canadian. The tree naturalizes itself readily and the region may not have been as remote as it seemed, for Abert himself mentions that ". . . this place is what is called the 'Spanish Crossing,' where the people of New Mexico pass with pack mules on their way to and from the Camanche country."

In traveling southward in winter by night train from New York and wakening in the Carolinas, the first plant that strikes the eye is the chinaberry tree—devoid of foliage, but covered with clusters of cream-colored, marble-like fruits. One such tree at least grows before every little Negro cabin along the railroad.

2. Abert refers to the stem, not the leaf, of the *Opuntia*, which (in this group) is flattened, each section much the shape and size of a ping-pong bat.

spines . . . when touched the joints would fly off,[3] adhere to the flesh, and cause much pain . . ."

On September 3 Abert crossed into Oldham County, in the Texas Panhandle and from that date until September 15 kept along the Canadian, crossing Oldham and Potter counties and entering Hutchinson County. On September 15 he turned south from the Canadian to examine, as directed, the upper waters of the Washita but, mistakenly, followed the upper waters of the North Fork of Red River for some [?] seventy miles. From September 17 to 23 he was on the upper waters of the Washita. On the 24th he turned north and camped again on the Canadian River. He appears to have entered Oklahoma at a point not far east of Antelope Hills ("Buttes") in Roger Mills County.

The "Journal" does not include many comments about plants during the above period but I quote a few:

On September 4, crossing a sandy waterless waste, their sufferings were ". . . greatly alleviated by the refreshing fruit of the plum tree . . . equal to any of the cultivated varieties . . . tasted in the United States." They made them into ". . . a nondescript pudding, which, had it not been for the fruit, one might liken to sailor's duff."

September 5: "The valley appeared full of grape vines, and the troublesome sand-burs covered the ground . . . We saw . . . an abundance of musquit, a species of 'leguminosæ' . . . thought by some persons to be the same as the 'acacia Arabica.' It is a thorny shrub, scarcely ever attaining the height of 5 feet. The legumes are long, sabre-form, cylindrical, and nearly white, filled with a solid substance of sweetish taste, from which the Camanches and Kioways manufacture a kind of flour. Gregg[4] states that in some of the fertile valleys of Chihuahua it attains the height of 30 and 40 feet . . . We saw here an abundance of the cardinal flower, ('lobelia cardinalis.') They looked most brilliantly as they glistened in bright scarlet array . . . We also found the cacti, 'argemone Mexicana,' and yucca, of the plains; the 'populus canadensis' and 'arundo canadensis,' near the streams; and in the ponds . . . the 'typhalatifolia,' 'nymphae lutea,' and 'sagittaria sagittifolia.' "

September 6: "Among the plants . . . most abundant, we noticed the artemisia and the 'cucurbita aurantia,' which was characteristic of the plains since leaving Bent's fort. We also found overgrowing the dry ponds or buffalo wallows, the 'myrtinia proboscidea.' St. Pierre describes it as peculiar to Vera Cruz; attributes to it traits so singular that we could not notice the plant without relating them. He says: 'I presume that often when the shores of Vera Cruz are overflowed by high tides, you must see fishes caught by this plant, for the stem of its pod is not easily broken off; its two crotchets pointed like fishing hooks are elastic and hard as thorn. Besides, when it is soaked in water,

3. Abert refers to the Opuntia group with terete stems—doubtless one of the chollas. See p. 953, *fn.* 7.

4. See p. 955, *fn.* 8.

its furrows, shaded with black, shine as if they were filled with globules of quicksilver. Now, the lustre of this light is a further bait to attract the fishes.' " [5]

September 10: "The cacti and musquit were most abundant . . . Whilst riding over a sandy waste we noticed a most delicious fragrance . . . All the sweetness arose from an unattractive little aster—a species of 'compositæ,' . . . growing in the barren sand . . ." . . ."

September 11: ". . . we stopped near a large grove of hackberry, 'celtis crossifolia,' and gathered quantities of the berries, which . . . repaid us for our trouble by the pleasantness of their flavor, which resembles that of tea."

The party was disappointed more than once, by finding that Indians had already gathered the plums, grapes and cactus fruits of the region. They came upon some petrified trees and the area seemed to be full of rock crystals and agate.

September 17: "This is the day which we look back upon . . . as the day of anxiety . . . we entered upon the famous table-land known to the Spaniards as 'el llano estacado.' This is the most extensive and continuous of the plains lying in the desert country, and gives rise not only to the Washita, but to all the main branches of the Red river, as well as the rivers of upper Texas, or their affluents. To reach the head water of the Washita, it was necessary to cross a portion of this dry and level tableau.[6] The only water . . . is contained in pools or lakes, often at great distances asunder, and with banks so low that the traveller . . . often passes them unnoticed. The sun was pouring down heat as heavy as clouds do rain . . ."

After some hours they finally ". . . descended into the breaks of the head waters of the Washita . . ." On the 19th (and until the 23rd) they kept along the Washita:

". . . the uplands were thickly spread over by a diminutive species of oak, commonly called 'shin oak,' not exceeding two feet in height . . . Our camp ground was covered with the sunflower, 'Heleantheæ,' among which we noticed a variety the flowers of which were sessile and axillary."

5. The quotation is doubtless from Jacques Henri Bernardin de Saint Pierre who wrote a popular book on plants which was translated into English by Henry Hunter and published at Worcester, England, in 1797, under the title *Botanical harmony delineated; or, applications of some general laws of nature to plants.* How much (for the fish story) Saint Pierre drew upon his imagination I do not know, but the long, hooked pods of *Martynia,* often called the unicorn plant, are in appearance much as he described them. Kearney and Peebles ("Flowering plants and ferns of Arizona," 835, 1942) state of the genus: "These plants are usually known in Arizona as devilsclaw. The young pods are sometimes eaten as a vegetable. The black designs in the baskets made by the Pima and other Arizona Indians are woven with the split mature pods of *M. parviflora.* The plants are regarded as somewhat of a pest on sheep ranges because the hooked beaks of the pods become entangled in the fleece." So serving nature's purpose of disseminating the plant's seeds! In gadget shops of our southwest one sees these curious pods transformed into imaginary birds and animals.

6. Surely an error for plateau.

On the 21st they found ". . . great quantities of the poke weed, 'Phytolacca de-candra,' the berries of which afford abundant food for the many varieties of birds . . ."

September 23: "The bottoms of the streams were very well timbered . . . we noticed the cotton-wood, elm, black locust, 'Robinia pseudo acacia;' black walnut, 'Juglans nigra;' and the coffee-nut tree, 'Gymnocladus Canadensis,' which contains, in its large legumes, a very palatable nut, which we collected and ate."

Back on the Canadian River on the 24th and, as noted, in Oklahoma, they followed that stream until they came to the confluence of its North Fork in McIntosh County, on October 18. During that period the "Journal" only rarely mentions plants.

September 27: "In some of the bottoms . . . we found the bur-oak, 'Quercus macro-carpa,' and collected many of the acorns; for they were new to many of us, and the singular beauty of the deeply-fringed edge of the cup which enveloped the acorn, and of the size of the fruit, could not but call forth our admiration."

Also noted on the 27th was the ". . . dense growth of oak commonly called the black-jack oak." Abert, on September 3, had mentioned that his party had started one forest fire: "The crackling of consuming vegetation . . . deepened the impression of this scene of grand sublimity." Now, on the 28th, they started another, and Abert, whose literary style inclines towards the "flowery," comments that he ". . . felt much grieved at this unnecessary destruction of vegetation, which a benign Providence had intended for the nourishment of many of his animated creatures." I might add that, on October 16, this same fire was raging and near at hand. There are not many references to plants along the Canadian:

October 8: "We saw . . . great quantities of the sumach, 'rhus glabrum.' This is much used by the Camanches in making their 'kinick-kinick' . . ."

October 9: "The thick woods which bordered the river were perfectly impenetrable, on account of the luxuriant growth of the greenbrier, 'smilax rotundifolia.' "

October 10: "The elm-trees . . . were in many places ornamented with beautiful clumps of the misletoe[7] . . . the plant was covered with a profusion of beautiful pearly berries . . . extremely mucilaginous . . . nutritious food for birds . . ."

It was on the 10th that the party passed ". . . the ruins of old Fort Holmes . . . Some of our people . . . mounted the chimney, and unfurled the American handkerchief, that it might float in the breeze."

October 11: "In some of the bottoms we found the persimmon, 'diosporus Virginia-nus.' The fruit was not quite ripe, therefore but few of those who were acquainted with its peculiar astringency could be induced to taste it."

7. The *Phoradendron* is now the state flower of Oklahoma.

October 13: "... we noticed many varieties of the oak, 'quercus obtusiloba,' 'Q. fer-ruginea,' and 'Q. alba.' A few specimens of hickory, the ash, and grape vines seemed to have found a grateful soil ..."

While examining a deserted Indian village on the 14th they "... found the bois d'arc, 'maclara aurantica,' the fruit of which is sometimes called the Osage orange ..." It was on the 16th that Abert remarks that the forest fire which they had started was still raging. They observed also the effects of "... one of those terrible hurricanes of frequent occurrence in the western world. The noblest trees of the forest had fallen from their high estates ..."

On October 18 the party crossed the North Fork of the Canadian and camped "about 2½ miles from its mouth." In order to take observations, Abert descended the stream to its confluence with the Arkansas River. On October 21 he followed the Arkansas to Fort Gibson, situated not far northeast of present Muskogee, Muskogee County. He had observed on the way (October 20) that they "... found some of the fruit of the pawpaw, 'annona triloba,' and black walnuts ...", and had seen ... among the sylva, the elm, and various species of the oak and hickory—among the latter, the bitternut hickory, 'juglans aurata' ...", as well as the "buttonwood and spicewood."

From the North Fork of the Canadian, Fitzpatrick had taken the "equipage" over another route and he arrived at Fort Gibson on the 22nd. The "Journal" states that, be-tween Fort Gibson and St. Louis, the road "... was literally lined with wagons of emi-grants to Texas ... until we arrived at St. Louis we continued daily to see hundreds of them."

Abert's party reached St. Louis on November 12. "On the 27th of November [1845] we had the honor to report to the chief of the bureau on our arrival at Washington city."

Abert's "Journal," describing the exploration of 1845 just recounted, included no record of the plants which he then collected.

In 1846 Abert, then attached to the command of William Hemsley Emory, was de-tailed to make a report "of such objects of natural history" as came under his observa-tion on an expedition between Fort Leavenworth, Kansas, and Bent's Fort, Colorado. Abert's "Notes" describing that journey were published in 1848 and included a list of his plants which had been identified by John Torrey. From the localities of collec-tion which are cited it is evident that—although most of the plants were gathered in 1846—some had been collected on the journey of 1845, or on the trip described in the present chapter. I shall refer to Torrey's list when discussing Abert's assignment of early 1846.

ABERT, JAMES WILLIAM. 1846. Journal of Lieutenant J. W. Abert, from Bent's Fort to St. Louis, in 1845. *U. S. 29th Cong., 1st Sess., Sen. Doc.* 8: No. 438, 1-75. Map.
Anonymous. 1897. James William Abert. *Nat. Cyclop. Am. Biog.* 4: 308.

BAILEY, LIBERTY HYDE. 1914. Melia. *In:* Bailey, L. H. Standard Cycl. Hort. 3 vols. New York. 2: 2024, 2025.

FRÉMONT, JOHN CHARLES. 1887. Memoirs of my life, by John Charles Frémont. Including in the narrative five journeys of western exploration, during the years 1842, 1843–4, 1845—6–7, 1848–9, 1853–4 . . . A retrospect of fifty years, covering the most eventful periods of modern American history . . . with maps and colored plates. Chicago. New York.

GEISER, SAMUEL WOOD. 1948. Naturalists of the frontier. ed. 2. Dallas. 270.

WARREN, GOUVENEUR KEMBLE. 1859. Memoir to accompany the map of the territory of the United States from the Mississippi River to the Pacific Ocean, giving a brief account of each of the exploring expeditions since A.D. 1800 . . . *U. S. War Dept. Rept. expl. surv. RR Mississippi Pacific* 11: 50, 51.

CHAPTER XLVI

WISLIZENUS REACHES SANTA FE BY WAY OF THE CIM-

ARRON DESERT AND DESCENDS THE VALLEY OF THE

RIO GRANDE ON HIS WAY TO MEXICO PROPER

IN 1912 the Missouri Historical Society of St. Louis published an English translation of a delightful little book entitled *A journey to the Rocky Mountains in
1839.* An included facsimile of an earlier title-page shows that it had been published in German in 1840 under the title: *Ein Ausflug nach der Felsen-Gebirgen im
Jahre 1839.* The author was Friedrich Adolph Wislizenus, a doctor of medicine.

The translation includes an account of the author's life written by his son and
namesake, Frederick A. Wislizenus. We are told that the father had been born ". . .
May 21, 1810, at Koenigsee, in Schwarzburg-Rudolstadt, a German principality of
duodecimo proportions . . ." and that, although a ministerial career had been planned
for the boy, his ". . . bent was for the natural sciences, and he entered the . . . University of Jena as a medical student. He studied successfully at Goettingen and Tuebingen,
till he fled for his life from Germany." After participating in attempts to break the
Napoleonic oppression and to liberate and unify Germany, Wislizenus finally reached
Switzerland, obtained his medical degree at the University of Zurich, emigrated to the
United States and, in 1836, settled down for three years to the practice of medicine in
Illinois. His savings were expended upon the trip of 1839 to the Rocky Mountains. He
traveled with a party of trappers and was able to observe the workings of the fur trade
at first hand.

Wislizenus' description of his journey contains little to indicate an interest in plants.
He mentions only a few specifically and only such as all travelers notice—the cotton-
wood, the sagebrush, the juniper, the buffalo grass, the *"Pomme blanche (Psoralea
esculenta)"* and such. The trip, which for the doctor was merely an interesting excursion, took him as far west as Fort Hall, Idaho, whence he returned homeward.

Ewan mentions that one may follow his route ". . . with some ease . . . since in the
main he followed the well-worn trails known to the trappers and adventurers . . ." and
he comments thus upon a portion of the excursion:

". . . On the return trip he passed through Browns Hole, the country of the Little
Snake River, and across the present Park Range to North Park, where mountain bison
were plentiful. Crossing the Medicine Bow Range, though of course only vaguely identified, and the medley of hills lying to the north of Cache la Poudre, he emerged from
the Front Range evidently at the exact point where J. C. Fremont . . . was subsequently
to enter the mountains in 1844. Visiting Fort St. Vrain he found the Indian troubles

941

too serious along the South Platte to risk that route to the Missouri. Accordingly, he set out southeastward for the South Platte-Arkansas Divide and continued on to Bent's Fort, from whence he struck out for St. Louis by the Arkansas route. This excursion of 1839 was of little scientific significance, lacking as he did instruments for records or equipment for coll[ection]s . . ."

When writing of Frémont's second expedition of 1843, the same author mentions that ". . . Pueblo, a trading station, was built in 1842 at the confluence of Fountain Creek but Fort Pueblo had been built by Dr. Frederick Adolphus Wislizenus, naturalist-explorer, in 1839 on the north bank of the Arkansas four miles above Bent's Fort." As authority for this he cited Chittenden (p. 969). Checking this—for I had found no record that Wislizenus "built" anything on his trip—the citation reads: "Wislizenus in 1839 found a small post called *Fort Pueblo* four miles above Bent's fort, 'inhabited principally by Mexicans and Frenchmen.' . . ."

For me the interest of Wislizenus' story lies in the skillful recounting of "the daily doings of the caravan" and, above all, in the picture of what had taken place—less than forty years after Lewis and Clark's crossing of the western wilderness—as the result of American exploration and exploitation and in the prediction of what the future held in store for that still-virgin country, for American settlement had advanced but little beyond the eastern margins of the country traversed by Wislizenus. The author's short "Postscript" ends thus:

"It is perhaps only a few years until the plow upturns the virgin soil, which is now only touched by the lightfooted Indian or the hoof of wild animals. Every decade will change the character of the country materially, and in a hundred years perhaps the present narratives of mountain life may sound like fairy tales."

Wislizenus believed that the weakening of Indian strength was in part attributable to the extermination of the buffalo: "The Indian and the buffalo are Siamese twins; both live and thrive only on one ground, that of the wilderness. Both will perish together." The rapid extermination of fur-bearing animals, beaver in particular, had been responsible for the decline of the fur trade. The British had been wiser and, certainly within their own territory, had protected the beaver:

"The Hudson's Bay Company has established more system in beaver trapping within its territories . . . In regions, however, on whose permanent possession the company does not count, it allows the trappers to do as they please. But if trapping is carried on in this ruthless fashion, in fifty years all the beavers there will have disappeared, as have those in the east, and the country will thereby lose a productive branch of commerce."

According to Chittenden, the great period of the American fur trade ended in 1843. Wislizenus wrote in 1839: "Another decade perhaps and the original trapper will have disappeared from the mountains." And we find this prediction in a chapter upon the Indian:

"There has been much fabulous talk about the Indian character. Some pose them as Roman heroes and unspoiled sons of Nature; others as cowards and the scum of humanity. The truth is between these extremes . . .

"The ultimate destiny of these wild tribes, now hunting unrestrained through the Far West of the United States, can be foretold almost to a certainty . . . Civilization, steadily pressing forward toward the West, has driven the Indians step by step before it. Where war with the whites and with each other was not enough to reduce their numbers, the result was brought about by disease and ardent spirits . . . Whole tribes . . . have entirely disappeared, leaving scarce a trace of their name behind . . . But the greatest danger threatens the Indians from the West; from the settlements on the Columbia. Along the Columbia River various Indian tribes have already perished; the rest live in entire dependence on the whites.

"So the waves of civilization will draw nearer and nearer from the East and from the West, till they cover the sandy plains, and cast their spray on the feet of the Rockies. The few fierce tribes who may have maintained themselves until that time in the mountains, may offer some resistance to the progress of the waves, but the swelling flood will at last rise higher and higher, till at last they are buried beneath it. The buffalo and the antelope will be buried with them; and the bloody tomahawk will be buried too. But for all that there will be no smoking of the pipe of peace; for the new generation with the virtues of civilization will bring also its vices. It will ransack the bowels of the mountains to bring to light the most precious of all metals,[1] which, when brought to the light, will arouse strife and envy and all ignoble passions, and the sons of civilization will be no happier than their red brethren who have perished."

Wislizenus predicted great things for the Columbia River! He was writing of the future of that region and of American stakes therein, and of the indifference of the United States government to the welfare of those of its citizens who had gone there to settle:

"The Americans, whose claims to the territory of the Columbia River are much better founded than those of the British, are now merely tolerated by the Hudson's Bay Company . . . While the Columbia is navigable for only a short distance, and its tributaries, on account of the numerous falls, are not suited for navigation, there is the better opportunity for mills and power plants; and better communications could easily be secured through the construction of canals. Moreover, the country is very suitable for agriculture and cattle raising. The interior trade and the coast trade with the Indians is very profitable. The intercourse with the Sandwich Islands, California, Russian America and Asia grows from year to year; and the trading vessels and whalers on the Pacific Ocean find here a safe base for action. In short, if any place on the western shore of North America seems designed by nature to be a western New

1. Written before Americans had demonstrated any appreciable interest in the possible existence of mineral wealth in the regions which they traversed.

York on the Pacific Ocean, it is this. The Straits of Juan de Fuca, somewhat further north, form a better harbor. It is said that a whole fleet could anchor there in safety. These straits also lie south of the forty-ninth parallel . . .”

Dr. Wislizenus had an interesting time on his holiday—as does a reader who accompanies him. Chittenden refers to the book as “. . . an exceedingly important contribution to our knowledge of the period and an interesting and ably written work.” In publishing it, the Missouri Historical Society made a valuable contribution, for it is a choice volume both in format and in content.

The son relates that, after his father's return from the journey of 1839, he entered into a medical partnership with George Engelmann and lived and practiced in St. Louis for about six years; he then could no longer “resist the longing for further explorations.” Possibly Engelmann had stimulated his interest in botany for certainly Dr. Wislizenus' next excursion produced some notable results in that field; not so great perhaps within the region of my story, although this was not neglected, but in present Mexico. I shall only follow his travels as far south as the El Paso crossing; on his trip down the Rio Grande valley from Santa Fe he seems to have been the first to collect plants.

Although Senate and House “Documents” have a forbidding appearance, stores of interesting matter are found within their covers. One such is the “Memoir of a tour to northern Mexico, connected with Col. Doniphan's expedition,[2] in 1846 and 1847, by A. Wislizenus, M.D.” which appeared as a Senate Miscellaneous Document in 1848. Included were three maps and a “Scientific Appendix” containing meteorological tables; also a “Botanical Appendix” of some twenty-eight pages contributed by George Engelmann. On January 13, 1848, five thousand copies of the “Memoir” were ordered printed for members of the Senate, “and 200 additional for Dr. Wislizenus.”

In a short preface Wislizenus mentions that his purpose in making the journey was scientific—his interest lay in geography, natural history and statistics, in compass and astronomical observations, in geology and in mines. He mentions also that he “. . . made a rich collection of quite new and undescribed plants.” He begged indulgence for his literary style: “I am a German by birth, and an American by choice . . . If the reader should . . . discover some Germanism in my English style, I hope he will not judge me with the severe criticism of an English grammarian, but with the philanthropic liberality of a citizen of the world.”

The “Memoir” mentions the great westward migrations which were taking place during this decade. War had been declared against Mexico on May 13, 1846, or before Wislizenus left the Missouri frontier; before his return to the United States he had participated in the conflict in Mexico.

Others of my collectors are soon to travel from Independence, Missouri, to Santa

2. Wislizenus had no official connection with the Doniphan expedition until early March of 1847. He had reached Mexico in early August, of 1846.

Fe, New Mexico, and there are to be many references to places encountered along the way, more than one of which are difficult to locate on the maps. I believe that Chittenden's outline of the route is as clear a one as can be found and, as a means of checking Wislizenus' itinerary and those of other travelers, it is invaluable. I shall, therefore, quote his record of some of the more important stopping places and their mileages. Many of the locations Chittenden discusses in detail but such data I omit, however interesting. The first halting place beyond Independence, Missouri, was the "Blue Camp," twenty miles away.

"*Blue Camp*, 20 miles . . . *Round Grove*, or *Lone Elm Tree*, 35 miles . . . *110-Mile Creek*, 100 miles . . . *Council Grove*, 150 miles . . . *Cottonwood Creek*, 192 miles . . . *Little Arkansas*, 234 miles . . . *Cow Creek*, 254 miles . . . *The Arkansas river*, 270 miles . . . *Walnut Creek*, 278 miles . . . *Ash Creek*, 297 miles . . . *Pawnee Fork*, 303 miles . . . *Coon Creek*, 336 miles . . . '*The Caches*,' 372 miles . . . five miles west of where Dodge City, Kansas, now stands . . . *The Ford of the Arkansas*, 392 miles . . . the regular crossing after 1829 . . . known as the Cimarron crossing . . . twenty miles above Dodge City, a little more than half way between Independence and Santa Fe . . . *Cimarron River, Lower Spring*, 450 miles . . . This part of the route was most dreaded of all. The distance, fifty-eight miles, required at least two, and more often three, days to traverse, and there was no water on the way . . . This part of the Trail passed near the localities now known by the names of Example, Ivanhoe, Conductor, and Zyonville in southwestern Kansas. The route then ascended the valley of the Cimarron for eighty-five miles . . . *Middle Spring of the Cimarron*, 486 miles . . . *Upper Spring*, 530 miles . . . *Cold Spring*, 535 miles. At this point the road left the valley of the Cimarron . . . *McNees' Creek*, 560 miles . . . *Rabbit Ear Creek*, 580 miles . . . It was near the head of Rabbit Ear Creek that Major Long passed in 1820 in his futile search for the Red river . . . *Round Mound*, 588 miles . . . *Rio Colorado*, 635 miles . . . Major Long discovered that it was the upper course of the Canadian . . . *Ocate Creek*, 641 miles . . . *Santa Clara Spring*, near *Wagon Mound*, 662 miles . . . here . . . the mountain branch [of the trail] from Bent's Fort rejoined the main trail . . . *Rio Mora*, 684 miles. Last of the Canadian waters . . . *Rio Gallinas*, 704 miles . . . *San Miguel*, 727 miles . . . *Pecos Village*, 750 miles . . . *Santa Fe*, 775 miles . . ."

The route outlined above was the one that Wislizenus was to follow. Turning back, however, to *The Ford of the Arkansas*, Chittenden specifies that "There was another, or *Lower Crossing*, seventeen miles *below* Dodge City. It was near the mouth of Mulberry creek at the extreme point of the large southern bend of the river." Furthermore, as Chittenden records, there was a "mountain branch of the Santa Fe Trail" and on this the "principal point" was "Bent's Fort, 530 miles from Independence . . . This branch of the Trail crossed the river very nearly where La Junta now stands, and thence ran south, crossing Raton Pass, and joined the main trail at Santa Clara Spring near Mora river. The mountain branch of the Santa Fe Trail has been closely followed by the 'Santa Fe [railroad] route' of the present day." Many of my collectors are to follow this mountain route into northern New Mexico rather than the waterless one by the Cimarron River. I have crossed Raton Pass both by railroad and by fine motor highway more than once.

Wislizenus' "Memoir" is dated from Independence, Missouri, May 9, 1846; I now follow that record.

1846

On May 4, 1846, he left St. Louis and on the 9th had arrived at ". . . the well known town" of Independence, ". . . the usual starting place of the companies going to Santa Fe, Oregon, or California . . . Seven years ago I had seen Independence as a small village . . . the great throng of emigrants to the 'far west,' and of Santa Fe traders . . . gives it quite a lively appearance . . . My own object was, to join the first large company destined for Santa Fe, and my enterprising countryman, Mr. A. Speyer,[3] . . . afforded me all the facilities for doing so." Since Speyer's caravan was not ready to leave, Wislizenus on May 14 went ahead to "Big Blue Camp" where he remained, waiting, until the 22nd. He describes it as "a charming spot," lying ". . . just on the western boundary line of the State of Missouri . . . at the very junction of civilization and wilderness . . . The scenery was enlivened by thousands of stock . . . and by the daily arrival of new wagons and prairie travellers . . ."

On May 22 the caravan started, ". . . 22 large wagons (each drawn by 10 mules,) several smaller vehicles, and 35 men." Wislizenus had provided himself ". . . with a small wagon on springs, to carry . . . baggage and instruments, and as a comfortable retreat in bad weather . . ." It traveled southwest, crossing from Missouri into Kansas at a point, according to Chittenden, just east of present Glenn, Kansas.

May 23: ". . . the whole plain was so covered with flowers, principally with the blue-sky *Tradescantia Virginica,* and the light-red *Phlox aristata,* that it resembled a vast carpet of green, interwoven with the most brilliant colors . . . At 'Lone Elm-tree' we halted at noon . . . How long the venerable elm-tree, that must have seen many ages, will still be respected by the traveller, I am unable to say; but I fear that its days are numbered . . ."

On May 26 they were at ". . . 110-*miles creek* . . . The name . . . refers to its distance from the old Fort Osage . . ." and on the 29th at Council Grove on the Neosho River, Morris County. ". . . Council Grove forms, as it were, a dividing point in the character of the country east and west of it." To the east was the prairie, to the west the country gradually ascended towards the Arkansas and the soil was to become ". . . dryer and more sandy, the vegetation scantier, timber and water more rare . . . the transition to the sandy plains on the other side of the Arkansas . . ."

3. According to M. G. Fulton, Albert Speyer, a Santa Fe trader, was a Prussian Jew, and on this trip was transporting two wagon loads of arms and ammunition which had been ordered in 1845 by the Governor of Chihuahua. We hear of him again at Santa Fe. For M. G. Fulton, *editor,* see p. 955, *fn.* 8.

At Cottonwood Creek (Marion County), on June 1, "The *Malva papaver,* with its violet flower, was here very common." On the Little Arkansas, reached June 3, Wislizenus found "For the first time . . . a prickly pear, or *cactus* . . . It was the *Opuntia vulgaris,* with its bright yellow flower . . ." They crossed the Little Arkansas, and were traversing McPherson County. On June 4, after passing "several 'Little Cow creeks' " they reached at noon "Big Cow Creek," and finally ". . . Camp Osage . . . the first camp near the Arkansas." They had crossed Rice County and encountered the Arkansas at the northern end of its great bend in Barton County. On the 5th they passed Walnut Creek (Barton County) and reached and crossed Ash Creek (Pawnee County) where the ". . . so-called buffalo grass, (*Sessleria dactyloidea*[4]) . . ." was noted.

The Wislizenus map shows that the party was keeping north of the Arkansas River. On June 6 it crossed Pawnee Fork (Pawnee County), on the 7th Little Coon and Big Coon creeks, and on the 8th encamped at the "Caches" (Ford County); on June 9 the Arkansas River was reached, and crossed on the 10th. Here, as noted above, there was a choice of routes to Santa Fe. The Speyer party took the "Cimarron crossing" and started on the Cimarron River route to Santa Fe. They reached the Cimarron in southwestern Kansas, on June 11, and followed it upstream until June 15. In so doing they had, from Kansas, crossed for a very short distance the extreme southeastern corner of Colorado, and the northwesternmost portion of the Oklahoma Panhandle. It was a journey through a nearly waterless desert and dreaded by all travelers. Chittenden recorded the location of the very few springs or water holes found along the trail.[5] In the Panhandle region of Oklahoma, in Cimarron County, the trail crossed from the Cimarron to the upper waters of the Canadian and followed that river upstream into Union County, northeastern New Mexico.

Before discussing Wislizenus' New Mexico journey I quote a few of his comments upon the route of June 10 to 17, from the "Cimarron crossing" of the Arkansas, to arrival at the New Mexico boundary. From the crossing, for fifteen miles "Our road lay through deep sand. Grass was very scanty, but there was quite an abundance of sand-plants; and the ground was so covered with the most variegated flowers, especially the gay *Gaillardia pulchella,* that it looked more like an immense flower garden than a sandy desert."

June 11: "Travelled about 18 miles . . . without seeing wood or water . . . The high plain between the Arkansas and Cimarron, whose elevation above the sea is about 3,000 feet, is the most desolate part of the whole Santa Fe road, and the first adventurers in Santa Fe trade stood many severe trials here. Within the distance of 66 miles, from the Arkansas to the lower springs of Cimarron, there is not one water-course or

4. This is now *Buchloe dactyloides* (Nutt.) Engelm.

5. In 1820 S. H. Long and the botanist Edwin James had traversed very similar desert regions somewhat farther south when, by what is now called Major Long's Creek, they crossed northeastern New Mexico and entered northwestern Texas. Neither participant had anything pleasant to report about this portion of their journey.

water pool to be depended upon in the dry season. The soil is generally dry and hard; the vegetation poor; scarcely anything grows there but short and parched buffalo grass and some cacti."

It was on this day that Wislizenus wrote at length upon "... the celebrated *'mirage,'* *(false ponds; fata morgana.)*[6] ... nowhere so common, so deceptive, and so well developed, as here." According to the map of the "Memoir," they camped in the triangle formed by the confluence of *"Sand creek"* with the Cimarron and followed the last for four days, on the 12th reaching "the *lower springs of Cimarron,*" and "the *middle springs*" on the 13th. Leaving them on the 14th, the camp was "... without water, but with tolerable grass, considering that we were on the Cimarron." The soil had become "... entirely sandy; different species of artemisia, those shrubs with bitter and terebinthine flavor, cover the whole plain; horn frogs, lizards, and rattlesnakes find a comfortable abode in the warm sand; thousands of grass-hoppers occupy all shrubs and plants, mosquitoes and buffalo gnats the air;—what a great place for settlements this would be!"

June 15: "... we ... encamped on the *crossing of Cimarron* ... On the road to-day we saw the skulls and bones of about 100 mules, which Mr. Speyer had lost there several years ago ..."

It was, probably, on June 17 that the present boundary between Oklahoma and New Mexico was crossed. Standley, who has outlined the Wislizenus route through New Mexico, begins the itinerary on that date: "Cold Spring to McNees Creek." [7]

Wislizenus was to remain in New Mexico from June 17 until August 7 when— after an extremely short passage through present El Paso County, Texas—he was to cross the Rio Grande River into Mexico proper on August 8. I now follow his New Mexican journey of seven weeks.

On June 18 Speyer's party crossed Cottonwood and Rabbit Ear creeks (shown on the map of the "Memoir" as upper tributaries of the "Rio Nutria or North Fork of Canadian r[iver]"); between the 18th and 21st crossed Rock and Whetstone creeks

6. Chittenden tells us that this phenomenon "... has received a variety of explanations ... Wislizenus holds that the true mirage always shows objects double, the lower erect by refraction through the stratum of air next to the ground, and the upper inverted by reflection against the surface of a different stratum some distance above. Whatever may be the true explanation the delusion is a perfect one, and its tantalizing effect upon the thirsty wayfarer was often more distressing than the thirst itself."

7. Josiah Gregg, Chittenden and others mention that the creek was named for one of two young men, McNees and Munroe, who were barbarously killed by Indians on its banks. But although this event is mentioned more than once, I have found McNees Creek only indicated by name on two maps, one Chittenden's, where it appears as a short stream rising in northeastern New Mexico and crossing into northwestern Oklahoma where it enters Rabbit Ear Creek which in turn flows into the North Fork of the Canadian. On the same map the trail to Santa Fe strikes McNees Creek just at the Oklahoma-New Mexico border and, from there, continuing southwestward soon reaches the upper waters of Rabbit Ear Creek, nearly north of Round Mound.

(on the same map shown as upper tributaries of a large unnamed stream—probably Ute Creek—flowing southward and nearly paralleling in direction the Canadian River of further west); on the 21st reached the main Canadian, and crossed a little north of the boundary between Colfax and Mora counties, for the party camped no great distance above the mouth of Ocate Creek which enters the Canadian from the northwest at that point; in other words the route had crossed Union County from its northeastern corner to near the middle of its western boundary and had then turned in a more southerly direction across the southeastern corner of Colfax County.

June 22: "We left the Colorado [the Canadian] this morning for the Ocaté creek . . . On Ocaté creek there are some pines, the first we have seen close to the road . . . We started in the afternoon for *Wagon mound* . . . a hail-storm . . . forced us to camp in the prairie . . ."

June 23: "Made this morning 12 miles to [towards] *Santa Clara* . . . The western mountains . . . are all covered thickly with pine timber. Some isolated mountains rise in the plains through which we travel. The road passes at the foot of the highest of them the so-called *Wagon mound,* which I ascended as far as the rocks would allow. On the Wagon mound I found for the first time a dry specimen of the *Opuntia arborescens,* (Eng) so common throughout Mexico, and whose porous stems are used in the south as torches . . . the caravan . . . camped on a spring near the Wagon mound, called the *Santa Clara.*"

Wagon Mound is in eastern Mora County. The caravan had been met by Mexican soldiers sent out from Santa Fe to escort it to that city. "The Mexicans reported that everything was quiet at Santa Fe, and that General [Manuel] Armijo was at the head of the government in New Mexico."

June 24: "Went in the morning but five miles, to *Wolf creek* . . . Pines, cedars, and sundry shrubs, grow along the creek; the grass and water are good . . . About eight miles from Wolf creek we reached the *Rio Mora* . . ."

They were still in Mora County, keeping considerably west of the Canadian River and encountering the various streams mentioned along their upper reaches.

June 25: "Made in the morning . . . as far as *Gallinas creek* . . . About a mile from the creek lies the small town of las Vegas, or Gallinas . . . In the afternoon we passed through town and turned [almost due west according to the map] into the mountains . . . we shall now travel mostly in narrow valleys, and through mountainous passes . . . so called *cañons.* Through such a cañon we travelled on that same afternoon . . . Two species of pine grow on the mountains, both of them undescribed yet. The one (*Pinus brachyptera,* Eng) is the most common pine of New Mexico, and the most useful for timber; the other (*Pinus edulis* Eng) or so-called *piñon,* contains in the cones seeds or small nuts, that are roasted and eaten. We encamped . . . about five miles from las Vegas."

On June 27 they passed through ". . . *San Miguel,* or [on] the Rio Pecos." To find water Wislizenus kept on for twenty miles, reaching two springs not far from "the *Rio Pecos,* opposite the old Pecos village." This was on the western border of San Miguel County. Reaching the village on the 28th, he found it deserted. Six years before the Indian inhabitants had left to join another tribe.

"From Pecos springs we went . . . over a very mountainous road, to *Cottonwood branch,* a small valley amidst high mountains, where oaks, maple *(Negundo fraxinifolio,)* common and bitter cottonwood *(Populus Canadensis* and *angustifolia)* grow, surrounded by pine trees. This is the highest point on the Santa Fe road; according to my barometrical measurements . . . 7,250 feet above the level of the sea."

They had crossed Glorieta Pass, in eastern Santa Fe County. The map shows that, from San Miguel they had turned northward along the west side of the Pecos River and had turned west at the old Pecos village. Daily marches were short. On the 29th they camped at ". . . the same spot where, some months after this narrative, Governor Armijo was encamped with his whole army, prepared for a battle[8] with General Kearny . . ." On June 30 they arrived at Santa Fe which from a distance had looked to Wislizenus ". . . more a prairie-dog village than a capital . . ."

News had reached the city that the battle of Palo Alto[9] had taken place. Before he had left the Missouri frontier Wislizenus had had word of ". . . the first skirmish, near Matamoras, that preceded this war . . ." but, under the conviction that the United States would finish the conflict "at a single blow," he had made his journey as planned.

"The people in Santa Fe appeared indifferent to the defeat at Palo Alto; no excitement prevailed; only General Armijo felt alarmed, because he had been informed that troops would be sent over the plains to occupy New Mexico. All the information we could really give him on that account was, that if they started at all, they could hardly reach New Mexico in less than two months hence.[1] In the meanwhile, General Armijo treated the traders as usual. After some bargaining, they agreed to pay $625 duty on each wagon; those who wanted to go into the interior received the usual passports from him, and everything went on as in perfect peace."

While the traders carried on their "mercantile business" Wislizenus occupied his time in making "some scientific observations," in acquiring information about the country and so on, all of which he incorporated into some nine pages—"Statistics of

8. W. H. Emory recounts the story of the Apache Cañon episode in his "Reconnoissance."

9. Zachary Taylor's force had invaded Mexico and had won battles at Palo Alto (May 8) and at Resaca de la Palma (May 9); both battlefields were in Texas, on the Gulf of Mexico, just north of the mouth of the Rio Grande, in Cameron County.

1. Kearny's Army of the West was to reach Santa Fe on August 18, or in six weeks; on Kearny's entry into the city Armijo had fled.

New Mexico"—of the "Memoir;" there is nothing of botanical interest therein. Wislizenus remained in Santa Fe until July 8, only one week.

The region of Santa Fe, which botanists the world over had long wished to see explored, was, in 1846, to receive the attention of a number of plant collectors. Only William Gambel had been there before Wislizenus, in 1841. In June, 1846, Wislizenus arrived. From August 18 to September 2 and from September 11 to 25, 1846, W. H. Emory was to be stationed there. And, more important, August Fendler was to reach Santa Fe in the autumn of 1846 and, from the spring of 1847 to the autumn of that year, was to collect plants intensively, within a radius of safety.

Speyer's caravan was encamped five miles west of Santa Fe, in Agua Fria, and on July 8 Wislizenus joined it:

"After a week, Mr. Speyer had finished his business[2] in Santa Fe, and resolved to go on to Chihuahua. No further news had during that time been received either from below or from the plains. In this state of uncertainty, I thought it best, instead of waiting idle in Santa Fe for the possible arrival of an army over the plains, to spend my time more usefully by extending my excursion as far as Chihuahua where, according to all accounts, everything was as quiet as in Santa Fe. Besides, I had a passport from Governor Armijo . . . securing my retreat in case of necessity."

The map of the "Memoir" shows that, from Santa Fe region, the route taken by the caravan led southwest to the Rio Grande, encountered near San Domingo and San Felipe in Sandoval County; from there—excepting for the transit of the Jornada del Muerto—it kept close to the east side of the Rio Grande until it crossed that river into Mexico proper. It started on its way on July 9, ". . . on the usual road, by Algodones, for the Rio del Norte." Wislizenus, anxious to visit the gold mines in the Sandia Mountains (". . . the old and new Placer, in a range of mountains southwest from Santa Fe . . .") left the party until the 11th when he went to Albuquerque to rejoin it. On leaving the mines he was "Loaded with specimens of gold ore . . ." The caravan had been delayed by rains and while waiting its arrival Wislizenus on the 12th moved to a nearby "rancho (small farm)," where, with the help of irrigation, were cultivated ". . . mostly maize, wheat, beans, and red pepper, (chile colorado)."

Finally, on July 17, they got started, passed "Sandival's hacienda," and camped on ". . . a sandy plain, covered with artemisia and similar shrubbery, but without grass." By the 19th they were near the southern boundary of Bernalillo County, camping ". . . nearly opposite to a pueblo on the other side, called *Isleta*." On the 20th, after passing ". . . a fine grove of cotton trees, called *bosque*, or *alamos de Pinos* . . ." they came to ". . . the *hacienda* of *Mariano Chavez's* widow." On the 21st they passed *"Otero's hacienda*, or *Peralta,"* went through *"Valencia,"* and a "long-stretched town,

---

2. According to M. G. Fulton, already quoted here, the fact that Speyer was transporting munitions destined for Mexico City seems to have been known and Speyer was desirous of getting "a certificate from the custom house" in New Mexico before his goods were confiscated by United States troops. For M. G. Fulton, *editor*, see p. 955, *fn.* 8.

Tomé," [3] and camped at its southern end. On the 22nd they camped at *"Casas Colorados,"* in Valencia County; on the 24th at *"Joyita,* a small town," in Socorro County; and on the 25th passed *"Joya . . .* another small town," which is La Joya of modern maps.

July 26: "Passed . . . through the town *Sabino* . . . our night'camp was five miles further, (near *Parida.)* The vegetable creation in the valley of the Rio del Norte, characterized principally by a great many sand plants, exhibits since a couple of days two specimens of shrub, which for their extension over the greatest part of Mexico, and their daily appearance hence, deserve a particular notice. The one is the so-called *mezquite,* a shrub belonging to the family of the mimmoseæ, and a species of algarobia.[4] It resembles in appearance our locust tree; it is very thorny; bears yellow flowers and long pods, with a pleasant sour taste . . . The mezquite . . . is no doubt the most common tree in the high plains of Mexico.[5] Pleased as I was with the first sight of the shrub, which I knew only by description, I soon got tired of it . . . The other new companion . . . is the *yucca,*[6] resembling the palm tree, and therefore commonly called palmilla . . . There are many species of this family . . . The first diminutive species of this plant, from two to three feet high, *(yucca angustifolia,)* I had seen on the Arkansas and near Santa Fe; but here a much larger species begins, which becomes every day now more common and taller . . ."

On July 28 they had arrived opposite to the town of Socorro, ". . . on the right bank of the Rio del Norte . . ." While Speyer was transacting business, Wislizenus with a guide visited old mines west of town, presumably in the Socorro or Magdalena Mountains. The mines were abandoned but the doctor found ". . . a new species of yucca, with large, oblong and edible fruits. The pulpy mass of the fruit tastes like paupau . . . For the first time, also, I saw here opuntias, with ripe, red fruits . . . sweet and refreshing . . ."

That evening the party camped a mile north of Lopez and on July 29 Wislizenus mentions that "To-day we have passed the last settlements above the much dreaded Jornada del Muerto." On the 30th they passed the " *'ruins of Valverde,'* (in prosaic translation, the mud walls of a deserted Mexican village,)" and on August 2 came to a ". . . camping place . . . known as *Fray Cristobal* . . ." at which point the Jornada began. It is explained that *"Fray Cristobal"* was ". . . generally the last camping place on or near the Rio del Norte before entering the *Jordana del Muerto* . . ." The map shows that the route avoided the bend in the Rio Grande (now Elephant Butte Reser-

3. Peralta, Valencia and Tomé are all in eastern Valencia County.

4. The carob *(Ceratonia)* is called algaroba. Wislizenus may refer to an acacia or to the mesquite.

5. The mesquite, *Prosopis juliflora* Torrey, is common in arid regions of our southwest, in both the Upper and Lower Sonoran zones.

6. Wislizenus was entering the range of *Yucca elata* Engelmann, a common species from this point southward in the Rio Grande valley.

voir) and kept to the west of the "Sierra Blanca" and "Organ" mountains; or, on modern maps, followed the valley to the west of the Obscura Mountains (north) and San Andreas Mountains (south). The northern end of the valley is in southern Socorro County, the southern in Sierra County—the valley runs northeast to southwest (or the reverse). Wislizenus refers to it as "This awful Jornada, a distance of about 90 miles . . ." In some seasons water was to be found during the passage, but in midsummer was not to be expected. A footnote supplies the name's origin:

"Jornada del Muerto, literally, the day's journey of the dead man, and refers to an old tradition that the first traveller who attempted to cross it in one day perished in it. The word Jornada (journey performed in one day) is especially applied in Mexico to wide tracts of country without water, which must for this reason be traversed in one day."

Stopping places along the route are mentioned: ". . . a place called *Laguna del Muerto,* another with ". . . a good spring . . . the so-called *Ojo del Muerto . . .*" (August 2); "Alamos," but the adjacent water-pool was dry, and "Barilla," with stagnant water (August 3); then a day without water (August 4); then they returned to the river and the camping place was called ". . . *Robledo . . .* From here to *Doñana,* the first small town again, it is about 12 miles." (August 5). Dona Ana, in the county of that name, was passed on the 6th. I presume they had completed the Jornada but have found no record of where it terminated to the south. Wislizenus records a discovery:

"Before reaching Doñana, I met on the road with the largest cactus of the kind that I have ever seen. It was an oval Echino cactus, with enormous fishhook-like prickles, measuring in height four feet, and in the largest circumference six feet eight inches. It had yellow flowers, and at the same time seed, both of which I took along with the ribs; but I really felt sorry that its size and weight prevented me from carrying the whole of this exquisite specimen with me. Dr. Engelmann, perceiving that it was a new, undescribed species, has done me the honor to call it after my name." [7]

The caravan continued down the Rio Grande on August 6 and on the 7th Wislizenus moved on ahead and during the night arrived ". . . near the *'upper crossing of the*

---

7. Wislizenus' "exquisite specimen" was a barrel cactus or visnaga. Although the name bestowed by Engelmann—*Echinocactus Wislizeni*—is scarcely euphonic, it can at least claim some significance.

Since we are entering a period in which the *Cactaceae* of our southwest are beginning to arouse interest —until Engelmann undertook to work upon the complicated group they had been largely neglected by American botanists—it may be of interest to tell how these well-armed plants are collected and prepared for the herbarium. Flowers, fruit, and clusters of the characteristic spines and prickles, are essentials and a photograph of the living plant is a valuable supplement. To handle any cactus—even the "cunning little ones"—requires extreme caution; I received valuable instruction in the art from the experienced Alice Eastwood who advised the use of an ordinary dinner-knife, fork and dessert-spoon; after some personal experience I added to my equipment a pair of short ice-tongs. These implements serve the following purposes: with the tongs one grasps the desired portion of the cactus and with the knife severs it from the plant; what has been cut off (still held in the tongs) is placed upon the sheet in which the specimen is to be preserved and held thereon with the fork while it is sliced into two parts with the knife and the interior pulp of each half scraped out with the spoon (so that the "specimen" will dry with a *certain amount*

Rio del Norte . . .' on both sides of the river rose mountains, which converge above el Paso, and confine the river for several miles to a narrow pass, hemmed in by precipitous rocks." On this day, states a footnote, they had stopped at the place which, ". . . according to all descriptions given me afterwards in relation to it, is the famous battleground, *Brazito,* where some months later Colonel Doniphan's regiment celebrated Christmas day by its first engagement with the enemy . . ."

Not far south of Brazito, meaning "little arm," the party crossed into El Paso County, Texas. The town of El Paso, in 1846, did not extend to the north side of the Rio Grande. On August 8 Wislizenus crossed the river into El Paso and into Mexico proper. All the way from Santa Fe to this crossing he had been in Mexican-held territory, and the "Memoir" frequently refers to it as Mexico.

Wislizenus' sojourn south of the Rio Grande was very different from what he had contemplated. Yet, despite restrictions imposed by the Mexican authorities, and despite his close contact with war activities, the doctor still pursued his scientific interests and made a collection of plants which was considered of great value by the professional botanists; actually his ten months in Mexico produced the most important collections of his entire trip.

On June 6, 1847, with American troops, he finally reached the mouth of the Rio Grande. From there he moved to Brazos on the little island of Santiago nine miles away, and on June 10 set sail for New Orleans where he arrived seven days later. From there he left for "St. Louis, my home," arriving in early July.

"After an absence of 14 months, I had travelled from Independence to Reynosa, on the Rio Grande, about 2,200 miles by land, and about 3,100 by water, and had been exposed to many privations, hardships, and dangers; but all of them I underwent, for the scientific purpose of my expedition, with pleasure, except for the unjust and arbitrary treatment from the government of the State of Chihuahua, which deprived me for six months of what I always valued the highest, my individual liberty . . ."

The six months referred to had been spent in "exile" at Cosihuiriachi, some ninety miles from Chihuahua. Wislizenus gave himself the satisfaction of saying what he thought of that place: "Gentle reader, whenever in the course of your life you should

---

of rapidity); what is left, the exterior skin as it were, with its characteristic and highly important armaments, is then strapped between blotters in the collecting press and, daily, until the "specimen" is brittle, the moisture-absorbent blotters must be changed. The entire process calls for HANDLE WITH CARE. While riding in my car, months after a collecting trip, Boston friends have complained of needles in the upholstery!

For the enlightenment of novices who contemplate collecting these plants I mention that the leaves of the opuntias are very small, deciduous affairs and not at all suggestive of leaves. And that what the unknowing usually mistake for leaves are actually joints or sections of stems; the opuntias, often mentioned in these pages, are divided—as a starting point—into two groups: those with flattened stems (the paddle-shaped portions are sections of the stem) and those with cylindrical stems; the chollas belong to this group, and of these the so-called "jumping cholla"—which does *not* leap but reaches the unwary by less active means—seems to be of most interest to the desert visitor.

feel tempted to pronounce a foreign, jaw-breaking word, or to visit a strange-looking, incomprehensible, awful place, I would recommend to your kind attention Cosihuiri-achi . . ."

The "Botanical appendix" of the Wislizenus "Memoir" was written by George Eng-elmann and, to the layman, his presentation of the subject is far more interesting than the classification according to orders, families, and so on usually adopted by the systematic botanists, for he discusses the plants as they occurred along the collector's route and relegates to footnotes the essential and orthodox descriptions of new species. But Engelmann is careful—"new" plants are described ". . . with the apprehension . . . that some . . . may have been published already from other sources, without my being aware of it." The "Botanical appendix" contains the common apology of the period, that there had been little "leisure" to do the task. It was undoubtedly true that Torrey, Gray and Engelmann were being deluged in these years with material from west of the Mississippi River, and I might add that Engelmann was a practising physician.

Some fourteen pages are devoted to the collections made by Wislizenus between Independence, Missouri, and his entry into Mexico proper. Eighteen new species and two new genera are described and some already-published descriptions revised on the basis of material gathered by Wislizenus and by other collectors such as Gregg[8] and

8. At this period no botanical paper concerned with New Mexican or Mexican plants fails to refer to Josiah Gregg whose *Commerce of the prairies*, published in 1844, may be called the *Baedeker* of the time. The first edition came out in New York, in two volumes, with maps and engravings, and later appeared in many editions. The book is replete with matters of historical and of geographical interest and one often turns to it for routes between Independence, Missouri, and Santa Fe, and southward down the Rio Grande valley. Many of the old place-names mentioned by Wislizenus, by Emory and by Abert, and difficult to locate on modern maps, are found in Gregg, often with full descriptions.

Unfortunately the book contains nothing substantial on botanical matters although a number of indigenous and cultivated plants are mentioned. Yet the author made many contributions in the form of collected plants which he either sent to Engelmann directly, or through an intermediary such as Wislizenus for the two men were in Mexico at the same time apparently and in communication with each other. Gregg corresponded with Engelmann who wrote in his introductory words to the "Botanical appendix":

"In examining the collections of Dr. Wislizenus, I have been materially aided by having it in my power to compare the plants which Dr. Josiah Gregg . . . has gathered between Chihuahua and the mouth of the Rio Grande, and particularly about Monterey and Saltillo, and a share of which, with great liberality, he has communicated to me. His and Dr. W[islizenus]'s collections together, form a very fine herbarium of those regions."

While this quotation applies in particular to Mexican plants, species found along portions of the Rio Grande valley did not all stop at the crossing and a flower found in New Mexico may have been supplemented by a fruit found across the border, giving a complete picture of the plant which Engelmann wished to describe. Many plants bear testimony to Gregg's work in the specific name *Greggii*.

Two valuable books about Josiah Gregg have been published by the University of Oklahoma Press, the first in 1941, the second in 1944: *Diary & letters of Josiah Gregg: Southwestern enterprises 1840–1847* and *Diary and letters of Josiah Gregg Excursions in Mexico and California 1847–1850*. Both were edited by M. G. Fulton.

We learn therefrom that Gregg died in California on February 25, 1850, while on an exploring trip between Humboldt Bay and the settlements on Trinity River, and that his death was due to starvation. It

Fendler (who had worked in New Mexico before the "Memoir" and "Botanical appendix" were published). I quote some of Engelmann's comments:

"The tour of Dr. Wislizenus encompassed, as it were, the valley of the Rio Grande and the whole of Texas[9] as a glance at the map will show. His plants partake, therefore, of the character of the floras of the widely different countries which are separated by this valley. Indeed, the flora of the valley of the Rio Grande connects the United States, the Californian, the Mexican, and the Texan floras, including species or genera, or families, peculiar to these countries."

The flora along Wislizenus' route is discussed sectionally; first from Independence to the crossing of the Arkansas River (June 10, 1846). In this region the plants collected ". . . are those well known as the inhabitants of our western plains. I mention among others, as peculiarly interesting to the botanist, or distinguished by giving a character to the landscape, in the order in which they were collected, *Tradescantia virginica, Phlox aristata, Oenothera missouriensis, serrulata, speciosa, &c., Pentstemon Cobaea, Astragalus caryocarpus,* (common as far west as Santa Fe,) *Delphinium azureum, Baptisia australis, Malva Papaver, Schrankia uncinata* and *angustata, Echinacea angustifolia, Aplopappus spinulosus, Gaura coccinea, Sida coccinea, Sophora sericea, Sesleria dactyloides, Hordeum pusillum, Engelmannia pinnatifida, Pyrrhopappus grandiflorus, Gaillardia pulchella, Argemone Mexicana* . . ."

Between the Arkansas and Cimarron rivers the plants were ". . . rarer, some of them known to us only through Dr. James, who accompanied Long's expedition to those

was a sad story and rendered more so by the fact that his last days were marked by the dissensions of a small group of men who, although in disagreement, still found it necessary to remain together for reasons of safety. He apparently retained an interest in plants to the end and, despite abuse from his companions, insisted on measuring the size of certain redwood trees seen along the route! Gregg could have been only about forty-four years of age at the time of his death—Fulton refers to him as eight years old in 1814.

Chittenden has only praise for Josiah Gregg and his *Commerce of the prairies,* referring to the author as a "distinguished trader" who "went to the plains for his health, and becoming enamored of the life went into business there." He notes that his book ". . . is discriminating, comprehensive and free from exaggeration. It is the classic of the Santa Fe Trail, well arranged and admirably written. Although limited in scope, it fills its particular niche so completely that it is entitled to rank as one of the great works of American history."

CHITTENDEN, HIRAM MARTIN. 1902. The American fur trade of the far west. A history of the pioneer trading posts and early fur companies of the Missouri valley and the Rocky Mountains and of the overland commerce with Santa Fe. Map. 3 vols. New York. 2: 544, *fn.* 3.

GHENT, WILLIAM JAMES. 1931. Josiah Gregg. *Dict. Am. Biog.* 7: 597.

GREGG, JOSIAH. 1844. Commerce of the prairies: or the journal of a Santa Fé trader, during eight expeditions across the great western prairies, and a residence of nearly nine years in northern Mexico. Maps. 2 vols. New York. 2: 544.

———— 1941. Diary & letters of Josiah Gregg. Southwestern enterprises 1840–1847. Edited by Maurice Garland Fulton with an introduction by Paul Horgan. 2 vols. Norman.

———— 1944. Diary & letters of Josiah Gregg. Excursions in Mexico & California. Edited by Maurice Garland Fulton with an introduction by Paul Horgan. 2 vols. Norman. 2: 367, 368.

9. Meaning, presumably, that, on his return journey, Wislizenus went by boat from Brazos on the island of Santiago through the Gulf of Mexico to New Orleans and thence to St. Louis.

regions in 1820." There was *"Cosmidium gracile,"* already described by Torrey and Gray; *"Cucumis ? perennis,"* of James, also found "by Mr. Lindheimer, in Texas." The ranges of certain species were being extended: *"Erysimum asperum"* had not been known to grow so far south, nor *"Rhus trilobata"* so far east. There was a new *"Talinum,* which I named *T. calycinum . . ."*

In the mountainous regions from Cedar Creek (June 17, 1846) to Santa Fe, Wislizenus had been in the region ". . . of the pines, and of the cacti." From his collections Engelmann described two new species of pine: ". . . the nut pine of New Mexico, (Piñon,) *Pinus edulis*[1] . . ." and *"Pinus brachyptera . . .* the most common pine of New Mexico and the most useful for timber." *Pinus brachyptera* is now considered a synonym of *Pinus ponderosa* Douglas, the yellow pine. Wislizenus had "overlooked" the limber pine, *Pinus flexilis,* found and first described by Edwin James. Later in his paper Engelmann refers to three more new pines found by Wislizenus in Mexico.

"Among the most remarkable plants . . . were the *Cactaceæ.* After having observed on the Arkansas, and northeast of it, nothing but an *opuntia* [June 3] . . . Dr. W. came at once, as soon as the mountain region and the pine woods commenced, on several . . . members of this curious family . . . On Waggon-mound [June 23] [he saw] the first (flowerless) specimens of a strange *opuntia* . . . with an erect, ligneous stem, and cylindrical,[2] horridly spinous, horizontal branches . . . I can give it no more appropriate name than *O[puntia] arborescens,* the *tree* cactus, or Foconoztle, as called by the Mexicans, according to Dr. Gregg . . . The first *Mammillaria* was also met at Waggon-mound . . . Mr. Fendler has collected the same species near Santa Fe."

From Wolf Creek (June 25, 1846) came the new *Geranium pentagynum.* And from prairies adjacent to the creek came ". . . the smallest of a tribe of cactaceæ . . . numerous species of which were found in the course of the journey south and southeast; several others have been discovered in Texas . . . I propose to distinguish them . . .

1. The piñon or nut pine offers an example of the complexities in classification and in nomenclature which were arising in the 1840's—and which were eventually to become worse! Sudworth wrote in 1917:

"The history of the piñon is closely connected with that of the single-leaf pine and shows that there has been much difference of judgment, which obtains even at the present time, regarding their botanical relationship, particularly their claims to specific rank. Single-leaf pine was found for the first time by Gen. John Charles Frémont in 1844 in southern California and was described in 1845 as a distinct species under the technical name *Pinus monophylla* Torrey. Similarly, the piñon, found first by Dr. Wislizenus in New Mexico in 1846, was botanically described in 1848 for the first time as a distinct species under the name *Pinus edulis* Engelmann . . ."

After discussing the various classifications, reclassifications and names suggested since the discoveries of Frémont and of Wislizenus were first described, Sudworth states that he prefers ". . . to consider the piñon [*Pinus edulis* Engelmann] and the single-leaf pine [*Pinus monophylla* Torrey and Frémont] as distinct species . . ."

On the other hand Alfred Rehder's *Manual of cultivated trees and shrubs* (1940) reduces both *Pinus monophylla* and *P. edulis* to varieties of *P. cembroides* of Zuccarini (the Mexican piñon or nut pine)!

2. See last paragraph of p. 953, *fn.* 7.

under the name *Echinocereus,* indicating their intermediate position between *Cereus* and *Echinocactus* . . ." The type was *E. viridiflorus.* Wislizenus had discovered two other species of *Echinocereus* (June 23), both of which Engelmann described as new: *E. triglochidiatus* and *E. coccineus.*

South of Santa Fe, along the Rio Grande, were found plants of the plains and of Texas; some had already been described by Torrey and Gray or by Torrey and Frémont. There was a new species of creosote bush from near "Olla," ("Joya," July 25)— *Larrea glutinosa*—". . . the first and most northern form of the shrubby *Zygophyllaceae,* more abundant farther south."

". . . near Sabino, an interesting bignoniaceous shrub was collected for the first time, undoubtedly the *Chilopsis* of Don, which farther south appears more abundantly. Its slightly twining branches, willow-like slender glutinous leaves, and large paler or darker red flowers, render it a very remarkable shrub. Dr. Gregg mentions it under the name of *'Mimbre,'* as one of the most beautiful shrubs of northern Mexico. The character given by Don, and that of Decandolle, appear defective, though I cannot doubt that both had our plant in view . . . I am enabled to correct those errors."

From near Albuquerque there was a new *Opuntia* with "short clavate joints, which make the name *O. clavata* most appropriate." "The famous desert, the Jornada del Muerto, furnished, as was to be expected, its quota of interesting plants." There was a new species of *Dithyrea,* a genus described by Harvey, *D. Wislizeni.* Also new, ". . . the gigantic *Echinocactus Wislizeni,* reminds us again that we are approaching the Mexican plateau . . . *E. Wislizeni* is . . . the third in size in this genus." There was as well a new *Mammillaria, M. macromeris,* found near "Doñana."

And it was Wislizenus who found for the first time the striking plant noted by all visitors to the deserts of the southwest, the ocotillo or coach-whip. It has many local names; Engelmann named it *Fouquieria splendens.* Wislizenus' specimen lacked both flowers and fruit, but he later collected flowers in Chihuahua and Gregg found the fruit, also in Mexico.

Towards El Paso Wislizenus had found ". . . a curious capparidaceous plant . . . l have named this new genus (in honor of its discoverer . . . the first naturalist, it is believed, who explored the regions betwen Santa Fe, Chihuahua and Saltillo) *Wislezenia!"* The type was *W. refracta,* ". . . from the upper crossing of the Rio Grande, near El Paso . . ." It must, therefore, have been gathered in what is now El Paso County, Texas.

But Wislizenus has now arrived at the Mexican border.

As late as 1902, in *The Silva of North America,* C. S. Sargent described a new cottonwood which he named *Populus Wislizeni,* noting that it had been ". . . discovered on the upper Rio Grande in July, 1846, by Dr. F. A. Wislizenus."

Many of the plants discussed in the "Botanical appendix" are referred to by Engelmann in later papers, as fuller material came under his scrutiny. Asa Gray had long been regarded as authority for the family of composites. Engelmann was now becoming authority for the *Cacti,* the *Yucca* and the century plants or *Agave;* for, in this decade when such men as Lindheimer, Wislizenus and Emory pursued their collecting in Texas, New Mexico and Arizona, and in the 1850's when collectors accompanied the United States government surveys along the Mexican boundary and crossed the trans-Mississippi west at many latitudes in search of suitable routes for the railroads which were to cross from the Mississippi River to the Pacific Ocean, knowledge of these genera was to grow by leaps and bounds as the great desert areas where they are most abundant and at their most vigorous were explored for the first time.

Wislizenus made one more trip, this time to California by way of Panama. Brewer records that he ". . . made a small collection in the State in 1851,[3] but his more extensive and better known collections were made earlier beyond our borders."

Dr. Wislizenus had the characteristic of perseverance. His son tells that, having fallen in love in 1848 with Miss Lucy Crane in the city of Washington, and having at that time been unsuccessful in his suit, he pursued her to Constantinople and won her hand in 1850.

In 1852 he settled permanently in St. Louis—"His days of wandering were over." For several years before his death he was totally blind, but did not lack for "willing readers to feed his active mind." A photograph of the old gentleman published by the Missouri Historical Society shows a happy face with, between the eyes, the deep furrows that suggest concentration.

3. Sargent wrote in 1895 of an oak which he found on this excursion: "Discovered by Frémont on the Sierra Nevada in the winter of 1844–45, *Quercus Wislizeni,* which was at first confounded with *Quercus agrifolia* [Née], was described from specimens gathered by Dr. F. A. Wislizenus in 1851 on the American fork of the Sacramento River." A. de Candolle in 1864 had noted the differences between these oaks and had named the undescribed species in honor of Wislizenus.

BANCROFT, HUBERT HOWE. 1889. The works of . . . 17 (Arizona and New Mexico 1530–1888): 464, *fn.* 32.

BREWER, WILLIAM HENRY. 1880. List of persons who have made botanical collections in California. *In:* Watson, Sereno. Geological survey of California. Botany of California 2: 557.

CHITTENDEN, HIRAM MARTIN. 1902. The American fur trade of the far west. A history of the pioneer trading posts and early fur companies of the Missouri valley and the Rocky Mountains and of the overland commerce with Santa Fe. Map. 3 vols. New York. 2: 535-542, 639, 758; 3: 969.

DUFFUS, ROBERT LUTHER. 1930. The Santa Fe trail. London. New York. Toronto. 189-191.

EMORY, WILLIAM HEMSLEY. 1848. Notes of a military reconnoissance, from Fort Leavenworth, in Missouri, to San Diego, in California, including part of the Arkansas, Del Norte, and Gila Rivers. By W. H. Emory, Brevet Major, Corps Topographical Engineers. Made in 1846–7, with the advanced guard of the "Army of the West." *U. S. 30th Cong., 1st Sess., Sen. Exec. Doc.* 3: No. 7, 1-126. Map.

ENGELMANN, GEORGE. 1848. Botanical appendix. *In:* Wislizenus, F. A. 1848. Memoir of ... 87-115. *Reprinted:* Trelease, William & Gray, Asa, *editors.* 1887. The botanical works of the late George Engelmann ... Cambridge, Massachusetts. 39-58.

EWAN, JOSEPH. 1943. Botanical explorers of Colorado III. John Charles Fremont. *Trail & Timberline* No. 291: 34.

―――― 1950. Rocky Mountain naturalists. Denver. 28, 338, 339.

GEISER, SAMUEL WOOD. 1948. Naturalists of the frontier. ed. 2. Dallas. 283.

GREGER, D. K. 1935. Dr. Frederick Adolphus Wislizenus (1810–1889). *Bull. Acad. Sci. St. Louis* 1: 21.

REHDER, ALFRED. 1940. Manual of cultivated trees and shrubs hardy in North America exclusive of the subtropical and warmer temperate regions. ed. 2. New York. 39.

SARGENT, CHARLES SPRAGUE. 1895. 1902. The Silva of North America. Boston. New York. 6: 94, *fn.* 1, 1895; 14: 71, 1902.

SPAULDING, PERLEY. 1909. A biographical history of botany at St. Louis, Missouri. *Pop. Sci. Monthly* 74: 244-246.

STANDLEY, PAUL CARPENTER. 1910. The type localities of plants first described from New Mexico. *Contr. U. S. Nat. Herb.* 13: 147, 148.

SUDWORTH, GEORGE BISHOP. 1917. The pine trees of the Rocky Mountain region. *U. S. Dept. Agric. Bull.* 460: 18, 19.

WARREN, GOUVENEUR KEMBLE. 1859. Memoir to accompany the map of the territory of the United States from the Mississippi River to the Pacific Ocean, giving a brief account of each of the exploring expeditions since A.D. 1800 ... *U. S. War Dept. Rept. expl. surv. RR Mississippi Pacific* 11: 53.

WISLIZENUS, FRIEDRICH ADOLPH. 1848. Memoir of a tour to northern Mexico, connected with Col. Doniphan's expedition, in 1846 and 1847, by A. Wislizenus, M.D. With a scientific appendix and three maps. *U. S. 30th Cong., 1st. Sess., Sen. Miscel. Doc.* No. 26: 1-86. Map. [For "Botanical appendix" see under Engelmann, George. 1848.]

―――― 1912. A journey to the Rocky Mountains in the year 1839. Translated from the German, with a sketch of the author's life, by Frederick A. Wislizenus, Esq. St. Louis. [The German edition was published in Missouri in 1840.]

CHAPTER XLVII

HARTWEG SPENDS SOME TWENTY MONTHS IN CALI-

FORNIA COLLECTING PLANTS FOR THE HORTICUL-

TURAL SOCIETY OF LONDON

KARL Theodor Hartweg, a plant collector whose name is famous in the annals of botany, arrived at Monterey, California, on June 7, 1846.

He had gone to Mexico in 1837 and had spent about seven years in that
country and elsewhere. William Jackson Hooker, in January, 1837, had advertised
the collections which he was expected to make on that journey in *The Companion to
the Botanical Magazine* and evidently thought highly of the man's abilities; he was
"happy to announce" that ". . . Mr. Theodore Hartweg embarked for Mexico in the
service of the Horticultural Society, to whom therefore all living plants, roots, and
seeds will be sent: but that useful Institution has generously allowed him to dispose
of dried specimens of plants on his own account, which he will do at the rate of £2
the hundred species. All applications, however, for shares must be made through the
Horticultural Society, by letters addressed either to Mr. Bentham, or to Mr. Lindley.
From the capital of Mexico, Mr. Hartweg will go to Guanaxato and proceed northward . . . keeping as much as he can to the Tierra fria.[1] He will remain in the country
two or three years, that is, if the state of it will admit of botanizing: but it is so disturbed, that he may probably have to take another direction and visit Bolivia, which
presents a yet more interesting field. Whichever way he goes, we are authorized in
anticipating great things from him."

Of Hartweg's life up to the time of his arrival in California in 1846, Charles Sprague
Sargent wrote in 1891:

"Karl Theodore[2] Hartweg (1812–1871) was a native of Carlsruhe, and the descendant of a long race of famous gardeners. At an early age he found employment in
the Jardin des Plantes in Paris, and afterwards in London in the garden of the Royal
Horticultural Society,[3] where his industry and intelligence soon attracted attention
and led to his being sent to Mexico by the society to collect plants and seeds. In 1836
Hartweg left England on this mission, passing seven years in Mexico, central and
western equatorial America, and in Jamaica, making important discoveries, including many coniferous trees of the Mexican highlands, and several orchids which he successfully introduced into cultivation. Hartweg returned to Mexico in 1845, and was
in California in 1846 and 1847, spending much of his time in Monterey and penetrating to the upper valley of the Sacramento River . . ."

1. The cold lands.   2. Hartweg's middle name usually appears in the Anglicized form.

3. Until 1866 this institution was called the Horticultural Society.

The California visit was described in a long paper entitled "Journal of a mission to California in search of plants. By Theodore Hartweg, in the service of the Horticultural Society," which was published in *The Journal of the Horticultural Society of London* in installments issued in 1846 (Part I, pp. 180-185); in 1847 (Part II, pp. 121-125; Part III, pp. 187-191); and in 1848 (Part IV, pp. 217-228). Despite the title, it was not until Part II that Hartweg begins his account of the California visit.

In the Horticultural Society's "Report from the Council to the anniversary meeting, May 1, 1846," it is stated that "When Mr. Fortune's[4] expedition was drawing to a close, the Council made arrangements for the despatch of a collector to California; and Mr. Hartweg has again sailed in that capacity. He left England on the 2nd of October, 1845,[5] reached Vera Cruz on the 13th of November, the city of Mexico (by way of Xalapa) on the 3rd of December; and his last letters are dated January 16th, from Tepic, a town near the port of S. Blas, on the Pacific side of Mexico . . . Short as the time is which has elapsed since Mr. Hartweg's departure, he has already sent home some useful seed and plants . . . The Council trust that when Mr. Hartweg shall have reached California he will reap a rich harvest of hardy plants, especially of the beautiful Zauschneria and the evergreen Castanea,[6] to which his attention has been most especially directed."

In a paragraph prefacing his "Journal" it is mentioned that the Council had instructed Hartweg regarding his route: after reaching Tepic, Mexico, he was to await the sailing of a ship bound for California; in California ". . . he was to spend one or two years, as might appear to himself most advisable. The following is the journal which Mr. Hartweg has kept, in pursuance of his instructions . . ."

Hartweg relates that he reached the roadstead of Vera Cruz "After a passage of forty-five days in one of the royal mail steam-packets . . . in the evening of the 13th of November, 1845 . . ."

In transcribing from Hartweg's paper I italicize the scientific plant names and break down some of his dense paragraphs in order to emphasize the dates.

1846

According to the "Journal" it was on March 14 that the collector ". . . finally left Tepic for San Blas . . . and embarked the following day on board of a small schooner for Mazatlan, where I arrived after a passage of five days." He notes that

4. Robert Fortune, who began his famous explorations in China in 1843, and who is said to have visited that country intermittently for eighteen years—according to Bretschneider "four times, 1843–45, 1848–51, 1853–56, 1861."

5. A manuscript letter from K. A. Geyer to W. J. Hooker, dated "Kew, Sept. 5th [18]45," states: "Mr. Hartweg will leave for Vera Cruz at the 16th of the month."

6. The reference is to the golden-leaved chestnut, or giant chinquapin, *Castanopsis chrysophylla* A.DC., which David Douglas had first discovered and which the botanists, Hooker in particular, were determined to obtain.

"... Mazatlan is now the most important port on the west coast of Mexico ... Upon making inquiries about merchant-vessels proceeding soon to Northern[7] California, I found to my consternation that no opportunity had offered for the last six months, nor was it likely there would be any for some time; but that the United States ship 'Portsmouth' would sail in a few days for Monterey ... I applied to Commodore Stoat [Sloat] for a passage ... but I was told by him in very few words that he could not serve me ... when a rupture between the United States and Mexico was hourly expected, he could not let his movements be known: thus wishing to keep the 'Portsmouth's' destination secret—her purpose being well known three weeks before she sailed.

"More successful was an application I made ... to Rear-Admiral Sir George Seymour of H.M.S. 'Collingwood,' who kindly allowed me a passage in H.M.S. 'Juno,' then proceeding to Monterey. Towards evening of the 11th of May, I went on board, and, sailing the following morning, we arrived at Monterey on Sunday, the 7th of June, after a passage of twenty-six days."

June 8: "... I delivered my letters of introduction, and the following morning I settled down in the quiet little town of Monterey.

"The verdant fields and pine-covered range of mountains at the back of the town form a pleasing contrast to the dried up vegetation about Mazatlan. The predominating trees are an evergreen oak *(Quercus californica),* forming a tree 30 feet high ... It occurs principally in low but dry situations. The higher parts are occupied by *Pinus insignis* ... 60 to 100 feet high ... This species is liable to vary much in the size of the leaves ... and in the cones ... These differences, which are too insignificant to establish even varieties of *Pinus insignis,* have given rise to the names *Pinus tuberculata* and *radiata,*[8] which were, according to Loudon, collected by the late Dr. Coulter near Monterey; that locality, no doubt, is Point Pinos, as it is the only habitat near Monterey where pines grow close to the beach; it is at the same place where I made the foregoing observations ...

"On the dry banks of ravines, to the north-east of the town, the Californian horse-chestnut *(Pavia californica)* is common ... Of shrubs I observed *Ceanothus thyrsiflorus* very common in the pine-woods ... *Sambucus,* No. 28: *Lonicera racemosa; Spiræa ariæfolia; Rhus* 3 sp.; *Caprifolium Douglasii,* No. 4; *Diplacus,* No. 65; *Garrya elliptica ... Lupinus arboreus* and *ornatus ..."* Many other plants are mentioned.

From 1839 to 1857,[9] George Bentham ("Georgius Bentham," for the entire volume is in Latin) published his *Plantæ Hartwegianæ:* "Sequunter stirpes Californiae annis

---

7. In distinction to Lower California.

8. *Pinus insignis* Douglas and *P. tuberculata* D. Don are now treated as synonyms of *P. radiata* D. Don.

9. In Bentham's *Plantae Hartwegianae* the plants of California appear as follows: pp. 294-308 (issued in 1848); pp. 309-332 (issued in 1849); and pp. 333-342, as well as the rest of the volume (issued in 1857).

1846 et 1847 lectae, imprimis circa Monterey et Sacramento." The first plant listed is "1625 (121*) . . ." and a footnote states that the numbers indicated between parentheses are cited in Hartweg's "Journal" published by the Horticultural Society. Since, as indicated above, the Hartweg numbers follow no numerical sequence (as do the numbers supplied by Bentham) it is necessary to search through his paper to locate a particular Hartweg number; and, although the Bentham index of "specierum et synonymorum" is helpful, Hartweg, unfortunately, does not assign names to all the plants included in his "Journal."

June 22: ". . . I left Monterey for the mission of Santa Cruz . . . across the bay, due north, of Monterey, and at a distance of sixty miles by land . . . Passing along the seashore over the plains, which present the same vegetation as about Monterey, we arrived in the afternoon at the mission, after a gallop of seven hours . . .

"The mountains of Santa Cruz are well wooded with *Taxodium sempervirens*,[1] called by the American settlers redwood or bastard cedar. In close forests it grows to an enormous size, averaging 200 feet in height, with a stem of 6 to 8 feet in diameter, which is straight as an arrow, and clear of branches up to 60 or 70 feet. One tree . . . termed by the Americans 'the giant of the forest,' is 270 feet high, with a stem measuring 55 feet in circumference at 6 feet from the ground . . .

"Some fine trees of *Abies Douglasii* are found in the mountains of Santa Cruz . . . thinly scattered among the redwood trees, with which they vie in size. The mountain oak (No. 84, *Castanea chrysophylla* ?)[2] also occurs here . . . a tree 50 feet high . . ."

July 2: ". . . I returned to Monterey, on board of an American bark, after a passage of four hours, and found that Commodore Sloat had arrived in the 'Savannah,' accompanied by two sloops of war. In consequence of a rupture between the United States' forces and the Mexicans, near Matamoras, wherein the latter were defeated, the American Commodore, on the 7th, landed a party of marines and seamen, and hoisted the American flag without opposition.

"The few days of absence produced a great change even in the vegetation; the fields and woods, which before were covered with flowers, are now gradually drying up from the total absence of rain during the summer months; even the bulbous plants had, during that time, shed their flowers and ripened their seeds.

"As yet I have not succeeded in procuring horses, Castro having taken all the available horses away, in order to mount the militia, with which he intended to have marched against the Americans . . . I cannot venture far away from Monterey . . . I

1. Jepson notes: "Doubtless one of the 'Santa Cruz Big Trees.' " These, as Hartweg states, are redwoods, in distinction to big trees. According to Sargent it was Hartweg who, in 1846, introduced the redwood into English gardens.

2. Was this the "evergreen Castanea" to which Hartweg's attention had been "most especially directed" by the Council of the Horticultural Society? or was it the "oak," *Quercus chrysolepis*?

might fall in with a party of country people, who could not be persuaded that a person would come all the way from London to look after weeds, which in their opinion are not worth picking up . . . I . . . confine my excursions within a few miles of the town.

"Crossing the wooded heights near Monterey I arrived at Carmel Bay, after an easy walk of two hours; here I found *Diervilla,* No. 47; *Cupressus macrocarpa,* No. 143, attaining the height of 60 feet, and a stem of 9 feet in circumference, with far-spreading branches, flat at the top like a full-grown cedar of Lebanon, which it closely resembles at a distance;[3] *Eschscholtzia crocea, E. californica* . . . [etc., etc.]

"Another excursion, which I made to the Rancho de Tularcitos, led over the mission of Carmel. This, like all other missions in California since their breaking up in 1836, is in a sad state of neglect; the buildings are fast falling to ruin, and the lands nearly in the same wilderness as the first settlers found them . . .

"Following up the narrow valley of the Carmel river I entered a beautiful wood of alders, willows, and plane-trees, some of the latter attaining the height of 80 feet, and 12 in circumference . . . On my return thence, over El Toro, a high mountain destitute of trees or shrubs . . . I found, on the north side, in a ravine, a few small trees of *Pinus Sabiniana,*[4] the highest of them not exceeding 30 feet . . ."

August 23: ". . . I embarked on board the bark 'Joven Guipuzcoana,' whose owner . . . invited me to take a trip with him up to the Bay of San Francisco. I gladly accepted . . . as I intended to visit the valley of the Sacramento river, where I possibly might procure horses, and return thence by land to Monterey."

August 24: ". . . we anchored off Santa Cruz, where the ship was to remain a day or two. I took advantage of this delay, and made an excursion to the mountains, in a different direction from that visited before. Passing through a copse wood, composed chiefly of *Pavia californica, Quercus californica, Ceanothus thyrsiflorus, Corylus* No. 85, *Rhus viride,* called 'Yedra,' and justly dreaded . . . for its poisonous properties, I entered a beautiful pine-wood. The leaves of this species of pine stand in threes . . . The trees rise to the height of 100 feet . . . This handsome species of pine, which appears to be new, I have named, in compliment to the late Secretary of the Society, George Bentham, Esq., *Pinus Benthamiana.*[5]

"Another kind of pine that I found within a few hundred yards of the foregoing species, is, probably, the doubtful and little known *Pinus californica*[6] . . ."

3. This is the first time, so far as I know, that a collector comments upon the Monterey cypress. Hartweg's name, *Cupressus macrocarpa,* is a *nomen nudum,* but was later adopted when the species was described by George Gordon. Hartweg's No. 143 was the type.

4. *Pinus Sabiniana,* described by David Douglas in 1833.

5. Hartweg was writing of the western yellow pine, *Pinus ponderosa* Douglas.

6. In 1892 J. G. Lemmon proposed the name *Pinus attenuata* because the names *P. californica* and *P. tuberculata* bestowed by David Don were untenable.

August 28: ". . . the bark got under weigh for Yerba Buena.[7] On the afternoon of the same day we encountered a strong north-west gale, which tore some sails and obliged us to put back to repair damages."

August 30: ". . . we sailed again, and . . . kept close in shore. The whole of the coast is destitute of trees or shrubs, with the exception of Point Año Nuevo, where some pines. or cypresses seem to grow."

September 2: ". . . we were opposite the narrow but safe entrance to the bay of San. Francisco; a large inland sea, divided into several branches, forming not only the principal port in California, but the largest and safest on the whole western coast of America. About noon we anchored off Yerba Buena, a small town, rising rapidly in importance.

"Two Wardian cases,[8] which were furnished by the Society for preserving such plants and seeds as will not carry otherwise, gave me some trouble in clearing them at the Customhouse. They were shipped in London for the Sandwich Islands, and thence to Yerba Buena, where they arrived under the Mexican government, without it being exactly known what they contained, nor who the owner was. Some miscreant, thinking he might profit by the occasion, denounced them as containing contraband goods; they were accordingly taken to the customhouse, where, upon examination, instead of silk stockings and printed calicos, they found 'two small greenhouses,' some kitchen-garden seeds, nails, &c. Soon after, by the change of government, they fell into the hands of the Americans, and having no papers to show that I was the owner, I had to

---

7. San Francisco.

8. In 1869 Asa Gray wrote an obituary notice of Nathaniel Bagshaw Ward for whom the cases were named. The invention had grown out of ". . . Mr. Ward's persistent endeavors to cultivate the plants he delighted in under the smoke and soot of the dingiest part of London, and which resulted in providing for the poor as well as the rich denizens of the smoky towns of the old world the inexpensive but invaluable luxury or comfort of being surrounded at all seasons with growing plants and fresh flowers . . . Equally important is the application of the Wardian case to the conveyance of living plants between distant countries. The writer well remembers the first case of growing plants sent to New York thirty-five years ago, which arrived as fresh and healthy as when they left London . . . So useful has this contrivance proved to be . . . that the Director of Kew Gardens 'feels safe in saying that a large proportion of the most valuable economic and other tropical plants now cultivated in England would, but for these cases, not yet have been introduced.' The earliest published account of the Wardian case was given by Mr. Ward in the form of a letter to . . . Sir William Hooker, and was printed in the 'Companion to the Botanical Magazine' for May, 1836 . . ."

Gray had written a review of the second edition (London) of a book (*On the growth of plants in closely glazed cases*) which Ward published on the subject in 1852, and included this interesting comment: ". . . Formerly only one plant in a thousand survived the voyage from China to England. Recently, availing himself of our author's discovery, Mr. [Robert] Fortune planted 250 species of plants in these cases in China, and landed 215 of them in England alive and healthful. The same person lately conveyed in this way 20,000 growing tea-plants, in safety and high health, from Shanghai to the Himalayas. In fact, this mode of conveyance is now universally adopted . . ."

In contrast with the little "glass frame" in which Menzies attempted to grow the seeds and plants acquired in the 1790's, Ward's cases must certainly have represented a great improvement!

send in a petition to the captain of the port, prove the property before the magistrate
. . . I finally received them from the commander of the place . . .

"The vegetation about Yerba Buena is poor; the sand-hills that surround the town,
and which extend for several miles into the interior, are but thinly covered with brushwood of oak (*Quercus californica, Ceanothus thyrsiflorus, Rhus,* 'Toyon,' [9] *Prunus*
No. 102, and *Baccharis* No. 123.).)"

On September 10 Hartweg ". . . went across the bay to Sausalito . . ." and on the
11th was ". . . joined at the mission of San Rafael by General Vallejo . . ." After enjoying the "hospitable board" of the General for three days Hartweg left the environs
of the Bay of San Francisco[1] and started on a trip which took him as far north as
Bodega Bay.

9. *Photinia arbutifolia* Lindl., the Christmas berry or Toyon.

1. Within a week of Hartweg's departure the British ship *Herald,* commanded by Berthold Seeman, spent
about five days (September 18-22) in the Bay of San Francisco. Seeman had spent some two months
(June 24–September 2) surveying many of the regions visited by Vancouver: the Straits of Juan de Fuca,
through which—as an indication of the changes which were taking place—the *Herald* had been ". . .
lugged . . . up about sixty or seventy miles . . ." by a steamboat, the waters about "Port Victoria," "Port
Discovery," "Protection Island," "Port Townshend," and so on. From these regions he had proceeded
southward along the coast and on the 18th of September ". . . ran into the Bay of San Francisco, about
which we had heard and read so much; but we were disappointed . . ."

Seeman seems to have been disappointed in practically everything related to California! The *Herald*
had hardly anchored in the Bay when he learned that ". . . the Americans were in possession of California . . ." The missions (his reflections thereon were aroused by a visit to the Mission of San Francisco)
had lost the "very shadow" of their "former fame" and their attempts to unite "religion and civilization"
had proven impracticable, and injurious rather than beneficial. There was "much fine land" on the banks
of the Sacramento and San Joaquin rivers, "but not equal to the speculator's hopes." "At the time of our
visit the gold had not been discovered, and San Francisco was extremely dull . . ."

Moving southward Seeman reached Monterey and later San Diego—where one ship, the *Pandora,* went
into the harbor and the *Herald* remained off ". . . the low, arid, and uninteresting shores. The land had a
denuded aspect . . . the only object to enliven the scene was the mission of San Diego . . ." At the
Coronados Islands—as California was left—everything was "dried up and withered" so that, after catching
"three rattlesnakes and a tarantula," Seeman seems to have been content to betake himself elsewhere.

This British surveying expedition was under the command of Captain Henry Kellett, "R.N., C.B.," and
was circumnavigating the globe. It had reached Monterey only to find ". . . the Americans in full occupation of the place . . ." After the reflection that California under Spanish and Mexican rule had never
risen from obscurity, Seeman comments that "It remains to be seen what the more enterprising and
energetic American will effect . . ."

Seeman wrote the *Narrative of the voyage of H.M.S. Herald* . . . which was published in 1853, in London. Chapter VII (volume I) records the visit to our northwest coast and Chapter VIII the one to California. He also wrote *The botany of the voyage of H.M.S. Herald* . . . , published from 1852 to 1857. This
contained one hundred very fine plates. Seeman expresses his ". . . sense of gratitude to Mr. W. Fitch, the
talented artist who supplied the drawings to most of the plates, and lithographed them in the spirited manner so peculiarly his own." For a full account of Walter Hood Fitch the reader should turn to Wilfrid
Blunt's remarkable book, *The art of botanical illustration.*

Two sections of *The botany* ("Flora of western Esquimaux-land" and "Flora of north-western Mexico") concern plant collections along the Pacific coast of North America. The first refers, very largely, to
plants from the far northwest; and the latter to plants from Lower California with numerous citations

On September 14 Hartweg went on to ". . . San Miguel, distant thirty miles . . . The face of the country about Sonoma and San Miguel[2] is perfectly level towards the bay, and capable of great agricultural improvements. Several species of oak (*Quercus,* Nos. 139, 140, and 141)[3] thrive well in the fine black vegetable mould, and are disposed into large irregular clumps, giving the country the appearance of an immense park . . . A ridge of mountains . . . at a short distance from San Miguel is thinly scattered over with oaks, and a few *Abies Douglasii* interspersed. No other kinds of pine occur here. In the shaded dells I found a *Viburnum, Euonymus,* and a large-leaved *Calycanthus* in seed.

"From San Miguel I went to Bodega, where the Russians a few years back had an establishment granted them by the Mexican Government . . . When their term expired, it was purchased by Captain S[utter], an American, who erected a steam saw-mill there, for which the redwood trees that cover the mountains supply him amply with material. This is the most northern limit of this magnificent tree, growing at intervals from the latitude of 32° N. up to the river Ross in 38° 15′.

"From Bodega I returned by way of San Rafael to Sausalito, passing over a beautifully undulated prairie, destitute of water or trees . . ."

On October 7 Hartweg records that he returned to Monterey ". . . in the bark 'Joven Guipuzcoana,' and made preparations for a trip to San Diego, her next destination, from whence I expected to return by land before the rains set in . . ." He was, however, obliged to give this up because of the disturbed state of the country. He records that at the beginning of November the periodical rains set in and terminate by the end of March—the heaviest rainfall occurring in January and February. Presumably he was unable to do much work during that period.

of Texas specimens credited to Berlandier, Drummond, Lindheimer and Wright, of New Mexican specimens credited to Wright, and of California specimens credited to Hartweg, Lobb and Coulter, and so on. As far as any Seeman specimens are concerned I cannot find that he made any along what is now our northwest coast (where he spent some two months) or in California. Knowledge of plant distribution was greatly on the increase at this time and I must believe that Hooker had much to do with the content of *The botany* (certainly as far as the sections mentioned are concerned). Seeman dedicated it to "Sir William Jackson Hooker, K.H., D.C.L. Oxon., F.R.A.S.A., and L.S. . . ." and so on, and mentions that he was "deeply indebted" to Hooker for "his generous encouragement and ready assistance."

BLUNT, WILFRID. 1950. The art of botanical illustration. London. 223-228, 268-282.

BOULGER, GEORGE SIMONDS. 1897. Berthold Seeman. *Dict. Nat. Biog.* 51: 194, 195.

SEEMAN, BERTHOLD. 1852-1857. The botany of the voyage of H.M.S. Herald, under the command of Captain Henry Kellett, R.N., C.B., during the years 1845-51 . . . with one hundred plates. London.

———— 1853. Narrative of the voyage of H.M.S. Herald during the years 1845-51, under the command of Captain Henry Kellett, R.N., C.B., being a circumnavigation of the globe, and three cruizes to the arctic regions in search of Sir John Franklin. 2 vols. London.

2. According to Jepson, this was the "Rancho de San Miguel, just above the present town of Santa Rosa," or in Sonoma County.

3. Jepson identifies these respectively as *Quercus lobata, Q. Douglasii* and *Q. Kelloggii;* the first collection of the last-named.

1847

"With January the rains set in unusually severe; the Salinas and other rivers . . . have now become impassable. The first indications of the returning spring I observed in the flowering of *Garrya elliptica, Berberis aquifolium, Ribes speciosum, R. malvaceum, Arctostaphylos,* Nos. 158, 159, and 160; *Vaccinium,* No. 157, a dwarf shrubby *Prunus* (No. 162), with white pendulous flowers, and *Ornithogalum,* No. 163.

"When the weather permitted it, I continued my rambles on foot in the mountains of Monterey, and discovered on the western declivity, within two miles of the seashore, a species of Pine which I had not found previously. The leaves are two in a sheath . . . The trees attain no great elevation, averaging twenty feet, rarely thirty . . . This species, which appears to be new, I have named, in compliment to Thomas Edgar, Esq., the Society's Treasurer, *Pinus Edgariana.* In the same locality with the above Pine, I observed a Cypress (*Cupressus,* No. 166)[4] with smaller cones than *C. macrocarpa,* of which it seems more than a variety, being a stunted shrub six to ten feet high.

"Returning by a different route, through a thick brushwood of *Arctostaphylos* and *Ceanothus,* I found on the steep acclivity, in a shaded dell, a *Rhododendron,* without seeds or flowers, forming a shrub five feet high, well beset with flower-buds, and *Castanea chrysophylla* in the same condition. This evergreen Chestnut forms a shrub three to eight feet high . . . From its situation, and habit in general, it may be expected, if I am fortunate enough to introduce it, to withstand the ordinary winters about the neighbourhood of London . . .

"In February, *Dodecatheon,* No. 170, appeared everywhere common . . ." Hartweg enumerates other plants, and among them, in ". . . sandy plains towards the river Salinas, the large, golden-flowered *Viola chrysantha, Nemophila insignis, Eschscholtzia crocea,* and *E. californica* were common."

At this point it was necessary for Hartweg to alter his plans somewhat, again because of unsettled conditions. He finally decided to visit the Sacramento valley ". . . where the settlers are all foreigners, and where I need not be under any apprehensions of disturbances in the lower country." Accordingly, on March 8, Hartweg left Monterey on "the American bark Tasso," and after five days arrived at Yerba Buena. Here a "few days' detention" made it possible for him to "examine the neighbourhood" and to add to his collection ". . . a white *Myosotis,* No. 190, *Liliacea,* No. 192, *Œnothera,* No. 194, a scarlet *Aquilegia,* No. 198, *Iris,* No. 204, and *Ribes echinatum* . . ."

On March 23 he ". . . embarked in a small launch with Mr. Cordua, who was pro-

4. This was, I believe, *Cupressus Goveniana,* later described by George Gordon. Sargent notes that it was introduced to English gardens by Hartweg in 1846, the specific name commemorating the services to horticulture of James Robert Gowen. He quotes Hooker to the effect that Gowen was "well known in the most respectable circles in London . . ."!

ceeding to his farm in the Sacramento Valley, and who kindly invited me to make his house my headquarters . . ." Late on the 24th they arrived at ". . . the Corte de Madera[5] . . . a wood-cutting establishment . . ." They left there the same night and on the morning of the 26th, after ". . . passing . . . through the straits of Carquinez into Suisun Bay . . .", entered the Sacramento River in the afternoon:

". . . the country is flat, presenting a boundless field of rushes as far as the eye can reach, bordered on both side by a distant ridge of mountains . . . a line of snow. The lowlands of the Sacramento are subject to inundations during the spring months, and are destitute of trees, with the exception of the banks . . . a belt of trees and shrubs, varying from thirty to two hundred yards in depth, extends along the banks and is chiefly composed of Oaks, *Platanus,* Willows, Poplars, Ash, *Negundo californicum, Pavia californica, Cornus,* a dwarf Birch, and a Grape-vine."

On March 31, after "a tedious process of warping up the launch against a strong current," they arrived at "the landing-place of Fort Sacramento," and disembarked at "the mouth of the American Fork, and following that stream about six miles . . . arrived at Mr. S[utter]'s." After a day and a half by horseback (April 1 and 2) they reached ". . . Mr. Cordua's farm, situate on the left bank of the Sacramento, at the junction of the Chuba with the Feather river, which twenty miles below falls into the Sacramento."

Hartweg found the vegetation "much earlier" than at the Bay of San Francisco, but he found "many old acquaintances" among the plants and enumerates some. On April 13 he started on a trip which was to take him "seventy miles higher up the valley":[6]

". . . Crossing Feather river . . . our course lay five and twenty miles along that river, through a beautiful wood of evergreen and deciduous Oaks . . .

"Leaving Feather river, we struck across a prairie for twenty miles: here immense fields of *Eschscholtzia crocea, E. californica,* and *Ranunculus,* No. 239,[7] presented themselves, each species growing by itself, which with the plants observed on Mr. Cordua's farm . . . produced a splendid effect . . .

"The prairies in the Sacramento valley are divided by small rivers, termed 'creeks' by the American settlers: these creeks generally have a border of Oaks . . . In the dry

5. According to Jepson, Corte Madera, Marin County.

6. Jepson wrote of this excursion: ". . . one northward seventy miles, to the Upper Sacramento Valley, which should have taken him to the neighborhood of the 40th parallel, about opposite what is now Tehama. There can be no doubt, it seems to us, however, that he went at least as far northward as the present town of Chico, and, it may be, further. From this point he made an excursion into the Sierra foot-hills, on which occasion *Ceanothus prostratus* was collected for the first time." See under April 14, below, where Hartweg cites "*Ceanothus,* No. 284, spreading on the ground."

7. According to Jepson: ". . . The Ranunculus mentioned here is the *R. canus* of Bentham, which has never been recollected. One is inclined, therefore, to venture the thought that the collector may have made a mistake concerning the locality or the abundance of the species. That Hartweg labeled his plants with the greatest care has, however, been abundantly proved by all subsequent explorations in the State."

beds . . . I observed plants, which nowhere are to be found on the prairie, the seeds of which have evidently been carried down from the mountains during the rains, as for example . . .[8]

"A four days' slow drive . . . brought us to the farm of my companion: the vegetation here differed in no respect from that already observed in the valley. An opportunity of visiting the mountains was afforded me[9] . . . I therefore joined a party of settlers who were going to the mountains . . . to find a site for a sawmill.

"On the first evening we encamped under a large oak, near Pine creek, a little mountain rivulet; here I found . . . evergreen and deciduous Oaks, and *Pinus Sabiniana.* This species of Pine . . . rises here to the height of fifty or sixty feet . . . The branches which in other Pines stand in whorls, are in this species quite irregular (except when young), which, combined with the paucity of its partly bent down, glaucous leaves, gives the tree a peculiar appearance."

April 14: "Early the following morning we ascended the gradual acclivity, and passed through a brushwood entirely composed of *Ceanothus,* No. 285. At noon we arrived at the edge of a noble Pine forest; a few moments rest[1] . . . enabled me to collect *Viola,* No. 287 . . . The species of Pine composing the forest is principally *Pinus Benthamiana,* with a few trees of *P. Lambertiana, Abies nobilis,* and a species of Thuja intermixed, No. 309, *Ceanothus,* No. 284 spreading on the ground, and *Cornus florida,* No. 297, were the only plants[2] observed in the pinewoods."

We are told that, by the end of April, ". . . the prairies in the Sacramento valley assumed a different aspect; two weeks ago they were a carpet of flowers, which have now disappeared, and a yellow, sickly tinge pervades the whole: such is the rapidity of vegetation under the cloudless sky of a tropical sun. Bulbous plants now make their appearance . . ." Hartweg notes that, now, being ". . . aware of the rapidity of Californian vegetation . . .", he ". . . lost no time in collecting such seeds as were worth taking . . ." and returned to his headquarters "by the beginning of May."

". . . Most kinds had . . . ripened their seeds, and it was with difficulty I found a few grains of the beautiful *Leptosiphon aureus,* and similar plants, which between their taller neighbors, had almost become invisible.

"An excursion to the 'Butes,'[3] an isolated group of mountains between the Sacra-

8. Hartweg's "Journal" abounds in specific citations of the plants which he observed at various points throughout all his travels, but I do not cite them all.

9. Hartweg refers to ". . . the hostile character of the mountain Indians towards the settlers . . ."

1. What Hartweg really means is that, *while others rested, he* had a rare opportunity to acquire plants!

2. *Ceanothus prostratus* Bentham and *Cornus Nuttallii* Audubon.

3. Jepson identifies these as ". . . the isolated group of mountains rising up out of the plains between the Feather and Sacramento Rivers, now known as the Marysville Buttes, from which he could see the great tule basins flooded with water."

mento and Feather rivers," was remunerative.[4] ". . . From the rocky summit of the Butes a beautiful view is obtained of the Sacramento valley . . . the lower country . . . presented an immense lake."

"Another excursion I made to the mountains led along the right bank of the Chuba [Yuba] river, over the now parched up prairie . . . fifteen miles brought me to the foot of the mountains. The lower range . . . is occupied by *Ceanothus* No. 285, a few live Oaks, and the *Pinus Sabiniana*. Following a small rivulet, I found there *Mentha* No. 348, *Labiata* No. 352, *Stenactis* No. 353, and a shrubby *Labiata* No. 355, with large white flowers, and *Collinsia tinctoria,* No. 354 . . . On a subsequent occasion, when I returned to this place, to procure seeds of it, my hands were stained yellow by the glandular hairs which cover the seedpods, from which circumstance I named it *Collinsia tinctoria*. Another interesting plant I found on this excursion is *Nemophila speciosa* . . . If the few seeds I procured should vegetate, it will prove a great acquisition to that handsome genus."

Since the Yuba River was much swollen, Hartweg was unable to get as high on the mountains as he wished. His "Journal" now tells of a trip which he made into the "Californian Mountains" with his host, Theodor Cordua. In 1933 the California Historical Society published in its *Quarterly* a document which, as a picture of the period, is extremely interesting: "The Memoirs of Theodor Cordua The pioneer of New Mecklenburg in the Sacramento Valley," edited and translated by Erwin G. Gudde. About a page and a half describe the trip which took Cordua and Hartweg to Bear Valley (Jepson's "Sierra foot-hills"). I quote from this record before giving Hartweg's version of the excursion:

"In the year of 1847, little of note occurred . . . In June . . . I was visited by Mr. Hartweg, a traveller for the London Botanical Society. With my guest I enjoyed frequent hunting and fishing trips and helped him to collect plants in the California mountains. We also visited the beautiful Bear Valley in these mountains. This valley of about the size of half a mile square forms a square, level meadow through which a brook flows slowly. This is the small Bear River which empties into the Feather River. The small mountain valley was covered with the most beautiful green grass and a mass of spring flowers, while in the shadowy canyons of the mountains the snow had not yet melted. In the center of the valley we found a small isolated fir which caused the greatest surprise to the botanist because this species was entirely unknown to him. In spite of a thorough search we could not find a second tree of its kind in the vicinity. At the end of the valley was a great block of granite, larger than the largest building of Europe. Attracted by this immense mass of rock, I called to my companion, 'Tell me, how did this giant stone get here?' Mr. Hartweg, who was already on the block looking for moss and other plants, and holding a beautiful flower in his hand triumphantly replied, 'But, please tell me first how this strange little flower gets here, where the

4. Some fifteen species are listed.

tiniest moss hardly finds nourishment?' From this rock we saw the Yuba River making its way into a canyon about five hundred feet deep . . . On the other side . . . a labyrinth of rugged mountains, whose valleys were shaded by proud coniferous growth, while their lofty peaks lay bald and deserted in the glaring sunlight. In the distance, high peaks covered with eternal snow appeared on the horizon. We remained several days on this beautiful high plateau, shot several deer, and Mr. Hartweg discovered many an unknown plant . . .

"The elevation of Bear Valley is estimated at 6,500 feet, that of the Emigrant Pass at 8,000 feet, and the highest peak of the Sierra Nevada from 11,000 to 12,000 feet . . . At the elevation of 5,000 feet the region of coniferous growth begins. We found silver fir with a circumference of twenty-five to thirty feet at the height of six feet from the ground; the lowest branches, however, were about one hundred and fifty to two hundred feet from the ground. The tallest trees usually stand in the impassable canyons and valleys.

"We were satisfied to admire these trees, but the bold Yankee who surmounts even unsurpassable difficulties will know how to make use of them. At an elevation of 5,000 feet down to the 2,000-foot level the region of coniferous and deciduous trees extends. Among the latter the oak is most common. Here forests are open and light. Here and there regions without trees are found, where the slopes of the mountains are partially covered with manzanitas. The Indians use the leaves of this shrub as tobacco and its small red fruit makes an agreeable cooling drink, a healthy nourishing mush, or a tasty cake when pounded into meal and baked. Large valleys with fine grass and clover cut through this region of the mountains in all directions. The lower region, in which prevails . . . reddish-brown clay, does not have many trees. Only here and there one finds thick thorny bushes and isolated crippled oaks and firs . . ."

I now turn back to the Hartweg "Journal." No specific dates are cited for a time but the author notes that, by the beginning of June, he set out with Cordua and an Indian, to visit, if possible, the ". . . snowy heights of the mountains, generally termed by emigrants from the United States the Californian Mountains.

"After crossing the Chuba river we struck across the prairie, and entered the mountains near Bear Creek, where we encamped . . . in a grove of *Pinus Sabiniana* and Oaks. The vegetation here differed in nothing from that observed on the right bank of the Chuba on a former visit. *Calochortus,* No. 306, which had been very common throughout the Sacramento Valley, was still in flower here, the white variety being more frequent than the yellow.

"Early the next morning we were *en route* again, passing through an interminable wood of *Pinus Sabiniana* and Oaks. Here I observed a pretty little *Allium,* No. 357, with purple flowers; *Asarum,* No. 364, *Viola,* 365; *V.* 366; *Polemonium* (?), No. 382; *Hosackia bicolor; Mimulus bicolor,* No. 376; and *M.* 377—the last two luxuriating in the sandy bed of dried up rivulets. Ascending the gradual acclivity, we left the region of *Pinus Sabiniana,* and entered that of *Pinus Benthamiana,* which seems to be the characteristic of the upper region. Some trees of this noble Pine attain an enormous size. The largest I measured were 28 feet in circumference, and 220 feet high. Of equal dimensions is *P. Lambertiana,* which however does not constitute masses by itself, but is thinly scattered among the former. The same is the case with *Thuja,* No. 402, which rises to the height of 130 feet, by 12 to 15 in circumference.

"Few plants occur in these Pine tracts . . .

"On the fourth day we reached Bear Valley, a beautiful little mountain valley surrounded by a lofty ridge of mountains, which is well wooded with *Pinus Benthamiana.* The north side of the valley was still covered with snow. On the south side . . . a few spring flowers had made their appearance, among which I observed *Pæonia californica,*[5] with brown petals edged with orange . . . A new species of Pine, *P.* No. 413, occurred in the valley, of which I only saw two trees of dwarf growth, probably stragglers from a more northern latitude . . . In general appearance the tree is not unlike a young Scotch fir. The cones . . . were open, and the seeds had fallen out.

"The upper end of the valley is bounded by a mass of granite, terminating in a precipice 800 feet in depth, below which the Chuba [Yuba] river is winding its way . . . In warm and sheltered situations, where the snow had melted, I observed an *Allium* . . . The more elevated parts above Bear Valley . . . were still several feet deep, covered with snow . . .

"Immediately upon my arrival at head-quarters, I proceeded once more to the Upper Sacramento Valley to collect such seeds as I could not procure before."[6]

On June 30 Hartweg notes that, having packed up his collections—sending some by water to San Francisco—he departed for Monterey, in company with an American whom he had engaged as guide:

". . . Towards evening . . . we arrived at the junction of the Feather river with the Sacramento; and passing, the following morning, our luggage over in a canoe, we swam the horses across, the distance from shore to shore being not less than 300 yards. We now continued our course over the prairie on the right bank of the Sacramento river for two days, and crossed again to the south side of a ferry-boat, at the Straits of Carquinez."

At this point the collector fell victim to a "kind of tertian fever" which left him in ". . . such a state of debility as scarcely to be able to sit on horseback." However, "From the Straits of Carquinez we passed along the Bay of San Francisco to the Pueblo of San José . . ." On July 8 they reached Monterey.

5. *Paeonia Brownii* (discovered, named, described, and beloved, by David Douglas); or, possibly, *P. californica* Nuttall which in recent years has been recognized as a distinct species.

6. Jepson notes that, before Hartweg, "A number of botanical collectors had been on the coast . . . but none except General Fremont and members of the Wilkes party had penetrated the interior, and none had collected in the High Sierras unless we except Fremont."

"Soon after my arrival (having, with the assistance of my little medicine-chest, cured myself), I continued my excursions about Monterey as far as my returning strength permitted, and collected such kinds of seeds as I thought worth preserving.

"Towards the end of July I went over to Santa Cruz for a similar purpose, and whilst visiting a family on their farm . . . I was again taken ill with fever and ague. In addition to the seeds which I collected in the Santa Cruz mountains last year, I found the evergreen Chestnut with ripe fruit. This shrub, of which I had been most anxious to procure seeds, attains the height of ten feet, and is of a pyramidal form . . ."

Under date of August 13 Hartweg mentions that he returned to Monterey but was again "laid up with fever and ague" from which he did not recover "until the beginning of September."

September 6: ". . . I went again over to Santa Cruz in quest of pine-cones, which were now ripening. The sorts I procured were *Abies Douglasii, Pinus californica,* and *P. Benthamiana.* The cones of the latter were unusually scarce this season and seem to have suffered from late spring frosts. A few cones were all I could preserve of this sort. They were smaller than those of the preceding year, and contained but few good seeds."

September 20: ". . . I again left Monterey for the southern parts, which, on account of the disturbed state of last year, I could not visit before . . . On the day of our starting we reached the mission of La Solidad . . . situate in the Salinas valley, and encamped . . . on the banks of the Salinas river, within a short distance of the mission."

September 21: "By sunrise . . . we were again on horseback, and leaving the main road on the right, we entered a mountain defile leading to the mission of San Antonio. Here I observed a shrubby *Arctostaphylos,* with large brown seeds . . .

"From San Antonio a range of mountains[7] extends along the coast, attaining a great elevation, which . . . I was assured on the western flank towards the sea is covered with large Pines. The lower region of this range, at the foot of which the mission is built, is thinly covered with the evergreen Californian Oak, a *Ceanothus, Cercocarpus,* a small-leaved shrubby *Fraxinus,* and *Pinus Sabiniana*—the latter at the time with ripe cones. An evergreen shrubby *Prunus,* called Islay, with a holly-like leaf, bearing a red fruit resembling the cherry-plum, grows also abundantly here . . . The kernel, after being roasted and made into gruel, is a favourite dish amongst the Indians. Having ascended the first ridge, we passed through thickets of *Arctostaphylos tomentosa* and *Ceanothus thyrsiflorus,* and entered a forest of *Pinus Lambertiana.* The cones of this noble pine are always hanging from the points of the branches, were . . . already open, and the seeds had fallen out. From cones that had been blown down, I picked out a few seeds.

7. The Santa Lucia Mountains.

"Descending the western flank of the great mountain range, I found at last the long-wished-for *Abies bracteata,*[8] occupying exclusively ravines. This remarkable Fir attains the height of 50 feet, with a stem from 12 to 15 feet in diameter, one-third of which is clear of branches, and the remainder forming an elongated tapering pyramid, of which the upper part, for three feet, is productive of cones. Having cut down some trees, I found to my regret that the cones were but half-grown, and had been frost-bitten. In more sheltered situations, towards the sea-shore, the same happened to be the case; and I was thus precluded all hope of introducing this remarkable Fir into Europe.

"Finding it impracticable to prosecute my journey to the south along the coast, from the numerous ravines which descend from the mountain range, I returned . . . to San Antonio, and crossed by the farm of El Piojo, where the ridge is less elevated. A small Pine wood, which became visible on our descent, extending along the beach, looked like an oasis in the desert . . . I found the wood to be composed of *Pinus insignis,*[9] with larger cones than those about Monterey . . . produced in less abundance. Following along the sea-shore for nine miles, we struck inland again, and arrived at the mission of San Luis Obispo, from whence we proceeded over a flat and uninteresting country to the mission of Santa Ines. The whole of this route is but poorly wooded by a few stunted Oaks. On the ascent to the mission of La Purissima, the monotony of the bare hills was somewhat relieved by a small forest of *Pinus Edgariana,*[1] which attains no larger size than those observed near Monterey.

"Previous to leaving Monterey I was told by several persons that a kind of thin-shelled pine-nut is occasionally brought for sale by the Indians to Santa Ines and Santa Barbara, without being able to learn any more respecting it. Upon making further inquiries at Santa Ines, I was told that the Indians bring them from a great distance, that the harvest of them was over, but that I might procure a few of the mission Indians. Proceeding to a hut . . . I bought a gallon of the fresh seeds: and inquiring about the size of the cones, the Indian handed me two, with the information that the trees are of a small size, when, judge my surprise, I recognised in them those of *Pinus Llaveana,*[2] which I had on former occasions found in several parts of Mexico.

"Seeing there was no prospect of enriching my collection of seeds by proceeding further to the south, I returned from Santa Ines to San Luis Obispo, near which mission[3] the late Dr. Coulter gives the station of *Pinus muricata,* and which seemed to have escaped my notice when first passing through that place. Upon a nearer examina-

8. This was *Abies venusta* Sargent, according to Jepson.

9. *Pinus radiata* D. Don, the Monterey pine.

1. *Pinus muricata* D. Don.

2. Presumably the piñon pine, *Pinus cembroides* var. *monophylla* Voss.

3. David Douglas had, in 1831, visited many of the missions to which Hartweg refers: Soledad (April 25), San Antonio (April 27), San Luis Obispo (May 3), La Purisima (May 5) and Santa Inez (May 6). In writing of Douglas I mentioned the location of these missions, and gave their names in full.

tion I found that on the 'Crusta,' or ascent from San Luis Obispo, only one kind of Pine is growing on the brow of the mountains, which may prove to be *P. macrocarpa*.[4]

"From San Luis we returned to San Antonio, over a flat and uninteresting road, and thence to Monterey, where we arrived on the 18th of October."

October 25: ". . . I again left Monterey, with my former guide, to visit the continuation of the San Antonio range of mountains, which . . . I attempted now by a different route . . ."

October 28: ". . . Following along the sea-coast over a succession of hills intersected by numerous deep ravines, we found our further progress impeded on the third day by the extreme steepness of the range. The only objects derived from this excursion were some fine cones of *Pinus macrocarpa*,[5] some measuring 15 inches in length . . .

"At the beginning of November we returned to Monterey; the rainy season being now close at hand, and having no more excursions to make, I prepared to return to Europe with my collection."

1848

There was but little traffic between California and the west coast of Mexico or Central America and Hartweg was obliged to wait until February 5, 1848, to obtain a passage; but on that date he embarked on ". . . the Hawaian schooner 'S.S.' bound for Mazatlan, and thence to the coast of Central America." Travel was difficult because of political disturbances but by various means he finally reached ". . . the settlement of San Juan de Nicaragua on the 21st of April." On the 24th he took passage ". . . on board the 'Severn,' one of the West India steamers, and arrived at Southampton, after a very fine passage, on the 3rd of June."

Of the many narratives which I have examined Hartweg's "Journal" is one of the clearest and most concise. He seems to have planned his excursions on the basis of advice as to where it was, or was not, wise to travel and, except for his illness, nothing occurred to interfere with his work. Moreover he seems to have collected a considerable number of valuable plants. It is, therefore, somewhat surprising to learn that his accomplishment was not considered altogether satisfactory to his sponsor, the Horticultural Society of London, and that he "left" the service of that institution.

Jepson's paper, "The explorations of Hartweg in America," published in 1897, states: "An enumeration of his plants from California, Mexico, Central and South America, and a systematic description of the new forms, was undertaken by George Bentham, and appeared in the volume so well known to West American botanists as

4. *Pinus Coulteri* D. Don, the Coulter or big-cone pine.

5. Sudworth ("Forest trees of the Pacific slope," *U. S. Dept. Agric., Forest Service*, 57, 1908) comments that "The horribly armed, extremely heavy cones . . . distinguish this pine from all its relatives and associates."

'Plantæ Hartwegianæ.' In this work, eighty-one species from his Californian collection were described as new, some of these being published also in De Candolle's Prodromus. His collection, in addition, furnished material for various contributions by Bentham and by Lindley to the 'Proceedings' and the 'Journal' of the London Horticultural Society and the 'Botanical Register.' Various numbers of the collection were at a later date published as new by Dr. Gray and Dr. Watson, and even within a year[6] a new species has been founded with a reference to a number of Hartweg as part of the type.

"It is said that the Society was not altogether satisfied with the result of his work in California, and blamed him especially for not securing seeds of *Abies venusta*.[7] So far as one may judge from his journal, the character and extent of the country he traversed, and the fruits of his collecting, he was altogether conscientious and faithful. However this may be, he left the service of the London Horticultural Society . . ."

Jepson supplies a ". . . list of the Californian plants first collected by Hartweg . . .", enumerating seventy-one species.

". . . Tracing his route on local maps by the aid of our present knowledge of the geographical distribution of the plants he collected and by the aid of landmarks described in his journal, we are enabled, through a comparison of the numbers of his journal, with those of his California collection in the Plantæ Hartwegianæ, to give for the first time the exact station of a number of his new plants, and a somewhat more definite locality for some others. It is very desirable in many instances that a full series of specimens of Hartweg's plants should be collected from the original station or region. Even at this late period there are few cases in which this has been done."

Bentham's enumeration of the plants collected by Hartweg in California had included about four hundred species and varieties (Bentham's numbers 1625 through 2042) and of these an approximate eighty—as Jepson noted—were designated as new. There were also new genera; as examples: *Chamaebatia* (known as mountain misery), and *Cycladenia* (a genus of the Dogbane Family). In the index of Bentham's paper we find approximately fifty plants—not all from California however—whose specific names, *Hartwegi, Hartwegiana* (or *Hartwegianum*), pay tribute to their collector. John Lindley named the genus *Hartwegia* in his honor; basing it upon an epiphytal orchid which, according to Sargent, Hartweg had first found in Central America [Mexico], "on the eastern declivities of Mount Orizaba . . ."

Sargent's short biographical sketch of Hartweg relates that, after returning to Europe, the collector ". . . was appointed by his friend, the Grand Duke of Baden, inspector of the ducal gardens at Schwetzingen, a position which he continued to fill for the rest of his life." Jepson states that he died on February 3, 1871.

It was presumably the editor of the *Beiträge zur botanischen Zeitung* (February 10,

6. Jepson's paper was read in May of 1893 but published only in 1897.

7. As he recorded in the "Journal," he had secured seeds, and could hardly be considered responsible for the fact that "the cones were but half-grown, and had been frost-bitten."

1843) who was responsible for one curious and scarcely credible comment which he attributed to William Jackson Hooker. It antedates the collector's visit to California of course. I translate:

"The well known botanist, Sir William Hooker, head of the Royal Gardens in Kew, made a very flattering remark to me about one of our fellow Germans. Speaking of travelers, he said: 'We have sent out a number of Englishmen, but none of them accomplished anything; the money was thrown away. The only exception was a German, Mr. Hartweg, who is at present in S. Jago, Peru.' This Mr. Hartweg, son of the Court Gardener in Karlsruhe, is a very able botanist and gardener and has worked for some time in Kew Gardens himself. The last four years he has been traveling in America . . ."

Possibly the comment, if made at all, may have been applicable to a particular field, but as supplied by the editor it refers to collectors in general. Hooker could scarcely have made such a sweeping condemnation—remembering Menzies, Drummond, and Douglas—if no others.

BENTHAM, GEORGE. 1848. 1849. 1857. Plantae Hartwegianae. Londoni. 294-342. [Pages 294-308 issued in 1848, 309-322 in 1849, 333-342 in 1857. Although the publication began in 1839 Hartweg's California plants were confined to these pages.]

BREWER, WILLIAM HENRY. 1880. List of persons who have made botanical collections in California. *In:* Watson, Sereno. Geological survey of California. Botany of California 2: 557.

CORDUA, THEODOR. 1933. The memoirs of Theodor Cordua, the pioneer of New Mecklenburg in the Sacramento valley. Edited and translated by Erwin G. Gudde. *Quart. Cal. Hist. Soc.* 12: 292, 293.

EASTWOOD, ALICE. 1939. Early botanical explorers on the Pacific coast and the trees they found there. *Quart. Cal. Hist. Soc.* 18: 342, 343.

GORDON, GEORGE. 1849. New plants, etc., from the Society's garden. *Jour. Hort. Soc. London* 4: 295-297.

GRAY, ASA. 1853. N. B. Ward, F.R.S., &c., On the growth of plants in closely glazed cases. Second edition.[8] London. *Am. Jour. Sci. Arts,* ser. 2, 16: 132, 133. *Reprinted:* Sargent, Charles Sprague, *compiler.* 1889. Scientific papers of Asa Gray. 2 vols. Boston. 1: 59-62.

———— 1869. Nathaniel Bagshaw Ward.[9] *Am. Jour. Sci. Arts.* ser. 2, 47: 141, 142. *Reprinted: Ibid.* 2: 349-350.

HARTWEG, THEODORE. 1846. 1847. 1848. Journal of a mission to California in search of

---

8. The first edition (London, 1842) was reviewed, presumably by one of the Sillimans, in the *Am. Jour. Sci. Arts* 3: 383-385 (1842). According to this review, "The earliest published account of Mr. Ward's new method of cultivating plants without open exposure to the air, was contained in a letter to Sir Wm. Hooker, published in the *Companion to the Botanical Magazine* for May, 1836 . . ."

9. Ward had died on June 4, 1868. Gray's obituary notice is preceded by another upon George Arnott Walker-Arnott, who died on June 17, 1868; Gray refers to his name as "The most distinguished . . . upon the list . . ." [of the "Botanical Necrology" for 1868].

plants. *Jour. Hort. Soc. London* 1: 180-185 (1846); 2: 121-125, 187-191 (1847); 3: 217-228 (1848).

HOOKER, WILLIAM JACKSON. 1837. Botanical information. *Comp. Bot. Mag.* 2: 184, 185.

JEPSON, WILLIS LINN. 1897. The explorations of Hartweg in America, *Erythea* 5: 31-35, 51-56.

——— 1925. A manual of the flowering plants of California. Berkeley.

LEMMON, JOHN GILL. 1892. Notes on cone-bearers of north-west America. I. *Gard. & Forest* 5: 64, 65.

——— Notes on west American Coniferae. III. Bibliography of two Californian pines. *Erythea* 1: 229-231.

SARGENT, CHARLES SPRAGUE. 1891. The Silva of North America. Boston. New York. 2: 34, *fn.* 3. [*See also* 4: 32 (1892); 8: 113, 142 (1895); 10: 104, 108, 143 (1896); 11: 140 (1897).]

CHAPTER XLVIII

THE YOUNGER ABERT EXAMINES THE "OBJECTS OF

NATURAL HISTORY" BETWEEN FORT LEAVENWORTH,

KANSAS, AND BENT'S FORT, COLORADO

IN an earlier chapter I wrote of an exploration which, under orders from John Charles Frémont, Lieutenant James William Abert made in 1845 into northern New Mexico and into portions of northwestern Oklahoma and Texas.

Early in 1846 Abert was given an assignment by William Hemsley Emory to examine the route from Fort Leavenworth, Kansas, to Bent's Fort, Colorado—a trip which covered a large portion of the trail followed by traders and emigrants bound for Santa Fe and beyond. In writing of Wislizenus I quoted from Chittenden's outline which covered much the same journey although Wislizenus started from Independence, Missouri, and at the Cimarron Crossing turned south for Santa Fe, whereas Abert, starting from Fort Leavenworth, struck the trail at 110-mile Creek and, when he arrived at the Cimarron Crossing, continued along the Arkansas River to Bent's Fort.

Abert's report upon his assignment was published in 1848, in Appendix VI of Emory's "Notes of a military reconnoissance," and Emory's included map shows the route taken by his subordinate officer. Abert's narrative—"Notes of Lieutenant J. W. Abert"—is prefaced by a letter which he wrote to Emory from "Washington City, *October* 8, 1847." It explains what his specific duty had been:

"I have the honor to submit, herewith, a report of such objects of natural history as came under my observation while I was attached to the topographical party, under your command, during the journey from Fort Leavenworth to Bent's Fort.

"The plants which were collected were submitted to the inspection of Dr. Torrey, to whom I am indebted for their names."

Abert's party was preceding General Kearny's "Army of the West" and, leaving Fort Leavenworth on June 27, arrived at Bent's Fort on July 29 or five days before Emory's contingent of the "Army" reached there on August 2. On the way, hearing that Kearny had been taken ill, Abert visited his encampment; the next day he himself was taken sick and, although he got to Bent's Fort, could proceed no farther for about two months. When Emory left Santa Fe for California on September 26, he left orders that Abert and his assistant William Guy Peck (who had also been ill and was unable to get beyond Santa Fe) should complete surveys which he himself had begun in New Mexico. I mention this assignment again. Emory's "Reconnoissance" explains that both Abert and Peck ". . . had but too recently returned from an exploring expedition in less favored climates . . ."—or from Frémont's assignment of 1845.

981

Abert's interest in "objects of natural history"—to judge by the content of his "Notes"—lay, rather obviously, in birds; Geiser refers to him as an "earnest student of ornithology." I shall, however, confine my citations mainly to his comments about plants. I have altered some of his paragraphs in order to condense. Here, as in his earlier "Report," Abert's literary style is somewhat "flowery" as his opening paragraph indicates:

"On the 27th of June, 1846, we set out from Fort Leavenworth.[1] The day was clear and bright; the woods were rejoiced with the voice of the mocking bird, and of the many little warblers that would join in the chorus of his song; the bluebird was there with his sprightly notes, and the meadow lark, perched on some tall mullein weed, caroled forth his song of love . . . we were well prepared to enjoy the beautiful scenes . . ."

Leaving Fort Leavenworth on June 27, 1846, Abert's route until August 26 or 27 lay through Kansas, at first nearly south. He records on the day of departure:

". . . Here are many varieties of useful timber: the hickory, the walnut, the linden, the ash, the hornbeam, the maple, the birch, and the beech, also the cotton wood; but, beyond the limits of the 'tall grass,' there is the cotton wood only.

"Five miles from Fort Leavenworth we passed a large butte, called 'Pilot Knob;' . . . we saw fine forests of timber, consisting chiefly of oak. Among the shrubs, we noticed the hazel, (corylus Americanus,) and the button bush, (cephalantus occidentalis;) among these the wild grape had twisted its tendrils . . . the wild rose was still in bloom, and mingled its pink flowers with the beautiful white clusters of the Jersey tea, (ceonothus Americanus.) The prairies were covered with tall stalks of the rattlesnake weed, (rudebeckia purpurea.) . . . After . . . twelve miles, we encamped . . . Here we noticed the white hickory, or downy hickory, (juglans pubescens,) the chestnut oak, (quercus prinus acuminata,) the spicewood, (laurus benzoin,) and, deep in the woods, the modest May apple, (podophyllum peltatum,) and bloodroot, (sanguinea canadensis.)"

They were moving south and southwest across Leavenworth County, bound for the Kansas River. On June 28 they came to Stranger Creek which, flowing south, enters the Kansas not far east of present Linwood. Abert refers to it as "a romantic little stream."

". . . The banks . . . are composed of rich loam that nourishes immense oaks and sycamores, (platanus occidentalis.) . . . We noticed a great quantity of the orange colored asclepias, (asclepias tuberosa,) around which gaudy butterflies were flitting. The low grounds . . . were covered with a prickly button-head rush, (eryingium aquaticum,) the roots of which, when candied over, formed the kissing comfits of Falstaff.[2]

1. This fort was situated on the west side of the Missouri River, in northeastern Leavenworth County, Kansas. The site had been selected in 1827 by Colonel Henry Leavenworth, for the purpose of protecting the old trail to Santa Fe.

2. The reference is to the *Merry Wives of Windsor*, act 5, scene 5. Falstaff is addressing Mistress Ford:

The woods were skirted by a dense growth of hazel, plum trees, and tangled grape vines . . ."

June 29: "The prairie was . . . rolling; the flat bottoms were covered with the rosin weed or polas plant, (silplicum [Silphium] laciniatum,) whose pennate-parted leaves have their lobes extending like fingers on each side of the mid rib. It is said that the planes of the leaves of this plant are coincident with the plane of the meridian; but those I have noticed must have been influenced by some local attraction that deranged their polarity. The orange colored asclepias, (A. tuberosa,) and the melanthium virginicum, a white-flowering bush, were also abundant. The timber on the ravines consisted of the white oak, (Q. alba,) black jack oak, (Q. ferruginea,) mulberry, (morus rubra,) walnut, (F. [*Juglans*] nigra,) the hickory, the red bud, (ericis [*Cercis*] canadensis.) The nettles (urtica canadensis) had grown to the height of 7 or 8 feet, all of which show the prodigal fertility of the soil . . ."

They were approaching ". . . the Kaw or Kansas River, . . ." and, crossing a high ridge, entered ". . . the Kansas bottom . . . overgrown with a tall grass (arundo phragmites) from 5 to 6 feet high . . . mingled with . . . the long-leafed willow and the cotton wood . . ." They had reached the Kansas opposite the mouth of Wakarusa ("Wakaroosa") Creek which enters from the south in Douglas County, to the east of present Lawrence, and crossed the river before camping, as shown on Emory's map, to the east of the creek. On June 30 they had a glimpse of "the Wakaroosa buttes" and their route lay between them; next they ". . . saw the 'divide' that separates the waters of the 'Wakaroosa' from those of the 'Alaris des cygnes,' or Osage; (as it is called near its mouth;) upon this divide the Santa Fé road is laid out. We soon saw the Oregon trail, which here unites with that of Santa Fé . . ." Abert's route now trended to the southwest, for a time along Wakarusa Creek, where he observed that ". . . The elder (sambucus pubescens) was still in bloom, and the orange asclepias still displaying its gaudy flowers, much to the delight of the brilliant butterflies that sported around it, and are so constantly found near it, that it is often called the butterfly plant . . . Along the margin of the creek I found a beautiful lily, (lilium tigrinum,) of a bright orange color, and beautifully dotted."

It was on July 1 that the party ". . . reached the broad trail of the traders from Independence, Missouri, to Santa Fé . . ." From here westward to the Cimarron Crossing all travelers took the same route apparently, and Abert's "Notes" record certain well-known points along the trail: first "110 mile creek" in Osage County; "Independence creek"; "Rock creek"; " 'Big John spring' "; "Council grove" and "Diamond spring" in Morris County; Cottonwood Fork ("a tributary of the Neosha") in

"Let the sky rain potatoes . . . hail kissing-comfits, and snow eringoes . . ." *Eryngium maritinum* is a European species, but Abert evidently had an American species of sea holly in mind.

According to *The National Cyclopedia of American Biography* (4: 395, 1887) Abert, in 1849, was ". . . elected instructor and assistant professor . . . in the department of English literature, *belles-lettres* and moral philosophy at West Point." He certainly seems to have had a knowledge of the classics.

Marion County; Turkey Creek and the Little Arkansas, both in McPherson County; and "Cow creek" in Rice County; many of these appear not only on Emory's map but on modern maps as well. I shall quote Abert's comments about some of the plants observed along the above way (italicizing the localities mentioned). Where, on July 1, the trail to Santa Fe was encountered, he ". . . collected some beautiful flowers, amongst which were the rudbeckia hirta, and the delicate bed straw, (galium tinctorum.) . . . The only trees . . . were some tall elms, (ulmus Amer.,) . . ."

July 2: ". . . we reached *110 mile creek* . . . Nigh the banks of the stream there was a low piece of ground covered with the purple monarda, (monarda allophylla)."

July 3: ". . . we reached *Independence creek,* so called by Colonel Frémont, in consequence of our encamping here on the 4th of July, one year previous . . . We encamped seven miles beyond Independence creek, in a ravine timbered with the elm, the cotton wood, the hickory and the oak."

Along the roadside that day Abert had gathered ". . . a plant called lamb's quarter, (chenopodium album,) the plantain weed, (plantago major,) and a beautiful sensitive plant, with a yellow flower, slightly resembling the violet, (cassia chamaecrista.)" On July 4, at a stream "composed of pools," the banks were ". . . heavily timbered with walnut, and we also noticed the buckeye, (pavia lutea,) and, skirting the stream, gooseberry bushes, (ribes triflorum,) and elder . . . we reached *Rock creek* . . . we pushed forward for *'Big John spring,'* which we reached at 5 o'clock. Here we luxuriated . . . reclining under the shade of a tall oak 'sub tegmine querci'[3] . . . We saw to-day two beautiful varieties of the evening primrose, (œnothera biennis,) the white and the yellow."

On July 5 they reached *"Council grove"* and next " *'Diamond spring' "*—where Abert ". . . procured . . . a beautiful white thistle, (cnicus acarna,) of delicious fragrance." At Cottonwood River on the 6th, he noticed ". . . thickets of the elder (S[ambucus] canadensis) in full bloom. The beautiful monarda (M. allophyla) covered the low portions of the banks." On the 7th they took a day of rest.

"Around our camp the ground looked golden with the different varieties of the golden rod, (solidago,) and along the stream we saw box elder, (acer negundo,) and extended thickets of plum bushes."

July 8: ". . . we were on the route for the *Turkey creeks;* they are three in number, and unite a few miles below the points where our road crosses them . . . We had now reached the short grass . . . and we saw little patches of the true buffalo grass, (sesleria dactyloides,) a short and curly grass, so unique in its general character that it catches the eye of the traveller . . . we observed little circular spots . . . where the buffalo once wallowed . . . These old wallows are now overgrown with plants that grow more luxu-

3. Is Abert paraphrasing Cicero's *sub tegmine caeli*—"the vault of heaven"?

riantly than on other portions of the prairie. There is the splendid coreopsis (coreopsis tinctoria) and the silver margined euphorbia; (euphorbia marginata;) . . . It is seldom, now, that the buffalo range this far . . . we . . . formed our camp on *Turkey creek*. Here . . . we found some beautiful plants with brilliant scarlet flowers (malva pedata) and roots which are eatable. We also obtained specimens of the pomme blanche, (psoralea esculenta,) . . ."

July 9: ". . . we reached the *Little Arkansas,* a tributary of the great river the name of which it bears . . . on its banks were some large elms and box elder; we also saw the common elder, (sambucus,) narrow leafed willow, and the grape, (vitis aestivalis,) the sorel (oxalis stricta) and lamb's quarter, (chenopodium album,) grew near the stream . . . We noticed to-day the pink sensitive plant (schrankia uncinata) of most delicious fragrance . . . with this plant, we also found a white variety, (darlingtonia brachypoda,) the flowers and leaves are smaller than the plant first mentioned, and has no odor."

July 10: "We collected some lamb's quarter[4] and had it cooked, and noticed along the road side the purslane, (portulaca oleracea:) this also would answer for the table of the prairie voyageur." [5]

On arrival at *Cow Creek* and near "the buttes" Abert collected "some beautiful Gaillardias of different species. Gaillardia amblyodon and G. pinnatifida we found abundant . . ." It was this day, July 11, that the party arrived at the Arkansas River and encamped, finding there a ". . . train of wagons, belonging to Messrs. Hoffman, of Baltimore." They had reached the river at the top of its great sweep to the north, in Barton County—the present city of Great Bend marks its apex. After reaching the Arkansas River, Abert followed its northern side, with only one minor deviation (July 12 to 18), to Bent's Fort. Still in Kansas, the "Notes" mention "Walnut creek"

4. This plant is so often mentioned by Abert—as well as by others of our collectors—that it arouses interest. *Chenopodium album* was first described by Linnaeus and is a European species; Gray's *New manual of botany* (1908) cites the common names of lamb's-quarters and pigweed, and states that it is introduced from Europe and naturalized "everywhere." In *A dictionary of English plant-names* Britten and Holland include a citation from Caleb Threlkeld (*Synopsis Stirpium Hibernicarum,* published in 1727) which records that, in Ireland, the plant is " 'sold in May by the country women by the name of *Lamb's quarter.*' " I have found no earlier reference to this curious name which may bear some relation to the shape of the leaf.

Anyone who is interested in increasing his list of eatable vegetables will find considerable entertainment in Fernald and Kinsey's *Edible wild plants of eastern North America* where the authors refer to the "common Pigweed" as a popular potherb, "highly prized by European peoples," but little used in this country where there exists "a spendthrift American prejudice against it because it is so common." They describe a variety of forms in which they offered the plant to their guests. They also mention that, in the Goosefoot Family *(Chenopodiaceae),* there are fourteen "other species" in addition to *Chenopodium album,* and they end with the "Caution" that certain of these should *not* be used as potherbs!

5. Again I refer the reader to Fernald and Kinsey (*l.c.,* 195, 196) for the edible desirability of the purslane, or "Pusley."

in Barton County; "Ash creek" in Pawnee County; and "Pawnee fork" in either Hodgeman County or in Ness County (for it was at this time that the party left the Arkansas River for a few days). Again I quote Abert's plant records:

July 12: "We left the Arkansas and marched to *Walnut creek* . . . We here noticed the prairie gourd (cucumis perennis) and the cactus, (cactus opuntia;) also the 'pinette de prairie,' or liatris pychnostachia, with a great abundance of the common sunflower, (helianthus annuus;) the bright scarlet malva (malva pedata) and the silver edged euphorbia, (E. marginata;) also the purslane, the convolvulus (ipomen leptophylla) rudbeckia hirta, and a species of cockle burr . . ."

This day they left Walnut Creek and ". . . entered upon vast plains of the buffalo grass, (sesleria. dactyloides.)" On July 13 they moved on to *Ash Creek* and finally reached *"Pawnee fork."*

". . .The country around was covered with the (cucumis perennis) prairie gourd . . . This creek is timbered with the elm, (ulmus Americana,) and the box elder, (aceo[6] negundo.) We frequently . . . noticed the purslane and the 'pinette de prairie;' in the low grounds the splendid coreopsis and the euphorbia were displaying their beauties; and on the uplands the prickly pear was seen in great abundance, but it had passed its bloom."

The water in Pawnee Fork was too high to cross on the 13th and the next two days were spent in making a raft which was ". . . built of the driest wood that we could find." Nonetheless, when the party crossed on the 16th it ". . . became water logged. The elm and box elder were the only trees we could get, and when green their specific gravity is but little less than that of water . . ." While waiting at the fork, Abert, on July 14, had one of the men ". . . dig up a root of the beautiful prairie convolvulus, (ipomea leptophylla.)"—the man's reaction to this task is not mentioned!

". . . This man worked for several hours, for the ground was extremely hard, so that he was at last obliged to tear it up, leaving much of the top [?tap] root behind. This root extended for about one foot . . . then it suddenly enlarged, forming a great tuber, 2 feet in length and 21 inches in circumference. The Cheyenne Indians told me that they eat it, that it has a sweet taste, and is good to cure the fever. They called it badger's food, and sometimes the man root, on account of its great size, for they say some of them are as large as a man. We also procured here the Mexican poppy, (argemone Mexicana;) noticed quantities of a willow brush, and several specimens of the tooth-ache tree (near zanthoxylum fraxinum.) . . . Thus far we have noticed several plants that have been so common that I have neglected to mention them. One is the lead plant, or tea plant, (amorpha canescens,) . . . the other is what our men call the prairie indigo, (baptisia leucantha,) it bears a large black cylindrical pod, filled with kidney-shaped seed."

6. *Acer.*

The day of the crossing Abert mentions another contingent of the "Army of the West":

". . . Colonel Doniphan came to our camp and informed us that General Kearny wished to see us . . . the general had some inquiries to make in regard to the route by the Smoky Hill fork; a route Lieutenant Peck and myself had travelled when we were attached to the command of Colonel Frémont[7] . . . the Arkansas river route much to be preferred."

July 16: ". . . we obtained a singular species of cactus, resembling roundish bodies covered with long protuberances, whose tips were crowned with stars of white spines, (near mammilarea sulcata.)"

July 17: "We have now entered that portion of the prairie that well deserves to be considered part of the great desert. The short, curly buffalo grass (sesleria dactyloides) is seen in all directions; the plain is dotted with cacti and thistle, (carduus lanceolatus,) while only in buffalo wallows one meets the silver margined euphorbia; and in the prairie dog villages, a species of asclepias, with truncated leaves."

At this time, and until arrival at Bent's Fort, Abert moved up the valley of the Arkansas River. His "Notes" do not record any landmarks, however.

July 18: ". . . we found the ground in many places covered with beautiful gallardias (g. amblyodon) and the eupatorium, whilst in the moist grounds we saw the curious dodder twining in its golden tendrils all the plants that grew around it . . ."

July 19: "Marching along the Arkansas bottom one is struck with the variety of swamp grasses. Here we find the triangular grass, (scirpus triguctio,) and mingled with it in great abundance the scouring rush (equisetum hyemale) and the beautiful liatris (liatris spicata.)"

It was on July 20 that Abert was informed that General Kearny was "very ill," and members of his own detachment soon became so. He noted on the 21st that ". . . we presented quite a sorry looking array of human faces . . . I was obliged to send for the doctor . . ." They pushed on that day, however, and reached the "Santa Fé crossing" where they camped. The reference is probably to the crossing at the ford of the Arkansas River which, according to Chittenden, was "the regular crossing after 1829 . . . known as the Cimarron crossing . . . twenty miles above Dodge City . . ." Here some persons turned off on the desert route which led southwest. Abert mentions that "At this place we obtained some beautiful purple lilies, (eustoma russeliana,) and Mr. Nourse brought me a psoralia, with a monosepalous calyx." Writes Abert:

"From the 21st of July until our arrival at Bent's fort, on the 29th, being all the time

7. The Frémont assignment of 1845 had begun at Bent's Fort, but perhaps Abert had followed the "Smoky Hill fork" on his way thither.

sick, I have no recollection of anything that transpired, except a drawing that I made of the sand rat . . . Of the plants that occur between the Arkansas crossing and Bent's fort, I cannot do better than refer to the list appended to this report, in which they are arranged in the family to which they belong, and, the locality mentioned in which they were obtained.

"As one approaches Bent's fort, he meets with many varieties of artemisia, with the obione canescens, and a plant which is extremely useful to the Mexicans as a substitute for soap, by them called the palmillo, by us Adams needle, or Spanish bayonet; its botanical name is the yucca angustifolia. We also have the prairie gourd, (cucumis perennis;) that is abundant also from Bent's fort to Santa Fé. We have the bartonia, several varieties of solanas, several varieties of œnothera, the martynia, the cleome, the salicornia, ipomea, and erigonums. Amongst the trees, several varieties of populus; amongst which are the populus canadensis and p. monolifera; several varieties of salix, and the plum and cherry."

Emory's map indicates that, on July 23, Abert and his men were not far from present Deerfield, Kearny County. About July 25 or 26 they must have crossed from Kansas into Colorado; on the 28th or 29th they were at the mouth of Purgatory River and on July 30 reached Bent's Fort. The journey covered by the "Notes" was ended. Abert remained at Bent's Fort for about two months, recuperating from his illness.

In writing of Abert's exploration of 1845 I mentioned that his "Journal" of that trip, published in 1846, contained no report upon the plants collected. Appended to Abert's "Notes" upon the journey of 1846 just described is a classified list of plants drawn up by John Torrey, supplying botanical names, localities of collection and occasional dates—little if anything more. From the localities cited, such as Raton Pass, the valley of the Timpas River, the headwaters of the Purgatory, and so on, it is obvious that they were collected on the journey of 1845 and that Torrey combined in one list the plants of both journeys. Most, however, were gathered on the trip narrated in this chapter. No new species are included. Abert's "Notes" and Torrey's list of plants were merely appended to the far more important Emory "Reconnoissance" published in 1848. Both Torrey and Engelmann contributed important botanical papers to the "Reconnoissance" and in the same year Engelmann contributed his "Botanical appendix" to Wislizenus' "Memoir." One may infer that in identifying the less important collections made by Abert, Torrey merely did so in order to add some interest to Emory's total accomplishment.

I mentioned in the beginning of this chapter that Abert and his assistant Peck were left instructions by Emory—when he left Santa Fe for California at the end of September, 1846—to complete surveys which he had begun in New Mexico. These two men started work on this assignment in October, 1846, and completed it in March, 1847. The results were published in the "Report of Lieut. J. W. Abert, of his examination of New Mexico, in the years 1846–'47." (*U. S. 30th Cong., 1st Sess., House Exec. Doc.* 4:

No. 41, 417-546, 1848.) Although this would seem to have been an interesting journey and although Abert mentions (once at least) collecting plants and refers many times to those which he observed, I find no record that anything was published upon the specimens which he obtained. For this reason I do not include the narrative of the journey.

ABERT, JAMES WILLIAM. 1848. Notes of Lieutenant J. W. Abert. *In:* Emory, W. H. 1848. Notes of ... Appendix 6: 386-405.

BRITTEN, JAMES & HOLLAND, ROBERT. 1878. A dictionary of English plant names. London. London Dialect Society. (*See* under: *Eringo.* 169.)

EMORY, WILLIAM HEMSLEY. 1848. Notes of a military reconnoissance, from Fort Leavenworth, in Missouri, to San Diego, California, including part of the Arkansas, Del Norte, and Gila Rivers. By Lieut. W. H. Emory. Made in 1846-7, with the advanced guard of the "Army of the West." *U. S. 30th Cong., 1st Sess., House Exec. Doc.* 4: No. 41, 1-614.

FERNALD, MERRITT LYNDON & KINSEY, ALFRED CHARLES. 1943. Edible wild plants of eastern North America, *Gray Herb. Special Publ.* 177-180, 182.

GEISER, SAMUEL WOOD. 1948. Naturalists of the frontier. ed. 2. Dallas. 270.

TORREY, JOHN. 1848. [List of plants collected by J. W. Abert in 1845, and from June 7– July 29, 1846, and named by Torrey.] *In:* Abert, J. W. 1848. Notes of ... (Appendix 6 of Emory, W. H. 1848. Notes of ... 406-414.) [The list of plants bears no title.]

CHAPTER XLIX

EMORY, ATTACHED TO KEARNY'S "ARMY OF THE

WEST," TAKES PART IN THE CAPTURE OF SANTA FE

AND, FOLLOWING THE GILA RIVER, CROSSES WEST-

ERN NEW MEXICO AND THE BREADTH OF ARIZONA

ON HIS WAY TO CALIFORNIA

ON June 5, 1846,[1] Lieutenant William Hemsley Emory of the Corps of Topographical Engineers received orders from his chief, Colonel J. J. Abert, to report to Colonel Stephen Watts Kearny. He learned that he was to get under way the next day.

"The column commanded by Colonel Kearny, to which we were attached, styled 'The Army of the West,' to march from Fort Leavenworth, was destined to strike a blow at the northern provinces of Mexico, more especially New Mexico and California . . . We left Washington on the 6th of June . . . As far as Santa Fé, I received the assistance of Lieutenants J. W. Abert and G. W. Peck[2] . . . both of whom had but too recently returned from an exploring expedition in less favored climates, and fell ill—the first at Bent's fort, and the last at Santa Fé. From Santa Fé to the Pacific, I was aided by First Lieutenant W. H. Warner . . . and Mr. Norman Bestor . . ."

In 1846 *Niles' National Register* reprinted (from the newspaper *Union*) an ". . . extract of an official journal of 1st Lieut. Emory . . . " entitled "General Kearney and the Army of the West." This paper covered the period beginning August 2 and ending with September 7, 1846. The content does not differ greatly from that of Emory's official report of 1848, but nonetheless Emory ends the second installment with the comment that "The more . . . I think of this journal . . . the more I am satisfied it is unfit for official use in its present state. Therefore, let it be considered as an unofficial record of passing, and often trivial events."

Emory's official report of the expedition was issued first in 1848 with the title: "Notes of a military reconnoissance, from Fort Leavenworth, in Missouri, to San Diego, in California, including part of the Arkansas, Del Norte, and Gila Rivers." In this chapter I cite from the Senate Executive Document No. 7, 30th Congress, 1st Session,

---

1. To bring into alignment the travels of our plant collectors, it was two days later that John Charles Frémont, on his expedition of 1845–1847, abandoned his scientific pursuits for military activities. The last station in his "Table of latitudes and longitudes" was taken at the "Buttes of the Sacramento," California, on June 7, 1846.

2. William Guy Peck.

which—on the authority of E. H. Whorf—I accept as the first issue of the "Reconnois-sance."[3] From the point of view of botany Emory's "Reconnoissance" was one of the most valuable reports issued to date. It was important from the historical aspect as well. F. L. Paxson comments that it ". . . ranks in importance with Frémont's vol-ume on the road to Oregon."

Despite the fact that the expedition had a military purpose, Emory turns casually from battles to botany, or *vice versa*. Will power is an asset if one is to read some of the narratives of my period, but the "Reconnoissance" is truly enjoyable; the author appreciated the comic and knew how to turn a phrase.

Up to this time the botanists and plant collectors had availed themselves of what-ever protection they could find, accompanying trading, exploring, even "sporting" expeditions; in the present decade we find them more than once traveling with mili-tary expeditions. Not until Kearny's "Army of the West" reached California did any carnage enter the picture; as De Voto states: "It looked simple when you studied it on the map. It turned out to be almost as simple as it looked."

About twenty years before Emory, two traders, Sylvester Pattie and his fifteen-year-old son, had entered California by way of the Gila River, but Emory never mentions—and undoubtedly knew nothing of—their trip and its sad ending. Only a few weeks before Emory, Wislizenus had descended the Rio Grande to the El Paso crossing. But, after Emory turned west from the Rio Grande valley (near the Mimbres Mountains of New Mexico), until he reached the confluence of the Gila and Colorado rivers, he was to cross virgin country as far as botanical exploration was concerned. And he was to find, therefore, a considerable number of new and important plants. Certainly Torrey and Engelmann were pleased with his accomplishments.

From Fort Leavenworth, Kansas, Emory traveled south to the Santa Fe trail and followed the trail to the northernmost point of the Great Bend of the Arkansas River (Barton County) and onward to Bent's Fort, situated on the south side of the Arkansas, about midway between the towns of Las Animas (Bent County) and La Junta (Otero County), Colorado; they were avoiding the desert route which turned off at the Cimar-ron Crossing (or about twenty miles above Dodge City, Ford County, Kansas). The first pages of the "Reconnoissance" describe some of the regional aspects of the coun-

3. Two papers have been published upon the different issues of Emory's report—one by John H. Barn-hart (1895), another by Frederick V. Coville (1896)—and bibliographical complexities appear to have been many. Coville stated that "It is clear that the first edition of the Emory report is the one issued as House Document No. 41, the second as Senate Document No. 7."

Associated with the twenty-one issues of Emory's report which are in the library of the Arnold Ar-boretum are letters and pages of notes summarizing the results of a "collation of seventy-two copies of Emory's Report by Edward H. Whorf." Believing that the facts therein presented, but never published, may be of interest, I have included them at the end of this chapter. The collator reached the conclusion that the first issue was *Senate Executive Document No. 7, 30th Congress, 1st Session*, of 416 pages, Wash-ington, Wendell and Van Benthuysen, 1848, ordered printed December 16, 1847; and that the second issue was *House Executive Document No. 41, 30th Congress, 1st Session*, of 614 pages, Washington, Wendell and Van Benthuysen, 1848, ordered printed February 9, 1848.

try between these two points and I shall quote the author's comments upon the flora; first, between Fort Leavenworth and the fork of Pawnee River (entering the Arkansas near Larned, Pawnee County, Kansas) and, second, between that point and Bent's Fort:

"Trees are to be seen only along the margins of the streams, and the general appearance of the country is that of vast, rolling fields, enclosed with colossal hedges. The growth along these streams, as they approach the eastern part of the sections under consideration, consists of ash, burr oak, black walnut, chesnut oak, black oak, long-leaved willow, sycamore, buck-eye, American elm, pig-nut hickory, hack-berry, and sumach; towards the west . . . the growth along the streams becomes almost exclusively cotton-wood. Council Grove creek forms an exception to this, as most of the trees enumerated above flourish in this vicinity, and render it, for that reason, a well-known halting-place for caravans, for the repairs of wagons, and the acquisition of spare axles.

"On the uplands the grass is luxuriant, and occasionally is found the wild tea (amorpha canescens,) and pilot weed (silphium lacinatum;) the low grounds abound in prickly rush, narrow leafed asclepias, white flowering indigo, flowering rush, spotted tulip, bed-straw, wild burgamot, spider wort, pink spider wort, pomme blanche, (psoralea esculenta,) scarlet malva, pilot weed, hazel, button bush, wild strawberry, cat-tail, and arrow-rush.

". . . near the meridian of Pawnee Fork . . . the country changes . . . until it merges into the arid, barren wastes described under that section. The transition is marked by the occurrence of cacti and other spinose plants, the first of which we saw in longitude 98°.

"Near the same meridian the buffalo grass was seen in small quantities . . ."

Turning to the section of country between the Pawnee River and Bent's Fort, we are told that

"The eye wanders in vain over these immense wastes in search of trees. Not one is to be seen. The principal growth is the buffalo grass, cacti in endless variety, though diminutive, yucca angustifolia, (soap plant,) the Darlingtonia brachyloba, schrankia uncinata, prairie gourd (cucurbita aurantia,) and very rarely that wonderful plant, the Ipomea leptophylla, called by the hunter man root, from the similarity of its root in size and shape to the body of a man. It is esculent, and served to sustain human life . . .

"Near the dry mouth of the Big Sandy creek, the yucca angustifolia, palmillo of the Spaniards, or soap plant, first made its appearance, and marked a new change in the soil and vegetation of the prairies.

"The narrow strip . . . the bottom land of the Arkansas . . . contains a luxuriant growth of grasses . . . The only tree of any magnitude . . . is the cotton-wood, (populus canadensis,) . . . About 35 miles before reaching Bent's Fort is found what is called the 'big timber.' . . . The 'big timber' is a thinly scattered growth of large cotton woods not more than three quarters of a mile wide, and three or four miles long . . .

"In addition to the grasses and cotton-wood mentioned, we find in the bottoms wild plum, wild cherry, willow, (salix longifolia,) summer grape, (vitis æstivalis,) cat-tail, (typha latifolia,) scouring rush, (equisetum hyemale,) . . . commelina angusti-folia, Mexican poppy, (argemone Mexicana,) monarda fistulosa, coreopsis tinctoria, psoralea esculenta, cassia chamærcrista, several varieties of solidego, œnothera, and helianthus; among which was the common sunflower."

At this point the "Reconnoissance" proper (headed "Notes") begins. It has been necessary to delete much that is interesting and, in order to condense, to alter some of Emory's paragraphing.

On August 2 Emory reached Bent's Fort; a "huge United States flag" was "flow-ing in the breeze" and a column of dust "marked the arrival of 'the Army of the West.' " Here Emory took his place with the staff. The expedition moved on the same day up the Arkansas River and, after five miles ". . . turned to the left . . . over an arid elevated plain for twenty miles . . ." and reached "the Timpas," a creek which en-ters the larger stream from the southwest, between Rocky Ford and La Junta, or in Otero County. Its course parallels that of the Purgatoire, or Purgatory, River lying further to the east.

"Along the Arkansas the principal growth consists of very coarse grass, and a few cotton-woods, willows, and euphorbia marginata. The plains are covered with very short grass, sesleria dactyloides, now burnt to cinder; artemisia, in abundance; Fre-montia vermicularis; yucca angustifolia, palmillo, of the Spaniards; verbena; eurotia lanata, and a few menzelia nuda."

On August 3 they ascended the Timpas for about seven miles finding the vegetation "similar" to that of the previous day except for the addition of ". . . an evergreen and a magnificent cactus three feet high, with round limbs shaped like a rope, three and a half inches in diameter, branching at right angles. It is said the Mexicans make hedges of it . . . On . . . hills we found cedar growing, very stunted; Missouri flax; sev-eral varieties of wild currants; a very stunted growth of plums; moss and cacti in great variety, but diminutive."

On the 4th, after following the valley of the Timpas for thirteen miles they crossed to the eastern side of that stream and began crossing ". . . the dividing line between the waters of the Timpas and those of the Purgatory, or Los Animos, of the Spaniards." The vegetation was the same except for the addition of ". . . a plant described by Dr. Torrey, as physalis perbalis, and one eriogonum tomentosum." It was on August 5, and presumably in Las Animas County, that they reached the valley of Purgatory River, ". . . called, by the mountain men, Picatoire, a corruption[4] of Purgatoire . . ." Here Emory observed that ". . . The blighted trunks of large cotton-wood and locust trees were seen for many miles . . . the cause of decay was not apparent. The growth of the

4. Still further corrupted at the present time to "Picket-wire," as I found at various gas stations in the region!

bottom . . . was black locust, the everlasting cotton-wood, willow, wild currants, hops, plum and grape, artemisia, clematis Virginiana, salix, in many varieties; and a species of angelica, but no fruit was on the bushes."

On August 6 the party began the ascent of Raton Pass, stopping within half a mile of the summit. Here, we are told, that ". . . Pine trees (pinus rigida) . . . obtain a respectable size . . . A few oaks, (quercus olivaformis,) big enough for axels, were found . . . we saw several clumps of the pinon, (pinus monophyllus.) . . . also the lamita in great abundance. It resembles the wild currant, and is, probably, one of its varieties; grows to the height of several feet, and bears a red berry, which is gathered, dried, pounded, and then mixed with sugar and water, making a very pleasant drink, resembling currant cordial . . .

"The view from our camp is inexpressibly beautiful, and reminds persons of the landscapes of Palestine[5] . . ."

Emory himself was reminded of ". . . the pass at the summit of the Boston and Albany railroad, but the scenery bolder, and less adorned with vegetation." Presumably, he was thinking of the Berkshires! On August 7 the expedition reached the summit and crossed into New Mexico, the boundary between that state and Colorado now passing through or near the top of Raton Pass. The party was to remain in New Mexico until October 23 when it crossed into Arizona. Emory wrote on the 7th:

"For two days our way was strewed with flowers . . . Among the flowers and shrubbery was the campanula rotundifolia, (hare bell,) sida coccinea, galium triflorum, the snowberry, eriogonum, geranium Frémontii, clematis virpuenna, ranunculus aquatilis, euphorbia marginata, linum perenne, malva pedata, lippia cuneifolia, and many pretty varieties of convolvulus."

After sixteen miles they reached ". . . the main branch of the Canadian . . ." seeing for the first time ". . . a few sprigs of the famous grama, Atheropogon oligostaclyum. The growth on to-day's march was piñon in small quantities, scrub oak, a few lamita bushes, and, on the Canadian, a few cotton-wood trees . . ." Here the Canadian River flows south through central Colfax County and Emory evidently followed its valley, crossing the river on the 9th and camping on its western side where he saw ". . . cacti in great abundance, and in every variety; also a plant which Dr. De Camp pointed out as being highly balsamic . . ." On the 10th they reached Vermejo Creek (in Mora County) and camped on the Little Cimarron. "The plants of to-day . . . were the Erysinum Arkansanum, lippa cuneifolia, myosotis glomerata . . . lytherus linearis, hypericum ellipticum, several verbenas, and several new varieties of oxybaphus, wild sage, and on the streams a few cotton-wood and willows."

Emory's party was, at this time, no great distance west of the regions crossed by Long and James, when, in 1820, they had traversed Union County and, by way of what is now called Major Long's Creek, had turned into Hartley County, Texas.

5. The very day I read this comment I received from my son, stationed at Tel Aviv, Israel, a letter which said how much the scenery reminded him of our southwest, of Arizona in particular.

On August 11 Ocate Creek (Mora County) is mentioned and Emory records that "matters" military ". . . are now becoming very interesting." From time to time they were capturing Mexicans, spies apparently, although they were by no means warlike— one presented ". . . the colonel with a fresh cream cheese." Emory mentions this day, for the first time, the presence of the famous guide Thomas Fitzpatrick.

On the 12th they crossed from ". . . the Ocaté to the valley of the Moro . . . Ten miles up the Moro is the Moro town . . . 200 houses." Mora, on the creek of the same name, is in southwestern Mora County. Here, near pools of cool water, was seen ". . . the wild liquorice (glycyrrhiza lepidota) . . ." growing plentifully. The next day they reached the junction of "the Moro and Sapillo," or Sapello, Creek which enters the Mora between San Miguel County (south) and Mora County (north); they were not far north of Las Vegas, San Miguel County, and were meeting American settlers, as well as Mexicans. One, escaped from Santa Fe, came to inform the "Army of the West" that Mexican forces were gathering. Writes Emory, "War now seems 'inevitable.' " On August 14 they came in sight of Las Vegas. "The country . . . was rolling, almost mountainous . . . Grass . . . was interspersed with malva pedata, lippia cunefolia, and several new species of geraniacæ, bartonia, and convolvulus." On the 15th Kearny addressed "Mr. Alcalde and the people of New Mexico," explaining that, on the orders of the government, he had come to take possession of their country and ". . . extend over it the laws of the United States." There were promises of protection against Indians, religious opinions were to be respected, and so on. The army then moved on to San Miguel where, again, ". . . General Kearny assembled the people and harangued them much in the same manner as at the Vegas." Nearing ". . . the ruins of the ancient town of Pecos . . .", on Pecos Creek, San Miguel County, the expedition was informed that the Mexican general, Armijo—who was thought to be assembling a force—had ". . . gone to hell, 'and the Cañon is all clear.' "

On August 18 the "Army of the West" was twenty-nine miles from Santa Fe and not a "hostile rifle or arrow" was between them and "the capital of New Mexico." A forced march was made to the town and, on the way, word came that Armijo had fled and that they would be received by the acting governor. Even on this day of days Emory records, ". . . the native potato in full bloom. The fruit was not quite as large as a wren's egg."

As Ghent observes: "The journey had taken less than fifty days. Not a drop of blood had been shed, and not a gun had been fired."

Emory's headquarters were at Santa Fe from August 19 until September 26. A fort was built and maps were drawn up. We are told much that is of interest about the famous old town; the population was then between two and four thousand and ". . . the inhabitants . . . the poorest people of any town in the province."

From September 2 to 11 Emory made a trip south, following the Rio Grande and taking him into Sandoval, Bernalillo and northeastern Valencia counties. He mentions the towns of San Felipe ("San Felippe") west of Santa Fe, of Bernalillo ("Bernallilo"), and of dining "sumptuously at Sandival's" (? Sandoval of recent maps). He refers to Sandia (". . . Zandia, an Indian town . . . at the base of a high mountain of the same name . . ."), to Isleta ("Isoletta"), and to "Tomé," southernmost point of the trip. Back at Santa Fe on September 11, he spent from the 15th to the 25th, "fitting out for California."

Neither Emory's assistant Abert (still at Bent's Fort because of illness) nor Peck (at Santa Fe, but on the sick list) were able to join the expedition. Emory, on September 14, wrote instructions as to their future duties in completing his own surveys of New Mexico. Orders were issued relative to the march on California and these mention that the battalion of Mormons "commanded by Captain Cook [Cooke]" was to follow the advance contingent; Colonel Doniphan's regiment was to remain in New Mexico until relieved and then effect a junction with "General Wool at Chihuahua." Emory's party now included Lieutenant Warner, topographical engineer, "J. M. Stanl[e]y, draughtsman," and Norman Bestor, "assistant."[6]

Leaving Santa Fe on September 26, Emory states that they marched over the ground which he had traveled and described between September 2 and 7; that they crossed to the west side of the Rio Grande del Norte at Albuquerque and camped until the 30th "a little more than half way between Albuquerque and Pardillas." "Feeling no desire to go over the same ground twice," he ". . . struck off on the table lands to the west and found them a succession of rolling sand hills, with obione canescens, franseria acanthocarpa, yerba del sapa of the Mexicans, and . . . at very long intervals, with scrub cedar, about as high as the boot-top." We are told that

"The plains and river bottoms were covered with much the same growth as that heretofore noted; to which may be added an erythera, a handsome little gentian-like plant, with deep rose-colored flowers, and a solanum, a kind of wild potato, with narrow leaves, which Dr. Torrey says is different from that in the United States."

On October 1 Emory made a trip across the Rio Grande to Tomé. "The air was elastic, and fragrant with the perfumes of the wild sage . . . Every thing was couleur de rose . . ." On the 2nd they camped opposite La Joya ["La Lloya"] in northern Socorro County, ". . . at the bend of the river Del Norte . . ." and ". . . found growing abundantly atriplex and salicornia . . . lycium in great abundance, senecis[7] longilobus, martynia proboscidea, (cuckold's horns,) and a small shrub like convolvulus."

On October 4 they found it necessary to mount to the tablelands lying west of the river, for they made travel somewhat easier.

"These plains . . . are of rolling sand hills, covered with obione, canescens, prosopis glandulosa, (romeria,) riddellia tegetina, paga-paga—an abundant shrubby

6. W. H. Brewer records that "Norman Bestor collected on Major W. H. Emory's expedition across the continent, joining the party at Santa Fé, and reaching San Diego December 12, 1846 . . ."

7. Government reports had to be issued promptly and more than once the botanists, who undoubtedly felt responsible for the plant names, complained that they had not seen proof. In the "Reconnoissance" one finds peculiar spellings and punctuation between generic and specific names is not uncommon.

plant, belonging to the family of the amaranths, but a genus not yet described— a new dieteria, a new fallugia, baileya multiradiata, abronia mellifera, and a few patches of grama. This last is the only nutriment the plains afford for horses and cattle; but mules and asses, when hard pressed, will eat the trato and romeria. The chamisa grows to a considerable height . . . To-day I eat, for the first time, the fruit of the prickly pear, the 'yerba de la vivera,' of the Mexicans . . . it tasted truly delicious, having the flavor of a lemon with crushed sugar.

"To the west, the hills of Pulvidera form an amphitheatre . . . Arrived at the town of Pulvidera[8] . . . as the name implies, covered with dust . . ."

On October 5, camp was at Socorro ("Secoro"), in the county of the same name. The town was prettily situated in the valley of the Rio Grande and had one hundred inhabitants.

"The growth on the sand plains to-day is chiefly iodeodonda and a little stunted acacia. The iodeodonda is a new plant, very offensive to the smell, and, when crushed, resembling kreosote.[9] Its usual growth is the height of a man on horseback, and is the only bush which mules will not eat when excessively hungry; besides this were varieties of ephedra, erythercea, helianthus petiolaris, and two well known and widely diffused grasses, the reed grass, and a short salty grass, uniola distichophylla."

On October 6 it was decided to follow the Rio Grande farther south before turning westward. On the 7th they passed the last settlement and in a few miles ". . . left the beaten road, which crosses to the east side of the river, and thenceforth a new road was to be explored. The land passed over to-day . . . is incomparably the best in New Mexico . . . One or two large white cedars were seen . . . and, in addition to the usual plants, was that rare one cevallia sinuata, guava parviflora, œnothera sinuata, and a species of wild liquorice, but with a root not sweet, like the European kind." By the 8th, the river valley was losing ". . . what little capacity for agriculture it possessed . . ." On the eastern side of the river ran the "Chihuahua road . . . and that part of it is the dreaded jornado . . . ninety miles without water." [1] The road on the western side ". . . led us through a great variety of vegetation, all totally different from that of the United States . . . First, there were cacti in endless variety and of gigantic size, our new and disagreeable friend, the larrea Mexicana, Fremontia vermicularis, obione canescens,

8. The reference is to the peak and town of Polvadera, Socorro County. The day before, the Apaches had made a brutal attack and the inhabitants had been saved only (and only to some extent) by the assistance of the people of "Lamitas" (? Limitar) situated two miles below. "These banditti will not long revel in scenes of plunder and violence. Yesterday Colonel Doniphan's regiment was directed to march into their country and destroy it."

9. A footnote has this to say of the creosote bush: "Since writing the above, the following extract of a note from Dr. Torrey was received in reference to this plant, which is so remarkable, and extends over so great a distance. 'The *iodeodonda* I find described in a late work by Moricand, entitled *"Plantes nouvelles ou rares d'Amerique."* It is described by him as a new genus, under the name larrea. It is well figured in his 48th plate as *Larrea Mexicana.* In its affinities it is allied to *guiacum.*' "

1. Wislizenus described its terrors in his "Memoir."

tessaria borealis, diotis lanata, franseria acanthocarpa, several varieties of mezquite, and ... peculiar to the ground passed over, were several compositæ, a species of malva convolvulus, an unknown shrub found in the beds of all deserted rivers; larger grama, as food for horses, nearly equal to oats, and dalea formosa, a much branc[h]ed shrub, three feet high, with beautiful purple flowers. The infinite variety of cacti could not be brought home for analysis, and this department of the Flora must be left to the enterprise of some traveller, with greater means of transportation than we possessed. A great many were sketched, but not with sufficient precision to classify them."

On October 9, the country being "broken," and the valley narrowing into a cañon, it was necessary to climb to the tablelands on the west side. The next three days the party remained in camp. They were in northeastern Sierra County (to the east of what is now Elephant Butte Reservoir). "Above and below us is a cañon, and on the eastern side of the river the Fra Cristobal shoots up to a great height." On October 13 the need of better grass for the animals necessitated moving a mile farther; it had been decided to abandon the wagons and they awaited the arrival of packsaddles. Emory wrote on the 14th that, with the sending back of the wagons, "... every man seemed to be greatly relieved. With me it was far otherwise. My chronometers and barometer, which rode ... safely, were now in constant danger ... All my endeavors, in the 24 hours allowed me in Washington to procure a pocket barometer, had failed ..."

It was on October 15 that they "... turned off from the Del Norte and took final leave of it ..." On the tableland where they now were "... the winter grama (a more delicate grass than summer grama) was in great abundance, but now dry and sun burnt. The other growth ... consisted of malva, senecio longilobus, small mezquite, fraxinus, (ash,) different from any in the United States; castilleja and datura."

They looked down on "... Paloma, (Pigeon creek.) " Palomas River, entering the Rio Grande from the northwest in Sierra County, flows not far south of present Hot Springs. In its valley "... grows cotton wood, a new variety of evergreen oak,[2] with leaves like the holly, a new variety of ash, and a new kind of black walnut, with fruit about half the size of ours. The oak was covered with round red balls, the size and color of apricots—the effect of disease or the sting of an insect."

On October 16 Emory and his party turned southwest[3] towards the Mimbres Mountains lying along the boundary between Sierra and Grant counties.

"We commenced the approach to the Mimbres mountains over a beautiful rolling country, traversed by small streams of pure water, fringed with a stunted growth of walnut, live oak and ash ... There were several new and beautiful varieties of cactus and the entamario (tessaica borealis) diotis lanata in great luxuriance; one a minia-

2. Sargent noted in *The Silva* (8: 103, 104, t. 297, 1905): "*Quercus Emoryi* was discovered on October 15, 1846, in the valley of Pigeon Creek in southern New Mexico, by Colonel W. H. Emory ..."

3. At much the same point where Frémont was to do so in 1849. When he had passed west of the Mimbres Mountains he continued southwest—probably along the road which Emory mentions as leading to "... Janos, a frontier town in Sonora ..." Emory moved almost due west.

ture tree, with a stalk six inches in diameter, a new species of dieteria like an aster, with fine purple flowers; aster hebecladus and three-leaved barberry (berberis trifoliolata.)"

On the 17th they reached a height of land between the Rio Grande valley and the Mimbres Mountains and then ascended an arroyo—with "sides and bank covered with a thick growth of stunted live oak"—leading to that range. Emory comments upon several ". . . new plants and shrubs, amongst which was the cercocarpus parvifolius, a curious rosaceous shrub, 'with a spiral, feathery tail, projecting from each calyx when the plant is in seed.' The spiral tailed or barbed seed-vessels fall when ripe, and, impelled by the wind, work into the ground by a gyratory motion. The cedar seen to-day was also very peculiar; in leaf resembling the common cedar of the States, but the body like the pine, except that its bark was much rougher. (For the rest of to-day's growth, see catalogue of plants for this date.)"

On the 18th they moved around the north end of "Ben Moore bluff," which Emory named for a friend, and then ". . . began to drop down into the valley of what is supposed to be an arm of the Mimbres, where there are some deserted copper mines."

These mines, situated at Santa Rita, Grant County, appear in the literature of botany under a number of names—Standley cites Santa Rita, Santa Rita del Cobre, and the Copper Mines;[4] he also notes that "Two localities in New Mexico are remarkable for the number of plants described from them, Santa Fe[5] and Santa Rita. The reason for this is the fact that the first extensive collections made in the Southwest were made largely at these two places." In other words, the copper mines (by whatever name) and Santa Fe are type localities of many plants. Emory records on the 18th:

"This afternoon I found the famous mezcal, (an agave,) about three feet in diameter, broad leaves, armed with teeth like a shark; the leaves arranged in concentric circles, and terminating in the middle of the plant in a perfect cone. Of this the Apaches make molasses, and cook it with horse meat. We also found to-day the dasylirion graminifolium, a plant with a long, narrow leaf, with sharp teeth on the margin, with a [flower] stalk eighteen feet high. According to Dr. Torrey, it has lately been 'described by Zuccarini,' who says 'four species of this genus are now known, all of them Mexican or Texan.' "

On October 19 Emory and his men moved on and three miles from the mines ". . .

4. Standley states that "The settlement is a very old one, the Spaniards having mined copper here for probably more than two hundred years ... Lieutenant Emory visited Santa Rita in 1847 [1846] and gathered a few plants in the vicinity ... In the ... Reconnoissance there is a drawing showing Santa Rita as it appeared at that time ..."

Cleland includes an interesting account of the history of the mines, some of it quoted from T. A. Rickard, *A history of American mining*. New York. 1932.

5. Standley refers to Santa Fe as ". . . historically the most interesting locality in the Southwest from a botanical standpoint, for here was made the first extensive collection of plants in the whole southwestern region ..." This was made by Fendler.

reached the 'divide,' ..." and began their descent from the mountains. On the 20th they were "but a few miles from the Gila," which Emory "... was no less desirous of seeing than the Del Norte." From his map it must have been encountered near the northern end of the Burro Mountains, perhaps by way of the Mangas Valley. After a conference with some Apaches[6] they reached and crossed the Gila River and followed its course.

"... We heard the fish playing in the water, and soon those who were disengaged were after them. At first it was supposed they were the mountain trout, but, being comparatively fresh from the hills of Maine, I soon saw the difference ... They are in great abundance."

On October 21 they followed the Gila, crossing and recrossing it "a dozen times" to avoid a deep canyon. According to Emory's map, they must have camped near Red Rock, in Grant County: "Under this date, in the catalogue of plants will be found many differing from those heretofore observed; amongst them the zanschneria [Zauschneria] Californica, also a new shrub with an edible nut, a grass allied to the grama, Adam's needle, artemisia cana, and many varieties of mezquite."

The map indicates that, on October 22, the party rounded the southern bend in the Gila River and camped near present Virden, Hidalgo County. Following the Gila River, they must have crossed from Hidalgo County, New Mexico, into Arizona,[7] for camp of October 23 appears to have been near present Duncan, Greenlee County. Here, on the 24th, the party "laid by to recruit"; it was not to leave Arizona until November 25.

For the remainder of October Emory's records are confusing and the rivers which he mentions acquire their modern identity only if one compares his map with a more recent one.

On October 25 Emory was eagerly looking for "the fabulous 'Casa Montezuma' " made famous "in olden times by the fables of Friar Marcos ..." This day, along the Gila, he found "... what was to us a very great vegetable curiosity, a cactus, 18 inches high, and 18 inches in its greatest diameter, containing 20 vertical volutes, armed with strong spines. When the traveller is parched with thirst, one of these, split open, will give sufficient liquid to afford relief ... These and the mezquite, acacia, prosopis odorata, and prosopis glandulosa, now form the principal growth. Under the name mezquite, the voyageur comprises all the acacia and prosopis family."

On the 26th Emory refers to three streams which enter the Gila from the north in

6. Wrote Emory: "They swore eternal friendship to the whites, and everlasting hatred to the Mexicans ... The road was open to the American now and forever. Carson, with a twinkle of his keen hazel eye, observed to me, 'I would not trust one of them.' ... One had a jacket made of a Henry Clay flag ... the acquisition, no doubt, cost one of our countrymen his life."

7. Botanical collecting in Arizona began with the advent of Emory's party although two other collectors may be said to have "set foot" in that state: Thomas Coulter, when he entered its extreme southwestern corner in 1832 near present Yuma; and Frémont, when he crossed its extreme northwestern corner while following the Virgin River from Nevada into Utah in 1844. Whether either man collected a single plant on these occasions I do not know.

this region: ". . . the Priete, the Azul and San Carlos rivers." [8] On the 27th he ". . . strolled a mile or two up the San Carlos . . ." (which I believe to have been Bonita Creek) and, at a spring near camp, "Several exquisite ferns were plucked . . . a new green-barked acacia, covering the plains above the river bed, but vegetation was . . . scarce; this is the first camp since leaving the Del Norte, in which we have not had good grass." Camp of October 28 must have been near present Safford, Graham County, with Mt. Graham lying southwest. On the tablelands Emory observed "Great quantities of green-barked acacia . . . also the chamiza, wild sage and mezquite; close to the river, cotton wood and willow. We found too, . . . the eriodictyon Californicum, several new grasses, and a sedge, very few of which have been seen on our journey."

Camp of the 29th must have been near present Geronimo, Graham County, with Mt. Turnbull to the west. On the 30th we are told that "Mount Turnbull . . . had been in view down the valley of the river for three days. To-day . . . we turned its base forming the northern terminus of the same chain, in which is Mt. Graham."

On October 31 Emory states that they reached "the San Francisco[9]" at about noon and "unsaddled" in order ". . . to look up a trail by which we could pass the formidable range of mountains through which the Gila cut its way, making a deep cañon impassable for the howitzers." Between this date and November 5 the expedition ceased to follow the Gila and made a detour through the Mescal Mountains, in the region dominated by El Capitan Mountain; they kept north and west of the Gila and returned to that stream near the point where the San Pedro River enters from the southeast, or near present Winkleman, Gila County, and Hayden, Pinal County. From there the route was again along the Gila. On November 11th they reached the Indian Pima Village ("Pimos village") situated at the so-called Gila Crossing in Maricopa County. From there to the Colorado River they kept to the south side of the Gila, leaving that stream only when they cut across country to avoid the Gila Bend. On Emory's map this great southward sweep of the river is scarcely noticeable. I resume Emory's itinerary at the San Carlos (his "San Francisco") River.

On November 1 he records climbing a peak which may have been El Capitan Mountain, in Gila County:

"No alternative seemed to offer but to pursue Carson's old trail sixty miles over a rough country, without water, and two, if not three days' journey . . . I took advantage of the early halt to ascend . . . a very high peak overhanging the camp, which I took to be the loftiest in the Piñon Lano range on the north side of the Gila . . . Near the top of this peak the mezcal grew in abundance, with the stalk of one 25 feet long we erected

8. Comparing his map with a modern one, his "Pierte" would seem to be the San Francisco River (entering the Gila near present Clifton, Greenlee County), his "Azul" would be the Eagle River (slightly further west), and his "San Carlos" would be Bonita Creek (entering the Gila in eastern Graham County).

9. Emory's "San Francisco" (the San Carlos River of the modern map) forms the boundary between Graham County (east) and Gila County (west) before entering what is now the San Carlos Reservoir, where the waters of the Gila River are held in check by the Coolidge Dam.

a flag-staff . . . At the point where we left the Gila, there stands a cactus six feet in circumference, and so high I could not reach half way to the top of it with the point of my sabre by many feet; and a short distance up the ravine is a grove of these or pita-haya, much larger than the one I measured, and with large branches. These plants bear a saccharine fruit much prized by the Indians and Mexicans. 'They are without leaves, the fruit growing to the boughs. The fruit resembles the burr of a chesnut and is full of prickles, but the pulp resembles that of a fig, only more soft and luscious.' In some it is white, in some red, and in others yellow, but always of exquisite taste.

"A new shrub bearing a delicious nutritious nut and in sufficient abundance to form an article of food for the Apaches. Mezcal and the fruit of the agave Americana, and for the first time arctostaphylos pungens. Two or three new shrubs and flowers."

The cactus was the famous sahuaro, one of Emory's most famous discoveries. In-serted at this point in the "Reconnoissance" is a plate bearing the name *Cereus gigan-teus* "Engelman, Appendix No. 2 Continued." The shrub with the "delicious nutri-tious nut" was doubtless the *Simmondsia,* or deernut.

On November 2 Emory mentions that the "mezcal" (mescal or *Agave*) flourished in the region and that they found ". . . several artificial craters, into which the Indians throw this fruit, with heated stones, to remove the sharp thorns and reduce it to its saccharine state." Near camp he observed ". . . a species of ash unknown in the United States, and the California plane tree, which is also distinct in species from our syca-more."

The excitement of November 3 seems to have been the visit of a female Apache who arrived in camp in a "gauze-like dress, trimmed with the richest and most costly Brus-sels lace, pillaged no doubt from some . . . belle of Sonora." She was in no way dis-turbed when the fastenings of this apparel broke as she rode about camp (demonstrat-ing the good points of a horse she wished to sell), but dextrously slipped it between her seat and the saddle and rode in a nude state about camp ". . . until at last, attaining the object of her ambition, a soldier's red flannel shirt, she made her adieu in that new costume."

On November 4 (the last day of the detour through the mountains) Emory men-tions finding in the ravines ". . . a luxuriant growth of sycamore, ash, cedar, pine, nut-wood, mezcal, and some walnut, the edible nut again, Adam's needle, small evergreen oak and cotton wood, and a gourd the cucumis perennis . . . We encamped in a grove of cacti of all kinds; amongst them the huge pitahaya, one of which was fifty feet high."

Back on the Gila on the 5th, the river ". . . now presents an inhospitable look . . ."; it was crossed and recrossed a number of times. They then traversed ". . . a ridge at the base of Saddle-Back mountain [Saddle Mountain, in eastern Pinal County] . . . and descended . . . to the San Pedro, running nearly north." As stated, this stream en-ters the Gila near present Winkleman. Its valley was ". . . covered with a dense growth of mezquite, (acacia prosopis,) cotton wood, and willow, through which it is hard to move without being unhorsed . . . For six miles we followed the Gila. The pitahaya [the giant cactus] and every other variety of cactus flourished in great luxuriance. The pitahaya, tall, erect, and columnar in its appearance, grew in every crevice from the base to the top of the mountains, and in one place I saw it growing nearly to its full dimensions from a crevice not much broader than the back of my sabre. These extraordinary looking plants seemed to seek the wildest and most unfrequented places ... The uplands covered as usual with mezquite, chimaza, ephydræ, the shrub with edible nut, and cactus, of this a new and beautiful variety ..."

The "Vegitation on the Gila" is portrayed (very poorly) in a plate inserted at this point in the narrative; however, it is possible to recognize two sorts of opuntia, a barrel cactus and the sahuaro.

Emory notes that they had crossed through the country of the Apaches, by way of ". . . the great highway leading from the mountain fastnesses into the plains of Santa Cruz, Santa Anna, and Tucsoon, frontier towns of Sonora ... Since the 1st November, we have been traversing ... the stronghold of these mountain robbers ..." Travel had been difficult and slow; on November 6 the party rested while the howitzers were being brought up.

". . . I was struck most forcibly with the fact that not one object in the whole view, animal, vegetable, or mineral, had any thing in common with the products of any State in the Union, with the single exception of the cotton wood ... In one view could be seen clustered, the larrea Mexicana, the cactus, (king) cactus, (chandelier) green wood acacia, chamiza, prosopis odorata, and a new variety of sedge, and then large open spaces of bare gravel. The only animals seen were lizzards, scorpions, and tarantulas."

On November 7 and 8 they moved slowly along the Gila. It was raining "in good earnest." On the 9th the Gila left the mountains and flowed ". . . off quietly ... into a wide plain, which extends south almost as far as the eye can reach.

"Upon this plain mezquite, chamiza, the green acacia, prosopis, artemisia, obione canescens, and petahaya, were the only vegetation ... After leaving the mountains all seemed for a moment to consider the difficulties of our journey at an end. The mules went off at a frolicsome pace, those which were loose contending with each other for precedence in the trail ... In overcoming one set of difficulties we were to encounter another ... we bade adieu to grass ..."

They must have visited the ruins of Casa Grande on the 10th—it is pictured in Emory's report—and that day were eight or nine miles from the Pima Village.[1] They arrived there the next day and, under a sheltering awning, did some trading:

". . . this place formed a perfect menagerie, into which crowded, with eager eyes, Pimos, Maricopas, Mexicans, French, Dutch, English and Americans. As I passed on

---

1. In "Arizona place names" (*Univ. of Arizona Bulletin* 6: 332, 1935) Will C. Barnes quotes J. Ross Brown as follows: " 'There were ten Pima and two Maricopa villages in this group known as the Pima Villages. They were scattered along the Gila for several miles ...' " Emory always refers to these settlements in the singular, as "Pimos village."

to take a peep . . . naked arms, hands, and legs protruded from the awning . . . The trade went merrily on, and the conclusion of each bargain was announced by a grunt and a joke, sometimes at the expense of the quartermaster, but oftener at that of the Pimos . . .

"To us it was a rare sight to be thrown in the midst of a large nation of what is termed wild Indians, surpassing many of the christian nations in agriculture, little behind them in the useful arts, and immeasurably before them in honesty and virtue . . . not a single instance of theft was reported . . . This peaceful and industrious race are in possession of a beautiful and fertile basin. Living remote from the civilized world, they are seldom visited by whites, and then only by those in distress, to whom they generously furnish horses and food."

Emory had only good to say of the Pimas and adds that the Maricopas were also a peaceful race—". . . all that has been said of the Pimos, is applicable to them . . ." The expedition camped on "the dividing line" between these two tribes. To the north ". . . there is a gap in the mountains through which the Salt river flows to meet the Gila . . ." At the Pima Village the party was some ten to twenty miles southwest of present Phoenix. Here, on November 13 and 14, Emory left the Gila and went southwest, striking the river again at or near the southern end of the great loop—the Gila Bend— which it makes around the mountains of the same name. It had been a thirsty trip:

". . . Five miles brought us into a grove of the pitahaya . . . After leaving the pitahaya, there is no growth except the larrea Mexicana, and occasionally, at long intervals, an acacia or ingra . . . after dark . . . dropped down in a dust hole near two large green-barked acacias. There was not a sprig of grass or a drop of water . . . the whole night the mules kept up a piteous cry for both . . . We went on briskly to the Gila, whose course, marked by the green cotton wood, could be easily traced. It looked much nearer than it really was. We reached it after making forty miles from our camp of yesterday."

After a day of rest on the 15th they again started down the Gila, keeping to its south side. The country on the 17th was dreary ". . . beyond description, covered with blocks of basalt, with a few intervals of dwarf growth of larrea . . .

Emory noted on the 18th that ". . . there was still a barrier of snow-clad mountains between ourselves and Monterey, which we must turn or scale." On the 19th everything was much the same as it had been for the past several days. On the 20th the ". . . table lands were of sand . . ." On the 21st the plains were ". . . now almost entirely of sand . . . covered sparcely with chamiza, larrea Mexicana, and a shrubby species of sage, (salvia.)"

It is sometimes hard to remember that Emory's expedition was a military one, but on November 22 they came upon a large party of Mexicans. "It was not [General José] Castro, as we expected, but a party of Mexicans with 500 horses from California, on their way to Sonora for the benefit of Castro." On the 23rd they took a day of rest to prepare the mules for the trip across the great desert which lay ahead. With two others Emory visited the junction of the Gila and Colorado rivers, only about a mile and a half beyond camp:

"The Gila comes into it nearly at right angles, and the point of junction . . . is the hard butte through which, with their united forces they cut a cañon, and then flow off due magnetic west . . . Near the junction, on the north side, are the remains of an old Spanish church, built near the beginning of the 17th century, by the renowned missionary, Father Kino . . ."

They captured a Mexican, finding in his wallet "the mail from California," and learned from these letters ". . . that a counter revolution had taken place in California, that the Americans were expelled from Santa Barbara, Puebla de los Angeles, and other places . . ." Although there was more news it all dated back to mid-October, so that Kearny and his party were ". . . left in great doubt as to the real state of affairs in California . . ." Nor could they extract anything from the wily captives except advice not to proceed to the Puebla. After a visit to the camp of the Mexicans the party moved on on the 24th and, after traveling a few miles over a sandy plain, ". . . descended into the wide bed of the Colorado, overgrown thickly with mezquite, willow and cotton wood." On the 25th they reached the ford, where ". . . the Colorado is 1,500 feet wide . . . The ford is narrow and circuitous, and a few feet to the right or left sets a horse afloat. This happened to my horse . . . The growth on the river bottom is cotton wood, willow of different kinds, equisetum hyemale, (scouring rush,) and a nutritious grass in small quantities."

It was now November 25 and Emory, with the "Army of the West," was in California.

Having crossed the Colorado, they ascended it ". . . three quarters of a mile" where they encountered "an immense sand drift . . . Prosopis glandulosa, wild sage, and ephedra compose the growth; the first is luxuriant." They stopped and tied the horses to the "mezquite trees (prosopis glandulosa,)" and Emory noted the inclination of the horses to eat its beans. He sent men to collect them and the few obtained "were eaten with avidity." On the 26th the course ". . . inclined a few degrees more to the north . . ." and after twenty-four miles they came to ". . . the Alamo or cotton wood." Emory's party had entered California in what is now Imperial County and had kept west and northwest to the Alamo River. Crossing from Mexico—not far east of the town of Calexico—this stream flows almost due north and enters the Salton Sea[2] which was, undoubtedly, the "lake" which they visited on November 27 and 28.

2. The volume *California* ("American Guide Series"), p. 460, states of the Salton Sea: ". . . a gourd-shaped body of water 20 miles long and from 8 to 14 miles, wide, was only a vast, sandy depression when discovered in 1853 by Professor W. P. Blake, who made the first governmental survey of Imperial Valley. In 1905 the Colorado River overflowed into the Imperial Valley . . . and poured into the Salton Sink, filling it to a depth of 83 feet and a length of 45 miles. When this flood was checked in 1907, it left the lowest area still filled with the present Salton Sea—a lake with no outlets. The present depth of the sea is kept

"Our course was a winding one, to avoid the sand-drifts. The Mexicans had informed us that the waters of the salt lake, some thirty or forty miles distant, were too salt to use, but other information led us to think the intelligence was wrong. We accordingly tried to reach it; about 3, p.m., we disengaged ourselves from the sand and went due (magnetic) west, over an immense level of clay detritus, hard and smooth as a bowling green.

"The desert was almost destitute of vegetation, now and then an ephedra, œnothera, or bunches of aristida were seen, and occasionally the level was covered with a growth of obione canescens, and a low bush with small oval plaited leaves, unknown . . .

"About 8 o'clock, as we approached the lake, the stench of dead animals confirmed the reports of the Mexicans, and put to flight all hopes of our being able to use the water.

"The basin of the lake, as well as I could judge at night, is about three-quarters of a mile long and half a mile wide. The water had receded to a pool, diminished to one-half its size, and the approach to it was through a thick soapy quagmire. It was wholly unfit for man or brute, and we studiously kept the latter from it, thinking that the use of it would but aggravate their thirst. . . .

"One or two of the men came in late and, rushing to the lake, threw themselves down and took many swallows before discovering their mistake; but the effect was not injurious except that it increased their thirst . . .

"A few mezquite trees and a chenopodiaceous shrub bordered the lake, and on these our mules munched till they had sufficiently refreshed themselves, when the call to saddle was sounded, and we groped silently our way in the dark . . . a heavy fog from the southwest . . . enveloped us for two or three hours . . . When the fog had entirely dispersed we found ourselves entering a gap in the mountains . . . it was not till 12 o'clock, m., that we struck the Cariso (cane) creek, within half a mile of one of its sources . . . We halted, having made fifty-four miles in two days, at the source, [was] a magnificent spring . . ."

Emory now moved around the southern end of the Salton Sea (for, like the Alamo River, Carrizo Creek enters the southern end of that body of water from the southwest) and was approaching the present boundary between Imperial and San Diego counties. He apparently turned up Vallecito Creek (a northern arm of Carrizo Creek) for he soon mentions Aqua Caliente, San Felipe and Warner's Ranch all in San Diego County. He wrote on November 29:

"The grass at the spring was anything but desirable . . . We followed the dry sandy bed of the Cariso nearly all day, at a snail's pace . . . A few miles from the spring

approximately constant, despite evaporation, by waters draining from the irrigation ditches into the New and Alamo Rivers, which empty into the southern end of the sea . . ."

Emory was not the first to "discover" the Salton Sea—Mexicans had told him of its existence—but his visit appears to antedate that of Professor Blake by about seven years. The small amount of water noted by Emory had evidently gone when Blake arrived.

called Ojo Grande, at the head of the creek, several scattered objects were seen projected against the cliffs, hailed by the Florida campaigners ... as old friends. They were cabbage trees,[3] and marked the locale of a spring and a small patch of grass. We found also to-day, in full bloom, the bronnia spinosa, a rare and beautiful plant; the plantago, new to our flora; a new species of eriogonum, very remarkable for its extremely numerous long hair-like fruit stalks and minute flowers. We rode for miles through thickets of the centennial plant, agave Americana, and found one in bloom. The sharp thorns terminating every leaf of this plant, were a great annoyance to our dismounted and wearied men whose legs were now almost bare. A number of these plants were cut by the soldiers, and the body of them used for food."

After a day of rest on November 30 Emory's party, on December 1, ascended a valley—presumably that of the San Felipe—"to its termination" and then "descended to the deserted Indian village of San Felippe."

"... we passed the summit which is said to divide the waters flowing into the Colorado from those flowing into the Pacific, but I think this is a mistake ... We are still to look for the glowing pictures drawn of California. As yet, barrenness and desolation hold their reign."

On the 2nd and 3rd they were crossing "... another 'divide,' and as we approached the summit the narrow valley leading to it was covered with timber and long grass. On both sides, the evergreen oak grew luxuriantly ... we saw in the distance the beautiful valley of Aqua Caliente ... where we expected to find the rancheria owned by an American named Warner ... suddenly it burst upon our view at the foot of the hill ... The rancheria was in charge of a young fellow from New Hampshire, named Marshall ... his employer was a prisoner to the Americans in San Diego ... the Mexicans were still in possession of the whole of the country except that port [San Diego], San Francisco, and Monterey ... we were near the heart of the enemy's stronghold ... To the south, down the valley of the Aqua Caliente, lay the road to San Diego ... Near the house is the source of the Aqua Caliente[4] ..."

Emory's comments upon the flora of necessity stop at this point, and any botanical interest in his report comes to an end. On December 4 the "Army of the West" moved to the "rancheria of San Isabel" and on the 5th to the "rancheria Santa Maria." On the 6th an engagement took place at San Pasqual,[5] situated not far east of Escondido, San Diego County; eighteen officers and men were killed and thirteen wounded (among

3. A name sometimes applied to the *Sabal* (palmetto), native of tropical regions. The Floridians in Emory's party may well have seen plants of the *Washingtonia*, or fan palm, which occurs in groves along the northern and western margins of the Colorado Desert of California. The fact that the plants seen grew in a moist place makes this seem probable.

4. Identical with Warner's Springs.

5. It was at San Pasqual (Pascual, or Pasquale) that, a few years later, the *Yucca* which is deservedly the pride of California was discovered by Arthur Schott while working on the surveys along the United States–Mexican boundary. Torrey proposed for it the name *Yucca Whipplei*.

these General Kearny). There were two more battles: one on January 8, 1847, at the ford of the San Gabriel River, another on January 9 (the Battle of the Mesa) on the elevated land between the "Rio San Gabriel and Rio San Fernando," or Los Angeles River. In both the "Californians" were defeated. After the enemy had surrendered "their dear City of the Angels," the place was fortified.

Emory now received special orders and went to San Diego; from there, on January 25, he sailed on the "prize brig Maled Adhel" and "coasted along the rocky and barren shore of Lower California." The final date in the "Reconnoissance" is January 20, 1847, and Emory closes with a comment upon floral matters concerned with southern California:

"The season of the year at which we visited the country was unfavorable to obtaining a knowledge of its botany. The vegetation, mostly deciduous, had gone to decay, and no flowers nor seeds were collected. The country generally, is entirely destitute of trees. Along the principal range of mountains are a few live oaks, sycamore, and pine . . . Wild oats every where cover the surface of the hills, and these, with the wild mustard and carrots, furnish good pasturage to the immense herds of cattle, which form the staple of California. Of the many fruits capable of being produced with success, by culture and irrigation, the grape is perhaps that which is brought nearest to perfection. Men experienced in growing it, and Europeans, pronounce the soil and climate of this portion of California, unequalled for the quality of the grape and the wine expressed from it."

Appendix 1 of the "Reconnoissance" includes a letter from Albert Gallatin who was writing an essay for the New York Botanical Society and was in a great hurry to obtain information upon many aspects of Emory's route. Emory answered him a week later and mentions his plant collections:

"When I left California, it was as a special envoy to the government, and on so short a notice that many of my collections and notes were left behind, with my assistants. Among the things so left, were the seed of the cotton. Most of the plants collected, however, were brought home. These will show a very complete history of the botany of the country. They are in the hands of Doctor Torrey, who is preparing an elaborate catalogue and drawings of those plants, heretofore unknown. This catalogue I should be very glad to place at the disposal of your society."

Appendix 2 includes Torrey's "catalogue"—more nearly a list and lacking any title—which is preceded by a letter to Emory from its author, dated "College of Physicians and Surgeons, *New York, February* 10, 1848."

"I have examined the interesting collection of plants . . . and herewith send you a list of them, as complete as my numerous engagements permit me to make at present. The route which you passed over is exceedingly rich in botanical treasures, as is evident from the number of new species and genera which you were enabled to make under great disadvantages, and in an expedition which was almost wholly military in its character. Most of the new plants which you found are only indicated, or, at most, very briefly described in the following list. A more full account of them will be given hereafter."

Preceding the Torrey list is also a letter from Emory to Torrey, dated July 22, 1847. It outlines the route taken between June 27 and December 12, 1846, and for ready reference is extremely valuable.[6]

Torrey's list included one new genus of the family *Compositae:* "Baileya, *n. gen. Harv. and Gr. ined.* Two other species of this unpublished genus, dedicated to that profound observer of nature, Professor Bailey of West Point, exist among the California plants collected by Coulter,[7] and will soon be described by Mr. Harvey and Dr. Gray . . . Very abundant along the Del Norte and in the dividing region between the waters of the Del Norte and those of the Gila . . . ."

There were, among Emory's specimens, a number which Torrey believed were novelties but which, because of insufficient material, etc., he did not designate as "new species." The following, however, he named and described as such:

"P[rosopis] (Strombocarpa) Emoryi . . . Found in fruit only; on the Gila river . . .

6. So valuable indeed that I quote its important dates:

"From the 27th of June to July 11th, we were traversing the country between Fort Leavenworth and the bend of the Arkansas, a rich rolling prairie . . .

"From July 11th to July 13th, followed the Arkansas to Pawnee fork . . . At this point the fertile soil ceases, except on the immediate margin of the streams.

"From 14th July to August 1st, we were in the valley of the Arkansas . . .

"From the 1st of August to the 8th, crossing the plain in a southerly direction and mounting the Raton mountain . . .

"From the 8th August to the 14th, in the valleys of the tributaries to the Canadian, and crossing the extensive plains between these valleys.

"From the 14th August to the 18th, ascending the great ridge between the head of the Canadian and the waters of the Del Norte, halting at Santa Fé . . . on a tributary of the Del Norte, about 15 miles distant from the Del Norte . . .

"From August 18th up to the 14th October, all the collections were made in New Mexico, in the valley of the Del Norte, or on the table lands adjacent, and between Santa Fé and . . . 230 miles below Santa Fé . . .

"From the 14th October to the 19th, we were crossing the great dividing ridge between the waters of the Del Norte and the waters of the Gila . . .

"From the 19th of October to the 22d November, we were following the course of the Gila river, occasionally forced into the mountains to avoid the cañons . . .

"From the 22d November to the 24th, we were on the Colorado of the west, traversing a low sandy bottom.

"From the 24th November to the 28th we were crossing the great desert of shifting sand in a course little north of west.

"On the 28th November, we encamped at the Cariso (Reed) creek or spring . . .

"From the 28th November we commenced to ascend the Cordilleras of California, (the continuation of which forms the peninsula of Lower California,) and reached the highest point of the route December 5th, 3,000 feet above the sea, and as many below the overhanging peaks. From that point we descended to San Diego, a seaport on the level of the sea . . . This point we reached December 12 . . . ."

According to Warren, Emory had crossed, on December 5, ". . . the summit of the Coast Range, through Warner's Pass . . ."

7. Thomas Coulter, whose collections went to Professor W. H. Harvey in Dublin.

"Spiraea Californica[8] . . . Grows on high mountains near the Gila . . .

"A[denostoma] sparsifolia . . . Cordilleras of California . . .

"D[ieteria] asteroides . . . Elevated land between the Del Norte and the waters of the Gila . . .

"Wyethia ovata . . . Torr. and Gr., ined. . . . Abundant on the western side of the Cordilleras of California . . .

"Ximenesia, n. sp. ? Valley of the Del Norte, and along the Gila . . .

"T[etradymia] (Polydymia) ramosissima . . . Hills bordering the Gila . . .

"Plantago, n. sp. ? Allied to P. gnaphaloides, Nutt. Great desert west of the Colorado, near the Cordilleras of California . . .

"Euploca grandiflora . . . On the Del Norte below Santa Fé . . .

"Phlox . . . likewise occurs in Texas, and will be described by Dr. Gray. It was found in various places on the tributaries of the Canadian.

"Fraxinus velutina . . . Grows in the region between the waters of the Del Norte and the Gila; also on the Mimbres . . .

"O[bione] polycarpa . . . Valley of the Gila . . .

"Eriogonum trichopes . . . Eastern slope of the Cordilleras of California[9] . . .

"E[riogonum] Abertianum . . . Very common in the region between the Del Norte and the Gila. Also found by Lieutenant Abert on the upper waters of the Arkansas. Just as I was sending these notes to the press, I received a visit from Mr. Nuttall, who informed me that a species allied to this was found by Mr. Gambel, in his late journey to California. He thinks its characters differ so much from all the Eriogona hitherto described, that he has constituted of it a new genus under the name of Eucycla. A full account of Mr. Gambel's plants, by Mr. Nuttall, will soon be published . . .

"Stillingia spinulosa . . . Abundant in the desert west of the Colorado . . .

"Quercus Emoryi . . . Common in the elevated country between the Del Norte and the Gila . . .

"Agave . . . Another species of Agave, or a very remarkable variety of the preceding [A. americana Linn.] was found in New Mexico, west of the Del Norte. It differs from

8. Sargent wrote in *Garden & Forest* in 1889: "Vauquelinia Torreyi Watson.—This Arizona tree was first described by Dr. Torrey (Emory's Rep., 140) as *Spiraea Californica*. Mr. Watson, in referring it to its right genus (*Proc. Am. Acad.*, xi., 147), dropped the specific name under which it was first described (for the reason, no doubt, that this tree has not been found within the actual limits of the State of California), and substituted for it the name of the first describer. Instances are not wanting in the annals of American botany of geographical specific names having been improperly applied; and unless all such names are to be corrected, it would be better to retain the earliest specific name for this plant, which would thus become *Vauquelinia Californica*."

This is the shrub or small tree known as the Arizona-rosewood, according to Kearney and Peebles' "Flowering plants and ferns of Arizona."

9. According to W. H. Brewer: ". . . In the report of the expedition Dr. Torrey enumerates one hundred and sixty species as especially worthy of mention, of which number less than thirty are cited as having been collected in California."

A. Americana in its much shorter and broader leaves, which are furnished with smaller marginal spines[1] . . .

"Chondrosium eriopodum . . . Abundant along the Del Norte, and in the region between that river and the waters of the Gila . . .

"Chondrosium fœneum . . . Uplands bordering the valley of the Del Norte . . ."

Torrey included with his list twelve fine plates illustrating the new genus *Baileya* and some of the new species and rarities in Emory's collection: *Dalea formosa, Fallugia paradoxa, Larrea mexicana, Zinnia grandiflora, Riddellia tagetina, Baileya multiflora, Arctostaphylos pungens, Fouquieria spinosa, Quercus Emoryi, Sesleria dactyloides, Ipomoea leptophylla,* and *Chondrosium foeneum.* I shall comment upon these plates later.

Following Torrey's list of plants is a letter from George Engelmann (dated "St. Louis, *February* 13, 1848") which serves as preface to his report upon Emory's *Cactaceae:*

". . . On the occasion of my report on the botany of Dr. Wislizenus's voyage, I have made a careful investigation of the cactaceæ, of which he brought home with him more than 20 species . . . I am now able to distinguish all the different genera of cactaceæ by their seed, and sometimes even the different sections of one genus. The small black shining seed sent me, belongs to a true *Cereus,* probably the plant which you mention under the name pitahaya, the larger opaque black seed is that of an *Echinocactus,* and the largest white seed is the seed of an Opuntia of the section *cylindraceæ.*

"I have ventured to describe some of your species from the drawing; my description, however, and the names given by me, must remain doubtful till we are able to obtain some more data to characterize the species. I have written it more for your information than for publication, but if you choose to append it to your published report, I have no objection to it . . ."

Engelmann advised some qualifications of Emory's statements and does not commit himself, irrevocably, about a single cactus. He enumerates twelve, proposing appropriate names for some: for the *Mammillaria* (collected October 18, 1846) the name *M. aggregata;* for another, "Rare," (collected October 26) *M. fasciculata;* for a third (collected November 4) *M. microcarpa.* One *Echinocactus* (collected October 26) he had already named for Wislizenus; for another (collected October 25) the name *E. Emoryi* was proposed. One *Cereus,* three feet high (collected November 21) had already been described as *C. Greggii,* from the collections of Gregg and of Wislizenus. For the giant cactus (figured and described by Emory in a letter) Engelmann

---

1. This plant was later described by Engelmann under the name *Agave Parryi.* Emory had observed it in the region of the copper mines (Santa Rita) on October 18. It is an extremely handsome plant and very abundant in the Pinos Altos to the north of Silver City, New Mexico, etc.

proposed the name *Cereus giganteus.*[2] For one *Opuntia* (undated), which was very abundant on the Del Norte and on the Gila, the name *O. microcarpa* was proposed; for another (collected October 28), common on the Gila, *O. violacea;* for another (October 22), *O. Stanleyi.* One (of November 3) was *O. arborescens,* already described from Wislizenus material. For another (of November 2), the name "O. Californica" was proposed; still another had been described from Wislizenus material as *O. vaginata.* "Nos. 13 to 15 are no Cacti." Period!! No. 13 was *Koeberlinia zuccarini,* collected in Mexico by Wislizenus and Gregg. No. 14 might be another species of the same genus. No. 15 was "entirely unknown" to Engelmann—might possibly be an *Agave* or an *Amaryllidaceous* plant.

Two plates, illustrating fourteen species (among these the three plants which were "no Cacti") are included.

The "Reconnoissance" includes more than a dozen delightful sketches which depict landscapes from the Santa Fe region westward, four fine drawings of Indians; two of plants (one of "Cereus giganteus," one of "Vegitation on the Gila" which includes four different cacti); and one of the "Gila trout" which, on October 20, Emory had noted was different from the Maine species. None mention the name of the artist, merely that of the lithographer, C. B. Graham. However the text of the "Reconnoissance" occasionally identifies "Stanly," Emory's draughtsman, with some of them.

The twelve fine drawings of plants which accompany the Torrey paper (Appendix 2) cite the same lithographer but, again, not the artist; they suggest in skill the work of Isaac Sprague but there is no indication that they were done by him.

The two plates accompanying Engelmann's letter (Appendix 2) bear no name, of either artist or lithographer. They are not to be compared with the extremely beautiful ones drawn by Paulus Roetter for Engelmann's later work, the "Cactaceæ of the boundary."

We know that John Mix Stanley, who gained repute as a painter of Indians, had joined Emory's party at Santa Fe on October 25, 1846. But his name appears rarely in the "Reconnoissance." Welcome, therefore, are comments by De Voto.

One appears in *The year of decision 1846* where, after paying tribute to the many men of the United States Army who ". . . at the head of small detachments, over three quarters of a century, traveled the American wilderness, making maps and recording observations of Indians, languages, religions, animals, trees, grasses, weather, rocks, ores, fossils, soils, and drainage . . . ," De Voto refers in particular to Emory and to Stanley:

"Even the *Dictionary of American Biography* can say of Lieutenant William H. Emory little more than that he served knowledge and diplomacy well, that his career was honorable for military virtues, and that his Civil War record was distinguished: a soldier's epitaph . . . He had [with him] . . . a handful of civilians, among them John

---

2. Although this was transferred by Britton and Rose to the genus *Carnegiea* in 1908, Engelmann's name is still used by some botanists.

Mix Stanley, a painter who would perhaps have a larger fame in indexes if the fire which burned down the Smithsonian Institution in 1851[3] had not destroyed all but five of the 151 paintings which he made of the West. It is without literary importance that this little group observed the weather and resources on the day's march and set them down, that Stanley's drawings of plants were the first made in this region, that Emory gave the region its first scientific scrutiny and made the first map, an exquisite map which is extremely useful still."

In *Across the wide Missouri* De Voto mentions Stanley a number of times and I quote one interesting paragraph:

". . . even if artists had been available who knew the West, books about it could not be counted upon to sell widely enough to cover the high cost of reproducing the pictures. Nuttall, Townsend, and other trained scientists beautifully sketched[4] the flora and fauna of the West, and nearly every Western expedition included someone capable of making accurate sketches of the scenery. But before 1850 a publisher of Western books was limited to woodcuts if he wanted to show a profit . . . True, there was lithography, but before 1850 only the national government was willing to pay for it. Of the four or five books which I have said could have visually instructed a Forty-Niner, three were government reports, Frémont's, Albert's [Abert's], and Emory's. They were all splendidly illustrated with lithographs, those in Emory beautifully drawn by John Mix Stanley, so beautifully drawn that some of them remain unsurpassed today."

According to Carey, writing in the *Dictionary of American biography*, William Hemsley Emory (1811–1887) had a distinguished career, not only during the Mexican War, but while commanding troops in the Indian Territory in 1861, when he ". . . was the only officer on the frontier who brought his entire command out of the insurrectionary country without the loss of a man. The troops thus saved from capture were of great importance beyond the consideration of numbers . . ." We are told also that, after considerable and honorable service during the Civil War, Emory ". . . commanded successively the Department of Washington, District of the Republican (1869–71), Department of the Gulf (1871–75), and was retired with the rank of brigadier-general, July 1, 1876, after a period of forty-five years of service . . ." Emory is referred to as ". . . a talented and skilful officer, always calm and dignified in bear-

3. This date is undoubtedly a slip; for, in *Across the wide Missouri* (p. 451), the date of the fire is correctly cited as 1865. The *Dictionary of American biography* states that Stanley's collection of paintings was ". . . deposited in the Smithsonian Institution in 1852. All but five of the pictures were destroyed by fire in 1865 . . ."

4. Perhaps De Voto refers to the sketches in Nuttall's *A journal of travels into the Arkansa territory* (1821). Nuttall mentions sketching the Cadron Hills and the copy of the book which I have seen includes five plates, evidently reproductions of pencil drawings—not, however, signed. These do not, to me, denote much ability as an artist but De Voto may have had others unknown to me in mind.

ing, courageous and firm. Though apparently stern in character he was really warm-hearted, sympathetic, and generous." To this I add that a general spirit of kindliness and of generosity may be read through the lines of his narrative. Interestingly enough, in 1838 Emory married a great-granddaughter of Benjamin Franklin.

Emory was to make further contributions to botanical knowledge when, from 1848 to 1853, he took part in the surveys along the boundary between the United States and Mexico.

Early in this chapter, I mentioned a collection of twenty-one issues of the Emory report which are in the library of the Arnold Arboretum. Associated therewith are three letters from Clarence W. Ayer, Librarian, Cambridge Public Library, to Charles E. Faxon, dated October 20 and 29, and November 4, 1910. From these it appeared that, on the breaking up of the "Whorf collection," these volumes had been presented to the Arboretum where, wrote Ayer, ". . . I . . . trust they will better serve the interests of science and fulfill the expressed wish of the donor by being kept together in the library of the Arnold Arboretum." Also associated with the collection are several pages of notes[5] which are explanatory of the differences between the different issues. Since two papers at least have been written upon the publication-dates of Emory's "Reconnoissance" I have felt that Mr. Whorf's notes might be of interest.

After the passage of more than forty years no one on the present staff of the Arboretum could tell me anything of Edward Henry Whorf's collection or of how he became interested in making the gift. On the chance that he might have been a Harvard graduate (correct), and that he might have had a son who went to Harvard (also correct), I telephoned Mr. Edward Webster Whorf (to me unknown) and discovered that he was a son of the donor and that his father was a nephew of Charles Edward Faxon, long associated with the Arboretum and the artist who made the drawings for the plates in *The Silva of North America*—the Faxon-Ayer letters were thus explainable!

The Edward H. Whorf paper is headed "Notes on the result of a collation of seventy-two copies of Emory's Report":

"While collating seventy-two copies of this book letters were assigned to the different issues and each copy was numbered. Letters A to M were assigned to Senate issues numbered 1 to 41 and N to Z to House issues numbered 1-31.

"Until the collating was nearly completed it was impossible to determine the exact order of the Senate issues or of the House issues, but the letters were assigned in what seemed to be the probable order. After completing the collating it was found that many copies to which letters had been assigned as separate issues were simply mixtures of two or more issues and such mixed copies were grouped and our letters assigned to each group.

"After all the copies obtainable had been collated the order of the two sets of issues

---

5. Of these notes there are two copies, one in longhand, one typewritten, and virtually the same; I shall cite from the typewritten copy which appears to be the final draft.

was determined with reasonable certainty, and Emory's manuscript notes in C 39 finally proved the order of the two distinct Emory texts and reversed what had at first been considered the probable order based upon the Emory notes in A 11.

"Senate C was the first issued and is the first edition. After the issue of Senate C the House issues printed from C forms and the Senate issues printed from A forms were being published at about the same time. Presumably House O came out before the Senate reprint A, and O is probably the second edition.

"Senate A came out while Emory was in Mexico and is probably the third edition. After that it seems impossible even to guess at the relative order of Senate and House issues.

"C and A are the two distinct Emory texts, all other issues having been printed mostly from those forms. But House P and QQ have some important changes in the re-set signatures 9 and 10.

"Only the House issues have the Abert and other reports[6] added, pp. 417-614.

"The large folding map, in cover pocket, was published with all the Senate issues, except the publisher's B and BB. It was not published with any of the House issues.

"The botanical plates were engraved at least twice by Weber, once by Graham and once by Endicott. In several cases the different engravings vary considerably, but they were used so indiscriminately in the various issues that it is impossible to tell anything about the order of the engravings from the order of the issues of the text, or vice versa.

"The principal variations in the botanical plates are in III, VI and VIII."

The issues in the Arboretum collection are described and differentiated as follows:

Senate Issues. Text

C. FIRST ISSUE. 30 Cong. 1 Sess. Executive, No. 7. 416 pp. Wash. Wendell and Van Benthuysen, 1848. Ordered printed, Dec. 16, 1847.

No. 39. (Arboretum Library.)

C is the original issue of the book as ordered by the Senate, Dec. 16, 1847, and was probably published before Emory went to Mexico[7] in April, 1848. After the Senate issue was printed the forms were used to print the House issues ordered by the House, Feb. 9, 1848.

6. The reference is to "Report of Lieut. J. W. Abert, of his examination of New Mexico, in the years 1846-'47" (*U. S. 30th Cong., 1st Sess., House Exec. Doc.* 4: No. 1, 417-516, 1848); the "other reports" are "Report of Lieut. Col. St. George Cooke . . ." and the "Journal of Captain A. R. Johnston . . ." It does *not* refer to the "Notes of Lieutenant J. W. Abert" which was issued as Appendix 6 of Emory's "Reconnoissance," in both Senate and House issues.

7. Included among the Whorf papers is a copy of a brief correspondence (June of 1905) between J. N. Rose, Associate Curator, Division of Plants, of the Smithsonian Institution, U. S. National Museum, and the War Department office, in regard to the period when Emory was absent in Mexico. A note from the Military Secretary states: "It is shown by the records that Lieutenant Colonel W. H. Emory, Maryland and District of Columbia Volunteers, Mexican War, joined his regiment at Jalapa, Mexico, in April, 1848, exact date not shown, and that he arrived at Fort McHenry, Maryland, from Mexico, July 17, 1848."

A. SECOND ISSUE. 30 Cong. 1 Sess. Executive, No. 7. 416 pp. Wash. Wendell and Van Benthuysen, 1848.

No. 11. Arboretum copy has copies of Emory's manuscript notes from the Rose copy.[8] (Arboretum Library.)

A was re-set with many changes and omissions, and was published while Emory was in Mexico, between April, 1848 and July 17, 1848.

After the issue was printed the type was double-leaded and a publisher's edition, BB, printed.

No. 38. (Arboretum Library.) This is a clean copy of A issue.

BB. THIRD ISSUE. New York, H. Long & Brother, 1848. 230 pp.

No. 1. (Arboretum Library.)

BB is a publisher's edition printed from the double-leaded A type. The early copies of BB, called B, have a part of one word dropped out on p. 167, which is supplied in the later BB copies.

D. FOURTH ISSUE. 30 Cong. 1 Sess. Executive, No. 7. 416 pp. Wash. Wendell and Van Benthuysen, 1848.

No. 12. (Arboretum Library.)

D is a mixture of A and C signatures, sig. 1 (having the t.p.) being always A, but few copies agreeing otherwise. Evidently a made up issue from left-over sheets of the two issues. As early as Jan. 17, 1849, this issue was coming out.

House Issues. Text

O. FIRST ISSUE. 30 Cong. 1 Sess. Ex. Doc. No. 41. 614 pp. Wash. Wendell and Van Benthuysen, 1848. Ordered printed, 9 Feb. 1848.

No. 3. (Arboretum Library.)

O is printed from Senate C forms, with heading of pp. changed from "[7]" to "Ex. Doc. No. 41.", except pp. 17-32 which have "[7]".

P. SECOND ISSUE. Same imprint, etc. 614 pp.

No. 2. (Arboretum Library.)

P is same as O except that signature 9 and 10 are re-set and some text is omitted in these signatures.

---

8. This copy bears on the flyleaf the following notation, signed by Edw. H. Whorf: "Boston, Aug. 20, 1905. All annotations in pencil are copied verbatim from Form A, #18. now in the possession of Mr. J. W. [N.] Rose, Associate Curator, U. S. Nat. Museum. The originals are in the handwriting of W. H. Emory." Also—and noted as from "[First Fly Leaf, Recto.]"—the notation: "This book was published while the Author was with his Regmt in Mexico. This corrected copy is presented by him to the Smithsonian. W. H. E."

Emory's corrections are many and appear to relate chiefly to plants and plant names. The Arboretum copy is inscribed with the name "R. Kimball" and the notation "Presented by Hon. I. P. Hale, U. S. S."

QQ. THIRD ISSUE. Same imprint, etc. 614 pp.

> No. 10. (Arboretum Library.)

> QQ is same as P except that sig. 2 is also re-set, and pp. 145-158 also have "[7]" at top of pages instead of "Ex. Doc. No. 41.".

Q. FOURTH ISSUE. Same imprint, etc. 416 pp.

> No. 1. (Arboretum Library.)

> Q is same as QQ except that pp. 129-144 have "[7]" instead of "Ex. Doc. 41", and pp. 145-158 have "Ex. Doc. No. 41" instead of "[7]". Also this issue has only the Emory 416 pp. like the Senate issues.

R. FIFTH ISSUE. Notes of Travel in California. 614 pp. Phila. George S. Appleton, 1849.

> No. 11. (Arboretum Library.)

> R is same text as O and was printed from same forms. In this copy pp. 159 and 160 are duplicated. In this issue the Gov't t.p. was cut out and a publisher's t.p. substituted.

BANCROFT, HUBERT HOWE. 1888. The works of . . . 17 (Arizona and New Mexico 1530–1888): 416, *fn.* 9, 417-419, 464.

BARNHART, JOHN HENDLEY. 1895. On the two editions of Emory's report, 1848. *Bull. Torrey Bot. Club* 22: 394, 395.

BREWER, WILLIAM HENRY. 1880. List of persons who have made botanical collections in California. *In:* Watson, Sereno, Geological survey of California. Botany of California 2: 556.

CAREY, WILLIAM F. 1931. William Hemsley Emory. *Dict. Am. Biog.* 6: 153, 154.

CLELAND, ROBERT GLASS. 1950. This reckless breed of men. The trappers and fur traders of the southwest. New York.

COVILLE, FREDERICK VERNON. 1896. Three editions of Emory's report, 1848. *Bull. Torrey Bot. Club* 23: 90-92.

DE VOTO, BERNARD. 1943. The year of decision 1846. Boston, 234, 360.

—— 1947. Across the wide Missouri. Illustrated with paintings by Alfred Jacob Miller, Charles Bodmer and George Catlin. Boston. 398.

EMORY, WILLIAM HEMSLEY. 1846. General Kearney and the Army of the West. [". . . extract of an unofficial journal of 1st Lieut. Emory . . ."]. *In: Niles Nat. Reg.* ser. 5, 21: 138-140, 174-175. [Reprinted from the newspaper *Union.*]

—— 1848. Notes of a military reconnoissance, from Fort Leavenworth, in Missouri, to San Diego, in California, including part of the Arkansas, Del Norte, and Gila Rivers. By W. H. Emory, Brevet Major, Corps Topographical Engineers. Made in 1846-7, with the advanced guard of the "Army of the West." *U. S. 30th Cong., 1st Sess., Sen. Exec. Doc. No. 7*, 1-126.

—— 1951. Lieutenant Emory Reports: A reprint of Lieutenant W. H. Emory's Notes of a military reconnoissance. Introduction and notes by Ross Calvin, Ph.D. University of New Mexico Press. Albuquerque. [Transcribed from 1848 edition published by H. Long & Brother, New York. The editor's introduction and footnotes are of special interest.]

ENGELMANN, GEORGE. 1848. [Letter from Engelmann to Emory, dated St. Louis, February 13, 1848, reviewing the Cactaceae collected on the Reconnoissance.] *In:* Emory, W. H., 1848. Notes of . . ., Appendix 2. 155-158. 2 plates. *Reprinted:* Trelease, William & Gray, Asa, *editors.* The botanical works of the late George Engelmann . . . Cambridge, Mass., 109-113. 2 plates.

―――― 1852. Notes on the Cereus giganteus of southeastern California, and other Californian Cactaceae. *Am. Jour. Sci. Arts ser.* 2, 14: 335-339. *Reprinted: Ibid.* 122-124.

GHENT, WILLIAM JAMES. 1931. The early far west. A narrative outline 1540–1850. New York. Toronto. 355.

PAXSON, FREDERIC LOGAN. 1924. History of the American frontier 1763–1893. Boston. New York. 359.

SARGENT, CHARLES SPRAGUE. 1889. Notes on some North American Trees.― VI. *Gard. & Forest* 2: 400.

―――― 1892. The Silva of North America. Boston. New York. 4: 59, 60, *fn.* 1.

STANDLEY, PAUL CARPENTER. 1910. The type localities of plants first described from New Mexico. *Contr. U. S. Nat. Herb.* 13: 145, 167, 170.

TORREY, JOHN. 1848. [Letter from Torrey to Emory, dated February 10, 1848 (p. 135), and letter from Emory to Torrey, dated July 22, 1847 (pp. 135, 136); these preface the "Appendix by Dr. Torrey" (his catalogue of plants which bears no title, pp. 137-155).] *In:* Emory, W. H. 1848. Notes of . . . Appendix 2. 135-156. 12 plates.

WARREN, GOUVENEUR KEMBLE. 1859. Memoir to accompany the map of the territory of the United States from the Mississippi River to the Pacific Ocean, giving a brief account of each of the exploring expeditions since A.D. 1800 . . . *U. S. War Dept. Rept. expl. surv. RR Mississippi Pacific* 11: 51, 52.

CHAPTER L

FENDLER COLLECTS PLANTS FOR FOUR MONTHS IN

THE NEIGHBORHOOD OF SANTA FE

IN 1836 there arrived on our eastern seaboard a young man of twenty-three, Augustus Fendler, a German, born at Gumbinnen in East Prussia. His efforts to acquire an education seem to have been hampered by family poverty, and the varied forms of remunerative work which he had undertaken had never been to his taste. He must have had somewhat the make-up of Thoreau, appreciating nature and not averse to solitude. From the east he went to St. Louis, where lamp making proved wearisome, and from there moved to New Orleans. The story of his life is supplied in "An autobiography and some reminiscences of August Fendler," edited by William M. Canby and published in *The Botanical Gazette* in 1885:

"Arrived at New Orleans, the talk about Texas induced F. to seek adventures still farther west. Embarked in a steamer and arrived at Galveston in January, 1839 . . . From Galveston he went to Houston, the capital of Texas. The government of Texas then granted to every emigrant a 'headright' of 320 acres of public land. F. applied for one and received it, but in order to have it selected and surveyed he was required, well armed, to join the surveyors, in order to strengthen their party against any premeditated attack from the wild Comanches . . . But, having no rifle, he could not join the surveying party, and hence had to leave his grant of land unselected."

For twelve months Fendler roamed the country alone, getting to the site of Austin. He eventually returned to Houston,[1] in time to observe the epidemic of yellow fever in the autumn of 1839.

"At last, dissatisfied with the country, nearly empty in purse, and broken down in health, he left for Illinois, where for some time he was engaged in teaching school . . . in 1841 . . . he was seized with an uncontrollable desire for the solitary life in the wild woods . . . for the independent life of a hermit . . . he came upon a small village called Wellington [Lafayette County] . . . on the banks of the Missouri river, three hundred miles above St. Louis. Here he learned that an uninhabited island . . . called Wolf's Island . . . was at his service . . . ."

He lived on this island for some time. In 1844 he made a visit to Germany: "At Koenigsberg he got acquainted with Ernst Meyer, Professor of Botany at the University, who first intimated to him that a certain number of sets of dried specimens of plants for the herbarium might be disposed of at a reasonable price, and advised him

---

1. Geiser mentions that he lived there "in obscurity." Also that he worked at market-gardening while in Texas.

to return to the Western United States to collect and send them on, for sale, to the Professor's address.

"On his return to . . . St. Louis, F. assiduously began to collect plants, and took the different species to Dr. Engelmann, who furnished him with their scientific names . . ."

Noting Fendler's "zeal" for this task, Engelmann recommended him to Asa Gray and thus the man's work as a collector of plants began.

The systematic botanists had been eager for a long time to have their collectors work at Santa Fe, and this desire was fulfilled[2] when Fendler got there on October 11, 1846, two months after the city's capture by the Americans. It was late in the season when he arrived but he remained through the winter and, from April to August of 1847 collected where conditions made it safe to do so. Standley describes the field:

"Santa Fe stands on a sandy plain which stretches westward for about 20 miles to the Rio Grande. To the east rise the high peaks of the Santa Fe Mountains, 3,600 to 3,900 meters high. The low foothills, like the mesa, are a poor collecting ground, covered with juniper, cedar, piñon, cactus, and low shrubs, with but a scanty mantle of herbaceous plants. But after one has gone 10 to 12 miles eastward into the mountains he finds a luxuriant vegetation. Santa Fe Creek, or River, as it is more commonly called, is a small stream which comes down from the mountains through Santa Fe Canyon and runs through the town . . . It was along this stream and on the plains about Santa Fe that Fendler got most of his plants. He went as far west as the valley of the Rio Grande . . . perhaps on account of the hostility of the Indians in 1847, he did not venture more than 12 miles or so into the mountains, while if he had gone farther he would have found hundreds of plants not in his collection. Fifteen or 20 miles away he would have found a subalpine flora that would have been rich in plants then undescribed."

Asa Gray, in 1849, published a famous paper on Fendler's collections: "Plantæ Fendlerianæ Novi-Mexicanæ: an account of a collection of plants made chiefly in the vicinity of Santa Fé, New Mexico, by Augustus Fendler; with descriptions of new species, critical remarks, and characters of other undescribed or little known plants from surrounding regions." Since the "surrounding" regions included such places as California, northern Washington and so on, Gray must have been using the qualifying word in an extremely broad sense! But by 1849 the botanists had ample material to provide them with corroborative or missing data upon puzzling genera and species and, moreover, were in a position to draw conclusions regarding plant distribution. The paper demonstrates that the botany of the trans-Mississippi west was, by this time, no simple[3] study.

2. William Gambel, sent out by Thomas Nuttall, had passed through Santa Fe in 1841 on his way to California, but seems to have done very little collecting of plants en route.

3. "Plantæ Fendlerianæ" offers an excellent example of why the Index Kewensis, citing as it does all literature references to new genera and species, is indispensable in a botanical reference library. For no one would turn to a paper on Fendler's New Mexican plants to find the first description of a plant from ". . .

Fendler's beautifully penned collection-records which are in the Gray Herbarium list 1,026 numbers, but do not provide an itinerary since the plants are arranged according to families and the dates and localities supplied are covering ones. For example his no. 1 bears the notation: "Santa Fe, 1st–19th July 1847. Moro River 15th August. Creek bottom land, near the water climbing to the tops of trees & shrubs Also the Rio de los Animas between Bent's Fort & Santa Fe. Fruit: 15th August–16th January."

Gray, in preparing his report, made notations on Fendler's manuscript, renumbered the plants when his own classification did not conform to Fendler's and entered the scientific names. He stopped his work in the midst of the *Compositae,* having enumerated a total of 462 species, so that more than one half of Fendler's numbers awaited treatment in the continuation which Gray contemplated writing. After Gray stopped his work, only a few plants on Fendler's manuscript are supplied with a name, and Gray's numerical notations soon cease.

About three pages of "Plantæ Fendlerianæ" are devoted to an outline of the collector's route from Fort Leavenworth, Kansas, to Santa Fe and his return journey to Fort Leavenworth. This data was supplied by Engelmann, and Gray notes that further information "as to the character and features of the country" was to be found in Wislizenus' "Memoir of a tour to northern Mexico . . ." in J. W. Abert's "Report of an Expedition to the Upper Arkansas and through the Country of the Camanche Indians, &c.",[4] and in Emory's "Notes of a military reconnoissance."

As outlined by Gray, Fendler's route westward from Fort Leavenworth to Bent's Fort, and thence by way of Raton Pass to the headwaters of the Canadian River and on to Santa Fe, was very similar to the one taken by Emory a few months earlier; and his eastward route after leaving Santa Fe was—although in reverse—much the one that Wislizenus had taken westward from the "Arkansas crossing" into extreme northeastern New Mexico and thence to Santa Fe. The two routes were alternates for the last portion of the Santa Fe Trail and, according to Gray, Fendler traveled both.

Canby tells us that Dr. Engelmann recommended Fendler ". . . to Dr. Asa Gray in 1846, during the war between Mexico and the United States. The latter being about to send troops to Santa Fe, New Mexico, Dr. Gray furnished F. with a letter of recommendation from the Secretary of War, by means of which he got free transportation for himself, his collections and luggage . . ."

Spokan and Coeur d'Aleine Mountains . . .", or of another from "Clear Water, on the Kooskooskee, Oregon . . ." collected by the missionary Spalding.

First published in 1895 under the editorship of Benjamin Daydon Jackson, under the direction of Joseph Dalton Hooker and with the assistance of other botanists, the book was in two volumes and bore the title: *Index Kewensis: an enumeration of the genera and species of flowering plants, from the time of Linnaeus to the year 1885 inclusive, together with their author's names, the works in which they were first published, their native countries and their synonyms.* By periodic installments the publication has now grown to a work of large size and of inestimable usefulness.

4. *See* p. 939: Abert, J. W. 1846. Journal of . . . (*U. S. 29th Cong., 1st Sess., Sen. Doc.* 8: No. 438).

With what troops he traveled I have been unable to discover. The date of his departure from Fort Leavenworth (August 10, 1846) and the date of his arrival at Santa Fe (October 11) suggest that it might have been with either the Second Regiment of Missouri Mounted Volunteers or with the Mormon Battalion which left Fort Leavenworth in broken contingents at about this time. De Voto states that the Mormon Battalion was ordered out from Fort Leavenworth on August 13 and that its advance party ". . . got to Santa Fe on October 9, the rest three days later." The arrival of the last contingent would, therefore, have been on October 11—the date of Fendler's first glimpse of the Santa Fe Mountains. But Gray states that Fendler went west by the Bent's Fort-Raton Pass route and De Voto states that ". . . the Battalion did not go near Bent's Fort. At the Arkansas crossing Smith took it down the shorter, thirstier route to Santa Fe, the Cimarron branch of the trail. Price's Second Missouri also took this route . . ."

I do not know, therefore, with what troops Fendler traveled.[5] Since Gray supplies all that is known of Fendler's trips to and from Santa Fe, and of his experiences while there, I shall quote from the "Plantæ Fendlerianæ":

". . . Mr. Fendler left Fort Leavenworth, on the Missouri, on the 10th of August, 1846, with a military train, he having been allowed by the Secretary of War a free transportation for himself, his luggage, and collections . . .

5. Since writing the above, Ewan has published his *Rocky Mountain naturalists* (1950). He states that James William Abert ". . . In 1846 . . . made a reconnaissance of New Mexico, leaving Fort Leavenworth, Kansas, on June 27, 1846, and returning in March, 1847, reaching Santa Fe via Santa Fe Trail. Augustus Fendler . . . set out with Abert." (p. 147)

And, again (p. 207): "It was evidently through the intervention of Asa Gray and George Engelmann . . . that permission was obtained for the transportation of Fendler and his luggage with the U. S. troops which took possession of Santa Fe, N. M., in 1846."

Kearny's "Army of the West" took Santa Fe in mid-August, 1846, and Fendler did not get there until mid-October. And, as far as Fendler's and Abert's dates are concerned, I find them to be as follows:

Abert left Fort Leavenworth on June 27, 1846, and reached Bent's Fort on July 29, remaining there for more than a month because of illness. He left Bent's Fort and forded the Arkansas River on September 9, crossed Raton Pass on September 17, and reached Santa Fe on October 2.

Fendler left Fort Leavenworth on August 10, 1846, and reached Bent's Fort on September 5. He crossed the Arkansas River (four miles above Bent's Fort) on September 25, and reached Santa Fe on October 11; Gray, as noted, states that he took the Raton Pass route.

These dates do not indicate that they traveled together, although they may have done so. Abert's account of his trip from Bent's Fort to Santa Fe does not mention Fendler by name although one incident might well refer to the botanist. It is found in the "Report of J. W. Abert, of his examination of New Mexico, in the years 1846—'47" (*U. S. 30th Cong., 1st Sess., House Exec. Doc.* 4: No. 41, 435). The date was September 12 and Abert was then two days out from Bent's Fort, while Fendler did not leave the fort until September 25. Abert, as I have had occasion to note elsewhere, was fond of quoting from the classics: "This afternoon a young German, who accompanied the ox wagons, entered my camp. I had seen him several times at Bent's Fort. On his approach, he greeted me with a salutation from Horace, 'quid agis, dulcessime rerum.' For some time I did not know in what language he had spoken, his pronunciation being so different from that of an American. He brought me a specimen of the horned lizard (agama cornuta) and a species of centipede."

"Mr. Fendler travelled the well-beaten track of the Santa Fé traders to the Arkansas, and then followed that river up to Bent's Fort, which he reached the 5th of September. On the 25th of September the Arkansas was crossed, four miles above Bent's Fort, and the westerly course was now changed to a southwestern direction. *Opuntia arborescens* was first observed in the barren region now traversed; and the shrubby *Atriplex* (No. 709) was the most characteristic and abundant plant . . .

". . . on September 27th, the base of the mountain chain was reached, which is an outlier of the Rocky Mountains, and attains in the Raton Mountains the elevation of eight thousand feet. West of these, in dim distance, the still higher Spanish Peaks[6] appear . . . Scattered pine-trees are here seen for the first time on the Rio de los Animos (or Purgatory River of the Anglo-Americans), which issues from the Raton Mountains . . . The sides of the Raton Mountains were studded with the tall *Pinus brachyptera,* Engelm. (831), and the elegant *Pinus concolor* (828).[7] Descending the mountains, the road led along their southeastern base, across the head-waters of the Canadian.

"On the 11th of October, Mr. Fendler obtained the first view of the valley of Santa Fé, and was disagreeably surprised by the apparent sterility of the region where his researches were to commence in the following season. The mountains rise probably to near nine thousand feet about the sea-level, two thousand feet above the town, but do not reach the line of perpetual snow, and are destitute, therefore, of strictly alpine plants. Their sides are studded with the two pines already mentioned, with *Pinus flexilis,*[8] &c.

"The Rio del Norte, twenty-five or thirty miles west of Santa Fé, is probably two thousand feet lower than the town . . . its peculiar flora is meagre . . .

"South and southwest of Santa Fé, a sterile, almost level plain extends for fifteen

6. Of Huerfano County, Colorado.

7. Nos. 831 and 828 are Fendler numbers. Engelmann's *P. brachyptera* is now *Pinus ponderosa,* the western yellow pine.

Since Fendler is credited with the discovery of "the elegant *Pinus concolor*"—actually *Abies concolor* (Gordon) Engelmann, the Colorado or white fir—I shall quote his manuscript record which indicates that he was a careful observer, indeed a collector after Gray's own heart:

"828. 1st November, 31st January, 4th March 1847. Santa Fe. In the higher part of the mountains about 5 or 6 miles East of Santa Fe. Especially on the Northern declivity of a sharp mountain ridge where they are to be found in great numbers with but very few other trees between them, while the southern declivity of the same ridge is occupied by a long leaved pine No. 831 to the exclusion of almost every other kind. The pistillate flower-branchlets are to be found on the top part of the tree, the staminate ones on the lower branches. Bark of the younger trees smooth and of a white color, on old trees only the summit and the branches are smooth and whitish. The terminal buds of the branches are furnished during the winter with a thick and transparent coat of a resinous substance. A kind of balsam or turpentine is elaborated in little cavities under the surface of the bark around the larger branches and the stem of the tree, by which the surface is raised into scattered nodulus. This turpentine is clear and of a very light color and has an agreeable odor. Trees 60-80 feet high, some of a very handsome pyramidal growth."

8. The limber pine had been discovered by Edwin James more than twenty-five years earlier and he had also described and named it.

miles, which offers few resources to the botanist[9] ... To the west and northwest of Santa Fé, a range of gravelly hills thinly covered with Cedar and the Nut-pine (830) offers a good botanizing ground in early spring. The valleys between these hills ... furnished some very interesting portions of Mr. Fendler's collection, and of Cactaceæ ...

"By far the richest and most interesting region about Santa Fé for the botanist ... is the valley of the Rio Chiquito (*little creek*) or Santa Fé Creek. It takes its origin about sixteen or eighteen miles northeast of the town, from a small mountain lake or pond, runs through a narrow, chasm-like valley, which widens about three miles from Santa Fé, and opens into the plain just where the town is built ... Most of the characteristic plants of the upper part of the creek and of the mountain-sides are those of the Rocky Mountains, or of allied forms; some of which ... have never before been met with in so low a latitude (under 36°).

"Mr. Fendler made his principal collections from the beginning of April to the beginning of August, 1847, in the region just described. At that time unforeseen obstacles[1] obliged him to leave the field of his successful researches. He quitted Santa Fé, August 9th, followed the usual road to Fort Leavenworth, which separates from the 'Bent's Fort road' at the Mora River, and unites with it again at the 'Crossing of the Arkansas.' The first part of the route from Santa Fé to Vegas leads through a mountainous, wooded country, of much botanical interest, crossing the water-courses of the Pecos, Ojo de Bernal, and Gallinas. From Vegas the road turns northeastwardly through an open prairie country, occasionally varied with higher hills, as far as the Round Mound ... The principal water-courses on this part of the route, all of which furnished remarkable species, were the Mora, Ocaté, Colorado (the head of the Canadian), and Rock Creek, all of which empty into the Canadian. Rabbit's Ear Creek and McNees Creek (the head-waters of the north fork of the Canadian) are east of the mountains altogether. From thence the Cimarron was reached, where the Cold Spring, Upper, Middle, and Lower Spring, and Sand Creek are interesting localities.[2] On September 4th, Mr. Fendler recrossed the Arkansas, and reached Fort Leavenworth on the 24th of that month."

A few extracts from Gray's letters to Engelmann are of interest at this point. They show why he thought so highly of Fendler, and they also indicate Gray's "driving capacity" as far as plant collectors were concerned. I cite volume and page in the Asa Gray *Letters* but omit dates:

9. Gray cites three *Opuntia*, a *Cereus* and the "Shrub Cedar" as the only plants, in addition to grasses, seen on these wide plains.

1. Undoubtedly the "... noncompliance with my repeated most fervent requests through Dr. Engelmann for an advance of some money ..." to which Fendler refers in a letter to Gray which I shall quote shortly.

2. Interesting to Gray, presumably, because of the fact that where a little moisture was present plants varied somewhat; interesting to travelers over this route because they were the only sources of water on an extremely arid route.

"... I wrote at once to Sullivant, telling him to forward fifty dollars to Fendler,—to take his pay in Mosses and Hepaticæ, and to give instructions about collecting these, his great favorites ... He is a capital fellow, and Fendler must be taught to collect Mosses for him ... Now Fendler has money enough to begin with. As soon as he is in the field, and shown by his first collections that he is deserving, I can get as much more money advanced for him, from other parties. If he only makes as good and handsome specimens as Lindheimer, all will be well. His collections should commence when he crosses the Arkansas; his first envoi should be the plants between that and Santa Fé, and be sent this fall, with seeds, cacti and bulbs, the former of every kind he can get ... Other live plants he had better not attempt now.

"His next collection must be at and around Santa Fé. But instruct him to get into high mountains, or as high as he can find, whenever he can. The mountains to the north of Santa Fé often rise to the snow-line, and are perfectly full of new things. But you can best judge what instructions to give him. We can sell just as many sets of plants as he will make good specimens of. But forty sets is about as many as he ought to make ..." (1: 341, 342)

"... Do you hear from Fendler? Hooker says that region, the mountains especially, is the best ground to explore in North America! There is a high mountain right back of Santa Fé. Fendler must ravish it." (1: 343)

"I got a parcel from New York ... containing ... a set of Fendler's from Santa Fé, up to Rosaceæ. The specimens are perfectly charming! so well made, so full and perfect. Better never were made ... I shall take them right up for study, and they are Rocky Mountain forms of vegetation entirely, so I can do so with ease and comfort ... If these come from the plains, what will the mountains yield? Fendler must go back, or a new collector, now that order is restored there.

"All Fendler's collection will sell at once,[3] no fear, such fine specimens and so many good plants. Pity that F. did not know enough to leave out some of the common plants, except two or three specimens for us, and bestow the same labor on the new plants around him." (1: 351, 352)

"... Now for Fendler himself. He ought to go back, and without delay. He has

3. Fendler's specimens did *not* sell at once! Canby mentions a Fendler letter written on August 11, 1853— or more than five years after Gray's—which "... states that many of the sets of his Santa Fe collections, which had been sent to Europe, were still unsold. Professor Gray's classic *Plantæ Fendlerianæ* has long since made these invaluable, and many botanists would gladly purchase them now."

I have yet to find a record indicating that a collector made a living wage from the sale of his "sets," or fascicles as they were called in botanical circles—the scheme seems to have been mainly advantageous to those who sent collectors into the field; not that they themselves profited thereby but their investment in the enterprise needed only to be small, if any; and it was customary for them to get the first (and presumably the best) set free of charge, doubtless as a recompense for determining the plants and distributing them to subscribers. How much effort they made to dispose of the sets apart from the notices which they inserted in botanical journals I do not know. They mention them in letters to their botanical friends certainly. But more than one collector records that his sets had not sold, and that long after he had been in the field.

gained much experience, and will now work to greater advantage. He makes unrivaled specimens, and with your farther instructions will collect so as to make more equable sets. If he will stay and bide his time he can get on to the mountains, and must try the higher ones, especially those near Taos.

"Let him stay two years, and if he is energetic he will reap a fine harvest for botany, and accumulate a pretty little sum for himself, and have learned a profession, for such that of a collector now is. Drummond made money quite largely.[4]

"I had rather Fendler would go north and west than south of Santa Fé. New Spain and Rocky Mountain botany is far more interesting to us than Mexican." (1: 355, 356)

Fendler did not return to Santa Fe. In the Gray Herbarium is a pitiful letter which he wrote Gray on July 25, 1848, or about ten months after his return from New Mexico. It was written from St. Louis and I quote it in its entirety:

"I have been very happy to learn from Dr. Engelmann that you were pleased with those specimens of my Santa Fe plants which have come to your hands. By your kindness and through your instrumentality I was enabled to start on my journey to Santa Fe in 1846.

"I also learned that you have expressed a desire to see me on a second expedition of this kind, but which desire I could not yet comply with for certain considerations.

"Not the dangers nor the risk of life, health, and property; not the many hardships and privations which are inseparably connected with such an undertaking are deterring me from entering upon it again;—on the contrary these botanical excursions were enough to make me passionately fond of herbarizing.

"It is rather the experience I have had since, which is of such a nature as to operate very unfavourably upon my mind. The noncompliance with my repeated most fervent requests through Dr. Engelmann for an advance of some money is also not much calculated to encourage me to a similar enterprise.

"When my pecuniary means at Santa Fe were nearly all exhausted, when I had to sacrifice one thing after another of my most necessary effects, to keep up a few days longer the scanty support of our lives, in order to collect something more of the vegetation of that region; I looked forward with the utmost confidence to those gentlemen in the East, who had induced me to go out to Santa Fe, and who would as I hoped, leave me not without assistance as soon as I should have returned with my collections to St. Louis. It was this hope that made me bear all the difficulties most cheerfully under the happy impression that the enjoyment of the fruits of my labour would soon compensate for all.

"But alas! it was to be otherwise.

"When I came back to St. Louis I had to assort about 17,000 plants, and to put them up in sets ready to be sent away. With this kind of work I was occupied till the beginning of April, during which time I could do nothing else to earn any thing to pay

---

4. If Drummond did so, it was certainly not during the sojourn in Texas with which I am familiar.

my current expenses, and I was therefore obliged to borrow money to keep from starving. Dr. Engelmann had been on such occasions so good as to advance little sums which amount now to about 80 Dollars (all that he was able to do).

"My brother who had accompanied me on my journey and assisted me as much as it was in his power, now found himself likewise sadly disappointed, and was for want of money obliged to enlist in the army.

"In regard to myself I also find how little I have gained after all by engaging in this botanical enterprise. By it I have expended about 200 Dollars of my own money; by it I gave up a business in which I was doing well, and in which I cannot at present engage again for want of means. To this may be added the sad prospect I have now before me, that by the time I shall receive some money, this money will be nearly all wanted to pay my debts which I had to contract during all this time.

"I have now endeavoured to show how I am situated here, how I am anxiously waiting day after day and week after week for some glad tidings from you; and I believe that these my simple statements will be sufficient to enlist your good will in my behalf.

"Very respectfully Your Obt Servant   Augustus Fendler."

A notation on this letter in Gray's handwriting reads: "Answered Aug 3d, enclosing check on C. River Bank for $100—as advance—" After paying his indebtedness of $80 to Engelmann, Fendler would have been richer by only $20 had Gray's "advance" reached him on July 25. But Gray did not get off his letter until August 3 and what Fendler's financial situation was when the money arrived is not disclosed.

Gray was undoubtedly busy. Not too busy, however, to plan for another trip after the second one to Santa Fe did not materialize. The new scheme is mentioned in Gray's letters. While the proposed region promised well, the lure of the gold fields was a trifle disquieting. The letter was written on January 24, 1849:

". . . In a few days I shall write to Marcy; send him the sheets of 'Plantæ Fendlerianæ,' and make a vigorous application for . . . aid . . . perhaps it might be almost as well for Fendler to go over with a party of emigrants directly to Mormon City . . . Fendler can do admirably well in that region, if he perseveres. But will he not take the gold-fever and leave us in the lurch? Will not living, etc., be very dear in Mormon City also? I fear it. I must leave much to your discretion. Only if you think Fendler has a strong tendency to gold-hunting (which few could resist) let him go. And afterwards, if he chooses to collect plants, very well. Few can withstand the temptation when fairly within the infected region, and we hear the Mormons have found gold also . . ." (1: 360, 361)

"I have just received from the secretary of war, Mr. Marcy, and inclose to you, what I think will procure all the facilities that Fendler can wish from United States troops. If, as I am informed, the secretary has no right to issue an order for rations to Fendler, he has certainly done the best thing by issuing a recommendation which will, if

the commander is favorably disposed, enable him to give all without any order . . ."
(1: 361)

Fendler started on this trip and Canby records the outcome in the "Autobiography":

"In the spring of 1849 F. started on another collecting expedition over the western plains. This time he intended to visit the Great Salt Lake region. To him the year 1849 proved most disastrous, for in crossing the plains he lost, in the Little Blue river, by a flood that came suddenly upon him, all his drying paper, besides many other things needful to his intended tour, as well as his principal means of transportation, so that he was forced to wait at Fort Kearney for a chance to return to St. Louis. Arrived at the latter place, he found that all his worldly goods, all his collections, all his books, and worse still, all the journals of his travels had been destroyed by the great conflagration that, at the same time, laid the best business quarter of St. Louis into ashes during his absence."

At the Gray Herbarium is a typewritten "Fendler Itinerary Compiled from specimen labels" by Dr. L. B. Smith and dated December 10, 1941. It indicates that, despite the unfortunate outcome of the 1849 journey, at least twenty-six specimens were saved. Starting from Fort Leavenworth, Kansas, Fendler evidently traveled across country by one of the several routes which, starting from the Missouri River, united on the Little Blue River not far from the present boundary between Kansas and Nebraska, and continued up that stream to Fort Kearny (starting-point for the Oregon-California trail) whence the road continued up the Platte River and westward. I quote the itinerary. The year was 1849:

June 6:	15 miles NW of Fort Leavenworth	Kansas
	17 miles NNW of Fort Leavenworth	"
June 7:	23 miles NNW of Fort Leavenworth	"
June 8:	30 miles NNW of Fort Leavenworth	"
June 10:	60 miles NW of Fort Leavenworth	"
June 11:	85 miles NW of Fort Leavenworth	"
June 12:	Big Nemahah River	Nebraska
June 13:	Little Nemahah River	"
	Big Nemahah River	"
June 18:	Big Sandy River (184 miles WNW of Fort Leavenworth)	"
June 19:	Big Sandy River	"
June 21:	Little Blue River	"
June 22:	Little Blue River to Big Platte River	"
June 23:	Little Blue River to Fort Kearney	"
June 24:	Fort Kearney (Big Platte River)	"
June 26:	Fort Kearney	"

June 30:	Fort Kearney	"
July 1:	Fort Kearney	"
July 6:	Little Blue River	"
July 7:	Big Sandy River	"
July 8:	Big Sandy River	"
July 10:	10 miles west of Big Blue River	Nebraska?
July 13:	Big Blue River and eastward to Fort Leavenworth	Nebraska or Kansas ?
July 20:	Wolf Creek to Fort Leavenworth	Kansas
July 22:	20 miles NW of Fort Leavenworth Independence Creek	"

The rest of Fendler's life was spent in regions beyond the locale of this story and I follow his activities no further. He died on the island of Trinidad in February, 1883, at the age of seventy. Asa Gray's biographical notice was commendatory:

". . . Fendler was a quick and keen observer and an admirable collector. He had much literary taste, and had formed a very good literary style in English, as his descriptive letters show. He was excessively diffident and shy, but courteous and most amiable, gentle, and delicately refined. Many species of his own discovery commemorate his name, as also a well-marked genus, a Saxifragaceous shrub, which is winning its way into ornamental cultivation." [5]

This was the genus *Fendlera*—the species *F. rupicola*—described in 1852 by Engelmann and Gray from a collection made in 1849 by Charles Wright. Of the genus Gray wrote Engelmann on February 23, 1852:

"I have carefully kept your flowering bit of Fendlera, ready to return it if Lindheimer does not get more, as I trust he will. It is the most interesting of North American genera, between Deutzia and Philadelphus, and shows plainly that both are saxifragaceous . . ." (2: 391)

A later collector, A. A. Heller, who was in the Santa Fe region in 1897, dedicated another genus to Fendler in 1898: *Fendlerella*—the species *F. utahensis*.

A multitude of specific names also bear him honor.

Standley includes a list of about one hundred and twenty-five plants which have been first described from collections made in the Santa Fe region—". . . most of them first collected by Fendler . . ." Of these an approximate fifty were described in "Plantæ Fendlerianæ"; others by Gray on later occasions; still others by such later men as Greene and Heller; still others by Standley himself, and so on.

Fendler had collected about Santa Fe for only a few months and without getting into the mountainous regions which offered the richest field for plants; but, in part

5. Wooton and Standley refer to this as a beautiful shrub and note: ". . . It has never been cultivated, so far as we can learn, but it is certainly as handsome as the commonly grown species of Philadelphus."

because he was in a virtually untouched botanical field, in part because of the fine quality of his collections, he left an indelible mark on the botanical history of New Mexico.

DE VOTO, BERNARD. 1943. The year of decision 1846. Boston. 322, 323, 327.

EWAN, JOSEPH. 1950. Rocky Mountain naturalists. Denver. 147, 207.

FENDLER, AUGUSTUS. 1885. An autobiography and some reminiscences of the late August Fendler. Edited by William Marriott Canby. *Bot. Gaz.* 10: 285-290; 301-304; 319-322.

GEISER, SAMUEL WOOD. 1948. Naturalists of the frontier. ed. 2. Dallas. 273.

GRAY, ASA. 1849. Plantae Fendlerianae Novi-Mexicanae: an account of a collection of plants made chiefly in the vicinity of Santa Fé, New Mexico, by Augustus Fendler; with descriptions of the new species, critical remarks, and characters of other undescribed or little known plants from surrounding regions. By Asa Gray, M.D. *Mem. Am. Acad. Arts Sci.* new ser. 4: 1-116.

———— 1884. Necrology, Augustus Fendler. *Bot. Gaz.* 9: 111, 112.

———— 1885. Augustus Fendler. *Am. Jour. Sci. Arts,* ser. 3, 29: 169-171. *Reprinted:* Sargent, Charles Sprague, *compiler.* 1889. Sci. Papers Asa Gray. 2 vols. Boston. New York. 2: 465-467.

———— 1893. Letters of Asa Gray. Edited by Jane Loring Gray. 2 vols. Boston. New York. 1: 341, 343, 355, 356, 360, 361; 2: 391.

SARGENT, CHARLES SPRAGUE. 1893. The Silva of North America. Boston. New York. 12: 123, *fn.* 7.

SPAULDING, PERLEY. 1909. A biographical history of botany at St. Louis, Missouri. *Pop. Sci. Monthly* 74: 240-243.

STANDLEY, PAUL CARPENTER. 1910. The type localities of plants first described from New Mexico. *Contr. U. S. Nat. Herb.* 13: 167-170.

SUDWORTH, GEORGE B. 1916. The spruce and balsam fir trees of the Rocky Mountain region. *U. S. Dept. Agric. Bull.* 327:33.

WOOTON, ELMER OTIS & STANDLEY, PAUL CARPENTER. 1915. Flora of New Mexico. *Contr. U. S. Nat. Herb.* 19: 300.

CHAPTER LI

PARRY ACCOMPANIES DAVID DALE OWEN ON HIS

GEOLOGICAL SURVEY OF WISCONSIN, IOWA AND

NEBRASKA

CHARLES Christopher Parry was born in England—at Admington, Glouces-
tershire—in 1823 but arrived in the United States—at Washington County,
New York—at the age of ten and remained in this country for the rest of his
life. After being graduated from Union College, Schenectady, where it is said that his
work in "medical botany" began, he took a medical degree at Columbia College and
then moved to Davenport, Iowa, where, according to Dr. Howard A. Kelly, he began
to practice his profession ". . . but soon discovered that his natural tastes ranged far
from disease and drew him to the treasures of wood and field. Thenceforward his life
story is interwoven with that of three of his spiritual kindred, Torrey, Gray and George
Engelmann, and presents, apart from the scientific side, a wonderful record of travel
and toil."

In 1878 Parry wrote of the beginnings of his botanical work thus:

"My earliest gatherings in the botanical field were begun in 1842, while residing in
the attractive floral district of North-Eastern New York, and continued more or less
actively for five years, while occupied in a course of medical studies . . . The last two
years of this period was especially memorable by being favored with the personal acquaintance of the distinguished American botanist, Dr. John Torrey, to whose assistance and encouragement, equally shared by nearly all active American botanists
of this generation, I am largely indebted for whatever success I may have attained.

"In the fall of 1846 I removed to Davenport, Iowa, and the season following (1847)
I was actively engaged in securing the flora of this district,[1] including a summer excur-
sion to Central Iowa, in the vicinity of the present State Capital, Des Moines, with a
United States land surveying party, under the charge of Lieut. J. Morehead." [2]

Jepson's account of Parry in the *Dictionary of American biography* mentions his
first important assignment, in 1848, when ". . . he served under David Dale Owen[3] in

1. In his "Biographical sketch of Dr. C. C. Parry," C. H. Preston notes that Parry "left a manuscript list"
of the wild flowers which he found about Davenport, "with the dates of finding."

2. Morehead does not appear in several important biographical dictionaries and I have been unable to
find—although it may exist—any account of his surveying work in the United States government documents. Nor is he mentioned in the index to the volume *Iowa* ("American Guide Series").

3. G. P. Merrill tells that David Dale Owen (1807–1860) was the third son of Robert Owen whose name
is associated with the colony at New Harmony, Indiana.

the geological survey of Wisconsin, Iowa and Minnesota . . .." Parry himself refers to it as Owen's ". . . geological survey of the North-West . . ." and states that he then made botanical collections ". . . along the course of the St. Peter's River [the Minnesota River] and up the St. Croix as far as Lake Superior. A list of the plants collected during this and the preceding season was included in Dr. Owen's report published in 1852."

Parry's "list of plants" was entitled "Systematic catalogue of plants of Wisconsin and Minnesota . . . made in connexion with the geological survey of the northwest, during the season of 1848." It appeared (Article V) in the Appendix to Owen's *Report of a geological survey of Wisconsin, Iowa, and Minnesota; and incidentally of a portion of Nebraska territory. Made under instructions from the United States Treasury Department,* which was published in Philadelphia in 1852—one volume of text and another of "Illustrations to the geological report . . ." This was an amplification and re-arrangement of a report which Owen had published in 1848.[4] It contained reports on specific regions written by J. W. Norwood, "Colonel Whittlesey," and B. F. Shumard, and a memoir on fossil mammals and reptiles by Joseph Leidy. Parry's name does not appear in any of these and I do not know to which contingent he was assigned. However, Owen's introduction to the volume mentions his presence with the survey:

"Being desirous to collect as much general scientific information as possible . . . I instructed the members of the corps, when not otherwise engaged, to record observations, and preserve specimens in those departments of natural history in which they were most proficient. Accordingly, Dr. Parry, who has a good knowledge of Botany, has reported to me his observations made in this branch of science, on the St. Peter's River, and the country lying between the Mississippi River and Lake Superior . . ."

Owen then records "a few of the general results, referring the Department for further information to Dr. Parry's own report."

"Where there is a lithological as well as a palæontological passage from one geological formation to another, there is a simultaneous change in the botany of the country. This is especially observable in the influence of the trap ranges. The vegetation superincumbent on that formation is so marked, that it may often serve to detect it when the rocks themselves are hidden from view.

"The drift-deposits of the St. Peter's support a peculiar growth, among which the following are the most striking: *Castilliga sessiliflora, Psoralea esculenta* (Pomme de prairie, or Bread root), *Oenothera surrulata, Oxytropis Lambertii, Lygodesmia junsea, Orthopogon oligostrachyum.* These contrast strongly with the plants which char-

---

4. "Report of a geological reconnoissance of the Chippewa land district of Wisconsin; and, incidentally, of a portion of the Kickapoo country, and of a part of Iowa and of the Minesota territory, made under instructions from the United States Treasury Department" (*U. S. 30th Cong., 1st Sess., Sen. Exec. Doc.* 7: No. 57, 1-72). Contains a geological report by J. G. Norwood, M.D. (73-129) and an Appendix (131-134).

acterize the drift-deposits occupying the height of land between the Mississippi and Lake Superior, which are *Pinus Banksiana, Vaccinium tenellum* (whortleberry), *Gaultheria procumbens* (wintergreen), and some species of *Lycopodiums,* proving a decided difference in the two regions, both in the composition of the beds of drift and the soil derived therefrom."

After a paragraph concerned with the flora of the shores of Lake Superior, Owen turns to still another aspect of Parry's work:

"Dr. Parry was instructed to collect as much information as possible with regard to the economical and medicinal applications of plants, used by the Indians. Several of their most important native articles of food, as he justly remarks, are found in the regions where we might least expect to find the means of subsistence; thus, the wild rice fringes the innumerable lakes and rivers of this northern Indian country, the cranberry delights in the irreclaimable marshes and bogs, and the huckleberry flourishes in the barren ridges. Most of the common articles of diet in use among the Indians, Dr. Parry has been able to refer both to their botanical terms and Indian names.

"On account of the peculiar interest attached to the cryptogamia in connexion with geology, I requested Dr. Parry to pay particular attention to that class of plants. He informs me that he has collected thirty-eight species of Ferns, and the allied orders Equisitacea and Lycopodiacea.[5]

"All doubtful specimens have been referred to Dr. Torrey, of New York, especially in the classification of grasses and sedges. The mosses collected have been submitted to Mr. Wm. S. Sullivant, of Ohio."

Parry's "Systematic catalogue"—sixteen quarto pages of *very* small type—includes no new species; it is merely a list of plants (already described) presented in botanical sequence; following the botanical name is the locality, usually a very general one, as well as the month of collection; then, information as to the uses of the plant and so on. Under the divisions—Exogens, Endogens and Ærogens[6]—a single paragraph is devoted to each family and its genera and species. Although saving of space, this form of presentation requires a good knowledge of botanical classification, is hard on the eyes, and at times is very confusing. The foreword to the enumeration is the interesting part of Parry's paper and, although it repeats some of Owen's comments quoted

5. The "Equisitacea" are the familiar plants called horsetails, common in moist places. The "Lycopodiacea" are the club mosses now much used (or much too much used) in Christmas wreaths and decorations; from their ripe spores a smooth, almost silky, dusting powder (Lycopodium powder) was once made and was considered a very superior article with which to powder babies! I am told, on good authority, that this has not been heard of in pharmacies for at least twenty years!

6. Is it, or is it *not,* clarifying to explain that "Exogens" are Dicotyledons (or plants whose stems form a ring of vessels near the surface); that "Endogens" are Monocotyledons (or plants which have their vessels scattered through the substance of the stem); and that—according to Parry's interpretation—"Ærogens" are the equisetums, the ferns and the mosses (or plants destitute of stamens, pistil and true seed) which are commonly called Cryptogams?

above, I shall cite therefrom. Parry states that his list embodies the observations made on the Owen "Geological Survey of the Northwest" of 1848, as well as "personal observations" made during the season when he worked with Morehead "in the State of Iowa," this being "properly comprised within the District of the Northwest . . . The precise region of country covered by these observations will be sufficiently indicated by the subjoined localities . . ."

It would be impossible (and useless) to attempt to reconstruct even a month-by-month itinerary from Parry's cited localities and occasional dates; but these indicate that he spent considerable time on the Minnesota (or St. Peter's) River, on the "Upper Mississippi," in central Iowa, and so on. Other plant collectors had preceded him in these regions: David Bates Douglass (with Schoolcraft in 1820) had collected near the headwaters of the Mississippi; also Douglass Houghton (on another Schoolcraft expedition, in 1832); Thomas Say (traveling with Long in 1823) had made a few collections on his way up the Minnesota River and down the Red River of the North, to Pembina, North Dakota; Karl Andreas Geyer (botanist in the employ of Nicollet in 1838 and 1839) had made a collection between "the Mississippi and Missouri Rivers" —as Torrey's title to the "Catalogue" of Geyer's plants explains: the country from Lake Superior westward into all these regions must have included many of the same plants, for we find more than one species recorded by Parry both from "St. Croix" (in the Wisconsin area) and from "St. Peter's" (far to the west).[7]

In the introductory words to his "Systematic catalogue" Parry states that "The order followed is the *Natural System* and the authority used, *'Torrey and Gray's North American Flora,'* and *'Gray's Botany of the Northern United States.'* " The author first discusses the relationship between the plants and the geology of certain regions, noting that this ". . . gives to the botany of a new country its chief interest, and makes a *suite* of native plants valuable *portable indices* of the country they inhabit, of its agricultural capacities, climate, and external features, affording a ready means of comparison or contrast with other countries. May they not, when enlarged experience has traced with more accuracy these relations, and especially when we keep in view the principle so much insisted on in geology, viz., to depend more on the *grouping* of specimens, and drawing nice distinctions, than in isolated examples,—may they not take the same rank to agriculture that fossils do to geology?

"This principle always has been in general application. By it the farmer naturally judges of the fertility or barrenness of unploughed fields, while to an experienced botanist, a complete suite of the plants of any country would convey a greater amount of interesting information, and impart more definite notions of a country, than can be drawn from any single source."

---

7. Persons interested in the flora of the regions mentioned might perhaps find, among the Parry collections and records preserved at Ames, in Iowa State College, some additional data thereon.

Parry then refers to the economic and medicinal uses to which the Indians put certain plants and ends with a discussion of the content of his paper and with various acknowledgements:

". . . Of the native vegetable productions of this region, several of the most useful . . . are connected with those features of the country which seem least desirable. Thus the excellent *cranberry* occupies its irreclaimable marshes; the delicious *huckleberry* its barren ridges; while the staple *wild rice* edges its innumerable lakes . . .

"With regard to the medicinal articles used, my information is less important, due . . . to . . . the difficulty of obtaining accurate information. Medicine, in the mind of the Indian, is always connected with superstitious observances . . . blended with his religious notions . . . an air of mystery is thrown over the subject, combining to render reliable information . . . difficult to obtain, and . . . good for nothing when obtained . . . I have contented myself with a single specification of alleged virtues, without taking the trouble to classify them."

"Particular attention has been given to the class of ferns, from their more intimate relation with geology, of which, including the allied orders, *Equisetaceæ* and *Lycopodiaceæ,* thirty-eight different species have been observed, including some of much interest.

"The class of *forest trees* having been designed for a special report,[8] they are merely included in their proper natural order . . .

"The number of plants comprised in this list is seven hundred and twenty-seven, included in one hundred and six natural orders; many of these have never before been referrd to this region.

"I am indebted to the distinguished botanist, Dr. John Torrey,—and what American botanist is not?—for the authentication of my doubtful specimens, partieularly in the class of *grasses* and *sedges.*

"My acknowledgments are also due to Mr. William S. Sullivan[t], of Ohio, for labelling my entire collection of *mosses* . . ."

In 1878 Parry presented to the Trustees of Davenport Academy of Natural Sciences, for "deposit in the botanical room of the Academy," his many collections and records. W. H. Leggett states that his botanical collections there deposited included ". . . thirty thousand specimens, the labor of thirty-six years." In making this gift Parry took occasion to refer briefly to another assignment which began in the year 1849:

"In 1849 I was appointed botanist to the Mexican boundary survey, going by way of the Isthmus of Panama to San Diego, California, which latter place was reached in July. In September of the same year I accompanied an astronomical party to the junction of the Gila and Colorado rivers, returning to San Diego in December. The important collections of this season were unfortunately lost in crossing the Isthmus of

---

8. I have found no record of such a report.

Panama while in charge of the late Gen. A. W. Whipple, being probably involved in a disastrous fire while stored in Panama awaiting transportation.

"In the subsequent year, 1850, this loss was partially made up by somewhat extensive collections in the vicinity of the Southern Boundary line, and including a land trip up the coast as far as Monterey."

Despite the fact that Parry's trip from San Diego to the confluence of the Gila and Colorado rivers took place between September and December of 1849—or within the last months of the 1840–1850 decade—I shall not discuss it here,[9] for it marks the beginning of a new era in botanical collecting in the trans-Mississippi west: the era of many government surveys, in the present instance the ones along the United States-Mexican boundary. It was, I believe, upon Parry's accomplishments with these surveys, and upon his later work in Colorado, that his fame as a plant collector is largely based.

By late 1849 the route into southern California by way of the Gila River which, when Emory followed it with Kearny's "Army of the West" in 1846, had been virtually untraveled, was already being used by the emigrants. One group which Whipple's party encountered near the mouth of the Gila included an interesting personage: "Arrived Colonel Collyer, collector of the port of San Francisco, escorted by Captain Thorne, with thirty dragoons. Under their protection is also a party of emigrants, commanded by Mr. Audubon the younger,[1] naturalist . . . this party . . . was suffering for the want of provisions."

The next day, October 19, Whipple recorded an accident—the individual involved would appear to have been with still another group of emigrants: "Brevet Captain Thorne, son of Mr. Herman Thorne, of New York, while superintending the transportation of his party across the Rio Colorado, just below the junction of the Rio Gila, was thrown into the river by the upsetting of his heavily-laden boat, and was drowned. The current of the river was so rapid that all exertions—even those of the Yuma Indians, the best swimmers in the world—were unavailing . . ."

9. The records of the journey appeared in a number of government documents:

Parry's duties are mentioned in William Hemsley Emory's "Report on the United States and Mexican Boundary Survey made under the direction of the Secretary of the Interior" (*U. S. 34th Cong., 1st Sess., Sen. Exec. Doc.* 20: pt. 1, No. 108, 4, 1857),—in chapter 1, Emory's "Personal account."

The route is described in Amiel Weeks Whipple's "Extract from a journal of an expedition from San Diego, California, to the Rio Colorado, from September 11 to December 11, 1849" (*U. S. 31st Cong., 2nd Sess., Sen. Exec. Doc.* 3: No. 19, 2-28, 1851).

Parry himself wrote a few pages about the trip—"Reconnoissance to the mouth of the Gila River, from San Diego, California, September 11 to December 10, 1849"—which were appended to the "Report of Lieut. Michler" which was included in the Emory "Report" cited above (chapter 7, 125-130).

1. John Woodhouse Audubon is said by Geiser to have ". . . returned to Texas in March of 1849 for a brief trip through the Rio Grande valley at Brownsville en route to California." Apparently he was reaching his destination on October 18, when Whipple records his presence.

In late 1849 Parry entered few regions which had not already been visited by collectors of plants: San Diego, point of departure for the journey, had been visited by Menzies (November 27–December 8, 1793), by Thomas Coulter (probably in late May or early June, 1832), by Nuttall (for some three weeks in April and May, 1836) and by Hinds (October, 1839). Coulter had also reached the confluence of the Gila and Colorado rivers, and Emory had traveled from that same point westward across southern California. The time was rapidly approaching when—in order to discover new plants—collectors must deviate from "the beaten track" and do intensive work in untouched regions. That this approach still offered much that was yet unknown is evidenced by the many new plants discovered in the course of the surveys of the 1850's.

After describing the fields in which he had worked, Parry mentioned the results of his labors and the disposition of his collections:

"From all these various sources collections, more or less complete, have accumulated on my hands, the great bulk being fortunately distributed far and wide to the different herbaria of America and Europe. An active correspondence with the principal American botanists during the past thirty years has added largely, in the way of exchanges, to the material for illustrating Western American Botany. Hoping . . . to see it safely deposited in some scientific institution in the West, where it properly belongs, I gladly avail myself of the invitation extended to me by the Trustees of the Davenport Academy of Natural Sciences."

Of Parry's overall accomplishment as a botanist, Sargent had this to say:

"No other botanist of his generation explored so many unexplored fields in North America or revealed so many undescribed North American plants, *Pinus aristata, Pinus Torreyana, Pinus Parryana, Picea pungens,* and *Picea Engelmannii* being among the trees which he added to our silva.

". . . one of the peaks of the Snowy Range of Colorado bears the name of this indefatigable and successful explorer, and *Parryella* of Gray . . . reminds American botanists how greatly they are indebted to his zeal, industry, and intelligence. His herbarium, gathered in the wanderings of forty-eight years, and containing duplicates of his discoveries, has been acquired by the Agricultural College of Iowa."

CANDOLLE, ALPHONSE DE. 1880. La phytographie; ou l'art de décrire les végétaux considérés sous différents points de vue. Paris. 439.

GEISER, SAMUEL WOOD. 1948. Naturalists of the frontier. ed. 2. Dallas. 270.

JEPSON, WILLIS LINN. 1934. Charles Christopher Parry. *Dict. Am. Biog.* 14: 261, 262.

KELLY, HOWARD ATWOOD. 1929. Some American medical botanists commemorated in our botanical nomenclature. New York. London. 180-186.

LEGGETT, WILLIAM HENRY. 1878. [Parry's herbarium.] *Bull. Torrey Bot. Club* 6: 280.

MERRILL, GEORGE P. 1934. David Dale Owen. *Dict. Am. Biog.* 14: 116, 117.

OWEN, DAVID DALE. 1852. Report of a geological survey of Wisconsin, Iowa, and Minnesota; and incidentally of a portion of Nebraska territory. Made under instructions from the United States Treasury Department. By David Dale Owen, United States Geologist. 2 vols. Philadelphia.

PARRY, CHARLES CHRISTOPHER. 1852. Systematic catalogue of plants of Wisconsin and Minnesota, by C. C. Parry, M.D., made in connexion with the geological survey of the northwest, during the season of 1848. *In:* Owen, D. D. 1852. Report of a geological survey of . . . 606-622.

——— 1878. [Letter] To the trustees of Davenport Academy of Natural Sciences. *Proc. Davenport Acad. Nat. Sci.* 2: 279-282.

PRESTON, C. H. 1897. Biographical sketch of Dr. C. C. Parry 1823–1890 . . . together with a list of papers published by Dr. C. C. Parry, prepared by Mrs. C. C. Parry. *Proc. Davenport Acad. Nat. Sci.* 6: 35-52.

SARGENT, CHARLES SPRAGUE. 1895. The Silva of North America. Boston. New York. 7: 130, *fn.* 2.

CHAPTER LII

AFTER A DISASTROUS ATTEMPT TO CROSS THE

MOUNTAINS OF COLORADO IN WINTER, FRÉMONT

REACHES THE JUNCTION OF THE GILA AND COLO-

RADO RIVERS BY WAY OF NEW MEXICO, MEXICO

PROPER AND SOUTHERN ARIZONA

JOHN Charles Frémont left no published day-by-day record of his fourth expedition, the last which he made before 1850. On this occasion he went of his own initiative, not as an army officer under government orders. His *Memoirs of my life,* published in 1887, ends before the journey of 1848–1849, but the included map shows the route followed in those years—shows it especially clearly from the time of his arrival in Taos, New Mexico, until he reached California, where our interest in his story ceases.

The first weeks of the expedition were disastrous ones. On arrival in Colorado in November, 1848, he was warned, both at Bent's Fort and at Pueblo, that snow was already exceptionally heavy, and common sense should have suggested that more was to be expected; but Frémont adhered to his intention of crossing the Rocky Mountains of southwestern Colorado instead of taking one of the alternate routes which, although slightly slower, promised greater safety for men and animals at that season. He had crossed the Sierras of California in 1844 despite warnings at Fort Vancouver that the season was too late and apparently was convinced of his own ability to accomplish the impossible. Whether his guide, "Old" Bill Williams, was responsible for the choice of route which precipitated the immediate disaster seems never to have been settled, but certain it is that the original mistake was Frémont's.

The Century Magazine published in 1891 portions of the diary of a member of the party which affords the only detailed and, in a sense, the only disinterested record written by a member of Frémont's party. It was entitled "Rough times in rough places. A personal narrative of the terrible experiences of Frémont's fourth expedition" and was ". . . made up of the records and diary of a member of the party, left at his death, and never before published." The writer was Micajah McGehee.[1] His brother, C. G. McGehee, of Woodville, Mississippi, was responsible for its publication. He omitted

---

1. He was the great-great grandfather of Stark Young, whose recent book *The Pavilion,* a record "Of people and times remembered, of stories and places," includes more than one reference to his ancestor.

some of the earlier portions of the diary as not "unique," and begins the citations at Pueblo on the Arkansas River.

After the Colorado episode was over Frémont supplied his wife and his father-in-law, Senator Thomas Hart Benton, with his side of the picture—in letters which are included in John Bigelow's *Memoir of the life and public services of John Charles Frémont* (1856). Mrs. Frémont reprinted those which she received in *A year of American travel* (1878). Except for the headings these letters include few dates and, in view of what transpired, dates are of no particular importance.

Although it is inconceivable, reading of what happened, that any collections of any sort should have been made during the first part of the journey, McGehee records that, the following spring of 1849, ". . . Bill Williams and Dr. Kern, with a company of Mexicans, went back into the mountains to recover some of the most valuable of the property[2] left by us, and were attacked and killed, either by the Indians or by the Mexicans who went out with them, we never could ascertain which." In view of the annihilation of this search party—which seems to have been moved by a superhuman interest in scientific pursuits—it is unlikely that any "specimens in natural history" were recovered. The "few plants" mentioned by Torrey in his "Plantæ Frémontianæ" may have been gathered later in the trip:

"There was still another journey to California made by that zealous traveller; the disastrous one commenced late in the year 1848. Even in this he gleaned a few plants, which, with all his other botanical collections, he kindly placed at my disposal."

The literature cited above seems to have represented the main source of information upon the early portions of the journey as it is described by Frémont's biographers, F. S. Dellenbaugh (1914) and Allan Nevins (1939). However, chapter 22 of Nevins' *Frémont. Pathmarker of the west,* includes some seven paragraphs from a hitherto unpublished source: "Years later, in preparation to write the second volume of his *Memoirs,* never published, Frémont jotted down some rough notes upon the adventure which are still kept in his manuscript." [3] These "rough notes," which offer little more than a check against the route given on the Frémont map and mentioned in his letters, begin at St. Louis and end at the junction of the Gila and Colorado rivers.

Frémont reached what is now the southernmost point of the boundary between Arizona and California in the spring of 1849, and my story of his journey ends at that point.

2. Frémont's biographer Nevins includes in "the property," ". . . Edward Kern's collection of specimens of natural history . . ."

3. Dellenbaugh had noted: "General Frémont did not complete the second volume of his *Memoirs,* but from his notes Mrs. Frémont and their younger son compiled it. The manuscript was sent to the Chicago firm that was to bring it out, but it was never printed. A rough draft is extant but I have been unable to consult it . . ." Whether the "rough draft" to which Dellenbaugh refers included the "rough notes" published by Nevins I do not know.

From St. Louis Frémont had proceeded to the Missouri frontier where he made his final preparations for the journey. He left Westport Landing, near Independence, on October 21, 1848. He then crossed Kansas following the Kansas River and reached Bent's Fort, Colorado. From there he wrote Benton on November 17 of what had transpired between Independence and that point:

"In order to avoid the chance of snow-storms upon the more exposed Arkansas road, I followed up the line of the Southern Kansas (the true Kansas River) and so far added something to geography . . . We find the Valley of the Kansas affords by far the most eligible approach to the Mountains. The whole valley soil is of very superior quality, well timbered, abundant grasses, and the route very direct. This line would afford continuous and good settlements certainly for 400 miles, and is therefore worthy of consideration in any plan of approach to the Mountains."

At Bent's Fort Frémont found that two of his old guides—Christopher, or "Kit," Carson and Thomas Fitzpatrick—were otherwise employed:

"We found our friend, Major Fitzpatrick, in the full exercise of his functions at a point about thirty miles below, in what is called the 'Big Timber,' [4] and surrounded by about 600 lodges of different nations . . . He is a most admirable agent, entirely educated for such a post . . . knowing how difficult this Indian question may become, I am particular in bringing Fitzpatrick's operations to your notice . . ."

Unable to obtain these experienced men, he engaged the services of Bill Williams,[5] also experienced but peculiar, and later wrote his wife that ". . . the error of our journey was in engaging this man . . ."

Bent's Fort was situated on the Arkansas River, about midway between La Junta, present Otero County, and Las Animas, Bent County. In the letter just quoted, Frémont optimistically described his future plans:

"We will ascend the Del Norte [the Rio Grande] to its head, descend on to the Colorado, and so across the Wahsatch mountains and the basin country somewhere near the 37th parallel, reaching the settled parts of California, near Monterey. There is, I think, a pass in the Sierra Nevada between the 37th and 38th, which I wish to examine . . . I think that I shall never cross the country again, except at Panama . . . Should we have reasonable success, we shall be in California early in January, say about the 8th . . ."

From Bent's Fort Frémont moved to Pueblo, a "mere hamlet," situated near the mouth of Fountain River (his "Fontaine-qui-bouit"), Pueblo County; he had last been there on June 29, 1844. Despite warnings from weatherwise, experienced trap-

---

4. On an earlier journey Frémont had mentioned camping, on June 14, 1843, ". . . on a little creek in the valley of the Republican, 256 miles by our travelling road from the mouth of the Kansas . . . at an elevation of 1,520 feet. That part of the river . . . is called by the Indians the *Big Timber*."

5. For William Sherley Williams (1787–1849) *see* Cleland, Robert Glass. 1950. *This reckless breed of men. The trappers and fur traders of the southwest.* New York.

pers, the party set out. According to McGehee they entered "the Rocky Mountains"—actually the Sangre de Christo range—on November 26 and crossed them through Roubidoux Pass; Frémont's map shows the passage into the San Luis Valley, Alamosa County. Nevins states that "They reached the Rio Grande at about the site of the present town of Monte Vista . . ." in Rio Grande County. The Continental Divide lay ahead. McGehee states that they found on December 17 that they could force their way no farther through "the main range of the Rocky Mountains proper."

Just where the crossing was attempted is not clear, nor where the party became, definitely, lost. But, when it was finally decided to make for safety, they were, according to McGehee, at an elevation of eleven thousand feet, the temperature was twenty degrees below zero, the snow four to thirty feet deep, and ". . . nearly the entire band of our one hundred mules had frozen to death." Not until December 22 were they even able to start back and head for the Rio Grande valley which they had left and down which they intended to make for the settlements.

Frémont wrote his wife that he had spent Christmas Day reading Blackstone! McGehee records the more important fact that on December 25 a party of four (selected from volunteers) was delegated to press on ahead and bring back relief to a designated point in the Rio Grande valley. One of those chosen was "Creutzfeldt," whom Nevins refers to as the botanist of the expedition. The name is spelled "Croitzfeldt" by McGehee and by Dellenbaugh, but Nevins' spelling[6] is undoubtedly the correct one. Bill Williams served as guide. Frémont was of the opinion that relief would arrive in sixteen days and, after moving camp to the rendezvous, they waited. Men were beginning to die of starvation and exposure. Wrote McGehee: ". . . one of the party, Proue, froze to death beside the trail; we passed and repassed his lifeless body, not daring to stop long enough in the intense cold to perform the useless rite of burial." After waiting the time allotted for the arrival of relief, Frémont and four men started on to see what had happened or to get aid (he wrote that he regarded this decision "as an inspiration"). According to McGehee, "He left an order, which we scarcely knew how to interpret, to the effect that we must finish packing the baggage to the river, and hasten on down as speedily as possible to the mouth of Rabbit River where we would meet relief, and that if we wished to see him we must be in a hurry about it, as he was going on to California."

6. This is the same "F. Creutzfeldt" who, in the capacity of botanist, accompanied Captain John W. Gunnison's ill-fated expedition into Utah in 1853; the party consisted of twelve men and Creutzfeldt and Gunnison were among the eight who were massacred and mutilated beyond recognition by a party of Piute Indians. The tragedy took place not far from Sevier Lake, no great distance from present Hinckley, Millard County; a monument now marks the spot. Frederick Brendel noted in *The American Naturalist* (1880) that Torrey and Gray, in volume two of their *Flora*, ". . . reported on the collection of plants made by F. Creutzfeldt, a German gardener from St. Louis, who was engaged as botanist under the command of Capt. Gunnison . . . He collected 124 species, with two new ones . . ." The episode is described in the Beckwith report (*U. S. 33rd Cong., 2nd Sess., Sen. Exec. Doc.*: No. 78).

Not until January 25, 1849, did aid reach the waiting men, by which time they had been reduced to the starvation point—some say that cannibalism had been resorted to—and a number had died.

Frémont overtook the relief party after he had been six days on his way—and twenty-two days after it had left the main camp; one had died and the body had been partially devoured, perhaps by wolves. The guide Williams and the botanist Creutz-feldt had survived. Frémont wrote his wife: "We found them . . . the most miserable objects I have ever seen. I did not recognize Creutzfeldt's features when Bracken-ridge brought him up to me and mentioned his name. They had been starving . . ." Ten days after Frémont had left the main party he reached one of the settlements and *sent back* assistance while he, himself, proceeded to Taos where he was cared for at the home of his friend Carson; he was ingenuous enough to write that he had enjoyed the hot chocolate provided!

By the time Frémont's party, or what remained of it, was reunited at Taos, a third of its members had died, eleven out of thirty-three. The famous Donner party, com-posed of novices, is said to have lost thirty-nine out of a party of eighty-seven, not far short of half its complement.

One would have expected the leader of an expedition to return to rescue his men. And one would have expected Frémont, when writing his wife or Senator Benton, to express some distress at what had occurred; but there is not the slightest evidence that he felt any! The day after his arrival at Taos, New Mexico, or on January 27, 1849, he wrote Mrs. Frémont that he hoped to get off "about Saturday" for California, which would have been on February 3. Since the survivors only reached Taos on February 6, this had to be put off—Frémont had expected them by "Wednesday evening, the 31st [of January]." That there should be any need for rest and recuperation did not enter the picture.

In a later letter to Benton, dated February 24, from Socorro, New Mexico, Frémont expressed the hope that his letter would be received ". . . in advance of exaggerated reports of the events which have delayed my journey . . ." and refers to the fact that he and his party ". . . were overtaken and surrounded by deep and impracticable snows in the Rocky Mountains . . ." The truth needed no exaggeration to make news for the public, and the snow had been there before Frémont started—even a "tenderfoot" might have predicted more at high altitudes at that season. One more sentence to his wife is hard to assimilate: he had encountered ". . . a persistence of misfortune, which no precaution had been adequate on my part to avert." As Nevins comments: "It is a pity that he did not detail some of his precautions."

I shall now follow Frémont's journey to the California border. Mrs. Frémont was traveling to the west coast by way of Panama and she describes her trip in *A year of American travel*. The day after reaching Taos Frémont wrote her:

"At the beginning of February (about Saturday) I shall set out for California, taking the southern route, by the *Rio Abejo,* the Paso del Norte, and the south side of the *Gila,* entering California at the *Agua Caliente,* thence to Los Angeles and immediately north. I shall break up my party here and take with me only a few men. The survey has been uninterrupted to this point,[7] and I shall carry it on consecutively . . ."

On January 29 he wrote her again—he may have had many admirable qualities, but a sense of humor was not among them: ". . . This is to be considered a poor country; mountainous, with severe winters . . ."! On February 6, still in Taos, he was ready to depart, before his party had rejoined him:

". . . I have engaged a Spaniard to furnish mules to take my little party with our baggage, as far down the Del Norte as Albuquerque . . . My road will take me down the Del Norte, about 160 miles below Albuquerque and then passes between this river and the head of the Gila to a little Mexican town called, I think Tusson. Thence to the mouth of the Gila and across the Colorado, direct to Agua Caliente, into California. I intend to make the journey rapidly . . ."

Everyone in Taos and in Santa Fe, where he spent two days, had been helpful in replenishing supplies and equipment. Frémont mentions to Benton the assistance rendered by army officers in particular—his father-in-law was an influential man. Bill Williams and some others of the original party "declined" (Nevins' choice of word) to accompany him. McGehee's account does not continue beyond Santa Fe, and it seems highly probable that Creutzfeldt proceeded no farther. There were twenty-five or thirty men in the new party and some sixty pack animals.

At this point in the story Nevins' biography includes three paragraphs of Frémont's "rough notes" already mentioned; unfortunately they supply no dates. In quoting from them I break down the paragraphs in which they are presented by Nevins (and doubtless in the original) and omit much. The map of Frémont's *Memoir* becomes of real assistance at this point. Frémont records:

". . . I resume journey, following down the Del Norte and intending to reach the Rio Grande [probably the Colorado River is intended] by a route south of the Gila River.

"The snows this season too heavy to insist on a direct route through the mountains.[8] Engage a New Mexican for guide . . ."

The party continued down the Rio Grande until it was considerably south of Socorro ("Socora" of the map), Socorro County; Frémont wrote Benton from there on February 24. It left the Rio Grande near the Palomas River ("Rio de los Palmono" of the map), which enters the larger stream not far south of Elephant Butte Reservoir, Sierra County, and then moved southwestward. The notes record:

7. A statement which hardly seems credible.

8. To "insist on" seems a curious phrase in this connection!

"Leave the river . . . Retreat into the Membres Mountains[9] . . . Travel along foot of mountains . . . The Indians [Apaches] go to Membres River with us . . ."

As planned before leaving Taos, Frémont was soon south of the Gila River, the upper waters of which flow southwest across Grant County. The map shows that he then crossed southeastern Arizona south of the Chiricahua Mountains and entered the Mexican province of Sonora, proceeding to Santa Cruz. He then turned north, down Santa Cruz River, re-entering Arizona near the town of Nogales, Santa Cruz County. The route lay in Arizona until, at the confluence of the Gila and Colorado rivers, he crossed into California. Some of this route is mentioned in the "rough notes" quoted by Nevins:[*]

"Follow down the Santa Cruz River—Tucson. Spring on the Santa Cruz—peach orchard—the ruined missions. River lost in the sand . . .

"Reach the Gila River. The Pimah village . . . Follow the river around the bend. Meet large party of Sonorans going to California. Their pleasure in meeting us. Their fear of Indians. They urge me to travel with them. I consent[1] . . . Reach the Gila River . . ."

The map shows that Frémont followed the Santa Cruz River, often subterraneous, from what is now the Arizona border northward to Tucson ("Tussone" of the map), probably passing the church of San Xavier del Bac ("the ruined missions"), for he went by way of Tubac; from there he turned northwest to the Gila River, passing to the west of the Catalina Mountains ("Sierra de la Catarina" on the map) and reached the Gila at the "Pimah village," [2] situated at the Gila Crossing. From there the Gila was followed around the Gila Bend Mountains and westward to its mouth in southwestern Yuma County. The "rough notes" end at this point. Frémont's "Rio Grande" was, again, the Colorado River:

"Determine position of the junction with the Rio Grande. Make bullboat—ferry women and children of the Sonorans across, with my party, and leave the bullboat for the men to complete their crossing."

These Sonorans, some twelve hundred men, women and children, were bound to the gold fields of California and from them Frémont first learned that gold had been discovered on Sutter's ranch. He made quick decisions. States Dellenbaugh:

"Believing that gold must also be found on his Mariposa lands he engaged twenty-eight of the men to dig for him on shares, when they should arrive there. He was to furnish the supplies, 'grubstake' them as it was later called, and they were to divide with him equally what was secured. These people were familiar with the work and hence would be valuable . . ."

9. The Mimbres Mountains, between Grant and Sierra counties.

1. Frémont's choice of words sometimes suggests a "superman" complex!

2. See p. 1003, fn. 1.

From this time onward Frémont's life—if one may judge from its fullness along other lines—was not devoted to scientific pursuits. He would seem to have reached the peak of his success at the end of his second expedition of 1843–1845, in the sense that, at its completion, his accomplishment had been indisputably important. From then onward much that he did ended in controversy at best, at times in outright failure. One wonders whether, had he limited his activities to the map making in which he unquestionably excelled, his stature might not in the end have been greater. To let imagination run riot, what a botanic garden he might have developed on his Mariposa property!

The speed characterizing all Frémont's expeditions was, of course, incompatible with the best collecting, in whatever field. Although I do not know the covering dates of his journey from Santa Fe to the California border there is every reason to suppose that it was as rapid as possible. On the journey just described (and ignoring the mid-winter trip into the Colorado mountains) Frémont entered very little country which had not been visited by other collectors: the botanist Coulter had reached the confluence of the Gila and Colorado rivers in 1832; Wislizenus had descended the Rio Grande from Santa Fe to the El Paso crossing in 1846; Emory had, in 1846–1847, descended the Rio Grande to the Socorro region, crossed to the Mimbres Mountains and thence to the Gila River and had followed that stream to its mouth. Therefore, after leaving Santa Fe, the only botanically untouched territory traversed by Frémont in 1849 lay from the Mimbres Mountains of New Mexico southwest across the southeastern corner of Arizona and (back in Arizona from Mexico) from the region of Nogales north to Tucson and thence northwest to the Gila Crossing at the Pima village. The chance of discovering new plants was, therefore, diminishing rapidly as collectors crossed, not always precisely the same routes, but similar plant-zones.

Other than the Torrey comment, quoted early in this chapter, I have found no reference to plants collected on the journey just described. It is possible that the Torrey Herbarium contains specimens which, because of dates or localities, might be associated therewith.

BIGELOW, JOHN. 1856. Memoir of the life and public services of John Charles Fremont, including an account of his explorations, discoveries and adventures on five successive expeditions across the North American continent; voluminous selections from his private and public correspondence; his defence before the court martial, and full reports of his principal speeches in the senate of the United States. New York.

BRENDEL, FREDERICK. 1880. Historical sketch of the science of botany in North America from 1840 to 1858. *Am. Naturalist* 14: 31.

DELLENBAUGH, FREDERICK SAMUEL. 1914. Frémont and '49; the story of a remarkable career and its relation to the exploration and development of our western territory, especially in California. Maps. New York. 381-421. [Map (opp. p. 68) shows routes of all Frémont's explorations.]

FRÉMONT, JESSIE BENTON. 1878. A year of American travel. New York. 69-81.

FRÉMONT, JOHN CHARLES. 1887. Memoirs of my life, by John Charles Frémont. Including in the narrative five journeys of western exploration, during the years 1842, 1843-4, 1845-6-7, 1848-9, 1853-4 ... A retrospect of fifty years, covering the most eventful periods of modern American history ... with maps and colored plates. Chicago. New York. [As to map only. The volume ends with the expedition of 1845-1847.]

GOODWIN, CARDINAL. 1922. The trans-Mississippi west (1803-1853). A history of its acquisition and settlement. New York. London. 243.

McGEHEE, MICAJAH. 1891. Rough times in rough places. A personal narrative of Frémont's fourth expedition. Edited by his brother, C. G. McGehee. *Century Mag.* 41: 771-780.

NEVINS, ALLAN. 1939. Frémont. Pathmarker of the west. New York. London. 341-372.

TORREY, JOHN. 1853. Plantae Frémontianae; or, descriptions of plants collected by Col. J. C. Frémont in California. *Smithsonian Contr. Knowledge* 6: 4.

CHAPTER LIII

IT IS POSSIBLE TO OUTLINE SOME OF TRÉCUL'S TRAV-

ELS IN NORTH AMERICA BY MEANS OF THE DATES AND

LOCALITIES CITED IN A NOTEBOOK ENUMERATING A

COLLECTION OF HIS SPECIMENS

IN *The Silva of North America* Charles Sprague Sargent wrote a short biographical account of a French collector about whom and about whose work in North America so little has been published that I quote the record in full, breaking down the author's single paragraph into three (each applicable to periods in the man's life):

"Auguste Adolph Lucien Trécul was born in Mondoubleau, near Vendôme, in France, on the 18th of January, 1818, where his father was a baker, and was educated in the primary school of his native place and in the college at Vendôme. On graduating from college he went to Paris to study pharmacy, and in 1841 was admitted as an assistant in the Paris hospitals, and began the study of natural history. A paper published in 1843 in the *Annales des Sciences Naturelles* on the Fruits of Prismatocarpus and on the Crucifers attracted the attention of the authorities of the Museum, who engaged him temporarily to assist in the arrangement of the herbarium. At this time Trécul prepared a monograph of the Artocarpeæ, published in the eighth volume of the third series of the Annales, and continued his studies upon the organs of plants, to which most of his attention as a botanist has been devoted.

"In 1847 he was sent by the Museum to North America to collect plants and animals, being also commissioned by the Minister of Agriculture and Commerce to study the esculent plants used by the Indians[1] of the western plains. Arriving in North America in 1848, he traveled through the region between the Mississippi River and the Rocky Mountains for nearly three years, and returned to France in the autumn of

---

1. Samuel Wood Geiser wrote that "... In 1849 Trécul visited Texas on his scientific mission to North America to study and collect farinaceous-rooted plants used for food by the Indians. Wright met him in Castroville in November. His unpublished reports are in the archives of the Museum of Natural History in Paris."

Only in May, 1954, did I learn of a naturalist and zoologist who, after visiting North America in 1841, was sent there a second time by the French government to acquire nutritive plants for introduction to the gardens of France—a somewhat similar mission, apparently, to the one undertaken by Trécul a year later. The man was Lamare-Picquot and I understand that Dr. Grace Lee Nute expects to publish upon his work and travels; these took him through the states of the upper Mississippi River valley, particularly into the western portion of the region known today as Minnesota. Dr. Nute tells me (*in litt.* May 17, 1954) that "No one as yet has solved the enigma of his first name ..."

1850. His collections made during the first year of his stay in America were lost in the wreck of the ship to which they had been intrusted; but those made in Texas and northern Mexico, where he passed the winter of 1849, reached France in good condition, and included living plants of *Ungnadia speciosa, Yucca Treculeana, Sophora secundiflora, Guaiacum augustifolium, Rhus virens,* and several species of Cactus.

"Since 1850 Trécul has devoted himself to morphology and physiology, and has published many papers on these subjects in the *Comptes Rendus de l'Academie des Sciences,* the *Journal de Pharmacie, Annales des Sciences Naturelles, Revue Horticole,* etc. In 1851 he delivered a course of lectures on botany before the Institute National Agronomique at Versailles; but since his return from America has occupied no official position."

Through the courtesy of those in charge of the Laboratoire de Phanérogamie of the Muséum National d'Histoire Naturelle, Paris, I learned that no Trécul journal was to be found but that there did exist a notebook of some two hundred pages, in Trécul's handwriting with a numerical list of his collections: ("Référence: Per K.g.15"). This notebook was sent me in photostat (a beautiful piece of work, done by the Centre de Documentation du Centre National de la Recherche Scientifique).

Headed "Plantes de l'Amérique septentrionale"—"Par M. A. Trécul" added in pencil—the manuscript lists in numerical order a total of 1,525 numbers, as well as many additional insertions differentiated by *bis* or by letters. It is neatly penned and only rarely is it difficult to decipher a word. There are two sets of entries: (1) a main set supplies, after each number, the plant's generic name (rarely more), cites with precision the locality where it was collected (and often the conditions under which it was growing) as well as the month, occasionally the day of the month, and nearly always the year in which it was gathered; (2) the second set of entries, appearing in a separate column, starts with Trécul's number *282* and continues, intermittently, to the end; it supplies a fuller determination for many species and the entries appear to have been added at a later time—probably after the specimens had been studied.

Because, at this late date, Trécul's collections are not particularly important, I decided to omit all plant names and, since so little is known about the man's travels in North America, merely outline his routes, following his numbers with their cited dates and localities. The manuscript suggested plain sailing!

After several attempts to carry out this plan I abandoned it; for it became evident that the notebook was not a field-record but one compiled at a later date, and the numbers added as the specimens were taken out in the herbarium.[2] For although there is

2. Dr. E. D. Merrill tells me that the Trécul photostat shows "... *not the slightest evidence* that the material was sorted out into families and genera before the specimens were numbered." This, apparently, was a very common practice in Europe, and in America, at the time and still persists occasionally, along with, so Dr. Merrill states, the bad practice of "... matching individual collections from different places 'by eye' and then adding the numbers regardless of localities so that in some Old World early numbered collections we often find two to several different species all under the same name and often localities remote from each other involved ..." In his note to me Dr. Merrill cites numerous examples.

a numerical sequence there is, certainly for the overall picture, no continuous geographical sequence. As an example: the first four hundred numbers, approximately, are from Kansas (July-October, 1848, and February and March, 1849); but, following these in numerical sequence, are collections made in June, 1848, from the Lake Michigan region, and then from eastern Missouri in June and July, 1848. The student does not learn from the record when and where Trécul landed in North America; on what date or from what point he left (from Texas possibly) for Europe; how he moved from one collecting ground to another, nor where he passed much of the winter of 1848–1849 and the months of June, July and August, 1849.

In its over-all picture the outline of Trécul's travels presented here is my interpretation of the notebook's records. Once in a particular region a series of numbers may for a time apply to that locality but even here the many insertions, out of sequence geographically, wreck what might otherwise be a clear presentation. I shall omit most of these confusing insertions and cite only enough numbers for one region to indicate the collector's presence at a particular spot at a particular time.[3]

1848

I believe that Trécul must have landed at some point in the eastern United States and, like others before him, have traveled west through Pennsylvania and thence, by way of the Great Lakes and the Illinois River, have reached St. Louis. For, although bewilderingly out of geographical sequence, there are collections from Pottsville *(704, 705)* and Minersville *(706-711)*, both in Pennsylvania, dated "Mai."[4] The Great Lakes are indicated by "Isle des Castor"—Beaver Island in northern Lake Michigan— *(419, 424, 427, 430)*, and by Racine *(420, 422, 423, 426)*, by Littleport *(421)* and by Sheboygan *(428, 429, 431)*. These only cite "Juin," but the later insertion *577* (clearly out of place in the geographical sequence where it appears) cites "l'ile du Castor dans le lac Michigan Juin 1848," thus supplying the year.

Trécul's first real collecting-ground appears to have been in eastern Missouri, in June and July, 1848. He was at "Montagne de fer," or Iron Mountain *(433-443, 462-465, 474-480,* etc.), and at Farmington *(460),* both localities in St. Francois County; at "Ste Geneviève" *(444-448, 450-454, 467-473,* etc.) in Ste. Genevieve County; at Potosi *(525, 526, 531-540,* etc.) and at Caledonia *(653, 654)* in Washington County; at Pilot Knob *(657)* in Iron County; and at Hillsboro *(546-551, 553-576,* etc.) in Jefferson County. In this numerical group is a specimen from Jefferson Barracks

---

3. The photostat of Trécul's document is in the Library of the Arnold Arboretum and available for reference to any one interested in a particular record of plant or route.

4. Of these only number 704 cites a year and unfortunately this is written across a vertical line and not quite clear. I believe that the year is 1848.

(545), and two specimens given the collector by George Engelmann *(578, 579)*, both undated. Trécul must have been at St. Louis more than once.

I place Trécul next in Kansas, from July to October, 1848[5] *(1-382)*. Then follows a period (October, 1848–February, 1849) when his activities are unaccounted for. Very likely he spent these months in Kansas, perhaps at Fort Scott, for nothing suggests that he left the state during that period; but of course he may have returned for the winter to St. Louis. Kansas collections begin again in February, 1849 *(383)*, and continue into March of 1849 *(418)*.

The only indication of how Trécul reached Kansas from, presumably, St. Louis, exists in specimens inserted among the Kansas records which are suggestive of a trip up the Missouri River to the Kansas border, at which point he must have turned south: ". . . entre Westport et West point, sur la frontière occidentale de l'état du Missouri. Juillet 1848" *(22)*, "Prairies a l'ouest de l'Etat du Missouri, près de la route de Santa Fé. Juillet." *(26)*; ". . . près du Kansas." *(45)*; "Sur les bords du Missouri . . ." *(107)*.

The Kansas collections begin on the Marais des Cygnes,[6] which Trécul refers to as the upper portion of the Osage River. Many specimens cite the Marais des Cygnes *(2-14, 17-21, 48-106, etc., etc.)*, the Osage River *(128, 129, 134, 135, 138, etc.)* and regions west of the state of Missouri *(31, 32, etc.)*. By August there were specimens from "Sugar creek au Neosho" *(146, 147, 149, 150, 152)*, from the "village Osage de Neosho" *(148, 155)*, from the Indian "Mission de Sugar creek, chez les Potowatomies" *(162, 174)* and from Fort Scott *(156, 254-256)*—all indicative that Trécul was working in the vicinity of the Missouri-Kansas border. By late August and September most of the numbers for some time cite the Neosho River *(185-206, 215-221, 223-238, 242-248, 261-268, etc.)* and it seems probable that his direction was northwestward; for soon there are citations from "l'Arkansas, le Petit Arkansas, et a l'ouest de ces rivières Septembre et Octobre 1848" *(284-349, 353-372, etc.)*. Just how far west Trécul traveled along these streams I do not know, perhaps to the center of the state, but he certainly did not reach the Rocky Mountains.[7] He must then have struck north for he was on the Grand Saline *(350-352, 376-382)*. With *382*, October, 1848, the Kansas records stop until, in February, 1849, they begin again "Près du Neosho" *(383)*.

5. The months during which Trécul was collecting in Kansas change from July *(46*, July 31) to August *(48)* and to September *(152)*.

6. The volume *Kansas* ("American Guide Series") recounts a legend—that the name was bestowed upon the stream by Evangeline (of Henry Wadsworth Longfellow fame) after she had been told the tale of two Indian lovers who were drowned in its waters and transformed into swans before the eyes of onlookers.

About ten years after Trécul's visit a massacre occurred on the Marais des Cygnes—situated north of Trading Post, Linn County—which is now commemorated by a monument and, as well, in a poem by John Greenleaf Whittier. Pro-slavery factions are said to have attacked persons holding different opinions.

7. This would seem to form a partial reply to Joseph Ewan's comment that ". . . How close Trécul approached the Rocky Mountains is not known from the brief published accounts."

1849

The remaining Kansas records are mainly from the Neosho, and as of March, 1849. Two mention what seems to be the "Petit Pawnie" *(407, 408),* and one the "Grand Pawnie, entre Neosho et Fort Scott, Mars" *(409),* and there are several Fort Scott citations *(411-417).* The last number from Kansas was from "Prairies du Neosho à la frontière de l'état du Missouri Mars 1849" *(418).*

In April, 1849, Trécul evidently descended the Mississippi River, collecting a few plants at various points. There are citations from regions adjacent to the river: from Illinois *(665-669);* from Missouri *(670-678);* from an island off Tennessee *(680, 681);* from Memphis *(682);* from Arkansas and the Arkansas River *(683-687);* from Natchez *(688-695);* Baton Rouge *(696-700);* Donaldsonville *(701-703);* and from Lake Pontchartrain *(712-726).*

But he seems to have returned north immediately, for May, 1849, was spent in the state of Illinois *(731-794).*

For June and until September, 1849, there are no records and I find nothing even to suggest where he spent his time. But with September the records begin again in Illinois *(795, 797)* and are followed by citations from "Chester, Etat du Missouri" *(798-814).* I have been unable to discover a town of that name in present Missouri, but Chester, Illinois, is situated at the mouth of the river Kaskaskia. Trécul carefully refers to Missouri as a state—it had been admitted to the Union in 1821.

It is to be hoped, perhaps even inferred, that Trécul enjoyed steamboating, for he seems now to have embarked on his third Mississippi River trip within six months (he had made one south in April and must have returned by the same route in that month or in May). Now, in September, there are citations from New Madrid *(816-823);* Memphis *(824-825);* Arkansas *(826-836);* Natchez *(837, 838);* Baton Rouge *(839-844);* Lake Pontchartrain *(845-864);* succeeded by a series from marshes ("Lieux maricageux" and "lieux humides") in Louisiana *(865-883)* and ending with one from New Orleans *(884).* He then moved to Biloxi, Mississippi *(885-1029),* where specimens are all dated September.

Trécul was now to journey into Texas, a trip which, with the exception of an extremely brief excursion into Mexico, was to extend from October, 1849, to February 23, 1850. As elsewhere, I shall only outline his route and cite only *some* of his collections from particular regions. The records at times are extremely confusing because of citations geographically out of sequence. The very fact that precise localities are always meticulously recorded (for Trécul seems always to mean just what he says) only adds to the complexity.

I believe that most, if not all, of the localities in Texas where Trécul made collections had already been visited by other botanists: some by Berlandier, some by Drummond; Wright had been in the Eagle Pass—Rio Grande region; Lindheimer at this time had his headquarters at New Braunfels and, according to Geiser, had been in 1844 at Port Lavaca and along the Guadalupe in Victoria County. Geiser has stated that Wright met Trécul at Castroville and one wonders whether Lindheimer and the Frenchman had any contact.

The first Texas collections cite Galveston and October, 1849 *(1030-1034, 1036-1039, 1044-1050)*; next Port Lavaca and Matagorda Bay *(1040-1043, 1051-1157)*.

Moving inland and northeastward from the Gulf of Mexico, Trécul made collections between Port Lavaca and Victoria and about the last-named *(1158-1193)*. Since Trécul seems to have followed the Guadalupe River all the way from the Gulf to New Braunfels, certain specimens mention that river in this series *(1179-1183, 1185, 1187, 1189-1191)*. Next the records cite Victoria to Gonzales and near the last *(1194-1253)*; next Gonzales to Seguin and near the last *(1256-1269)*. The month changes in this series: *1262* is dated October 31, 1849, and *1263* November 1, 1849; succeeding numbers for some time (or to *1442*) are dated November. From Seguin, Trécul moved on to New Braunfels *(1270-1277)* and collected about that town *(1286-1296, 1298-1331)*. Some numbers mention the Guadalupe and one the Comal *(1298)*.

Numbers *1332-1338, 1340-1349* suggest that, from New Braunfels, Trécul may have made a trip to Seguin and to Gonzales and back, for specimens are cited as from Seguin to Gonzales, others as from Gonzales to Seguin,[8] and several cite the "rio San Marcos." These are followed by others from New Braunfels *(1350-1367)*.

From New Braunfels Trécul went to San Antonio de Bexar *(1368-1377)* and collected about that town *(1378-1402)*. In this sequence the "rio de San Antonio" is cited *(1382)*, also nearby "Sulphur Springs" *(1386, 1387, 1389, 1392-1394, 1396-1401)* and "Salado creek" *(1389)*.

The next records cite Castroville *(1403-1429)* and in the sequence are mentioned: the "rio Medina" *(1416)* on which Castroville is situated; the "Ruisseau appelé Quihi" *(1416 bis-1419)*; and what appears to read "la rivière Lionne" *(1421)*, presumably Leon Creek although, if out of geographical sequence, it might be Leona River.

Trécul was bound for Eagle Pass on the Rio Grande del Norte and, on the way, he collected at various streams: at the "rio Seco" *(1430-1437, 1439, 1440)*, at the "rivière de la Lionne" *(1441, 1453-1457)*, which in this sequence was probably Leona River, and at the Nueces or between that stream and the Rio Grande del Norte *(1442-1445, 1447-1452)*. Beginning with *1442* all numbers in this series are dated December, 1849.

Records from the Rio Grande valley cite Eagle Pass *(1458-1463, 1488)*, which Trécul locates "30 milles au dessus de Presidio del Rio Grande," or merely refer to situations "près du rio Grande" *(1489-1490, 1492, etc.)*. In this series *1488* and upward are dated January, 1850. Number *1490* specifies January 7.

---

8. It is of course immaterial from the point of view of locality of collection whether Trécul was facing north or south when the specimen was gathered! Trécul appears, however, to be so careful in his citations that his differentiation suggests the trip mentioned above.

In December, 1849, and into January, 1850, Trécul, from Eagle Pass, made a trip into Coahuila, Mexico, to near San Fernando, a small town about twenty miles southwest of Eagle Pass *(1464-1469); 1465* is dated December 30, 1849; *1466, 1468* are dated December 31, 1849, and *1469* is dated January 1, 1850.

1850

A few more Mexican collections from Coahuila and "Texas occidental" are cited later but without precise locality *(1475-1483)* and are dated January, 1850. There is no reason to suppose, from any record in his notebook, that Trécul left the vicinity of Eagle Pass when on the Texas side of the Rio Grande. One specimen *(1484)* cites "la rivière du Diable," but the plant is stated to have been brought from that river.

From Eagle Pass Trécul must have returned to New Braunfels, and by much the way he had come. He was at Castroville on January 15 *(1493),* and at New Braunfels on January 24 *(1494).* There are other collections from New Braunfels dated January *(1497-1501).*

In February, 1850, he was on his way from Bastrop to San Felipe de Austin ("Bastrop à San Felipe de Austin" on the "rio Colorado") *(1502, 1504),* and at San Felipe de Austin *(1505-1524).* Of this series, some numbers *(1505, 1512-1515, 1518, 1519, 1521, 1524)* cite the "rio Brazos," which in one instance Trécul calls the "rio Brazos de Dios."

This brings us to the last number in the notebook, *1525,* which is from "Prairies salées à Quintana Texas. 23 février 1850." Quintana, Brazoria County, is at the mouth of the Brazos River and Trécul may have sailed from Galveston not far away.

Sargent, as quoted at the opening of this chapter, mentions that Trécul had included living plants among his collections, of which several were species of *Cacti.* The notebook mentions nine which were sent as living plants ("Envoyé vivant"), all collected in January, 1850, all from the region of Coahuila, Mexico. They included three *Mammillaria (1476, 1481, 1482),* two *Echinocactus (1477, 1478),* three *Opuntia (1479, 1480, 1483)* and a cactus, undetermined *(1475).* Also sent as a living plant was a *Mammillaria* which grew from the regions about Gonzales to the Rio Grande *(1474).*

When I was studying the yuccas of the southwestern United States I was sent, by Professor Humbert of the Paris Museum, a photograph of a yucca which bore on one label (just a scrap of paper) the number *1966* and the notation "Acaule ou plutôt à tige souterraine. Sur les collines de New-Braunfels Texas 5 9ième [the ninth month or November] 1849." Another printed label is inscribed "Voyage de Mr Trécul, dans l'Amérique septentrionale 1848-1850." It was clearly *Yucca rupicola* Scheele.

Trécul's notebook cites three yuccas: (1) *1441,* from "près de la rivière Lionne," and dated November, 1849; it was identified in the notebook as *Yucca angustifolia*

Pursh, and the citation differs entirely from the one supplied with *1966* above. (2) *1496* is the number of the specimen which Carrière used as type of *Yucca Treculeana*. (3) *1472* grew from New Braunfels to the Rio Grande, was collected in January, 1850, and was shipped as a living plant. It bears no determination. None of these, on the label evidence, could have been confused with Trécul's number *1966*.

Either *1966* is misnumbered or—what seems to me highly probable—Trécul collected many more plants on his journey of 1848-1850 than he recorded in his notebook which stopped, as noted, with number *1525;* he may have intended a continuation of his "Plantes de l'Amérique septentrionale," but never have been able to accomplish it. The Museum authorities mention only one such record. Trécul had been at New Braunfels in November, 1849, as his earlier numbers *(1286-1296, 1298-1331)* record, so that 1966 of the same date and locality would—like many of his notebook records—merely be out of numerical sequence had he continued listing his plants.

EWAN, JOSEPH. 1950. Rocky Mountain naturalists. Denver. 324.

GEISER, SAMUEL WOOD. 1948. Naturalists of the frontier. Dallas. ed. 2. 282.

SARGENT, CHARLES SPRAGUE. 1896. The Silva of North America. Boston. New York. 10: 10, *fn.* 2.

TRÉCUL, AUGUSTE ADOLPH LUCIEN. [Undated.] Plantes de l'Amérique septentrionale. [Manuscript in Muséum National d'Histoire Naturelle, Paris. Photostat in Library of Arnold Arboretum, Harvard University. ca. 200 pp.]

CHAPTER LIV

WRIGHT MAKES A FINE COLLECTION OF PLANTS BE-

TWEEN SAN ANTONIO, TEXAS, AND EL PASO, THEN LO-

CATED ON THE MEXICAN SIDE OF THE RIO GRANDE

CHARLES Wright, renowned plant collector, did much of his most important work when attached to the United States-Mexican Boundary Survey in the decade of 1850–1860, or at a later period than the one covered by my story. But he had arrived in eastern Texas as early as 1837 and, because of his interest in botany, had collected plants wherever his work as a teacher and surveyor had taken him. He is said to have begun a correspondence with Asa Grày in 1844, soon thereafter sending him specimens. In 1849, at the behest of Gray and taking up collecting as a profession, Wright crossed Texas from Galveston to San Antonio. From there he traveled to El Paso and back, finding many new plants in regions hitherto unexplored by botanists.

Wright was born at Wethersfield, Connecticut, on October 29, 1811, and died there on August 11, 1885. Gray wrote an obituary notice, published in 1886 in the *American Journal of Science and Arts,* which supplies an over-all picture of his life. This records his graduation from Yale College in 1835—where his "fondness for botany was developed"—and states that, immediately after, he went to teach in a family living at Natchez, Mississippi: "The opportunity of gratifying this predilection in an inviting region may have determined his acceptance ..." This employment did not last long. Gray wrote:

"At this time there was a flow of immigration into Texas, then an independent republic ... Wright joining in it, in the spring of 1837 made his way from the Mississippi to the Sabine, and over the border, chiefly on foot, botanizing as he went. Making his headquarters for two or three years at a place called Zarvala,[1] on the Neches, he occupied himself with land-surveying, explored the surrounding country ... and inured himself to the various hardships of a frontier life at this period. When the business of surveying fell off he took again to teaching; and in the year 1844 he opened a botanical correspondence with the present writer, sending an interesting collection of the plants of eastern Texas to Cambridge. In 1845 he went to Rutersville in Fayette County, and for a year or two he was a teacher in a so-called college[2] at that place, or

1. In Jasper County. *See* Geiser, 1948, p. 181.

2. Writes Geiser: "... perhaps a natural remark from the direction of Cambridge ..."! He adds: "The College, whatever its shortcomings, must be viewed as a courageous attempt to overcome the limitations of the frontier. It was the first institution of higher learning to open in Texas—and although its life was

in private families there and at Austin, devoting all his leisure to his favorite avocation. In the summer of 1847–8 he had an opportunity of carrying his botanical explorations farther south and west. His friend, Dr. Veitch,[3] whom he had known in eastern Texas, raised a company of volunteers for the Mexican war, then going on (Texas having been annexed to the United States), and gave Mr. Wright a position with moderate pay and light duties. This took him to Eagle Pass on the Mexican frontier, where he botanized on both sides of the river. He returned to the north in the autumn of that year with his botanical collections, and passed the ensuing winter [of 1848–1849] in Connecticut and at Cambridge.

"In the spring of 1849 . . . Wright returned to Texas, and, at the beginning of the summer, with some difficulty obtained leave to accompany a small body of United States troops which was sent across the unexplored country from San Antonio to El Paso on the Rio Grande . . ."

This epitome of Wright's activities up to the spring of 1849 is amplified in Samuel Wood Geiser's paper on Charles Wright, first published in the *Southwest Review* in 1930 and afterwards incorporated, with some changes and additions, in the two editions (1937 and 1948) of his *Naturalists of the frontier*. I quote what he says of the progress of botanical knowledge in Texas when Wright arrived there from Natchez in the spring of 1837:

". . . Wright was destined to make Texas his home for the next fifteen years—if indeed, during his productive years as a botanist, Wright can be said to have had a home. It was a fruitful period for the advancement of knowledge of the botany of the Southwest. Lindheimer . . . was just then beginning to collect plants in the region of Houston and San Felipe. South of San Antonio and as far as Laredo and Matamoras, Jean Louis Berlandier had made, in 1828-34, his extensive collection for DeCandolle. Only a few years later Lindheimer would make with Roemer a collection of plants from the banks of the Guadalupe in Comal County that would show the world the richness of the Texan flora. Five years before Wright's arrival, Thomas Drummond had gathered seven hundred and fifty species of plants for Hooker in the Austin Grant. But all of this had been done in central and 'southwestern' Texas. No collecting of note, besides Dr. M. C. Leavenworth's slight efforts, had been done in eastern Texas. Hence Asa Gray, busily engaged in various magisterial works on American botany, received with something like delight a letter from Wright in eastern Texas in 1844 . . ."

After picturing the conditions facing Wright when he began his life in Texas—". . . a region where life and property were never secure . . ."—Geiser states that ". . . Altogether, it was quite too much for a quiet, studious-minded man. There was but one bright spot in Wright's situation: his friendship for a physician, Dr. John A. Veatch,

short, it accomplished under frontier conditions a task that I doubt Harvard or Yale would have attempted. It never was large . . . But it was a noble dream whose fruition was commensurate with the possibilities of time and place."

3. Dr. John A. Veatch.

with whom he shared a love of botany." After his period of teaching at Rutersville College—where his work represented "some of the first science field-work done in the schools of Texas" and which lasted from July, 1845, until late 1847—and after teaching in Austin from November, 1847 to July, 1848, he set out about ". . . the middle of July . . . from San Antonio for Eagle Pass to join his friend Veatch,[4] whose company was posted there guarding the Mexican frontier. During July and August . . . while holding a commissary position in Captain Veatch's company, Wright botanized on both sides of the Rio Grande, and returned to San Antonio about the middle of September." He then went east for the winter, returning to Texas in the spring of 1849. Gray had arranged that Wright should make a trip and wrote Torrey of the plan on February 26, 1849. This communication is included in the *Letters of Asa Gray*, edited by his wife:

"Having determined on an expedition for Wright, you may be sure I was not going to be altogether disappointed. Accordingly I have got one all arranged . . . thus far everything has wonderfully conspired to favor it. Wright . . . takes first vessel for Galveston . . . to reach Austin and Fredricksburg in time to accompany the troops that are about to be sent up, by a new road, across a new country, to El Paso, in New Mexico. Look on the map (Wislizenus) and you will see the region we mean him to explore this summer; the hot valley of Rio del Norte, early in the season, the mountains east, and especially those west in summer. He will probably stay two years and get to Taos[5] and Spanish Peaks [of Colorado] this year or next . . ." (1:362, 363)

Geiser, evidently basing his facts upon a letter which Wright wrote Gray from Austin on May 26, 1849, outlines the collector's movements from the time he left New York until he reached San Antonio. I quote a part:

"Wright left New York . . . about the first of April, 1849, and arrived in Galveston the twenty-fourth of that month . . . He . . . proceeded from Galveston to Houston,

4. Geiser refers to John Allen Veatch as ". . . a native of Kentucky who came to Texas about 1836 . . . an amateur botanist and surveyor in Vehlein's Grant (*c.* 1837–45) . . . also an explorer in Lower California . . . his plants were described by Albert Kellogg . . ."

Kellogg named a number of species after Veatch. And Gray named the genus *Veatchia* (now *Pachycornus*) in his honor. Whether any Texan plants bear his name I do not know. Veatch went to Lower California when its flora was largely unknown and brought to the attention of botanists the remarkable plants of "Cerros," or Cedros, Island, which lies off the west coast of the peninsula and where he was the first to collect. Writing of the island he refers to the ". . . bristling cacti and fearful rattlesnakes—the two prominent products of the land . . ."

Veatch also collected in the state of California, for plants are credited to him by Kellogg; for some reason he is not mentioned in Brewer's enumeration of those who worked there. Although Kellogg does not give the dates of Veatch's collections, the man's publications appeared in the 1850's and 1860's and his plants were presumably of that period. Veatch wrote mainly on geological subjects; one of his articles —on the mud volcanoes, or Sales, of the Colorado Desert—appeared in several periodicals.

5. Gray had finally succeeded in his determination to get Fendler to Santa Fe in 1846, but had *not* succeeded in his determination that Fendler should go back again—and the collector explained the reasons for his refusal very explicitly. But Gray seems never to have abandoned his objectives and Wright finally collected in New Mexico in 1851.

and from Houston west across the Brazos bottoms of Waller and Austin counties to Rutersville, which he reached on May 12 . . . during the ten days from May 15 to May 25 he followed the Colorado from Lagrange to Austin; May 27 and 28 he spent in the neighborhood of San Marcos and on the road to New Braunfels; on the twenty-ninth he was at New Braunfels and on the Guadalupe, and he arrived in San Antonio on the thirtieth of May. He had timed his arrival well, for the baggage train which he was to accompany left San Antonio the next day."

Wright's letter told of a difficult but, nonetheless, somewhat fruitful journey, for he mentions that, between Galveston and Austin, he had ". . . collected some 250 species probably which will furnish some 2000 specimens some few rather bad—injured by rain and most of which we had a plenty. They are mostly old and known plants but will do for exchanges . . ." According to Ivan M. Johnston's "Field notes of Charles Wright," the collectors' numbers *1* through *319* were gathered between April and June 1, 1849—or after his arrival at Galveston Island, Galveston County, until he left San Antonio, Bexar County, to begin his westward journey; the route, generally speaking, had led from Galveston northwestward to Austin, Travis County, and thence southwest to San Antonio. The regions of collection—more specifically cited by Johnston—were as follows:

April 28: nos. *1* through *30,* Galveston Island, Galveston County. May 5: nos. *31* through *55,* region west of Houston, western Harris County. May 6 and 7 (in part): nos. *56* through *107,* near and along the Brazos, Waller County. May 7 (in part), 8, 9, and 10: nos. *108* through *185,* from west of the Brazos to Rutersville, crossing Waller, Austin and Fayette counties. May 12, 15, 18: nos. *186* through *227,* Rutersville and thence to Austin, crossing western Fayette, Bastrop and eastern Travis counties. May 23-25, 27-30: nos. *228* through *319,* Austin to San Antonio, crossing southern Travis, Hays, Comal and northern Bexar counties.

Gray's introductory words to his treatise on Wright's collections of 1849 explain the auspices under which the journey to El Paso was to be made. Why, over the years, the United States army should have been expected to supply Gray's plant collectors with food, free transportation and much else in addition to protection, is never explained, but it is obvious that the military responsible for such expeditions cared less than nothing about promoting botanical knowledge:

"Mr. Charles Wright . . . in the spring of 1849 . . . proceeded to San Antonio, purposing to avail himself of the opportunity afforded by the movement of a body of United States troops from this place across the country to El Paso, in Southern New Mexico, to investigate the natural history, and especially the botany, of this hitherto untrodden region. A recommendation from the War Department, that all proper facilities be furnished to Mr. Wright . . . procured for him only the free transportation of his paper for preserving specimens, and of the collections he was enabled to make. This favor he owes to the kindness of Captain French, the quartermaster of the expedition, to whom and to Major Henry and Major Van Horn, Mr. Wright desires to express his thanks . . ." [6]

Wright, who was undoubtedly better informed than his employer in Cambridge as to what the contemplated journey involved in a number of respects, had asked Gray for a horse and wagon but did not get it and made the trip—"a distance of 675 miles" and taking "one hundred and five days," according to Geiser—on foot. Although written too late to alter the setup Wright gave himself the satisfaction of saying what he thought; his letter is dated "Quihi June 2nd/49," or before the trip had well begun, and is in the Gray Herbarium collection.

"My dear Dr

"I wrote you so recently that if I were not full I would keep silence But steam is so high that if I do not blow off fearful consequences may follow.

"Yesterday morning we had a violent norther cold and accompanied with rain after which and when ready to start my baggage, paper &c was distributed about into three or four waggons It was so packed that it was not much injured This morning about daylight we had another more severe accompanied with hail My collections were nearly all wet and I have had no time to dry them   so they will be much damaged My paper is nearly all wet I should not wonder if we had another storm tonight

"Now these are misfortunes attendant on my dependent situation and I can not prevent them   The officers care nothing about my affairs and the waggoners have a little curiosity to gratify by looking on while I change my plants and care no more about it or rather would be pleased if they were sunk in the river and their load would be lightened

"You will recollect I suppose a suggestion made to you that I should be equipped with a waggon and horse from which you dissented instancing the labors of other botanists who had made large collections But I venture to say that Drummond did not attempt to save 12-15 specimens of each species or if he did he had an art which I do not possess

"The outfit which I proposed seemed to you perhaps large but I am sincerely of the opinion that the entire cost of the outfit might have been *clear* saved *the present year* I would rather have a horse and carriage and ten dollars in my pocket than have five hundred as I am so far as it facilitates my operations I have money in my pocket but it does me no good I can buy nothing with it I sit uninvited and see others eating and it is a severe trial to my feelings to *thrust* myself among them The men have their rations and often none to spare and how I am to get along to El Paso I know not If

6. With future trips under army auspices in mind Gray was being diplomatic, no more, for he wrote Wright on January 2, 1850: "As to French's cool request that you should give away two sets of your plants to Smithsonian & National Instit. as a token of your gratitude to *him*. I do think it beats all . . . I will see him in — — — — — Texas before I make any such disposition. For tho in the preface it is right & proper to speak gratefully of his & Baker's good offices—yet in fact French has been ungenerous—has done *nothing* to what he *might* have done . . ."

I had consulted my own feelings alone I should have stopped at San Antonio and turned back But you & Mr Lowell[7] had expectations which would not have been realized and I felt reluctant to disappoint you You wrote to me of working like a dog I know how you live—then call your situation dog-paradise and mine hog- and ass-paradise combined and you *may* realize my situation—sleep all night if you can in the rain and walk 12-15 miles next day in the mud and then overhall a huge package of soaked plants and dry them in the heat of the clouds

"I have been now three or four days in such a state of uncertainty about the possibility of going on that I have no enjoyment and today I have not saved a specimen—have merely collected some seeds as I walked along the way As for studying the plants I have not attempted it so long that I have almost forgotten how I have been vexed enough to cry or swear when thinking that I have the pleasing prospect of being dependent for six months on a parcel of men who call me a fool and wish me at the bottom of the sea

"There is a man who is bound for California, in our company—provided with a carriage and mules provisions and cooking utensils—independent as a wood sawyer and dependent on others only for safety against enemies If I had such a one my expenses would be very trifling I could collect twice as many specimens of twice as many species and twice as well preserved I could attend to them at any time I pleased in wet or dry weather and have the assurance that the rain at least could be prevented from coming to them I could also take them to Houston or other seaport and put them on shipboard *myself* and then I would know they would depend for their forwarding on no careless agent

"I am fully resolved that this season will close my botanical travels on horseback or on foot if I can not operate to better advantage I'll give it up and turn my attention to some thing else

"I can now only *hope* that when Capt. French arrives in camp by [my] situation will be improved by an appointment or in some other way . . . What I wrote about money I recall as I know not what to do with that which I have. You now know my sent[i]-ment on the mode of botanizing in *this* country & if you wish to continue it on *my* plan I am ready to do all I can . . .

"Affectionately Yours Charles Wright"

Whether Wright's entirely justifiable protest brought about a more satisfactory arrangement on his later expeditions I do not know, nor have I discovered Gray's reply to the above letter, if he wrote one!

Gray's introductory words to his treatise on Wright's plants includes the following account of the journey:

". . . The train left the frontier settlement of Castroville about the first of June, and reached El Paso early in September. The remainder of the month was devoted to

---

7. John Amory Lowell of Boston, subscriber to Wright's contemplated "sets."

making collections in the vicinity of that interesting station.[8] Finding that much time would necessarily be lost in passing the long winter in New Mexico, Mr. Wright retraced his steps, and accompanied his rich collections back to Texas by the return train, leaving El Paso in October, and reaching San Antonio late in November . . .

"A proper account of the topography and physical character of the region traversed by the United States troops in their march from Texas to New Mexico will doubtless be published, before the printing of this memoir is completed. It is therefore unnecessary for me to attempt to compile any such account from Mr. Wright's disjoined and necessarily imperfect memoranda."

This statement is dated "May, 1850," and Gray's paper was issued in 1852. As he surmised, Captain S. G. French, Assistant Quartermaster, presented his "Report"—". . . in relation to the route over which the government train moved from San Antonio to El Paso del Norte . . . in pursuance of orders . . . dated May 30, 1849 . . ."—with promptitude and it was published in 1850 (*U. S. 31st Cong., 1st Sess., Sen. Exec. Doc.* 14: No. 64, 40-54). French gives a picture of the sparsely settled country and of traveling conditions but, after the first four days, supplies few dates and little that is precise about the route. In 1849, 1850 and 1851 the War Department was sending a number of exploratory, topographical and road-building expeditions into western Texas and French's train—transporting "government stores"—seems to have served several contingents. Reports upon some of these parties—which Warren calls "Reconnoissances in Texas"—were published, like French's, in Senate Document No. 64. The included map ("Reconnoissances of routes *from* San Antonio de Bexar to *El Paso del Norte*") does not show French's (and Wright's) route but is nonetheless helpful. Warren's "Memoir" contains a brief resumé of these various expeditions and to some extent clarifies their complexities, although when the "Memoir" was published in 1859 not all the army reports had been issued.

In addition to French's "Report" and Warren's "Memoir" there are two papers which are based upon the more precise data supplied in Wright's specimen-labels and field-lists. The first of these papers, published in 1935 is by Geiser: "Charles Wright's 1849 botanical collecting-trip from San Antonio to El Paso; with type-localities for new species." This is a valuable supplement to the same author's *Naturalists of the frontier*, where very little is said about Wright's trip of 1849, for it outlines the collector's itinerary with dates and modernized place-names.[9]

The second paper, unpublished but available in the Library of the Gray Herbarium,

8. On August 8, 1846, Wislizenus, having descended the Rio Grande from Santa Fe, had crossed to El Paso which consisted at that time of a few houses on what is now the Mexican side of the river. Hurrying on with Speyer's caravan, he could have made few collections if any. Wright appears to have been the first botanist to collect on both sides of the Rio Grande in the El Paso region.

9. The "Geological map of Texas," prepared by J. A. Udden, C. L. Baker and Emil Böse, published by the University of Texas in 1916, I found extremely helpful as an adjunct to Geiser's paper.

is Ivan M. Johnston's "Field notes of Charles Wright for 1849 and 1851-52[1] relating to collections from Texas, New Mexico, Arizona, and adjacent Sonora and Chihuahua . . . with commentary," dated February, 1940. The author supplies the plant names and localities given in Wright's field-lists (numbers *320* through *1404*) with, so he informs me, some editing. The running commentary upon the route contains many extracts from Geiser's aforementioned paper, but incorporates as well considerable information taken from Wright's notes and labels. To fully appreciate the complexity of Johnston's task one should understand the treatment accorded Wright's personal labels and field-notes by Asa Gray when he distributed the man's "sets" to subscribers and published upon them. Gray's "Plantæ Wrightianæ Texano-Neo-Mexicanæ: an account of a collection of plants made by Charles Wright, A.M., in an expedition from Texas to New Mexico, in the summer and autumn of 1849", Part I, was issued in March of 1852. In his introductory words Gray states that "The numbers prefixed to the names are those under which the specimens are distributed." Since most botanists concerned with the plants indigenous to our southwestern states must sooner or later study Wright's material, Johnston's comments upon Gray's methods should be welcome:

". . . The . . . *field-numbers* assigned the collections by Wright . . . were disregarded by Gray when he made up Wright's plants into sets. The numbers on the labels of Wright specimens in herbaria belong to a new series assigned them by Gray, the set of *distribution-numbers*. Not only did Gray ignore Wright's field-numbers but he also frequently united and distributed under a single distribution-number, two or even more of the collections which Wright had collected under different field-numbers, frequently at distant stations, and at different seasons. If Gray thought two or more of Wright's collections represented the same species and if there was any advantage in uniting them, he did so regularly without scruples.

"Many of the plants distributed by Gray can not be definitely associated with any of the numbers given in Wright's field-lists. Since Wright had a good knowledge of genera and families of Texan plants, his field identification and the geographical data given by Gray in the *Plantae Wrightianae* help one to make a reasonably good guess as to the precise identity of many collections. In the Gray Herbarium it is common to find, obscurely pencilled in the corner of the familiar blue label of the Wright collection, numbers which refer to the field-lists . . . In the pocket of other collections there is frequently found a small bit of paper bearing Wright's original field-number, in the script of the collector, probably representing the tag which Wright placed with each collection as he assembled it in the field . . . Gray . . . as a general rule made little or no effort to preserve Wright's field-ticket, or to note the proper field-number on the herbarium-label of these collections . . .

1. My comments thereon relate only to Johnston's data for the year 1849.

". . . Gray had two different herbarium-labels, one for the collection of 1849, the other for the collection of 1851–52. Field-numbers found associated with the collections of 1849 are easily located in the list for 1849 . . . The collections of 1849 were distributed in 1850, and though representing 1404 numbers in Wright's list, were in Gray's distribution accommodated by numbers 1-828 . . .

"The extent to which Gray united collections is well shown by his treatment of the collections of 1849. In that year Wright made 1404 separate collections. These, supplemented by some of Wright's earlier collections (made about Eagle Pass in 1848), made 828 collections when distributed by Gray. This of course does not take into account the unicates, which Gray retained and did not number, but, even so, it indicates that about half of the specimens Gray distributed must be a mixture of two or more collections, made at different places and at different times by Wright . . ."

In following Wright's journey of 1849 I shall utilize all the aforementioned sources, but depend upon Geiser's paper of 1935 for the interpretation of the route. Johnston, in his running commentary, assigns each of Wright's specimen-numbers to a particular county. Changing this form of presentation, I record the route under county-headings, followed by the total number of specimens which Johnston assigns thereto.

Wright's letter to Gray already quoted mentioned the storms marking the first days of the journey—no single factor produces *more* trouble for a plant collector than moisture in any form! French mentions the weather:

". . . The day fixed for . . . departure proved exceedingly unfavorable; the rain fell in torrents, which, added to those that had fallen a few days previous rendered the roads extremely bad . . . The following day a violent thunder-storm arose early in the morning, and the command remained in camp. On the morning of the 3rd . . . before the tents were pitched, again the rains began to fall. The prairies were now inundated —the roads so bad that it was with difficulty the company teams, overloaded as they were, could move . . ."

Wright had started from San Antonio, Bexar County, on the afternoon of May 31 and had joined French's train ". . . on present Medio Creek on the way to Castroville . . ." which, according to Geiser, was a fifteen-mile journey.

Medina County
Wright nos. *320* through *347*

On June 1 they reached Castroville, about twenty-five miles from San Antonio and situated, according to French, "on the west bank of the Medina river." He mentions that the town had ". . . 500 inhabitants, mostly German emigrants. The place presents but few signs of improvement, and idleness and poverty are more visible than industry and wealth; houses are falling to decay, and the rich lands lie uncultivated." Ten miles beyond at Quihi—where on June 2 Wright wrote his letter of protest to Gray—

was another German settlement, "a branch of the main one at Castroville," which boasted only "a few miserably rude huts." June 3 and 4 were spent in getting to the Hondo eleven miles away. On the 5th they made nine miles to the Seco, after crossing "a 'hog-wallow' that we found nearly impassable." Geiser states that they camped ". . . at a crossing two miles above present d'Hanis, then an embryo settlement of Germans . . ."

Uvalde County
Wright nos. 348 through 549

On June 6, having crossed "Rancheros . . . Creek, a branch of the Sabinal" and having reached and crossed the Sabinal, they came to the Frio River, remaining there the next day. Writes French:

"Thus far the road over which we had travelled is known as Wool's, or the Presidio road, and extends to the Rio Grande. But at the crossing of the Rrio Frio, the road to El Paso leaves it, and commences its course over the hitherto untrodden prairie. Bearing a more northernly course, it strikes the head-waters of the Leona above the site of the military post."

The Wool or Presidio road led to Eagle Pass in Maverick County. French's train was bound for Camp Inge which, so Geiser states, was ". . . near a rocky conical hill or mound now known as 'Inge Mountain', about two miles south of present Uvalde . . ." French was beginning to think better of the country along the Leona:

"The lands on this stream will vie in fertility with any portion of Texas . . . No part of the State offers greater inducements to the agriculturist, and as a grazing country it is unrivalled . . . The post is located on the left bank, above the Presidio crossing, near a rocky conical hill or mound . . . a beautiful site for a military station . . ."

They remained at Camp Inge from June 8 to 18 and then, on the 19th and 20th, moved twenty miles from the head of the Leona to the head of Turkey Creek. Geiser notes that the route taken ". . . probably followed the present course of the railroad to the neighborhood of present Obi Hill, and then followed the route later taken by the Uvalde-Cline-Brackett road. Camp was probably made near present Asphalt Mountain, at the head of Turkey Creek, in western Uvalde County. A spring made the head of Turkey Creek . . . heavy rains . . . made the roads difficult; and the progress of the train was stopped for nine days (June 21-29)."

This camp must have been close to the boundary between Uvalde and Kinney counties, in the vicinity of present Cline.

Kinney County
Wright nos. 550 through 598

Geiser records that the route taken on June 29 and 30 ". . . very closely anticipated the course of the present U. S. Highway No. 90, between Cline and Brackettville . . .":

". . . a march of thirteen miles brought Wright's company to the bed of Arenosa Creek ['Live Oak Creek']; and three miles farther on Elm Creek . . . Seven miles beyond . . . the road headed the beautiful Las Moras Springs . . . These . . . lie in present Brackettville. They . . . give rise to Las Moras Creek, which flows into the Rio Grande. Here the train camped for two days (June [July] 1-2).

"From Brackettville, on June [July] 3, the company set out for the springs at the head of the San Felipe. After seven miles' travel . . . they crossed Pinto Creek . . . at the crossing . . . Here they camped for the night."

French's name for the Pinto was "Piedra Pinta." He notes that on July 4 they came to "the Zoquete, a small stream that flows through a bed of rushes." Also spelled "Zacate" in the "Report," it is identified by Geiser as Mud Creek and as more nearly a "mud-hole." Three miles beyond the road crossed "Arroyo Pedro" (Sycamore Creek according to Geiser) and nine miles from San Felipe Creek. This stream now forms the boundary between Kinney County (east) and Valverde County (west).

Valverde County
Wright nos. *599* through *804*

On July 7 the party reached San Felipe Creek, remaining through the 21st. Geiser locates the camp ". . . about a half-mile below the beautiful San Felipe Spring, near present Del Rio . . . The flow-off of this, with that of other springs, forms San Felipe Creek, which flows into the Rio Grande, five miles to the south." French notes that they ". . . moved on the 22d, and thus continued the march to El Paso. Eleven miles distant is the San Pedro River." This was Devils River. Geiser states that it was crossed ". . . at a ford near the point where Sells Creek enters the river from the west. The road ascended the tableland beyond the Devils River by the arroyo of Sells Creek; and the company encamped . . . at a spring two-and-a-half miles from the mouth of the creek. From now on, until the wagon-train left the valley of the Devils River at Johnson's Draw, progress was slow, averaging only six miles a day. The route . . . fairly approximated, north of present Comstock, that of the Comstock-Ozona[2] highway. They approached the Devils River, and crossed to the left bank at a point about three miles south of the Thirtieth Parallel; and between that point and present Juno subsequently crossed and recrossed the Devils River . . .

"At the 'summer head' of the Devils River, about a mile below present Juno, the party, . . . turned up Johnson's Draw, and followed a westerly and northerly course by the . . . Juno-Ozona road . . ."

This was on August 2 and Geiser states that the train left the present Juno-Ozona road ". . . about seven miles north of the Valverde-Crockett County line."

2. This highway, Route 163, runs almost north through Juno, Valverde County, and eventually reaches Ozona in eastern Crockett County.

Crockett County
Wright made no collections between August 2 and August 11

According to Geiser, they now turned westwardly, ". . . up a ravine, from Johnson's Draw . . . descending by another ravine opening into Howard's Creek, they crossed the latter at Howard's Springs . . . distant forty-one miles from the Devils River . . ." He states that on the day, August 2, on which the train left Devils River, Wright ". . . fell ill with malaria, and was completely incapacitated for any work until the Pecos was crossed, and for some distance beyond. As the expedition took it, the distance between the Devils River and the Pecos was eighty-one miles. The region between the two rivers had an abundant and rich flora; and it was unfortunate that Wright could not have explored that region . . ."

Wright's no. *803* of August 2 was from Valverde County. His no. *804* is dated August 11; and his no. *805, et seq.,* from "Prairies of the Pecos," are dated August 12, according to Johnston's "Field notes." Johnston states that "The area not botanized includes most of the Johnson Creek (Johnson Draw) and Howard Creek and Live Oak Creek, all in Crockett County . . ."

Pecos County
Wright nos. *805* through *886*

Geiser states that the Pecos was reached on August 12, and that, until the 16th, the party ascended that stream ". . . to a point approximately 35 miles below Horsehead Crossing, and near the present point Iraan . . ." This is in easternmost Pecos County.[3] After the crossing, on August 16, ". . . the road lay in a westerly direction over a plain margined with broken ridges. Twelve miles from the river, the road turns over a gentle hill, and at . . . 18-20 miles from the Pecos, entered the valley of an intermittent creek, called in Wright's day 'Escondido Creek,' but now known as Mule Creek. Eight miles further along a good road . . . they came upon Escondido Springs . . . The general trend was westerly, over a good road . . . to Comanche Spring . . ." This Geiser identifies with present Fort Stockton. They camped here on the 19th and on the 20th at Leon Springs. On the 21st they camped ". . . somewhere not far from present Barilla Spring, forty miles or so west of Leon Springs, at the opening of the wide valley down which the Limpia Creek intermittently flows to join a north-flowing tributary of the Pecos[4] . . ."

3. Johnston takes issue with French's comment (p. 45) that ". . . 'For the distance of near forty miles the route lies up the east bank of the river . . .' " and states that ". . . French's map, Emory's map, and other old maps, all show the road on the *west* side of the Pecos. The road crossed the Pecos at the Ferry (7 miles above camp in lower Live Oak Creek, i.e. near Ft. Lancaster) at a point near present Sheffield." Sheffield is in the easternmost corner of Pecos County.

4. ? Paisano Creek.

Jeff Davis County
Wright nos. *887* through *1058*

On August 21, states Geiser, the ". . . stream-bed of the Limpia was struck fifteen miles up the valley . . . Wright's company ascended this stream until they reached the Cañon and Pass of the Limpia ('Wild Rose Pass') in the Davis Mountains. Through this lovely pass[5] . . . the company journeyed from the 24th to the 26th of August . . . On the 26th of August the train left the Limpia Pass . . . skirted the Davis Mountains, on their way from the Limpia to the Van Horn Flats." This took from the 26th to the 28th. While the party was still on the Limpia French refers to a halt at "Painted Camp" —Johnston's paper places this "at the head of the Limpia . . . a mile or two above [or west of] present Fort Davis." French next mentions "Smith's Run, twenty-six miles from the Limpia," identified by Johnston with "Rock Creek," and "18 miles east and a bit south of Valentine." Geiser, citing no dates from August 28 until September 3, records that the party now ". . . followed the Van Horn trough up the Chispa or Wildhorse Creek as far as the Van Horn Wells, in present Culbertson County. The Tierra Viega and Van Horn Mountains lay close at hand [or to the west] on their way up."

Culberson County
? Wright nos. *1059* through *1093*

On August 29 and 30, according to Johnston, Wright ". . . may have been collecting near Van Horn . . ."

Hudspeth County
Wright nos. ? *1094* through *1158* or *1173*

Following for a time what is now the line of the Southern Pacific Railroad, French's train reached and passed, states Geiser, ". . . Eagle Mountain, with Eagle Spring . . . From this point, the road bore to the left [which was southwest] between Eagle Mountain and Devil Ridge, crossing the present Quitman Arroyo. About twenty-three miles from Eagle Spring, the road entered a deep rugged cañon in the Quitman Mountains by which, after a course of eight or nine miles, the train entered (September 3) upon the valley of the Rio Grande, opposite present Las Banderas, Mexico. From this point, the course led up the valley of the Rio Grande to El Paso, which was reached the twelfth of the month . . ."

They reached the Rio Grande, according to Johnston, at a point six miles below "old Fort Quitman." This was situated on the north side of the river about midway along the southern boundary of Hudspeth County. If the daily mileage averaged about the same and if it took from September 3 to 12 to reach El Paso from the point where the party entered the valley of the Rio Grande—a distance estimated by French as

5. *See* Rogers McVaugh's "Wild Rose Pass . . ." (1948).

eighty miles—the Hudspeth-El Paso county line (which falls about halfway between Fort Quitman and El Paso) should have been reached on September 6 or 7.

El Paso County
Wright nos. *1159* (or *1174*) through *1189*

The specimens cited above would seem to have been collected *before* Wright's arrival at El Paso, reached on September 12.

French states that the town of El Paso was then ". . . wholly situated in Mexico[6] —there being, excepting the villages on the island, but three houses on the American side."

French's contingent remained at El Paso for one month, or through October 10. During that time Wright evidently moved about and his nos. *1190* (dated September 12) through *1293* (of October 10) were made between date of arrival and departure. Such towns as Socorro, San Elizario and Yselta were visited, all on the American side of the Rio Grande and all in present El Paso County and within a radius of about twenty miles of their camp.

Geiser's paper of 1935 ends, as far as itinerary is concerned, with Wright's arrival at El Paso. Between October 11 (departure from El Paso) and November 13 (arrival at San Antonio)—and there were thirteen days, eight in October and five in November, when he collected nothing—the Wright records cite only about one hundred and ten numbers (? *1297* through *1404*). The westward journey had consumed one hundred and five days; French states that the eastward one took "forty-one days," but my arithmetic would make it even shorter, certainly as far as Wright's participation is concerned. The trip may have been too rapid to make collecting possible; moreover it was late in the season and—from the crossing of the Pecos River eastward—he was in country where he had already worked on the way west. It is even possible that he did not feel like collecting—the weather was getting cold!

French reports that, in returning to San Antonio, they took ". . . the Fredericksburg or upper road to the point where it crosses the Pecos. Leaving El Paso, the road bears an easterly course for thirty miles to the Waco tanks." "Waco [or Hueco] tanks" is the first of a number of localities cited by French which serve to mark the route. Next mentioned are " 'Ojo del Cuerpo'," the Guadalupe Mountains, Delaware Creek and the Pecos River (followed for one hundred and seventy-five miles before reaching the Horsehead Crossing, where the party had last been on August 16 and which, ac-

---

6. Standley wrote in 1910 (and by "Wright's labels" he means Gray's labels):
"Wright's labels for his collection of 1849 read: 'Collected on a journey from San Antonio, Texas, to El Paso, New Mexico.' Now, New Mexico at that time was a term which included parts of Chihuahua, Texas, Arizona, in addition to its present territory. The El Paso thus referred to is not in New Mexico and it is not El Paso, Texas, but the present town of Ciudad Juárez, Chihuahua . . . Even now it is often impossible for the best informed to tell whether some of Wright's specimens came from New Mexico, from Texas, or from Mexico . . ."

cording to Geiser, was "near present point Iraan," in Pecos County). The route is shown clearly on the map of Senate Document No. 164 and, given that the latitudes and longitudes of this map are correct, the route between El Paso and Horsehead Crossing is not too difficult to follow, let us say approximately:

El Paso County
Wright nos. ? *1297* through *1308*

After crossing the Rio Grande the French train went northeastward to Hueco Tanks, situated in the northeastern corner of El Paso County, not far south of latitude 32° and not far west of longitude 106°, in the Hueco Mountains. Continuing northeast, it soon crossed latitude 32° and entered southern Otero County, New Mexico.

Otero County, New Mexico
? Wright nos. *1309* through *1329*

The direction taken across Otero County[7] was northeast, southeast and then east to a point west of the Sacramento Mountains. From there the route reëntered Texas in northeastern Hudspeth County, at a point to the west of longitude 105° and at latitude 32°, or in the region of French's (and the map's) " 'Ojo del Cuerpo' " and of the map's "Salt Lagunes." Johnston's interpretation of this portion of the route differs from mine (above), for he indicates that Wright crossed northern Hudspeth County without entering New Mexico. New Mexico was crossed from October 13 to 16 inclusive, or so I believe.

Hudspeth County

If this county was crossed, it must have been in its extreme northeastern corner and the transit a quick and short one.

Culberson County
Wright nos. *1330* through *1340*

After crossing the Guadalupe Mountains the party reached the headwaters of Delaware Creek and followed that stream across northern Culberson County east and northeast to near latitude 32° and longitude 104° when it left the creek and turned southeast and, crossing the Culberson-Reeves county line, reached the Pecos River at a point not far from present Angeles or Orla in the northernmost corner of Reeves County.

From their notations, supplied in the Johnston paper, Wright's nos. *1341* through

---

7. Wooton wrote in 1906: "Practically all the plants of the 1849 collection were gathered in what is now the State of Texas. Wright may have entered what is now New Mexico on the return journey, between October 12 and October 20, but this is quite doubtful, since the party apparently went down the river [the Rio Grande] to San Elezario and then turned eastward." This is incorrect, the Rio Grande was *not* descended on the return journey.

1345 were collected on October 20, between the Guadalupe Mountains (of Culberson County) and the Pecos (reached in Reeves County). Johnston notes that, between October 21 and November 2, the party with which Wright traveled was in the Pecos valley, "in Reeves, Pecos and Crockett counties." It is doubtless impossible to be more precise. For the French train descended the Pecos for the eastern length of Reeves County, and of Pecos County to the point where, on August 16, it had turned away from that stream to ascend Mule Creek on its way west towards Escondido Springs.

Wright's nos. *1346* through *1355*, dated October 25, were all from the "Valley of Pecos," as was his no. *1356*, of October 28. His nos. *1357* through *1367*, collected from October 29 to 31 inclusive, were gathered along or near the Pecos—probably on its western side and before the crossing; for his nos. *1368* through *1374*, dated November 2 are, by distinction, cited from "east of" the Pecos, and doubtless from Crockett County. Wright nos. *1375* through *1377*, dated November 3, were gathered "3 days march from the Pecos;" nos. *1378* through *1403*, dated November 5 through 11, came from Devils River or nearby—with the exception of no. *1403* which was from "San Felipe Creek," in southern Valverde County. Wright's last collection, no. *1404*, dated November 13, was from the "Nueces bottom," and presumably from Uvalde County.

From these records it is clear that, from El Paso to the crossing of the Pecos River in southeastern Pecos County, Wright's route had been farther north than the one taken westward. From the crossing to San Antonio his westward and eastward routes had been the same. It seems to me possible that, at the crossing, he may have separated from French's contingent and joined the one headed by Colonel J. E. Johnston in whose "Report," included in Senate Document No. 64, it is stated that, ". . . in the middle of October, the winter set in with such severity, that I thought the lives of our mules depended on turning southward. The men, also, were equipped for summer. Therefore, instead of crossing the Pecos we marched down it to the southern road which was followed to San Antonio . . ." That Wright was in touch with Colonel Johnston is suggested by a notation on Wright's no. *1321* of October 15: "on rocks in mountains east of El Paso, presented by Col. Johnson . . ." It is not clear, from a statement made by French in the last paragraph of his "Report," whether he *took,* or merely described, another route from the crossing:

"From there [the crossing] . . . the return route joins the one over which we went to El Paso near the point where it turns off from the Pecos for Escondido Springs. The time occupied in returning with a small train of about thirty-five wagons was forty-one days. The upper route from the Horse-head crossing on the Pecos continues, by way of the Concho and San Saba rivers, &c., to Fredericksburg, and thence to the seaboard, either by San Antonio or Austin . . . Such were the routes we travelled . . ."

The final part of Geiser's paper of 1935 is headed "An incomplete list of the new species collected with type-localities" and is stated to have been compiled ". . . from the *Plantae Wrightianae* of Asa Gray ... the numbers ... copied from Charles Wright's Manuscript *Field List* of his 1849 expedition ... preserved in the Gray Herbarium ..." The list cites under Wright's own numbers sixty-seven new species and three new varieties.

Gray's "Plantæ Wrightianæ," a systematic arrangement, only extends through the family *Compositae*. Johnston wrote:

"... Following the sequence of Bentham and Hooker, Gray treated in the *Plantae Wrightianae,* I and II,[8] the *Dicotyledonae* families up to and including the *Compositae.*[9] Gray appears to have considered that he had extracted all of value from the field-list for those collections treated in the *Plantae Wrightianae,* and as a general rule made little or no effort to preserve Wright's field-ticket, or to note the proper field-number on the herbarium-label of those collections. It is on the specimens belonging to the families not treated in either of the two volumes of the *Plantae Wrightianae* that Gray usually saved the field-tag or hastily pencilled (sometimes incorrectly) the field-numbers ..."

Part I of Gray's treatise included descriptions of numerous species collected by Wright in 1851—for Part I was not published until 1852. Also described for the first time are collections made by Gregg and other collectors with whose names we are familiar. And long footnotes provide classifications of plant-groups contributed by such authorities as George Engelmann and John Torrey. Seven genera, including the handsome shrubs *Mortonia* and *Fendlera,* and more than eighty species which Gray considered new are described from Wright's 1849 collections; some of these are figured in beautiful plates made from the drawings of Isaac Sprague. About twenty species (twelve new) which are listed in the index bear Wright's name. Other botanists are similarly honored.

On January 2, 1850, Gray wrote Wright from Cambridge that he had received a letter "... & at the same time your *Catalogue* (postage $1.40), which you had better have put in the boxes to come with the plants as it can be no use till they come ..." On February 10: "... Yesterday, without any previous notice, I was greatly surprised by receiving *through express from New York,* your collection of 8 bundles & 2 boxes (Express $10.00) ..." Wright, poor fellow, had probably considered it unwise to put all his eggs in one basket and, considering that shipping facilities from Texas could have been none too good or too frequent in 1849, his choice of "express" and his total expenditure of $11.40 seems not extravagant!

The best collection of Wright specimens is, of course, in the Gray Herbarium. On May 30, 1850, Gray wrote him regarding the distribution of the "sets":

---

8. Part II of Gray's treatise describes Wright's collections made in 1851 and 1852.

9. Standley states: "... When we look for the diagnoses of new species described in the remaining groups, we can find them in no one or two papers, but must look for them in dozens of places scattered through our botanical literature." Although Standley was writing of New Mexican plants, his statement has a general application.

". . . I[n] th[e] distribution of your sets the highest number is *828*. *This* runs through my set Lowell's an[d] Smithsonian's. The Bentham's which I pay for is *680*, Harvey's *690*, Hooker's *650*, Fielding's *644*, B. D. Greene's *631* (As he advanced 50$ that leaves only 13$, coming from him), British Museum *519*—Webb *453* (these two last are not engaged but I hope they will take them, I shall carry them abroad for th[e] purpose) A set I have offered to Durand contains *498*. As he bought a former Texan collection from you which contains many of th[e] same things and has also Lind-heimer's, I have written to him offering this set for 40$. Should he decline I shall try to dispose of it abroad for 50$ but there will be some expenses. In th[e] distribution I have used those collected between Galveston and San Antonio where they held out well, also I have used th[e] Rio Grande plants[1] and others [of] th[e] old collection. Here after do not send those collected in Texas or any formerly collected as they de-teriorate your sets   Every one of our subscribers have them from Drummonds and Lindheimers before   It was not well thought of to send that Texan bundle, as they are of no use, except that I have used a few to eke out other sets. For the rest—Torrey's set *433* specimens, when the specimens run short his are always scraps; those go against his paper etc. To Engelmann I have sent a number of supernumeraries and plants which I want him to study. DeCandolles set amounts to *410* specimens this I have offered to him as you proposed in exchange for the Prodromus I shall bundle up the other Texan plants remaining here and send to DeCandolle with them   I have not time to label them now but must do that at intervals of leisure while I am at Geneva this summer   The above enumeration takes all your collection Supernumeraries of interesting plants are thrown into my set from which specimens can be taken here after for yourself   If Durand, Webb, and the Brit[ish] Museum take the sets saved for them they will all be off our hands at once   I sent by private hand some of the earlier numbers of the set to Bentham and have got th[e] returns of the Leguminosae, which I asked him to elaborate; they have much novelty and interest   I have little time to reply to your letter of April 10th . . . I have sent him [Sullivant] all the mosses of the last collection, very few indeed; also I have sent all the Lichens down to Tuckermans room; he is still abroad . . ."

To collect plants for Asa Gray was no sinecure—defined in one dictionary as an "office of profit or honour without duties attached"—but it is obvious from the above that to distribute a man's sets involved a vast amount of "fussy" detail over and above the time-consuming work of classifying the plants. Johnston states that "Wright ap-parently received ten cents a specimen." Gray's recompense must have lain in the in-tangibles of accomplishment and accomplishment's rewards.

As stated in Gray's letter last quoted, one set of Wright's plants went to the Smith-sonian Institution. W. G. Farlow's chapter on "Botany" (included in *The Smith-sonian Institution 1846–1896. The history of its first half century,* edited by G. B.

---

1. The Eagle Pass collections made by Wright in 1848.

Goode and published in Washington in 1897) has considerable to tell of the early assistance rendered by that organization to botanical research, towards the publication of important papers (some of which have been mentioned in my story) and towards botanical exploration. Gray's "Plantæ Wrightianæ" seems to have been the first botanical treatise issued by the Smithsonian.

"The earliest reference to botanical work undertaken under the direction of the Institution is to be found in 1848[2] where mention is made of some drawings and engravings of a paper on the botany of Oregon, for which a small advance had been made, and in the same volume it is said that a 'report on the forest trees of North America,' by Professor Asa Gray, is in progress. The paper on the botany of Oregon refers apparently to the work on the plants of the Wilkes expedition, of which the part by Gray appeared in 1854 and that by Torrey in 1873. The Report of 1849 again mentions the 'report on the forest trees of North America . . .'[3]

"In 1849 the Institution contributed $150 towards defraying the expenses of the botanist, Charles Wright,[4] on an expedition to El Paso, Texas, and in 1852 appeared the first contribution to phænogamic knowledge, entitled 'Plantæ Wrightianæ Texano-Neo-Mexicanæ,' containing a description by Gray of the plants collected by Wright, together with many of those collected by Wislizenus in the valley of the Rio Grande and Chihuahua, and by Dr. Gregg in the same district and northern part of Mexico. A second part of the 'Plantæ Wrightianæ' appeared the following year, both parts with illustrations by Mr. Sprague. In rapid succession appeared three other 'Contributions to Knowledge' by Professor John Torrey . . . Professor Torrey published in 1854, in the sixth volume of the 'Contributions to Knowledge,' a monograph entitled 'Plantæ Frémontianæ,' in which he gave an account of twelve of the most characteristic genera and species collected by Frémont in California, including the new genera Spraguea, Fremontia, Coelogyne, Emplectocladus, Carpenteria, and Sarcodes, the type of the latter being the then remarkable but now familiar snow-plant of the Sierras, *S. sanguinea*. In the same volume are two other important papers[5] by Torrey . . ."

After mentioning other papers published by the Institution, all of a later period

2. Smithsonian Report, 1848, p. 16.

3. Farlow goes on to tell of the various delays which took place in regard to the publication of this paper and of the final abandonment of the project in 1884: "In 1891, the twenty-three beautifully colored plates which had been drawn by Mr. [Isaac] Sprague between 1849 and 1859 were issued by the Institution without text; and thus what was, as originally planned, to have been the first botanical publication of the Institution, in reality formed its last quarto publication on a botanical subject."

4. In the same volume from which I am quoting is a chapter by Frederick William True on "Exploration work of the Smithsonian Institution" which states: "In the direction of botanical explorations, the first aid rendered by the Institution took the form of a small appropriation for the expense of an expedition to Texas, in 1849, by Charles Wright, under the direction of Asa Gray. The results of this expedition were published in the 'Contributions' in 1852 and 1853."

5. Torrey's "Observations on the Batis maritima of Linnaeus," and his paper on the *Darlingtonia californica* discovered by Brackenridge.

than the one considered here, Farlow turns to the ". . . first step taken by the Institution towards the formation of a national herbarium . . ."

"In consideration of the $150 subscribed towards defraying the expenses of Mr. Wright on his botanical trip to El Paso, the Institution was entitled to a full set of all the plants he collected. At about the same date, a set of the plants collected by Fendler in 1847 in the vicinity of Santa Fé was purchased, and it was proposed, further, to assist him by the purchase of a set of the collections he might make in the future. The policy of the Institution in regard to giving aid to collectors and receiving in return sets of the plants collected was expressed in the Report for 1849 in the following words: 'By coöperating in this way with individuals and institutions, we are enabled, at a small expense, materially to advance the cause of science.' The Report for 1851, referring again to the sets of Wright and Fendler, states that these sets, together with plants collected by Lindheimer, 'form the nucleus of an important and authentic North American herbarium.' . . . Other additions were from time to time reported, the most important being a set of Doctor Berlandier's Texas plants in 1855, and the unique set of ferns collected by Brackenridge on the Wilkes Exploring Expedition in 1862 . . ."

FARLOW, WILLIAM GILSON. 1897. Botany. *In:* The Smithsonian Institution 1846–1896. The history of its first half century. Edited by George Brown Goode. Washington. 697-710.

FRENCH, SAMUEL GIBBS. 1850. Report. *In:* Reports of the Secretary of War, with reconnoissances of routes from San Antonio to El Paso, by Brevet Lt. Col. J. E. Johnston; Lieutenant W. F. Smith; Lieutenant F. T. Bryan; Lieutenant N. H. Michler; and Captain S. G. French, of Q'rmaster Dep't. *U. S. 31st Cong., 1st Sess., Sen. Exec. Doc.* 14: No. 64, 40-54. Map.

GEISER, SAMUEL WOOD. 1930. Charles Wright. *Southwest Review* 15: 342-378. *Reprinted* (with revisions): 1937. Naturalists of the frontier. Dallas. 215-252, 336. *Ibid.* 1948. ed. 2, 172–198, 283.

——— 1935. Charles Wright's 1849 botanical collecting trip from San Antonio to El Paso; with type-localities for new species. *Field & Lab.* 4: 23-32.

——— 1936. Charles Wright. *Dict. Am. Biog.* 20: 545-546.

——— 1942. [Corrections of biographical statements concerning John Allen Veatch.] *Southwestern Hist. Quart.* 46: 169-173.

——— 1948. John A. Veatch. *In:* Naturalists of the frontier. ed. 2. Dallas. 282.

GRAY, ASA. 1852. Plantae Wrightianae Texano-Neo-Mexicanae: an account of a collection of plants made by Charles Wright, A.M., in an expedition from Texas to New Mexico, in the summer and autumn of 1849. Part I. *Smithsonian Contr. Knowledge* 3: Art. 5, 1-129.

——— 1886. Charles Wright. *Am. Jour. Sci. Arts* ser. 3, 31: 12-17. *Reprinted:* Sargent, Charles Sprague, *compiler.* 1889. Scientific papers of Asa Gray. 2 vols. Boston. New York. 2: 468-474.

——— 1893. Letters of Asa Gray. Edited by Jane Loring Gray. 2 vols. Boston. New York. 1: 362, 363.

JOHNSTON, IVAN MURRAY. 1940. Field notes of Charles Wright for 1849 and 1851–52, relating to collections from Texas, New Mexico, Arizona, and adjacent Sonora and Chihuahua. A copy with commentary by I. M. Johnston. February 1940. 1-48. [Unpublished manuscript in Gray Herbarium.]

JOHNSTON, J. E. 1850. Report of Brevet Lieutenant Colonel J. E. Johnston ... dated December 28, 1849. *U. S. 31st Cong., 1st Sess., Sen. Doc.* 14: No. 64, 26-29.

McVAUGH, ROGERS. 1948. Wild Rose Pass, an obscure locality and its actual position. *Wrightia* 1: 207-213.

SARGENT, CHARLES SPRAGUE. 1890. The Silva of North America. Boston. New York. 1: 94, *fn.* 3.

STANDLEY, PAUL CARPENTER. 1910. The type localities of plants first described from New Mexico. *Contr. U. S. Nat. Herb.* 13: 143-146.

TRUE, FREDERICK WILLIAM. 1897. Exploration work of the Smithsonian Institution. *In:* The Smithsonian Institution 1846–1896. The history of its first half century. Edited by George Brown Goode. Washington. 462-463.

VEATCH, JOHN ALLEN. 1860. About Cerros Island. *Hesperian* 3: 529-534.

WARREN, GOUVENEUR KEMBLE. 1859. Memoir to accompany the map of the territory of the United States from the Mississippi River to the Pacific Ocean, giving a brief account of each of the exploring expeditions since A.D. 1800 ... *U. S. War Dept. Rept. expl. surv. RR Mississippi Pacific* 11: 58-60.

WINKLER, CHARLES HERMAN. 1915. The botany of Texas, an account of botanical investigations in Texas and adjoining territory. *Univ. Texas Bull.* 18: 9-10.

WOOTON, ELMER OTIS. 1906. Southwestern localities visited by Charles Wright. *Bull. Torrey Bot. Club* 33: 561-566.

CHAPTER LV

STANSBURY DESCRIBES THE CARAVANS MOVING WEST-

WARD OVER THE OREGON TRAIL AND REPORTS UPON

THE RECENTLY ESTABLISHED STATE OF DESERET

IN the years 1849—the last of my story—and 1850, Howard Stansbury of the Corps of Topographical Engineers of the United States Army made a journey to Salt Lake City and back. The content of a collection of plants made on this occasion must, in largest part, have been gathered on Stansbury's return trip and his exploration might appropriately, perhaps, be considered under a later decade. But in another sense it forms a fitting termination to the present story. My plant collectors, from the start, have participated in the historical picture and Stansbury describes at first hand, and extremely well, two important events: the migrations of citizens of the United States to the far west, and the establishment of a great colony—the State of Deseret—well towards the west of a region which in 1790, when my story opened, was an unknown wilderness to residents on the eastern seaboard. Much of the content of this chapter is devoted, therefore, to Stanbury's depiction of the caravans and of the progress of the colony on Great Salt Lake, begun only about two years before his advent.

Stansbury's report, "Exploration and survey of the valley of the Great Salt Lake of Utah, including a reconnoissance of a new route through the Rocky Mountains," was published in 1851 (*U. S. 32nd Cong., Special Sess., Sen. Exec. Doc.* 2: No. 3, 1-267) and is a model of clarity as to routes; and, to make clarity even clearer, its Appendix A includes a "Table of distances measured along the route traveled by the expedition in 1849." [1] The maps are fine—"Drawn by Lieut. Gunnison and Charles Preuss" —and the included lithographs, which, as usual, leave the name of the artist to the imagination, are well drawn and pleasing.

For most of his way west Stansbury traveled the already-well-worn trail to Oregon. His first chapter covers the journey from Fort Leavenworth, Kansas, to Fort Kearny,[2] Nebraska; chapter two its continuation to Fort Laramie in southeastern Wyoming; chapter three its progress to Fort Bridger in southwestern Wyoming; and chapter four the rest of the way to Great Salt Lake. The journey began on May 31, 1849, and "the City of the Great Salt Lake" was reached on August 23. In the autumn of 1849 there was an "Exploration of a route from Great Salt Lake City to Fort Hall [in southern

---

1. Although the title does not mention the year 1850, the substance covers both the "Outward journey," of 1849 and the "Homeward journey" of 1850.

2. For spelling of name see p. 758, *fn.* 2.

Idaho], and a reconnoissance of Cache Valley [northern Utah]," as well as a "Reconnoissance of the deserts around the western shores of the Great Salt Lake." The winter of 1849–1850 was spent in the vicinity of the State of Deseret and in the spring and early summer of 1850 the shores of Great Salt Lake and some of its islands were surveyed. The return journey began on August 27, 1850, and ended on November 6 of that year at Fort Leavenworth. Stansbury occasionally mentions a few plants in his narrative but not often.

Even before leaving Fort Leavenworth the expedition had been joined by ". . . a small party of emigrants for California, who desired to travel in our company for the sake of protection, and who continued with us as far as Salt Lake City . . ." Females—"of the species"—are rarely mentioned in my narrative[3] but Stansbury refers to one, for the first and only time, in this connection:

". . . This proved a fortunate arrangement, since we thereby secured the society of an excellent and intelligent lady, who not only, by her cheerfulness and vivacity, beguiled the tedium of many a monotonous and wearisome hour, but, by her fortitude and patient endurance of exposure and fatigue, set an example worthy the imitation of the ruder sex."

We are told that the route ". . . has been travelled by thousands of people, both before and since we passed over it . . ." Cholera had for some time "been raging" on the Missouri and ". . . fearful rumours of its prevalence and fatality among the emigrants on the route daily reached us from the plains . . ." One member of the party died within twenty-four hours on the first day's journey. On the second day (June 1) they passed ". . . the travelling-train of a Mr. Allen . . . about twenty-five ox-teams, bound for the land of gold. They had been on the spot several days, detained by sickness." One member of his party had died of cholera the day before and two more were ill. Four men from the Allen camp were already starting back, "on foot, with their effects on their backs, frightened by the danger and disgusted already with the trip." On June 7 they met a ". . . Mr. Brulet, a French trader, from Fort Laramie, with a large train of wagons, laden with packs of buffalo-robes, bound for St. Louis.

"He had been forty days on the road, and had met not less than four thousand wagons, averaging four persons to a wagon. This large number of emigrants appeared to him to be getting along rather badly, from their want of experience as to the proper mode of travelling on the prairies, to which cause much of the suffering experienced on these plains is doubtless to be ascribed . . . In the course of the morning passed the fresh grave of a poor fellow whose last resting-place had been partially disturbed by

3. Unless it be the wife of Henry Harmon Spalding, no women participated in the work of plant collecting in my period. The fact brings to mind Professor Sargent's story of a botanical excursion of which he was a member. It was a "mixed" party—ladies and gentlemen—and they had been obliged to camp out unexpectedly for the night in a one-room cabin. After the candles had been put out one of the gentlemen was heard to inquire: "Am I permitted to remove my spectacles?"

the wolves . . . It was an affecting object, and no good omen of what might be looked for, should any of us fall by the way . . ."

June 8: ". . . a small party, with a single wagon, drove into camp . . . They . . . had proceeded within sixty miles of Fort Kearny, but had quarrelled, and become disgusted with the trip and with each other . . . Those persons were on their return to St. Louis. They gave discouraging accounts of matters ahead. Wagons, they said, could be bought, upon the route of emigration, for from ten to fifteen dollars apiece, and provisions for almost nothing at all . . ."

June 9, at camp on the Big Blue River: ". . . We found the trees and stumps . . . carved all over with the names of hundreds of emigrants . . . the dates of their passing, the state of their health and spirits . . . Such a record, in the midst of a wide solitude like this, could not but make a strong and cheering impression on every new-comer, who thus suddenly found himself, as it were, in the midst of a great company of friends and fellow-travellers . . . We had passed six graves already during the day. Melancholy accompaniments they are of a road silent and solitary at best . . . a small party of emigrants . . . had lost most of their cattle on the journey; and the father of three of them having died on the road, they, in conformity with his dying wishes, were now on their return to the settlements . . ."

June 10: ". . . a species of mallow and *Œnothera* occurred on the bottoms of the streams, with *Digitalis* and *Loasa nitida*. Phlox once abundant is becoming scarce."

June 12: ". . . We have been in company with multitudes of emigrants the whole day. The road has been lined to a long extent by their wagons, whose white covers, glittering in the sunlight, resembled, at a distance, ships upon the ocean. We passed a company from Boston, consisting of seventy persons, one hundred and forty pack and riding mules, a number of riding horses, and a drove of cattle for beef. The expedition, as might be expected, and as is too generally the case, was badly conducted: the mules were overloaded, and the manner of securing and arranging the packs elicited many a sarcastic criticism from our party, most of whom were old and experienced mountain-men, with whom the making up of a pack and the loading of a mule amounted to a science. We passed also an old Dutchman, with an immense wagon, drawn by six yoke of cattle, and loaded with household furniture. Behind, followed a covered cart containing his wife, driving herself, and a host of babies—the whole bound to the land of promise, of the distance to which . . . they seemed to have not the remotest idea. To the tail of the wagon was attached a large chicken-coop, full of fowls; two milch-cows followed, and next came an old mare, upon the back of which was perched a little, brown-faced, barefooted girl, not more than seven years old, while a small sucking colt brought up the rear. We had occasion to see this old gentleman and his caravan frequently afterward, as we passed and repassed each other, from time to time, on the road. The last we saw of him was on the Sweetwater, engaged in sawing his wagon into two parts, for the purpose of converting it into two carts, and in disposing of everything he could sell or give away, to lighten his load.

"*Œnothera*, with its bright yellow flowers, was frequent, both to-day and yester-day, with *Amorpha* and *Artemisia*. The prairie-rose is becoming quite abundant."

June 13, on Little Walnut Creek: ". . . The wind blew a hurricane, the rain fell in tor-rents, while the thunder and lightening were terrible and incessant . . . Our men were exposed to all its fury for several hours . . . Our poor mules having been ticketed . . . seemed dejected, tired and hollow . . . altogether the camp seemed weary and dis-pirited . . . the men . . . engaged in drying their bedding, which had been thoroughly soaked . . . An immense number of black beetles and other insects swarmed around the camp last evening . . . they annoyed us beyond measure, and could be heard all night, pattering against the tents . . ."

June 14: ". . . The aloe and the prickly-pear were found in the sand-hills, as were the *Commelina* and the *saxifrage*. The prairie-rose, *Amorpha*, *Œnothera*, and *Arte-misia* abound. A blue lupine and a white mallow were gathered."

June 18: ". . . The grass is generally very abundant and *prêle* (the common scouring-rush) is found in great plenty. Our mules ate it with avidity . . . The valley of the Little Blue has not presented any great novelty in the way of flowers. The only new plants met with have been a lupine, the flower of which, of a bright purple, rises di-rectly from the root; the plant is totally leafless. A splendid variety of the mallow, of a bright carmine colour, its trailing stems sending up flowers in little patches of a few yards square, presented a rich and beautiful appearance. The aloe occurred in some places in abundance; and there were a few cacti, and a species of a leguminous plant . . . having a flower of a pale purple colour, resembling the vetch; also a species of pale blue digitalis . . .

"Archambault, our guide, told me that the last time he had passed this spot, the whole of the immense plain, as far as the eye could reach, was black with herds of buf-falo. Now, not so much as one is to be seen; they have fled before the advancing tide of emigration. Driven from their ancient and long-loved haunts, these aboriginal herds, confined within still narrowing bounds, seem destined to final extirpation at the hand of man . . ."

On June 19 they reached Fort Kearny, then under the command of ". . . Colonel Bonneville, whose adventures among the Rocky Mountains are so well known to the world. He received us courteously . . ." Here Stansbury was able to buy a little "spring-carriage"—discarded by some emigrants—for the use of Gunnison who had been ill and could not ride:

". . . Such abandonments are very common; most of these sanguine and adventur-ous companies, by the time they get thus far . . . find out that they have started on their journey with more than they can contrive to carry . . . to lighten their load, most of them dispose of every thing they can possibly spare, and at almost any price. Flour and bacon, for example, have been sold as low as one cent per pound; and many, being unable to sell even at that price, had used their meat for fuel. The pack company from Boston, which had passed us on the route, and which we found encamped here on our arrival, left before our departure. As they had been entirely unaccustomed to the operation of packing, their mules, as was to be expected, were in a most terrible condition, with galled backs and sides that made one shudder to behold. The proper mode of arranging the load of these suffering animals is an art taught only by experience. These people, though belonging to a race famous for foresight and calculation, had, like others from less thrifty and managing portions of the Union, been selling and giving away all they could dispense with . . ."

Between Fort Kearny and Fort Laramie little is said about the emigrants. There are more references to plants seen along the way. Encamped on the banks of the Platte, "fifty-six miles above Fort Kearny," the soil was found to be richer and contained more clay. On June 24 ". . . The plants seen were *Tradescantia,* the purple mallow, (the root of which resembles the parsnip, and it is used by the Indians for food,) the small yellow *Œnothera,* and a pretty, small stellate-flowered plant. Over large portions of the bottom, no flowers were met with; on the high ground, red mallow, *Mimosa, Linum,* a white *Mimulus,* and a sort of larkspur. The aloe was flowering in abundance on the face of some very steep bluffs."

On June 26 Stansbury mentions "*Carduus, Cactus* with a large sickly-looking yellow flower, *Amorpha, Tradescantia,* a small sunflower, and a species of milk-plant . . . The *Amorpha* is beginning to bloom. The vetch, with its purple clusters . . . seems of a different species from that seen heretofore, and has not so much foliage." That day the pack train from Boston was overtaken. They reported having seen a small herd of one hundred buffalo, and on the 27th Stansbury's hunters killed their first.

". . . Some idea may be formed of the great digestibility of this species of food, as well as of the enormous quantities devoured at a single meal, from the fact that the regular daily allowance or ration for one employee in the Fur Company's service is eight pounds, the whole of which is often consumed. It is true, however, that an old mountaineer seldom eats anything else. If he can get a cup of strong coffee, with plenty of sugar, and as much buffalo-meat as he can devour, he is perfectly happy and content, never feeling the want either of bread or vegetables."

After a halt of six days at Fort Laramie the Stansbury party moved on towards Fort Bridger. At Fort Laramie the emigrants were less than half way on their journey. Many had learned—by the bitter lesson of experience—many things which they should have been told by the well-informed before they started on their way.

July 19: ". . . We passed to-day the nearly consumed fragments of about a dozen wagons that had been broken up and burned by their owners; and near them was piled up, in one heap, from six to eight hundred weight of bacon, thrown away for want of means to transport it farther. Boxes, bonnets, trunks, wagon-wheels, whole wagon-bodies, cooking utensils, and, in fact, almost every article of household furniture, were found from place to place along the prairie, abandoned for the same reason . . ."

July 21: "The road, as usual, was strewn with fragments of broken and burnt wagons, trunks, and immense quantities of white beans . . . thrown away by the sackful, their owners having become tired of carrying them farther, or afraid to consume them from danger of the cholera. The commanding officer at Fort Kearny had forbidden their issue at that post on this account. Stoves, gridirons, moulding-planes and carpenters' tools of all sorts, were to be had at every step for the mere trouble of picking them up."

July 22: ". . . A considerable change has taken place in the flora as the country begins to ascend. Since leaving Fort Laramie, a variety of geranium has been frequent upon the borders of the streams. A small-leaved Œnothera, white, and the blue Digitalis, were also found. On the north side of the ridge, some plants were seen which we had not met with before; Azalea; a small white Œnothera, on a tall stem, with flowers not more than a line and a-half in diameter; two species of Potentilla, yellow, and two or three varieties of Campanula."

July 27: "To-day we find additional and melancholy evidence of the difficulties encountered by those who are ahead of us . . . we passed eleven wagons that had been broken up, the spokes of the wheels taken to make pack-saddles, and the rest burned or otherwise destroyed. The road has been literally strewn with articles that have been thrown away. Bar-iron and steel, large blacksmiths' anvils and bellows, crow-bars, drills, augers, gold-washers, chisels, axes, lead, trunks, spades, ploughs, large grindstones, baking-ovens, cooking-stoves without number, kegs, barrels, harness, clothing, bacon, and beans, were found along the road in pretty much the order . . . enumerated. The carcasses of eight oxen, lying in one heap . . . explained a part of the trouble. I recognised the trunks of some of the passengers who had accompanied me from St. Louis to Kansas . . . an excellent rifle was found in the river, thrown there by some desperate emigrant . . . In the course of this one day the relics of seventeen wagons and the carcasses of twenty-seven dead oxen have been seen. Day's march, twenty-four miles."

August 6: "I witnessed, at the Pacific Springs, an instance of no little ingenuity on the part of some emigrant. Immediately alongside of the road was what purported to be a grave, prepared with more than usual care, having a headboard on which was painted the name and age of the deceased, the time of his death, and the part of the country from which he came. I afterward ascertained that this was only a ruse to conceal the fact that the grave . . . had been made a safe receptacle for divers casks of brandy, which the owner could carry no farther. He afterward sold his liquor to some traders farther on, who, by his description of its locality, found it without difficulty."

On August 11 the Stansbury expedition reached Fort Bridger[4] and were received "... with great kindness and lavish hospitality by the proprietor, Major James Bridger, one of the oldest mountain-men in this entire region, who has been engaged in the Indian trade, here, and upon the heads of the Missouri and Columbia, for the last thirty years."

At Fort Bridger the trail to Oregon turned northwest to Fort Hall and Stansbury's comments upon the migrations cease. Many diaries and reminiscences written by participants in these migrations have been published. Stansbury, who was obviously highly intelligent, and thoroughly experienced in the art of overland travel, wrote as an observer rather than as a participant and his descriptions are factual and dispassionate. I have read none that are more convincing. I have quoted but a fraction of what he relates.

With the assistance of Bridger, Stansbury was now to investigate a new route to the head of Great Salt Lake—what he calls a " 'cut-off' to the travel for either Oregon or California." They turned southwest and reached their destination on August 29. I turn now to Stansbury's account of the remarkable accomplishments of those of Mormon faith.

"Before reaching Great Salt Lake City, I had heard from various sources that much uneasiness was felt by the Mormon community at my anticipated coming among them. I was told that they would never permit any survey of their country to be made; while it was darkly hinted that if I persevered in attempting to carry it on, my life would scarce be safe ... I at once called on Brigham Young, the president of the Mormon church and the governor of the commonwealth ... explained to him the views of the Government in directing an exploration and survey of the lake, assuring him that these were the sole objects of the expedition ... The impression was that a survey was to be made of their country in the same manner that other public lands are surveyed, for the purpose of dividing it into townships and sections, and of thus establishing and recording the claims of the Government to it, and thereby anticipating any claim the Mormons might set up from their previous occupation ... it must be remembered that these people are exasperated by the wrongs and persecutions they had previously suffered in Illinois and Missouri ... they had ... fled to these far distant wilds, that they might enjoy undisturbed the religious liberty which had been practically denied them ... now they supposed themselves to be followed up by the Federal Government with the view of driving them out from even this solitary spot, where they had hoped they should at length be permitted to set up their habitation in peace ..."

When the true purpose of Stansbury's visit was understood all objections were withdrawn and he notes that it was a pleasure to acknowledge "... the warm interest manifested and efficient aid rendered, as well by the president as by all the leading men

4. Fort Bridger, built by James Bridger in 1843, belongs to the period of the great migrations and Stansbury, in 1849, seems to have been the first and only one of my plant collectors to refer to this famous trading post.

of the community, both in our personal welfare and in the successful prosecution of the work." I omit Stansbury's account of his explorations in the autumn of 1849 and turn at once to his comments upon the Mormons and upon their settlement:

"The founding, within the space of three years, of a large and flourishing community, upon a spot so remote from the abodes of man, so completely shut out by natural barriers from the rest of the world, so entirely unconnected by watercourses with either of the oceans that wash the shores of this continent—a country offering no advantages of inland navigation or of foreign commerce, but, on the contrary, isolated by vast uninhabitable deserts, and only to be reached by long, painful, and often hazardous journeys by land—presents an anomaly so very peculiar, that it deserves more than a passing notice. In this young and progressive country of ours, where cities grow up in a day, and states spring into existence in a year, the successful planting of a colony, where the natural advantages have been such as to hold out the promise of adequate reward to the projectors, would have excited no surprise; but the success of an enterprise under circumstances so at variance with all our preconceived ideas of its probability, may well be considered as one of the most remarkable incidents of the present age." [5]

After describing the origins of the Mormon faith and the persecutions to which its adherents had been subjected wherever they went, Stansbury takes pains—he does so more than once—to vouch for their patriotism. It had been demonstrated by their readiness to supply a military contingent—the "Mormon Battalion"—for service in the Mexican War: this people had hardly started for Great Salt Lake when ". . . an officer of the United States Government presented himself, with a requisition for five hundred men to serve in the war with Mexico. This demand, though sudden and unexpected, was promptly and patriotically complied with; but in consequence the expedition was broken up for the season . . ." The advance guard of the Mormons had reached Great Salt Lake on July 21, 1847, and the main body a few days later:

". . . A piece of ground was selected, consecrated by prayer, broken up, and planted; and thus, in 1847, was formed the nucleus of what, in 1850, was admitted as a Territory of the Union, and which bids fair ere long to present itself at the door of the national legislature for admission as one of the States of the confederacy.

"In a short time . . . ground was surveyed and laid out into streets and squares for a large city; a fort or enclosure was erected, of houses made of logs and sun-dried brick, opening into a large square . . . In October [of 1847] . . . an addition of between three and four thousand was made to their number, by the emigration of such as had been left behind . . . ploughing and planting continued throughout the whole winter and into July following, by which time a line of fence had been constructed,

5. H. H. Bancroft wrote in his *History of Nevada, Colorado, and Wyoming* (p. 65): ". . . Much notoriety was given to Frémont's explorations, and less to a greater movement—that of the Latter-day Saints, who founded a city two thirds of the way across the continent, and in so doing forestalled the necessity about to arise for such a station at such a place . . ."

enclosing upward of six thousand acres of land, laid down in crops, besides a large tract of pasture land ... This year (1848,) a small grist-mill was erected, and two saw-mills nearly completed ... The following winter and spring, a settlement was commenced on the banks of Weber River ... Upon Ogden Creek ... a city has since (1850) been laid out, and called Ogden City, and is already surrounded by a flourishing agricultural population. In the autumn, another large immigration arrived under the president, Brigham Young ... Building and agriculture were prosecuted with renewed vigour. Numerous settlements continued to be made wherever water could be found for irrigation ...."

Still other settlements had already been started. In March of 1849, a convention was held and a constitution adopted for the State of Deseret—an essential step until such time as the federal government should see fit to admit the colony as a state or territory. Writes Stansbury: "... Such is a brief sketch of the origin and progress of this colony, and the condition in which we found it upon our arrival in August, 1849."

"A city had been laid out upon a magnificent scale, being nearly four miles in length and three in breadth; the streets at right angles to each other ... one hundred and thirty-two feet wide, with sidewalks of twenty feet; the blocks ... divided into eight lots, each of ... an acre and a-quarter of ground. By an ordinance of the city, each house is to be placed twenty feet back from the front line of the lot, the intervening space being designed for shrubbery and trees. The site for the city is most beautiful ...

"The facilities for beautifying this admirable site are manifold. The irrigating canals, which flow before every door, furnish abundance of water for the nourishment of shade-trees, and the open space between each building ... when planted with shrubbery and adorned with flowers, will make this one of the lovely spots between the Mississippi and the Pacific. One of the most unpleasant characteristics of the whole country, after leaving the Blue River, is the entire absence of trees from the landscape ... The studding ... of this beautiful city with noble trees,[6] will render it, by contrast with the surrounding regions, a second 'Diamond of the Desert,'...."

The Mormons, we are told, had recruited converts "... in most of the countries of Europe, as well as in the Sandwich Islands, and even here in our own country, with all of whom it is made a cardinal point to 'gather in the mountains.'" To make the journey of these recruits possible, a fund had been established:

"Measures are being taken to open a southern route, by which the converts coming from abroad may cross the Isthmus of Panama, and, landing at San Diego, may thence reach the land of promise by a comparatively short and easy transit, without being subjected to the hazard of a sickly voyage up the Mississippi, or to the tedious and expensive journey across the plains. In the mean while, preparations are industriously

6. In regions of the southwest far removed from northern Utah, one can recognize a Mormon settlement from afar by the trees, usually fine old poplars. The planting of trees seems to have been considered as essential as the erection of a dwelling.

making in the valley for the reception and immediate accommodation of the coming tide, by the building of houses, sowing large quantities of grain, the erection of mills, the establishment of manufactures, the importation of labour-saving machinery, and the establishment upon a solid basis of the means of education . . .''

As we follow Stansbury's picture of what a journey across the continent involved, in difficulties of transportation alone, it reads like a fairy tale that such a prodigious enterprise should have been launched and should have attained within less than three years the proportions and potentials for future development that it had. But Stansbury's "Exploration" contains no suggestion of the exaggerated or of the fictional.

Appendix D of the "Exploration" includes a sixteen-page paper, "Catalogue of plants collected by the expedition," by John Torrey. There are nine fine plates but who made the drawings for them is not stated. Torrey's biographer Rodgers mentions that Torrey "placed a German at work drawing plants for a report on Stansbury's expedition," but does not disclose his name. The plates included two new species named for Stansbury—*Cowania*[7] *Stansburiana* and *Monothrix Stansburiana*. *Monothrix* was described as a new genus.[8] The greater part of the plants listed in the "Catalogue" were gathered in 1850, rather than in 1849.

According to Rodgers, Torrey was disappointed that the number of plants collected was smaller than he had anticipated,[9] and annoyed that he had not seen the proof sheets of his paper, but Stansbury, "like other officers of the time seeking to hurry the publication of their reports," did not submit them for his inspection and corrections.

In his introduction to the "Exploration" Stansbury refers to the fact that, "In the Department of Natural Science . . . I was not successful in securing the services of a competent assistant . . ." Rodgers comments: "It was not that Stansbury did not have a

7. The man who drove me on many collecting trips in the southwest became familiar with the scientific names of certain plants and occasionally used them to some effect. Once, wishing to collect some specimens under a tremendous oak, I was hesitating because of a cow of menacing appearance and with an aversion to moving. "That tree," said my driver, "is certainly a Cowania!"

8. According to Coville: ". . . In the second edition [of Stanbury's report, a *House Document* published in 1853] the genus Monothrix is discarded, the name given for this plant being *Laphamia Stansburii* (Gray, Plant. Wright. 101. 1852) . . ." Coville discusses the matter of priority, concluding that "It is evident that Monothrix does not have priority over Laphamia."

9. Rodgers quotes a letter which Torrey wrote W. D. Brackenridge—disclosing his excitement on learning of the collection:

" 'I see by the papers that an officer (whose name I have forgotten) has returned from the Salt Lake Valley with a large collection of objects of natural history—among the rest, the *plants of the region!* Now you known what a passion I have for *dried specimens* especially from the *far west*, as I have spent so much time in studying its Botany. Now I wish you would put me on the track of these plants, that I may obtain them for description. If they are sent to me, I will prepare a report on them for the officer who conducted the exploration—i.e. unless he has already made another disposition of them. As Dr. Gray is now absent in Europe, & will not return for another six months—there are no botanists except Dr. Engelmann & myself who have the means of giving them a proper examination—for we alone have extensive herbaria of Western plants & the books required for determining new collections . . .' "

competent naturalist. It was that Blake,[1] his naturalist, deserted his post to go to the California gold mines."

1. I have found no mention of the delinquent in Stansbury's report but Rodgers identifies him in his index as "W. P. Blake." William Phipps Blake is best known as a geologist; but he did collect plants on the railroad survey led by R. S. Williamson which, in 1854–1856, explored passes in the Sierra Nevada and in the Coast Range. Ewan notes that Blake's "... 'Observations on the mineral resources of the Rocky Mountain chain, etc.' ... was much sought during the gold rush days in the central Rocky Mountains ... Inexplicably his date of birth is variously given by biographers, as June 21, 1825, June 1, 1826, or even 1828!"

BANCROFT, HUBERT HOWE. 1890. The works of ... 25 (Nevada, Colorado, & Wyoming 1540–1888): 65.

———— 1889. The works of ... 26 (Utah 1540–1886) 463-467, *fn.* 53, 54, 55.

COVILLE, FREDERICK VERNON. 1896. Three editions of Stansbury's Report. *Bull. Torrey Bot. Club* 23: 137-139.

EWAN, JOSEPH. 1950. Rocky Mountain naturalists. Denver. 166.

GHENT, WILLIAM JAMES. 1935. Howard Stansbury. *Dict. Am. Biog.* 17: 516.

RODGERS, ANDREW DENNY 3RD. 1942. John Torrey. A story of North American botany. Princeton. 215, 218.

STANSBURY, HOWARD. 1851. Exploration and survey of the valley of the Great Salt Lake in Utah, including a reconnoissance of a new route through the Rocky Mountains. By Howard Stansbury, Captain Corps of Topographical Engineers, U. S. Army. *U. S. 32nd Cong., Spec. Sess., Sen. Ex. Doc.* 2: No. 3, 1-267. Maps and plates.

TORREY, JOHN. 1851. Catalogue of plants collected by the expedition. By Professor Torrey. *In:* Stansbury, H. 1851. Exploration and ... Appendix D. Botany. 9 plates. 381-397.

CHAPTER LVI

NOTES ON AUGUST FITCH, GEORGE GIBBS, ALBERT KELLOGG, WILLIAM LOBB, JAMES McNAB AND JACOB DAVIS BABCOCK STILLMAN

PERSONS interested in the plant collectors of the region and of the period of my story will doubtless point out "the forgotten man." This seems inevitable. I mention a few such possible candidates for inclusion but only to indicate that of some I have been able to learn very little and that the accomplishments of others appear to have been associated with a later decade than the last considered here. I have not, however, examined herbarium material nor searched the files for letters which might—but more probably do not—contain important facts about their activities *before 1850*. Most of the men mentioned are cited in W. H. Brewer's "List of persons who have made botanical collections in California" as having arrived on the Pacific coast near the end of the decade of 1840–1850. I refer to them alphabetically.

AUGUST FITCH

Brewer mentions that "Rev. A. Fitch travelled extensively in California about 1846 to 1849, and sent several collections to Dr. Torrey, including some species of much interest. Most of the specimens were from the southern half of the State." Turning, therefore, to the New York Botanical Garden as the probable source of information, I learned from Dr. H. W. Rickett (*in litt.*, March 1, 1951) that "There seems to be very little information available on Fitch . . . except that . . . [he] collected in California and sent specimens to Torrey. The date of Fitch's collections seems to have been 1846 to 49 . . . We have a few letters from . . . [Fitch] in the Torrey correspondence dealing with specimens which . . . [he] sent him." Except for the "few letters"—upon which I took no action—this differs little from Brewer's statement.

Rodgers' biography of John Torrey includes a chapter on "The Pacific railroad and other surveys"—concerned with the period after 1850—in which the author mentions that Torrey was ". . . overwhelmed with the abundance of materials that had come to him for study . . . Plants kept coming from isolated explorers in California such as Dr. G. W. Hulse[1] and a Rev. Fitch . . ." This comment, and Brewer's, are the only published references to Fitch's collecting which I have discovered.

---

1. Brewer records that "Dr. G. E. Hulse collected some plants in the northern part of the State of California about 1850 and sent his specimens to Dr. Torrey." Despite a difference in middle initial the reference is undoubtedly to Gilbert White Hulse. I have mentioned elsewhere that ". . . Dr. G. W. Hulse, of New Orleans . . . in May, 1851 . . ." had been sent by Torrey to obtain flowers of the *Darlingtonia*, a pitcher

However, the "Rev. August Fitch" wrote two short articles: (1) "The shrubs, flowers and plants of California," "Communicated to the Institute [of the City of New York] . . ." and published in its *Transactions* in 1853 (pp. 75-84); and (2) "Trees of California and Oregon," published in the same journal and in the same year (pp. 239-244). In a final paragraph in the latter Fitch states that he had planned to include ". . . California's flowers, shrubs, &c. . . ." and to present its agricultural "capabilities" but feared that his paper was ". . . already lengthened beyond your patience, and must leave these subjects for future papers." The editors did some abridging, but Fitch must have been modest for his paper was only five small pages long. The editors mention by way of introduction that "The following communication from the Rev. A. Fitch, (recently from California and Oregon,) to Hon. James Tallmadge, President of the American Institute, was read, and the *galls,* the *hyssopifolia,* and the *laurus California,* with their leaves and prints, exhibited to the club. Gen. Tallmadge's note to the club was read. He presents the specimens to the club."

The two Fitch articles were on the "amateurish" side and I surmise that that may be one reason why Torrey took no part in their publication. They do indicate where Fitch did his collecting: but, since few scientific names are mentioned, one cannot be sure of the plant in many instances! The article (1) above refers frequently to regions "around San Francisco," "about Monterey, on the hill sides," to "the valley of San Jose," "the banks of the San Joaquin . . . about Stockton and Sacramento," "along the Mercedes River," "about Trinity river," to "hills in the mineral region," "on the burial ground near Sacramento," and so on. Article (2) refers to much the same collecting fields.

If Frémont's claim—that he bestowed the name Golden Gate—is true, as seems to be the case, then Fitch drew upon his imagination when, writing of "the golden Ranunculus [?]," he states that "The prevalence of this gold-colored flower on the hills has given to the entrance of San Francisco harbor the appelation of 'golden gate.' "

While it is possible that the "Rev. Fitch" may not have qualified as a "botanist," he evidently kept his eyes open on his travels and I found some of his comments interesting. Writing of the madroña, *Arbutus menziesii* Pursh, he states:

". . . This tree has been held in such estimation that, under the Spanish law, a heavy fine was imposed upon its destroyer. It has become rather a rare tree, as I have met with it in but few localities. Magnificent specimens may be met with in the valleys of San Jose and Sonoma: a few small specimens are to be found in Happy Valley, San Francisco. There have been some large trees here as would appear from the stumps, but they have been prostrated by the axe of civilization. This beautiful tree I think will stand the climate of the middle states, as I have found it high up on the Columbia and Wilametic rivers in Oregon . . . also in the cold regions of Nevada."

plant, which Brackenridge of the United States Exploring Expedition had collected on October 2, 1841, on the Sacramento River, and that Hulse had been successful in his mission.

BREWER, WILLIAM HENRY. 1880. List of persons who have made botanical collections in California. *In:* Watson, Sereno. Geological survey of California. Botany of California 2: 556.

FITCH, AUGUST. 1853. The shrubs, flowers and plants of California. *Trans. Am. Instit. New York,* 75-84.

———— 1853. Trees of California and Oregon. *Ibid.* 239-244.

RODGERS, ANDREW DENNY 3RD. 1942. John Torrey. A story of North American botany. Princeton. 244.

GEORGE GIBBS

Brewer records that "Dr. George Gibbs was on this coast from 1848 until 1860. From 1849 to 1854 he was most of the time in this State [California], where he made collections, mostly near Columbia. Later he made more extensive collections in Oregon and Washington, which were in part incorporated with the collections of Drs. Cooper and Suckley and with those of the Northern Boundary Survey."

This appears to have been the same George Gibbs who, according to Bancroft, arrived in Oregon in the autumn of 1849 to fill the post of "deputy-collector," at a time when the United States government was extending its revenue laws in that region. Bancroft wrote:

"Gibbs, who came with the rifle regiment[2] was employed in various positions on the Pacific coast for several years. He became interested in philology and published a *Dictionary of the Chinook Jargon,* and other matter concerning the native races, as well as the geography and geology of the west coast. In *Suckley and Cooper's Natural History* it is said that he spent two years in southern Oregon, near the Klamath; that in 1853 he joined McClellan's surveying party, and afterwards made explorations with I. I. Stevens in Washington. In 1859 he was still employed as geologist of the northwest boundary survey with Kennerly. He was for a short time collector of customs at Astoria. He went from there to Puget Sound, where he applied himself to the habits, languages, and traditions of the natives, which study enabled him to make some valuable contributions to the Smithsonian Institution. Mr. Gibbs died at New Haven, Conn., May 11, 1873. 'He was a man of fine scholarly attainments,' says the *Olympia Pacific Tribune,* May 17, 1873, 'and ardently devoted to science and polite literature. He was something of a wag withal, and on several occasions, in conjunction with the late Lieut. Derby (John Phoenix[3]) and others, perpetrated "sells" that obtained a world wide publicity. His friends were many, warm, and earnest.' "

2. This "rifle regiment" was commanded by Brevet-Colonel W. W. Loring, and left Fort Leavenworth in May of 1849. Bancroft states on another occasion that it was accompanied by "George Gibbs, deputy collector at Astoria."

3. This was George Horatio Derby (1823–1861), who wrote under the pseudonym of John Phoenix. He graduated from West Point and after the Mexican War was assigned to exploring and surveying work in the far west. The "sells" were humorous sketches which appeared in the 1850's in the periodical *Pioneer* of San Francisco, in two series, under the titles "Phoenixiana" and "The Squibob papers." These are said to

Bancroft's reference quoted above to *"Suckley and Cooper's Natural History"* doubtless refers *in part* to Suckley's paper on mammals[4] and *in part* to his two papers ("Plants collected in Washington Territory" and "Plants collected west of the Cascade Mountains during 1854–'55") which were issued as part of J. C. Cooper's "Catalogue of plants collected in the Washington Territory" in 1869. In his "Catalogue" Cooper acknowledges that he was ". . . indebted to Mr. Gibbs for much assistance in collecting . . ." and many of the plants mentioned in the papers are credited to Gibbs.

Except for the Brewer reference first quoted, Gibbs' collecting seems to have been done after 1850, or later than the period of my story.

BANCROFT, HUBERT HOWE. 1888. The works of . . . 30 (Oregon II 1848–1888): 81, *fn.* 33, 104, *fn.* 8.

BREWER, WILLIAM HENRY. 1880. List of persons who have made botanical collections in California. *In:* Watson, Sereno. Geological survey of California. Botany of California 2: 556.

COOPER, JAMES GRAHAM. 1869. Catalogue of plants collected in Washington Territory. *U. S. War Dept. Rept. expl. surv. RR Mississippi Pacific* 12: 13-39.

GRINNELL, JOSEPH. 1930. James Graham Cooper. *Dict. Am. Biog.* 4: 406, 407.

MONAGHAN, FRANK. 1931. George Gibbs. *Dict. Am. Biog.* 7: 245, 246.

ALBERT KELLOGG

In a "Biographical notice of Dr. Albert Kellogg" the author, Edward L. Greene, refers to his subject as ". . . the first botanist who became a resident of California . . ." and states that he had been in New England when gold was discovered in that state and had joined a party bound for the mining regions:

". . . This company reached Sacramento, by way of the Straits of Magellan, as early as the eighth of August, 1849. Landings had been made on Tierra del Fuego, and at several points along the southern coasts of South America, and the botanist had eagerly availed himself of every opportunity for collecting plant specimens. This collection which would in after years have been of individual interest, as well as of scientific value, perished by a flood at Sacramento, not long after the arrival."

Greene mentions further that "After three or four years at Sacramento and in the mining districts above that place, Dr. Kellogg took up his residence in San Francisco . . ." It seems doubtful to me that he did any plant collecting of any importance between the time of his arrival at the gold fields in August, 1849, and the year 1850, when my story terminates.

have influenced other writers such as Mark Twain. See *The Columbia Encyclopedia* (13th printing, 1942, p. 490). I found both series rather dull, for the humor is largely based on local and contemporaneous matters. Although the authority mentioned above gives *Deadham*, Massachusetts, as Derby's birthplace, Dedham seems more probable! Boston, as well as the west coast comes in for a number of "digs." Both series came out in book form.

4. The Library of Congress cards cite "Cooper, J. G., Suckley, G., and Gibbs, G. Mammals. v. 12."

Referring to still earlier years, Greene states that Kellogg turned from medicine to the botanical work which interested him more, after "an extended tour southwestward" made in company with John James Audubon: "This journey landed our botanist eventually in San Antonio, Texas, in the autumn of 1845 ... Returning from Texas to his native state [Connecticut], he appears soon to have been upon his way to new fields of study, and other parts of the Mississippi river region. There is uncertainty about the length of time occupied in these journeyings ..." Sargent makes much the same statement. As does Jepson: ". . . He traveled widely through the Southern states and the Mississippi Valley, gratifying his taste for natural history, and on one occasion fell in with John James Audubon, and accepted his invitation to accompany him on a journey to Texas . . ." Kellogg's travels in the Mississippi River valley might bear investigating although Greene, who seems to have been the sole authority on the collector's early life, states that there is "uncertainty about the length of time" he spent there. I have found no reference to Kellogg in my reading on Audubon and believe that reliance may be placed upon Geiser's conclusions as to their association:

". . . Professor W. L. Jepson (*DAB*, *s.v.* 'Albert Kellogg') states that Kellogg came to Texas with the elder Audubon. This would have been in 1837. My evidence, to me conclusive, seems to show that neither in 1837, in 1845–6, nor in 1849 (all the possible dates) did Kellogg accompany any of the Audubons to Texas. The statement, based by Jepson on Edward L. Greene's published sketch of Kellogg, seems thus to be in error."

One thing is certain: that very little has been published about Kellogg's work as a collector before the year 1850—whether in the Mississippi River region, in Texas (where he may have gone even if not in famous company), or in California. Brewer states that Kellogg ". . . came to San Francisco in 1849 . . ."

BREWER, WILLIAM HENRY. 1880. List of those who have made botanical collections in California. *In:* Watson, Sereno. Geological survey of California. Botany of California 2: 556.

GEISER, SAMUEL WOOD. 1948. Naturalists of the frontier. ed. 2, Dallas. 276.

GREENE, EDWARD LEE. 1887. Biographical notice of Dr. Albert Kellogg. *Pittonia* 1: 145-151.

JEPSON, WILLIS LINN. 1933. Albert Kellogg. *Dict. Am. Biog.* 10: 300, 301.

SARGENT, CHARLES SPRAGUE. 1895. The Silva of North America. Boston. New York. 8: 120, *fn.* 4.

WILLIAM LOBB

In *A manual of the Coniferae* James Veitch and Sons in 1881 published a sketch of William Lobb. They state that he was born in 1809 in Cornwall: "The place is unknown,[5] nor is anything known of his early life." After working in the horticultural

5. Britten and Boulger mention that he was born at ". . . Perran-ar-worthal, E. Cornwall, 1899 [1809] . . ." and died at San Francisco in 1863. They report that his plants are at Kew.

establishment of Stephen Davey of Redruth, he entered the service of the Veitch firm and went on two collecting trips to South America (in 1840–1844, and in [?] 1845–1848). Next he was commissioned to make a trip to California.

"... The wonderful Conifers discovered by Douglas in California and Oregon, were then very scarce in England, and young plants of most of the important species could scarcely be bought for money. Hartweg had succeeded in sending consignments of cones and seeds to the Horticultural Society of London, three years previous, but the plants raised from them were distributed to fellows only. It was therefore decided that Lobb should proceed to California with a view of obtaining seeds of all the most important kinds known, and to discover others, if possible. He landed at San Francisco in the summer of 1849, and at once made arrangements for exploring southern California. Lobb's experience as a collector, his indomitable perseverance and courage, which was deterred by no danger, no toil, or no privation, enabled him to surmount difficulties and accomplish enterprises during the succeeding seven years of his collecting excursions through California and Oregon, which were scarcely equalled by Douglas himself . . ."

After enumerating some of the conifers which Lobb sent back (in the form of seeds and cones) to his employers in 1850–1851, in 1852, and in 1853, the *Manual* relates that in 1853 he explored the Sierra Nevada ". . . whither he was led by reports of the discovery of trees of extraordinary magnitude, and which he had the good fortune to find, and to secure the first cones and seeds of the Wellingtonia[6] received in England. He brought home with him two living plants, which were afterwards planted out in our Exeter Nursery, where they survived for three or four years . . ." Lobb, we are told, returned to California in the autumn of 1854 and up to the end of 1856 continued to send his employers plants and seeds. Although his work for the Veitches ended in 1857, he remained in California until his death in 1863, and was buried "in Lone Mountain Cemetery (now Laurel Hill)" [7]

Such was Lobb's assiduity that it seems unlikely that he wasted any opportunities to collect plants during the first five months of his sojourn in California in 1849. How-

6. The Rt. Rev. J. W. Hunkin, Bishop of Truro, writing of the two Lobb brothers—for William had a brother Thomas who also collected for the Veitch firm, but in the Far East—states in this connection: "*Sequoia gigantea* was called Wellingtonia by Lindley. This is generally taken to be after the Duke of Wellington, who had died 14 September, 1852. But it is noted that the Lobb family were then living in a cottage at Wellington Place, near Devoran (so called after its owner, a Mr. John Wellington); and it is possible that this circumstance may have had something to do with the choice of name." Fortunately, since it eliminates any possibility of confusion, Lindley explicitly states that the generic name was bestowed in honor of the Duke. (*See Gard. Chron.* 819-820, 1853. Also Rehder, Alfred, 1949. *Bibliography of cultivated trees and shrubs* ... p. 42, where the unsigned article of the *Gard. Chron.* is attributed to Lindley.)

7. We are told that ". . . his grave was located by the California Botanical Club several years ago, and the remains were moved to a new location and the grave put in good condition." I cite from Alice Eastwood and, although she fails to mention the "moving spirit" in this connection, I venture the guess that she was responsible for initiating this token of respect.

ever, I do not believe that it is possible, or even important, to distinguish those months from his over-all visit of 1849–1851.

The Veitch *Manual* expresses regret that not a single plant "worth mentioning" ". . . will perpetuate his name, or . . . keep in remembrance his great achievements." [8]

BREWER, WILLIAM HENRY. 1880. List of persons who have made botanical collections in California. *In:* Watson, Sereno. Geological survey of California. Botany of California 2: 557.

BRITTEN, JAMES & BOULGER, GEORGE SIMONDS. 1931. A biographical index of deceased British and Irish botanists. ed. 2. London. 191.

EASTWOOD, ALICE. 1939. Early botanical explorers on the Pacific coast and the trees they found there. *Quart. Cal. Hist. Soc.* 18: 343.

HUNKIN, J. W. 1942. William and Thomas Lobb: two Cornish plant collectors. *Jour. Royal Hort. Soc.* 67: 48-51.

PARRY, CHARLES CHRISTOPHER. 1883. Early botanical explorers of the Pacific coast. *Overland Monthly,* ser. 2, 2: 415, 416.

SARGENT, CHARLES SPRAGUE. 1896. The Silva of North America. Boston. New York. 10: 60, *fn.* 3.

VEITCH, JAMES & SONS. 1881. A manual of the Coniferae, containing a general review of the order; a synopsis of the hardy kinds cultivated in Great Britain; their place and use in horticulture, etc., etc. London. 258-261.

JAMES McNAB

Some time having been spent in verifying the fact that James McNab[9] did *not* extend his travels in North America to the west of the Mississippi River it is perhaps worth while to mention his trip if only to make that fact clear.

Sargent supplies a short biographical account of the man when writing of *Cupressus Macnabiana* A. Murray, which had been discovered by William Murray in 1854 at the southern base of Mount Shasta and states that the name "commemorates the horticultural and botanical labors of James McNab, the distinguished curator of the Edinburgh Botanic Garden." His father, William McNab, had held that post before him, according to Britten and Boulger. The same authorities note that he had been born at

---

8. The authors comment, in this connection, upon the custom, which was then prevalent, of naming genera and species in honor of persons who had not the slightest connection with the particular plant involved— "It is not for us to call into question the strictness of precedence in botanical nomenclature so much insisted on. It is enough to state the fact . . ." They then relate that Sargent had sent them seeds ". . . of a Ribes from Vancouver Island, labelled *Ribes Lobbi* (Gray), but figured and described in the *Botanical Magazine,* Tab. 4931, under the name of *R. subvestitum.*" While they felt that this was "truly an *amende honorable* on the part of the distinguished American botanist," they retained their opinion that this "pretty shrub" could hardly bear comparison with the "noble Thuia with which British horticulturists associate Lobb's name." The reference is to the canoe cedar, *Thuja gigantea* Nuttall, now *T. plicata* D. Don. Regrettable as it may be, the name *Thuja Lobbii* is now followed by "Hort."

9. I use the spelling adopted by Britten and Boulger, viz., *McNab.* The name appears in the literature as Macnab, MacNab, M'Nab and McNab.

Richmond, Surrey, on April 25, 1810, and died at Edinburgh on November 19, 1878. Sargent states that he ". . . learned the theory and practice of horticulture under his father at Edinburgh . . . in 1834 [he] visited America, and traveled extensively in Canada and the United States, where he made a large herbarium and collections of living plants and seeds. An account of the interesting plants which he gathered during his journey was published in the *Edinburgh Philosophical Journal* of 1835 and in the early volumes of the *Transactions of the Botanical Society of Edinburgh* . . ."

The *Edinburgh New Philosophical Journal* mentioned by Sargent contains an article entitled "Account of some of the rarer plants observed during an excursion in the United States and the Canadas in 1834. Communicated by Mr James Macnab." The plant stations cited prove that "Macnab" visited "Upper Canada," the Great Lakes (Canadian side), New Jersey and its pine barrens, the Pittsburgh region, the vicinity of the Ohio River and so on—all east of the region of my story.

McNab, in a footnote to his paper, mentions that his journey ". . . was undertaken chiefly for the gratification of my valued friend Mr Brown,[1] formerly of the Perth Nurseries, whose ardent love of science led him to America, that he might have the opportunity of seeing, in a state of nature, those forest trees and shrubs, to the rearing of which, in early life, his attention had been devoted . . . The tour was strictly private, and not (as some of the public journals seem to have imagined) under the auspices of any public institution."

An obituary notice of McNab, published in the *Transactions and Proceedings of the Botanical Society of Edinburgh,* credits him with having described "one or more new species" in his paper on the "rarer" plants observed in North America. Several plates in *The Botanical Magazine* and in *The British Flower Garden,* edited by Robert Sweet, are said to have been drawn by McNab.

Anonymous. 1879. Obituary notices. *Trans. Proc. Bot. Soc. Edinburgh* 13: 381, 382.
Britten, James & Boulger, George Simonds. 1931. A biographical index of deceased British and Irish botanists. ed. 2. London. 48, 201, 202.
McNab, James. 1835. Account of some of the rarer plants observed during an excursion in the United States and the Canadas in 1834. Communicated by Mr James Macnab. *Edinburgh New Philos. Jour.* 19: 56-64.
Sargent, Charles Sprague. 1896. The Silva of North America. Boston. New York. 10: 110, *fn.* 1.

JACOB DAVIS BABCOCK STILLMAN

Brewer noted briefly that "Dr. J. D. B. Stillman collected plants in 1849 near Sacramento, and in 1850 between Marysville and Long Bar. The specimens are in the Torrey Herbarium." Dr. Rickett wrote me (March 1, 1951) that "There seems to be little information available on . . . Stillman except that . . . he collected in California and sent specimens to Torrey . . . Stillman was born at Schenectady 21 F 1819 and died

1. Britten and Boulger identify him as Robert Brown (1767?–1845).

in 1888. We have a few letters . . . [from him] in the Torrey correspondence dealing with specimens which . . . [he] sent him." Stillman's name does not appear in the index of Rodgers' biography of Torrey.

Stillman wrote a short article entitled "Footprints of early California discoverers" which was published in the *Overland Monthly* in 1869, and which tells of "early travelers, other than Spanish, in California." Mentioned are many of the men to whom I have called attention here: "La Perouse," Vancouver, von Langsdorff, von Kotzebue, Douglas, Thomas Coulter, Beechey, Belcher, De Mofras, Wilkes, and finally Seeman of the *Herald*. Stillman's short paper ends at the point where another somewhat similar paper, written by C. C. Parry and published in 1869 in the *Overland Monthly,* begins its narrative. Stillman seems to have been familiar with the old published records and his article is useful as a brief summary.

He also wrote a story entitled "St. Jo" which was published, like his first, in the *Overland Monthly,* in the same volume which contained what are presumably the first printings[2] of Francis Bret Harte's "Miggles," and "The outcasts of Poker Flat." It must have been based to some extent upon his experiences as a medical man in the gold fields. St. Jo, the hero of the story—who might have been born in Gascony, in Buncombe or in both!—made life so miserable for those to whom he attached himself that he was finally forced into a duel. Stillman, "being the only surgeon," made a show of preparations, getting out his field case of instruments and spreading them on the table:

". . . I laid out the saw, placed the tourniquet in a convenient place, examined the edges of the long catlings and laid them down in a row; bandages were laid out with needles and thread. A pail of water with sponges was placed on the ground by the side of a bottle of Stoughton bitters, the best substitute to be had for spirits . . . and every preparation made for prompt assistance . . . as it seemed that both [contestants] would need it . . ."

After St. Jo—who was allowed to think that he had killed his adversary and was to be hung—had been well frightened, he was permitted to escape. When found, two days later, he was dead. At the inquest Stillman, as an expert, was called upon to give his opinion ". . . as to the cause of death. In accordance with that opinion a verdict was rendered of—*Frightened to death*." A far cry from botany and plant collecting! Stillman has the distinction of being the only one of my collectors to have written in humorous vein!

BREWER, WILLIAM HENRY. 1880. List of persons who have made botanical collections in California. *In:* Watson, Sereno. Geological survey of California. Botany of California 2: 556.

STILLMAN, JACOB DAVIS BABCOCK. 1869. St. Jo. *Overland Monthly* 2: 170-175.

———— 1869. Footprints of early California discoverers. *Overland Monthly* 2: 256-263.

2. Merle Johnson's *American first editions* . . . (ed. 3. New York. 1936, p. 207) states that no formal bibliography of Bret Harte's writings exists.

A Selected List of Plants Mentioned in the Text

The reasons for limiting my citations of plants and for presenting their names in a separate list should be explained.

In the Introduction (p. xxxii) I comment that to identify in the text all of the plants mentioned in the early narratives would be to enter the province of the botanical manual and the realm of synonymy. To include them all in an index would be but to enter the same fields in a different form. And it soon becomes apparent—if one attempts instead to limit the citations by some predetermined criterion, whether economic, horticultural, aesthetic or other— that there is no obvious point at which to stop the enumeration; as from the pebble dropped in the pond, the emanating circles grow larger and larger.

During much, certainly, of the period covered in this book, the plant names used by the collectors and to a lesser extent by the scientific botanists followed no stabilized pattern as far as newly discovered plants were concerned. For want of anything better the men in the field employed descriptive phrases or had recourse to colloquial names; misapplied the Latin names of plants with which they were familiar to others which to them appeared to be the same; employed Latin epithets (at times misspelled) which subsequently, because of priority or other rulings, came to be regarded as synonyms; there existed spelling variants since outmoded; the genus *Pinus* was interpreted to include a number of conifers, such as *Abies*, *Picea* and so on, now treated as distinct genera. These and other usages of the period involving plant names, if scattered through an index without apparent connection the one with the other, would serve no useful purpose.

Employing the form of presentation adopted here, it is possible to make the list elastic— as small or as large as seems desirable—and, by consolidating under a now-accepted generic and/or specific name the nomenclatorial usages mentioned above, to present a fairly coherent picture of any one plant as it appeared in the literature of the day.

I include in the list plants which (like Collignon's Sand Verbena or Née's two Oaks) have an historical significance, or (like the Salal) a sentimental connotation; ones which (like the California Poppy or the Blue Columbine) are now popular the world over; others which (like the Creosote Bush, the Mesquite, the Narrow-leaved Poplar or the Western Yellow Pine) are widespread; plants (like the two Sequoias, the Giant Cactus or the Joshua Tree) before which all persons pause; still others which (like the Osage Orange) interested the early travelers, or (like the Golden-leaved Chestnut) the taxonomists were determined to obtain. I even include a few introduced plants such as the Filaree of carpeting propensity, the Pride of India met in an unexpected region, or the rampant Mustard noted by early visitors to California. The list, therefore, represents my personal selection, from many hundreds of plants, of a mere handful which seem to possess a general or specific interest. Certain it is that no two persons would compile identical ones.

Latin names now in general use and colloquial names appear in Roman type. Latin names now regarded as synonyms, as spelling variants, as misidentifications, or which represent still-controversial reclassifications, appear in *italic type*.

INDEX

INDEX

1106 INDEX

fn. 4, 821; plants in plates of, 830-834; visits Texas, 834; his Missouri River journey, 834-839; comments of, on Catlin, 838-839; interest of, in collections of Nuttall and Townsend, 839-840; visit of, to Washington, D. C., 841; his reported trip with Kellogg to Texas, 1092. *See also:* Isaac Sprague.

Audubon, John Woodhouse, 1036 & *fn.* 1

Audubon, Lucy, 837, 839, 841

Austin, Stephen, 207, *fn.* 7, 372, 498 & *fn.* 2

Austin, Wright near, 1059

Auxumné (Merced) River, 920

Ayer, Clarence W., 1014

Azul River. *See:* Eagle River.

B

BACON, Leonard, 91, *fn.* 3

Bad (or Teton) River, 119, 663

Bagley, Clarence B., 465, 466 & *fn.* 1

Bailey, Liberty H., 935, *fn.* 1

Bailey, Professor (West Point), 1009

Baily, James, 50, *fn.* 7

Baird, S. F., 379, 838

Baker, Joseph, 31

Baker, Mount, named, 31 & *fn.* 3, 45, *fn.* 7

Baker's Bay, named, 45, *fn.* 7; Scouler at, 287, 293; Douglas at, 310, 311, 314, 321; Brackenridge at, 706

Balcones Escarpment, 381

Baldwin, Hannah Webster, 197 & *fn.* 6, 199, 200

Baldwin, William, xvi, xxix, 104, 147, *fns.* 1, 2, 197, *fn.* 6, 213, 244, *fn.* 6, 387; journey of, with Long expedition (1819), 188-196; instructions, letters, notes, collections, publications of, 189-200. *See also:* Darlington, W., and James.

Balfour, Alastair Norman, 301, *fn.* 9

Balfour, Frederick R. S., 301, *fn.* 9; on Menzies, 26, *fn.* 3, 29, *fn.* 8, 41, *fn.* 8, 57-58; on Douglas, 309, *fn.* 3, 313, *fn.* 3, 320, *fns.* 2, 3, 334, *fn.* 2, 406, *fn.* 3, 407-408, 411, *fn.* 7, 422, 423

Balfour, Graham, 682, *fn.* 6

Ball, John, 510, *fn.* 4

Bancroft, Hubert Howe, on: John Groem, 17, *fn.* 4; Malaspina, 18, 45; von Langsdorff, 87; Douglas, 305, *fn.* 1, 323, *fn.* 6, 399-400 & *fn.* 6, 401, 402, 405, 406, 415; Duhaut-Cilly, 360, 363; U. S.—Texas boundary dispute, 367-368; Wyeth, 387, *fn.* 1, 509, 510; Barnston, 394, *fn.* 3; on wreck of *Isabel*, 395, *fn.* 4; W. E. P. Hartnell, 400, *fn.* 5; Deppe, 426, *fn.* 9; Coulter, 428, *fn.* 1, 429, 435; Tolmie, 464, 476-479 *passim;* Fort Hall, 602; Townsend, 615; Russians in California, 679; De Mofras, 717, *fn.* 6; Gambel, 731 & *fn.* 2, 732, 733, *fn.* 7, 742, *fn.* 6; Frémont, 754; Applegate Cut-off, 813, *fns.* 1, 2; Lüders, 860, *fn.* 5; Childs, 867, *fn.* 6; Mormons, 1084, *fn.* 5; Gibbs, 1090 & *fn.* 2

Bandinel, James, 303, 308, 419

Banks, Joseph, xxii, xxiii, 26 & *fn.* 3, 28, 29 & *fn.* 8, 42, 47, 53, 57, 58, 108, 131, *fn.* 1, 135

Banks, Peter, 795 & *fn.* 5, 797, *fn.* 8

Bannock Pass, 805

Bannock River, McLeod on, 631; Frémont on, 855, 856

Baranoff, Alexander A. von, 90 & *fn.* 2, 155

Barclay, Mr., 637, 656

Barnhart, John H., on fictitious botanists, 15, *fn.* 1; on Douglass, 252, *fn.* 4; on Agate, 707 & *fn.* 8; on *Darlingtonia*, 715, *fn.* 4, 722, *fn.* 1; on W. D. Brackenridge, 724, 727; on Emory report, 991, *fn.* 3

Barnston, George, and Douglas, 315, *fn.* 7, 316, *fn.* 1, 396, *fn.* 5, 397-399, 405-412 *passim,* 424

Bartlett, H. H., 687, 724, 725

Barton, Benjamin Smith, 68, 71, 72, 110, 127, 199; contract of, with Clark, 72-75; and Nuttall, 142-144, 146-148, 182, *fn.* 3

Bartram, William, 174, 176, 379, *fn.* 6

Basalt Point, 33

Bass, Seth, 831, *fn.* 3

D

E

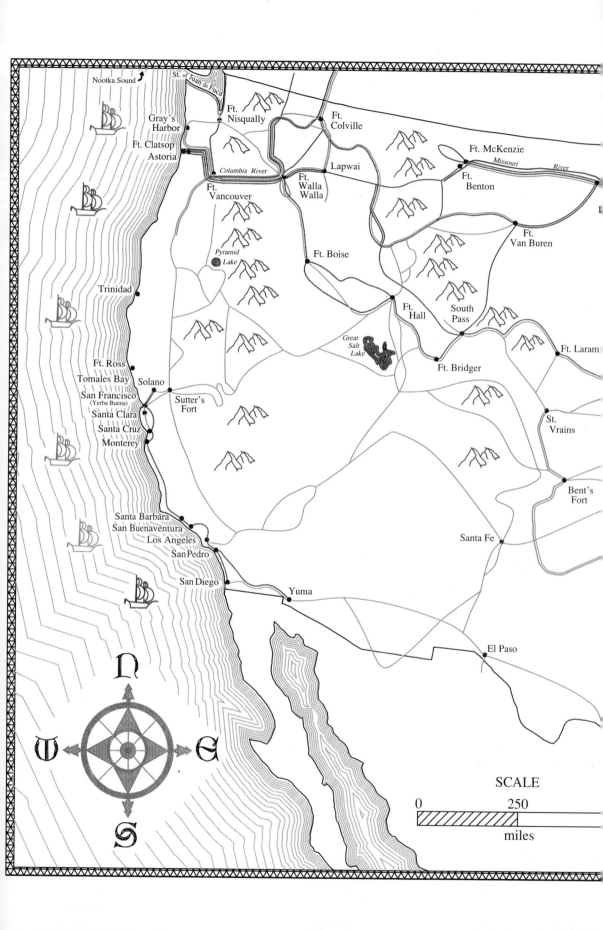

Nootka Sound
St. of Juan de Fuca
Ft. Nisqually
Gray's Harbor
Ft. Colville
Ft. McKenzie
Missouri   *River*
Ft. Clatsop
Astoria
Columbia River
Lapwai
Ft. Benton
Ft. Vancouver
Ft. Walla Walla
Ft. Boise
Ft. Van Buren
Pyramid
Lake
Trinidad
Great
Salt
Lake
Ft. Hall
South Pass
Ft. Laram
Ft. Ross
Tomales Bay
Ft. Bridger
San Francisco
(Yerba Buena)
Solano
Sutter's Fort
St. Vrains
Santa Clara
Santa Cruz
Monterey
Santa Barbara
San Buenaventura
Los Angeles
San Pedro
Bent's Fort
Santa Fe
San Diego
Yuma
El Paso

N
W   E
S

SCALE
0   250
miles